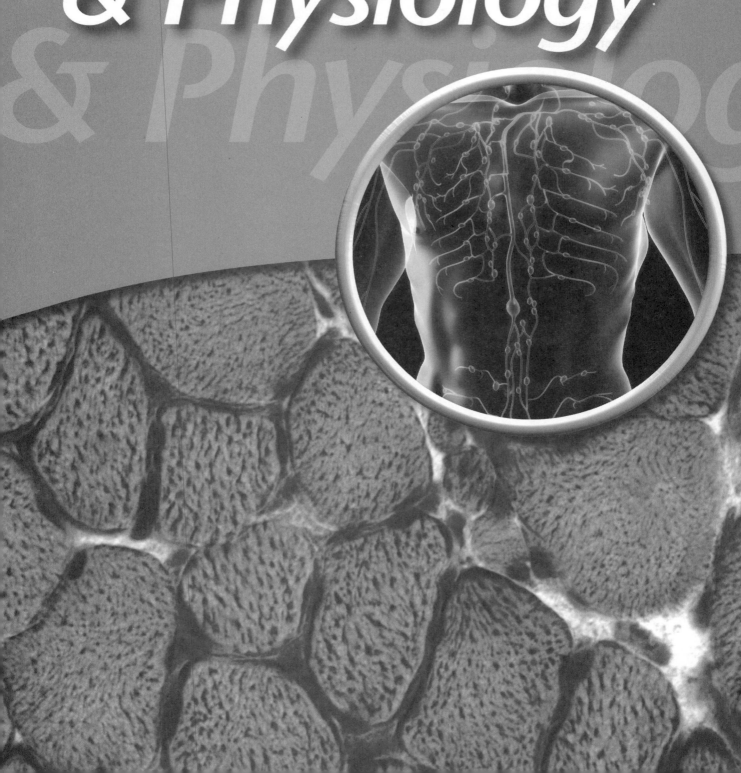

BIOZONE

Biology Modular W...

Anatomy
& Physiology

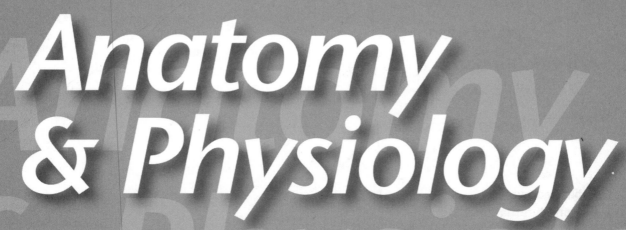

The Biozone Writing Team

Tracey Greenwood

Richard Allan

Artwork Support

Will Robinson

Bardoe Besselaar

Published by:
Biozone International Ltd
109 Cambridge Road, Hamilton 3216, New Zealand

Second printing with corrections
Printed by REPLIKA PRESS PVT LTD using paper
produced from renewable and waste materials

Distribution Offices:

United Kingdom & Europe	**Biozone Learning Media (UK) Ltd**, UK
	Telephone: +44 1283-553-257
	Fax: +44 1283-553-258
	Email: sales@biozone.co.uk
	Website: www.biozone.co.uk

USA, Canada, South America, Africa	**Biozone International Ltd**, New Zealand
	Telephone: +64 7-856-8104
	Freefax: 1-800717-8751 (USA-Canada only)
	Fax: +64 7-856-9243
	Email: sales@biozone.co.nz
	Website: www.biozone.co.nz

Asia & Australia	**Biozone Learning Media Australia**, Australia
	Telephone: +61 7-5535-4896
	Fax: +61 7-5508-2432
	Email: sales@biozone.com.au
	Website: www.biozone.com.au

Front cover photographs:
Muscle fibers of the quadriceps. Image ©2006 Armando Villalta from iStock Photos
www.istockphoto.com

Artist's 3-D render of the human lymphatic system. ©2006 Sebastian Kaulitzki from
iStock Photos www.istockphoto.com

Biology Modular Workbook Series

The Biozone *Biology Modular Workbook Series* has been developed to meet the demands of customers with a need for a flexible modular resource. As with Biozone's popular Student Course Workbooks, these resources provide a collection of visually engaging activities, which encourage independent thought and facilitate differential learning. The workbooks are divided into a series of chapters, each comprising an introductory section with detailed learning objectives and useful resources, and a series of write-on activities ranging from paper practicals and data handling exercises, to questions requiring short essay style answers. Page tabs identifying **"Related activities"** in the workbook and **"Web links"** help students to locate related material within the workbook and identify web links and activities (including video clips and animations) that will enhance their understanding of the topic. Web links are accessed directly from Biozone's website as described in the introductory section of the workbook. During the development of this series, we have taken the opportunity to improve the design and content, while retaining the basic philosophy of a student-friendly resource spanning the gulf between textbook and study guide. With its unique, highly visual presentation, it is possible to engage and challenge students, increase their motivation and empower them to take control of their learning.

Anatomy and Physiology

This title in the *Biology Modular Workbook Series* provides students with a set of comprehensive guidelines and highly visual worksheets through which to explore anatomy and physiology. *Anatomy & Physiology* is the ideal companion for students of the life sciences, encompassing the basic principles of cell biology, functional anatomy, physiological processes, and the impact of disease and aging on the body's systems. Homeostasis is the unifying theme throughout, and contextual examples provide the student with a relevant framework for their knowledge. This workbook comprises twelve chapters corresponding to each of the eleven body systems and an introductory chapter covering basic cells and tissues. The material is explained through a series of activities, usually of one or two pages, each of which explores a specific concept (e.g. joints or cell signaling). Model answers (on CD-ROM) accompany each order free of charge. *Anatomy & Physiology* is a student-centered resource and is to be complemented by the **Anatomy & Physiology Presentation Media CD-ROM** (to be released in 2010). Students completing the activities, in concert with their other classroom and practical work, will consolidate existing knowledge and develop and practise skills that they will use throughout their course. This workbook may be used in the classroom or at home as a supplement to a standard textbook. Some activities are introductory in nature, while others may be used to consolidate and test concepts already covered by other means. Biozone has a commitment to produce a cost-effective, high quality resource, which acts as a student's companion throughout their biology study. Please do not photocopy from this workbook; we cannot afford to provide single copies to schools and continue to develop, update, and improve the material they contain.

Acknowledgements & Photo Credits

Biozone's authors also acknowledge the generosity of all those who have kindly provided information and photographs for this first edition: • J. Armstrong • Prof. John Bath, California State Poly (photographs of vertebral bones) • Dan Butler • David Fankhauser, Uni. of Cincinnati, Clermont College • Dept of Biological Science, University of Delaware (endothelial tight junction) • Dr David Wells, AgResearch • Wellington Harrier Club • Dr D. Cooper, University of California, San Francisco • Helen Hall • Ed Uthman • Danny Wann, Carl Albert State College • Clinical Cases • L. Howard and K Connolly, Dartmouth College • Wadsworth Center, New York State Dept of Health. We also acknowledge photographs from **Wikimedia Commons** and the following contributors: • Stevenfruitsmaak • Lyn Bry (Mad Science) • Bill Rhodes • Georgetown University Hospital • Dan Ferber • pan Pavel Recyl • UC Regents Davis Campus • Michael Berry • Wbensmith Coded credits are: **BF**: Brian Finerran (University of Canterbury), **CA**: Clipart.com, **DS**: Digital Stock, **EII**: Education Interactive Imaging, **RCN**: Ralph Cocklin, **Eyewire**: Eyewire Inc © 1998-2001, www.eyewire.com, **JDG**: John Green, **CDC**: Centers for Disease Control and Prevention, Atlanta, USA, **RA**: Richard Allan, **TG**: Tracey Greenwood, **WMU**: Waikato Microscope Unit, **WIKI**: Wikimedia Commons

Royalty free images, purchased by Biozone International Ltd, are used throughout this manual and do not have on-page credits. They have been obtained from: istockphotos (www.istockphoto.com) • Corel Corporation from various titles in their Professional Photos CD-ROM collection; ©Hemera Technologies Inc, 1997-2001; © 2005 JupiterImages Corporation www.clipart.com; PhotoDisc®, Inc. USA, www.photodisc.com; and ©Digital Vision. 3D models were created using Poser IV, Curious Labs and Bryce 5.5.

Also in this series:

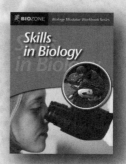

Skills in Biology

Health & Disease

Microbiology & Biotechnology

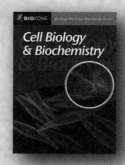

Cell Biology & Biochemistry

For other titles in this series go to:
www.thebiozone.com/modular.html

Contents

Activity is marked: ☐ to be done;   ☑ when completed

How to Use this Workbook

Anatomy & Physiology is designed to provide you with a resource that will make acquiring knowledge and skills in this area easier and more enjoyable. Anatomy and physiology is a interesting and rewarding area to study, particularly as the subject matter is so personally relevant to each of us. It is a subject with a great deal of terminology, so be sure to concentrate on understanding basic principles and use this workbook in conjunction with your textbook where necessary. *Anatomy & Physiology* provides an engaging way in which to reinforce and extend the ideas developed by your teacher. It is **not a textbook** and its aim is to complement your text and enable you to consolidate your knowledge and explore relevant contexts. *Anatomy & Physiology* provides the following resources in each chapter. You should refer back to them as you work through each set of worksheets.

Guidance Provided for Each Topic

Learning objectives:

These provide you with a map of the chapter content. Completing the learning objectives relevant to your course will help you to satisfy the knowledge requirements of your syllabus. Your teacher may decide to leave out points or add to this list.

Chapter content:

The upper panel of the header identifies the general content of the chapter. The lower panel provides a brief summary of the chapter content.

Key words:

Key words are displayed in **bold** type in the learning objectives and should be used to create a glossary as you study each topic. From your teacher's descriptions and your own reading, write your own definition for each word.

Note: Only the terms relevant to your selected learning objectives should be used to create your glossary. Free glossary worksheets are also available from our web site.

Use the check boxes to mark objectives to be completed.
Use a **dot** to be done (•).
Use a **tick** when completed (✓).

Periodical articles:

Ideal for those seeking more depth or the latest research on a specific topic. Articles are sorted according to their suitability for student or teacher reference. Visit your school, public, or university library for these articles.

Supplementary texts:

References to supplementary texts suitable for use with this workbook are provided. Chapter references are provided as appropriate. The details of these are provided on page 9, together with other resources information.

Internet addresses:

Access our database of links to more than **1000** web sites (updated regularly) relevant to the topics covered. Go to Biozone's web site: **www.thebiozone.com** and link directly to listed sites using the *BioLinks* button.

Supplementary resources
Biozone's Presentation MEDIA are noted where appropriate.

Activity Pages

The activities and exercises make up most of the content of this workbook. They are designed to reinforce the concepts you have learned about in the topic. Your teacher may use the activity pages to introduce a topic for the first time, or you may use them to revise ideas already covered. They are excellent for use in the classroom, and as homework exercises and revision. In most cases, the activities should not be attempted until you have carried out the necessary background reading from your textbook. As a self-check, model answers for each activity are provided on CD-ROM with each order of workbooks.

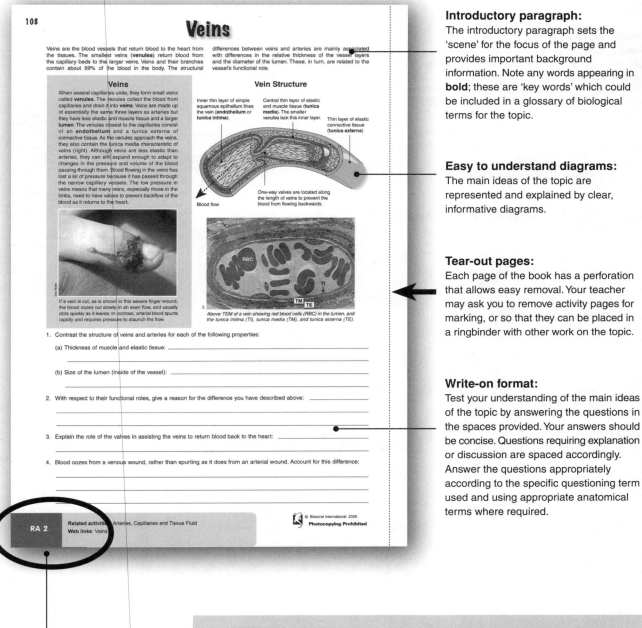

Introductory paragraph:
The introductory paragraph sets the 'scene' for the focus of the page and provides important background information. Note any words appearing in **bold**; these are 'key words' which could be included in a glossary of biological terms for the topic.

Easy to understand diagrams:
The main ideas of the topic are represented and explained by clear, informative diagrams.

Tear-out pages:
Each page of the book has a perforation that allows easy removal. Your teacher may ask you to remove activity pages for marking, or so that they can be placed in a ringbinder with other work on the topic.

Write-on format:
Test your understanding of the main ideas of the topic by answering the questions in the spaces provided. Your answers should be concise. Questions requiring explanation or discussion are spaced accordingly. Answer the questions appropriately according to the specific questioning term used and using appropriate anatomical terms where required.

Activity code:
Activities are coded to help you in identifying the type of activities and the skills they require. Most activities require some basic knowledge recall, but will usually build on this to include applying the knowledge to explain observations or predict outcomes. The least difficult questions generally occur early in the activity, with more challenging questions towards the end of the activity.

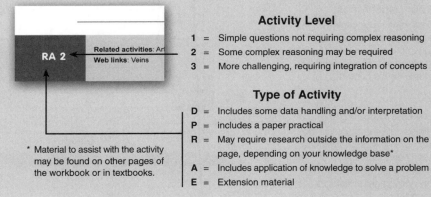

* Material to assist with the activity may be found on other pages of the workbook or in textbooks.

Activity Level
1 = Simple questions not requiring complex reasoning
2 = Some complex reasoning may be required
3 = More challenging, requiring integration of concepts

Type of Activity
D = Includes some data handling and/or interpretation
P = includes a paper practical
R = May require research outside the information on the page, depending on your knowledge base*
A = Includes application of knowledge to solve a problem
E = Extension material

Introduction

Chapter Summary and Contexts

Each chapter (or in some cases two chapters) in *Anatomy & Physiology* is preceded by a two page summary of homeostatic interactions and contextual examples. The first of the two pages provides an overview of how the specific body system (in this case, the respiratory system) interacts with the other body systems to maintain homeostasis. A lower panel summarizes the general functions of the system (for example, the respiratory

system's functional role is in gas exchange). Homeostasis is a unifying theme throughout *Anatomy & Physiology*. The second of the two pages continues this theme by showing how the selected body system can be affected by disease, aging, and exercise, and how our current medical knowledge can be applied to disorders of homeostasis. These contexts provide a relevant framework for understanding the subject matter.

A Contextual Framework

Interacting Systems:

The purpose of this page is to summarize the interactions of the body system under study (in this case the respiratory system) with all other body systems in turn. This summary describes the way in which systems work together to maintain homeostasis.

Most systems are treated singly, although those systems that operate very closely (e.g. muscular and skeletal) are mapped together.

The intersecting regions of the center panel of the context map highlight topics of focus within each context. These are specifically addressed within the workbook.

Interacting Systems

The Respiratory System

Cardiovascular system
- Blood transports respiratory gases.
- The carbonate-bicarbonate system in blood contributes to blood buffering.

Urinary system
- Kidneys dispose of the waste products of respiratory metabolism (other than CO_2 which is breathed out).

Reproductive system
- Pregnancy has significant effects on breathing mediated largely through sex hormones. The enlarged uterus (its position at full term shown by dotted line) pushes abdominal organs up and outwards and compromises the functioning of the diaphragm.

Skeletal system
- Bones enclose and protect the lungs and bronchi from damage.
- Expansion and elastic recoil of the ribcage produce the volume changes necessary for inhalation/exhalation (breathing).

Integumentary system
- Skin forms a surface barrier protecting the organs of the respiratory system.

Nervous system
- Control centers in the medulla and the pons regulate the rate and depth of breathing.
- Sensory feedback to the respiratory control centers is provided by stretch receptors in the bronchioles and chemoreceptors in the aorta and carotid arteries.

Lymphatic system and immunity
- Immune system provides general protection against pathogens; specifically the tonsils protect against pharyngeal and upper respiratory tract infections.
- Lymphatic system helps to maintain blood volume required for efficient transport of respiratory gases.

Endocrine system
- Epinephrine (from the adrenal medulla atop the kidneys) acts as a bronchodilator.
- Testosterone promotes enlargement of the larynx at puberty in males.
- Cortisol has role in lung maturation and a number of hormones directly or indirectly influence breathing rates.

General Functions and Effects
The respiratory system provides an interface for gas exchange ... ultimately responsible for providing all the cells of the body with ... dioxide produced as a result of cellular respiration. These resp...

		Stem cell therapy
Asthma		Gene therapy
Lung cancer		X rays
Chronic bronchitis		Vaccination
Emphysema		
Asbestosis		

Taking a Breath
The Respiratory System

The respiratory system can be affected by disease and undergoes changes associated with training and aging.

Medical technologies can be used to diagnose and treat respiratory disorders. Exercise and lifestyle management can prevent some respiratory diseases.

- Lung cancer
- Chronic bronchitis
- Emphysema

- VO_2max
- Ventilation efficiency
- Ventilation rhythm

Disease

A summary of some of the diseases affecting the body system. These provide a good context for examining departures from homeostasis.

Disease		Medicine & Technology	
Symptoms of disease	• Chest pain • Excessive mucus production • Coughing, sneezing • Difficulty breathing, cyanosis	Diagnosis of disorders	• Chest X-ray and CT scans • Pulmonary function tests • Sputum cultures and biopsy • DNA tests and screening
Infectious respiratory diseases	• Bacterial pneumonia • Pulmonary tuberculosis • Influenza	Preventing and treating diseases of the respiratory system	• Drug therapies (e.g. antibiotics) • Vaccination • Surgery (e.g. transplants) • Radiotherapy • Gene or cell therapy (e.g. CF) • Behavior modification
Non-infectious respiratory diseases	• Asthma • Fibrosis (scarring) • Smoking-related diseases • Inherited diseases (e.g. CF)		

- Asthma
- Lung cancer
- Chronic bronchitis
- Emphysema
- Asbestosis

- Stem cell therapy
- Gene therapy
- X rays
- Vaccination

Taking a Breath
The Respiratory System

The respiratory system can be affected by disease and undergoes changes associated with training and aging.

Medical technologies can be used to diagnose and treat respiratory disorders. Exercise and lifestyle management can prevent some respiratory diseases.

- Lung cancer
- Chronic bronchitis
- Emphysema

- VO_2max
- Ventilation efficiency
- Ventilation rhythm

Exercise can delay or reverse some age-related changes to respiratory function

| Aging and the respiratory system | • Decline in respiratory capacity
• Decline in aerobic capacity (VO_2max)
• Increased incidence of chronic respiratory disease | Effects of exercise on the respiratory system | • Increased rate and depth of breathing
• Increased aerobic capacity (VO_2max)
• Increased respiratory efficiency
• Improved oxygen loading-unloading
• Better diaphragmatic performance |

| The Effects of Aging | Exercise |

TAKING A BREATH

Medicine and Technology

A summary of how medicine and technology are used to study the chosen body system, and how new technologies can be used to diagnose and treat specific diseases. An awareness of how technology is applied in a medical context is essential to the study of anatomy and physiology.

Four-panel focus:

Each of the four panels on this page focuses on one context to which you can apply your knowledge and understanding of the topic's content.

Exercise

Exercise has different effects on different body systems. Some of the physiological effects of exercise to be considered in the workbook are summarized here.

The Effects of Aging

Degenerative changes in the body system are summarized in this panel. The effects of aging provide another context for considering disruptions to homeostasis.

Resources Information

Your set textbook should always be a starting point for information, but there are also many other resources available. A list of readily available resources is provided below. Access to the publishers of these resources can be made directly from Biozone's web site through our resources hub: **www.thebiozone.com/resource-hub.html**. Please note that our listing of any product in this workbook does not denote Biozone's endorsement of it.

Comprehensive Texts

We have listed two commonly used texts for anatomy and physiology below. There are many other excellent texts available also.

Marieb, E. N. & K. Hoehn, 2008
Human Anatomy & Physiology, 7 ed., 1296 pp.
Publisher: Benjamin Cummings.
ISBN: 978-0321559111
Comments: *Well illustrated and clearly explained coverage of the human body. 'The Essentials of Human Anatomy and Physiology', also by Marieb is a popular option for a one-semester course.*

Shier, D.N, ad J.L Butler, 2009
Hole's Essentials of Human Anatomy & Physiology, 10 ed. 640 pp.
Publisher: Mcgraw-Hill
ISBN: 978-0077221355
Comments: *Designed for a one-semester A&P course, this text assumes no prior knowledge and supports core material with clinical examples.*

Supplementary Texts

Morton, D. & J.W. Perry, 1997
Photo Atlas for Anatomy & Physiology, 160 p.
Publisher: Brooks Cole. **ISBN**: 0-534-51716-1
Comments: *An excellent photographic guide to lab work. Includes a good section on histology, with representations of tissue types and clear images for comparison with textbook drawings.*

Rowett, H.G.Q, 1999
Basic Anatomy & Physiology, 4 ed. 132 pp.
Publisher: John Murray
ISBN: 0-7195-8592-9
Comments: *A revision of a well established reference book for the basics of human anatomy and physiology. Accurate coverage of required AS/A2 content with clear, informative diagrams.*

Tobin, A.J. and R.E Morel, 1997
Asking About Cells, 698 pp (paperback)
Publisher: Thomson Brooks/Cole
ISBN: 0-030-98018-6
Comments: *An introduction to cell biology, cellular processes and specialisation, DNA and gene expression, and inheritance. The focus is on presenting material through inquiry.*

Periodicals, Magazines, & Journals

Biological Sciences Review: *An informative quarterly publication for biology students.* Enquiries:
UK: Philip Allan Publishers **Tel**: 01869 338652
Fax: 01869 338803 **E-mail**: sales@philipallan.co.uk
Australasia: **Tel**: 08 8278 5916, **E-mail**: rjmorton@adelaide.on.net

New Scientist: *Widely available weekly magazine with research summaries and features.* Enquiries: Reed Business Information Ltd, 51 Wardour St. London WIV 4BN **Tel**: (UK and intl):+44 (0) 1444 475636 **E-mail**: ns.subs@qss-uk.com *or subscribe from their web site.*

Scientific American: *A monthly magazine containing specialist features. Articles range in level of reading difficulty and assumed knowledge.* Subscription enquiries: 415 Madison Ave. New York. NY10017-1111 **Tel**: (outside North America): 515-247-7631 **Tel**: (US& Canada): 800-333-1199

School Science Review: *A quarterly journal which includes articles, reviews, and news on current research and curriculum development. Free to Ordinary Members of the ASE or available on subscription.* Enquiries: **Tel**: 01707 28300 **Email**: info@ase.org.uk *or visit their web site.*

The American Biology Teacher: *The peer-reviewed journal of the NABT. Published nine times a year and containing information and activities relevant to biology teachers.* Contact: NABT, 12030 Sunrise Valley Drive, #110, Reston, VA 20191-3409 **Web**: www.nabt.org

Biology Dictionaries

A good dictionary is useful when dealing with the terminology of human biology. Some of the titles available are listed below. Link to the relevant publisher via Biozone's resources hub: **www.thebiozone.com > resources > dictionaries**

Hale, W.G. **Collins: Dictionary of Biology** 4 ed. 2005, 528 pp. Collins.
ISBN: 0-00-720734-4.
Updated to take in the latest developments in biology and now internet-linked. (§ This latest edition is currently available only in the UK. The earlier edition, ISBN: 0-00-714709-0, is available though amazon.com in North America).

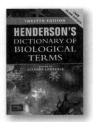

Henderson, I.F, W.D. Henderson, and E. Lawrence. **Henderson's Dictionary of Biological Terms**, 1999, 736 pp. Prentice Hall.
ISBN: 0582414989
An updated edition, rewritten for clarity, and reorganized for ease of use. An essential reference and the dictionary of choice for many.

Thain, M. **Penguin Dictionary of Human Biology**, 2009, 768 pp. Penguin Global.
ISBN: 978-0140514827
An essential dictionary for those studying human biology, medicine, or nursing. With a focus on human anatomy and physiology, this dictionary (in paperback) would be a good choice for students of the life sciences.

Using the Internet

The internet is a powerful resource for locating information. There are several key areas of Biozone's web site that may be of interest to you. Go to the **BioLinks** area to browse through the hundreds of web sites hosted by other organizations. These sites provide a supplement to the activities provided in our workbooks and have been selected on the basis of their accurate, current, and relevant content. We have also provided links to biology-related **podcasts** and **RSS newsfeeds**. These provide regularly updated information about new discoveries in biology; perfect for those wanting to keep abreast of changes in this dynamic field.

The BIOZONE website: www.thebiozone.com

The current internet address (URL) for the web site is displayed here. You can type a new address directly into this space.

Use Google to search for web sites of interest. The more precise your search words are, the better the list of results. EXAMPLE: If you type in "biotechnology", your search will return an overwhelmingly large number of sites, many of which will not be useful to you. Be more specific, e.g. "biotechnology medicine DNA uses".

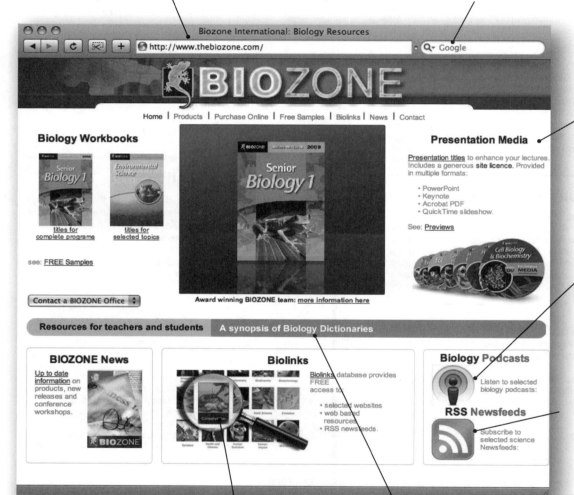

Find out about our superb **Presentation Media**. These slide shows are designed to provide in-depth, highly accessible illustrative material and notes on specific areas of biology.

Podcasts: Access the latest news as audio files (mp3) that may be downloaded to your ipod (mp3 player) or played directly off your computer.

RSS Newsfeeds: Read about the latest news and major new discoveries in biology directly from Biozone's web site.

News: Find out about product announcements, shipping dates, and workshops and trade displays by Biozone at teachers' conferences around the world.

Access the **BioLinks** database of web sites related to each major area of biology.

The **Resource Hub** provides links to the supporting resources referenced in the workbook. These resources include comprehensive and supplementary texts, biology dictionaries, computer software, videos, and science supplies.

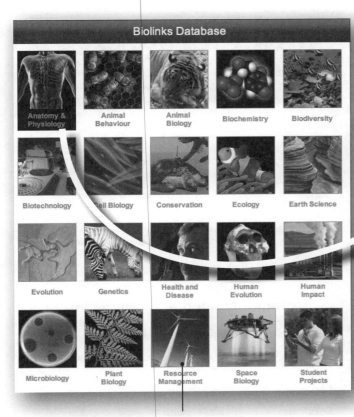

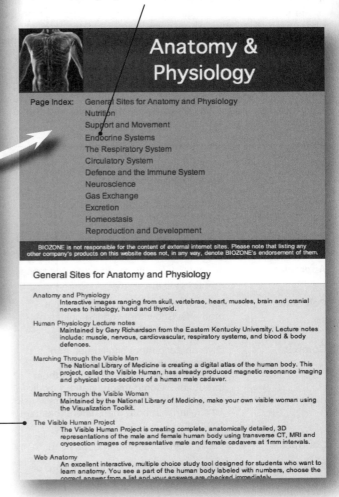

Index of sub-topics on this page. Click on these to jump down to the desired section.

Click on each topic to see a list of all related biology links. Each topic has relevant subtopics to make searching easier and each link has a brief description.

Click on the link to access the named site. The brief description tells you how the site may be of interest, as well as any country specific bias, if this is relevant.

Weblinks:

Go to: **www.thebiozone.com/weblink/AnaPhy-2269.html**

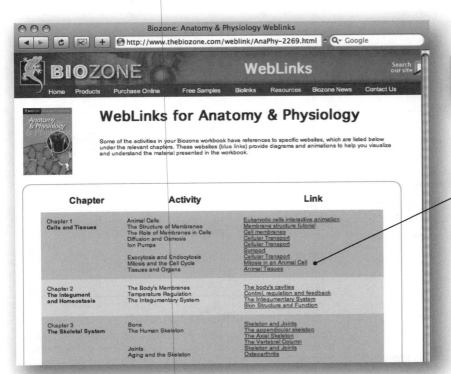

Throughout this workbook, some pages make reference to additional or alternative activities, as well as web sites that have particular relevance to the activity. See example of page reference below:

RA 2	Related activities: Arteries, Capillaries and Tissue Fluid Web links: Veins

Web Links: These provide links to **external web sites** with supporting information for the activity against which they are listed. The sites have been specifically chosen for their clear presentation, accuracy, and appeal. They are largely restricted to brief but helpful explanatory animations and video clips and can be supplemented with online text-based material available though the Biozone's Biolinks area.

Terms in Anatomy and Physiology

The study of anatomy and physiology requires a good mental map of the body's different regions and a sound understanding of the terms used to describe the location of structures on the body. As well as this, when answering questions, you should have a good knowledge of what specific questioning terms mean and phrase your answers accordingly. The following short guide, which is by no means exhaustive, lists some of the basic questioning terms, and directional and regional terms that you will come across when you study anatomy and physiology. Refer to as frequently as you see fit.

Commonly used questioning terms

The following terms are frequently used when asking questions in examinations and assessments. You should have a clear understanding of each of the following terms and use this to answer questions appropriately.

Account for: Provide a satisfactory explanation or reason for an observation.

Analyse: Interpret data to reach stated conclusions.

Annotate: Add **brief** notes to a diagram, drawing, or graph.

Appreciate: To understand the meaning or relevance of a particular situation.

Calculate: Find an answer using mathematical methods. Show the working unless instructed not to.

Compare: Give an account of similarities and differences between two or more items, referring to both (or all) of them throughout. Comparisons can be given using a table. Comparisons generally ask for similarities more than differences (see contrast).

Contrast: Show differences. Set in opposition.

Define: Give the precise meaning of a word or phrase as concisely as possible.

Describe: Give an account, with all the relevant information.

Discuss: Give an account including, where possible, a range of arguments, assessments of the relative importance of various factors, or comparison of alternative hypotheses.

Distinguish: Give the difference(s) between two or more different items.

Evaluate: Assess the implications and limitations.

Explain: Give a clear account including causes, reasons, or mechanisms.

Identify: Find an answer from a number of possibilities.

Illustrate: Give concrete examples. Explain clearly by using comparisons or examples.

Interpret: Comment upon, give examples, describe relationships. Describe, then evaluate.

List: Give a sequence of names or other brief answers with no elaboration. Each one should be clearly distinguishable from the others.

Measure: Find a value for a quantity.

Outline: Give a brief account or summary.

Predict: Give an expected result.

Solve: Obtain an answer using algebraic and/or numerical methods.

State: Give a specific name, value, or other answer. No supporting argument or calculation is necessary.

Suggest: Propose a hypothesis or other possible explanation.

Summarize: Give a brief, condensed account. Include conclusions and avoid unnecessary details.

Commonly used anatomical terms

Abdominal: Anterior body trunk inferior to ribs.

Anterior: Toward or at the front of the body, (ventral).

Cephalic: Head region.

Cervical: Neck region.

Deep: Away from the body surface, (internal).

Distal: Farther from the point of attachment of a limb to the body trunk.

Dorsal: Toward or at the back side of the body, (posterior).

Frontal plane: Divides the body into front and back portions.

Inferior: Toward the lower part of the body.

Lateral: Away from the midline of the body.

Medial: Toward or at the midline of the body.

Median plane: A vertical plane through the midline of the body; divides the body into right and left halves, (midsagittal).

Pelvic: Area overlying the pelvis.

Posterior: Toward or at the back side of the body, (dorsal).

Proximal: The point of attachment of a limb to the body trunk.

Pubic: Genital region.

Superior: Toward the head or upper part of the body.

Superficial: Body surface, (external).

Thoracic: Chest region.

Transverse plane: Divides the body into top and bottom portions.

Ventral: Toward or at the front of the body, (anterior).

In Conclusion

You should familiarise yourself with this list of terms. The aim is to become familiar with interpreting questions and answering them appropriately using the correct anatomical terminology.

 © Biozone International 2009

Concepts in Anatomy & Physiology

Cells and Tissues

| Cell structure and function | • Cellular membranes and organelles
• Cellular transport processes
• Cell division and specialization |

| Tissues are made up of cells with different roles | • Epithelial tissues
• Connective tissues
• Muscle tissue
• Nervous tissue |

| Organs are made up of different tissues. Organ systems have different roles. | • Exchanges with the environment
• Support and movement
• Control and coordination
• Internal transport
• Internal defense
• Reproduction and development
• Excretion and fluid balance |

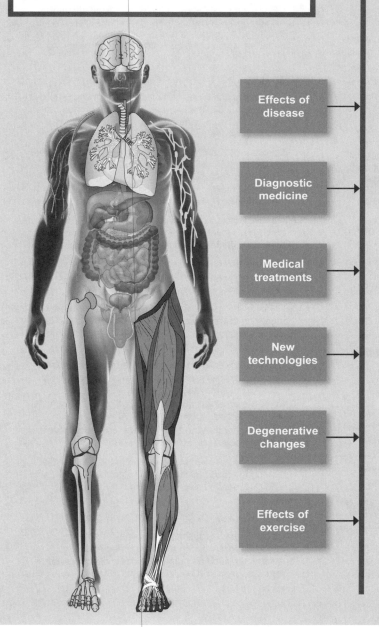

Effects of disease

Diagnostic medicine

Medical treatments

New technologies

Degenerative changes

Effects of exercise

The Integumentary System
The skin and its accessory structures

The Skeletal System
The bones, cartilage, and ligaments

The Muscular System
Smooth, cardiac, and skeletal muscle

The Nervous System
Neurons, glial cells, sensory receptors, and sense organs

The Endocrine System
Endocrine glands and hormones, including the hypothalamus

The Cardiovascular System
The heart, blood vessels, and blood

The Lymphatic System
The lymphoid tissues and organs, the leukocytes

The Digestive System
The digestive tract and accessory organs, including the liver

The Respiratory System
The lungs and associated structures

The Urinary System
The kidneys, bladder, and accessory structures

The Reproductive System
The sex organs and associated structures

Cells and Tissues

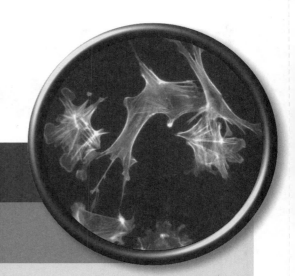

Understanding the structural and functional components of the human body

Cells and cellular organelles, passive and active transport processes in cells, biological organization, cell division and differentiation, tissues, organs, and organ systems

Learning Objectives

☐ 1. Compile your own glossary from the **KEY WORDS** displayed in **bold type** in the learning objectives below.

Basic Biochemistry *(review)*

☐ 2. With reference to your textbook, review your understanding of basic biochemistry, including:

   (a) The common **elements** found in living things and where these occur in cells.

   (b) The difference between an atom and an ion.

   (c) The difference between ionic bonds and covalent bonds and the importance of **covalent bonding** in carbon-based compounds.

   (d) The difference between organic and inorganic compounds and the role of each in the body.

   (e) The functional role of **macromolecules** in the body.

   (f) The structure of water, including reference to the polar nature of the water molecule, the nature of the bonding within the molecule, and the role of **hydrogen bonding** between water molecules.

   (g) The physical properties of water that are important in biological systems and the various biological roles of water: *e.g. metabolic role, as a solvent, as a lubricant, as a coolant, as a transport medium.*

   (h) The role of inorganic ions in the human body.

   (i) The general properties of enzymes and their role in regulating cell metabolism.

Features of Cells *(pages 16-24)*

☐ 3. Describe and interpret drawings and photographs of typical **cells** as seen using light and electron microscopy. Describe the role of the following:

- **nucleus, nuclear envelope, nucleolus**
- **mitochondria, nucleolus**
- rough/smooth **endoplasmic reticulum, ribosomes,**
- **plasma membrane**
- **Golgi apparatus, lysosomes, peroxisomes**
- **cytoplasm, cytoskeleton** (of **microtubules**), **centrioles, cilia** (if present)

☐ 4. Describe the interrelationship between the organelles involved in the production and secretion of proteins.

☐ 5. PRACTICAL: Demonstrate an ability to correctly use a light microscope to locate material and focus images. Identify the steps required for preparing a **temporary mount** for viewing with a compound light microscope. Understand why **stains** are useful in the preparation of specimens. If required, use simple **staining techniques** to show specific features of cells.

☐ 6. Draw a simple labeled diagram of the structure of the **plasma membrane** (cell surface membrane), clearly identifying the arrangement of the lipids and proteins.

☐ 7. Describe and explain the current **fluid mosaic model** of membrane structure, including the terms **lipid bilayer** and **partially permeable membrane**. Explain the roles of **phospholipids, cholesterol, glycolipids, proteins,** and **glycoproteins** in membrane structure. Explain how the **hydrophobic** and **hydrophilic** properties of phospholipids help to maintain membrane structure. Appreciate that the plasma membrane is essentially no different to the membranes of organelles.

☐ 8. Describe the functions of membranes (including the **plasma membrane**) in cells, identifying their role in the structure of organelles and in regulating the transport of materials within cells, and into and out of cells.

Cellular Transport *(pages 25-28)*

☐ 9. Summarize the types of movements that occur across membranes. Outline the role of proteins in membranes as receptors and carriers in membrane transport. Define: **passive transport, concentration gradient**.

☐ 10. Describe **diffusion** and **osmosis**, identifying the types of substances moving in each case. Describe **facilitated diffusion** (facilitated transport) involving carrier or channel proteins. Identify when and where facilitated diffusion might occur in a cell.

☐ 11. Identify factors determining the rate of diffusion. Explain how **Fick's law** provides a framework for determining maximum diffusion rates across cell surfaces.

☐ 12. Suggest why cell size is limited by the rate of diffusion. Discuss the significance of **surface area to volume ratio** to cells.

☐ 13. Distinguish between passive and **active transport** mechanisms. Understand the principles involved in active transport, clearly identifying the involvement of protein molecules and energy.

☐ 14. Describe the following active transport mechanisms, including examples of when and where each type of transport mechanism might occur:

   (a) **Ion-exchange pumps**, e.g. sodium-potassium pumps and proton pumps

   (b) Vesicular transport, e.g. **exocytosis, endocytosis, phagocytosis,** and **pinocytosis**.

☐ 15. Identify the mechanisms involved in the transport of: water, fatty acids, glucose, amino acids, O_2, CO_2, and ions (e.g. mineral and metal ions).

Cell Division *(pages 29-31, 212, 215, 228)*

☐ 16. Using diagrams, describe the behavior of **chromosomes** during a mitotic **cell cycle** in eukaryotes such as humans. Include reference to: **mitosis, growth** (G_1 and G_2), and DNA replication (S).

17. Recognize and describe the following events in mitosis: **prophase**, **metaphase**, **anaphase**, and **telophase**. With respect to mitosis in animal cells, understand the term **cytokinesis**, and distinguish between nuclear division and division of the cytoplasm.

18. Describe the role of mitosis in growth and repair. Recognize the importance of **daughter nuclei** with chromosomes identical in number and type. Recognize that mitotic cell division is a necessary prelude to **cellular differentiation**.

19. Explain how **carcinogens** can upset the normal controls regulating cell division. Define the terms: **cancer**, **tumor suppressor genes**, **oncogenes**. Describe factors that increase the chances of cancerous growth and **metastasis**.

20. Understand the functional role of programmed cell death (**apoptosis** or PCD) and describe the situations in which apoptosis plays a crucial role, e.g. during development or when DNA is damaged beyond repair.

21. Contrast mitosis and **meiosis** with respect to the functional role and the end products of each type of division (also see spermatogenesis and oogenesis).

Tissues and Organs *(pages 32-35)*

22. Describe how a **zygote** undergoes **mitotic cell division** and differentiation to produce an adult. With reference to specific examples, explain what is meant by **differentiation** and **specialized cell**.

23. Recognize the hierarchy of organization in multicellular organisms (including humans). Appreciate the role of cooperation between cells, **tissues**, **organs**, and **organ systems** in the structure and function of human body. Recognize that each step in the hierarchy of organization is associated with the emergence of properties (e.g. cognition) not present at simpler levels of organization.

24. Recognize structural and functional diversity in the cells that make up the tissues of the of the human body. identify and describe cells that:
    - connect body parts (e.g. fibroblasts, erythrocytes)
    - cover or line body organs (e.g. epithelial cells)
    - defend against disease (e.g. white blood cells)

- store nutrients (e.g. fat cells)
- are involved in reproduction (e.g. sperm, oocyte)
- communicate (e.g. neurons)
- move body parts (e.g. muscle cells)

25. With reference to specific examples, explain how cells are organized into **tissues**. Note that tissues frequently contain extracellular material.

26. Recognize the characteristic features and functional roles of the four main tissue types in humans. In each case, describe where each tissue type is found and its suitability for that location:

    (a) Epithelial tissue (epithelium):
        simple or stratified epithelium
    - squamous, cuboidal, columnar
    - glandular epithleium
    (b) Connective tissue:
    - bone (compact, spongy)
    - cartilage (elastic, hyaline)
    - dense connective tissue (e.g. tendons, ligaments)
    - loose connective tissue (e.g. areolar, adipose)
    - blood
    (c) Muscle tissue:
    - skeletal, smooth, cardiac
    (d) Nervous tissue

27. Briefly describe the components and function of the eleven organ systems of the body, recognizing the way in which they interact to maintain the body's steady state (**homeostasis**):

    - integumentary system
    - skeletal system
    - muscular system
    - nervous system
    - endocrine system
    - cardiovascular system
    - lymphatic system
    - respiratory system
    - digestive system
    - urinary system
    - reproductive system

Supplementary Texts

See page 9 for additional details of these texts:

■ Morton, D. and Perry, J.W, 1997. **Photo Atlas for Anatomy and Physiology**, (Brooks/Cole).

■ Rowett, H.G.Q., 1999. **Basic Anatomy & Physiology**, (John Murray), pp. 2-6.

■ Tobin, A.J. and Morel, R.E., 1997. **Asking About Cells**, (Thomson Brooks/Cole), as required.

Periodicals

See page 9 for details of publishers of periodicals:

STUDENT'S REFERENCE

■ **Getting in and Out** Biol. Sci. Rev., 20(3), Feb. 2008, pp. 14-16. *A good account of diffusion including some common misunderstandings.*

■ **Border Control** New Scientist, 15 July 2000 (Inside Science). *The role of the plasma membrane in cell function: membrane structure and transport, and the role of membrane receptors.*

■ **Lysosomes: The Cell's Recycling Centres** Biol. Sci. Rev., 17(2) Nov. 2004, pp. 21-23. *The nature and role of lysosomes: small membrane-bound organelles found in all eukaryotic cells.*

■ **Lysosomes and their Versatile and Potentially Fatal Membranes** Biol. Sci. Rev., 17(3) Feb. 2005, pp. 14-16. *The critical importance of the lysosome membrane.*

■ **The Cell Cycle and Mitosis** Biol. Sci. Rev., 14(4) April 2002, pp. 37-41. *Cell growth and division, key stages in the cell cycle, and the complex control over different stages of mitosis.*

■ **Rebels without a Cause** New Scientist, 13 July 2002, (Inside Science). *The causes of cancer: the uncontrolled division of cells that results in tumour formation. Breast cancer is a case example.*

■ **Light Microscopy** Biol. Sci. Rev., 13(1) Sept. 2000, pp. 36-38. *An excellent account of the basis and various techniques of light microscopy.*

■ **Transmission Electron Microscopy** Biol. Sci. Rev., 13(2) Nov. 2000, pp. 32-35. *The techniques and applications of TEM. Includes a diagram comparing features of TEM and light microscopy.*

■ **Scanning Electron Microscopy** Biol. Sci. Rev., 20(1) Sept. 2007, pp. 38-41. *An excellent account of the techniques and applications of SEM. Includes details of specimen preparation and recent advancements in the technology.*

Internet

See pages 10-11 for details of how to access **Bio Links** from our web site: www.thebiozone.com From Bio Links, access sites under the topics:

BIOCHEMISTRY: • Basic biochemistry • Chapter 7: Metabolism and Biochemistry • Energy and enzymes • Enzymes • Reactions and enzymes

CELL BIOLOGY: General sites: • Kimball's biology pages > **Microscopy:** • Histology • University of Delaware: Histology > **Cell Structure and Transport:** • Animal cells • Aquaporins • CELLS alive! • Nanoworld • Passive transport (animation) • Various cellular animations ... *and many others* > **Cell Division:** • The cell cycle and mitosis tutorial • Comparison of meiosis and mitosis • Mitosis vs meiosis (animation) • How cancer grows (animation)

Presentation MEDIA to support this topic: **Anatomy & Physiology**

Anatomy & Physiology

Cell Processes

All of the organelles and other structures in the cell have specific functions. The cell can be compared to a factory with an assembly line. Organelles in the cell provide the equivalent of the power supply, assembly line, packaging department, repair and maintenance, transport system, and the control centre.

The sum total of all the processes occurring in a cell is known as **metabolism**. Some of these processes store energy in molecules (**anabolism**) while others release that stored energy (**catabolism**). A summary of the major processes that take place in a cell are illustrated below.

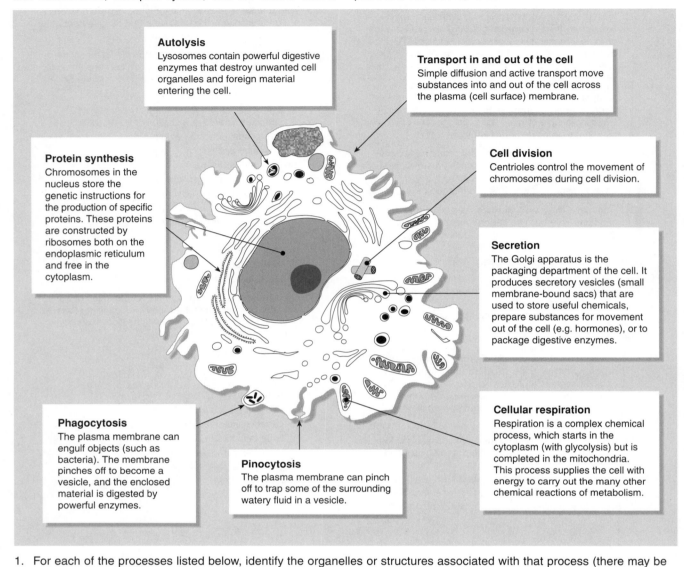

Autolysis
Lysosomes contain powerful digestive enzymes that destroy unwanted cell organelles and foreign material entering the cell.

Transport in and out of the cell
Simple diffusion and active transport move substances into and out of the cell across the plasma (cell surface) membrane.

Protein synthesis
Chromosomes in the nucleus store the genetic instructions for the production of specific proteins. These proteins are constructed by ribosomes both on the endoplasmic reticulum and free in the cytoplasm.

Cell division
Centrioles control the movement of chromosomes during cell division.

Secretion
The Golgi apparatus is the packaging department of the cell. It produces secretory vesicles (small membrane-bound sacs) that are used to store useful chemicals, prepare substances for movement out of the cell (e.g. hormones), or to package digestive enzymes.

Phagocytosis
The plasma membrane can engulf objects (such as bacteria). The membrane pinches off to become a vesicle, and the enclosed material is digested by powerful enzymes.

Pinocytosis
The plasma membrane can pinch off to trap some of the surrounding watery fluid in a vesicle.

Cellular respiration
Respiration is a complex chemical process, which starts in the cytoplasm (with glycolysis) but is completed in the mitochondria. This process supplies the cell with energy to carry out the many other chemical reactions of metabolism.

1. For each of the processes listed below, identify the organelles or structures associated with that process (there may be more than one associated with a process):

   (a) Secretion: _____

   (b) Respiration: _____

   (c) Pinocytosis: _____

   (d) Phagocytosis: _____

   (e) Protein synthesis: _____

   (f) Cell division: _____

   (g) Autolysis: _____

   (h) Transport in/out of cell: _____

2. (a) Explain what is meant by **metabolism** and describe an example of a metabolic process: _____

   _____

   _____

   (b) Identify one catabolic process in the diagram above and explain your choice: _____

   _____

   (c) Identify one anabolic process in the diagram above and explain your choice: _____

   _____

Related activities: Basic Cell Structure

Basic Cell Structure

Cells have a similar basic structure, although they may vary tremendously in size, shape, and function. Certain features are common to almost all cells, including their three main regions: a **nucleus** (usually located near the center of the cell), surrounded by a watery **cytoplasm**, which is itself enclosed by the **plasma membrane**. Animal cells do not have a regular shape, and some (such as the phagocytic white blood cells) are quite mobile. The diagram below illustrates the basic ultrastructure of an **intestinal epithelial cell**. It contains organelles common to most relatively unspecialized human cells. The intestine is lined with these columnar epithelial cells. They are taller than they are wide, with the nucleus close to the base and hairlike projections (**microvilli**) on their free surface. Microvilli increase the surface area of the cell, greatly increasing the capacity for absorption.

Structures and Organelles in an Intestinal Epithelial Cell

Cells and Tissues

Mitochondrion (pl. mitochondria): 1.5 µm X 2–8 µm. Ovoid organelle bounded by a double membrane. They are the cell's energy transformers, and convert chemical energy into ATP.

Transverse section through a mitochondrion

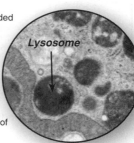

Each epithelial cell has many small projections, called microvilli, discernible in this photograph as a fuzzy brush border (arrowed).

Lysosome: A sac bounded by a single membrane. Lysosomes are pinched off from the Golgi and contain and transport enzymes that break down foreign material. Lysosomes show little internal structure but often contain fragments of degraded material.

Lysosome

Peroxisomes: Self-replicating organelles containing oxidative enzymes, which function to rid the body of toxic substances. They are distinguished from lysosomes by the crystalline core.

Golgi apparatus (above): A series of flattened, disc-shaped sacs, stacked one on top of the other and connected with the ER. The Golgi stores, modifies, and packages proteins. It 'tags' proteins so that they go to their correct destination.

Cytoplasm: A watery solution containing dissolved substances, enzymes, and the cell organelles and structures.

Ribosomes: These small (20 nm) structures manufacture proteins. Ribosomes are made of ribosomal RNA and protein. They may be free in the cytoplasm or associated with the surface of the endoplasmic reticulum.

Rough ER: Endoplasmic reticulum with ribosomes attached to its surface. It is where the proteins destined for transport are synthesized.

Rough endoplasmic reticulum showing ribosomes (dark spots)

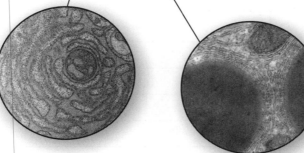

Plasma membrane: 3-10 nm thick phospholipid bilayer with associated proteins and lipids.

Tight junction: impermeable junction binding neighboring cells together.

Nucleus (below): 5 µm diameter. A large organelle containing most of the cell's DNA. Within the nucleus, the **nucleolus** (n) is a dense structure of crystalline protein and nucleic acid involved in ribosome synthesis.

Nuclear pore: A hole in the nuclear membrane. It allows communication between the nucleus and the rest of the cell.

Centrioles: Microtubular structures associated with nuclear division. Under a light microscope, they appear as small, featureless particles, 0.25 µm diameter.

Endoplasmic reticulum (ER): Comprises a network of tubules and flattened sacs. ER is continuous with the plasma membrane and the nuclear membrane. **Smooth ER**, as shown here, is a site for lipid and carbohydrate metabolism, including hormone synthesis.

Related activities: Cell Structures & Organelles, Cell Processes
Web links: Eukaryotic Cells Interactive Animation

A 2

18

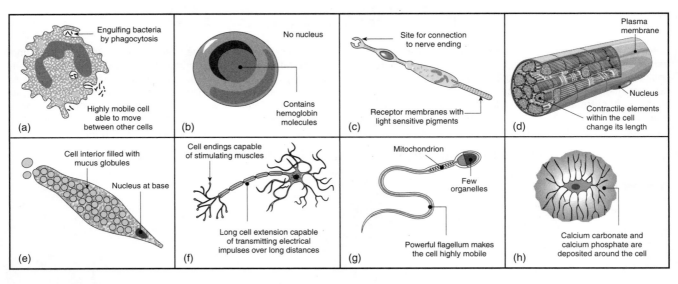

1. Explain what you understand by the term generalized cell: _____

_____

_____

2. Each of the cells (a) to (h) exhibits **specialized features** specific to its **functional role** in the body. For each, describe one specialized feature of the cell and its purpose:

(a) Phagocytic white blood cell: _____

_____

(b) Red blood cell (erythrocyte) _____

_____

(c) Rod cell of the retina: _____

_____

(d) Skeletal muscle fiber (part of): _____

_____

(e) Intestinal globlet cell: _____

_____

(f) Motor neuron: _____

_____

(g) Spermatozoon: _____

_____

(h) Osteocyte: _____

_____

3. Discuss how the shape and size of a specialized cell, as well as the number and types of organelles it has, is related to its functional role. Use examples to illustrate your answer:

_____

_____

_____

_____

_____

The Structure of Membranes

All cells have a **plasma membrane** forming the outer limit of the cell. Cellular membranes are also found inside eukaryotic cells as part of membranous **organelles**, such as the endoplasmic reticulum. Present day knowledge of membrane structure has been built up as a result of many observations and experiments. The now-accepted model of membrane structure is the **fluid-mosaic model** (below). The plasma membrane is more than just a passive envelope; it is a dynamic structure actively involved in cellular activities. Specializations of the plasma membrane, including microvilli and membrane junctions (e.g. desmosomes and tight junctions), are particularly numerous in epithelial cells, which line hollow organs such as the small intestine.

The Fluid Mosaic Model

The currently accepted model for membrane structure, the **fluid mosaic model**, satisfies the observed properties of cellular membranes. In this model there is a double layer of phospholipid molecules (fats) which are arranged with their hydrophobic 'tails' facing inwards. This self-orientating property allows cellular membranes to reseal themselves when disrupted. The double layer of lipids is quite fluid, within which the proteins move quite freely. The mobile proteins are thought to have a number of functions, including roles in active transport.

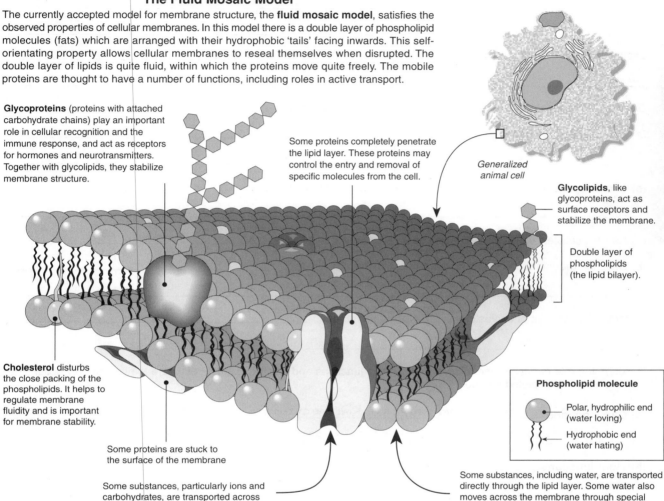

Glycoproteins (proteins with attached carbohydrate chains) play an important role in cellular recognition and the immune response, and act as receptors for hormones and neurotransmitters. Together with glycolipids, they stabilize membrane structure.

Some proteins completely penetrate the lipid layer. These proteins may control the entry and removal of specific molecules from the cell.

Generalized animal cell

Glycolipids, like glycoproteins, act as surface receptors and stabilize the membrane.

Double layer of phospholipids (the lipid bilayer).

Cholesterol disturbs the close packing of the phospholipids. It helps to regulate membrane fluidity and is important for membrane stability.

Some proteins are stuck to the surface of the membrane

Some substances, particularly ions and carbohydrates, are transported across the membrane via the channel proteins.

Phospholipid molecule

Polar, hydrophilic end (water loving)

Hydrophobic end (water hating)

Some substances, including water, are transported directly through the lipid layer. Some water also moves across the membrane through special protein channels called **aquaporins**.

Membrane Specializations

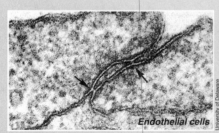

Endothelial cells

Tight junctions bind the membranes of neighboring cells together to form a virtually impermeable barrier to fluid. Tight junctions prevent molecules passing through the spaces between cells.

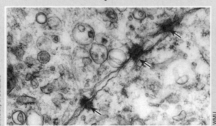

Desmosomes (arrowed) are anchoring junctions that allow cell-to-cell adhesion. Desmosomes help to resist shearing forces in tissues subjected to mechanical stress (such as skin cells).

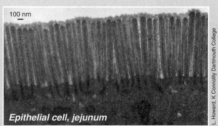

100 nm

Epithelial cell, jejunum

Microvilli are microscopic protrusions of the plasma membrane that increase the surface area of cells. Microvilli are involved in a wide variety of functions, including absorption (e.g. in the intestine).

1. (a) Explain how phospholipids organize themselves into a bilayer in an aqueous environment: _____

_____

_____

_____

Related activities: The Role of Membranes in Cells
Web links: Membrane Structure Tutorial, Animal Cell Junctions

RA 2

(b) Explain how the fluid mosaic model accounts for the observed properties of cellular membranes:

_____

_____

_____

_____

_____

_____

2. Explain how the membrane surface area is increased within cells and organelles: _____

_____

3. Discuss the importance of each of the following to cellular function:

   (a) High membrane surface area: _____

   _____

   (b) Channel proteins and carrier proteins in the plasma membrane: _____

   _____

4. (a) Name a cellular organelle that possesses a membrane: _____

   (b) Describe the membrane's purpose in this organelle: _____

   _____

5. Describe the purpose of cholesterol in the plasma membrane: _____

   _____

6. Describe the role of each of the following membrane junctions and give an example of where they commonly occur. The first example is completed for you:

   (a) **Gap junctions**: Communicating junctions linking the cytoplasm of neighboring cells. They allow rapid passage of signals between cells, e.g. electrical messages in cardiac muscle cells.

   (b) **Tight junctions**: _____

   _____

   (c) **Desmosomes**: _____

   _____

7. Explain why tight junctions are especially abundant in epithlelial cells, e.g. in the skin and intestine: _____

   _____

   _____

8. Use the symbol for a phospholipid molecule (below) to draw a **simple labelled diagram** to show the structure of a plasma membrane (include features such as lipid bilayer and various kinds of proteins):

The Role of Membranes in Cells

Many of the important structures and organelles in cells are composed of, or are enclosed by, membranes. These include the endoplasmic reticulum, mitochondria, nucleus, Golgi apparatus, lysosomes, peroxisomes, and the plasma membrane itself. All membranes within eukaryotic cells share the same basic structure as the plasma membrane that encloses the entire cell.

They perform a number of critical functions in the cell: serving to compartmentalize regions of different function within the cell, controlling the entry and exit of substances, and fulfilling a role in recognition and communication between cells. Some of these roles are described below and electron micrographs of the organelles involved are on the following page.

Cells and Tissues

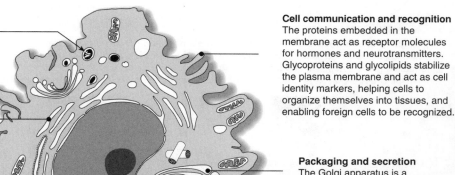

Isolation of enzymes Membrane-bound lysosomes contain enzymes for the destruction of wastes and foreign material. Peroxisomes are the site for destruction of the toxic and reactive molecule, hydrogen peroxide (formed as a result of some cellular reactions).

Role in lipid synthesis
The smooth ER is the site of lipid and steroid synthesis.

Containment of DNA
The nucleus is surrounded by a nuclear envelope of two membranes, forming a separate compartment for the cell's genetic material.

Role in protein and membrane synthesis
Some protein synthesis occurs on free ribosomes, but much occurs on membrane-bound ribosomes on the rough endoplasmic reticulum. Here, the protein is synthesized directly into the space within the ER membranes. The rough ER is also involved in membrane synthesis, growing in place by adding proteins and phospholipids.

Entry and export of substances The plasma membrane may take up fluid or solid material and form membrane-bound vesicles (or larger vacuoles) within the cell. Membrane-bound transport vesicles move substances to the inner surface of the cell where they can be exported from the cell by exocytosis.

Cell communication and recognition
The proteins embedded in the membrane act as receptor molecules for hormones and neurotransmitters. Glycoproteins and glycolipids stabilize the plasma membrane and act as cell identity markers, helping cells to organize themselves into tissues, and enabling foreign cells to be recognized.

Packaging and secretion
The Golgi apparatus is a specialized membrane-bound organelle which produces lysosomes and compartmentalizes the modification, packaging and secretion of substances such as proteins and hormones.

Transport processes
Channel and carrier proteins are involved in selective transport across the plasma membrane. Cholesterol in the membrane can help to prevent ions or polar molecules from passing through the membrane (acting as a plug).

Energy transfer The reactions of cellular respiration (and photosynthesis in plants) take place in the membrane-bound energy transfer systems occurring in mitochondria and chloroplasts respectively. See the example explained below.

Compartmentation within Membranes

Membranes play an important role in separating regions within the cell (and within organelles) where particular reactions occur. Specific enzymes are therefore often located in particular organelles. Reaction rates are controlled by controlling the rate at which substrates enter the organelle. This regulates the availability of the raw materials required for the metabolic reactions.

Example: *The enzymes involved in cellular respiration are arranged in different parts of the mitochondria. The various reactions are localized and separated by membrane systems.*

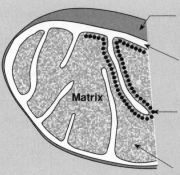

Amine oxidases and other enzymes on the outer membrane surface

Adenylate kinase and other phosphorylases between the membranes

Respiratory assembly enzymes embedded in the membrane (ATPase)

Many soluble enzymes of the Krebs cycle floating in the matrix, as well as enzymes for fatty acid degradation.

Matrix

Cross-section of a mitochondrion

1. Discuss the various functional roles of membranes in cells: _____

_____

_____

_____

_____

Related activities: The Structure of Membranes, Cell Structures & Organelles
Web links: Cell Membranes

A 2

Functional Roles of Membranes in Cells

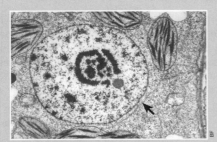

The **nuclear membrane**, which surrounds the nucleus, regulates the passage of genetic information to the cytoplasm and may also protect the DNA from damage.

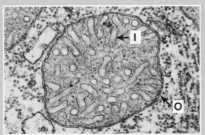

Mitochondria have an outer membrane (**O**) which controls the entry and exit of materials involved in aerobic respiration. Inner membranes (**I**) provide attachment sites for enzyme activity.

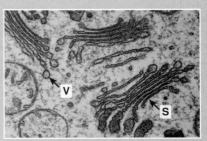

The **Golgi apparatus** comprises stacks of membrane-bound sacs (**S**). It is involved in packaging materials for transport or export from the cell as secretory vesicles (**V**).

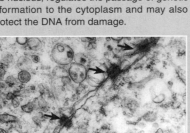

The **plasma membrane** surrounds the cell. In this photo, intercellular junctions called **desmosomes**, which connect neighbouring cells, are indicated with arrows.

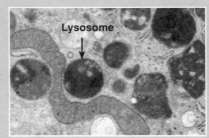

Lysosomes are membrane-bound organelles containing enzymes capable of digesting worn-out cellular structures and foreign material. They are abundant in phagocytes.

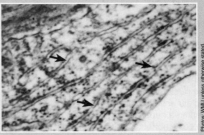

This EM shows stacks of rough endoplasmic reticulum (arrows). The membranes are studded with ribosomes, which synthesize proteins into the intermembrane space.

2. Match each of the following organelles with the correct description of its functional role in the cell:

   peroxisome, rough endoplasmic reticulum, lysosome, smooth endoplasmic reticulum, mitochondrion, Golgi apparatus

   (a) Active in synthesis, sorting, and secretion of cell products: _____

   (b) Digestive organelle where macromolecules are hydrolyzed: _____

   (c) Organelle where most cellular respiration occurs and most ATP is generated: _____

   (d) Active in membrane synthesis and synthesis of secretory proteins: _____

   (e) Active in lipid and hormone synthesis and secretion: _____

   (f) Small organelle responsible for the destruction of toxic substances: _____

3. Explain the importance of membrane systems and organelles in providing compartments within the cell:

   _____

   _____

4. (a) Explain why non-polar (lipid-soluble) molecules diffuse more rapidly through membranes than polar molecules:

   _____

   _____

   (b) Explain the implications of this to the transport of substances into the cell through the plasma membrane:

   _____

   _____

5. Identify three substances that need to be transported **into** all kinds of human cells, in order for them to survive:

   (a) _____ (b) _____ (c) _____

6. Identify two substances that need to be transported **out** of all kinds of human cells, in order for them to survive:

   (a) _____ (b) _____

Cell Structures and Organelles

This activity requires you to summarize information about the components of a typical eukaryotic cell. Complete the table using the list provided and by referring to other pages in this chapter. Fill in the final column with either 'YES' or 'NO'. The first has been completed for you as a guide and the log scale of measurements (next page) illustrates the relative sizes of some cells and cell structures. **List of components**: nucleus, ribosome, centrioles, mitochondrion, lysosome and vacuole (given), endoplasmic reticulum, Golgi apparatus, plasma membrane (given), cell cytoskeleton, flagella or cilia (given), cellular junctions (given).

Cells and Tissues

Cell Component	Details	Visible under light microscope
(a) 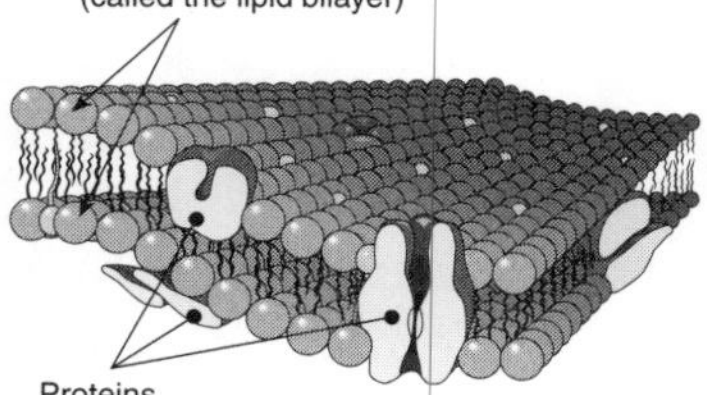 Double layer of phospholipids (called the lipid bilayer), Proteins	Name: Plasma (cell surface) membrane Location: Surrounding the cell Function: Gives the cell shape and protection. It also regulates the movement of substances into and out of the cell.	YES (but not at the level of detail shown in the diagram)
(b) 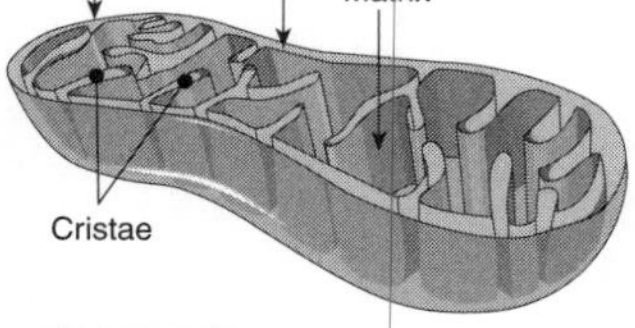 Outer membrane, Inner membrane, Matrix, Cristae	Name: Location: Function:	
(c) 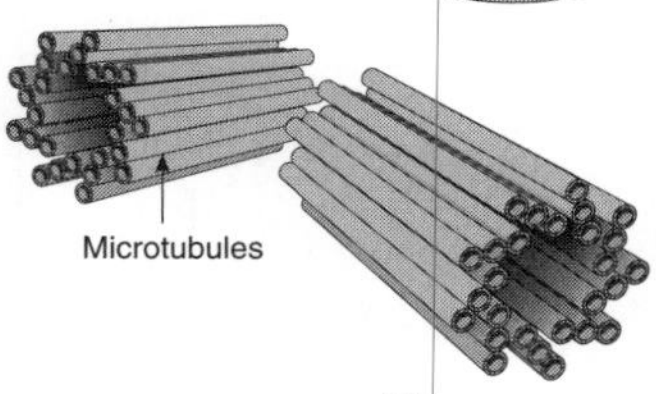 Microtubules	Name: Location: Function:	
(d) 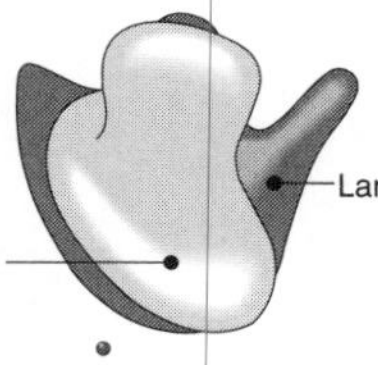 Large subunit, Small subunit	Name: Location: Function:	
(e) 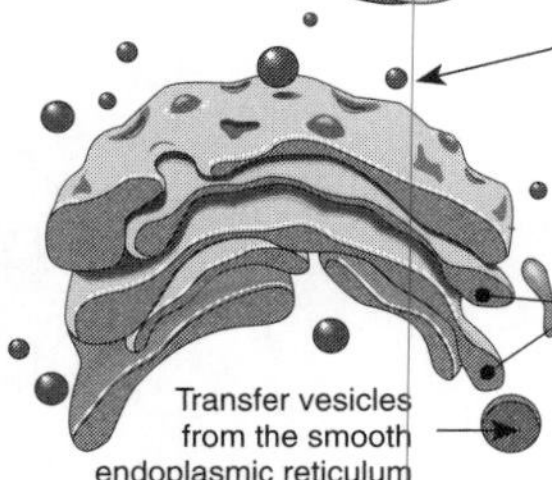 Secretory vesicles budding off, Cisternae, Transfer vesicles from the smooth endoplasmic reticulum	Name: Location: Function:	
(f) 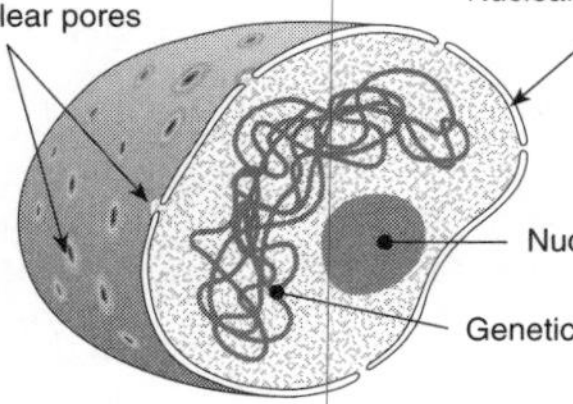 Nuclear pores, Nuclear membrane, Nucleolus, Genetic material	Name: Location: Function:	
(g) 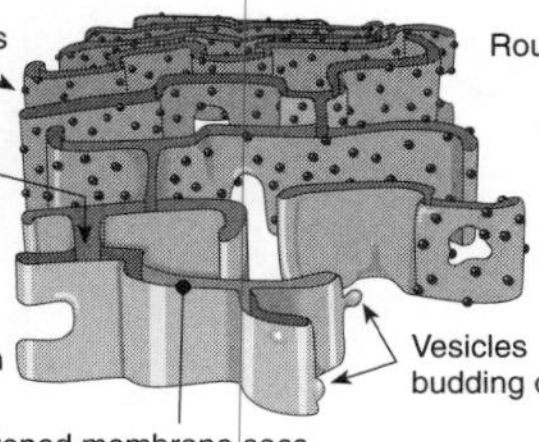 Ribosomes, Rough, Transport pathway, Smooth, Vesicles budding off, Flattened membrane sacs	Name: Location: Function:	

Related activities: Basic Cell Structure
Web links: Eukaryotic Cells Interactive Animation

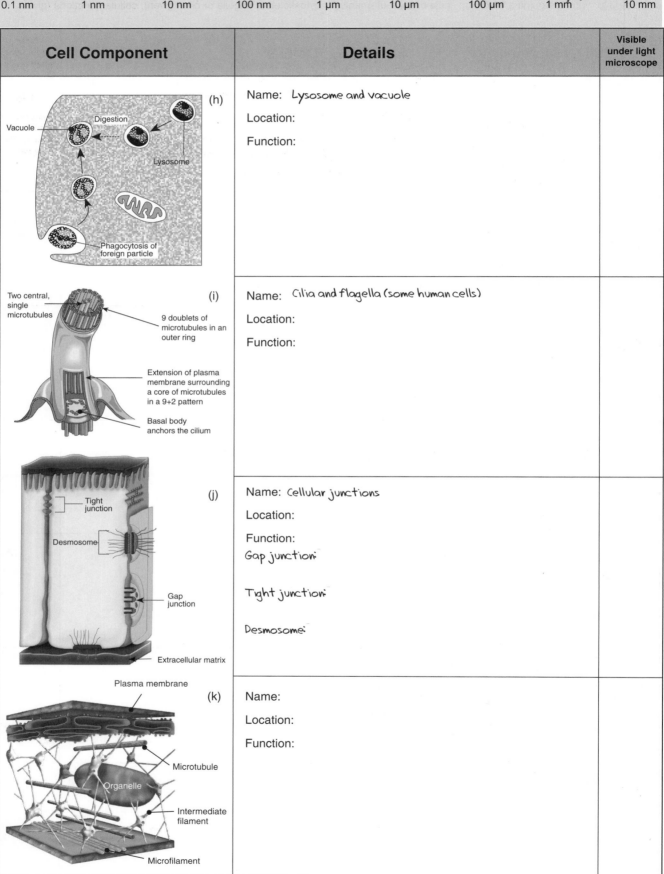

Cell Component	Details	Visible under light microscope
(h)	Name: Lysosome and vacuole Location: Function:	
(i)	Name: Cilia and flagella (some human cells) Location: Function:	
(j)	Name: Cellular junctions Location: Function: Gap junction: Tight junction: Desmosome:	
(k)	Name: Location: Function:	

Passive Transport Processes

The molecules that make up substances are constantly moving in a random way. This random motion causes molecules to disperse from areas of high to low concentration (i.e. down a **concentration gradient**). This process is called **diffusion**. All types of diffusion, including **osmosis**, are **passive transport** processes, in that they require no expenditure of energy. Diffusion occurs freely across membranes, as long as the membrane is permeable to that molecule (**partially permeable membranes** allow the passage of some molecules but not others). Simple diffusion may occur directly across the lipid bilayer, whereas facilitated diffusion utilizes transmembrane proteins to assist the diffusion of specific molecules. **Filtration** is also a passive process, in which fluid pressure is used to push substances through a membrane or capillary wall. Filtration obeys the same rules of movement as diffusion, but the gradient involved is a pressure gradient rather than a concentration gradient.

Diffusion of Molecules Along Concentration Gradients

Diffusion is the movement of molecules (and ions) from regions of high to low concentration, with the end result being that the molecules become evenly distributed. In biological systems, diffusion often occurs across **partially permeable membranes**. Each type of diffusing molecule (gas, solvent, solute) moves along its own **concentration gradient**. Various factors (right) determine the rate at which this occurs.

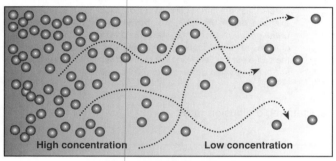

Factors affecting rates of diffusion for any given diffusing molecule

- **Concentration gradient**: The greater the concentration gradient, the higher the diffusion rate.
- **Surface area**: The larger the area across which diffusion occurs, the greater the rate of diffusion.
- **Barriers to diffusion**: Thicker barriers slow diffusion rate. Pores in a barrier enhance diffusion.
- **Temperature**: Diffusion rates are higher at higher temperatures (within the body this is a negilible factor).

High concentration   **Low concentration**

Concentration gradient

If molecules are free to move, they move from high to low concentration until they are evenly dispersed.

These factors are expressed in **Fick's law**, which governs the rate of diffusion of substances across membranes. It is described by:

$$\frac{\text{Surface area of membrane} \quad \times \quad \text{Difference in concentration across the membrane}}{\text{Length of the diffusion path (thickness of the membrane)}}$$

Diffusion Through Membranes

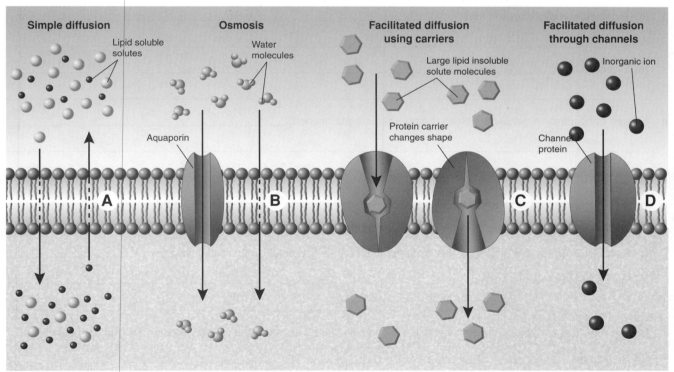

Simple diffusion — Lipid soluble solutes — A

Osmosis — Water molecules — Aquaporin — B

Facilitated diffusion using carriers — Large lipid insoluble solute molecules — Protein carrier changes shape — C

Facilitated diffusion through channels — Inorganic ion — Channel protein — D

A: Some molecules (e.g. gases and lipid soluble molecules) diffuse directly across the plasma membrane. Two-way diffusion is common in biological systems, e.g. at the alveolar surface of the lung, CO_2 diffuses out and oxygen diffuses into the blood.

B: Osmosis describes the diffusion of water across a partially permeable membrane (in this case, the plasma membrane). Some water can diffuse directly through the lipid bilayer, but movement is also aided by specific protein channels called **aquaporins**.

C: In **carrier-mediated facilitated diffusion**, a lipid-insoluble molecule is aided across the membrane by a transmembrane carrier protein specific to the molecule being transported (e.g. glucose transport into red blood cells).

D: Small polar molecules and ions diffuse rapidly across the membrane by **channel-mediated facilitated diffusion**. Protein channels create hydrophilic pores that allow some solutes, usually inorganic ions, to pass through.

Related activities: Absorption and Transport
Web links: Cellular Transport

Cellular Tonicity and Osmotic Pressure

In the study of physiology, it is important to understand the consequences of changes to the solute concentrations of cellular environments. The tendency of a solution to 'pull' water into it is called the **osmotic pressure** and it is directly related to the concentration of solutes in the solution. The higher the solute concentration, the greater the osmotic pressure and the greater the tendency of water to move into the solution (see previous page on factors affecting diffusion rates). In biology, **relative tonicity** (isotonic, hypotonic, or hypertonic) is used describe the difference in osmotic pressure between solutions. Only solutes that cannot cross the plasma membrane affect tonicity.

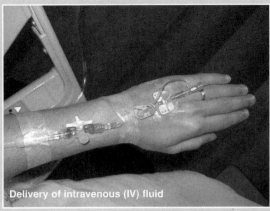

Delivery of intravenous (IV) fluid

Michael Berry (Wikipedia)

Tonicity of solution relative to the cytosol	Extracellular environment (solution)	Intracellular environment (cytosol)	Consequence to a cell in the solution
Isotonic	Equal osmotic environment		Normal shape and form
Hypotonic	Lower solute concentration	Higher solute concentration	Water enters cell causing the cell to burst (cell **lysis**)
Hypertonic	Higher solute concentration	Lower solute concentration	Water leaves cell causing shrinkage (**crenation**)

The relative tonicity of cells can be used to predict the consequences of changes in solute concentration either side of a partially permeable membrane (i.e. the plasma membrane surrounding each of the body's cells). Such predictions are important in a practical sense as, for example, during the delivery of **intravenous fluid** to patients (*intravenous means within vein*). To prevent life-threatening changes to cell volumes, intravenous (IV) fluids must present the same osmotic environment as the blood cells they will be surrounding when delivered (e.g. 0.9% saline solution).

1. Describe two properties of an exchange surface that would facilitate rapid diffusion rates:

   (a) _____   (b) _____

2. Describe two biologically important features of diffusion:

   (a) _____

   (b) _____

3. Describe how facilitated diffusion is achieved for:

   (a) Small polar molecules and ions: _____

   _____

   (b) Glucose: _____

   _____

4. Explain concentration gradients across membranes are maintained: _____

   _____

5. Explain the role of aquaporins in the rapid movement of water through some cells: _____

   _____

6. Fluid replacements are usually provided for heavily perspiring athletes after endurance events.

   (a) Identify the preferable tonicity of these replacement drinks (isotonic, hypertonic, or hypotonic): _____

   (b) Give a reason for your answer: _____

   _____

7. Describe what would happen to a patient's red blood cells if they were treated with an intravenous drip containing:

   (a) Pure water: _____

   (b) A hypertonic solution: _____

   (c) A hypotonic solution: _____

8. The malarial parasite lives in human blood. Relative to the tonicity of the blood, the parasite's cell contents would be hypotonic / isotonic / hypertonic (circle the correct answer).

Ion Pumps

Diffusion alone cannot supply the cell's entire requirements for molecules (and ions) in all situations; sometimes molecules must be moved in a certain direction according to the specific needs of the cell. The movement of molecules against their concentration gradient requires **energy expenditure** and is achieved through **active transport** mechanisms (ion pumps and cytosis). **Ion pumps** are specific transmembrane carrier proteins that harness the energy of ATP to move molecules from a low to a high concentration. When ATP transfers a phosphate group to the carrier protein, the protein changes its shape, moving the bound

molecule across the membrane. Three types of membrane pump are described below. The sodium-potassium pump (center) is almost universal in animal cells. The concentration gradient created by ion pumps such as this and the proton pump (left) is frequently coupled to the transport of other larger molecules, as shown below right. Note that glucose enters most cells by facilitated diffusion (i.e. passively) but moves by active transport into the intestinal epithelial cells. In this way, uptake of glucose from ingested food continues despite what might be highly fluctuating glucose levels in the intestines.

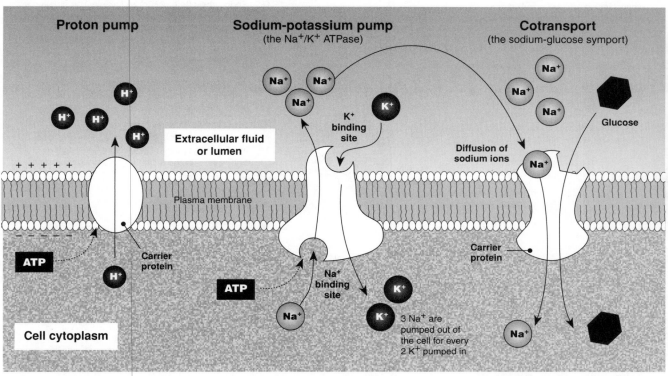

Cells and Tissues

Proton pumps
ATP driven proton pumps use energy to remove hydrogen ions (H$^+$) from inside the cell to the outside. This creates a large difference in the proton concentration either side of the membrane, with the inside of the plasma membrane being negatively charged. This potential difference can be coupled to the transport of other molecules.

Sodium-potassium pump
The sodium-potassium pump is a specific protein in the membrane that uses energy in the form of ATP to exchange sodium ions (Na$^+$) for potassium ions (K$^+$) across the membrane. The unequal balance of Na$^+$ and K$^+$ across the membrane creates large concentration gradients that can be used to drive transport of other substances (e.g. cotransport of glucose).

Cotransport (coupled transport)
A gradient in sodium ions drives the active transport of **glucose** in intestinal epithelial cells. The specific transport protein couples the return of Na$^+$ down its concentration gradient to the transport of glucose into the intestinal epithelial cell. A low intracellular concentration of Na$^+$ (and therefore the concentration gradient) is maintained by a sodium-potassium pump.

1. Explain why the ATP is required for membrane pump systems to operate: _____

_____

2. (a) Explain what is meant by cotransport: _____

_____

   (b) Explain how cotransport is used to move glucose into the intestinal epithelial cells: _____

_____

_____

   (c) Explain what happens to the glucose that is transported into the intestinal epithelial cells: _____

_____

3. Describe two consequences of the extracellular accumulation of sodium ions: _____

_____

_____

Related activities: Transmission of Nerve Impulses, Absorption & Transport
Web links: Cellular Transport, Symport

A 2

Exocytosis and Endocytosis

Most cells carry out **cytosis**: a form of **active transport** involving infolding or outfolding of the plasma membrane. Cells are able to do this because of the flexibility of the plasma membrane. Cytosis results in the bulk transport of materials into or out of the cell and is achieved through the localized activity of microfilaments and microtubules in the cell cytoskeleton. Engulfment of material is termed **endocytosis.** Endocytosis typically occurs in certain white blood cells of the human defense system (e.g. neutrophils, macrophages). **Exocytosis** is the reverse of endocytosis and involves the release of material from vesicles or vacuoles that have fused with the plasma membrane. Exocytosis is typical of cells that export material (secretory cells).

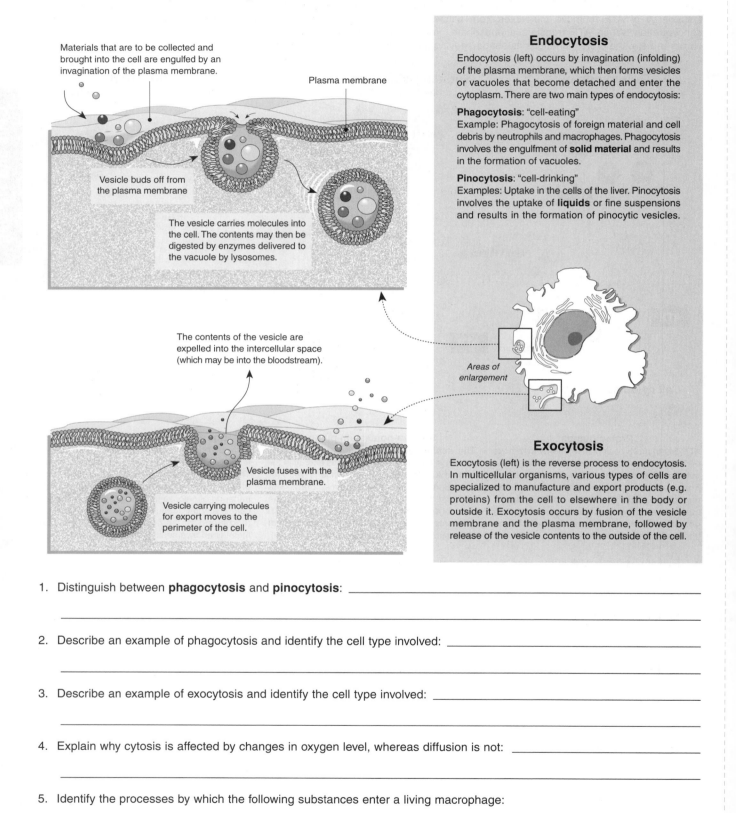

Materials that are to be collected and brought into the cell are engulfed by an invagination of the plasma membrane.

Plasma membrane

Vesicle buds off from the plasma membrane

The vesicle carries molecules into the cell. The contents may then be digested by enzymes delivered to the vacuole by lysosomes.

The contents of the vesicle are expelled into the intercellular space (which may be into the bloodstream).

Vesicle fuses with the plasma membrane.

Vesicle carrying molecules for export moves to the perimeter of the cell.

Areas of enlargement

Endocytosis

Endocytosis (left) occurs by invagination (infolding) of the plasma membrane, which then forms vesicles or vacuoles that become detached and enter the cytoplasm. There are two main types of endocytosis:

Phagocytosis: "cell-eating"
Example: Phagocytosis of foreign material and cell debris by neutrophils and macrophages. Phagocytosis involves the engulfment of **solid material** and results in the formation of vacuoles.

Pinocytosis: "cell-drinking"
Examples: Uptake in the cells of the liver. Pinocytosis involves the uptake of **liquids** or fine suspensions and results in the formation of pinocytic vesicles.

Exocytosis

Exocytosis (left) is the reverse process to endocytosis. In multicellular organisms, various types of cells are specialized to manufacture and export products (e.g. proteins) from the cell to elsewhere in the body or outside it. Exocytosis occurs by fusion of the vesicle membrane and the plasma membrane, followed by release of the vesicle contents to the outside of the cell.

1. Distinguish between **phagocytosis** and **pinocytosis**: _____

_____

2. Describe an example of phagocytosis and identify the cell type involved: _____

_____

3. Describe an example of exocytosis and identify the cell type involved: _____

_____

4. Explain why cytosis is affected by changes in oxygen level, whereas diffusion is not: _____

_____

5. Identify the processes by which the following substances enter a living macrophage:

   (a) Oxygen: _____   (c) Water: _____

   (b) Cellular debris: _____   (d) Glucose: _____

RA 2   **Related activities**: The Role of Membranes in Cells, The Action of Phagocytes,
Passive Transport Processes   **Web links**: Cellular Transport

Mitosis and the Cell Cycle

Mitosis is part of the **cell cycle** in which an existing cell (the parent cell) divides into two (the **daughter cells**). In humans and other multicellular organisms, mitosis has a role in growth and development, and in repairing damaged cells and tissues. Although mitosis is part of a continuous cell cycle, it is divided into stages (below). The example below illustrates the cell cycle in a generic animal cell. Unlike meiosis, which is involved in gamete production, mitosis does not reduce the chromosome number in the daughter cells, which are identical to the parent cell. The two types of cell division are compared on the following page. Note that in animal cells, **cytokinesis** (with formation of a cleavage furrow) is usually well underway by the end of telophase.

Cells and Tissues

The Cell Cycle and Stages of Mitosis

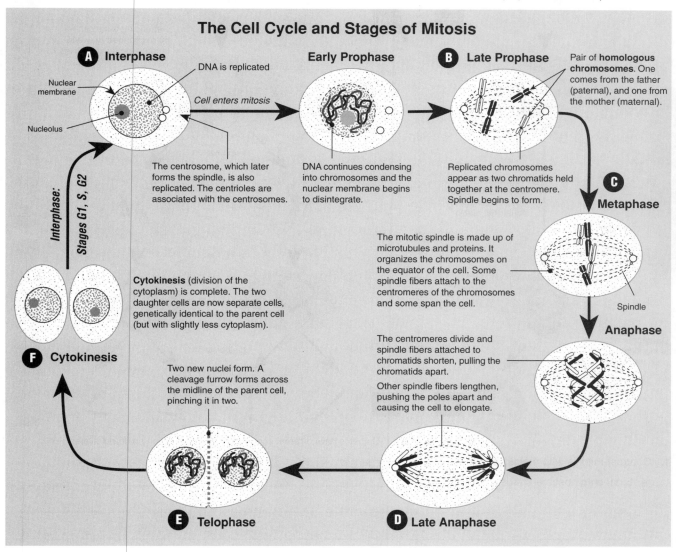

A Interphase

DNA is replicated

Nuclear membrane

Nucleolus

Cell enters mitosis

Early Prophase

B Late Prophase

Pair of **homologous chromosomes**. One comes from the father (paternal), and one from the mother (maternal).

The centrosome, which later forms the spindle, is also replicated. The centrioles are associated with the centrosomes.

DNA continues condensing into chromosomes and the nuclear membrane begins to disintegrate.

Replicated chromosomes appear as two chromatids held together at the centromere. Spindle begins to form.

C Metaphase

The mitotic spindle is made up of microtubules and proteins. It organizes the chromosomes on the equator of the cell. Some spindle fibers attach to the centromeres of the chromosomes and some span the cell.

Spindle

Interphase: *Stages G1, S, G2*

F Cytokinesis

Cytokinesis (division of the cytoplasm) is complete. The two daughter cells are now separate cells, genetically identical to the parent cell (but with slightly less cytoplasm).

Two new nuclei form. A cleavage furrow forms across the midline of the parent cell, pinching it in two.

Anaphase

The centromeres divide and spindle fibers attached to chromatids shorten, pulling the chromatids apart.

Other spindle fibers lengthen, pushing the poles apart and causing the cell to elongate.

E Telophase

D Late Anaphase

The Cell Cycle Overview

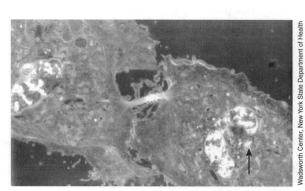

S Phase
Chromosome replication (DNA synthesis)

Second Gap Phase
The chromosomes begin condensing.

G2

S

The Cell Cycle

M

Mitosis
Nuclear division

C

G1

Cytokinesis
Division of the cytoplasm and separation of the two cells. Cytokinesis is distinct from nuclear division.

First Gap Phase
Cell growth and development

Animal cell cytokinesis (above) begins shortly after the sister chromatids have separated in anaphase of mitosis. A contractile ring of microtubular elements assembles in the middle of the cell, next to the plasma membrane, constricting it to form a **cleavage furrow**. In an energy-using process, the cleavage furrow moves inwards, forming a region of abscission where the two cells will separate. In the photograph above, an arrow points to a centrosome, which is still visible near the nucleus.

Wadsworth Center, New York State Department of Health

Related activities: Apoptosis: Programed Cell Death
Web links: Mitosis in an Animal Cell

A 2

Mitosis
(growth and repair)

Meiosis
(gamete formation)

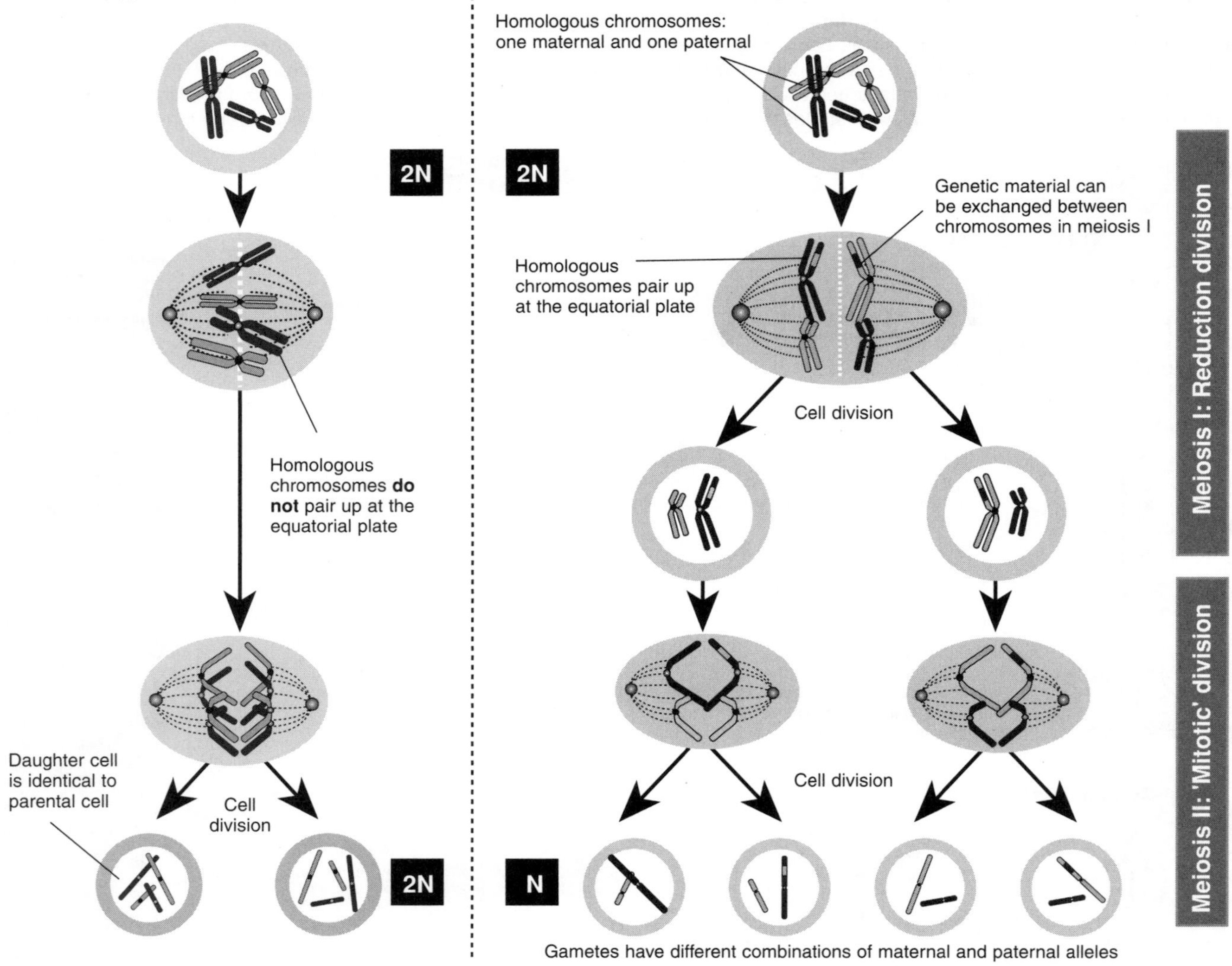

Homologous chromosomes: one maternal and one paternal

2N

2N

Genetic material can be exchanged between chromosomes in meiosis I

Homologous chromosomes pair up at the equatorial plate

Homologous chromosomes **do not** pair up at the equatorial plate

Cell division

Meiosis I: Reduction division

Daughter cell is identical to parental cell

Cell division

2N

Cell division

Meiosis II: 'Mitotic' division

N

Cell division

Gametes have different combinations of maternal and paternal alleles

1. Contrast mitosis with meiosis in terms of:

(a) Final chromosome status: _____

_____

_____

(b) Biological role in humans: _____

_____

_____

2. State two important changes that chromosomes must undergo before cell division can take place: _____

_____

3. Summarize stages of the cell cycle by describing what is happening at points **A-F** in the diagram on the previous page:

A. _____

B. _____

C. _____

D. _____

E. _____

F. _____

Cancer: Cells out of Control

Normal cells do not live forever. Under certain circumstances, cells are programed to die, particularly during development. Cells that become damaged beyond repair will normally undergo this programed cell death (called **apoptosis** or **cell suicide**). Cancer cells evade this control and become immortal, continuing to divide regardless of any damage incurred. **Carcinogens** are agents capable of causing cancer. Roughly 90% of carcinogens are also mutagens, i.e. they damage DNA. Chronic exposure to carcinogens accelerates the rate at which dividing cells make errors. Susceptibility to cancer is also influenced by genetic make-up. Any one or a number of cancer-causing factors (including defective genes) may interact to induce cancer.

Cancer: Cells out of Control

Cancerous transformation results from changes in the genes controlling normal cell growth and division. The resulting cells become immortal and no longer carry out their functional role. Two types of gene are normally involved in controlling the cell cycle: **proto-oncogenes**, which start the cell division process and are essential for normal cell development, and **tumor-suppressor** genes, which switch off cell division. In their normal form, both kinds of genes work as a team, enabling the body to perform vital tasks such as repairing defective cells and replacing dead ones. However mutations in these genes can disrupt these checks and balances. Proto-oncogenes, through mutation, can give rise to **oncogenes**; genes that lead to uncontrollable cell division. Mutations to tumor-suppressor genes initiate most human cancers. The best studied tumor-suppressor gene is **p53**, which encodes a protein that halts the cell cycle so that DNA can be repaired before division.

The panel, right, shows the mutagenic action of some selected carcinogens on four of five codons of the **p53 gene**.

Features of Cancer Cells

The diagram right shows a single **lung cell** that has become cancerous. It no longer carries out the role of a lung cell, and instead takes on a parasitic lifestyle, taking from the body what it needs in the way of nutrients and contributing nothing in return. The rate of cell division is greater than in normal cells in the same tissue because there is no *resting phase* between divisions.

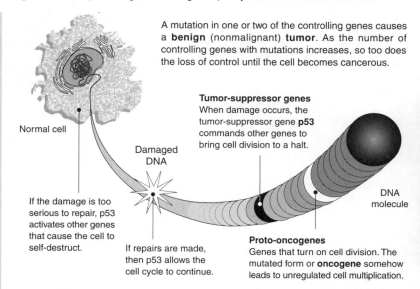

A mutation in one or two of the controlling genes causes a **benign** (nonmalignant) **tumor**. As the number of controlling genes with mutations increases, so too does the loss of control until the cell becomes cancerous.

Normal cell

Damaged DNA

Tumor-suppressor genes
When damage occurs, the tumor-suppressor gene **p53** commands other genes to bring cell division to a halt.

DNA molecule

If the damage is too serious to repair, p53 activates other genes that cause the cell to self-destruct.

If repairs are made, then p53 allows the cell cycle to continue.

Proto-oncogenes
Genes that turn on cell division. The mutated form or **oncogene** somehow leads to unregulated cell multiplication.

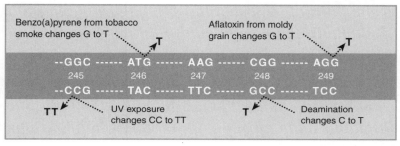

Benzo(a)pyrene from tobacco smoke changes G to T

Aflatoxin from moldy grain changes G to T

```
--GGC ------ ATG ------ AAG ------ CGG ------ AGG
   245          246          247          248          249
--CCG ------ TAC ------ TTC ------ GCC ------ TCC
```

UV exposure changes CC to TT

Deamination changes C to T

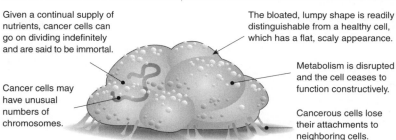

Given a continual supply of nutrients, cancer cells can go on dividing indefinitely and are said to be immortal.

The bloated, lumpy shape is readily distinguishable from a healthy cell, which has a flat, scaly appearance.

Cancer cells may have unusual numbers of chromosomes.

Metabolism is disrupted and the cell ceases to function constructively.

Cancerous cells lose their attachments to neighboring cells.

1. Explain how cancerous cells differ from normal cells: _____

_____

2. Explain how the cell cycle is normally controlled, including reference to the role of **tumor-suppressor genes**:

_____

_____

_____

_____

3. With reference to the role of **oncogenes**, explain how the normal controls over the cell cycle can be lost:

_____

_____

_____

Related activities: : Apoptosis: Programed Cell Death, Breast Cancer

Levels of Organization

Organization and the emergence of novel properties in complex systems are two of the defining features of living organisms. Organisms are organized according to a hierarchy of structural levels (below), each level building on the one below it. At each level, novel properties emerge that were not present at the simpler level. Hierarchical organization allows specialized cells to group together into tissues and organs to perform a particular function. This improves efficiency of function in the organism.

In the spaces provided below, assign each of the examples listed to one of the levels of organization as indicated.

1. Examples: *blood, bone, brain, cardiac muscle, cartilage, epinephrine (adrenaline), collagen, DNA, heart, leukocyte, lysosome, pancreas, mast cell, nervous system, phospholipid, reproductive system, ribosomes, neuron, Schwann cell, spleen, squamous epithelium, astrocyte, respiratory system, muscular system, peroxisome, ATP, collagen, testis, liver.*

(a) Chemical level: _____

_____

_____

_____

(b) Organelles: _____

_____

_____

_____

(c) Cells: _____

_____

_____

_____

(d) Tissues: _____

_____

_____

_____

(e) Organs: _____

_____

_____

_____

(f) Organ system: _____

_____

_____

_____

2. State the name given to the microscopic study of cells and tissues:

_____

CHEMICAL LEVEL
Atoms and molecules form the most basic level of organization. This level includes all the chemicals essential for maintaining life, e.g. water, ions, fats, carbohydrates, amino acids, proteins, and nucleic acids.

ORGANELLE LEVEL
Many diverse molecules may associate together to form complex, highly specialized structures within cells called **cellular organelles**, e.g. mitochondria, Golgi apparatus, endoplasmic reticulum, chloroplasts.

CELLULAR LEVEL
Cells are the basic structural and functional units of an organism. Each cell type has a different structure and function (the result of cellular differentiation during development).
Human examples *include: epithelial cells, osteoblasts, muscle fibers.*

TISSUE LEVEL
Tissues are composed of groups of cells of similar structure that perform a particular, related function.
Human examples *include: epithelial tissue, bone, muscle.*

ORGAN LEVEL
Organs are structures of definite form and structure, comprising two or more tissues with related functions.
Human examples *include: heart, lungs, brain, stomach, kidney.*

ORGAN SYSTEM LEVEL
In animals, organs form parts of even larger units known as **organ systems**. An organ system is an association of organs with a common function, e.g. digestive system, cardiovascular system, and the urinogenital system. In all, 11 organ systems make up the body, or **organism**.

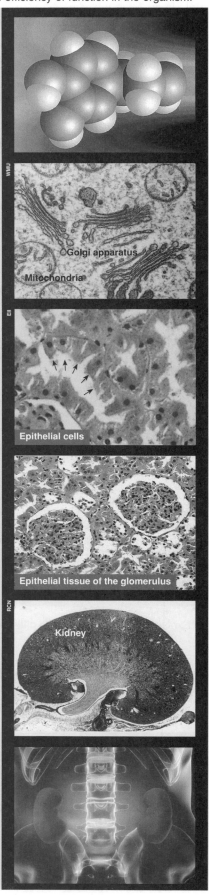

Tissues and Organs

The microscopic study of tissues is called **histology**. The cells of a tissue, and their associated extracellular substances, e.g. collagen, are grouped together to perform particular functions. Tissues improve the efficiency of operation because they enable tasks to be shared amongst various specialized cells. **Animal tissues** can be divided into four broad groups: **epithelial tissues**, **connective tissues**, **muscle**, and **nervous tissues**. Organs usually consist of several types of tissue. The heart mostly consists of cardiac muscle tissue, but also has epithelial tissue, which lines the heart chambers to prevent leaking, connective tissue for strength and elasticity, and nervous tissue, in the form of neurons, which direct the contractions of the cardiac muscle. The features of some of the more familiar tissues of the human body are described below.

Epithelial tissues line internal and external surfaces (e.g. blood vessels, ducts, gut lining) and protect the underlying structures from wear and tear, infection, and pressure.

Features of epithelial tissues
- Epithelium always has one free surface called the **apical surface**. On the lower **basal surface**, the epithelial cells are anchored on a **basement membrane** of collagen fibers held together by a carbohydrate-based glue.
- Except for glandular epithelium, epithelial cells form fitted continuous sheets, held in place by desmosomes and tight junctions.
- Epithelial tissues are **avascular**, i.e. they have no blood supply and rely on diffusion from underlying capillaries.
- Epithelia are classified as **simple** (single layered) or **stratified** (two or more layers), and the cells may be **squamous** (flat), **cuboidal**, or **columnar** (rectangular). Thus at least two adjectives describe any particular epithelium (e.g. stratified cuboidal).
- **Pseudostratified epithelium** is a type of simple epithelium that appears layered because the cells are of different heights. All cells rest on the basement membrane.
- **Transitional epithelium** is a type of stratified epithelium which is capable of considerable stretching. It lines organs such as the urinary bladder.
- Epithelia may be modified, e.g. ciliated or specialized for secretion, absorption, or filtration.
- Glandular epithelium is specialized for secretion. Epithelia may be also be ciliated e.g. in the respiratory tract or specialized for absorption or filtration.

Epithelial Tissue

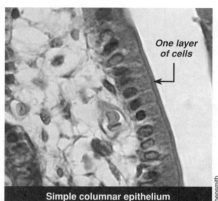

Simple columnar epithelium

The simple epithelium of the gastrointestinal tract is easily recognized by the regular column-like cells. it is specialized for secretion and absorption.

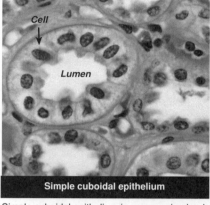

Simple cuboidal epithelium

Simple cuboidal epithelium is common in glands and their ducts and also lines the kidney tubules (above) and the surface of the ovaries.

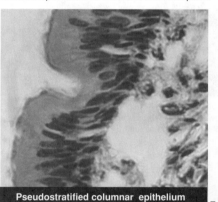

Pseudostratified columnar epithelium

This type of epithelium lines most of the respiratory tract (above). Mucus produced by goblet cells in the epithelium traps dust particles.

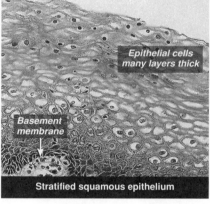

Stratified squamous epithelium

Stratified epithelium is more durable than simple epithelium because it has several layers. It has a protective role, e.g. in the vagina above.

Muscle Tissue

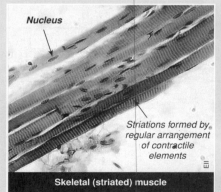

Skeletal (striated) muscle

Muscle tissue consists of very highly specialized cells called fibers, held together by connective tissue. Muscle tissues are specialized to contract. Skeletal muscle (above) brings about voluntary movements. Note the multinucleated cells.

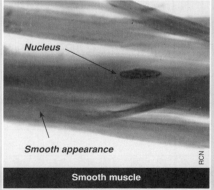

Smooth muscle

The spindle shaped cells of **smooth muscle** have only one nucleus per cell. The contractile elements are not regularly arranged so the tissue appears smooth. Smooth muscle is responsible for involuntary movements (e.g. in the gut wall).

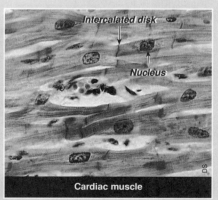

Cardiac muscle

Cardiac muscle is found only in the heart. It has striations, but the cells are short with only one nucleus, and they are held together by specialized intercalated disks with gap junctions to allow rapid communication between the cells.

Related activities: Levels of Organization
Web links: Animal Tissues

RA 2

Connective tissue

Connective tissue is the major supporting tissue of the body. It is made of living cells widely dispersed in a semi-fluid extracellular matrix. Connective tissues bind other structures together and provide support and protection against damage, infection, or heat loss. Most connective tissues have a plentiful blood supply, although tendons and ligaments are poorly vascularized and cartilage is avascular. Connective tissues range from very hard to fluid: classified as bone, cartilage, dense connective tissue (e.g. tendons, ligaments), loose connective tissue, and blood.

Nervous tissue

Nervous tissue makes up the structures of the nervous system. It contains densely packed nerve cells (neurons), specialized for transmitting electrochemical impulses. Neurons are usually associated with supporting cells (**neuroglia**) and connective tissue containing blood vessels.

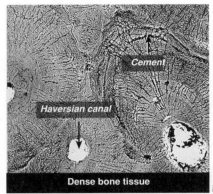

Dense bone tissue

Bone is the hardest connective tissue and consists of bone cells surrounded by a hard matrix.

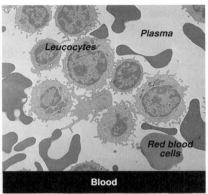

Blood

Blood is a liquid tissue, comprising cells floating in a liquid matrix, which includes soluble fibers.

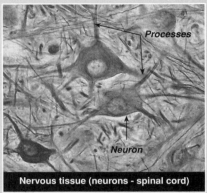

Nervous tissue (neurons - spinal cord)

Neurons have long processes which allow nerve impulses to be transmitted over long distances.

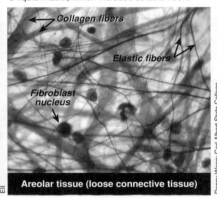

Fibrocartilage of intervertebral disc

Cartilage is more flexible than bone and forms supporting structures in the skeleton. Fibrocartilage forms the cushiony disks between vertebrae.

Areolar tissue (loose connective tissue)

Loose connective tissues have more cells and fewer fibers than harder connective tissues. Areolar tissue helps to hold internal organs in position.

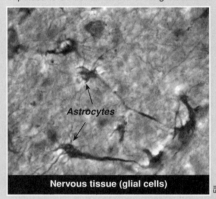

Nervous tissue (glial cells)

Astrocytes (astroglia) provide physical and metabolic support to the neurons of the CNS and help maintain the composition of the extracellular fluid.

1. (a) Describe the basic components of a tissue: _____

   _____

   (b) Explain how the development of tissues improves functional efficiency: _____

   _____

2. Describe the general functional role of each of the following broad tissue types:

   (a) Epithelial tissue: _____ (c) Muscle tissue: _____

   (b) Nervous tissue: _____ (d) Connective tissue: _____

3. Describe the particular features that contribute to the particular functional role of each of the following tissue types:

   (a) Muscle tissue: _____

   _____

   (b) Nervous tissue: _____

   _____

   (c) Connective tissue: _____

   _____

   (d) Epithelial tissue: _____

   _____

Human Organ Systems

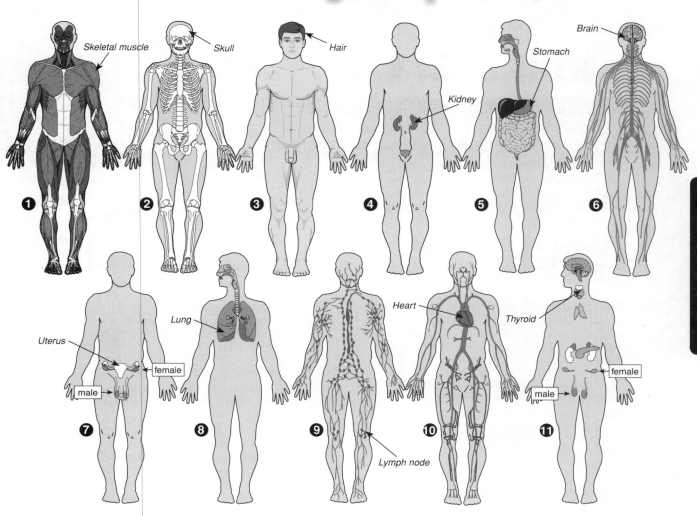

Skeletal muscle · Skull · Hair · Kidney · Stomach · Brain · Uterus · female · male · Lung · Heart · Thyroid · female · male · Lymph node

① ② ③ ④ ⑤ ⑥ ⑦ ⑧ ⑨ ⑩ ⑪

There are 11 organ systems in the human body, each comprising a number of components with specific functions. Identify each organ system (1-11) pictured and state its basic functional role in a few words. Some labels are given as clues and you can use the web links, your textbook or other chapters in this workbook to help you. In addition, identify the main components of organ system #11, briefly state the primary role of each organ, and identify which organs are specific to each gender:

1. _____

2. _____

3. _____

4. _____

5. _____

6. _____

7. _____

8. _____

9. _____

10. _____

11. _____

_____

_____

_____

The Integument and Homeostasis

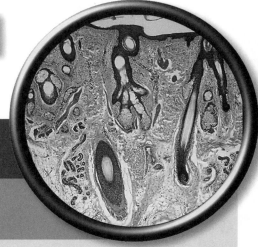

The structure and function of the skin and body membranes. Contributions to homeostasis.

Principles of homeostasis, organ systems and homeostasis, diagnostic medicine, the integumentary system and thermoregulation.

Learning Objectives

☐ 1. Compile your own glossary from the **KEY WORDS** displayed in **bold type** in the learning objectives below.

Body Cavities and Membranes *(pages 12, 38)*

☐ 2. Review terms used in relation to the position of anatomical parts. You should be familiar with the correct way in which to refer to the location of body parts, including: anterior/ventral, posterior/dorsal, superior, inferior, lateral, distal, proximal, medial, superficial. Use these terms when referring to body parts in your discussions and responses to questions.

☐ 3. Describe the location of the internal body cavities: the **dorsal body cavity** (=the cranial cavity + spinal cavity), and the **ventral body cavity** (= the thoracic cavity + the abdominal cavity + the pelvic cavity). Identify the organs associated with each of these cavities.

☐ 4. Describe the features, location, and functional role of the body's membranes. Recognize:
(a) Epithelial membranes
- cutaneous membranes (the skin)
- mucous membranes
- serous membranes
(b) Connective tissue membranes
- synovial membranes

Homeostasis *(pages 37, 39-40, 45)*

☐ 5. Explain the need for **homeostasis**. Identify the factors that need to be regulated to maintain a steady state. Explain the role of **negative feedback** in maintaining homeostasis, identifying how they stabilize systems against excessive change.

☐ 6. Recognize the mechanisms (hormonal and nervous) by which homeostasis is achieved. Describe the contribution of the body's organ systems in collectively maintaining homeostasis in the face of environmental fluctuations and changing metabolic demands.

☐ 7. Describe how body temperature is regulated in humans, including the role of:
(a) the **hypothalamus**
(b) the **autonomic nervous system**
(c) the circulatory system (including the blood)
(d) the skin and its associated sensory receptors

The Integumentary System *(pages 43-44, 90)*

☐ 8. Describe the functions of the integumentary system (**integument**) and identify structural features associated with these functions.

☐ 9. Describe the structure of the skin recognizing the **epidermis** and **dermis**. Appreciate that the **hypodermis** (or subcutaneous tissue) is not considered part of the skin but is closely associated with it. Describe features of the skin's two tissues:

(a) **Epidermis:** Five layered, non-vascular outer-most region of the skin. Recognize the keratinized region and pigment-producing melanocytes.
(b) **Dermis:** Dense connective tissue containing dermal papillae, hair follicles, blood vessels, sweat and sebaceous glands, and sensory receptors.

Medical Imaging and Diagnosis *(pages 41-42)*

☐ 10. Discuss the technologies available for medical diagnosis and identify their advantages and disadvantages. You could include reference to any of the following: **x-ray imaging** and **radionuclide scanning**, computer imaging techniques (e.g. **MRI, CT**, and **PET**), **endoscopy**, **electrocardiography**, and **biosensors**. In each case, describe the basic principle of the technique and situations appropriate to its use.

Supplementary Texts

See page 9 for additional details of these texts:

■ Morton, D. and Perry, J.W, 1997. **Photo Atlas for Anatomy and Physiology**, (Brooks/Cole).

■ Rowett, H.G.Q., 1999. **Basic Anatomy & Physiology**, (John Murray), pp. 80-83.

Presentation MEDIA
to support this topic:
Anatomy & Physiology

Periodicals

See page 9 for details of publishers of periodicals:

STUDENT'S REFERENCE

■ **Temperature Regulation** Biol. Sci. Rev., 17(2) Nov. 2004, pp. 2-7. *Principles of thermoregulation, the body's hypothalamic thermostat, and the role of the skin in detecting and regulating temperature.*

■ **Homeostasis** Biol. Sci. Rev., 12(5) May 2000, pp. 2-5. *Homeostasis: what it is, how it is achieved through feedback mechanisms, the involvement of the autonomic nervous system, and adaptations for homeostasis in different environments.*

■ **The Autonomic Nervous System** Biol. Sci. Rev., 18(3) Feb. 2006, pp. 21-25. *A thorough account of the structure and roles of the ANS, including its role in thermoregulation.*

■ **The Color Code** New Scientist, 10 March 2007, pp. 34-37. *Researchers are developing a better understanding of the basis of genetic differences in skin color and how they evolved.*

■ **Skin, Scabs and Scars** Biol. Sci. Rev., 17(3) Feb. 2005, pp. 2-6. *The many roles of skin, including its importance in wound healing and the processes involved in its repair when damaged.*

Internet

See pages 10-11 for details of how to access **Bio Links** from our web site: **www.thebiozone.com** From Bio Links, access sites under the topics:

ANATOMY & PHYSIOLOGY > Homeostasis:
• BBC schools - Homeostasis • Homeostasis: general principles • Homeostasis lecture notes • Physiological homeostasis • Thermoregulation

Principles of Homeostasis

Homeostasis the relative physiological constancy of the body, despite external fluctuations. Homeostasis of the internal environment is an essential feature of complex animals and it is the job of the body's **organ systems** to maintain it, even as they make necessary exchanges with the environment. Homeostatic control systems have three functional components: a receptor to detect change, a control centre, and an effector to direct an appropriate response. In **negative feedback** systems, movement away from a steady state triggers a mechanism to counteract further change in that direction. Using negative feedback systems, the body counteracts disturbances and restores the steady state. **Positive feedback** is also used in physiological systems, but to a lesser extent since positive feedback leads to the response escalating in the same direction.

Organ systems maintain a constant internal environment that provides for the needs of all the body's cells, making it possible for animals to move through different and often highly variable external environments. This representation of a mammal shows how organ systems permit exchanges with the environment. The exchange surfaces of organ systems are usually internal, but may be connected to the environment via openings on the body surface.

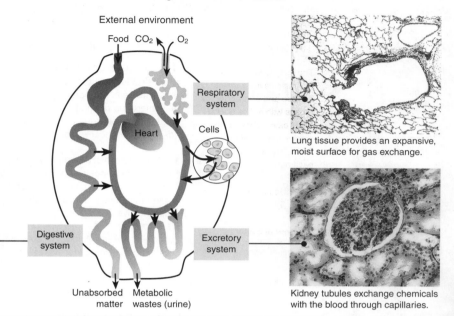

The finger-like villi of the small intestine greatly expand the surface area for nutrient absorption.

Lung tissue provides an expansive, moist surface for gas exchange.

Kidney tubules exchange chemicals with the blood through capillaries.

Negative Feedback and Control Systems

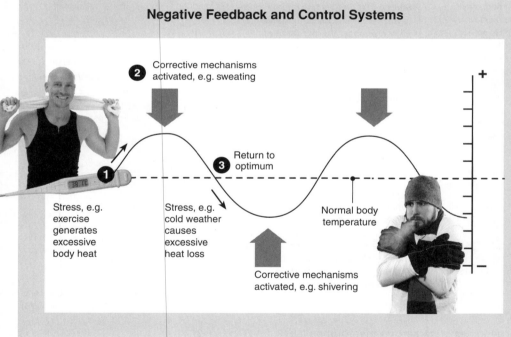

Corrective mechanisms activated, e.g. sweating

Return to optimum

Normal body temperature

Stress, e.g. exercise generates excessive body heat

Stress, e.g. cold weather causes excessive heat loss

Corrective mechanisms activated, e.g. shivering

❶ A stress or disturbance, e.g. exercise, takes the internal environment away from optimum.

❷ Stress is detected by receptors and corrective mechanisms (e.g. sweating) are activated.

❸ The corrective mechanisms act to restore optimum.

Negative feedback acts to counteract any departures from a steady physiological state. The diagram shows how a stress (disturbance) is counteracted by corrective mechanisms in the case of body temperature.

In contrast to negative feedback, positive feedback will push physiological levels out of the normal range. While it is inherently unstable, it has a useful role at certain times, e.g. during childbirth.

The Integument & Homeostasis

1. Describe the three main components of a regulatory control system in the human body: _____

_____

_____

2. Explain how negative feedback mechanisms maintain homeostasis in a variable environment: _____

_____

_____

_____

Related activities: Maintaining Homeostasis

A 1

Body Membranes and Cavities

The study of anatomy and physiology requires a basic understanding of anatomical terms, including the directional (e.g. distal) and regional (e.g. pelvic) terms used to describe the position of body parts, the location of the body's **cavities** (dorsal and ventral), and the way in which the body's **membranes** line those cavities and protect the organs within. A membrane is a thin layer of tissue that covers a structure or lines a cavity. The body's membranes fall into two broad categories: **epithelial** **membranes** (the skin, mucosa, and serosa), and **synovial membranes**, which lack epithelium. Membranes line and cover the internal and external surfaces of the body, protecting and, in some cases, lubricating them. Epithelial membranes are formed from epithelium and the connective tissue on which it rests. Whereas the skin (**cutaneous membrane**) is exposed to air and is a dry membrane, **mucous membranes** (mucosa) and **serous membranes** (serosa) are moist and bathed in secretions.

Body Cavities

Cranial cavity

Spinal cavity

The body is divided into two main cavities, which enclose and protect the organs within. The **dorsal body cavity** contains the cranial and spinal cavity. These are continuous with each other.

Thoracic cavity

Abdominal cavity (contains the abdominal organs)

Pelvic cavity (contains pelvic organs)

The **ventral body cavity** contains the thoracic, abdominal, and pelvic cavities. The thoracic cavity lies above the diaphragm and contains the lungs (within their pleural cavities) separated by the mediastinum within which lies the heart, thymus, esophagus, trachea, and blood vessels.

Fig. 1: Location of dorsal and ventral body cavities in the human body. A knee joint shows a typical location of connective tissue (synovial) membranes.

Body Membranes

Cutaneous membrane
- The **skin** forms a protective covering over the surface of the body.
- It is made up of an epidermis of stratified squamous epithelium and an underlying dermis of connective tissue.

Mucous membranes (muscosa)
- The **mucosa** lines all body cavities that open to the exterior, i.e. the hollow organs of the respiratory, digestive, urinary, and reproductive tracts.
- It is composed of some type of simple epithelium (e.g. columnar or squamous) resting on loose connective tissue.
- The epithelium of mucosae is often absorptive or secretory.
- Many of them, but not all, produce mucus.

Serous membranes (serosa)
- Serous membranes line internal body cavities that are closed to the exterior.
- They are made of a thin layer of squamous epithelium resting on a thin layer of loose connective tissue.
- They occur in pairs: the **parietal layer** lines the body wall and the **visceral layer** lines the organ within that cavity.
- The membranes are separated by a thin film of **serous fluid**.
- They are named according to location in the body: **peritoneum** (abdomen), **pleura** (lungs), **pericardium** (heart).

Synovial membranes
- Synovial membranes line the capsules around joints and secrete a lubricating synovial fluid (see the activity '*Joints*').
- They are composed of connective tissue and contain no epithelial cells.
- The provide a smooth surface and cushion moving structures.

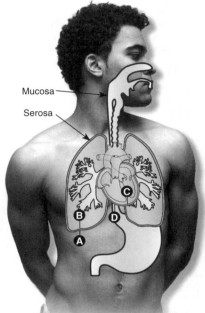

Mucosa

Serosa

Fig. 2: Location of the mucous and serous membranes in the thorax.

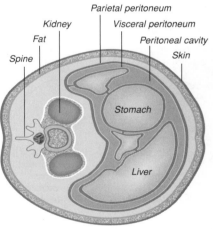

Parietal peritoneum

Kidney

Fat

Spine

Visceral peritoneum

Peritoneal cavity

Skin

Stomach

Liver

Fig. 3: Relationship of parietal and visceral peritoneal membranes in the abdomen (TS)

1. Use the information given above to name the **serous membranes** labeled A-D in Fig. 2:

   (a) A: _____   (c) C: _____

   (b) B: _____   (d) D: _____

2. (a) Describe the general role of epithelial membranes in the body: _____

   _____

   _____

   (b) Explain how epithelial membranes differ from synovial membranes: _____

   _____

Related activities: Terms in Anatomy and Physiology, Joints
Web links: The Body's Cavities, Quizlet: Body Cavities and Membranes

Maintaining Homeostasis

The various organ systems of the body act to maintain homeostasis through a combination of hormonal and nervous mechanisms. In everyday life, the body must regulate respiratory gases, protect itself against agents of disease (pathogens), maintain fluid and salt balance, regulate energy and nutrient supply, and maintain a constant body temperature. All these must be coordinated and appropriate responses made to incoming stimuli. In addition, the body must be able to repair itself when injured and be capable of reproducing (leaving offspring).

The Integument & Homeostasis

Regulating Respiratory Gases

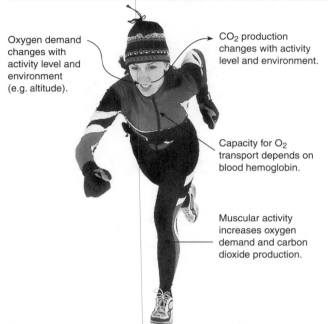

Oxygen demand changes with activity level and environment (e.g. altitude).

CO$_2$ production changes with activity level and environment.

Capacity for O$_2$ transport depends on blood hemoglobin.

Muscular activity increases oxygen demand and carbon dioxide production.

Oxygen must be delivered to all cells and carbon dioxide (a waste product of cellular respiration) must be removed. **Breathing** brings in oxygen and expels CO$_2$, and the cardiovascular and lymphatic systems circulate these respiratory gases (the oxygen mostly bound to hemoglobin). The rate of breathing is varied according to oxygen demands (as detected by CO$_2$ levels in the blood).

Coping with Pathogens

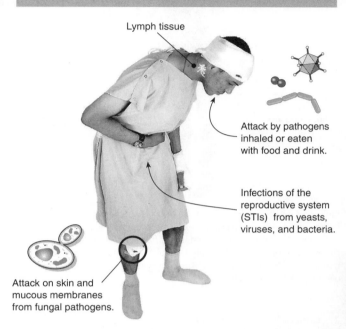

Lymph tissue

Attack by pathogens inhaled or eaten with food and drink.

Infections of the reproductive system (STIs) from yeasts, viruses, and bacteria.

Attack on skin and mucous membranes from fungal pathogens.

All of us are under constant attack from pathogens (disease causing organisms). The body has a number of mechanisms that help to prevent the entry of pathogens and limit the damage they cause if they do enter the body. The skin, the digestive system, and the immune system are all involved in the body's defense, while the cardiovascular and lymphatic systems circulate the cells and antimicrobial substances involved.

Maintaining Nutrient Supply

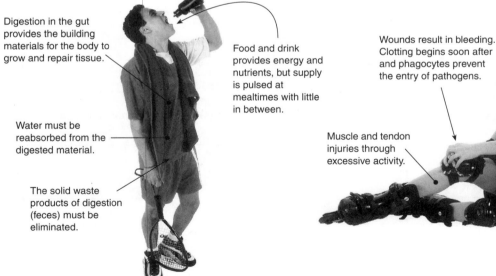

Digestion in the gut provides the building materials for the body to grow and repair tissue.

Food and drink provides energy and nutrients, but supply is pulsed at mealtimes with little in between.

Water must be reabsorbed from the digested material.

The solid waste products of digestion (feces) must be eliminated.

Food and drink must be taken in to maintain the body's energy supplies. The digestive system makes these nutrients available, and the cardiovascular system distributes them throughout the body. Food intake is regulated largely through nervous mechanisms while hormones control the regulation of cellular uptake of glucose.

Repairing Injuries

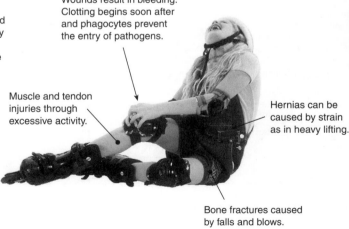

Wounds result in bleeding. Clotting begins soon after and phagocytes prevent the entry of pathogens.

Muscle and tendon injuries through excessive activity.

Hernias can be caused by strain as in heavy lifting.

Bone fractures caused by falls and blows.

Damage to body tissues triggers the **inflammatory response** and white blood cells move to the injury site. The inflammatory response is started (and ended) by chemical signals (e.g. from histamine and prostaglandins) released when tissue is damaged. The cardiovascular and lymphatic systems distribute the cells and molecules involved.

Related activities: Principles of Homeostasis

RA 2

Maintaining Fluid and Ion Balance

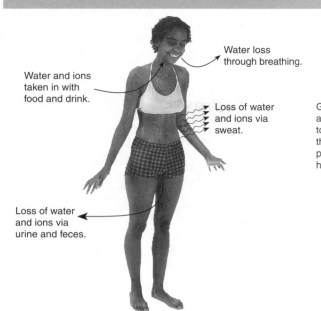

Water loss through breathing.

Water and ions taken in with food and drink.

Loss of water and ions via sweat.

Loss of water and ions via urine and feces.

Coordinating Responses

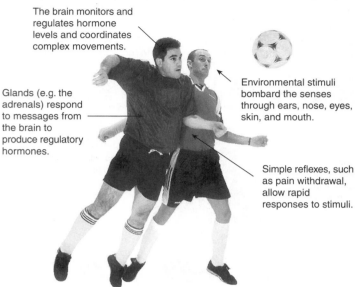

The brain monitors and regulates hormone levels and coordinates complex movements.

Glands (e.g. the adrenals) respond to messages from the brain to produce regulatory hormones.

Environmental stimuli bombard the senses through ears, nose, eyes, skin, and mouth.

Simple reflexes, such as pain withdrawal, allow rapid responses to stimuli.

Fluid and electrolyte balance in the body is maintained by the kidneys (although the skin is also important). Osmoreceptors monitor blood volume and bring about the release of regulatory hormones; the kidneys regulate reabsorption of water and sodium from blood in response to levels of the hormones ADH and aldosterone. The cardiovascular and lymphatic system distribute fluids around the body.

The body is constantly bombarded by stimuli from the environment. The brain sorts these stimuli into those that require a response and those that do not. Responses are coordinated via nervous or hormonal controls. Simple nervous responses (reflexes) act quickly. Hormones, which are distributed by the cardiovascular and lymphatic systems, take longer to produce a response and the response is more prolonged.

1. Describe two mechanisms that operate to restore homeostasis after infection by a pathogen:

   (a) _____

   _____

   (b) _____

   _____

2. Describe two mechanisms by which responses to stimuli are brought about and coordinated:

   (a) _____

   _____

   (b) _____

   _____

3. Explain two ways in which water and ion balance are maintained. Name the organ(s) and any hormones involved:

   (a) _____

   _____

   (b) _____

   _____

4. Explain two ways in which the body regulates its respiratory gases during exercise:

   (a) _____

   _____

   (b) _____

   _____

Diagnostic Medicine

Sophisticated **medical imaging** has provided the means to look in detail at the tissues and organs of the body, making it possible to accurately and rapidly diagnose disorders and therefore treat people more effectively for medical problems. As well as imaging techniques, simpler methods, such as blood tests, are also widely used for diagnostic purposes.

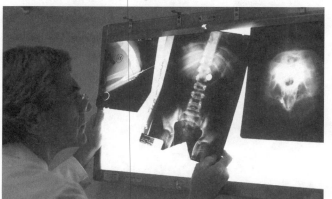

X-ray imaging

X-rays are a form of electromagnetic radiation that can pass through tissues and expose photographic film. The X-rays are absorbed by dense body tissues (e.g. bone) which appear as white areas, but they pass easily through less dense tissues (e.g. muscle), which appear dark. X-rays are used to identify fractures or abnormalities in bone. X-ray technology is also used in conjunction with computer imaging techniques (see below).

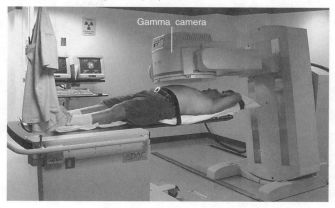

Gamma camera

Radionuclide scanning

Radionuclide scanning involves introducing a radioactive substance (the radionuclide) into the body, where it is taken up in different amounts by different tissues (e.g. radioactive iodine is taken up preferentially by the thyroid). The radiation emitted by the tissues that take up the radionuclide is detected by a gamma camera. Radionuclide scanning provides better detail of function than other techniques, but gives less anatomical detail.

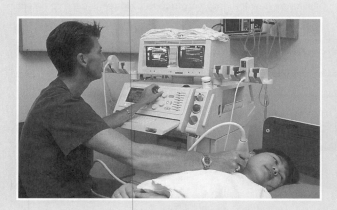

Diagnostic uses of ultrasound

Ultrasound is a diagnostic tool used to visualize internal structures without surgery or X-rays. Ultrasound imaging is based on the fact that tissues of different densities reflect sound waves differently. Sound waves are directed towards a structure (e.g. uterus, heart, kidney, liver) and the reflected sound waves are recorded. An image of the internal structures is analyzed by computer and displayed on a screen.

Echocardiography uses ultrasound to investigate heart disorders such as congenital heart disease and valve disorders. The liver and other abdominal organs can also be viewed with ultrasound for diagnosis of disorders such as cirrhosis, cysts, blockages, and tumors. Ultrasound scans of the uterus are commonly used during pregnancy to indicate placental position and aspects of foetal growth and development. This information aids better pregnancy management.

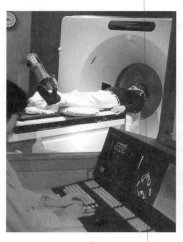

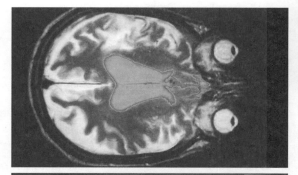
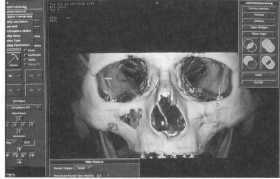

Computer imaging techniques

Computers are used extensively to examine the soft tissues of the body for diagnostic purposes. The photos directly above show **magnetic resonance imaging** (MRI), which uses computer analysis of high frequency radio waves to map out variations in tissue density, especially of the central nervous system (above, far right). In **computerized tomography** (CT) scans, a series of X-rays is made through an organ and the picture from each X-ray slice is reconstructed (using computer software) into a 3-D image (e.g. the skull, right). Such images can be used to detect abnormalities such as tumors.

The Integument & Homeostasis

Related activities: Aging and the Skeleton, Monoclonal Antibodies, Type 1 Diabetes Mellitus, Type 2 Diabetes Mellitus

A 2

Endoscopy

An **endoscope** is an illuminated tube comprising fiber-optic cables with lenses attached. Endoscopy can be used for a visual inspection of the inside of organs (or any body cavity) to look for blockages or damage. Endoscopes can also be fitted with devices to remove foreign objects, temporarily stop bleeding, remove tissue samples (biopsy), and remove polyps or growths.

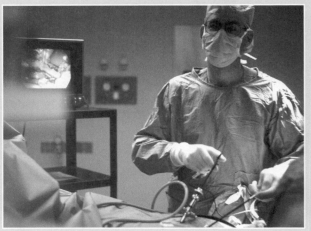

Laparoscopy is the endoscopic examination of the organs in the abdominal cavity, and is used during simple surgical operations (e.g. tubal ligation). Endoscopic examination of the stomach is called gastroscopy.

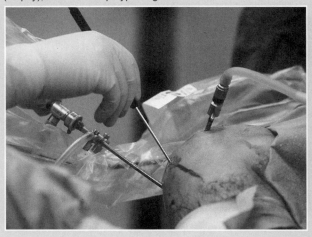

Arthroscopy is used for inspecting joints, usually knee joints (above), while the patient is under a general anesthetic. Using very small incisions, damaged cartilage can be removed from the joint using other instruments.

Biosensors

Biosensors are electronic monitoring devices that use biological material to detect the presence or concentration of a particular substance. Because of their specificity and sensitivity, enzymes are ideally suited for use in biosensors. The example below illustrates how the enzyme **glucose oxidase** is used to detect blood sugar level in diabetics.

The enzyme, *glucose oxidase* is immobilized in a semi-conducting silicon chip. It catalyses the conversion of glucose (from the blood sample) to gluconic acid.

Hydrogen ions from the gluconic acid cause a movement of electrons in the silicon which is detected by a transducer. The strength of the electric current is directly proportional to the blood glucose concentration.

Results are shown on a liquid crystal display.

| Plastic sleeve | Membrane permeable to glucose → | Biological recognition layer | Transducer | Amplifier | 932 |

The signal is amplified

---

1. Describe the basic principle of the scanning technology behind each of the following computer imaging techniques:

   (a) Computerised Tomography (CT): _____

   _____

   (b) Magnetic Resonance Imaging (MRI): _____

   _____

2. Describe the benefits of using computer imaging techniques such as MRI or CT: _____

   _____

3. Explain how radionuclide scanning differs from X-rays: _____

   _____

4. Describe the benefits of surgery using endoscopy over conventional open surgery: _____

   _____

   _____

5. Describe the basic principle of a biosensor: _____

   _____

   _____

The Integumentary System

The skin, or **cutaneous membrane**, and its associated structures (hair, sweat glands, nails) collectively make the **integumentary system**. The skin is the body's largest organ. Unlike other epithelial membranes, it is a dry membrane, made up of an outer **epidermis** and underlying **dermis**. The subcutaneous tissue beneath the dermis, which is largely fat (a loose connective tissue) is called the **hypodermis**. It is not part of the skin, but it does anchor the skin to underlying organs, thereby insulating and protecting them. The homeostatic interactions of the skin with other body systems are described below (highlighted panels).

Endocrine system
- Estrogens help to maintain skin hydration.
- Androgens activate the sebaceous glands and help to regulate the growth of hair.
- Changes in skin pigmentation are associated with hormonal changes during pregnancy and puberty.

Respiratory system
- Provides oxygen to the cells of the skin via gas exchange with the blood.
- Removes carbon dioxide (gaseous metabolic waste) from the cells of the skin via gas exchange with the blood.

Cardiovascular system
- Blood vessels transport O_2 and nutrients to the skin and remove wastes (via the blood).
- The skin prevents fluid loss and acts as a reservoir for blood.
- Dilation and constriction of blood vessels is important thermoregulatory mechanism.
- The blood supplies substances required for functioning of the skin's glands.

Digestive system
- Skin synthesizes vitamin D, which is required for absorption of calcium from the gut.
- Digestive system provides nutrients for growth, repair, and maintenance of the skin (delivered via the cardiovascular system).

Skeletal system
- Skin absorbs ultraviolet light and produces a vitamin D precursor. Vitamin D is involved in calcium and phosphorus metabolism, and is needed for normal calcium absorption and deposition of calcium salts in bone.

Nervous system
- Many sensory organs (e.g. Pacinian corpuscle) and simple receptors (e.g. thermoreceptors), are located in the skin.
- The nervous system regulates blood vessel dilation and sweat gland secretion.
- CNS interprets sensory information from the skin's sensory receptors.
- Nervous stimulation of arrector pili muscles causes erection of hair (thermoregulatory response).

Lymphatic system and immunity
- Tissue fluid bathes and nourishes skin cells. Lymphatic vessels prevent edema by collecting and returning tissue fluid to the general circulation.

Urinary system
- Vitamin D synthesis begins in the skin but final activation of vitamin D occurs in the kidneys.
- Urination controlled by a voluntary sphincter in the urethra.

Muscular system
- Muscular activity generates heat, which is dissipated via an increase in blood flow to the skin's surface.
- Muscular activity and increased blood flow increases secretion from the skin's glands (e.g. sweating).

Reproductive system
- Mammary glands, which are modified sweat glands, nourish the infant in lactating women.
- Skin stretches during pregnancy to accommodate enlargement of the uterus.
- Changes in skin pigmentation are associated with pregnancy and puberty.

Fingernail

Hairy skin

Fingernails (inset, right) and **toenails** are modifications of the epidermis that protect the ends of the digits. Nails are formed from the horny layer of the epidermis (stratum corneum) that contains hard keratin, and they grow from the basal layer of cells in an area called the nail matrix.

Hair is found over almost all of the body. Each hair has a shaft, which protrudes above the skin's surface, and a root and hair bulb beneath. The root and shaft are made of dead, keratinized epithelial cells in three layers (a central medulla, thick inner cortex, and outer cuticle). As with skin, melanin is responsible for the color of the hair.

General Functions and Effects on all Systems
The skin is the body's largest organ and the integumentary system covers and provides physical and chemical protection for most parts of all other body systems. It has a critical role in thermoregulation and in the absorption of sunlight and synthesis of a vitamin D precursor.

The Integument & Homeostasis

Related activities: Tissues & Organs, Human Organ Systems, Skin Senses
Web links: The Integumentary System, Skin Structure and Function

RA 2

The Structure of the Skin

The skin is made up of two kinds of tissue resting on a underlying layer of fat.

- The upper **epidermis** of **stratified squamous epithelium**. There are up to five layers of cells in the epidermis, with the regenerative **basal layer** furthest from the surface. The basal layer also houses the melanocytes which contain the pigment **melanin** that gives skin a dark color. Cells divide at the basal layer and migrate towards the surface, becoming flatter and increasingly **keratinized**. These outermost layers vary in thickness, depending on the skin type. Like all epithelial tissue, the skin's epidermis lacks blood vessels.

- The lower **dermis** of dense connective tissue containing collagen (strength), elastic fibers (flexibility) and reticular fibers (support). The dermis also contains the skin's sensory receptors (discussed in a later chapter). The upper papillary region is uneven and has small projections which indent the epidermis above. The lower reticular layer contains the blood vessels, sweat and sebaceous (oil) glands, and deep dermal sensory receptors.

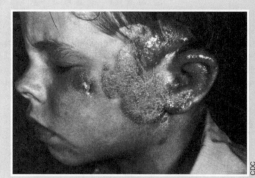

Ringworm is a fungal infection of the skin. It can affect any area of the body surface and causes raised ring-like scaly lesions, which may be itchy and inflamed.

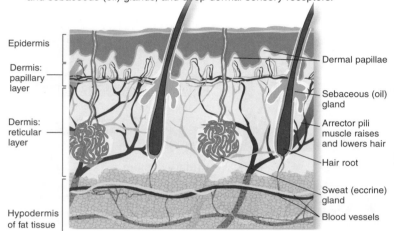

Epidermis

Dermis: papillary layer

Dermis: reticular layer

Hypodermis of fat tissue

Dermal papillae

Sebaceous (oil) gland

Arrector pili muscle raises and lowers hair

Hair root

Sweat (eccrine) gland

Blood vessels

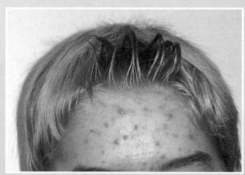

Acne is a common skin condition in adolescents in which the sebaceous glands become infected and **pimples** appear on the skin. Acne can be extremely severe, causing deep abcesses and scarring.

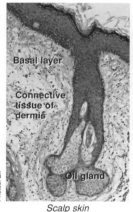

Basal layer

Connective tissue of dermis

Oil gland

Scalp skin

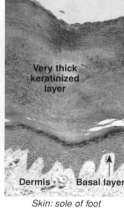

Very thick keratinized layer

Dermis   Basal layer

Skin: sole of foot

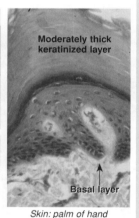

Moderately thick keratinized layer

Basal layer

Skin: palm of hand

There are up to five layers of cells in the epidermis. The thickness of the layers, particularly the outermost heavily keratinized layer, varies depending on where the skin is. Keratin protects the deeper cell layers, and skin subjected to regular wear and tear is heavily keratinized.

A **tattoo** is a permanent marking made by inserting ink into the layers of skin. The pigment remains stable in the upper layer of the dermis, trapped within fibroblasts but, in the long term, it tends to migrate deeper into the dermis and the tattoo degrades.

1. Describe two homeostatic functions of the skin:

   (a) _____

   (b) _____

2. Explain why a tattoo stays permanently in the skin: _____

   _____

3. Identify the location (dermis or epidermis) of each of the following and identify its role as part of the skin:

   (a) Basal layer (stratum basale): _____

   (b) Outermost keratinized layer (stratum corneum): _____

   (c) Sweat glands: _____

   (d) Collagen fibers: _____

Thermoregulation

In humans (and other placental mammals), the temperature regulation center of the body is in the **hypothalamus**. In humans, it has a 'set point' temperature of 36.7°C. The hypothalamus responds directly to changes in core temperature and to nerve impulses from receptors in the skin. It then coordinates appropriate nervous and hormonal responses to counteract the changes and restore normal body temperature. Like a thermostat, the hypothalamus detects a return to normal temperature and the corrective mechanisms are switched off (negative feedback). Toxins produced by pathogens, or substances released from some white blood cells, cause the set point to be set to a higher temperature. This results in fever and is an important defense mechanism in the case of infection.

Counteracting Heat Loss

Heat promoting center* in the hypothalamus monitors fall in skin or core temperature below 35.8°C and coordinates responses that generate and conserve heat. These responses are mediated primarily through the **sympathetic nerves** of the autonomic nervous system.

Thyroxine (together with epinephrine) **increases metabolic rate**.

Under conditions of *extreme* cold, epinephrine and thyroxine increase the energy releasing activity of the liver. Under normal conditions, the liver is thermally neutral.

Muscular activity (including *shivering*) produces internal heat.

Erector muscles of hairs contract to raise hairs and increase insulating layer of air. Blood flow to skin decreases (**vasoconstriction**).

Factors causing heat loss
- Wind chill factor accelerates heat loss through conduction.
- Heat loss due to temperature difference between the body and the environment.
- The rate of heat loss from the body is increased by being wet, by inactivity, dehydration, inadequate clothing, or shock.

Factors causing heat gain
- Gain of heat directly from the environment through radiation and conduction.
- Excessive fat deposits make it harder to lose the heat that is generated through activity.
- Heavy exercise, especially with excessive clothing.

*NOTE: The heat promoting center is also called the cold centre and the heat losing center is also called the hot centre. We have used the terminology descriptive of the activities promoted by the center in each case.

Counteracting Heat Gain

Heat losing center* in the hypothalamus monitors any rise in skin or core temperature above 37.5°C and coordinates responses that increase heat loss. These responses are mediated primarily through the **parasympathetic nerves** of the autonomic nervous system.

Sweating increases. Sweat cools by evaporation.

Muscle tone and **metabolic rate** are decreased. These mechanisms reduce the body's heat output.

Blood flow to skin (**vasodilation**) increases. This increases heat loss.

Erector muscles of hairs relax to flatten hairs and decrease insulating air layer.

The Integument & Homeostasis

The Skin and Thermoregulation

Thermoreceptors in the dermis (probably free nerve endings) detect changes in skin temperature outside the normal range and send nerve impulses to the hypothalamus, which mediates a response. Thermoreceptors are of two types: **hot thermoreceptors** detect a rise in skin temperature above 37.5°C while the **cold thermoreceptors** detect a fall below 35.8°C. Temperature regulation by the skin involves **negative feedback** because the output is fed back to the skin receptors and becomes part of a new stimulus-response cycle.

Note that the thermoreceptors detect the temperature change, but the hair erector muscles and blood vessels are the **effectors** for mediating a response.

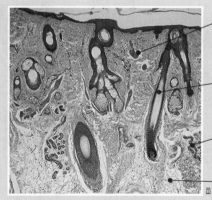

Cross section through the skin of the scalp.

Blood vessels in the dermis dilate (vasodilation) or constrict (vasoconstriction) to respectively promote or restrict heat loss.

Hairs raised or lowered to increase or decrease the thickness of the insulating air layer between the skin and the environment.

Sweat glands produce sweat in response to parasympathetic stimulation from the hypothalamus. Sweat cools through evaporation.

Fat in the subdermal layers insulates the organs against heat loss.

1. State two mechanisms by which body temperature could be reduced after intensive activity (e.g. hard exercise):

   (a) _____ (b) _____

2. Briefly state the role of the following in regulating internal body temperature:

   (a) The hypothalamus: _____

   (b) The skin: _____

   (c) Nervous input to effectors: _____

   (d) Hormones: _____

Muscular and Skeletal Systems

Endocrine system

- The skeleton protects the endocrine organs especially in the pelvis, chest, and brain.
- Bone takes up and releases calcium in response to hormones.
- Androgens and growth hormone promote muscle strength and increase in mass.

Respiratory system

- Skeleton encloses and protects lungs
- Flexible ribcage enables ventilation of the lungs for exchange of gases (O_2/CO_2).
- Diaphragm and intercostals produce volume changes in breathing.

Cardiovascular system

- Heart and blood vessels transport O_2, nutrients, and waste products to all the body.
- Bone marrow produces red blood cells
- Bone matrix stores calcium, which is required for muscle contraction

Digestive system

- Skeleton provides some protection and support for the abdominal organs.
- Digestive system provides nutrients for growth, repair, and maintenance of muscle and connective tissues.

Skeletal system

- Muscular activity maintains bone strength and helps determine bone shape.
- Muscles pull on bones to create movement.

Integumentary system

- Skin absorbs and produces precursor of vitamin D, which is involved in calcium and phosphorus metabolism.
- Skin covers and protects the muscle tissue.

Nervous system

- The skeleton protects the CNS.
- Bone acts as a store of calcium ions required for nerve function.
- Innervation of bone and joint capsules provides sensation and positional awareness.
- Muscular activity is dependent on innervation.

Lymphatic system and immunity

- Stem cells in the bone marrow give rise to the lymphocytes involved in the immune response.

Urinary system

- The skeleton protects the pelvic organs.
- Final activation of vitamin D, which is involved in calcium and phosphorus metabolism, occurs in the kidneys.
- Urination controlled by a voluntary sphincter in the urethra.

Reproductive system

- The skeleton protects the reproductive organs.
- Reproductive (sex) hormones influence skeletal development.

Muscular system

- Skeleton acts as a system of levers for muscular activity.
- Bone provides a store of calcium for muscle contraction.

General Functions and Effects on all Systems

The skeletal system provides bony protection for the internal organs, especially the brain and spinal cord, and the lungs, heart, and pelvic organs. The muscular system acts with the skeletal system to generate voluntary movements. Smooth and cardiac muscle provide motility for involuntary activity.

Disease

| Symptoms of disease | • Pain (moderate to severe)
• Inflammation
• Limitations in function |

| Disorders of the bones and joints | • Growth disorders
• Trauma (fractures and sprains)
• Infection
• Tumors
• Degenerative diseases |

| Diseases of the skeletal muscles | • Inherited diseases
• Fibrosis (scarring)
• Strains, tears, and cramps
• Denervation and atrophy |

Medicine & Technology

| Diagnosis of disorders | • Blood tests
• Bone scans
• Medical imaging techniques
• Athroscopy |

| Treatment of injury | • Surgery
• Physical and drug therapies
• Prosthetics and orthotics |

| Treatment of inherited disorders | • Surgery
• Radiotherapy (for cancers)
• Physical and drug therapies
• Prosthetics and orthotics
• Gene therapy |

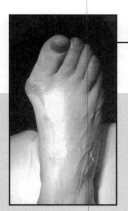

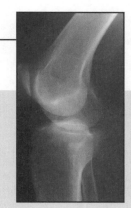

• Osteomalacia
• Osteoarthritis
• Osteoporosis
• Sarcomas
• Muscular dystrophy

• Joint replacement
• Grafts
• Genetic counselling
• X-rays
• MRI

Support & Movement

The Musculoskeletal System

The musculoskeletal system can be affected by disease and undergoes changes associated with aging.

Medical technologies and exercise can be used to diagnose, treat, and delay the onset of musculoskeletal disorders.

• Osteoarthritis
• Osteoporosis
• Muscular atrophy

• Muscle fatigue
• Fast vs slow twitch
• Aerobic training
• Anaerobic training

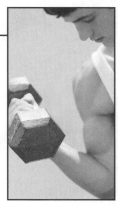

| Aging and the bones, joints, and muscles | • Bone loss
• Loss of muscle mass
• Accumulated trauma
• Increased incidence of cancers |

| Effects of exercise on bones, joints, and muscles | • Increased bone density
• Increased lean muscle mass
• Changes in flexibility & joint mobility
• Changes in fiber type & recruitment
• Changes in oxidative capacity |

The Effects of Aging

Exercise

The Skeletal System

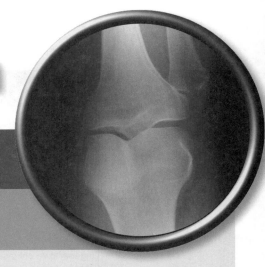

Understanding the structure and function of the skeleton and the basis of movement

Bone structure and formation, components of the skeleton, the structure and operation of joints, degenerative changes in the skeleton

Learning Objectives

☐ 1. Compile your own glossary from the **KEY WORDS** displayed in **bold type** in the learning objectives below.

Bone and Cartilage *(pages 49-50)*

☐ 2. Describe the tissues that contribute to the structural and functional components of the skeleton: **bone**, cartilage (e.g. **hyaline cartilage**), dense connective tissue (**tendons** and **ligaments**), and reticular connective tissue (in bone marrow).

☐ 3. Describe the various functions of bone with respect to: support, protection, movement, mineral storage, and blood cell production. Discuss the ways in which bone tissue contributes to the body's homeostasis.

☐ 4. Recognize **bone** as a type of **connective tissue**. Describe the composition of **bone**, including reference to the **osteocytes** (within lacunae) within a **matrix** of calcium salts and collagen fibers. Distinguish between the features of **compact** (cortical) **bone** and **spongy** (cancellous) **bone**, and describe where these occur.

☐ 5. Using examples, classify bones according to their size and shape as follows: **long bones**, **short bones**, **flat bones**, and **irregular bones**. Identify any differences in the structure of these different types.

☐ 6. Using a diagram, describe the gross structure of a long bone, including the features conferring strength and shock absorption. Indicate the shaft (**diaphysis**), **periosteum**, and the **epiphyses** and their associated **articular cartilage**. Describe the location of yellow and red bone marrow and explain their functions.

☐ 7. Describe the ultrastructure of compact bone, identifying the **periosteum**, osteoblasts, **osteocytes**, matrix, lacunae, and **Haversian canals**.

☐ 8. Describe **ossification** (bone formation), explaining the role of the **osteoblasts** and the process by which hyaline cartilage is replaced with hard bone. Describe how bone is **remodeled** during growth and repaired in response to injury (e.g. **fracture**).

☐ 9. Explain the role of hormones in regulating bone growth and repair, including the role of **growth hormone** and sex hormones in the growth of long bones, and the role of **parathyroid hormone** (PTH) and **calcitonin** in calcium metabolism and bone remodeling.

The Skeleton *(pages 46-47, 51-55)*

☐ 10. Identify the components of the **skeleton** and describe the features of its two functional regions: the **axial skeleton** and the **appendicular skeleton**. Describe the basic structure and functions of the skull, backbone, and ribcage, and the **pectoral** and **pelvic** girdles.

☐ 11. Describe the role of **joints** (articulations) in the skeleton. Classify joints structurally (e.g. **fibrous**, **cartilaginous**, **synovial**) and functionally (based on the amount of movement permitted by the joint).

☐ 12. Describe the structure and function of a typical **synovial joint** (e.g. the elbow or knee joint), including reference to: **cartilage**, **synovial fluid**, **tendons**, **ligaments**, and named **bones**.

☐ 13. Describe the degenerative changes in the skeleton that occur with increasing age, including a reduction in the rate of bone remodeling, accelerated rates of bone loss, **osteroporosis**, and **osteoarthritis**.

See page 9 for additional details of these texts:

■ Morton, D. and Perry, J.W, 1997. **Photo Atlas for Anatomy and Physiology**, (Brooks/Cole).

■ Rowett, H.G.Q., 1999. **Basic Anatomy & Physiology**, (John Murray), pp. 8-37.

See page 9 for details of publishers of periodicals:

STUDENT'S REFERENCE

■ **Age - Old Story** New Scientist, 23 Jan. 1999, (Inside Science). *The processes involved in aging. An accessible, easy-to-read, but thorough account.*

■ **Adolescence - Hormones Rule OK?** Biol. Sci. Rev., 19(3) Feb. 2007, pp. 2-6. *The role of sex hormones in maturation, including their role in bone growth and skeletal development.*

■ **Aching Joints and Breaking Bones** Biol. Sci. Rev., 20(2) Nov. 2007, pp. 10-13. *As people age they can suffer problems with their bones and joints, including osteoporosis and osteoarthritis.*

TEACHER'S REFERENCE

■ **Teaching Musculoskeletal Anatomy** American Biology Teacher, 62(3), March 2000, pp. 198-201. *Students create their own string model of bone and muscle anatomy to better understand the relationship between structure and function in the skeletal and muscular systems.*

■ **Vascular Spaces in Compact Bone** American Biology Teacher, 65(9), Nov. 2003, pp. 701-707. *A simple technique to better illustrate the 3-D arrangement of spaces in bone.*

■ **Cell Defenses and the Sunshine Vitamin** Scientific American, Nov. 2007, pp. 36-44. *Vitamin D has a critical role in bone formation, but is also involved in immune system responses*

See pages 10-11 for details of how to access **Bio Links** from our web site: **www.thebiozone.com** From Bio Links, access sites under the topics:

ANATOMY & PHYSIOLOGY > **General sites**:
• Anatomy & Physiology • Human physiology lecture notes • WebAnatomy > **Support and Movement**: • Bone and joint sources • The living skeleton • The structure and function of bones

Presentation MEDIA to support this topic:
Anatomy & Physiology

Bone

The skeleton is formed from two hard connective tissues: **bone** and **cartilage**. It has five basic functions: it provides **support** and **protection** for soft tissues and organs, it allows for **movement**, and it is involved in **storing minerals** (calcium and phosphorus) and **blood cell formation**. Bones also have a role in the conduction of sound in the middle ear. Although bone is a hard tissue, it is dynamic and is continually **remodeled** and repaired according to needs and in response to blood calcium levels and the pull of gravity and muscles. Hormones are also involved: the maturation of bones is influenced by thyroid hormones, adrenal androgens, and gonadal sex steroids. Bones have a simple gross structure, typified by the long bone such as the humerus, but a complex internal structure of **lamellae**. Most bones of the skeleton are formed from hyaline cartilage by a process of **ossification** (bone formation) and they grow by **bone remodeling**. Bone remodeling is also important in bone repair.

The Skeletal System

Ossification and Bone Growth

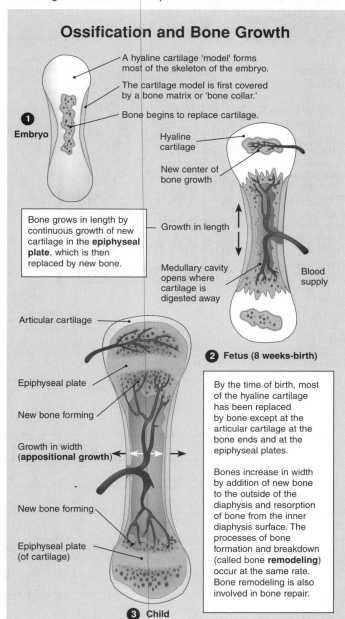

A hyaline cartilage 'model' forms most of the skeleton of the embryo.

The cartilage model is first covered by a bone matrix or 'bone collar.'

Bone begins to replace cartilage.

1 Embryo

Hyaline cartilage

New center of bone growth

Bone grows in length by continuous growth of new cartilage in the **epiphyseal plate**, which is then replaced by new bone.

Growth in length

Medullary cavity opens where cartilage is digested away

Blood supply

2 Fetus (8 weeks-birth)

Articular cartilage

Epiphyseal plate

New bone forming

Growth in width (**appositional growth**)

New bone forming

Epiphyseal plate (of cartilage)

By the time of birth, most of the hyaline cartilage has been replaced by bone except at the articular cartilage at the bone ends and at the epiphyseal plates.

Bones increase in width by addition of new bone to the outside of the diaphysis and resorption of bone from the inner diaphysis surface. The processes of bone formation and breakdown (called bone **remodeling**) occur at the same rate. Bone remodeling is also involved in bone repair.

3 Child

Mature Long Bone

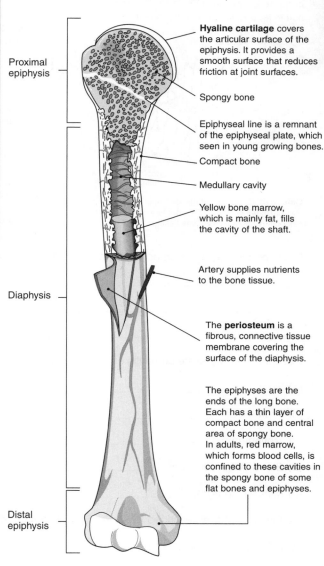

Proximal epiphysis

Hyaline cartilage covers the articular surface of the epiphysis. It provides a smooth surface that reduces friction at joint surfaces.

Spongy bone

Epiphyseal line is a remnant of the epiphyseal plate, which seen in young growing bones.

Compact bone

Medullary cavity

Yellow bone marrow, which is mainly fat, fills the cavity of the shaft.

Diaphysis

Artery supplies nutrients to the bone tissue.

The **periosteum** is a fibrous, connective tissue membrane covering the surface of the diaphysis.

The epiphyses are the ends of the long bone. Each has a thin layer of compact bone and central area of spongy bone. In adults, red marrow, which forms blood cells, is confined to these cavities in the spongy bone of some flat bones and epiphyses.

Distal epiphysis

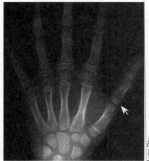

This x-ray shows the epiphyseal plates of a child's hand, seen as separate from the longer bones.

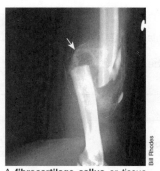

A **fibrocartilage callus** or tissue mass (indicated) begins the repair process on a fractured humerus.

Red bone marrow is stored in the cavities of spongy bone. Here it is being extracted for transplant.

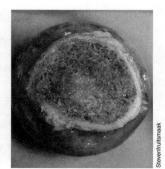

A section of a femur head shows outer compact bone surrounding inner spongy bone and marrow.

The Ultrastructure of Bone

The cells that produce the bone are called **osteoblasts**. They secrete the matrix of calcium phosphate and collagen fibers that forms the rigid bone. Once mature and embedded within the matrix, the bone cells are called **osteocytes**. Dense bone has a very regular structure, composed of repeating units called **Haversian systems**. Each Haversian system has concentric rings of hard material enclosing the bone cells. Haversian canals running through the bone contain blood vessels and nerves so that the bone cells can be supplied with oxygen and nutrients, and wastes can be removed.

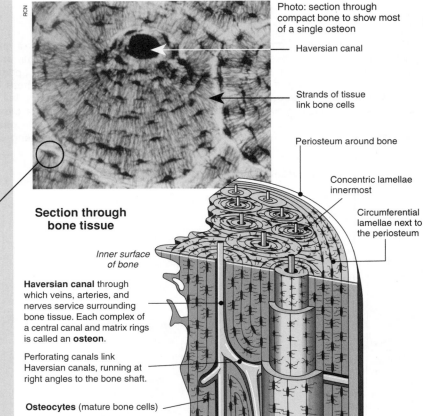

Photo: section through compact bone to show most of a single osteon

Haversian canal

Strands of tissue link bone cells

Periosteum around bone

Concentric lamellae innermost

Circumferential lamellae next to the periosteum

Section through bone tissue

Inner surface of bone

Haversian canal through which veins, arteries, and nerves service surrounding bone tissue. Each complex of a central canal and matrix rings is called an **osteon**.

Perforating canals link Haversian canals, running at right angles to the bone shaft.

Osteocytes (mature bone cells)

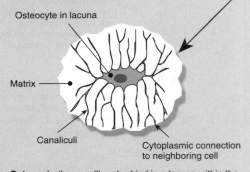

Osteocyte in lacuna

Matrix

Canaliculi

Cytoplasmic connection to neighboring cell

Osteocyte (bone cell) embedded in a lacuna within the matrix (mainly calcium phosphate and collagen). The osteocytes maintain the bone tissue.

1. Describe the way in which bones grow in length and distinguish this from appositional growth:

   _____

   _____

   _____

2. Describe how the skeleton fulfills each of the following functional roles:

   (a) Support: _____

   (b) Protection: _____

   (c) Movement: _____

   (d) Blood cell production: _____

   (e) Mineral storage: _____

3. Identify the feature described by each of the following definitions:

   (a) A feature of bones that are still increasing in length: _____

   (b) The long shaft of a mature bone: _____

   (c) Fibrous, connective tissue membrane covering the surface of the bone shaft: _____

   (d) The end of a long bone, covered in articular hyaline cartilage: _____

4. Distinguish between osteoblasts and osteocytes and describe the role of each: _____

   _____

   _____

5. Explain the function of the Haversian canal system in hard bone tissue: _____

   _____

   _____

The Human Skeleton

The human skeleton consists of two main divisions: the **axial skeleton**, comprising the **skull**, **rib cage**, and **spine**, and the **appendicular skeleton**, made up of the limbs and the pectoral and pelvic girdles. As well as being identified by their location, bones are also described by their shape (e.g. irregular, flat, long, or short), which is related to their functional position in the skeleton. Most of the bones of the upper and lower limbs, for example, are long bones. Bones also have features such as processes, holes (foramina, *sing.* **foramen**), and depressions (**fossae**), associated with nerves, blood vessels, ligaments, and muscles. Understanding the basic organization of the skeleton, the particular features associated with its component bones, and the nature of skeletal articulations (**joints**) is essential to understanding how the movement of body parts is achieved.

WORD LIST:
phalanges, humerus, patella, scapula, tibia, clavicle, sternum, lumbar vertebra, femur, phalanges, cranium, sacrum, metacarpals, rib, ilium, fibula, carpals, tarsals, metatarsals, facial bones

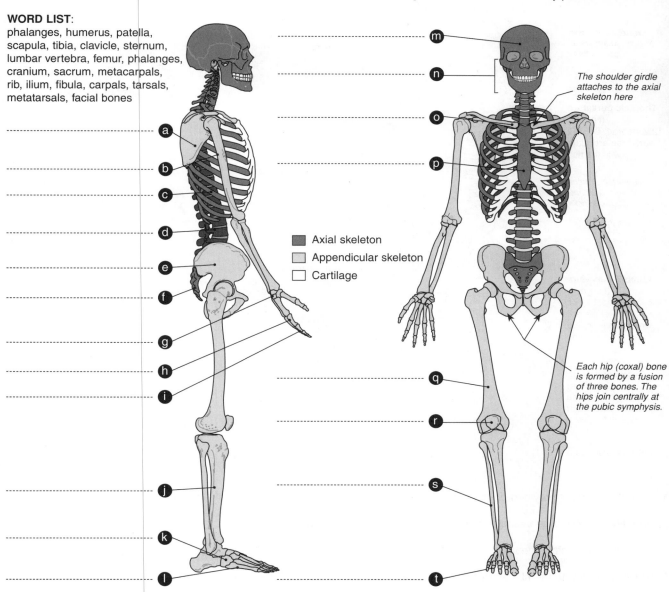

The shoulder girdle attaches to the axial skeleton here

- Axial skeleton
- Appendicular skeleton
- Cartilage

Each hip (coxal) bone is formed by a fusion of three bones. The hips join centrally at the pubic symphysis.

Bone Shapes

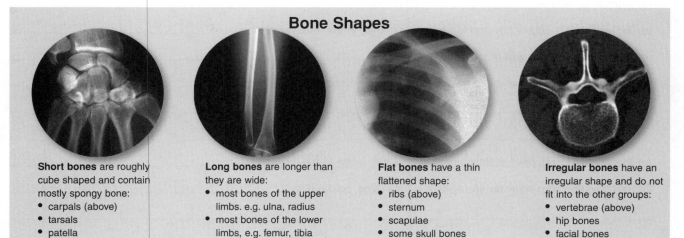

Short bones are roughly cube shaped and contain mostly spongy bone:
- carpals (above)
- tarsals
- patella

Long bones are longer than they are wide:
- most bones of the upper limbs. e.g. ulna, radius
- most bones of the lower limbs, e.g. femur, tibia

Flat bones have a thin flattened shape:
- ribs (above)
- sternum
- scapulae
- some skull bones

Irregular bones have an irregular shape and do not fit into the other groups:
- vertebrae (above)
- hip bones
- facial bones

Related activities: Joints, Aging and the Skeleton   **Web links**: The Vertebral Column, The Axial Skeleton, The Appendicular Skeleton

RA 2

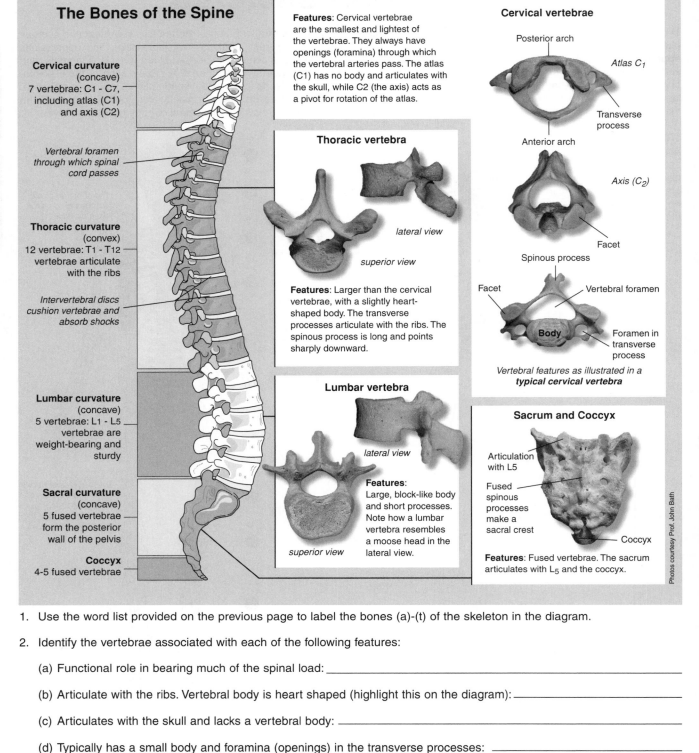

The Bones of the Spine

Cervical curvature
(concave)
7 vertebrae: C1 - C7,
including atlas (C1)
and axis (C2)

*Vertebral foramen
through which spinal
cord passes*

Thoracic curvature
(convex)
12 vertebrae: T1 - T12
vertebrae articulate
with the ribs

*Intervertebral discs
cushion vertebrae and
absorb shocks*

Lumbar curvature
(concave)
5 vertebrae: L1 - L5
vertebrae are
weight-bearing and
sturdy

Sacral curvature
(concave)
5 fused vertebrae
form the posterior
wall of the pelvis

Coccyx
4-5 fused vertebrae

Features: Cervical vertebrae are the smallest and lightest of the vertebrae. They always have openings (foramina) through which the vertebral arteries pass. The atlas (C1) has no body and articulates with the skull, while C2 (the axis) acts as a pivot for rotation of the atlas.

Thoracic vertebra

lateral view

superior view

Features: Larger than the cervical vertebrae, with a slightly heart-shaped body. The transverse processes articulate with the ribs. The spinous process is long and points sharply downward.

Lumbar vertebra

lateral view

Features:
Large, block-like body and short processes. Note how a lumbar vertebra resembles a moose head in the lateral view.

superior view

Cervical vertebrae

Posterior arch

Atlas C$_1$

Transverse process

Anterior arch

Axis (C$_2$)

Facet

Spinous process

Facet

Vertebral foramen

Body

Foramen in transverse process

Vertebral features as illustrated in a **typical cervical vertebra**

Sacrum and Coccyx

Articulation with L5

Fused spinous processes make a sacral crest

Coccyx

Features: Fused vertebrae. The sacrum articulates with L$_5$ and the coccyx.

Photos courtesy Prof. John Bath

1. Use the word list provided on the previous page to label the bones (a)-(t) of the skeleton in the diagram.

2. Identify the vertebrae associated with each of the following features:

   (a) Functional role in bearing much of the spinal load: _____

   (b) Articulate with the ribs. Vertebral body is heart shaped (highlight this on the diagram): _____

   (c) Articulates with the skull and lacks a vertebral body: _____

   (d) Typically has a small body and foramina (openings) in the transverse processes: _____

   (e) Forms the posterior wall of the bony pelvis: _____

3. (a) Describe the function of the shoulder (pectoral) girdle and name its components: _____
   _____
   _____

   (b) Identify the single point of attachment of shoulder girdle to the axial skeleton: _____

4. Explain how and why the male and female pelves (*sing.* pelvis) differ: _____
   _____
   _____

Joints

Bones are too rigid to bend without damage. To allow movement, the skeletal system consists of many bones held together at **joints** by flexible connective tissues called **ligaments**. All movements of the skeleton occur at joints: points of contact between bones, or between cartilage and bones. Joints may be classified structurally as fibrous, cartilaginous, or synovial based on whether fibrous tissue, cartilage, or a joint cavity separates

the bones of the joint. Each of these joint types allows a certain degree of movement. **Fibrous joints**, such as the sutures of the skull, generally allow little or no movement. **Cartilaginous joints** (e.g the pubic **symphysis**) generally allow slight movement, while **synovial joints** enable free movement in one or more planes (see table overleaf). Bones are made to move about a joint by the force of muscles acting upon them.

Cartilaginous Joints

Here, the bone ends are connected by cartilage. Most allow limited movement although some (e.g. between the first ribs and the sternum) are immovable.

Immovable Fibrous Joints

The bones are connected by fibrous tissue. In some (e.g. sutures of the skull), the bones are tightly bound by connective tissue fibers and there is no movement.

Synovial Joints

These allow free movement in one or more planes. The articulating bone ends are separated by a joint cavity containing lubricating synovial fluid (see overleaf).

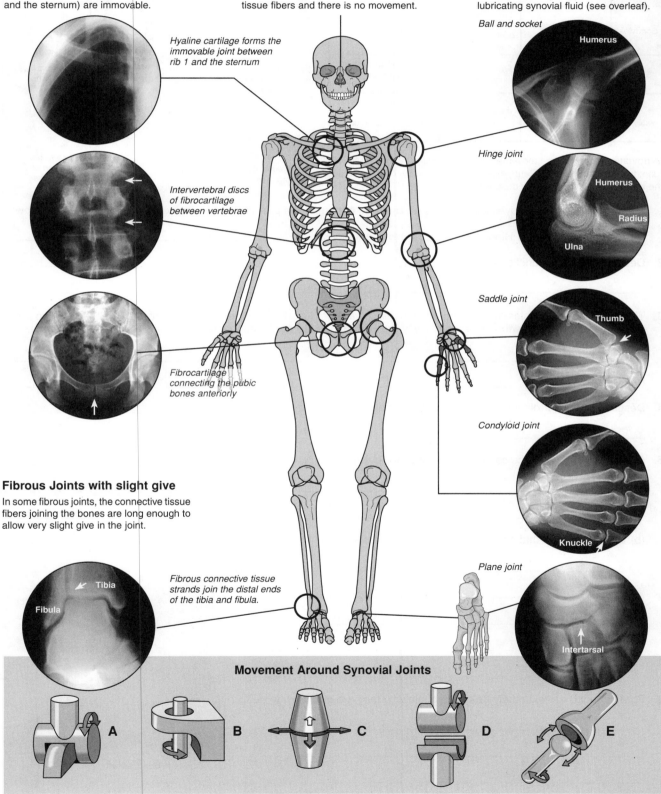

Hyaline cartilage forms the immovable joint between rib 1 and the sternum

Intervertebral discs of fibrocartilage between vertebrae

Fibrocartilage connecting the pubic bones anteriorly

Ball and socket

Humerus

Hinge joint

Humerus

Radius

Ulna

Saddle joint

Thumb

Condyloid joint

Knuckle

Plane joint

Intertarsal

Fibrous Joints with slight give

In some fibrous joints, the connective tissue fibers joining the bones are long enough to allow very slight give in the joint.

Tibia

Fibula

Fibrous connective tissue strands join the distal ends of the tibia and fibula.

Movement Around Synovial Joints

A   B   C   D   E

Related activities: Body Membranes and Cavities, The Mechanics of Locomotion, Autoimmune Diseases **Web links**: Skeleton and Joints

RA 2

Classification of Synovial Joints

Joint	Structure	Movement	Examples
Plane (gliding)	Flat articular surfaces	Short gliding movements	- intercarpal joints - intertarsal joints
Hinge	Cylindrical end of one bone fits into a trough-like surface on the other bone	Angular movement in one plane	- elbow (humerus-ulna) - ankle joint - joints between the phalanges
Pivot	Round end of one bone fits into a sleeve or ring of bone	Rotation around the long axis	- joint between radius and ulna - joint between atlas and axis
Condyloid	Two oval articular surfaces	Free movement in two planes (side to side and back and forth)	- knuckles of the hands
Saddle	One convex and one concave articular surface	Free movement in two planes	- joint between carpals and metacarpals in the thumb
Ball and socket	Spherical head of one bone fits into a round socket on the other	Free movement in all axes	- shoulder joint - hip joint (hip/femur)

Structure of a Synovial Joint

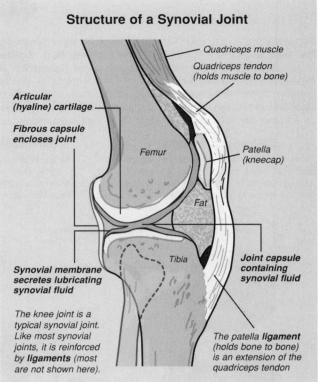

The knee joint is a typical synovial joint. Like most synovial joints, it is reinforced by **ligaments** (most are not shown here).

The patella **ligament** (holds bone to bone) is an extension of the quadriceps tendon

Synovial joints allow free movement of body parts in varying directions (one, two or three planes). Some examples of typical synovial joints and the movements they allow are described above.

1. Classify each of the synovial joint models (**A-E**) at the bottom of the previous page, according to the descriptors below:

   (a) Pivot: _____ (b) Hinge: _____ (c) Ball-and-socket: _____ (d) Saddle: _____ (e) Gliding: _____

2. Describe the basic function of joints: _____

   _____

3. The skull bones of babies at birth and early in infancy are not fused and some areas (the **fontanelles**) have still to be converted to bone. Describe two reasons why the skull bones are not fused into sutures until around 2 years of age:

   (a) _____

   (b) _____

4. Describe the major difference between a fibrous and a cartilaginous joint: _____

   _____

5. (a) Describe the features common to most synovial joints: _____

   _____

   _____

   (b) Explain the role that synovial fluid and cartilage play in the structure and function of a synovial joint:

   _____

   _____

   _____

   _____

   _____

A **sprain** occurs when the ligaments or tendons reinforcing a joint are stretched excessively or torn. The sprain is accompanied by bruising and swelling.

6. Sprains (right) are extremely painful. Explain why they are also slow to heal:

   _____

   _____

   _____

Bruise around sprain

Aging and the Skeleton

After reaching physical maturity the body undergoes **degenerative changes** known as **senescence** or **aging**. The process is characterized by increasing homeostatic imbalance, and increasing susceptibility to stress and disease. Aging occurs as a result of cells dying and renewal rates slowing or stopping. This applies to bone as well as other tissues; the rate of bone remodelling slows and bone resorption rates begin to exceed rates of deposition. Consequently, there is a loss on skeletal strength and an increased tendency for bones to fracture. Note that the degenerative diseases of the skeleton described below (**osteoporosis** and **osteoarthritis**) are related to but not caused by aging. There are many people well into old age who have no signs of skeletal disease.

Osteoarthritis

Osteoarthritis (OA) a chronic degenerative disease aggravated by mechanical stress on bone joints. It is characterized by the degeneration of cartilage and the formation of bony outgrowths (**bone spurs** or osteophytes) around the edges of the eroded cartilage. This leads to pain, stiffness, inflammation, and full or partial loss of joint function. OA occurs in almost all people over the age of 60 and affects three times as many women as men. Weight bearing joints such as those in the knee, foot, hips, and spine are the most commonly affected. Although there is no cure, the symptoms can be greatly relieved by painkillers, anti-inflammatory drugs, and exercises to maintain joint mobility.

Loss of lubricating fluid and cartilage

Spur

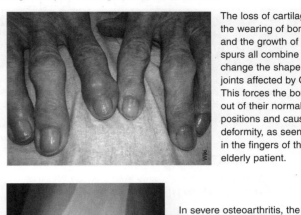

The loss of cartilage, the wearing of bone, and the growth of spurs all combine to change the shape of joints affected by OA. This forces the bones out of their normal positions and causes deformity, as seen in the fingers of this elderly patient.

In severe osteoarthritis, the cartilage can become so thin that it no longer covers the thickened bone ends. The bone ends touch, rub against each other, and start to wear away, as shown in this X-ray of an osteoarthritic knee (left). X rays are the commonly used diagnostic method for OA as the features associated with the disease show up clearly.

Osteoporosis

Osteoporosis is an age-related disorder where bone mass decreases, and there is a loss of height and an increased tendency for bones to break (**fracture**). Women are at greater risk of developing the disease than men because their skeletons are lighter and their estrogen levels fall after menopause (estrogen provides some protection against bone loss). Younger women with low hormone levels and/or low body weight are also affected. Osteoporosis affects the whole skeleton, but especially the spine, hips, and legs.

Loss of height

Hunching of spine

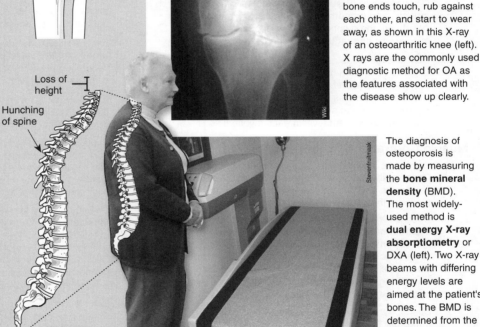

Osteoporosis affecting the spine

The diagnosis of osteoporosis is made by measuring the **bone mineral density** (BMD). The most widely-used method is **dual energy X-ray absorptiometry** or DXA (left). Two X-ray beams with differing energy levels are aimed at the patient's bones. The BMD is determined from the absorption of each beam by bone.

1. Describe the main reason for age-related loss of mass:

   _____

   _____

2. Describe how the structural changes in an osteoathritic joint relate to loss of function: _____

   _____

   _____

3. Explain why loss of bone mineral density is associated with increased risk of bone fracture: _____

   _____

   _____

Related activities: Aging and the Endocrine System
Web links: Osteoarthritis

RA 2

The Muscular System

Understanding the structure and function of muscles and the basis of movement

Muscle structure: smooth, skeletal, and cardiac muscle. The ultrastructure of skeletal muscle, muscle contraction, muscle tone and posture, energy and muscle fatigue

Learning Objectives

☐ 1. Compile your own glossary from the **KEY WORDS** displayed in **bold type** in the learning objectives below.

Muscle Structure *(pages 46-47, 57-59)*

☐ 2. Describe the distinguishing features of muscle tissue and its various roles. Describe the features of the three types of muscle tissue: **cardiac muscle**, **skeletal muscle**, and **smooth muscle**. Include reference to the cell shape, the number of nuclei, appearance under microscopy, and regulation of contraction.

☐ 3. Describe the ultrastructure of **skeletal muscle**:
   • The organization of the muscle into bundles of **fibers**.
   • The organization of the individual fibers (muscle cells) into bundles of smaller **myofibrils**.
   • The composition and arrangement of the thick and thin **(myo)filaments** within each myofibril.

Muscle Contraction *(pages 46-47, 57-59)*

☐ 4. Recognize a **sarcomere** as a complete contractile unit. Describe the relationship between the structure of the sarcomere and the distribution of **actin** and **myosin** (thin and thick filaments) within the myofibril.

☐ 5. Describe the **sliding-filament hypothesis** of **muscle contraction**. Identify the role of **calcium ions** (Ca^{2+}), the **troponin-tropomyosin** complex, and **ATP** in the cycle of **actin/myosin bridge** (cross-bridge) formation. Recognize that contraction of a muscle fiber is an **all-or-nothing** event, but the reactions of the muscle as a whole may be graded (see #8).

☐ 6. Describe the structure and function of the **neuro-muscular junction**, including the role of **acetylcholine** (Ach), the **T-tubules**, and the **sarcoplasmic reticulum**.

☐ 7. Describe the structural differences between **fast twitch** and **slow twitch** muscle fibers. Explain the physiological basis of these differences.

The Functioning Muscle *(pages 60-65)*

☐ 8. Recognize graded responses in muscles as a whole and describe the two ways in which these are achieved:
   (a) By changing the frequency of stimulation (to tetany)
   (b) By recruitment of motor units.

☐ 9. Identify sources of energy for muscle contraction and describe ATP production in muscle during exercise of varying intensity and duration. In terms of energy yield and waste products, compare **aerobic** and **anaerobic** pathways as sources of ATP for muscle contraction.

☐ 10. Explain **muscle fatigue** and relate it to the increase in **blood lactate**, depletion of carbohydrate supplies, and decreased pH. Explain how these changes provide the stimulus for increased breathing (and heart) rates.

☐ 11. Explain what is meant by **oxygen debt**. Describe the ultimate fate of blood lactate and explain how the oxygen debt is repaid after intense exercise.

☐ 12. Explain the effects of regular exercise on muscle, contrasting the effects of **endurance** (aerobic) **exercise** and **resistance** (or isometric) **exercise**. Describe the effects of inactivity on muscle and identify some of the common causes of this.

☐ 13. Understand how joints, together with **antagonistic muscles**, permit particular movements of specific parts of the skeleton. With reference to muscles and their activity, define the terms **origin** and **insertion** Distinguish between **isotonic contraction** and **isometric contraction** and explain how **muscle tone** is achieved and maintained.

☐ 14. Describe examples of movement of parts of the skeleton in terms of **antagonistic muscle action**. Understand the use of the terms **extension**, **flexion**, **rotation**, **abduction**, **adduction**, and **circumduction** when describing the movement of body parts. Identify the role of reflex inhibition in the movement of antagonistic muscle pairs.

Supplementary Texts

See page 9 for additional details of these texts:

■ Morton, D. and Perry, J.W, 1997. **Photo Atlas for Anatomy and Physiology**, (Brooks/Cole).

■ Rowett, H.G.Q., 1999. **Basic Anatomy & Physiology**, (John Murray), pp. 38-59.

Anatomy & Physiology

Presentation MEDIA to support this topic: **Anatomy & physiology**

Periodicals

See page 9 for details of publishers of periodicals:

STUDENT'S REFERENCE
■ **Human Muscle: Structure and Function** Biol. Sci. Rev., 19(4) April 2007, pp. 25-29. *The structure and function of muscle (including skeletal muscle) in humans, the physiology of contraction and the mechanics of locomotion.*

TEACHER'S REFERENCE
■ **Acting Out Muscle Contraction** The Am. Biology Teacher, 65(2), Feb. 2003, pp. 128-132. *An experiment to demonstrate muscle contraction.*

■ **Demonstrating the Stretch Reflex** The Am. Biology Teacher, 62(7), Sept., 2000, pp. 503-507. *Demonstrating muscular coordination and control.*

Internet

See pages 10-11 for details of how to access **Bio Links** from our web site: **www.thebiozone.com** From Bio Links, access sites under the topics:

ANATOMY & PHYSIOLOGY > **General sites**: • Anatomy & Physiology • Human physiology lecture notes • WebAnatomy > **Support and Movement**: • Anatomy of skeletal muscle • Energy systems: Part 1 • Exercise physiology • Muscle spindle and stretch reflexes • Muscle structure and function • Muscles

Muscle Structure and Function

There are three kinds of muscle tissue: **skeletal, cardiac**, and **smooth** muscle, each with a distinct structure. The muscles used for posture and locomotion are skeletal (voluntary) muscles and are largely under conscious control. Their distinct appearance is the result of the regular arrangement of contractile elements within the muscle cells. Muscle fibers are innervated by the branches of motor neurons, each of which terminates in a specialized cholinergic synapse called the **neuromuscular junction** (or motor end plate). A motor neuron and all the fibers it innervates (which may be a few or several hundred) are called a **motor unit**.

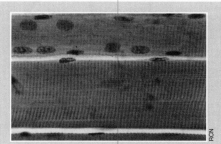
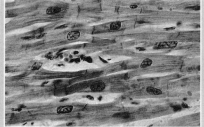

Skeletal muscle
Also called striated or striped muscle. It has a banded appearance under high power microscope. Sometimes called voluntary muscle because it is under conscious control. The cells are large with many nuclei at the edge of each cell.

Cardiac muscle
Specialized striated muscle that does not fatigue. Cells branch and connect with each other to assist the passage of nerve impulses through the muscle. Cardiac muscle is not under conscious control (it is involuntary).

Smooth muscle
Also called involuntary muscle because it is not under conscious control. Contractile filaments are irregularly arranged so the contraction is not in one direction as in skeletal muscle. Cells are spindle shaped with one central nucleus.

Structure of Skeletal Muscle

Skeletal muscle is organized into bundles of muscle cells or **fibers**. Each fiber is a single cell with many nuclei and each fiber is itself a bundle of smaller **myofibrils** arranged lengthwise. Each myofibril is in turn composed of two kinds of **myofilaments** (thick and thin), which overlap to form light and dark bands. It is the alternation of these light and dark bands which gives skeletal muscle its striated or striped appearance. The **sarcomere**, bounded by the dark Z lines, forms one complete contractile unit.

Longitudinal section of a sarcomere

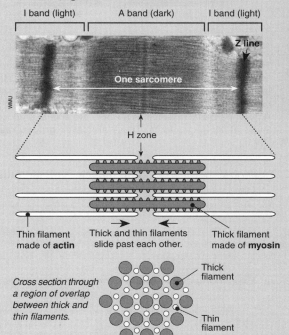

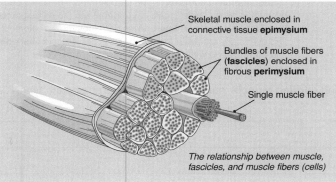

The relationship between muscle, fascicles, and muscle fibers (cells)

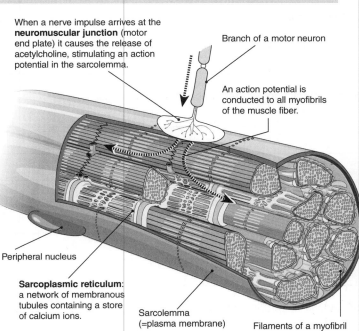

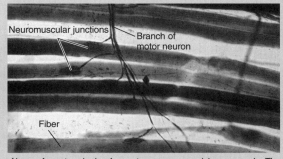

Above: Axon terminals of a motor neuron supplying a muscle. The branches of the axon terminate on the sarcolemma of a fiber at regions called the neuromuscular junction. Each fiber receives a branch of an axon, but one axon may supply many muscle fibers.

The Muscular System

Related activities: The Sliding Filament Theory, Chemical Synapses, Integration at Synapses **Web links**: Muscle Structure and Contraction

RA 2

The Banding Pattern of Myofibrils

Within a myofibril, the thin filaments, held together by the **Z lines**, project in both directions. The arrival of an action potential sets in motion a series of events that cause the thick and thin filaments to slide past each other. This is called **contraction** and it results in shortening of the muscle fiber and is accompanied by a visible change in the appearance of the myofibril: the I band and the sarcomere shorten and H zone shortens or disappears (below).

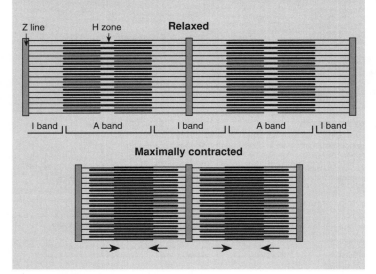

Z line   H zone   **Relaxed**

I band | A band | I band | A band | I band

Maximally contracted

The response of a single muscle fiber to stimulation is to contract maximally or not at all; its response is referred to as the **all-or-none law** of muscle contraction. If the stimulus is not strong enough to produce an action potential, the muscle fiber will not respond. However skeletal muscles as a whole are able to produce varying levels of contractile force. These are called **graded responses**.

When Things Go Wrong

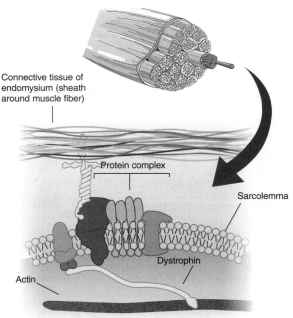

Connective tissue of endomysium (sheath around muscle fiber)

Protein complex

Sarcolemma

Dystrophin

Actin

Duchenne's muscular dystrophy is an X-linked disorder caused by a mutation in the gene DMD, which codes for the protein **dystrophin**. The disease causes a rapid deterioration of muscle, eventually leading to loss of function and death. It is the most prevalent type of muscular dystrophy and affects only males. Dystrophin is an important structural component within muscle tissue and it connects muscles fibers to the extracellular matrix through a protein complex on the sarcolemma. The absence of dystrophin allows excess calcium to penetrate the sarcolemma (the fiber's plasma membrane). This damages the sarcolemma, and eventually results in the death of the cell. Muscle fibers die and are replaced with adipose and connective tissue.

1. Distinguish between **smooth muscle**, **striated muscle**, and **cardiac muscle**, summarizing the features of each type:

   _____

   _____

   _____

2. (a) Explain the cause of the banding pattern visible in striated muscle: _____

   _____

   _____

   (b) Explain the change in appearance of a myofibril during contraction with reference to the following:

   The I band: _____

   The H zone: _____

   The sarcomere: _____

3. Describe the purpose of the connective tissue sheaths surrounding the muscle and its fascicles: _____

   _____

4. Explain what is meant by the all-or-none response of a muscle fiber: _____

   _____

5. Explain why the inability to produce **dystrophin** leads to a loss of muscle function: _____

   _____

The Sliding Filament Theory

The previous activity described how muscle contraction is achieved by the thick and thin muscle filaments sliding past one another. This sliding is possible because of the structure and arrangement of the thick and thin filaments. The ends of the thick myosin filaments are studded with heads or **cross bridges** that can link to the thin filaments next to them. The thin filaments contain the protein actin, but also a regulatory protein complex. When the cross bridges of the thick filaments connect to the thin filaments, a shape change moves one filament past the other. Two things are necessary for cross bridge formation: calcium ions, which are released from the **sarcoplasmic reticulum** when the muscle receives an action potential, and ATP, which is hydrolyzed by ATPase enzymes on the myosin. When cross bridges attach and detach in sarcomeres throughout the muscle cell, the cell shortens. Although a muscle fiber responds to an action potential by contracting maximally, skeletal muscles as a whole can produce varying levels of contractile force. These **graded responses** are achieved by changing the frequency of stimulation (**frequency summation**) and by changing the number and size of motor units recruited (**multiple fiber summation**). Maximal contractions of a muscle are achieved when nerve impulses arrive at the muscle at a rapid rate and a large number of motor units are active at once.

The Sliding Filament Theory

Muscle contraction requires calcium ions (Ca^{2+}) and energy (in the form of ATP) in order for the thick and thin filaments to slide past each other. The steps are:

1. The binding sites on the **actin** molecule (to which myosin 'heads' will locate) are blocked by a complex of two protein molecules: tropomyosin and troponin.

2. Prior to muscle contraction, ATP binds to the heads of the myosin molecules, priming them in an erect high energy state. Arrival of an action potential causes a release of Ca^{2+} from the sarcoplasmic reticulum. The Ca^{2+} binds to the troponin and causes the blocking complex to move so that the myosin binding sites on the actin filament become exposed.

3. The heads of the cross-bridging myosin molecules attach to the binding sites on the actin filament. Release of energy from the hydrolysis of ATP accompanies the cross bridge formation.

4. The energy released from ATP hydrolysis causes a change in shape of the myosin **cross bridge**, resulting in a bending action (*the power stroke*). This causes the actin filaments to slide past the myosin filaments towards the centre of the sarcomere.

5. (Not illustrated). Fresh ATP attaches to the myosin molecules, releasing them from the binding sites and repriming them for a repeat movement. They become attached further along the actin chain as long as ATP and Ca^{2+} are available.

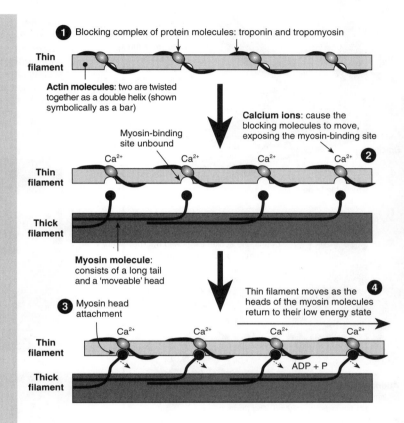

1 Blocking complex of protein molecules: troponin and tropomyosin

Thin filament

Actin molecules: two are twisted together as a double helix (shown symbolically as a bar)

Calcium ions: cause the blocking molecules to move, exposing the myosin-binding site

Myosin-binding site unbound

Ca^{2+} Ca^{2+} Ca^{2+} Ca^{2+} **2**

Thin filament

Thick filament

Myosin molecule: consists of a long tail and a 'moveable' head

3 Myosin head attachment

4 Thin filament moves as the heads of the myosin molecules return to their low energy state

Ca^{2+} Ca^{2+} Ca^{2+} Ca^{2+}

Thin filament

Thick filament

ADP + P

The Muscular System

1. Match the following chemicals with their functional role in muscle movement (draw a line between matching pairs):

(a) Myosin
• Bind to the actin molecule in a way that prevents myosin head from forming a cross bridge

(b) Actin
• Supplies energy for the flexing of the myosin 'head' (power stroke)

(c) Calcium ions
• Has a moveable head that provides a power stroke when activated

(d) Troponin-tropomyosin
• Two protein molecules twisted in a helix shape that form the thin filament of a myofibril

(e) ATP
• Bind to the blocking molecules, causing them to move and expose the myosin binding site

2. Describe the two ways in which a muscle as a whole can produce contractions of varying force:

(a) _____

_____

(b) _____

_____

3. (a) Identify the two things necessary for cross bridge formation: _____

(b) Explain where each of these comes from: _____

_____

Related activities: Muscle Structure and Function
Web links: Muscle Cell Contraction

A 3

The Mechanics of Movement

We are familiar with the many different bodily movements achievable through the action of muscles. Contractions in which the length of the muscle shortens in the usual way are called **isotonic contractions**: the muscle shortens and movement occurs. When a muscle contracts against something immovable and does not shorten the contraction is called **isometric**. Skeletal muscles are attached to bones by tough connective tissue structures called **tendons**. They always have at least two attachments: the **origin** and the **insertion**. They create movement of body parts when they contract across **joints**. The type and degree of movement achieved depends on how much movement the joint allows and where the muscle is located in relation to the joint. Some common types of body movements are described below (left panel). Because muscles can only pull and not push, most body movements are achieved through the action of opposing sets of muscles (below, right panel).

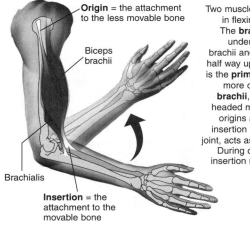

Origin = the attachment to the less movable bone

Biceps brachii

Brachialis

Insertion = the attachment to the movable bone

Two muscles are involved in flexing the forearm. The **brachialis**, which underlies the biceps brachii and has an origin half way up the humerus, is the **primer mover**. The more obvious **biceps brachii**, which is a two headed muscle with two origins and a common insertion near the elbow joint, acts as the synergist. During contraction, the insertion moves towards the origin.

The Action of Antagonistic Muscles

The skeleton works as a system of levers. The joint acts as a **fulcrum**, the muscles exert the **force**, and the weight of the bone being moved represents the **load**. The flexion (bending) and extension (unbending) of limbs is caused by the action of **antagonistic muscles**; muscles that work in pairs and whose actions oppose each other. Every coordinated movement in the body requires the application of muscle force. This is accomplished by the action of agonists, antagonists, and synergists. The opposing action of agonists and antagonists also produces muscle tone. Note that either muscle in an antagonistic pair can act as the **prime mover**, depending on the movement (e.g. flexion or extension).

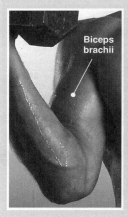

Biceps brachii

Agonists or prime movers: muscles that are primarily responsible for the movement and produce most of the force required.

Antagonists: muscles that oppose the prime mover. They may also play a protective role by preventing over-stretching of the prime mover.

Synergists: muscles that assist the prime movers and may be involved in fine-tuning the direction of the movement.

During flexion of the forearm (left) the **brachialis** muscle acts as the prime mover and the **biceps brachii** is the synergist. The antagonist, the **triceps brachii** at the back of the arm, is relaxed. During extension, their roles are reversed.

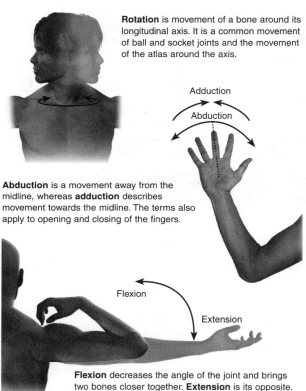

Rotation is movement of a bone around its longitudinal axis. It is a common movement of ball and socket joints and the movement of the atlas around the axis.

Adduction

Abduction

Abduction is a movement away from the midline, whereas **adduction** describes movement towards the midline. The terms also apply to opening and closing of the fingers.

Flexion

Extension

Flexion decreases the angle of the joint and brings two bones closer together. **Extension** is its opposite. Extension more than 180° is called **hyperextension**.

Quadriceps

Hamstrings

Movement of the leg is achieved through the action of several large groups of muscles, collectively called the **quadriceps** and the **hamstrings**.

The hamstrings are actually a collection of three muscles, which act together to flex the leg. The quadriceps at the front of the thigh (a collection of four large muscles) opposes the motion of the hamstrings and extends the leg. When the prime mover contracts forcefully, the antagonist also contracts very slightly. This stops overstretching and allows greater control over thigh movement.

1. Using appropriate terminology, explain how antagonistic muscles act together to raise and lower a limb:

    _____

    _____

2. (a) Identify the insertion for the biceps brachii during flexion of the forearm: _____

    (b) Identify the insertion of the brachialis muscle during flexion of the forearm: _____

    (c) Given its insertion, describe the forearm movement during which the biceps brachialis is the prime mover:

    _____

Related activities: Joints
Web links: Muscles in Action

Energy for Muscle Contraction

Exercise places an immediate demand on the body's energy supply systems. During exercise, the metabolic rate of the muscles increases by up to 20 times and the body's systems must respond appropriately to maintain homeostasis. The ability to exercise for any given length of time depends on maintaining adequate supplies of ATP to the muscles. There are three energy systems to do this: the **ATP-CP system**, the **glycolytic system**, and the **oxidative system**. The ultimate sources of energy for ATP generation in muscle via these systems are glucose, and stores of glycogen and triglycerides. Prolonged intense exercise utilizes the oxidative system, and relies on a constant supply of oxygen to the tissues. The VO2 is the amount of oxygen (expressed as a volume) used by muscles during a specified interval for cell metabolism and energy production. **VO2max** is the maximum volume of oxygen that can be delivered and used per minute and therefore represents an individual's upper limit of aerobic metabolism. VO2max is used as a measure of fitness, and is high in trained athletes. At some percentage of VO2max (the **anaerobic threshold**) the body is unable to meet its energy demands aerobically and an **oxygen debt** is incurred.

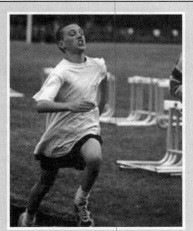

CP provides enough energy to fuel about 10 s of maximum effort (e.g. a 100 m race).

The ATP-CP System

The simplest of the energy systems is the **ATP-CP system**. CP or **creatine phosphate** is a high energy compound that stores energy sufficient for brief periods of muscular effort. Energy released from the breakdown of CP is not used directly to accomplish cellular work. Instead it rebuilds ATP to maintain a relatively constant supply. This process is anaerobic, occurs very rapidly, and is accomplished without any special structures in the cell. CP levels decline steadily as it is used to replenish depleted ATP levels. The ability of the ATP-CP system to maintain energy levels is limited to 3-15 seconds during an all out sprint. Beyond this, the muscle must rely on other processes for ATP generation.

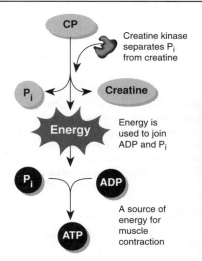

Rugby and other field sports demand brief intense efforts with recovery in-between.

The Glycolytic System

ATP can also be provided by glycolysis: the first phase of cellular respiration. The ATP yield from glycolysis is low (only net 2ATP per molecule of glucose), but it produces ATP rapidly and does not require oxygen. The fuel for the glycolytic system is glucose in the blood, or glycogen, which is stored in the muscle or liver and broken down to glucose-6-phosphate. Glycolysis provides ATP for exercise for just a few minutes. Its main limitation is that it causes lactic acid ($C_3H_6O_8$) to accumulate in the tissues. Indirectly, it is the accumulation of lactic acid that gives the feeling of muscle fatigue. The lactic acid must transported to the liver and respired aerobically. The extra oxygen needed for this is the **oxygen debt**.

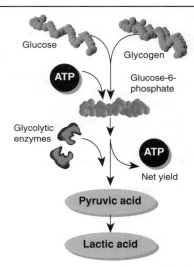

Prolonged aerobic effort (e.g. distance running) requires a sustained ATP supply.

The Oxidative System

In the oxidative system, glucose is completely broken down to yield (about) 36 molecules of ATP. This process uses oxygen and takes place within the mitochondria. Aerobic metabolism has a high energy yield and is the primary method of energy production during sustained high activity. It is reliant on a continued supply of oxygen and therefore on the body's ability to deliver oxygen to the muscles. The fuels for aerobic respiration are glucose, stored glycogen, or stored **triglycerides**. Triglycerides provide free fatty acids, which are oxidized in the mitochondria by the successive removal of two-carbon fragments (a process called β-oxidation). These two carbon units enter the Krebs cycle as acetyl coenzyme A (acetyl CoA).

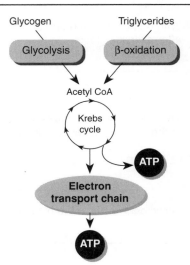

Oxygen Uptake During Exercise and Recovery

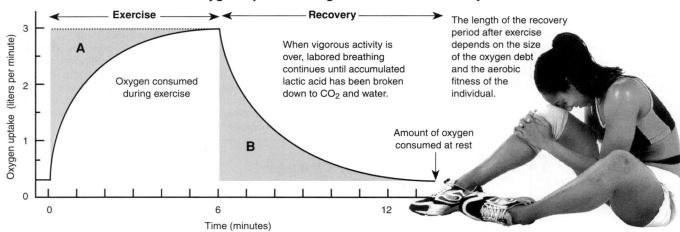

The graph above illustrates the principle of oxygen debt. In the graph, the energy demands of aerobic exercise require 3 liters of oxygen per minute. The rate of oxygen uptake increases immediately exercise starts, but the full requirement is not met until six minutes later. The **oxygen deficit** is the amount of oxygen needed (for aerobic energy supply) but not supplied by breathing. During the first six minutes, the energy is supplied largely from anaerobic pathways: the ATP-CP and glycolytic systems (previous page). After exercise, the oxygen uptake per minute does not drop immediately to its resting level. Extra oxygen is taken in despite the drop in energy demand (the **oxygen debt**). The oxygen debt is used to replace oxygen reserves in the body, restore creatine phosphate, and break down the lactic acid (through various intermediates) to CO_2 and water.

1. Explain why the supply of energy through the glycolytic system is limited: _____

_____

_____

2. Summarize the features of the three energy systems in the table below:

	ATP-CP system	Glycolytic system	Oxidative system
ATP supplied by:			
Duration of ATP supply:			

3. Study the graph and explanatory paragraph above, then identify and describe what is represented by:

   (a) The shaded region **A**: _____

   _____

   (b) The shaded region **B**: _____

   _____

4. With respect to the graph above, explain why the rate of oxygen uptake does not immediately return to its resting level after exercise stops:

_____

_____

5. The rate of oxygen uptake increases immediately exercise starts. Explain how the oxygen supply from outside the body to the cells is increased during exercise:

_____

_____

6. Lactic acid levels in the blood continue to rise for a time after exercise has stopped. Explain why this occurs:

_____

_____

Muscle Tone and Posture

Even when we consciously relax a muscle, a few of its fibers at any one time will be involuntarily active. This continuous and passive partial contraction of the muscles is responsible for **muscle tone** and is important in maintaining **posture**. The contractions are not visible but they are responsible for the healthy, firm appearance of muscle. The amount of muscle contraction is monitored by sensory receptors in the muscle called **muscle spindle organs**. These provide the sensory information necessary to adjust movement as required. Abnormally low muscle tone (**hypotonia**) can arise as a result of traumatic or degenerative nerve damage, so that the muscle no longer receives the innervation it needs to contract. The principal treatment for these disorders is physical therapy to help the person compensate for the neuromuscular disability.

We are usually not aware of the skeletal muscles that maintain posture, although they work almost continuously making fine adjustments to maintain body position. Both posture and functional movements of the body are highly dependent on the strength of the body's core (the muscles in the pelvic floor, belly, and mid and lower back). The core muscles stabilize the thorax and the pelvis and lack of core strength is a major contributor to postural problems and muscle imbalances.

Physical therapy is a branch of health care concerned with maintaining or restoring functional movement throughout life. Loss of muscle tone and strength can develop as a result of aging, disease, or trauma. As a result of not being used, muscles will **atrophy**, losing both mass and strength. Although the type of physical therapy depends on the problem, it usually includes therapeutic exercise to help restore mobility and strength, and prevent or slow down the loss of muscle tissue.

The Role of the Muscle Spindle

Changes in length of a muscle are monitored by the **muscle spindle organ**, a stretch receptor located within skeletal muscle, parallel to the muscle fibers themselves. The muscle spindle is stimulated in response to sustained or sudden stretch on the central region of its specialized intrafusal fibers. Sensory information from the muscle spindle is relayed to the spinal cord. The motor response brings about adjustments to the degree of stretch in the muscle. These adjustments help in the coordination and efficiency of muscle contraction. Muscle spindles are important in the maintenance of muscle tone, postural reflexes, and movement control, and are concentrated in muscles that exert fine control over movement.

Motor nerves send impulses to adjust the degree of contraction in the intrafusal and extrafusal fibers.

Sensory nerves monitor stretch in the non-contractile region of the spindle and send impulses to the spinal cord.

Striated appearance of contractile elements

Nucleus of muscle fiber

The **muscle spindle organ** comprises special **intrafusal fibers** which lie parallel to the muscle fibers within a lymph-filled capsule. Only the regions near the end can contract.

The spindle is surrounded by the muscle fibers of the skeletal muscle.

The Muscular System

1. (a) Explain what is meant by muscle tone: _____

   (b) Explain how this is achieved: _____
   _____

2. (a) Explain the role of the muscle spindle organ: _____
   _____

   (b) With reference to the following, describe how the structure of the muscle spindle organ is related to its function:

   Intrafusal fibers lie parallel to the extrafusal fibers: _____
   _____

   Sensory neurons are located in the non-contractile region of the organ: _____
   _____

   Motor neurons synapse in the extrafusal fibers and the contractile region of the intrafusal fibers: _____
   _____

Muscle Fatigue

Muscle fatigue refers to the decline in a muscle's ability to maintain force in a prolonged or repeated contraction. It is a normal result of vigorous exercise but the reasons for it are complex. During strenuous activity, the oxygen demand of the working muscle exceeds supply and the muscle metabolizes anaerobically. This leads to the metabolic changes (including lactic acid accumulation) that result in fatigue. The lactic acid (lactate) produced during muscle contraction diffuses into the bloodstream and is carried to the liver where it is reconverted to glucose via the **Cori cycle**. This requires energy from aerobic respiration and accounts for the **oxygen debt** which causes the continued heavy breathing after exercise has stopped. During moderate intensity exercise, the rate at which lactate enters the blood from the muscles is equal to its rate of removal by the liver. This steady-state situation is suitable for endurance training, because the lactic acid doesn't accumulate in the muscle tissue.

At Rest

- Muscles produce a surplus of ATP
- This extra energy is stored in CP (creatine phosphate) and glycogen

During Moderate Activity

- ATP requirements are met by the aerobic metabolism of glycogen and lipids

During Peak Activity

- Effort is limited by ATP. ATP production is ultimately limited by availability of oxygen.
- During short-term, intense activity, ATP is increased through anaerobic metabolism of glycogen (glycolysis).
- A by-product of this is lactic acid, which lowers tissue pH and affects cellular activity.
- When **fatigued**, the muscle can no longer contract fully.

Lactic Acid and Muscle Fatigue

Lactic acid is a by-product of ATP production through anaerobic metabolism when oxygen demand exceeds supply. Lactic acid accumulation in the muscle causes a fall in tissue pH and inhibits the activity of the key enzymes involved in ATP production. This decline in ATP supply limits muscular performance during peak activity (see graph below). Together with the effects of ATP and creatine phosphate breakdown (accumulating phosphate (P_i) for example), lactic acid buildup also acts to slow the release of calcium into the T tubules and affects the ion pumps which move calcium ions back into the sarcoplasmic reticulum. This contributes to fatigue because calcium is a key component in muscle contraction.

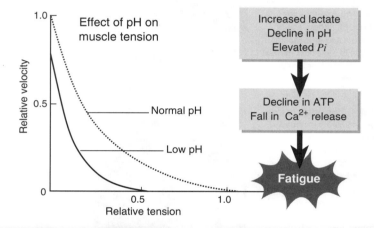

Effect of pH on muscle tension

Relative velocity — Relative tension

Normal pH

Low pH

Increased lactate
Decline in pH
Elevated P_i

→

Decline in ATP
Fall in Ca^{2+} release

→

Fatigue

Short term maximal exertion (sprint)
- Lactic acid build-up lowers pH ● Depletion of creatine phosphate ● Buildup of phosphate (P_i) affects the sensitivity of the muscle to Ca^{2+}

Mixed aerobic and anaerobic (5 km race)
- Lactate accumulation in the muscle ● Build-up of ADP and P_i ● Decline in Ca^{2+} release affects the ability of the muscle to contract

Extended sub-maximal effort (marathon)
- Depletion of all energy stores (glycogen, lipids, amino acids) leads to a failure of Ca^{2+} release
- Repetitive overuse damages muscle fibers.

1. Explain the mechanism by which lactic acid accumulation leads to muscle fatigue: _____

_____

2. Identify the two physiological changes in the muscle that ultimately result in a decline in muscle performance:

   (a) _____   (b) _____

3. Suggest why the reasons for fatigue in a long distance race are different to those in a 100 m sprint: _____

_____

_____

Related activities: Energy and Exercise, Muscle Physiology and Performance

Muscle Physiology and Performance

The overall effect of **aerobic training** on muscle is improved oxidative function and better endurance. Regardless of the type of training, some of our ability to perform different types of activity depends on our genetic make-up. This is particularly true of aspects of muscle physiology, such as the relative proportions of different fiber types in the skeletal muscles. Muscle fibers are primarily of two types: **fast twitch** (FT) or **slow twitch** (ST). Fast twitch fibers predominate during anaerobic, explosive activity, whereas slow twitch fibers predominate during endurance activity. In the table below, note the difference in the degree to which the two fiber types show fatigue (a decrease in the capacity to do work). Training can increase fiber size and, to some extent, the makeup of the fiber, but not the proportion of ST to FT, which is genetically determined.

The Effects of Aerobic Training on Muscle Physiology

Improved oxidation of glycogen. Training increases the capacity of skeletal muscle to generate ATP aerobically.

An increased capacity of the muscle to oxidize fats. This allows muscle and liver glycogen to be used at a slower rate. The body also becomes more efficient at mobilizing free fatty acids from adipose tissue for use as fuel.

Increased myoglobin content. Myoglobin stores oxygen in the muscle cells and aids oxygen delivery to the mitochondria. Endurance training increases muscle myoglobin by 75%-80%.

Increase in lean muscle mass and decrease in body fat. Trained endurance athletes typically have body fat levels of 15-19% (women) or 6-18% (men), compared with 26% (women) and 15% (men) for non-athletes.

The size of **slow twitch fibers** increases. This change in size is associated with increased aerobic capacity.

An increase in the size and density of mitochondria in the skeletal muscles and an increase in the activity and concentration of Krebs cycle enzymes.

An increase in the number of capillaries surrounding each muscle fiber. Endurance trained men have 5%-10% more capillaries in their muscles than sedentary men.

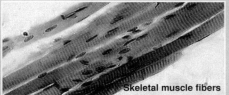

Skeletal muscle fibers

Fast vs Slow Twitch Muscle

Feature	Fast twitch	Slow twitch
Color	White	Red
Diameter	Large	Small
Contraction rate	Fast	Slow
ATP production	Fast	Slow
Metabolism	Anaerobic	Aerobic
Rate of fatigue	Fast	Slow
Power	High	Low

There are two basic types of muscle fibers. Slow twitch (type I) muscle and fast twitch (type II) muscle fiber. Human muscles contain a genetically determined mixture of both slow and fast fiber types. On average, we have about 50% slow and 50% fast fibers in most of the muscles used for movement.

The slow twitch fibers contain more mitochondria and myoglobin than fast twitch fibers, which makes them more efficient at using oxygen to generate ATP without lactate acid build up. In this way, they can fuel repeated and extended muscle contractions such as those required for endurance events like a marathon.

Both fiber types generally produce the same force per contraction, but fast twitch fibers produce that force at a higher rate, so fast fibers are important when there is a limited time in which to generate maximal force (as in a sprint)

1. Explain three ways in which aerobic (endurance) training improves the oxidative function of muscle:

   (a) _____

   _____

   (b) _____

   _____

   (c) _____

   _____

2. Contrast the properties of fast and slow twitch skeletal muscle fibers, identifying how these properties contribute to their performance in different conditions:

   _____

   _____

   _____

The Muscular System

Related activities: Exercise and Blood Flow, Muscle Fatigue

RA 2

Nervous and Endocrine Systems

Endocrine system

- The hypothalamus controls the activity of the anterior pituitary gland.
- Sympathetic NS stimulates the release of "fight or flight" hormones from the adrenal medulla.

Cardiovascular system

- ANS regulates heart rate and blood pressure
- Cardiovascular system supplies nervous and endocrine tissues with O_2 and removes wastes
- Hormones influence blood volume and pressure, and heart activity.
- The hormones erythropoietin (EPO) stimulates production of red blood cells.

Respiratory system

- Nervous system regulates breathing
- Epinephrine (adrenaline) dilates bronchioles and increases the rate and depth of breathing
- Respiratory system enables gas exchange, providing O_2 and removing CO_2
- Angiotensin II, which increases blood pressure, is activated in the lung capillaries.

Skeletal system

- Nerves supply bones.
- The skeleton protects the nervous and endocrine systems and provides Ca^{2+} for nerve function.
- Parathyroid hormone regulates blood calcium.
- Growth hormone, and thyroid and sex hormones regulate skeletal growth and development.

Integumentary system

- Sebaceous glands are influenced by sex hormones.
- Sympathetic NS activity regulates sweat glands and blood vessels in the skin (**thermoregulation**).

Nervous system

- Hormones (e.g. growth hormone, thyroid hormones, and sex hormones) influence the growth, development, and activity of the nervous system.

Lymphatic system and immunity

- Glucocorticoids (e.g. coritsol) depress the immune response
- Maturation of lymphocytes occurs in response to hormones from the thymus.
- Nervous system innervates and regulates immune system function.

Urinary system

- Final activation of vitamin D occurs in the kidneys (bioactive vitamin D is classed as a steroid hormone).
- Bladder emptying in regulated by both the autonomic NS and voluntary activity (control of the urethral sphincter).
- Autonomic NS regulates renal blood pressure

Digestive system

- Digestion is regulated by the autonomic NS, by local GI hormones, and by adrenal hormones (e.g. epinephrine).
- Bioactive vitamin D is needed to absorb calcium from the diet.

Reproductive system

- Autonomic NS activity regulates erectile tissue in males and females and ejaculation in males.
- Testosterone underlies sexual drive.
- Hormones from the hypothalamus, gonads, and pituitary regulate the development and functioning of the reproductive system.

Muscular system

- Somatic NS controls skeletal muscle activity.
- Muscular activity promotes release of catecholamine hormones from the adrenal medulla.
- Growth hormone is required for normal muscle development.
- Thyroid hormones and catecholamines influence muscle metabolism.

General Functions and Effects on all Systems

The nervous system regulates all the visceral and motor functions of the body, integrating with the endocrine system to provide both short term and longer term responses to stimuli. The endocrine system produces hormones that activate and regulate homeostatic functions, growth, and development.

Disease

Symptoms of disease	• Loss of function or control • Loss of voluntary control • Failure to develop normally
Diseases of the nervous system	• Inherited (e.g. Huntington's disease) • Trauma (e.g. paraplegia) • Infection (e.g.encephalitis, meningitis) • Autoimmune (multiple sclerosis) • Tumors • Degenerative diseases
Diseases of the endocrine system	• Inherited disease (e.g. congenital adrenal hyperplasia) • Cancers (e.g. thyroid cancer) • Autoimmune damage • Dietary related diseases

Medicine & Technology

Diagnosis of disorders	• Genetic testing and screening • Genetic counselling • Medical imaging techniques • Blood tests
Treatment of injury	• Surgery • Physical therapies • Drug therapies
Treatment of inherited disorders	• Dietary management • Physical and drug therapies • Radiotherapy (for cancers) • Stem cell therapy and transplants • Gene therapy (e.g. for Huntington's)

- Multiple sclerosis
- Type 1 & 2 diabetes
- Goiter (left)
- GH deficiency

- Brain scans (right)
- Genetic counselling
- Dietary modification
- Transplants
- Cell therapy

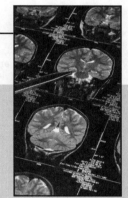

Integration & Control

Nervous & Endocrine Systems

The endocrine and nervous systems are closely linked. They can be affected by disease and undergo marked changes associated with aging.

Nervous and endocrine disorders may respond to exercise and medical treatment.

- Alzheimer's disease
- Hearing loss
- Changes in vision
- Menopause

- Control of diabetes
- Improved cognitive function (brain)
- Improved autonomic NS function

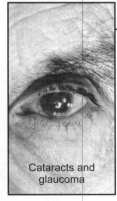

Cataracts and glaucoma

Photo: pan Pavel Rycl

The Effects of Aging

Effects of aging on nervous and endocrine function	• Sensory impairment • Loss of neurons (Alzheimer's) • Increased risks of falls • Increased risks of cancers • Decreased levels of some hormones

Exercise

Effects of exercise on nervous and endocrine function	• Improved coordination • Reduced risk of memory loss • Improved blood glucose management • Reduced risk of type 2 diabetes

INTEGRATION & CONTROL

The Nervous System

Understanding the role of the nervous system in regulating and coordinating the body's activities

Nervous tissue, nervous system organization, nerve impulses, synaptic transmission, sensory perception: pain, pressure, touch, hearing, balance, vision, taste, smell

Learning Objectives

☐ 1. Compile your own glossary from the **KEY WORDS** displayed in **bold type** in the learning objectives below.

Nervous System Organization *(page 66-67, 70-76)*

☐ 2. Review the structure and function of nervous tissue including the general features of:
(a) **Neurons**: sensory, motor and interneurons
(b) Supporting **glial cells** of the CNS, e.g. astrocytes
(c) Supporting cells of the PNS: e.g. **Schwann cells**

☐ 3. Describe the three overlapping functions of the nervous system: **monitoring** (via sensory input), **integration**, and **response** (via motor output). Identify the parts of the nervous system associated with each of these functions. Appreciate the close relationship between the nervous and endocrine systems in coordinating the body's responses and maintaining homeostasis.

☐ 4. Describe the basic structure and organization of the human **nervous system**, distinguishing between the **central nervous system** (CNS) and the **peripheral nervous system** (PNS). Describe the functional role of these two major divisions of the nervous system.

☐ 5. In more detail than 2(a), describe the anatomy of a typical motor neuron, including reference to the **cell body**, processes (**dendrites** and **axons**), **myelin sheath**, and **axon terminals**. Identify **nodes of Ranvier** on diagrams of myelinated fibers.

☐ 6. Draw the gross structure of the **brain**, and describe the functional role of its main regions including:
- **cerebellum**
- **brain stem: midbrain**, **pons**, **medulla oblongata**
- **diencephalon**: including the thalamus, **hypothalamus** and **pituitary gland**
- **cerebral hemispheres**

☐ 7. Recognize and describe the organization and functions of the two divisions of the PNS:
(a) **Sensory division**
(b) **Motor division**, which is divided into:
- **autonomic nervous system** (involuntary)
- **somatic nervous system** (voluntary)

☐ 8. Recognize the components of the autonomic nervous system (**ANS**): the **sympathetic** and **parasympathetic neurons** (nervous systems). Describe their roles and recognize that they have generally **antagonistic** effects.

☐ 9. Describe some of the effects of the sympathetic and parasympathetic systems, e.g. in the control of bladder emptying, pupil diameter, or heart rate.

Neuron Structure and Function *(pages 75-81)*

☐ 10. Describe the structure and function of different types of **neurons**: **motor (effector)** and **sensory neurons**, and **interneurons** (relay neurons or association neurons).

☐ 11. With reference to both somatic and autonomic reflexes, describe the adaptive value of **reflexes**. Using an annotated diagram, describe a simple spinal **reflex arc** involving three neurons. Outline examples of reflexes: the **pain withdrawal reflex** and one other spinal reflex, and the pupil reflex and one other **cranial reflex**. Discuss diagnostic uses of reflexes, e.g. the use of the pupillary reflex as a test for brain death.

☐ 12. Explain how the **resting potential** of a neuron is established. Include reference to the movement of Na^+ and K^+, the differential permeability of the membrane, and the generation of an **electrochemical gradient**.

☐ 13. Describe the generation of the **nerve impulse** with reference to the change in **membrane permeability** of the nerve leading to **depolarization**.

☐ 14. Describe how an **action potential** is propagated along a myelinated nerve by **saltatory conduction**. Include reference to the **all-or-nothing** nature of the impulse, and role of **myelin** and the **nodes of Ranvier**.

☐ 15. Describe impulse conduction in a **non-myelinated** nerve. Appreciate the difference in speed of conduction between myelinated and non-myelinated fibers.

☐ 16. Identify the role of synapses in the nervous system. Describe the basic features of a **cholinergic synapse** (as seen using electron microscopy).

☐ 17. With reference to **acetylcholine** and **norepinephrine** (also called **noradrenaline**), recognize that the synapses of the PNS are classified according to the **neurotransmitter** involved.

☐ 18. Explain the principles of synaptic transmission, e.g. at a **cholinergic synapse**. Include reference to the arrival of the **action potential** at the **presynaptic terminal**, the role of Ca^{2+} and **neurotransmitter**, the depolarization of the **post-synaptic neuron**, and subsequent removal of the neurotransmitter. Recognize the **neuromuscular junction** as a specialized cholinergic synapse.

☐ 19. Explain how presynaptic neurons can encourage or inhibit postsynaptic transmission. Appreciate the role of synapses in **unidirectionality** and **integration** through **summation** and **inhibition**.

☐ 20. Describe the effects of neurotransmitters and **psycho-active** drugs on the nervous system and behavior, as illustrated by pain and the action of **endorphins** and **enkephalins** or by **Parkinson's disease** and the action of psychoactive drugs and **dopamine**.

Neurological Disease *(pages 81)*

☐ 21. Describe the degenerative structural and physiological changes associated with **Alzheimer's disease** (AD). Describe the symptoms of the disease and relate these to the formation of amyloid plaques in the brain tissue.

Sensory Perception *(pages 63, 84-91)*

☐ 22. Define the terms: **stimulus** and **response**. Distinguish between a **sensory receptor** and a **sense organ**. Identify internal and external **stimuli** and describe examples of sensory receptors that respond to these.

☐ 23. Explain how sensory receptors receive and respond to stimuli. Appreciate that they act as **biological transducers**, converting different forms of energy into nerve impulses. With reference to specific examples, explain the following features of sensory receptors:
- They respond only to specific stimuli (**stimulation**).
- They respond by producing changes in membrane potential which may result in a nerve impulse.
- The strength of receptor response is proportional to the stimulus strength.

☐ 24. Describe the structure and function of some simple sensory receptors, including nerve endings (pain), **Meissner's corpuscle** (touch), **Pacinian corpuscle** (pressure), and **muscle spindle** (proprioception).

☐ 25. Describe **chemical senses** (gustation and olfaction):
- Structure and function of the taste buds.
- Structure and function of the **olfactory receptors**.

☐ 26. With respect to sensory receptors, explain what is meant by the **threshold** for stimulation and explain its significance. Understand that sensory receptors show **sensory adaptation** and explain its adaptive value.

Hearing and Balance *(page 89)*

☐ 27. Recognize the senses of equilibrium and hearing are forms of **mechanoreception**. Recognize that both senses rely on **sensory hair cells** that respond to movement of endolymph.

☐ 28. Describe the structure of the human **ear** and explain the function of each of its main regions, distinguishing between the structures involved with **equilibrium** and those involved with **hearing**. Include
- **External ear**: pinna, auditory canal, eardrum
- **Middle ear**: auditory ossicles, oval window, round window, auditory tube.

- **Inner ear**: the **cochlea** (hearing), **vestibular apparatus** (equilibrium).

☐ 29. Explain the mechanisms by which **static equilibrium** (movement with respect to gravity) and **dynamic equilibrium** (angular movement) are detected.

☐ 30. Describe the anatomy of the cochlea, identifying the **cochlear duct** and **organ of Corti**, **vestibular canal**, **tympanic canal**, and hair cells. Explain the nature of sound and describe how it is received by the ear and transmitted to the hair cells in the organ of Corti. Explain how the sense cells produce a signal for interpretation. Identify the regions of the brain involved in processing auditory information.

Vision *(pages 85-88)*

☐ 31. Using a diagram, describe the basic structure of the human **eye**, including: sclera, cornea, iris, pupil, lens, suspensory ligaments, choroid, aqueous humour, vitreous humour, retina. Describe the function of the each of these structures, including reference to how light is received by the eye and focused on the retina.

☐ 32. Describe **accommodation** of the eye for near and far vision. Recognise different defects of vision (myopia, hyperopia) and describe measures for their correction.

☐ 33. Annotate a diagram of the **retina**, identifying the **rods** and **cones**, and the direction of light movement.

☐ 34. Describe the function of the **rods** and **cones**, including:
- The role of the photosensitive pigments.
- Vision in bright and dim light.
- The basis of monochromatic and trichromatic vision.
- The neural basis for differences in **sensitivity** and **acuity** between rods and cones.
- The structural and functional differences between the **blind spot** and **central fovea**.

☐ 35. Describe the response of the photoreceptor cells to light, outlining how action potentials are generated in the **optic nerve** and how the signal is interpreted. Identify the regions of the brain involved in processing visual information.

The Nervous System

Supplementary Texts

See page 9 for additional details of these texts:

■ Morton, D. and Perry, J.W, 1997. **Photo Atlas for Anatomy and Physiology**, (Brooks/Cole).

■ Rowett, H.G.Q., 1999. **Basic Anatomy & Physiology**, (John Murray), pp. 60-79.

Periodicals

See page 9 for details of publishers of periodicals:

STUDENT'S REFERENCE
■ **Neuroscience** Biol. Sci. Rev., 19(1) Sept. 2006, pp. 32-35. *How the brain processes information, brain disorders and the future of neuroscience.*

■ **The Autonomic Nervous System** Biol. Sci. Rev., 18(3) Feb. 2006, pp. 21-25. *A thorough account of the structure and roles of the ANS.*

Presentation MEDIA to support this topic: **Anatomy & Physiology**

■ **Refractory Period** Biol. Sci. Rev., 20(4) April 2008, pp. 7-9. *The nature and purpose of the refractory period in response stimuli. The biological principles involved are discussed with in the context of the refractory period of the human heart.*

■ **Make a Move: How the Brain Controls our Movements** Biol. Sci. Rev., 12 (5) May 2000, pp. 28-32. *An excellent overview of the way in which the brain regulates muscular activity.*

■ **Color Vision and Color Blindness** Biol. Sci. Rev., 19(3) Feb. 2007, pp. 28-32. *Color is a quality our brains give to light based on the different sensitivities of three types of cones in the retina to different wavelengths of light.*

■ **What's Your Poison** Bio. Sci. Rev. 16(2) Nov. 2003, pp. 33-37. *The action of naturally derived poisons on synaptic transmission. This account includes an account of toxins and their actions.*

■ **The Nervous System** (series) New Scientist, 10 June 1989, 11 Nov. 1989, 29 June 1991 (Inside Science). *Nervous system structure and function.*

■ **A Pacinian Corpuscle** Biol. Sci. Rev., 12(3) Jan. 2000, pp. 33-34. *An account of the structure and operation of a common pressure receptor.*

■ **Before Your Very Eyes** New Scientist 15 March 1997 (Inside Science). *An excellent, highly readable account of eye structure, and the perception and processing of visual information.*

■ **Making the Connection** Biol. Sci. Rev.,13(3) Jan. 2001, pp. 10-13. *The central nervous system, neurotransmitters, and synapses.*

■ **Remodelling the Eye** Biol. Sci. Rev., 17(2) Nov. 2004, pp. 9-12. *An account of the structure of the human eye and disorders of vision.*

TEACHER'S REFERENCE
■ **How Neurons Work: An Analogy and Demonstration Using a Sparkler & a Frying Pan** The Am. Biology Teacher, 68(7), Sept. 2006, pp. 412-417. *Simulating how neurons generate and propagate action potentials.*

■ **Seeking the Neural Code** Scientific American, Dec. 2006, pp. 48-55. *Deciphering the complex neural processing responsible for our thoughts, emotions, and sensations.*

■ **Windows on the Mind** Scientific American, August 2007, pp. 40-47. *Human eyes at rest move imperceptibly in ways that are essential for seeing. These movements engender visibility when a person's gaze is fixed.*

Internet

See pages 10-11 for details of how to access **Bio Links** from our web site: **www.thebiozone.com** From Bio Links, access sites under the topics:

ANATOMY & PHYSIOLOGY > **General sites:**
• Anatomy & Physiology • Human physiology lecture notes • WebAnatomy > **Neuroscience:**
• Anatomy of the human ear • Basic neural processes • Information on drugs of abuse • Neuroscience for kids • Seeing, hearing, smelling the world • The effects of LSD on the human brain • The human eye • Virtual tour of the ear: hearing mechanism • The secret of the brain

Nervous Regulatory Systems

An essential feature of living organisms is their ability to coordinate their activities. In mammals, such as humans, detecting and responding to environmental change, and regulating the internal environment (**homeostasis**) are brought about by two coordinating systems: the nervous and endocrine systems. Although these two systems are quite different structurally, they frequently interact to coordinate behavior and physiology. The nervous system contains cells called **neurons** (or nerve cells). Neurons are specialized to transmit information in the form of electrochemical impulses (action potentials). The nervous system is a signaling network with branches carrying information directly to and from specific target tissues. Impulses can be transmitted over considerable distances and the response is very precise and rapid. Whilst it is extraordinarily complex, comprising millions of neural connections, its basic plan (below, left) is quite simple, structured around reception of sensory input, integration or processing of the information, and formulation of a response.

Coordination by the Nervous System

The vertebrate nervous system consists of the central nervous system (brain and spinal cord), and the nerves and receptors outside it (peripheral nervous system). Sensory input to receptors comes via stimuli. Information about the effect of a response is provided by feedback mechanisms so that the system can be readjusted. The basic organization of the nervous system can be simplified into a few key components: the sensory receptors, a central nervous system processing point, and the effectors which bring about the response (below):

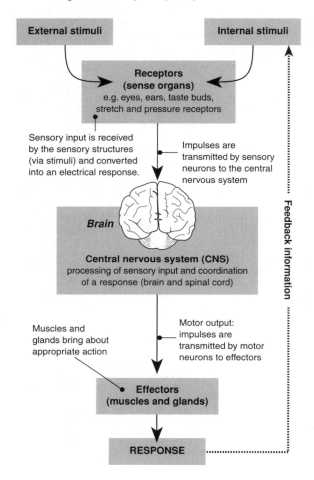

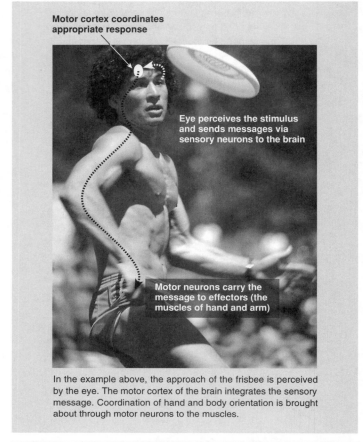

Motor cortex coordinates appropriate response

Eye perceives the stimulus and sends messages via sensory neurons to the brain

Motor neurons carry the message to effectors (the muscles of hand and arm)

In the example above, the approach of the frisbee is perceived by the eye. The motor cortex of the brain integrates the sensory message. Coordination of hand and body orientation is brought about through motor neurons to the muscles.

Comparison of nervous and hormonal control

	Nervous control	Hormonal control
Communication	Impulses across synapses	Hormones in the blood
Speed	Very rapid (within a few milliseconds)	Relatively slow (over minutes, hours, or longer)
Duration	Short term and reversible	Longer lasting effects
Target pathway	Specific (through nerves) to specific cells	Hormones broadcast to target cells everywhere
Action	Causes glands to secrete or muscles to contract	Causes changes in metabolic activity

1. Identify the three basic components of a nervous system and explain how they function to maintain homeostasis:

    (a) _____

    (b) _____

    (c) _____

2. Describe two differences between nervous control and endocrine (hormonal) control of body systems:

    (a) _____

    (b) _____

Related activities: The Nervous System, Hormonal Regulatory Systems
Web links: Nervous System Animation

The Nervous System

The **nervous system** is the body's control and communication center. It has three broad functions: detecting stimuli, interpreting them, and initiating appropriate responses. Its basic structure is outlined below. Further detail is provided in the following pages.

The Human Nervous System

The **central nervous system** (**CNS**) comprises the brain and spinal cord. The spinal cord is a cylinder of nervous tissue extending from the base of the brain down the back, protected by the spinal column. It transmits messages to and from the brain, and controls spinal reflexes.

The **peripheral nervous system**, or **PNS**, (right, far right) comprises all the nerves and sensory receptors outside the central nervous system.

Legend:
- ■ Brain (see below)
- ▨ Spinal cord
- ▨ Peripheral nerves

Below: cross sections through the spinal cord to show entry and exit of neurons.

Sensory neurons enter the spinal cord by the **dorsal root**.

Gray matter

Motor neurons leave the spinal cord by the **ventral root**.

White matter (myelinated nerves)

The **spinal cord** has an H shaped central area of gray matter, comprising nerve cell bodies, dendrites, and synapses around a central canal filled with cerebrospinal fluid. The area of white matter contains the nerve fibers.

The Peripheral Nervous System (PNS)

The PNS comprises **sensory** and **motor divisions**. Peripheral nerves all enter or leave the CNS, either from the spinal cord (the spinal nerves) or the brain (cranial nerves). They can be **sensory** (from sensory receptors), **motor** (running to a muscle or gland), or **mixed** (containing sensory and motor neurons). Cranial nerves are numbered in roman numerals, I-XII. They include the vagus (X), a mixed nerve with an important role in regulating bodily functions, including heart rate and digestion.

Sensory Division

Sensory nerves arise from **sensory receptors** (left) and carry messages to the central nervous system for processing.

The sensory system keeps the central nervous system aware of the external and internal environments. This division includes the familiar sense organs such as ears, eyes (A), and taste buds (B) as well as internal receptors that monitor internal state (e.g. thirst, hunger, body position, movement, pain).

Motor Division

Motor nerves carry impulses from the CNS to **effectors**: muscles (left) and glands. The motor division comprises two parts:

Somatic nervous system: the neurons that carry impulses to voluntary (skeletal) muscles (C).

Autonomic nervous system: regulates visceral functions over which there is generally no conscious control, e.g. heart rate, gut peristalsis involving smooth muscle (D), pupil reflex, and sweating.

1. Identify and briefly describe the three main functions of the nervous system:

    (a) _____

    (b) _____

    (c) _____

2. In the human nervous system, briefly explain the structure and role of each of the following:

    (a) The central nervous system: _____

    _____

    (b) The peripheral nervous system: _____

    _____

3. Explain the significance of the separation of the motor division of the PNS into somatic and autonomic divisions:

    _____

    _____

    _____

Related activities: Nervous Regulatory Systems, The Autonomic Nervous System

A 2

The Nervous System

The Autonomic Nervous System

The **autonomic nervous system** (ANS) regulates involuntary visceral functions through **reflexes**. Although most autonomic nervous system activity is beyond our conscious control, voluntary control over some basic reflexes (such as bladder emptying) can be learned. Most visceral effectors have dual innervation, receiving fibers from both branches of the ANS. These two branches, the **parasympathetic** and **sympathetic** divisions, have broadly opposing actions on the organs they control (excitatory or inhibitory). Nerves in the parasympathetic division release acetylcholine. This neurotransmitter is rapidly deactivated at the synapse and its effects are short lived and localized. Most sympathetic postganglionic nerves release norepinephrine (noradrenaline), which enters the bloodstream and is deactivated slowly. Hence, sympathetic stimulation tends to have more widespread and long lasting effects than parasympathetic stimulation. Aspects of ANS structure and function are illustrated below. The arrows indicate nerves to organs or ganglia (concentrations of nerve cell bodies).

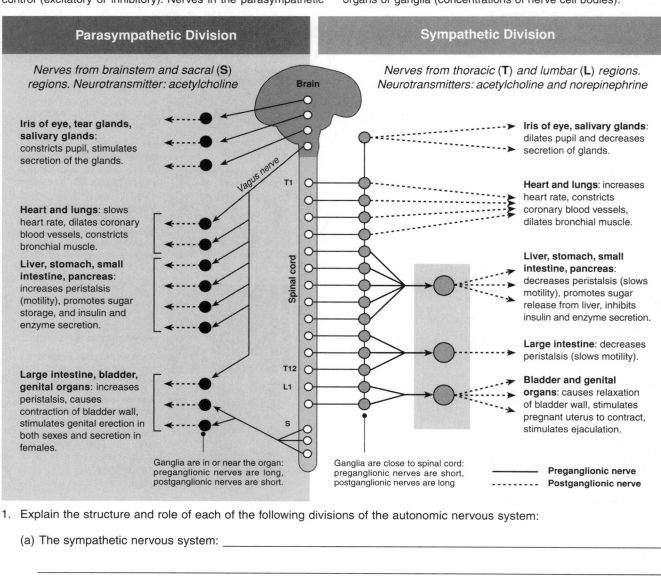

1. Explain the structure and role of each of the following divisions of the autonomic nervous system:

   (a) The sympathetic nervous system: _____

   _____

   (b) The parasympathetic nervous system: _____

   _____

2. Using an example (e.g. pupil reflex or control of heart rate), describe the role of reflexes in the functioning of the autonomic nervous system:

   _____

   _____

3. With reference to the emptying of the bladder, explain how a reflex activity can be modified by learning:

   _____

   _____

   _____

Related activities: The Nervous System, Reflexes

The Human Brain

The brain is one the largest organs in the body. It is protected by the skull, the **meninges**, and the **cerebrospinal fluid** (CSF). The brain is the body's control center. It receives a constant flow of sensory information, but responds only to what is important at the time. Some responses are very simple (e.g. cranial reflexes), whilst others require many levels of processing. The human brain is noted for its large, well developed cerebral region, with its prominent folds (**gyri**) and grooves (**sulci**). Each cerebral hemisphere has an outer region of gray matter and an inner region of white matter, and is divided into four lobes by deep sulci or fissures. These lobes: temporal, frontal, occipital, and parietal, correspond to the bones of the skull under which they lie.

Primary Structural Regions of the Brain

Cerebrum: Divided into two cerebral hemispheres. Many, complex roles. It contains sensory, motor, and association areas, and is involved in memory, emotion, language, reasoning, and sensory processing.

Ventricles: Cavities containing the CSF, which absorbs shocks and delivers nutritive substances.

Thalamus is the main relay center for all sensory messages that enter the brain, before they are transmitted to the cerebrum.

Hypothalamus controls the autonomic nervous system and links nervous and endocrine systems. Regulates appetite, thirst, body temperature, and sleep.

Midbrain
Pons
Medulla

Cerebellum coordinates body movements, posture, and balance.

Brainstem: Relay center for impulses between the rest of the brain and the spinal cord. Controls breathing, heartbeat, and the coughing and vomiting reflexes.

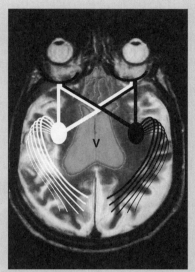

MRI scan of the brain viewed from above. The visual pathway has been superimposed on the image. Note the crossing of some sensory neurons to the opposite hemisphere and the fluid filled ventricles (V) in the center.

Sensory and Motor Regions in the Cerebrum

Primary somatic sensory area receives sensations from receptors in the skin, muscles and viscera, allowing recognition of pain, temperature, or touch. Sensory information from receptors on one side of the body crosses to the opposite side of the cerebral cortex where conscious sensations are produced. The size of the sensory region for different body parts depends on the number of receptors in that particular body part.

Visual areas within the occipital lobe receive, interpret, and evaluate visual stimuli. In vision, each eye views both sides of the visual field but the brain receives impulses from left and right visual fields separately (see photo caption above). The visual cortex combines the images into a single impression or **perception** of the image.

Primary motor area controls muscle movement. Stimulation of a point one side of the motor area results in muscular contraction on the opposite side of the body.

Primary gustatory area interprets sensations related to taste.

Sulci (grooves)
Gyri (elevated folds)

Language areas: The motor speech area (Broca's area) is concerned with speech production. The sensory speech area (Wernicke's area) is concerned with speech recognition and coherence.

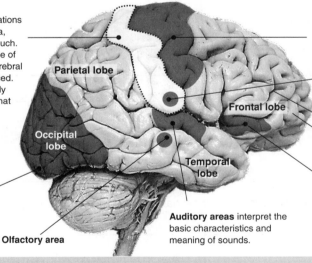

Parietal lobe
Frontal lobe
Occipital lobe
Temporal lobe
Auditory areas
Olfactory area

Auditory areas interpret the basic characteristics and meaning of sounds.

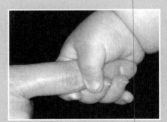

Touch is interpreted in the primary somatic sensory area. The fingertips and the lips have a relatively large amount of area devoted to them.

Humans rely heavily on vision. The importance of this **special sense** in humans is indicated by the large occipital region of the brain.

The olfactory tract connects the olfactory bulb with the cerebral hemispheres where olfactory information is interpreted.

The endothelial tight junctions of the capillaries supplying the brain form a protective **blood-brain barrier** against toxins and infection.

Dan Ferber

The Nervous System

Related activities: The Nervous System, Aging and the Brain
Web links: Inside the Brain: An Interactive Tour, Brain Anatomy

A 2

The Ventricles and CSF

The delicate nervous tissue of the brain and spinal cord is protected against damage by the **bone** of the skull and vertebral column, the membranes overlying the brain (the **meninges**), and the watery but nutritive **cerebrospinal fluid** (CSF), which lies between the inner two of the meningeal layers.

The meninges are collectively three membranes: a tough double-layered outer **dura mater**, a web-like middle **arachnoid mater**, and an inner delicate **pia mater** that adheres to the surface of the brain. The CSF is formed from the blood by clusters of capillaries on the roof of each of the brain's ventricles (choroid plexuses). The CSF is constantly circulated through the ventricles of the brain (and into the spinal cord), returning to the blood via specialized projections of the middle meningeal layer (the arachnoid).

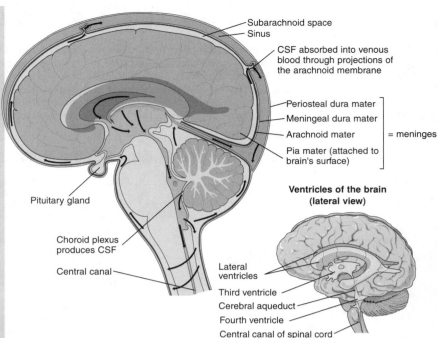

Subarachnoid space
Sinus
CSF absorbed into venous blood through projections of the arachnoid membrane
Periosteal dura mater
Meningeal dura mater
Arachnoid mater
Pia mater (attached to brain's surface)
= meninges
Pituitary gland
Choroid plexus produces CSF
Central canal

Ventricles of the brain (lateral view)

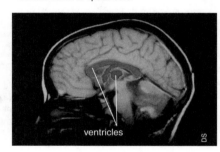

Lateral ventricles
Third ventricle
Cerebral aqueduct
Fourth ventricle
Central canal of spinal cord

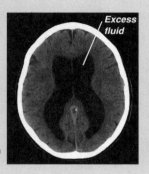

Excess fluid

If the passages that normally allow the CSF to exit the brain become blocked, the CSF accumulates within the brain's ventricles causing a condition called hydrocephalus

The accumulated fluid can be seen in this MRI scan.

MRI scanning is a powerful technique to visualize the structure and function of the body. It provides much greater contrast between the different soft tissues than computerized tomography (CT) does, making it especially useful in neurological (brain) imaging, especially for indicating the presence of tumors or fluid, and showing up abnormalities in blood supply. In the scan pictured right, the fluid within the lateral and third ventricles is clearly visible.

ventricles

DS

1. For each of the following bodily functions, identify the region(s) of the brain involved in its control:

   (a) Breathing and heartbeat: _____

   (b) Memory and emotion: _____

   (c) Posture and balance: _____

   (d) Autonomic functions: _____

   (e) Visual processing: _____

   (f) Body temperature: _____

   (g) Language: _____

   (h) Muscular movement: _____

2. Explain how the brain is protected against physical damage and infection: _____

   _____

   _____

   _____

   _____

3. (a) Describe where CSF is produced and how the CSF returns to the blood: _____

   _____

   (b) Explain the consequences of blocking this return flow of CSF: _____

   _____

The Cells of Nervous Tissue

Nervous tissue is made up of two main cell types: **neurons**, which are specialized to transmit nerve impulses, and supporting cells, which are collectively called **neuroglia**. Neurons (right panel) have a recognizable structure with a cell body and armlike **processes** (**dendrites** and **axons**). Most long neurons in the PNS are also supported by a fatty insulating sheath of myelin. Supporting cells are more variable and are specialized to perform different roles. Some, such as astrocytes, help to form the **blood-brain barrier**, while others line the cavities of the brain and spinal cord. Although they appear similar in structure to neurons, glial cells are unable to transmit nerve impulses. The features of some of these cell types are described below.

Supporting Glial Cells

Glial cells are of four basic types, each with a particular function in supporting the neurons of the CNS.

Astrocyte connects capillary and neuron

Capillary

Neuron

Processes cling to capillary walls

Astrocytes are the most abundant supportive cells in nervous tissue. They anchor neurons to capillaries and support the blood-brain barrier by restricting the passage of certain substances. They are also important in the repair of the brain and spinal cord following injury.

Cilia circulate CSF   CSF

Nervous tissue

Ependymal cells are the epithelial cells lining the ventricles in the brain and the central canal of the spinal cord. The surfaces of these cuboidal cells are covered in cilia and microvilli, which circulate and absorb CSF. Specialized ependymal cells and capillaries together form the choroid plexuses, which produce the CSF.

Microglial cell

Spiderlike processes

Neuron

Neuron

Microglia are the defense cells of nervous tissue. Antibodies are too large to cross the blood-brain barrier, so the phagocytic microglia must be able to recognize and dispose of foreign material and debris.

Oligodendrocytes produce insulating myelin sheaths around the axons of neurons in the CNS. A single oligodendrocyte can extend to wrap around up to 50 axons.

This image shows an oligodendrocyte genetically altered to fluoresce.

WIKI

Neuron Types

Sensory neuron
Transmits impulses from sensory receptors to other neurons.

Axon: A long extension of the cell transmits the nerve impulse to another neuron or to an **effector** (e.g. muscle). Axons may be very long and, in the peripheral nervous system, many are myelinated.

Sense organ (pressure receptor) in the skin.

Axon branches

Cell body containing the organelles to keep the neuron alive and functioning.

Node of Ranvier
Axon surrounded by myelin sheath.

Motor neuron
Transmits impulses from the CNS to muscles or glands.

Dendrites

Axon branches have synaptic knobs at each end. These release neurotransmitters, which transmit the impulse between neurons or between a neuron and a muscle cell.

Cell body

Impulse direction

Myelin sheath

Axon hillock (generation of action potential)

Relay neuron
Also called association or interneurons. Located in the CNS and carry impulses from sensory to motor neurons (as in reflexes).

Dendrites: Bushy extensions of the cell body, specialized to receive stimuli.

Axon

Axon branches

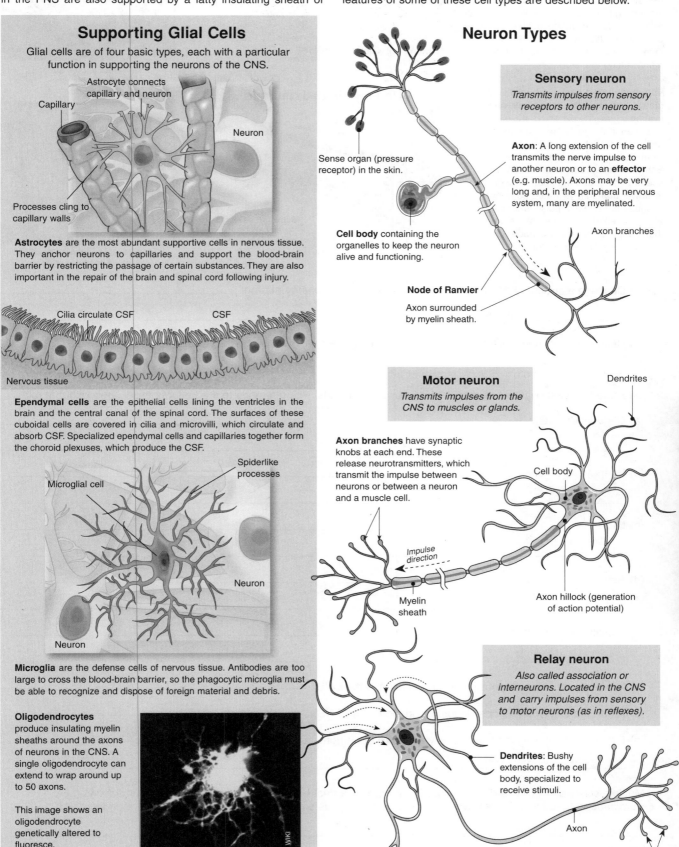

The Nervous System

Related activities: The Nervous System, Nervous Regulatory Systems, Autoimmune Diseases   **Web links**: Unipolar and Multipolar Neurons

RA 2

Where conduction speed is important, the axons of neurons are sheathed within a lipid and protein rich substance called **myelin**. Myelin is produced by **oligodendrocytes** in the central nervous system (CNS) and by **Schwann cells** in the peripheral nervous system (PNS). Myelin acts as an insulator, increasing the speed at which nerve impulses travel because it prevents ion flow across the neuron membrane and forces the current to "jump" along the axon.

Non-myelinated axons are relatively more common in the central nervous system where the distances travelled are less than in the PNS. Here, the axons are encased within the cytoplasmic extensions of oligodendrocytes or Schwann cells, rather than within a myelin sheath. The speed of impulse conduction is slower than in myelinated neurons because the nerve impulse is propagated along the entire axon membrane, rather than jumping from node to node as occurs in myelinated neurons.

Myelinated Neurons

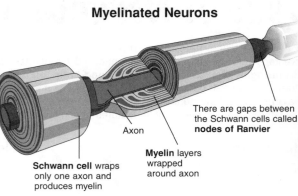

Axon

Schwann cell wraps only one axon and produces myelin

Myelin layers wrapped around axon

There are gaps between the Schwann cells called **nodes of Ranvier**

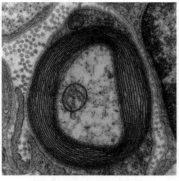

TEM cross section through a myelinated axon

Non-myelinated Neurons

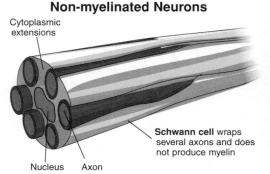

Cytoplasmic extensions

Nucleus   Axon

Schwann cell wraps several axons and does not produce myelin

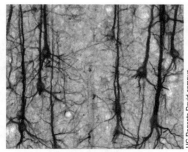

Unmyelinated pyramidal neurons of the cerebral cortex

1. (a) Describe a structural difference between a motor and a sensory neuron: _____

_____

(b) Describe a functional difference between a motor and a relay or interneuron: _____

_____

2. Describe the functional role of each of the following glial cells, with reference to the features associated with that role:

(a) Oligodendrocytes: _____

_____

(b) Ependymal cells: _____

_____

(c) Microglia: _____

_____

(d) Astrocytes: _____

_____

3. (a) Explain the function of myelination in neurons: _____

_____

(b) Name the cell type responsible for myelination in the CNS: _____

(c) Name the cell type responsible for myelination in the PNS: _____

(d) Explain why myelination is a typically a feature of neurons in the peripheral nervous system: _____

_____

4. Multiple sclerosis (MS) is a disease involving progressive destruction of the myelin sheaths around axons (see *Autoimmune Diseases*). Explain why MS impairs nervous system function even though axons are undamaged:

_____

_____

Reflexes

A reflex is an automatic response to a stimulus involving a small number of neurons and a central nervous system (CNS) processing point (usually the spinal cord, but sometimes the brain stem). This type of circuit is called a **reflex arc**. Reflexes permit rapid responses to stimuli. They are classified according to the number of CNS synapses involved; **monosynaptic reflexes** involve only one CNS synapse (e.g. knee jerk reflex), **polysynaptic reflexes** involve two or more (e.g. pain withdrawal reflex). Both are spinal reflexes. The pupil reflex (opening and closure of the pupil) is an example of a cranial reflex.

Pain Withdrawal: A Polysynaptic Reflex Arc

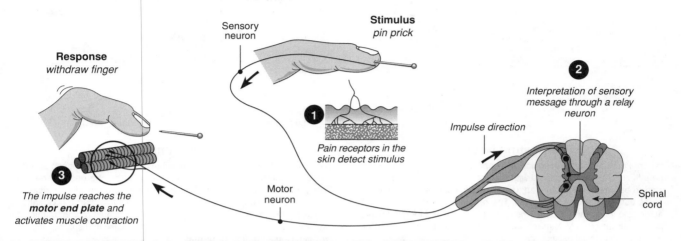

Response
withdraw finger

Sensory neuron

Stimulus
pin prick

2 Interpretation of sensory message through a relay neuron

Impulse direction

1 Pain receptors in the skin detect stimulus

3 The impulse reaches the **motor end plate** and activates muscle contraction

Motor neuron

Spinal cord

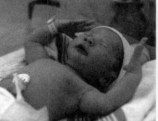

Normal newborns exhibit a number of primitive reflexes in response to particular stimuli. These reflexes disappear within a few months of birth as the child develops. Primitive reflexes include the grasp reflex (above left) and the startle or Moro reflex (right) in which a sudden noise will cause the infant to throw out its arms, extend the legs and head, and cry. The rooting and sucking reflexes are other examples of primitive reflexes.

The pupillary light reflex refers to the rapid expansion or contraction of the pupils in response to the intensity of light falling on the retina. It is a polysynaptic cranial reflex and can be used to test for brain death.

The patella (knee jerk) reflex is a simple deep tendon reflex that is used to test the function of the femoral nerve and spinal cord segments L2-L4. It helps to maintain posture and balance when walking.

All photos: istock

The Nervous System

1. Explain why higher reasoning or conscious thought are not necessary or desirable features of reflex behaviors:

_____

2. Distinguish between a spinal reflex and a cranial reflex and give an example of each: _____

_____

_____

3. Distinguish between a monosynaptic and a polysynaptic reflex arc and give an example of each: _____

_____

_____

4. (a) With reference to specific examples, describe the adaptive value of primitive reflexes in newborns:

_____

(b) Explain why newborns are tested for the presence of these reflexes: _____

_____

5. Describe the adaptive value of cranial reflexes such as the pupillary light reflex and the blink reflex:

_____

_____

Related activities: The Nervous System, Chemical Synapses
Web links: Parasympathetic Eye Response, Knee Jerk Reflex

RA 1

Transmission of Nerve Impulses

Neurons, like all cells, contain ions or charged atoms. Those of special importance include sodium (Na^+), potassium (K^+), and negatively charged proteins. Neurons are **electrically excitable** cells: a property that results from the separation of ion charge either side of the neuron membrane. They may exist in either a resting or stimulated state. When stimulated, neurons produce electrical impulses that are transmitted along the axon. These impulses are transmitted between neurons across junctions called **synapses**. Synapses enable the transmission of impulses rapidly all around the body.

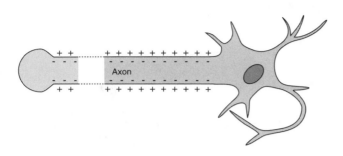

Impulse travels in this direction

Na^+ Na^+

Axon

Area of impulse

Next area to be stimulated

Area returning to resting state

The Resting Neuron

When a neuron is not transmitting an impulse, the inside of the cell is negatively charged compared with the outside of the cell. The cell is said to be electrically polarized, because the inside and the outside of the cell are oppositely charged. The potential difference (voltage) across the membrane is called the resting potential and for most nerve cells is about -70 mV. Nerve transmission is possible because this membrane potential exists.

The Nerve Impulse

When a neuron is stimulated, the distribution of charges on each side of the membrane changes. For a millisecond, the charges reverse. This process, called **depolarization**, causes a burst of electrical activity to pass along the axon of the neuron. As the charge reversal reaches one region, local currents depolarize the next region. In this way the impulse spreads along the axon. An impulse that spreads this way is called an **action potential**.

The Action Potential

The depolarization described above can be illustrated as a change in membrane potential (in millivolts). In order for an action potential to be generated, the stimulation must be strong enough to reach the **threshold** potential; this is the potential (voltage) at which the depolarization of the membrane becomes "unstoppable" and the action potential is generated. The action potential is **all or none** in its generation. Either the **threshold** is reached and the action potential is generated or the nerve does not fire. The resting potential is restored by the movement of potassium ions (K^+) out of the cell. During this **refractory period**, the nerve cannot respond.

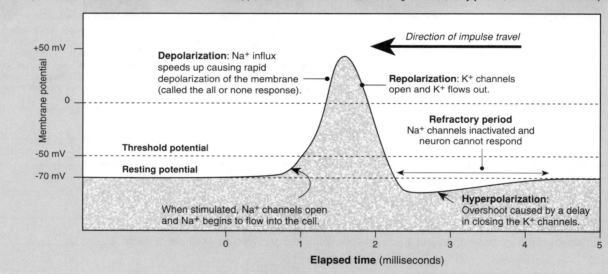

Direction of impulse travel

Depolarization: Na^+ influx speeds up causing rapid depolarization of the membrane (called the all or none response).

Repolarization: K^+ channels open and K^+ flows out.

Refractory period Na^+ channels inactivated and neuron cannot respond

Threshold potential

Resting potential

When stimulated, Na^+ channels open and Na^+ begins to flow into the cell.

Hyperpolarization: Overshoot caused by a delay in closing the K^+ channels.

Membrane potential: +50 mV, 0, -50 mV, -70 mV

Elapsed time (milliseconds): 0 1 2 3 4 5

1. Explain how an action potential is able to pass along a nerve: _____

_____

2. Explain how the refractory period influences the direction in which an impulse will travel: _____

_____

3. Action potentials themselves are indistinguishable from each other. Explain how the nervous system is able to interpret the impulses correctly and bring about an appropriate response:

_____

_____

Related activities: Chemical Synapses
Web links: Nerve Action Potential, Neurobiology

Chemical Synapses

Action potentials are transmitted between neurons across **synapses**: junctions between the end of one axon and the dendrite or cell body of a receiving neuron. Electrical synapses, where cells are electrically coupled through gap junctions between cells, occur in heart muscle and in the cerebral cortex, but they are relatively uncommon elsewhere. Most synapses in the nervous system are **chemical synapses**. In these, the axon terminal is a swollen knob, and a small gap, the **synaptic cleft**, separates it from the receiving neuron. The synaptic knobs are filled with tiny packets of chemicals called **neurotransmitters**.

Nerve transmission involves the diffusion of the neurotransmitter across the cleft, where it interacts with the receiving membrane and causes an electrical response. The response of a receiving (post-synaptic) cell to the arrival of a neurotransmitter depends on the nature of the cell itself, on its location in the nervous system, and on the neurotransmitter involved. Synapses that release acetylcholine (ACh) are termed **cholinergic**. In the example below, ACh results in membrane depolarization and an action potential (an excitatory response). Unlike electrical synapses, transmission at chemical synapses is always unidirectional.

A Cholinergic Synapse

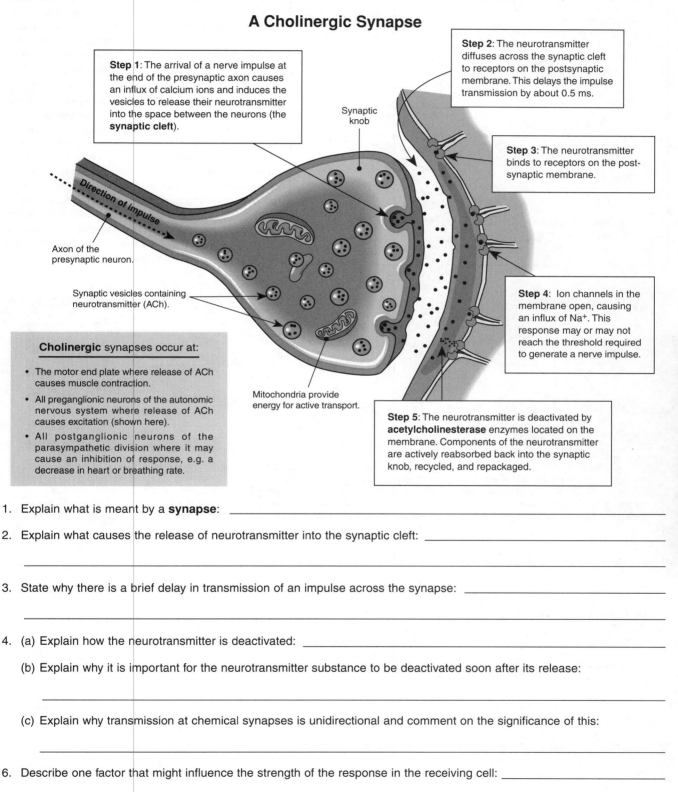

Step 1: The arrival of a nerve impulse at the end of the presynaptic axon causes an influx of calcium ions and induces the vesicles to release their neurotransmitter into the space between the neurons (the **synaptic cleft**).

Step 2: The neurotransmitter diffuses across the synaptic cleft to receptors on the postsynaptic membrane. This delays the impulse transmission by about 0.5 ms.

Step 3: The neurotransmitter binds to receptors on the post-synaptic membrane.

Step 4: Ion channels in the membrane open, causing an influx of Na^+. This response may or may not reach the threshold required to generate a nerve impulse.

Step 5: The neurotransmitter is deactivated by **acetylcholinesterase** enzymes located on the membrane. Components of the neurotransmitter are actively reabsorbed back into the synaptic knob, recycled, and repackaged.

Synaptic knob

Axon of the presynaptic neuron.

Direction of impulse

Synaptic vesicles containing neurotransmitter (ACh).

Mitochondria provide energy for active transport.

Cholinergic synapses occur at:

- The motor end plate where release of ACh causes muscle contraction.
- All preganglionic neurons of the autonomic nervous system where release of ACh causes excitation (shown here).
- All postganglionic neurons of the parasympathetic division where it may cause an inhibition of response, e.g. a decrease in heart or breathing rate.

1. Explain what is meant by a **synapse**: _____

2. Explain what causes the release of neurotransmitter into the synaptic cleft: _____
   _____

3. State why there is a brief delay in transmission of an impulse across the synapse: _____
   _____

4. (a) Explain how the neurotransmitter is deactivated: _____

   (b) Explain why it is important for the neurotransmitter substance to be deactivated soon after its release:
   _____

   (c) Explain why transmission at chemical synapses is unidirectional and comment on the significance of this:
   _____

6. Describe one factor that might influence the strength of the response in the receiving cell: _____
   _____

The Nervous System

Integration at Synapses

Synapses play a pivotal role in the ability of the nervous system to respond appropriately to stimulation and to adapt to change. The nature of synaptic transmission allows the **integration** (interpretation and coordination) of inputs from many sources. These inputs need not be just excitatory (causing depolarization). Inhibition results when the neurotransmitter released causes negative chloride ions (rather than sodium ions) to enter the postsynaptic neuron. The postsynaptic neuron then becomes more negative inside (hyperpolarized) and an action potential is less likely to be generated. At synapses, it is the sum of **all** inputs (excitatory and inhibitory) that leads to the final response in a postsynaptic cell. Integration at synapses makes possible the various responses we have to stimuli. It is also the most probable mechanism by which learning and memory are achieved.

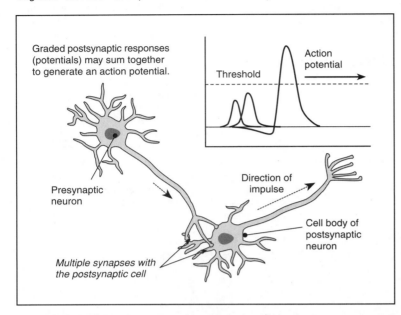

Graded postsynaptic responses (potentials) may sum together to generate an action potential.

Threshold

Action potential

Presynaptic neuron

Direction of impulse

Cell body of postsynaptic neuron

Multiple synapses with the postsynaptic cell

Synapses and Summation

Nerve transmission across chemical synapses has several advantages, despite the delay caused by neurotransmitter diffusion. Chemical synapses transmit impulses in one direction to a precise location and, because they rely on a limited supply of neurotransmitter, they are subject to fatigue (inability to respond to repeated stimulation). This protects the system against overstimulation.

Synapses also act as centers for the **integration** of inputs from many sources. The response of a postsynaptic cell is often graded; it is not strong enough on its own to generate an action potential. However, because the strength of the response is related to the amount of neurotransmitter released, subthreshold responses can sum to produce a response in the post-synaptic cell. This additive effect is termed **summation**. Summation can be **temporal** or **spatial** (below). A neuromuscular junction (photo below) is a specialized form of synapse between a motor neuron and a skeletal muscle fiber. Functionally, it is similar to any excitatory cholinergic synapse.

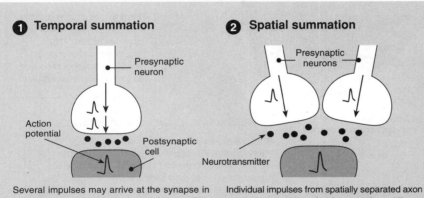

1 Temporal summation

Presynaptic neuron

Action potential

Postsynaptic cell

Several impulses may arrive at the synapse in quick succession from a single axon. The individual responses are so close together in time that they sum to reach threshold and produce an action potential in the postsynaptic neuron.

2 Spatial summation

Presynaptic neurons

Neurotransmitter

Individual impulses from spatially separated axon terminals may arrive **simultaneously** at different regions of the same postsynaptic neuron. The responses from the different places sum to reach threshold and produce an action potential.

3 Neuromuscular junction

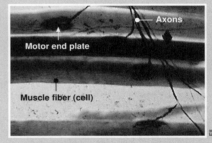

Axons

Motor end plate

Muscle fiber (cell)

The arrival of an impulse at the neuromuscular junction causes the release of acetylcholine from the synaptic knobs. This causes the muscle cell membrane (sarcolemma) to depolarize, and an action potential is generated in the muscle cell.

1. Explain the purpose of nervous system integration: _____

_____

2. (a) Explain what is meant by **summation**: _____

_____

   (b) In simple terms, distinguish between temporal and spatial summation: _____

_____

_____

3. Describe two ways in which a neuromuscular junction is similar to any excitatory cholinergic synapse:

   (a) _____

   (b) _____

Drugs at Synapses

Synapses in the peripheral nervous system are classified according to the neurotransmitter they release; **cholinergic** synapses release **acetylcholine (Ach)** while **adrenergic** synapses release **epinephrine** (adrenaline) or **norepinephrine** (noradrenaline). The effect produced by these neurotransmitters depends, in turn, on the type of receptors present on the postsynaptic membrane. Ach receptors are classified as nicotinic or muscarinic according of their response to nicotine or muscarine (a fungal toxin). Adrenergic receptors are also of two types, alpha (α) or beta (β), classified according to their particular responses to specific chemicals. **Drugs** exert their effects on the nervous system by mimicking (**agonists**) or blocking (**antagonists**) the action of neurotransmitters at synapses. Because of the small amounts of chemicals involved in synaptic transmission, drugs that affect the activity of neurotransmitters, or their binding sites, can have powerful effects even in small doses.

Drugs at Cholinergic Synapses

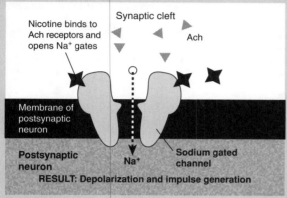

Nicotine acts as a **direct agonist** at nicotinic synapses. Nicotine binds to and activates acetylcholine (Ach) receptors on the postsynaptic membrane. This opens sodium gates, leading to a sodium influx and membrane depolarization. Some agonists work indirectly at the synapse by preventing Ach breakdown. Such drugs are used to treat elderly patients with Alzheimer's disease.

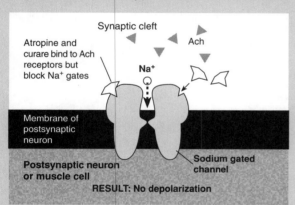

Atropine and **curare** act as antagonists at some cholinergic synapses. These molecules compete with Ach for binding sites on the postsynaptic membrane, and block sodium influx so that impulses are not generated. If the postsynaptic cell is a muscle cell, muscle contraction is prevented. In the case of curare, this causes death by flaccid paralysis.

Drugs at Adrenergic Synapses

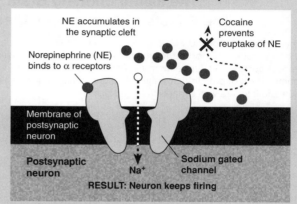

Under normal circumstances, the continued activity of the neurotransmitter norepinephrine (NE) at the synapse is prevented by reuptake of NE by the presynaptic neuron. **Cocaine** and **amphetamine** drugs act indirectly as agonists by preventing this reuptake. This action allows NE to linger at the synapse and continue to exert its effects.

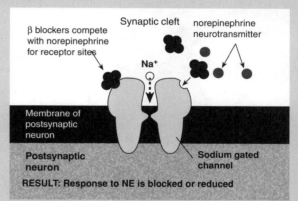

Therapeutic drugs called **beta** (β) **blockers** act as direct antagonists at adrenergic synapses (sympathetic nervous system). They compete for the adrenergic β receptors on the postsynaptic membrane and block impulse transmission. Beta blockers are prescribed primarily to treat hypertension and heart disorders because they slow heart rate and reduce the force of contraction.

The Nervous System

1. Providing an example of each, outline two ways in which drugs can act at a cholinergic synapse:

    (a) _____

    (b) _____

2. Providing an example, outline one way in which drugs can operate at adrenergic synapses: _____
    _____

3. Explain why atropine and curare are described as direct antagonists: _____
    _____

4. Suggest why curare (carefully administered) is used during abdominal surgery: _____
    _____

Aging and the Brain

Alzheimer's disease (AD) is a disabling neurological disorder affecting about 5% of the population over 65. Although its causes are largely unknown, people with a family history of Alzheimer's have a greater risk, implying that a genetic factor is involved. Some of the cases of Alzheimer's with a familial (inherited) pattern involve a mutation of the gene for amyloid precursor protein (APP), found on chromosome 21 and nearly all people with Down syndrome (trisomy 21) who live into their 40s develop the disease. The gene for the protein apoE, which has an important role in lipid transport, degeneration and regulation in nervous tissue, is also a risk factor that may be involved in modifying the age of onset. Sufferers of Alzheimer's have trouble remembering recent events and they become confused and forgetful. In the later stages of the disease, people with Alzheimer's become very disorientated, lose past memories, and may become paranoid and moody. Dementia and loss of reason occur at the end stages of the disease. The effects of the disease are irreversible and it has no cure.

The Malfunctioning Brain: The Effects of Alzheimer's Disease

Alzheimer's is associated with accelerated loss of neurons, particularly in regions of the brain that are important for memory and intellectual processing, such as the cerebral cortex and hippocampus. The disease has been linked to abnormal accumulations of protein-rich **amyloid** plaques and tangles, which invade the brain tissue and interfere with synaptic transmission.

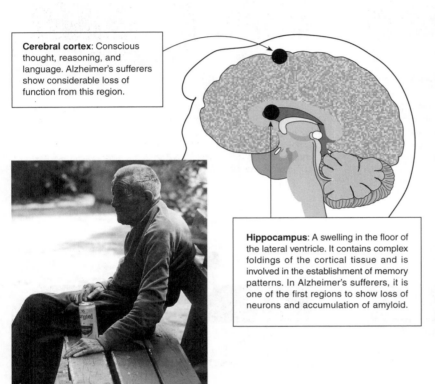

Cerebral cortex: Conscious thought, reasoning, and language. Alzheimer's sufferers show considerable loss of function from this region.

Hippocampus: A swelling in the floor of the lateral ventricle. It contains complex foldings of the cortical tissue and is involved in the establishment of memory patterns. In Alzheimer's sufferers, it is one of the first regions to show loss of neurons and accumulation of amyloid.

It is not uncommon for Alzheimer's sufferers to wander and become lost and disorientated.

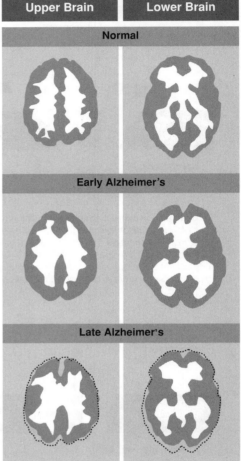

The brain scans above show diminishing brain function in certain areas of the brain in Alzheimer's sufferers. Note, particularly in the two lower scans, how much the brain has shrunk (original size indicated by the dotted line). Light areas indicate brain activity.

1. Describe the biological basis behind the degenerative changes associated with Alzheimer's disease:

2. Describe the evidence for the Alzheimer's disease having a genetic component in some cases:

3. Some loss of neuronal function occurs normally as a result of aging. Identify the features distinguishing Alzheimer's disease from normal age related loss of neuronal function:

Detecting Changing States

A **stimulus** is any physical or chemical change in the environment capable of provoking a response in an organism. All organisms respond to stimuli in order to survive. This response is adaptive; it acts to maintain the organism's state of homeostasis. Stimuli may be either external (outside the organism) or internal (within its body). Some of the stimuli to which humans respond are described below, together with the sense organs that detect and respond to these stimuli. Note that sensory receptors respond only to specific stimuli, so the sense organs we possess determine how we perceive the world.

Hair cells in the vestibule of the inner ear respond to **gravity** by detecting the rate of change and direction of the head and body. Other hair cells in the cochlea of the inner ear detect **sound** waves. The sound is directed and amplified by specialized regions of the outer and middle ear (pinna, canal, middle ear bones).

Photoreceptor cells in the eyes detect color, intensity, and movement of **light**.

Olfactory receptors in the nose detect airborne **chemicals**. The human nose has about 5 million of these receptors, a bloodhound nose has more than 200 million. The taste buds of the tongue detect dissolved chemicals (gustation). Tastes are combinations of five basic sensations: sweet, salt, sour, bitter, and savory (umami receptor).

Chemoreceptors in certain blood vessels, e.g. carotid arteries, monitor carbon dioxide levels (and therefore pH) of the blood. Breathing and heart rate increase or decrease (as appropriate) to adjust blood composition.

Baroreceptors in the walls of some arteries, e.g. aorta, monitor blood pressure. Heart rate and blood vessel diameter are adjusted accordingly.

Pressure deforms the skin surface and stimulates sensory receptors in the dermis. These receptors are especially abundant on the lips and fingertips.

Proprioreceptors (stretch receptors) in the muscles, tendons, and joints monitor limb position, **stretch**, and **tension**. The muscle spindle is a stretch receptor that monitors the state of muscle contraction and enables muscle to maintain its length.

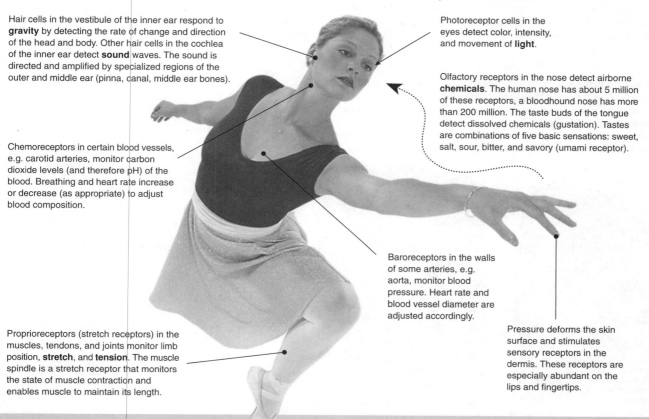

 Pain and temperature are detected by simple nerve endings in the skin. Deep tissue injury is sometimes felt on the skin as referred pain.

 Humans rely heavily on their hearing when learning to communicate; without it, speech and language development are more difficult.

 Breathing and heart rates are regulated in response to sensory input from chemoreceptors.

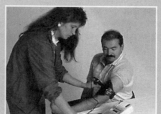 Baroreceptors and osmoreceptors act together to keep blood pressure and volume within narrow limits.

The Nervous System

1. Provide a concise definition of a **stimulus**: _____

2. Using humans as an example, discuss the need for communication systems to respond to changes in the environment: _____

3. (a) Name one internal stimulus and its sensory receptor: _____

(b) Describe the role of this sensory receptor in contributing to **homeostasis**: _____

Related activities: The Basis of Sensory Perception

A 2

The Basis of Sensory Perception

Sensory receptors are specialized to detect stimuli and respond by producing an electrical discharge. In this way they act as **biological transducers**, converting the energy from a stimulus into an electrochemical signal. Stimulation of a sensory receptor cell results in an electrical impulse with specific properties. The frequency of impulses produced by the receptor cell encodes information about the strength of the stimulus; a stronger stimulus produces more frequent impulses. Sensory receptors also show **sensory adaptation** and will cease responding to a stimulus of the same intensity. The simplest sensory receptors consist of a single sensory neuron (e.g. free nerve endings). More complex sense cells form synapses with their sensory neurons (e.g. taste buds). Sensory receptors are classified according to the stimuli to which they respond (for example, photoreceptors respond to light). The response of a simple **mechanoreceptor**, the Pacinian corpuscle, to a stimulus (pressure) is described below.

The Pacinian Corpuscle

Pacinian corpuscles are pressure receptors occurring in deep subcutaneous tissues all over the body. They are relatively large and simple in structure, consisting of a sensory nerve ending (dendrite) surrounded by a capsule of layered connective tissue. Pressure deforms the capsule, stretching the nerve ending and leading to a localized depolarization. Once a **threshold** value is reached, an **action potential** propagates along the axon.

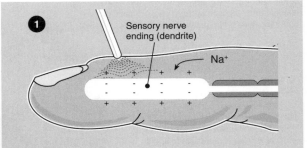

Deforming the corpuscle leads to an increase in the permeability of the nerve to sodium. Na^+ diffuses into the nerve ending creating a localized depolarization. This depolarization is called a **generator potential**.

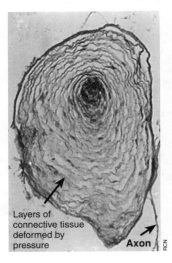

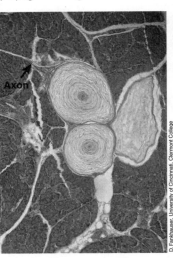

Pacinian corpuscle (above, left), illustrating the distinctive layers of connective tissue. The photograph on the right shows corpuscles grouped together in the pancreas. Pacinian corpuscles are rapidly adapting receptors; they fire at the beginning and end of a stimulus, but do not respond to unchanging pressure.

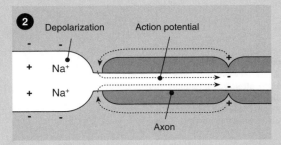

A volley of **action potentials** is triggered once the generator potential reaches or exceeds a **threshold value**. These action potentials are conducted along the sensory axon. A strong stimulus results in a high frequency of impulses.

1. Explain why sensory receptors are termed 'biological transducers': _____

_____

_____

2. Explain the significance of linking the magnitude of a sensory response to stimulus intensity: _____

_____

_____

3. Explain the physiological importance of sensory adaptation: _____

_____

4. (a) Describe the properties of a generator potential: _____

_____

   (b) Suggest why a simple mechanoreceptor, such as the Pacinian corpuscle, does not fire action potentials unless a stimulus of threshold value is reached:

_____

_____

_____

Related activities: The Physiology of Vision, Skin Senses, Hearing, Taste and Smell Web links: Neuron Information Coding and Transfer

The Structure of the Eye

The eye is a complex and highly sophisticated sense organ specialized to detect light. The adult eyeball is about 25 mm in diameter. Only the anterior one-sixth of its total surface area is exposed; the rest lies recessed and protected by the **orbit** into which it fits. The eyeball is protected and given shape by a fibrous tunic. The posterior part of this structure is the **sclera** (the white of the eye), while the anterior transparent portion is the **cornea**, which covers the colored iris.

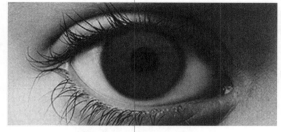

Forming a Visual Image

Before light can reach the photoreceptor cells of the retina, it must pass through the cornea, aqueous humor, pupil, lens, and vitreous humor. For vision to occur, light reaching the photoreceptor cells must form an image on the retina. This requires **refraction** of the incoming light, **accommodation** of the lens, and **constriction** of the pupil.

The anterior of the eye is concerned mainly with **refracting** (bending) the incoming light rays so that they focus on the retina (below left). Most refraction occurs at the cornea. The lens adjusts the degree of refraction to produce a sharp image. **Accommodation** (below right) adjusts the eye for near or far objects. Constriction of the pupil narrows the diameter of the hole through which light enters the eye, preventing light rays entering from the periphery.

The point at which the nerve fibers leave the eye as the optic nerve, is the **blind spot** (the point at which there are no photoreceptor cells). Nerve impulses travel along the optic nerves to the visual processing areas in the cerebral cortex. Images on the retina are inverted and reversed by the lens but the brain interprets the information it receives to correct for this image reversal.

The Structure and Function of the Human Eye

The human eye is essentially a three layered structure comprising an outer fibrous layer (the sclera and cornea), a middle vascular layer (the choroid, ciliary body, and iris), and inner **retina** (neurons and **photoreceptor cells**). The shape of the eye is maintained by the fluid filled cavities (aqueous and vitreous humors), which also assist in light refraction. Eye color is provided by the pigmented iris. The iris also regulates the entry of light into the eye through the contraction of circular and radial muscles.

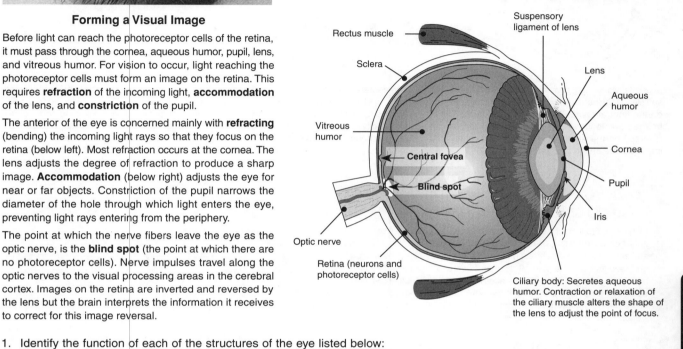

Ciliary body: Secretes aqueous humor. Contraction or relaxation of the ciliary muscle alters the shape of the lens to adjust the point of focus.

1. Identify the function of each of the structures of the eye listed below:

    (a) Cornea: _____

    (b) Ciliary body: _____

    (c) Iris: _____

2. (a) The first stage of vision involves forming an image on the retina. In simple terms, explain what this involves:

    (b) Explain how accommodation is achieved: _____

The lens of the eye has two convex surfaces (biconvex). When light enters the eye, the lens bends the incoming rays towards each other so that they intersect at the focal point on the central fovea of the retina. By altering the curvature of the lens, the focusing power of the eye can be adjusted. This adjustment of the eye for near or far vision is called **accommodation** and it is possible because of the elasticity of the lens. For some people, the shape of the eyeball or the lens prevents convergence of the light rays on the central fovea, and images are focused in front of, or behind, the retina. Such visual defects (below) can be corrected with specific lenses. As we age, the lens loses some of its elasticity and, therefore, its ability to accommodate. This inability to focus on nearby objects due to loss of lens elasticity is a natural part of **aging** and is called **far sight**.

Normal vision

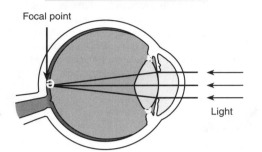

In normal vision, light rays from an object are bent sufficiently by the cornea and lens, and converge on the central fovea. A clear image is formed. Images are focused upside down and mirror reversed on the retina. The brain automatically interprets the image as right way up.

Accommodation for near and distant vision

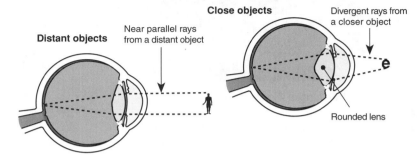

The degree of refraction occurring at each surface of the eye is precise. The light rays reflected from an object 6 m or more away are nearly parallel to one another. Those reflected from near objects are divergent. The light rays must be refracted differently in each case so that they fall exactly on the central fovea. This is achieved through adjustment of the shape of the lens (accommodation). Accommodation from distant to close objects occurs by rounding the lens to shorten its focal length, since the image distance to the object is essentially fixed.

Short sightedness (myopia)

Myopia (top row, right) results from an elongated eyeball or a thickened lens. Left uncorrected, distant objects are focused in front of the retina and appear blurred. To correct myopic vision, concave (negative) lenses are used to move the point of focus backward to the retina. Myopia is not necessarily genetic, nor is it necessarily caused by excessive close work, as was once thought, although myopia does seem to be more prevalent amongst those living in very confined spaces (e.g. people working and living in submarines).

Long sightedness (hypermetropia)

Long sightedness (bottom row, right) results from a shortened eyeball or from a lens that is too thin. Left uncorrected, light is focused behind the retina and near objects appear blurred. Mild or moderate hypermetropia, which occurs naturally in young children, may be overcome by **accommodation**. In more severe cases, corrective lenses are used to bring the point of focus forward to produce a clear image. This is achieved using a convex (positive) lens.

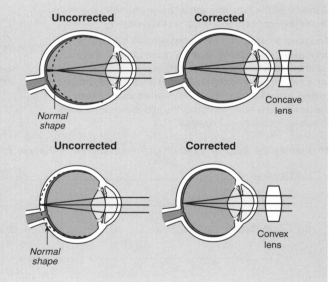

3. (a) Describe the function of the pupil: _____

   (b) Suggest why control of pupil diameter would be under reflex control: _____

   _____

   _____

4. With respect to formation of the image, describe what is happening in:

   (a) Short sighted people: _____

   (b) Long sighted people: _____

5. In general terms, describe how the use of lenses corrects the following problems associated with vision:

   (a) Myopia: _____

   (b) Hypermetropia: _____

The Physiology of Vision

Vision involves two stages: formation of the image on the retina and generation and conduction of nerve impulses. When light reaches the retina, it is absorbed by the photosensitive pigments associated with the membranes of the photoreceptor cells (the rods and cones). The pigment molecules are altered by the absorption of light in such a way as to lead to the generation of nerve impulses. It is these impulses that are conducted via nerve fibers to the occipital lobe in the cerebral cortex. Some of nerve fibers from each eye cross over to the opposite side of the brain so each side of the brain receives visual input from both eyes.

The Structure and Arrangement of Photoreceptor Cells in the Retina

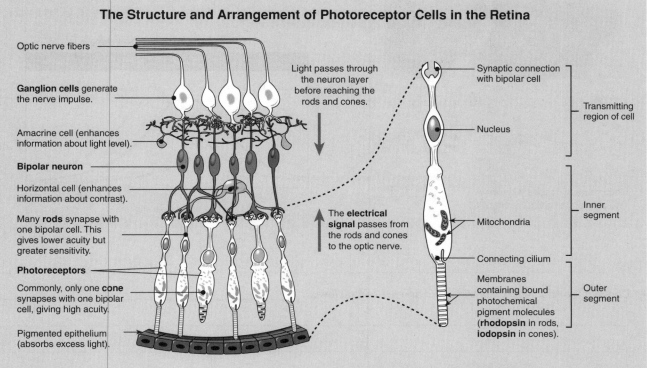

Arrangement of photoreceptors and neurons in the retina

Structure of a rod photoreceptor cell

The photoreceptor cells of the mammalian retina are the **rods** and **cones**. Rods are specialized for vision in dim light, whereas cones are specialized for color vision and high visual acuity. Cone density and visual acuity are greatest in the **central fovea** (rods are absent here). After an image is formed on the retina, light impulses must be converted into nerve impulses. The first step is the development of **generator potentials** by the rods and cones. Light induces structural changes in the **photochemical pigments** (or photopigments) of the rod and cone membranes. The generator potential that develops from the pigment breakdown in the rods and cones is different from the generator potentials that occur in other types of sensory receptors because stimulation results in a **hyperpolarization** rather than a depolarization (in other words, there is a net loss of Na$^+$ from the photoreceptor cell). Once generator potentials have developed, the graded changes in membrane conductance spread through the photoreceptor cell. Each photoreceptor makes synaptic connection with a bipolar neuron, which transmits the potentials to the **ganglion cells**. The ganglion cells become **depolarized** and initiate nerve impulses which pass through the optic chiasma and eventually to the visual areas of the cerebral cortex. The frequency and pattern of impulses in the optic nerve conveys information about the changing visual field.

The Nervous System

1. Complete the table below, comparing the features of the **photoreceptor cells**, the rods and cones:

Feature	Rod cells	Cone cells
Visual pigment(s):	_____	_____
Visual acuity:	_____	_____
Overall function:	_____	_____

2. Identify the three major types of neuron making up the retina and describe their basic function:

   (a) _____

   (b) _____

   (c) _____

3. Identify two types of accessory neurons in the retina and describe their basic function:

   (a) _____

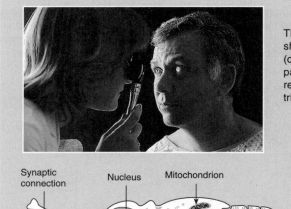

The Basis of Trichromatic Vision

There are three classes of **cones**, each with a maximal response in either short (blue), intermediate (green) or long (yellow-green) wavelength light (coded B, G, and R below). The yellow-green cone is also sensitive to the red part of the spectrum and is often called the red cone (R). The differential responses of the cones to light of different wavelengths provides the basis of trichromatic color vision.

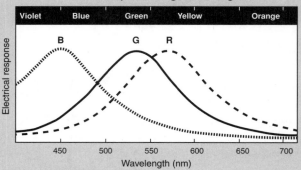

Cone response to light wavelengths

Synaptic connection   Nucleus   Mitochondrion

Membranes containing bound **iodopsin** pigment molecules.

Each **cone** synapses with only one bipolar cell giving high acuity.

(b) _____

4. Contrast the structure of the blind spot and the central fovea: _____

_____

_____

5. Explain the differences in acuity and sensitivity between rod and cone cells: _____

_____

_____

_____

6. (a) Explain clearly what is meant by the term photochemical pigment (photopigment): _____

_____

_____

(b) Identify two photopigments and their location: _____

_____

7. In your own words, explain how light is able to produce a nerve impulse in the ganglion cells: _____

_____

_____

_____

_____

_____

8. Explain the physiological basis for color vision in humans: _____

_____

_____

_____

Hearing

In humans and other mammals, the receptors for detecting sound waves are organized into hearing organs called **ears**. Sound is produced by the vibration of particles in a medium and it travels in waves that can pass through solids, liquids, or gases. The distance between wave 'crests' determines the frequency (pitch) of the sound. The absolute size (amplitude) of the waves determines the intensity or loudness of the sound. Sound

reception in humans is the role of **mechanoreceptors**: tiny **hair cells** in the cochlea of the inner ear. The hair cells are very sensitive and are easily damaged by prolonged exposure to high intensity sounds. Gradual hearing loss with age is often caused by the cumulative loss of sensory hair cell function, especially at the higher frequencies. The ear also houses the vestibular apparatus, which is sensitive to the body's equilibrium.

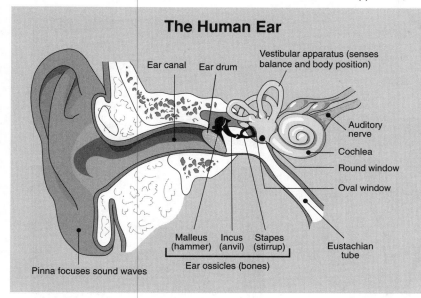

The Human Ear

In mammals, sound waves are converted to pressure waves in the inner ear. The ears of mammals use mechanoreceptors (sensory hair cells) to change the pressure waves into nerve impulses. The mammalian ear contains not only the organ of hearing, the **cochlea**, but all the specialized structures associated with gathering, directing, and amplifying the sound. The cochlea is a tapered, coiled tube, divided lengthwise into **three fluid filled canals**. The cochlea is shown below, unrolled to indicate the way in which sound waves are transmitted through the canals to the sensory cells. The mechanisms involved in hearing are outlined in a simplified series of steps. In mammals, the inner ear is also associated with the organ for detecting balance and position (the vestibular apparatus), although this region is not involved in hearing.

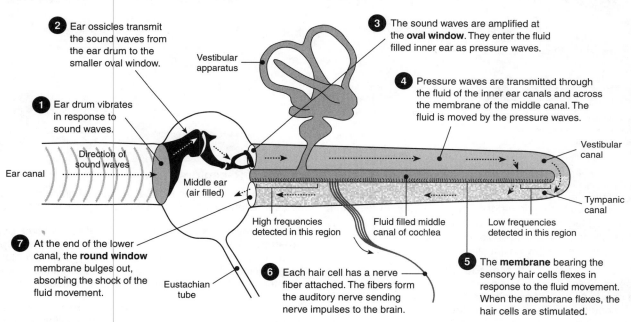

The Nervous System

1. In a short sentence, outline the role of each of the following in the reception and response to sound:

   (a) The ear drum: _____

   (b) The ear ossicles: _____

   (c) The oval window: _____

   (d) The sensory hair cells: _____

   (e) The auditory nerve: _____

2. Explain the significance of the inner ear being fluid filled: _____

Related activities: The Basis of Sensory Perception
Web links: Hearing, Sound Waves and the Cochlea

A 2

Skin Senses

The skin is an important sensory organ, with receptors for pain, pressure, touch, and temperature. While some are specialized receptor structures, many are simple unmyelinated nerves. Tactile (touch) and pressure receptors are **mechanoreceptors** and are stimulated by mechanical distortion. In the **Pacinian corpuscle**, the layers of tissue comprising the sensory structure are pushed together with pressure, stimulating the axon. Views of human skin and its structures are illustrated below. Human skin is fairly uniform in its basic structure, but the density and distribution of glands, hairs, and receptors varies according to the region of the body. **Meissner's corpuscles**, for example, are concentrated in areas sensitive to light touch and, in hairy skin, tactile receptors are clustered into specialized epithelial structures called touch domes or hair disks.

Tactile receptors
Meissner's corpuscles are rapidly adapting nerve endings providing information on light touch. Other more slowly adapting receptors provide information on texture. Both are in the superficial layers of the skin.

Pain receptors are free nerve endings that respond if damaged. They are found in the skin and in almost every tissue of the body.

Hair

Skin surface

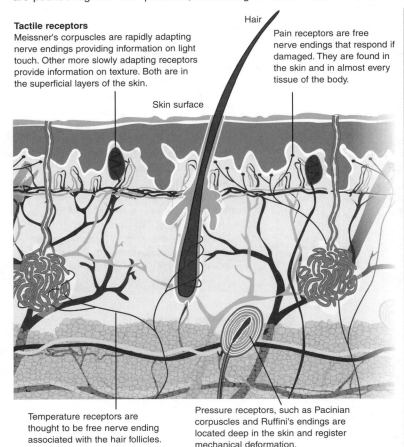

Temperature receptors are thought to be free nerve ending associated with the hair follicles.

Pressure receptors, such as Pacinian corpuscles and Ruffini's endings are located deep in the skin and register mechanical deformation.

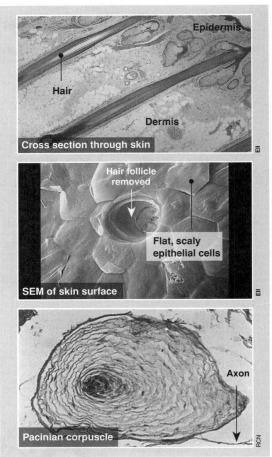

Cross section through skin

Epidermis

Hair

Dermis

Hair follicle removed

Flat, scaly epithelial cells

SEM of skin surface

Pacinian corpuscle

Axon

Testing the Distribution of Touch Receptors

The receptors in the skin are more concentrated in some parts of the body than in others. The distribution of receptors can be tested by using the **two point touch test**. This involves finding the smallest distance at which someone can correctly distinguish two point stimuli.

Method
Repeatedly touch your lab partner's skin lightly with either one or two points of fine scissors or tweezers. Your partner's eyes should be closed. At each touch, they should report the sensation as "one" or "two", depending on whether they perceive one or two touches.

Begin with the scissor points far apart and gradually reduce the separation until only about 8 in 10 reports are correct.

This separation distance (in mm) is called the **two point threshold**. When the test subject can feel only one receptor (when there are two) it means that only one receptor is being stimulated. A large two point threshold indicates a low receptor density, a low one indicates a high receptor density.

Repeat this exercise for: the forearm, the back of the hand, the palm of the hand, the fingertip, and the lips, and then complete the table provide below:

Area of skin	Two point threshold (in mm)
Forearm	
Back of hand	
Palm of hand	
Fingertip	
Lips	

1. Name the region with the greatest number of touch receptors:

   _____

2. Name the region with the least number of touch receptors:

   _____

3. Explain why there is a difference between these two regions:

   _____

   _____

   _____

Related activities: The Basis of Sensory Perception, The Integumentary System
Web links: Touch Receptors

Taste and Smell

Chemosensory receptors are responsible for our sense of smell (**olfaction**) and taste (**gustation**). The receptors for smell and taste both respond to chemicals, either carried in the air (smell) or dissolved in a fluid (taste). In humans and other mammals, these are located in the nose and tongue respectively.

Each receptor type is basically similar: they are collections of receptor cells equipped with chemosensory microvilli or cilia. When chemicals stimulate their membranes, the cells respond by producing nerve impulses that are transmitted to the appropriate region of the cerebral cortex for interpretation.

Taste (Gustation)

The organs of taste are the **taste buds**, which are located on the tongue. Most of the taste buds on the tongue are located on raised protrusions of the tongue surface called **papillae**. Each bud is flask-like in shape, with a pore opening to the surface of the tongue enabling molecules and ions dissolved in saliva to reach the receptor cells inside. Each taste bud is an assembly of 50-150 taste cells. These connect with nerves that send messages to the gustatory region of the brain. There are five basic taste sensations. **Salty** and **sour** operate through ion channels, while **sweet**, **bitter**, and **umami** (savory) operate through membrane signaling proteins. These taste senations are found on all areas of the tongue although some regions are more sensitive than others.

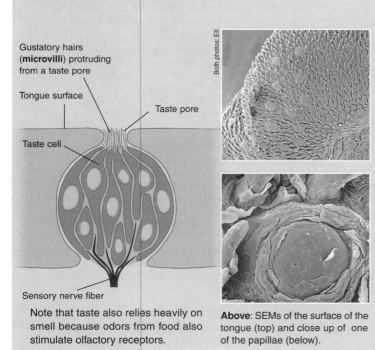

Gustatory hairs (**microvilli**) protruding from a taste pore

Tongue surface

Taste pore

Taste cell

Sensory nerve fiber

Both photos: Ell

Note that taste also relies heavily on smell because odors from food also stimulate olfactory receptors.

Above: SEMs of the surface of the tongue (top) and close up of one of the papillae (below).

Smell (Olfaction)

In humans, the receptors for smell are located at the top of the nasal cavity. The receptors are specialized hair cells that detect airborne molecules and respond by sending nerve impulses to the olfactory centre of the brain. Unlike taste receptors, olfactory receptors can detect many different odors. However, they quickly adapt to the same smell and will cease to respond to it. This phenomenon is called **sensory adaptation**.

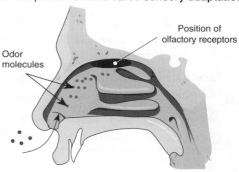

Position of olfactory receptors

Odor molecules

Detail of olfactory membrane

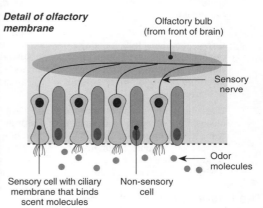

Olfactory bulb (from front of brain)

Sensory nerve

Odor molecules

Sensory cell with ciliary membrane that binds scent molecules

Non-sensory cell

The Nervous System

1. Describe the basic mechanism by which chemical sense operates: _____

_____

_____

2. Take a deep breath of a non-toxic, pungent substance such as perfume. Take a sniff of the substance at 10 second intervals for about 1 minute. Make a record of how strongly you perceived the smell to be at each time interval. Use a scale of 1 to 6: **1.** *Very strong;* **2.** *Quite strong;* **3.** *Noticeable;* **4.** *Weak;* **5.** *Very faint;* **6.** *Could not detect.*

Time	Strength	Time	Strength	Time	Strength
10 s	_____	30 s	_____	50 s	_____
20 s	_____	40 s	_____	1 min	_____

3. (a) Explain what happened to your sense of smell over the time period: _____

   (b) State the term that describes this phenomenon: _____

   (c) Describe the adaptive advantage of this phenomenon: _____

   _____

Related activities: The Basis of Sensory Perception
Web links: Sense of Taste, Sense of Smell, Olfactory Receptor Stimulation

A 1

The Endocrine System

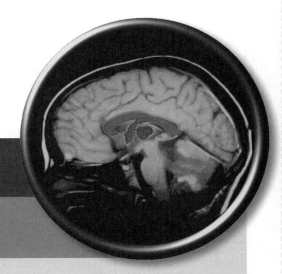

Understanding how the endocrine system operates to regulate bodily processes

Hormonal regulation and cell signaling, location and function of the endocrine organs, the hypothalamus and pituitary, regulation of blood glucose, diabetes mellitus, menopause

Learning Objectives

☐ 1. Compile your own glossary from the **KEY WORDS** displayed in **bold type** in the learning objectives below.

The Basics of Endocrine Control *(pages 93-95)*

☐ 2. Appreciate the interdependence and general roles of the two regulatory systems (hormonal and nervous) with which humans achieve **homeostasis**.

☐ 3. Define the terms **endocrine gland**, **hormone**, and **target cell** (tissue or organ). Describe the general organization of the **endocrine system** and outline the general role of hormones in the maintenance of homeostasis.

☐ 4. Explain how hormones bring about their effects (see also #5) and why they have wide-ranging physiological effects. Describe the stimuli for endocrine gland activity: **hormonal**, **humoral**, and **neural**. Explain the role of **negative feedback mechanisms** in regulating hormone levels.

☐ 5. Describe the types of **cell signaling** and identify the involvement of signaling molecules in a **signal transduction** pathway. Describe examples of types of signaling molecules. Distinguish between the mode of action of steroid hormones (direct activation of genes) and peptide hormones (action by second messenger).

The Endocrine System *(pages 96-103)*

☐ 6. Identify the major endocrine glands and tissues of the body. Name the hormones produced and their functions, and explain how their release is regulated.

☐ 7. Describe the functional relationship between the **hypothalamus** and the **pituitary gland**. Discuss the central role of the hypothalamus in regulating the secretions of other endocrine glands, as illustrated by the role of the hypothalamus in the **stress response**.

☐ 8. Distinguish between the anterior and posterior pituitary and identify the position and role of the portal vein and **neurosecretory cells**. Describe the function and regulation of the pituitary hormones, as follows:
   (a) **Anterior pituitary** (adenohypophysis): GH, TSH, ACTH, MSH, LH, FSH, and prolactin
   (b) **Posterior pituitary** (neurohypophysis) ADH and oxytocin

Case study: Blood glucose regulation

☐ 9. Describe the factors that lead to variation in blood glucose levels and understand the normal range over which blood glucose levels fluctuate. Describe the general structure of the pancreas and outline its role as both an exocrine and an **endocrine organ**. Explain how the regulation of **blood glucose** level is achieved in humans, including reference to:

   (a) Negative feedback mechanisms.
   (b) The hormones **insulin** and **glucagon**.
   (c) The role of the liver in glucose-glycogen conversions.
   (d) Causes and effects of type 1 and type 2 diabetes

Aging and the endocrine system

☐ 10. Describe degenerative changes occurring in endocrine function, with particular reference to **menopause**. Describe the physiological changes associated with menopause and relate these to increased risk of some diseases, e.g. **osteoporosis** and heart disease.

Supplementary Texts

See page 9 for additional details of these texts:
■ Morton, D. and Perry, J.W., 1997. **Photo Atlas for Anatomy and Physiology**, (Brooks/Cole).
■ Rowett, H.G.Q., 1999. **Basic Anatomy & Physiology**, (John Murray), pp. 121-123.

Periodicals

See page 9 for details of publishers of periodicals:
STUDENT'S REFERENCE
■ **Adolescence - Hormones Rule OK?** Biol. Sci. Rev., 19(3) Feb. 2007, pp. 2-6. *The physical and physiological changes during puberty are coordinated by the hypothalamus and mediated by the sex hormones and by growth hormone.*

■ **Glucose Center Stage** Biol. Sci. Rev., 19(2) Nov. 2006, pp. 14-17. *Insulin and glucagon are responsible for the tight regulation of glucose in the blood and the metabolism of glucose in the liver.*
■ **Thyroxine** Biol. Sci. Rev., 12(2) Nov. 1999, pp. 19-21. *A good account of the structure of the thyroid, the physiological roles of its hormones, and their regulation through negative feedback.*
■ **Growth Hormone** Biol. Sci. Rev., 12 (4) March 2000, pp. 26-28. *The consequences of growth hormone deficiencies in humans.*
■ **Menopause - Design Fault, or By Design** Biol. Sci. Rev., 14(1) Sept. 2001, pp. 2-6. *An excellent synopsis of the basic biology of menopause.*

TEACHER'S REFERENCE
■ **Pregnancy Tests** Scientific American, Nov. 2000, pp. 92-93. *Pregnancy tests: how they work and the role of HCG in signalling pregnancy.*
■ **New Bull's-Eye for Drugs** Sci. American, Oct. 2005, pp. 32-39. *New drugs target the receptors on the cell membrane involved in signal cascades.*
■ **Are Your Cells Pregnant?** The Am. Biology Teacher, 63(7), Sept. 2001, pp. 514-517. *An experiment to investigate the principles of hormone release from endocrine cells.*

Internet

See pages 10-11 for details of how to access **Bio Links** from our web site: **www.thebiozone.com** From Bio Links, access sites under the topics:

ANATOMY & PHYSIOLOGY > General sites: • Anatomy & Physiology • Human physiology lecture notes • WebAnatomy > **Endocrine System:** • Hormones and their effects • Drag and drop hormone match • The endocrine system • Tour of the endocrine system • Endocrine topics **Homeostasis** • BBC schools - Homeostasis • Physiological homeostasis

Presentation MEDIA to support this topic: **Anatomy & Physiology**

Hormonal Regulatory Systems

The endocrine system regulates the body's processes by releasing chemical messengers called **hormones** into the bloodstream. Hormones are potent chemical regulators: they are produced in minute quantities yet can have a large effect on metabolism. The endocrine system is made up of endocrine cells (organized into endocrine glands), and the hormones they produce. Unlike exocrine glands (e.g. sweat and salivary glands), endocrine glands are **ductless glands**, secreting hormones directly into the bloodstream rather than through a duct or tube. Some organs (e.g. the pancreas) have both endocrine and exocrine regions, but these are structurally and functionally distinct. The stimulus for hormone release may be **hormonal** (e.g. action of hypothalamic releasing hormones), a **humoral** trigger (such as high blood glucose), or **neural** (e.g. sympathetic nervous system stimulation). The basics of hormonal action and regulation are described below.

The Mechanism of Hormone Action

Endocrine cells produce hormones and secrete them into the bloodstream where they are distributed throughout the body. Although hormones are broadcast throughout the body, they affect only specific target cells. These target cells have receptors on the plasma membrane which recognize and bind the hormone (see inset, below right). The binding of hormone and receptor triggers the response in the target cell. Cells are unresponsive to a hormone if they do not have the appropriate receptors.

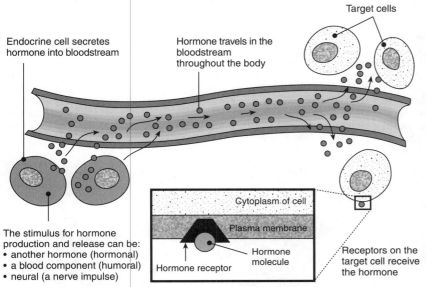

Target cells

Endocrine cell secretes hormone into bloodstream

Hormone travels in the bloodstream throughout the body

Cytoplasm of cell

Plasma membrane

Hormone receptor

Hormone molecule

Receptors on the target cell receive the hormone

The stimulus for hormone production and release can be:
• another hormone (hormonal)
• a blood component (humoral)
• neural (a nerve impulse)

Antagonistic Hormones

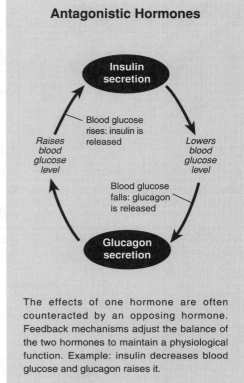

Insulin secretion

Blood glucose rises: insulin is released

Raises blood glucose level

Lowers blood glucose level

Blood glucose falls: glucagon is released

Glucagon secretion

The effects of one hormone are often counteracted by an opposing hormone. Feedback mechanisms adjust the balance of the two hormones to maintain a physiological function. Example: insulin decreases blood glucose and glucagon raises it.

1. (a) Explain what is meant by **antagonistic hormones** and describe an example of how two such hormones operate:

_____

_____

Example: _____

_____

(b) Explain the role of feedback mechanisms in adjusting hormone levels (explain using an example if this is helpful):

_____

_____

2. Explain how a hormone can bring about a response in target cells even though all cells may receive the hormone:

_____

_____

3. Explain why hormonal control differs from nervous system control with respect to the following:

(a) The speed of hormonal responses is slower: _____

_____

(b) Hormonal responses are generally longer lasting: _____

_____

Related activities: Nervous Regulatory Systems, Control of Blood Glucose
Web links: Control of Endocrine Activity

RA 1

The Endocrine System

Cell Signaling

Cells use **signals** (chemical messengers) to gather information about, and respond to, changes in their cellular environment and for communication between cells. The signaling and response process is called the **signal transduction pathway**, and often involves a number of enzymes and molecules in a **signal cascade** which causes a large response in the target cell. Cell signaling pathways are categorized primarily on the distance over which the signal molecule travels to reach its target cell, and generally fall into three categories. The **endocrine** pathway involves the transport of hormones over large distances through the circulatory system. During **paracrine** signaling, the signal travels an intermediate distance to act upon neighboring cells. **Autocrine** signaling involves a cell producing and reacting to its own signal. These three pathways are illustrated below.

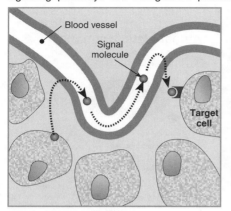

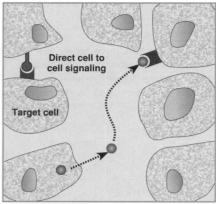

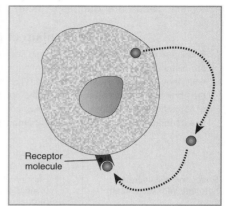

Endocrine signaling: Hormone signals are released by ductless endocrine glands and carried long distances through the body by the circulatory system to the target cells. Examples include sex hormones, growth factors, and neurohormones such as dopamine.

Paracrine signaling: Signals released from a cell act upon target cells within the immediate vicinity. The chemical messenger can be transferred through the extracellular fluid (e.g. at synapses) or directly between cells, which is important during embryonic development.

Autocrine signaling: Cells produce and react to their own signals. In vertebrates, when a foreign antibody enters the body, some T-cells (lymphocytes) produce a growth factor to stimulate their own production. The increased number of T-cells helps to fight the infection.

Signaling Receptors and Signaling Molecules

Examples of cell signaling molecules

Insulin like growth factor 1 (IGF-1)

Progesterone

Structure of a transmembrane receptor

Labels: Extracellular domain, Outside of cell, Cell membrane, Cell cytosol, Intracellular domain

The binding sites of cell receptors are very specific; they only bind certain **ligands** (signal molecules). This stops them from reacting to every signal bombarding the cell. Receptors fall into two main categories :

▶ **Cytoplasmic receptors** Cytoplasmic receptors, located within the cell cytoplasm, bind ligands which are able to cross the plasma membrane unaided.

▶ **Transmembrane receptors** These span the cell membrane and bind ligands which cannot cross the plasma membrane on their own. They have an extra-cellular domain outside the cell, and an intracellular domain within the cell cytosol. Ion channels, protein kinases and G-protein linked receptors are examples of transmembrane receptors (see diagram on left).

1. Briefly describe the three types of cell signaling:

   (a) _____

   (b) _____

   (c) _____

2. Identify the components that all three cell signaling types have in common: _____

   _____

   _____

© Biozone International 2009

Related activities: Hormonal Regulatory Systems, Signal Transduction
Web links: Hormones, Receptors, and Target Cells

Signal Transduction

Once released, a hormone is carried in the blood to affect specific target cells. Water soluble hormones are carried free in the blood, whilst steroid and thyroid hormones are carried bound to plasma proteins. Target cells have receptors to bind the hormone, initiating a cascade of reactions which results in a specific target cell response (e.g. protein synthesis, change in membrane permeability, enzyme activation, or secretion). **Peptide hormones** operate by interacting with transmembrane receptors and activating a second messenger system (e.g. cyclic AMP). **Steroid hormones** enter the cell to interact directly with intracellular cytoplasmic receptors. Once the target cell responds, the response is recognized by the hormone-producing cell through a feedback signal and the hormone is degraded.

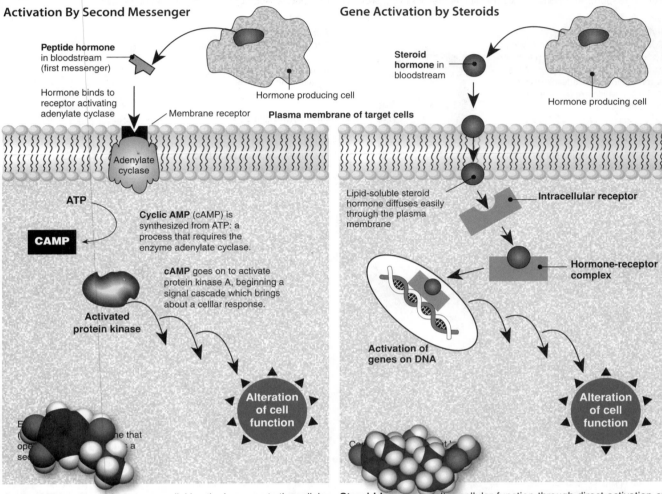

Activation By Second Messenger

Peptide hormone in bloodstream (first messenger)

Hormone binds to receptor activating adenylate cyclase

Membrane receptor

Hormone producing cell

Plasma membrane of target cells

Adenylate cyclase

ATP

CAMP

Cyclic AMP (cAMP) is synthesized from ATP: a process that requires the enzyme adenylate cyclase.

cAMP goes on to activate protein kinase A, beginning a signal cascade which brings about a cellular response.

Activated protein kinase

Alteration of cell function

Gene Activation by Steroids

Steroid hormone in bloodstream

Hormone producing cell

Lipid-soluble steroid hormone diffuses easily through the plasma membrane

Intracellular receptor

Hormone-receptor complex

Activation of genes on DNA

Alteration of cell function

Cyclic AMP is a **second messenger** linking the hormone to the cellular response. Cellular concentration of cAMP increases markedly once a hormone binds and the cascade of enzyme-driven reactions is initiated.

Steroid hormones alter cellular function through direct activation of genes. Once inside the target cell, steroids bind to intracellular receptor sites, creating hormone-receptor complexes that activate specific genes.

1. Describe the two mechanisms by which a hormone can bring about a cellular response:

   (a) _____

   _____

   (b) _____

   _____

2. State in what way these two mechanisms are alike: _____

   _____

   _____

3. Explain how a very small amount of hormone is able to exert a disproportionately large effect on a target cell:

   _____

   _____

   _____

   _____

The Endocrine System

Related activities: Hormonal Regulatory Systems, Cell Signaling
Web links: Signal Transduction, Peptide Hormone Action

A 3

The Endocrine System

The endocrine glands are distributed throughout the body, frequently associated with the organs of other body systems. Under appropriate stimulation they secrete **hormones**, which are carried in the blood to exert a specific metabolic effect on target cells. After exerting their effect, hormones are broken down and excreted from the body. Although a hormone circulates in the blood, only the targets respond. Hormones may be amino acids, peptides, proteins (often modified), fatty acids, or steroids. Some basic features of the human endocrine system are explained below. The hypothalamus, although part of the brain and not strictly an endocrine gland, contains neurosecretory cells, and links the nervous and endocrine systems. The hypothalamus, together with the pituitary, adrenal, and thyroid glands, form a central axis of endocrine control that regulates much of the body's metabolic activity.

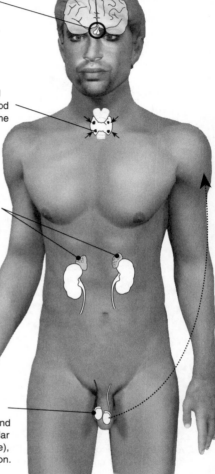

Hypothalamus
Coordinates nervous and endocrine systems. Secretes releasing hormones, which regulate the hormones of the anterior pituitary. Produces oxytocin and ADH, which are released from the posterior pituitary.

Pineal
This small gland in the brain secretes melatonin, which regulates the sleep-wake cycle. Melatonin secretion follows a circadian rhythm and coordinates reproductive hormones too.

Thyroid gland
Secretes thyroxine, an iodine containing hormone needed for normal growth and development. Thyroxine stimulates metabolism and growth via protein synthesis.

Pancreatic islets
Specialized α and β endocrine cells in the pancreas produce glucagon and insulin. Together, these control blood sugar levels.

Ovaries (in females)
The ovaries produce estrogen and progesterone. These hormones control and maintain female characteristics, stimulate the menstrual cycle, maintain pregnancy, and prepare the mammary glands for lactation.

Pituitary gland
The pituitary is located below the hypothalamus. It secretes at least nine hormones that regulate the activities of other endocrine glands.

Parathyroid glands
On the surface of the thyroid, they secrete PTH (parathyroid hormone), which regulates blood calcium levels and promotes the release of calcium from bone. High levels of calcium in the blood inhibit PTH secretion.

Adrenal glands
The adrenal medulla produces epinephrine (adrenaline) and norepinephrine (noradrenaline) responsible for the fight or flight response. The adrenal cortex produces various steroid hormones, including cortisol (response to stress) and aldosterone (sodium regulation).

Testes (in males)
The testes of males produce testosterone, which controls and maintains "maleness" (muscular development and deeper voice), and promotes sperm production.

1. Explain how a hormone is different from a neurotransmitter: _____

_____

2. Using ruled lines, connect each of the following endocrine glands with its correct role in the body

   (a) Pituitary gland   The hormone from this gland regulates the levels of calcium in the blood

   (b) Ovaries   Master gland secreting at least nine hormones, including growth hormone and TSH

   (c) Pineal gland   Produces hormones involved in the regulation of metabolic rate

   (d) Parathyroid glands   Secretes melatonin to regulate sleep patterns and cycles of reproductive hormones

   (e) Thyroid   Produce estrogen and progesterone in response to hormones from the pituitary

3. Review the three types of stimuli for hormone release and describe a specific example of each:

   (a) Hormonal stimulus: _____

   (b) Humoral stimulus: _____

   (c) Neural stimulus: _____

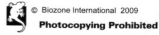

The Hypothalamus and Pituitary

The **hypothalamus** is located below the thalamus, just above the brain stem and the pituitary gland, with which it has a close structural and functional relationship. Information comes to the hypothalamus through sensory pathways from sensory receptors. On the basis of this information, the hypothalamus controls and integrates many basic physiological activities (e.g. temperature regulation, food and fluid intake, and sleep), including the reflex activity of the **autonomic nervous system**.

One of the most important functions of the hypothalamus is to link the nervous system to the endocrine system (via the pituitary). The hypothalamus contains **neurosecretory cells**. These are specialized secretory neurons, which are both nerve cells and endocrine cells. They produce hormones (usually peptides) in the cell body, which are packaged into droplets and transported along the axon. At the axon terminal, the **neurohormone** is released into the blood in response to nerve impulses.

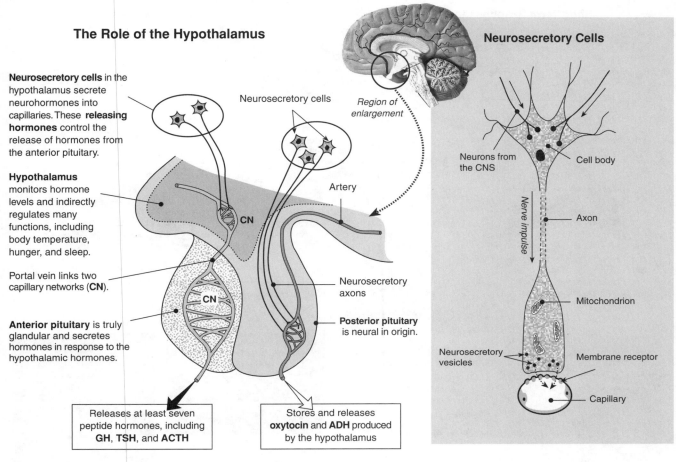

The Role of the Hypothalamus

Neurosecretory cells in the hypothalamus secrete neurohormones into capillaries. These **releasing hormones** control the release of hormones from the anterior pituitary.

Hypothalamus monitors hormone levels and indirectly regulates many functions, including body temperature, hunger, and sleep.

Portal vein links two capillary networks (**CN**).

Anterior pituitary is truly glandular and secretes hormones in response to the hypothalamic hormones.

Neurosecretory cells

Region of enlargement

Artery

CN

CN

Neurosecretory axons

Posterior pituitary is neural in origin.

Releases at least seven peptide hormones, including **GH**, **TSH**, and **ACTH**

Stores and releases **oxytocin** and **ADH** produced by the hypothalamus

Neurosecretory Cells

Neurons from the CNS

Cell body

Nerve impulse

Axon

Mitochondrion

Neurosecretory vesicles

Membrane receptor

Capillary

1. (a) Explain how the anterior and posterior pituitary differ with respect to their relationship to the hypothalamus:

   _____

   _____

   (b) Explain how these differences relate to the nature of the hormonal secretions for each region: _____

   _____

   _____

2. Describe the role of the neurohormones released by the hypothalamus: _____

   _____

3. Explain why the adrenal and thyroid glands atrophy if the pituitary gland ceases to function: _____

   _____

   _____

4. Although the anterior pituitary is often called the master gland, the hypothalamus could also claim that title. Explain:

   _____

   _____

Related activities: Hormones of the Pituitary, The Stress Response

RA 2

The Endocrine System

The Stress Response

The interactions of the **hypothalamus, pituitary** and **adrenal glands** together constitute the hypothalamic-pituitary-adrenal axis, which is responsible for controlling the body's reactions to stress and regulating many of the body's processes, including digestion, immune function, mood, sexuality, and energy storage and expenditure. The **stress response** is triggered through sympathetic stimulation of the central medulla region of the adrenal glands, in what is popularly know as the **fight or flight syndrome**. This stimulation causes the release of catecholamines (epinephrine and norepinephrine). These hormones help to prepare the body to cope with short-term stressful situations. Continued stress results in release of glucocorticoids (especially **cortisol**) from the outer cortex of the adrenals. These hormones help the body to resist longer term stress. Their secretion is a normal part of what is called the general adaptation syndrome, but continued unrelieved stress can be damaging, or even fatal.

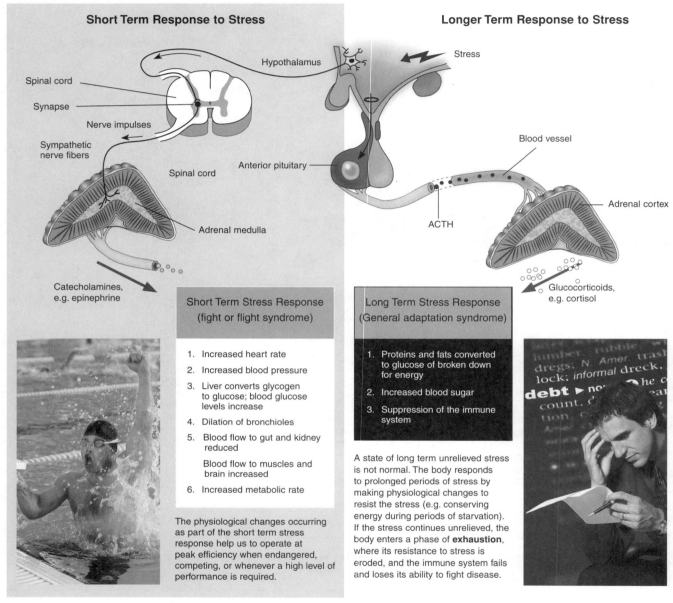

Short Term Response to Stress

Spinal cord
Synapse
Nerve impulses
Sympathetic nerve fibers
Spinal cord
Hypothalamus
Adrenal medulla
Catecholamines, e.g. epinephrine

Longer Term Response to Stress

Stress
Blood vessel
Anterior pituitary
ACTH
Adrenal cortex
Glucocorticoids, e.g. cortisol

Short Term Stress Response
(fight or flight syndrome)

1. Increased heart rate
2. Increased blood pressure
3. Liver converts glycogen to glucose; blood glucose levels increase
4. Dilation of bronchioles
5. Blood flow to gut and kidney reduced

   Blood flow to muscles and brain increased
6. Increased metabolic rate

The physiological changes occurring as part of the short term stress response help us to operate at peak efficiency when endangered, competing, or whenever a high level of performance is required.

Long Term Stress Response
(General adaptation syndrome)

1. Proteins and fats converted to glucose of broken down for energy
2. Increased blood sugar
3. Suppression of the immune system

A state of long term unrelieved stress is not normal. The body responds to prolonged periods of stress by making physiological changes to resist the stress (e.g. conserving energy during periods of starvation). If the stress continues unrelieved, the body enters a phase of **exhaustion**, where its resistance to stress is eroded, and the immune system fails and loses its ability to fight disease.

1. Explain how the body's short term response to stress is adaptive: _____

_____

_____

2. (a) Describe features of the long term stress response that help to maintain activity through the period of the stress:

    _____

    _____

   (b) Describe how these responses could be damaging if the stress is unrelieved for prolonged periods (e.g. months):

    _____

    _____

Hormones of the Pituitary

The **pituitary gland** (or hypophysis) is a tiny endocrine gland, about the size of a pea, hanging from the inferior surface of the hypothalamus. It has two regions or lobes, each with different structure and origin. The **posterior pituitary** is neural (nervous) in origin and is essentially an extension of the hypothalamus. Its neurosecretory cells have their cell bodies in the hypothalamus, and release oxytocin and ADH directly into the bloodstream in response to nerve impulses. The **anterior pituitary** is connected to the hypothalamus by blood vessels and receives releasing and inhibiting hormones (factors) from the hypothlamus via a capillary network. These releasing hormones regulate the secretion of the anterior pituitary's hormones.

ANTERIOR PITUITARY

The anterior pituitary releases at least seven **peptide hormones** (below) into the blood from simple secretory cells. The release of these hormones is regulated by releasing and inhibiting hormones from the hypothalamus.

POSTERIOR PITUITARY

The posterior pituitary develops as an extension of the hypothalamus. The release of its two hormones, oxytocin and antidiuretic hormone, occurs directly as a result of nervous input to the hypothalamus.

Hypothalamus

Hormones from the anterior pituitary

Hormones from the posterior pituitary

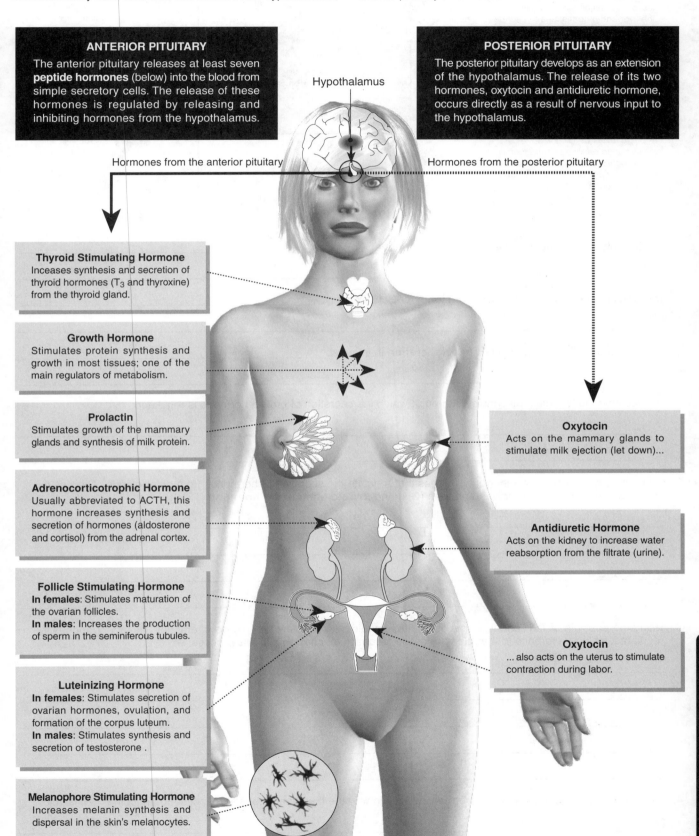

Thyroid Stimulating Hormone
Inceases synthesis and secretion of thyroid hormones (T_3 and thyroxine) from the thyroid gland.

Growth Hormone
Stimulates protein synthesis and growth in most tissues; one of the main regulators of metabolism.

Prolactin
Stimulates growth of the mammary glands and synthesis of milk protein.

Adrenocorticotrophic Hormone
Usually abbreviated to ACTH, this hormone increases synthesis and secretion of hormones (aldosterone and cortisol) from the adrenal cortex.

Follicle Stimulating Hormone
In females: Stimulates maturation of the ovarian follicles.
In males: Increases the production of sperm in the seminiferous tubules.

Luteinizing Hormone
In females: Stimulates secretion of ovarian hormones, ovulation, and formation of the corpus luteum.
In males: Stimulates synthesis and secretion of testosterone .

Melanophore Stimulating Hormone
Increases melanin synthesis and dispersal in the skin's melanocytes.

Oxytocin
Acts on the mammary glands to stimulate milk ejection (let down)...

Antidiuretic Hormone
Acts on the kidney to increase water reabsorption from the filtrate (urine).

Oxytocin
... also acts on the uterus to stimulate contraction during labor.

The Endocrine System

Related activities: The Endocrine System, The Hypothalamus and Pituitary, The Stress Response

RA 2

Effects of Growth Hormone

Growth hormone (GH) is released in response to GHRH (growth hormone releasing hormone) from the hypothalamus. GH acts both directly and indirectly to affect metabolic activities associated with growth.

GH directly stimulates metabolism of fat, but its major role is to stimulate the liver and other tissues to secrete IGF-I (Insulin-like Growth Factor) and through this stimulate bone and muscle growth. GH secretion is regulated is via negative feedback:

- *High levels of IGF-1 suppress GHRH secretion from the hypothalamus.*

- *High levels of IGF-1 also stimulate release of somatostatin from the hypothalamus, which also suppresses GH secretion (not shown).*

GHRH Hypothalamus

Anterior pituitary

GH

Fat

Directly stimulates utilization of fat

IGF-1

Liver

IGF-1

Bone

Stimulates the proliferation of chondrocytes and bone growth

Stimulates muscle growth through protein synthesis and proliferation of myoblasts

Muscle

1. (a) Describe the metabolic effects of growth hormone: _____

_____

(b) Predict the effect of chronic **GH deficiency** of GH in infancy: _____

_____

(c) Predict the effect of chronic **GH hypersecretion** in infancy: _____

_____

(d) Describe the two main mechanisms through which the secretion of growth hormone is regulated:

_____

_____

2. "The pituitary releases a number of hormones that regulate the secretion of hormones from other glands". Discuss this statement with reference to growth hormone (GH) and **thyroid stimulating hormone** (TSH):

_____

_____

_____

_____

3. Using the example of TSH and its target tissue (the thyroid), explain how the release of anterior pituitary hormones is regulated. Include reference to the role of negative feedback mechanisms in this process:

_____

_____

_____

_____

_____

4. Iodine is needed to produce thyroid hormones. Explain why the thyroid enlarges in response to an iodine deficiency:

_____

_____

Control of Blood Glucose

The endocrine portion of the **pancreas** (the α and β cells of the **islets of Langerhans**) produces two hormones, **insulin** and **glucagon**, which maintain blood glucose at a steady state through **negative feedback**. Insulin promotes a decrease in blood glucose by promoting cellular uptake of glucose and synthesizing glycogen. **Glucagon** promotes an increase in blood glucose through the breakdown of glycogen and the synthesis of glucose from amino acids. When normal blood glucose levels are restored, negative feedback stops hormone secretion. Regulating blood glucose to within narrow limits allows energy to be available to cells as needed. Extra energy is stored as glycogen or fat, and is mobilized to meet energy needs as required. The liver is pivotal in these carbohydrate conversions. One of the consequences of a disruption to this system is the disease **diabetes mellitus**. In type 1 diabetes, the insulin-producing β cells are destroyed as a result of autoimmune activity and insulin is not produced. In type 2 diabetes, the pancreatic cells produce insulin, but the body's cells become increasingly resistant to it.

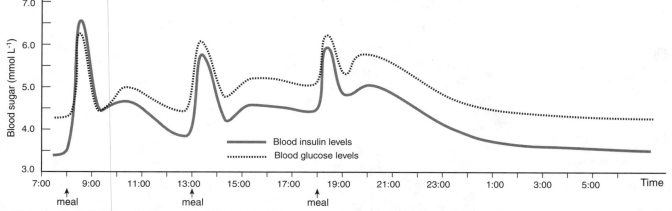

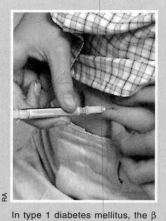

In type 1 diabetes mellitus, the β cells of the pancreas are destroyed and insulin must be delivered to the bloodstream by injection. Type 2 diabetics produce insulin, but their cells do not respond to it.

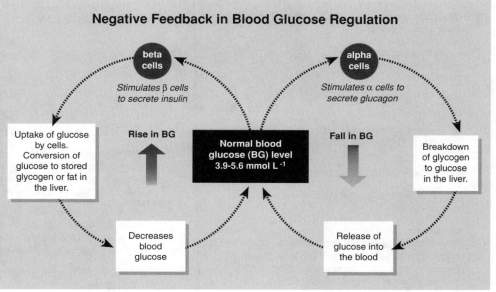

Negative Feedback in Blood Glucose Regulation

1. (a) Identify the stimulus for the release of insulin: _____

   (b) Identify the stimulus for the release of glucagon: _____

   (c) Explain how glucagon brings about an increase in blood glucose level: _____

   _____

   (d) Explain how insulin brings about a decrease in blood glucose level: _____

   _____

2. Explain the pattern of fluctuations in blood glucose and blood insulin levels in the graph above:

   _____

   _____

   _____

   _____

3. Identify the mechanism regulating insulin and glucagon secretion (humoral, hormonal, neural): _____

The Endocrine System

Related activities: The Liver's Homeostatic Role, Stomach and Small Intestine, Type 1 Diabetes Mellitus

A 2

Type 1 Diabetes Mellitus

Diabetes is a general term for a range of disorders sharing two common symptoms: production of large amounts of urine and excessive thirst. There may be other symptoms, depending on the type of diabetes. **Diabetes mellitus** is the most common form of diabetes and is characterized by high blood sugar. **Type 1** has a juvenile onset while **type 2** affects adults.

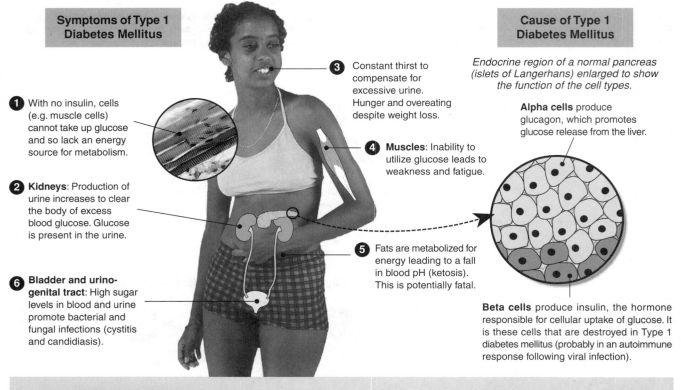

Symptoms of Type 1 Diabetes Mellitus

1 With no insulin, cells (e.g. muscle cells) cannot take up glucose and so lack an energy source for metabolism.

2 Kidneys: Production of urine increases to clear the body of excess blood glucose. Glucose is present in the urine.

6 Bladder and urino-genital tract: High sugar levels in blood and urine promote bacterial and fungal infections (cystitis and candidiasis).

3 Constant thirst to compensate for excessive urine. Hunger and overeating despite weight loss.

4 Muscles: Inability to utilize glucose leads to weakness and fatigue.

5 Fats are metabolized for energy leading to a fall in blood pH (ketosis). This is potentially fatal.

Cause of Type 1 Diabetes Mellitus

Endocrine region of a normal pancreas (islets of Langerhans) enlarged to show the function of the cell types.

Alpha cells produce glucagon, which promotes glucose release from the liver.

Beta cells produce insulin, the hormone responsible for cellular uptake of glucose. It is these cells that are destroyed in Type 1 diabetes mellitus (probably in an autoimmune response following viral infection).

Type 1 Diabetes mellitus (insulin dependent)

Incidence: About 10-15% of all diabetics.

Age at onset: Early; often in childhood (often called juvenile onset diabetes).

Symptoms: Symptoms are severe. Insulin deficiency accelerates fat breakdown and leads to a number of metabolic complications: hyperglycemia (high blood sugar), excretion of glucose in the urine, increased urine production, excessive thirst and hunger, weight loss, and ketosis.

Cause: Absolute deficiency of insulin due to lack of insulin production (pancreatic beta cells are destroyed in an autoimmune reaction). There is a genetic component but usually a childhood viral infection triggers the development of the disease. Mumps, coxsackie, and rubella are implicated.

Treatments

Present treatments: Regular insulin injections combined with dietary management to keep blood sugar levels stable. Blood glucose is monitored regularly with testing kits to guard against sudden falls in blood glucose (hypoglycemia).

Until recently, insulin was extracted from dead animals. Now, genetically engineered yeast or bacterial cells containing the gene for human insulin are grown in culture, providing abundant, low cost insulin, without the side effects associated with animal insulin.

New treatments: Cell therapy involves the transplant of insulin producing islet cells. To date, approximately 400 patients worldwide have received islet cell transplants from donor pancreases, with varying degrees of success. Cell therapy promises to be a practical and effective way to provide sustained relief for Type 1 diabetics.

Future treatments: In the future, gene therapy, where the gene for insulin is inserted into the diabetic's cells, may be possible.

1. Describe the **symptoms** for type 1 diabetes mellitus and relate these to the physiological cause of the disease:

_____

_____

_____

_____

2. Summarize the **treatments** for type 1 diabetes mellitus (list key words/phrases only):

(a) Present treatment: _____

(b) New treatments: _____

(c) Future treatments: _____

Related activities: Control of Blood Glucose, Type 2 Diabetes Mellitus

Aging and the Endocrine System

Despite age-related changes, the endocrine system functions well in most older people. However, some endocrine changes do occur because of normal cellular damage accumulating as a result of the aging process and genetically programmed cellular changes. Aging produces changes in hormone production and secretion, hormone metabolism (how quickly excess hormones are broken down and leave the body, for example, through urination), levels of circulating hormones, target tissue response to hormones, and biological rhythms such as sleep and the menstrual cycle.

The pituitary gland shrinks and becomes more fibrous in old age. Production and secretion of growth hormone declines markedly, and this is related to a decrease in lean muscle mass, an increase in fat mass, and a decrease in bone density.

Melatonin secretion from the pineal declines in the elderly, leading to a disruption in the normal sleep patterns. Melatonin supplements are used to relieve some of the symptoms associated with menopause.

Aging is frequently associated with failure in other organs. Cardiovascular and renal problems become more common for example. The liver and kidneys are primarily responsible for clearing hormones from the bloodstream. Several clearance processes become altered or slowed in individuals who have chronic heart, liver, or kidney disorders.

Unlike women, men remain fertile well into old age, although fertility declines somewhat and there is an increase in sperm abnormalities.

Increasing age is correlated with increased rates of type 2 diabetes and associated problems such as poor peripheral circulation. With aging, the target cell response time becomes slower, especially in people who might be at risk for this disorder.

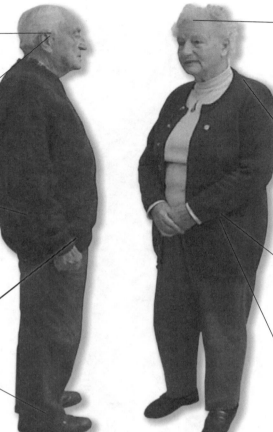

Declining levels of circulating testosterone affects sex drive in both men and women. Age-related thinning and loss of hair is also related to hormonal changes as circulating levels of a testosterone derivative (DHT) increase and damage the hair follicles. In women, hair loss and thinning is the result of DHT sensitivity increasing as estrogen levels decline.

In post menopausal women, low estrogen levels increase the risk of cardiovascular disease and accelerate bone loss, resulting in **osteoporosis**, hunching of the spine, and increased risk of fractures.

In **menopause**, the ovaries stop responding to FSH and LH from the anterior pituitary. Ovarian hormone production of estrogen and progesterone slows and stops, and the menstrual cycle becomes at first irregular and then stops altogether. At this point, a woman is no longer fertile.

Aging is associated with an increase in the number of aberrant cells and an increased incidence of cancers as cellular damage accumulates. Tumors of the endocrine organs are more common in old age.

Menopause is a stage of life and not a disease. The symptoms of approaching menopause, which include hot flushes of the skin, mood swings, and irregular menstrual periods, can be alleviated with attention to a healthy diet and lifestyle. Menopause is determined retrospectively, after 12 months without menstruation.

Physical exercise can help slow or reverse some of the physical changes that occur with aging as a result of hormonal changes. Regular exercise helps to regulate body weight and insulin levels, reduces insulin resistance, improves cardiovascular efficiency and muscle strength, and slows the loss of bone mass.

Production and secretion of growth hormone (GH) is maximal during adolescence and declines thereafter. When GH is administered to adults with GH deficiency, body fat declines and muscle mass and bone density increase. However, GH does not reverse all indications of aging and can even increase mortality in the elderly.

1. Describe two consequences of declining output of growth hormone in old age:

   (a) _____

   (b) _____

2. Explain why the levels of LH and FSH increase in post-menopausal women: _____

   _____

   _____

3. Identify one process that accelerates in elderly women as a result of loss of estrogen: _____

Related activities: Principles of Homeostasis, The Hypothalamus and Pituitary, Hormones of the Pituitary

RA 2

The Endocrine System

Cardiovascular and Lymphatic Systems

Endocrine system

- Lymph and blood distribute hormones.
- A number of hormones, including epinephrine and ADH, influence blood pressure.
- The thymus (a lymphoid organ with endocrine function) produces hormones for the development of the lymphatic organs and T cell maturation.
- Blood pressure and fluid volume are regulated via the renin-angiotensin-aldosterone system.

Cardiovascular system

- Lymphatic vessels pick up leaked tissue fluid and proteins and returns them to general circulation.
- Spleen destroys RBCs, removes cellular debris from the blood, and stores iron
- Blood is the source of lymph and circulates antibodies and immune system cells.

Digestive system

- Products of digestion are transported in the blood and lymph.
- Lymphoid nodules in the intestinal wall protect against invasion by pathogens.
- Gastric acidity destroys pathogens.

Skeletal system

- Bones are the site of blood cell production (hematopoiesis) for both cardiovascular and lymphatic systems.
- Bones protect cardiovascular and lymphatic organs and provide a store of calcium, which is needed for regulating blood volume.

Integumentary system

- Blood vessels in the skin provide a reservoir of blood and are a site of heat loss.
- Skin provides a physical and chemical barrier to pathogens.

Nervous system

- The nervous system innervates lymphatic organs and vessels. The brain helps to regulate the immune response.
- The autonomic ANS regulates cardiac output; the sympathetic division regulates blood pressure and distribution.

Lymphatic system and immunity

- The lymphatic system returns tissue fluid and proteins to the cardiovascular system.
- Lymph forms from blood. Blood and lymph transport antibodies and immune cells.

Respiratory system

- Gas exchange loads O_2 and unloads CO_2 to and from the blood. Respiration aids return of venous blood and lymph to the heart.
- IgA secreted by plasma cells protects the respiratory mucosa against pathogens.
- The pharynx houses lymphoid tissues.

Urinary system

- Urinary system eliminates wastes transported in the blood and maintains water, electrolyte, and acid-base balance.
- The urinary system helps to regulate blood pressure and volume by altering urine output and via release of renin.
- Urine flushes some pathogens from the body.

Reproductive system

- Slightly acidic vaginal secretions in women prevent the growth of bacteria and fungi.
- Estrogen maintains cardiovascular health in women (and in men as a result of conversion from testosterone).

Muscular system

- Activity of skeletal muscles aids return of blood and lymph to the heart.
- Muscles protect superficial lymph nodes.
- Aerobic activity improves cardiovascular health and efficiency.

General Functions and Effects on all Systems

The cardiovascular system delivers oxygen and nutrients to organs and tissues and removes carbon dioxide and other waste products of metabolism. Lymphatic vessels pick up leaked tissue fluid and proteins and return them to general circulation. The immune system protects the body's tissues and organs against pathogens.

Disease

Symptoms of disease	• Pain (moderate to severe) • Loss of function (e.g. cardiac arrest) • Swelling and immunodeficiency • Tissue loss and gangrene
Diseases of the heart and blood vessels	• Congenital heart disorders • Atherosclerosis and heart disease • Peripheral vascular disease & stroke • Cancers (e.g. leukemia)
Lymphatic system disorders	• Cancers (e.g. lymphomas) • Infectious diseases (e.g. HIV) • Inherited disorders (e.g. SCID) • Autoimmune disorders

Medicine & Technology

Diagnosis of disorders	• Electrocardiography • X-rays, MRI and CT scans • Ultrasound • Blood and DNA tests
Treatment of CVD	• Surgery • Diet and lifestyle management • Drug therapies
Treatment of immune disorders	• Surgery • Radiotherapy (for cancers) • Drug and physical therapies • Immunotherapy (desensitization) • Gene and cell therapy (SCID)

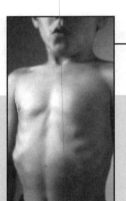

• Asthma
• Allergy
• Atherosclerosis
• Myocardial infarction
• HIV

• ECGs
• Coronary bypass
• Valve replacement
• Asthma treatment
• Organ transplants

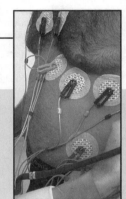

Internal Transport

The Cardiovascular and Lymphatic Systems

The cardiovascular and lymphatic systems are linked in a communication network extending throughout the body.

Exercise and medical therapies can alleviate the symptoms of some age and disease related changes to these systems.

• CVD risk factors
• Cardiac pacemakers
• Stroke
• Cancer
• Rheumatoid arthritis

Effects of exercise on:
• stroke volume
• cardiac output
• pulse rate
• blood pressure
• venous return

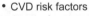

Exercise can delay or reverse some age-related changes to cardiovascular function

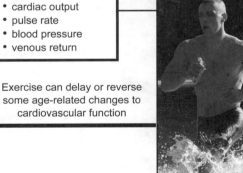

Cpl. Earnest J. Barnes

Effects of aging on heart and immune function	• Higher rates of autoimmune disease • Increased susceptibility to disease • Decline in cardiovascular performance • Higher risk of stroke and CVD • Increased risk of hypertension

Effects of exercise on heart and immune function	• Improved immune function • Improved cardiovascular performance • Increase in cardiac output • Reduced risk of CVD and stroke • Lowered blood pressure

The Effects of Aging

Exercise

The Cardiovascular System

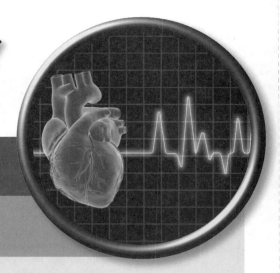

Understanding the structure and function of the cardiovascular system

Structure and organization of the cardiovascular system, blood and blood vessels, heart structure and the cardiac cycle, aerobic exercise and cardiovascular health, heart disease

Learning Objectives

☐ 1. Compile your own glossary from the **KEY WORDS** displayed in **bold type** in the learning objectives below.

The Cardiovascular System *(pages 104, 107-114)*

☐ 2. Describe the primary components of the human **cardiovascular system**, identifying the location of the **heart** and major vessels. Distinguish between the **pulmonary circulation** and the **systemic circulation**.

☐ 3. Explain the relationship between the structure and functional role of **arteries**, **capillaries**, and **veins**. Draw labeled diagrams to illustrate the important features of these comparisons.

☐ 4. Explain the importance of **capillaries**. Draw a diagram to show the relative positions of blood vessels in a capillary network and their relationship to the **lymphatic vessels**. Explain the functional relationship between **blood**, **lymph**, **plasma**, and **tissue fluid**, including reference to how the tissue fluid is formed.

☐ 5. Recognize the transport functions of the lymphatic system, identifying how lymph is returned to the blood circulatory system. Recognize the lymphatic system as a network of vessels that parallels the blood system.

☐ 6. Describe the composition of **blood**, including the functional role of each of the following:

Non-cellular components: **plasma** (water, mineral ions, blood proteins, hormones, nutrients, urea, vitamins).
Cellular components: **erythrocytes**, **leukocytes** (**lymphocytes**, **monocytes**, **granulocytes**), **platelets**.

Heart Structure and Function *(pages 115-120)*

☐ 7. Use a diagram to describe the internal and external gross structure of the **heart**, including **atria**, **ventricles**, **atrioventricular valves**, **semilunar valves**, major vessels (**aorta**, **vena cava**, **pulmonary artery** and **vein**) and coronary circulation. Relate the differences in the thickness of the heart chambers to their functions.

☐ 8. Explain the events of the **cardiac cycle**, relating stages in the cycle (**atrial systole**, **ventricular systole**, and **diastole**) to the maintenance of blood flow through the heart. Analyze data showing pressure and volume changes in the left atrium, left ventricle, and the aorta during the cardiac cycle. Understand the terms **systolic** and **diastolic blood pressure**.

☐ 9. Explain the mechanisms controlling heartbeat, including the role of the **sinoatrial node** (SAN), the **atrioventricular** (AV) **node**, and the conducting fibers in the ventricular walls (**bundle of His**, **Purkinje fibers**).

☐ 10. Describe the extrinsic regulation of **heart rate** through **autonomic nerves**. Identify the role of the **medulla**, **baroreceptors**, and **chemoreceptors** in the response of the heart to changing demands.

Cardiovascular Health *(pages 118-122)*

☐ 11. Explain how **aerobic exercise** affects the cardiovascular system, including reference to **resting pulse**, **heart rate**, **stroke volume**, and **cardiac output**.

☐ 12. Recognize **cardiovascular disease** as a term that includes a range of diseases. Describe **atherosclerosis** and its relationship to **myocardial infarction**. Describe **risk factors** in the development of CVD.

See page 9 for additional details of these texts:

■ Morton, D. and Perry, J.W, 1997. **Photo Atlas for Anatomy and Physiology**, (Brooks/Cole).

■ Rowett, H.G.Q., 1999. **Basic Anatomy & Physiology**, (John Murray), pp. 99-113.

See page 9 for details of publishers of periodicals:

STUDENT'S REFERENCE

■ **Keeping Pace - Cardiac Muscle and Heartbeat** Biol. Sci. Rev., 19(3), Feb. 2007, pp. 21-24. *Cardiac muscle cells can generate electrical activity like nerve impulses, and these impulses produce a smooth contraction of the muscle.*

■ **Blood Pressure** Biol. Sci. Rev., 12(5) May 2000, pp. 9-12. *Blood pressure: its control, measurement, and significance to diagnosis.*

■ **Cunning Plumbing** New Scientist, 6 February 1999, pp. 32-37. *The arteries can actively respond to changes in blood flow, spreading the effects of mechanical stresses to avoid extremes.*

■ **Venous Disease** Biol. Sci. Rev., 19(3), Feb. 2007, pp. 15-17. *Valves in the deep veins of the legs assist venous return and circulation is compromised when they become damaged.*

■ **Mending Broken Hearts** National Geographic, 211(2), Feb. 2007, pp. 40-65. *Assessing susceptibility to heart disease may be the key to treating the disease more effectively.*

TEACHER'S REFERENCE

■ **The Search for Blood Substitutes** Scientific American, Feb. 1998, pp. 60-65. *Replicating the blood's unique properties in a blood substitute.*

■ **Atherosclerosis: The New View** Scientific American, May 2002, pp. 28-37. *An excellent account of the pathological development and rupture of plaques in atherosclerosis.*

■ **Breaking Out of the Box** The Am. Biology Teacher, 63(2), Feb. 2001, pp. 101-115. *Using a web-based activity to investigate the cardiac cycle.*

See pages 10-11 for details of how to access **Bio Links** from our web site: **www.thebiozone.com** From Bio Links, access sites under the topics:

ANATOMY & PHYSIOLOGY > **General sites:**
• Anatomy & Physiology • Human physiology lecture notes • WebAnatomy > **Circulatory System:** • American Heart Association • Heart disease • How the heart works • NOVA online: Cut to the heart • The circulatory system • The heart: A virtual exploration • The matter of the human heart

Presentation MEDIA to support this topic:
Anatomy & Physiology

Arteries

In all vertebrates, including humans, arteries are the blood vessels that carry blood away from the heart to the capillaries within the tissues. The large arteries that leave the heart divide into medium-sized (distributing) arteries. Within the tissues and organs, these distribution arteries branch to form very small vessels called **arterioles**, which deliver blood to capillaries. Arterioles lack the thick layers of arteries and consist only of an endothelial layer wrapped by a few smooth muscle fibers at intervals along their length. Resistance to blood flow is altered by contraction (**vasoconstriction**) or relaxation (**vasodilation**) of the blood vessel walls, especially in the arterioles. Vasoconstriction increases resistance and leads to an increase in blood pressure whereas vasodilation has the opposite effect. This mechanism is important in regulating the blood flow into tissues.

Arteries

Arteries have an elastic, stretchy structure that gives them the ability to withstand the high pressure of blood being pumped from the heart. At the same time, they help to maintain pressure by having some contractile ability themselves (a feature of the central muscle layer). Arteries nearer the heart have more elastic tissue, giving greater resistance to the higher blood pressures of the blood leaving the left ventricle. Arteries further from the heart have more muscle to help them maintain blood pressure. Between heartbeats, the arteries undergo elastic recoil and contract. This tends to smooth out the flow of blood through the vessel.

Arteries comprise three main regions (right):

1. A thin inner layer of epithelial cells called the **endothelium** lines the artery.

2. A central layer (the **tunica media**) of elastic tissue and smooth muscle that can stretch and contract.

3. An outer connective tissue layer (the **tunica externa**) has a lot of elastic tissue.

(a)

(b)

(c)

(d)

Artery Structure

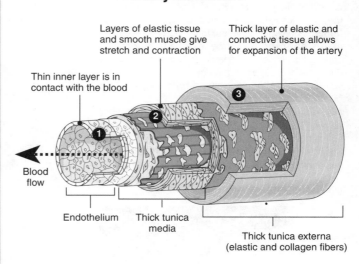

Layers of elastic tissue and smooth muscle give stretch and contraction

Thick layer of elastic and connective tissue allows for expansion of the artery

Thin inner layer is in contact with the blood

Blood flow

Endothelium

Thick tunica media

Thick tunica externa (elastic and collagen fibers)

Cross section through a large artery

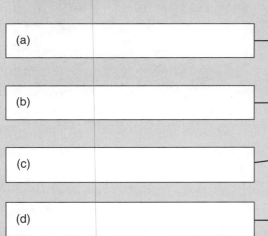

1. Using the diagram to help you, label the photograph of the cross section through an artery (above).

2. (a) Explain why the walls of arteries need to be thick with a lot of elastic tissue: _____

_____

(b) Explain why arterioles lack this elastic tissue layer: _____

_____

3. Explain the purpose of the smooth muscle in the artery walls: _____

_____

4. (a) Describe the effect of vasodilation on the diameter of an arteriole: _____

(b) Describe the effect of vasodilation on blood pressure: _____

Related activities: Heart Function
Web links: Arteries

A 1

Veins

Veins are the blood vessels that return blood to the heart from the tissues. The smallest veins (**venules**) return blood from the capillary beds to the larger veins. Veins and their branches contain about 59% of the blood in the body. The structural differences between veins and arteries are mainly associated with differences in the relative thickness of the vessel layers and the diameter of the lumen. These, in turn, are related to the vessel's functional role.

Veins

When several capillaries unite, they form small veins called **venules**. The venules collect the blood from capillaries and drain it into **veins**. Veins are made up of essentially the same three layers as arteries but they have less elastic and muscle tissue and a larger **lumen**. The venules closest to the capillaries consist of an **endothelium** and a tunica externa of connective tissue. As the venules approach the veins, they also contain the tunica media characteristic of veins (right). Although veins are less elastic than arteries, they can still expand enough to adapt to changes in the pressure and volume of the blood passing through them. Blood flowing in the veins has lost a lot of pressure because it has passed through the narrow capillary vessels. The low pressure in veins means that many veins, especially those in the limbs, need to have valves to prevent backflow of the blood as it returns to the heart.

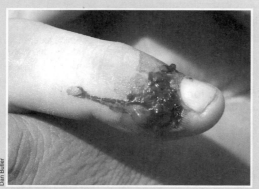

If a vein is cut, as is shown in this severe finger wound, the blood oozes out slowly in an even flow, and usually clots quickly as it leaves. In contrast, arterial blood spurts rapidly and requires pressure to staunch the flow.

Vein Structure

Inner thin layer of simple squamous epithelium lines the vein (**endothelium** or **tunica intima**).

Central thin layer of elastic and muscle tissue (**tunica media**). The smaller venules lack this inner layer.

Thin layer of elastic connective tissue (**tunica externa**)

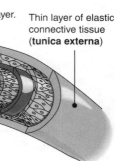

One-way valves are located along the length of veins to prevent the blood from flowing backwards.

Blood flow

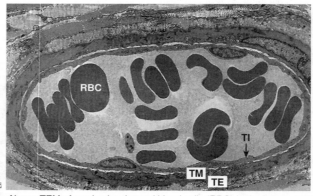

Above: TEM of a vein showing red blood cells (RBC) in the lumen, and the tunica intima (TI), tunica media (TM), and tunica externa (TE).

1. Contrast the structure of veins and arteries for each of the following properties:

   (a) Thickness of muscle and elastic tissue: _____

   _____

   (b) Size of the lumen (inside of the vessel): _____

   _____

2. With respect to their functional roles, give a reason for the difference you have described above: _____

   _____

   _____

3. Explain the role of the valves in assisting the veins to return blood back to the heart: _____

   _____

4. Blood oozes from a venous wound, rather than spurting as it does from an arterial wound. Account for this difference:

   _____

   _____

   _____

Related activities: Arteries, Capillaries and Tissue Fluid
Web links: Veins

Capillaries and Tissue Fluid

Capillaries are very small vessels that connect arterial and venous circulation and allow efficient exchange of nutrients and wastes between the blood and tissues. Capillaries form networks or beds and are abundant where metabolic rates are high. The fluid that leaks out of the capillaries has an essential role in bathing the tissues. The movement of fluid into and out of capillaries depends on the balance between the blood (hydrostatic) pressure (HP) and the solute concentrations at each end of a capillary bed. Not all the fluid is returned to the capillaries and this extra fluid must be returned to the general circulation. This is the role of the **lymphatic system**; a system of vessels that parallels the system of arteries and veins. The lymphatic system also has a role in internal defense, and in transporting the lipids that are absorbed from the digestive tract.

Exchanges in Capillaries

Blood passes from the arterioles into capillaries: small blood vessels with a diameter of only 4-10 μm. (red blood cells only just squeeze through). The only tissue present is an **endothelium** of squamous epithelial cells. Capillaries form networks of vessels that penetrate all parts of the body. They are so numerous that most cells are within 25 μm of a capillary.

Exchange of materials between the body's cells and the blood takes place in capillaries. Blood pressure causes fluid to leak from capillaries through small gaps where the endothelial cells join. This fluid bathes the tissues, supplying nutrients and oxygen, and removing wastes (right).The density of capillaries in a tissue is an indication of that tissue's metabolic activity. For example, cardiac muscle relies heavily on oxidative metabolism. It has a high demand for blood flow and is well supplied with capillaries. Smooth muscle is far less active than cardiac muscle, relies more on anaerobic metabolism, and does not require such an extensive blood supply.

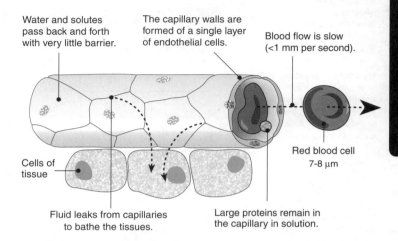

Water and solutes pass back and forth with very little barrier.

The capillary walls are formed of a single layer of endothelial cells.

Blood flow is slow (<1 mm per second).

Red blood cell 7-8 μm

Cells of tissue

Fluid leaks from capillaries to bathe the tissues.

Large proteins remain in the capillary in solution.

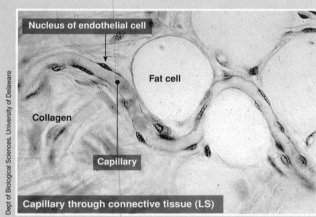

Nucleus of endothelial cell

Fat cell

Collagen

Capillary

Dept of Biological Sciences, University of Delaware

Capillary through connective tissue (LS)

Capillaries are found near almost every cell in the body. In many places, the capillaries form extensive branching networks. In most tissues, blood normally flows through only a small portion of a capillary network when the metabolic demands of the tissue are low. When the tissue becomes active, the entire capillary network fills with blood.

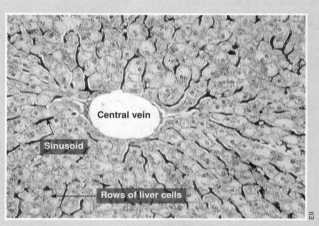

Central vein

Sinusoid

Rows of liver cells

Microscopic blood vessels in some dense organs, such as the liver (above), are called **sinusoids**. They are wider than capillaries and follow a more convoluted path through the tissue. Instead of the usual endothelial lining, they are lined with phagocytic cells. Like capillaries, sinusoids transport blood from arterioles to venules.

1. Describe the structure of a capillary, contrasting it with the structure of a vein and an artery:

_____

_____

_____

_____

2. Sinusoids provide a functional replacement for capillaries in some organs:

   (a) Describe how sinusoids differ structurally from capillaries: _____

   _____

   (b) Describe in what way capillaries and sinusoids are similar: _____

   _____

Related activities: The Lymphatic System, Arteries, Veins
Web links: Microcirculation

RA 2

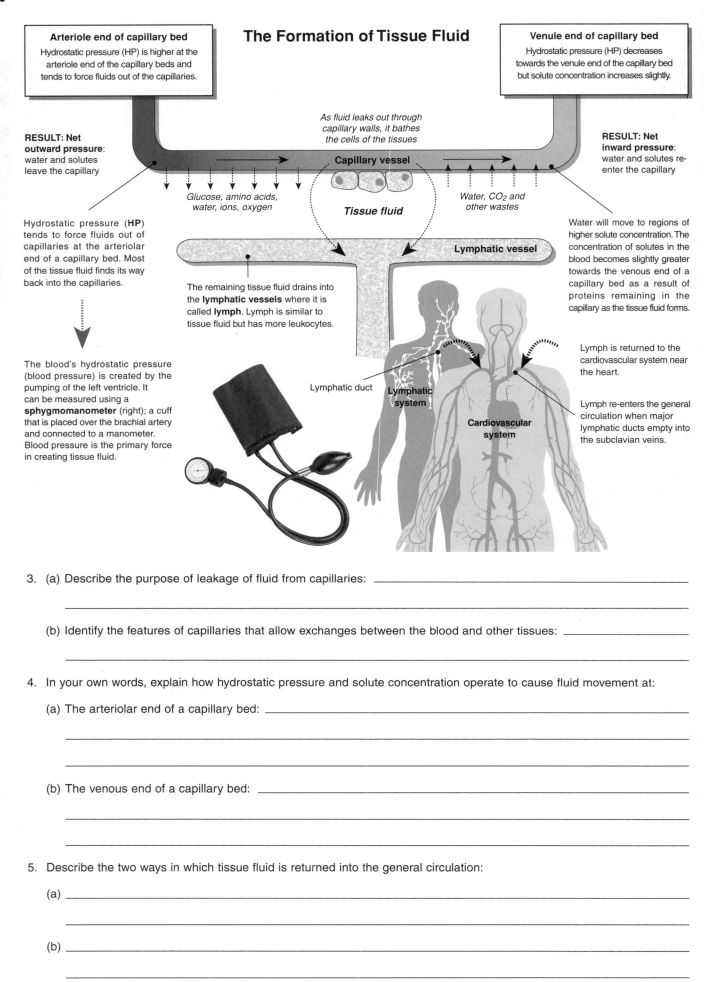

The Formation of Tissue Fluid

Arteriole end of capillary bed
Hydrostatic pressure (HP) is higher at the arteriole end of the capillary beds and tends to force fluids out of the capillaries.

Venule end of capillary bed
Hydrostatic pressure (HP) decreases towards the venule end of the capillary bed but solute concentration increases slightly.

As fluid leaks out through capillary walls, it bathes the cells of the tissues

RESULT: Net outward pressure: water and solutes leave the capillary

Capillary vessel

RESULT: Net inward pressure: water and solutes re-enter the capillary

Glucose, amino acids, water, ions, oxygen

Water, CO_2 and other wastes

Tissue fluid

Hydrostatic pressure (**HP**) tends to force fluids out of capillaries at the arteriolar end of a capillary bed. Most of the tissue fluid finds its way back into the capillaries.

Lymphatic vessel

Water will move to regions of higher solute concentration. The concentration of solutes in the blood becomes slightly greater towards the venous end of a capillary bed as a result of proteins remaining in the capillary as the tissue fluid forms.

The remaining tissue fluid drains into the **lymphatic vessels** where it is called **lymph**. Lymph is similar to tissue fluid but has more leukocytes.

The blood's hydrostatic pressure (blood pressure) is created by the pumping of the left ventricle. It can be measured using a **sphygmomanometer** (right); a cuff that is placed over the brachial artery and connected to a manometer. Blood pressure is the primary force in creating tissue fluid.

Lymphatic duct

Lymphatic system

Cardiovascular system

Lymph is returned to the cardiovascular system near the heart.

Lymph re-enters the general circulation when major lymphatic ducts empty into the subclavian veins.

3. (a) Describe the purpose of leakage of fluid from capillaries: _____

_____

(b) Identify the features of capillaries that allow exchanges between the blood and other tissues: _____

_____

4. In your own words, explain how hydrostatic pressure and solute concentration operate to cause fluid movement at:

(a) The arteriolar end of a capillary bed: _____

_____

_____

(b) The venous end of a capillary bed: _____

_____

_____

5. Describe the two ways in which tissue fluid is returned into the general circulation:

(a) _____

_____

(b) _____

_____

Blood

Blood is a complex liquid tissue comprising cellular components suspended in plasma. It makes up about 8% of body weight. If a blood sample is taken, the cells can be separated from the plasma by centrifugation. The cells (formed elements) settle as a dense red pellet below the transparent, straw-colored **plasma**. Blood performs many functions. It transports nutrients, respiratory gases, hormones, and wastes and has a role in thermoregulation through the distribution of heat. Blood also defends against infection and its ability to clot protects against blood loss. The examination of blood is also useful in diagnosing disease. The cellular components of blood are normally present in particular specified ratios. A change in the morphology, type, or proportion of different blood cells can therefore be used to indicate a specific disorder or infection (see the next page).

Non-Cellular Blood Components

The non-cellular blood components form the plasma. Plasma is a watery matrix of ions and proteins and makes up 50-60% of the total blood volume.

Water
The main constituent of blood and lymph.
Role: Transports dissolved substances. Provides body cells with water. Distributes heat and has a central role in thermoregulation. Regulation of water content helps to regulate blood pressure and volume.

Mineral ions
Sodium, bicarbonate, magnesium, potassium, calcium, chloride.
Role: Osmotic balance, pH buffering, and regulation of membrane permeability. They also have a variety of other functions, e.g. Ca^{2+} is involved in blood clotting.

Plasma proteins
7-9% of the plasma volume.
Serum albumin
Role: Osmotic balance and pH buffering, Ca^{2+} transport.
Fibrinogen and prothrombin
Role: Take part in blood clotting.
Immunoglobulins
Role: Antibodies involved in the immune response.
α-globulins
Role: Bind/transport hormones, lipids, fat soluble vitamins.
β-globulins
Role: Bind/transport iron, cholesterol, fat soluble vitamins.
Enzymes
Role: Take part in and regulate metabolic activities.

Substances transported by non-cellular components
Products of digestion
Examples: sugars, fatty acids, glycerol, and amino acids.
Excretory products
Example: urea
Hormones and vitamins
Examples: insulin, sex hormones, vitamins A and B_{12}.
Importance: These substances occur at varying levels in the blood. They are transported to and from the cells dissolved in the plasma or bound to plasma proteins.

Cellular Blood Components

The cellular components of the blood (also called the formed elements) float in the plasma and make up 40-50% of the total blood volume.

Erythrocytes (red blood cells or RBCs)
5-6 million per mm^3 blood; 38-48% of total blood volume.
Role: RBCs transport oxygen (O_2) and a small amount of carbon dioxide (CO_2). The oxygen is carried bound to hemoglobin (Hb) in the cells. Each Hb molecule can bind four molecules of oxygen.

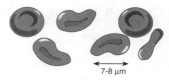

7-8 μm

Platelets
Small, membrane bound cell fragments derived from bone marrow cells; about 1/4 the size of RBCs.
0.25 million per mm^3 blood.
Role: To start the blood clotting process.

2 μm

Leukocytes (white blood cells)
5-10 000 per mm^3 blood
2-3% of total blood volume.
Role: Involved in internal defense. There are several types of white blood cells (see below).

Lymphocytes
T and B cells.
24% of the white cell count.
Role: Antibody production and cell mediated immunity.

Neutrophils
Phagocytes.
70% of the white cell count.
Role: Engulf foreign material.

Eosinophils
Rare leukocytes; normally 1.5% of the white cell count.
Role: Mediate allergic responses such as hayfever and asthma.

Basophils
Rare leukocytes; normally 0.5% of the white cell count.
Role: Produce heparin (an anti-clotting protein), and histamine. Involved in inflammation.

Related activities: Gas Transport in Humans, The Body's Defenses, Thermoregulation, Blood Clotting and Defense

A 2

The Examination of Blood

Different types of microscopy give different information about blood. A SEM (right) shows the detailed external morphology of the blood cells. A fixed smear of a blood sample viewed with a light microscope (far right) can be used to identify the different blood cell types present, and their ratio to each other. Determining the types and proportions of different white blood cells in blood is called a **differential white blood cell count**. Elevated counts of particular cell types indicate allergy or infection.

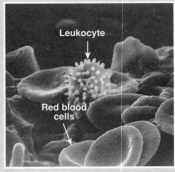

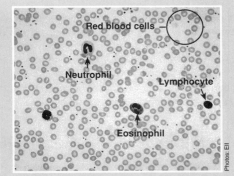

SEM of red blood cells and a leukocytes. **Light microscope** view of a fixed blood smear.

1. For each of the following blood functions, identify the component (or components) of the blood responsible and state how the function is carried out (the mode of action). The first one is done for you:

   (a) **Temperature regulation**. *Blood component:* Water component of the plasma

   Mode of action: Water absorbs heat and dissipates it from sites of production (e.g. organs)

   (b) **Protection against disease**. *Blood component:* _____

   Mode of action: _____

   (c) **Communication between cells, tissues, and organs**. *Blood component:* _____

   Mode of action: _____

   (d) **Oxygen transport**. *Blood component:* _____

   Mode of action: _____

   (e) **CO_2 transport**. *Blood components:* _____

   Mode of action: _____

   (f) **Buffer against pH changes**. *Blood components:* _____

   Mode of action: _____

   (g) **Nutrient supply**. *Blood component:* _____

   Mode of action: _____

   (h) **Tissue repair**. *Blood components:* _____

   Mode of action: _____

   (i) **Transport of hormones, lipids, and fat soluble vitamins**. *Blood component:* _____

   Mode of action: _____

2. Identify a feature that distinguishes red and white blood cells: _____

3. Explain two physiological advantages of red blood cell structure (lacking nucleus and mitochondria):

   (a) _____

   (b) _____

4. Suggest what each of the following results from a differential white blood cell count would suggest:

   (a) Elevated levels of eosinophils (above the normal range): _____

   (b) Elevated levels of neutrophils (above the normal range): _____

   (c) Elevated levels of basophils (above the normal range): _____

   (d) Elevated levels of lymphocytes (above the normal range): _____

Hematopoiesis

Hematopoiesis (also called hemopoiesis) refers to the formation of blood cells. All cellular blood components are derived from **hematopoietic stem cells** (HSCs). In a healthy adult person, approximately 10^{11}-10^{12} new blood cells are produced every day in order to maintain homeostasis of the peripheral circulation. Before birth, hematopoiesis occurs in aggregates of blood cells in the yolk sac, then the liver, and eventually the bone marrow. In normal adults, HSCs reside in the marrow and have the ability to give rise to all of the different mature blood cell types. Like all stem cells, HSCs are able to differentiate into other cell types, and they are self renewing. When they proliferate, at least some

of their daughter cells remain as HSCs, so the pool of stem cells is not depleted. The other daughters of HSCs (myeloid and lymphoid progenitor cells), can each commit to any of the alternative differentiation pathways that lead to the production of one or more types of blood cells, but cannot self renew. Blood cells are divided into lineages. **Erythroid cells** are the oxygen carrying red blood cells. **Lymphoid cells** are the white blood cells of the adaptive immune system. They are derived from common lymphoid progenitors. **Myeloid cells** are derived from common myeloid progenitors, and are involved in many diverse roles within the body's defense system.

Stem Cells and Blood Cell Production

New blood cells are produced in the red bone marrow, which becomes the main site of blood production after birth, taking over from the fetal liver. All types of blood cells develop from a single cell type: called a **multipotent stem cell** or hemocytoblast. These cells are capable of mitosis and of differentiation into 'committed' precursors of each of the main types of blood cell.

Each of the different cell lines is controlled by a specific **growth factor**. When a stem cell divides, one of its daughters remains a stem cell, while the other becomes a precursor cell, either a **lymphoid cell** or **myeloid cell**. These cells continue to mature into the various type of blood cells, developing their specialized features and characteristic roles as they do so.

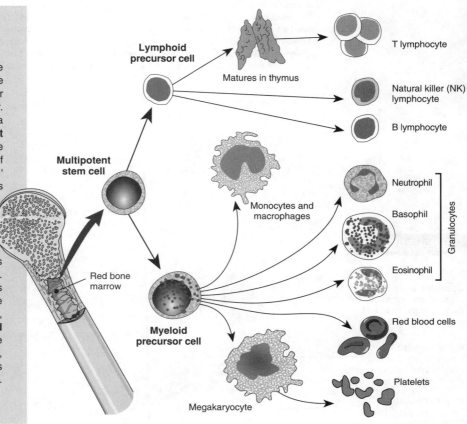

1. Describe the two defining features of stem cells:

   (a) _____

   (b) _____

2. Identify the blood cell types arising from each of the progenitor cell types:

   (a) Myeloid progenitor cells: _____

   _____

   (b) Lymphoid progenitor cells: _____

3. (a) Using an example, explain the purpose of stem cells in an adult: _____

   _____

   (b) Identify where else in the body, apart from the red bone marrow, you might find stem cells: _____

   (c) Explain why blood cells are constantly being produced, when some other cells (e.g. neurons) are not:

   _____

   _____

The Human Transport System

The blood vessels of the circulatory system form a vast network of tubes that carry blood away from the heart, transport it to the tissues of the body, and then return it to the heart. The arteries, arterioles, capillaries, venules, and veins are organized into specific routes to circulate the blood throughout the body. The figure below shows a number of the basic **circulatory routes** through which the blood travels. Humans, like all mammals have a double circulatory system: a **pulmonary circulation**, which carries blood between the heart and lungs, and a **systemic circulation**, which carries blood between the heart and the rest of the body. The systemic circulation has many subdivisions. Two important subdivisions are the coronary (cardiac) circulation, which supplies the heart muscle, and the **hepatic portal circulation**, which runs from the gut to the liver.

Schematic Overview of the Human Circulatory System

Deoxygenated blood (colored gray below) travels to the right side of the heart via the vena cavae. The heart pumps the deoxygenated blood to the lungs where it releases carbon dioxide and receives oxygen. The oxygenated blood (colored white below) travels via the pulmonary vein back to the heart from where it is pumped to all parts of the body. The **venous system** (figure, left) returns blood from the capillaries to the heart. The **arterial system** (figure right) carries blood from the heart to the capillaries. **Portal systems** carry blood between two capillary beds.

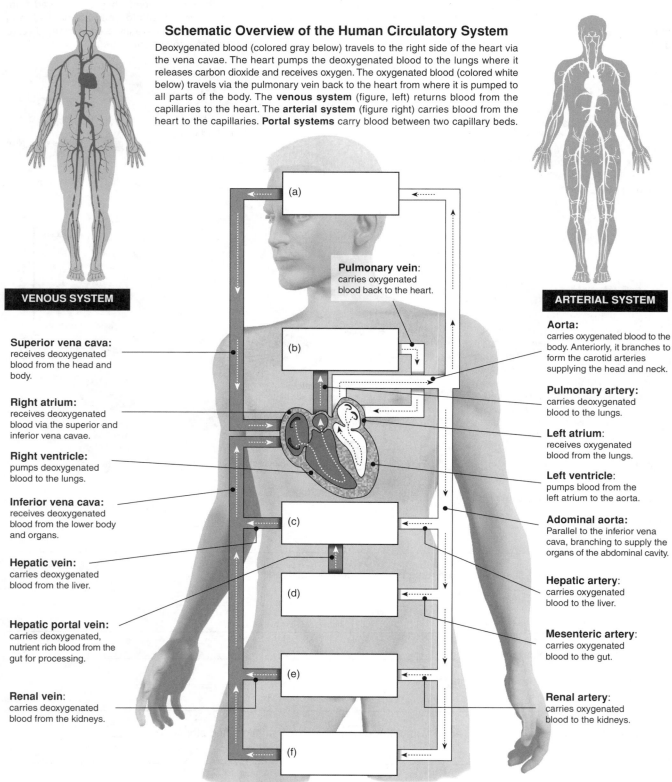

VENOUS SYSTEM

Superior vena cava:
receives deoxygenated blood from the head and body.

Right atrium:
receives deoxygenated blood via the superior and inferior vena cavae.

Right ventricle:
pumps deoxygenated blood to the lungs.

Inferior vena cava:
receives deoxygenated blood from the lower body and organs.

Hepatic vein:
carries deoxygenated blood from the liver.

Hepatic portal vein:
carries deoxygenated, nutrient rich blood from the gut for processing.

Renal vein:
carries deoxygenated blood from the kidneys.

Pulmonary vein:
carries oxygenated blood back to the heart.

ARTERIAL SYSTEM

Aorta:
carries oxygenated blood to the body. Anteriorly, it branches to form the carotid arteries supplying the head and neck.

Pulmonary artery:
carries deoxygenated blood to the lungs.

Left atrium:
receives oxygenated blood from the lungs.

Left ventricle:
pumps blood from the left atrium to the aorta.

Adominal aorta:
Parallel to the inferior vena cava, branching to supply the organs of the abdominal cavity.

Hepatic artery:
carries oxygenated blood to the liver.

Mesenteric artery:
carries oxygenated blood to the gut.

Renal artery:
carries oxygenated blood to the kidneys.

1. Complete the diagram above by labeling the boxes with the organs or structures they represent.

The Human Heart

The heart is at the center of the human cardiovascular system. It is a hollow, muscular organ, weighing on average 342 grams. Each day it beats over 100 000 times to pump 3780 litres of blood through 100 000 kilometers of blood vessels. It comprises a system of four muscular chambers (two **atria** and two **ventricles**) that alternately fill and empty of blood, acting as a double pump.

The left side pumps blood to the body tissues and the right side pumps blood to the lungs. The heart lies between the lungs, to the left of the body's midline, and it is surrounded by a double layered **pericardium** of tough fibrous connective tissue. The pericardium prevents over-distension of the heart and anchors the heart within the **mediastinum**.

Human Heart Structure

(sectioned, anterior view)

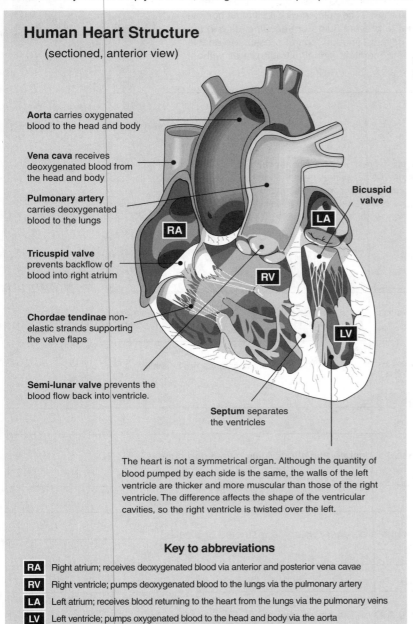

Aorta carries oxygenated blood to the head and body

Vena cava receives deoxygenated blood from the head and body

Pulmonary artery carries deoxygenated blood to the lungs

Tricuspid valve prevents backflow of blood into right atrium

Chordae tendinae non-elastic strands supporting the valve flaps

Semi-lunar valve prevents the blood flow back into ventricle.

Bicuspid valve

Septum separates the ventricles

The heart is not a symmetrical organ. Although the quantity of blood pumped by each side is the same, the walls of the left ventricle are thicker and more muscular than those of the right ventricle. The difference affects the shape of the ventricular cavities, so the right ventricle is twisted over the left.

Key to abbreviations

RA Right atrium; receives deoxygenated blood via anterior and posterior vena cavae

RV Right ventricle; pumps deoxygenated blood to the lungs via the pulmonary artery

LA Left atrium; receives blood returning to the heart from the lungs via the pulmonary veins

LV Left ventricle; pumps oxygenated blood to the head and body via the aorta

Top view of a heart in section, showing valves

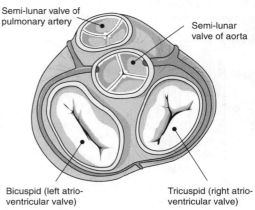

Semi-lunar valve of pulmonary artery

Semi-lunar valve of aorta

Bicuspid (left atrio-ventricular valve)

Tricuspid (right atrio-ventricular valve)

Posterior view of heart

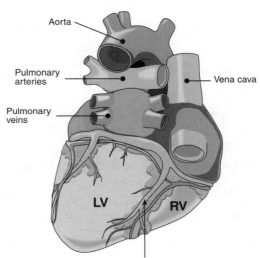

Aorta

Pulmonary arteries

Pulmonary veins

Vena cava

LV

RV

Coronary arteries: The high oxygen demands of the heart muscle are met by a dense capillary network. Coronary arteries arise from the aorta and spread over the surface of the heart supplying the cardiac muscle with oxygenated blood. Deoxygenated blood is collected by cardiac veins and returned to the right atrium via a large coronary sinus.

1. In the schematic diagram of the heart, below, label the four chambers and the main vessels entering and leaving them. The arrows indicate the direction of blood flow. Use large coloured circles to mark the position of each of the four valves.

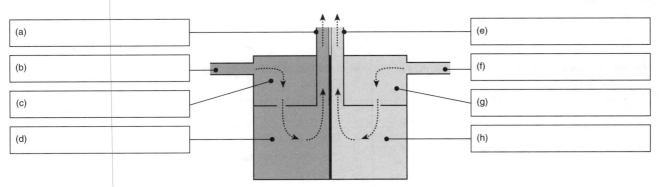

(a)
(b)
(c)
(d)
(e)
(f)
(g)
(h)

Related activities: Arteries, Veins, Capillaries and Tissue Fluid
Web links: Anatomy of the Heart, How the Heart Works

RA 2

Pressure Changes and the Asymmetry of the Heart

aorta, 100 mg Hg

The heart is not a symmetrical organ. The left ventricle and its associated arteries are thicker and more muscular than the corresponding structures on the right side. This asymmetry is related to the necessary pressure differences between the pulmonary (lung) and systemic (body) circulations (not to the distance over which the blood is pumped per se). The graph below shows changes in blood pressure in each of the major blood vessel types in the systemic and pulmonary circuits (the horizontal distance not to scale). The pulmonary circuit must operate at a much lower pressure than the systemic circuit to prevent fluid from accumulating in the alveoli of the lungs. The left side of the heart must develop enough "spare" pressure to enable increased blood flow to the muscles of the body and maintain kidney filtration rates without decreasing the blood supply to the brain.

Blood pressure during contraction (systole)

The greatest fall in pressure occurs when the blood moves into the capillaries, even though the distance through the capillaries represents only a tiny proportion of the total distance traveled.

Blood pressure during relaxation (diastole)

Pressure (mm Hg)

120
100
80
60
40
20
0

radial artery, 98 mg Hg

arterial end of capillary, 30 mg Hg

aorta   arteries   **A**   capillaries   **B**   veins   vena cava   pulmonary arteries   **C**   **D**   venules   pulmonary veins

Systemic circulation
horizontal distance not to scale

Pulmonary circulation
horizontal distance not to scale

2. Explain the purpose of the valves in the heart: _____

_____

3. The heart is full of blood. Suggest two reasons why, despite this, it needs its own blood supply:

(a) _____

(b) _____

4. Predict the effect on the heart if blood flow through a coronary artery is restricted or blocked: _____

_____

_____

5. Identify the vessels corresponding to the letters **A-D** on the graph above:

A: _____   B: _____   C: _____   D: _____

6. (a) Explain why the pulmonary circuit must operate at a lower pressure than the systemic system: _____

_____

_____

(b) Relate this to differences in the thickness of the wall of the left and right ventricles of the heart: _____

_____

_____

7. Explain what you are recording when you take a pulse: _____

_____

Control of Heart Activity

When removed from the body, cardiac muscle continues to beat. This indicates that the origin of the heartbeat is **myogenic**; the contractions arise as an intrinsic property of the cardiac muscle itself. The heartbeat is regulated by a special conduction system consisting of the pacemaker (**sinoatrial node**) and specialized conduction fibers called **Purkinje fibers**. The pacemaker sets a basic rhythm for the heart, but this rate is influenced by the cardiovascular control center in the medulla in response to sensory information from pressure receptors in the walls of the heart and blood vessels, and by higher brain functions. Changing the rate and force of heart contraction is the main mechanism for controlling cardiac output in order to meet changing demands.

The Cardiovascular System

Generation of the Heartbeat

The basic rhythmic heartbeat is **myogenic**. The nodal cells (SAN and atrioventricular node) spontaneously generate rhythmic action potentials without neural stimulation. The normal resting rate of self-excitation of the SAN is about 50 beats per minute.

The amount of blood ejected from the left ventricle per minute is called the **cardiac output**. It is determined by the **stroke volume** (the volume of blood ejected with each contraction) and the **heart rate** (number of heart beats per minute).

Cardiac output

= **stroke volume** × **heart rate**

Cardiac muscle responds to stretching by contracting more strongly. The greater the blood volume entering the ventricle, the greater the force of contraction. This relationship is known as **Starling's Law.**

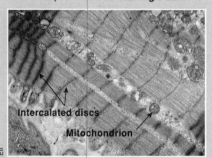

Intercalated discs

Mitochondrion

TEM of cardiac muscle showing branched fibers (muscle cells). Each fiber has one or two nuclei and many large mitochondria. **Intercalated discs** are specialized regions between neighboring cells that support synchronized contraction of the muscle. They contain **gap junctions**, specialized electrical synapses that allow very rapid spread of nerve impulses through the heart muscle.

Sinoatrial node (SAN) is also called the **pacemaker**. It is a mass of specialized muscle cells near the opening of the superior vena cava. The pacemaker initiates the cardiac cycle, spontaneously generating action potentials that cause the atria to contract. The SAN sets the basic pace of the heart rate, although this rate is influenced by hormones and impulses from the autonomic nervous system.

Atrioventricular node (AVN) at the base of the atrium briefly delays the impulse to allow time for the atrial contraction to finish before the ventricles contract.

Bundle of His (atrioventricular bundle) containing Purkinje tissue. A tract of conducting fibers that distribute the action potentials over the ventricles causing ventricular contraction.

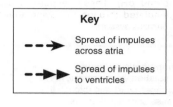

Key

- - - → Spread of impulses across atria

- - -▶▶ Spread of impulses to ventricles

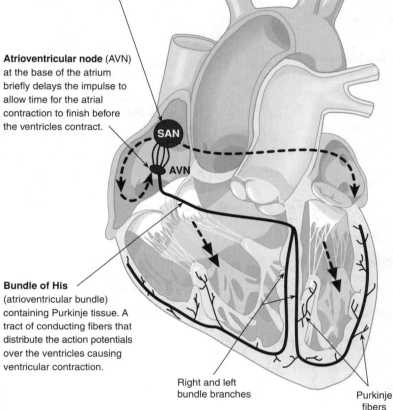

SAN

AVN

Right and left bundle branches

Purkinje fibers

1. Identify the role of each of the following in heart activity:

   (a) The sinoatrial node: _____

   (b) The atrioventricular node: _____

   (c) The bundle of His: _____

2. Explain the significance of the delay in impulse conduction at the AVN: _____

   _____

3. (a) Calculate the **cardiac output** when stroke volume is 70 cm³ and the heart rate is 70 beats per minute:

   _____

   (b) Trained endurance athletes have a very high cardiac output. Suggest how this is achieved: _____

   _____

Related activities: Review of the Human Heart, Exercise and Blood Flow

RDA 2

Autonomic Nervous System
Control of Heartbeat

Cardiovascular control center in the medulla of the brain comprises the accelerator center (speeds up heart rate and force of contraction) and the inhibitory center (decreases heart rate and force of contraction).

The accelerator center also responds directly to epinephrine in the blood and to changes in blood composition (low blood pH). These responses are mediated through the sympathetic nervous system.

Cerebral hemispheres may send impulses (e.g. in sexual arousal).

Hypothalamus may send impulses (e.g. anger or alarm).

The carotid reflex: Pressure receptors in the carotid sinus detect stretch caused by increased arterial flow (blood flow leaving the heart). They send impulses to the inhibitory center to mediate decrease in heart rate via the vagus nerve (parasympathetic stimulation).

Sympathetic nervous stimulation via the cardiac nerve increases heart rate through the release of norepinephrine.

Parasympathetic nervous stimulation via the vagus nerve decreases heart rate through the release of acetylcholine.

The aortic reflex: Pressure receptors in the aorta detect stretch caused by increased arterial flow. They send impulses to the inhibitory center to mediate decrease in heart rate via the vagus nerve.

The Bainbridge reflex: Pressure receptors in the vena cava and atrium respond to stretch caused by increased venous return by sending impulses to the accelerator center, mediating an increase in heart rate.

●······▶ Parasympathetic motor nerve (vagus)

●– –▶ Sympathetic motor nerve (cardiac nerve)

●———▶ Sensory nerve

4. (a) With respect to the heart beat, explain what is meant by **myogenic**: _____

_____

(b) Describe the evidence for the myogenic nature of the heart beat: _____

_____

5. During heavy exercise, heart rate increases. Describe the mechanisms that are involved in bringing about this increase:

_____

_____

6. (a) Identify a stimulus for a decrease in heart rate: _____

(b) Explain how this change in heart rate is brought about: _____

_____

7. Identify two pressure receptors involved in control of heart rate and state what they respond to:

(a) _____

(b) _____

8. Guarana is a chemical found in many energy drinks. A group of students designed an experiment to test whether guarana stimulates a cardiovascular response. The test subjects had their pulses recorded before and after drinking an energy drink containing a known amount of guarana.

(a) Suggest two reasons why the test subjects may respond in different ways: _____

_____

(b) Describe a suitable control for this experiment: _____

_____

The Cardiac Cycle

The heart pumps with alternate contractions (**systole**) and relaxations (**diastole**). The **cardiac cycle** refers to the sequence of events of a heartbeat and involves three main stages: atrial systole, ventricular systole, and complete cardiac diastole. Pressure changes within the heart's chambers generated by the cycle of contraction and relaxation are responsible for blood movement and cause the heart valves to open and close, preventing the backflow of blood. The noise of the blood when the valves open and close

produces the heartbeat sound (**lubb-dupp**). The heart beat occurs in response to electrical impulses, which can be recorded as a trace, called an **electrocardiogram** or **ECG**. The ECG pattern is the result of the different impulses produced at each phase of the cardiac cycle, and each part is identified with a letter code. An ECG provides a useful method of monitoring changes in heart rate and activity and detection of heart disorders. The electrical trace is accompanied by volume and pressure changes (below).

The Cardiac Cycle

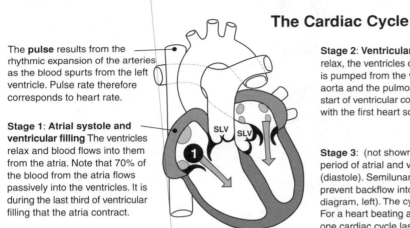

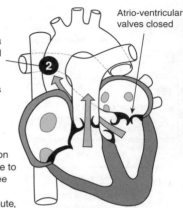

Atrio-ventricular valves closed

The **pulse** results from the rhythmic expansion of the arteries as the blood spurts from the left ventricle. Pulse rate therefore corresponds to heart rate.

Stage 1: **Atrial systole and ventricular filling** The ventricles relax and blood flows into them from the atria. Note that 70% of the blood from the atria flows passively into the ventricles. It is during the last third of ventricular filling that the atria contract.

Stage 2: **Ventricular systole** The atria relax, the ventricles contract, and blood is pumped from the ventricles into the aorta and the pulmonary artery. The start of ventricular contraction coincides with the first heart sound.

Stage 3: (not shown) There is a short period of atrial and ventricular relaxation (diastole). Semilunar valves (**SLV**) close to prevent backflow into the ventricles (see diagram, left). The cycle begins again. For a heart beating at 75 beats per minute, one cardiac cycle lasts about 0.8 seconds.

Heart during ventricular filling

Heart during ventricular contraction

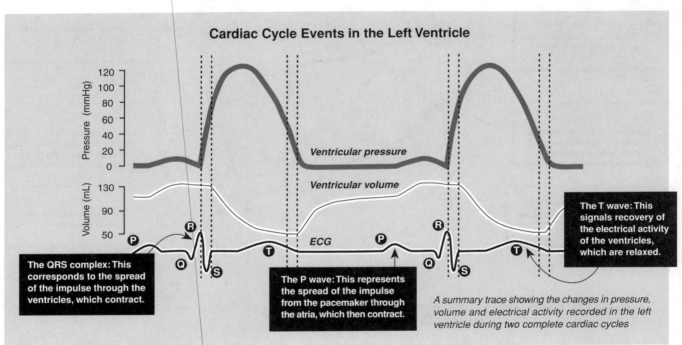

Cardiac Cycle Events in the Left Ventricle

Ventricular pressure

Ventricular volume

ECG

The QRS complex: This corresponds to the spread of the impulse through the ventricles, which contract.

The P wave: This represents the spread of the impulse from the pacemaker through the atria, which then contract.

The T wave: This signals recovery of the electrical activity of the ventricles, which are relaxed.

A summary trace showing the changes in pressure, volume and electrical activity recorded in the left ventricle during two complete cardiac cycles

1. Identify each of the following phases of an ECG by its international code:

   (a) Excitation of the ventricles and ventricular systole: _____

   (b) Electrical recovery of the ventricles and ventricular diastole: _____

   (c) Excitation of the atria and atrial systole: _____

2. Suggest the physiological reason for the period of electrical recovery experienced each cycle (the T wave):

   _____

3. Using the letters indicated, mark the points on the trace above corresponding to each of the following:

   (a) E: Ejection of blood from the ventricle

   (b) AVC: Closing of the atrioventricular valve

   (c) FV: Filling of the ventricle

   (d) AVO: Opening of the atrioventricular valve

Related activities: The Human Heart
Web links: Electrocardiogram, Cardiac Cycle Animation

RA 2

Review of the Human Heart

A circulatory system is required to transport materials because diffusion is too inefficient and slow to supply all the cells of the body adequately. The circulatory system in humans transports nutrients, respiratory gases, wastes, and hormones, aids in regulating body temperature and maintaining fluid balance, and has a role in internal defence. The circulatory system comprises a network of vessels, a circulatory fluid (blood), and a heart. This activity summarizes key features of the structure and function of the human heart. The necessary information can be found in earlier activities in this topic.

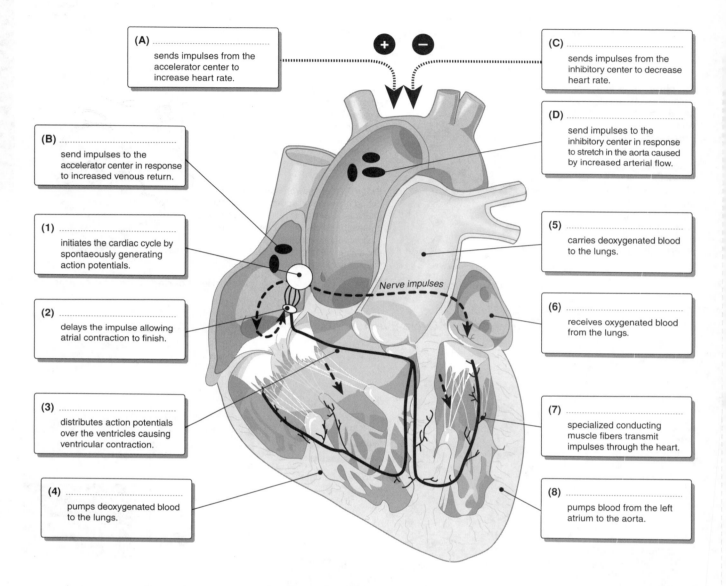

(A) .... sends impulses from the accelerator center to increase heart rate.

(C) .... sends impulses from the inhibitory center to decrease heart rate.

(B) .... send impulses to the accelerator center in response to increased venous return.

(D) .... send impulses to the inhibitory center in response to stretch in the aorta caused by increased arterial flow.

(1) .... initiates the cardiac cycle by spontaneously generating action potentials.

(5) .... carries deoxygenated blood to the lungs.

(2) .... delays the impulse allowing atrial contraction to finish.

(6) .... receives oxygenated blood from the lungs.

(3) .... distributes action potentials over the ventricles causing ventricular contraction.

(7) .... specialized conducting muscle fibers transmit impulses through the heart.

(4) .... pumps deoxygenated blood to the lungs.

(8) .... pumps blood from the left atrium to the aorta.

Nerve impulses

1. On the diagram above, label the identified components of heart structure and intrinsic control (**1-8**), and the components involved in extrinsic control of heart rate (**A-D**).

2. An **ECG** is the result of different impulses produced at each phase of the **cardiac cycle** (the sequence of events in a heartbeat). For each electrical event indicated in the ECG below, describe the corresponding event in the cardiac cycle:

A ---
The spread of the impulse from the pacemaker (sinoatrial node) through the atria.

B ---
The spread of the impulse through the ventricles.

C ---
Recovery of the electrical activity of the ventricles.

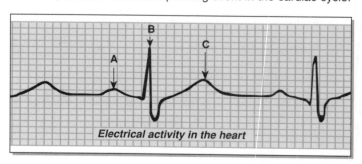

Electrical activity in the heart

3. Describe one treatment that may be indicated when heart rhythm is erratic or too slow: _____
_____

Related activities: The Human Heart, Control of Heart Activity, The Cardiac Cycle

Cardiovascular Disease

Cardiovascular disease (CVD) is a term describing all diseases involving the heart and blood vessels. It includes coronary heart disease (CHD), atherosclerosis, hypertension (high blood pressure), peripheral vascular disease, stroke, and congenital heart disorders. CVD is responsible for 20% of all deaths worldwide and is the principal cause of deaths in developed countries. In westernized countries, deaths due to CVD have been declining since the 1970s due to better prevention and treatment. Despite this, CVD is still one of the leading causes of mortality,. The continued prevalence of CVD is of considerable public health concern, particularly as many of the **risk factors** involved, such as cigarette smoking, obesity, and high blood cholesterol, are controllable. Uncontrollable risk factors include advancing age, gender, and heredity.

Cardiovascular Diseases

Atherosclerosis: Atherosclerosis is a disease of the arteries caused by **atheroma** (fatty deposits) on the inner walls of the arteries. An atheroma is made up of cells (mostly macrophages) or cell debris, with associated fatty acids, cholesterol, calcium, and varying amounts of fibrous connective tissue. The lining of the arteries degenerates due to the accumulation of fat and plaques. Atheroma weakens the arterial walls and eventually restricts blood flow through the arteries, increasing the risk of **aneurysm** and **thrombosis** (blood clot formation). Complications arising as a result of atherosclerosis include **infarction**, stroke, and gangrene.

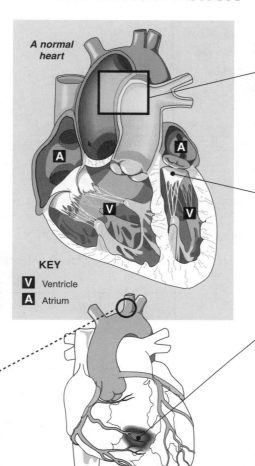

A normal heart

KEY

V	Ventricle
A	Atrium

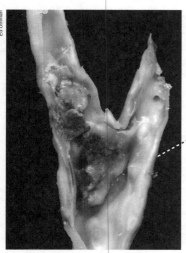

Atherosclerotic plaque in the carotid artery. Plaque material can detach from the artery wall and enters the circulation, increasing the risk of thrombosis.

Restricted supply of blood to heart muscle resulting in myocardial infarction

Aortic aneurysm: Ballooning of the wall of the aorta. Atheroma increases the risk of aneurysm in arteries because it weakens the artery wall. Aneurysms usually result from generalized heart disease and high blood pressure.

Valve defects: Unusual heart sounds (murmurs) can result when a valve (often the mitral valve) does not close properly, allowing blood to bubble back into the atria. Valve defects may be congenital (present at birth) but they can also occur as a result of rheumatic fever.

Myocardial infarction (*heart attack*): Occurs when an area of the heart is deprived of blood supply resulting in tissue damage or death. It is the major cause of death in developed countries. Symptoms of infarction include a sudden onset of chest pain, breathlessness, nausea, and cold clammy skin. Damage to the heart may be so severe that it leads to heart failure and even death (myocardial infarction is fatal within 20 days in 40 to 50% of all cases).

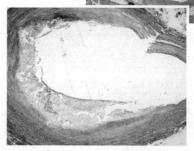

Normal unobstructed coronary artery above, and a coronary artery (left) with moderately severe atheroma. Note the formation of the plaque on the inside surface of the artery. Plaques obstruct blood flow through the artery.

Cholesterol and Risk of CVD

Cholesterol is a sterol lipid found in all animal tissues as part of cellular membranes. It is transported within complex spherical particles called **lipoproteins**. One form of cholesterol-transporting molecule is called **high density lipoprotein** or **HDL**. HDL helps remove cholesterol from the bloodstream by transporting it to the liver. Another form of lipoprotein, called **LDL** (**low density lipoprotein**) deposits cholesterol onto the walls of blood vessels to form **plaques**.

Abnormally high concentrations of LDL and lower concentrations of functional HDL are strongly associated with CVD because these promote development of atheroma in arteries. This disease process leads to a myocardial infarction. It is the **LDL:HDL ratio**, rather than total cholesterol *per se*, that best indicates the risk of cardiovascular disease, and the risk profile is different for men and women (tables right). The LDL:HDL ratio is mostly genetically determined but can be changed by body build, diet, and exercise regime.

	Ratio of LDL to HDL	
Risk	**Men**	**Women**
Very low (half average)	1.0	1.5
Average risk	3.6	3.2
Moderate risk (2X average risk)	6.3	5.0
High (3X average risk)	8.0	6.1

Related activities: Correcting Heart Problems, Organ Transplants, Smoking and the Lungs **Web links**: LifeBeat Online

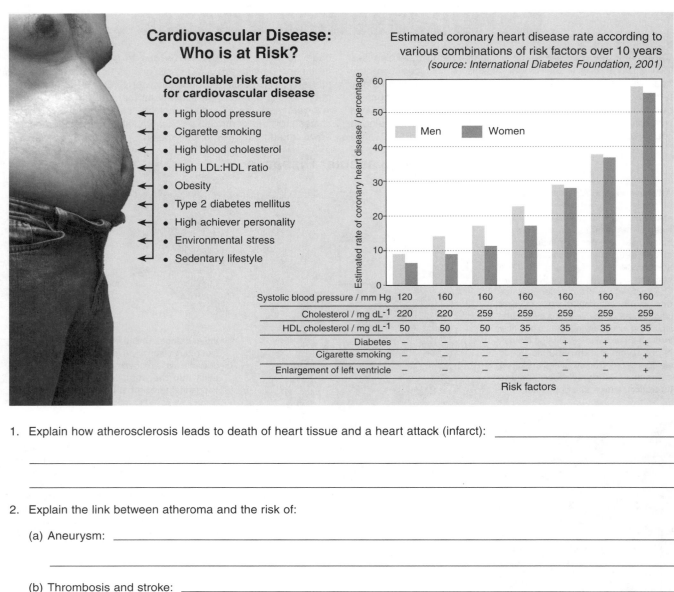

Cardiovascular Disease: Who is at Risk?

Controllable risk factors for cardiovascular disease

- High blood pressure
- Cigarette smoking
- High blood cholesterol
- High LDL:HDL ratio
- Obesity
- Type 2 diabetes mellitus
- High achiever personality
- Environmental stress
- Sedentary lifestyle

Estimated coronary heart disease rate according to various combinations of risk factors over 10 years
(source: International Diabetes Foundation, 2001)

Men   Women

Systolic blood pressure / mm Hg	120	160	160	160	160	160	160
Cholesterol / mg dL^{-1}	220	220	259	259	259	259	259
HDL cholesterol / mg dL^{-1}	50	50	50	35	35	35	35
Diabetes	–	–	–	–	+	+	+
Cigarette smoking	–	–	–	–	–	+	+
Enlargement of left ventricle	–	–	–	–	–	–	+

Risk factors

1. Explain how atherosclerosis leads to death of heart tissue and a heart attack (infarct): _____

_____

_____

2. Explain the link between atheroma and the risk of:

   (a) Aneurysm: _____

   _____

   (b) Thrombosis and stroke: _____

   _____

3. (a) Distinguish between controllable and uncontrollable risk factors in the development of CVD: _____

   _____

   (b) Suggest why some of the controllable risk factors often occur together: _____

   _____

   _____

   (c) Evaluate the evidence supporting the observation that patients with several risk factors are at higher risk of CVD:

   _____

   _____

4. (a) Explain the link between high LDL:HDL ratio and the risk of cardiovascular disease: _____

   _____

   _____

   (b) Explain why this ratio is more important to medical practitioners than total blood cholesterol *per se*:

   _____

   _____

Correcting Heart Problems

Over the last few decades the death rates from CVD have slowly declined, despite an increase in its prevalence. This reduction in mortality has been achieved partly through better management and treatment of the disease. Medical technology now provides the means to correct many heart problems, even if only temporarily. Some symptoms of CVD, arising as a result of blockages to the coronary arteries, are now commonly treated using techniques such as coronary bypass surgery and angioplasty. Other cardiac disorders, such as disorders of heartbeat, are frequently treated using cardiac pacemakers. Valve defects, which are often congenital, can be successfully corrected with surgical valve replacement. The latest technology, still in its trial phase, involves non-surgical replacement of aortic valves. The procedure, known as percutaneous (through the skin) heart valve replacement, will greatly reduce the trauma associated with correcting these particular heart disorders.

Coronary Bypass Surgery

This is a now commonly used surgery to bypass blocked coronary arteries with blood vessels from elsewhere in the body (e.g. leg vein or mammary artery). Sometimes, double or triple bypasses are performed.

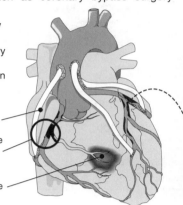

Bypass vessel

Blocked coronary artery; the tissue it normally supplies with blood dies

Dying heart tissue

Angioplasty

Angioplasty (right) is an alternative procedure used for some patients with coronary artery disease. A balloon tipped catheter is placed via the aorta into the coronary artery. The balloon is inflated to reduce the blockage of the artery and later removed. Heparin (an anticlotting agent) is given to prevent the formation of blood clots. The death rate from complications is about 1%.

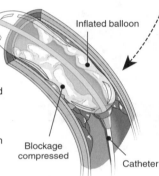

Inflated balloon

Blockage compressed

Catheter

Heart Valve Replacement

Heart valves can be replaced with either **biological** (tissue) **valves** or **synthetic valves**. Tissue valves are sourced from animal (e.g. pig) or human donors. They last only 7-10 years, but there are relatively few blood clotting and tissue rejection problems associated with them. For these reasons, they are often used in older patients. Synthetic ball or disc valves are constructed from non-biological materials. They last a long time but tend to create blood clots (raising the risk of stroke). They are used on younger patients, who must take long-term anti-clotting drugs.

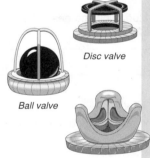

Disc valve

Ball valve

Tissue valve

Cardiac Pacemakers

A cardiac pacemaker is sometimes required to maintain an effective heart rate in cases where the heart beats irregularly or too slowly. Pacemakers provide regular electrical stimulation of the heart muscle so that it contracts and relaxes with a normal rhythm. They stand by until the heart rate falls below a pre-set rate. **Temporary pacemakers** are often used after cardiac surgery or heart attacks, while **permanent pacemakers** are required for patients with ongoing problems. Pacemakers allow a normal (even strenuous) lifestyle.

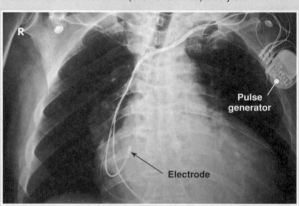

R

Pulse generator

Electrode

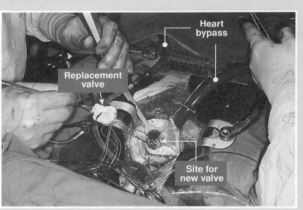

Heart bypass

Replacement valve

Site for new valve

Above: Valve replacement operation in progress. The valve can be seen threaded up and ready for placement. Two large tubes bypass the heart so that circulation to the lungs and rest of the body is maintained.

1. Describe the problems associated with the use of each of the following types of replacement heart valve:

    (a) Tissue valves: _____

    (b) Synthetic valves: _____

2. Suggest why tissue valves are usually a preferred option for use in elderly patients: _____

    _____

3. Explain why patients who have undergone coronary bypass surgery or angioplasty require careful supervision of their diet and lifestyle following the operation, even though their problem has been alleviated:

    _____

    _____

    _____

Related activities: Cardiovascular Disease

A 2

Exercise and Blood Flow

Exercise promotes health by improving the rate of blood flow back to the heart (venous return). This is achieved by strengthening all types of muscle and by increasing the efficiency of the heart.

During exercise blood flow to different parts of the body changes in order to cope with the extra demands of the muscles, the heart and the lungs.

1. The following table gives data for the **rate** of blood flow to various parts of the body at rest and during strenuous exercise. **Calculate** the **percentage** of the total blood flow that each organ or tissue receives under each regime of activity.

Organ or tissue	At rest		Strenuous exercise	
	cm^3 min^{-1}	% of total	cm^3 min^{-1}	% of total
Brain	700	14	750	4.2
Heart	200		750	
Lung tissue	100		200	
Kidneys	1100		600	
Liver	1350		600	
Skeletal muscles	750		12 500	
Bone	250		250	
Skin	300		1900	
Thyroid gland	50		50	
Adrenal glands	25		25	
Other tissue	175		175	
TOTAL	5000	100	17 800	100

2. Explain how the body increases the rate of blood flow during exercise: _____

3. (a) State approximately how many times the total rate of blood flow increases between rest and exercise: _____

   (b) Explain why the increase is necessary: _____

4. (a) Identify which organs or tissues show no change in the rate of blood flow with exercise: _____

   (b) Explain why this is the case: _____

5. (a) Identify the organs or tissues that show the most change in the rate of blood flow with exercise: _____

   (b) Explain why this is the case: _____

The Effects of Aerobic Training

The body has an immediate response to an exercise bout but also, over time, responds to the stress of repeated exercise (or **training**) by adapting and improving both its capacity for exercise and the efficiency with which it performs. Regular, intense exercise causes predictable physiological changes in muscular, cardiovascular, and respiratory performance. The heart adapts so that it can pump more blood per stroke. This increase in **cardiac output** is brought about not only by an increase in heart rate, but also by a training-induced increase in the stroke volume. The circulatory system also adapts to repeated exercise, with changes in blood flow facilitating an increased flow of blood to the muscles and skin and an increased rate of gas exchange. The pulmonary system adjusts accordingly, with greater efficiencies in ventilation rate and breathing rhythm.

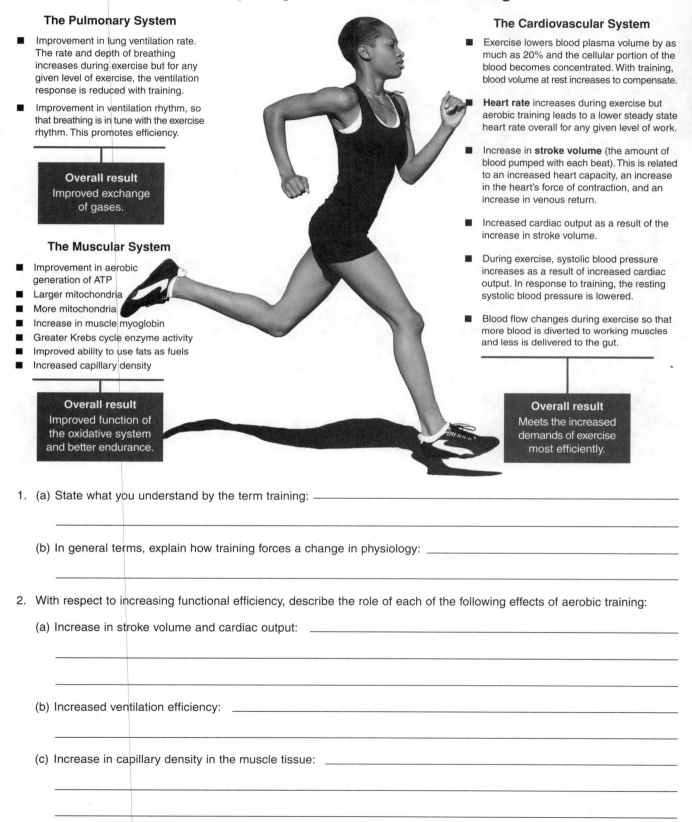

The Physiological Effects of Aerobic Training

The Pulmonary System

- Improvement in lung ventilation rate. The rate and depth of breathing increases during exercise but for any given level of exercise, the ventilation response is reduced with training.
- Improvement in ventilation rhythm, so that breathing is in tune with the exercise rhythm. This promotes efficiency.

Overall result
Improved exchange of gases.

The Muscular System

- Improvement in aerobic generation of ATP
- Larger mitochondria
- More mitochondria
- Increase in muscle myoglobin
- Greater Krebs cycle enzyme activity
- Improved ability to use fats as fuels
- Increased capillary density

Overall result
Improved function of the oxidative system and better endurance.

The Cardiovascular System

- Exercise lowers blood plasma volume by as much as 20% and the cellular portion of the blood becomes concentrated. With training, blood volume at rest increases to compensate.
- **Heart rate** increases during exercise but aerobic training leads to a lower steady state heart rate overall for any given level of work.
- Increase in **stroke volume** (the amount of blood pumped with each beat). This is related to an increased heart capacity, an increase in the heart's force of contraction, and an increase in venous return.
- Increased cardiac output as a result of the increase in stroke volume.
- During exercise, systolic blood pressure increases as a result of increased cardiac output. In response to training, the resting systolic blood pressure is lowered.
- Blood flow changes during exercise so that more blood is diverted to working muscles and less is delivered to the gut.

Overall result
Meets the increased demands of exercise most efficiently.

The Cardiovascular System

1. (a) State what you understand by the term training: _____

_____

(b) In general terms, explain how training forces a change in physiology: _____

_____

2. With respect to increasing functional efficiency, describe the role of each of the following effects of aerobic training:

(a) Increase in stroke volume and cardiac output: _____

_____

_____

(b) Increased ventilation efficiency: _____

_____

(c) Increase in capillary density in the muscle tissue: _____

_____

_____

Endurance refers to the ability to carry out sustained activity. Muscular strength and short term muscular endurance allows sprinters to run fast for a short time or body builders and weight lifters to lift an immense weight and hold it. Cardiovascular and respiratory endurance refer to the body as a whole: the ability to endure a high level of activity over a prolonged period. This type of endurance is seen in marathon runners, and long distance swimmers and cyclists.

Sprint-focused sports demand quite different training to that required for endurance sports, and the physiologies and builds of the athletes are quite different. A body builder ready for a competition would be ill-equipped to complete a 90 km cycle race!

2008's Mr Olympia winner, Dexter Jackson: high muscular development and endurance.

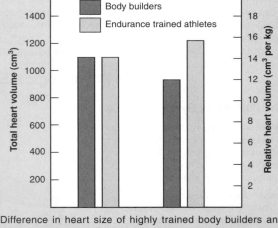

Difference in heart size of highly trained body builders and endurance athletes. Total heart volume is compared to heart volume as related to body weight. Average weights as follows: Body builders = 90.1 kg. Endurance athletes = 68.7 kg.

Weightlifters have high muscular strength and short term muscular endurance; they can lift extremely heavy weights and hold them for a short time. Typical sports requiring these attributes are sprinting, weight lifting, body building, boxing, and wrestling.

Distance runners have very good cardiovascular, respiratory, and muscular endurance; they sustain high intensity exercise for a long time. Typical sports needing overall endurance are distance running, cycling, and swimming (triathletes combine all three).

3. Explain why heart size increases with endurance activity: _____

_____

_____

4. In the graph above right, explain why the relative heart volume of endurance athletes is greater than that of body builders, even though their total heart volumes are the same:

_____

_____

5. Heart stroke volume increases with endurance training. Explain how this increases the efficiency of the heart as a pump:

_____

6. The **resting pulse** is much lower in trained athletes compared with non-active people. Explain the health benefits of a lower resting pulse:

_____

_____

The Lymphatic System & Immunity

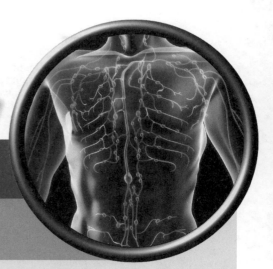

Understanding the role of the lymphatic system in transport and internal defense

Structure and organization of the lymphatic system, self recognition, non-specific defense, immune system, disease and the immune system, stem cell technology, monoclonal antibodies, gene therapy

Learning Objectives

☐ 1. Compile your own glossary from the **KEY WORDS** displayed in **bold type** in the learning objectives below.

Recognizing Self and Non-self *(pages 129-130)*

☐ 2. Explain how a body is able to distinguish between self and non-self and comment on the importance of this. Appreciate the nature of **major histocompatibility complex (MHC)** and its role in self-recognition and in determining tissue compatibility in transplant recipients.

☐ 3. Explain the basis of the **Rh** and **ABO blood group systems** in humans. Explain the consequences of blood type incompatibility in **blood transfusions**.

☐ 4. Discuss the physiological basis of **transplant rejection** and suggest how it may be avoided. Explain why it is so difficult to find compatible tissue and organ donors and suggest how this problem may be solved in the future.

Defense Mechanisms *(pages 104-105)*

Blood clotting *(page 133)*

☐ 5. Describe the process of **blood clotting**, including the role of **clotting factors**, **thrombin**, and **fibrin**. Appreciate the role of blood clotting in the resistance of the body to infection by sealing off damage and restricting invasion of the tissues by microorganisms.

Non-specific defenses *(pages 131-132, 134-136)*

☐ 6. Explain what is meant by a **non-specific defense mechanism**. Distinguish between first and second lines of defense. Describe the nature and role of each of the following in protecting against pathogens:
(a) Skin (including sweat and sebum production).
(b) Mucus-secreting and ciliated membranes.
(c) Body secretions (tears, urine, saliva, gastric juice).
(d) Natural anti-bacterial and anti-viral proteins such as **interferon** and **complement.**
(e) The **inflammatory response**, **fever**, and cell death.
(f) **Phagocytosis** by phagocytes. Recognize the term phagocyte as referring to any of a number of phagocytic leukocytes (e.g. macrophages).

Specific defenses *(pages 131, 137, 139-144)*

☐ 7. Identify the role of **specific resistance** in the body's resistance to infection. Explain how the **immune response** involves recognition and response to foreign material. Explain the significance of the immune system having both **specificity** and **memory**.

☐ 8. Describing examples, distinguish between **naturally acquired** and **artificially acquired immunity**, and between **active** and **passive immunity**. Compare the duration of the immunity gained by active and passive means and explain the difference.

☐ 9. Describe the basic structure of the **lymphatic system**, identifying lymphoid tissues and organs. Describe the role of the lymphatic system in producing and transporting leukocytes.

The Immune System *(pages 139-144)*

☐ 10. Distinguish between **cell-mediated immunity** and **humoral (antibody-mediated) immunity**, identifying the specific white blood cells involved in each case.

☐ 11. Recall that other types of white blood cells are involved in non-specific defense mechanisms.

☐ 12. Explain the role of the **thymus** in the immune response. Describe the nature, origin, and role of **macrophages** (a type of phagocyte). Appreciate the role of macrophages in processing and presenting foreign antigens and in stimulating lymphocyte activity.

☐ 13. Explain the origin and maturation of **B lymphocytes** (cells) and **T lymphocytes** (cells). Describe and distinguish between the activities of the B and T lymphocytes in the immune response.

☐ 14. With reference to immune system function, outline the principle of challenge and response. Outline **clonal selection** and the basis of **immunological memory**. Explain how the immune system is able to respond to the large and unpredictable range of potential antigens.

☐ 15. Appreciate that **self-tolerance** occurs during development as a result of the selective destruction of the B cells that react to self-antigens.

Cell-mediated immunity

☐ 16. T cells are responsible for **cell-mediated immunity**. Describe how T cells recognize **specific** foreign antigens. Describe the roles of named T cells, including **cytotoxic** (killer) **T cells** (T_C) and **helper T cells** (T_H). Identify the organisms against which these T cells act.

☐ 17. Appreciate the role of T lymphocytes in the rejection of transplanted tissues and organs.

Humoral immunity

☐ 18. Define the terms: **antibody** (immunoglobulin), and **antigen**. Name some common antigens and explain their role in provoking a specific immune response.

☐ 19. Describe the structure of an antibody identifying the constant and variable regions, and the antigen binding site. Relate the structure of antibodies to their function.

☐ 20. Describe the activation and differentiation of B-cells. Explain antibody production, including how B cells bring about **humoral** (antibody-mediated) **immunity** to specific antigens. Describe how antibodies inactivate antigens and facilitate their destruction.

☐ 21. Contrast the roles of **plasma cells** and **memory cells**. Discuss the role of **immunological memory** in long term immunity.

☐ 22. Outline the principle of **vaccination**. Explain what is meant by a **primary** and a **secondary response** to infection and identify the role of these, and the immune system memory, in the success of vaccines.

☐ 23. Appreciate that **immunization** involves the production of immunity by artificial means and that **vaccination** usually refers to immunization by inoculation. Know that these terms are frequently used synonymously.

Immune Dysfunction *(pages 138, 145-146, 153)*

☐ 24. Explain what is meant by an **autoimmune disease** and describe some common examples. Identify common triggers for autoimmune diseases and suggest how they can be managed.

☐ 25. With reference to **asthma** or **hayfever**, outline the role of the immune system in allergic reactions, including the role of **histamine** in these allergies. Identify common triggers for allergies in susceptible people.

☐ 26. Provide examples of hypersensitivity reactions that differ in their speed of onset (rapid vs delayed).

☐ 27. With reference to HIV replication and T cell population, discuss the effects of **HIV** (*Human Immunodeficiency Virus*) on immune system function. Include reference to both the long and short term effects of HIV infection.

Medical Applications *(pages 147-152, 154-155)*

Monoclonal antibodies

☐ 28. Explain how **monoclonal antibodies** are produced the principles by which they work. Describe the role of monoclonal antibodies in the diagnosis and treatment of disease, e.g. Herceptin for the treatment of cancer.

Stem cell technology and transplants

☐ 29. Discuss the use of **stem cell therapy** and **transplant technology** (tissue and organ transplants) in the treatment of disease. Describe problems of **immunocompatibility** between tissue and organ donors and explain how this can be overcome. Explain how stem cell therapy has the potential to provide fully immune-compatible tissues for transplant recipients.

Gene therapy

☐ 30. Describe how **gene therapy** has been used to restore immune system function in patients with the heritable immunodeficiency disease, SCID (*Severe Combine Immune Deficiency*). Explain the basis of **gene therapy**, identifying the techniques involved. Explain why SCID was a good candidate for gene therapy trials and identify advantages of the therapy (relative to alternatives) and any problems.

See page 9 for additional details of these texts:

■ Morton, D. and Perry, J.W, 1997. **Photo Atlas for Anatomy and Physiology**, (Brooks/Cole).

■ Rowett, H.G.Q., 1999. **Basic Anatomy & Physiology**, (John Murray), pp. 114-116.

See page 9 for details of publishers of periodicals:

STUDENT'S REFERENCE

■ **Looking Out for Danger: How White Blood Cells Protect Us** Biol. Sci. Rev., 19 (4) April 2007, pp. 34-37. *Types of lymphocytes and they work together to protect against infection.*

■ **Red Blood Cells** Biol. Sci. Rev., 11(2) Nov. 1998, pp. 2-4. *The function of red blood cells, including their role in antigenic recognition.*

■ **Beware! Allergens** New Scientist, 22 Jan. 2000 (Inside Science). *The allergic response: sensitization and the role of the immune system.*

■ **Anaphylactic Shock** Biol. Sci. Rev., 19(2) Nov. 2006, pp. 11-13. *Describes anaphylactic shock, a severe allergic reaction caused by a massive overreaction of the body's immune system.*

■ **Skin, Scabs and Scars** Biol. Sci. Rev., 17(3) Feb. 2005, pp. 2-6. *The many roles of skin, including its importance in wound healing and the processes involved in its repair when damaged.*

■ **Antibodies** Biol. Sci. Rev., 11(3) Jan. 1999, pp. 34-35. *The operation of the immune system and the production of antibodies (including procedures for producing monoclonal antibodies).*

■ **Misery for all Seasons** National Geographic, 209(5) May 2006, pp. 116-135. *The causes, effects, and prevention of common allergies.*

■ **Lymphocytes - The Heart of the Immune System** Biol. Sci. Rev., 12 (1) September 1999 pp. 32-35. *An excellent account of the role of lymphocytes in the immune response (includes the types and actions of different lymphocytes).*

■ **HIV Focus** New Scientist, 8 Feb. 2003, pp. 33-44. *A special issue covering HIV research: why the immune system responds in different ways in different individuals, new hope in vaccine development, and use of protective microbiocides.*

■ **AIDS** Biol. Sci. Rev., 20(1), Sept. 2007, pp. 10-12. *The rapid evolution of HIV explains how it can continue to evade the immune system and become resistant to new drug therapies.*

■ **Immunotherapy** Biol. Sci. Rev., 15(1), Sept. 2002, pp. 39-41. *Medical research is uncovering ways in which our immune system can be used in developing vaccines for cancer.*

TEACHER'S REFERENCE

■ **Training Killers: Dendritic Cells** Biol. Sci. Rev., 19(4), April 2007, pp. 14-17. *Dendritic cells initiate the adaptive immune response, the second line of defense, by seeking out pathogens and training other cells to destroy them. T cells and B cells both play a part in the response of the adaptive immune system so that both intracellular and extracellular pathogens can be targeted. It is hoped that dendritic cells can be used to target cancer and prevent autoimmune diseases.*

■ **Monoclonals as Medicines** Biol. Sci. Rev., 18(4) April 2006, pp. 38-40. *The use of monoclonal antibodies in therapeutic and diagnostic medicine.*

■ **Genetic Vaccines** Scientific American, July 1999, pp. 34-41. *This excellent article includes a description of how the vaccines work and a table of specific diseases treatable by this method.*

■ **Filthy Healthy** New Scientist, 12 Jan 2008, pp. 34-37. *Exposure to dirt and infections may aid our long term immunity, and help fight cancer.*

■ **Grown to Order** New Scientist, 3 May 2008, pp. 40-43. *Stem cells offer the potential for immune compatible regeneration of organs and limbs.*

■ **Super Immune** New Scientist, 24 Feb. 2007, pp. 42-45. *Immunotherapy - genetic engineering of the immune system - could give us the power to fight off most devastating diseases. Recent trials have successfully targeted melanoma.*

■ **New Predictors of Disease** Scientific American, March 2007, pp. 54-61. *In autoimmune diseases, the immune system mistakenly manufactures antibodies that target the body's tissues. Screening for these could help predict disease progression in those at risk.*

■ **Peacekeepers of the Immune System** Scientific American, Oct. 2006, pp. 34-41. *Regulatory T cells suppress immune activity and combat autoimmunity.*

■ **The Long Arm of the Immune System** Sci. American, Nov. 2002, pp. 34-41. *The role of dendritic cells, a class of leukocyte with a role in activating the immune system (good extension).*

■ **Cell Defenses and the Sunshine Vitamin** Scientific American, Nov. 2007, pp. 36-44. *Vitamin D is active throughout the body, influencing immune system responses and cell defenses. A lack of vitamin D is associated with autoimmune disorders, cancers, and lack of disease resistance.*

■ **Preparing for Battle** Scientific American, Feb. 2001, pp. 68-69. *Preparation and mode of action of the influenza vaccine.*

■ **Exploring the Innate Immune System** The Am. Biology Teacher, Feb. 2008, pp. 103-108. *Using complement-mediated cell lysis to demonstrate the operation of the of the adaptive immune response.*

■ **ABO-Rh Blood-Typing Model: a Problem Solving Activity** The Am. Biology Teacher, March 2005, pp. 158-162. *Using an easy to make blood-typing model kit to practise problem solving skills.*

See pages 10-11 for details of how to access **Bio Links** from our web site: **www.thebiozone.com** From Bio Links, access sites under the topics:

ANATOMY & PHYSIOLOGY > **General sites**: • Anatomy & Physiology • Human physiology lecture notes • WebAnatomy > **Defense and the Immune System**: • Blood group antigens • Immune defense against microbial pathogens • Autoimmune disease • Monoclonal antibodies • Inducible defenses against pathogens • Microbiology and immunology • Primary immunodeficiency diseases • The immune system: An overview • Understanding the immune system • Constitutive defenses against pathogens

Presentation MEDIA to support this topic: **Anatomy & Physiology**

Anatomy & Physiology

Targets for Defense

In order for the body to present an effective defense against pathogens, it must first be able to recognize its own tissues (self) and ignore the body's normal microflora (e.g. the bacteria of the skin and gastrointestinal tract). In addition, the body needs to be able to deal with abnormal cells which, if not eliminated, may become cancerous. Failure of self/non-self recognition can lead to autoimmune disorders, in which the immune system mistakenly attacks its own tissues. The body's ability to recognize its own molecules has implications for procedures such as tissue grafts, organ transplants, and blood transfusions. Incompatible tissues (identified as foreign) are attacked by the body's immune system (**rejected**). Even a healthy pregnancy involves suppression of specific features of the self recognition system, allowing the mother to tolerate a nine month gestation with the fetus.

The Body's Natural Microbiota

After birth, normal and characteristic microbial populations begin to establish themselves on and in the body. A typical human body contains 1×10^{13} body cells, yet harbors 1×10^{14} bacterial cells. These microorganisms establish more or less permanent residence but, under normal conditions, do not cause disease. In fact, this normal microflora can benefit the host by preventing the overgrowth of harmful pathogens. They are not found throughout the entire body, but are located in certain regions.

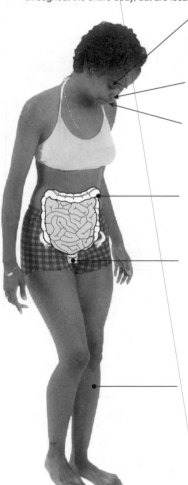

Eyes: The conjuctiva, a continuation of the skin or mucous membrane, contains a similar microbiota to the skin.

Nose and throat: Harbors a variety of microorganisms, e.g. *Staphylococcus spp.*

Mouth: Supports a large and diverse microbiota. It is an ideal microbial environment; high in moisture, warmth, and nutrient availability.

Large intestine: Contains the body's largest resident population of microbes because of its available moisture and nutrients.

Urinary and genital systems: The lower urethra in both sexes has a resident population; the vagina has a particular acid-tolerant population of microbes because of the low pH nature of its secretions.

Skin: Skin secretions prevent most of the microbes on the skin from becoming residents.

Distinguishing Self from Non-Self

The human immune system achieves self-recognition through the **major histocompatibility complex** (MHC). This is a cluster of tightly linked genes on chromosome 6 in humans. These genes code for protein molecules (MHC antigens) that are attached to the surface of body cells. They are used by the immune system to recognize its own or foreign material. **Class I MHC** antigens are located on the surface of virtually all human cells, but **Class II MHC** antigens are restricted to macrophages and the antibody-producing B-lymphocytes.

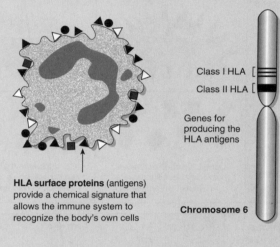

HLA surface proteins (antigens) provide a chemical signature that allows the immune system to recognize the body's own cells

Tissue Transplants

The MHC is responsible for the rejection of tissue grafts and organ transplants. Foreign MHC molecules are antigenic, causing the immune system to respond in the following way:

- T cells directly lyse the foreign cells
- Macrophages are activated by T cells and engulf foreign cells
- Antibodies are released that attack the foreign cell
- The complement system injures blood vessels supplying the graft or transplanted organ

To minimize this rejection, attempts are made to match the MHC of the organ donor to that of the recipient as closely as possible.

The Lymphatic System & Immunity

1. Explain why it is healthy to have a natural population of microbes on and inside the body: _____

2. (a) Explain the nature and purpose of the **major histocompatibility complex** (MHC): _____

(b) Explain the importance of such a self-recognition system: _____

3. Identify two situations when the body's recognition of 'self' is undesirable: _____

Blood Group Antigens

Blood groups classify blood according to the different marker proteins on the surface of red blood cells (RBCs). These marker proteins act as **antigens** and affect the ability of RBCs to provoke an immune response. The **ABO blood group** is the most important blood typing system in medical practice, because of the presence of anti-A and anti-B antibodies in nearly all people who lack the corresponding red cell antigens (these antibodies are carried in the plasma and are present at birth). If a patient is to receive blood from a blood donor, that blood must be compatible otherwise the red blood cells of the donated blood will clump together (agglutinate), break apart, and block capillaries. There is a small margin of safety in certain blood group combinations, because the volume of donated blood is usually relatively small and the donor's antibodies are quickly diluted in the plasma. In practice, blood is carefully matched, not only for ABO types, but for other types as well. Although human RBCs have more than 500 known antigens, fewer than 30 (in 9 blood groups) are regularly tested for when blood is donated for transfusion. The blood groups involved are: *ABO, Rh, MNS, P, Lewis, Lutheran, Kell, Duffy,* and *Kidd.* The ABO and rhesus (Rh) are the best known. Although blood typing has important applications in medicine, it can also be used to rule out individuals in cases of crime (or paternity) and establish a list of potential suspects (or fathers).

	Blood Type A	Blood Type B	Blood Type AB	Blood Type O
Antigens present on the **red blood cells**	antigen *A*	antigen *B*	antigens *A* and *B*	Neither antigen *A* nor *B*
Anti-bodies present in the **plasma**	Contains **anti-B** antibodies; but no antibodies that would attack its own antigen *A*	Contains **anti-A** antibodies; but no antibodies that would attack its own antigen *B*	Contains neither **anti-A** nor **anti-B** antibodies	Contains both **anti-A** and **anti-B** antibodies

Blood type	Frequency in US Rh+	Rh−	Antigen	Antibody	Can donate blood to:	Can receive blood from:
A	34%	6%	A	anti-B	A, AB	A, O
B	9%	2%				
AB	3%	1%				
O	38%	7%				

1. Complete the table above to show the antibodies and antigens in each blood group, and donor/recipient blood types:

2. In a hypothetical murder case, blood from both the victim and the murderer was left at the scene. There were five suspects under investigation:

    (a) Describe what blood typing could establish about the guilt or innocence of the suspects: _____

    _____

    (b) Identify what a blood typing could not establish: _____

    _____

    (c) Suggest how the murderer's identity could be firmly established (assuming that s/he was one of the five suspects): _____

    _____

    (d) Explain why blood typing is not used forensically to any great extent: _____

    _____

3. Explain why the discovery of the ABO system was such a significant medical breakthrough: _____

    _____

Related activities: Antibodies, Blood
Web links: Blood Typing Game

The Body's Defenses

If microorganisms never encountered resistance from our body defenses, we would be constantly ill and would eventually die of various diseases. Fortunately, in most cases our defenses prevent this from happening. Some of these defenses are designed to keep microorganisms from entering the body. Other defenses remove the microorganisms if they manage to get inside. Further defenses attack the microorganisms if they remain inside the body. The ability to ward off disease through the various defense mechanisms is called **resistance**. The lack of resistance, or vulnerability to disease, is known as **susceptibility**. One form of

defense is referred to as **non-specific resistance**, and includes defenses that protect us from any pathogen. This includes a first line of defense such as the physical barriers to infection (skin and mucous membranes) and a second line of defense (phagocytes, inflammation, fever, and antimicrobial substances). **Specific resistance** is a third line of defense that forms the **immune response** and targets specific pathogens. Specialized cells of the immune system, called lymphocytes, produce specific proteins called antibodies which are produced against specific antigens.

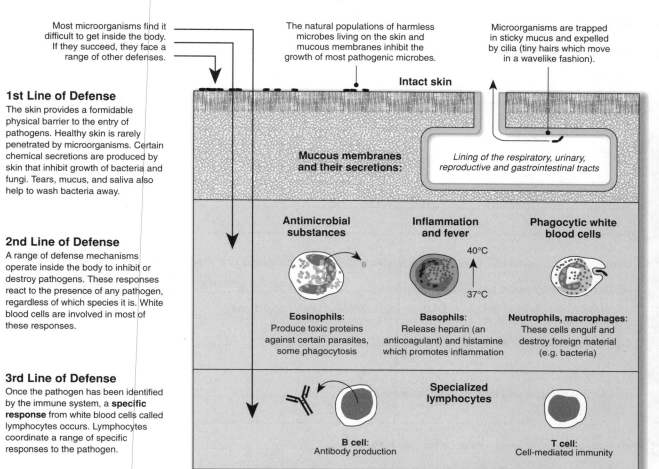

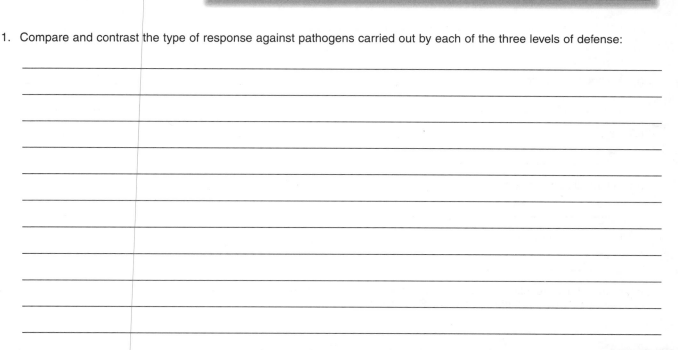

1st Line of Defense

The skin provides a formidable physical barrier to the entry of pathogens. Healthy skin is rarely penetrated by microorganisms. Certain chemical secretions are produced by skin that inhibit growth of bacteria and fungi. Tears, mucus, and saliva also help to wash bacteria away.

2nd Line of Defense

A range of defense mechanisms operate inside the body to inhibit or destroy pathogens. These responses react to the presence of any pathogen, regardless of which species it is. White blood cells are involved in most of these responses.

3rd Line of Defense

Once the pathogen has been identified by the immune system, a **specific response** from white blood cells called lymphocytes occurs. Lymphocytes coordinate a range of specific responses to the pathogen.

Most microorganisms find it difficult to get inside the body. If they succeed, they face a range of other defenses.

The natural populations of harmless microbes living on the skin and mucous membranes inhibit the growth of most pathogenic microbes.

Microorganisms are trapped in sticky mucus and expelled by cilia (tiny hairs which move in a wavelike fashion).

Intact skin

Mucous membranes and their secretions:

Lining of the respiratory, urinary, reproductive and gastrointestinal tracts

Antimicrobial substances

Eosinophils:
Produce toxic proteins against certain parasites, some phagocytosis

Inflammation and fever
40°C
37°C

Basophils:
Release heparin (an anticoagulant) and histamine which promotes inflammation

Phagocytic white blood cells

Neutrophils, macrophages:
These cells engulf and destroy foreign material (e.g. bacteria)

Specialized lymphocytes

B cell:
Antibody production

T cell:
Cell-mediated immunity

The Lymphatic System & Immunity

1. Compare and contrast the type of response against pathogens carried out by each of the three levels of defense:

    _____

    _____

    _____

    _____

    _____

    _____

    _____

    _____

    _____

    _____

    _____

Related activities: Targets for Defense, The Action of Phagocytes, Fever, Inflammation, The Immune System   **Web links**: Immunoanimations

RA 2

2. Distinguish between specific and non-specific resistance: _____

_____

_____

_____

_____

3. Describe features of the different types of white blood cells and explain how these relate to their role in the second line of defense:

_____

_____

_____

_____

_____

_____

_____

_____

_____

_____

_____

4. Describe the functional role of each of the following defense mechanisms (the first one has been completed for you):

   (a) Skin (including sweat and sebum production): _Skin helps to prevent direct entry of pathogens_

   into the body. Sebum slows growth of bacteria and fungi.

   (b) Phagocytosis by white blood cells: _____

   _____

   (c) Mucus-secreting and ciliated membranes: _____

   _____

   (d) Body secretions: tears, urine, saliva, gastric juice: _____

   _____

   (e) Natural antimicrobial proteins (e.g. interferon): _____

   _____

   (f) Antibody production: _____

   _____

   (g) Fever: _____

   _____

   (h) Cell-mediated immunity: _____

   _____

   (i) The inflammatory response: _____

   _____

5. Infection with HIV results in the progressive destruction of T lymphocytes. Suggest why this leads to an increasing number of opportunistic infections in AIDS sufferers:

_____

Blood Clotting and Defense

Apart from its transport role, **blood** has a role in the body's defense against infection and **hemostasis** (the prevention of bleeding and maintenance of blood volume). The tearing or puncturing of a blood vessel initiates **clotting**. Clotting is normally a rapid process that seals off the tear, preventing blood loss and the invasion of bacteria into the site. Clot formation is triggered by the release of clotting factors from the damaged cells at the site of the tear or puncture. A hardened clot forms a scab, which acts to prevent further blood loss and acts as a mechanical barrier to the entry of pathogens.

Blood Clotting

When tissue is wounded, the blood quickly coagulates to prevent further blood loss and maintain the integrity of the circulatory system. For external wounds, clotting also prevents the entry of pathogens. Blood clotting involves a cascade of reactions involving at least twelve clotting factors in the blood. The end result is the formation of an insoluble network of fibers, which traps red blood cells and seals the wound.

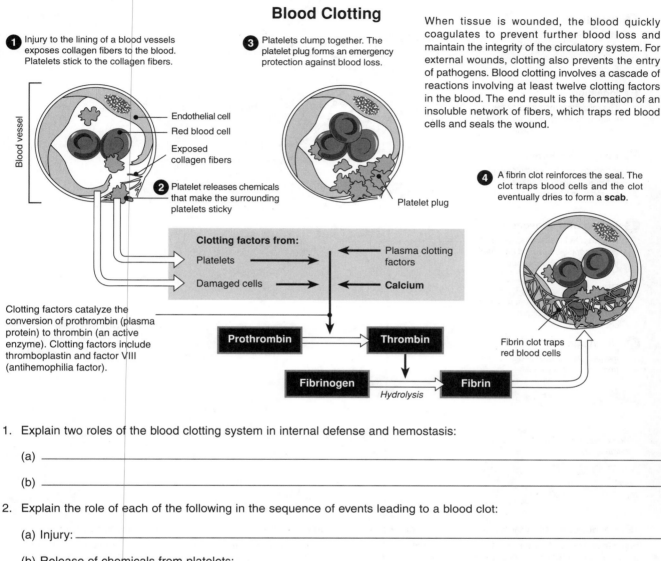

1 Injury to the lining of a blood vessels exposes collagen fibers to the blood. Platelets stick to the collagen fibers.

3 Platelets clump together. The platelet plug forms an emergency protection against blood loss.

Endothelial cell
Red blood cell
Exposed collagen fibers

2 Platelet releases chemicals that make the surrounding platelets sticky

Platelet plug

4 A fibrin clot reinforces the seal. The clot traps blood cells and the clot eventually dries to form a **scab**.

Blood vessel

Clotting factors from:
Platelets ⟶ ⟵ Plasma clotting factors
Damaged cells ⟶ ⟵ **Calcium**

Clotting factors catalyze the conversion of prothrombin (plasma protein) to thrombin (an active enzyme). Clotting factors include thromboplastin and factor VIII (antihemophilia factor).

Prothrombin ⟹ **Thrombin**

Fibrinogen ⟹ **Fibrin**
Hydrolysis

Fibrin clot traps red blood cells

1. Explain two roles of the blood clotting system in internal defense and hemostasis:

    (a) _____

    (b) _____

2. Explain the role of each of the following in the sequence of events leading to a blood clot:

    (a) Injury: _____

    (b) Release of chemicals from platelets: _____

    _____

    (c) Clumping of platelets at the wound site: _____

    _____

    (d) Formation of a fibrin clot: _____

    _____

3. (a) Explain the role of clotting factors in the blood in formation of the clot: _____

    _____

    (b) Explain why these clotting factors are not normally present in the plasma: _____

    _____

4. (a) Name one inherited disease caused by the absence of a clotting factor: _____

    (b) Name the clotting factor involved: _____

The Action of Phagocytes

Human cells that ingest microbes and digest them by the process of **phagocytosis** are called **phagocytes**. All are types of white blood cells. During many kinds of infections, especially bacterial infections, the total number of white blood cells increases by two to four times the normal number. The ratio of various white blood cell types changes during the course of an infection.

How a Phagocyte Destroys Microbes

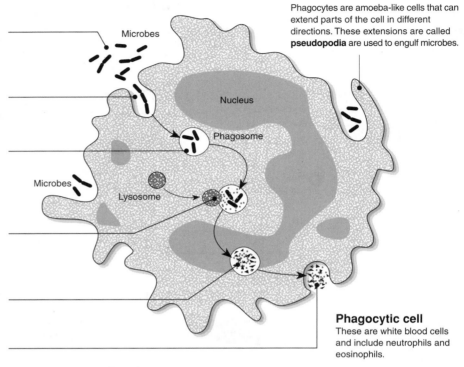

1 Detection
Phagocyte detects microbes by the chemicals they give off (chemotaxis) and sticks the microbes to its surface.

2 Ingestion
The microbe is engulfed by the phagocyte wrapping pseudopodia around it to form a vesicle.

3 Phagosome forms
A phagosome (phagocytic vesicle) is formed, which encloses the microbes in a membrane.

4 Fusion with lysosome
Phagosome fuses with a lysosome (which contains powerful enzymes that can digest the microbe).

5 Digestion
The microbes are broken down by enzymes into their chemical constituents.

6 Discharge
Indigestible material is discharged from the phagocyte cell.

Phagocytes are amoeba-like cells that can extend parts of the cell in different directions. These extensions are called **pseudopodia** are used to engulf microbes.

Labels: Microbes, Nucleus, Phagosome, Microbes, Lysosome

Phagocytic cell
These are white blood cells and include neutrophils and eosinophils.

The Interaction of Microbes and Phagocytes

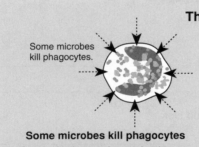

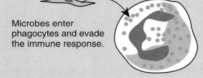

Some microbes kill phagocytes.

Microbes enter phagocytes and evade the immune response.

Dormant microbes may hide inside phagocytes.

Some microbes kill phagocytes
Some microbes produce toxins that can actually kill phagocytes, e.g. toxin-producing staphylococci and the dental plaque-forming bacteria *Actinobacillus*.

Microbes evade immune system
Some microbes can evade the immune system by entering phagocytes. The microbes prevent fusion of the lysosome with the phagosome and multiply inside the phagocyte, almost filling it. Examples include *Chlamydia*, *Mycobacterium tuberculosis*, *Shigella*, and malarial parasites.

Dormant microbes hide inside
Some microbes can remain dormant inside the phagocyte for months or years at a time. Examples include the microbes that cause brucellosis and tularemia.

1. Identify the white blood cells capable of phagocytosis: _____

_____

2. Describe how a blood sample from a patient may be used to determine whether they have a microbial infection (without looking for the microbes themselves):

_____

_____

3. Explain how some microbes are able to overcome phagocytic cells and use them to their advantage:

_____

_____

Related activities: The Body's Defenses, Blood
Web links: Phagocytosis and Bacterial Pathogens

Inflammation

Damage to the body's tissues can be caused by physical agents (e.g. sharp objects, heat, radiant energy, or electricity), microbial infection, or chemical agents (e.g. gases, acids and bases). The damage triggers a defensive response called **inflammation**. It is usually characterized by four symptoms: pain, redness, heat and swelling. The inflammatory response is beneficial and has the following functions: (1) to destroy the cause of the infection and remove it and its products from the body; (2) if this fails, to limit the effects on the body by confining the infection to a small area; (3) replacing or repairing tissue damaged by the infection. The process of inflammation can be divided into three distinct stages. These are described below.

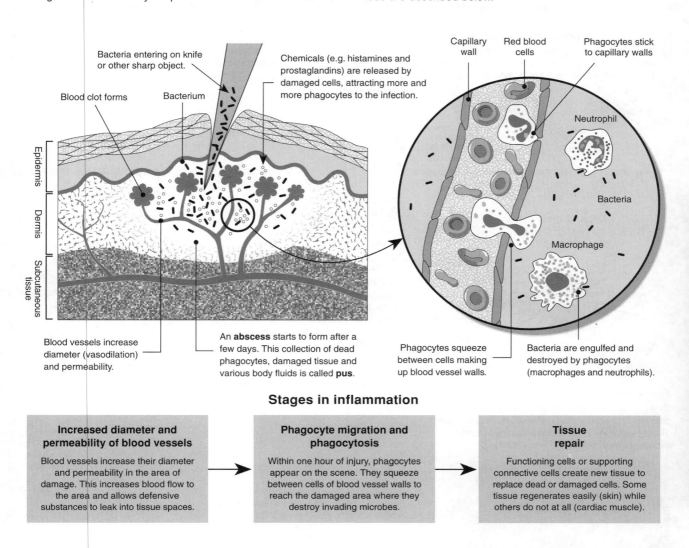

Stages in inflammation

Increased diameter and permeability of blood vessels	Phagocyte migration and phagocytosis	Tissue repair
Blood vessels increase their diameter and permeability in the area of damage. This increases blood flow to the area and allows defensive substances to leak into tissue spaces.	Within one hour of injury, phagocytes appear on the scene. They squeeze between cells of blood vessel walls to reach the damaged area where they destroy invading microbes.	Functioning cells or supporting connective cells create new tissue to replace dead or damaged cells. Some tissue regenerates easily (skin) while others do not at all (cardiac muscle).

1. Outline the three stages of inflammation and identify the beneficial role of each stage:

    (a) _____

    (b) _____

    (c) _____

2. Describe two features of phagocytes important in the response to microbial invasion: _____

3. State the role of histamines and prostaglandins in inflammation: _____

4. Explain why pus forms at the site of infection: _____

Related activities: The Body's Defenses, The Action of Phagocytes
Web links: Inflammation and Healing

A 1

The Lymphatic System & Immunity

Fever

To a point, fever is beneficial, because it assists a number of the defense processes. The release of the protein **interleukin-1** helps to reset the thermostat of the body to a higher level and increases production of T cells (lymphocytes). High body temperature also intensifies the effect of **interferon** (an antiviral protein) and may inhibit the growth of some bacteria and viruses. Because high temperatures speed up the body's **metabolic** **reactions**, it may promote more rapid tissue repair. Fever also increases heart rate so that white blood cells are delivered to sites of infection more rapidly. The normal body temperature range for most people is 36.2 to 37.2°C. Fevers of less than 40°C do not need treatment for **hyperthermia**, but excessive fever requires prompt attention (particularly in children). Death usually results if body temperature rises above 44.4 to 45.5°C.

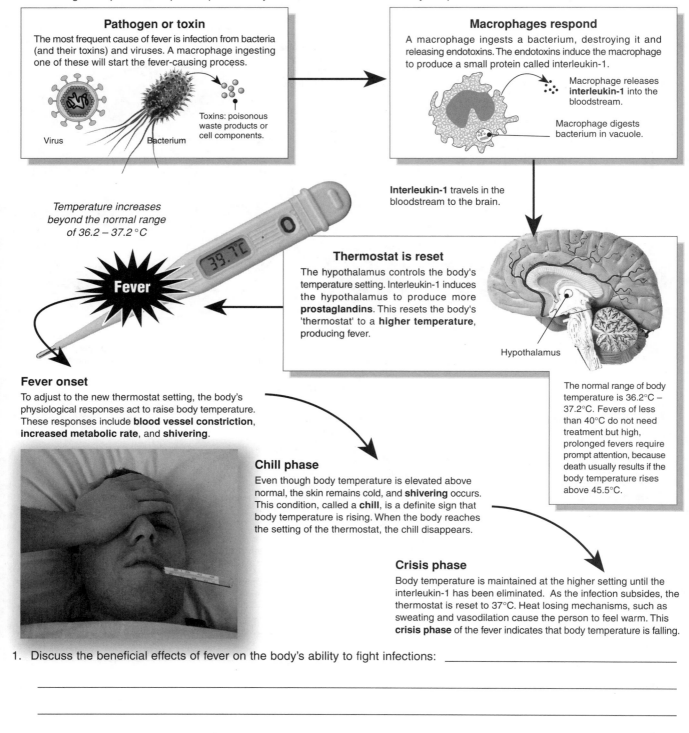

Pathogen or toxin
The most frequent cause of fever is infection from bacteria (and their toxins) and viruses. A macrophage ingesting one of these will start the fever-causing process.

Virus

Bacterium

Toxins: poisonous waste products or cell components.

Macrophages respond
A macrophage ingests a bacterium, destroying it and releasing endotoxins. The endotoxins induce the macrophage to produce a small protein called interleukin-1.

Macrophage releases **interleukin-1** into the bloodstream.

Macrophage digests bacterium in vacuole.

Interleukin-1 travels in the bloodstream to the brain.

Temperature increases beyond the normal range of 36.2 – 37.2 °C

39.7°C

Fever

Thermostat is reset
The hypothalamus controls the body's temperature setting. Interleukin-1 induces the hypothalamus to produce more **prostaglandins**. This resets the body's 'thermostat' to a **higher temperature**, producing fever.

Hypothalamus

The normal range of body temperature is 36.2°C – 37.2°C. Fevers of less than 40°C do not need treatment but high, prolonged fevers require prompt attention, because death usually results if the body temperature rises above 45.5°C.

Fever onset
To adjust to the new thermostat setting, the body's physiological responses act to raise body temperature. These responses include **blood vessel constriction**, **increased metabolic rate**, and **shivering**.

Chill phase
Even though body temperature is elevated above normal, the skin remains cold, and **shivering** occurs. This condition, called a **chill**, is a definite sign that body temperature is rising. When the body reaches the setting of the thermostat, the chill disappears.

Crisis phase
Body temperature is maintained at the higher setting until the interleukin-1 has been eliminated. As the infection subsides, the thermostat is reset to 37°C. Heat losing mechanisms, such as sweating and vasodilation cause the person to feel warm. This **crisis phase** of the fever indicates that body temperature is falling.

1. Discuss the beneficial effects of fever on the body's ability to fight infections: _____

_____

_____

_____

2. Summarize the key steps of how the body's thermostat is set at a higher level by infection: _____

_____

_____

_____

Related activities: The Body's Defenses

The Lymphatic System

Fluid leaks out from capillaries and forms the tissue fluid, which is similar in composition to plasma but lacks large proteins. This fluid bathes the tissues, supplying them with nutrients and oxygen, and removing wastes. Some of the tissue fluid returns directly into the capillaries, but some drains back into the blood circulation through a network of lymph vessels. This fluid, called **lymph**, is similar to tissue fluid, but contains more leukocytes.

Apart from its circulatory role, the lymphatic system also has an important function in the immune response. Lymph nodes are the primary sites where pathogens and other foreign substances are destroyed. A lymph node that is fighting an infection becomes swollen and hard as the leukocytes reproduce rapidly to increase their numbers. The thymus, spleen, and bone marrow also contribute leukocytes to the lymphatic and circulatory systems.

Tonsils: Tonsils (and adenoids) comprise a collection of large lymphatic nodules at the back of the throat. They produce lymphocytes and antibodies and are well-placed to protect against invasion of pathogens.

Thymus gland: The thymus is a two-lobed organ located close to the heart. It is prominent in infants and diminishes after puberty to a fraction of its original size. Its role in immunity is to help produce **T cells** that destroy invading microbes directly or indirectly by producing various substances.

Spleen: The oval spleen is the largest mass of lymphatic tissue in the body, measuring about 12 cm in length. It stores and releases blood in case of demand (e.g. in cases of bleeding), produces mature **B cells**, and destroys bacteria by phagocytosis.

Bone marrow: Bone marrow produces red blood cells and many kinds of leukocytes: monocytes (and macrophages), neutrophils, eosinophils, basophils, and lymphocytes (B cells and T cells).

Lymphatic vessels: When tissue fluid is picked up by lymph capillaries, it is called **lymph**. The lymph is passed along lymphatic vessels to a series of lymph nodes. These vessels contain one-way valves that move the lymph in the direction of the heart until it is reintroduced to the blood at the subclavian veins.

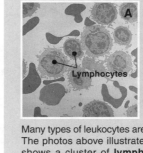

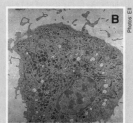

Photos: Ell

Many types of leukocytes are involved in internal defense. The photos above illustrate examples of leukocytes. **A** shows a cluster of **lymphocytes**. **B** shows a single **macrophage**: large, phagocytic cells that develop from monocytes and move from the blood to reside in many organs and tissues, including the spleen and lymph nodes.

Lymph node

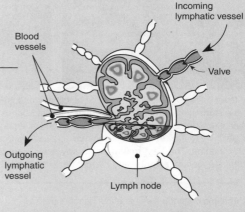

Lymph nodes are oval or bean-shaped structures, scattered throughout the body, usually in groups, along the length of lymphatic vessels. As lymph passes through the nodes, it filters foreign particles (including pathogens) by trapping them in fibers. Lymph nodes are also a "store" of **lymphocytes**, which may circulate to other parts of the body. Once trapped, macrophages destroy the foreign substances by phagocytosis. T cells may destroy them by releasing various products, and/or B cells may release antibodies that destroy them.

1. Briefly describe the composition of lymph: _____

   _____

2. Discuss the various roles of lymph: _____

   _____

   _____

3. Describe one role of each of the following in the lymphatic system:

   (a) Lymph nodes: _____

   (b) Bone marrow: _____

Allergies and Hypersensitivity

Sometimes the immune system may overreact, or react to the wrong substances instead of responding appropriately. This is termed **hypersensitivity** and the immunological response leads to tissue damage rather than immunity. Hypersensitivity reactions occur after a person has been **sensitized** to an antigen. In some cases, this causes only localized discomfort, as in the case of hayfever. More generalized reactions (such as anaphylaxis from insect venom or drug injections), or localized reactions that affect essential body systems (such as asthma), can cause death through asphyxiation and/or circulatory shock.

Hypersensitivity

A person becomes **sensitized** when they form antibodies to harmless substances in the environment such as pollen or spores (steps 1-2 right). These substances, termed **allergens**, act as antigens to induce antibody production and an allergic response. Once a person is sensitized, the antibodies respond to further encounters with the allergen by causing the release of **histamine** from mast cells (steps 4-5). It is histamine that mediates the symptoms of hypersensitivity reactions such as hay fever and asthma. These symptoms include wheezing and airway constriction, inflammation, itching and watering of the eyes and nose, and/or sneezing.

Eyewire

The Basis of Hypersensitivity

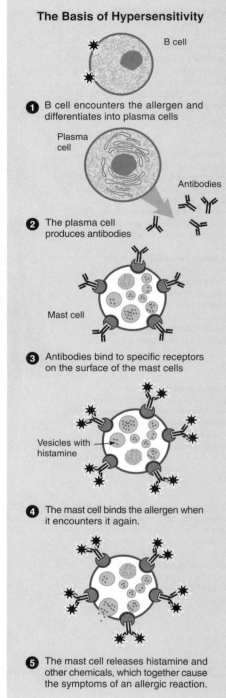

B cell

1 B cell encounters the allergen and differentiates into plasma cells

Plasma cell

Antibodies

2 The plasma cell produces antibodies

Mast cell

3 Antibodies bind to specific receptors on the surface of the mast cells

Vesicles with histamine

4 The mast cell binds the allergen when it encounters it again.

5 The mast cell releases histamine and other chemicals, which together cause the symptoms of an allergic reaction.

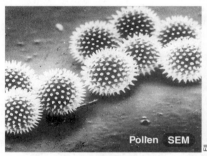
Pollen SEM

Ragweed

Hay fever (allergic rhinitis) is an allergic reaction to airborne substances such as dust, molds, pollens, and animal fur or feathers. Allergy to wind-borne pollen is the most common, and certain plants (e.g. ragweed and privet) are highly allergenic. There appears to be a genetic susceptibility to hay fever, as it is common in people with a family history of eczema, hives, and/or asthma. The best treatment for hay fever is to avoid the allergen, although anti-histamines, decongestants, and steroid nasal sprays will assist in alleviating symptoms.

Asthma is a common disease affecting more than 15 million people in the US. It usually occurs as a result of an allergic reaction to allergens such as house dust and the feces of house dust mites, pollen, and animal dander. As with all hypersensitivity reactions, it involves the production of histamines from mast cells (far right). The site of the reaction is the respiratory bronchioles where the histamine causes constriction of the airways, accumulation of fluid and mucus, and inability to breathe. During an attack, sufferers show labored breathing with overexpansion of the chest cavity (photo, right).

Asthma attacks are often triggered by environmental factors such as cold air, exercise, air pollutants, and viral infections. Recent evidence has also indicated the involvement of a bacterium: *Chlamydia pneumoniae*, in about half of all cases of asthma in susceptible adults.

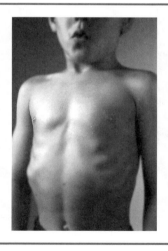

1. Explain the role of histamine in hypersensitivity responses: _____

_____

2. Explain what is meant by becoming **sensitized** to an allergen: _____

_____

3. Explain the effect of **bronchodilators** and explain why they are used to treat asthma: _____

_____

The Immune System

The efficient internal defense provided by the immune system is based on its ability to respond specifically against a foreign substance and its ability to hold a memory of this response. There are two main components of the immune system: the humoral and the cell-mediated responses. They work separately and together to protect us from disease. The **humoral immune response** is associated with the serum (non-cellular part of the blood) and involves the action of **antibodies** secreted by B cell lymphocytes. Antibodies are found in extracellular fluids including lymph, plasma, and mucus secretions. The humoral response protects the body against circulating viruses, and bacteria and their toxins. The **cell-mediated immune response** is associated with the production of specialized lymphocytes called **T cells**. It is most effective against bacteria and viruses located within host cells, as well as against parasitic protozoa, fungi, and worms. This system is also an important defense against cancer, and is responsible for the rejection of transplanted tissue. Both B and T cells develop from stem cells located in the liver of fetuses and the bone marrow of adults. T cells complete their development in the thymus, whilst the B cells mature in the bone marrow.

Lymphocytes and their Functions

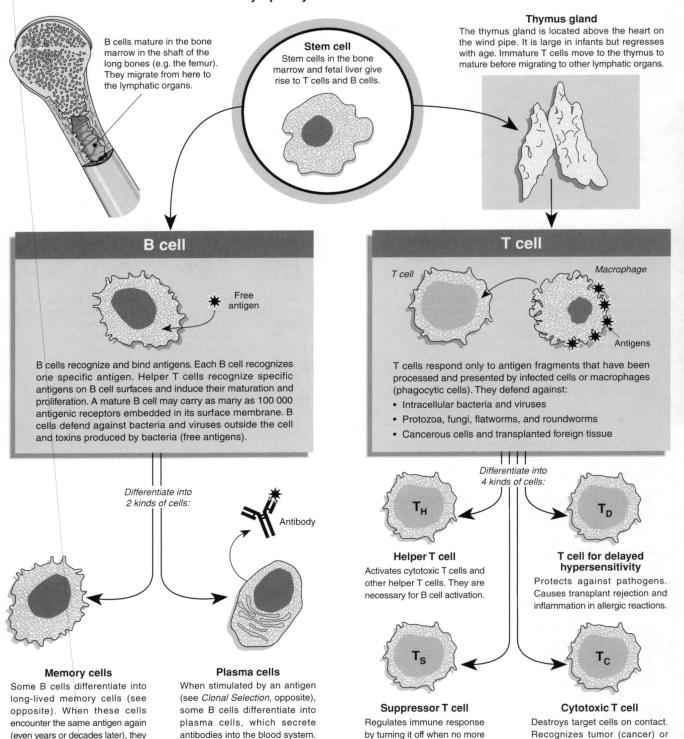

B cells mature in the bone marrow in the shaft of the long bones (e.g. the femur). They migrate from here to the lymphatic organs.

Stem cell
Stem cells in the bone marrow and fetal liver give rise to T cells and B cells.

Thymus gland
The thymus gland is located above the heart on the wind pipe. It is large in infants but regresses with age. Immature T cells move to the thymus to mature before migrating to other lymphatic organs.

B cell

Free antigen

B cells recognize and bind antigens. Each B cell recognizes one specific antigen. Helper T cells recognize specific antigens on B cell surfaces and induce their maturation and proliferation. A mature B cell may carry as many as 100 000 antigenic receptors embedded in its surface membrane. B cells defend against bacteria and viruses outside the cell and toxins produced by bacteria (free antigens).

T cell

T cell   Macrophage

Antigens

T cells respond only to antigen fragments that have been processed and presented by infected cells or macrophages (phagocytic cells). They defend against:
- Intracellular bacteria and viruses
- Protozoa, fungi, flatworms, and roundworms
- Cancerous cells and transplanted foreign tissue

Differentiate into 2 kinds of cells:

Antibody

Differentiate into 4 kinds of cells:

T_H

T_D

Helper T cell
Activates cytotoxic T cells and other helper T cells. They are necessary for B cell activation.

T cell for delayed hypersensitivity
Protects against pathogens. Causes transplant rejection and inflammation in allergic reactions.

T_S

T_C

Memory cells
Some B cells differentiate into long-lived memory cells (see opposite). When these cells encounter the same antigen again (even years or decades later), they rapidly differentiate into antibody-producing plasma cells.

Plasma cells
When stimulated by an antigen (see *Clonal Selection*, opposite), some B cells differentiate into plasma cells, which secrete antibodies into the blood system. The antibodies then inactivate the circulating antigens.

Suppressor T cell
Regulates immune response by turning it off when no more antigen is present.

Cytotoxic T cell
Destroys target cells on contact. Recognizes tumor (cancer) or virus infected cells by their surface (antigens and MHC markers).

The Lymphatic System & Immunity

Related activities: The Lymphatic System, Allergies and Hypersensitivity
Web links: Introducing... Specific Immunity, The Immune System Overview

A 2

The immune system has the ability to respond to the large and unpredictable range of potential antigens encountered in the environment. The diagram below explains how this ability is based on **clonal selection** after antigen exposure. The example illustrated is for B cell lymphocytes. In the same way, a T cell stimulated by a specific antigen will multiply and develop into different types of T cells. Clonal selection and differentiation of lymphocytes provide the basis for **immunological memory**.

Five (a-e) of the many, randomly generated B cells. Each one can recognize only one specific antigen.

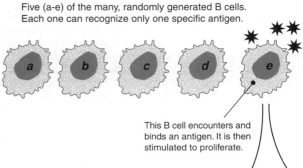

This B cell encounters and binds an antigen. It is then stimulated to proliferate.

Clonal Selection Theory

Millions of randomly generated B cells form during development. Collectively, they can recognize many antigens, including those that have never been encountered. Each B cell makes antibodies corresponding to the specific antigenic receptor on its surface. The receptor reacts only to that specific antigen. When a B cell encounters its antigen, it responds by proliferating and producing many clones all with the same kind of antibody. This is called **clonal selection** because the antigen selects the B cells that will proliferate.

Memory cells

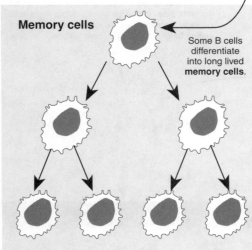

Some B cells differentiate into long lived **memory cells**.

Plasma cells

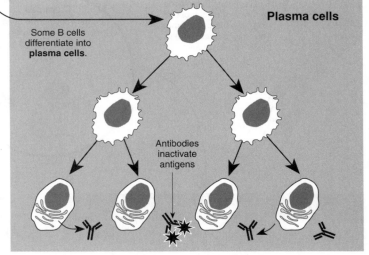

Some B cells differentiate into **plasma cells**.

Antibodies inactivate antigens

Some B cells differentiate into long lived **memory cells**. These are retained in the lymph nodes to provide future immunity (**immunological memory**). In the event of a second infection, B-memory cells react more quickly and vigorously than the initial B-cell reaction to the first infection.

Plasma cells secrete antibodies specific to the antigen that stimulated their development. Each plasma cell lives for only a few days, but can produce about 2000 antibody molecules per second. Note that during development, any B cells that react to the body's own antigens are selectively destroyed in a process that leads to **self tolerance** (acceptance of the body's own tissues).

1. State the general action of the two major divisions in the immune system:

   (a) Humoral immune system: _____

   (b) Cell-mediated immune system: _____

2. Identify the origin of B cells and T cells (before maturing): _____

3. (a) State where B cells mature: _____ (b) State where T cells mature: _____

4. Briefly describe the function of each of the following cells in the immune system response:

   (a) Memory cells: _____

   (b) Plasma cells: _____

   (c) Helper T cells: _____

   (d) Suppressor T cells: _____

   (e) Delayed hypersensitivity T cells: _____

   (f) Cytotoxic T cells: _____

5. Explain the basis of **immunological memory**: _____

   _____

   _____

   _____

Antibodies

Antibodies and antigens play key roles in the response of the immune system. Antigens are foreign molecules that are able to bind to antibodies (or T cell receptors) and provoke a specific immune response. Antigens include potentially damaging microbes and their toxins (see below) as well as substances such as pollen grains, blood cell surface molecules, and the surface proteins on transplanted tissues. **Antibodies** (also called immunoglobulins) are proteins that are made in response to antigens. They are secreted into the plasma where they circulate and can recognize, bind to, and help to destroy antigens. There are five classes of **immunoglobulins**. Each plays a different

role in the immune response (including destroying protozoan parasites, enhancing phagocytosis, protecting mucous surfaces, and neutralizing toxins and viruses). The human body can produce an estimated 100 million antibodies, recognizing many different antigens, including those it has never encountered. Each type of antibody is highly specific to only one particular antigen. The ability of the immune system to recognize and ignore the antigenic properties of its own tissues occurs early in development and is called **self-tolerance**. Exceptions occur when the immune system malfunctions and the body attacks its own tissues, causing an **autoimmune disorder**.

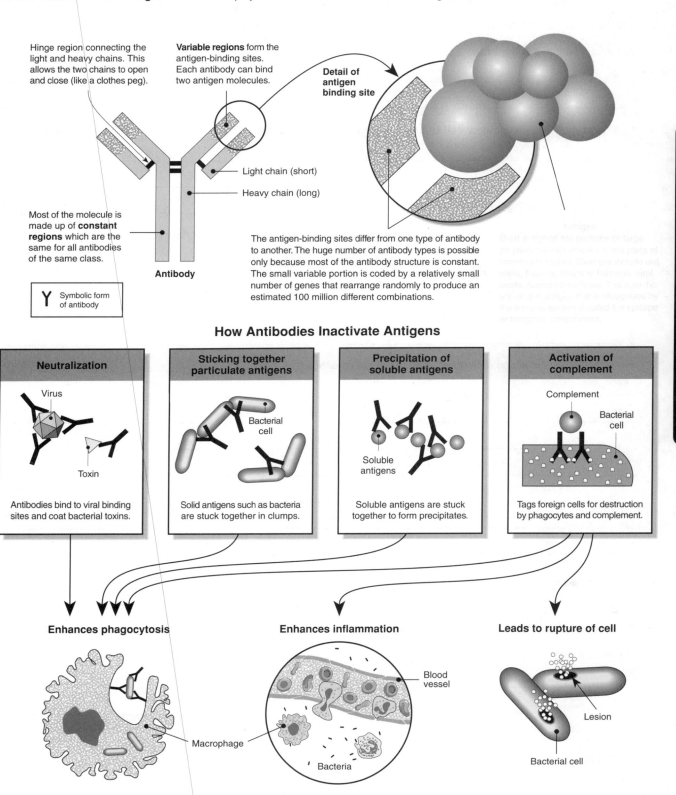

Hinge region connecting the light and heavy chains. This allows the two chains to open and close (like a clothes peg).

Variable regions form the antigen-binding sites. Each antibody can bind two antigen molecules.

Detail of antigen binding site

Light chain (short)

Heavy chain (long)

Most of the molecule is made up of **constant regions** which are the same for all antibodies of the same class.

Antibody

Y Symbolic form of antibody

The antigen-binding sites differ from one type of antibody to another. The huge number of antibody types is possible only because most of the antibody structure is constant. The small variable portion is coded by a relatively small number of genes that rearrange randomly to produce an estimated 100 million different combinations.

How Antibodies Inactivate Antigens

Neutralization
Virus

Toxin

Antibodies bind to viral binding sites and coat bacterial toxins.

Sticking together particulate antigens
Bacterial cell

Solid antigens such as bacteria are stuck together in clumps.

Precipitation of soluble antigens
Soluble antigens

Soluble antigens are stuck together to form precipitates.

Activation of complement
Complement

Bacterial cell

Tags foreign cells for destruction by phagocytes and complement.

Enhances phagocytosis

Macrophage

Enhances inflammation

Blood vessel

Bacteria

Leads to rupture of cell

Lesion

Bacterial cell

Related activities: The Immune System, Acquired Immunity
Web links: The Humoral Response, How Lymphocytes Produce Antibodies

RA 2

1. Distinguish between an antibody and an antigen: _____

   _____

   _____

   _____

   _____

   _____

   _____

2. It is necessary for the immune system to clearly distinguish foreign cells and proteins from those made by the body.

   (a) Explain why this is the case: _____

   _____

   _____

   (b) In simple terms, explain how **self tolerance** develops (see the activity *The Immune System* if you need help):

   _____

   _____

   (c) Name the type of disorder that results when this recognition system fails: _____

   (d) Describe two examples of disorders that are caused in this way, identifying what happens in each case:

   _____

   _____

   _____

   _____

3. Discuss the ways in which antibodies work to inactivate antigens: _____

   _____

   _____

   _____

   _____

   _____

   _____

   _____

   _____

   _____

4. Explain how antibody activity enhances or leads to:

   (a) Phagocytosis: _____

   _____

   (b) Inflammation: _____

   _____

   (c) Bacterial cell lysis: _____

   _____

Acquired Immunity

We have natural or **innate resistance** to certain illnesses, including most diseases of other animal species. **Acquired immunity** refers to the protection we develop during our lifetime against microbes and foreign substances. Immunity can be acquired either passively or actively. **Active immunity** develops when a person is exposed to foreign substances or to microorganisms (e.g. through infection) and the immune system

responds. **Passive immunity** is acquired when antibodies are transferred from one person to another. Recipients do not make the antibodies themselves and the effect lasts only as long as the antibodies are present (usually several weeks or months). Either type of immunity may also be **naturally acquired** through natural exposure to microbes, or **artificially acquired** as a result of medical treatment.

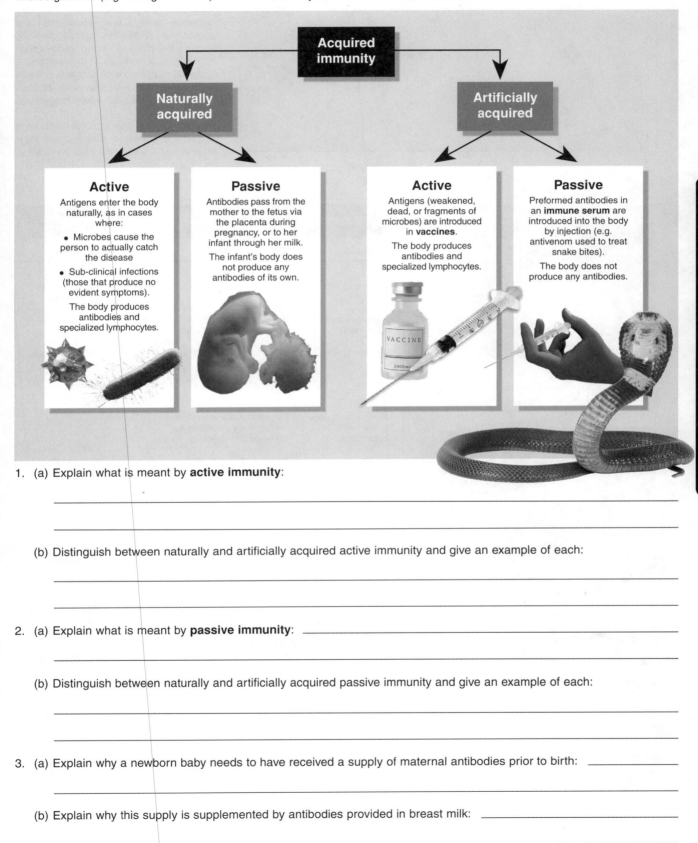

The Lymphatic System & Immunity

1. (a) Explain what is meant by **active immunity**:

   _____

   _____

   (b) Distinguish between naturally and artificially acquired active immunity and give an example of each:

   _____

   _____

2. (a) Explain what is meant by **passive immunity**: _____

   _____

   (b) Distinguish between naturally and artificially acquired passive immunity and give an example of each:

   _____

   _____

3. (a) Explain why a newborn baby needs to have received a supply of maternal antibodies prior to birth: _____

   _____

   (b) Explain why this supply is supplemented by antibodies provided in breast milk: _____

   _____

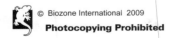
Related activities: Antibodies
Web links: Steps in Vaccine Development

A 2

Primary and Secondary Responses to Antigens

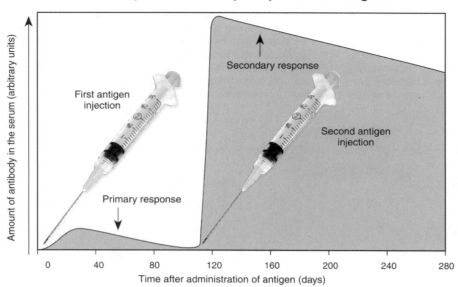

When the B cells encounter antigens and produce antibodies, the body develops **active immunity** against that antigen.

The initial response to antigenic stimulation, caused by the sudden increase in B cell clones, is called the **primary response**. Antibody levels as a result of the primary response peak a few weeks after the response begins and then decline. However, because the immune system develops an immunological memory of that antigen, it responds much more quickly and strongly when presented with the same antigen subsequently (the **secondary response**).

This forms the basis of immunization programs where one or more booster shots are provided following the inital vaccination.

Vaccines to protect against common diseases are administered at various stages during childhood according to an immunization schedule.

While most vaccinations are given in childhood, adults may be vaccinated against specific diseases (e.g. tuberculosis) if they are in a high risk group or if they are traveling to a region in the world where a disease is prevalent.

Selected Vaccines Used To Prevent Diseases In Humans

Disease	Type of vaccine	Recommendation
Diphtheria	Purified diphtheria toxoid	From early childhood and every 10 years for adults
Meningococcal meningitis	Purified polysaccharide of *Neisseria menigitidis*	For people with substantial risk of infection
Whooping cough	Killed cells or fragments of *Bordetella pertussis*	Children prior to school age
Tetanus	Purified tetanus toxoid	14-16 year olds with booster every 10 years
Meningitis caused by *Hemophilus influenzae* b	Polysaccharide from virus conjugated with protein to enhance effectiveness	Early childhood
Influenza	Killed virus (vaccines using genetically engineered antigenic fragments are also being developed)	For chronically ill people, especially with respiratory diseases, or for healthy people over 65 years of age
Measles	Attenuated virus	Early childhood
Mumps	Attenuated virus	Early childhood
Rubella	Attenuated virus	Early childhood; for females of child-bearing age who are not pregnant

*Vaccine development is an important part of public health. Immunization programs have been behind the eradication of some debilitating human diseases, such as smallpox. Many childhood diseases for which immunization programs exist are kept at a low level because of the phenomenon of **herd immunity**. If a large proportion of the population is immune, those that are not immunized may be protected because the disease becomes uncommon.*

4. (a) Describe two differences between the primary and secondary responses to antigen presentation:

_____

_____

(b) Explain why the secondary response is so different from the primary response: _____

_____

_____

_____

HIV and the Immune System

AIDS (Acquired Immune Deficiency Syndrome) first appeared in the news in 1981, with cases being reported in Los Angeles, in the US. By 1983, the pathogen causing the disease had been identified as a retrovirus that selectively infects **helper T cells**. The disease causes a massive deficiency in the immune system due to infection with **HIV** (human immunodeficiency virus). HIV is a **retrovirus** and is able to splice its genes into the host cell's chromosome. As yet, there is no cure or vaccine, and the disease has taken the form of a **pandemic**, spreading to all parts of the globe and killing more than a million people each year. HIV is transmitted in blood, vaginal secretions, semen, breast milk, and across the placenta. In developed countries, blood transfusions are no longer a likely source of infection because blood is tested for HIV antibodies. Historically, transmission of HIV in developed countries has been primarily through intravenous drug use and homosexual activity, but heterosexual transmission is increasing.

Capsid
Protein coat that protects the nucleic acids (RNA) within.

Viral envelope
A piece of the cell membrane budded off from the last human host cell.

Nucleic acid
Two identical strands of RNA contain the genetic blueprint for making more HIV viruses.

Reverse transcriptase
Two copies of this important enzyme convert the RNA into DNA once inside a host cell.

Surface proteins
These spikes allow HIV to attach to receptors on the host cells (T cells and macrophages).

The structure of HIV

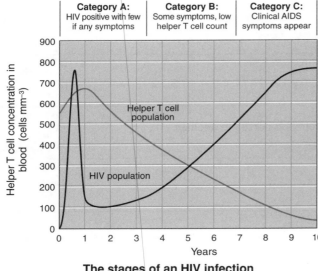

Category A: HIV positive with few if any symptoms	Category B: Some symptoms, low helper T cell count	Category C: Clinical AIDS symptoms appear

Helper T cell concentration in blood (cells mm^{-3})

Helper T cell population

HIV population

Years

The stages of an HIV infection

AIDS is actually only the end stage of an HIV infection. Shortly after the initial infection, HIV antibodies appear within the blood. The progress of infection has three clinical categories shown on the graph above.

HIV/AIDS

Individuals affected by the human immunodeficiency virus (HIV) may have no symptoms, while medical examination may detect swollen lymph glands. Others may experience a short-lived illness when they first become infected (resembling infectious mononucleosis). The range of symptoms resulting from HIV infection is huge, and is not the result of the HIV infection directly. The symptoms arise from an onslaught of secondary infections that gain a foothold in the body due to the suppressed immune system (due to the few helper T cells). These infections are from normally rare fungal, viral, and bacterial sources. Full blown AIDS can also feature some rare forms of cancer. Some symptoms are listed below:

Fever, lymphoma (cancer) and toxoplasmosis of the brain, dementia.

Eye infections (*Cytomegalovirus*).

Skin inflammation (dermatitis) particularly affecting the face.

Oral thrush (*Candida albicans*) of the esophagus, bronchi, and lungs.

A variety of opportunistic infections, including: chronic or persistent *Herpes simplex*, tuberculosis (TB), pneumocystis pneumonia, shingles, shigellosis and salmonellosis.

Diarrhea caused by *Isospora* or *Cryptosporidium*.

Marked weight loss.
A number of autoimmune diseases, especially destruction of platelets.

Kaposi's sarcoma: a highly aggressive malignant skin tumor consisting of blue-red nodules, usually start at the feet and ankles, spreading to the rest of the body later, including respiratory and gastrointestinal tracts.

The Lymphatic System & Immunity

1. Explain why the HIV virus has such a devastating effect on the human body's ability to fight disease:

_____

_____

_____

2. Consult the graph above showing the stages of HIV infection (remember, HIV infects and destroys helper T cells).

(a) Describe how the virus population changes with the progression of the disease: _____

_____

_____

Related activities: Patterns of Disease, Epidemiology of AIDS, The Control of Disease, The Immune System   **Web links**: HIV Interactive Animation

RDA 2

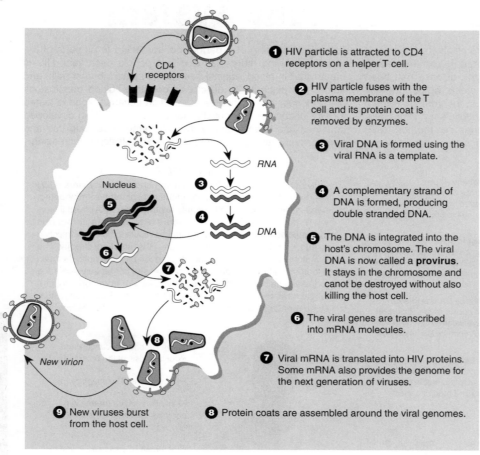

① HIV particle is attracted to CD4 receptors on a helper T cell.

CD4 receptors

② HIV particle fuses with the plasma membrane of the T cell and its protein coat is removed by enzymes.

RNA

Nucleus

③ Viral DNA is formed using the viral RNA is a template.

④ A complementary strand of DNA is formed, producing double stranded DNA.

DNA

⑤ The DNA is integrated into the host's chromosome. The viral DNA is now called a **provirus**. It stays in the chromosome and canot be destroyed without also killing the host cell.

⑥ The viral genes are transcribed into mRNA molecules.

New virion

⑦ Viral mRNA is translated into HIV proteins. Some mRNA also provides the genome for the next generation of viruses.

⑨ New viruses burst from the host cell.

⑧ Protein coats are assembled around the viral genomes.

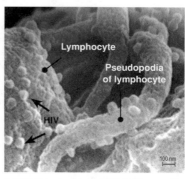

A SEM shows spherical HIV-1 virions on the surface of a human lymphocyte.

Lymphocyte

Pseudopodia of lymphocyte

HIV

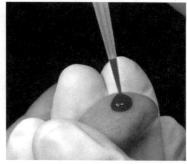

Diagnosis of HIV is possible using a simple antibody-based test on a blood sample.

(b) Describe how the helper T cells respond to the infection: _____

_____

_____

3. Name three common ways in which HIV can be transmitted from one person to another:

(a) _____

(b) _____

(c) _____

4. Explain what is meant by the term **HIV positive**: _____

_____

_____

5. In the years immediately following the discovery of the HIV pathogen, there was a sudden appearance of AIDS cases amongst **hemophiliacs** (people with an inherited blood disorder). State why this group was being infected with HIV:

_____

_____

6. Explain the significance of the formation of a provirus, both for the virus and for the human host: _____

_____

_____

_____

_____

_____

Organ and Tissue Transplants

Transplant surgery involves the replacement of a diseased tissue or organ with a healthy, living substitute. The tissue or organ is usually taken from a person who has just died, although some (e.g. blood, kidneys, bone marrow) can be taken from living donors. Around the world, more than 100 000 major organs have been transplanted, mostly in the past few decades. About 80% of patients are alive and well one year after the transplantation, and most survive at least 5 years. There are two major problems associated with organ transplants. One is the lack of donors and the second is **tissue rejection**. The **major histocompatibility complex** (MHC) is responsible for encoding proteins (antigens) on the surface of all cells. Cells from donor tissue have different antigens to those of the recipient. When the new tissue is transplanted, the recipient's immune system launches an attack to kill the foreign material, which often results in the rejection or death of the transplanted tissue. The use of **immunosuppressant drugs** and improved **tissue-typing** can decrease the MHC response, and this along with better techniques for organ preservation and transport has increased the success of organ transplants. Attempts to carry out transplants of organs from other species into humans (**xenotransplantation**) have not been very successful due to rejection, although there are hopes that genetically modified pigs may be used to produce organs engineered to be compatible with human recipients. With the advent of **tissue engineering** and **stem cell** technology, researchers are rapidly moving towards their goal of creating semi-synthetic, living organs that may be used as human replacement parts.

Organ Transplants

Currently, there are five organs that are routinely transplanted (below). In addition to organs, whole hand transplants and partial or whole face transplants are now possible.

Face: Facial reconstructions began initially with reconnection of a patient's own facial components damaged in accidents. More recently, medical techniques have developed so that partial reconstructions (usually of nose and mouth) are possible using facial material from a dead donor. In 2008, a French medical team completed the world's first full face transplant. These types of transplants require careful connection of blood vessels, skin, muscles, and bone, tendons, and other connective tissues. Performing face transplants also involves addressing a number of ethical concerns.

Heart (H): These transplants are carried out after a patient suffers heart failure due to heart attack, viral infections of the heart, or congenital, irreparable defects.

Lungs (Ls): Replace organs damaged by cystic fibrosis or emphysema. Typically, lungs are transplanted together, but single lung transplants and heart-lung transplants are also possible.

Hands: In 1999 the first successful hand transplant was performed on Matthew Scott in the USA. A year and a half after the operation, he could sense temperature, pressure and pain, and could write, turn the pages of a newspaper, tie shoelaces and throw a baseball. Twenty six successful hand transplants have been undertaken since.

Liver (Li): Substitute for a liver destroyed by cirrhosis, congenital defects, or hepatitis.

Pancreas (P): Restores insulin production in Type I diabetics (caused by autoimmune destruction of the insulin producing cells of the pancreas).

Kidneys (K): Used in cases of renal failure, diabetes, high blood pressure, inherited illnesses, and infection. The failing kidneys are usually left *in situ* and the transplant (Kt) is placed in a location different from the original kidney.

Tissue Transplants

A large number of tissues are currently used in transplant procedures. An estimated 200 patients can potentially benefit from the organs and tissues donated from a single body.

Cornea: Transplants can restore impaired vision.

Dental powder: This tissue is prepared to help rebuild defects in the mandible (which supports the teeth).

Jaw: The mandible is used in facial reconstruction.

Ear bones: The three bones of the inner ear can be transplanted to improve some forms of deafness.

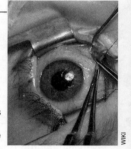
WIKI

Pericardium: The pericardium surrounding the heart is made of tough tissue that can be used to cover the brain after surgery. Transplants of the brain coverings themselves are no longer performed because of the risk of transmitting prion infections.

Blood and blood vessels: Blood transfusions are transplants of blood tissue. Blood vessels, mostly veins, can be transplanted to reroute blood around blockages, such as this atherosclerotic plaque (right) in the body.

Plaque

WIKI

Hip joints: Joints can be reconstructed by transplanting the head of the femur.

Bone marrow: Marrow is extracted from living donors and used to help people with a wide variety of illnesses, such as leukaemia.

Bones: Long bones of the arms and legs can be used in limb reconstruction; ribs can be used for spinal fusions and facial repair.

Cartilage and ligaments: Orthopaedic surgeons use these materials to rebuild ankle, knee, hip, elbow and shoulder joints.

Skin: Skin can be used as a temporary covering for burn injuries until the patient's own skin grows back.

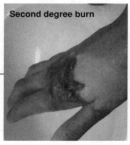
Second degree burn

WIKI

The Lymphatic System & Immunity

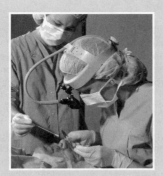  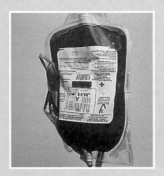 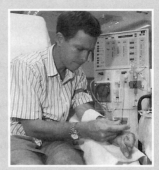

Corneal transplants can be used to restore sight in patients with damaged vision. The cornea naturally has a poor blood supply so rejection is less of a problem than with some other tissues

For many amputees, being fitted with an artificial limb is the first step towards mobility. In the future, such prostheses may be replaced with limb transplants, in much the same way as current hand transplants.

Transplants of whole blood, blood plasma, platelets, and other blood components are crucial to many medical procedures. The donor blood is carefully typed to ensure compatibility with the recipient.

Many patients with kidney failure rely on regular dialysis in order to function. This is expensive and inconvenient, and carries health risks. Such patients are usually waiting for a kidney transplant.

1. Describe three major technical advances that have improved the success rate of organ transplantation:

   (a) _____

   (b) _____

   (c) _____

2. (a) Explain the basis for organ and tissue rejection: _____

   _____

   _____

   (b) Discuss the role of **tissue typing** and **immunosuppressant drugs** in reducing or preventing this response:

   _____

   _____

   _____

   (c) Describe one of the major undesirable side-effects of using immunosuppressant drugs in transplant recipients:

   _____

3. Describe how tissue engineering is being developed in response to the shortage of donor organs:

   _____

   _____

   _____

4. Describe the ethical issues associated with organ and tissue transplants. Consider costs, benefits, source of tissue, and criteria for choosing recipients. If required, debate the issue, or develop your arguments as a separate report:

   _____

   _____

   _____

   _____

   _____

   _____

   _____

   _____

Stem Cell Technology

Stem cells are undifferentiated cells. They are characterized by two features. The first is **self renewal**, which enables them to undergo numerous cycles of cell division while maintaining an unspecialized state. The second is **potency**, which is the ability to differentiate into specialized cells. Stem cells have the ability to develop and form all the tissues of the body. The best source of these is from very early embryos. **Embryonic stem cells** are **pluripotent** and can differentiate into cells derived from the three germ layers. Some adult tissues (e.g. bone marrow) also contain stem cells. **Adult stem cells**, are **multipotent**, they can develop into several cell types, but are generally limited to

certain cell lineages such as blood and heart cells. Because of these properties, human stem cells have many potential applications. These include studies of human development and gene regulation, testing the safety and effectiveness of new drugs and vaccines, monoclonal antibody production, and supplying cells to treat diseased and damaged tissue. These technologies require a disease-free and plentiful supply of cells of specific types. Therapeutic **stem cell cloning** is still in its very early stages. The recent decision of President Barack Obama's government to lift the ban on stem cell research has been welcomed by many as a first step in developing new therapies.

Embryonic Stem Cells

Embryonic stem cells (ESC) are stem cells taken from **blastocysts** (below). Blastocysts are embryos which are about five days old and consist of a hollow ball containing 50-150 cells. In general, ESCs come from embryos which have been fertilised *in vitro* at fertilisation clinics and have then been donated for research.

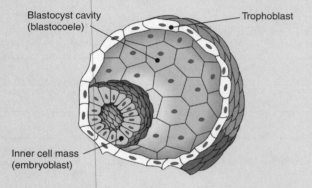

Blastocyst cavity (blastocoele)

Trophoblast

Inner cell mass (embryoblast)

Cells derived from the inner cell mass are **pluripotent**; they can become any cells of the body, with the exception of placental cells. When grown *in vitro*, without any stimulation for differentiation, these cells retain their pluripotency through multiple cell divisions. As a consequence of this, **ESC therapies** have potential use in regenerative medicine and tissue replacement. By manipulating the culture conditions, scientists are able to select and control the type of cells grown (e.g. heart cells). This ability could allow for specific cell types to be grown to treat specific diseases or replace damaged tissue. However, no ESC treatments have been approved to date due to ethical issues.

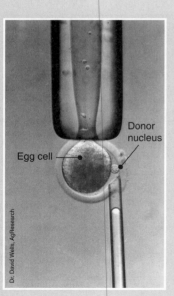

Egg cell
Donor nucleus
Dr. David Wells, AgResearch

Poor **histocompatibility**, in which the recipient's immune system rejects foreign cells, is one of the major difficulties with transplantation. Stem cell cloning (also called **therapeutic cloning**) provides a way around this problem. Stem cell cloning produces cells that have been derived from the recipient and are therefore histocompatible. Such an approach would mean immunosuppressant drugs would no longer be required. Diseases such as leukaemia and Parkinson's could be treated using this technique.

Embryonic Stem Cell Cloning

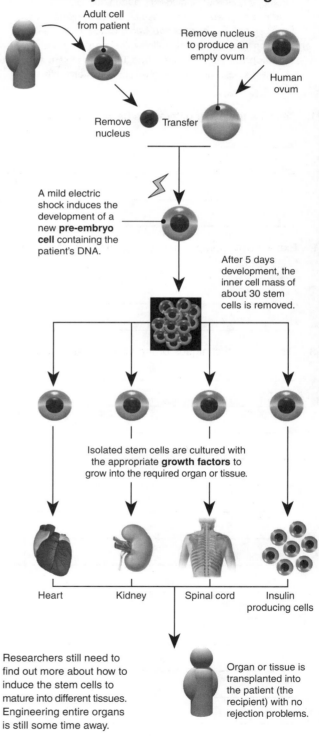

Adult cell from patient

Remove nucleus to produce an empty ovum

Human ovum

Remove nucleus

Transfer

A mild electric shock induces the development of a new **pre-embryo cell** containing the patient's DNA.

After 5 days development, the inner cell mass of about 30 stem cells is removed.

Isolated stem cells are cultured with the appropriate **growth factors** to grow into the required organ or tissue.

Heart | Kidney | Spinal cord | Insulin producing cells

Researchers still need to find out more about how to induce the stem cells to mature into different tissues. Engineering entire organs is still some time away.

Organ or tissue is transplanted into the patient (the recipient) with no rejection problems.

The Lymphatic System & Immunity

Related activities: Gene Therapy, Hematopoiesis
Web links: Stem Cells in the Spotlight

RA 3

Adult Stem Cells

Adult stem cells (ASC) are undifferentiated cells found in several types of tissues in adults and children (e.g. brain, bone marrow, skin, and liver), and in umbilical cord blood. Unlike ESCs, they are multipotent and can only differentiate into a limited number of cell types, usually related to the tissue of origin. In the body, the main role of ASC is to replace damaged or dying cells. There are fewer ethical issues associated with using ASC for therapeutic purposes, and for this reason ASC are already used to treat a number of diseases including leukemia and other blood disorders.

In many countries, including Canada, the US, and UK, parents of newborns can have blood from the umbilical cord stored in a **cord blood bank**. Cord blood is a rich source of multipotent cells which can be used for autologous (self) transplants to treat a range of diseases. ASC are also termed somatic stem cells.

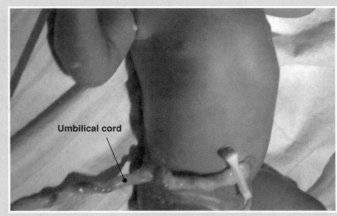

Umbilical cord

Cells obtained from umbilical cord blood (above) or bone marrow could be used to treat patients with a variety of diseases including leukemia, lymphomas, anemia, and a range of congenital diseases. Multipotent stem cells from marrow or cord blood give rise to the precursor cells for red blood cells, all white blood cell types, and platelets.

1. Describe the major differences between embryonic stem cells (ESC) and adult stem cells (ASC):

   _____

   _____

   _____

   _____

2. Explain why therapies using ASC are less controversial than therapies using ESC: _____

   _____

   _____

   _____

3. Discuss the techniques or the applications of therapeutic stem cell cloning (including ethical issues where relevant):

   _____

   _____

   _____

   _____

   _____

   _____

   _____

   _____

   _____

   _____

   _____

   _____

   _____

   _____

Monoclonal Antibodies

A **monoclonal antibody** is an artificially produced antibody that neutralizes only one specific protein (antigen). Monoclonal antibodies are produced in the laboratory by stimulating the production of B-cell in mice injected with the antigen. These B-cells produce an antibody against the antigen. When isolated and made with immortal tumor cells, they can be cultured indefinitely in a suitable growing medium (as illustrated below). Monoclonal antibodies are useful for three reasons: they are totally uniform (i.e. clones), they can be produced in large quantities, and they are highly specific. The uses of antibodies produced by this method have ranged from use as diagnostic tools to treatments for infections and cancer. The therapeutic use of monoclonal antibodies has been somewhat limited because they are currently produced by non-human cells and the immune systems of some people have reacted against these foreign proteins (remember that antibodies are proteins). It is hoped in the future to produce monoclonal antibodies derived from human cells, which will probably cause fewer reactions.

Making Monoclonal Antibodies

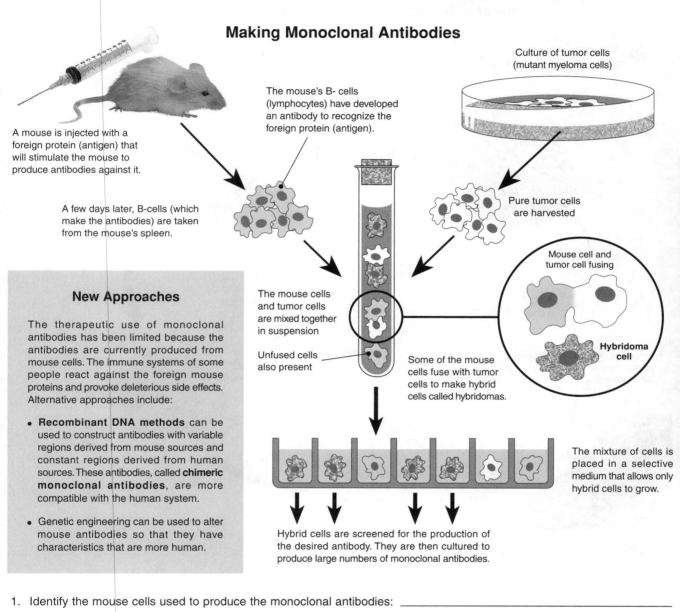

Culture of tumor cells (mutant myeloma cells)

A mouse is injected with a foreign protein (antigen) that will stimulate the mouse to produce antibodies against it.

The mouse's B- cells (lymphocytes) have developed an antibody to recognize the foreign protein (antigen).

A few days later, B-cells (which make the antibodies) are taken from the mouse's spleen.

Pure tumor cells are harvested

Mouse cell and tumor cell fusing

Hybridoma cell

The mouse cells and tumor cells are mixed together in suspension

Unfused cells also present

Some of the mouse cells fuse with tumor cells to make hybrid cells called hybridomas.

New Approaches

The therapeutic use of monoclonal antibodies has been limited because the antibodies are currently produced from mouse cells. The immune systems of some people react against the foreign mouse proteins and provoke deleterious side effects. Alternative approaches include:

- **Recombinant DNA methods** can be used to construct antibodies with variable regions derived from mouse sources and constant regions derived from human sources. These antibodies, called **chimeric monoclonal antibodies**, are more compatible with the human system.

- Genetic engineering can be used to alter mouse antibodies so that they have characteristics that are more human.

The mixture of cells is placed in a selective medium that allows only hybrid cells to grow.

Hybrid cells are screened for the production of the desired antibody. They are then cultured to produce large numbers of monoclonal antibodies.

The Lymphatic System & Immunity

1. Identify the mouse cells used to produce the monoclonal antibodies: _____

2. Describe the characteristic of tumor cells that allows an ongoing culture of antibody-producing lymphocytes to be made:

   _____

   _____

3. Compare the method of producing monoclonal antibodies using mice with the alternative methods now available:

   _____

   _____

   _____

   _____

Herceptin is the patented name of a **monoclonal antibody** for the targeted treatment of breast cancer. This drug (chemical name Trastuzumab) recognizes and is specific to the receptor proteins on the outside of cancer cells that are produced by the **HER2 gene**. The HER2 (**H**uman **E**pidermal growth factor **R**eceptor **2**) gene codes for proteins on the cell surface that signal to the cell when it should divide. Cancerous cells contain 20-30% more of the HER2 gene than non-cancerous cells and this causes **over-expression** of the HER2 gene, producing large amounts of HER2 protein.

The overexpression causes the cell to divide more often than normal, producing a tumor. Cancerous cells are designated **HER2+** indicating receptor protein over-expression. The immune system fails to destroy these cells because they are not recognized as being abnormal. Herceptin's role is to recognize and bind to the HER2 protein on the surface of the cancerous cell. The immune system can then identify the antibodies as foreign and destroy the cell. The antibody also has the effect of blocking the cell's signaling pathway and thus stops the cell from dividing.

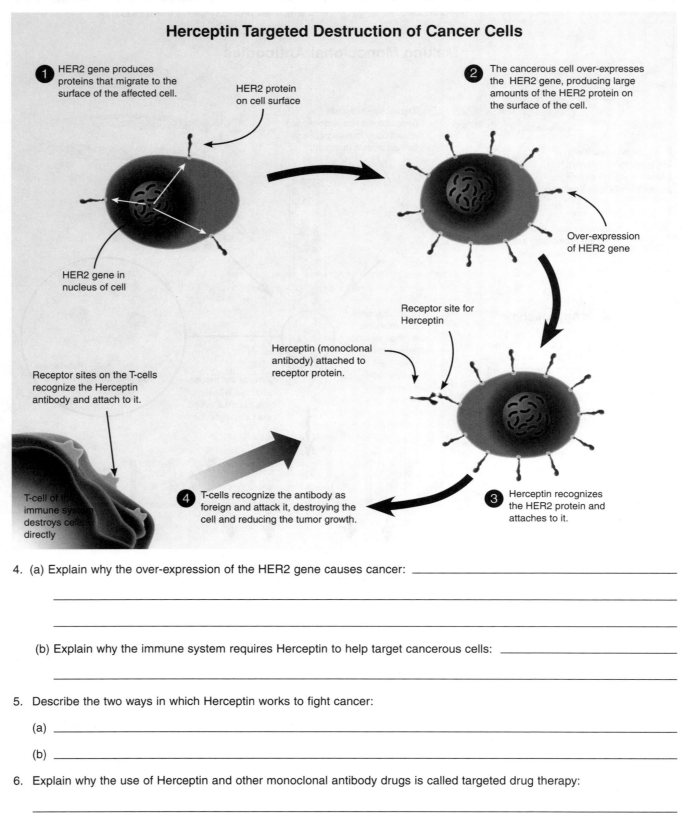

Herceptin Targeted Destruction of Cancer Cells

1 HER2 gene produces proteins that migrate to the surface of the affected cell.

HER2 protein on cell surface

HER2 gene in nucleus of cell

2 The cancerous cell over-expresses the HER2 gene, producing large amounts of the HER2 protein on the surface of the cell.

Over-expression of HER2 gene

Receptor site for Herceptin

Herceptin (monoclonal antibody) attached to receptor protein.

Receptor sites on the T-cells recognize the Herceptin antibody and attach to it.

T-cell of the immune system destroys cells directly

4 T-cells recognize the antibody as foreign and attack it, destroying the cell and reducing the tumor growth.

3 Herceptin recognizes the HER2 protein and attaches to it.

4. (a) Explain why the over-expression of the HER2 gene causes cancer: _____

_____

_____

(b) Explain why the immune system requires Herceptin to help target cancerous cells: _____

_____

5. Describe the two ways in which Herceptin works to fight cancer:

(a) _____

(b) _____

6. Explain why the use of Herceptin and other monoclonal antibody drugs is called targeted drug therapy:

_____

_____

Gene Therapy

Gene therapy refers to the application of gene technology to treat disease by correcting or replacing faulty genes. It was first envisioned as a treatment, or even a cure, for genetic disorders, but it could also be used to treat a wide range of diseases, including those that resist conventional treatments. Although varying in detail, all gene therapies are based around the same technique. Normal (non-faulty) DNA containing the correct gene is inserted into a vector, which is able to transfer the DNA into the patient's cells in a process called **transfection**. The vector may be a virus, liposome, or any one of a variety of other molecular transporters. The vector is introduced into a sample of the patient's cells and

these are cultured to amplify the correct gene. The cultured cells are then transferred back to the patient. The use of altered stem cells instead of mature somatic cells has so far achieved longer lasting results in many patients. The treatment of somatic cells or stem cells may be **therapeutic** but the changes are not inherited. Germline therapy would enable genetic changes to be passed on. To date there have been limited successes with gene therapy because transfection of targeted cells is inefficient and side effects can be severe or even fatal. However there have been a small number of successes, including the treatment of one form of SCID, a genetic disease affecting the immune system.

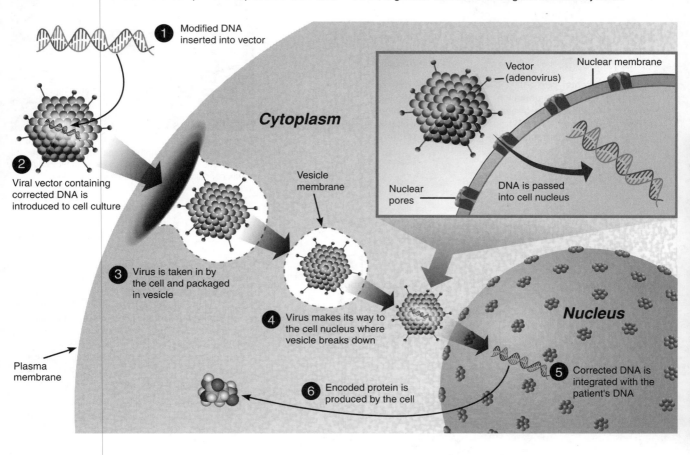

1. Modified DNA inserted into vector

2. Viral vector containing corrected DNA is introduced to cell culture

Cytoplasm

Vector (adenovirus)

Nuclear membrane

Nuclear pores

DNA is passed into cell nucleus

Vesicle membrane

3. Virus is taken in by the cell and packaged in vesicle

4. Virus makes its way to the cell nucleus where vesicle breaks down

Plasma membrane

6. Encoded protein is produced by the cell

Nucleus

5. Corrected DNA is integrated with the patient's DNA

The Lymphatic System & Immunity

1. (a) Describe the general principle of gene therapy: _____

_____

 (b) Describe the medical areas where gene therapy might be used: _____

_____

2. Explain the significance of transfecting **germline cells** rather than **somatic** (body) cells: _____

_____

3. Describe the purpose of **amplifying the gene** in gene therapy: _____

_____

4. Explain why CF and SCID are both considered to be good candidates for treatment using gene therapy:

_____

_____

_____

_____

Related activities: Stem Cell Technology
Web links: Gene Therapy, Gene Therapy Primer

RA 2

Treating SCID using Gene Therapy

The most common form of **SCID** (Severe Combined Immune Deficiency) is **X-linked SCID**, which results from mutations to a gene on the X chromosome encoding the **common gamma chain**, a protein forming part of a receptor complex for numerous types of leukocytes. A less common form of the disease, (**ADA-SCID**) is caused by a defective gene that codes for the enzyme adenosine deaminase (ADA).

Both of these types of SCID lead to immune system failure. A common treatment for SCID is bone marrow transplant, but this is not always successful and runs the risks of infection from unscreened viruses. **Gene therapy** appears to hold the best chances of producing a cure for SCID because the mutation affects only one gene whose location is known. DNA containing the corrected gene is placed into a **gutted retrovirus** and introduced to a sample of the patient's **bone marrow**. The treated cells are then returned to the patient.

In some patients with ADA-SCID, treatment was so successful that supplementation with purified ADA was no longer required. The treatment carries risks though. In early trials, two of ten treated patients developed leukemia when the corrected gene was inserted next to a gene regulating cell growth.

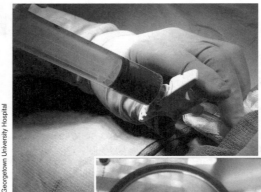

Samples of bone marrow being extracted prior to treatment with gene therapy.

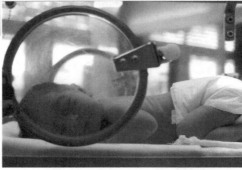

Detection of SCID is difficult for the first months of an infant's life due to the mother's antibodies being present in the blood. Suspected SCID patients must be kept in sterile conditions at all times to avoid infection.

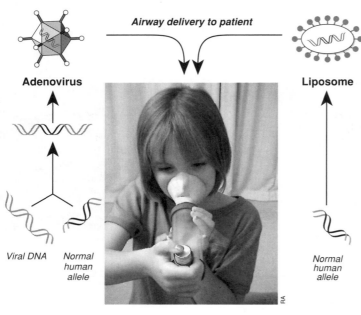

Airway delivery to patient

Adenovirus

Liposome

Viral DNA *Normal human allele*

Normal human allele

An **adenovirus** that normally causes colds is genetically modified to make it safe and to carry the normal (unmutated) CFTR ('cystic fibrosis') gene.

Liposomes are tiny fat globules. Normal CF genes are enclosed in liposomes, which fuse with plasma membranes and deliver the genes into the cells.

Gene Therapy - Potential Treatment for Cystic Fibrosis?

Cystic fibrosis (CF) is caused by a mutation to the gene coding for a chloride ion channel important in creating sweat, digestive juices, and mucus. The dysfunction results in abnormally thick, sticky mucus that accumulates in the lungs and intestines. The identification and isolation of the CF gene in 1989 meant that scientists could look for ways in which to correct the genetic defect rather than just treating the symptoms using traditional therapies.

The main target of CF gene therapy is the lung, because the progressive lung damage associated with the disease is eventually lethal.

In trials, normal genes were isolated and inserted into patients using vectors such as **adenoviruses** and **liposomes**, delivered via the airways (left). The results of trials were disappointing: on average, there was only a 25% correction, the effects were short lived, and the benefits were quickly reversed. Alarmingly, the adenovirus used in one of the trials led to the death of one patient.

Source: Cystic Fibrosis Trust, UK.

5. (a) Describe the difference between X-linked SCID and ADA-SCID: _____

_____

(b) In preparation for receiving the treated cells, the recipient's marrow is often depleted first. Explain why: _____

_____

6. Discuss the potential risks associated with gene therapy: _____

_____

_____

_____

Autoimmune Diseases

Any of numerous disorders, including **rheumatoid arthritis**, insulin dependent (type 1) **diabetes mellitus**, and **multiple sclerosis**, are caused by an individual's immune system reaction to their own cells or tissues. The immune system normally distinguishes self from non-self. Some lymphocytes are capable of reacting against self, but these are generally suppressed. **Autoimmune diseases** occur when there is some interruption of the normal control process, allowing lymphocytes to escape from suppression, or when there is an alteration in some body tissue so that it is no longer recognized as self. The exact mechanisms behind autoimmune malfunctions are not fully understood but pathogens or drugs may play a role in triggering an autoimmune response in someone who already has a genetic predisposition. The reactions are similar to those that occur in allergies, except that in autoimmune disorders, the hypersensitivity response is to the body itself, rather than to an outside substance.

Multiple Sclerosis

MS is a progressive inflammatory disease of the central nervous system in which scattered patches of **myelin** (white matter) in the brain and spinal cord are destroyed. Myelin is the fatty connective tissue sheath surrounding conducting axons and its destruction results in the symptoms of MS: numbness, tingling, muscle weakness and **paralysis**.

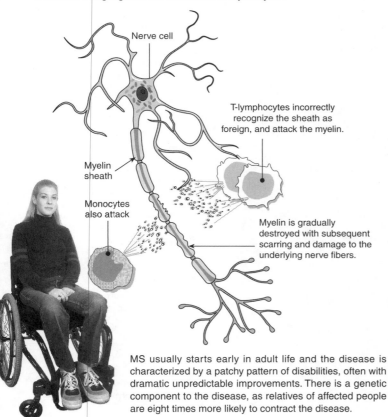

Nerve cell

T-lymphocytes incorrectly recognize the sheath as foreign, and attack the myelin.

Myelin sheath

Monocytes also attack

Myelin is gradually destroyed with subsequent scarring and damage to the underlying nerve fibers.

MS usually starts early in adult life and the disease is characterized by a patchy pattern of disabilities, often with dramatic unpredictable improvements. There is a genetic component to the disease, as relatives of affected people are eight times more likely to contract the disease.

Other Immune System Disorders

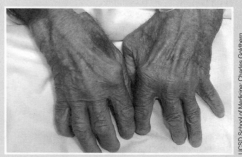

Rheumatoid arthritis is a type of joint inflammation, usually in the hands and feet, which results in destruction of cartilage and painful, swollen joints. The disease often begins in adulthood, but can also occur in children or the elderly. Rheumatoid arthritis affects more women than men and is treated with anti-inflammatory and immunosuppressant drugs, and physiotherapy.

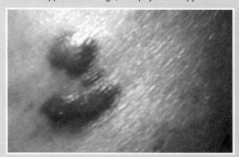

Lacking a sufficient immune response is called **immune deficiency**, and may be either **congenital** (present at birth) or **acquired** as a result of drugs, cancer, or infectious agents (e.g. HIV infection). HIV causes AIDS, which results in a steady destruction of the immune system. Sufferers then succumb to opportunistic infections and rare cancers such as Kaposi s sarcoma (above).

1. Explain the basis of the following autoimmune diseases:

   (a) Multiple sclerosis: _____

   _____

   (b) Rheumatoid arthritis: _____

   _____

2. Suggest why autoimmune diseases are difficult to treat effectively: _____

   _____

   _____

3. Explain why sufferers of immune deficiencies, such as AIDS, develop a range of debilitating infections:

   _____

   _____

The Lymphatic System & Immunity

The Respiratory System

Cardiovascular system

- Blood transports respiratory gases.
- The carbonate-bicarbonate system in blood contributes to blood buffering.
- Blood proteins (e.g. immunoglobulins) also contribute to blood buffering.

Urinary system

- Kidneys dispose of the waste products of respiratory metabolism (other than CO_2 which is breathed out).

Reproductive system

- Pregnancy has significant effects on breathing mediated largely through sex hormones. The enlarged uterus (its position at full term shown by dotted line) pushes abdominal organs up and outwards and compromises the functioning of the diaphragm.

Skeletal system

- Bones enclose and protect the lungs and bronchi from damage.
- Expansion and elastic recoil of the ribcage produce the volume changes necessary for inhalation/exhalation (breathing).

Integumentary system

- Skin forms a surface barrier protecting the organs of the respiratory system.

Nervous system

- Control centers in the medulla and the pons regulate the rate and depth of breathing.
- Sensory feedback to the respiratory control centers is provided by stretch receptors in the bronchioles and chemoreceptors in the aorta and carotid arteries.

Lymphatic system and immunity

- Immune system provides general protection against pathogens; specifically the tonsils protect against pharyngeal and upper respiratory tract infections.
- Lymphatic system helps to maintain blood volume required for efficient transport of respiratory gases.

Endocrine system

- Epinephrine (from the adrenal medulla atop the kidneys) acts as a bronchodilator.
- Testosterone promotes enlargement of the larynx at puberty in males.
- Cortisol has role in lung maturation and a number of hormones directly or indirectly influence breathing rates.

Digestive system

- Digestive system provides the nutrients required by the respiratory system.

Muscular system

- Diaphragm and intercostal muscles (with the ribcage) produce the volume changes necessary for breathing.
- Regular aerobic exercise improves the efficiency of the respiratory system.

General Functions and Effects on all Systems

The respiratory system provides an interface for gas exchange with the external environment. It is ultimately responsible for providing all the cells of the body with oxygen and disposing of waste carbon dioxide produced as a result of cellular respiration. These respiratory gases are transported in the blood.

Disease

Symptoms of disease	• Chest pain • Excessive mucus production • Coughing, sneezing • Difficulty breathing, cyanosis
Infectious respiratory diseases	• Bacterial pneumonia • Pulmonary tuberculosis • Influenza
Non-infectious respiratory diseases	• Asthma • Fibrosis (scarring) • Smoking-related diseases • Inherited diseases (e.g. CF)

Medicine & Technology

Diagnosis of disorders	• Chest X-ray and CT scans • Pulmonary function tests • Sputum cultures and biopsy • DNA tests and screening
Preventing and treating diseases of the respiratory system	• Drug therapies (e.g. antibiotics) • Vaccination • Surgery (e.g. transplants) • Radiotherapy • Gene or cell therapy (e.g. CF) • Behavior modification

CDC

• Asthma
• Lung cancer
• Chronic bronchitis
• Emphysema
• Asbestosis

• Stem cell therapy
• Gene therapy
• X rays
• Vaccination

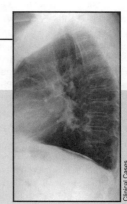

Clinical Cases

Taking a Breath
The Respiratory System

The respiratory system can be affected by disease and undergoes changes associated with training and aging.

Medical technologies can be used to diagnose and treat respiratory disorders. Exercise and lifestyle management can prevent some respiratory diseases.

istock

• Lung cancer
• Chronic bronchitis
• Emphysema

• VO$_2$max
• Ventilation efficiency
• Ventilation rhythm

Wintec Academy of Sport

Exercise can delay or reverse some age-related changes to respiratory function

Aging and the respiratory system	• Decline in respiratory capacity • Decline in aerobic capacity (VO$_2$max) • Increased incidence of chronic respiratory disease

Effects of exercise on the respiratory system	• Increased rate and depth of breathing • Increased aerobic capacity (VO$_2$max) • Increased respiratory efficiency • Improved oxygen loading -unloading • Better diaphragmatic performance

The Effects of Aging

Exercise

The Respiratory System

Understanding the structure and function of the respiratory system

Structure and organization of the respiratory system, the mechanics and control of breathing, gas transport, gas exchange at altitude and during exercise, respiratory disease

Learning Objectives

☐ 1. Compile your own glossary from the **KEY WORDS** displayed in **bold type** in the learning objectives below.

The Basics of Gas Exchange *(pages 156-157)*

☐ 2. Distinguish between **cellular respiration** and **gas exchange** and explain why it is necessary to exchange materials with the environment.

☐ 3. Name the gases involved in respiration and explain how these **respiratory gases** are exchanged across gas exchange surfaces. Describe the essential features of an efficient gas exchange surface. With reference to **Fick's law**, explain the significance of these features.

Gas Exchange in Humans *(pages 160-164)*

☐ 4. Describe the basic structure, location, adaptations, and function of the gas exchange surfaces and related structures, including: **trachea**, **bronchi**, **bronchioles**, **lungs**, **alveoli** (and **alveolar epithelium**). Explain how these features contribute to efficient gas exchange.

☐ 5. In more detail than #4 above, describe the gross structure of the lungs and their relationship to the thoracic wall. Include reference to the **visceral pleura**, **parietal pleura**, **pleural fluid**, and **pleural space**.

☐ 6. Describe the distribution of the following tissues and cells in the **trachea**, **bronchi**, and **bronchioles**: **cartilage**, **ciliated epithelium**, **goblet cells**, and **smooth muscle cells**. Describe the function of the **cartilage**, **cilia**, **goblet cells**, **smooth muscle**, and **elastic fibers** in the gas exchange system.

☐ 7. Describe the features of the alveoli that are adaptations to their functional role in gas exchange, including reference to the total surface area provided, the thin layer of squamous epithlelial cells, moist lining, and capillary network.

☐ 8. Recognize the relationship between gas exchange surfaces (the alveoli) and the blood vessels in the lung tissue. Draw a simple diagram of an **alveolus** (air sac) to illustrate the movement of O_2 and CO_2, into and out of the blood in the surrounding capillary.

☐ 9. In more detail than #7 above, describe the structure of the **respiratory membrane**, identifying its components and commenting on the extent of the air-blood barrier it represents and the role of the **alveolar macrophages**.

☐ 10. Explain the mechanism of ventilation (**breathing**), including reference to the following:

    (a) The role of the **diaphragm**, internal and external **intercostal muscles**, and the abdominal muscles in changing the air pressure in the lungs.

    (b) The role of negative **intrapleural pressure** as the major factor preventing lung collapse.

    (c) The distinction between **inspiration** (inhalation) as an active process and **expiration** (exhalation) as a passive process (during normal, quiet breathing).

    (d) The difference between quiet and forced breathing.

    (e) The role of **surfactant** in lung function.

    (f) The composition of inhaled and exhaled air.

☐ 11. Explain how ventilation is measured in humans using a **spirometer**. Explain how the **breathing rate** and **pulmonary ventilation rate** (PV) are calculated and expressed. Explain what is meant by the **dead space air**, and comment on its significance to the exchange of gases. Explain the terms: **tidal volume**, **vital capacity**, and **residual volume**, and provide some typical values for these. Describe how each of these is affected by strenuous exercise.

Gas Transport in Humans *(pages 165-166)*

☐ 12. Describe the general role of respiratory pigments in the transport and delivery of oxygen to the tissues. Include reference to **myoglobin**, **fetal hemoglobin,** and **adult hemoglobin**, and comment on any differences in the oxygen carrying capacity of these.

☐ 13. Define the term **partial pressure** and understand its significance with respect to gas transport.

☐ 14. Describe the transport of oxygen in relation to the **oxygen-hemoglobin dissociation curve**. Explain the oxygen dissociation curves of adult and fetal **hemoglobin** and **myoglobin** and identify the significance of the differences described.

☐ 15. Describe the effect of pH (CO_2 level) on the oxygen-hemoglobin dissociation curve (the **Bohr effect**) and explain its significance.

☐ 16. Explain how CO_2 is carried in the blood, including the action and role of the following: **carbonic anhydrase**, **chloride shift**, plasma proteins as blood buffers, and hydrogen-carbonate (bicarbonate) ions.

Control of Breathing *(pages 167-168)*

☐ 17. Explain how basic rhythm of breathing is controlled through the activity of the **respiratory center** in the medulla and its output via the **phrenic nerves** and the **intercostal nerves**.

☐ 18. Explain the role of the **stretch receptors** and the **vagus nerve** in ending inspiration during normal breathing. Identify this control as the **inflation reflex**.

☐ 19. Identify influences on the respiratory center and comment on how these reflect changes in the body's demand for oxygen. Include reference to the activity of the **carotid** and **aortic bodies** (chemoreceptors in the carotid arteries and the aorta).

☐ 20. Distinguish between **involuntary** and **voluntary control** of breathing.

□ 21. Explain how and why **breathing rate** varies to meet changing metabolic demands. Your explanation should include reference to the changes in blood composition that occur with exercise and the physiological effects of these changes. Understanding the mechanisms by which ventilation rate is adjusted to meet the demands of exercise requires a basic understanding of the mechanisms controlling breathing (see #11-14).

Gas Exchange at High Altitude *(page 169)*

□ 22. Explain the problem of gas exchange at **high altitude**. Describe the short and long term effects of high altitude on human physiology, and explain the way the body **acclimatizes**. Include reference to blood composition (red blood cell number, blood viscosity), density of capillaries, and breathing and heart rates. Comment on the significance of these changes to the maintenance of adequate rates of exchange at altitude.

Gas Exchange & Training *(pages 125, 164, 167)*

□ 23. Explain the immediate and long term adjustments made to cope with the respiratory demands of strenuous exercise and aerobic training, including:
(a) Immediate: increase in rate and depth of breathing
(b) Long term: improved respiratory efficiency, improvements in aerobic capacity as measured by VO_2 max (maximum uptake of oxygen (in mL) per kilogram of bodyweight per minute ($mL^{-1}kg^{-1}min^{-1}$).

Respiratory Disease *(pages 170-173)*

□ 24. Explain how both pathogens (e.g. *Mycobacterium tuberculosis*) and environmental factors (e.g. tobacco smoking and pollution) contribute to lung diseases with specific, recognizable symptoms.

□ 25. Distinguish between **obstructive pulmonary diseases** (passage blockage) and **restrictive pulmonary diseases** (scarring). Describe the characteristics of some obstructive pulmonary diseases including **chronic bronchitis**, **emphysema**, and **asthma**. Explain the effect of each on lung function.

□ 26. Outline the link between **tobacco smoking** and respiratory disease, including the incidence of lung cancer and chronic obstructive pulmonary diseases such as chronic bronchitis and emphysema. Identify the components of tobacco smoke including those associated with addiction and those associated with lung damage.

□ 27. Describe the effects of lung cancer on the lung tissue, identifying detrimental changes to the alveolar epithelium, lung capacity, blood vessel walls, and composition of gases in the lung.

□ 28. Analyze and/or interpret data related to:
• the effect of pollution and smoking on the incidence of lung disease.
• specific risk factors associated with the incidence of lung disease.

Supplementary Texts

See page 9 for additional details of these texts:
■ Morton, D. and Perry, J.W, 1997. **Photo Atlas for Anatomy and Physiology**, (Brooks/Cole).
■ Rowett, H.G.Q., 1999. **Basic Anatomy & Physiology**, (John Murray), pp. 95-98.

Periodicals

See page 9 for details of publishers of periodicals:

STUDENT'S REFERENCE

■ **Lungs and the Control of Breathing** Bio. Sci. Rev. 14(4) April 2002, pp. 2-5. *The mechanisms, control, and measurement of breathing in humans. This article includes good, clear diagrams and useful summaries of important points.*

■ **Gas Exchange in the Lungs** Bio. Sci. Rev. 16(1) Sept. 2003, pp. 36-38. *The structure and function of the alveoli of the lungs, with an account of respiratory problems and diseases such as respiratory distress syndrome and emphysema.*

■ **Getting in and Out** Biol. Sci. Rev., 20(3), Feb. 2008, pp. 14-16. *An excellent account of diffusion and factors governing diffusion rates. Some of the adaptations for diffusion across gas exchange surfaces are also discussed.*

■ **Red Blood Cells** Bio. Sci. Rev. 11(2) Nov. 1998, pp. 2-4. *The structure and function of red blood cells, including details of oxygen transport.*

■ **Fetal Hemoglobin** Biol. Sci. Rev., 16(1) Sept. 2003, pp. 15-17. *The complex structure of hemoglobin molecules: the molecule in red blood cells that delivers oxygen to the tissues.*

■ **Humans with Altitude** New Scientist, 2 Nov. 2002, pp. 36-39. *The short term adjustments and (evolutionary) adaptations of those living at altitude.*

■ **The White Plague** New Scientist (Inside Science), 9 Nov. 2002. *The causes and nature of TB, its global incidence, and a discussion of the implications of drug resistance to TB treatment.*

■ **Tuberculosis** Biol. Sci. Rev., 14(1) Sept. 2001, pp. 30-33. *Despite vaccination, TB has become more common recently. Why has it returned?*

■ **Asbestos and the Lung** Biol. Sci. Rev., 20(2) Nov. 2007, pp. 2-5. *The health effects of asbestos, specifically the damage it causes to lungs.*

■ **Smoking** Biol. Sci. Rev., 10(1) Sept. 1997, pp. 14-16. *Smoking related diseases and the effects of tobacco smoking on human physiology.*

■ **Dust to Dust** New Scientist, 21 Sept. 2002 (Inside Science). *A supplement that concentrates on dust pollution, but also examines the impact of this pollutant on respiratory health.*

■ **Environmental Lung Disease** New Scientist, 23 September 1995 (Inside Science). *An excellent supplement on lung disorders, with good diagrams illustrating lung functioning and gas transport.*

TEACHER'S REFERENCE

■ **The Effect of Hyperventilation on the Ability to Hold one's Breath** The Am. Biology Teacher, 59(4), April, 1997, pp. 229-231. *Investigating the effects of hyperventilation and blood CO_2 and O_2 level on respiratory physiology.*

■ **An Improved Model for Demonstrating the Mechanism of Breathing** The Am. Biology Teacher, 60(7), Sept. 1998, pp. 528-530. *A how-to-do-it example with instructions on how to make a jar and balloon model to demonstrate lung function.*

■ **Everest - Altitude and the Death Zone** National Geographic, 203(5), May 2003, pp. 30-33. *An account of the changes that occur to the human body at extreme altitudes. At 26,000 feet on Mount Everest, climbers suffer frostbite, dehydration, and the effects of oxygen deprivation on the organs, including the brain, lungs, and heart.*

■ **Gene Therapy for Cystic Fibrosis** Bio. Sci. Rev. 11(1) Feb. 1998, pp. 37-40. *The causes and effects of cystic fibrosis, including the physiology of how CF causes changes to ion transport across membranes. Could gene therapy offer therapy or even a cure for people affected by the disease?*

Internet WWW

See pages 10-11 for details of how to access **Bio Links** from our web site: www.thebiozone.com From Bio Links, access sites under the topics:

ANATOMY & PHYSIOLOGY > General sites:
• Anatomy & Physiology • Human physiology lecture notes • WebAnatomy **> Gas Exchange:**
• Asthma • **Lesson 11:** The respiratory system • The respiratory system

HEALTH & DISEASE > Non-infectious disease:
• Chemicals & human health • Cancer Group Institute Information Center • NCI's CancerNet

Presentation MEDIA to support this topic:
Anatomy & Physiology

The Respiratory System

Breathing in Humans

In mammals, the mechanism of breathing (ventilation) provides a continual supply of fresh air to the lungs and helps to maintain a large diffusion gradient for respiratory gases across the gas exchange surface. Oxygen must be delivered regularly to supply the needs of respiring cells. Similarly, carbon dioxide, which is produced as a result of cellular metabolism, must be quickly eliminated from the body. Adequate lung ventilation is essential to these exchanges. The cardiovascular system participates by transporting respiratory gases to and from the cells of the body. The volume of gases exchanged during breathing varies according to the physiological demands placed on the body (e.g. by exercise). These changes can be measured using spirometry.

Inspiration (inhalation or breathing in)

During quiet breathing, inspiration is achieved by increasing the space (therefore decreasing the pressure) inside the lungs. Air then flows into the lungs in response to the decreased pressure inside the lung. Inspiration is always an active process involving muscle contraction.

1a External intercostal muscles contract causing the ribcage to expand and move up.

1b Diaphragm contracts and moves down.

2 Thoracic volume increases, lungs expand, and the pressure inside the lungs decreases.

3 Air flows into the lungs in response to the pressure gradient.

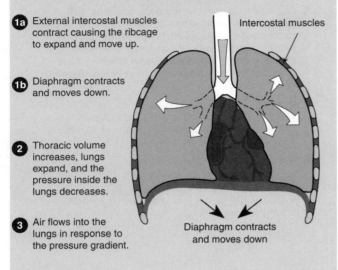

Intercostal muscles

Diaphragm contracts and moves down

Expiration (exhalation or breathing out)

During quiet breathing, expiration is achieved passively by decreasing the space (thus increasing the pressure) inside the lungs. Air then flows passively out of the lungs to equalise with the air pressure. In active breathing, muscle contraction is involved in bringing about both inspiration and expiration.

1 In **quiet breathing**, external intercostal muscles and diaphragm relax. Elasticity of the lung tissue causes recoil.

In **forced breathing**, the internal intercostals and abdominal muscles also contract to increase the force of the expiration.

2 Thoracic volume decreases and the pressure inside the lungs increases.

3 Air flows passively out of the lungs in response to the pressure gradient.

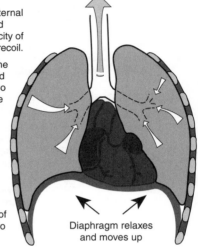

Diaphragm relaxes and moves up

1. Explain the purpose of breathing: _____

_____

2. (a) Describe the sequence of events involved in quiet breathing: _____

_____

_____

(b) Explain the essential difference between this and the situation during heavy exercise or forced breathing:

_____

_____

3. Identify what other gas is lost from the body in addition to carbon dioxide: _____

4. Explain the role of the elasticity of the lung tissue in normal, quiet breathing: _____

_____

_____

5. Breathing rate is regulated through the medullary respiratory center in response to demand for oxygen. The trigger for increased breathing rate is a drop in blood pH. Suggest why this is an appropriate trigger to increase breathing rate:

_____

_____

_____

Related activities: Measuring Lung Function, Gas Transport in Humans
Web links: Respiratory Basics Learning Activity

The Human Respiratory System

The paired lungs of mammals, including humans, are located within the thorax and are connected to the outside air by way of a system of tubular passageways: the trachea, bronchi, and bronchioles. Ciliated, mucus secreting epithelium lines this system of tubules, trapping and removing dust and pathogens before they reach the gas exchange surfaces. Each lung is divided into a number of lobes, each receiving its own bronchus.

Each bronchus divides many times, terminating in the respiratory bronchioles from which arise 2-11 alveolar ducts and numerous **alveoli** (air sacs). These provide a very large surface area (around 70 m²) for the exchange of respiratory gases by diffusion between the alveoli and the blood in the capillaries. The details of this exchange across the **respiratory membrane** are described on the next page.

Morphology of the Gas Exchange System

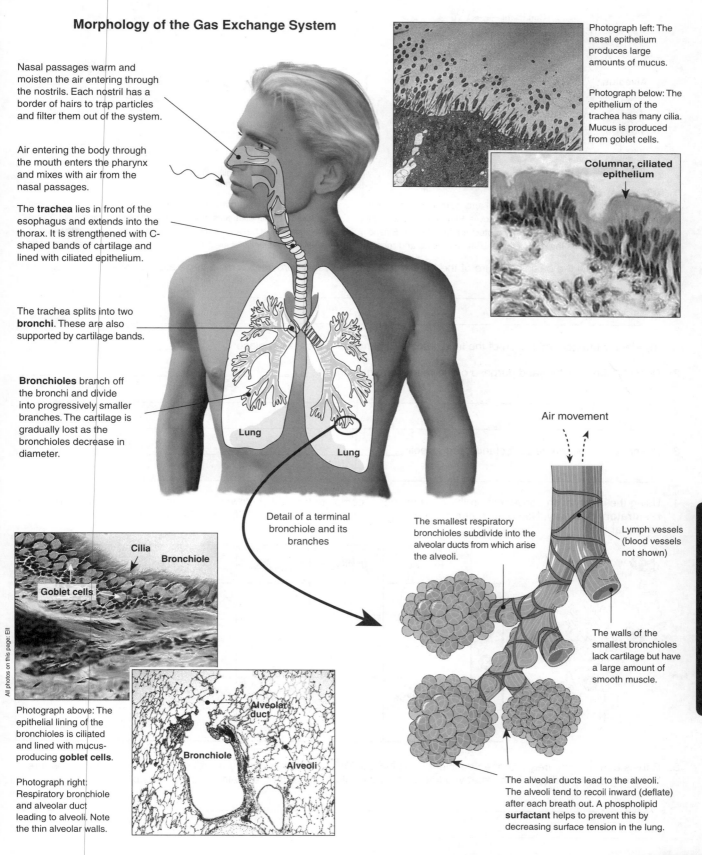

Nasal passages warm and moisten the air entering through the nostrils. Each nostril has a border of hairs to trap particles and filter them out of the system.

Air entering the body through the mouth enters the pharynx and mixes with air from the nasal passages.

The **trachea** lies in front of the esophagus and extends into the thorax. It is strengthened with C-shaped bands of cartilage and lined with ciliated epithelium.

The trachea splits into two **bronchi**. These are also supported by cartilage bands.

Bronchioles branch off the bronchi and divide into progressively smaller branches. The cartilage is gradually lost as the bronchioles decrease in diameter.

Lung

Lung

Photograph left: The nasal epithelium produces large amounts of mucus.

Photograph below: The epithelium of the trachea has many cilia. Mucus is produced from goblet cells.

Columnar, ciliated epithelium

Air movement

Detail of a terminal bronchiole and its branches

The smallest respiratory bronchioles subdivide into the alveolar ducts from which arise the alveoli.

Lymph vessels (blood vessels not shown)

The walls of the smallest bronchioles lack cartilage but have a large amount of smooth muscle.

Cilia
Bronchiole

Goblet cells

Photograph above: The epithelial lining of the bronchioles is ciliated and lined with mucus-producing **goblet cells**.

Photograph right: Respiratory bronchiole and alveolar duct leading to alveoli. Note the thin alveolar walls.

Alveolar duct

Bronchiole

Alveoli

The alveolar ducts lead to the alveoli. The alveoli tend to recoil inward (deflate) after each breath out. A phospholipid **surfactant** helps to prevent this by decreasing surface tension in the lung.

All photos on this page: EII

The Respiratory System

Related activities: Gas Transport in Humans

RA 2

An Alveolus

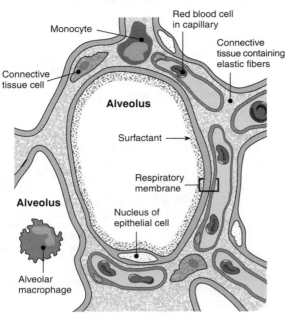

The diagram above illustrates the physical arrangement of the alveoli to the capillaries through which the blood moves. Phagocytic monocytes and macrophages are also present to protect the lung tissue. Elastic connective tissue gives the alveoli their ability to expand and recoil.

The Respiratory Membrane

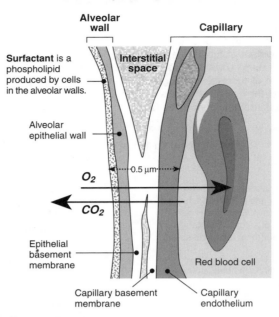

The **respiratory membrane** is the term for the layered junction between the alveolar epithelial cells, the endothelial cells of the capillary, and their associated basement membranes (thin, collagenous layers that underlie the epithelial tissues). Gases move freely across this membrane.

1. (a) Explain how the basic structure of the human respiratory system provides such a large area for gas exchange:

    _____

    _____

    (b) Identify the general region of the lung where exchange of gases takes place: _____

2. Describe the structure and purpose of the respiratory membrane: _____

    _____

    _____

3. Describe the role of the surfactant in the alveoli: _____

    _____

4. Using the information above and on the previous page, complete the table below summarizing the **histology of the respiratory pathway**. Name each numbered region and use a tick or cross to indicate the presence or absence of particular tissues.

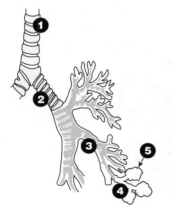

	Region	Cartilage	Ciliated epithelium	Goblet cells (mucus)	Smooth muscle	Connective tissue
1						✓
2						
3		gradually lost				
4	Alveolar duct		✗	✗		
5					very little	

5. Babies born prematurely are often deficient in surfactant. This causes respiratory distress syndrome; a condition where breathing is very difficult. From what you know about the role of surfactant, explain the symptoms of this syndrome:

    _____

    _____

Measuring Lung Function

Changes in lung volume can be measured using a technique called **spirometry**. Total adult lung capacity varies between 4 and 6 litres (L or dm³) and is greater in males. The **vital capacity**, which describes the volume exhaled after a maximum inspiration, is somewhat less than this because of the residual volume of air that remains in the lungs even after expiration. The exchange between fresh air and the residual volume is a slow process and

the composition of gases in the lungs remains relatively constant. Once measured, the tidal volume can be used to calculate the **pulmonary ventilation rate** or PV, which describes the amount of air exchanged with the environment per minute. Measures of respiratory capacity provide one way in which a reduction in lung function can be assessed (for example, as might occur as result of disease or an obstructive lung disorder such asthma).

Determining changes in lung volume using spirometry

The apparatus used to measure the amount of air exchanged during breathing and the rate of breathing is a **spirometer** (also called a respirometer). A simple spirometer consists of a weighted drum, containing oxygen or air, inverted over a chamber of water. A tube connects the air-filled chamber with the subject's mouth, and soda lime in the system absorbs the carbon dioxide breathed out. Breathing results in a trace called a spirogram, from which lung volumes can be measured directly.

During inspiration
Air is removed from the chamber, the drum sinks, and an upward deflection is recorded on the paper on the rotating drum.

During expiration
Air is added to the chamber, the drum rises, and a downward deflection is recorded.

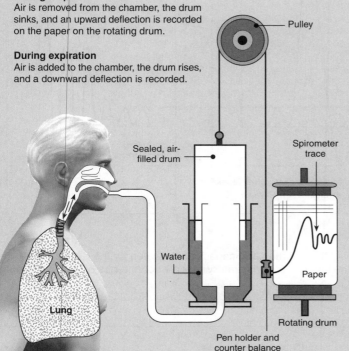

Lung Volumes and Capacities

The air in the lungs can be divided into volumes. Lung capacities are combinations of volumes.

DESCRIPTION OF VOLUME	Vol / L
Tidal volume (TV) Volume of air breathed in and out in a single breath	0.5
Inspiratory reserve volume (IRV) Volume breathed in by a maximum inspiration at the end of a normal inspiration	3.3
Expiratory reserve volume (ERV) Volume breathed out by a maximum effort at the end of a normal expiration	1.0
Residual volume (RV) Volume of air remaining in the lungs at the end of a maximum expiration	1.2

DESCRIPTION OF CAPACITY	
Inspiratory capacity (IC) = TV + IRV Volume breathed in by a maximum inspiration at the end of a normal expiration	3.8
Vital capacity (VC) = IRV + TV + ERV Volume that can be exhaled after a maximum inspiration.	4.8
Total lung capacity (TLC) = VC + RV The total volume of the lungs. Only a fraction of TLC is used in normal breathing	6.0

PRIMARY INDICATORS OF LUNG FUNCTION
Forced expiratory volume in 1 second (FEV_1)
The volume of air that is maximally exhaled in the first second of exhalation.

Forced vital capacity (FVC)
The total volume of air that can be forcibly exhaled after a maximum inspiration.

The Respiratory System

1. Describe how each of the following might be expected to influence values for lung volumes and capacities obtained using spirometry:

   (a) Height: _____

   (b) Gender: _____

   (c) Age: _____

2. A percentage decline in FEV_1 and FVC (to <80% of normal) are indicators of impaired lung function, e.g in asthma:

   (a) Explain why a forced volume is a more useful indicator of lung function than tidal volume:

   _____

   (b) Asthma is treated with drugs to relax the airways. Suggest how spirometry could be used during asthma treatment:

   _____

Related activities: The Human Respiratory System
Web links: Respiratory Basics Learning Activity

DA 2

Respiratory gas	Approximate percentages of O_2 and CO_2		
	Inhaled air	Air in lungs	Exhaled air
O_2	21.0	13.8	16.4
CO_2	0.04	5.5	3.6

Above: The percentages of respiratory gases in air (by volume) during normal breathing. The percentage volume of oxygen in the alveolar air (in the lung) is lower than that in the exhaled air because of the influence of the **dead air volume** (the air in the spaces of the nose, throat, larynx, trachea and bronchi). This air (about 30% of the air inhaled) is unavailable for gas exchange.

Left: During exercise, the breathing rate, tidal volume, and PV increase up to a maximum (as indicated below).

Spirogram for a male during quiet and forced breathing, and during exercise

Lung volume (L) — y-axis (0 to 6)

Inspiratory reserve volume (IRV) = 3.3 L

A
B
C
D E F G

Resting   Exercise

PV = breathing rate X tidal volume
$L\ min^{-1}$ = breaths min^{-1} X L

Time

3. Using the definitions given on the previous page, identify the volumes and capacities indicated by the letters **A-F** on the spirogram above. For each, indicate the volume (vol) in liters (L). The inspiratory reserve volume has been identified:

(a) A: _____ Vol: _____ (d) D: _____ Vol: _____

(b) B: _____ Vol: _____ (e) E: _____ Vol: _____

(c) C: _____ Vol: _____ (f) F: _____ Vol: _____

4. Explain what is happening in the sequence indicated by the letter **G**: _____

_____

5. Calculate PV when breathing rate is 15 breaths per minute and tidal volume is 0.4 L: _____

6. (a) Describe what would happen to PV during strenuous exercise: _____

(b) Explain how this is achieved: _____

_____

7. The table above gives approximate percentages for respiratory gases during breathing. Study the data and then:

(a) Calculate the difference in CO_2 between inhaled and exhaled air: _____

(b) Explain where this 'extra' CO_2 comes from: _____

(c) Explain why the dead air volume raises the oxygen content of exhaled air above that in the lungs: _____

_____

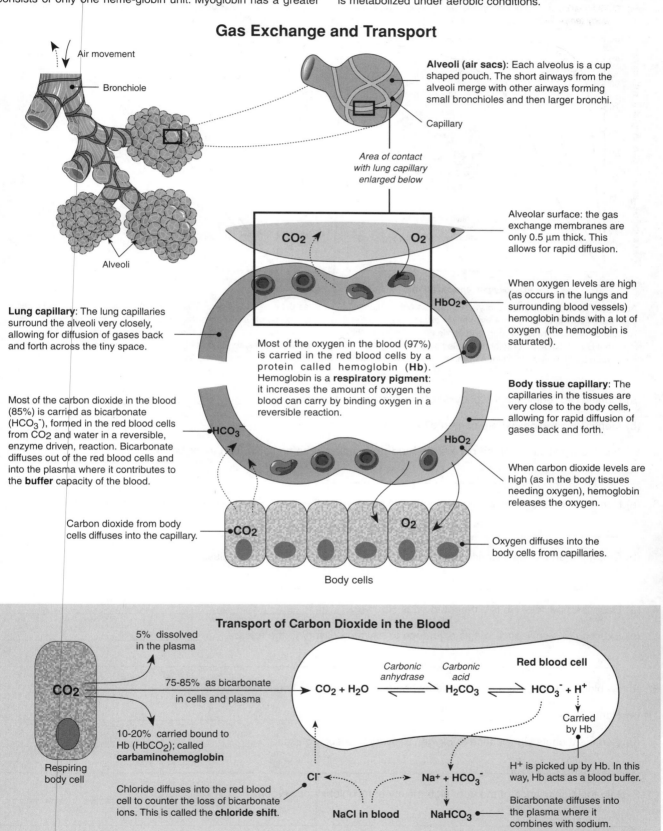

Gas Transport in Humans

The transport of respiratory gases around the body is the role of the blood and its respiratory pigments. Oxygen is transported throughout the body chemically bound to the respiratory pigment **hemoglobin** inside the red blood cells. In the muscles, oxygen from hemoglobin is transferred to and retained by **myoglobin**, a molecule that is chemically similar to hemoglobin except that it consists of only one heme-globin unit. Myoglobin has a greater

affinity for oxygen than hemoglobin and acts as an oxygen store within muscles, releasing the oxygen during periods of prolonged or extreme muscular activity. If the myoglobin store is exhausted, the muscles are forced into oxygen debt and must respire anaerobically. The waste product of this, lactic acid, accumulates in the muscle and is transported (as lactate) to the liver where it is metabolized under aerobic conditions.

Gas Exchange and Transport

Alveoli (air sacs): Each alveolus is a cup shaped pouch. The short airways from the alveoli merge with other airways forming small bronchioles and then larger bronchi.

Alveolar surface: the gas exchange membranes are only 0.5 µm thick. This allows for rapid diffusion.

When oxygen levels are high (as occurs in the lungs and surrounding blood vessels) hemoglobin binds with a lot of oxygen (the hemoglobin is saturated).

Lung capillary: The lung capillaries surround the alveoli very closely, allowing for diffusion of gases back and forth across the tiny space.

Most of the oxygen in the blood (97%) is carried in the red blood cells by a protein called hemoglobin (**Hb**). Hemoglobin is a **respiratory pigment**: it increases the amount of oxygen the blood can carry by binding oxygen in a reversible reaction.

Body tissue capillary: The capillaries in the tissues are very close to the body cells, allowing for rapid diffusion of gases back and forth.

Most of the carbon dioxide in the blood (85%) is carried as bicarbonate (HCO_3^-), formed in the red blood cells from CO_2 and water in a reversible, enzyme driven, reaction. Bicarbonate diffuses out of the red blood cells and into the plasma where it contributes to the **buffer** capacity of the blood.

When carbon dioxide levels are high (as in the body tissues needing oxygen), hemoglobin releases the oxygen.

Carbon dioxide from body cells diffuses into the capillary.

Oxygen diffuses into the body cells from capillaries.

Body cells

Transport of Carbon Dioxide in the Blood

5% dissolved in the plasma

75-85% as bicarbonate in cells and plasma

10-20% carried bound to Hb ($HbCO_2$); called **carbaminohemoglobin**

Respiring body cell

Chloride diffuses into the red blood cell to counter the loss of bicarbonate ions. This is called the **chloride shift**.

$$CO_2 + H_2O \rightleftharpoons H_2CO_3 \rightleftharpoons HCO_3^- + H^+$$

Carbonic anhydrase — Carbonic acid — Red blood cell

Carried by Hb

H^+ is picked up by Hb. In this way, Hb acts as a blood buffer.

Cl^- ← NaCl in blood → $Na^+ + HCO_3^-$ → $NaHCO_3$

Bicarbonate diffuses into the plasma where it combines with sodium.

The Respiratory System

Related activities: The Human Respiratory System, Respiratory Pigments

A 2

Oxygen does not easily dissolve in blood, but is carried in chemical combination with hemoglobin (Hb) in red blood cells. The most important factor determining how much oxygen is carried by Hb is the level of oxygen in the blood. The greater the oxygen tension, the more oxygen will combine with Hb. This relationship can be illustrated with an oxygen-hemoglobin dissociation curve as shown below (Fig. 1). In the lung capillaries, (high O_2), a lot of oxygen is picked up and bound by Hb. In the tissues, (low O_2), oxygen is released. In skeletal muscle, myoglobin picks up oxygen from hemoglobin and therefore serves as an oxygen store when oxygen tensions begin to fall. The release of oxygen is enhanced by the **Bohr effect** (Fig. 2).

Respiratory Pigments and the Transport of Oxygen

Fig. 1: Dissociation curves for hemoglobin and myoglobin at normal body temperature for fetal and adult human blood.

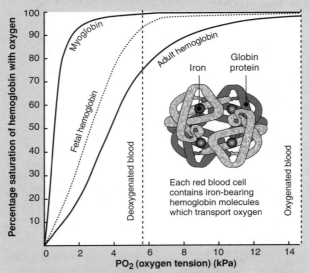

Fig. 2: Oxygen-hemoglobin dissociation curves for human blood at normal body temperature at different blood pH.

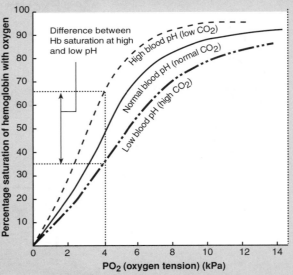

As oxygen level increases, more oxygen combines with hemoglobin (Hb). Hb saturation remains high, even at low oxygen tensions. Fetal Hb has a high affinity for oxygen and carries 20-30% more than maternal Hb. Myoglobin in skeletal muscle has a very high affinity for oxygen and will take up oxygen from hemoglobin in the blood.

As pH increases (lower CO_2), more oxygen combines with Hb. As the blood pH decreases (higher CO_2), Hb binds less oxygen and releases more to the tissues (**the Bohr effect**). The difference between Hb saturation at high and low pH represents the amount of oxygen released to the tissues.

1. (a) Identify two regions in the body where oxygen levels are very high: _____

   (b) Identify two regions where carbon dioxide levels are very high: _____

2. Explain the significance of the **reversible binding** reaction of hemoglobin (Hb) to oxygen: _____

   _____

3. (a) Hemoglobin saturation is affected by the oxygen level in the blood. Describe the nature of this relationship:

   _____

   (b) Comment on the significance of this relationship to oxygen delivery to the tissues: _____

   _____

4. (a) Describe how fetal Hb is different to adult Hb: _____

   (b) Explain the significance of this difference to oxygen delivery to the fetus: _____

   _____

5. At low blood pH, less oxygen is bound by hemoglobin and more is released to the tissues:

   (a) Name this effect: _____

   (b) Comment on its significance to oxygen delivery to respiring tissue: _____

   _____

6. Explain the significance of the very high affinity of myoglobin for oxygen: _____

   _____

7. Identify the two main contributors to the buffer capacity of the blood: _____

Control of Breathing

The basic rhythm of breathing is controlled by the **respiratory center**, a cluster of neurons located in the medulla oblongata. This rhythm is adjusted in response to the physical and chemical changes that occur when we carry out different activities. Although the control of breathing is involuntary, we can exert some degree of conscious control over it. The diagram below illustrates these controls.

The Control of Breathing

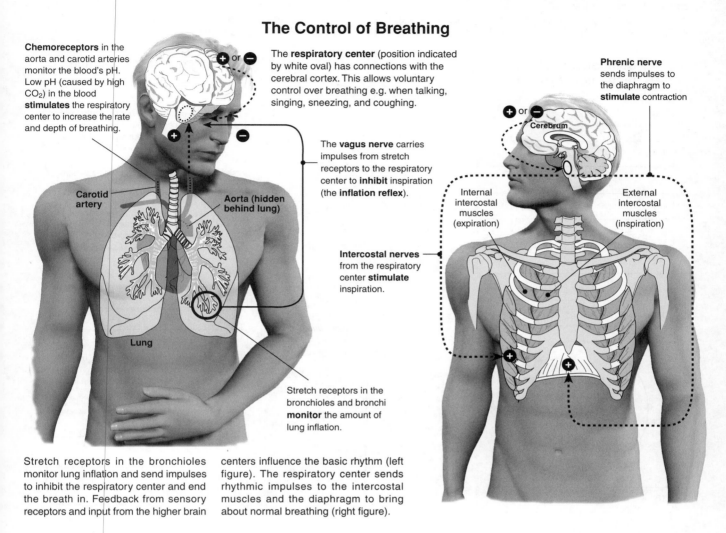

Chemoreceptors in the aorta and carotid arteries monitor the blood's pH. Low pH (caused by high CO_2) in the blood **stimulates** the respiratory center to increase the rate and depth of breathing.

The **respiratory center** (position indicated by white oval) has connections with the cerebral cortex. This allows voluntary control over breathing e.g. when talking, singing, sneezing, and coughing.

Phrenic nerve sends impulses to the diaphragm to **stimulate** contraction

Cerebrum

The **vagus nerve** carries impulses from stretch receptors to the respiratory center to **inhibit** inspiration (the **inflation reflex**).

Carotid artery

Aorta (hidden behind lung)

Internal intercostal muscles (expiration)

External intercostal muscles (inspiration)

Intercostal nerves from the respiratory center **stimulate** inspiration.

Lung

Stretch receptors in the bronchioles and bronchi **monitor** the amount of lung inflation.

Stretch receptors in the bronchioles monitor lung inflation and send impulses to inhibit the respiratory center and end the breath in. Feedback from sensory receptors and input from the higher brain centers influence the basic rhythm (left figure). The respiratory center sends rhythmic impulses to the intercostal muscles and the diaphragm to bring about normal breathing (right figure).

1. Explain how the basic rhythm of breathing is controlled: _____

   _____

2. Describe the role of each of the following in the regulation of breathing:

   (a) Phrenic nerve: _____

   (b) Intercostal nerves: _____

   (c) Vagus nerve: _____

   (d) Inflation reflex: _____

3. (a) Describe the effect of low blood pH on the rate and depth of breathing: _____

      _____

   (b) Explain how this effect is mediated: _____

      _____

   (c) Suggest why blood pH is a good mechanism by which to regulate breathing rate: _____

      _____

      _____

Related activities: Gas Transport in Humans

A 2

The Respiratory System

Review of Lung Function

The respiratory system in humans (and other air breathing vertebrates) includes the lungs and the system of tubes through which the air reaches them. Breathing (ventilation) provides a continual supply of fresh air to the lungs and helps to maintain a large diffusion gradient for respiratory gases across the gas exchange surface. The basic rhythm of breathing is controlled by the respiratory center in the medulla of the hindbrain. The volume of gases exchanged during breathing varies according to the physiological demands placed on the body. These changes can be measured using spirometry. The following activity summarizes the key features of respiratory system structure and function. The stimulus material can be found in earlier exercises in this topic.

Components of the respiratory system

(a)

(b)

(c)

(d)

(e)

(f)

(g)

The control of breathing

(i) .... controls the rate and depth of breathing. It also has connections with the cerebral cortex that allow voluntary control over breathing (e.g. when talking, singing, sneezing, and coughing).

(ii) .... carries impulses from stretch receptors to the respiratory center to **inhibit** inspiration (the **inflation reflex**).

(iii) .... from the respiratory center, **stimulate** inspiration.

(iv) .... in the aorta and carotid arteries, monitor blood pH. Low pH (caused by high CO_2) in the blood stimulates an increase in the rate and depth of breathing.

(v) .... in the bronchioles and bronchi, **monitor** the amount of lung inflation.

(vi) .... sends impulses to the diaphragm to **stimulate** contraction.

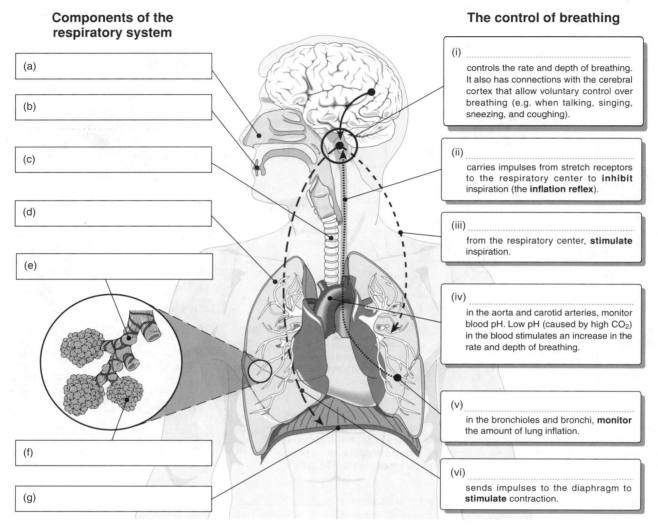

1. On the diagram above, label the components of the respiratory system (a-g) and the components that control the rate of breathing (i - vi).

2. Identify the volumes and capacities indicated by the letters A - E on the diagram of a spirogram below.

A = ...

B = ...

C = ...

D = ...

E = ...

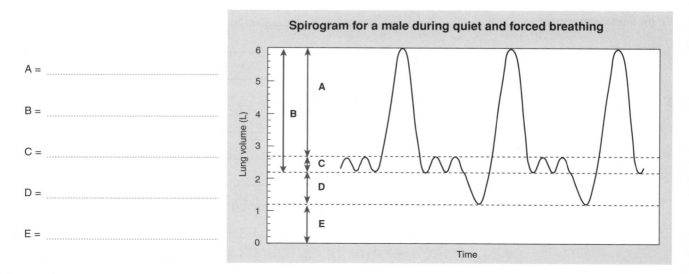

Spirogram for a male during quiet and forced breathing

Related activities: Breathing in Humans, The Human Respiratory System, Measuring Lung Function, Control of Breathing

Respiratory Diseases

Respiratory diseases are diseases of the respiratory system, including diseases of the lung, bronchial tubes, trachea, and upper respiratory tract. Respiratory diseases include mild and self-limiting diseases such as the common cold, to life-threatening infections such as tuberculosis. One in six people in the US is affected by some form of chronic lung disease, the most common being asthma and chronic obstructive pulmonary disease (including emphysema and chronic bronchitis). Non-infectious respiratory diseases are categorized according to whether they prevent air reaching the alveoli (**obstructive**) or whether they affect the gas exchange tissue itself (**restrictive**).

Such diseases have different causes and different symptoms (below) but all are characterized by difficulty in breathing and the end result is similar in that gas exchange rates are too low to meet metabolic requirements. Non-infectious respiratory diseases are strongly correlated with certain behaviors and are made worse by exposure to air pollutants. Obstructive diseases, such as emphysema, are associated with an inflammatory response of the lung to noxious particles or gases, most commonly tobacco smoke. In contrast, scarring (**fibrosis**) of the lung tissue underlies restrictive lung diseases such as **asbestosis** and **silicosis**. Such diseases are often called occupational lung diseases.

Chronic bronchitis
Excess mucus blocks airway, leading to inflammation and infection

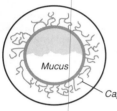

Mucus

Capillary

Asthma
Thickening of bronchiole wall and muscle hypertrophy. Bronchioles narrow.

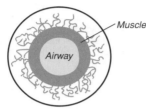

Muscle

Airway

Emphysema
Destruction of capillaries and structures supporting the small airways and lung tissue

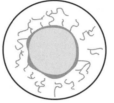

Cross sections through a bronchiole with various types of obstructive lung disease

A peak flow meter is a small, hand-held device used to monitor a person's ability to breathe out air. It measures the airflow through the bronchi and thus the degree of obstruction in the airways.

Obstructive lung disease
– passage blockage –

In obstructive lung diseases, a blockage prevents the air getting to the gas exchange surface.

The flow of air may be obstructed because of constriction of the airways (as in **asthma**), excess mucus secretion (as in **chronic bronchitis**), or because of reduced lung elasticity, which causes alveoli and small airways to collapse (as in **emphysema**). Shortness of breath is a symptom in all cases and chronic bronchitis is also associated with a persistent cough.

Chronic bronchitis and emphysema often occur together and are commonly associated with cigarette smoking, but can also occur with chronic exposure to air pollution.

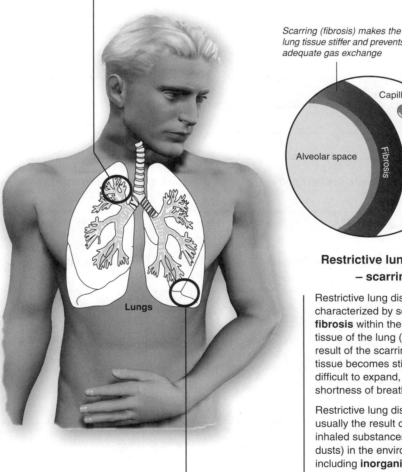

Lungs

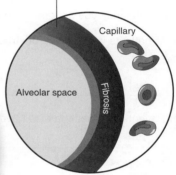

Scarring (fibrosis) makes the lung tissue stiffer and prevents adequate gas exchange

Capillary

Alveolar space

Fibrosis

Restrictive lung disease
– scarring –

Restrictive lung diseases are characterized by scarring or **fibrosis** within the gas exchange tissue of the lung (above). As a result of the scarring, the lung tissue becomes stiffer and more difficult to expand, leading to shortness of breath.

Restrictive lung diseases are usually the result of exposure to inhaled substances (especially dusts) in the environment, including **inorganic dusts** such as silica, asbestos, or coal dust, and **organic dusts**, such as those from bird droppings or moldy hay. Like most respiratory diseases, the symptoms are exacerbated by poor air quality (such as occurs in smoggy cities).

SEM of asbestos fibers. Asbestos has different toxicity depending on the type. Some types are very friable, releasing fibers into the air, where they can be easily inhaled.

USGS

The Respiratory System

Related activities: Measuring Lung Function, Allergies and Hypersensitivity, Diseases Caused by Smoking **Web links**: What is Asthma?

RA 2

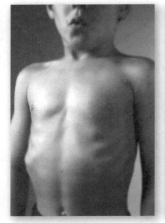

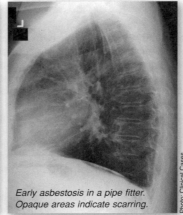

Early asbestosis in a pipe fitter. Opaque areas indicate scarring.

Photo: Clinical Cases

Asthma is a common disease affecting millions of people worldwide (20 million in the US alone). Asthma is the result of a hypersensitive reaction to allergens such as house dust or pollen, but attacks can be triggered by environmental factors such as cold air, exercise, or air pollutants. During an attack, sufferers show labored breathing with overexpansion of the chest cavity (above left). Asthma is treated with drugs that help to expand the airways (bronchodilators). These are usually delivered via a nebulizer or inhaler (above).

Asbestosis is a restrictive lung disease caused by breathing in asbestos fibers. The tiny fibers make their way into the alveoli where they cause damage and lead to scarring. Other occupational lung diseases include silicosis (exposure to silica dust) and coal workers' pneumoconiosis.

Chronic bronchitis is accompanied by a persistent, productive cough, where sufferers attempt to cough up the sputum or mucus which accumulates in the airways. Chronic bronchitis is indicated using **spirometry** by a reduced FEV_1/FVC ratio that is not reversed with bronchodilator therapy.

1. Distinguish between obstructive and restrictive lung diseases, and provide some examples:

   _____

   _____

   _____

   _____

2. Physicians may use spirometry to diagnosis certain types of respiratory disease. Explain the following typical results:

   (a) In patients with chronic obstructive pulmonary disease, the FEV_1 / FVC ratio declines (to <70% of normal):

   _____

   _____

   (b) Patients with asthma also have a FEV_1 / FVC ratio of <70%, but this improves following use of bronchodilators:

   _____

   _____

   (c) In patients with restrictive lung disease, both FEV_1 and FVC are low but the FEV_1 / FVC ratio is normal to high:

   _____

   _____

3. Describe the mechanisms by which restrictive lung diseases reduce lung function and describe an example:

   _____

   _____

   _____

4. Suggest why many restrictive lung diseases are also classified as occupational lung diseases: _____

   _____

   _____

5. Describe the role of histamine in the occurrence of an asthma attack: _____

   _____

   _____

Smoking and the Lungs

Tobacco smoking has been accepted as a major health hazard only relatively recently in historical terms, despite its practice in Western countries for more than 400 years, and much longer elsewhere. Cigarettes became popular at the end of World War I because they were cheap, convenient, and easier to smoke than pipes and cigars. They remain popular for the further reason that they are more addictive than other forms of tobacco. The milder smoke can be more readily inhaled, allowing **nicotine** (a powerful addictive poison) to be quickly absorbed into the bloodstream. **Lung cancer** is the most widely known and most harmful effect of smoking; 98% of cases are associated with cigarette smoking. Symptoms include chest pain, breathlessness, and coughing up blood. Tobacco smoking is also directly associated with coronary artery disease, emphysema, chronic bronchitis, peripheral vascular disease, and stroke. The damaging components of cigarette smoke include tar, carbon monoxide, nitrogen dioxide, and nitric oxide. Many of these harmful chemicals occur in greater concentrations in sidestream smoke (as occurs as a result of **passive smoking**) than in mainstream smoke (inhaled) due to the presence of a filter in the cigarette.

Long term effects of tobacco smoking

Smoking damages the arteries of the brain and may result in a **stroke**.

All forms of tobacco-smoking increase the risk of **mouth cancer**, **lip cancer**, and **cancer of the throat** (pharynx).

Lung cancer is the best known harmful effect of smoking.

In a young man who smokes 20 cigarettes a day, the risk of **coronary artery disease** is increased by about three times over that of a nonsmoker.

Smoking leads to severe constriction of the arteries supplying blood to the extremities and leads to **peripheral vascular disease**.

Short term effects of tobacco smoking

- Reduction in capacity of the lungs.
- Increase in muscle tension and a decrease in steadiness of the hands.
- Raised blood pressure (10-30 points).
- Very sharp rise in carbon monoxide levels in the lungs contributing to breathlessness.
- Increase in pulse rate by up to 20 beats per minute.
- Surface blood vessel constriction drops skin temperature by up to 5°C.
- Dulling of appetite as well as the sense of smell and taste.

How smoking damages the lungs

Non-smoker

Normal alveoli arrangement

Thin layer of mucus

Cilia

Cells lining airways

Smoker

Coalesced alveoli

Extra mucus produced

Smoke particles

Cancerous cell

Smoke particles indirectly destroy the walls of the lung's alveoli.

Cavities lined by heavy black tar deposits.

SPECIMEN A-73-309 DATE

Gross pathology of lung tissue from a patient with emphysema. Tobacco tar deposits can be seen. Tar contains at least 17 known carcinogens.

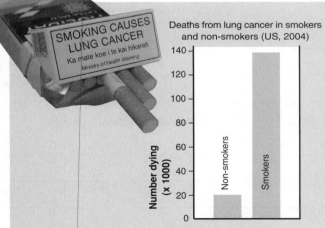

SMOKING CAUSES LUNG CANCER
Ka mate koe i te kai hikareti
Ministry of Health Warning

Deaths from lung cancer in smokers and non-smokers (US, 2004)

Number dying (x 1000): Non-smokers, Smokers
140, 120, 100, 80, 60, 40, 20, 0

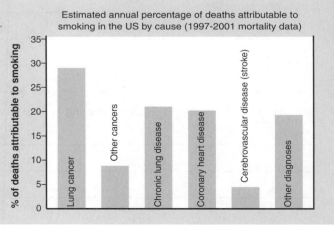

Estimated annual percentage of deaths attributable to smoking in the US by cause (1997-2001 mortality data)

% of deaths attributable to smoking: Lung cancer, Other cancers, Chronic lung disease, Coronary heart disease, Cerebrovascular disease (stroke), Other diagnoses
35, 30, 25, 20, 15, 10, 5, 0

The Respiratory System

Related activities: Gas Transport In Humans, Cardiovascular Disease
Web links: CDC: The Health Consequences of Smoking

RDA 2

Components of Cigarette Smoke

Particulate Phase

Nicotine: a highly addictive alkaloid

Tar: composed of many chemicals

Benzene: carcinogenic hydrocarbon

Gas Phase

Carbon monoxide: a poisonous gas

Ammonia: a pungent, colorless gas

Formaldehyde: a carcinogen

Hydrogen cyanide: a highly poisonous gas

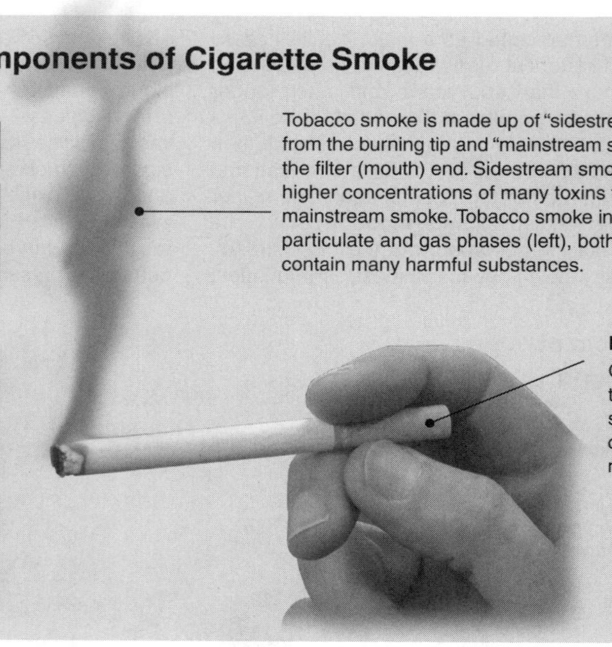

Tobacco smoke is made up of "sidestream smoke" from the burning tip and "mainstream smoke" from the filter (mouth) end. Sidestream smoke contains higher concentrations of many toxins than mainstream smoke. Tobacco smoke includes both particulate and gas phases (left), both of which contain many harmful substances.

Filter
Cellulose acetate filters trap some of the tar and smoke particles. They cool the smoke slightly, making it easier to inhale.

1. Discuss the physical changes to the lung that result from long-term smoking:

_____

_____

_____

_____

2. Determine the physiological effect of each of the following constituents of tobacco smoke when inhaled:

(a) Tar: _____

_____

(b) Nicotine: _____

_____

(c) Carbon monoxide: _____

_____

3. Describe the symptoms of the following diseases associated with long-term smoking:

(a) Emphysema: _____

_____

(b) Chronic bronchitis: _____

_____

(c) Lung cancer: _____

_____

4. Evaluate the evidence linking cigarette smoking to increased incidence of respiratory and cardiovascular diseases:

_____

_____

_____

_____

The Effects of High Altitude

The air at high altitudes contains less oxygen than the air at sea level. Air pressure decreases with altitude so the pressure (therefore amount) of oxygen in the air also decreases. Sudden exposure to an altitude of 2000 m would make you breathless on exertion and above 7000 m most people would become unconscious. The effects of altitude on physiology are related to this lower oxygen availability. Humans and other animals can make some physiological adjustments to life at altitude; this is called acclimatization. Some of the changes to the cardiovascular and respiratory systems to high altitude are outlined below.

Mountain Sickness

Altitude sickness or mountain sickness is usually a mild illness associated with trekking to altitudes of 5000 meters or so. Common symptoms include headache, insomnia, poor appetite and nausea, vomiting, dizziness, tiredness, coughing and breathlessness. The best way to avoid mountain sickness is to ascend to altitude slowly (no more than 300 m per day above 3000 m). Continuing to ascend with mountain sickness can result in more serious illnesses: accumulation of fluid on the brain (cerebral edema) and accumulation of fluid in the lungs (pulmonary edema). These complications can be fatal if not treated with oxygen and a rapid descent to lower altitude.

Physiological Adjustment to Altitude

Effect	Minutes	Days	Weeks
Increased heart rate	←——————→		
Increased breathing		←——————→	
Concentration of blood		←———→	
Increased red blood cell production			←———————→
Increased capillary density			←———→

The human body can make adjustments to life at altitude. Some of these changes take place almost immediately: breathing and heart rates increase. Other adjustments may take weeks (see above). These responses are all aimed at improving the rate of supply of oxygen to the body's tissues. When more permanent adjustments to physiology are made (increased blood cells and capillary networks) heart and breathing rates can return to normal.

People who live permanently at high altitude, e.g. Tibetans, Nepalese, and Peruvian Indians, have physiologies adapted (genetically, through evolution) to high altitude. Their blood volumes and red blood cell counts are high, and they can carry heavy loads effortlessly despite a small build. In addition, their metabolism uses oxygen very efficiently.

Llamas, vicunas, and Bactrian camels are well suited to high altitude life. Vicunas and llamas, which live in the Andes, have high blood cell counts and their red blood cells live almost twice as long as those in humans. Their hemoglobin also picks up and offloads oxygen more efficiently than the hemoglobin of most mammals.

1. (a) Describe the general effects of high altitude on the body: _____

    _____

   (b) Name the general term given to describe these effects: _____

2. (a) Name one short term physiological adaptation that humans make to high altitude: _____

    _____

   (b) Explain how this adaptation helps to increase the amount of oxygen the body receives: _____

    _____

3. (a) Describe one longer term adaptation that humans can make to living at high altitude: _____

    _____

   (b) Explain how this adaptation helps to increase the amount of oxygen the body receives: _____

    _____

Related activities: Control of Breathing

A 2

The Respiratory System

The Digestive System

Respiratory system

- Respiratory system provides O_2 to the organs of the digestive system and disposes of CO_2 produced by cellular respiration.

Cardiovascular system

- Digestive system absorbs iron required for synthesis of hemoglobin and water for maintenance of blood volume.
- Hepatic portal system transports nutrient-rich blood from substantial parts of the gastrointestinal tract to the liver. Ultimately the cardiovascular system distributes nutrients throughout the body.
- The liver produces angiotensinogen, a precursor of the protein angiotensin, which is involved in the system regulating blood pressure and fluid volume.
- Blood distributes hormones of the digestive tract.

Endocrine system

- The liver removes hormones from circulation and prevents their continued activity.
- Pancreas contains endocrine cells that produce hormones for regulating blood sugar.
- Local hormones (e.g. gastrin from the stomach, cholecystokinin and secretin from the intestinal mucosa) help to regulate digestive function, including secretion of digestive juices and gut motility.

Skeletal system

- Digestive system absorbs calcium needed for bone maintenance, growth, and repair.
- Skeletal system protects some of the digestive organs from major damage.
- Bone acts as a storage depot for some nutrients (e.g. calcium).

Integumentary system

- Digestive system provides fats for insulation in dermal and subcutaneous tissues.
- Skin provides external covering to protect the digestive organs.
- The skin synthesizes vitamin D, which is required for absorption of calcium from the gut.

Nervous system

- The feeding center of the hypothalamus stimulates hunger. The satiety center suppresses the feeding center's activity after eating.
- Autonomic NS activity regulates much of gut function. Generally, parasympathetic stimulation increases and sympathetic stimulation decreases gut activity.
- There are reflex and voluntary controls over defecation.

Lymphatic system and immunity

- The lymphatic vessels of the small intestine (the lacteals) drain fat-laden lymph from the gut to the liver.
- Acidic gastric secretions destroy pathogens (non-specific defense).
- Lymphoid tissues in the gut mesenteries and intestinal wall house macrophages and leukocytes that protect against infection.

Urinary system

- Kidneys excrete toxins and the breakdown products of hormones which have been metabolized by the liver.
- Final activation of vitamin D, which is involved in calcium and phosphorus metabolism, occurs in the kidneys.

Reproductive system

- The digestive system provides nutrients required both for normal growth and repair, and the extra nutrition required to support pregnancy and lactation.

Muscular system

- Liver removes and metabolizes lactic acid produced by intense muscular activity.
- Calcium absorbed in the gut as part of the diet is required for muscle contraction.
- Activity of skeletal muscles increases the motility of the gastrointestinal tract, aiding passage of food through the gut.

General Functions and Effects on all Systems

The digestive system is responsible for the physical and chemical digestion and absorption of ingested food. Ultimately, it provides the nutrients required by all body systems for energy metabolism, growth, repair, and maintenance of tissues. Some nutrients may be stored (e.g. in bone, liver, and adipose tissue).

Disease

Symptoms of disease	• Pain (moderate to severe) • Bleeding or change in bowel function • Gastric reflux, nausea or vomiting • Nutritional deficiencies
Infectious diseases of the digestive system	• Cholera • Viral hepatitis • Bacterial food poisoning • Viral gastroenteritis
Non-infectious disorders of the digestive system	• Bowel cancer • Appendicitis • Inflammatory bowel diseases • Food allergies or intolerance • Cirrhosis of the liver

Medicine & Technology

Diagnosis of disorders	• Endoscopy and colonoscopy • Gastrointestinal biopsy • MRI scans and barium enema X-ray • Blood tests
Preventing and treating diseases of the digestive system	• Drug therapies • Surgery • Radiotherapy • Dietary management • Behavior modification

- Appendicitis
- Lactose intolerance
- Celiac disease
- Salmonellosis
- Cholera

- Endoscopy
- MRI scanning
- Appendectomy
- Diet for health

Eating to Live

The Digestive System

The digestive system provides for the energy and nutritional needs of all the body's systems.

While the digestive system is fairly robust against degenerative changes, gastrointestinal disorders are common. Gut function is improved by moderate exercise

- Constipation
- Gastric emptying
- Bowel cancer

- GI blood flow
- Sports nutrition
- Carbo-loading
- Nutrition & recovery

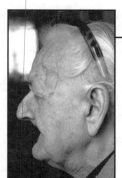

Effects of aging on the digestive system	• Increased risk of bowel cancers • Slower passage of food, constipation • Fibrosis of some organs (pancreas) • Decline in gastric emptying rate • Reduced gastric capacity

Effects of exercise on the digestive system	• Reduced blood flow to gut (short term) • Decreased intestinal transit time • Improved digestive function (long term) • GI upset in highly trained athletes

The Effects of Aging

Exercise

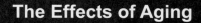

The Digestive System

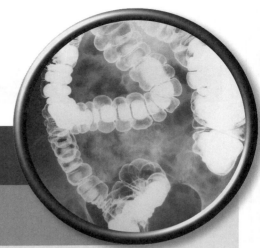

Understanding the structure and function of the alimentary canal and associated structures

Histology, structure and organization of the alimentary canal: dentition, digestive secretions, stomach and intestines, feces formation and egestion. Liver structure and function. Diet and disease.

Learning Objectives

☐ 1. Compile your own glossary from the **KEY WORDS** displayed in **bold type** in the learning objectives below.

The Alimentary Canal (pages 174-175, 178-182)

☐ 2. Describe the overall function of the **digestive system**. Recognize the stages in processing food: **ingestion**, **digestion**, **absorption**, and **egestion**.

☐ 3. Annotate a diagram of the human **alimentary canal** (digestive tract), identifying the main regions and associated organs and glands.

☐ 4. Describe the structure and function of the **mouth**, **pharynx**, and **esophagus**. Describe deciduous and permanent **dentition**, and the basic anatomy of a **tooth**. Describe the composition and function of **saliva**.

☐ 5. Explain how food is moved through the digestive tract by **peristalsis**, including reference to the musculature of the gut wall. Describe the mechanism of **swallowing**.

☐ 6. Describe the structure and function of the **stomach**, including the composition and function of **gastric juice**.

☐ 7. Describe the structure and function of the **small intestine**, including identifying differences in specific regions (**duodenum**, **jejunum**, and **ileum**).

☐ 8. Describe the exocrine role of the **pancreas**, including the composition and function of **pancreatic juice**.

☐ 9. List the enzymes or enzyme groups produced by the organs of the digestive tract, and identify their substrates and the products of digestion.

The Liver (pages 186-188)

☐ 10. Describe the structure and histology of the **liver**. Describe the liver's role in digestion, including the composition, function, and secretion of **bile**.

☐ 11. Describe the central role of the liver in **metabolism**, including reference to storage of nutrients, synthesis of plasma proteins and cholesterol, detoxification, and breakdown of hemoglobin.

Absorption and Transport (pages 181-185)

☐ 12. Distinguish between **absorption** and **assimilation**. Identify where each of the following is absorbed: water, small molecules (alcohol, glucose), breakdown products of carbohydrate, protein, and fat digestion.

☐ 13. Describe the structure of an **intestinal villus** (including blood and lymph vessels) and relate the structure to its role in absorption. Explain the ultrastructure of an epithelial cell of a villus, including: **microvilli**, **pinocytotic vesicles**, and **tight junctions**.

☐ 14. Describe the mechanisms by which nutrients are absorbed in the ileum, including the roles of **diffusion**, **facilitated diffusion**, and **active transport** (including **endocytosis**). Explain the role of **micelles** and **chylomicrons** in lipid absorption and transport.

☐ 15. List the materials that are not absorbed and explain the role of the large intestine and rectum in feces formation and **egestion**. Describe the mechanism of **defecation**.

☐ 16. Outline the control of digestive secretions by nerves and hormones, using gastric secretion as an example. Outline the role of the appetite control center in the brain in regulating food intake.

Diet, Nutrition and Disease (pages 177, 189-193)

☐ 17. Understand aspects of balanced human nutrition, including the consequences of various types of malnutrition (e.g. mineral and vitamin deficiencies, starvation, and obesity).

☐ 18. Recognize **type 2 diabetes mellitus** as a disease with a hereditary component, associated with excessive weight gain and **insulin resistance**, and explain how it can be managed (in part) through dietary modification.

☐ 19. Describe **cholera** as an example of a **bacterial infection** affecting normal ion and water transport in the digestive tract. Explain why diarrheal diseases, such as cholera, are treated with **oral rehydration**.

Supplementary Texts

See page 9 for additional details of these texts:

■ Morton, D. and Perry, J.W, 1997. **Photo Atlas for Anatomy and Physiology**, (Brooks/Cole).

■ Rowett, H.G.Q., 1999. **Basic Anatomy & Physiology**, (John Murray), pp. 84-94.

Anatomy & Physiology

Presentation MEDIA to support this topic: **Anatomy & Physiology**

Periodicals

See page 9 for details of publishers of periodicals:

STUDENT'S REFERENCE

■ **Feast and Famine** Scientific American, Sept. 2007, (special issue). *Coverage of the most recent developments in health and nutrition science.*

■ **Metabolic Powerhouse** New Scientist, 11 Nov. 2000 (Inside Science). *The myriad roles of the liver in metabolic processes, including discussion of amino acid and glucose metabolism.*

■ **The Pancreas and Pancreatitis** Biol. Sci. Rev., 13(5) May 2001, pp. 2-6. *The structure and role of the pancreas, including acinar cell secretion.*

■ **Diarrhea, Digestion, and Dehydration** Biol. Sci. Rev., 14(1), Sept. 2001, pp. 7-9. *The causes and consequences of bacterial-induced diarrhea.*

Internet

See pages 10-11 for details of how to access **Bio Links** from our web site: **www.thebiozone.com** From Bio Links, access sites under the topics:

ANATOMY & PHYSIOLOGY > Nutrition: • Your digestive system and how it works • Human anatomy online • Constituents of human milk • Pathophysiology of the digestive system • Large intestine • Structure of the pancreas

The Human Diet

Nutrients are required for metabolism, tissue growth and repair, and as an energy source. Good nutrition (provided by a **balanced diet**) is recognized as a key factor in good health. Conversely poor nutrition (malnutrition) may cause ill-health or **deficiency diseases**. A diet refers to the quantity and nature of the food eaten. While not all foods contain all the representative nutrients, we can obtain the required balance of different nutrients by eating a wide variety of foods. In a recent overhaul of previous dietary recommendations, the health benefits of monounsaturated fats (such as olive and canola oils), fish oils, and whole grains have been recognized, and people are being urged to reduce their consumption of highly processed foods and saturated (rather than total) fat. Those on diets that restrict certain food groups (e.g. vegans) must take care to balance their intake of foods to ensure an adequate supply of protein and other nutrients (e.g. iron and B vitamins). Dietary information, including **Recommended Daily Amounts** (RDAs) for energy and nutrients, is provided to consumers through the food labeling. Such information helps individuals to assess their nutrient and energy intake and adjust their diet accordingly.

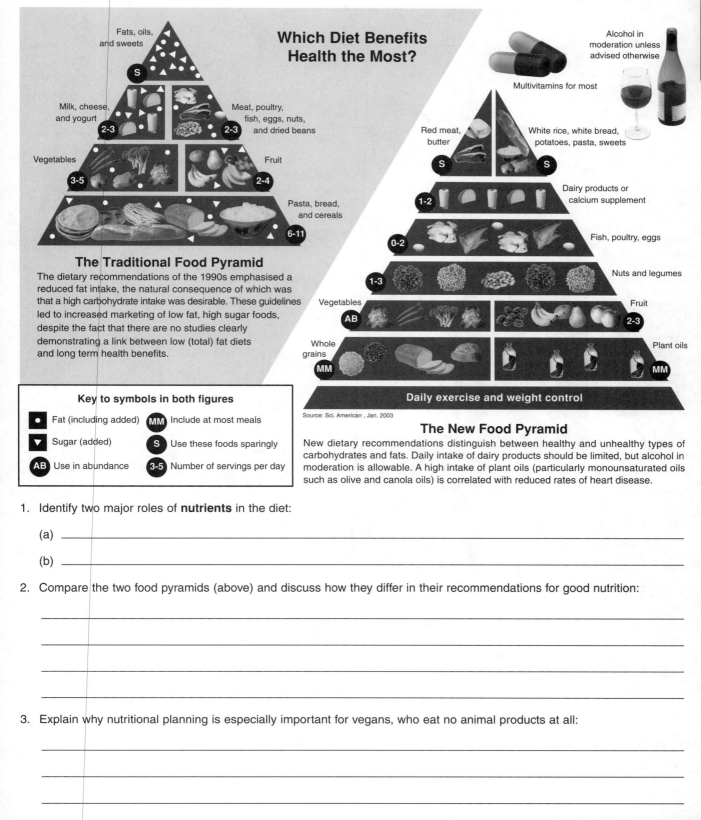

Which Diet Benefits Health the Most?

The Traditional Food Pyramid

The dietary recommendations of the 1990s emphasised a reduced fat intake, the natural consequence of which was that a high carbohydrate intake was desirable. These guidelines led to increased marketing of low fat, high sugar foods, despite the fact that there are no studies clearly demonstrating a link between low (total) fat diets and long term health benefits.

Key to symbols in both figures

- Fat (including added)
- ▼ Sugar (added)
- **AB** Use in abundance
- **MM** Include at most meals
- **S** Use these foods sparingly
- **3-5** Number of servings per day

Source: Sci. American, Jan. 2003

The New Food Pyramid

New dietary recommendations distinguish between healthy and unhealthy types of carbohydrates and fats. Daily intake of dairy products should be limited, but alcohol in moderation is allowable. A high intake of plant oils (particularly monounsaturated oils such as olive and canola oils) is correlated with reduced rates of heart disease.

1. Identify two major roles of **nutrients** in the diet:

   (a) _____

   (b) _____

2. Compare the two food pyramids (above) and discuss how they differ in their recommendations for good nutrition:

   _____

   _____

   _____

   _____

3. Explain why nutritional planning is especially important for vegans, who eat no animal products at all:

   _____

   _____

   _____

The Mouth and Pharynx

The mouth (**oral cavity**), is formed by the cheeks, hard and soft palate, and tongue (below right). The teeth are very hard structures, specialized for chewing food (**mastication**). The tongue, which moves food around, and the salivary glands, which produce saliva, act with the teeth to begin digestion. The structure of a tooth is illustrated by a section through a molar (below, right). The oral cavity is divided into quadrants and the number of teeth in each quadrant given by a **dental formula**. There are 32 adult (permanent) teeth, organized as shown below left. The basic structure of a tooth is described in the inset below.

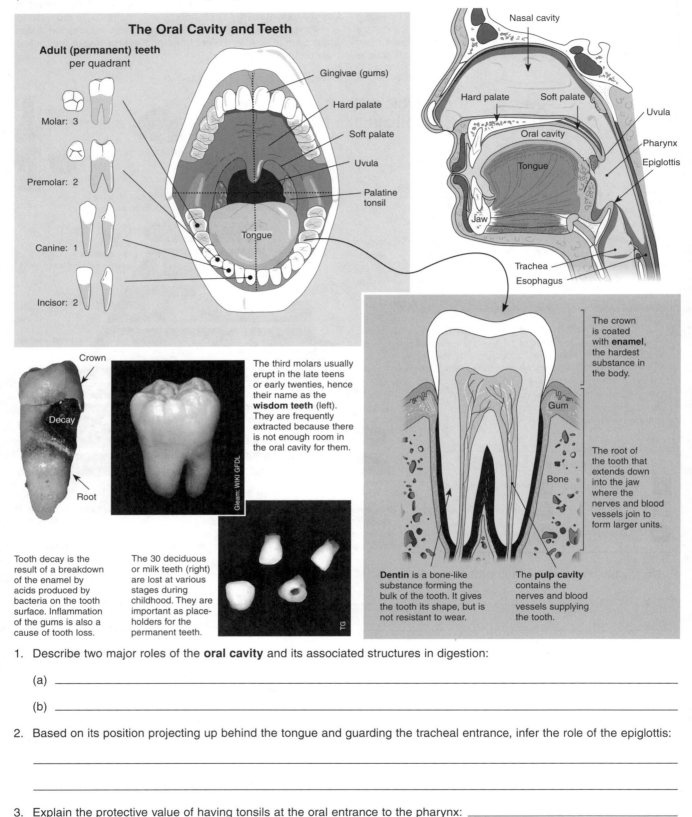

The Oral Cavity and Teeth

Adult (permanent) teeth
per quadrant

Molar: 3
Premolar: 2
Canine: 1
Incisor: 2

Gingivae (gums)
Hard palate
Soft palate
Uvula
Palatine tonsil
Tongue

Nasal cavity
Hard palate
Soft palate
Uvula
Oral cavity
Pharynx
Epiglottis
Tongue
Jaw
Trachea
Esophagus

Crown
Decay
Root

The third molars usually erupt in the late teens or early twenties, hence their name as the **wisdom teeth** (left). They are frequently extracted because there is not enough room in the oral cavity for them.

Gleam: WIKI GFDL

The crown is coated with **enamel**, the hardest substance in the body.

Gum

Bone

The root of the tooth that extends down into the jaw where the nerves and blood vessels join to form larger units.

Tooth decay is the result of a breakdown of the enamel by acids produced by bacteria on the tooth surface. Inflammation of the gums is also a cause of tooth loss.

The 30 deciduous or milk teeth (right) are lost at various stages during childhood. They are important as place-holders for the permanent teeth.

TG

Dentin is a bone-like substance forming the bulk of the tooth. It gives the tooth its shape, but is not resistant to wear.

The **pulp cavity** contains the nerves and blood vessels supplying the tooth.

1. Describe two major roles of the **oral cavity** and its associated structures in digestion:

   (a) _____

   (b) _____

2. Based on its position projecting up behind the tongue and guarding the tracheal entrance, infer the role of the epiglottis:

   _____

   _____

3. Explain the protective value of having tonsils at the oral entrance to the pharynx: _____

   _____

   _____

Related activities: The Human Digestive Tract
Web links: Swallowing

The Human Digestive Tract

It is estimated that an adult consumes about 20 000 kg of food between the ages of 18 and 38 years; about a metric tonne a year. Although babies grow rapidly from birth, growth is not the most significant reason for our ongoing eating. Our bodies require a constant source of energy for the vast number of biochemical reactions that constitute **metabolism**. Food provides the source of this energy. Humans have a tube-like digestive tract (gut), which runs through the body from the mouth to the anus and includes specialized regions for storage and absorption. It has various organs and glands functionally and structurally associated with it, including the liver, gall bladder, and pancreas. The digestive tract prepares food for use by the body's cells through five activities: eating (**ingestion**), movement of food, **digestion** (physical and chemical breakdown), **absorption**, and **elimination**.

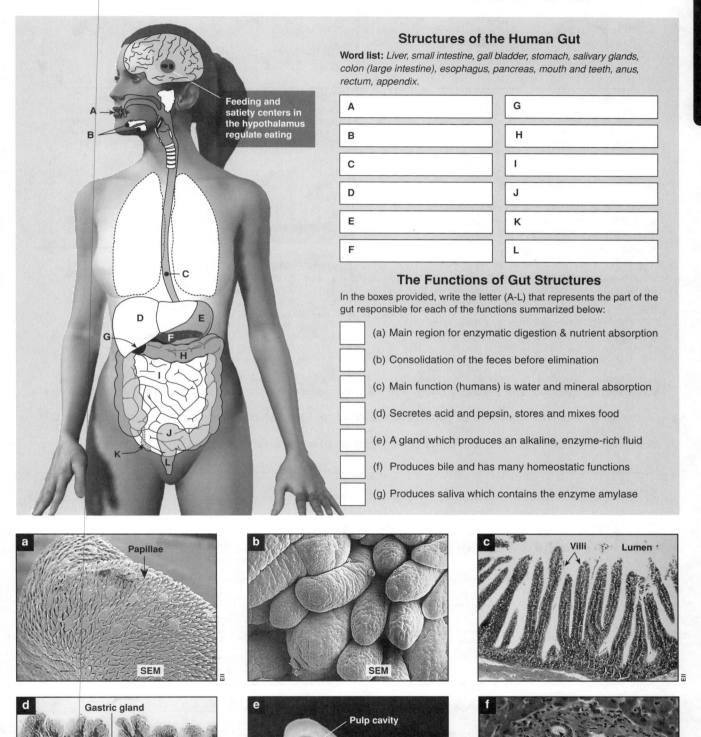

Feeding and satiety centers in the hypothalamus regulate eating

Structures of the Human Gut

Word list: *Liver, small intestine, gall bladder, stomach, salivary glands, colon (large intestine), esophagus, pancreas, mouth and teeth, anus, rectum, appendix.*

A | G
B | H
C | I
D | J
E | K
F | L

The Functions of Gut Structures

In the boxes provided, write the letter (A-L) that represents the part of the gut responsible for each of the functions summarized below:

(a) Main region for enzymatic digestion & nutrient absorption

(b) Consolidation of the feces before elimination

(c) Main function (humans) is water and mineral absorption

(d) Secretes acid and pepsin, stores and mixes food

(e) A gland which produces an alkaline, enzyme-rich fluid

(f) Produces bile and has many homeostatic functions

(g) Produces saliva which contains the enzyme amylase

a Papillae — SEM

b SEM

c Villi — Lumen

d Gastric gland

e Pulp cavity — Enamel

f Bile ducts

Related activities: Stomach and Small Intestine, The Large Intestine
Web links: Digestion Animation

RA 2

Processes in a Tube Gut

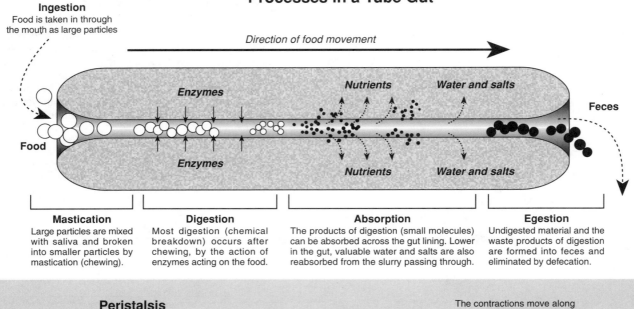

Ingestion
Food is taken in through the mouth as large particles

Direction of food movement

Enzymes

Enzymes

Food

Nutrients

Water and salts

Nutrients

Water and salts

Feces

Mastication	**Digestion**	**Absorption**	**Egestion**
Large particles are mixed with saliva and broken into smaller particles by mastication (chewing).	Most digestion (chemical breakdown) occurs after chewing, by the action of enzymes acting on the food.	The products of digestion (small molecules) can be absorbed across the gut lining. Lower in the gut, valuable water and salts are also reabsorbed from the slurry passing through.	Undigested material and the waste products of digestion are formed into feces and eliminated by defecation.

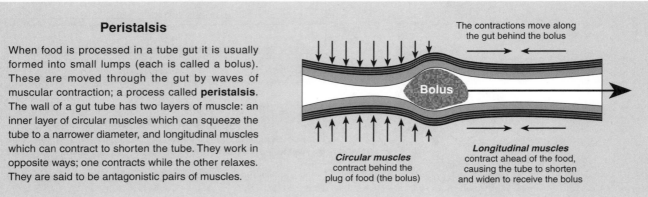

Peristalsis

When food is processed in a tube gut it is usually formed into small lumps (each is called a bolus). These are moved through the gut by waves of muscular contraction; a process called **peristalsis**. The wall of a gut tube has two layers of muscle: an inner layer of circular muscles which can squeeze the tube to a narrower diameter, and longitudinal muscles which can contract to shorten the tube. They work in opposite ways; one contracts while the other relaxes. They are said to be antagonistic pairs of muscles.

The contractions move along the gut behind the bolus

Bolus

Circular muscles
contract behind the plug of food (the bolus)

Longitudinal muscles
contract ahead of the food, causing the tube to shorten and widen to receive the bolus

1. In the spaces provided on the diagram (previous page), identify the parts **A-L** (choose from the word list provided). Match each of the **functions** described (a)-(g) with the letter representing the corresponding structure on the diagram.

2. On the same diagram, mark with lines and labels: anal sphincter (**AS**), pyloric sphincter (**PS**), cardiac sphincter (**CS**).

3. Identify the region of the gut illustrated by the photographs (a)-(f) on the previous page:

   (a) _____   (b) _____

   (c) _____   (d) _____

   (e) _____   (f) _____

4. Explain how a **bolus** of food is moved through a tube gut: _____

   _____

   _____

   _____

5. Describe an advantage of having a gut where food moves in only one direction: _____

   _____

6. Identify one factor that would influence the length and specialization in the gut: _____

7. Briefly describe how the following processes are involved in processing food:

   (a) Mastication: _____

   _____

   (b) Absorption: _____

   _____

Stomach and Small Intestine

Digestion in the gut depends on both the physical movement of the food and its enzymatic breakdown into constituent components. Most digestion occurs in the stomach and small intestine. The digestive enzymes involved may be bound to the surfaces of the intestinal epithelial cells or occur as components of the secretions of digestive glands (e.g. pancreas). The structure and functions of the stomach and small intestines, and their enzymic secretions are shown on this and the next page.

The Stomach and Organs of the Small Intestine

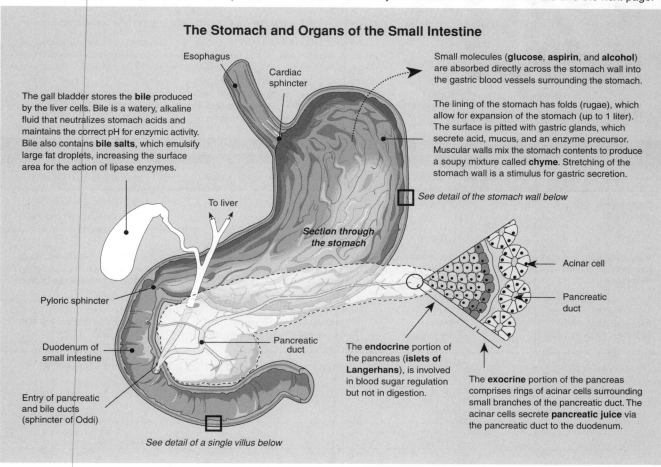

The gall bladder stores the **bile** produced by the liver cells. Bile is a watery, alkaline fluid that neutralizes stomach acids and maintains the correct pH for enzymic activity. Bile also contains **bile salts**, which emulsify large fat droplets, increasing the surface area for the action of lipase enzymes.

Esophagus

Cardiac sphincter

Small molecules (**glucose**, **aspirin**, and **alcohol**) are absorbed directly across the stomach wall into the gastric blood vessels surrounding the stomach.

The lining of the stomach has folds (rugae), which allow for expansion of the stomach (up to 1 liter). The surface is pitted with gastric glands, which secrete acid, mucus, and an enzyme precursor. Muscular walls mix the stomach contents to produce a soupy mixture called **chyme**. Stretching of the stomach wall is a stimulus for gastric secretion.

See detail of the stomach wall below

To liver

Section through the stomach

Pyloric sphincter

Acinar cell

Pancreatic duct

Duodenum of small intestine

Pancreatic duct

The **endocrine** portion of the pancreas (**islets of Langerhans**), is involved in blood sugar regulation but not in digestion.

The **exocrine** portion of the pancreas comprises rings of acinar cells surrounding small branches of the pancreatic duct. The acinar cells secrete **pancreatic juice** via the pancreatic duct to the duodenum.

Entry of pancreatic and bile ducts (sphincter of Oddi)

See detail of a single villus below

Detail of a Single Villus From Intestinal Wall

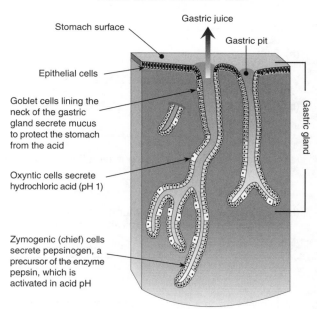

Epithelial cells on the tip of the villus are brushed off as a result of regular wear and tear.

The intestinal enzymes are bound to surfaces of the epithelial cells

Epithelial cells divide and migrate toward the tip of the villus to replace lost and worn cells.

Crypt of Lieberkühn: tubular exocrine gland that secretes alkaline fluid

Brunner's gland produces mucus which empties into the crypt of Lieberkühn

Goblet cells in the epithelium produce mucus

Columnar epithelium

Capillary network

Lymph vessel

Detail of the Stomach Wall

Gastric juice

Stomach surface

Gastric pit

Epithelial cells

Goblet cells lining the neck of the gastric gland secrete mucus to protect the stomach from the acid

Oxyntic cells secrete hydrochloric acid (pH 1)

Zymogenic (chief) cells secrete pepsinogen, a precursor of the enzyme pepsin, which is activated in acid pH

Gastric gland

1. Describe the two important roles of gut movements: _____

_____

_____

Related activities: The Human Digestive Tract, The Large Intestine
Web links: Acid Secretion in the Stomach, Interactive Digestion Quiz

A 2

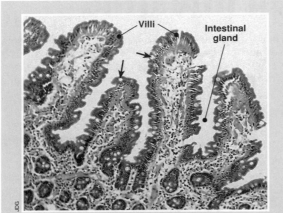

Villi | Intestinal gland

Intestinal villi and microvilli

The photograph (left) shows a section through the ileum with the **intestinal villi** and **intestinal glands** (crypts of Lieberkühn) indicated. The intestinal glands secrete mucus and alkaline fluid. **Epithelial cells** lining the surface of the villi are regularly worn off and replaced by new cells migrating from the base of the intestinal glands. Each epithelial cell has many **microvilli** (microscopic projections called the brush border) which further increase the intestinal surface area.

Enzymes bound to the microvilli surfaces of the epithelial cells (peptidases, maltase, lactase, and sucrase) break down small peptides and carbohydrate molecules into their constituent parts. The breakdown products (monosaccharides, amino acids) are then absorbed into the underlying blood and lymph vessels. **Mucous cells** (white spots arrowed) produce mucus to protect the epithelial cells from enzymatic digestion. The **blood vessels** transport nutrients to the liver. **Lymph vessels** transport the products of fat digestion.

Enzyme secretions of the gut and their role in digestion

Secretion and source	Site of action	Active enzyme	Substrate and products	Control of secretion
Gastric juice: stomach	Stomach	Pepsin	Protein ⟶ peptides	Reflex stimulation, stretching of the stomach wall, and the hormone **gastrin**.
Pancreatic juice: pancreas (exocrine region only)	Duodenum	Pancreatic amylase Trypsin Chymotrypsin Pancreatic lipase	Starch ⟶ maltose Protein ⟶ peptides Protein ⟶ peptides Fats ⟶ fatty acids + glycerol	Control of pancreatic secretions is via release of the hormones **secretin** and **cholecystokinin**.
Intestinal juice and enzymes: small intestine	Small intestine	Maltase Peptidases	Maltose ⟶ glucose Polypeptides ⟶ amino acids	Reflex action and contact with intestinal wall.

2. Discuss the digestive and storage role of the stomach in humans, identifying important structures and secretions:

_____

_____

_____

_____

_____

3. Identify two sites for enzyme secretion in the gut, give an example of an enzyme produced there, and state its role:

(a) Site: _____ Enzyme: _____

Enzyme's role: _____

(b) Site: _____ Enzyme: _____

Enzyme's role: _____

4. (a) Suggest why the pH of the gut secretions varies at different regions in the gut: _____

_____

(b) Explain why it is necessary for protein-digesting enzymes (e.g. trypsin, chymotrypsin, and pepsin) to be secreted in an inactive form and then activated after release:

_____

_____

5. Explain why alcohol exerts its effects more rapidly when the stomach is empty (rather than full): _____

_____

6. Explain the role of sphincter muscles in the digestive tract: _____

_____

The Large Intestine

After most of the nutrients have been absorbed in the small intestine, the remaining fluid contents pass into the large intestine (appendix, cecum, and colon). The fluid comprises undigested or undigestible food, bacteria, dead cells sloughed off from the gut wall, mucus, bile, ions, and a large amount of water. In humans and other omnivores, the large intestine is concerned mainly with the reabsorption of water and electrolytes. Infection or disease can cause an increase in gut movements, resulting in insufficient reabsorption of water and diarrhea. Sluggish gut movements cause the reabsorption of too much water and the feces become hard and difficult to pass, a condition known as constipation. The semi-solid waste material (feces) passes from the **colon** to the rectum, where it is stored and consolidated before being expelled (egested). Egestion of feces is controlled by the activity of two sphincters in the **anus**, one voluntary and the other under involuntary reflex control.

Structure of the Large Intestine

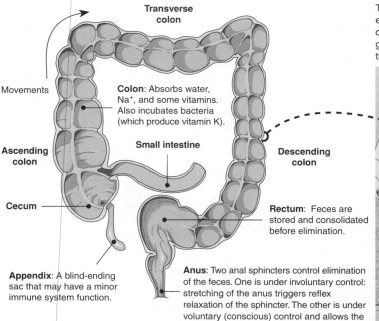

Transverse colon

Movements

Colon: Absorbs water, Na⁺, and some vitamins. Also incubates bacteria (which produce vitamin K).

Small intestine

Ascending colon

Descending colon

Cecum

Rectum: Feces are stored and consolidated before elimination.

Appendix: A blind-ending sac that may have a minor immune system function.

Anus: Two anal sphincters control elimination of the feces. One is under involuntary control: stretching of the anus triggers reflex relaxation of the sphincter. The other is under voluntary (conscious) control and allows the reflex activity to be modified.

A single crypt from the intestinal wall

The lining of the large intestine consists of a simple columnar epithelium. The epithelium is not folded into villi, but instead contains tubular glands called crypts, containing numerous goblet cells. The goblet cells produce mucus, which lubricates the colon wall and aids formation of the feces.

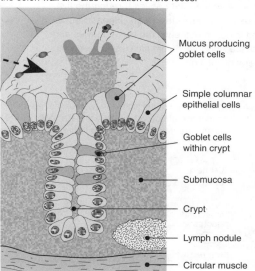

Mucus producing goblet cells

Simple columnar epithelial cells

Goblet cells within crypt

Submucosa

Crypt

Lymph nodule

Circular muscle

Appendicitis

Obstruction of the appendix by fecal matter or some other cause can lead to an inflammation called *appendicitis*. Appendicitis usually develops rapidly with little warning over a period of 6-12 hours. The usual symptom is abdominal pain, accompanied by nausea, vomiting and a slight fever. When severe, it can be life threatening. Acute appendicitis is treated by surgical removal of the appendix (**appendectomy**). The entire procedure usually takes about one hour and is performed in one of two ways: through what is called an open operation or through the laparoscopic technique.

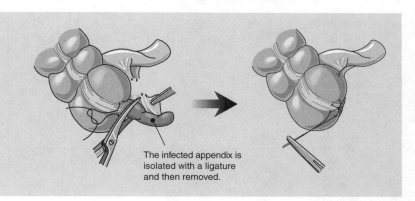

The infected appendix is isolated with a ligature and then removed.

1. Outline the main function of the large intestine: _____

2. Suggest why the lining of the large intestine consists of crypts as opposed to villi like projections: _____

_____

3. The photograph below shows a cross section through the colon wall. Using the diagram of the single crypt (above right) label the features indicated using the following word list: *circular muscle, submucosa, lymph nodule, epithelial cells.*

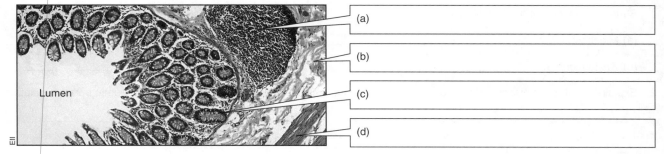

Lumen

(a)

(b)

(c)

(d)

Absorption and Transport

All the chemical and physical processes of digestion from the mouth to the small intestine are aimed at the breakdown of food molecules into forms that can pass through intestinal lining into the underlying blood and lymph vessels. These breakdown products include monosaccharides, amino acids, fatty acids, glycerol, and glycerides. Passage of these molecules from the gut into the blood or lymph is called **absorption**. After absorption, nutrients are transported directly or indirectly to the liver for storage or processing. Some of the features of nutrient absorption and transport are shown below. For simplicity, all nutrients are shown in the lumen of the intestine, even though some nutrients are digested on the epithelial cell surfaces.

The Hepatic Portal System

The liver obtains oxygenated blood from the hepatic artery, but it also receives deoxygenated blood containing newly absorbed nutrients via the hepatic portal vein. The **hepatic portal system** refers to all the blood flow from the digestive organs that passes through the liver before returning to the heart. Hepatic portal blood is rich in nutrients: the liver monitors and processes this load before the blood passes into general circulation.

Absorption: Most of the simple molecules that are the final products of food breakdown are absorbed by the epithelial cells of the villi into the blood vessels and are transported directly to the liver where they are processed.

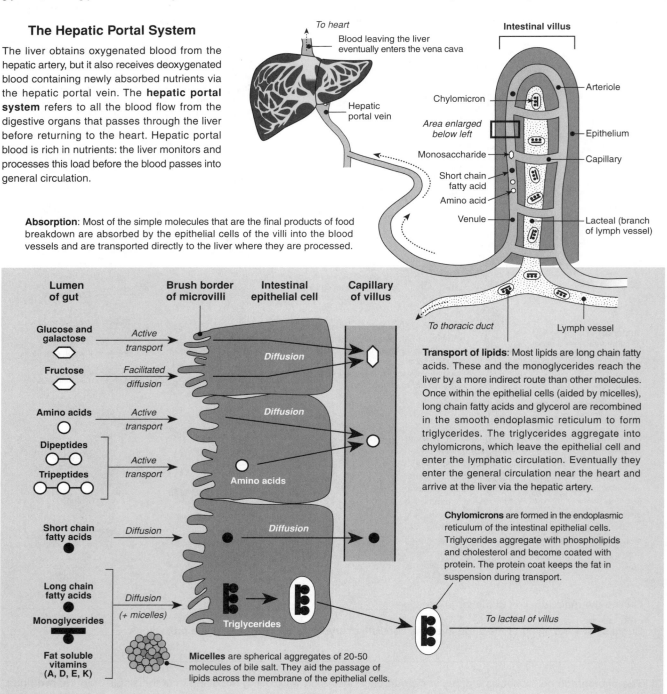

Transport of lipids: Most lipids are long chain fatty acids. These and the monoglycerides reach the liver by a more indirect route than other molecules. Once within the epithelial cells (aided by micelles), long chain fatty acids and glycerol are recombined in the smooth endoplasmic reticulum to form triglycerides. The triglycerides aggregate into chylomicrons, which leave the epithelial cell and enter the lymphatic circulation. Eventually they enter the general circulation near the heart and arrive at the liver via the hepatic artery.

Chylomicrons are formed in the endoplasmic reticulum of the intestinal epithelial cells. Triglycerides aggregate with phospholipids and cholesterol and become coated with protein. The protein coat keeps the fat in suspension during transport.

Micelles are spherical aggregates of 20-50 molecules of bile salt. They aid the passage of lipids across the membrane of the epithelial cells.

1. State the function of the following in fat digestion:

   (a) Micelles: _____

   (b) Chylomicrons: _____

2. Explain why it is important that venous blood from the gut is transported first to the liver via the hepatic portal circulation:

   _____

   _____

Related activities: Stomach and Small Intestine, Transport System in Humans, The Liver's Homeostatic Role

A 2

The Control of Digestion

The majority of digestive juices are secreted only when there is food in the gut and both nervous and hormonal mechanisms are involved in coordinating and regulating this activity appropriately. The digestive system is innervated by branches of the **autonomic nervous system** (sympathetic and parasympathetic stimulation).

Hormonal regulation is achieved through the activity of several hormones: **gastrin**, **secretin**, and **cholecystokinin** (formerly called **pancreozymin**). These are released into the bloodstream in response to nervous or chemical stimuli and influence the activity of gut and associated organs.

Hormonal and Nervous Control of Digestion

Salivation is entirely under nervous control. Some saliva is secreted continuously in response to **parasympathetic stimulation** via the vagus nerve. The presence of food in the mouth stimulates the salivary glands (and stomach) to increase their secretions. This response operates through a simple cranial reflex via the vagus nerve. The smell, sight, and thought of food also stimulates salivary (and gastric) secretion. These stimuli involve higher brain activity and learning (a conditioned reflex).

The feeding center of the brain

The **feeding center** in the hypothalamus is constantly active. It monitors metabolites in the blood and stimulates hunger when these metabolites reach low levels. After a meal, a neighboring region of the hypothalamus, **the satiety center**, suppresses the activity of the feeding center for a period of time. Impulses from these two centers travel via the vagus nerve to stimulate the secretion of particular digestive hormones.

The secretions and muscular activity of the gut are regulated by both nervous and hormonal mechanisms. **Parasympathetic stimulation** of the stomach and pancreas via the **vagus nerve** increases their secretion. Sympathetic stimulation has the opposite effect.

The entry of food into the small intestine, especially fat and gastric acid, stimulates the cells of the intestinal mucosa to secrete the hormones **cholecystokinin** (CCK) and **secretin**.

The presence of food in the stomach causes it to stretch. This mechanical stimulus results in secretion of the hormone **gastrin** from cells in the mucosa of the stomach. This activity is mediated through a simple **reflex**.

Cholecystokinin circulates in the blood and stimulates the pancreas to increase its secretion of enzyme-rich fluid. CCK also stimulates the release of bile into the intestine from the gall bladder. **Secretin** stimulates the pancreas to increase its secretion of alkaline fluid. This fluid neutralizes the acid entering the intestine. Secretin also stimulates the production of bile from the liver cells. **Both secretin and CCK** stimulate the secretion of intestinal juice but inhibit gastric secretion and general motility of the gastrointestinal tract.

Gastrin is secreted in response to eating food (particularly protein). Gastrin is released into the bloodstream where it acts back on the stomach to increase gastric secretion and motility. Gastrin also increases the motility of the gastrointestinal tract in general, and this helps to propel food through the gut.

Vagus nerve

Gastrin

CCK and secretin

1. Describe the role of each of the following stimuli in the control of digestion, identifying both the response and its effect:

    (a) Presence of food in the mouth: _____

    _____

    (b) Presence of fat and acid in the small intestine: _____

    _____

    (c) Stretching of the stomach by the presence of food: _____

    _____

2. Outline the role of the vagus nerve in regulating digestive activity: _____

    _____

Related activities: Stomach & Small Intestine, The Liver's Role in Digestion
Web links: Three Phases of Gastric Secretion, CCK Causes Bile Release

A 3

The Liver's Role in Digestion

The liver is a large organ, weighing about 1.4 kg, and is well supplied with blood. It carries out several hundred different functions and has a pivotal role in maintaining homeostasis. Its role in the digestion of food centers around the production of the alkaline fluid, **bile**, which is secreted at a rate of 0.8-1.0 liter per day. It is also responsible for processing absorbed nutrients, which arrive at the liver via the hepatic portal system. These functions are summarized below.

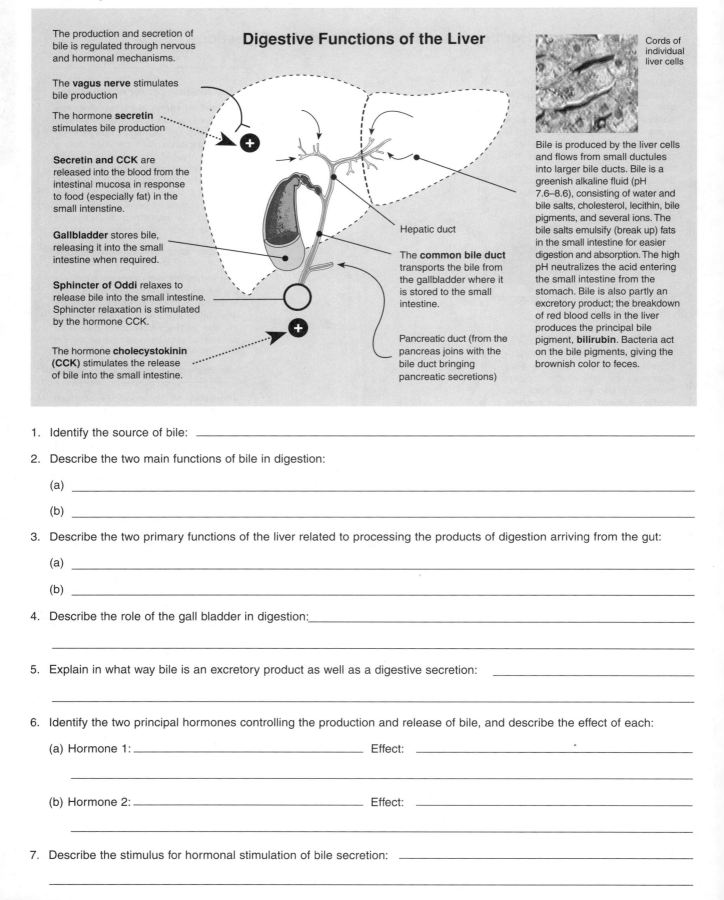

Digestive Functions of the Liver

The production and secretion of bile is regulated through nervous and hormonal mechanisms.

The **vagus nerve** stimulates bile production

The hormone **secretin** stimulates bile production

Secretin and CCK are released into the blood from the intestinal mucosa in response to food (especially fat) in the small intestine.

Gallbladder stores bile, releasing it into the small intestine when required.

Sphincter of Oddi relaxes to release bile into the small intestine. Sphincter relaxation is stimulated by the hormone CCK.

The hormone **cholecystokinin (CCK)** stimulates the release of bile into the small intestine.

Hepatic duct

The **common bile duct** transports the bile from the gallbladder where it is stored to the small intestine.

Pancreatic duct (from the pancreas joins with the bile duct bringing pancreatic secretions)

Cords of individual liver cells

Bile is produced by the liver cells and flows from small ductules into larger bile ducts. Bile is a greenish alkaline fluid (pH 7.6–8.6), consisting of water and bile salts, cholesterol, lecithin, bile pigments, and several ions. The bile salts emulsify (break up) fats in the small intestine for easier digestion and absorption. The high pH neutralizes the acid entering the small intestine from the stomach. Bile is also partly an excretory product; the breakdown of red blood cells in the liver produces the principal bile pigment, **bilirubin**. Bacteria act on the bile pigments, giving the brownish color to feces.

1. Identify the source of bile: _____

2. Describe the two main functions of bile in digestion:

    (a) _____

    (b) _____

3. Describe the two primary functions of the liver related to processing the products of digestion arriving from the gut:

    (a) _____

    (b) _____

4. Describe the role of the gall bladder in digestion: _____

    _____

5. Explain in what way bile is an excretory product as well as a digestive secretion: _____

    _____

6. Identify the two principal hormones controlling the production and release of bile, and describe the effect of each:

    (a) Hormone 1: _____ Effect: _____

    _____

    (b) Hormone 2: _____ Effect: _____

    _____

7. Describe the stimulus for hormonal stimulation of bile secretion: _____

    _____

Related activities: The Control of Digestion, The Liver's Homeostatic Role

The Liver's Homeostatic Role

The liver, located just below the diaphragm and making up 3-5% of body weight, is the largest homeostatic organ. It performs a vast number of functions including production of bile, storage and processing of nutrients, and detoxification of poisons and metabolic wastes. The liver has a unique **double blood supply** and up to 20% of the total blood volume flows through it at any one time.

This rich vascularization makes it the central organ for regulating activities associated with the blood and circulatory system. In spite of the complexity of its function, the liver tissue and the liver cells themselves are structurally relatively simple. Features of liver structure and function are outlined below. The histology of the liver in relation to its role is described on the next page.

Homeostatic Functions of the Liver

The liver is one of the largest and most complex organs in the body. It has a central role as an organ of homeostasis and performs many functions, particularly in relation to the regulation of blood composition. General functions of the liver are outlined below. Briefly summarized, the liver:

1. Secretes bile, important in emulsifying fats in digestion.
2. Metabolizes amino acids, fats, and carbohydrates (below).
3. Synthesizes glucose from non-carbohydrate sources when glycogen stores are exhausted (gluconeogenesis).
4. Stores iron, copper, and some vitamins (A, D, E, K, B_{12}).
5. Converts unwanted amino acids to urea (urea cycle).
6. Manufactures heparin and plasma proteins (e.g. albumin).
7. Detoxifies poisons or turns them into less harmful forms.
8. Some liver cells phagocytose worn-out blood cells.
9. Synthesizes cholesterol from acetyl coenzyme A.

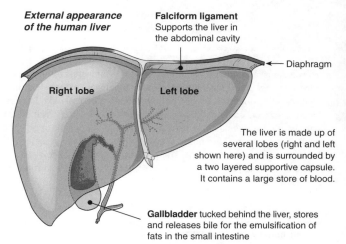

External appearance of the human liver

Falciform ligament Supports the liver in the abdominal cavity

Diaphragm

Right lobe

Left lobe

The liver is made up of several lobes (right and left shown here) and is surrounded by a two layered supportive capsule. It contains a large store of blood.

Gallbladder tucked behind the liver, stores and releases bile for the emulsification of fats in the small intestine

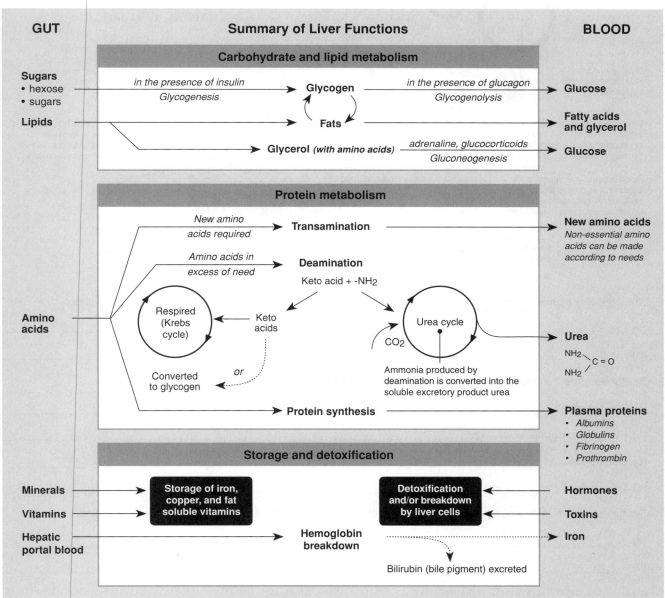

GUT | **Summary of Liver Functions** | **BLOOD**

Carbohydrate and lipid metabolism

Sugars
• hexose
• sugars

in the presence of insulin — Glycogenesis → **Glycogen** → in the presence of glucagon — Glycogenolysis → **Glucose**

Lipids → **Fats**

→ **Glycerol** *(with amino acids)* → adrenaline, glucocorticoids — Gluconeogenesis → **Glucose**

Fatty acids and glycerol

Protein metabolism

New amino acids required → **Transamination** → **New amino acids** *Non-essential amino acids can be made according to needs*

Amino acids in excess of need → **Deamination**

Keto acid + -NH_2

Amino acids

Respired (Krebs cycle) ← Keto acids → Urea cycle → **Urea**

CO_2

$$\begin{matrix} NH_2 \\ \\ NH_2 \end{matrix} \bigg\rangle C = O$$

Converted to glycogen *or*

Ammonia produced by deamination is converted into the soluble excretory product urea

→ **Protein synthesis** → **Plasma proteins**
• Albumins
• Globulins
• Fibrinogen
• Prothrombin

Storage and detoxification

Minerals → **Storage of iron, copper, and fat soluble vitamins**

Vitamins

Detoxification and/or breakdown by liver cells ← **Hormones**

← **Toxins**

Hepatic portal blood → **Hemoglobin breakdown** ----→ **Iron**

Bilirubin (bile pigment) excreted

Related activities: The Liver's Role in Digestion
Web links: Fatty Acid Metabolism

RA 2

The Internal Structure of the Liver

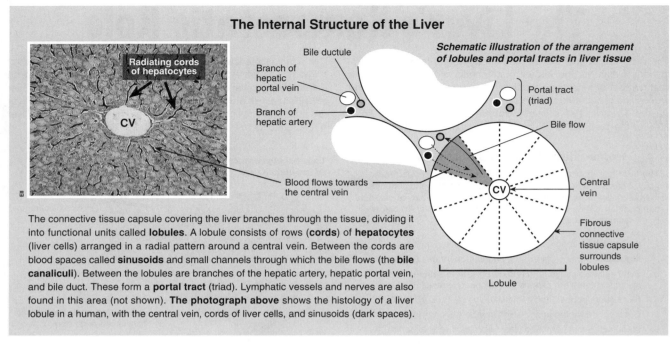

Bile ductule

Branch of hepatic portal vein

Branch of hepatic artery

Blood flows towards the central vein

Radiating cords of hepatocytes

CV

Schematic illustration of the arrangement of lobules and portal tracts in liver tissue

Portal tract (triad)

Bile flow

Central vein

Fibrous connective tissue capsule surrounds lobules

Lobule

The connective tissue capsule covering the liver branches through the tissue, dividing it into functional units called **lobules**. A lobule consists of rows (**cords**) of **hepatocytes** (liver cells) arranged in a radial pattern around a central vein. Between the cords are blood spaces called **sinusoids** and small channels through which the bile flows (the **bile canaliculi**). Between the lobules are branches of the hepatic artery, hepatic portal vein, and bile duct. These form a **portal tract** (triad). Lymphatic vessels and nerves are also found in this area (not shown). **The photograph above** shows the histology of a liver lobule in a human, with the central vein, cords of liver cells, and sinusoids (dark spaces).

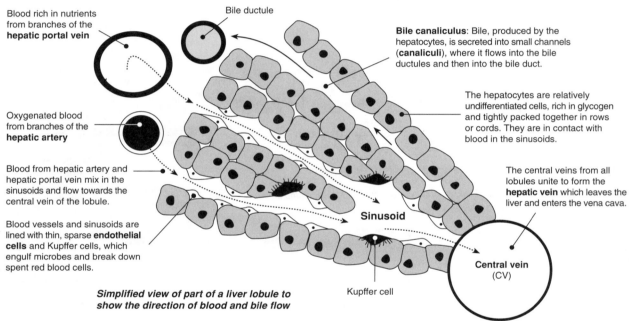

Blood rich in nutrients from branches of the **hepatic portal vein**

Bile ductule

Bile canaliculus: Bile, produced by the hepatocytes, is secreted into small channels (**canaliculi**), where it flows into the bile ductules and then into the bile duct.

The hepatocytes are relatively undifferentiated cells, rich in glycogen and tightly packed together in rows or cords. They are in contact with blood in the sinusoids.

Oxygenated blood from branches of the **hepatic artery**

Blood from hepatic artery and hepatic portal vein mix in the sinusoids and flow towards the central vein of the lobule.

Blood vessels and sinusoids are lined with thin, sparse **endothelial cells** and Kupffer cells, which engulf microbes and break down spent red blood cells.

Sinusoid

The central veins from all lobules unite to form the **hepatic vein** which leaves the liver and enters the vena cava.

Central vein (CV)

Kupffer cell

Simplified view of part of a liver lobule to show the direction of blood and bile flow

1. State the two sources of blood supply to the liver, describing the primary physiological purpose of each supply:

   (a) Supply 1: _____ Purpose: _____

   (b) Supply 2: _____ Purpose: _____

2. Briefly describe the role of the following structures in liver tissue:

   (a) Bile canaliculi: _____

   (b) Phagocytic Kupffer cells: _____

   (c) Central vein: _____

   (d) Sinusoids: _____

3. Briefly explain three important aspects of **either** protein metabolism **or** carbohydrate metabolism in the liver:

   (a) _____

   (b) _____

   (c) _____

Cholera is an acute intestinal infection caused by the bacterium *Vibrio cholerae*. The bacterium produces an **enterotoxin** which binds to membrane receptors on the small intestine, opening the ion channels and increasing permeability of the mucosal epithelium to chloride ions. According to the principles of osmosis, water follows the salt across the membrane resulting in copious, painless, watery **diarrhea** that can lead to severe dehydration, kidney failure, and death within hours if left untreated. Cholera can be prevented by hygienic disposal of human feces, provision of an adequate supply of safe drinking water, safe food handling and preparation (e.g. preventing contamination of food and water), and effective general hygiene (e.g. hand washing with soap). Once contracted, the only treatment for cholera is the administration of **oral rehydration solutions (ORS)** to prevent dehydration or death. In severe cases the rehydration solution is administered intravenously, and the patient may be prescribed antibiotics to reduce the infection time. With prompt and appropriate ORS treatment, the fatality rate from cholera infection is less than 1%.

Development of Oral Rehydration Solutions

Many scientific disciplines have been involved in developing modern ORS. Key discoveries include:

1950s: Physiologists first noted that glucose and sodium were transported together across the intestinal epithelium.

1960s: The first ORS formulations were developed to treat severe diarrhea. In addition to electrolytes, they also contained glucose, which had been proven to increase water reabsorption.

The discovery that the cholera enterotoxin was responsible for fluid loss (diarrhea) by interfering with membrane cAMP activity and G-proteins.

Current: The development of low osmolarity solutions which use alternative carbohydrate sources such as rice, instead of sugars to minimize diarrheal effect.

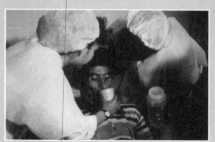

CDC

Administering ORS to a cholera patient

Vibrio cholerae

Oral Rehydration Solutions

Diarrhea causes water and electrolytes to be lost from the body, causing dehydration and electrolyte imbalance. This in turn can alter osmotic gradients in the body, affecting hydration, blood pH, and nerve and muscle function. Drinking water alone to treat diarrhea is ineffective for two reasons: during bouts of diarrhea the large intenstine is losing rather than absorbing water, and secondly, electrolyte loss is not addressed. Instead, **oral rehydation solutions** (ORS) are prescribed. Modern ORS are simple and inexpensive, and can be administered with no medical training. They contain water and salts in specific ratios designed to replenish fluids and electrolytes. Carbohydrates, such as glucose or sucrose, are added to enhance electrolyte absorption in the intestinal tract. Although the presence of sugars can increase the rate of diarrhea, they still have an overall benefit because they increase fluid replacement and improve patient hydration.

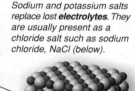

*Sodium and potassium salts replace lost **electrolytes**. They are usually present as a chloride salt such as sodium chloride, NaCl (below).*

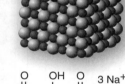

ORAL REHYDRATION SALTS

Each sachet contains the equivalent of:
Sodium Chloride .... 3.5 g.
Potassium Chloride .... 1.5 g.
Sodium Citrate, dihydrate .. 2.9 g.
Glucose Anhydrous .... 20.0 g.

DIRECTIONS
Dissolve in ONE LITRE of drinking water.

To be taken orally-
Infants - over a 24 hour period
Children - over an 8 to 24 hour period,
according to age or as otherwise
directed under medical supervision.

CAUTION: DO NOT BOIL SOLUTION

Carbohydrates, such as sucrose or glucose (above) increase water and electrolyte absorption.

Sodium bicarbonate or sodium citrate (right) help maintain homeostatic blood pH and revert metabolic acidosis which occurs if blood pH falls below 7.35.

3 Na⁺

1. Identify the pathogen that causes cholera: _____

2. Describe why severe diarrhea caused by cholera infection can be so dangerous if not treated quickly:

    _____

    _____

    _____

3. Briefly describe why ORS are more effective in treating the symptoms of cholera than water alone:

    _____

    _____

4. Explain why a patient taking an ORS with glucose might feel that their symptoms were worsening and stop treatment:

    _____

    _____

5. Discuss some of the ethical issues associated with trialing new ORS formulations on humans: _____

    _____

    _____

Related activities: Absorption and Transport, Passive Transport Processes

Malnutrition and Obesity

Malnutrition describes an imbalance between what someone eats and what is required to remain healthy. In economically developed areas of the world, most (but not all) forms of malnutrition are the result of poorly balanced nutrient intakes rather than a lack of food *per se*. Amongst the most common of these is **obesity**, as indicated by BMI values in excess of 30 (below). Although some genetic and hormonal causes are known, obesity is commonly the result of excessive energy intake, usually associated with a highly processed diet, high in fat and sugar. In addition, incidental physical activity is declining: we drive more, use labor-saving machines, and exercise less. Obesity is a risk factor in a number of chronic diseases, including hypertension, cardiovascular disease, and type 2 diabetes. Paradoxically, obesity in developed countries is more common in poorly educated, lower socio-economic groups than amongst the wealthy, who often have more options in terms of food choices.

Obesity and Malnutrition

In adults, the exact level of obesity is determined by reference to the Body Mass Index (BMI). A score of 30+ on the BMI indicates mild obesity, while those with severe or morbid obesity have BMIs of 40+. Child obesity is based on BMI-for-age, and is assessed in relation to the weight of other children of a similar age and gender. Central or abdominal obesity, now classified as an independent risk factor for some serious diseases, refers to excessive fat around the abdomen. While the explanation for excessive body fat is simple (energy in exceeds energy out), a complex of biological and socio-economic factors are implicated in creating the problems of modern obesity.

Obesity Prevalance

In 2000, an estimated 20.4% of adult Americans were obese and a further 36.7% overweight, leaving fewer than half of adult Americans not at risk for health problems related to excess weight. The prevalence of obesity (BMI > 30) continues to be a health concern for adults, children, and adolescents in the US and other developed countries, as indicated by self reported BMI assessment data below.

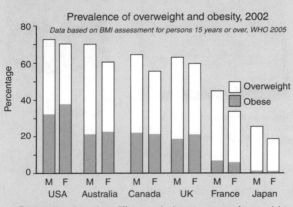

Prevalence of overweight and obesity, 2002

Data based on BMI assessment for persons 15 years or over, WHO 2005

Recent data show clear differences in the percentage of overweight and obese adults between Australia, Canada, the USA, and the UK (which are similar) and Japan and France, where obesity is much less common. These differences are attributable largely to customary dietary differences between these nations.

Health Effects of Obesity

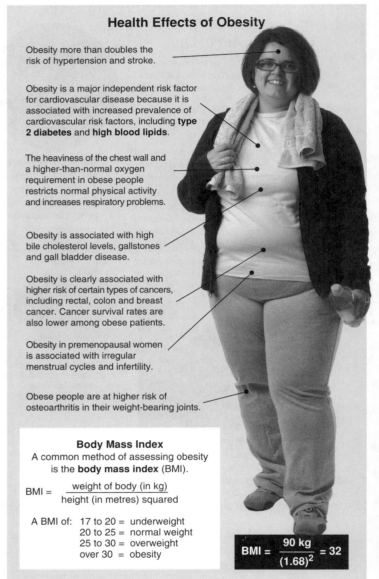

Obesity more than doubles the risk of hypertension and stroke.

Obesity is a major independent risk factor for cardiovascular disease because it is associated with increased prevalence of cardiovascular risk factors, including **type 2 diabetes** and **high blood lipids**.

The heaviness of the chest wall and a higher-than-normal oxygen requirement in obese people restricts normal physical activity and increases respiratory problems.

Obesity is associated with high bile cholesterol levels, gallstones and gall bladder disease.

Obesity is clearly associated with higher risk of certain types of cancers, including rectal, colon and breast cancer. Cancer survival rates are also lower among obese patients.

Obesity in premenopausal women is associated with irregular menstrual cycles and infertility.

Obese people are at higher risk of osteoarthritis in their weight-bearing joints.

Body Mass Index
A common method of assessing obesity is the **body mass index** (BMI).

$$BMI = \frac{\text{weight of body (in kg)}}{\text{height (in metres) squared}}$$

A BMI of: 17 to 20 = underweight
20 to 25 = normal weight
25 to 30 = overweight
over 30 = obesity

$$BMI = \frac{90 \text{ kg}}{(1.68)^2} = 32$$

---

1. (a) Explain why obesity is regarded as a form of malnutrition: _____

   _____

   (b) Describe the two basic energy factors that determine how a person's weight will change: _____

   _____

2. Using the BMI, calculate the minimum and maximum weight at which a 1.85 m tall man would be considered:

   (a) Overweight: _____   (b) Obese: _____

3. BMI is routinely used to assess healthy weight. Explain why BMI might sometimes not be a reliable in this respect:

   _____

Related activities: Deficiency Diseases, Cardiovascular Disease, Type 2 Diabetes

Deficiency Diseases

Malnutrition is the general term for nutritional disorders resulting from not having enough food (**starvation**) or not enough of the right food (**deficiency**). Children are the most at risk from malnutrition because they are growing rapidly and are more susceptible to disease. Malnutrition is a key factor in the deaths of 6 million children each year and, in developing countries, dietary deficiencies are a major problem. In these countries, malnutrition usually presents as energy and protein deficiencies. Specific vitamin and mineral deficiencies in adults are associated with specific disorders, e.g. **scurvy** (vitamin C), **rickets** (vitamin D),

pellagra (niacin), **pernicious anemia** (vitamin B$_{12}$), or **anemia** (iron). Deficiency diseases are rare in developed countries and are usually limited to people with very restricted diets, intestinal disorders, or drug and alcohol problems, although some deficiencies, e.g. iron deficient anemia, are much more common than others. The ideas on what constitutes a balanced diet have changed somewhat in the last decade. Previous recommendations emphasized a reduced fat intake, but more recent advice emphasizes the benefits of 'healthy fats' and whole grains in preference to highly processed food of any kind.

Vitamin Deficiencies

Vitamin A (found in animal livers, eggs, and dairy products) is essential for the production of light-absorbing pigments in the eye and for the formation of cell structures. Symptoms of **vitamin A deficiency** include loss of night vision, inflammation of the eye, corneal damage, and the presence of Bitot's spots (foamy patches on the white of the eye, arrowed).

Vitamin C Deficiency causes a disease known as scurvy, once the scourge of sailors but now rare in developed countries. Inadequate vitamin C intake disturbs the body's normal production of collagen, a protein in connective tissue that holds body structures together. This results in poor wound healing, rupture of small blood vessels (visible bleeding in the skin), swollen gums, and loose teeth.

Vitamin B$_{12}$ (found primarily in meat, but also in eggs and dairy products) is required for nucleic acid and protein metabolism, and for the maturation of red blood cells. It is essential for proper growth and for the proper nervous system function. Deficiency results in pernicious anemia, poor appetite, weight loss, growth failure, fatigue, brain damage, nervousness, muscle tics, depression, spinal cord degeneration, and lack of balance.

Lack of vitamin D in children produces the disease rickets. In adults a similar disease is called osteomalacia. Suffers typically show skeletal deformities (e.g. bowed legs) because inadequate amounts of phosphorus and calcium are incorporated into the bones. Vitamin D is produced by the skin when exposed to sunlight and it is vital for the absorption of calcium from the diet.

Common Mineral Deficiencies

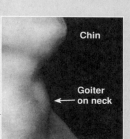

Iodine
Iodine is essential for the production of thyroid hormones, which control growth, metabolic rate, and development. Shortage of iodine in the diet may lead to **goiter** (enlargement of the thyroid). Iodine deficiency is also responsible for some cases of thyroid underactivity.

Iron
Anemia results from lower than normal levels of hemoglobin in red blood cells. Iron from the diet is required to produce hemoglobin. People most at risk include women during **pregnancy** and those with an inadequate dietary intake. Symptoms include fatigue, fainting, breathlessness, and heart palpitations.

Zinc
Zinc is found in red meat, poultry, fish, whole grain cereals and breads, legumes, and nuts. It is important for enzyme activity, production of insulin, making of sperm, and perception of taste. A deficiency in zinc causes growth retardation, a delay in puberty, muscular weakness, dry skin, and a delay in wound healing.

Calcium
Calcium is required for enzyme function, formation of bones and teeth, blood clotting, and muscular contraction. Calcium deficiency causes poor bone growth and structure, increasing the tendency of bones to fracture and break. It also results in muscular spasms and poor blood clotting.

All photos CDC unless indicated otherwise

Related activities: The Human Diet, Malnutrition and Obesity

RA 1

Protein and Energy Deficiencies

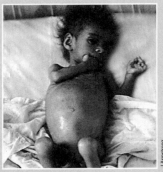

Marasmus is the most common form of deficiency disease. It is a severe protein and energy malnutrition that usually occurs in famine or starvation conditions. Children suffering from marasmus are stunted and extremely emaciated. They have loose folds of skin on the limbs and buttocks, due to the loss of fat and muscle tissue. Sufferers have no resistance to disease and common infections are typically fatal.

Kwashiorkor is a severe type of protein-energy deficiency in young children (1-3 years old), occurring mainly in poor rural areas in the tropics. Kwashiorkor occurs when a child is suddenly weaned on to a diet that is low in calories, protein, and certain essential micronutrients. Children have stunted growth, low resistance to infection, oedema (accumulation of fluid in the tissues), and are inactive, apathetic and weak.

Alcohol Abuse and Nutritional Deficiency

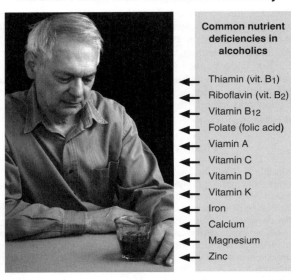

Common nutrient deficiencies in alcoholics

- Thiamin (vit. B_1)
- Riboflavin (vit. B_2)
- Vitamin B_{12}
- Folate (folic acid)
- Viamin A
- Vitamin C
- Vitamin D
- Vitamin K
- Iron
- Calcium
- Magnesium
- Zinc

People who regularly consume excessive alcohol are at increased risk of nutritional deficiencies. Even when food intake is adequate, alcohol interferes with the metabolism of food by affecting digestion, storage, utilization, and excretion of nutrients. Alcohol damages the cells lining the small intestine and impairs absorption of nutrients. For example, alcohol inhibits fat absorption, impairs the digestion of proteins, and interferes with glucose metabolism.

1. Using examples, distinguish between **malnutrition** and **starvation**: _____
   _____
   _____

2. Suggest why young children, pregnant women, and athletes are among the most susceptible to dietary deficiencies:
   _____
   _____
   _____

3. (a) Explain why a lack of iron leads to the symptoms of anemia (fatigue and breathlessness): _____
   _____

   (b) Explain why iron deficiency is relatively more common in women of child-bearing age than in men: _____
   _____

4. Using the example of **iodine**, explain how artificial dietary supplementation can be achieved and discuss its benefits:
   _____
   _____

5. Suggest why a zinc deficiency is associated with muscular weakness and a delay in puberty: _____
   _____
   _____

6. Explain why alcoholics are likely to be deficient in fat soluble vitamins (A, D, K) even when food intake is adequate:
   _____
   _____

Type 2 Diabetes Mellitus

Diabetes is a general term for a range of disorders sharing two common symptoms: production of large amounts of urine and excessive thirst. **Diabetes mellitus** is the most common form of diabetes and is characterized by **hyperglycemia** (high blood sugar). **Type 1** is characterized by a complete lack of insulin production and usually begins in childhood, while **type 2** is more typically a disease of older, overweight people whose cells develop a resistance to insulin uptake. Both types are chronic, incurable conditions and are managed differently. Type 1 is treated primarily with insulin injection, whereas type 2 sufferers manage their disease through diet and exercise in an attempt to limit the disease's long term detrimental effects.

Symptoms of Type 2 Diabetes Mellitus

a Symptoms may be mild at first. The body's cells do not respond appropriately to the insulin that is present and blood glucose levels become elevated. Normal blood glucose level is 60-110 mgL^{-1}. In diabetics, fasting blood glucose level is 126 mgL^{-1} or higher.

b Symptoms occur with varying degrees of severity:

▶ Cells are starved of fuel. This can lead to increased appetite and overeating and may contribute to an existing obesity problem.

▶ Urine production increases to rid the body of the excess glucose. Glucose is present in the urine and patients are frequently very thirsty.

▶ The body's inability to use glucose properly leads to muscle weakness and fatigue, irritability, frequent infections, and poor wound healing.

c Uncontrolled elevated blood glucose eventually results in damage to the blood vessels and leads to:

▶ coronary artery disease
▶ peripheral vascular disease
▶ retinal damage, blurred vision and blindness
▶ kidney damage and renal failure
▶ persistent ulcers and gangrene

Risk Factors

Obesity: BMI greater than 27. Distribution of weight is also important.

Age: Risk increases with age, although the incidence of type 2 diabetes is increasingly reported in obese children.

Sedentary lifestyle: Inactivity increases risk through its effects on bodyweight.

Family history: There is a strong genetic link for type 2 diabetes. Those with a family history of the disease are at greater risk.

Ethnicity: Certain ethnic groups are at higher risk of developing of type 2 diabetes.

High blood pressure: Up to 60% of people with undiagnosed diabetes have high blood pressure.

High blood lipids: More than 40% of people with diabetes have abnormally high levels of cholesterol and similar lipids in the blood.

Treating Type 2 Diabetes

Diabetes is not curable but can be managed to minimize the health effects:

▶ Regularly check blood glucose level
▶ Manage diet to reduce fluctuations in blood glucose level
▶ Take regular exercise
▶ Reduce weight
▶ Reduce blood pressure
▶ Reduce or stop smoking
▶ Take prescribed anti-diabetic drugs
▶ In time, insulin therapy may be required

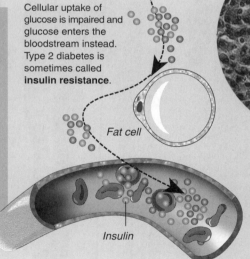

Cellular uptake of glucose is impaired and glucose enters the bloodstream instead. Type 2 diabetes is sometimes called **insulin resistance**.

Fat cell

Insulin

The **beta cells** of the pancreatic islets (above) produce insulin, the hormone responsible for the cellular uptake of glucose. In type 2 diabetes, the body's cells do not utilize the insulin properly.

1. Distinguish between type 1 and type 2 diabetes, relating the differences to the different methods of treatment:

_____

_____

2. Explain what dietary advice you would give to a person diagnosed with type 2 diabetes: _____

_____

_____

3. Explain why the increase in type 2 diabetes is considered epidemic in the developed world: _____

_____

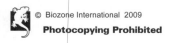
Related activities: Control of Blood Glucose
Web links: Cellular Mechanisms of Diabetes

RA 2

The Urinary System

Respiratory system

- Respiratory system provides O_2 to the urinary system and disposes of CO_2 produced by cellular respiration.
- An enzyme in the cells of the lung capillaries converts angiotensin I to angiotensin II (involved in regulation of glomerular filtration).

Cardiovascular system

- Regulation of salt and water balance in the kidney is important in regulation of blood pressure.
- Regulation of blood composition of Na^+, K^+, and Ca^{2+} helps maintain heart function.
- Arterial blood pressure is the driving force for glomerular filtration.
- Heart muscle cells secrete a peptide (ANP) in response to high blood pressure. ANP results in greater excretion of Na^+ and water from the kidney.
- Blood distributes the hormones that influence renal function (e.g. ADH and aldosterone).

Digestive system

- The final activation of vitamin D occurs in the kidneys. Bioactive vitamin D is required for calcium absorption in the gut.
- The liver synthesizes most of the body's urea, which is then excreted via the kidneys.
- The digestive system provides nutrients for maintenance and health of the urinary organs.

Skeletal system

- Bones of the ribcage provide some protection for the kidneys.
- Erythropoietin from the kidneys promotes the formation of red blood cells in the bone marrow.

Integumentary system

- The skin provides an external protective barrier.
- Final activation of vitamin D (synthesized in the skin) occurs in the kidneys.
- Skin is a site of water loss.

Nervous system

- Renal regulation of the Na^+, K^+, and Ca^{2+} content in the extracellular fluid is necessary for nerve function.
- Micturition (urination) is controlled by voluntary and reflex nervous activity.
- Sympathetic NS activity triggers the renin-angiotensin system for regulation of blood volume.

Lymphatic system and immunity

- The lymphatic vessels return leaked fluid to the general circulation and help to maintain the blood volume/pressure required for kidney function.
- The immune system protects the urinary organs from infection and cancer.

Endocrine system

- Kidneys produce erythropoietin, a hormone which promotes the formation of red blood cells in the bone marrow.
- Regulation of salt and water balance by the kidneys maintains the blood volume necessary for hormone transport.
- ADH, ANP, aldosterone and other hormones interact to regulate reabsorption of water and electrolytes in the kidney.

Reproductive system

- Urinogenital systems are closely aligned. Urethra has an excretory function in both sexes and a reproductive function in males, for the passage of semen.

Muscular system

- Renal regulation of the Na^+, K^+, and Ca^{2+} content in the extracellular fluid is necessary for muscle function.
- Muscles of the pelvic floor and external urethral sphincter are involved in voluntary control of micturition.
- Creatinine, which is a break-down product of creatine phosphate in muscle, is produced at a fairly constant rate by the body and must be excreted by the kidneys.

General Functions and Effects on all Systems

The urinary system (kidneys and associated structures) is responsible for disposing of nitrogenous wastes toxins, and metabolic breakdown products. The kidneys maintain the fluid, electrolytes, and acid-base balance of the body fluids, which is essential for the proper functioning of all body systems.

Disease

Symptoms of disease
- Pain (moderate to severe)
- Abnormal urine composition or volume
- Abnormal electrolyte balance
- Abnormal fluid balance

Diseases and disorders of the urinary system
- Kidney stones
- Hereditary disorders
- Nephrotic syndrome
- Congenital diseases (malformations)
- Bladder and kidney cancer
- Chronic kidney disease (CKD)
- Incontinence

Medicine & Technology

Diagnosis of disorders
- MRI scans
- Kidney biopsy
- Urine tests
- Blood tests

Prevention of urinary system disorders
- Control of hypertension
- Control of diabetes
- Control of weight
- Behavior to control risk of UTIs

Treatment of urinary system disorders
- Drug therapy (e.g. antibiotics)
- Physical therapy (e.g for pelvic floor)
- Surgery (e.g. removal of kidney stones)
- Transplant (e.g. of donor kidney)
- Renal dialysis

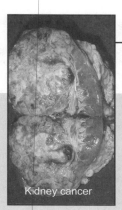
Kidney cancer

- Kidney stones
- CKD & renal failure
- Polycystic kidney disease

- MRI scans
- Urine analysis
- Kidney transplants
- Renal dialysis

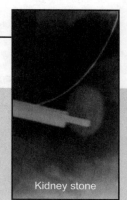
Kidney stone

Waste Removal

The Urinary System

The urinary system had a primary role in excretion of nitrogenous wastes, and in fluid and electrolyte balance.

Degenerative changes in kidney and bladder function can be severe and debilitating. Renal dialysis and transplants are options for sufferers of kidney disease.

- Prostate enlargement
- Poor renal function
- Incontinence

- Effects of exercise on health
- Creatine metabolism
- Dehydration

Effects of aging on the urinary system
- Lower number of functional nephrons
- Reduction in glomerular filtration rate
- Reduced response to ADH
- Prostate enlargement (males)
- Loss of bladder/sphincter muscle tone

Effects of exercise on the urinary system
- Lowered blood pressure, therefore...
- Reduced risk of chronic kidney disease
- Increased rates of creatinine excretion
- (Rarely) exercise-induced renal failure (dehydration or electrolyte imbalance)

The Effects of Aging

Exercise

The Urinary System

Understanding the structure and function of the urinary system

Structure and function of the urinary system: fluid budgets, nephron structure and function, regulation of fluid and electrolytes, maintenance of blood pH. Renal failure: dialysis and kidney transplants.

Learning Objectives

☐ 1. Compile your own glossary from the **KEY WORDS** displayed in **bold type** in the learning objectives below.

The Urinary System *(pages 194, 197-98, 200-202)*

☐ 2. Describe the main functions of the **kidney** in the regulation of body fluids. On an annotated diagram, identify the structures of the urinary system: **kidneys, ureters, renal blood vessels, bladder, urethra**.

☐ 3. Describe the gross structure of the kidney to include the **cortex, medulla,** and **renal pelvis**. Interpret features of the histology of the kidney from sections viewed with a light microscope.

☐ 4. Using a labelled diagram, describe the structure and arrangement of a **nephron** and its associated blood vessels in relation to its function in producing urine. Include reference to **glomerulus**, proximal and distal **convoluted tubules**, and **collecting duct**.

☐ 5. Explain concisely how the kidney nephron produces urine. Include reference to:
(a) The process of **ultrafiltration** in the **glomerulus**
(b) The ultrastructure of the glomerulus and **renal capsule** in relation to ultrafiltration.
(c) The **selective reabsorption** of water and solutes in the **proximal convoluted tubule**.
(d) The ultrastructure of the proximal convoluted tubule in relation to reabsorption.
(e) The role of the **loop of Henle** and the **counter-current multiplier system** in creating and maintaining the ionic (salt) gradient in the kidney.
(f) The role of the ionic gradient in the kidney in producing a concentrated urine and in fluid balance.

☐ 6. Describe **micturition** (urination or voiding), including reference to the involvement of urethral sphincters and voluntary and involuntary controls over micturition.

Fluid and Electrolyte Balance *(pages 203-205)*

☐ 7. Describe the main fluid compartments in the body: the **intracellular fluid** and the **extracellular fluid** (comprising interstitial fluid and plasma). Recognize that exchanges with the environment require that there be constant adjustments of these fluid compartments in order to maintain homeostasis.

☐ 8. Explain how blood composition (electrolyte balance) and volume are regulated. Include reference to:
(a) The role of **osmoreceptors** in the hypothalamus
(b) Release of antidiuretic hormone (**ADH**) from the **posterior pituitary** and its action on the kidney.
(c) The regulation of ADH output.
(d) The role of **aldosterone** in promoting sodium reabsorption in the kidney.
(e) The regulation of aldosterone release through the renin-angiotensin mechanism.

☐ 9. Recognize the requirement for maintenance of blood pH at pH 7.35-7.45. Describe how the acid-base balance of the body fluids is maintained through:
(a) Bicarbonate buffer system in the blood
(b) Respiratory system controls
(c) Renal mechanisms

Urinary System *(pages 195, 199, 206-207)*

☐ 10. Describe the characteristics of healthy urine and describe the use of urine analysis (**urinalysis**) in the diagnosis of renal disease.

☐ 11. Describe some common causes of renal failure and describe its effects. Explain the treatment of renal failure through **renal dialysis**. Outline the principles involved in dialysis and describe the structure and action of kidney dialysis machines.

☐ 12. Describe kidney transplants as a long term solution to renal failure. Explain the advantages and problems associated with kidney transplants.

Supplementary Texts

See page 9 for additional details of these texts:
■ Morton, D. and Perry, J.W, 1997. **Photo Atlas for Anatomy and Physiology**, (Brooks/Cole).
■ Rowett, H.G.Q., 1999. **Basic Anatomy & Physiology**, (John Murray), pp. 117-120.

Periodicals

See page 9 for details of publishers of periodicals:

STUDENT'S REFERENCE
■ **Metabolic Powerhouse** New Scientist, 11 November 2000 (Inside Science). *The myriad roles of the liver in metabolic processes, including its role in amino acid (nitrogen) metabolism.*

■ **The Kidney** Biol. Sci. Rev., 16(2) Nov. 2003, pp. 2-6. *The structure of the kidneys, and their essential role in regulating extracellular fluid volume, blood pressure, acid-base balance, and metabolic waste products such as urea.*

■ **Urea: A Product of Excess Dietary Protein** Biol. Sci. Rev., 17(4) April 2005, pp. 6-8. *An account of how and why urea is formed; nitrogen balance, and urea and the nitrogen cycle.*

Presentation MEDIA to support this topic: **Anatomy & Physiology**

Internet

See pages 10-11 for details of how to access **Bio Links** from our web site: **www.thebiozone.com** From Bio Links, access sites under the topics:

ANATOMY & PHYSIOLOGY > **General sites**:
• Anatomy & Physiology • Human physiology lecture notes • WebAnatomy > **Excretion**: • The excretory system • The kidney • Urinary system > **Homeostasis**: • BBC schools - homeostasis

Waste Products in Humans

In humans and other mammals, a number of organs are involved in the excretion of the waste products of metabolism: mainly the kidneys, lungs, skin, and gut. The liver is a particularly important organ in the initial treatment of waste products, particularly the breakdown of hemoglobin and the formation of urea from ammonia. Excretion should not be confused with the elimination or egestion of undigested and unabsorbed food material from the gut. Note that the breakdown products of hemoglobin (blood pigment) are excreted in bile and pass out with the feces, but they are not the result of digestion.

CO_2 Water

Lungs
Excretion of carbon dioxide (CO_2) with some loss of water.

Skin
Excretion of water, CO_2, hormones, salts and ions, and small amounts of urea as sweat.

Liver
Produces urea from ammonia in the urea cycle. Breakdown of hemoglobin in the liver produces the bile pigments e.g. bilirubin.

Gut
Excretion of bile pigments in the feces. Also loses water, salts, and carbon dioxide.

Bladder
Storage of urine before it is expelled to the outside.

All cells
All the cells that make up the body carry out cellular respiration; they break down glucose to release energy and produce the waste products, carbon dioxide and water.

Excretion In Humans

In mammals, the kidney is the main organ of excretion, although the skin, gut, and lungs also play important roles. As well as ridding the body of nitrogenous wastes, the kidneys are also able to excrete many unwanted poisons and drugs that are taken in from the environment. Usually these are ingested with food or drink, or inhaled. As long as these are not present in toxic amounts, they can usually be slowly eliminated from the body.

Kidney
Filtration of the blood to remove urea. Unwanted ions, particularly hydrogen (H^+) and potassium (K^+), and some hormones are also excreted by the kidneys. Some poisons and drugs (e.g. penicillin) are also excreted by active secretion into the urine. Water is lost in excreting these substances and extra water may be excreted if necessary.

Substance	Origin*	Organ(s) of excretion
Carbon dioxide		
Water		
Bile pigments		
Urea		
Ions (K^+, HCO_3^-, H^+)		
Hormones		
Poisons		
Drugs		

* Origin refers to from where in the body each substance originates

1. Complete the table above summarizing the origin of excretory products and the main organ(s) of excretion for each.

2. Explain the role of the liver in excretion, even though it is not primarily an organ of excretion: _____

_____

_____

3. Tests for pregnancy are sensitive to an excreted substance in the urine. Suggest what type of substance this might be:

_____

4. In people suffering renal failure, the kidneys cease to produce filtrate. Based on your knowledge of the central role of the kidneys in fluid and electrolyte balance, as well as nitrogen excretion, describe the typical symptoms of kidney failure:

_____

_____

_____

_____

_____

The Urinary System

Related activities: Acid-Base Balance, Kidney Dialysis

A 3

Water Budget in Humans

We cannot live without water for more than about 100 hours and adequate water is a requirement for physiological function and health. Body water content varies between individuals and through life, from above about 90% of total weight as a fetus to 74% as an infant, 60% as a child, and around 50-59% in adults, depending on gender and age. Gender differences (males usually have a higher water content than females) are the result of differing fat levels. Water intake and output are highly variable but closely matched to less than 0.1% over an extended period. Typical values for water gains and losses, as well as daily water transfers are given below. Men need more water than women due to their higher (on average) fat-free mass and energy expenditure. Infants and young children need more water in proportion to their body weight as they cannot concentrate their urine as efficiently as adults. They also have a greater surface area relative to weight, so water losses from the skin are greater.

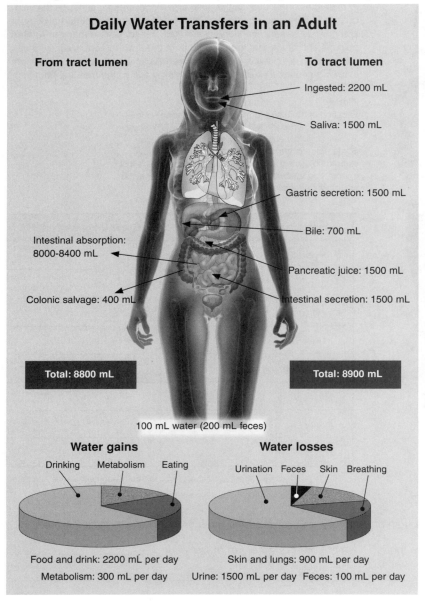

Daily Water Transfers in an Adult

From tract lumen

To tract lumen
- Ingested: 2200 mL
- Saliva: 1500 mL
- Gastric secretion: 1500 mL
- Bile: 700 mL

Intestinal absorption: 8000-8400 mL

- Pancreatic juice: 1500 mL
- Intestinal secretion: 1500 mL

Colonic salvage: 400 mL

Total: 8800 mL   **Total: 8900 mL**

100 mL water (200 mL feces)

Water gains
Drinking Metabolism Eating

Food and drink: 2200 mL per day
Metabolism: 300 mL per day

Water losses
Urination Feces Skin Breathing

Skin and lungs: 900 mL per day
Urine: 1500 mL per day Feces: 100 mL per day

About 63% of our daily requirement for water is met through drinking fluids, 25% is obtained from food, and the remaining 12% comes from metabolism (the oxidation of glucose to ATP, CO_2, and water).

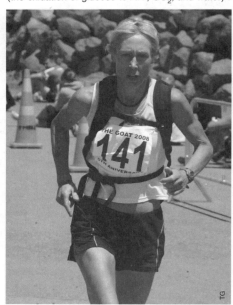

Typically, we lose 60% of body water through urination, 36% through the skin and lungs, and 4% in feces. Losses through the skin and from the lungs (breathing) average about 900 mL per day or more during heavy exercise. These are called **insensible losses**.

1. Explain how metabolism provides water for the body's activities: _____

2. Describe four common causes of physiological dehydration:

    (a) _____ (c) _____

    (b) _____ (d) _____

3. Some recent sports events have received media coverage because athletes have collapsed after excessive water intakes. This condition, called **hyponatremia** or water intoxication, causes nausea, confusion, diminished reflex activity, stupor, and eventually coma. From what you know of fluid and electrolyte balances in the body, explain these symptoms:

Related activities: Control of Urine Output, Fluid and Electrolyte Balance

Urine Analysis

Urine is the liquid waste product of the body. It contains water, electrolytes, and other waste metabolites which are filtered out of the blood by the kidneys. **Urine analysis** (urinalysis) is used as a medical diagnostic tool for a wide range of metabolic disorders. In addition, urine analysis can be used to detect the presence of illicit (non-prescription) drugs and for diagnosing pregnancy.

Diagnostic Urinalysis

A urinalysis (UA) is an array of tests performed on urine. It is one of the most common methods of medical diagnosis, as most tests are quick and easy to perform, they are non-invasive, and well understood diagnostically.

A typical urinalysis usually includes a **macroscopic analysis**, a **dipstick chemical analysis**, in which the test results can be read as color changes, and a **microscopic analysis**, which involves centrifugation of the sample and examination for crystals, blood cells, or microbial contamination.

MACROSCOPIC URINALYSIS
The first part of a urinalysis is direct visual observation. Normal, fresh urine is pale to dark yellow or amber in color and clear.

Turbidity or cloudiness may be caused by excessive cellular material or protein in the urine. A red or red-brown (abnormal) color could be from a food dye, eating fresh beets, a drug, or the presence of either hemoglobin or myoglobin. If the sample contained many red blood cells, it would be cloudy as well as red, as in this sample indicating hematuria (blood in the urine).

DIPSTICK URINALYSIS
Commonly dipstick tests indicate:
Urine pH: normal range is 4.5-8.0.
Specific gravity: Normal is 1.002 - 1.035 Specific gravity measures urine density, or the ability of the kidney to concentrate or dilute the urine over that of plasma.
Protein: Normal total protein excretion does not exceed 10 mg per 100 ml in any single specimen. More than 150 mg per day is defined as proteinuria.
Glucose: Less than 0.1% of glucose filtered by the glomerulus normally appears in urine. Excess sugar in urine generally indicates diabetes mellitus.
Ketones: Ketones in the urine result from diabetic ketosis or some other form of calorie deprivation (starvation).
Nitrite: Nitrites indicate that bacteria may be present in significant numbers.
Leukocyte esterase: A positive leukocyte esterase test results from the presence of whole or lysed white blood cells (indicating infection).

Testing For Anabolic Steroids

Anabolic steroids are synthetic steroids related to the male sex hormone **testosterone** (right). They work by increasing protein synthesis within cells, causing tissue, especially skeletal muscle, to build mass. They are used legitimately in medicine to stimulate bone growth and appetite, induce male puberty, and treat chronic wasting conditions. Misuse of anabolic steroids can have many adverse effects including elevated blood pressure, cardiovascular disease, and alteration of cholesterol ratios.

Steroids increase muscle mass and physical strength, and are used illegally by some athletes to gain an unfair advantage over their competitors. Anabolic steroid use is banned by most major sporting bodies, but many athletes continue to use them illegally. Athletes are routinely tested for the presence of **performance enhancing drugs**, including anabolic steroids.

Anabolic steroids break down into known metabolites which are excreted in the urine. The presence of specific metabolites indicates which substance has been used by the athlete. Some steroid metabolites stay in the urine for weeks or months after being taken, while others are eliminated quite rapidly. Athletes using anabolic steroids can escape detection by stopping use of the drugs prior to competition. This allows the body time to break down and eliminate the components, and the drug use goes undetected.

The Urinary System

1. Explain why urinalysis is a frequently used diagnostic technique for many common disorders: _____
_____

2. Explain why the pH of normal urine (4.5-8.0) is much more variable than the pH of the blood (pH 7.35-7.45): __
_____

3. Identify what each of the following might indicate in a urine sample:

(a) Cloudy, red colour: _____ (b) Positive leukocyte esterase test: _____

4. Explain why athletes exploiting illegal drugs might withhold them for a period before competition: _____
_____
_____

Related activities: Blood, Acid-Base Balance, The Hormones of Pregnancy

A 2

The Urinary System

The mammalian urinary system consists of the kidneys and bladder, and their associated blood vessels and ducts. The kidneys have a plentiful blood supply from the renal artery. The blood plasma is filtered by the kidneys to form urine. Urine is produced continuously, passing along the ureters to the bladder, a hollow muscular organ lined with smooth muscle and stretchable epithelium. Each day the kidneys filter about 180 liters of plasma. Most of this is reabsorbed, leaving a daily urine output of about 1 liter. By adjusting the composition of the fluid excreted, the kidneys help to maintain the body's internal chemical balance. The ability of the kidneys to concentrate the urine improves with maturity. It is less efficient in infants and young children, who produce relatively larger amounts of dilute urine and are also more susceptible to dehydration.

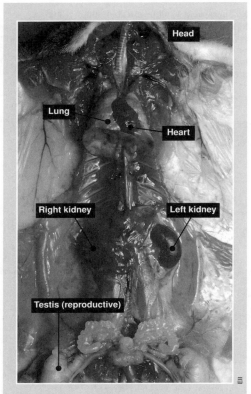

The kidneys of **rats** (above), humans, and other mammals are distinctive, bean shaped organs that lie at the back of the abdominal cavity to either side of the spine. The kidneys lie outside the peritoneum of the abdominal cavity and are partly protected by the lower ribs. Each kidney is surrounded by three layers of tissue. The innermost renal capsule is a smooth fibrous membrane that acts as a barrier against trauma and infection. The two outer layers comprise fatty tissue and fibrous connective tissue. These act to protect the kidney and anchor it firmly in place.

The Human Urinary System

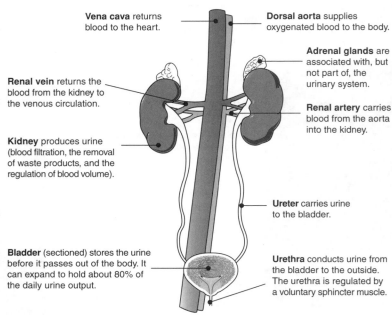

Vena cava returns blood to the heart.

Dorsal aorta supplies oxygenated blood to the body.

Adrenal glands are associated with, but not part of, the urinary system.

Renal vein returns the blood from the kidney to the venous circulation.

Renal artery carries blood from the aorta into the kidney.

Kidney produces urine (blood filtration, the removal of waste products, and the regulation of blood volume).

Ureter carries urine to the bladder.

Bladder (sectioned) stores the urine before it passes out of the body. It can expand to hold about 80% of the daily urine output.

Urethra conducts urine from the bladder to the outside. The urethra is regulated by a voluntary sphincter muscle.

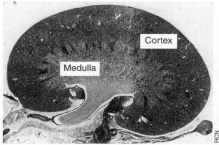

The very precise alignment of the nephrons (the filtering elements of the kidney) and their associated blood vessels gives the kidney tissue a striated appearance, as seen in this cross section.

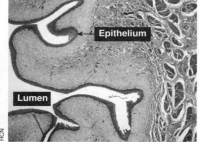

Transitional epithelium lines the bladder. This type of epithelium is layered, or stratified, and can be stretched without the outer cells breaking apart from each other.

1. Identify the components of the urinary system and describe their functions: _____

_____

_____

_____

2. Calculate the percentage of the plasma reabsorbed by the kidneys: _____

3. The kidney receives blood at a higher pressure than other organs. Explain why this is the case: _____

_____

_____

4. Explain the purpose of the fatty connective tissue surrounding the kidneys: _____

_____

The Physiology of the Kidney

The functional unit of the kidney, the **nephron**, is a selective filter element, comprising a renal tubule and its associated blood vessels. Filtration, i.e. forcing fluid and dissolved substances through a membrane by pressure, occurs in the first part of the nephron, across the membranes of the capillaries and the glomerular capsule. The passage of water and solutes into the nephron and the formation of the glomerular filtrate depends on

the pressure of the blood entering the afferent arteriole (below). If it increases, filtration rate increases; when it falls, glomerular filtration rate also falls. This process is so precisely regulated that, in spite of fluctuations in arteriolar pressure, glomerular filtration rate per day stays constant. After formation of the initial filtrate, the **urine** is modified through secretion and tubular reabsorption according to physiological needs at the time.

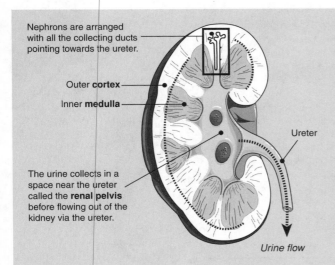

Nephrons are arranged with all the collecting ducts pointing towards the ureter.

Outer **cortex**

Inner **medulla**

Ureter

The urine collects in a space near the ureter called the **renal pelvis** before flowing out of the kidney via the ureter.

Urine flow

Internal Structure of the Human Kidney

Human kidneys are about 100-120 mm long and 25 mm thick. The functional unit of the kidney is the **nephron**. The other parts of the urinary system are primarily passageways and storage areas. The inner tissue of the kidney appears striated, because the nephrons and their surrounding blood vessels are aligned. It is this precise alignment of the nephrons in the kidney that makes it possible to fit in all the filtering units required.

Each kidney contains more than 1 million nephrons. They are **selective filter elements**, which regulate blood composition and pH, and excrete wastes and toxins. The initial urine is formed by **filtration** in the glomerulus. Plasma is filtered through three layers: the capillary wall, and the basement membrane and epithelium of Bowman's capsule. The epithelium comprises very specialized epithelial cells called **podocytes**. The filtrate is modified as it passes through the tubules of the nephron and the final urine passes out the ureter.

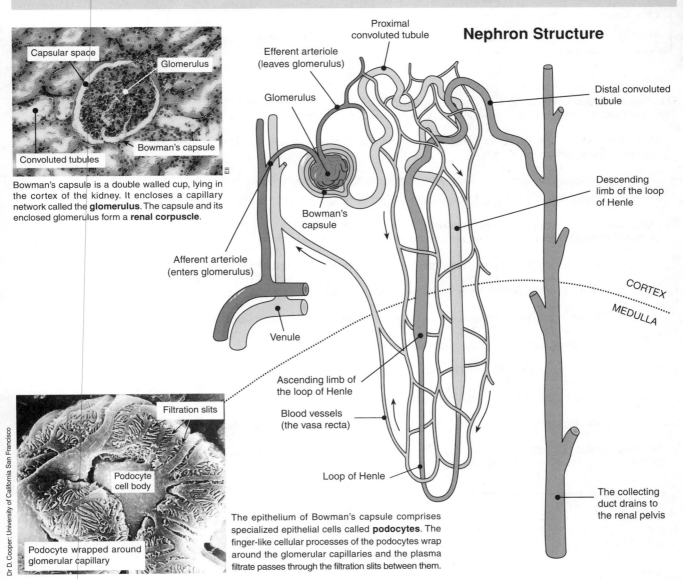

Capsular space

Glomerulus

Convoluted tubules

Bowman's capsule

Bowman's capsule is a double walled cup, lying in the cortex of the kidney. It encloses a capillary network called the **glomerulus**. The capsule and its enclosed glomerulus form a **renal corpuscle**.

Nephron Structure

Proximal convoluted tubule

Efferent arteriole (leaves glomerulus)

Glomerulus

Bowman's capsule

Afferent arteriole (enters glomerulus)

Venule

Distal convoluted tubule

Descending limb of the loop of Henle

CORTEX

MEDULLA

Ascending limb of the loop of Henle

Blood vessels (the vasa recta)

Loop of Henle

The collecting duct drains to the renal pelvis

Dr D. Cooper: University of California San Francisco

Filtration slits

Podocyte cell body

Podocyte wrapped around glomerular capillary

The epithelium of Bowman's capsule comprises specialized epithelial cells called **podocytes**. The finger-like cellular processes of the podocytes wrap around the glomerular capillaries and the plasma filtrate passes through the filtration slits between them.

Related activities: The Urinary System   **Web links**: Interactive Kidney Quiz, Kidney Vascular System, The Juxtaglomerular Apparatus

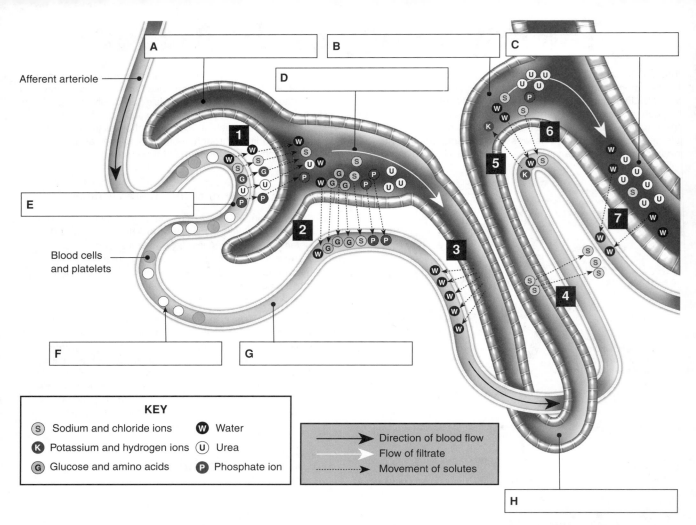

The diagram above presents an overview of the structures and processes involved in the formation of urine in the kidney. The structures involved are labelled with letters (**A-H**), while the major processes are identified with numbers (**1-7**).

1. Using the word list provided below, identify each of the structures marked with a letter. Write the name of the structure in the space provided on the diagram.

   distal convoluted tubule, efferent arteriole, glomerulus, Bowman's capsule,
   proximal convoluted tubule, loop of Henle, large blood proteins, collecting duct

2. Match each of the processes (identified on the diagram with numbers 1-7) to the correct summary of the process provided below. Write the process number next to the appropriate sentence.

   .... Active transport of salt (Na+ and Cl-) from the ascending limb of the loop of Henle.

   .... Filtration through the membranes of the glomerulus. Glucose, water, and ions pass through.

   .... Reabsorption of glucose and ions by active transport. Water follows by diffusion.

   .... Reabsorption of water by osmosis from the descending limb of the loop of Henle.

   .... Active secretion (into the filtrate) of H+ and K+ (NH3 also diffuses into the filtrate).

   .... Concentration of the urine by osmotic withdrawal of water from the flitrate.

   .... Reabsorption of Na+ and Cl- by active transport and water by osmosis.

3. There is marked gradient in salt concentration in the extracellular fluid of the medulla, produced by the transport of salt out of the filtrate. Explain the purpose of this salt gradient:

   _____

   _____

   _____

   _____

Control of Urine Output

Variations in salt and water intake, and in the environmental conditions to which we are exposed, contribute to fluctuations in blood volume and composition. The primary role of the kidneys is to regulate blood volume and composition (including the removal of nitrogenous wastes), so that homeostasis is maintained. This is achieved through varying the volume and composition of the urine. Two hormones, **antidiuretic hormone** (ADH) and **aldosterone**, are involved in the process.

Control of Urine Output

Osmoreceptors in the **hypothalamus** detect a fall in the concentration of water in the blood. They stimulate **neurosecretory cells** in the hypothalamus to synthesize and secrete the hormone ADH (antidiuretic hormone).

ADH passes from the hypothalamus to the posterior pituitary where it is released into the blood. ADH increases the permeability of the kidney collecting duct to water so that more water is reabsorbed and urine volume decreases.

Brain

ADH ACTS ON KIDNEY

Factors inhibiting ADH release
- Low solute concentration
  - High blood volume
  - Low blood sodium levels
- High fluid intake
- Alcohol consumption

ADH levels decrease → Water reabsorption decreases. Urine output increases.

Factors causing ADH release
- High solute concentration
  - Low blood volume
  - High blood sodium levels
- Low fluid intake
- Nicotine and morphine

ADH levels increase → Water reabsorption increases. Urine output decreases.

Factors causing release of aldosterone
Low blood volumes also stimulate secretion of aldosterone from the adrenal cortex. This is mediated through a complex pathway involving the hormone renin from the kidney.

Aldosterone → Sodium reabsorption increases, water follows, blood volume restored.

The Urinary System

1. (a) **Diabetes insipidus** is a disease caused by a lack of ADH. Based on what you know of the role of ADH in regulating urine volumes, describe the symptoms of this disease:

_____

_____

(b) Suggest how this disorder might be treated: _____

_____

2. Explain why alcohol consumption (especially to excess) causes dehydration and thirst: _____

_____

_____

3. Explain how negative feedback mechanisms operate to regulate blood volume and urine output: _____

_____

_____

4. **Diuretics** are drugs that increase urine volume. Many work by inhibiting the active transport of sodium and chloride in the nephron. Explain how this would lead to an increase in urine volume:

_____

_____

© Biozone International 2009
Photocopying Prohibited

Related activities: Water Budget in a Human, The Physiology of the Kidney, The Endocrine System, Hormones of the Pituitary

RA 2

Fluid and Electrolyte Balance

The body's fluid and electrolyte balance is critical to metabolic function. Water makes up around 60% of the body and is found within two main fluid compartments. The **intracellular fluid** makes up 60-65% of the water in the body and is found within the body's cells. The **extracellular fluid** makes up the rest of the body's water and can be divided into **intravascular fluid** (mostly blood) and the **extravascular fluid** (interstitial fluid around the cells). Electrolytes in the body fluids are responsible for maintaining osmotic gradients and permitting ion exchanges. For example, in the blood plasma, electrolytes help to maintain the blood volume by keeping water moving into the capillaries. When electrolyte (mostly Na^+) levels fall, water moves out of the capillaries and into the tissues. This causes blood volume and pressure to fall and plasma to thicken. Two hormones are involved in regulating blood volume: **ADH**, which promotes water reabsorption in the kidney collecting ducts, and **aldosterone**. Aldosterone promotes sodium reabsorption in the kidney tubules and the most important mechanism for regulating its release is the **renin-angiotensin system (RAS)**. The RAS is mediated by the **juxtaglomerular (JG) apparatus** in the renal tubules.

The Renin-Angiotensin System

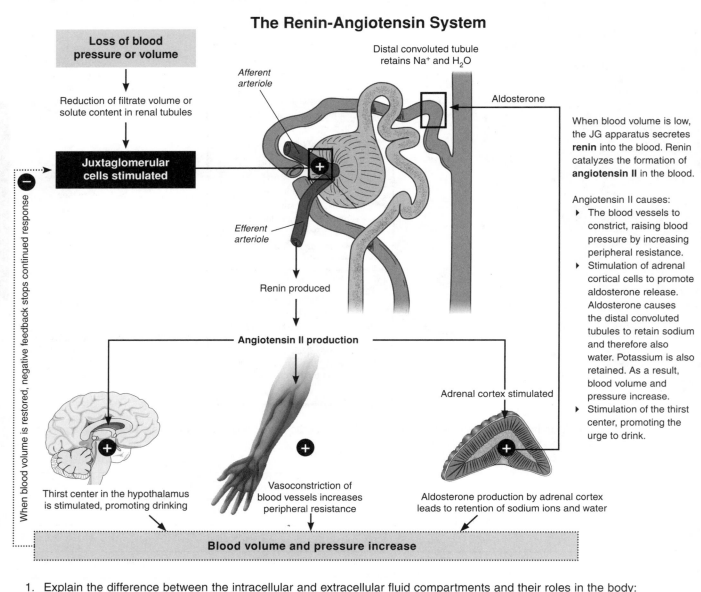

1. Explain the difference between the intracellular and extracellular fluid compartments and their roles in the body:

_____

_____

2. (a) Describe two situations that could cause a fall in blood volume: _____

_____

(b) Explain how the renin-angiotensin system responds to this loss of blood volume: _____

_____

_____

_____

Related activities: Control of Urine Output, Kidney Transplants
Web links: The Juxtaglomerular Apparatus

Acid-Base Balance

The pH of the body's fluids must be maintained within a very narrow range (pH 7.35-7.45). The products of metabolic activity are generally acidic and could alter pH considerably without a buffer system to counteract pH changes. The carbonic acid-bicarbonate buffer works throughout the body to maintain the pH of blood plasma close to 7.40. The body maintains the buffer by eliminating either the acid (carbonic acid) or the base (bicarbonate ions). The blood buffers, the lungs, and the kidneys represent the three defense systems against disturbances of pH homeostasis. Changes in carbonic acid concentration can be effected within seconds through increased or decreased respiration. The renal system, although acting more slowly, can permanently eliminate metabolic acids and regulate the levels of alkaline substances, controlling pH by either excreting or retaining bicarbonate ions.

The Blood Buffer System

Strong base neutralized to weak base

OH^- → HCO_3^-

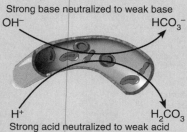

H^+ → H_2CO_3
Strong acid neutralized to weak acid

A buffer is able to resist changes to the pH of a fluid when either an acid or base is added to it. The bicarbonate ion (HCO_3^-) and its acid, carbonic acid (H_2CO_3), work in the following way:

$$H^+ + HCO_3^- \rightleftharpoons H_2CO_3$$

$$H_2CO_3 \rightleftharpoons H^+ + HCO_3^-$$

If a strong acid (such as HCl) is added to the system a weak acid is formed and thus the pH falls only slightly. Note that the blood also contains proteins, which contain basic and acidic groups that may act either as H^+ acceptors or donors to help maintain blood pH.

The Respiratory System

Signal to brain $CO_2 + H_2O \rightleftharpoons H_2CO_3$

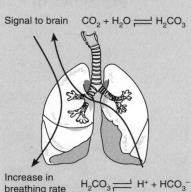

Increase in breathing rate $H_2CO_3 \rightleftharpoons H^+ + HCO_3^-$

Carbon dioxide (CO_2) in the blood, an end-product of cellular respiration, forms carbonic acid (H_2CO_3) which dissociates to form H^+ and bicarbonate (HCO_3^-). This means that as CO_2 rises in the blood so too does the H^+ concentration. **Chemoreceptors** in the brain detect the rise in H^+ ions and increase the rate of breathing to expel the CO_2. Low levels of CO_2 have the effect of depressing the respiratory system so that H^+ builds up and the pH is once again restored.

The Renal System

Rise in pH stimulates: Fall in pH stimulates:

Removal HCO_3^- Removal H^+

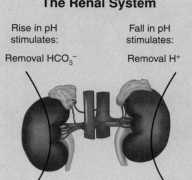

Gain H^+ Gain HCO_3^-

Recall that a net loss of HCO_3^- effectively results in the gain of H^+.

When blood pH rises, bicarbonate is excreted (lost from the body) and H^+ is retained by the tubule cells. Conversely, when blood pH falls, bicarbonate is reabsorbed and H^+ is actively secreted. Urine pH can normally vary from 4.5 to 8.0, reflecting the ability of the renal tubules to excrete or retain ions to maintain the homeostasis of blood pH.

1. Explain why the blood must be kept at a pH between 7.35 and 7.45: _____

_____

2. A drop in the blood pH to below 7.35 is called metabolic acidosis, even though the blood might still be at pH >7 and not strictly acidic. Describe how metabolic acidosis might arise:

_____

3 (a) Describe how the blood buffer system maintains blood pH: _____

_____

(b) Explain the effects of adding a base (e.g. ingestion of alkaline substances) to the system: _____

_____

4. (a) Describe the respiratory response to excess H^+ in the blood: _____

_____

(b) Explain where these H^+ ions come from: _____

(c) Describe how **respiratory acidosis** might arise: _____

_____

5. Explain the role of the renal system in maintaining the pH of the blood: _____

_____

_____

The Urinary System

Kidney Dialysis

A dialysis machine is a machine designed to remove wastes from the blood. It is used when the kidneys fail, or when blood acidity, urea, or potassium levels increase much above normal. In kidney dialysis, blood flows through a system of tubes composed of partially permeable membranes. Dialysis fluid (dialyzate) has a composition similar to blood except that the concentration of wastes is low. It flows in the opposite direction to the blood on the outside of the dialysis tubes. Consequently, waste products like urea diffuse from the blood into the dialysis fluid, which is constantly replaced. The dialysis fluid flows at a rate of several 100 cm^3 per minute over a large surface area. For some people dialysis is an ongoing procedure, but for others dialysis just allows the kidneys to rest and recover from injury or the effects of drugs or other metabolic disturbance.

Principles of Kidney Dialysis

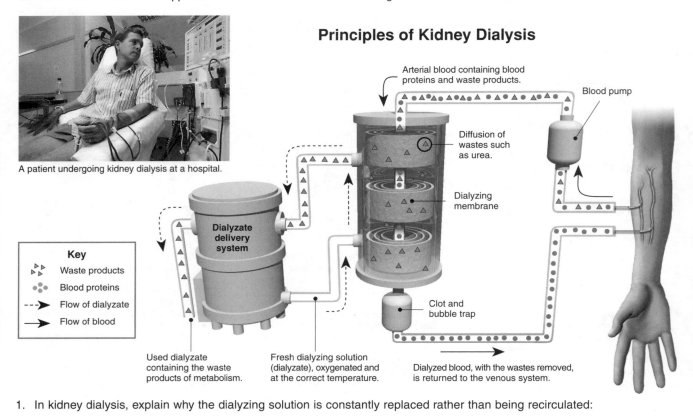

A patient undergoing kidney dialysis at a hospital.

Key
- Waste products
- Blood proteins
- Flow of dialyzate
- Flow of blood

Dialyzate delivery system

Arterial blood containing blood proteins and waste products.

Blood pump

Diffusion of wastes such as urea.

Dialyzing membrane

Clot and bubble trap

Used dialyzate containing the waste products of metabolism.

Fresh dialyzing solution (dialyzate), oxygenated and at the correct temperature.

Dialyzed blood, with the wastes removed, is returned to the venous system.

1. In kidney dialysis, explain why the dialyzing solution is constantly replaced rather than being recirculated:

_____

_____

2. Explain why ions such as potassium and sodium, and small molecules like glucose do not diffuse rapidly from the blood into the dialyzing solution along with the urea:

_____

_____

3. Explain why the urea passes from the blood into the dialyzing solution: _____

_____

4. Describe the general transport process involved in dialysis: _____

_____

5. Give a reason why the dialyzing solution flows in the opposite direction to the blood: _____

_____

_____

6. Explain why a clot and bubble trap is needed after the blood has been dialyzed but before it re-enters the body:

_____

_____

_____

Related activities: Passive Transport Processes, Organ Transplants, Kidney Transplants Web links: Dialysis Animation

Kidney Transplants

Kidney failure (also called renal failure) arises when the kidneys fail to function adequately and filtrate formation decreases or stops. In cases of renal failure, normal blood volume levels and electrolyte balances are not maintained, and waste products build up in the body. Kidney failure is classified as **acute** (rapid onset) or **chronic** (developing over a period of months or years). There are many causes of kidney failure including decreased blood supply, drug overdose, chemotherapy, infection, and poorly controlled diabetic or hypertensive conditions. Recovery from acute renal failure is possible, but chronic renal damage can not be reversed. If kidney deterioration is ignored, the kidneys will fail completely. In some cases diet and medication can be used to treat kidney failure, but when the damage is extensive, **kidney dialysis** or a **kidney transplant** are required to keep the patient alive.

Renal Failure

Kidney (renal) failure is indicated by levels of **serum creatinine**, as well as by kidney size on ultrasound and the presence of anemia (chronic kidney disease generally leads to anemia and small kidney size). Creatinine is a break-down product of creatine phosphate in muscle, and is usually produced at a fairly constant rate by the body (depending on muscle mass). It is chiefly filtered out of the blood by the kidneys, although a small amount is actively secreted by the kidneys into the urine. A rise in blood creatinine levels is observed only with marked damage to functioning nephrons.

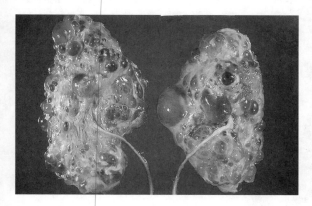

Acute renal failure (ARF) is characterized by decreased urine production (<400mL per day), and commonly arises because of low blood volume (blood loss), dehydration, or widespread infection. In contrast, chronic renal failure, which develops over months or years, is commonly the result of poorly controlled diabetes, poorly controlled high blood pressure, or **polycystic kidney disease**, a genetic disorder characterized by the growth of numerous cysts in the kidneys (above).

Kidney Transplants

Transplantation of a healthy kidney from an organ donor is the preferred treatment for end-stage kidney failure. The organ is usually taken from a person who has just died, although kidneys can also be taken from living donors. The failed organs are left in place and the new kidney transplanted into the lower abdomen. Provided recipients comply with medical requirements (e.g. correct diet and medication) over 85% of kidney transplants are successful.

There are two major problems associated with kidney transplants: lack of donors and tissue rejection. Cells from donor tissue have different antigens to that of the recipient, and are not immunocompatible. Tissue-typing and the use of immunosuppressant drugs helps to decrease organ rejection rates. In the future, xenotransplants of genetically modified organs from other species may help to solve both the problems of supply and immune rejection.

Creatinine

Creatinine levels in both blood and urine is used to calculate the creatinine clearance (CrCl), which reflects the glomerular filtration rate (GFR). The GFR is a clinically important measurement of renal function and more accurate than serum creatinine alone, since serum creatinine only rises when nephron function is very impaired.

1. Distinguish between acute and chronic renal failure and contrast their causes: _____

   _____

   _____

2. (a) Explain why a rise in blood (serum) levels of creatinine would indicate a failure of nephron function: _____

   _____

   (b) Explain why a creatinine clearance is a more accurate indicator of renal function than a serum creatinine test alone:

   _____

   _____

3. Describe some of the advantages and disadvantages of kidney transplantation over a life-time of kidney dialysis:

   _____

   _____

   _____

The Reproductive System

Cardiovascular system

- Estrogens lower blood HDL cholesterol levels and promote cardiovascular health in premenopausal women.
- Pregnancy places extra demands on the cardiovascular system. Blood volume increase 40-50% during pregnancy.
- Local vasodilation is responsible for aspects of the sexual response in men and women.
- Blood transports sex hormones to target tissues.

Respiratory system

- Respiratory system provides O_2 to the reproductive system and disposes of CO_2 produced by cellular respiration.
- Vital capacity and breathing rate increase in pregnancy. Enlarged uterus impairs descent of the diaphragm and can cause shortness of breath late in pregnancy.

Endocrine system

- Reproductive hormones from the ovaries (in females) and testes (in males) are responsible for the development of secondary sexual characteristics. They are regulated via feedback mechanisms to the hypothalamic-pituitary axis.
- Gonadotropins (e.g. LH and FSH) help to regulate gonadal function.
- Placental hormones maintain pregnancy.

Skeletal system

- Sex hormones are responsible for secondary sexual characteristics associated with the skeleton: in males, androgens masculinize the skeleton (broad shoulders and expanded chest) and increase bone density; in females, estrogen causes pelvic widening and maintains bone mass.
- Bony pelvis protect some reproductive organs.

Integumentary system

- In lactating women, milk from mammary glands nourishes the infant.
- Androgens activate oil glands and lubricate skin and hair. Estrogen increases skin hydration and increases skin pigmentation in pregnancy.
- Sex hormones are responsible for secondary sexual characteristics associated with the integument, e.g. appearance of pubic hair and changes in fat distribution associated with male and female body shape.

Nervous system

- Sex hormones masculinize or feminize the brain and influence sex drive.
- The neurohormone, GnRH from the hypothalamus regulates the timing of puberty.
- Reflexes regulate aspects of the sexual response (e.g. orgasm).

Lymphatic system and immunity

- Immune cells protect against pathogens. Regulatory T cells important in the immune tolerance to the developing fetus.
- Lymphatic (and blood) vessels transport sex hormones.
- Maternal antibodies pass to the fetus *in-utero* and are present in breast milk, providing passive immunity.
- Increased abdominal pressure in pregnancy impairs lymphatic return leading to edema.

Digestive system

- Digestive organs are crowded in late pregnancy and constipation and heartburn are common.
- Increased hormone levels result in nausea and vomiting in early pregnancy.

Urinary system

- Increased frequency and urgency of urination in pregnancy as a result of pressure on the bladder and pelvic floor.
- Enlargement of prostate (usually) in older men can impede urination.
- Kidneys dispose of nitrogenous waste and maintain fluid and electrolyte balance of mother and fetus in pregnancy.
- Urethra provides passage for semen.

Muscular system

- Androgens promote an increase in muscle mass in post-pubertal males.
- Pelvic floor muscles provide support for the reproductive organs and are involved in aspects of the sexual response.
- Abdominal, uterine, and pelvic floor muscles are active in childbirth.

General Functions and Effects on all Systems

The reproductive system in adults is responsible for reproduction, i.e. the production of gametes and offspring. Unlike other body systems, which are functioning almost continuously since birth, the reproductive system is quiescent until puberty, at which time it begins development towards maturity.

Disease

Symptoms of disease	• Pain (moderate to severe) • Abnormal bleeding • Infertility
Disorders and diseases of the male reproductive system	• Sexually transmitted infections • Cancers (e.g. prostate cancer) • Congenital abnormalities • Functional disorders (e.g. erectile dysfunction, premature ejaculation)
Disorders and diseases of the female reproductive system	• Sexually transmitted infections • Cancers (e.g. cervical cancer) • Congenital abnormalities • Functional disorders (e.g. infertility, ectopic pregnancy, endometriosis)

Medicine & Technology

Diagnosis of disorders	• MRI scan and ultrasound • Semen analysis • Laparoscopy • Blood and DNA (genetic) tests
Treating reproductive disorders	• Drug therapy (e.g. antibiotics) • Hormone therapy • Surgery (e.g. hysterectomy)
Treatment of infertility	• Assisted reproductive technologies • Hormone therapy (e.g. clomiphene) • Laparoscopic surgery
Contraception	• Physical (barrier) methods • Hormonal (e.g. oral contraceptive pill) • Surgical (e.g. vasectomy)

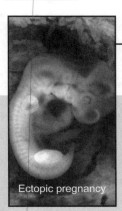

Ectopic pregnancy

• Infertility
• Breast cancer
• Menstrual cycle
• Ectopic pregnancy

• IVF and GIFT
• Ultrasound
• HRT
• Oral contraception
• Pregnancy testing

The Human Life Span
The Reproductive System

The reproductive system undergoes marked changes associated with aging and the end of fertility. Disease may affect it both directly and indirectly.

Medical technologies are used to detect, diagnose and treat reproductive disorders and control fertility.

• Menopause
• Decline in hormone levels
• Decline in fertility

Excessive exercise may lead to:
• Exercise induced amenorrhea
• GNRH depression

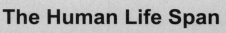

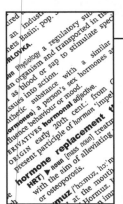

Aging and the reproductive system	• Decline in sperm production and erectile function • Cessation of menses (women) • Thinning and prolapse of organs

Effects of exercise on the reproductive system	• Improved muscle tone • Reduced risk of reproductive cancers • Heavy endurance exercise can lead to hormonal and menstrual irregularities • Exercise beneficial during pregnancy

The Effects of Aging

Exercise

Reproduction and Development

Understanding the structure and function of the urinary system

Structure and function of the urinary system: fluid budgets, nephron structure and function, regulation of fluid and electrolytes, maintenance of blood pH. Renal failure: dialysis and kidney transplants.

Learning Objectives

☐ 1. Compile your own glossary from the **KEY WORDS** displayed in **bold type** in the learning objectives below.

☐ 2. Recall the role of **meiosis** in producing **gametes**, identifying the significance of the reduction division. Explain how **fertilization** restores the diploid chromosome number in the zygote.

The Reproductive System *(pages 99, 212, 215-16)*

☐ 3. Using an annotated diagram, describe the structure and function of the reproductive system in males and females, including the external **genitalia**. Describe the dual roles of the **primary sex organs** (gonads) in males and females.

☐ 4. Identify major sex hormones in males and females and briefly describe their various roles, including reference to **estrogen**, **progesterone**, and **testosterone**. Recognize the general role of pituitary **gonadotropic hormones** in regulating gonadal function.

☐ 5. Draw the structure of **testis** tissue as seen using a light microscope. Include reference to **seminiferous tubules**, blood capillaries, and **interstitial cells**. Draw a seminiferous tubule with adjacent interstitial cells in XS. Indicate the **Sertoli cells** and developing sperm.

☐ 6. Outline the processes involved in **spermatogenesis** and their location. Include reference to mitosis, cell growth, meiosis, and cell differentiation. Outline the role of follicle stimulating hormone (**FSH**), luteinizing hormone (**LH**), and **testosterone** in spermatogenesis.

☐ 7. Outline the role of the **epididymis**, **seminal vesicle**, and **prostate gland** in the production of **semen**.

☐ 8. Draw the structure of the ovary as seen using a light microscope. Include reference to the following: **primary oocytes**, **zona pellucida**, **Graafian follicles**.

☐ 9. Outline the processes involved in **oogenesis**, including reference to mitosis, cell growth, meiosis, the unequal division of the cytoplasm, and the degeneration of the **polar bodies**. If required, outline the role of **hormones** in **gametogenesis** in females.

☐ 10. Compare spermatogenesis and oogenesis with respect to the number of gametes formed and the timing of the gamete formation and release. Draw the structure of mature **sperm** and **egg**, relating specific structural features to their functional role.

The Menstrual Cycle *(pages 213-214)*

☐ 11. Explain the main features of the human **menstrual cycle** including the development of the ovarian **follicles** and **corpora lutea**, the cyclical changes to the **uterine endometrium**, and **menstruation**. Include to:
  • Days 1-5: Menstrual phase
  • Days 6-14: Proliferative phase
  • Days 15-18: Secretory phase

☐ 12. Relate the changes in the menstrual cycle to the changes in the hormones regulating the cycle: **progesterone**, **estrogen**, **FSH**, and **LH**. Emphasize the role of **feedback control** in the menstrual cycle.

Human Development *(pages 217-221, 225, 227)*

☐ 13. Appreciate how **fertilization** in humans is dependent on the timing of gamete transfer. In general terms, describe the events in fertilization.

☐ 14. In more detail than above, describe fertilization, including reference to the timing and significance of the **acrosome reaction**, **penetration** of the egg membrane by the sperm, and the **cortical reaction**.

☐ 15. Describe early embryonic **development** up to the **implantation** of the **blastocyst** (including **cleavage**).

☐ 16. Describe the main events in embryonic **development** between implantation and 5-8 weeks, including the early development of nervous and circulatory systems. Describe the role of programmed cell death (**apoptosis**) in the development of the fetus.

☐ 17. Explain the role of **human chorionic gonadotropin** (also called human chorionic gonadotrophin or HCG) in early pregnancy. Identify the source of this hormone. Describe the use of monoclonal antibodies for detecting HCG in pregnancy tests.

☐ 18. Draw a diagram of the uterus during **pregnancy** and identify the following: **uterus**, **placenta**, **umbilical cord**, **embryonic membranes**, **amniotic fluid**, **fetus**. Recognize pregnancy as the period of **gestation**.

☐ 19. Describe the role of the amniotic sac and amniotic fluid in supporting and protecting the fetus. Appreciate the role of **placenta** in the exchange of materials between the maternal and fetal blood.

☐ 20. Describe the structure and function of the **placenta**. Explain how the placenta is maintained during **pregnancy** and describe its functions in relation to the development of the embryo.

☐ 21. Describe the effects of pregnancy on maternal physiology, including both anatomical and hormonal changes, and nutritional requirements.

☐ 22. Recognize **labor** as the series of events that expel the infant from the uterus. Outline the process of **parturition** (**birth**) and its control, including the role of **oxytocin** and **prostaglandins**, and decline in levels of **progesterone**. Explain how parturition is accomplished through **positive feedback** mechanisms. Outline the events in each of the three phases of labor: dilation of the cervix, expulsion, and delivery of the placenta.

☐ 23. Describe growth and physical development from birth to adult, including reference to changes in body proportions between newborn, 2 year old, 5 year old, adolescent, and adult. Relate these changes to differential growth in the skeleton.

24. Explain what is meant by **puberty** and describe the physical changes associated with it in both males and females. Identify **primary** and **secondary sexual characteristics** in males and females and explain the role of **sex hormones** (**testosterone** and **estrogen**) in human development.

The Breast and Lactation *(pages 226, 229)*

25. Describe the structure of the functioning mammary gland in women, identifying the **areola**, **nipple**, **alveolar glands**, and **lactiferous ducts**.

26. Explain the importance of **lactation** to early nutrition in infants. Describe the composition of human milk and discuss its beneficial nutritional and protective properties. Describe lactation, including the function and regulation of **prolactin** and **oxytocin**.

27. Describe some of the causes of **breast cancer** and describe risk factors in its development. Recognize the involvement of environmental, hereditary, and biological factors in the development of breast cancer. Describe changes in the breast tissue occurring as a result of cancer and outline some of the diagnostic and treatment options available.

Aging *(pages 55, 103)*

28. Explain the biological basis of **aging** and discuss the physiological changes that occur as a result of it, including its effects on the reproductive system of males and females.

29. Describe the causes **menopause** in women and describe how it is defined. Recognize menopause as a normal life stage and not a disorder. Describe the

hormonal changes that occur prior to menopause (peri-menopause) and explain why **post-menopausal** women are at higher risk (than pre-menopausal women) of certain diseases, including heart disease and osteoporosis.

30. Describe some of the **degenerative diseases** of aging that affect body systems other than the reproductive system. Recognize that the development of degenerative diseases is often the result of a decline in the function of several body systems that normally operate together to maintain homeostasis. Examples could include: **Alzheimer's disease**, **osteoarthritis**, **osteoporosis**, **cataracts**, or **hypermetropia**.

Reproductive Technology *(pages 222-224)*

31. Explain what is meant by **contraception**. Describe four methods of contraception, including at least one method from each of the following: **mechanical**, **chemical**, and **behavioral**. identify where in the cycle between gamete production and fertilization each of the methods operates and comment on its effectiveness in preventing pregnancy. Recognize that contraception represent an ethical concern for some.

32. Recognize the role of reproductive technologies in treating **infertility**. Identify reasons for infertility in males and females and discuss some of the options available to enhance fertility 9incluing any ethical concerns raised by the procedures). Include reference to any of the following:
    (a) *In-vitro* fertilization (IVF)
    (b) GIFT (**gamete intrafallopian transfer**)
    (c) **Artificial insemination** (AI)

Supplementary Texts

See page 9 for additional details of these texts:

■ Morton, D. and Perry, J.W. 1997. **Photo Atlas for Anatomy and Physiology**, (Brooks/Cole).

■ Rowett, H.G.Q., 1999. **Basic Anatomy & Physiology**, (John Murray), pp. 125-131.

Periodicals

See page 9 for details of publishers of periodicals:

■ **The Biology of Milk** Biol. Sci. Rev., 16(3) Feb. 2004, pp. 2-6. *The production and composition of milk, its role in mammalian biology, and the physiological processes controlling its release.*

■ **Spermatogenesis** Biol. Sci. Rev., 15(4) April 2003, pp. 10-14. *The process and control of sperm production in humans, with a discussion of the possible reasons for male infertility.*

■ **The Great Escape** New Scientist, 15 Sept. 2001, (Inside Science). *How the fetus is accepted by the mother's immune system during pregnancy.*

Presentation MEDIA to support this topic:
Anatomy & physiology

■ **Measuring Female Hormones in Saliva** Biol. Sci. Rev., 13(3) Jan. 2001, pp. 37-39. *The female reproductive system, and the complex hormonal control of the female menstrual cycle.*

■ **Adolescence - Hormones Rule OK?** Biol. Sci. Rev., 19(3) Feb. 2007, pp. 2-6. *The hormonal changes bringing about reproductive maturity.*

■ **The Placenta** Biol. Sci. Rev., 12 (4) March 2000, pp. 2-5. *The structure and function of the human placenta (includes prenatal diagnoses).*

■ **Menopause - Design Fault, or By Design** Biol. Sci. Rev., 14(1) Sept. 2001, pp. 2-6. *An excellent synopsis of the basic biology of menopause.*

■ **Aching Joints and Breaking Bones** Biol. Sci. Rev., 20(2) Nov. 2007, pp.10-13. *As people age, they suffer problems with their bones and joints, particularly, osteoarthritis and osteoporosis.*

■ **How to Live to be 100... and Enjoy it** New Scientist, 3 June 2006, pp. 35-45. *Nine things that seem to lead to a longer happier life and why.*

■ **Age - Old Story** New Scientist, 23 Jan. 1999, (Inside Science). *The processes involved in aging. An accessible, easy-to-read, but thorough account.*

TEACHER'S REFERENCE

■ **Pregnancy Tests** Scientific American, Nov. 2000, pp. 92-93. *Pregnancy tests: how they work and the role of HCG in signalling pregnancy.*

■ **The Evolution of Human Birth** Sci. American, Nov. 2001, pp. 60-65. *An examination of the unique aspects of human reproduction and how they arose.*

■ **The Timing of Birth** Scientific American, March 1999, pp. 50-57. *A hormone found in the human placenta influences the timing of delivery.*

■ **Let me Out** New Scientist, 10 Jan. 1998, pp. 24-29. *Fetus and mother are in conflict and it is the fetus that determines the timing of birth.*

■ **Male Contraception** Biol. Sci. Rev., 13(2) Nov. 2000, pp. 6-9. *A new contraceptive technology involves the inhibition of spermatogenesis in males.*

Internet WWW

See pages 10-11 for details of how to access **Bio Links** from our web site: **www.thebiozone.com** From Bio Links, access sites under the topics:

ANATOMY & PHYSIOLOGY > **General sites**:
• Anatomy & Physiology • Human physiology lecture notes • WebAnatomy > **Reproduction and Development**: • Anatomical travelogue • Atlanta Reproductive Health Center • Effects of drugs in pregnancy • Fertility UK: Physiology • Hormones of the reproductive system • Menstrual cycle and pregnancy • UNSW embryology

BIOTECHNOLOGY > **Applications in Biotechnology** > **Reproductive Biotechnology**
• Assisted reproductive technologies • Atlanta Reproductive Health Center • Fertility NZ • Frequently asked questions about infertiity • London Fertility Centre

Reproduction & Development

Female Reproductive System

The female reproductive system in mammals produces eggs, receives the penis and sperm during sexual intercourse, and houses and nourishes the young. Female reproductive systems in mammals are similar in their basic structure (uterus, ovaries etc.) but the shape of the uterus and the form of the placenta during pregnancy vary. The human system is described below.

Oogenesis

Oogenesis is the process by which mature ova (egg cells) are produced by the ovary. Oogonia are formed in the female embryo and undergo repeated mitotic divisions to form the primary oocyte. These remain in prophase of meiosis I throughout childhood. At this stage, all the eggs a female will ever have are present, but they remain in this resting phase until puberty. At puberty, meiosis resumes. Eggs are released, arrested in metaphase of meiosis II. This second division is only completed upon fertilization.

Oogonium

Growth (mitotic cell division)

Primary oocyte (2N)

1st meiotic division (meiosis I)

Completed in the fetus

Secondary oocyte (N)

First polar body (N)

2nd meiotic division (meiosis II)

Completed in the adult

Mature ovum (N)   Second polar body (N)   Additional polar bodies (do not always form)

The Female Reproductive System

(a)   (b)   (c)

Spine

Colon

Bladder

(d)

Pubis

(e)

(f)

Anus

Labia

Urethra

Side view of reproductive organs

Ovulation and Implantation

The unfertilized egg lives only for a day or so. It travels along the **fallopian tube**, where fertilization may occur if sperm are present.

A

Eggs or ova are produced by the **ovaries** and are released at ovulation.

If the egg is fertilized it will become implanted in the lining of the **uterus**. If it is not fertilized the prepared lining is shed, passing out through the vagina in a process called menstruation.

Fertilization occurs in the fallopian tube, after which it passes down to the uterus.

Front view of uterus and associated structures

1. The female human reproductive system and associated structures are illustrated above. Using the word list, identify the labeled parts. **Word list**: *ovary, uterus (womb), vagina, fallopian tube (oviduct), cervix, clitoris.*

2. In a few words or a short sentence, state the function of each of the structures labeled (a) - (d) in the above diagram:

   (a) _____

   (b) _____

   (c) _____

   (d) _____

3. (a) Name the organ labeled (**A**) in the diagram: _____

   (b) Name the event associated with this organ that occurs every month: _____

   (c) Name the process by which mature ova are produced: _____

4. (a) Name the stage in meiosis at which the oocyte is released from the ovary: _____

   (b) State when in the reproductive process meiosis II is completed: _____

Related activities: The Menstrual Cycle, Mitosis and the Cell Cycle

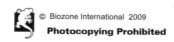

The Menstrual Cycle

In contrast to non-primate mammals, which have a breeding season, humans and other primates are sexually receptive throughout the year and may mate at any time, although fertilization of the ovum is most likely to occur during a relatively restricted period around the time of ovulation. Like all placental mammals, their uterine lining thickens in preparation for pregnancy. However, unlike other mammals, primates shed this lining as a discharge through the vagina if fertilization does not occur. This event, called **menstruation**, characterizes the human reproductive or **menstrual cycle**. In human females, the menstrual cycle starts from the first day of bleeding and lasts for about 28 days. It involves a predictable series of changes that occur in response to hormones. The cycle is divided into three phases (see below), defined by the events in each phase.

The Menstrual Cycle

Luteinizing hormone (LH) and follicle stimulating hormone (FSH): These hormones from the anterior pituitary have numerous effects. FSH stimulates the development of the ovarian follicles resulting in the release of estrogen. Estrogen levels peak, stimulating a surge in LH and triggering ovulation.

Hormone levels: Of the follicles that begin developing in response to FSH, usually only one (the Graafian follicle) becomes dominant. In the first half of the cycle, estrogen is secreted by this developing Graafian follicle. Later, the Graafian follicle develops into the corpus luteum (below right) which secretes large amounts of progesterone (and smaller amounts of estrogen).

The corpus luteum: The Graafian follicle continues to grow and then (around day 14) ruptures to release the egg (ovulation). LH causes the ruptured follicle to develop into a corpus luteum (yellow body). The corpus luteum secretes progesterone which promotes full development of the uterine lining, maintains the embryo in the first 12 weeks of pregnancy, and inhibits the development of more follicles.

Menstruation: If fertilization does not occur, the corpus luteum breaks down. Progesterone secretion declines, causing the uterine lining to be shed (menstruation). If fertilization occurs, high progesterone levels maintain the thickened uterine lining. The placenta develops and nourishes the embryo completely by 12 weeks.

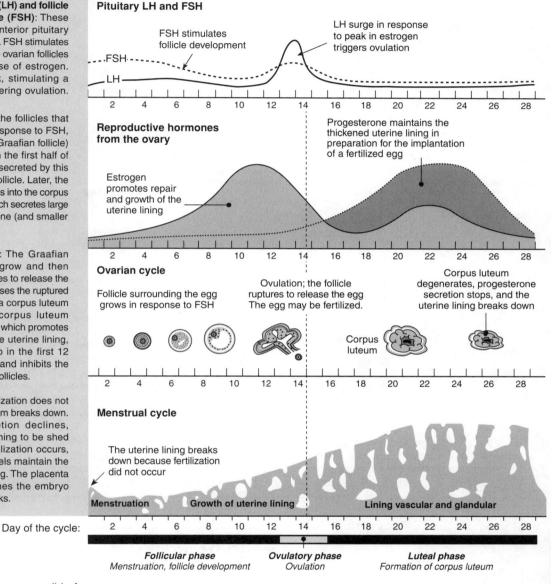

1. Name the hormone responsible for:

   (a) Follicle growth: _____   (b) Ovulation: _____

2. Each month, several ovarian follicles begin development, but only one (the Graafian follicle) develops fully:

   (a) Name the hormone secreted by the developing follicle: _____

   (b) State the role of this hormone during the follicular phase: _____

   (c) Suggest what happens to the follicles that do not continue developing: _____

3. (a) Identify the principal hormone secreted by the corpus luteum: _____

   (b) State the purpose of this hormone: _____

4. State the hormonal trigger for menstruation: _____

Related activities: Control of the Menstrual Cycle
Web links: The Menstrual Cycle Animation, Ovarian and Uterine Cycle

A 2

Reproduction & Development

Control of the Menstrual Cycle

The female menstrual cycle is regulated by the interplay of several reproductive hormones. The main control centers for this regulation are the **hypothalamus** and the **anterior pituitary gland**. The hypothalamus secretes GnRH (gonadotrophin releasing hormone), a hormone that is essential for normal gonad function in males and females. GnRH is transported in blood vessels to the anterior pituitary where it brings about the release of two hormones: follicle stimulating hormone (FSH) and luteinizing hormone (LH). It is these two hormones that induce the cyclical changes in the ovary and uterus. Regulation of blood hormone levels during the menstrual cycle is achieved through **negative feedback** mechanisms. The exception to this is the mid cycle surge in LH (see previous page) which is induced by the rapid increase in estrogen secreted by the developing follicle.

Control of the Menstrual Cycle

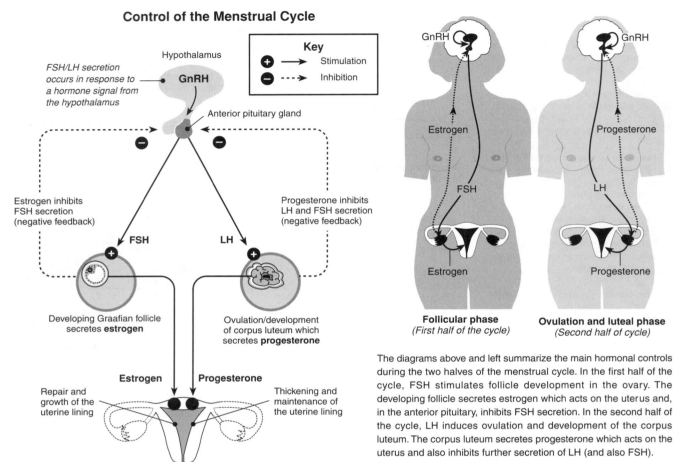

The diagrams above and left summarize the main hormonal controls during the two halves of the menstrual cycle. In the first half of the cycle, FSH stimulates follicle development in the ovary. The developing follicle secretes estrogen which acts on the uterus and, in the anterior pituitary, inhibits FSH secretion. In the second half of the cycle, LH induces ovulation and development of the corpus luteum. The corpus luteum secretes progesterone which acts on the uterus and also inhibits further secretion of LH (and also FSH).

1. Using the information above and on the previous page, complete the table below summarizing the role of hormones in the control of the menstrual cycle. To help you, some of the table has been completed:

Hormone	Site of secretion	Main effects and site of action during the menstrual cycle
GnRH		
		Stimulates the growth of ovarian follicles
LH		
		At high levels, stimulates LH surge. Promotes growth and repair of the uterine lining.
Progesterone		

2. Briefly explain the role of negative feedback in the control of hormone levels in the menstrual cycle:

_____

_____

3. **FSH** and **LH** (called ICSH or interstitial cell stimulating hormone in males) also play a central role in male reproduction. Refer to the activity *Male Reproductive System* and state how these two hormones are involved **in male reproduction**:

_____

Related activities: The Menstrual Cycle, Male Reproductive System, Sexual Development

Male Reproductive System

The reproductive role of the male is to produce the sperm and deliver them to the female. When a sperm combines with an egg, it contributes half the genetic material of the offspring and, in humans and other mammals, determines its sex. The reproductive structures in human males (shown below) are in many ways typical of other mammals.

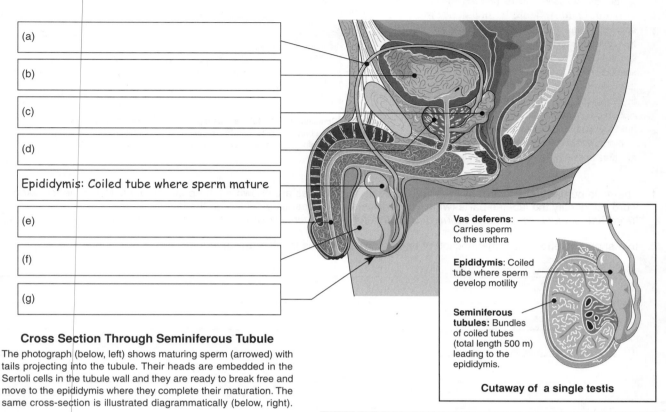

(a)

(b)

(c)

(d)

Epididymis: Coiled tube where sperm mature

(e)

(f)

(g)

Vas deferens: Carries sperm to the urethra

Epididymis: Coiled tube where sperm develop motility

Seminiferous tubules: Bundles of coiled tubes (total length 500 m) leading to the epididymis.

Cutaway of a single testis

Cross Section Through Seminiferous Tubule

The photograph (below, left) shows maturing sperm (arrowed) with tails projecting into the tubule. Their heads are embedded in the Sertoli cells in the tubule wall and they are ready to break free and move to the epididymis where they complete their maturation. The same cross-section is illustrated diagrammatically (below, right).

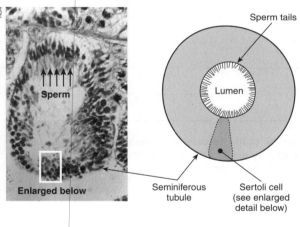

Sperm

Enlarged below

Sperm tails

Lumen

Seminiferous tubule

Sertoli cell (see enlarged detail below)

Spermatogenesis

Spermatogenesis is the process by which mature spermatozoa (sperm) are produced in the testis. In humans, they are produced at the rate of about 120 million per day. Spermatogenesis is regulated by the hormones **FSH** (from the anterior pituitary) and testosterone (secreted from the testes in response to **ICSH** (LH) from the anterior pituitary). Spermatogonia, in the outer layer of the seminiferous tubules, multiply throughout reproductive life. Some of them divide by meiosis into spermatocytes, which produce spermatids. These are transformed into mature sperm by the process of spermiogenesis in the seminiferous tubules of the testis. Full sperm motility is achieved in the epididymis.

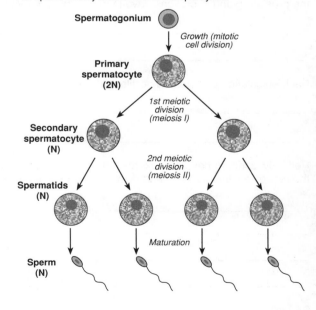

Spermatogonium

Growth (mitotic cell division)

Primary spermatocyte (2N)

1st meiotic division (meiosis I)

Secondary spermatocyte (N)

2nd meiotic division (meiosis II)

Spermatids (N)

Maturation

Sperm (N)

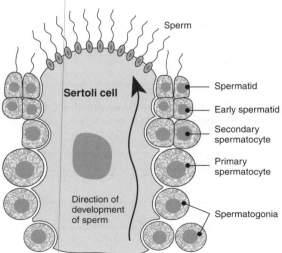

Sertoli cell

Sperm

Spermatid

Early spermatid

Secondary spermatocyte

Primary spermatocyte

Direction of development of sperm

Spermatogonia

Reproduction & Development

Related activities: Mitosis and the Cell Cycle

RA 3

Sperm Structure

Mature spermatozoa (sperm) are produced by a process called spermatogenesis in the testes (see description of the process on the previous page). Meiotic division of spermatocytes produces spermatids which then differentiate into mature sperm. Sperm are quite simple in structure because their sole purpose is to swim to the egg and donate their genetic material. They are composed of three regions: headpiece, midpiece, and tail. Sperm do not live long (only about 48 hours), but they swim quickly and there are so many of them (millions per ejaculation) that some are able to reach the egg to fertilize it.

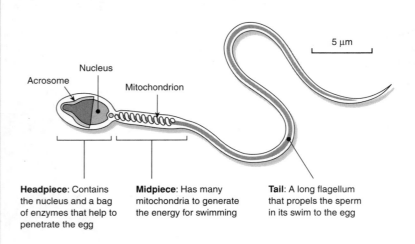

5 μm

Acrosome

Nucleus

Mitochondrion

Headpiece: Contains the nucleus and a bag of enzymes that help to penetrate the egg

Midpiece: Has many mitochondria to generate the energy for swimming

Tail: A long flagellum that propels the sperm in its swim to the egg

1. The male human reproductive system and associated structures are shown on the previous page. Using the following word list identify the labeled parts (write your answers in the spaces provided on the diagram).
   Word list: *bladder, scrotal sac, sperm duct (vas deferens), seminal vesicle, testis, urethra, prostate gland*

2. In a short sentence, state the function of each of the structures labeled (a)-(g) in the diagram on the previous page:

   (a) _____

   (b) _____

   (c) _____

   (d) _____

   (e) _____

   (f) _____

   (g) _____

3. (a) Name the process by which mature sperm are formed: _____

   (b) Name the hormones regulating this process: _____

   (c) State where most of this process occurs: _____

   (d) State where the process is completed: _____

4. The secretions of the prostate gland (which make up a large proportion of the seminal fluid produced in an ejaculation) are of alkaline pH, while the secretions of the vagina are normally slightly acidic. With this information, explain the role the prostate gland secretions have in maintaining the viability of sperm deposited in the vagina.

   _____

   _____

5. Each ejaculation of a healthy, fertile male contains 100-400 million sperm. Suggest why so many sperm are needed:

   _____

   _____

6. Recently, concern has been expressed about the level of synthetic estrogen-like chemicals in the environment. Explain the reason for this concern and discuss evidence in support of the claim that these chemicals lower male fertility:

   _____

   _____

   _____

   _____

Fertilization and Early Growth

When an egg cell is released from the ovary it is arrested in metaphase of meiosis II and is termed a secondary oocyte. **Fertilization** occurs when a sperm penetrates an egg cell at this stage and the sperm and egg nuclei unite to form the zygote. Fertilization is always regarded as time 0 in a period of gestation (pregnancy) and has five distinct stages (below). After fertilization, the zygote begins its **development** i.e. its growth and differentiation into a multicellular organism (see next page).

Fertilization (Time 0)

The stages in fertilization are represented below in a numbered sequence (1-5)

1. Capacitation

The surface of the sperm cell undergoes changes that are essential to enabling the acrosome reaction and sperm entry.

2. The Acrosome Reaction

Enzymes from the acrosome (an enzyme-filled bag at the tip of the sperm) are released and digest a pathway through the follicle cells (not shown) and the jelly-like zona pellucida surrounding the egg cell (secondary oocyte).

3. Fusion of Sperm Head

The plasma membranes of the sperm and egg fuse, and the nucleus of the sperm enters the egg cytoplasm. Fusion causes a sudden membrane depolarization that acts as a "fast block" to further sperm entry. The fusion of the two plasma membranes also triggers the completion of meiosis II in the egg cell and induces the cortical reaction (below).

4. The Cortical Reaction

The fusion of the two plasma membranes induces a permanent change in the egg surface that prevents further sperm entry. Cortical granules in the egg cytoplasm release their contents into the space between the plasma membrane and the vitelline layer. Substances released from the granules raise and harden the vitelline layer to form a slow (permanent) block to further sperm entry.

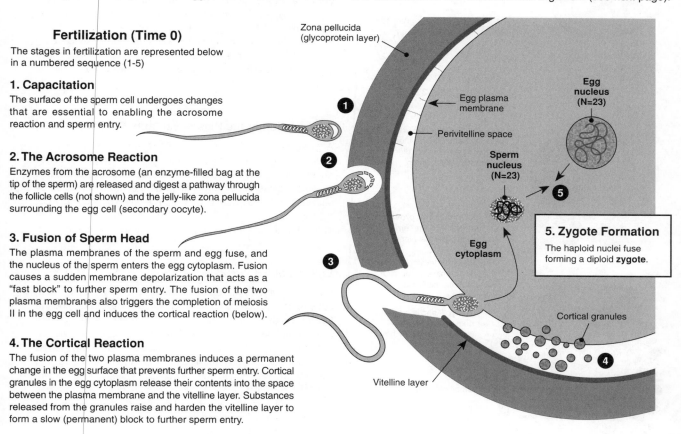

Zona pellucida (glycoprotein layer)
Egg plasma membrane
Perivitelline space
Egg nucleus (N=23)
Sperm nucleus (N=23)
Egg cytoplasm

5. Zygote Formation
The haploid nuclei fuse forming a diploid **zygote**.

Cortical granules
Vitelline layer

---

1. Briefly describe the significant events (and their importance) occurring at each of the following stages of fertilization:

    (a) Capacitation: _____
    _____

    (b) The acrosome reaction: _____
    _____

    (c) Fusion of egg and sperm plasma membranes: _____
    _____

    (d) The cortical reaction: _____
    _____

    (e) Fusion of egg and sperm nuclei: _____
    _____

2. Explain the significance of the blocks that prevent entry of more than one sperm into the egg (polyspermy):
    _____
    _____

3. (a) Explain why the egg cell, when released from the ovary, is termed a secondary oocyte: _____
    _____

    (b) State at which stage, its meiotic division is completed: _____

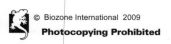

Reproduction & Development

Related activities: Female Reproductive System
Web links: Fertilization

A 2

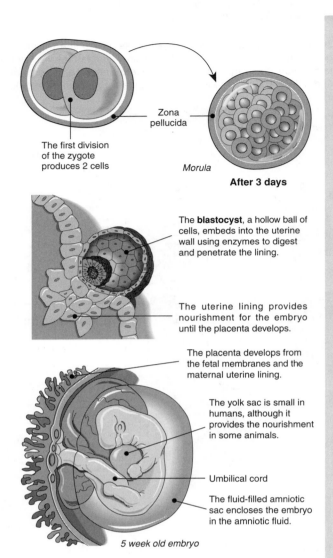

Zona pellucida

The first division of the zygote produces 2 cells

Morula

After 3 days

The **blastocyst**, a hollow ball of cells, embeds into the uterine wall using enzymes to digest and penetrate the lining.

The uterine lining provides nourishment for the embryo until the placenta develops.

The placenta develops from the fetal membranes and the maternal uterine lining.

The yolk sac is small in humans, although it provides the nourishment in some animals.

Umbilical cord

The fluid-filled amniotic sac encloses the embryo in the amniotic fluid.

5 week old embryo

Early Growth and Development

Cleavage and Development of the Morula

Immediately after fertilization, rapid cell division takes place. These early cell divisions are called **cleavage** and they increase the number of cells, but not the size of the zygote. The first cleavage is completed after 36 hours, and each succeeding division takes less time. After 3 days, successive cleavages have produced a solid mass of cells called the **morula**, (left) which is still about the same size as the original zygote.

Implantation of the Blastocyst (after 6-8 days)

After several days in the uterus, the morula develops into the blastocyst. It makes contact with the uterine lining and pushes deeply into it, ensuring a close maternal-fetal contact. Blood vessels provide early nourishment as they are opened up by enzymes secreted by the blastocyst. The embryo produces **HCG** (human chorionic gonadotropin), which prevents degeneration of the corpus luteum and signals that the woman is pregnant.

Embryo at 5-8 Weeks

Five weeks after fertilization, the embryo is only 4-5 mm long, but already the central nervous system has developed and the heart is beating. The embryonic membranes have formed; the amnion encloses the embryo in a fluid-filled space, and the allanto-chorion forms the fetal portion of the placenta. From two months the embryo is called a fetus. It is still small (30-40 mm long), but the limbs are well formed and the bones are beginning to harden. The face has a flat, rather featureless appearance with the eyes far apart. Fetal movements have begun and brain development proceeds rapidly. The placenta is well developed, although not fully functional until 12 weeks. The umbilical cord, containing the fetal umbilical arteries and vein, connects fetus and mother.

4. State what contribution the sperm and egg cell make to each of the following:

(a) The nucleus of the zygote: Sperm contribution: _____   Egg contribution: _____

(b) The cytoplasm of the zygote: Sperm contribution: _____   Egg contribution: _____

5. Explain what is meant by cleavage and comment on its significance to the early development of the embryo:

_____

_____

6. (a) Explain the importance of implantation to the early nourishment of the embryo: _____

_____

(b) Identify the fetal tissues that contribute to the formation of the placenta: _____

(c) Suggest a purpose of the amniotic sac and comment on its importance to the developing embryo: _____

_____

(d) Suggest why the heart is one of the very first structures to develop in the embryo: _____

_____

7. State why the fetus is particularly prone to damage from drugs towards the end of the first trimester (2-3 months):

_____

_____

The Placenta

As soon as an embryo embeds in the uterine wall it begins to obtain nutrients from its mother and increase in size. At two months, when the major structures of the adult are established, it is called a fetus. It is entirely dependent on its mother for nutrients, oxygen, and elimination of wastes. The placenta is the specialized organ that performs this role, enabling exchange between fetal and maternal tissues, and allowing a prolonged period of fetal growth and development within the protection of the uterus. The placenta also has an endocrine role, producing hormones that enable the pregnancy to be maintained.

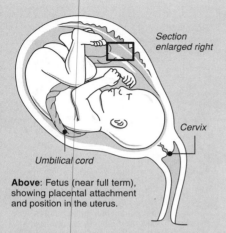

Section enlarged right

Cervix

Umbilical cord

Above: Fetus (near full term), showing placental attachment and position in the uterus.

Below: Photograph shows a 14 week old fetus. Limbs are fully formed, many bones are beginning to ossify, and joints begin to form. Facial features are becoming more fully formed.

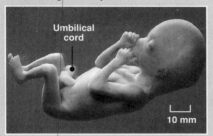

Umbilical cord

10 mm

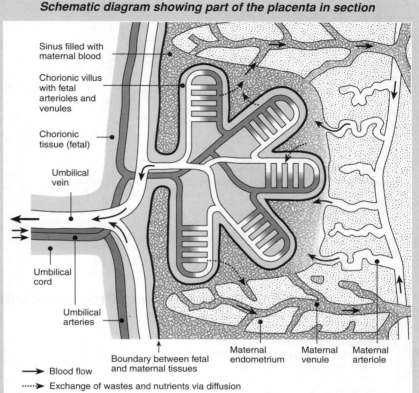

Schematic diagram showing part of the placenta in section

Sinus filled with maternal blood

Chorionic villus with fetal arterioles and venules

Chorionic tissue (fetal)

Umbilical vein

Umbilical cord

Umbilical arteries

Boundary between fetal and maternal tissues

Maternal endometrium

Maternal venule

Maternal arteriole

→ Blood flow

·····➤ Exchange of wastes and nutrients via diffusion

The placenta is a disc-like organ, about the size of a dinner plate and weighing about 1 kg. It develops when fingerlike projections of the fetal chorion (the chorionic villi) grow into the endometrium of the uterus. The villi contain the numerous capillaries connecting the fetal arteries and vein. They continue invading the maternal tissue until they are bathed in the maternal blood sinuses. The maternal and fetal blood vessels are in such close proximity that oxygen and nutrients can diffuse from the maternal blood into the capillaries of the villi. From the villi, the nutrients circulate in the umbilical vein, returning to the fetal heart. Carbon dioxide and other wastes leave the fetus through the umbilical arteries, pass into the capillaries of the villi, and diffuse into the maternal blood. Note that fetal blood and maternal blood do not mix: the exchanges occur via diffusion through thin walled capillaries.

1. In simple terms, describe the basic structure of the human placenta: _____

_____

_____

_____

2. The umbilical cord contains the fetal arteries and vein. Describe the status of the blood in each type of fetal vessel:

   (a) Fetal arteries: Oxygenated and containing nutrients / Deoxygenated and containing nitrogenous wastes (delete one)

   (b) Fetal vein: Oxygenated and containing nutrients / Deoxygenated and containing nitrogenous wastes (delete one)

3. Teratogens are substances that may cause malformations in embryonic development (e.g. nicotine, alcohol):

   (a) Give a general explanation why substances ingested by the mother have the potential to be harmful to the fetus:

   _____

   (b) Explain why cigarette smoking is so harmful to fetal development: _____

   _____

Reproduction & Development

Related activities: Fertilization and Early Growth

A 2

The Hormones of Pregnancy

Human reproductive physiology occurs in a cycle (the menstrual cycle) which follows a set pattern and is regulated by the interplay of several hormones. Control of hormone release is brought about through feedback mechanisms: the levels of the female reproductive hormones, estrogen and progesterone, regulate the secretion of the pituitary hormones that control the ovarian cycle (see earlier pages). Pregnancy interrupts this cycle and maintains the corpus luteum and the placenta as endocrine organs with the specific role of maintaining the developing fetus for the period of its development. During the last month of pregnancy the peptide hormone oxytocin induces the uterine contractions that will expel the baby from the uterus.

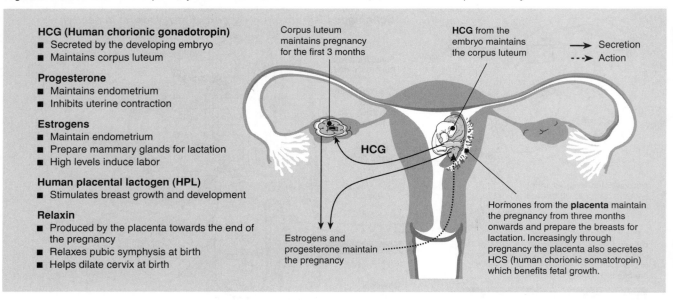

Hormonal Changes During Pregnancy, Birth, and Lactation

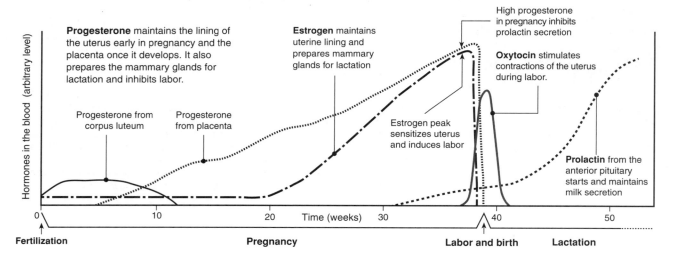

During the first 12-16 weeks pregnancy, the **corpus luteum** secretes enough progesterone to maintain the uterine lining and sustain the developing embryo. After this, the placenta takes over as the primary endocrine organ of pregnancy. **Progesterone** and **estrogen** from the placenta maintain the uterine lining, inhibit the development of further ova (eggs), and prepare the breast tissue for **lactation** (milk production). At the end of pregnancy, the placenta loses competency, progesterone levels fall, and high estrogen levels trigger the onset of labor. The estrogen peak coincides with an increase in oxytocin, which stimulates uterine contractions in a postive feedback loop: the contractions and the increasing pressure of the cervix from the infant stimulate release of more oxytocin, and more contractions and so on, until the infant exits the birth canal. After birth, the secretion of prolactin increases. Prolactin maintains lactation during the period of infant nursing.

1. (a) Explain why the corpus luteum is the main source of progesterone in early pregnancy:

   (b) Name the hormones responsible for maintaining pregnancy:

2. (a) Identify two hormones involved in **labor** and explain their roles:

   (b) Describe two physiological factors in initiating labor:

Related activities: The Menstrual Cycle, Control of the Menstrual Cycle, Birth and Lactation

Detecting Pregnancy using Monoclonal Antibodies

Pregnancy testing determines whether a woman is pregnant or not. The most frequently used methods involve detecting **chemical markers** in urine or blood, but obstetric ultrasonography is also used to detect pregnancy. When a woman becomes pregnant, a hormone called **human chorionic gonadotropin** (HCG) is released from the placenta, accumulates in the blood, and is excreted in the urine.

Blood tests can be used to confirm pregnancy 48 hours after fertilization, but they are very expensive and time consuming. By comparison, urine analyses are quick and inexpensive. If used correctly they are 97% accurate. HCG is detectable only after implantation has occurred (6-12 days after fertilization), so false negative results can be obtained if the test is performed too early in the pregnancy.

Monoclonal antibodies are also used in other home testing kits, such as those for detecting ovulation time (below).

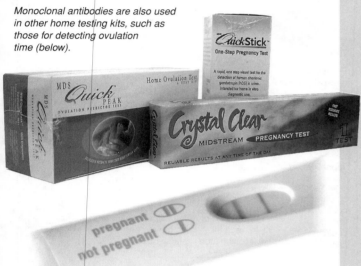

How home pregnancy detection kits work

HCG is a glycoprotein, so antibodies can be produced against it and used to determine if a woman is pregnant. The test area of the dipstick (below) contains two types of antibodies: free monoclonal antibodies and capture monoclonal antibodies, bound to the substrate in the test window.

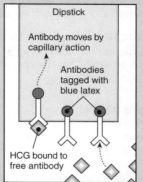

The free antibodies are specific for HCG and are color-labeled. HCG in the urine of a pregnant woman binds to the free antibodies on the surface of the dipstick. The antibodies then travel up the dipstick by capillary action.

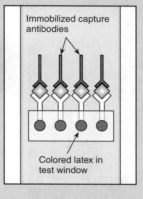

The capture antibodies are specific for the HCG-free antibody complex. The HCG-free antibody complexes traveling up the dipstick are bound by the immobilized **capture antibodies**, forming a sandwich. The color labeled antibodies then create a visible color change in the test window.

3. Explain why prolactin secretion increases markedly after birth: _____

_____

4. Explain why the placenta is regarded as an endocrine organ, even though it is a temporary structure:

_____

_____

5. (a) Labor is one of the few physiological situations involving a positive feedback loop. Explain its role in this situation:

_____

_____

(b) Describe another situation involving mother and infant in which positive feedback is important: _____

_____

_____

6. (a) Describe the principle behind the use of monoclonal antibodies in a home pregnancy test: _____

_____

_____

(b) Explain why a urine tests for pregnancy are so widely used and describe a precaution with their use:

_____

_____

Reproduction & Development

Contraception

Humans have many ways in which to manage their own reproduction. They may choose to prevent or assist fertilization of an egg by a sperm (conception). **Contraception** refers to the use of methods or devices that prevent conception. There are many contraceptive methods available including physical barriers (such as condoms) that prevent egg and sperm ever meeting. The most effective methods (excluding sterilization) involve chemical interference in the normal female cycle so that egg production is inhibited. This is done by way of **oral contraceptives** (below, left) or hormonal implants. If taken properly, oral contraceptives are almost 100% effective at preventing pregnancy. The placement of their action in the normal cycle of reproduction (from gametogenesis to pregnancy) is illustrated below. Other contraceptive methods are included for comparison.

Hormonal Contraception

The most common method by which to prevent conception using hormones is by using an oral contraceptive pill (OCP). These may be **combined OCPs**, or low dose mini pills.

Combined oral contraceptive pills (OCPs)

These pills exploit the feedback controls over hormone secretion normally operating during a menstrual cycle. They contain combinations of synthetic **estrogens** and **progesterone**. They are taken daily for 21 days, and raise the levels of these hormones in the blood so that FSH secretion is inhibited and no ova develop. Sugar pills are taken for 7 days; long enough to allow menstruation to occur but not long enough for ova to develop. Combined OCPs can be of two types:

Monophasic pills (left): Hormones (**H**) are all at one dosage level. Sugar pills (**S**) are usually larger and differently colored.

Triphasic pills (right): The hormone dosage increases in stages (**1,2,3**), mimicking the natural changes in a menstrual cycle.

Mini-pill (progesterone only)

The mini-pill contains 28 days of low dose progesterone; generally too low to prevent ovulation. The pill works by thickening the cervical mucus and preventing endometrial thickening. The mini-pill is less reliable than combined pills and must be taken at a regular time each day. However, it is safer for older women and those who are breastfeeding.

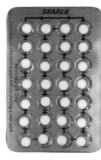

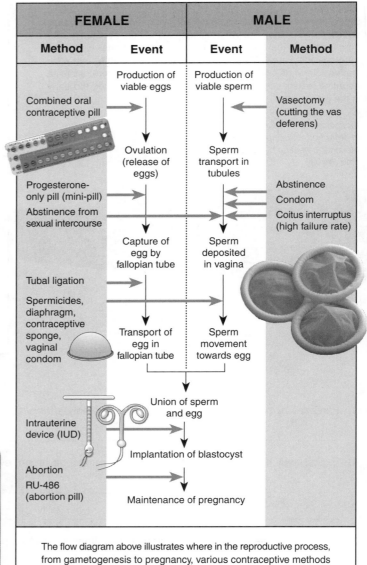

FEMALE		MALE	
Method	**Event**	**Event**	**Method**
	Production of viable eggs	Production of viable sperm	
Combined oral contraceptive pill			Vasectomy (cutting the vas deferens)
	Ovulation (release of eggs)	Sperm transport in tubules	
Progesterone-only pill (mini-pill)			Abstinence
Abstinence from sexual intercourse			Condom
			Coitus interruptus (high failure rate)
	Capture of egg by fallopian tube	Sperm deposited in vagina	
Tubal ligation			
Spermicides, diaphragm, contraceptive sponge, vaginal condom	Transport of egg in fallopian tube	Sperm movement towards egg	
	Union of sperm and egg		
Intrauterine device (IUD)			
	Implantation of blastocyst		
Abortion RU-486 (abortion pill)			
	Maintenance of pregnancy		

The flow diagram above illustrates where in the reproductive process, from gametogenesis to pregnancy, various contraceptive methods operate. Note the early action of hormonal contraceptives.

1. Explain briefly how the **combined oral contraceptive pill** acts as a contraceptive: _____

   _____

   _____

2. Contrast the mode of action of OCPs with that of the mini-pill, giving reasons for the differences: _____

   _____

   _____

3. Suggest why oral contraceptives offer such effective control over conception: _____

   _____

   _____

Treating Female Infertility

Failure to ovulate is a common cause of female infertility. In most cases, the cause is hormonal, although sometimes the ovaries may be damaged or not functioning normally. Female infertility may also arise through damage to the fallopian tubes as a result of infection or scarring. These cases are usually treated with hormones, followed by **IVF**. Most treatments for female infertility involve the use of synthetic female hormones, which stimulate ovulation, boost egg production, and induce egg release.

Treating Female Infertility

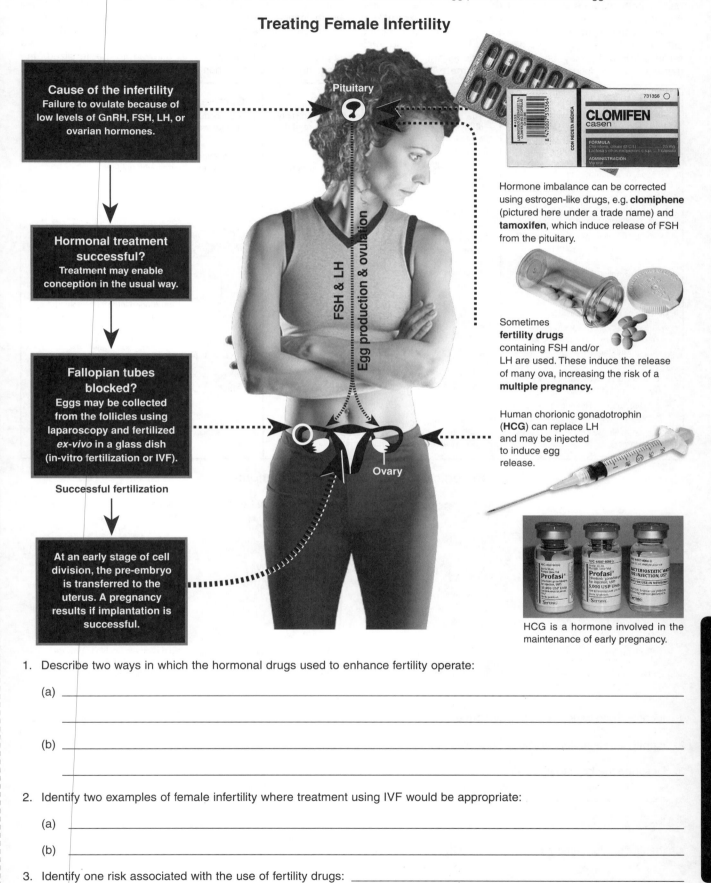

Cause of the infertility
Failure to ovulate because of low levels of GnRH, FSH, LH, or ovarian hormones.

Hormonal treatment successful?
Treatment may enable conception in the usual way.

Fallopian tubes blocked?
Eggs may be collected from the follicles using laparoscopy and fertilized *ex-vivo* in a glass dish (in-vitro fertilization or IVF).

Successful fertilization

At an early stage of cell division, the pre-embryo is transferred to the uterus. A pregnancy results if implantation is successful.

Pituitary

FSH & LH

Egg production & ovulation

Ovary

CLOMIFEN
casen

Hormone imbalance can be corrected using estrogen-like drugs, e.g. **clomiphene** (pictured here under a trade name) and **tamoxifen**, which induce release of FSH from the pituitary.

Sometimes **fertility drugs** containing FSH and/or LH are used. These induce the release of many ova, increasing the risk of a **multiple pregnancy.**

Human chorionic gonadotrophin (**HCG**) can replace LH and may be injected to induce egg release.

Profasi

HCG is a hormone involved in the maintenance of early pregnancy.

1. Describe two ways in which the hormonal drugs used to enhance fertility operate:

   (a) _____

   _____

   (b) _____

2. Identify two examples of female infertility where treatment using IVF would be appropriate:

   (a) _____

   (b) _____

3. Identify one risk associated with the use of fertility drugs: _____

Related activities: Control of the Menstrual Cycle, Human Reproductive Technology

A 2

Reproduction & Development

Human Reproductive Technology

Infertility may result from a disturbance of any of the factors involved in fertilization or embryonic development. Female infertility may be due to a failure to ovulate, requiring stimulation of the ovary, with or without hormone therapy. For couples with one or both partners incapable of providing suitable gametes, it may be possible for them to receive eggs and/or sperm from donors. **Artificial insemination** (AI) may be used to introduce selected sperm from the male partner or from a donor. **In vitro fertilization** (IVF) may be used for patients with irreparable damage to the fallopian tubes. The **gamete intrafallopian transfer** (GIFT) technique is now widely accepted as a form of treatment for patients with one or more functioning fallopian tubes. **Surrogate mothers** may be used to 'incubate' the fetus in cases where the woman is incapable of sustaining a pregnancy. **Fertility drugs** may be used to treat ovulation failure, as well as to induce the production of many eggs for use in IVF or GIFT. Such drugs stimulate the pituitary gland and may induce the simultaneous release of numerous eggs; an event called superovulation. If each egg is allowed to be fertilized, the resulting embryos may then be frozen after 24-72 hours culture.

Causes of Infertility

Infertility is a common problem (as many as one in six couples require help from a specialist). The cause of the infertility may be inherited, due to damage caused by an infectious disease, or psychological.

Causes of male infertility:
- *Penis:* Fails to achieve or maintain erection; abnormal ejaculation.
- *Testes:* Too few sperm produced or sperm are abnormally shaped, have impaired motility, or too short lived.
- *Vas deferens:* Blockage or structural abnormality may impede passage of sperm.

Causes of female infertility:
- *Fallopian tubes:* Blockage may prevent sperm from reaching egg; one or both tubes may be damaged (disease) or absent (congenital).
- *Ovaries:* Eggs may fail to mature or may not be released.
- *Uterus:* Abnormality or disorder may prevent implantation of the egg.
- *Cervix:* Antibodies in cervical mucus may damage or destroy the sperm.

In Vitro Fertilization (IVF)

The woman is given hormone therapy (fertility drugs) causing a number of eggs to mature at the same time (superovulation).

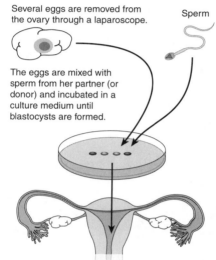

Several eggs are removed from the ovary through a laparoscope.

Sperm

The eggs are mixed with sperm from her partner (or donor) and incubated in a culture medium until blastocysts are formed.

The blastocyst(s) is then implanted in the mother's uterus and the pregnancy is allowed to continue normally.

Gamete Intrafallopian Transfer (GIFT)

A procedure for assisting conception, suitable only for women with healthy fallopian tubes.

Using a needle for aspiration, under laparoscopic or ultrasound guidance, the eggs are removed from the ovary.

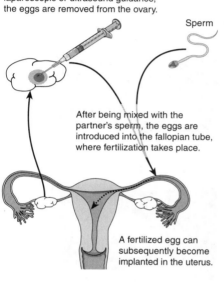

Sperm

After being mixed with the partner's sperm, the eggs are introduced into the fallopian tube, where fertilization takes place.

A fertilized egg can subsequently become implanted in the uterus.

Biological Origins of Gamete Donations

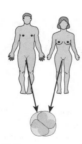

Both partners provide gametes for IVF or GIFT (they donate their own gametes).

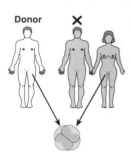

Male partner unable to provide sperm; sperm from male donor.

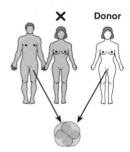

Female partner unable to provide eggs; egg from female donor.

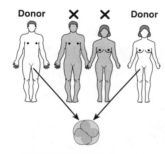

Both partners unable to provide gametes; sperm and egg obtained from donors.

1. Identify three causes of infertility for men: _____

2. Identify three causes of infertility for women: _____

3. Describe the key stages of **IVF**: _____

_____

_____

4. Identify the fundamental way in which **GIFT** differs from IVF: _____

_____

Related activities: Treating Female Infertility
Web links: In Vitro Fertilization

Birth and Lactation

A human pregnancy (the period of **gestation**) lasts, on average, about 38 weeks after fertilization. It ends in labor, the birth of the baby, and expulsion of the placenta. During pregnancy, progesterone maintains the placenta and inhibits contraction of the uterus. At the end of a pregnancy, increasing estrogen levels overcome the influence of progesterone and labor begins. Prostaglandins, factors released from the placenta, and the physiological state of the baby itself are also involved in triggering the actual timing of labor onset. Labor itself comprises three stages (below), and ends with the delivery of the placenta. After birth, the mother provides nutrition for the infant through **lactation**: the production and release of milk from mammary glands. Breast milk provides infants with a complete, easily digested food for the first 4-6 months of life. All breast milk contains maternal antibodies, which give the infant protection against infection while its own immune system develops.

Birth and the Stages of Labor

Delivery of the Baby: The End of Stage 2

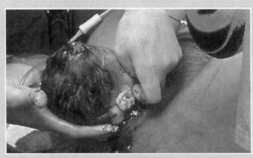

Delivery of the head. This baby is face forward. The more usual position for delivery is face to the back of the mother.

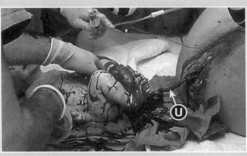

Full delivery of the baby. Note the umbilical cord (U), which supplies oxygen until the baby's breathing begins.

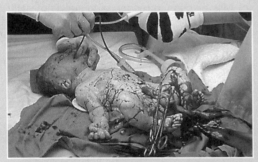

Post-birth check of the baby. The baby is still attached to the placenta and the airways are being cleared of mucus.

Stage 1: Dilation

Duration: 2-20 hours

The time between the onset of labor and complete opening (dilation) of the cervix. The amniotic sac may rupture at this stage, releasing its fluid. The hormone **oxytocin** stimulates the uterine contractions necessary to dilate the cervix and expel the baby. It is these uterine contractions that give the pain of labor, most of which is associated with this first stage.

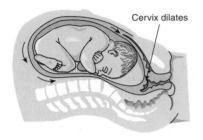

Cervix dilates

Stage 2: Expulsion

Duration: 2-100 minutes

The time from full dilation of the cervix to delivery. Strong, rhythmic contractions of the uterus pass in waves (arrows), and push the baby to the end of the vagina, where the head appears.

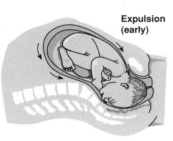

Expulsion (early)

As labor progresses, the time between each contraction shortens. Once the head is delivered, the rest of the body usually follows very rapidly. Delivery completes stage 2.

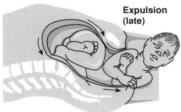

Expulsion (late)

Stage 3: Delivery of placenta

Time: 5-45 minutes after delivery

The third or **placental stage**, refers to the expulsion of the placenta from the uterus. After the placenta is delivered, the placental blood vessels constrict to stop bleeding.

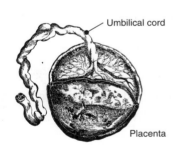

Umbilical cord

Placenta

1. Name the three stages of birth, and briefly state the main events occurring in each stage:

    (a) Stage 1: _____

    (b) Stage 2: _____

    (c) Stage 3: _____

2. (a) Name the hormone responsible for triggering the onset of labor: _____

    (b) Describe two other factors that might influence the timing of labor onset: _____

    _____

    _____

Related activities: The Hormones of Pregnancy

A 2

Reproduction & Development

Lactation and its Control

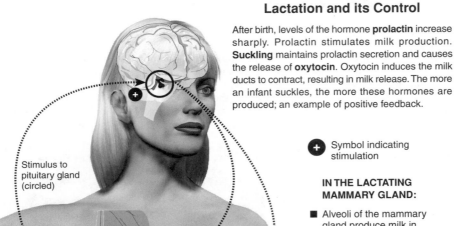

After birth, levels of the hormone **prolactin** increase sharply. Prolactin stimulates milk production. **Suckling** maintains prolactin secretion and causes the release of **oxytocin**. Oxytocin induces the milk ducts to contract, resulting in milk release. The more an infant suckles, the more these hormones are produced; an example of positive feedback.

Stimulus to pituitary gland (circled)

Prolactin

Oxytocin

Alveolus

Mammary duct

+ Symbol indicating stimulation

IN THE LACTATING MAMMARY GLAND:

■ Alveoli of the mammary gland produce milk in response to prolactin.

■ Contraction of the mammary ducts ejects milk to the nipple in a reflex letdown (induced by oxytocin).

■ Suckling stimulates secretion of prolactin from the antior pituitary and oxytocin from the posterior pituitary.

It is essential to establish breast feeding soon after birth, as this is when infants exhibit the strong reflexes that enable them to learn to suckle effectively. The first formed milk, colostrum, has very little sugar, virtually no fat, and is rich in maternal antibodies. Breast milk that is produced later has a higher fat content, and its composition varies as the nutritional needs of the infant change during growth.

3. Explain why the umbilical cord continues to supply blood to the baby for a short time after delivery: _____

_____

4. For each of the following processes, state the primary controlling hormone and its site of production:

(a) Uterine contraction during labor: Hormone: _____ Site of production: _____

(b) Production of milk: Hormone: _____ Site of production: _____

(c) Milk ejection in response to suckling: Hormone: _____ Site of production: _____

5. State which hormone inhibits prolactin secretion during pregnancy: _____

6. Describe two benefits of breast feeding to the health of the infant:

(a) _____

(b) _____

7. (a) Describe the nutritional differences between the first formed milk (colostrum) and the milk that is produced later:

_____

_____

(b) Suggest a reason for these differences: _____

_____

8. Explain why the nutritional composition of breast milk might change during a six-month period of breast feeding:

_____

9. Infants exhibit marked growth spurts at six weeks and three months of age. At these times, their caloric (energy intake) requirements also increase sharply. With reference to what you know about the control of lactation, suggest how a breast-feeding mother could continue to provide for the increased energy requirements of her infant:

_____

_____

Growth and Development

Development describes the process of growing to maturity, from zygote to adult. After birth, development continues rapidly and is marked by specific stages recognized by the set of physical and cognitive skills present. Obvious physical changes include the elongation of the bones, increasing ossification of cartilage, and changes to the proportions of the body. These proportional changes are the result of **allometric growth** (differential growth rates) and occur concurrently with motor, intellectual, and emotional and social development. These changes lead the child to increasing independence and maturity.

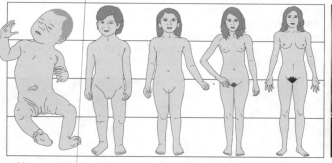

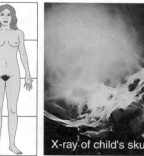

| Newborn | 2 years | 5 years | 15 years | Adult |

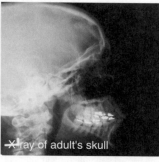

X-ray of child's skull   X-ray of adult's skull

At birth the cranium is very large in comparison to the face and the skull makes up around one quarter of the infant's height. During early life, the face continues to grow outward, reducing the relative proportions of the cranium, while at adulthood the size of the skull in proportion to the body is much less.

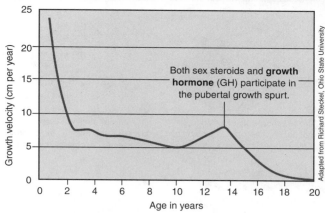

Both sex steroids and **growth hormone** (GH) participate in the pubertal growth spurt.

Growth velocity (cm per year)

Age in years

Adapted from Richard Steckel, Ohio State University

By 6 weeks old, a human baby is usually able to hold its head up if placed on its stomach. At 3 months the infant will exercise limbs aimlessly but by 5 months is able to grasp objects and sit up. The infant may be able to crawl by 8 months and walk by 12 months. It is more or less independent by two years and undergoes changes to adulthood at around 11 years of age (puberty).

Babies are effectively born premature so that they complete much of their early development in the first two to three years of life. The rate of growth declines slowly through childhood, but increases again to a peak in puberty (the growth spurt). By 20 years of age the cartilage in the long bones has been replaced by bone and growth stops.

1. Describe the most noticeable change in body proportion from birth to adulthood: _____

    _____

2. Describe the changes that occur in the first period of rapid growth in humans: _____

    _____

    _____

3. Describe the changes that occur in the second period of rapid growth in humans: _____

    _____

4. Answer the following questions with respect to the graph depicting growth rates 0-20 years (above, right):

    (a) Describe what happens to growth velocity in the first two years of an infant's life: _____

    _____

    (b) Describe what the graph infers about the rate of growth in the period before birth: _____

    _____

    (c) Identify the age range (in years) marking the pubertal growth spurt: _____

5. Relate the changes in physical development to the changes occurring in the mental development of an infant:

    _____

    _____

Reproduction & Development

Related activities: Bone, Sexual Development

RA 2

Apoptosis: Programed Cell Death

Apoptosis or programed cell death (PCD) is a normal and necessary mechanism in multicellular organisms to trigger the death of a cell. Apoptosis has a number of crucial roles in the body, including the maintenance of adult cell numbers, and defence against damaged or dangerous cells, such as virus-infected cells and cells with DNA damage. Apoptosis also has a role in "sculpting" embryonic tissue during its development, e.g. in the formation of fingers and toes in a developing human embryo. Programed cell death involves an orderly series of biochemical events that result in set changes in cell morphology and end in cell death. The process is carried out in such a way as to safely dispose of cell remains and fragments. This is in contrast to another type of cell death, called **necrosis**, in which traumatic damage to the cell results in spillage of cell contents. Apoptosis is tightly regulated by a balance between the factors that promote cell survival and those that trigger cell death. An imbalance between these regulating factors leads to defective apoptotic processes and is implicated in an extensive variety of diseases. For example, low rates of apoptosis result in uncontrolled proliferation of cells and cancers.

Stages in Apoptosis

Apoptosis is a normal cell suicide process in response to particular cell signals. It characterized by an overall compaction (shrinking) of the cell and its nucleus, and the orderly dissection of chromatin by endonucleases. Death is finalized by a rapid engulfment of the dying cell by phagocytosis. The cell contents remain membrane-bound and there is no inflammation.

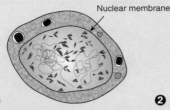

Nuclear membrane

Chromatin

1 The cell shrinks and loses contact with neighboring cells. The chromatin condenses and begins to degrade.

2 The nuclear membrane degrades. The cell loses volume. The chromatin clumps into **chromatin bodies**.

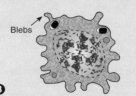

Blebs

Organelle

Nucleus

3 **Zeiosis**: The plasma membrane forms bubble like **blebs** on its surface.

4 The nucleus collapses, but many membrane-bound organelles are unaffected.

Apoptotic body

5 The nucleus breaks up into spheres and the DNA breaks up into small fragments.

6 The cell breaks into numerous **apoptotic bodies**, which are quickly resorbed by phagocytosis.

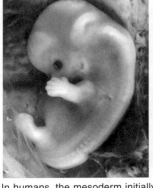
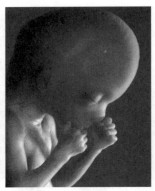

Ed Uthman

In humans, the mesoderm initially formed between the fingers and toes is removed by apoptosis. 41 days after fertilization (top left), the digits of the hands and feet are webbed, making them look like small paddles. Apoptosis selectively destroys this superfluous webbing and, later in development, each of the digits can be individually seen (right).

Regulating Apoptosis

Apoptosis is a complicated and tightly controlled process, distinct from cell necrosis (uncontrolled cell death), when the cell contents are spilled. Apoptosis is regulated through both:

Positive signals, which prevent apoptosis and allow a cell to function normally. They include:
▶ interleukin-2
▶ bcl-2 protein and growth factors

Interleukin-2 is a positive signal for cell survival. Like other signalling molecules, it binds to surface receptors on the cell to regulate metabolism.

Negative signals (death activators), which trigger the changes leading to cell death. They include:
▶ inducer signals generated from within the cell itself in response to stress, e.g. DNA damage or cell starvation.
▶ signalling proteins and peptides such as lymphotoxin.

1. The photograph (right) depicts a condition called syndactyly. Explain what might have happened during development to result in this condition:

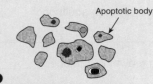

_____

_____

2. Describe one difference between apoptosis and necrosis: _____

_____

3. Describe two situations, other than digit formation in development, in which apoptosis plays a crucial role:

(a) _____

_____

(b) _____

_____

Related activities: Mitosis and the Cell Cycle, Cancer: Cells out of Control
Web links: Apoptosis: The Dance of Death

Breast Cancer

After non-melanocytic skin cancer, breast cancer is the most commonly diagnosed cancer amongst US women, and it is the most common cause of cancer death after lung cancer. In the USA, breast cancer has an **incidence rate** (2008) of approximately 13%, although this rate is not projected to increase. Fewer than 1% of breast cancer cases occur in men. Female sex hormones are implicated in the development of many breast cancer. The incidence of the disease is higher in women who began menstruation early and/or whose menopause was late. Women who have no children or who had their first child when they were in their late 20s or 30s are also at higher risk. There is also a definite familial (heredity) factor in 5-10% of cases. A high fat diet is also implicated. In Japan, where a low fat diet is typical, the disease is rare. Yet Japanese women living in the United States and eating a higher-fat American diet have the same rate of breast cancer as American women generally.

Characteristics of breast cancer

The incidence of breast cancer increases with age; it is almost unknown before age 25 but the incidence rises sharply in women aged 25-44. Most deaths from breast cancer occur because the disease has already spread beyond the breast when first detected.

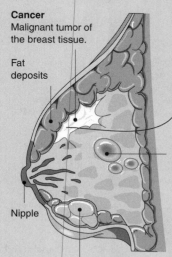

Cancer
Malignant tumor of the breast tissue.

Fat deposits

Nipple

Cancer cells may detach from the tumor mass and enter the blood vessels and lymph ducts where they can form secondary tumors. This process is called **metastasis**.

Cysts, which are fluid-filled sacs, are not associated with cancer.

Fibroadenoma
One of a number of common benign tumors.

Cross-section through a female breast, showing the various types of tissue masses present.

Breast self examination

Regular self-examinations to detect lumps in the breast tissue are recommended (left). Regular breast self examination may be the first step in detecting abnormal changes in the breast tissue.

Mammography

Mammography, which involves a breast X-ray, can detect tumors less than 15 mm in diameter (too small to be detected by a physical breast exam).

In mammography, the breast is compressed between the X-ray plate below and a plastic cover screen above. Several views are then taken.

A **biopsy** is performed if there is a chance that a lump may be malignant. A sample of the affected tissue is taken with a hollow needle and examined under a microscope. If there are cancerous cells present, X-rays, ultrasound scanning, and blood tests can determine if the disease has spread to other parts of the body, such as the bones or the liver. Treatment may then follow.

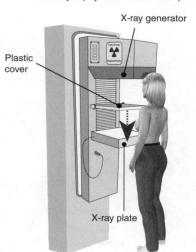

X-ray generator

Plastic cover

X-ray plate

Treatment

Surgical removal of the tumor achieves a cure (as defined by survival for 20 years after treatment) in one third of women with early breast cancer. Studies have shown that survival is not improved by radical surgery such as radical mastectomy (1). Less radical procedures (2 and 3) are now frequently recommended, combined with **radiotherapy** or **anticancer drugs**, such as tamoxifen.

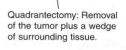

Lumpectomy: only the area of cancerous tissue is removed.

Radical mastectomy: removal of the entire breast, chest muscle, associated lymph nodes, and fat and skin.

Quadrantectomy: Removal of the tumor plus a wedge of surrounding tissue.

1. Describe three factors associated with increased risk of developing breast cancer and suggest why they increase risk:

_____

_____

2. Suggest in what way breast self examination may be deficient in detecting early breast cancers: _____

_____

3. Describe a possible treatment for early breast cancer involving a small isolated tumor: _____

_____

4. State the evidence for a link between diet and increased risk of developing breast cancer: _____

_____

_____

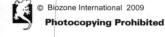

Related activities: Apoptosis: Programed Cell Death, Cancer: Cells out of Control

Reproduction & Development

A 2

Sexual Development

Like many animals, humans differentiate into the male or female sex by the action of a combination of different hormones. The hormones testosterone (in males), and estrogen and progesterone (in females), are responsible for puberty (the onset of sexual maturity), the maintenance of gender differences, and the production of gametes. In females, estrogen and progesterone also regulate the menstrual cycle, and ensure the maintenance of pregnancy and nourishment of young.

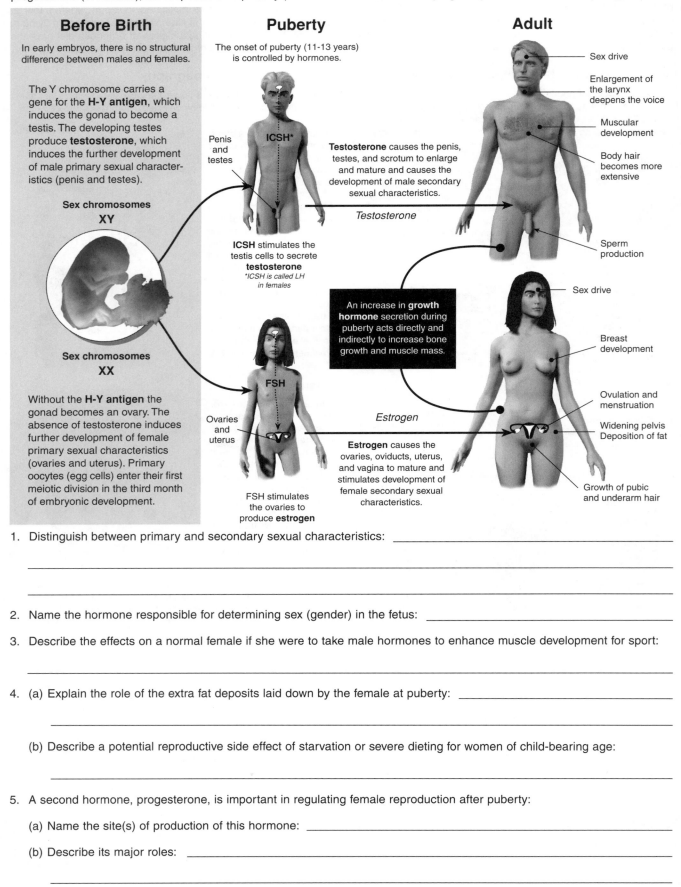

Before Birth

In early embryos, there is no structural difference between males and females.

The Y chromosome carries a gene for the **H-Y antigen**, which induces the gonad to become a testis. The developing testes produce **testosterone**, which induces the further development of male primary sexual characteristics (penis and testes).

Sex chromosomes XY

Sex chromosomes XX

Without the **H-Y antigen** the gonad becomes an ovary. The absence of testosterone induces further development of female primary sexual characteristics (ovaries and uterus). Primary oocytes (egg cells) enter their first meiotic division in the third month of embryonic development.

Puberty

The onset of puberty (11-13 years) is controlled by hormones.

Penis and testes

ICSH*

ICSH stimulates the testis cells to secrete **testosterone**
*ICSH is called LH in females

Testosterone

Testosterone causes the penis, testes, and scrotum to enlarge and mature and causes the development of male secondary sexual characteristics.

An increase in **growth hormone** secretion during puberty acts directly and indirectly to increase bone growth and muscle mass.

Ovaries and uterus

FSH

FSH stimulates the ovaries to produce **estrogen**

Estrogen

Estrogen causes the ovaries, oviducts, uterus, and vagina to mature and stimulates development of female secondary sexual characteristics.

Adult

Sex drive

Enlargement of the larynx deepens the voice

Muscular development

Body hair becomes more extensive

Sperm production

Sex drive

Breast development

Ovulation and menstruation

Widening pelvis Deposition of fat

Growth of pubic and underarm hair

1. Distinguish between primary and secondary sexual characteristics: _____

_____

_____

2. Name the hormone responsible for determining sex (gender) in the fetus: _____

3. Describe the effects on a normal female if she were to take male hormones to enhance muscle development for sport:

_____

4. (a) Explain the role of the extra fat deposits laid down by the female at puberty: _____

_____

(b) Describe a potential reproductive side effect of starvation or severe dieting for women of child-bearing age:

_____

5. A second hormone, progesterone, is important in regulating female reproduction after puberty:

(a) Name the site(s) of production of this hormone: _____

(b) Describe its major roles: _____

_____

Related activities: Hormones of the Pituitary, Control of the Menstrual Cycle, The Hormones of Pregnancy

© Biozone International 2009
Photocopying Prohibited

Index

Rational (Reciprocal) Function

$f(x) = \dfrac{1}{x}$

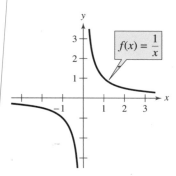

Domain: $(-\infty, 0) \cup (0, \infty)$
Range: $(-\infty, 0) \cup (0, \infty)$
No intercepts
Decreasing on $(-\infty, 0)$ and $(0, \infty)$
Odd function
Origin symmetry
Vertical asymptote: y-axis
Horizontal asymptote: x-axis

Exponential Function

$f(x) = a^x, \ a > 1$

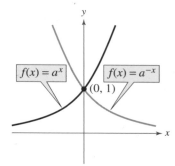

Domain: $(-\infty, \infty)$
Range: $(0, \infty)$
Intercept: $(0, 1)$
Increasing on $(-\infty, \infty)$
    for $f(x) = a^x$
Decreasing on $(-\infty, \infty)$
    for $f(x) = a^{-x}$
Horizontal asymptote: x-axis
Continuous

Logarithmic Function

$f(x) = \log_a x, \ a > 0, \ a \neq 1$

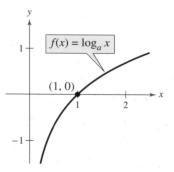

Domain: $(0, \infty)$
Range: $(-\infty, \infty)$
Intercept: $(1, 0)$
Increasing on $(0, \infty)$
Vertical asymptote: y-axis
Continuous
Reflection of graph of $f(x) = a^x$
    in the line $y = x$

Sine Function

$f(x) = \sin x$

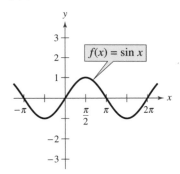

Domain: $(-\infty, \infty)$
Range: $[-1, 1]$
Period: 2π
x-intercepts: $(n\pi, 0)$
y-intercept: $(0, 0)$
Odd function
Origin symmetry

Cosine Function

$f(x) = \cos x$

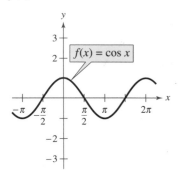

Domain: $(-\infty, \infty)$
Range: $[-1, 1]$
Period: 2π
x-intercepts: $\left(\dfrac{\pi}{2} + n\pi, 0\right)$
y-intercept: $(0, 1)$
Even function
y-axis symmetry

Tangent Function

$f(x) = \tan x$

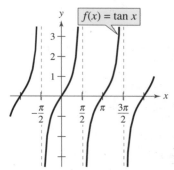

Domain: all $x \neq \dfrac{\pi}{2} + n\pi$
Range: $(-\infty, \infty)$
Period: π
x-intercepts: $(n\pi, 0)$
y-intercept: $(0, 0)$
Vertical asymptotes:
    $x = \dfrac{\pi}{2} + n\pi$
Odd function
Origin symmetry

Cosecant Function

$f(x) = \csc x$

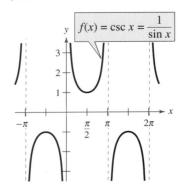

Domain: all $x \neq n\pi$
Range: $(-\infty, -1] \cup [1, \infty)$
Period: 2π
No intercepts
Vertical asymptotes: $x = n\pi$
Odd function
Origin symmetry

Secant Function

$f(x) = \sec x$

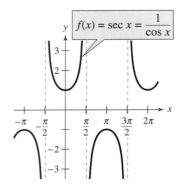

Domain: all $x \neq \dfrac{\pi}{2} + n\pi$

Range: $(-\infty, -1] \cup [1, \infty)$
Period: 2π
y-intercept: $(0, 1)$
Vertical asymptotes:

$$x = \frac{\pi}{2} + n\pi$$

Even function
y-axis symmetry

Cotangent Function

$f(x) = \cot x$

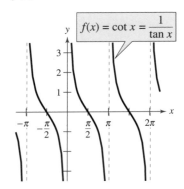

Domain: all $x \neq n\pi$
Range: $(-\infty, \infty)$
Period: π

x-intercepts: $\left(\dfrac{\pi}{2} + n\pi, 0\right)$

Vertical asymptotes: $x = n\pi$
Odd function
Origin symmetry

Inverse Sine Function

$f(x) = \arcsin x$

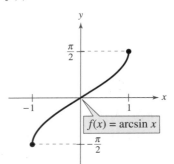

Domain: $[-1, 1]$

Range: $\left[-\dfrac{\pi}{2}, \dfrac{\pi}{2}\right]$

Intercept: $(0, 0)$
Odd function
Origin symmetry

Inverse Cosine Function

$f(x) = \arccos x$

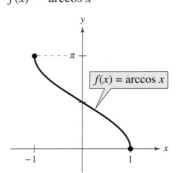

Domain: $[-1, 1]$
Range: $[0, \pi]$

y-intercept: $\left(0, \dfrac{\pi}{2}\right)$

Inverse Tangent Function

$f(x) = \arctan x$

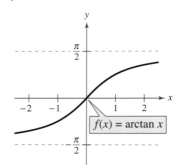

Domain: $(-\infty, \infty)$

Range: $\left(-\dfrac{\pi}{2}, \dfrac{\pi}{2}\right)$

Intercept: $(0, 0)$
Horizontal asymptotes:

$$y = \pm\frac{\pi}{2}$$

Odd function
Origin symmetry

Mat 180: Precalculus

Ron Larson

With the assistance of David C. Falvo

Tatiana Serrão
MAT - 180 - 604
Spring 11

(201) 800 - 2192

CENGAGE
Learning™

Australia • Brazil • Japan • Korea • Mexico • Singapore • Spain • United Kingdom • United States

CENGAGE
Learning™

Mat 180: Precalculus

Precalculus: A Concise Course, 2e
Ron Larson

© 2011, 2007 Brooks/Cole, Cengage Learning. All rights reserved.

Executive Editors:
 Maureen Staudt
 Michael Stranz

Senior Project Development Manager:
 Linda deStefano

Marketing Specialist:
 Courtney Sheldon

Senior Production/Manufacturing Manager:
 Donna M. Brown

PreMedia Manager:
 Joel Brennecke

Sr. Rights Acquisition Account Manager:
 Todd Osborne

Cover Image:
 Getty Images*

*Unless otherwise noted, all cover images used by Custom
Solutions, a part of Cengage Learning, have been supplied
courtesy of Getty Images with the exception of the Earthview
cover image, which has been supplied by the National
Aeronautics and Space Administration (NASA).

For product information and technology assistance, contact us at
Cengage Learning Customer & Sales Support, 1-800-354-9706

For permission to use material from this text or product,
submit all requests online at **cengage.com/permissions**
Further permissions questions can be emailed to
permissionrequest@cengage.com

This book contains select works from existing Cengage Learning resources and
was produced by Cengage Learning Custom Solutions for collegiate use. As such,
those adopting and/or contributing to this work are responsible for editorial
content accuracy, continuity and completeness.

Compilation © 2010 Cengage Learning

ISBN-13: 978-1-111-46709-8

ISBN-10: 1-111-46709-9

Cengage Learning
5191 Natorp Boulevard
Mason, Ohio 45040
USA
Cengage Learning is a leading provider of customized learning solutions with
office locations around the globe, including Singapore, the United Kingdom,
Australia, Mexico, Brazil, and Japan. Locate your local office at:
international.cengage.com/region.

Cengage Learning products are represented in Canada by Nelson Education, Ltd.
For your lifelong learning solutions, visit **www.cengage.com/custom.**
Visit our corporate website at **www.cengage.com.**

Printed in the United States of America

Custom Table of Contents

Contents

- *New Capstone Exercises*   *Capstones* are conceptual problems that synthesize key topics and provide students with a better understanding of each section's concepts. Capstone exercises are excellent for classroom discussion or test prep, and teachers may find value in integrating these problems into their reviews of the section.

- *New Chapter Summaries*   The *Chapter Summary* now includes an explanation and/or example of each objective taught in the chapter.

- *Revised Exercise Sets*   The exercise sets have been carefully and extensively examined to ensure they are rigorous and cover all topics suggested by our users. Many new skill-building and challenging exercises have been added.

For the past several years, we've maintained an independent website—**CalcChat.com**—that provides free solutions to all odd-numbered exercises in the text. Thousands of students using our textbooks have visited the site for practice and help with their homework. For the Second Edition, we were able to use information from CalcChat.com, including which solutions students accessed most often, to help guide the revision of the exercises.

I hope you enjoy the Second Edition of *Precalculus: A Concise Course*. As always, I welcome comments and suggestions for continued improvements.

Ron Larson

Acknowledgments

I would like to thank the many people who have helped me prepare the text and the supplements package. Their encouragement, criticisms, and suggestions have been invaluable.

Thank you to all of the instructors who took the time to review the changes in this edition and to provide suggestions for improving it. Without your help, this book would not be possible.

Reviewers

Chad Pierson, *University of Minnesota-Duluth*; Sally Shao, *Cleveland State University*; Ed Stumpf, *Central Carolina Community College*; Fuzhen Zhang, *Nova Southeastern University*; Dennis Shepherd, *University of Colorado, Denver*; Rhonda Kilgo, *Jacksonville State University*; C. Altay Özgener, *Manatee Community College Bradenton*; William Forrest, *Baton Rouge Community College*; Tracy Cook, *University of Tennessee Knoxville*; Charles Hale, *California State Poly University Pomona*; Samuel Evers, *University of Alabama*; Seongchun Kwon, *University of Toledo*; Dr. Arun K. Agarwal, *Grambling State University*; Hyounkyun Oh, *Savannah State University*; Michael J. McConnell, *Clarion University*; Martha Chalhoub, *Collin County Community College*; Angela Lee Everett, *Chattanooga State Tech Community College*; Heather Van Dyke, *Walla Walla Community College*; Gregory Buthusiem, *Burlington County Community College*; Ward Shaffer, *College of Coastal Georgia*; Carmen Thomas, *Chatham University*

My thanks to David Falvo, The Behrend College, The Pennsylvania State University, for his contributions to this project. My thanks also to Robert Hostetler, The Behrend College, The Pennsylvania State University, and Bruce Edwards, University of Florida, for their significant contributions to the previous edition of this text.

I would also like to thank the staff at Larson Texts, Inc. who assisted with proofreading the manuscript, preparing and proofreading the art package, and checking and typesetting the supplements.

On a personal level, I am grateful to my spouse, Deanna Gilbert Larson, for her love, patience, and support. Also, a special thanks goes to R. Scott O'Neil. If you have suggestions for improving this text, please feel free to write to me. Over the past two decades I have received many useful comments from both instructors and students, and I value these comments very highly.

Ron Larson

Ron Larson

Supplements

Supplements for the Instructor

Annotated Instructor's Edition   This AIE is the complete student text plus point-of-use annotations for the instructor, including extra projects, classroom activities, teaching strategies, and additional examples. Answers to even-numbered text exercises, Vocabulary Checks, and Explorations are also provided.

Complete Solutions Manual   This manual contains solutions to all exercises from the text, including Chapter Review Exercises and Chapter Tests.

Instructor's Companion Website   This free companion website contains an abundance of instructor resources.

PowerLecture™ with ExamView®   The CD-ROM provides the instructor with dynamic media tools for teaching Precalculus. PowerPoint® lecture slides and art slides of the figures from the text, together with electronic files for the test bank and a link to the Solution Builder, are available. The algorithmic ExamView allows you to create, deliver, and customize tests (both print and online) in minutes with this easy-to-use assessment system. Enhance how your students interact with you, your lecture, and each other.

Solutions Builder   This is an electronic version of the complete solutions manual available via the PowerLecture and Instructor's Companion Website. It provides instructors with an efficient method for creating solution sets to homework or exams that can then be printed or posted.

Supplements for the Student

Student Companion Website   This free companion website contains an abundance of student resources.

Instructional DVDs   Keyed to the text by section, these DVDs provide comprehensive coverage of the course—along with additional explanations of concepts, sample problems, and applications—to help students review essential topics.

Student Study and Solutions Manual   This guide offers step-by-step solutions for all odd-numbered text exercises, Chapter and Cumulative Tests, and Practice Tests with solutions.

Premium eBook   The Premium eBook offers an interactive version of the textbook with search features, highlighting and note-making tools, and direct links to videos or tutorials that elaborate on the text discussions.

Enhanced WebAssign   Enhanced WebAssign is designed for you to do your homework online. This proven and reliable system uses pedagogy and content found in Larson's text, and then enhances it to help you learn Precalculus more effectively. Automatically graded homework allows you to focus on your learning and get interactive study assistance outside of class.

Functions and Their Graphs

1

In Mathematics

Functions show how one variable is related to another variable.

In Real Life

Functions are used to estimate values, simulate processes, and discover relationships. For instance, you can model the enrollment rate of children in preschool and estimate the year in which the rate will reach a certain number. Such an estimate can be used to plan measures for meeting future needs, such as hiring additional teachers and buying more books. (See Exercise 113, page 64.)

Jose Luis Pelaez/Getty Images

IN CAREERS

There are many careers that use functions. Several are listed below.

- Financial analyst
  Exercise 95, page 51

- Biologist
  Exercise 73, page 91

- Tax preparer
  Example 3, page 104

- Oceanographer
  Exercise 83, page 112

1.1 RECTANGULAR COORDINATES

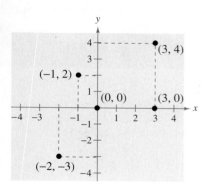

The Cartesian Plane

Just as you can represent real numbers by points on a real number line, you can represent ordered pairs of real numbers by points in a plane called the **rectangular coordinate system,** or the **Cartesian plane,** named after the French mathematician René Descartes (1596–1650).

The Cartesian plane is formed by using two real number lines intersecting at right angles, as shown in Figure 1.1. The horizontal real number line is usually called the **x-axis,** and the vertical real number line is usually called the **y-axis.** The point of intersection of these two axes is the **origin,** and the two axes divide the plane into four parts called **quadrants.**

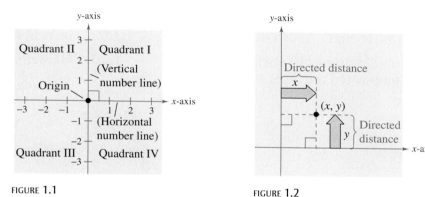

FIGURE 1.1

FIGURE 1.2

Each point in the plane corresponds to an **ordered pair** (x, y) of real numbers x and y, called **coordinates** of the point. The **x-coordinate** represents the directed distance from the y-axis to the point, and the **y-coordinate** represents the directed distance from the x-axis to the point, as shown in Figure 1.2.

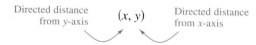

The notation (x, y) denotes both a point in the plane and an open interval on the real number line. The context will tell you which meaning is intended.

| Example 1 | Plotting Points in the Cartesian Plane

Plot the points $(-1, 2)$, $(3, 4)$, $(0, 0)$, $(3, 0)$, and $(-2, -3)$.

Solution

To plot the point $(-1, 2)$, imagine a vertical line through -1 on the x-axis and a horizontal line through 2 on the y-axis. The intersection of these two lines is the point $(-1, 2)$. The other four points can be plotted in a similar way, as shown in Figure 1.3.

CHECK***Point***   Now try Exercise 7.

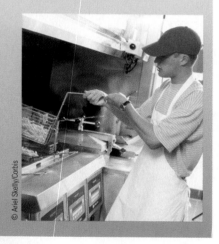

FIGURE 1.3

The beauty of a rectangular coordinate system is that it allows you to *see* relationships between two variables. It would be difficult to overestimate the importance of Descartes's introduction of coordinates in the plane. Today, his ideas are in common use in virtually every scientific and business-related field.

Example 2   Sketching a Scatter Plot

From 1994 through 2007, the numbers N (in millions) of subscribers to a cellular telecommunication service in the United States are shown in the table, where t represents the year. Sketch a scatter plot of the data.   (Source: CTIA-The Wireless Association)

Solution

To sketch a *scatter plot* of the data shown in the table, you simply represent each pair of values by an ordered pair (t, N) and plot the resulting points, as shown in Figure 1.4. For instance, the first pair of values is represented by the ordered pair $(1994, 24.1)$. Note that the break in the t-axis indicates that the numbers between 0 and 1994 have been omitted.

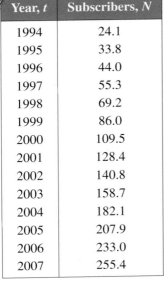

Year, t	Subscribers, N
1994	24.1
1995	33.8
1996	44.0
1997	55.3
1998	69.2
1999	86.0
2000	109.5
2001	128.4
2002	140.8
2003	158.7
2004	182.1
2005	207.9
2006	233.0
2007	255.4

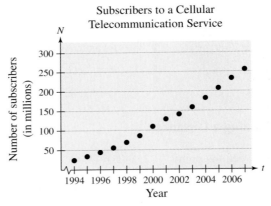

FIGURE 1.4

CHECK **Point**   Now try Exercise 25.

In Example 2, you could have let $t = 1$ represent the year 1994. In that case, the horizontal axis would not have been broken, and the tick marks would have been labeled 1 through 14 (instead of 1994 through 2007).

TECHNOLOGY

The scatter plot in Example 2 is only one way to represent the data graphically. You could also represent the data using a bar graph or a line graph. If you have access to a graphing utility, try using it to represent graphically the data given in Example 2.

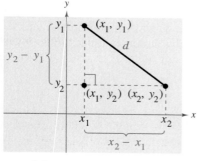

$a^2 + b^2 = c^2$

a

c

b

FIGURE **1.5**

The Pythagorean Theorem and the Distance Formula

The following famous theorem is used extensively throughout this course.

Pythagorean Theorem

For a right triangle with hypotenuse of length *c* and sides of lengths *a* and *b*, you have $a^2 + b^2 = c^2$, as shown in Figure 1.5. (The converse is also true. That is, if $a^2 + b^2 = c^2$, then the triangle is a right triangle.)

Suppose you want to determine the distance *d* between two points (x_1, y_1) and (x_2, y_2) in the plane. With these two points, a right triangle can be formed, as shown in Figure 1.6. The length of the vertical side of the triangle is $|y_2 - y_1|$, and the length of the horizontal side is $|x_2 - x_1|$. By the Pythagorean Theorem, you can write

$$d^2 = |x_2 - x_1|^2 + |y_2 - y_1|^2$$

$$d = \sqrt{|x_2 - x_1|^2 + |y_2 - y_1|^2} = \sqrt{(x_2 - x_1)^2 + (y_2 - y_1)^2}.$$

This result is the **Distance Formula.**

FIGURE **1.6**

The Distance Formula

The distance *d* between the points (x_1, y_1) and (x_2, y_2) in the plane is

$$d = \sqrt{(x_2 - x_1)^2 + (y_2 - y_1)^2}.$$

Example 3   Finding a Distance

Find the distance between the points $(-2, 1)$ and $(3, 4)$.

Algebraic Solution

Let $(x_1, y_1) = (-2, 1)$ and $(x_2, y_2) = (3, 4)$. Then apply the Distance Formula.

$$d = \sqrt{(x_2 - x_1)^2 + (y_2 - y_1)^2} \qquad \text{Distance Formula}$$

$$= \sqrt{[3 - (-2)]^2 + (4 - 1)^2} \qquad \begin{matrix}\text{Substitute for}\\ x_1, y_1, x_2, \text{ and } y_2.\end{matrix}$$

$$= \sqrt{(5)^2 + (3)^2} \qquad \text{Simplify.}$$

$$= \sqrt{34} \qquad \text{Simplify.}$$

$$\approx 5.83 \qquad \text{Use a calculator.}$$

So, the distance between the points is about 5.83 units. You can use the Pythagorean Theorem to check that the distance is correct.

$$d^2 \overset{?}{=} 3^2 + 5^2 \qquad \text{Pythagorean Theorem}$$

$$\left(\sqrt{34}\right)^2 \overset{?}{=} 3^2 + 5^2 \qquad \text{Substitute for } d.$$

$$34 = 34 \qquad \text{Distance checks. } \checkmark$$

Graphical Solution

Use centimeter graph paper to plot the points $A(-2, 1)$ and $B(3, 4)$. Carefully sketch the line segment from *A* to *B*. Then use a centimeter ruler to measure the length of the segment.

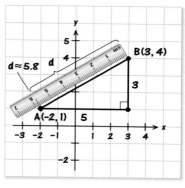

FIGURE **1.7**

The line segment measures about 5.8 centimeters, as shown in Figure 1.7. So, the distance between the points is about 5.8 units.

CHECK**Point**   Now try Exercise 31.

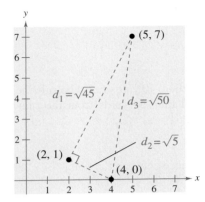

FIGURE **1.8**

![Algebra Help]

You can review the techniques for evaluating a radical in Appendix A.2.

Example 4   Verifying a Right Triangle

Show that the points $(2, 1)$, $(4, 0)$, and $(5, 7)$ are vertices of a right triangle.

Solution

The three points are plotted in Figure 1.8. Using the Distance Formula, you can find the lengths of the three sides as follows.

$$d_1 = \sqrt{(5 - 2)^2 + (7 - 1)^2} = \sqrt{9 + 36} = \sqrt{45}$$

$$d_2 = \sqrt{(4 - 2)^2 + (0 - 1)^2} = \sqrt{4 + 1} = \sqrt{5}$$

$$d_3 = \sqrt{(5 - 4)^2 + (7 - 0)^2} = \sqrt{1 + 49} = \sqrt{50}$$

Because

$$(d_1)^2 + (d_2)^2 = 45 + 5 = 50 = (d_3)^2$$

you can conclude by the Pythagorean Theorem that the triangle must be a right triangle.

CHECK *Point*   Now try Exercise 43.

The Midpoint Formula

To find the **midpoint** of the line segment that joins two points in a coordinate plane, you can simply find the average values of the respective coordinates of the two endpoints using the **Midpoint Formula.**

The Midpoint Formula

The midpoint of the line segment joining the points (x_1, y_1) and (x_2, y_2) is given by the Midpoint Formula

$$\text{Midpoint} = \left(\frac{x_1 + x_2}{2}, \frac{y_1 + y_2}{2} \right).$$

For a proof of the Midpoint Formula, see Proofs in Mathematics on page 122.

Example 5   Finding a Line Segment's Midpoint

Find the midpoint of the line segment joining the points $(-5, -3)$ and $(9, 3)$.

Solution

Let $(x_1, y_1) = (-5, -3)$ and $(x_2, y_2) = (9, 3)$.

$$\text{Midpoint} = \left(\frac{x_1 + x_2}{2}, \frac{y_1 + y_2}{2} \right) \qquad \text{Midpoint Formula}$$

$$= \left(\frac{-5 + 9}{2}, \frac{-3 + 3}{2} \right) \qquad \text{Substitute for } x_1, y_1, x_2, \text{ and } y_2.$$

$$= (2, 0) \qquad \text{Simplify.}$$

The midpoint of the line segment is $(2, 0)$, as shown in Figure 1.9.

CHECK *Point*   Now try Exercise 47(c).

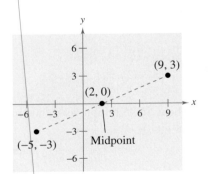

FIGURE **1.9**

Applications

Example 6 Finding the Length of a Pass

A football quarterback throws a pass from the 28-yard line, 40 yards from the sideline. The pass is caught by a wide receiver on the 5-yard line, 20 yards from the same sideline, as shown in Figure 1.10. How long is the pass?

Solution

You can find the length of the pass by finding the distance between the points (40, 28) and (20, 5).

$$d = \sqrt{(x_2 - x_1)^2 + (y_2 - y_1)^2} \qquad \text{Distance Formula}$$

$$= \sqrt{(40 - 20)^2 + (28 - 5)^2} \qquad \text{Substitute for } x_1, y_1, x_2, \text{ and } y_2.$$

$$= \sqrt{400 + 529} \qquad \text{Simplify.}$$

$$= \sqrt{929} \qquad \text{Simplify.}$$

$$\approx 30 \qquad \text{Use a calculator.}$$

So, the pass is about 30 yards long.

CHECK Point Now try Exercise 57.

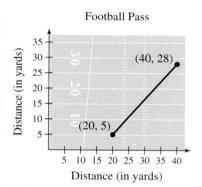

Football Pass

FIGURE **1.10**

In Example 6, the scale along the goal line does not normally appear on a football field. However, when you use coordinate geometry to solve real-life problems, you are free to place the coordinate system in any way that is convenient for the solution of the problem.

Example 7 Estimating Annual Revenue

Barnes & Noble had annual sales of approximately $5.1 billion in 2005, and $5.4 billion in 2007. Without knowing any additional information, what would you estimate the 2006 sales to have been? (Source: Barnes & Noble, Inc.)

Solution

One solution to the problem is to assume that sales followed a linear pattern. With this assumption, you can estimate the 2006 sales by finding the midpoint of the line segment connecting the points (2005, 5.1) and (2007, 5.4).

$$\text{Midpoint} = \left(\frac{x_1 + x_2}{2}, \frac{y_1 + y_2}{2} \right) \qquad \text{Midpoint Formula}$$

$$= \left(\frac{2005 + 2007}{2}, \frac{5.1 + 5.4}{2} \right) \qquad \text{Substitute for } x_1, x_2, y_1 \text{ and } y_2.$$

$$= (2006, 5.25) \qquad \text{Simplify.}$$

So, you would estimate the 2006 sales to have been about $5.25 billion, as shown in Figure 1.11. (The actual 2006 sales were about $5.26 billion.)

CHECK Point Now try Exercise 59.

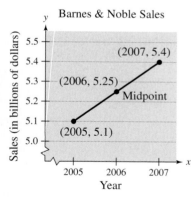

Barnes & Noble Sales

FIGURE **1.11**

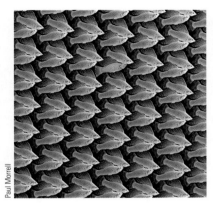

Paul Morrell

Much of computer graphics, including this computer-generated goldfish tessellation, consists of transformations of points in a coordinate plane. One type of transformation, a translation, is illustrated in Example 8. Other types include reflections, rotations, and stretches.

Example 8   Translating Points in the Plane

The triangle in Figure 1.12 has vertices at the points $(-1, 2)$, $(1, -4)$, and $(2, 3)$. Shift the triangle three units to the right and two units upward and find the vertices of the shifted triangle, as shown in Figure 1.13.

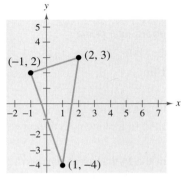

FIGURE **1.12**

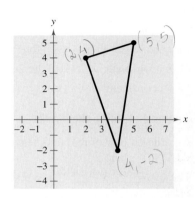

FIGURE **1.13**

Solution

To shift the vertices three units to the right, add 3 to each of the *x*-coordinates. To shift the vertices two units upward, add 2 to each of the *y*-coordinates.

Original Point	*Translated Point*
$(-1, 2)$	$(-1 + 3, 2 + 2) = (2, 4)$
$(1, -4)$	$(1 + 3, -4 + 2) = (4, -2)$
$(2, 3)$	$(2 + 3, 3 + 2) = (5, 5)$

CHECK*Point*   Now try Exercise 61.

The figures provided with Example 8 were not really essential to the solution. Nevertheless, it is strongly recommended that you develop the habit of including sketches with your solutions—even if they are not required.

CLASSROOM DISCUSSION

Extending the Example   Example 8 shows how to translate points in a coordinate plane. Write a short paragraph describing how each of the following transformed points is related to the original point.

Original Point	*Transformed Point*
(x, y)	$(-x, y)$
(x, y)	$(x, -y)$
(x, y)	$(-x, -y)$

1.1   EXERCISES

See www.CalcChat.com for worked-out solutions to odd-numbered exercises.

VOCABULARY

1. Match each term with its definition.

(a) x-axis
(b) y-axis
(c) origin
(d) quadrants
(e) x-coordinate
(f) y-coordinate

(i) point of intersection of vertical axis and horizontal axis
(ii) directed distance from the x-axis
(iii) directed distance from the y-axis
(iv) four regions of the coordinate plane
(v) horizontal real number line
(vi) vertical real number line

In Exercises 2–4, fill in the blanks.

2. An ordered pair of real numbers can be represented in a plane called the rectangular coordinate system or the _Cartesian_ plane.

3. The _Distance Formula_ is a result derived from the Pythagorean Theorem.

4. Finding the average values of the representative coordinates of the two endpoints of a line segment in a coordinate plane is also known as using the _Midpoint Formula_.

SKILLS AND APPLICATIONS

In Exercises 5 and 6, approximate the coordinates of the points.

5.

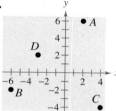

6.

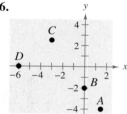

In Exercises 7–10, plot the points in the Cartesian plane.

7. $(-4, 2), (-3, -6), (0, 5), (1, -4)$

8. $(0, 0), (3, 1), (-2, 4), (1, -1)$

9. $(3, 8), (0.5, -1), (5, -6), (-2, 2.5)$

10. $\left(1, -\frac{1}{3}\right), \left(\frac{3}{4}, 3\right), (-3, 4), \left(-\frac{4}{3}, -\frac{3}{2}\right)$

In Exercises 11–14, find the coordinates of the point.

11. The point is located three units to the left of the y-axis and four units above the x-axis.

12. The point is located eight units below the x-axis and four units to the right of the y-axis.

13. The point is located five units below the x-axis and the coordinates of the point are equal.

14. The point is on the x-axis and 12 units to the left of the y-axis.

In Exercises 15–24, determine the quadrant(s) in which (x, y) is located so that the condition(s) is (are) satisfied.

15. $x > 0$ and $y < 0$

16. $x < 0$ and $y < 0$

17. $x = -4$ and $y > 0$

18. $x > 2$ and $y = 3$

19. $y < -5$

20. $x > 4$

21. $x < 0$ and $-y > 0$

22. $-x > 0$ and $y < 0$

23. $xy > 0$

24. $xy < 0$

In Exercises 25 and 26, sketch a scatter plot of the data shown in the table.

25. NUMBER OF STORES The table shows the number y of Wal-Mart stores for each year x from 2000 through 2007. (Source: Wal-Mart Stores, Inc.)

Year, x	Number of stores, y
2000	4189
2001	4414
2002	4688
2003	4906
2004	5289
2005	6141
2006	6779
2007	7262

26. METEOROLOGY The table shows the lowest temperature on record y (in degrees Fahrenheit) in Duluth, Minnesota for each month x, where $x = 1$ represents January. (Source: NOAA)

Month, x	Temperature, y
1	−39
2	−39
3	−29
4	−5
5	17
6	27
7	35
8	32
9	22
10	8
11	−23
12	−34

In Exercises 27–38, find the distance between the points.

27. $(6, -3), (6, 5)$ **28.** $(1, 4), (8, 4)$

29. $(-3, -1), (2, -1)$ **30.** $(-3, -4), (-3, 6)$

31. $(-2, 6), (3, -6)$ **32.** $(8, 5), (0, 20)$

33. $(1, 4), (-5, -1)$ **34.** $(1, 3), (3, -2)$

35. $\left(\frac{1}{2}, \frac{4}{3}\right), (2, -1)$ **36.** $\left(-\frac{2}{3}, 3\right), \left(-1, \frac{5}{4}\right)$

37. $(-4.2, 3.1), (-12.5, 4.8)$

38. $(9.5, -2.6), (-3.9, 8.2)$

In Exercises 39–42, (a) find the length of each side of the right triangle, and (b) show that these lengths satisfy the Pythagorean Theorem.

39.
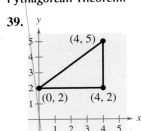

40.
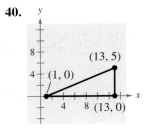

41.
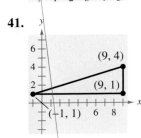

42.
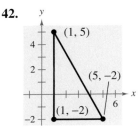

In Exercises 43–46, show that the points form the vertices of the indicated polygon.

43. Right triangle: $(4, 0), (2, 1), (-1, -5)$

44. Right triangle: $(-1, 3), (3, 5), (5, 1)$

45. Isosceles triangle: $(1, -3), (3, 2), (-2, 4)$

46. Isosceles triangle: $(2, 3), (4, 9), (-2, 7)$

In Exercises 47–56, (a) plot the points, (b) find the distance between the points, and (c) find the midpoint of the line segment joining the points.

47. $(1, 1), (9, 7)$ **48.** $(1, 12), (6, 0)$

49. $(-4, 10), (4, -5)$ **50.** $(-7, -4), (2, 8)$

51. $(-1, 2), (5, 4)$ **52.** $(2, 10), (10, 2)$

53. $\left(\frac{1}{2}, 1\right), \left(-\frac{5}{2}, \frac{4}{3}\right)$ **54.** $\left(-\frac{1}{3}, -\frac{1}{3}\right), \left(-\frac{1}{6}, -\frac{1}{2}\right)$

55. $(6.2, 5.4), (-3.7, 1.8)$ **56.** $(-16.8, 12.3), (5.6, 4.9)$

57. FLYING DISTANCE An airplane flies from Naples, Italy in a straight line to Rome, Italy, which is 120 kilometers north and 150 kilometers west of Naples. How far does the plane fly?

58. SPORTS A soccer player passes the ball from a point that is 18 yards from the endline and 12 yards from the sideline. The pass is received by a teammate who is 42 yards from the same endline and 50 yards from the same sideline, as shown in the figure. How long is the pass?

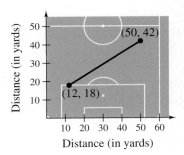

SALES In Exercises 59 and 60, use the Midpoint Formula to estimate the sales of Big Lots, Inc. and Dollar Tree Stores, Inc. in 2005, given the sales in 2003 and 2007. Assume that the sales followed a linear pattern. (Source: Big Lots, Inc.; Dollar Tree Stores, Inc.)

59. Big Lots

Year	Sales (in millions)
2003	$4174
2007	$4656

60. Dollar Tree

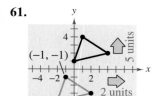

Year	Sales (in millions)
2003	$2800
2007	$4243

In Exercises 61–64, the polygon is shifted to a new position in the plane. Find the coordinates of the vertices of the polygon in its new position.

61.

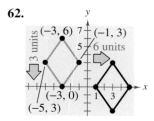

62.

63. Original coordinates of vertices: $(-7, -2), (-2, 2),$ $(-2, -4), (-7, -4)$

Shift: eight units upward, four units to the right

64. Original coordinates of vertices: $(5, 8), (3, 6), (7, 6),$ $(5, 2)$

Shift: 6 units downward, 10 units to the left

RETAIL PRICE In Exercises 65 and 66, use the graph, which shows the average retail prices of 1 gallon of whole milk from 1996 to 2007. (Source: U.S. Bureau of Labor Statistics)

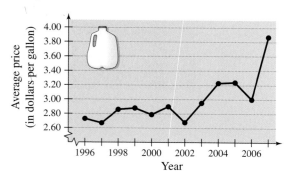

65. Approximate the highest price of a gallon of whole milk shown in the graph. When did this occur?

66. Approximate the percent change in the price of milk from the price in 1996 to the highest price shown in the graph.

67. ADVERTISING The graph shows the average costs of a 30-second television spot (in thousands of dollars) during the Super Bowl from 2000 to 2008. (Source: Nielson Media and TNS Media Intelligence)

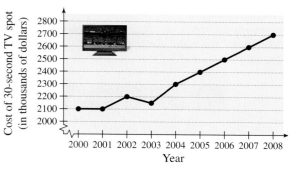

FIGURE FOR **67**

(a) Estimate the percent increase in the average cost of a 30-second spot from Super Bowl XXXIV in 2000 to Super Bowl XXXVIII in 2004.

(b) Estimate the percent increase in the average cost of a 30-second spot from Super Bowl XXXIV in 2000 to Super Bowl XLII in 2008.

68. ADVERTISING The graph shows the average costs of a 30-second television spot (in thousands of dollars) during the Academy Awards from 1995 to 2007. (Source: Nielson Monitor-Plus)

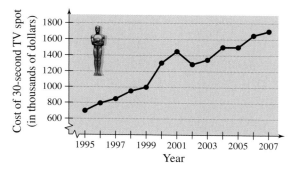

(a) Estimate the percent increase in the average cost of a 30-second spot in 1996 to the cost in 2002.

(b) Estimate the percent increase in the average cost of a 30-second spot in 1996 to the cost in 2007.

69. MUSIC The graph shows the numbers of performers who were elected to the Rock and Roll Hall of Fame from 1991 through 2008. Describe any trends in the data. From these trends, predict the number of performers elected in 2010. (Source: rockhall.com)

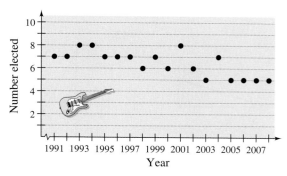

70. LABOR FORCE   Use the graph below, which shows the minimum wage in the United States (in dollars) from 1950 to 2009.   (Source: U.S. Department of Labor)

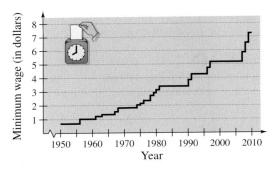

(a) Which decade shows the greatest increase in minimum wage?

(b) Approximate the percent increases in the minimum wage from 1990 to 1995 and from 1995 to 2009.

(c) Use the percent increase from 1995 to 2009 to predict the minimum wage in 2013.

(d) Do you believe that your prediction in part (c) is reasonable? Explain.

71. SALES   The Coca-Cola Company had sales of $19,805 million in 1999 and $28,857 million in 2007. Use the Midpoint Formula to estimate the sales in 2003. Assume that the sales followed a linear pattern.   (Source: The Coca-Cola Company)

72. DATA ANALYSIS: EXAM SCORES   The table shows the mathematics entrance test scores x and the final examination scores y in an algebra course for a sample of 10 students.

x	22	29	35	40	44	48	53	58	65	76
y	53	74	57	66	79	90	76	93	83	99

(a) Sketch a scatter plot of the data.

(b) Find the entrance test score of any student with a final exam score in the 80s.

(c) Does a higher entrance test score imply a higher final exam score? Explain.

73. DATA ANALYSIS: MAIL   The table shows the number y of pieces of mail handled (in billions) by the U.S. Postal Service for each year x from 1996 through 2008.   (Source: U.S. Postal Service)

Year, x	Pieces of mail, y
1996	183
1997	191
1998	197
1999	202
2000	208
2001	207
2002	203
2003	202
2004	206
2005	212
2006	213
2007	212
2008	203

TABLE FOR 73

(a) Sketch a scatter plot of the data.

(b) Approximate the year in which there was the greatest decrease in the number of pieces of mail handled.

(c) Why do you think the number of pieces of mail handled decreased?

74. DATA ANALYSIS: ATHLETICS   The table shows the numbers of men's M and women's W college basketball teams for each year x from 1994 through 2007.   (Source: National Collegiate Athletic Association)

Year, x	Men's teams, M	Women's teams, W
1994	858	859
1995	868	864
1996	866	874
1997	865	879
1998	895	911
1999	926	940
2000	932	956
2001	937	958
2002	936	975
2003	967	1009
2004	981	1008
2005	983	1036
2006	984	1018
2007	982	1003

(a) Sketch scatter plots of these two sets of data on the same set of coordinate axes.

(b) Find the year in which the numbers of men's and women's teams were nearly equal.

(c) Find the year in which the difference between the numbers of men's and women's teams was the greatest. What was this difference?

EXPLORATION

75. A line segment has (x_1, y_1) as one endpoint and (x_m, y_m) as its midpoint. Find the other endpoint (x_2, y_2) of the line segment in terms of $x_1, y_1, x_m,$ and y_m.

76. Use the result of Exercise 75 to find the coordinates of the endpoint of a line segment if the coordinates of the other endpoint and midpoint are, respectively,

   (a) $(1, -2), (4, -1)$ and (b) $(-5, 11), (2, 4)$.

77. Use the Midpoint Formula three times to find the three points that divide the line segment joining (x_1, y_1) and (x_2, y_2) into four parts.

78. Use the result of Exercise 77 to find the points that divide the line segment joining the given points into four equal parts.

   (a) $(1, -2), (4, -1)$ (b) $(-2, -3), (0, 0)$

79. **MAKE A CONJECTURE** Plot the points $(2, 1)$, $(-3, 5)$, and $(7, -3)$ on a rectangular coordinate system. Then change the sign of the x-coordinate of each point and plot the three new points on the same rectangular coordinate system. Make a conjecture about the location of a point when each of the following occurs.

   (a) The sign of the x-coordinate is changed.

   (b) The sign of the y-coordinate is changed.

   (c) The signs of both the x- and y-coordinates are changed.

80. **COLLINEAR POINTS** Three or more points are *collinear* if they all lie on the same line. Use the steps below to determine if the set of points $\{A(2, 3), B(2, 6), C(6, 3)\}$ and the set of points $\{A(8, 3), B(5, 2), C(2, 1)\}$ are collinear.

   (a) For each set of points, use the Distance Formula to find the distances from A to B, from B to C, and from A to C. What relationship exists among these distances for each set of points?

   (b) Plot each set of points in the Cartesian plane. Do all the points of either set appear to lie on the same line?

   (c) Compare your conclusions from part (a) with the conclusions you made from the graphs in part (b). Make a general statement about how to use the Distance Formula to determine collinearity.

TRUE OR FALSE? In Exercises 81 and 82, determine whether the statement is true or false. Justify your answer.

81. In order to divide a line segment into 16 equal parts, you would have to use the Midpoint Formula 16 times.

82. The points $(-8, 4)$, $(2, 11)$, and $(-5, 1)$ represent the vertices of an isosceles triangle.

83. **THINK ABOUT IT** When plotting points on the rectangular coordinate system, is it true that the scales on the x- and y-axes must be the same? Explain.

84. **CAPSTONE** Use the plot of the point (x_0, y_0) in the figure. Match the transformation of the point with the correct plot. Explain your reasoning. [The plots are labeled (i), (ii), (iii), and (iv).]

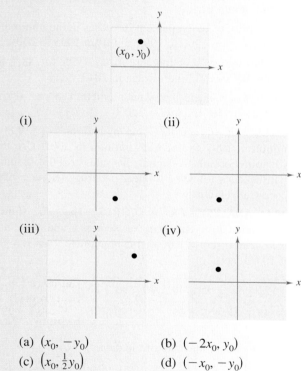

   (a) $(x_0, -y_0)$ (b) $(-2x_0, y_0)$
   (c) $\left(x_0, \frac{1}{2}y_0\right)$ (d) $(-x_0, -y_0)$

85. **PROOF** Prove that the diagonals of the parallelogram in the figure intersect at their midpoints.

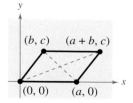

1.2 GRAPHS OF EQUATIONS

What you should learn

- Sketch graphs of equations.
- Find x- and y-intercepts of graphs of equations.
- Use symmetry to sketch graphs of equations.
- Find equations of and sketch graphs of circles.
- Use graphs of equations in solving real-life problems.

Why you should learn it

The graph of an equation can help you see relationships between real-life quantities. For example, in Exercise 87 on page 23, a graph can be used to estimate the life expectancies of children who are born in 2015.

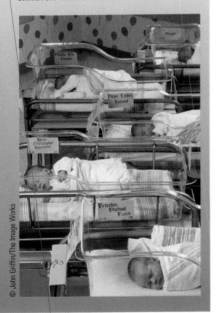

Algebra Help

When evaluating an expression or an equation, remember to follow the Basic Rules of Algebra. To review these rules, see Appendix A.1.

The Graph of an Equation

In Section 1.1, you used a coordinate system to represent graphically the relationship between two quantities. There, the graphical picture consisted of a collection of points in a coordinate plane.

Frequently, a relationship between two quantities is expressed as an **equation in two variables.** For instance, $y = 7 - 3x$ is an equation in x and y. An ordered pair (a, b) is a **solution** or **solution point** of an equation in x and y if the equation is true when a is substituted for x and b is substituted for y. For instance, $(1, 4)$ is a solution of $y = 7 - 3x$ because $4 = 7 - 3(1)$ is a true statement.

In this section you will review some basic procedures for sketching the graph of an equation in two variables. The **graph of an equation** is the set of all points that are solutions of the equation.

Example 1   Determining Solution Points

Determine whether (a) $(2, 13)$ and (b) $(-1, -3)$ lie on the graph of $y = 10x - 7$.

Solution

a.

$y = 10x - 7$	Write original equation.
$13 \overset{?}{=} 10(2) - 7$	Substitute 2 for x and 13 for y.
$13 = 13$	$(2, 13)$ is a solution. ✓

The point $(2, 13)$ *does* lie on the graph of $y = 10x - 7$ because it is a solution point of the equation.

b.

$y = 10x - 7$	Write original equation.
$-3 \overset{?}{=} 10(-1) - 7$	Substitute -1 for x and -3 for y.
$-3 \neq -17$	$(-1, -3)$ is not a solution.

The point $(-1, -3)$ *does not* lie on the graph of $y = 10x - 7$ because it is *not* a solution point of the equation.

CHECK*Point* Now try Exercise 7.

The basic technique used for sketching the graph of an equation is the **point-plotting method.**

Sketching the Graph of an Equation by Point Plotting

1. If possible, rewrite the equation so that one of the variables is isolated on one side of the equation.

2. Make a table of values showing several solution points.

3. Plot these points on a rectangular coordinate system.

4. Connect the points with a smooth curve or line.

Example 2   Sketching the Graph of an Equation

Sketch the graph of

$$y = 7 - 3x.$$

Solution

Because the equation is already solved for y, construct a table of values that consists of several solution points of the equation. For instance, when $x = -1$,

$$y = 7 - 3(-1)$$

$$= 10$$

which implies that $(-1, 10)$ is a solution point of the graph.

x	$y = 7 - 3x$	(x, y)
-1	10	$(-1, 10)$
0	7	$(0, 7)$
1	4	$(1, 4)$
2	1	$(2, 1)$
3	-2	$(3, -2)$
4	-5	$(4, -5)$

From the table, it follows that

$$(-1, 10), (0, 7), (1, 4), (2, 1), (3, -2), \text{ and } (4, -5)$$

are solution points of the equation. After plotting these points, you can see that they appear to lie on a line, as shown in Figure 1.14. The graph of the equation is the line that passes through the six plotted points.

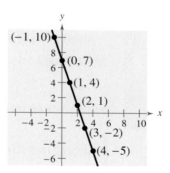

FIGURE **1.14**

CHECK *Point*   Now try Exercise 15.

| Example 3 | **Sketching the Graph of an Equation** |

Sketch the graph of

$$y = x^2 - 2.$$

Solution

Because the equation is already solved for y, begin by constructing a table of values.

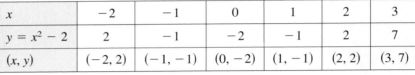

x	-2	-1	0	1	2	3
$y = x^2 - 2$	2	-1	-2	-1	2	7
(x, y)	$(-2, 2)$	$(-1, -1)$	$(0, -2)$	$(1, -1)$	$(2, 2)$	$(3, 7)$

Next, plot the points given in the table, as shown in Figure 1.15. Finally, connect the points with a smooth curve, as shown in Figure 1.16.

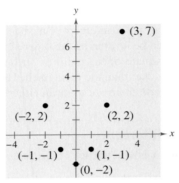

FIGURE 1.15

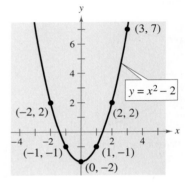

FIGURE 1.16

CHECK*Point*   Now try Exercise 17.

The point-plotting method demonstrated in Examples 2 and 3 is easy to use, but it has some shortcomings. With too few solution points, you can misrepresent the graph of an equation. For instance, if only the four points

$$(-2, 2), (-1, -1), (1, -1), \text{ and } (2, 2)$$

in Figure 1.15 were plotted, any one of the three graphs in Figure 1.17 would be reasonable.

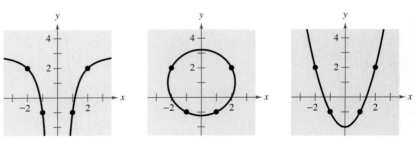

FIGURE 1.17

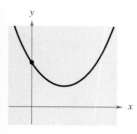

No *x*-intercepts; one *y*-intercept

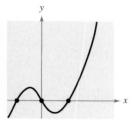

Three *x*-intercepts; one *y*-intercept

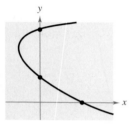

One *x*-intercept; two *y*-intercepts

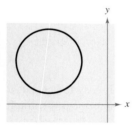

No intercepts

FIGURE **1.18**

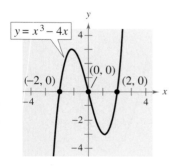

FIGURE **1.19**

TECHNOLOGY

To graph an equation involving *x* and *y* on a graphing utility, use the following procedure.

1. Rewrite the equation so that *y* is isolated on the left side.

2. Enter the equation into the graphing utility.

3. Determine a *viewing window* that shows all important features of the graph.

4. Graph the equation.

Intercepts of a Graph

It is often easy to determine the solution points that have zero as either the *x*-coordinate or the *y*-coordinate. These points are called **intercepts** because they are the points at which the graph intersects or touches the *x*- or *y*-axis. It is possible for a graph to have no intercepts, one intercept, or several intercepts, as shown in Figure 1.18.

Note that an *x*-intercept can be written as the ordered pair $(x, 0)$ and a *y*-intercept can be written as the ordered pair $(0, y)$. Some texts denote the *x*-intercept as the *x*-coordinate of the point $(a, 0)$ [and the *y*-intercept as the *y*-coordinate of the point $(0, b)$] rather than the point itself. Unless it is necessary to make a distinction, we will use the term *intercept* to mean either the point or the coordinate.

Finding Intercepts

1. To find *x*-intercepts, let *y* be zero and solve the equation for *x*.

2. To find *y*-intercepts, let *x* be zero and solve the equation for *y*.

Example 4   Finding *x*- and *y*-Intercepts

Find the *x*- and *y*-intercepts of the graph of $y = x^3 - 4x$.

Solution

Let $y = 0$. Then

$$0 = x^3 - 4x = x(x^2 - 4)$$

has solutions $x = 0$ and $x = \pm 2$.

   x-intercepts: $(0, 0), (2, 0), (-2, 0)$

Let $x = 0$. Then

$$y = (0)^3 - 4(0)$$

has one solution, $y = 0$.

   y-intercept: $(0, 0)$ See Figure 1.19.

CHECK*Point* Now try Exercise 23.

Symmetry

Graphs of equations can have **symmetry** with respect to one of the coordinate axes or with respect to the origin. Symmetry with respect to the x-axis means that if the Cartesian plane were folded along the x-axis, the portion of the graph above the x-axis would coincide with the portion below the x-axis. Symmetry with respect to the y-axis or the origin can be described in a similar manner, as shown in Figure 1.20.

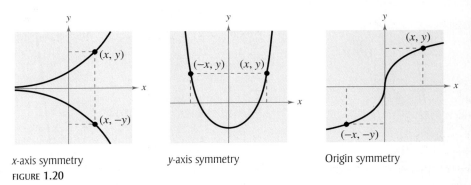

x-axis symmetry

y-axis symmetry

Origin symmetry

FIGURE **1.20**

Knowing the symmetry of a graph *before* attempting to sketch it is helpful, because then you need only half as many solution points to sketch the graph. There are three basic types of symmetry, described as follows.

Graphical Tests for Symmetry

1. A graph is **symmetric with respect to the x-axis** if, whenever (x, y) is on the graph, $(x, -y)$ is also on the graph.

2. A graph is **symmetric with respect to the y-axis** if, whenever (x, y) is on the graph, $(-x, y)$ is also on the graph.

3. A graph is **symmetric with respect to the origin** if, whenever (x, y) is on the graph, $(-x, -y)$ is also on the graph.

You can conclude that the graph of $y = x^2 - 2$ is symmetric with respect to the y-axis because the point $(-x, y)$ is also on the graph of $y = x^2 - 2$. (See the table below and Figure 1.21.)

x	-3	-2	-1	1	2	3
y	7	2	-1	-1	2	7
(x, y)	$(-3, 7)$	$(-2, 2)$	$(-1, -1)$	$(1, -1)$	$(2, 2)$	$(3, 7)$

Algebraic Tests for Symmetry

1. The graph of an equation is symmetric with respect to the x-axis if replacing y with $-y$ yields an equivalent equation.

2. The graph of an equation is symmetric with respect to the y-axis if replacing x with $-x$ yields an equivalent equation.

3. The graph of an equation is symmetric with respect to the origin if replacing x with $-x$ and y with $-y$ yields an equivalent equation.

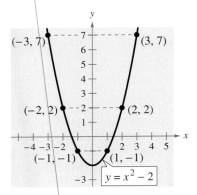

FIGURE **1.21** y-axis symmetry

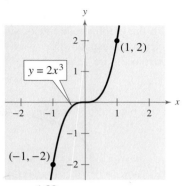

FIGURE **1.22**

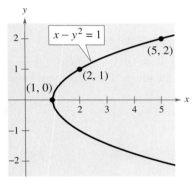

FIGURE **1.23**

Algebra Help

In Example 7, $|x - 1|$ is an absolute value expression. You can review the techniques for evaluating an absolute value expression in Appendix A.1.

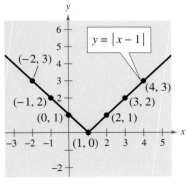

FIGURE **1.24**

Example 5 Testing for Symmetry

Test $y = 2x^3$ for symmetry with respect to both axes and the origin.

Solution

x-axis:	$y = 2x^3$	Write original equation.
	$-y = 2x^3$	Replace y with $-y$. Result is *not* an equivalent equation.
y-axis:	$y = 2x^3$	Write original equation.
	$y = 2(-x)^3$	Replace x with $-x$.
	$y = -2x^3$	Simplify. Result is *not* an equivalent equation.
Origin:	$y = 2x^3$	Write original equation.
	$-y = 2(-x)^3$	Replace y with $-y$ and x with $-x$.
	$-y = -2x^3$	Simplify.
	$y = 2x^3$	Equivalent equation

Of the three tests for symmetry, the only one that is satisfied is the test for origin symmetry (see Figure 1.22).

CHECK*Point* Now try Exercise 33.

Example 6 Using Symmetry as a Sketching Aid

Use symmetry to sketch the graph of $x - y^2 = 1$.

Solution

Of the three tests for symmetry, the only one that is satisfied is the test for *x*-axis symmetry because $x - (-y)^2 = 1$ is equivalent to $x - y^2 = 1$. So, the graph is symmetric with respect to the *x*-axis. Using symmetry, you only need to find the solution points above the *x*-axis and then reflect them to obtain the graph, as shown in Figure 1.23.

CHECK*Point* Now try Exercise 49.

Example 7 Sketching the Graph of an Equation

Sketch the graph of $y = |x - 1|$.

Solution

This equation fails all three tests for symmetry and consequently its graph is not symmetric with respect to either axis or to the origin. The absolute value sign indicates that y is always nonnegative. Create a table of values and plot the points, as shown in Figure 1.24. From the table, you can see that $x = 0$ when $y = 1$. So, the y-intercept is $(0, 1)$. Similarly, $y = 0$ when $x = 1$. So, the x-intercept is $(1, 0)$.

x	-2	-1	0	1	2	3	4
$y = |x - 1|$	3	2	1	0	1	2	3
(x, y)	$(-2, 3)$	$(-1, 2)$	$(0, 1)$	$(1, 0)$	$(2, 1)$	$(3, 2)$	$(4, 3)$

CHECK*Point* Now try Exercise 53.

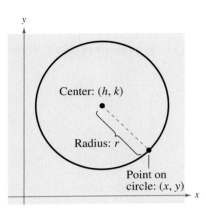

FIGURE 1.25

Throughout this course, you will learn to recognize several types of graphs from their equations. For instance, you will learn to recognize that the graph of a second-degree equation of the form

$$y = ax^2 + bx + c$$

is a parabola (see Example 3). The graph of a **circle** is also easy to recognize.

Circles

Consider the circle shown in Figure 1.25. A point (x, y) is on the circle if and only if its distance from the center (h, k) is r. By the Distance Formula,

$$\sqrt{(x - h)^2 + (y - k)^2} = r.$$

By squaring each side of this equation, you obtain the **standard form of the equation of a circle.**

Standard Form of the Equation of a Circle

The point (x, y) lies on the circle of **radius** r and **center** (h, k) if and only if

$$(x - h)^2 + (y - k)^2 = r^2.$$

> ⚠️ **WARNING / CAUTION**
>
> Be careful when you are finding h and k from the standard equation of a circle. For instance, to find the correct h and k from the equation of the circle in Example 8, rewrite the quantities $(x + 1)^2$ and $(y - 2)^2$ using subtraction.
>
> $$(x + 1)^2 = [x - (-1)]^2,$$
> $$(y - 2)^2 = [y - (2)]^2$$
>
> So, $h = -1$ and $k = 2$.

From this result, you can see that the standard form of the equation of a circle *with its center at the origin*, $(h, k) = (0, 0)$, is simply

$$x^2 + y^2 = r^2.$$ Circle with center at origin

Example 8   Finding the Equation of a Circle

The point $(3, 4)$ lies on a circle whose center is at $(-1, 2)$, as shown in Figure 1.26. Write the standard form of the equation of this circle.

Solution

The radius of the circle is the distance between $(-1, 2)$ and $(3, 4)$.

$$r = \sqrt{(x - h)^2 + (y - k)^2}$$ Distance Formula

$$= \sqrt{[3 - (-1)]^2 + (4 - 2)^2}$$ Substitute for x, y, h, and k.

$$= \sqrt{4^2 + 2^2}$$ Simplify.

$$= \sqrt{16 + 4}$$ Simplify.

$$= \sqrt{20}$$ Radius

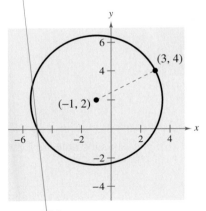

FIGURE 1.26

Using $(h, k) = (-1, 2)$ and $r = \sqrt{20}$, the equation of the circle is

$$(x - h)^2 + (y - k)^2 = r^2$$ Equation of circle

$$[x - (-1)]^2 + (y - 2)^2 = \left(\sqrt{20}\right)^2$$ Substitute for h, k, and r.

$$(x + 1)^2 + (y - 2)^2 = 20.$$ Standard form

CHECK **Point** Now try Exercise 73.

Application

In this course, you will learn that there are many ways to approach a problem. Three common approaches are illustrated in Example 9.

A *Numerical Approach:* Construct and use a table.
A *Graphical Approach:* Draw and use a graph.
An *Algebraic Approach:* Use the rules of algebra.

Example 9   Recommended Weight

The median recommended weight y (in pounds) for men of medium frame who are 25 to 59 years old can be approximated by the mathematical model

$$y = 0.073x^2 - 6.99x + 289.0, \quad 62 \le x \le 76$$

where x is the man's height (in inches). (Source: Metropolitan Life Insurance Company)

a. Construct a table of values that shows the median recommended weights for men with heights of 62, 64, 66, 68, 70, 72, 74, and 76 inches.

b. Use the table of values to sketch a graph of the model. Then use the graph to estimate *graphically* the median recommended weight for a man whose height is 71 inches.

c. Use the model to confirm *algebraically* the estimate you found in part (b).

Solution

a. You can use a calculator to complete the table, as shown at the left.

b. The table of values can be used to sketch the graph of the equation, as shown in Figure 1.27. From the graph, you can estimate that a height of 71 inches corresponds to a weight of about 161 pounds.

Height, x	Weight, y
62	136.2
64	140.6
66	145.6
68	151.2
70	157.4
72	164.2
74	171.5
76	179.4

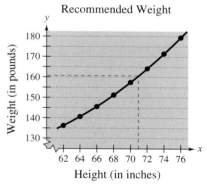

FIGURE 1.27

c. To confirm algebraically the estimate found in part (b), you can substitute 71 for x in the model.

$$y = 0.073(71)^2 - 6.99(71) + 289.0 \approx 160.70$$

So, the graphical estimate of 161 pounds is fairly good.

CHECK*Point* Now try Exercise 87.

1.2   EXERCISES

See www.CalcChat.com for worked-out solutions to odd-numbered exercises.

VOCABULARY: Fill in the blanks.

1. An ordered pair (a, b) is a _____ of an equation in x and y if the equation is true when a is substituted for x, and b is substituted for y.

2. The set of all solution points of an equation is the _____ of the equation.

3. The points at which a graph intersects or touches an axis are called the _____ of the graph.

4. A graph is symmetric with respect to the _____ if, whenever (x, y) is on the graph, $(-x, y)$ is also on the graph.

5. The equation $(x - h)^2 + (y - k)^2 = r^2$ is the standard form of the equation of a _____ with center _____ and radius _____.

6. When you construct and use a table to solve a problem, you are using a _____ approach.

SKILLS AND APPLICATIONS

In Exercises 7–14, determine whether each point lies on the graph of the equation.

Equation	*Points*	
7. $y = \sqrt{x + 4}$	(a) $(0, 2)$	(b) $(5, 3)$
8. $y = \sqrt{5 - x}$	(a) $(1, 2)$	(b) $(5, 0)$
9. $y = x^2 - 3x + 2$	(a) $(2, 0)$	(b) $(-2, 8)$
10. $y = 4 - |x - 2|$	(a) $(1, 5)$	(b) $(6, 0)$
11. $y = |x - 1| + 2$	(a) $(2, 3)$	(b) $(-1, 0)$
12. $2x - y - 3 = 0$	(a) $(1, 2)$	(b) $(1, -1)$
13. $x^2 + y^2 = 20$	(a) $(3, -2)$	(b) $(-4, 2)$
14. $y = \frac{1}{3}x^3 - 2x^2$	(a) $\left(2, -\frac{16}{3}\right)$	(b) $(-3, 9)$

In Exercises 15–18, complete the table. Use the resulting solution points to sketch the graph of the equation.

15. $y = -2x + 5$

x	-1	0	1	2	$\frac{5}{2}$
y					
(x, y)					

16. $y = \frac{3}{4}x - 1$

x	-2	0	1	$\frac{4}{3}$	2
y					
(x, y)					

17. $y = x^2 - 3x$

x	-1	0	1	2	3
y					
(x, y)					

18. $y = 5 - x^2$

x	-2	-1	0	1	2
y					
(x, y)					

In Exercises 19–22, graphically estimate the x- and y-intercepts of the graph. Verify your results algebraically.

19. $y = (x - 3)^2$

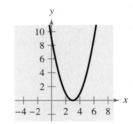

20. $y = 16 - 4x^2$

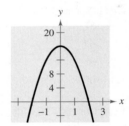

21. $y = |x + 2|$

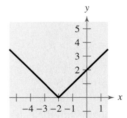

22. $y^2 = 4 - x$

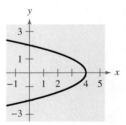

In Exercises 23–32, find the x- and y-intercepts of the graph of the equation.

23. $y = 5x - 6$

24. $y = 8 - 3x$

25. $y = \sqrt{x + 4}$

26. $y = \sqrt{2x - 1}$

27. $y = |3x - 7|$

28. $y = -|x + 10|$

29. $y = 2x^3 - 4x^2$

30. $y = x^4 - 25$

31. $y^2 = 6 - x$

32. $y^2 = x + 1$

In Exercises 33–40, use the algebraic tests to check for symmetry with respect to both axes and the origin.

33. $x^2 - y = 0$

34. $x - y^2 = 0$

35. $y = x^3$

36. $y = x^4 - x^2 + 3$

37. $y = \dfrac{x}{x^2 + 1}$

38. $y = \dfrac{1}{x^2 + 1}$

39. $xy^2 + 10 = 0$

40. $xy = 4$

In Exercises 41–44, assume that the graph has the indicated type of symmetry. Sketch the complete graph of the equation. To print an enlarged copy of the graph, go to the website *www.mathgraphs.com*.

41.

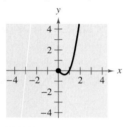

y-axis symmetry

42.

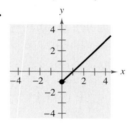

x-axis symmetry

43.

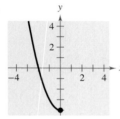

Origin symmetry

44.

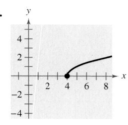

y-axis symmetry

In Exercises 45–56, identify any intercepts and test for symmetry. Then sketch the graph of the equation.

45. $y = -3x + 1$

46. $y = 2x - 3$

47. $y = x^2 - 2x$

48. $y = -x^2 - 2x$

49. $y = x^3 + 3$

50. $y = x^3 - 1$

51. $y = \sqrt{x - 3}$

52. $y = \sqrt{1 - x}$

53. $y = |x - 6|$

54. $y = 1 - |x|$

55. $x = y^2 - 1$

56. $x = y^2 - 5$

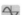 In Exercises 57–68, use a graphing utility to graph the equation. Use a standard setting. Approximate any intercepts.

57. $y = 3 - \frac{1}{2}x$

58. $y = \frac{2}{3}x - 1$

59. $y = x^2 - 4x + 3$

60. $y = x^2 + x - 2$

61. $y = \dfrac{2x}{x - 1}$

62. $y = \dfrac{4}{x^2 + 1}$

63. $y = \sqrt[3]{x} + 2$

64. $y = \sqrt[3]{x + 1}$

65. $y = x\sqrt{x + 6}$

66. $y = (6 - x)\sqrt{x}$

67. $y = |x + 3|$

68. $y = 2 - |x|$

In Exercises 69–76, write the standard form of the equation of the circle with the given characteristics.

69. Center: $(0, 0)$; Radius: 4

70. Center: $(0, 0)$; Radius: 5

71. Center: $(2, -1)$; Radius: 4

72. Center: $(-7, -4)$; Radius: 7

73. Center: $(-1, 2)$; Solution point: $(0, 0)$

74. Center: $(3, -2)$; Solution point: $(-1, 1)$

75. Endpoints of a diameter: $(0, 0)$, $(6, 8)$

76. Endpoints of a diameter: $(-4, -1)$, $(4, 1)$

In Exercises 77–82, find the center and radius of the circle, and sketch its graph.

77. $x^2 + y^2 = 25$

78. $x^2 + y^2 = 16$

79. $(x - 1)^2 + (y + 3)^2 = 9$

80. $x^2 + (y - 1)^2 = 1$

81. $\left(x - \frac{1}{2}\right)^2 + \left(y - \frac{1}{2}\right)^2 = \frac{9}{4}$

82. $(x - 2)^2 + (y + 3)^2 = \frac{16}{9}$

83. **DEPRECIATION** A hospital purchases a new magnetic resonance imaging (MRI) machine for \$500,000. The depreciated value y (reduced value) after t years is given by $y = 500{,}000 - 40{,}000t$, $0 \le t \le 8$. Sketch the graph of the equation.

84. **CONSUMERISM** You purchase an all-terrain vehicle (ATV) for \$8000. The depreciated value y after t years is given by $y = 8000 - 900t$, $0 \le t \le 6$. Sketch the graph of the equation.

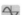 **85.** **GEOMETRY** A regulation NFL playing field (including the end zones) of length x and width y has a perimeter of $346\frac{2}{3}$ or $\frac{1040}{3}$ yards.

(a) Draw a rectangle that gives a visual representation of the problem. Use the specified variables to label the sides of the rectangle.

(b) Show that the width of the rectangle is $y = \dfrac{520}{3} - x$ and its area is $A = x\left(\dfrac{520}{3} - x\right)$.

(c) Use a graphing utility to graph the area equation. Be sure to adjust your window settings.

(d) From the graph in part (c), estimate the dimensions of the rectangle that yield a maximum area.

(e) Use your school's library, the Internet, or some other reference source to find the actual dimensions and area of a regulation NFL playing field and compare your findings with the results of part (d).

The symbol 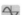 indicates an exercise or a part of an exercise in which you are instructed to use a graphing utility.

86. GEOMETRY A soccer playing field of length x and width y has a perimeter of 360 meters.

(a) Draw a rectangle that gives a visual representation of the problem. Use the specified variables to label the sides of the rectangle.

(b) Show that the width of the rectangle is $y = 180 - x$ and its area is $A = x(180 - x)$.

(c) Use a graphing utility to graph the area equation. Be sure to adjust your window settings.

(d) From the graph in part (c), estimate the dimensions of the rectangle that yield a maximum area.

(e) Use your school's library, the Internet, or some other reference source to find the actual dimensions and area of a regulation Major League Soccer field and compare your findings with the results of part (d).

87. POPULATION STATISTICS The table shows the life expectancies of a child (at birth) in the United States for selected years from 1920 to 2000. (Source: U.S. National Center for Health Statistics)

Year	Life Expectancy, y
1920	54.1
1930	59.7
1940	62.9
1950	68.2
1960	69.7
1970	70.8
1980	73.7
1990	75.4
2000	77.0

A model for the life expectancy during this period is

$y = -0.0025t^2 + 0.574t + 44.25, \quad 20 \le t \le 100$

where y represents the life expectancy and t is the time in years, with $t = 20$ corresponding to 1920.

 (a) Use a graphing utility to graph the data from the table and the model in the same viewing window. How well does the model fit the data? Explain.

(b) Determine the life expectancy in 1990 both graphically and algebraically.

(c) Use the graph to determine the year when life expectancy was approximately 76.0. Verify your answer algebraically.

(d) One projection for the life expectancy of a child born in 2015 is 78.9. How does this compare with the projection given by the model?

(e) Do you think this model can be used to predict the life expectancy of a child 50 years from now? Explain.

88. ELECTRONICS The resistance y (in ohms) of 1000 feet of solid copper wire at 68 degrees Fahrenheit can be approximated by the model

$y = \dfrac{10{,}770}{x^2} - 0.37, \quad 5 \le x \le 100$

where x is the diameter of the wire in mils (0.001 inch). (Source: American Wire Gage)

(a) Complete the table.

x	5	10	20	30	40	50
y						

x	60	70	80	90	100
y					

(b) Use the table of values in part (a) to sketch a graph of the model. Then use your graph to estimate the resistance when $x = 85.5$.

(c) Use the model to confirm algebraically the estimate you found in part (b).

(d) What can you conclude in general about the relationship between the diameter of the copper wire and the resistance?

EXPLORATION

89. THINK ABOUT IT Find a and b if the graph of $y = ax^2 + bx^3$ is symmetric with respect to (a) the y-axis and (b) the origin. (There are many correct answers.)

90. CAPSTONE Match the equation or equations with the given characteristic.

(i) $y = 3x^3 - 3x$ (ii) $y = (x + 3)^2$

(iii) $y = 3x - 3$ (iv) $y = \sqrt[3]{x}$

(v) $y = 3x^2 + 3$ (vi) $y = \sqrt{x + 3}$

(a) Symmetric with respect to the y-axis

(b) Three x-intercepts

(c) Symmetric with respect to the x-axis

(d) $(-2, 1)$ is a point on the graph

(e) Symmetric with respect to the origin

(f) Graph passes through the origin

1.3 LINEAR EQUATIONS IN TWO VARIABLES

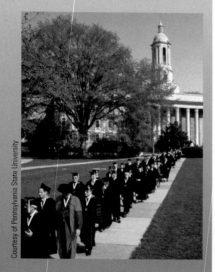

Using Slope

The simplest mathematical model for relating two variables is the **linear equation in two variables** $y = mx + b$. The equation is called *linear* because its graph is a line. (In mathematics, the term *line* means *straight line*.) By letting $x = 0$, you obtain

$$y = m(0) + b \qquad \text{Substitute 0 for } x.$$

$$= b.$$

So, the line crosses the y-axis at $y = b$, as shown in Figure 1.28. In other words, the y-intercept is $(0, b)$. The steepness or slope of the line is m.

$$y = mx + b$$

Slope ——⌐   ⌐—— y-Intercept

The **slope** of a nonvertical line is the number of units the line rises (or falls) vertically for each unit of horizontal change from left to right, as shown in Figure 1.28 and Figure 1.29.

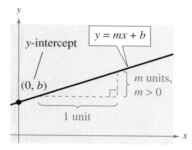

Positive slope, line rises.

FIGURE 1.28

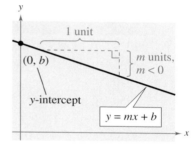

Negative slope, line falls.

FIGURE 1.29

A linear equation that is written in the form $y = mx + b$ is said to be written in **slope-intercept form.**

> ### The Slope-Intercept Form of the Equation of a Line
>
> The graph of the equation
>
> $$y = mx + b$$
>
> is a line whose slope is m and whose y-intercept is $(0, b)$.

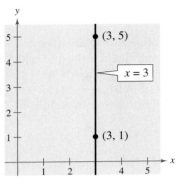

FIGURE **1.30**   Slope is undefined.

Once you have determined the slope and the *y*-intercept of a line, it is a relatively simple matter to sketch its graph. In the next example, note that none of the lines is vertical. A vertical line has an equation of the form

$$x = a.$$ Vertical line

The equation of a vertical line cannot be written in the form $y = mx + b$ because the slope of a vertical line is undefined, as indicated in Figure 1.30.

| **Example 1** | **Graphing a Linear Equation** |

Sketch the graph of each linear equation.

a. $y = 2x + 1$

b. $y = 2$

c. $x + y = 2$

Solution

a. Because $b = 1$, the *y*-intercept is $(0, 1)$. Moreover, because the slope is $m = 2$, the line *rises* two units for each unit the line moves to the right, as shown in Figure 1.31.

b. By writing this equation in the form $y = (0)x + 2$, you can see that the *y*-intercept is $(0, 2)$ and the slope is zero. A zero slope implies that the line is horizontal—that is, it doesn't rise *or* fall, as shown in Figure 1.32.

c. By writing this equation in slope-intercept form

$$x + y = 2$$ Write original equation.

$$y = -x + 2$$ Subtract *x* from each side.

$$y = (-1)x + 2$$ Write in slope-intercept form.

you can see that the *y*-intercept is $(0, 2)$. Moreover, because the slope is $m = -1$, the line *falls* one unit for each unit the line moves to the right, as shown in Figure 1.33.

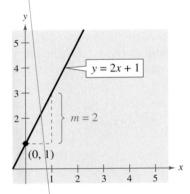

When *m* is positive, the line rises.
FIGURE **1.31**

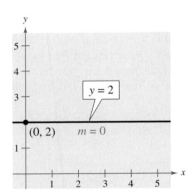

When *m* is 0, the line is horizontal.
FIGURE **1.32**

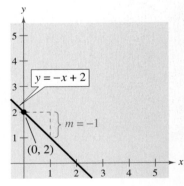

When *m* is negative, the line falls.
FIGURE **1.33**

CHECK***Point***   Now try Exercise 17.

Finding the Slope of a Line

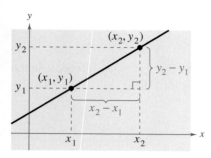

Given an equation of a line, you can find its slope by writing the equation in slope-intercept form. If you are not given an equation, you can still find the slope of a line. For instance, suppose you want to find the slope of the line passing through the points (x_1, y_1) and (x_2, y_2), as shown in Figure 1.34. As you move from left to right along this line, a change of $(y_2 - y_1)$ units in the vertical direction corresponds to a change of $(x_2 - x_1)$ units in the horizontal direction.

$$y_2 - y_1 = \text{the change in } y = \text{rise}$$

and

$$x_2 - x_1 = \text{the change in } x = \text{run}$$

The ratio of $(y_2 - y_1)$ to $(x_2 - x_1)$ represents the slope of the line that passes through the points (x_1, y_1) and (x_2, y_2).

$$\text{Slope} = \frac{\text{change in } y}{\text{change in } x}$$

$$= \frac{\text{rise}}{\text{run}}$$

$$= \frac{y_2 - y_1}{x_2 - x_1}$$

The Slope of a Line Passing Through Two Points

The **slope** m of the nonvertical line through (x_1, y_1) and (x_2, y_2) is

$$m = \frac{y_2 - y_1}{x_2 - x_1}$$

where $x_1 \neq x_2$.

When this formula is used for slope, the *order of subtraction* is important. Given two points on a line, you are free to label either one of them as (x_1, y_1) and the other as (x_2, y_2). However, once you have done this, you must form the numerator and denominator using the same order of subtraction.

$$m = \frac{y_2 - y_1}{x_2 - x_1} \qquad m = \frac{y_1 - y_2}{x_1 - x_2} \qquad m = \frac{y_2 - y_1}{x_1 - x_2}$$

Correct   Correct   Incorrect

For instance, the slope of the line passing through the points $(3, 4)$ and $(5, 7)$ can be calculated as

$$m = \frac{7 - 4}{5 - 3} = \frac{3}{2}$$

or, reversing the subtraction order in both the numerator and denominator, as

$$m = \frac{4 - 7}{3 - 5} = \frac{-3}{-2} = \frac{3}{2}.$$

Example 2   Finding the Slope of a Line Through Two Points

Find the slope of the line passing through each pair of points.

a. $(-2, 0)$ and $(3, 1)$ **b.** $(-1, 2)$ and $(2, 2)$

c. $(0, 4)$ and $(1, -1)$ **d.** $(3, 4)$ and $(3, 1)$

Solution

a. Letting $(x_1, y_1) = (-2, 0)$ and $(x_2, y_2) = (3, 1)$, you obtain a slope of

$$m = \frac{y_2 - y_1}{x_2 - x_1} = \frac{1 - 0}{3 - (-2)} = \frac{1}{5}. \qquad \text{See Figure 1.35.}$$

b. The slope of the line passing through $(-1, 2)$ and $(2, 2)$ is

$$m = \frac{2 - 2}{2 - (-1)} = \frac{0}{3} = 0. \qquad \text{See Figure 1.36.}$$

c. The slope of the line passing through $(0, 4)$ and $(1, -1)$ is

$$m = \frac{-1 - 4}{1 - 0} = \frac{-5}{1} = -5. \qquad \text{See Figure 1.37.}$$

d. The slope of the line passing through $(3, 4)$ and $(3, 1)$ is

$$m = \frac{1 - 4}{3 - 3} = \frac{-3}{0}. \qquad \text{See Figure 1.38.}$$

Because division by 0 is undefined, the slope is undefined and the line is vertical.

Algebra Help

To find the slopes in Example 2, you must be able to evaluate rational expressions. You can review the techniques for evaluating rational expressions in Appendix A.4.

Study Tip

In Figures 1.35 to 1.38, note the relationships between slope and the orientation of the line.

a. Positive slope: line rises from left to right

b. Zero slope: line is horizontal

c. Negative slope: line falls from left to right

d. Undefined slope: line is vertical

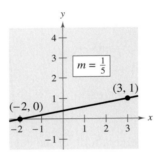

FIGURE **1.35**

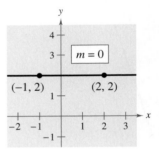

FIGURE **1.36**

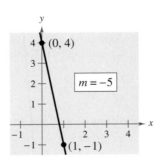

FIGURE **1.37**

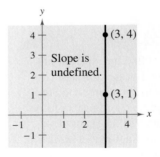

FIGURE **1.38**

CHECK *Point*   Now try Exercise 31.

Writing Linear Equations in Two Variables

If (x_1, y_1) is a point on a line of slope m and (x, y) is *any other* point on the line, then

$$\frac{y - y_1}{x - x_1} = m.$$

This equation, involving the variables x and y, can be rewritten in the form

$$y - y_1 = m(x - x_1)$$

which is the **point-slope form** of the equation of a line.

Point-Slope Form of the Equation of a Line

The equation of the line with slope m passing through the point (x_1, y_1) is

$$y - y_1 = m(x - x_1).$$

The point-slope form is most useful for *finding* the equation of a line. You should remember this form.

Example 3   Using the Point-Slope Form

Find the slope-intercept form of the equation of the line that has a slope of 3 and passes through the point $(1, -2)$.

Solution

Use the point-slope form with $m = 3$ and $(x_1, y_1) = (1, -2)$.

$y - y_1 = m(x - x_1)$	Point-slope form
$y - (-2) = 3(x - 1)$	Substitute for m, x_1, and y_1.
$y + 2 = 3x - 3$	Simplify.
$y = 3x - 5$	Write in slope-intercept form.

The slope-intercept form of the equation of the line is $y = 3x - 5$. The graph of this line is shown in Figure 1.39.

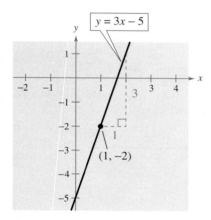

$y = 3x - 5$

$(1, -2)$

FIGURE **1.39**

CHECK *Point*   Now try Exercise 51.

The point-slope form can be used to find an equation of the line passing through two points (x_1, y_1) and (x_2, y_2). To do this, first find the slope of the line

$$m = \frac{y_2 - y_1}{x_2 - x_1}, \quad x_1 \neq x_2$$

and then use the point-slope form to obtain the equation

$$y - y_1 = \frac{y_2 - y_1}{x_2 - x_1}(x - x_1). \qquad \text{Two-point form}$$

This is sometimes called the **two-point form** of the equation of a line.

Study Tip

When you find an equation of the line that passes through two given points, you only need to substitute the coordinates of one of the points in the point-slope form. It does not matter which point you choose because both points will yield the same result.

Parallel and Perpendicular Lines

Slope can be used to decide whether two nonvertical lines in a plane are parallel, perpendicular, or neither.

Parallel and Perpendicular Lines

1. Two distinct nonvertical lines are **parallel** if and only if their slopes are equal. That is, $m_1 = m_2$.

2. Two nonvertical lines are **perpendicular** if and only if their slopes are negative reciprocals of each other. That is, $m_1 = -1/m_2$.

Example 4   Finding Parallel and Perpendicular Lines

Find the slope-intercept forms of the equations of the lines that pass through the point $(2, -1)$ and are (a) parallel to and (b) perpendicular to the line $2x - 3y = 5$.

Solution

By writing the equation of the given line in slope-intercept form

$$2x - 3y = 5 \qquad \text{Write original equation.}$$

$$-3y = -2x + 5 \qquad \text{Subtract } 2x \text{ from each side.}$$

$$y = \tfrac{2}{3}x - \tfrac{5}{3} \qquad \text{Write in slope-intercept form.}$$

you can see that it has a slope of $m = \tfrac{2}{3}$, as shown in Figure 1.40.

a. Any line parallel to the given line must also have a slope of $\tfrac{2}{3}$. So, the line through $(2, -1)$ that is parallel to the given line has the following equation.

$$y - (-1) = \tfrac{2}{3}(x - 2) \qquad \text{Write in point-slope form.}$$

$$3(y + 1) = 2(x - 2) \qquad \text{Multiply each side by 3.}$$

$$3y + 3 = 2x - 4 \qquad \text{Distributive Property}$$

$$y = \tfrac{2}{3}x - \tfrac{7}{3} \qquad \text{Write in slope-intercept form.}$$

b. Any line perpendicular to the given line must have a slope of $-\tfrac{3}{2}$ (because $-\tfrac{3}{2}$ is the negative reciprocal of $\tfrac{2}{3}$). So, the line through $(2, -1)$ that is perpendicular to the given line has the following equation.

$$y - (-1) = -\tfrac{3}{2}(x - 2) \qquad \text{Write in point-slope form.}$$

$$2(y + 1) = -3(x - 2) \qquad \text{Multiply each side by 2.}$$

$$2y + 2 = -3x + 6 \qquad \text{Distributive Property}$$

$$y = -\tfrac{3}{2}x + 2 \qquad \text{Write in slope-intercept form.}$$

CHECK *Point*   Now try Exercise 87.

Notice in Example 4 how the slope-intercept form is used to obtain information about the graph of a line, whereas the point-slope form is used to write the equation of a line.

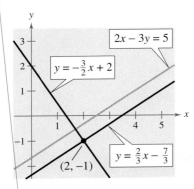

FIGURE **1.40**

TECHNOLOGY

On a graphing utility, lines will not appear to have the correct slope unless you use a viewing window that has a square setting. For instance, try graphing the lines in Example 4 using the standard setting $-10 \leq x \leq 10$ and $-10 \leq y \leq 10$. Then reset the viewing window with the square setting $-9 \leq x \leq 9$ and $-6 \leq y \leq 6$. On which setting do the lines $y = \tfrac{2}{3}x - \tfrac{5}{3}$ and $y = -\tfrac{3}{2}x + 2$ appear to be perpendicular?

Applications

In real-life problems, the slope of a line can be interpreted as either a *ratio* or a *rate*. If the *x*-axis and *y*-axis have the same unit of measure, then the slope has no units and is a **ratio**. If the *x*-axis and *y*-axis have different units of measure, then the slope is a **rate** or **rate of change.**

Example 5   Using Slope as a Ratio

The maximum recommended slope of a wheelchair ramp is $\frac{1}{12}$. A business is installing a wheelchair ramp that rises 22 inches over a horizontal length of 24 feet. Is the ramp steeper than recommended?   (Source: *Americans with Disabilities Act Handbook*)

Solution

The horizontal length of the ramp is 24 feet or 12(24) = 288 inches, as shown in Figure 1.41. So, the slope of the ramp is

$$\text{Slope} = \frac{\text{vertical change}}{\text{horizontal change}} = \frac{22 \text{ in.}}{288 \text{ in.}} \approx 0.076.$$

Because $\frac{1}{12} \approx 0.083$, the slope of the ramp is not steeper than recommended.

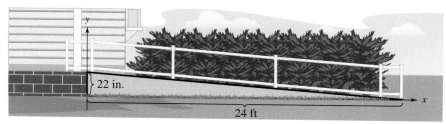

FIGURE **1.41**

CHECKPoint   Now try Exercise 115.

Example 6   Using Slope as a Rate of Change

A kitchen appliance manufacturing company determines that the total cost in dollars of producing *x* units of a blender is

$$C = 25x + 3500. \qquad \text{Cost equation}$$

Describe the practical significance of the *y*-intercept and slope of this line.

Solution

The *y*-intercept (0, 3500) tells you that the cost of producing zero units is $3500. This is the *fixed cost* of production—it includes costs that must be paid regardless of the number of units produced. The slope of *m* = 25 tells you that the cost of producing each unit is $25, as shown in Figure 1.42. Economists call the cost per unit the *marginal cost*. If the production increases by one unit, then the "margin," or extra amount of cost, is $25. So, the cost increases at a rate of $25 per unit.

CHECKPoint   Now try Exercise 119.

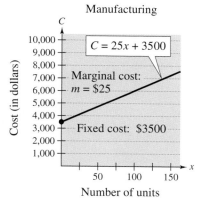

FIGURE **1.42**   Production cost

Most business expenses can be deducted in the same year they occur. One exception is the cost of property that has a useful life of more than 1 year. Such costs must be *depreciated* (decreased in value) over the useful life of the property. If the *same amount* is depreciated each year, the procedure is called *linear* or *straight-line depreciation*. The *book value* is the difference between the original value and the total amount of depreciation accumulated to date.

Example 7   Straight-Line Depreciation

A college purchased exercise equipment worth $12,000 for the new campus fitness center. The equipment has a useful life of 8 years. The salvage value at the end of 8 years is $2000. Write a linear equation that describes the book value of the equipment each year.

Solution

Let V represent the value of the equipment at the end of year t. You can represent the initial value of the equipment by the data point $(0, 12{,}000)$ and the salvage value of the equipment by the data point $(8, 2000)$. The slope of the line is

$$m = \frac{2000 - 12{,}000}{8 - 0} = -\$1250$$

which represents the annual depreciation in *dollars per year*. Using the point-slope form, you can write the equation of the line as follows.

$$V - 12{,}000 = -1250(t - 0) \qquad \text{Write in point-slope form.}$$

$$V = -1250t + 12{,}000 \qquad \text{Write in slope-intercept form.}$$

The table shows the book value at the end of each year, and the graph of the equation is shown in Figure 1.43.

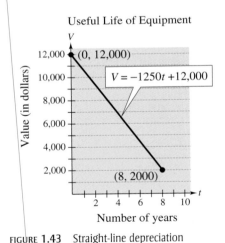

Useful Life of Equipment

$V = -1250t +12{,}000$

FIGURE **1.43**   Straight-line depreciation

Year, t	Value, V
0	12,000
1	10,750
2	9500
3	8250
4	7000
5	5750
6	4500
7	3250
8	2000

CHECK*Point*   Now try Exercise 121.

In many real-life applications, the two data points that determine the line are often given in a disguised form. Note how the data points are described in Example 7.

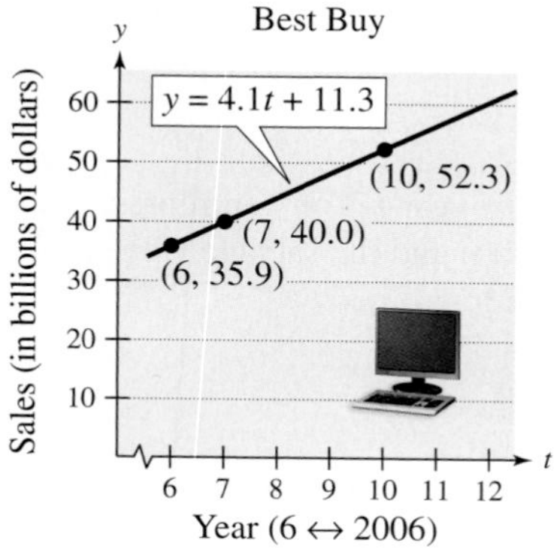

Example 8   Predicting Sales

The sales for Best Buy were approximately $35.9 billion in 2006 and $40.0 billion in 2007. Using only this information, write a linear equation that gives the sales (in billions of dollars) in terms of the year. Then predict the sales for 2010.   (Source: Best Buy Company, Inc.)

Solution

Let $t = 6$ represent 2006. Then the two given values are represented by the data points $(6, 35.9)$ and $(7, 40.0)$. The slope of the line through these points is

$$m = \frac{40.0 - 35.9}{7 - 6}$$

$$= 4.1.$$

Using the point-slope form, you can find the equation that relates the sales y and the year t to be

$y - 35.9 = 4.1(t - 6)$ Write in point-slope form.

$y = 4.1t + 11.3.$ Write in slope-intercept form.

According to this equation, the sales for 2010 will be

$y = 4.1(10) + 11.3 = 41 + 11.3 = \$52.3$ billion. (See Figure 1.44.)

CHECK Point   Now try Exercise 129.

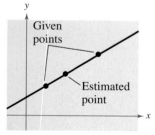

Best Buy

$y = 4.1t + 11.3$

$(10, 52.3)$

$(7, 40.0)$

$(6, 35.9)$

Sales (in billions of dollars)

Year ($6 \leftrightarrow 2006$)

FIGURE **1.44**

The prediction method illustrated in Example 8 is called **linear extrapolation.** Note in Figure 1.45 that an extrapolated point does not lie between the given points. When the estimated point lies between two given points, as shown in Figure 1.46, the procedure is called **linear interpolation.**

Because the slope of a vertical line is not defined, its equation cannot be written in slope-intercept form. However, every line has an equation that can be written in the **general form**

$Ax + By + C = 0$ General form

where A and B are not both zero. For instance, the vertical line given by $x = a$ can be represented by the general form $x - a = 0$.

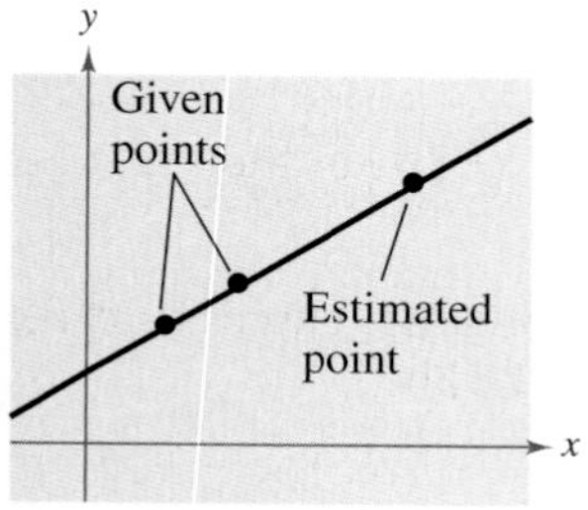

Given points

Estimated point

Linear extrapolation
FIGURE **1.45**

Given points

Estimated point

Linear interpolation
FIGURE **1.46**

Summary of Equations of Lines

1. General form: $Ax + By + C = 0$

2. Vertical line: $x = a$

3. Horizontal line: $y = b$

4. Slope-intercept form: $y = mx + b$

5. Point-slope form: $y - y_1 = m(x - x_1)$

6. Two-point form: $y - y_1 = \dfrac{y_2 - y_1}{x_2 - x_1}(x - x_1)$

1.3   EXERCISES

See www.CalcChat.com for worked-out solutions to odd-numbered exercises.

VOCABULARY

In Exercises 1–7, fill in the blanks.

1. The simplest mathematical model for relating two variables is the _____ equation in two variables $y = mx + b$.

2. For a line, the ratio of the change in y to the change in x is called the _____ of the line.

3. Two lines are _____ if and only if their slopes are equal.

4. Two lines are _____ if and only if their slopes are negative reciprocals of each other.

5. When the x-axis and y-axis have different units of measure, the slope can be interpreted as a _____.

6. The prediction method _____ _____ is the method used to estimate a point on a line when the point does not lie between the given points.

7. Every line has an equation that can be written in _____ form.

8. Match each equation of a line with its form.

   (a) $Ax + By + C = 0$ (i) Vertical line
   (b) $x = a$ (ii) Slope-intercept form
   (c) $y = b$ (iii) General form
   (d) $y = mx + b$ (iv) Point-slope form
   (e) $y - y_1 = m(x - x_1)$ (v) Horizontal line

SKILLS AND APPLICATIONS

In Exercises 9 and 10, identify the line that has each slope.

9. (a) $m = \frac{2}{3}$
   (b) m is undefined.
   (c) $m = -2$

10. (a) $m = 0$
    (b) $m = -\frac{3}{4}$
    (c) $m = 1$

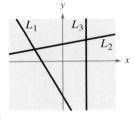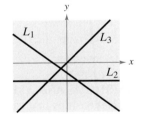

In Exercises 11 and 12, sketch the lines through the point with the indicated slopes on the same set of coordinate axes.

	Point	Slopes
11.	$(2, 3)$	(a) 0 (b) 1 (c) 2 (d) -3
12.	$(-4, 1)$	(a) 3 (b) -3 (c) $\frac{1}{2}$ (d) Undefined

In Exercises 13–16, estimate the slope of the line.

13.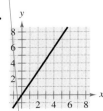

14.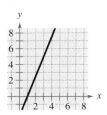

15.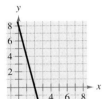

16.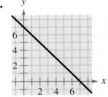

In Exercises 17–28, find the slope and y-intercept (if possible) of the equation of the line. Sketch the line.

17. $y = 5x + 3$
18. $y = x - 10$
19. $y = -\frac{1}{2}x + 4$
20. $y = -\frac{3}{2}x + 6$
21. $5x - 2 = 0$
22. $3y + 5 = 0$
23. $7x + 6y = 30$
24. $2x + 3y = 9$
25. $y - 3 = 0$
26. $y + 4 = 0$
27. $x + 5 = 0$
28. $x - 2 = 0$

In Exercises 29–40, plot the points and find the slope of the line passing through the pair of points.

29. $(0, 9), (6, 0)$
30. $(12, 0), (0, -8)$
31. $(-3, -2), (1, 6)$
32. $(2, 4), (4, -4)$
33. $(5, -7), (8, -7)$
34. $(-2, 1), (-4, -5)$
35. $(-6, -1), (-6, 4)$
36. $(0, -10), (-4, 0)$
37. $\left(\frac{11}{2}, -\frac{4}{3}\right), \left(-\frac{3}{2}, -\frac{1}{3}\right)$
38. $\left(\frac{7}{8}, \frac{3}{4}\right), \left(\frac{5}{4}, -\frac{1}{4}\right)$
39. $(4.8, 3.1), (-5.2, 1.6)$
40. $(-1.75, -8.3), (2.25, -2.6)$

In Exercises 41–50, use the point on the line and the slope m of the line to find three additional points through which the line passes. (There are many correct answers.)

41. $(2, 1), \quad m = 0$
42. $(3, -2), \quad m = 0$
43. $(5, -6), \quad m = 1$
44. $(10, -6), \quad m = -1$
45. $(-8, 1), \quad m$ is undefined.
46. $(1, 5), \quad m$ is undefined.
47. $(-5, 4), \quad m = 2$
48. $(0, -9), \quad m = -2$
49. $(7, -2), \quad m = \frac{1}{2}$
50. $(-1, -6), \quad m = -\frac{1}{2}$

In Exercises 51–64, find the slope-intercept form of the equation of the line that passes through the given point and has the indicated slope m. Sketch the line.

51. $(0, -2), \quad m = 3$
52. $(0, 10), \quad m = -1$
53. $(-3, 6), \quad m = -2$
54. $(0, 0), \quad m = 4$
55. $(4, 0), \quad m = -\frac{1}{3}$
56. $(8, 2), \quad m = \frac{1}{4}$
57. $(2, -3), \quad m = -\frac{1}{2}$
58. $(-2, -5), \quad m = \frac{3}{4}$
59. $(6, -1), \quad m$ is undefined.
60. $(-10, 4), \quad m$ is undefined.
61. $\left(4, \frac{5}{2}\right), \quad m = 0$
62. $\left(-\frac{1}{2}, \frac{3}{2}\right), \quad m = 0$
63. $(-5.1, 1.8), \quad m = 5$
64. $(2.3, -8.5), \quad m = -2.5$

In Exercises 65–78, find the slope-intercept form of the equation of the line passing through the points. Sketch the line.

65. $(5, -1), (-5, 5)$
66. $(4, 3), (-4, -4)$
67. $(-8, 1), (-8, 7)$
68. $(-1, 4), (6, 4)$
69. $\left(2, \frac{1}{2}\right), \left(\frac{1}{2}, \frac{5}{4}\right)$
70. $\left(1, 1\right), \left(6, -\frac{2}{3}\right)$
71. $\left(-\frac{1}{10}, -\frac{3}{5}\right), \left(\frac{9}{10}, -\frac{9}{5}\right)$
72. $\left(\frac{3}{4}, \frac{3}{2}\right), \left(-\frac{4}{3}, \frac{7}{4}\right)$
73. $(1, 0.6), (-2, -0.6)$
74. $(-8, 0.6), (2, -2.4)$
75. $(2, -1), \left(\frac{1}{3}, -1\right)$
76. $\left(\frac{1}{5}, -2\right), (-6, -2)$
77. $\left(\frac{7}{3}, -8\right), \left(\frac{7}{3}, 1\right)$
78. $(1.5, -2), (1.5, 0.2)$

In Exercises 79–82, determine whether the lines are parallel, perpendicular, or neither.

79. $L_1: y = \frac{1}{3}x - 2$
$L_2: y = \frac{1}{3}x + 3$
80. $L_1: y = 4x - 1$
$L_2: y = 4x + 7$
81. $L_1: y = \frac{1}{2}x - 3$
$L_2: y = -\frac{1}{2}x + 1$
82. $L_1: y = -\frac{4}{5}x - 5$
$L_2: y = \frac{5}{4}x + 1$

In Exercises 83–86, determine whether the lines L_1 and L_2 passing through the pairs of points are parallel, perpendicular, or neither.

83. $L_1: (0, -1), (5, 9)$
$L_2: (0, 3), (4, 1)$
84. $L_1: (-2, -1), (1, 5)$
$L_2: (1, 3), (5, -5)$

85. $L_1: (3, 6), (-6, 0)$
$L_2: (0, -1), \left(5, \frac{7}{3}\right)$
86. $L_1: (4, 8), (-4, 2)$
$L_2: (3, -5), \left(-1, \frac{1}{3}\right)$

In Exercises 87–96, write the slope-intercept forms of the equations of the lines through the given point (a) parallel to the given line and (b) perpendicular to the given line.

87. $4x - 2y = 3, \quad (2, 1)$
88. $x + y = 7, \quad (-3, 2)$
89. $3x + 4y = 7, \quad \left(-\frac{2}{3}, \frac{7}{8}\right)$
90. $5x + 3y = 0, \quad \left(\frac{7}{8}, \frac{3}{4}\right)$
91. $y + 3 = 0, \quad (-1, 0)$
92. $y - 2 = 0, \quad (-4, 1)$
93. $x - 4 = 0, \quad (3, -2)$
94. $x + 2 = 0, \quad (-5, 1)$
95. $x - y = 4, \quad (2.5, 6.8)$
96. $6x + 2y = 9, \quad (-3.9, -1.4)$

In Exercises 97–102, use the *intercept form* to find the equation of the line with the given intercepts. The intercept form of the equation of a line with intercepts $(a, 0)$ and $(0, b)$ is

$$\frac{x}{a} + \frac{y}{b} = 1, \quad a \neq 0, \ b \neq 0.$$

97. x-intercept: $(2, 0)$
y-intercept: $(0, 3)$
98. x-intercept: $(-3, 0)$
y-intercept: $(0, 4)$
99. x-intercept: $\left(-\frac{1}{6}, 0\right)$
y-intercept: $\left(0, -\frac{2}{3}\right)$
100. x-intercept: $\left(\frac{2}{3}, 0\right)$
y-intercept: $(0, -2)$
101. Point on line: $(1, 2)$
x-intercept: $(c, 0)$
y-intercept: $(0, c), \quad c \neq 0$
102. Point on line: $(-3, 4)$
x-intercept: $(d, 0)$
y-intercept: $(0, d), \quad d \neq 0$

GRAPHICAL ANALYSIS In Exercises 103–106, identify any relationships that exist among the lines, and then use a graphing utility to graph the three equations in the same viewing window. Adjust the viewing window so that the slope appears visually correct—that is, so that parallel lines appear parallel and perpendicular lines appear to intersect at right angles.

103. (a) $y = 2x$ (b) $y = -2x$ (c) $y = \frac{1}{2}x$
104. (a) $y = \frac{2}{3}x$ (b) $y = -\frac{3}{2}x$ (c) $y = \frac{2}{3}x + 2$
105. (a) $y = -\frac{1}{2}x$ (b) $y = -\frac{1}{2}x + 3$ (c) $y = 2x - 4$
106. (a) $y = x - 8$ (b) $y = x + 1$ (c) $y = -x + 3$

In Exercises 107–110, find a relationship between x and y such that (x, y) is equidistant (the same distance) from the two points.

107. $(4, -1), (-2, 3)$
108. $(6, 5), (1, -8)$
109. $\left(3, \frac{5}{2}\right), (-7, 1)$
110. $\left(-\frac{1}{2}, -4\right), \left(\frac{7}{2}, \frac{5}{4}\right)$

111. SALES   The following are the slopes of lines representing annual sales y in terms of time x in years. Use the slopes to interpret any change in annual sales for a one-year increase in time.

(a) The line has a slope of $m = 135$.

(b) The line has a slope of $m = 0$.

(c) The line has a slope of $m = -40$.

112. REVENUE   The following are the slopes of lines representing daily revenues y in terms of time x in days. Use the slopes to interpret any change in daily revenues for a one-day increase in time.

(a) The line has a slope of $m = 400$.

(b) The line has a slope of $m = 100$.

(c) The line has a slope of $m = 0$.

113. AVERAGE SALARY   The graph shows the average salaries for senior high school principals from 1996 through 2008. (Source: Educational Research Service)

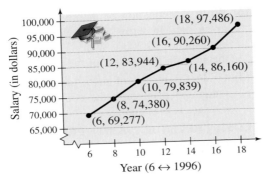

(a) Use the slopes of the line segments to determine the time periods in which the average salary increased the greatest and the least.

(b) Find the slope of the line segment connecting the points for the years 1996 and 2008.

(c) Interpret the meaning of the slope in part (b) in the context of the problem.

114. SALES   The graph shows the sales (in billions of dollars) for Apple Inc. for the years 2001 through 2007. (Source: Apple Inc.)

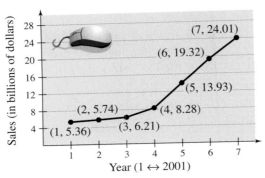

(a) Use the slopes of the line segments to determine the years in which the sales showed the greatest increase and the least increase.

(b) Find the slope of the line segment connecting the points for the years 2001 and 2007.

(c) Interpret the meaning of the slope in part (b) in the context of the problem.

115. ROAD GRADE   You are driving on a road that has a 6% uphill grade (see figure). This means that the slope of the road is $\frac{6}{100}$. Approximate the amount of vertical change in your position if you drive 200 feet.

116. ROAD GRADE   From the top of a mountain road, a surveyor takes several horizontal measurements x and several vertical measurements y, as shown in the table (x and y are measured in feet).

x	300	600	900	1200	1500	1800	2100
y	-25	-50	-75	-100	-125	-150	-175

(a) Sketch a scatter plot of the data.

(b) Use a straightedge to sketch the line that you think best fits the data.

(c) Find an equation for the line you sketched in part (b).

(d) Interpret the meaning of the slope of the line in part (c) in the context of the problem.

(e) The surveyor needs to put up a road sign that indicates the steepness of the road. For instance, a surveyor would put up a sign that states "8% grade" on a road with a downhill grade that has a slope of $-\frac{8}{100}$. What should the sign state for the road in this problem?

RATE OF CHANGE   In Exercises 117 and 118, you are given the dollar value of a product in 2010 and the rate at which the value of the product is expected to change during the next 5 years. Use this information to write a linear equation that gives the dollar value V of the product in terms of the year t. (Let $t = 10$ represent 2010.)

	2010 Value	*Rate*
117.	$2540	$125 decrease per year
118.	$156	$4.50 increase per year

119. DEPRECIATION   The value V of a molding machine t years after it is purchased is

$$V = -4000t + 58,500, \quad 0 \le t \le 5.$$

Explain what the V-intercept and the slope measure.

120. COST   The cost C of producing n computer laptop bags is given by

$$C = 1.25n + 15,750, \quad 0 < n.$$

Explain what the C-intercept and the slope measure.

121. DEPRECIATION   A sub shop purchases a used pizza oven for $875. After 5 years, the oven will have to be replaced. Write a linear equation giving the value V of the equipment during the 5 years it will be in use.

122. DEPRECIATION   A school district purchases a high-volume printer, copier, and scanner for $25,000. After 10 years, the equipment will have to be replaced. Its value at that time is expected to be $2000. Write a linear equation giving the value V of the equipment during the 10 years it will be in use.

123. SALES   A discount outlet is offering a 20% discount on all items. Write a linear equation giving the sale price S for an item with a list price L.

124. HOURLY WAGE   A microchip manufacturer pays its assembly line workers $12.25 per hour. In addition, workers receive a piecework rate of $0.75 per unit produced. Write a linear equation for the hourly wage W in terms of the number of units x produced per hour.

125. MONTHLY SALARY   A pharmaceutical salesperson receives a monthly salary of $2500 plus a commission of 7% of sales. Write a linear equation for the salesperson's monthly wage W in terms of monthly sales S.

126. BUSINESS COSTS   A sales representative of a company using a personal car receives $120 per day for lodging and meals plus $0.55 per mile driven. Write a linear equation giving the daily cost C to the company in terms of x, the number of miles driven.

127. CASH FLOW PER SHARE   The cash flow per share for the Timberland Co. was $1.21 in 1999 and $1.46 in 2007. Write a linear equation that gives the cash flow per share in terms of the year. Let $t = 9$ represent 1999. Then predict the cash flows for the years 2012 and 2014.   (Source: The Timberland Co.)

128. NUMBER OF STORES   In 2003 there were 1078 J.C. Penney stores and in 2007 there were 1067 stores. Write a linear equation that gives the number of stores in terms of the year. Let $t = 3$ represent 2003. Then predict the numbers of stores for the years 2012 and 2014. Are your answers reasonable? Explain. (Source: J.C. Penney Co.)

129. COLLEGE ENROLLMENT   The Pennsylvania State University had enrollments of 40,571 students in 2000 and 44,112 students in 2008 at its main campus in University Park, Pennsylvania.   (Source: *Penn State Fact Book*)

(a) Assuming the enrollment growth is linear, find a linear model that gives the enrollment in terms of the year t, where $t = 0$ corresponds to 2000.

(b) Use your model from part (a) to predict the enrollments in 2010 and 2015.

(c) What is the slope of your model? Explain its meaning in the context of the situation.

130. COLLEGE ENROLLMENT   The University of Florida had enrollments of 46,107 students in 2000 and 51,413 students in 2008.   (Source: University of Florida)

(a) What was the average annual change in enrollment from 2000 to 2008?

(b) Use the average annual change in enrollment to estimate the enrollments in 2002, 2004, and 2006.

(c) Write the equation of a line that represents the given data in terms of the year t, where $t = 0$ corresponds to 2000. What is its slope? Interpret the slope in the context of the problem.

(d) Using the results of parts (a)–(c), write a short paragraph discussing the concepts of *slope* and *average rate of change*.

131. COST, REVENUE, AND PROFIT   A roofing contractor purchases a shingle delivery truck with a shingle elevator for $42,000. The vehicle requires an average expenditure of $6.50 per hour for fuel and maintenance, and the operator is paid $11.50 per hour.

(a) Write a linear equation giving the total cost C of operating this equipment for t hours. (Include the purchase cost of the equipment.)

(b) Assuming that customers are charged $30 per hour of machine use, write an equation for the revenue R derived from t hours of use.

(c) Use the formula for profit

$$P = R - C$$

to write an equation for the profit derived from t hours of use.

(d) Use the result of part (c) to find the break-even point—that is, the number of hours this equipment must be used to yield a profit of 0 dollars.

132. RENTAL DEMAND A real estate office handles an apartment complex with 50 units. When the rent per unit is $580 per month, all 50 units are occupied. However, when the rent is $625 per month, the average number of occupied units drops to 47. Assume that the relationship between the monthly rent p and the demand x is linear.

(a) Write the equation of the line giving the demand x in terms of the rent p.

(b) Use this equation to predict the number of units occupied when the rent is $655.

(c) Predict the number of units occupied when the rent is $595.

133. GEOMETRY The length and width of a rectangular garden are 15 meters and 10 meters, respectively. A walkway of width x surrounds the garden.

(a) Draw a diagram that gives a visual representation of the problem.

(b) Write the equation for the perimeter y of the walkway in terms of x.

(c) Use a graphing utility to graph the equation for the perimeter.

(d) Determine the slope of the graph in part (c). For each additional one-meter increase in the width of the walkway, determine the increase in its perimeter.

134. AVERAGE ANNUAL SALARY The average salaries (in millions of dollars) of Major League Baseball players from 2000 through 2007 are shown in the scatter plot. Find the equation of the line that you think best fits these data. (Let y represent the average salary and let t represent the year, with $t = 0$ corresponding to 2000.) (Source: Major League Baseball Players Association)

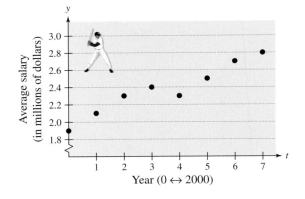

Year (0 ↔ 2000)

135. DATA ANALYSIS: NUMBER OF DOCTORS The numbers of doctors of osteopathic medicine y (in thousands) in the United States from 2000 through 2008, where x is the year, are shown as data points (x, y). (Source: American Osteopathic Association)

(2000, 44.9), (2001, 47.0), (2002, 49.2), (2003, 51.7), (2004, 54.1), (2005, 56.5), (2006, 58.9), (2007, 61.4), (2008, 64.0)

(a) Sketch a scatter plot of the data. Let $x = 0$ correspond to 2000.

(b) Use a straightedge to sketch the line that you think best fits the data.

(c) Find the equation of the line from part (b). Explain the procedure you used.

(d) Write a short paragraph explaining the meanings of the slope and y-intercept of the line in terms of the data.

(e) Compare the values obtained using your model with the actual values.

(f) Use your model to estimate the number of doctors of osteopathic medicine in 2012.

136. DATA ANALYSIS: AVERAGE SCORES An instructor gives regular 20-point quizzes and 100-point exams in an algebra course. Average scores for six students, given as data points (x, y), where x is the average quiz score and y is the average test score, are (18, 87), (10, 55), (19, 96), (16, 79), (13, 76), and (15, 82). [*Note:* There are many correct answers for parts (b)–(d).]

(a) Sketch a scatter plot of the data.

(b) Use a straightedge to sketch the line that you think best fits the data.

(c) Find an equation for the line you sketched in part (b).

(d) Use the equation in part (c) to estimate the average test score for a person with an average quiz score of 17.

(e) The instructor adds 4 points to the average test score of each student in the class. Describe the changes in the positions of the plotted points and the change in the equation of the line.

EXPLORATION

TRUE OR FALSE? In Exercises 137 and 138, determine whether the statement is true or false. Justify your answer.

137. A line with a slope of $-\frac{5}{7}$ is steeper than a line with a slope of $-\frac{6}{7}$.

138. The line through $(-8, 2)$ and $(-1, 4)$ and the line through $(0, -4)$ and $(-7, 7)$ are parallel.

139. Explain how you could show that the points $A(2, 3)$, $B(2, 9)$, and $C(4, 3)$ are the vertices of a right triangle.

140. Explain why the slope of a vertical line is said to be undefined.

141. With the information shown in the graphs, is it possible to determine the slope of each line? Is it possible that the lines could have the same slope? Explain.

(a)   (b)

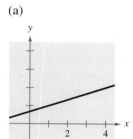   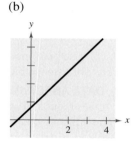

142. The slopes of two lines are -4 and $\frac{5}{2}$. Which is steeper? Explain.

143. Use a graphing utility to compare the slopes of the lines $y = mx$, where $m = 0.5$, 1, 2, and 4. Which line rises most quickly? Now, let $m = -0.5$, -1, -2, and -4. Which line falls most quickly? Use a square setting to obtain a true geometric perspective. What can you conclude about the slope and the "rate" at which the line rises or falls?

144. Find d_1 and d_2 in terms of m_1 and m_2, respectively (see figure). Then use the Pythagorean Theorem to find a relationship between m_1 and m_2.

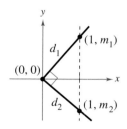

145. THINK ABOUT IT Is it possible for two lines with positive slopes to be perpendicular? Explain.

146. CAPSTONE Match the description of the situation with its graph. Also determine the slope and y-intercept of each graph and interpret the slope and y-intercept in the context of the situation. [The graphs are labeled (i), (ii), (iii), and (iv).]

(i)   (ii)

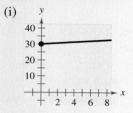   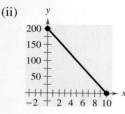

(iii)   (iv)

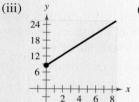   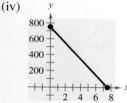

(a) A person is paying $20 per week to a friend to repay a $200 loan.

(b) An employee is paid $8.50 per hour plus $2 for each unit produced per hour.

(c) A sales representative receives $30 per day for food plus $0.32 for each mile traveled.

(d) A computer that was purchased for $750 depreciates $100 per year.

PROJECT: BACHELOR'S DEGREES To work an extended application analyzing the numbers of bachelor's degrees earned by women in the United States from 1996 through 2007, visit this text's website at *academic.cengage.com*. (Data Source: U.S. National Center for Education Statistics)

1.4   FUNCTIONS

What you should learn

- Determine whether relations between two variables are functions.
- Use function notation and evaluate functions.
- Find the domains of functions.
- Use functions to model and solve real-life problems.
- Evaluate difference quotients.

Why you should learn it

Functions can be used to model and solve real-life problems. For instance, in Exercise 100 on page 52, you will use a function to model the force of water against the face of a dam.

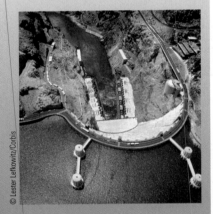

Introduction to Functions

Many everyday phenomena involve two quantities that are related to each other by some rule of correspondence. The mathematical term for such a rule of correspondence is a **relation.** In mathematics, relations are often represented by mathematical equations and formulas. For instance, the simple interest I earned on $1000 for 1 year is related to the annual interest rate r by the formula $I = 1000r$.

The formula $I = 1000r$ represents a special kind of relation that matches each item from one set with *exactly one* item from a different set. Such a relation is called a **function.**

Definition of Function

A **function** f from a set A to a set B is a relation that assigns to each element x in the set A exactly one element y in the set B. The set A is the **domain** (or set of inputs) of the function f, and the set B contains the **range** (or set of outputs).

To help understand this definition, look at the function that relates the time of day to the temperature in Figure 1.47.

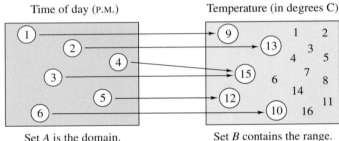

Set A is the domain.
Inputs: 1, 2, 3, 4, 5, 6

Set B contains the range.
Outputs: 9, 10, 12, 13, 15

FIGURE 1.47

This function can be represented by the following ordered pairs, in which the first coordinate (x-value) is the input and the second coordinate (y-value) is the output.

$$\{(1, 9°), (2, 13°), (3, 15°), (4, 15°), (5, 12°), (6, 10°)\}$$

Characteristics of a Function from Set *A* to Set *B*

1. Each element in A must be matched with an element in B.

2. Some elements in B may not be matched with any element in A.

3. Two or more elements in A may be matched with the same element in B.

4. An element in A (the domain) cannot be matched with two different elements in B.

Functions are commonly represented in four ways.

Four Ways to Represent a Function

1. *Verbally* by a sentence that describes how the input variable is related to the output variable

2. *Numerically* by a table or a list of ordered pairs that matches input values with output values

3. *Graphically* by points on a graph in a coordinate plane in which the input values are represented by the horizontal axis and the output values are represented by the vertical axis

4. *Algebraically* by an equation in two variables

To determine whether or not a relation is a function, you must decide whether each input value is matched with exactly one output value. If any input value is matched with two or more output values, the relation is not a function.

Example 1   Testing for Functions

Determine whether the relation represents y as a function of x.

a. The input value x is the number of representatives from a state, and the output value y is the number of senators.

b.

Input, x	Output, y
2	11
2	10
3	8
4	5
5	1

c.

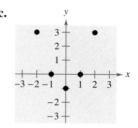

FIGURE **1.48**

Solution

a. This verbal description *does* describe y as a function of x. Regardless of the value of x, the value of y is always 2. Such functions are called *constant functions*.

b. This table *does not* describe y as a function of x. The input value 2 is matched with two different y-values.

c. The graph in Figure 1.48 *does* describe y as a function of x. Each input value is matched with exactly one output value.

CHECK *Point* ▶ Now try Exercise 11.

Representing functions by sets of ordered pairs is common in *discrete mathematics*. In algebra, however, it is more common to represent functions by equations or formulas involving two variables. For instance, the equation

$$y = x^2 \qquad \text{\scriptsize y is a function of x.}$$

represents the variable y as a function of the variable x. In this equation, x is

the **independent variable** and y is the **dependent variable.** The domain of the function is the set of all values taken on by the independent variable x, and the range of the function is the set of all values taken on by the dependent variable y.

Example 2   Testing for Functions Represented Algebraically

Which of the equations represent(s) y as a function of x?

a. $x^2 + y = 1$ **b.** $-x + y^2 = 1$

Solution

To determine whether y is a function of x, try to solve for y in terms of x.

a. Solving for y yields

$$x^2 + y = 1 \qquad \text{Write original equation.}$$
$$y = 1 - x^2. \qquad \text{Solve for } y.$$

To each value of x there corresponds exactly one value of y. So, y is a function of x.

b. Solving for y yields

$$-x + y^2 = 1 \qquad \text{Write original equation.}$$
$$y^2 = 1 + x \qquad \text{Add } x \text{ to each side.}$$
$$y = \pm\sqrt{1 + x}. \qquad \text{Solve for } y.$$

The $\pm$ indicates that to a given value of x there correspond two values of y. So, y is not a function of x.

CHECK*Point*   Now try Exercise 21.

Function Notation

When an equation is used to represent a function, it is convenient to name the function so that it can be referenced easily. For example, you know that the equation $y = 1 - x^2$ describes y as a function of x. Suppose you give this function the name "f." Then you can use the following **function notation.**

Input	*Output*	*Equation*
x	$f(x)$	$f(x) = 1 - x^2$

The symbol $f(x)$ is read as *the value of f at x* or simply *f of x*. The symbol $f(x)$ corresponds to the y-value for a given x. So, you can write $y = f(x)$. Keep in mind that f is the *name* of the function, whereas $f(x)$ is the *value* of the function at x. For instance, the function given by

$$f(x) = 3 - 2x$$

has *function values* denoted by $f(-1)$, $f(0)$, $f(2)$, and so on. To find these values, substitute the specified input values into the given equation.

For $x = -1,$ $\quad f(-1) = 3 - 2(-1) = 3 + 2 = 5.$

For $x = 0,$ $\quad\quad f(0) = 3 - 2(0) = 3 - 0 = 3.$

For $x = 2,$ $\quad\quad f(2) = 3 - 2(2) = 3 - 4 = -1.$

Although f is often used as a convenient function name and x is often used as the independent variable, you can use other letters. For instance,

$$f(x) = x^2 - 4x + 7, \quad f(t) = t^2 - 4t + 7, \quad \text{and} \quad g(s) = s^2 - 4s + 7$$

all define the same function. In fact, the role of the independent variable is that of a "placeholder." Consequently, the function could be described by

$$f(\quad) = (\quad)^2 - 4(\quad) + 7.$$

> **⚠ WARNING / CAUTION**
>
> In Example 3, note that $g(x + 2)$ is not equal to $g(x) + g(2)$. In general, $g(u + v) \neq g(u) + g(v)$.

Example 3 Evaluating a Function

Let $g(x) = -x^2 + 4x + 1$. Find each function value.

a. $g(2)$ **b.** $g(t)$ **c.** $g(x + 2)$

Solution

a. Replacing x with 2 in $g(x) = -x^2 + 4x + 1$ yields the following.

$$g(2) = -(2)^2 + 4(2) + 1 = -4 + 8 + 1 = 5$$

b. Replacing x with t yields the following.

$$g(t) = -(t)^2 + 4(t) + 1 = -t^2 + 4t + 1$$

c. Replacing x with $x + 2$ yields the following.

$$g(x + 2) = -(x + 2)^2 + 4(x + 2) + 1$$
$$= -(x^2 + 4x + 4) + 4x + 8 + 1$$
$$= -x^2 - 4x - 4 + 4x + 8 + 1$$
$$= -x^2 + 5$$

CHECK Point Now try Exercise 41.

A function defined by two or more equations over a specified domain is called a **piecewise-defined function.**

Example 4 A Piecewise-Defined Function

Evaluate the function when $x = -1, 0,$ and 1.

$$f(x) = \begin{cases} x^2 + 1, & x < 0 \\ x - 1, & x \geq 0 \end{cases}$$

Solution

Because $x = -1$ is less than 0, use $f(x) = x^2 + 1$ to obtain

$$f(-1) = (-1)^2 + 1 = 2.$$

For $x = 0$, use $f(x) = x - 1$ to obtain

$$f(0) = (0) - 1 = -1.$$

For $x = 1$, use $f(x) = x - 1$ to obtain

$$f(1) = (1) - 1 = 0.$$

CHECK Point Now try Exercise 49.

| **Example 5**   **Finding Values for Which $f(x) = 0$** |

Find all real values of x such that $f(x) = 0$.

a. $f(x) = -2x + 10$

b. $f(x) = x^2 - 5x + 6$

Solution

For each function, set $f(x) = 0$ and solve for x.

a. $-2x + 10 = 0$ — Set $f(x)$ equal to 0.

$\qquad -2x = -10$ — Subtract 10 from each side.

$\qquad x = 5$ — Divide each side by -2.

So, $f(x) = 0$ when $x = 5$.

b. $x^2 - 5x + 6 = 0$ — Set $f(x)$ equal to 0.

$\qquad (x - 2)(x - 3) = 0$ — Factor.

$\qquad x - 2 = 0 \implies x = 2$ — Set 1st factor equal to 0.

$\qquad x - 3 = 0 \implies x = 3$ — Set 2nd factor equal to 0.

So, $f(x) = 0$ when $x = 2$ or $x = 3$.

CHECK**Point** Now try Exercise 59.

| **Example 6**   **Finding Values for Which $f(x) = g(x)$** |

Find the values of x for which $f(x) = g(x)$.

a. $f(x) = x^2 + 1$ and $g(x) = 3x - x^2$

b. $f(x) = x^2 - 1$ and $g(x) = -x^2 + x + 2$

Solution

a. $x^2 + 1 = 3x - x^2$ — Set $f(x)$ equal to $g(x)$.

$\qquad 2x^2 - 3x + 1 = 0$ — Write in general form.

$\qquad (2x - 1)(x - 1) = 0$ — Factor.

$\qquad 2x - 1 = 0 \implies x = \frac{1}{2}$ — Set 1st factor equal to 0.

$\qquad x - 1 = 0 \implies x = 1$ — Set 2nd factor equal to 0.

So, $f(x) = g(x)$ when $x = \dfrac{1}{2}$ or $x = 1$.

b. $x^2 - 1 = -x^2 + x + 2$ — Set $f(x)$ equal to $g(x)$.

$\qquad 2x^2 - x - 3 = 0$ — Write in general form.

$\qquad (2x - 3)(x + 1) = 0$ — Factor.

$\qquad 2x - 3 = 0 \implies x = \frac{3}{2}$ — Set 1st factor equal to 0.

$\qquad x + 1 = 0 \implies x = -1$ — Set 2nd factor equal to 0.

So, $f(x) = g(x)$ when $x = \dfrac{3}{2}$ or $x = -1$.

CHECK**Point** Now try Exercise 67.

Algebra Help

To do Examples 5 and 6, you need to be able to solve equations. You can review the techniques for solving equations in Appendix A.5.

The Domain of a Function

The domain of a function can be described explicitly or it can be *implied* by the expression used to define the function. The **implied domain** is the set of all real numbers for which the expression is defined. For instance, the function given by

$$f(x) = \frac{1}{x^2 - 4}$$ Domain excludes x-values that result in division by zero.

has an implied domain that consists of all real x other than $x = \pm 2$. These two values are excluded from the domain because division by zero is undefined. Another common type of implied domain is that used to avoid even roots of negative numbers. For example, the function given by

$$f(x) = \sqrt{x}$$ Domain excludes x-values that result in even roots of negative numbers.

is defined only for $x \geq 0$. So, its implied domain is the interval $[0, \infty)$. In general, the domain of a function *excludes* values that would cause division by zero *or* that would result in the even root of a negative number.

TECHNOLOGY

Use a graphing utility to graph the functions given by $y = \sqrt{4 - x^2}$ and $y = \sqrt{x^2 - 4}$. What is the domain of each function? Do the domains of these two functions overlap? If so, for what values do the domains overlap?

Example 7   Finding the Domain of a Function

Find the domain of each function.

a. f: $\{(-3, 0), (-1, 4), (0, 2), (2, 2), (4, -1)\}$ **b.** $g(x) = \frac{1}{x + 5}$

c. Volume of a sphere: $V = \frac{4}{3}\pi r^3$ **d.** $h(x) = \sqrt{4 - 3x}$

Solution

a. The domain of f consists of all first coordinates in the set of ordered pairs.

Domain $= \{-3, -1, 0, 2, 4\}$

b. Excluding x-values that yield zero in the denominator, the domain of g is the set of all real numbers x except $x = -5$.

c. Because this function represents the volume of a sphere, the values of the radius r must be positive. So, the domain is the set of all real numbers r such that $r > 0$.

d. This function is defined only for x-values for which

$$4 - 3x \geq 0.$$

By solving this inequality, you can conclude that $x \leq \frac{4}{3}$. So, the domain is the interval $\left(-\infty, \frac{4}{3}\right]$.

CHECK Point   Now try Exercise 73.

Algebra Help

In Example 7(d), $4 - 3x \geq 0$ is a linear inequality. You can review the techniques for solving a linear inequality in Appendix A.6.

In Example 7(c), note that the domain of a function may be implied by the physical context. For instance, from the equation

$$V = \frac{4}{3}\pi r^3$$

you would have no reason to restrict r to positive values, but the physical context implies that a sphere cannot have a negative or zero radius.

$$\frac{h}{r} = 4$$

FIGURE **1.49**

Applications

Example 8   The Dimensions of a Container

You work in the marketing department of a soft-drink company and are experimenting with a new can for iced tea that is slightly narrower and taller than a standard can. For your experimental can, the ratio of the height to the radius is 4, as shown in Figure 1.49.

a. Write the volume of the can as a function of the radius r.
b. Write the volume of the can as a function of the height h.

Solution

a. $V(r) = \pi r^2 h = \pi r^2 (4r) = 4\pi r^3$ Write V as a function of r.

b. $V(h) = \pi \left(\dfrac{h}{4}\right)^2 h = \dfrac{\pi h^3}{16}$ Write V as a function of h.

CHECK *Point*   Now try Exercise 87.

Example 9   The Path of a Baseball

A baseball is hit at a point 3 feet above ground at a velocity of 100 feet per second and an angle of 45°. The path of the baseball is given by the function

$$f(x) = -0.0032x^2 + x + 3$$

where x and $f(x)$ are measured in feet. Will the baseball clear a 10-foot fence located 300 feet from home plate?

Algebraic Solution

When $x = 300$, you can find the height of the baseball as follows.

$f(x) = -0.0032x^2 + x + 3$ Write original function.

$f(300) = -0.0032(300)^2 + 300 + 3$ Substitute 300 for x.

$\quad\quad = 15$ Simplify.

When $x = 300$, the height of the baseball is 15 feet, so the baseball will clear a 10-foot fence.

CHECK *Point*   Now try Exercise 93.

Graphical Solution

Use a graphing utility to graph the function $y = -0.0032x^2 + x + 3$. Use the *value* feature or the *zoom* and *trace* features of the graphing utility to estimate that $y = 15$ when $x = 300$, as shown in Figure 1.50. So, the ball will clear a 10-foot fence.

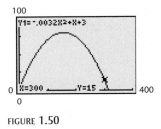

FIGURE **1.50**

In the equation in Example 9, the height of the baseball is a function of the distance from home plate.

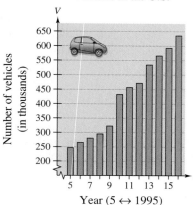

Number of Alternative-Fueled
Vehicles in the U.S.

FIGURE **1.51**

Example 10   Alternative-Fueled Vehicles

The number V (in thousands) of alternative-fueled vehicles in the United States increased in a linear pattern from 1995 to 1999, as shown in Figure 1.51. Then, in 2000, the number of vehicles took a jump and, until 2006, increased in a different linear pattern. These two patterns can be approximated by the function

$$V(t) = \begin{cases} 18.08t + 155.3, & 5 \le t \le 9 \\ 34.75t + 74.9, & 10 \le t \le 16 \end{cases}$$

where t represents the year, with $t = 5$ corresponding to 1995. Use this function to approximate the number of alternative-fueled vehicles for each year from 1995 to 2006. (Source: Science Applications International Corporation; Energy Information Administration)

Solution

From 1995 to 1999, use $V(t) = 18.08t + 155.3$.

245.7	263.8	281.9	299.9	318.0
1995	1996	1997	1998	1999

From 2000 to 2006, use $V(t) = 34.75t + 74.9$.

422.4	457.2	491.9	526.7	561.4	596.2	630.9
2000	2001	2002	2003	2004	2005	2006

CHECK Point   Now try Exercise 95.

Difference Quotients

One of the basic definitions in calculus employs the ratio

$$\frac{f(x + h) - f(x)}{h}, \quad h \ne 0.$$

This ratio is called a **difference quotient,** as illustrated in Example 11.

Example 11   Evaluating a Difference Quotient

For $f(x) = x^2 - 4x + 7$, find $\dfrac{f(x + h) - f(x)}{h}$.

Solution

$$\frac{f(x + h) - f(x)}{h} = \frac{[(x + h)^2 - 4(x + h) + 7] - (x^2 - 4x + 7)}{h}$$

$$= \frac{x^2 + 2xh + h^2 - 4x - 4h + 7 - x^2 + 4x - 7}{h}$$

$$= \frac{2xh + h^2 - 4h}{h} = \frac{h(2x + h - 4)}{h} = 2x + h - 4, \quad h \ne 0$$

CHECK Point   Now try Exercise 103.

The symbol $\int$ indicates an example or exercise that highlights algebraic techniques specifically used in calculus.

You may find it easier to calculate the difference quotient in Example 11 by first finding $f(x + h)$, and then substituting the resulting expression into the difference quotient, as follows.

$$f(x + h) = (x + h)^2 - 4(x + h) + 7 = x^2 + 2xh + h^2 - 4x - 4h + 7$$

$$\frac{f(x + h) - f(x)}{h} = \frac{(x^2 + 2xh + h^2 - 4x - 4h + 7) - (x^2 - 4x + 7)}{h}$$

$$= \frac{2xh + h^2 - 4h}{h} = \frac{h(2x + h - 4)}{h} = 2x + h - 4, \quad h \neq 0$$

Summary of Function Terminology

Function: A **function** is a relationship between two variables such that to each value of the independent variable there corresponds exactly one value of the dependent variable.

Function Notation: $y = f(x)$

  f is the *name* of the function.

  y is the **dependent variable.**

  x is the **independent variable.**

  $f(x)$ is the *value of the function at x.*

Domain: The **domain** of a function is the set of all values (inputs) of the independent variable for which the function is defined. If x is in the domain of f, f is said to be *defined* at x. If x is not in the domain of f, f is said to be *undefined* at x.

Range: The **range** of a function is the set of all values (outputs) assumed by the dependent variable (that is, the set of all function values).

Implied Domain: If f is defined by an algebraic expression and the domain is not specified, the **implied domain** consists of all real numbers for which the expression is defined.

CLASSROOM DISCUSSION

Everyday Functions   In groups of two or three, identify common real-life functions. Consider everyday activities, events, and expenses, such as long distance telephone calls and car insurance. Here are two examples.

a. The statement, "Your happiness is a function of the grade you receive in this course" is *not* a correct mathematical use of the word "function." The word "happiness" is ambiguous.

b. The statement, "Your federal income tax is a function of your adjusted gross income" *is* a correct mathematical use of the word "function." Once you have determined your adjusted gross income, your income tax can be determined.

Describe your functions in words. Avoid using ambiguous words. Can you find an example of a piecewise-defined function?

1.4   EXERCISES

See www.CalcChat.com for worked-out solutions to odd-numbered exercises.

VOCABULARY: Fill in the blanks.

1. A relation that assigns to each element x from a set of inputs, or _____, exactly one element y in a set of outputs, or _____, is called a _____.

2. Functions are commonly represented in four different ways, _____, _____, _____, and _____.

3. For an equation that represents y as a function of x, the set of all values taken on by the _____ variable x is the domain, and the set of all values taken on by the _____ variable y is the range.

4. The function given by

$$f(x) = \begin{cases} 2x - 1, & x < 0 \\ x^2 + 4, & x \geq 0 \end{cases}$$

   is an example of a _____ function.

5. If the domain of the function f is not given, then the set of values of the independent variable for which the expression is defined is called the _____ _____.

6. In calculus, one of the basic definitions is that of a _____ _____, given by $\dfrac{f(x + h) - f(x)}{h}$, $h \neq 0$.

SKILLS AND APPLICATIONS

In Exercises 7–10, is the relationship a function?

7. Domain   Range

$-2 \longrightarrow 5$
$-1 \qquad 6$
$0 \qquad 7$
$1 \qquad 8$
2

8. Domain   Range

$-2 \longrightarrow 3$
$-1 \qquad 4$
$0 \longrightarrow 5$
1
2

9.   Domain   Range

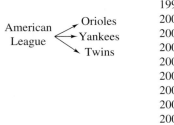

National League → Cubs, Pirates, Dodgers

American League → Orioles, Yankees, Twins

10. Domain   Range
(Year)   (Number of North Atlantic tropical storms and hurricanes)

1999 — 10
2000 — 12
2001 — 15
2002 — 16
2003 — 21
2004 — 27
2005
2006
2007
2008

In Exercises 11–14, determine whether the relation represents y as a function of x.

11.

Input, x	-2	-1	0	1	2
Output, y	-8	-1	0	1	8

12.

Input, x	0	1	2	1	0
Output, y	-4	-2	0	2	4

13.

Input, x	10	7	4	7	10
Output, y	3	6	9	12	15

14.

Input, x	0	3	9	12	15
Output, y	3	3	3	3	3

In Exercises 15 and 16, which sets of ordered pairs represent functions from A to B? Explain.

15. $A = \{0, 1, 2, 3\}$ and $B = \{-2, -1, 0, 1, 2\}$
   (a) $\{(0, 1), (1, -2), (2, 0), (3, 2)\}$
   (b) $\{(0, -1), (2, 2), (1, -2), (3, 0), (1, 1)\}$
   (c) $\{(0, 0), (1, 0), (2, 0), (3, 0)\}$
   (d) $\{(0, 2), (3, 0), (1, 1)\}$

16. $A = \{a, b, c\}$ and $B = \{0, 1, 2, 3\}$
   (a) $\{(a, 1), (c, 2), (c, 3), (b, 3)\}$
   (b) $\{(a, 1), (b, 2), (c, 3)\}$
   (c) $\{(1, a), (0, a), (2, c), (3, b)\}$
   (d) $\{(c, 0), (b, 0), (a, 3)\}$

CIRCULATION OF NEWSPAPERS   In Exercises 17 and 18, use the graph, which shows the circulation (in millions) of daily newspapers in the United States. (Source: Editor & Publisher Company)

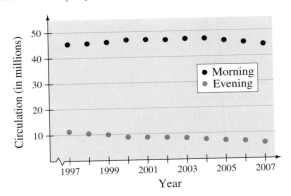

17. Is the circulation of morning newspapers a function of the year? Is the circulation of evening newspapers a function of the year? Explain.

18. Let $f(x)$ represent the circulation of evening newspapers in year x. Find $f(2002)$.

In Exercises 19–36, determine whether the equation represents y as a function of x.

19. $x^2 + y^2 = 4$ **20.** $x^2 - y^2 = 16$

21. $x^2 + y = 4$ **22.** $y - 4x^2 = 36$

23. $2x + 3y = 4$ **24.** $2x + 5y = 10$

25. $(x + 2)^2 + (y - 1)^2 = 25$

26. $(x - 2)^2 + y^2 = 4$

27. $y^2 = x^2 - 1$ **28.** $x + y^2 = 4$

29. $y = \sqrt{16 - x^2}$ **30.** $y = \sqrt{x + 5}$

31. $y = |4 - x|$ **32.** $|y| = 4 - x$

33. $x = 14$ **34.** $y = -75$

35. $y + 5 = 0$ **36.** $x - 1 = 0$

In Exercises 37–52, evaluate the function at each specified value of the independent variable and simplify.

37. $f(x) = 2x - 3$
   (a) $f(1)$ (b) $f(-3)$ (c) $f(x - 1)$

38. $g(y) = 7 - 3y$
   (a) $g(0)$ (b) $g\left(\frac{7}{3}\right)$ (c) $g(s + 2)$

39. $V(r) = \frac{4}{3}\pi r^3$
   (a) $V(3)$ (b) $V\left(\frac{3}{2}\right)$ (c) $V(2r)$

40. $S(r) = 4\pi r^2$
   (a) $S(2)$ (b) $S\left(\frac{1}{2}\right)$ (c) $S(3r)$

41. $g(t) = 4t^2 - 3t + 5$
   (a) $g(2)$ (b) $g(t - 2)$ (c) $g(t) - g(2)$

42. $h(t) = t^2 - 2t$
   (a) $h(2)$ (b) $h(1.5)$ (c) $h(x + 2)$

43. $f(y) = 3 - \sqrt{y}$
   (a) $f(4)$ (b) $f(0.25)$ (c) $f(4x^2)$

44. $f(x) = \sqrt{x + 8} + 2$
   (a) $f(-8)$ (b) $f(1)$ (c) $f(x - 8)$

45. $q(x) = 1/(x^2 - 9)$
   (a) $q(0)$ (b) $q(3)$ (c) $q(y + 3)$

46. $q(t) = (2t^2 + 3)/t^2$
   (a) $q(2)$ (b) $q(0)$ (c) $q(-x)$

47. $f(x) = |x|/x$
   (a) $f(2)$ (b) $f(-2)$ (c) $f(x - 1)$

48. $f(x) = |x| + 4$
   (a) $f(2)$ (b) $f(-2)$ (c) $f(x^2)$

49. $f(x) = \begin{cases} 2x + 1, & x < 0 \\ 2x + 2, & x \geq 0 \end{cases}$
   (a) $f(-1)$ (b) $f(0)$ (c) $f(2)$

50. $f(x) = \begin{cases} x^2 + 2, & x \leq 1 \\ 2x^2 + 2, & x > 1 \end{cases}$
   (a) $f(-2)$ (b) $f(1)$ (c) $f(2)$

51. $f(x) = \begin{cases} 3x - 1, & x < -1 \\ 4, & -1 \leq x \leq 1 \\ x^2, & x > 1 \end{cases}$
   (a) $f(-2)$ (b) $f\left(-\frac{1}{2}\right)$ (c) $f(3)$

52. $f(x) = \begin{cases} 4 - 5x, & x \leq -2 \\ 0, & -2 < x < 2 \\ x^2 + 1, & x \geq 2 \end{cases}$
   (a) $f(-3)$ (b) $f(4)$ (c) $f(-1)$

In Exercises 53–58, complete the table.

53. $f(x) = x^2 - 3$

x	-2	-1	0	1	2
$f(x)$					

54. $g(x) = \sqrt{x - 3}$

x	3	4	5	6	7
$g(x)$					

55. $h(t) = \frac{1}{2}|t + 3|$

t	-5	-4	-3	-2	-1
$h(t)$					

56. $f(s) = \dfrac{|s - 2|}{s - 2}$

s	0	1	$\frac{3}{2}$	$\frac{5}{2}$	4
$f(s)$					

57. $f(x) = \begin{cases} -\frac{1}{2}x + 4, & x \le 0 \\ (x - 2)^2, & x > 0 \end{cases}$

x	-2	-1	0	1	2
$f(x)$					

58. $f(x) = \begin{cases} 9 - x^2, & x < 3 \\ x - 3, & x \ge 3 \end{cases}$

x	1	2	3	4	5
$f(x)$					

In Exercises 59–66, find all real values of x such that $f(x) = 0$.

59. $f(x) = 15 - 3x$

60. $f(x) = 5x + 1$

61. $f(x) = \dfrac{3x - 4}{5}$

62. $f(x) = \dfrac{12 - x^2}{5}$

63. $f(x) = x^2 - 9$

64. $f(x) = x^2 - 8x + 15$

65. $f(x) = x^3 - x$

66. $f(x) = x^3 - x^2 - 4x + 4$

In Exercises 67–70, find the value(s) of x for which $f(x) = g(x)$.

67. $f(x) = x^2, \quad g(x) = x + 2$

68. $f(x) = x^2 + 2x + 1, \quad g(x) = 7x - 5$

69. $f(x) = x^4 - 2x^2, \quad g(x) = 2x^2$

70. $f(x) = \sqrt{x} - 4, \quad g(x) = 2 - x$

In Exercises 71–82, find the domain of the function.

71. $f(x) = 5x^2 + 2x - 1$

72. $g(x) = 1 - 2x^2$

73. $h(t) = \dfrac{4}{t}$

74. $s(y) = \dfrac{3y}{y + 5}$

75. $g(y) = \sqrt{y - 10}$

76. $f(t) = \sqrt[3]{t + 4}$

77. $g(x) = \dfrac{1}{x} - \dfrac{3}{x + 2}$

78. $h(x) = \dfrac{10}{x^2 - 2x}$

79. $f(s) = \dfrac{\sqrt{s - 1}}{s - 4}$

80. $f(x) = \dfrac{\sqrt{x + 6}}{6 + x}$

81. $f(x) = \dfrac{x - 4}{\sqrt{x}}$

82. $f(x) = \dfrac{x + 2}{\sqrt{x - 10}}$

In Exercises 83–86, assume that the domain of f is the set $A = \{-2, -1, 0, 1, 2\}$. Determine the set of ordered pairs that represents the function f.

83. $f(x) = x^2$

84. $f(x) = (x - 3)^2$

85. $f(x) = |x| + 2$

86. $f(x) = |x + 1|$

87. GEOMETRY Write the area A of a square as a function of its perimeter P.

88. GEOMETRY Write the area A of a circle as a function of its circumference C.

89. MAXIMUM VOLUME An open box of maximum volume is to be made from a square piece of material 24 centimeters on a side by cutting equal squares from the corners and turning up the sides (see figure).

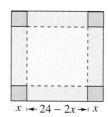 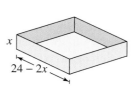

(a) The table shows the volumes V (in cubic centimeters) of the box for various heights x (in centimeters). Use the table to estimate the maximum volume.

Height, x	1	2	3	4	5	6
Volume, V	484	800	972	1024	980	864

(b) Plot the points (x, V) from the table in part (a). Does the relation defined by the ordered pairs represent V as a function of x?

(c) If V is a function of x, write the function and determine its domain.

90. MAXIMUM PROFIT The cost per unit in the production of an MP3 player is $60. The manufacturer charges $90 per unit for orders of 100 or less. To encourage large orders, the manufacturer reduces the charge by $0.15 per MP3 player for each unit ordered in excess of 100 (for example, there would be a charge of $87 per MP3 player for an order size of 120).

(a) The table shows the profits P (in dollars) for various numbers of units ordered, x. Use the table to estimate the maximum profit.

Units, x	110	120	130	140
Profit, P	3135	3240	3315	3360

Units, x	150	160	170
Profit, P	3375	3360	3315

(b) Plot the points (x, P) from the table in part (a). Does the relation defined by the ordered pairs represent P as a function of x?

(c) If P is a function of x, write the function and determine its domain.

91. GEOMETRY A right triangle is formed in the first quadrant by the x- and y-axes and a line through the point $(2, 1)$ (see figure). Write the area A of the triangle as a function of x, and determine the domain of the function.

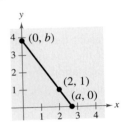

FIGURE FOR **91**

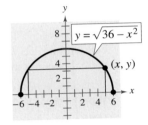

FIGURE FOR **92**

92. GEOMETRY A rectangle is bounded by the x-axis and the semicircle $y = \sqrt{36 - x^2}$ (see figure). Write the area A of the rectangle as a function of x, and graphically determine the domain of the function.

93. PATH OF A BALL The height y (in feet) of a baseball thrown by a child is

$$y = -\frac{1}{10}x^2 + 3x + 6$$

where x is the horizontal distance (in feet) from where the ball was thrown. Will the ball fly over the head of another child 30 feet away trying to catch the ball? (Assume that the child who is trying to catch the ball holds a baseball glove at a height of 5 feet.)

94. PRESCRIPTION DRUGS The numbers d (in millions) of drug prescriptions filled by independent outlets in the United States from 2000 through 2007 (see figure) can be approximated by the model

$$d(t) = \begin{cases} 10.6t + 699, & 0 \le t \le 4 \\ 15.5t + 637, & 5 \le t \le 7 \end{cases}$$

where t represents the year, with $t = 0$ corresponding to 2000. Use this model to find the number of drug prescriptions filled by independent outlets in each year from 2000 through 2007. (Source: National Association of Chain Drug Stores)

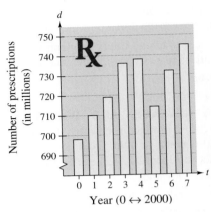

FIGURE FOR **94**

95. MEDIAN SALES PRICE The median sale prices p (in thousands of dollars) of an existing one-family home in the United States from 1998 through 2007 (see figure) can be approximated by the model

$$p(t) = \begin{cases} 1.011t^2 - 12.38t + 170.5, & 8 \le t \le 13 \\ -6.950t^2 + 222.55t - 1557.6, & 14 \le t \le 17 \end{cases}$$

where t represents the year, with $t = 8$ corresponding to 1998. Use this model to find the median sale price of an existing one-family home in each year from 1998 through 2007. (Source: National Association of Realtors)

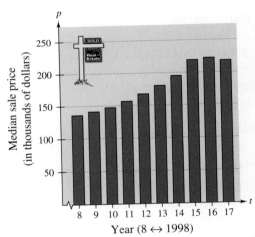

96. POSTAL REGULATIONS A rectangular package to be sent by the U.S. Postal Service can have a maximum combined length and girth (perimeter of a cross section) of 108 inches (see figure).

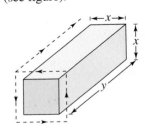

(a) Write the volume V of the package as a function of x. What is the domain of the function?

(b) Use a graphing utility to graph your function. Be sure to use an appropriate window setting.

(c) What dimensions will maximize the volume of the package? Explain your answer.

97. COST, REVENUE, AND PROFIT A company produces a product for which the variable cost is $12.30 per unit and the fixed costs are $98,000. The product sells for $17.98. Let x be the number of units produced and sold.

(a) The total cost for a business is the sum of the variable cost and the fixed costs. Write the total cost C as a function of the number of units produced.

(b) Write the revenue R as a function of the number of units sold.

(c) Write the profit P as a function of the number of units sold. (*Note:* $P = R - C$)

98. AVERAGE COST The inventor of a new game believes that the variable cost for producing the game is $0.95 per unit and the fixed costs are $6000. The inventor sells each game for $1.69. Let x be the number of games sold.

(a) The total cost for a business is the sum of the variable cost and the fixed costs. Write the total cost C as a function of the number of games sold.

(b) Write the average cost per unit $\overline{C} = C/x$ as a function of x.

99. TRANSPORTATION For groups of 80 or more people, a charter bus company determines the rate per person according to the formula

Rate $= 8 - 0.05(n - 80), \quad n \geq 80$

where the rate is given in dollars and n is the number of people.

(a) Write the revenue R for the bus company as a function of n.

(b) Use the function in part (a) to complete the table. What can you conclude?

n	90	100	110	120	130	140	150
$R(n)$							

100. PHYSICS The force F (in tons) of water against the face of a dam is estimated by the function $F(y) = 149.76\sqrt{10}\,y^{5/2}$, where y is the depth of the water (in feet).

(a) Complete the table. What can you conclude from the table?

y	5	10	20	30	40
$F(y)$					

(b) Use the table to approximate the depth at which the force against the dam is 1,000,000 tons.

(c) Find the depth at which the force against the dam is 1,000,000 tons algebraically.

101. HEIGHT OF A BALLOON A balloon carrying a transmitter ascends vertically from a point 3000 feet from the receiving station.

(a) Draw a diagram that gives a visual representation of the problem. Let h represent the height of the balloon and let d represent the distance between the balloon and the receiving station.

(b) Write the height of the balloon as a function of d. What is the domain of the function?

102. E-FILING The table shows the numbers of tax returns (in millions) made through e-file from 2000 through 2007. Let $f(t)$ represent the number of tax returns made through e-file in the year t. (Source: Internal Revenue Service)

Year	Number of tax returns made through e-file
2000	35.4
2001	40.2
2002	46.9
2003	52.9
2004	61.5
2005	68.5
2006	73.3
2007	80.0

(a) Find $\dfrac{f(2007) - f(2000)}{2007 - 2000}$ and interpret the result in the context of the problem.

(b) Make a scatter plot of the data.

(c) Find a linear model for the data algebraically. Let N represent the number of tax returns made through e-file and let $t = 0$ correspond to 2000.

(d) Use the model found in part (c) to complete the table.

t	0	1	2	3	4	5	6	7
N								

(e) Compare your results from part (d) with the actual data.

(f) Use a graphing utility to find a linear model for the data. Let $x = 0$ correspond to 2000. How does the model you found in part (c) compare with the model given by the graphing utility?

In Exercises 103–110, find the difference quotient and simplify your answer.

103. $f(x) = x^2 - x + 1$, $\dfrac{f(2 + h) - f(2)}{h}$, $h \neq 0$

104. $f(x) = 5x - x^2$, $\dfrac{f(5 + h) - f(5)}{h}$, $h \neq 0$

105. $f(x) = x^3 + 3x$, $\dfrac{f(x + h) - f(x)}{h}$, $h \neq 0$

106. $f(x) = 4x^2 - 2x$, $\dfrac{f(x + h) - f(x)}{h}$, $h \neq 0$

107. $g(x) = \dfrac{1}{x^2}$, $\dfrac{g(x) - g(3)}{x - 3}$, $x \neq 3$

108. $f(t) = \dfrac{1}{t - 2}$, $\dfrac{f(t) - f(1)}{t - 1}$, $t \neq 1$

109. $f(x) = \sqrt{5x}$, $\dfrac{f(x) - f(5)}{x - 5}$, $x \neq 5$

110. $f(x) = x^{2/3} + 1$, $\dfrac{f(x) - f(8)}{x - 8}$, $x \neq 8$

In Exercises 111–114, match the data with one of the following functions

$f(x) = cx$, $g(x) = cx^2$, $h(x) = c\sqrt{|x|}$, and $r(x) = \dfrac{c}{x}$

and determine the value of the constant c that will make the function fit the data in the table.

111.

x	-4	-1	0	1	4
y	-32	-2	0	-2	-32

112.

x	-4	-1	0	1	4
y	-1	$-\frac{1}{4}$	0	$\frac{1}{4}$	1

113.

x	-4	-1	0	1	4
y	-8	-32	Undefined	32	8

114.

x	-4	-1	0	1	4
y	6	3	0	3	6

EXPLORATION

TRUE OR FALSE? In Exercises 115–118, determine whether the statement is true or false. Justify your answer.

115. Every relation is a function.

116. Every function is a relation.

117. The domain of the function given by $f(x) = x^4 - 1$ is $(-\infty, \infty)$, and the range of $f(x)$ is $(0, \infty)$.

118. The set of ordered pairs $\{(-8, -2), (-6, 0), (-4, 0), (-2, 2), (0, 4), (2, -2)\}$ represents a function.

119. THINK ABOUT IT Consider

$$f(x) = \sqrt{x - 1} \quad \text{and} \quad g(x) = \dfrac{1}{\sqrt{x - 1}}.$$

Why are the domains of f and g different?

120. THINK ABOUT IT Consider $f(x) = \sqrt{x - 2}$ and $g(x) = \sqrt[3]{x - 2}$. Why are the domains of f and g different?

121. THINK ABOUT IT Given $f(x) = x^2$, is f the independent variable? Why or why not?

122. CAPSTONE

(a) Describe any differences between a *relation* and a *function*.

(b) In your own words, explain the meanings of *domain* and *range*.

In Exercises 123 and 124, determine whether the statements use the word *function* in ways that are mathematically correct. Explain your reasoning.

123. (a) The sales tax on a purchased item is a function of the selling price.

(b) Your score on the next algebra exam is a function of the number of hours you study the night before the exam.

124. (a) The amount in your savings account is a function of your salary.

(b) The speed at which a free-falling baseball strikes the ground is a function of the height from which it was dropped.

The symbol ▮ indicates an example or exercise that highlights algebraic techniques specifically used in calculus.

1.5 ANALYZING GRAPHS OF FUNCTIONS

The Graph of a Function

In Section 1.4, you studied functions from an algebraic point of view. In this section, you will study functions from a graphical perspective.

The **graph of a function** f is the collection of ordered pairs $(x, f(x))$ such that x is in the domain of f. As you study this section, remember that

$x = $ the directed distance from the y-axis

$y = f(x) = $ the directed distance from the x-axis

as shown in Figure 1.52.

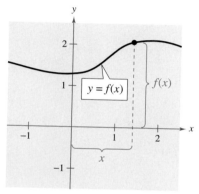

FIGURE 1.52

Example 1   Finding the Domain and Range of a Function

Use the graph of the function f, shown in Figure 1.53, to find (a) the domain of f, (b) the function values $f(-1)$ and $f(2)$, and (c) the range of f.

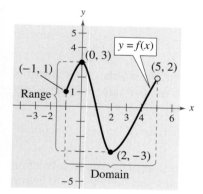

FIGURE 1.53

Solution

a. The closed dot at $(-1, 1)$ indicates that $x = -1$ is in the domain of f, whereas the open dot at $(5, 2)$ indicates that $x = 5$ is not in the domain. So, the domain of f is all x in the interval $[-1, 5)$.

b. Because $(-1, 1)$ is a point on the graph of f, it follows that $f(-1) = 1$. Similarly, because $(2, -3)$ is a point on the graph of f, it follows that $f(2) = -3$.

c. Because the graph does not extend below $f(2) = -3$ or above $f(0) = 3$, the range of f is the interval $[-3, 3]$.

CHECK**Point**   Now try Exercise 9.

The use of dots (open or closed) at the extreme left and right points of a graph indicates that the graph does not extend beyond these points. If no such dots are shown, assume that the graph extends beyond these points.

By the definition of a function, at most one y-value corresponds to a given x-value. This means that the graph of a function cannot have two or more different points with the same x-coordinate, and no two points on the graph of a function can be vertically above or below each other. It follows, then, that a vertical line can intersect the graph of a function at most once. This observation provides a convenient visual test called the **Vertical Line Test** for functions.

> ## Vertical Line Test for Functions
>
> A set of points in a coordinate plane is the graph of y as a function of x if and only if no *vertical* line intersects the graph at more than one point.

Example 2   Vertical Line Test for Functions

Use the Vertical Line Test to decide whether the graphs in Figure 1.54 represent y as a function of x.

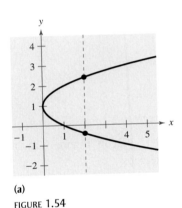

(a)

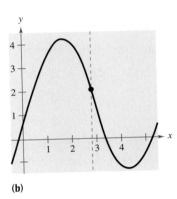

(b)

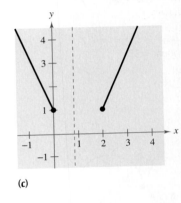

(c)

FIGURE 1.54

Solution

a. This *is not* a graph of y as a function of x, because you can find a vertical line that intersects the graph twice. That is, for a particular input x, there is more than one output y.

b. This *is* a graph of y as a function of x, because every vertical line intersects the graph at most once. That is, for a particular input x, there is at most one output y.

c. This *is* a graph of y as a function of x. (Note that if a vertical line does not intersect the graph, it simply means that the function is undefined for that particular value of x.) That is, for a particular input x, there is at most one output y.

CHECK *Point*   Now try Exercise 17.

TECHNOLOGY

Most graphing utilities are designed to graph functions of x more easily than other types of equations. For instance, the graph shown in Figure 1.54(a) represents the equation $x - (y - 1)^2 = 0$. To use a graphing utility to duplicate this graph, you must first solve the equation for y to obtain $y = 1 \pm \sqrt{x}$, and then graph the two equations $y_1 = 1 + \sqrt{x}$ and $y_2 = 1 - \sqrt{x}$ in the same viewing window.

Algebra Help

To do Example 3, you need to be able to solve equations. You can review the techniques for solving equations in Appendix A.5.

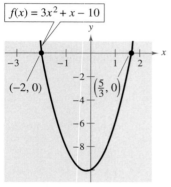

$f(x) = 3x^2 + x - 10$

Zeros of f: $x = -2, x = \frac{5}{3}$

FIGURE **1.55**

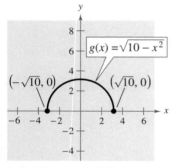

Zeros of g: $x = \pm\sqrt{10}$

FIGURE **1.56**

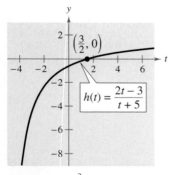

Zero of h: $t = \frac{3}{2}$

FIGURE **1.57**

Zeros of a Function

If the graph of a function of x has an x-intercept at $(a, 0)$, then a is a **zero** of the function.

> ### Zeros of a Function
>
> The **zeros of a function** f of x are the x-values for which $f(x) = 0$.

Example 3   Finding the Zeros of a Function

Find the zeros of each function.

a. $f(x) = 3x^2 + x - 10$ **b.** $g(x) = \sqrt{10 - x^2}$ **c.** $h(t) = \dfrac{2t - 3}{t + 5}$

Solution

To find the zeros of a function, set the function equal to zero and solve for the independent variable.

a.

$3x^2 + x - 10 = 0$	Set $f(x)$ equal to 0.
$(3x - 5)(x + 2) = 0$	Factor.
$3x - 5 = 0$ ⟹ $x = \frac{5}{3}$	Set 1st factor equal to 0.
$x + 2 = 0$ ⟹ $x = -2$	Set 2nd factor equal to 0.

The zeros of f are $x = \frac{5}{3}$ and $x = -2$. In Figure 1.55, note that the graph of f has $\left(\frac{5}{3}, 0\right)$ and $(-2, 0)$ as its x-intercepts.

b.

$\sqrt{10 - x^2} = 0$	Set $g(x)$ equal to 0.
$10 - x^2 = 0$	Square each side.
$10 = x^2$	Add x^2 to each side.
$\pm\sqrt{10} = x$	Extract square roots.

The zeros of g are $x = -\sqrt{10}$ and $x = \sqrt{10}$. In Figure 1.56, note that the graph of g has $\left(-\sqrt{10}, 0\right)$ and $\left(\sqrt{10}, 0\right)$ as its x-intercepts.

c.

$\dfrac{2t - 3}{t + 5} = 0$	Set $h(t)$ equal to 0.
$2t - 3 = 0$	Multiply each side by $t + 5$.
$2t = 3$	Add 3 to each side.
$t = \dfrac{3}{2}$	Divide each side by 2.

The zero of h is $t = \frac{3}{2}$. In Figure 1.57, note that the graph of h has $\left(\frac{3}{2}, 0\right)$ as its t-intercept.

CHECK*Point*   Now try Exercise 23.

Increasing and Decreasing Functions

The more you know about the graph of a function, the more you know about the function itself. Consider the graph shown in Figure 1.58. As you move from *left to right*, this graph falls from $x = -2$ to $x = 0$, is constant from $x = 0$ to $x = 2$, and rises from $x = 2$ to $x = 4$.

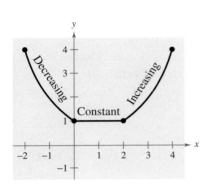

FIGURE **1.58**

Increasing, Decreasing, and Constant Functions

A function f is **increasing** on an interval if, for any x_1 and x_2 in the interval, $x_1 < x_2$ implies $f(x_1) < f(x_2)$.

A function f is **decreasing** on an interval if, for any x_1 and x_2 in the interval, $x_1 < x_2$ implies $f(x_1) > f(x_2)$.

A function f is **constant** on an interval if, for any x_1 and x_2 in the interval, $f(x_1) = f(x_2)$.

Example 4   Increasing and Decreasing Functions

Use the graphs in Figure 1.59 to describe the increasing or decreasing behavior of each function.

Solution

a. This function is increasing over the entire real line.

b. This function is increasing on the interval $(-\infty, -1)$, decreasing on the interval $(-1, 1)$, and increasing on the interval $(1, \infty)$.

c. This function is increasing on the interval $(-\infty, 0)$, constant on the interval $(0, 2)$, and decreasing on the interval $(2, \infty)$.

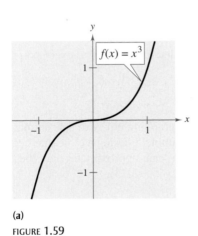
(a)

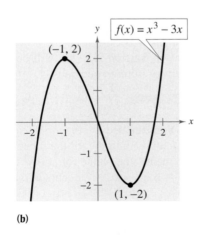
(b)

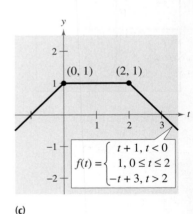
(c)

FIGURE **1.59**

CHECK*Point* Now try Exercise 41.

To help you decide whether a function is increasing, decreasing, or constant on an interval, you can evaluate the function for several values of x. However, calculus is needed to determine, for certain, all intervals on which a function is increasing, decreasing, or constant.

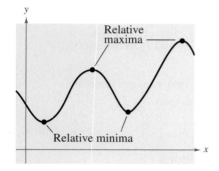

FIGURE **1.60**

The points at which a function changes its increasing, decreasing, or constant behavior are helpful in determining the **relative minimum** or **relative maximum** values of the function.

Definitions of Relative Minimum and Relative Maximum

A function value $f(a)$ is called a **relative minimum** of f if there exists an interval (x_1, x_2) that contains a such that

$$x_1 < x < x_2 \quad \text{implies} \quad f(a) \leq f(x).$$

A function value $f(a)$ is called a **relative maximum** of f if there exists an interval (x_1, x_2) that contains a such that

$$x_1 < x < x_2 \quad \text{implies} \quad f(a) \geq f(x).$$

Figure 1.60 shows several different examples of relative minima and relative maxima. In Section 2.1, you will study a technique for finding the *exact point* at which a second-degree polynomial function has a relative minimum or relative maximum. For the time being, however, you can use a graphing utility to find reasonable approximations of these points.

Example 5 Approximating a Relative Minimum

Use a graphing utility to approximate the relative minimum of the function given by $f(x) = 3x^2 - 4x - 2$.

Solution

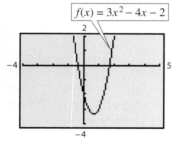

FIGURE **1.61**

The graph of f is shown in Figure 1.61. By using the *zoom* and *trace* features or the *minimum* feature of a graphing utility, you can estimate that the function has a relative minimum at the point

$(0.67, -3.33).$ Relative minimum

Later, in Section 2.1, you will be able to determine that the exact point at which the relative minimum occurs is $\left(\frac{2}{3}, -\frac{10}{3}\right)$.

CHECK*Point* Now try Exercise 57.

You can also use the *table* feature of a graphing utility to approximate numerically the relative minimum of the function in Example 5. Using a table that begins at 0.6 and increments the value of x by 0.01, you can approximate that the minimum of $f(x) = 3x^2 - 4x - 2$ occurs at the point $(0.67, -3.33)$.

TECHNOLOGY

If you use a graphing utility to estimate the *x*- and *y*-values of a relative minimum or relative maximum, the *zoom* feature will often produce graphs that are nearly flat. To overcome this problem, you can manually change the vertical setting of the viewing window. The graph will stretch vertically if the values of Ymin and Ymax are closer together.

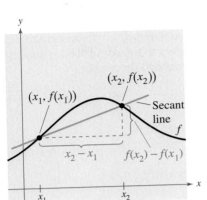

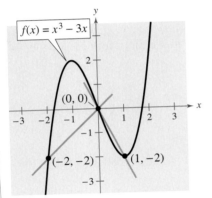

FIGURE **1.62**

FIGURE **1.63**

Average Rate of Change

In Section 1.3, you learned that the slope of a line can be interpreted as a *rate of change*. For a nonlinear graph whose slope changes at each point, the **average rate of change** between any two points $(x_1, f(x_1))$ and $(x_2, f(x_2))$ is the slope of the line through the two points (see Figure 1.62). The line through the two points is called the **secant line,** and the slope of this line is denoted as m_{sec}.

$$\text{Average rate of change of } f \text{ from } x_1 \text{ to } x_2 = \frac{f(x_2) - f(x_1)}{x_2 - x_1}$$

$$= \frac{\text{change in } y}{\text{change in } x}$$

$$= m_{sec}$$

| **Example 6**   **Average Rate of Change of a Function**

Find the average rates of change of $f(x) = x^3 - 3x$ (a) from $x_1 = -2$ to $x_2 = 0$ and (b) from $x_1 = 0$ to $x_2 = 1$ (see Figure 1.63).

Solution

a. The average rate of change of f from $x_1 = -2$ to $x_2 = 0$ is

$$\frac{f(x_2) - f(x_1)}{x_2 - x_1} = \frac{f(0) - f(-2)}{0 - (-2)} = \frac{0 - (-2)}{2} = 1. \qquad \text{Secant line has positive slope.}$$

b. The average rate of change of f from $x_1 = 0$ to $x_2 = 1$ is

$$\frac{f(x_2) - f(x_1)}{x_2 - x_1} = \frac{f(1) - f(0)}{1 - 0} = \frac{-2 - 0}{1} = -2. \qquad \text{Secant line has negative slope.}$$

CHECK**Point**   Now try Exercise 75.

| **Example 7**   **Finding Average Speed**

The distance s (in feet) a moving car is from a stoplight is given by the function $s(t) = 20t^{3/2}$, where t is the time (in seconds). Find the average speed of the car (a) from $t_1 = 0$ to $t_2 = 4$ seconds and (b) from $t_1 = 4$ to $t_2 = 9$ seconds.

Solution

a. The average speed of the car from $t_1 = 0$ to $t_2 = 4$ seconds is

$$\frac{s(t_2) - s(t_1)}{t_2 - t_1} = \frac{s(4) - s(0)}{4 - (0)} = \frac{160 - 0}{4} = 40 \text{ feet per second.}$$

b. The average speed of the car from $t_1 = 4$ to $t_2 = 9$ seconds is

$$\frac{s(t_2) - s(t_1)}{t_2 - t_1} = \frac{s(9) - s(4)}{9 - 4} = \frac{540 - 160}{5} = 76 \text{ feet per second.}$$

CHECK**Point**   Now try Exercise 113.

Even and Odd Functions

In Section 1.2, you studied different types of symmetry of a graph. In the terminology of functions, a function is said to be **even** if its graph is symmetric with respect to the y-axis and to be **odd** if its graph is symmetric with respect to the origin. The symmetry tests in Section 1.2 yield the following tests for even and odd functions.

Tests for Even and Odd Functions

A function $y = f(x)$ is **even** if, for each x in the domain of f,

$$f(-x) = f(x).$$

A function $y = f(x)$ is **odd** if, for each x in the domain of f,

$$f(-x) = -f(x).$$

Example 8   Even and Odd Functions

a. The function $g(x) = x^3 - x$ is odd because $g(-x) = -g(x)$, as follows.

$$g(-x) = (-x)^3 - (-x) \qquad \text{Substitute } -x \text{ for } x.$$

$$= -x^3 + x \qquad \text{Simplify.}$$

$$= -(x^3 - x) \qquad \text{Distributive Property}$$

$$= -g(x) \qquad \text{Test for odd function}$$

b. The function $h(x) = x^2 + 1$ is even because $h(-x) = h(x)$, as follows.

$$h(-x) = (-x)^2 + 1 \qquad \text{Substitute } -x \text{ for } x.$$

$$= x^2 + 1 \qquad \text{Simplify.}$$

$$= h(x) \qquad \text{Test for even function}$$

The graphs and symmetry of these two functions are shown in Figure 1.64.

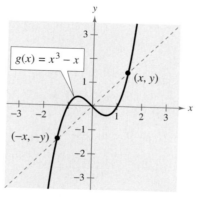

(a) **Symmetric to origin: Odd Function**

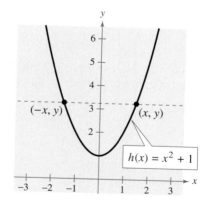

(b) **Symmetric to y-axis: Even Function**

FIGURE **1.64**

CHECK**Point**   Now try Exercise 83.

1.5   EXERCISES

See www.CalcChat.com for worked-out solutions to odd-numbered exercises.

VOCABULARY: Fill in the blanks.

1. The graph of a function f is the collection of _____ _____ $(x, f(x))$ such that x is in the domain of f.

2. The _____ _____ _____ is used to determine whether the graph of an equation is a function of y in terms of x.

3. The _____ of a function f are the values of x for which $f(x) = 0$.

4. A function f is _____ on an interval if, for any x_1 and x_2 in the interval, $x_1 < x_2$ implies $f(x_1) > f(x_2)$.

5. A function value $f(a)$ is a relative _____ of f if there exists an interval (x_1, x_2) containing a such that $x_1 < x < x_2$ implies $f(a) \geq f(x)$.

6. The _____ _____ _____ _____ between any two points $(x_1, f(x_1))$ and $(x_2, f(x_2))$ is the slope of the line through the two points, and this line is called the _____ line.

7. A function f is _____ if, for each x in the domain of f, $f(-x) = -f(x)$.

8. A function f is _____ if its graph is symmetric with respect to the y-axis.

SKILLS AND APPLICATIONS

In Exercises 9–12, use the graph of the function to find the domain and range of f.

9.

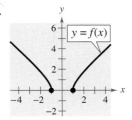

10.

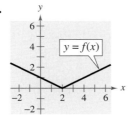

11.

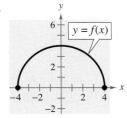

12.

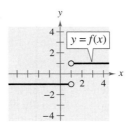

In Exercises 13–16, use the graph of the function to find the domain and range of f and the indicated function values.

13. (a) $f(-2)$ (b) $f(-1)$
   (c) $f\left(\frac{1}{2}\right)$ (d) $f(1)$

14. (a) $f(-1)$ (b) $f(2)$
   (c) $f(0)$ (d) $f(1)$

15. (a) $f(2)$ (b) $f(1)$
   (c) $f(3)$ (d) $f(-1)$

16. (a) $f(-2)$ (b) $f(1)$
   (c) $f(0)$ (d) $f(2)$

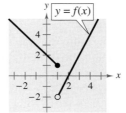

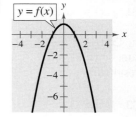

In Exercises 17–22, use the Vertical Line Test to determine whether y is a function of x. To print an enlarged copy of the graph, go to the website *www.mathgraphs.com*.

17. $y = \frac{1}{2}x^2$

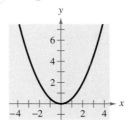

18. $y = \frac{1}{4}x^3$

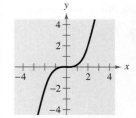

19. $x - y^2 = 1$

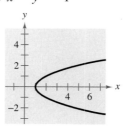

20. $x^2 + y^2 = 25$

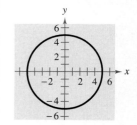

21. $x^2 = 2xy - 1$

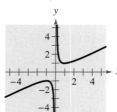

22. $x = |y + 2|$

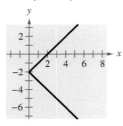

In Exercises 23–32, find the zeros of the function algebraically.

23. $f(x) = 2x^2 - 7x - 30$

24. $f(x) = 3x^2 + 22x - 16$

25. $f(x) = \dfrac{x}{9x^2 - 4}$

26. $f(x) = \dfrac{x^2 - 9x + 14}{4x}$

27. $f(x) = \frac{1}{2}x^3 - x$

28. $f(x) = x^3 - 4x^2 - 9x + 36$

29. $f(x) = 4x^3 - 24x^2 - x + 6$

30. $f(x) = 9x^4 - 25x^2$

31. $f(x) = \sqrt{2x} - 1$

32. $f(x) = \sqrt{3x + 2}$

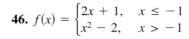 In Exercises 33–38, (a) use a graphing utility to graph the function and find the zeros of the function and (b) verify your results from part (a) algebraically.

33. $f(x) = 3 + \dfrac{5}{x}$

34. $f(x) = x(x - 7)$

35. $f(x) = \sqrt{2x + 11}$

36. $f(x) = \sqrt{3x - 14} - 8$

37. $f(x) = \dfrac{3x - 1}{x - 6}$

38. $f(x) = \dfrac{2x^2 - 9}{3 - x}$

In Exercises 39–46, determine the intervals over which the function is increasing, decreasing, or constant.

39. $f(x) = \frac{3}{2}x$

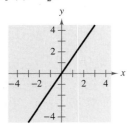

40. $f(x) = x^2 - 4x$

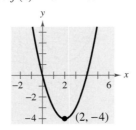

41. $f(x) = x^3 - 3x^2 + 2$

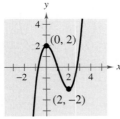

42. $f(x) = \sqrt{x^2 - 1}$

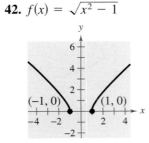

43. $f(x) = |x + 1| + |x - 1|$

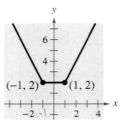

44. $f(x) = \dfrac{x^2 + x + 1}{x + 1}$

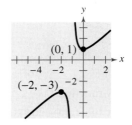

45. $f(x) = \begin{cases} x + 3, & x \le 0 \\ 3, & 0 < x \le 2 \\ 2x + 1, & x > 2 \end{cases}$

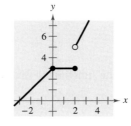

46. $f(x) = \begin{cases} 2x + 1, & x \le -1 \\ x^2 - 2, & x > -1 \end{cases}$

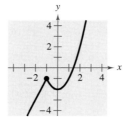

 In Exercises 47–56, (a) use a graphing utility to graph the function and visually determine the intervals over which the function is increasing, decreasing, or constant, and (b) make a table of values to verify whether the function is increasing, decreasing, or constant over the intervals you identified in part (a).

47. $f(x) = 3$

48. $g(x) = x$

49. $g(s) = \dfrac{s^2}{4}$

50. $h(x) = x^2 - 4$

51. $f(t) = -t^4$

52. $f(x) = 3x^4 - 6x^2$

53. $f(x) = \sqrt{1 - x}$

54. $f(x) = x\sqrt{x + 3}$

55. $f(x) = x^{3/2}$

56. $f(x) = x^{2/3}$

In Exercises 57–66, use a graphing utility to graph the function and approximate (to two decimal places) any relative minimum or relative maximum values.

57. $f(x) = (x - 4)(x + 2)$
58. $f(x) = 3x^2 - 2x - 5$
59. $f(x) = -x^2 + 3x - 2$
60. $f(x) = -2x^2 + 9x$
61. $f(x) = x(x - 2)(x + 3)$
62. $f(x) = x^3 - 3x^2 - x + 1$
63. $g(x) = 2x^3 + 3x^2 - 12x$
64. $h(x) = x^3 - 6x^2 + 15$
65. $h(x) = (x - 1)\sqrt{x}$
66. $g(x) = x\sqrt{4 - x}$

In Exercises 67–74, graph the function and determine the interval(s) for which $f(x) \geq 0$.

67. $f(x) = 4 - x$
68. $f(x) = 4x + 2$
69. $f(x) = 9 - x^2$
70. $f(x) = x^2 - 4x$
71. $f(x) = \sqrt{x - 1}$
72. $f(x) = \sqrt{x + 2}$
73. $f(x) = -(1 + |x|)$
74. $f(x) = \frac{1}{2}(2 + |x|)$

In Exercises 75–82, find the average rate of change of the function from x_1 to x_2.

Function	x-Values
75. $f(x) = -2x + 15$	$x_1 = 0, x_2 = 3$
76. $f(x) = 3x + 8$	$x_1 = 0, x_2 = 3$
77. $f(x) = x^2 + 12x - 4$	$x_1 = 1, x_2 = 5$
78. $f(x) = x^2 - 2x + 8$	$x_1 = 1, x_2 = 5$
79. $f(x) = x^3 - 3x^2 - x$	$x_1 = 1, x_2 = 3$
80. $f(x) = -x^3 + 6x^2 + x$	$x_1 = 1, x_2 = 6$
81. $f(x) = -\sqrt{x - 2} + 5$	$x_1 = 3, x_2 = 11$
82. $f(x) = -\sqrt{x + 1} + 3$	$x_1 = 3, x_2 = 8$

In Exercises 83–90, determine whether the function is even, odd, or neither. Then describe the symmetry.

83. $f(x) = x^6 - 2x^2 + 3$
84. $h(x) = x^3 - 5$
85. $g(x) = x^3 - 5x$
86. $f(t) = t^2 + 2t - 3$
87. $h(x) = x\sqrt{x + 5}$
88. $f(x) = x\sqrt{1 - x^2}$
89. $f(s) = 4s^{3/2}$
90. $g(s) = 4s^{2/3}$

In Exercises 91–100, sketch a graph of the function and determine whether it is even, odd, or neither. Verify your answers algebraically.

91. $f(x) = 5$
92. $f(x) = -9$
93. $f(x) = 3x - 2$
94. $f(x) = 5 - 3x$
95. $h(x) = x^2 - 4$
96. $f(x) = -x^2 - 8$

97. $f(x) = \sqrt{1 - x}$
98. $g(t) = \sqrt[3]{t - 1}$
99. $f(x) = |x + 2|$
100. $f(x) = -|x - 5|$

In Exercises 101–104, write the height h of the rectangle as a function of x.

101.
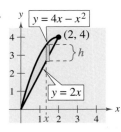
$y = -x^2 + 4x - 1$

102.
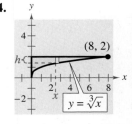
$(1, 3)$
$y = 4x - x^2$

103.
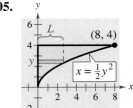
$y = 4x - x^2$
$(2, 4)$
$y = 2x$

104.
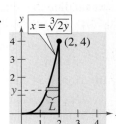
$(8, 2)$
$y = \sqrt[3]{x}$

In Exercises 105–108, write the length L of the rectangle as a function of y.

105.
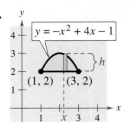
$(8, 4)$
$x = \frac{1}{2}y^2$

106.
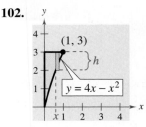
$x = \sqrt[3]{2y}$
$(2, 4)$

107.
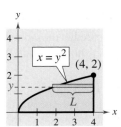
$x = y^2$
$(4, 2)$

108.
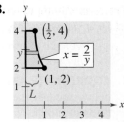
$\left(\frac{1}{2}, 4\right)$
$x = \frac{2}{y}$
$(1, 2)$

109. ELECTRONICS The number of lumens (time rate of flow of light) L from a fluorescent lamp can be approximated by the model

$$L = -0.294x^2 + 97.744x - 664.875, \quad 20 \leq x \leq 90$$

where x is the wattage of the lamp.

(a) Use a graphing utility to graph the function.

(b) Use the graph from part (a) to estimate the wattage necessary to obtain 2000 lumens.

110. DATA ANALYSIS: TEMPERATURE The table shows the temperatures y (in degrees Fahrenheit) in a certain city over a 24-hour period. Let x represent the time of day, where $x = 0$ corresponds to 6 A.M.

Time, x	Temperature, y
0	34
2	50
4	60
6	64
8	63
10	59
12	53
14	46
16	40
18	36
20	34
22	37
24	45

A model that represents these data is given by

$$y = 0.026x^3 - 1.03x^2 + 10.2x + 34, \quad 0 \le x \le 24.$$

(a) Use a graphing utility to create a scatter plot of the data. Then graph the model in the same viewing window.

(b) How well does the model fit the data?

(c) Use the graph to approximate the times when the temperature was increasing and decreasing.

(d) Use the graph to approximate the maximum and minimum temperatures during this 24-hour period.

(e) Could this model be used to predict the temperatures in the city during the next 24-hour period? Why or why not?

111. COORDINATE AXIS SCALE Each function described below models the specified data for the years 1998 through 2008, with $t = 8$ corresponding to 1998. Estimate a reasonable scale for the vertical axis (e.g., hundreds, thousands, millions, etc.) of the graph and justify your answer. (There are many correct answers.)

(a) $f(t)$ represents the average salary of college professors.

(b) $f(t)$ represents the U.S. population.

(c) $f(t)$ represents the percent of the civilian work force that is unemployed.

112. GEOMETRY Corners of equal size are cut from a square with sides of length 8 meters (see figure).

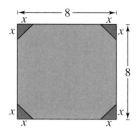
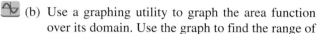

(a) Write the area A of the resulting figure as a function of x. Determine the domain of the function.

 (b) Use a graphing utility to graph the area function over its domain. Use the graph to find the range of the function.

(c) Identify the figure that would result if x were chosen to be the maximum value in the domain of the function. What would be the length of each side of the figure?

113. ENROLLMENT RATE The enrollment rates r of children in preschool in the United States from 1970 through 2005 can be approximated by the model

$$r = -0.021t^2 + 1.44t + 39.3, \quad 0 \le t \le 35$$

where t represents the year, with $t = 0$ corresponding to 1970. (Source: U.S. Census Bureau)

 (a) Use a graphing utility to graph the model.

(b) Find the average rate of change of the model from 1970 through 2005. Interpret your answer in the context of the problem.

114. VEHICLE TECHNOLOGY SALES The estimated revenues r (in millions of dollars) from sales of in-vehicle technologies in the United States from 2003 through 2008 can be approximated by the model

$$r = 157.30t^2 - 397.4t + 6114, \quad 3 \le t \le 8$$

where t represents the year, with $t = 3$ corresponding to 2003. (Source: Consumer Electronics Association)

 (a) Use a graphing utility to graph the model.

(b) Find the average rate of change of the model from 2003 through 2008. Interpret your answer in the context of the problem.

 PHYSICS In Exercises 115–120, (a) use the position equation $s = -16t^2 + v_0t + s_0$ to write a function that represents the situation, (b) use a graphing utility to graph the function, (c) find the average rate of change of the function from t_1 to t_2, (d) describe the slope of the secant line through t_1 and t_2, (e) find the equation of the secant line through t_1 and t_2, and (f) graph the secant line in the same viewing window as your position function.

115. An object is thrown upward from a height of 6 feet at a velocity of 64 feet per second.

$t_1 = 0, t_2 = 3$

116. An object is thrown upward from a height of 6.5 feet at a velocity of 72 feet per second.

$t_1 = 0, t_2 = 4$

117. An object is thrown upward from ground level at a velocity of 120 feet per second.

$t_1 = 3, t_2 = 5$

118. An object is thrown upward from ground level at a velocity of 96 feet per second.

$t_1 = 2, t_2 = 5$

119. An object is dropped from a height of 120 feet.

$t_1 = 0, t_2 = 2$

120. An object is dropped from a height of 80 feet.

$t_1 = 1, t_2 = 2$

EXPLORATION

TRUE OR FALSE? In Exercises 121 and 122, determine whether the statement is true or false. Justify your answer.

121. A function with a square root cannot have a domain that is the set of real numbers.

122. It is possible for an odd function to have the interval $[0, \infty)$ as its domain.

123. If f is an even function, determine whether g is even, odd, or neither. Explain.

(a) $g(x) = -f(x)$ (b) $g(x) = f(-x)$

(c) $g(x) = f(x) - 2$ (d) $g(x) = f(x - 2)$

124. THINK ABOUT IT Does the graph in Exercise 19 represent x as a function of y? Explain.

THINK ABOUT IT In Exercises 125–130, find the coordinates of a second point on the graph of a function f if the given point is on the graph and the function is (a) even and (b) odd.

125. $\left(-\frac{3}{2}, 4\right)$ **126.** $\left(-\frac{5}{3}, -7\right)$

127. $(4, 9)$ **128.** $(5, -1)$

129. $(x, -y)$ **130.** $(2a, 2c)$

131. WRITING Use a graphing utility to graph each function. Write a paragraph describing any similarities and differences you observe among the graphs.

(a) $y = x$ (b) $y = x^2$ (c) $y = x^3$

(d) $y = x^4$ (e) $y = x^5$ (f) $y = x^6$

132. CONJECTURE Use the results of Exercise 131 to make a conjecture about the graphs of $y = x^7$ and $y = x^8$. Use a graphing utility to graph the functions and compare the results with your conjecture.

133. Use the information in Example 7 to find the average speed of the car from $t_1 = 0$ to $t_2 = 9$ seconds. Explain why the result is less than the value obtained in part (b) of Example 7.

134. Graph each of the functions with a graphing utility. Determine whether the function is *even*, *odd*, or *neither*.

$f(x) = x^2 - x^4$

$g(x) = 2x^3 + 1$

$h(x) = x^5 - 2x^3 + x$

$j(x) = 2 - x^6 - x^8$

$k(x) = x^5 - 2x^4 + x - 2$

$p(x) = x^9 + 3x^5 - x^3 + x$

What do you notice about the equations of functions that are odd? What do you notice about the equations of functions that are even? Can you describe a way to identify a function as odd or even by inspecting the equation? Can you describe a way to identify a function as neither odd nor even by inspecting the equation?

135. WRITING Write a short paragraph describing three different functions that represent the behaviors of quantities between 1998 and 2009. Describe one quantity that decreased during this time, one that increased, and one that was constant. Present your results graphically.

136. CAPSTONE Use the graph of the function to answer (a)–(e).

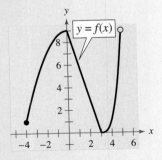

(a) Find the domain and range of f.

(b) Find the zero(s) of f.

(c) Determine the intervals over which f is increasing, decreasing, or constant.

(d) Approximate any relative minimum or relative maximum values of f.

(e) Is f even, odd, or neither?

1.6   A LIBRARY OF PARENT FUNCTIONS

© Getty Images

Linear and Squaring Functions

One of the goals of this text is to enable you to recognize the basic shapes of the graphs of different types of functions. For instance, you know that the graph of the **linear function** $f(x) = ax + b$ is a line with slope $m = a$ and y-intercept at $(0, b)$. The graph of the linear function has the following characteristics.

- The domain of the function is the set of all real numbers.
- The range of the function is the set of all real numbers.
- The graph has an x-intercept of $(-b/m, 0)$ and a y-intercept of $(0, b)$.
- The graph is increasing if $m > 0$, decreasing if $m < 0$, and constant if $m = 0$.

Example 1   Writing a Linear Function

Write the linear function f for which $f(1) = 3$ and $f(4) = 0$.

Solution

To find the equation of the line that passes through $(x_1, y_1) = (1, 3)$ and $(x_2, y_2) = (4, 0)$, first find the slope of the line.

$$m = \frac{y_2 - y_1}{x_2 - x_1} = \frac{0 - 3}{4 - 1} = \frac{-3}{3} = -1$$

Next, use the point-slope form of the equation of a line.

$y - y_1 = m(x - x_1)$ Point-slope form

$y - 3 = -1(x - 1)$ Substitute for x_1, y_1, and m.

$y = -x + 4$ Simplify.

$f(x) = -x + 4$ Function notation

The graph of this function is shown in Figure 1.65.

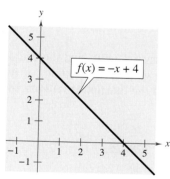

$f(x) = -x + 4$

FIGURE 1.65

CHECK *Point*   Now try Exercise 11.

There are two special types of linear functions, the **constant function** and the **identity function.** A constant function has the form

$$f(x) = c$$

and has the domain of all real numbers with a range consisting of a single real number c. The graph of a constant function is a horizontal line, as shown in Figure 1.66. The identity function has the form

$$f(x) = x.$$

Its domain and range are the set of all real numbers. The identity function has a slope of $m = 1$ and a y-intercept at $(0, 0)$. The graph of the identity function is a line for which each x-coordinate equals the corresponding y-coordinate. The graph is always increasing, as shown in Figure 1.67.

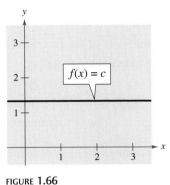

FIGURE **1.66**

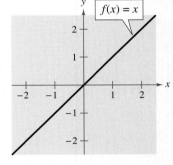

FIGURE **1.67**

The graph of the **squaring function**

$$f(x) = x^2$$

is a U-shaped curve with the following characteristics.

- The domain of the function is the set of all real numbers.
- The range of the function is the set of all nonnegative real numbers.
- The function is even.
- The graph has an intercept at $(0, 0)$.
- The graph is decreasing on the interval $(-\infty, 0)$ and increasing on the interval $(0, \infty)$.
- The graph is symmetric with respect to the y-axis.
- The graph has a relative minimum at $(0, 0)$.

The graph of the squaring function is shown in Figure 1.68.

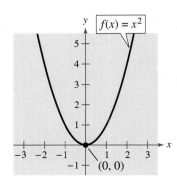

FIGURE **1.68**

Cubic, Square Root, and Reciprocal Functions

The basic characteristics of the graphs of the **cubic, square root,** and **reciprocal functions** are summarized below.

1. The graph of the *cubic* function $f(x) = x^3$ has the following characteristics.
- The domain of the function is the set of all real numbers.
- The range of the function is the set of all real numbers.
- The function is odd.
- The graph has an intercept at $(0, 0)$.
- The graph is increasing on the interval $(-\infty, \infty)$.
- The graph is symmetric with respect to the origin.

The graph of the cubic function is shown in Figure 1.69.

2. The graph of the *square root* function $f(x) = \sqrt{x}$ has the following characteristics.
- The domain of the function is the set of all nonnegative real numbers.
- The range of the function is the set of all nonnegative real numbers.
- The graph has an intercept at $(0, 0)$.
- The graph is increasing on the interval $(0, \infty)$.

The graph of the square root function is shown in Figure 1.70.

3. The graph of the *reciprocal* function $f(x) = \dfrac{1}{x}$ has the following characteristics.
- The domain of the function is $(-\infty, 0) \cup (0, \infty)$.
- The range of the function is $(-\infty, 0) \cup (0, \infty)$.
- The function is odd.
- The graph does not have any intercepts.
- The graph is decreasing on the intervals $(-\infty, 0)$ and $(0, \infty)$.
- The graph is symmetric with respect to the origin.

The graph of the reciprocal function is shown in Figure 1.71.

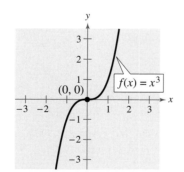

Cubic function
FIGURE **1.69**

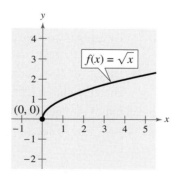

Square root function
FIGURE **1.70**

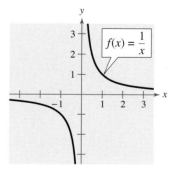

Reciprocal function
FIGURE **1.71**

Step and Piecewise-Defined Functions

Functions whose graphs resemble sets of stairsteps are known as **step functions.** The most famous of the step functions is the **greatest integer function,** which is denoted by $[\![x]\!]$ and defined as

$$f(x) = [\![x]\!] = \textit{the greatest integer less than or equal to } x.$$

Some values of the greatest integer function are as follows.

$$[\![-1]\!] = (\text{greatest integer} \le -1) = -1$$

$$[\![-\tfrac{1}{2}]\!] = (\text{greatest integer} \le -\tfrac{1}{2}) = -1$$

$$[\![\tfrac{1}{10}]\!] = (\text{greatest integer} \le \tfrac{1}{10}) = 0$$

$$[\![1.5]\!] = (\text{greatest integer} \le 1.5) = 1$$

The graph of the greatest integer function

$$f(x) = [\![x]\!]$$

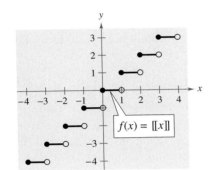

FIGURE **1.72**

has the following characteristics, as shown in Figure 1.72.

- The domain of the function is the set of all real numbers.
- The range of the function is the set of all integers.
- The graph has a y-intercept at $(0, 0)$ and x-intercepts in the interval $[0, 1)$.
- The graph is constant between each pair of consecutive integers.
- The graph jumps vertically one unit at each integer value.

| **Example 2** | **Evaluating a Step Function** |

Evaluate the function when $x = -1, 2,$ and $\tfrac{3}{2}$.

$$f(x) = [\![x]\!] + 1$$

Solution

For $x = -1$, the greatest integer ≤ -1 is -1, so

$$f(-1) = [\![-1]\!] + 1 = -1 + 1 = 0.$$

For $x = 2$, the greatest integer ≤ 2 is 2, so

$$f(2) = [\![2]\!] + 1 = 2 + 1 = 3.$$

For $x = \tfrac{3}{2}$, the greatest integer $\le \tfrac{3}{2}$ is 1, so

$$f\left(\tfrac{3}{2}\right) = [\![\tfrac{3}{2}]\!] + 1 = 1 + 1 = 2.$$

You can verify your answers by examining the graph of $f(x) = [\![x]\!] + 1$ shown in Figure 1.73.

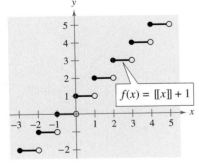

FIGURE **1.73**

CHECK *Point*   Now try Exercise 43.

Recall from Section 1.4 that a piecewise-defined function is defined by two or more equations over a specified domain. To graph a piecewise-defined function, graph each equation separately over the specified domain, as shown in Example 3.

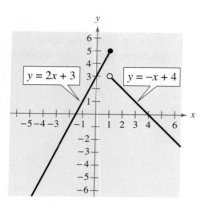

$y = 2x + 3$

$y = -x + 4$

FIGURE 1.74

| **Example 3** **Graphing a Piecewise-Defined Function**

Sketch the graph of

$$f(x) = \begin{cases} 2x + 3, & x \le 1 \\ -x + 4, & x > 1 \end{cases}.$$

Solution

This piecewise-defined function is composed of two linear functions. At $x = 1$ and to the left of $x = 1$ the graph is the line $y = 2x + 3$, and to the right of $x = 1$ the graph is the line $y = -x + 4$, as shown in Figure 1.74. Notice that the point $(1, 5)$ is a solid dot and the point $(1, 3)$ is an open dot. This is because $f(1) = 2(1) + 3 = 5$.

CHECK**Point** Now try Exercise 57.

Parent Functions

The eight graphs shown in Figure 1.75 represent the most commonly used functions in algebra. Familiarity with the basic characteristics of these simple graphs will help you analyze the shapes of more complicated graphs—in particular, graphs obtained from these graphs by the rigid and nonrigid transformations studied in the next section.

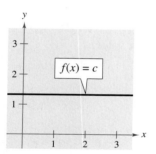

$f(x) = c$

(a) Constant Function

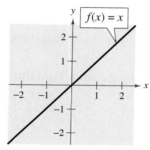

$f(x) = x$

(b) Identity Function

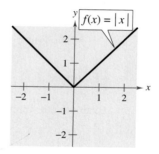

$f(x) = |x|$

(c) Absolute Value Function

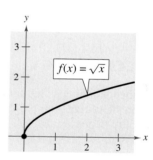

$f(x) = \sqrt{x}$

(d) Square Root Function

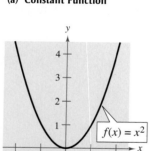

$f(x) = x^2$

(e) Quadratic Function

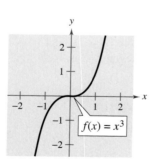

$f(x) = x^3$

(f) Cubic Function

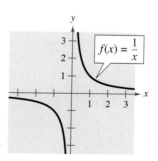

$f(x) = \dfrac{1}{x}$

(g) Reciprocal Function

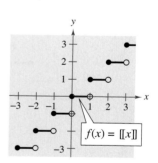

$f(x) = [\![x]\!]$

(h) Greatest Integer Function

FIGURE 1.75

1.6   EXERCISES

See www.CalcChat.com for worked-out solutions to odd-numbered exercises.

VOCABULARY

In Exercises 1–9, match each function with its name.

1. $f(x) = [\![x]\!]$

2. $f(x) = x$

3. $f(x) = 1/x$

4. $f(x) = x^2$

5. $f(x) = \sqrt{x}$

6. $f(x) = c$

7. $f(x) = |x|$

8. $f(x) = x^3$

9. $f(x) = ax + b$

(a) squaring function

(b) square root function

(c) cubic function

(d) linear function

(e) constant function

(f) absolute value function

(g) greatest integer function

(h) reciprocal function

(i) identity function

10. Fill in the blank: The constant function and the identity function are two special types of _____ functions.

SKILLS AND APPLICATIONS

In Exercises 11–18, (a) write the linear function f such that it has the indicated function values and (b) sketch the graph of the function.

11. $f(1) = 4, f(0) = 6$

12. $f(-3) = -8, f(1) = 2$

13. $f(5) = -4, f(-2) = 17$

14. $f(3) = 9, f(-1) = -11$

15. $f(-5) = -1, f(5) = -1$

16. $f(-10) = 12, f(16) = -1$

17. $f\left(\frac{1}{2}\right) = -6, f(4) = -3$

18. $f\left(\frac{2}{3}\right) = -\frac{15}{2}, f(-4) = -11$

In Exercises 19–42, use a graphing utility to graph the function. Be sure to choose an appropriate viewing window.

19. $f(x) = 0.8 - x$

20. $f(x) = 2.5x - 4.25$

21. $f(x) = -\frac{1}{6}x - \frac{5}{2}$

22. $f(x) = \frac{5}{6} - \frac{2}{3}x$

23. $g(x) = -2x^2$

24. $h(x) = 1.5 - x^2$

25. $f(x) = 3x^2 - 1.75$

26. $f(x) = 0.5x^2 + 2$

27. $f(x) = x^3 - 1$

28. $f(x) = 8 - x^3$

29. $f(x) = (x - 1)^3 + 2$

30. $g(x) = 2(x + 3)^3 + 1$

31. $f(x) = 4\sqrt{x}$

32. $f(x) = 4 - 2\sqrt{x}$

33. $g(x) = 2 - \sqrt{x + 4}$

34. $h(x) = \sqrt{x + 2} + 3$

35. $f(x) = -1/x$

36. $f(x) = 4 + (1/x)$

37. $h(x) = 1/(x + 2)$

38. $k(x) = 1/(x - 3)$

39. $g(x) = |x| - 5$

40. $h(x) = 3 - |x|$

41. $f(x) = |x + 4|$

42. $f(x) = |x - 1|$

In Exercises 43–50, evaluate the function for the indicated values.

43. $f(x) = [\![x]\!]$

    (a) $f(2.1)$ (b) $f(2.9)$ (c) $f(-3.1)$ (d) $f\left(\frac{7}{2}\right)$

44. $g(x) = 2[\![x]\!]$

    (a) $g(-3)$ (b) $g(0.25)$ (c) $g(9.5)$ (d) $g\left(\frac{11}{3}\right)$

45. $h(x) = [\![x + 3]\!]$

    (a) $h(-2)$ (b) $h\left(\frac{1}{2}\right)$ (c) $h(4.2)$ (d) $h(-21.6)$

46. $f(x) = 4[\![x]\!] + 7$

    (a) $f(0)$ (b) $f(-1.5)$ (c) $f(6)$ (d) $f\left(\frac{5}{3}\right)$

47. $h(x) = [\![3x - 1]\!]$

    (a) $h(2.5)$ (b) $h(-3.2)$ (c) $h\left(\frac{7}{3}\right)$ (d) $h\left(-\frac{21}{3}\right)$

48. $k(x) = \left[\!\left[\frac{1}{2}x + 6\right]\!\right]$

    (a) $k(5)$ (b) $k(-6.1)$ (c) $k(0.1)$ (d) $k(15)$

49. $g(x) = 3[\![x - 2]\!] + 5$

    (a) $g(-2.7)$ (b) $g(-1)$ (c) $g(0.8)$ (d) $g(14.5)$

50. $g(x) = -7[\![x + 4]\!] + 6$

    (a) $g\left(\frac{1}{8}\right)$ (b) $g(9)$ (c) $g(-4)$ (d) $g\left(\frac{3}{2}\right)$

In Exercises 51–56, sketch the graph of the function.

51. $g(x) = -[\![x]\!]$

52. $g(x) = 4[\![x]\!]$

53. $g(x) = [\![x]\!] - 2$

54. $g(x) = [\![x]\!] - 1$

55. $g(x) = [\![x + 1]\!]$

56. $g(x) = [\![x - 3]\!]$

In Exercises 57–64, graph the function.

57. $f(x) = \begin{cases} 2x + 3, & x < 0 \\ 3 - x, & x \geq 0 \end{cases}$

58. $g(x) = \begin{cases} x + 6, & x \leq -4 \\ \frac{1}{2}x - 4, & x > -4 \end{cases}$

59. $f(x) = \begin{cases} \sqrt{4 + x}, & x < 0 \\ \sqrt{4 - x}, & x \geq 0 \end{cases}$

60. $f(x) = \begin{cases} 1 - (x - 1)^2, & x \leq 2 \\ \sqrt{x - 2}, & x > 2 \end{cases}$

61. $f(x) = \begin{cases} x^2 + 5, & x \leq 1 \\ -x^2 + 4x + 3, & x > 1 \end{cases}$

62. $h(x) = \begin{cases} 3 - x^2, & x < 0 \\ x^2 + 2, & x \geq 0 \end{cases}$

63. $h(x) = \begin{cases} 4 - x^2, & x < -2 \\ 3 + x, & -2 \leq x < 0 \\ x^2 + 1, & x \geq 0 \end{cases}$

64. $k(x) = \begin{cases} 2x + 1, & x \leq -1 \\ 2x^2 - 1, & -1 < x \leq 1 \\ 1 - x^2, & x > 1 \end{cases}$

In Exercises 65–68, (a) use a graphing utility to graph the function, (b) state the domain and range of the function, and (c) describe the pattern of the graph.

65. $s(x) = 2\left(\frac{1}{4}x - \left[\!\left[\frac{1}{4}x\right]\!\right]\right)$

66. $g(x) = 2\left(\frac{1}{4}x - \left[\!\left[\frac{1}{4}x\right]\!\right]\right)^2$

67. $h(x) = 4\left(\frac{1}{2}x - \left[\!\left[\frac{1}{2}x\right]\!\right]\right)$

68. $k(x) = 4\left(\frac{1}{2}x - \left[\!\left[\frac{1}{2}x\right]\!\right]\right)^2$

69. DELIVERY CHARGES The cost of sending an overnight package from Los Angeles to Miami is $23.40 for a package weighing up to but not including 1 pound and $3.75 for each additional pound or portion of a pound. A model for the total cost C (in dollars) of sending the package is $C = 23.40 + 3.75[\![x]\!]$, $x > 0$, where x is the weight in pounds.

(a) Sketch a graph of the model.

(b) Determine the cost of sending a package that weighs 9.25 pounds.

70. DELIVERY CHARGES The cost of sending an overnight package from New York to Atlanta is $22.65 for a package weighing up to but not including 1 pound and $3.70 for each additional pound or portion of a pound.

(a) Use the greatest integer function to create a model for the cost C of overnight delivery of a package weighing x pounds, $x > 0$.

(b) Sketch the graph of the function.

71. WAGES A mechanic is paid $14.00 per hour for regular time and time-and-a-half for overtime. The weekly wage function is given by

$$W(h) = \begin{cases} 14h, & 0 < h \leq 40 \\ 21(h - 40) + 560, & h > 40 \end{cases}$$

where h is the number of hours worked in a week.

(a) Evaluate $W(30)$, $W(40)$, $W(45)$, and $W(50)$.

(b) The company increased the regular work week to 45 hours. What is the new weekly wage function?

72. SNOWSTORM During a nine-hour snowstorm, it snows at a rate of 1 inch per hour for the first 2 hours, at a rate of 2 inches per hour for the next 6 hours, and at a rate of 0.5 inch per hour for the final hour. Write and graph a piecewise-defined function that gives the depth of the snow during the snowstorm. How many inches of snow accumulated from the storm?

73. REVENUE The table shows the monthly revenue y (in thousands of dollars) of a landscaping business for each month of the year 2008, with $x = 1$ representing January.

Month, x	Revenue, y
1	5.2
2	5.6
3	6.6
4	8.3
5	11.5
6	15.8
7	12.8
8	10.1
9	8.6
10	6.9
11	4.5
12	2.7

A mathematical model that represents these data is

$$f(x) = \begin{cases} -1.97x + 26.3 \\ 0.505x^2 - 1.47x + 6.3 \end{cases}$$

(a) Use a graphing utility to graph the model. What is the domain of each part of the piecewise-defined function? How can you tell? Explain your reasoning.

(b) Find $f(5)$ and $f(11)$, and interpret your results in the context of the problem.

(c) How do the values obtained from the model in part (a) compare with the actual data values?

EXPLORATION

TRUE OR FALSE? In Exercises 74 and 75, determine whether the statement is true or false. Justify your answer.

74. A piecewise-defined function will always have at least one x-intercept or at least one y-intercept.

75. A linear equation will always have an x-intercept and a y-intercept.

76. CAPSTONE For each graph of f shown in Figure 1.75, do the following.

(a) Find the domain and range of f.

(b) Find the x- and y-intercepts of the graph of f.

(c) Determine the intervals over which f is increasing, decreasing, or constant.

(d) Determine whether f is even, odd, or neither. Then describe the symmetry.

1.7 TRANSFORMATIONS OF FUNCTIONS

Transtock Inc./Alamy

Shifting Graphs

Many functions have graphs that are simple transformations of the parent graphs summarized in Section 1.6. For example, you can obtain the graph of

$$h(x) = x^2 + 2$$

by shifting the graph of $f(x) = x^2$ *upward* two units, as shown in Figure 1.76. In function notation, h and f are related as follows.

$$h(x) = x^2 + 2 = f(x) + 2 \qquad \text{Upward shift of two units}$$

Similarly, you can obtain the graph of

$$g(x) = (x - 2)^2$$

by shifting the graph of $f(x) = x^2$ to the *right* two units, as shown in Figure 1.77. In this case, the functions g and f have the following relationship.

$$g(x) = (x - 2)^2 = f(x - 2) \qquad \text{Right shift of two units}$$

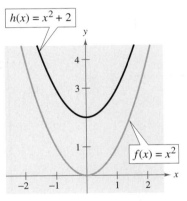

FIGURE **1.76**

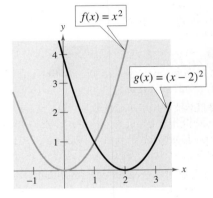

FIGURE **1.77**

The following list summarizes this discussion about horizontal and vertical shifts.

Vertical and Horizontal Shifts

Let c be a positive real number. **Vertical and horizontal shifts** in the graph of $y = f(x)$ are represented as follows.

1. Vertical shift c units *upward:* $h(x) = f(x) + c$

2. Vertical shift c units *downward:* $h(x) = f(x) - c$

3. Horizontal shift c units to the *right:* $h(x) = f(x - c)$

4. Horizontal shift c units to the *left:* $h(x) = f(x + c)$

Some graphs can be obtained from combinations of vertical and horizontal shifts, as demonstrated in Example 1(b). Vertical and horizontal shifts generate a *family of functions*, each with the same shape but at different locations in the plane.

Example 1   Shifts in the Graphs of a Function

Use the graph of $f(x) = x^3$ to sketch the graph of each function.

a. $g(x) = x^3 - 1$ **b.** $h(x) = (x + 2)^3 + 1$

Solution

a. Relative to the graph of $f(x) = x^3$, the graph of

$$g(x) = x^3 - 1$$

is a downward shift of one unit, as shown in Figure 1.78.

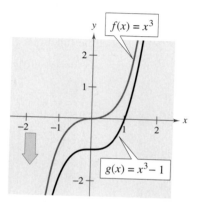

FIGURE **1.78**

b. Relative to the graph of $f(x) = x^3$, the graph of

$$h(x) = (x + 2)^3 + 1$$

involves a left shift of two units and an upward shift of one unit, as shown in Figure 1.79.

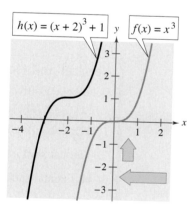

FIGURE **1.79**

CHECK*Point*   Now try Exercise 7.

In Figure 1.79, notice that the same result is obtained if the vertical shift precedes the horizontal shift *or* if the horizontal shift precedes the vertical shift.

Reflecting Graphs

The second common type of transformation is a **reflection.** For instance, if you consider the x-axis to be a mirror, the graph of

$$h(x) = -x^2$$

is the mirror image (or reflection) of the graph of

$$f(x) = x^2,$$

as shown in Figure 1.80.

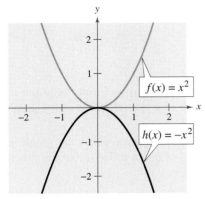

FIGURE **1.80**

Reflections in the Coordinate Axes

Reflections in the coordinate axes of the graph of $y = f(x)$ are represented as follows.

1. Reflection in the x-axis: $h(x) = -f(x)$

2. Reflection in the y-axis: $h(x) = f(-x)$

| **Example 2**   **Finding Equations from Graphs**

The graph of the function given by

$$f(x) = x^4$$

is shown in Figure 1.81. Each of the graphs in Figure 1.82 is a transformation of the graph of f. Find an equation for each of these functions.

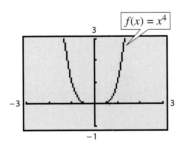

FIGURE **1.81**

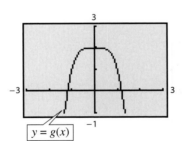

(a)

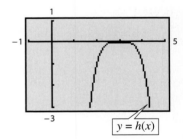

(b)

FIGURE **1.82**

Solution

a. The graph of g is a reflection in the x-axis *followed by* an upward shift of two units of the graph of $f(x) = x^4$. So, the equation for g is

$$g(x) = -x^4 + 2.$$

b. The graph of h is a horizontal shift of three units to the right *followed by* a reflection in the x-axis of the graph of $f(x) = x^4$. So, the equation for h is

$$h(x) = -(x - 3)^4.$$

CHECK*Point*   Now try Exercise 15.

Example 3   Reflections and Shifts

Compare the graph of each function with the graph of $f(x) = \sqrt{x}$.

a. $g(x) = -\sqrt{x}$ **b.** $h(x) = \sqrt{-x}$ **c.** $k(x) = -\sqrt{x+2}$

Algebraic Solution

a. The graph of g is a reflection of the graph of f in the x-axis because

$$g(x) = -\sqrt{x}$$
$$= -f(x).$$

b. The graph of h is a reflection of the graph of f in the y-axis because

$$h(x) = \sqrt{-x}$$
$$= f(-x).$$

c. The graph of k is a left shift of two units followed by a reflection in the x-axis because

$$k(x) = -\sqrt{x+2}$$
$$= -f(x+2).$$

Graphical Solution

a. Graph f and g on the same set of coordinate axes. From the graph in Figure 1.83, you can see that the graph of g is a reflection of the graph of f in the x-axis.

b. Graph f and h on the same set of coordinate axes. From the graph in Figure 1.84, you can see that the graph of h is a reflection of the graph of f in the y-axis.

c. Graph f and k on the same set of coordinate axes. From the graph in Figure 1.85, you can see that the graph of k is a left shift of two units of the graph of f, followed by a reflection in the x-axis.

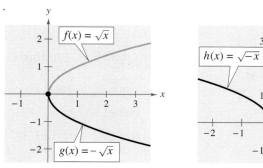

FIGURE **1.83**   FIGURE **1.84**

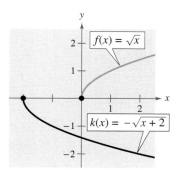

FIGURE **1.85**

CHECK*Point*   Now try Exercise 25.

When sketching the graphs of functions involving square roots, remember that the domain must be restricted to exclude negative numbers inside the radical. For instance, here are the domains of the functions in Example 3.

Domain of $g(x) = -\sqrt{x}$: $x \geq 0$

Domain of $h(x) = \sqrt{-x}$: $x \leq 0$

Domain of $k(x) = -\sqrt{x+2}$: $x \geq -2$

Nonrigid Transformations

Horizontal shifts, vertical shifts, and reflections are **rigid transformations** because the basic shape of the graph is unchanged. These transformations change only the *position* of the graph in the coordinate plane. **Nonrigid transformations** are those that cause a *distortion*—a change in the shape of the original graph. For instance, a nonrigid transformation of the graph of $y = f(x)$ is represented by $g(x) = cf(x)$, where the transformation is a **vertical stretch** if $c > 1$ and a **vertical shrink** if $0 < c < 1$. Another nonrigid transformation of the graph of $y = f(x)$ is represented by $h(x) = f(cx)$, where the transformation is a **horizontal shrink** if $c > 1$ and a **horizontal stretch** if $0 < c < 1$.

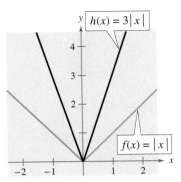

FIGURE **1.86**

| **Example 4**   **Nonrigid Transformations** |

Compare the graph of each function with the graph of $f(x) = |x|$.

a. $h(x) = 3|x|$ **b.** $g(x) = \frac{1}{3}|x|$

Solution

a. Relative to the graph of $f(x) = |x|$, the graph of

$$h(x) = 3|x| = 3f(x)$$

is a vertical stretch (each y-value is multiplied by 3) of the graph of f. (See Figure 1.86.)

b. Similarly, the graph of

$$g(x) = \tfrac{1}{3}|x| = \tfrac{1}{3}f(x)$$

is a vertical shrink $\left(\text{each } y\text{-value is multiplied by } \tfrac{1}{3}\right)$ of the graph of f. (See Figure 1.87.)

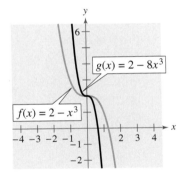

FIGURE **1.87**

CHECK**Point**   Now try Exercise 29.

| **Example 5**   **Nonrigid Transformations** |

Compare the graph of each function with the graph of $f(x) = 2 - x^3$.

a. $g(x) = f(2x)$ **b.** $h(x) = f\!\left(\tfrac{1}{2}x\right)$

Solution

a. Relative to the graph of $f(x) = 2 - x^3$, the graph of

$$g(x) = f(2x) = 2 - (2x)^3 = 2 - 8x^3$$

is a horizontal shrink $(c > 1)$ of the graph of f. (See Figure 1.88.)

b. Similarly, the graph of

$$h(x) = f\!\left(\tfrac{1}{2}x\right) = 2 - \left(\tfrac{1}{2}x\right)^3 = 2 - \tfrac{1}{8}x^3$$

is a horizontal stretch $(0 < c < 1)$ of the graph of f. (See Figure 1.89.)

CHECK**Point**   Now try Exercise 35.

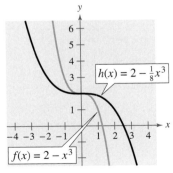

FIGURE **1.88**

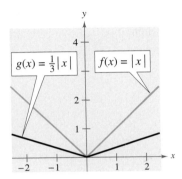

FIGURE **1.89**

1.7   EXERCISES

See www.CalcChat.com for worked-out solutions to odd-numbered exercises.

VOCABULARY

In Exercises 1–5, fill in the blanks.

1. Horizontal shifts, vertical shifts, and reflections are called _____ transformations.

2. A reflection in the x-axis of $y = f(x)$ is represented by $h(x) =$ _____, while a reflection in the y-axis of $y = f(x)$ is represented by $h(x) =$ _____.

3. Transformations that cause a distortion in the shape of the graph of $y = f(x)$ are called _____ transformations.

4. A nonrigid transformation of $y = f(x)$ represented by $h(x) = f(cx)$ is a _____ _____ if $c > 1$ and a _____ _____ if $0 < c < 1$.

5. A nonrigid transformation of $y = f(x)$ represented by $g(x) = cf(x)$ is a _____ _____ if $c > 1$ and a _____ _____ if $0 < c < 1$.

6. Match the rigid transformation of $y = f(x)$ with the correct representation of the graph of h, where $c > 0$.
   (a) $h(x) = f(x) + c$ (i) A horizontal shift of f, c units to the right
   (b) $h(x) = f(x) - c$ (ii) A vertical shift of f, c units downward
   (c) $h(x) = f(x + c)$ (iii) A horizontal shift of f, c units to the left
   (d) $h(x) = f(x - c)$ (iv) A vertical shift of f, c units upward

SKILLS AND APPLICATIONS

7. For each function, sketch (on the same set of coordinate axes) a graph of each function for $c = -1, 1,$ and 3.
   (a) $f(x) = |x| + c$
   (b) $f(x) = |x - c|$
   (c) $f(x) = |x + 4| + c$

8. For each function, sketch (on the same set of coordinate axes) a graph of each function for $c = -3, -1, 1,$ and 3.
   (a) $f(x) = \sqrt{x} + c$
   (b) $f(x) = \sqrt{x - c}$
   (c) $f(x) = \sqrt{x - 3} + c$

9. For each function, sketch (on the same set of coordinate axes) a graph of each function for $c = -2, 0,$ and 2.
   (a) $f(x) = [\![x]\!] + c$
   (b) $f(x) = [\![x + c]\!]$
   (c) $f(x) = [\![x - 1]\!] + c$

10. For each function, sketch (on the same set of coordinate axes) a graph of each function for $c = -3, -1, 1,$ and 3.
    (a) $f(x) = \begin{cases} x^2 + c, & x < 0 \\ -x^2 + c, & x \geq 0 \end{cases}$
    (b) $f(x) = \begin{cases} (x + c)^2, & x < 0 \\ -(x + c)^2, & x \geq 0 \end{cases}$

In Exercises 11–14, use the graph of f to sketch each graph. To print an enlarged copy of the graph, go to the website *www.mathgraphs.com*.

11. (a) $y = f(x) + 2$
    (b) $y = f(x - 2)$
    (c) $y = 2f(x)$
    (d) $y = -f(x)$
    (e) $y = f(x + 3)$
    (f) $y = f(-x)$
    (g) $y = f\left(\frac{1}{2}x\right)$

12. (a) $y = f(-x)$
    (b) $y = f(x) + 4$
    (c) $y = 2f(x)$
    (d) $y = -f(x - 4)$
    (e) $y = f(x) - 3$
    (f) $y = -f(x) - 1$
    (g) $y = f(2x)$

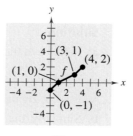

FIGURE FOR 11   **FIGURE FOR 12**

13. (a) $y = f(x) - 1$
    (b) $y = f(x - 1)$
    (c) $y = f(-x)$
    (d) $y = f(x + 1)$
    (e) $y = -f(x - 2)$
    (f) $y = \frac{1}{2}f(x)$
    (g) $y = f(2x)$

14. (a) $y = f(x - 5)$
    (b) $y = -f(x) + 3$
    (c) $y = \frac{1}{3}f(x)$
    (d) $y = -f(x + 1)$
    (e) $y = f(-x)$
    (f) $y = f(x) - 10$
    (g) $y = f\left(\frac{1}{3}x\right)$

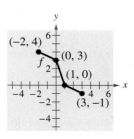

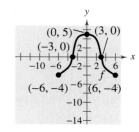

FIGURE FOR 13   FIGURE FOR 14

15. Use the graph of $f(x) = x^2$ to write an equation for each function whose graph is shown.

(a)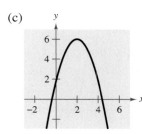

(b)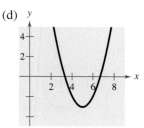

(c)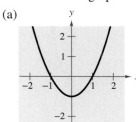

(d)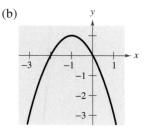

16. Use the graph of $f(x) = x^3$ to write an equation for each function whose graph is shown.

(a)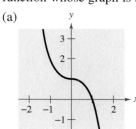

(b)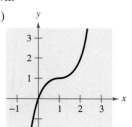

(c)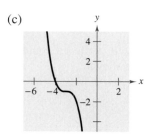

(d)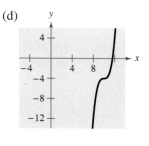

17. Use the graph of $f(x) = |x|$ to write an equation for each function whose graph is shown.

(a)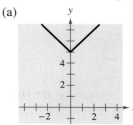

(b)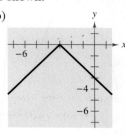

(c)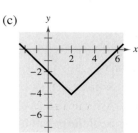

(d)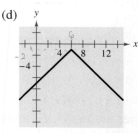

18. Use the graph of $f(x) = \sqrt{x}$ to write an equation for each function whose graph is shown.

(a)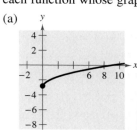

(b)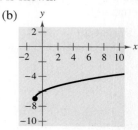

(c)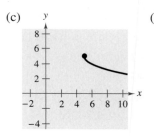

(d)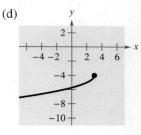

In Exercises 19–24, identify the parent function and the transformation shown in the graph. Write an equation for the function shown in the graph.

19.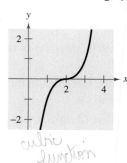

cubic function

$f(x) = (x-2)^3$

Horizontal shift

20.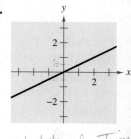

identity function

$f(x) = \frac{1}{2}x$

non rigid transformation (vertical shrink)

21. *greatest integer*

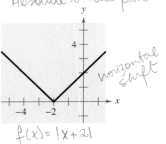

22. $f(x) = [x] + 4$ *vertical shift*

square root function

23. *x-axis reflection and vertical shift*

$f(x) = -\sqrt{x} + 1$

24. *Absolute value function*

horizontal shift

$f(x) = |x + 2|$

In Exercises 25–54, g is related to one of the parent functions described in Section 1.6. (a) Identify the parent function f. (b) Describe the sequence of transformations from f to g. (c) Sketch the graph of g. (d) Use function notation to write g in terms of f.

25. $g(x) = 12 - x^2$
26. $g(x) = (x - 8)^2$
27. $g(x) = x^3 + 7$
28. $g(x) = -x^3 - 1$
29. $g(x) = \frac{2}{3}x^2 + 4$
30. $g(x) = 2(x - 7)^2$
31. $g(x) = 2 - (x + 5)^2$
32. $g(x) = -(x + 10)^2 + 5$
33. $g(x) = 3 + 2(x - 4)^2$
34. $g(x) = -\frac{1}{4}(x + 2)^2 - 2$
35. $g(x) = \sqrt{3x}$
36. $g(x) = \sqrt{\frac{1}{4}x}$
37. $g(x) = (x - 1)^3 + 2$
38. $g(x) = (x + 3)^3 - 10$
39. $g(x) = 3(x - 2)^3$
40. $g(x) = -\frac{1}{2}(x + 1)^3$
41. $g(x) = -|x| - 2$
42. $g(x) = 6 - |x + 5|$
43. $g(x) = -|x + 4| + 8$
44. $g(x) = |-x + 3| + 9$
45. $g(x) = -2|x - 1| - 4$
46. $g(x) = \frac{1}{2}|x - 2| - 3$
47. $g(x) = 3 - [\![x]\!]$
48. $g(x) = 2[\![x + 5]\!]$
49. $g(x) = \sqrt{x - 9}$
50. $g(x) = \sqrt{x + 4} + 8$
51. $g(x) = \sqrt{7 - x} - 2$
52. $g(x) = -\frac{1}{2}\sqrt{x + 3} - 1$
53. $g(x) = \sqrt{\frac{1}{2}x} - 4$
54. $g(x) = \sqrt{3x} + 1$

In Exercises 55–62, write an equation for the function that is described by the given characteristics.

55. The shape of $f(x) = x^2$, but shifted three units to the right and seven units downward

56. The shape of $f(x) = x^2$, but shifted two units to the left, nine units upward, and reflected in the x-axis

57. The shape of $f(x) = x^3$, but shifted 13 units to the right

58. The shape of $f(x) = x^3$, but shifted six units to the left, six units downward, and reflected in the y-axis

59. The shape of $f(x) = |x|$, but shifted 12 units upward and reflected in the x-axis

60. The shape of $f(x) = |x|$, but shifted four units to the left and eight units downward

61. The shape of $f(x) = \sqrt{x}$, but shifted six units to the left and reflected in both the x-axis and the y-axis

62. The shape of $f(x) = \sqrt{x}$, but shifted nine units downward and reflected in both the x-axis and the y-axis

63. Use the graph of $f(x) = x^2$ to write an equation for each function whose graph is shown.

(a) (b)

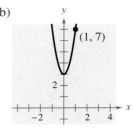

64. Use the graph of $f(x) = x^3$ to write an equation for each function whose graph is shown.

(a) (b)

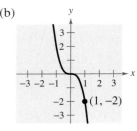

65. Use the graph of $f(x) = |x|$ to write an equation for each function whose graph is shown.

(a) (b)

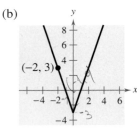

66. Use the graph of $f(x) = \sqrt{x}$ to write an equation for each function whose graph is shown.

(a) (b)

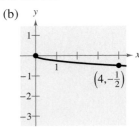

In Exercises 67–72, identify the parent function and the transformation shown in the graph. Write an equation for the function shown in the graph. Then use a graphing utility to verify your answer.

67.
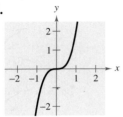

68.
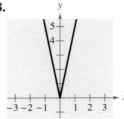

69.
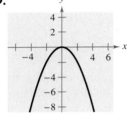

70.
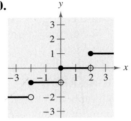

71.
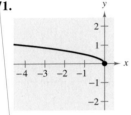

72.
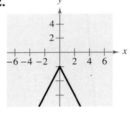

GRAPHICAL ANALYSIS In Exercises 73–76, use the viewing window shown to write a possible equation for the transformation of the parent function.

73.
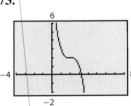

74.
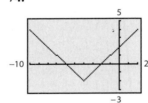

75.
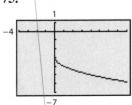

76.
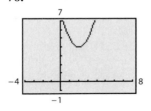

GRAPHICAL REASONING   In Exercises 77 and 78, use the graph of f to sketch the graph of g. To print an enlarged copy of the graph, go to the website *www.mathgraphs.com*.

77.
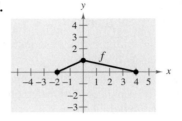

(a) $g(x) = f(x) + 2$ (b) $g(x) = f(x) - 1$
(c) $g(x) = f(-x)$ (d) $g(x) = -2f(x)$
(e) $g(x) = f(4x)$ (f) $g(x) = f\left(\frac{1}{2}x\right)$

78.
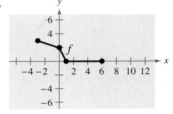

(a) $g(x) = f(x) - 5$ (b) $g(x) = f(x) + \frac{1}{2}$
(c) $g(x) = f(-x)$ (d) $g(x) = -4f(x)$
(e) $g(x) = f(2x) + 1$ (f) $g(x) = f\left(\frac{1}{4}x\right) - 2$

79. MILES DRIVEN   The total numbers of miles M (in billions) driven by vans, pickups, and SUVs (sport utility vehicles) in the United States from 1990 through 2006 can be approximated by the function

$$M = 527 + 128.0\sqrt{t}, \quad 0 \le t \le 16$$

where t represents the year, with $t = 0$ corresponding to 1990.   (Source: U.S. Federal Highway Administration)

(a) Describe the transformation of the parent function $f(x) = \sqrt{x}$. Then use a graphing utility to graph the function over the specified domain.

(b) Find the average rate of change of the function from 1990 to 2006. Interpret your answer in the context of the problem.

(c) Rewrite the function so that $t = 0$ represents 2000. Explain how you got your answer.

(d) Use the model from part (c) to predict the number of miles driven by vans, pickups, and SUVs in 2012. Does your answer seem reasonable? Explain.

80. **MARRIED COUPLES** The numbers N (in thousands) of married couples with stay-at-home mothers from 2000 through 2007 can be approximated by the function

$$N = -24.70(t - 5.99)^2 + 5617, \quad 0 \le t \le 7$$

where t represents the year, with $t = 0$ corresponding to 2000. (Source: U.S. Census Bureau)

 (a) Describe the transformation of the parent function $f(x) = x^2$. Then use a graphing utility to graph the function over the specified domain.

(b) Find the average rate of the change of the function from 2000 to 2007. Interpret your answer in the context of the problem.

(c) Use the model to predict the number of married couples with stay-at-home mothers in 2015. Does your answer seem reasonable? Explain.

EXPLORATION

TRUE OR FALSE? In Exercises 81–84, determine whether the statement is true or false. Justify your answer.

81. The graph of $y = f(-x)$ is a reflection of the graph of $y = f(x)$ in the x-axis.

82. The graph of $y = -f(x)$ is a reflection of the graph of $y = f(x)$ in the y-axis.

83. The graphs of

$$f(x) = |x| + 6$$

and

$$f(x) = |-x| + 6$$

are identical.

84. If the graph of the parent function $f(x) = x^2$ is shifted six units to the right, three units upward, and reflected in the x-axis, then the point $(-2, 19)$ will lie on the graph of the transformation.

85. **DESCRIBING PROFITS** Management originally predicted that the profits from the sales of a new product would be approximated by the graph of the function f shown. The actual profits are shown by the function g along with a verbal description. Use the concepts of transformations of graphs to write g in terms of f.

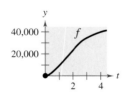

(a) The profits were only three-fourths as large as expected.

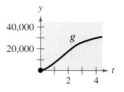

(b) The profits were consistently $10,000 greater than predicted.

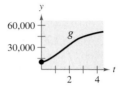

(c) There was a two-year delay in the introduction of the product. After sales began, profits grew as expected.

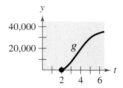

86. **THINK ABOUT IT** You can use either of two methods to graph a function: plotting points or translating a parent function as shown in this section. Which method of graphing do you prefer to use for each function? Explain.

(a) $f(x) = 3x^2 - 4x + 1$

(b) $f(x) = 2(x - 1)^2 - 6$

87. The graph of $y = f(x)$ passes through the points $(0, 1)$, $(1, 2)$, and $(2, 3)$. Find the corresponding points on the graph of $y = f(x + 2) - 1$.

 88. Use a graphing utility to graph f, g, and h in the same viewing window. Before looking at the graphs, try to predict how the graphs of g and h relate to the graph of f.

(a) $f(x) = x^2$, $g(x) = (x - 4)^2$,
   $h(x) = (x - 4)^2 + 3$

(b) $f(x) = x^2$, $g(x) = (x + 1)^2$,
   $h(x) = (x + 1)^2 - 2$

(c) $f(x) = x^2$, $g(x) = (x + 4)^2$,
   $h(x) = (x + 4)^2 + 2$

89. Reverse the order of transformations in Example 2(a). Do you obtain the same graph? Do the same for Example 2(b). Do you obtain the same graph? Explain.

90. **CAPSTONE** Use the fact that the graph of $y = f(x)$ is increasing on the intervals $(-\infty, -1)$ and $(2, \infty)$ and decreasing on the interval $(-1, 2)$ to find the intervals on which the graph is increasing and decreasing. If not possible, state the reason.

(a) $y = f(-x)$ (b) $y = -f(x)$ (c) $y = \frac{1}{2}f(x)$

(d) $y = -f(x - 1)$ (e) $y = f(x - 2) + 1$

1.8 COMBINATIONS OF FUNCTIONS: COMPOSITE FUNCTIONS

What you should learn

- Add, subtract, multiply, and divide functions.
- Find the composition of one function with another function.
- Use combinations and compositions of functions to model and solve real-life problems.

Why you should learn it

Compositions of functions can be used to model and solve real-life problems. For instance, in Exercise 76 on page 91, compositions of functions are used to determine the price of a new hybrid car.

© Jim West/The Image Works

Arithmetic Combinations of Functions

Just as two real numbers can be combined by the operations of addition, subtraction, multiplication, and division to form other real numbers, two *functions* can be combined to create new functions. For example, the functions given by $f(x) = 2x - 3$ and $g(x) = x^2 - 1$ can be combined to form the sum, difference, product, and quotient of f and g.

$$f(x) + g(x) = (2x - 3) + (x^2 - 1)$$
$$= x^2 + 2x - 4 \qquad \text{Sum}$$
$$f(x) - g(x) = (2x - 3) - (x^2 - 1)$$
$$= -x^2 + 2x - 2 \qquad \text{Difference}$$
$$f(x)g(x) = (2x - 3)(x^2 - 1)$$
$$= 2x^3 - 3x^2 - 2x + 3 \qquad \text{Product}$$
$$\frac{f(x)}{g(x)} = \frac{2x - 3}{x^2 - 1}, \qquad x \neq \pm 1 \qquad \text{Quotient}$$

The domain of an **arithmetic combination** of functions f and g consists of all real numbers that are common to the domains of f and g. In the case of the quotient $f(x)/g(x)$, there is the further restriction that $g(x) \neq 0$.

Sum, Difference, Product, and Quotient of Functions

Let f and g be two functions with overlapping domains. Then, for all x common to both domains, the *sum*, *difference*, *product*, and *quotient* of f and g are defined as follows.

1. *Sum:* $(f + g)(x) = f(x) + g(x)$

2. *Difference:* $(f - g)(x) = f(x) - g(x)$

3. *Product:* $(fg)(x) = f(x) \cdot g(x)$

4. *Quotient:* $\left(\dfrac{f}{g}\right)(x) = \dfrac{f(x)}{g(x)}, \quad g(x) \neq 0$

Example 1   Finding the Sum of Two Functions

Given $f(x) = 2x + 1$ and $g(x) = x^2 + 2x - 1$, find $(f + g)(x)$. Then evaluate the sum when $x = 3$.

Solution

$$(f + g)(x) = f(x) + g(x) = (2x + 1) + (x^2 + 2x - 1) = x^2 + 4x$$

When $x = 3$, the value of this sum is

$$(f + g)(3) = 3^2 + 4(3) = 21.$$

CHECK*Point*   Now try Exercise 9(a).

Example 2 Finding the Difference of Two Functions

Given $f(x) = 2x + 1$ and $g(x) = x^2 + 2x - 1$, find $(f - g)(x)$. Then evaluate the difference when $x = 2$.

Solution

The difference of f and g is

$$(f - g)(x) = f(x) - g(x) = (2x + 1) - (x^2 + 2x - 1) = -x^2 + 2.$$

When $x = 2$, the value of this difference is

$$(f - g)(2) = -(2)^2 + 2 = -2.$$

CHECK Point Now try Exercise 9(b).

Example 3 Finding the Product of Two Functions

Given $f(x) = x^2$ and $g(x) = x - 3$, find $(fg)(x)$. Then evaluate the product when $x = 4$.

Solution

$$(fg)(x) = f(x)g(x) = (x^2)(x - 3) = x^3 - 3x^2$$

When $x = 4$, the value of this product is

$$(fg)(4) = 4^3 - 3(4)^2 = 16.$$

CHECK Point Now try Exercise 9(c).

In Examples 1–3, both f and g have domains that consist of all real numbers. So, the domains of $f + g$, $f - g$, and fg are also the set of all real numbers. Remember that any restrictions on the domains of f and g must be considered when forming the sum, difference, product, or quotient of f and g.

Example 4 Finding the Quotients of Two Functions

Find $(f/g)(x)$ and $(g/f)(x)$ for the functions given by $f(x) = \sqrt{x}$ and $g(x) = \sqrt{4 - x^2}$. Then find the domains of f/g and g/f.

Solution

The quotient of f and g is

$$\left(\frac{f}{g}\right)(x) = \frac{f(x)}{g(x)} = \frac{\sqrt{x}}{\sqrt{4 - x^2}}$$

and the quotient of g and f is

$$\left(\frac{g}{f}\right)(x) = \frac{g(x)}{f(x)} = \frac{\sqrt{4 - x^2}}{\sqrt{x}}.$$

Study Tip

Note that the domain of f/g includes $x = 0$, but not $x = 2$, because $x = 2$ yields a zero in the denominator, whereas the domain of g/f includes $x = 2$, but not $x = 0$, because $x = 0$ yields a zero in the denominator.

The domain of f is $[0, \infty)$ and the domain of g is $[-2, 2]$. The intersection of these domains is $[0, 2]$. So, the domains of f/g and g/f are as follows.

Domain of f/g: $[0, 2)$ Domain of g/f: $(0, 2]$

CHECK Point Now try Exercise 9(d).

Composition of Functions

Another way of combining two functions is to form the **composition** of one with the other. For instance, if $f(x) = x^2$ and $g(x) = x + 1$, the composition of f with g is

$$f(g(x)) = f(x + 1)$$
$$= (x + 1)^2.$$

This composition is denoted as $f \circ g$ and reads as "f composed with g."

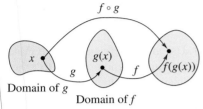

$f \circ g$

x

$g(x)$

g

$f(g(x))$

f

Domain of g

Domain of f

FIGURE **1.90**

Definition of Composition of Two Functions

The **composition** of the function f with the function g is

$$(f \circ g)(x) = f(g(x)).$$

The domain of $f \circ g$ is the set of all x in the domain of g such that $g(x)$ is in the domain of f. (See Figure 1.90.)

Example 5   Composition of Functions

Given $f(x) = x + 2$ and $g(x) = 4 - x^2$, find the following.

a. $(f \circ g)(x)$ **b.** $(g \circ f)(x)$ **c.** $(g \circ f)(-2)$

Solution

a. The composition of f with g is as follows.

$$(f \circ g)(x) = f(g(x)) \qquad \text{Definition of } f \circ g$$
$$= f(4 - x^2) \qquad \text{Definition of } g(x)$$
$$= (4 - x^2) + 2 \qquad \text{Definition of } f(x)$$
$$= -x^2 + 6 \qquad \text{Simplify.}$$

b. The composition of g with f is as follows.

$$(g \circ f)(x) = g(f(x)) \qquad \text{Definition of } g \circ f$$
$$= g(x + 2) \qquad \text{Definition of } f(x)$$
$$= 4 - (x + 2)^2 \qquad \text{Definition of } g(x)$$
$$= 4 - (x^2 + 4x + 4) \qquad \text{Expand.}$$
$$= -x^2 - 4x \qquad \text{Simplify.}$$

Note that, in this case, $(f \circ g)(x) \neq (g \circ f)(x)$.

c. Using the result of part (b), you can write the following.

$$(g \circ f)(-2) = -(-2)^2 - 4(-2) \qquad \text{Substitute.}$$
$$= -4 + 8 \qquad \text{Simplify.}$$
$$= 4 \qquad \text{Simplify.}$$

CHECK**Point**   Now try Exercise 37.

Study Tip

The following tables of values help illustrate the composition $(f \circ g)(x)$ given in Example 5.

x	0	1	2	3
$g(x)$	4	3	0	−5

$g(x)$	4	3	0	−5
$f(g(x))$	6	5	2	−3

x	0	1	2	3
$f(g(x))$	6	5	2	−3

Note that the first two tables can be combined (or "composed") to produce the values given in the third table.

Example 6   Finding the Domain of a Composite Function

Find the domain of $(f \circ g)(x)$ for the functions given by

$$f(x) = x^2 - 9 \qquad \text{and} \qquad g(x) = \sqrt{9 - x^2}.$$

Algebraic Solution

The composition of the functions is as follows.

$$(f \circ g)(x) = f(g(x))$$

$$= f\left(\sqrt{9 - x^2}\right)$$

$$= \left(\sqrt{9 - x^2}\right)^2 - 9$$

$$= 9 - x^2 - 9$$

$$= -x^2$$

From this, it might appear that the domain of the composition is the set of all real numbers. This, however, is not true. Because the domain of f is the set of all real numbers and the domain of g is $[-3, 3]$, the domain of $f \circ g$ is $[-3, 3]$.

Graphical Solution

You can use a graphing utility to graph the composition of the functions $(f \circ g)(x)$ as $y = \left(\sqrt{9 - x^2}\right)^2 - 9$. Enter the functions as follows.

$$y_1 = \sqrt{9 - x^2} \qquad y_2 = y_1^2 - 9$$

Graph y_2, as shown in Figure 1.91. Use the *trace* feature to determine that the x-coordinates of points on the graph extend from -3 to 3. So, you can graphically estimate the domain of $f \circ g$ to be $[-3, 3]$.

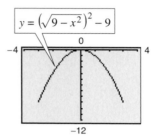

FIGURE **1.91**

CHECK *Point*   Now try Exercise 41.

In Examples 5 and 6, you formed the composition of two given functions. In calculus, it is also important to be able to identify two functions that make up a given composite function. For instance, the function h given by $h(x) = (3x - 5)^3$ is the composition of f with g, where $f(x) = x^3$ and $g(x) = 3x - 5$. That is,

$$h(x) = (3x - 5)^3 = [g(x)]^3 = f(g(x)).$$

Basically, to "decompose" a composite function, look for an "inner" function and an "outer" function. In the function h above, $g(x) = 3x - 5$ is the inner function and $f(x) = x^3$ is the outer function.

Example 7   Decomposing a Composite Function

Write the function given by $h(x) = \dfrac{1}{(x - 2)^2}$ as a composition of two functions.

Solution

One way to write h as a composition of two functions is to take the inner function to be $g(x) = x - 2$ and the outer function to be

$$f(x) = \frac{1}{x^2} = x^{-2}.$$

Then you can write

$$h(x) = \frac{1}{(x - 2)^2} = (x - 2)^{-2} = f(x - 2) = f(g(x)).$$

CHECK *Point*   Now try Exercise 53.

Application

Example 8   Bacteria Count

The number N of bacteria in a refrigerated food is given by

$$N(T) = 20T^2 - 80T + 500, \quad 2 \le T \le 14$$

where T is the temperature of the food in degrees Celsius. When the food is removed from refrigeration, the temperature of the food is given by

$$T(t) = 4t + 2, \quad 0 \le t \le 3$$

where t is the time in hours. (a) Find the composition $N(T(t))$ and interpret its meaning in context. (b) Find the time when the bacteria count reaches 2000.

Solution

a. $N(T(t)) = 20(4t + 2)^2 - 80(4t + 2) + 500$

$$= 20(16t^2 + 16t + 4) - 320t - 160 + 500$$

$$= 320t^2 + 320t + 80 - 320t - 160 + 500$$

$$= 320t^2 + 420$$

The composite function $N(T(t))$ represents the number of bacteria in the food as a function of the amount of time the food has been out of refrigeration.

b. The bacteria count will reach 2000 when $320t^2 + 420 = 2000$. Solve this equation to find that the count will reach 2000 when $t \approx 2.2$ hours. When you solve this equation, note that the negative value is rejected because it is not in the domain of the composite function.

CHECK*Point* Now try Exercise 73.

CLASSROOM DISCUSSION

Analyzing Arithmetic Combinations of Functions

a. Use the graphs of f and $(f + g)$ in Figure 1.92 to make a table showing the values of $g(x)$ when $x = 1, 2, 3, 4, 5,$ and 6. Explain your reasoning.

b. Use the graphs of f and $(f - h)$ in Figure 1.92 to make a table showing the values of $h(x)$ when $x = 1, 2, 3, 4, 5,$ and 6. Explain your reasoning.

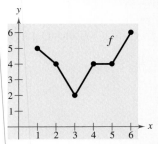
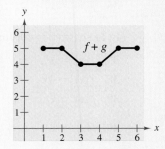
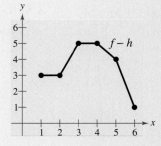

FIGURE 1.92

1.8   EXERCISES

See www.CalcChat.com for worked-out solutions to odd-numbered exercises.

VOCABULARY: Fill in the blanks.

1. Two functions f and g can be combined by the arithmetic operations of _____, _____, _____, and _____ to create new functions.

2. The _____ of the function f with g is $(f \circ g)(x) = f(g(x))$.

3. The domain of $(f \circ g)$ is all x in the domain of g such that _____ is in the domain of f.

4. To decompose a composite function, look for an _____ function and an _____ function.

SKILLS AND APPLICATIONS

In Exercises 5–8, use the graphs of f and g to graph $h(x) = (f + g)(x)$. To print an enlarged copy of the graph, go to the website *www.mathgraphs.com*.

5.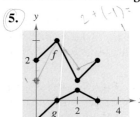

6.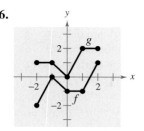

7.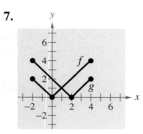

8.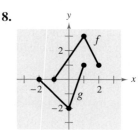

In Exercises 9–16, find (a) $(f + g)(x)$, (b) $(f - g)(x)$, (c) $(fg)(x)$, and (d) $(f/g)(x)$. What is the domain of f/g?

9. $f(x) = x + 2, \quad g(x) = x - 2$

10. $f(x) = 2x - 5, \quad g(x) = 2 - x$

11. $f(x) = x^2, \quad g(x) = 4x - 5$

12. $f(x) = 3x + 1, \quad g(x) = 5x - 4$

13. $f(x) = x^2 + 6, \quad g(x) = \sqrt{1 - x}$

14. $f(x) = \sqrt{x^2 - 4}, \quad g(x) = \dfrac{x^2}{x^2 + 1}$

15. $f(x) = \dfrac{1}{x}, \quad g(x) = \dfrac{1}{x^2}$

16. $f(x) = \dfrac{x}{x + 1}, \quad g(x) = x^3$

In Exercises 17–28, evaluate the indicated function for $f(x) = x^2 + 1$ and $g(x) = x - 4$.

17. $(f + g)(2)$

18. $(f - g)(-1)$

19. $(f - g)(0)$

20. $(f + g)(1)$

21. $(f - g)(3t)$

22. $(f + g)(t - 2)$

23. $(fg)(6)$

24. $(fg)(-6)$

25. $(f/g)(5)$

26. $(f/g)(0)$

27. $(f/g)(-1) - g(3)$

28. $(fg)(5) + f(4)$

In Exercises 29–32, graph the functions f, g, and $f + g$ on the same set of coordinate axes.

29. $f(x) = \frac{1}{2}x, \quad g(x) = x - 1$

30. $f(x) = \frac{1}{3}x, \quad g(x) = -x + 4$

31. $f(x) = x^2, \quad g(x) = -2x$

32. $f(x) = 4 - x^2, \quad g(x) = x$

 GRAPHICAL REASONING In Exercises 33–36, use a graphing utility to graph f, g, and $f + g$ in the same viewing window. Which function contributes most to the magnitude of the sum when $0 \le x \le 2$? Which function contributes most to the magnitude of the sum when $x > 6$?

33. $f(x) = 3x, \quad g(x) = -\dfrac{x^3}{10}$

34. $f(x) = \dfrac{x}{2}, \quad g(x) = \sqrt{x}$

35. $f(x) = 3x + 2, \quad g(x) = -\sqrt{x + 5}$

36. $f(x) = x^2 - \frac{1}{2}, \quad g(x) = -3x^2 - 1$

In Exercises 37–40, find (a) $f \circ g$, (b) $g \circ f$, and (c) $g \circ g$.

37. $f(x) = x^2, \quad g(x) = x - 1$

38. $f(x) = 3x + 5, \quad g(x) = 5 - x$

39. $f(x) = \sqrt[3]{x - 1}, \quad g(x) = x^3 + 1$

40. $f(x) = x^3, \quad g(x) = \dfrac{1}{x}$

In Exercises 41–48, find (a) $f \circ g$ and (b) $g \circ f$. Find the domain of each function and each composite function.

41. $f(x) = \sqrt{x + 4}, \quad g(x) = x^2$

42. $f(x) = \sqrt[3]{x - 5}, \quad g(x) = x^3 + 1$

43. $f(x) = x^2 + 1$, $\quad g(x) = \sqrt{x}$

44. $f(x) = x^{2/3}$, $\quad g(x) = x^6$

45. $f(x) = |x|$, $\quad g(x) = x + 6$

46. $f(x) = |x - 4|$, $\quad g(x) = 3 - x$

47. $f(x) = \dfrac{1}{x}$, $\quad g(x) = x + 3$

48. $f(x) = \dfrac{3}{x^2 - 1}$, $\quad g(x) = x + 1$

In Exercises 49–52, use the graphs of f and g to evaluate the functions.

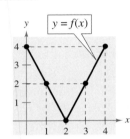
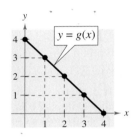

49. (a) $(f + g)(3)$ (b) $(f/g)(2)$

50. (a) $(f - g)(1)$ (b) $(fg)(4)$

51. (a) $(f \circ g)(2)$ (b) $(g \circ f)(2)$

52. (a) $(f \circ g)(1)$ (b) $(g \circ f)(3)$

In Exercises 53–60, find two functions f and g such that $(f \circ g)(x) = h(x)$. (There are many correct answers.)

53. $h(x) = (2x + 1)^2$ **54.** $h(x) = (1 - x)^3$

55. $h(x) = \sqrt[3]{x^2 - 4}$ **56.** $h(x) = \sqrt{9 - x}$

57. $h(x) = \dfrac{1}{x + 2}$ **58.** $h(x) = \dfrac{4}{(5x + 2)^2}$

59. $h(x) = \dfrac{-x^2 + 3}{4 - x^2}$ **60.** $h(x) = \dfrac{27x^3 + 6x}{10 - 27x^3}$

61. STOPPING DISTANCE   The research and development department of an automobile manufacturer has determined that when a driver is required to stop quickly to avoid an accident, the distance (in feet) the car travels during the driver's reaction time is given by $R(x) = \frac{3}{4}x$, where x is the speed of the car in miles per hour. The distance (in feet) traveled while the driver is braking is given by $B(x) = \frac{1}{15}x^2$.

(a) Find the function that represents the total stopping distance T.

(b) Graph the functions R, B, and T on the same set of coordinate axes for $0 \le x \le 60$.

(c) Which function contributes most to the magnitude of the sum at higher speeds? Explain.

62. SALES   From 2003 through 2008, the sales R_1 (in thousands of dollars) for one of two restaurants owned by the same parent company can be modeled by

$$R_1 = 480 - 8t - 0.8t^2, \quad t = 3, 4, 5, 6, 7, 8$$

where $t = 3$ represents 2003. During the same six-year period, the sales R_2 (in thousands of dollars) for the second restaurant can be modeled by

$$R_2 = 254 + 0.78t, \quad t = 3, 4, 5, 6, 7, 8.$$

(a) Write a function R_3 that represents the total sales of the two restaurants owned by the same parent company.

(b) Use a graphing utility to graph R_1, R_2, and R_3 in the same viewing window.

63. VITAL STATISTICS   Let $b(t)$ be the number of births in the United States in year t, and let $d(t)$ represent the number of deaths in the United States in year t, where $t = 0$ corresponds to 2000.

(a) If $p(t)$ is the population of the United States in year t, find the function $c(t)$ that represents the percent change in the population of the United States.

(b) Interpret the value of $c(5)$.

64. PETS   Let $d(t)$ be the number of dogs in the United States in year t, and let $c(t)$ be the number of cats in the United States in year t, where $t = 0$ corresponds to 2000.

(a) Find the function $p(t)$ that represents the total number of dogs and cats in the United States.

(b) Interpret the value of $p(5)$.

(c) Let $n(t)$ represent the population of the United States in year t, where $t = 0$ corresponds to 2000. Find and interpret

$$h(t) = \dfrac{p(t)}{n(t)}.$$

65. MILITARY PERSONNEL   The total numbers of Navy personnel N (in thousands) and Marines personnel M (in thousands) from 2000 through 2007 can be approximated by the models

$$N(t) = 0.192t^3 - 3.88t^2 + 12.9t + 372$$

and

$$M(t) = 0.035t^3 - 0.23t^2 + 1.7t + 172$$

where t represents the year, with $t = 0$ corresponding to 2000.   (Source: Department of Defense)

(a) Find and interpret $(N + M)(t)$. Evaluate this function for $t = 0$, 6, and 12.

(b) Find and interpret $(N - M)(t)$ Evaluate this function for $t = 0$, 6, and 12.

66. SPORTS   The numbers of people playing tennis T (in millions) in the United States from 2000 through 2007 can be approximated by the function

$$T(t) = 0.0233t^4 - 0.3408t^3 + 1.556t^2 - 1.86t + 22.8$$

and the U.S. population P (in millions) from 2000 through 2007 can be approximated by the function $P(t) = 2.78t + 282.5$, where t represents the year, with $t = 0$ corresponding to 2000. (Source: Tennis Industry Association, U.S. Census Bureau)

(a) Find and interpret $h(t) = \dfrac{T(t)}{P(t)}$.

(b) Evaluate the function in part (a) for $t = 0$, 3, and 6.

BIRTHS AND DEATHS   In Exercises 67 and 68, use the table, which shows the total numbers of births B (in thousands) and deaths D (in thousands) in the United States from 1990 through 2006. (Source: U.S. Census Bureau)

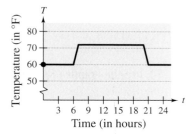

Year, t	Births, B	Deaths, D
1990	4158	2148
1991	4111	2170
1992	4065	2176
1993	4000	2269
1994	3953	2279
1995	3900	2312
1996	3891	2315
1997	3881	2314
1998	3942	2337
1999	3959	2391
2000	4059	2403
2001	4026	2416
2002	4022	2443
2003	4090	2448
2004	4112	2398
2005	4138	2448
2006	4266	2426

The models for these data are

$$B(t) = -0.197t^3 + 8.96t^2 - 90.0t + 4180$$

and

$$D(t) = -1.21t^2 + 38.0t + 2137$$

where t represents the year, with $t = 0$ corresponding to 1990.

67. Find and interpret $(B - D)(t)$.

68. Evaluate $B(t)$, $D(t)$, and $(B - D)(t)$ for the years 2010 and 2012. What does each function value represent?

69. GRAPHICAL   REASONING   An   electronically controlled thermostat in a home is programmed to lower the temperature automatically during the night. The temperature in the house T (in degrees Fahrenheit) is given in terms of t, the time in hours on a 24-hour clock (see figure).

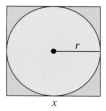

(a) Explain why T is a function of t.

(b) Approximate $T(4)$ and $T(15)$.

(c) The thermostat is reprogrammed to produce a temperature H for which $H(t) = T(t - 1)$. How does this change the temperature?

(d) The thermostat is reprogrammed to produce a temperature H for which $H(t) = T(t) - 1$. How does this change the temperature?

(e) Write a piecewise-defined function that represents the graph.

70. GEOMETRY   A   square   concrete   foundation   is prepared as a base for a cylindrical tank (see figure).

(a) Write the radius r of the tank as a function of the length x of the sides of the square.

(b) Write the area A of the circular base of the tank as a function of the radius r.

(c) Find and interpret $(A \circ r)(x)$.

71. RIPPLES   A pebble is dropped into a calm pond, causing ripples in the form of concentric circles. The radius r (in feet) of the outer ripple is $r(t) = 0.6t$, where t is the time in seconds after the pebble strikes the water. The area A of the circle is given by the function $A(r) = \pi r^2$. Find and interpret $(A \circ r)(t)$.

72. POLLUTION   The spread of a contaminant is increasing in a circular pattern on the surface of a lake. The radius of the contaminant can be modeled by $r(t) = 5.25\sqrt{t}$, where r is the radius in meters and t is the time in hours since contamination.

(a) Find a function that gives the area A of the circular leak in terms of the time t since the spread began.

(b) Find the size of the contaminated area after 36 hours.

(c) Find when the size of the contaminated area is 6250 square meters.

73. BACTERIA COUNT   The number N of bacteria in a refrigerated food is given by

$$N(T) = 10T^2 - 20T + 600, \quad 1 \le T \le 20$$

where T is the temperature of the food in degrees Celsius. When the food is removed from refrigeration, the temperature of the food is given by

$$T(t) = 3t + 2, \quad 0 \le t \le 6$$

where t is the time in hours.

(a) Find the composition $N(T(t))$ and interpret its meaning in context.

(b) Find the bacteria count after 0.5 hour.

(c) Find the time when the bacteria count reaches 1500.

74. COST   The weekly cost C of producing x units in a manufacturing process is given by $C(x) = 60x + 750$. The number of units x produced in t hours is given by $x(t) = 50t$.

(a) Find and interpret $(C \circ x)(t)$.

(b) Find the cost of the units produced in 4 hours.

(c) Find the time that must elapse in order for the cost to increase to $15,000.

75. SALARY   You are a sales representative for a clothing manufacturer. You are paid an annual salary, plus a bonus of 3% of your sales over $500,000. Consider the two functions given by $f(x) = x - 500,000$ and $g(x) = 0.03x$. If x is greater than $500,000, which of the following represents your bonus? Explain your reasoning.

(a) $f(g(x))$ (b) $g(f(x))$

76. CONSUMER AWARENESS   The suggested retail price of a new hybrid car is p dollars. The dealership advertises a factory rebate of $2000 and a 10% discount.

(a) Write a function R in terms of p giving the cost of the hybrid car after receiving the rebate from the factory.

(b) Write a function S in terms of p giving the cost of the hybrid car after receiving the dealership discount.

(c) Form the composite functions $(R \circ S)(p)$ and $(S \circ R)(p)$ and interpret each.

(d) Find $(R \circ S)(20,500)$ and $(S \circ R)(20,500)$. Which yields the lower cost for the hybrid car? Explain.

EXPLORATION

TRUE OR FALSE?   In Exercises 77 and 78, determine whether the statement is true or false. Justify your answer.

77. If $f(x) = x + 1$ and $g(x) = 6x$, then

$$(f \circ g)(x) = (g \circ f)(x).$$

78. If you are given two functions $f(x)$ and $g(x)$, you can calculate $(f \circ g)(x)$ if and only if the range of g is a subset of the domain of f.

In Exercises 79 and 80, three siblings are of three different ages. The oldest is twice the age of the middle sibling, and the middle sibling is six years older than one-half the age of the youngest.

79. (a) Write a composite function that gives the oldest sibling's age in terms of the youngest. Explain how you arrived at your answer.

(b) If the oldest sibling is 16 years old, find the ages of the other two siblings.

80. (a) Write a composite function that gives the youngest sibling's age in terms of the oldest. Explain how you arrived at your answer.

(b) If the youngest sibling is two years old, find the ages of the other two siblings.

81. PROOF   Prove that the product of two odd functions is an even function, and that the product of two even functions is an even function.

82. CONJECTURE   Use examples to hypothesize whether the product of an odd function and an even function is even or odd. Then prove your hypothesis.

83. PROOF

(a) Given a function f, prove that $g(x)$ is even and $h(x)$ is odd, where $g(x) = \frac{1}{2}[f(x) + f(-x)]$ and $h(x) = \frac{1}{2}[f(x) - f(-x)]$.

(b) Use the result of part (a) to prove that any function can be written as a sum of even and odd functions. [*Hint:* Add the two equations in part (a).]

(c) Use the result of part (b) to write each function as a sum of even and odd functions.

$$f(x) = x^2 - 2x + 1, \quad k(x) = \frac{1}{x+1}$$

84. CAPSTONE   Consider the functions $f(x) = x^2$ and $g(x) = \sqrt{x}$.

(a) Find f/g and its domain.

(b) Find $f \circ g$ and $g \circ f$. Find the domain of each composite function. Are they the same? Explain.

1.9 INVERSE FUNCTIONS

Inverse Functions

Recall from Section 1.4 that a function can be represented by a set of ordered pairs. For instance, the function $f(x) = x + 4$ from the set $A = \{1, 2, 3, 4\}$ to the set $B = \{5, 6, 7, 8\}$ can be written as follows.

$$f(x) = x + 4: \ \{(1, 5), (2, 6), (3, 7), (4, 8)\}$$

In this case, by interchanging the first and second coordinates of each of these ordered pairs, you can form the **inverse function** of f, which is denoted by f^{-1}. It is a function from the set B to the set A, and can be written as follows.

$$f^{-1}(x) = x - 4: \ \{(5, 1), (6, 2), (7, 3), (8, 4)\}$$

Note that the domain of f is equal to the range of f^{-1}, and vice versa, as shown in Figure 1.93. Also note that the functions f and f^{-1} have the effect of "undoing" each other. In other words, when you form the composition of f with f^{-1} or the composition of f^{-1} with f, you obtain the identity function.

$$f(f^{-1}(x)) = f(x - 4) = (x - 4) + 4 = x$$

$$f^{-1}(f(x)) = f^{-1}(x + 4) = (x + 4) - 4 = x$$

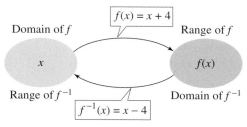

FIGURE 1.93

Example 1   Finding Inverse Functions Informally

Find the inverse function of $f(x) = 4x$. Then verify that both $f(f^{-1}(x))$ and $f^{-1}(f(x))$ are equal to the identity function.

Solution

The function f *multiplies* each input by 4. To "undo" this function, you need to *divide* each input by 4. So, the inverse function of $f(x) = 4x$ is

$$f^{-1}(x) = \frac{x}{4}.$$

You can verify that both $f(f^{-1}(x)) = x$ and $f^{-1}(f(x)) = x$ as follows.

$$f(f^{-1}(x)) = f\left(\frac{x}{4}\right) = 4\left(\frac{x}{4}\right) = x \qquad f^{-1}(f(x)) = f^{-1}(4x) = \frac{4x}{4} = x$$

CHECK *Point*   Now try Exercise 7.

Definition of Inverse Function

Let f and g be two functions such that

$$f(g(x)) = x \qquad \text{for every } x \text{ in the domain of } g$$

and

$$g(f(x)) = x \qquad \text{for every } x \text{ in the domain of } f.$$

Under these conditions, the function g is the **inverse function** of the function f. The function g is denoted by f^{-1} (read "f-inverse"). So,

$$f(f^{-1}(x)) = x \qquad \text{and} \qquad f^{-1}(f(x)) = x.$$

The domain of f must be equal to the range of f^{-1}, and the range of f must be equal to the domain of f^{-1}.

Do not be confused by the use of -1 to denote the inverse function f^{-1}. In this text, whenever f^{-1} is written, it *always* refers to the inverse function of the function f and *not* to the reciprocal of $f(x)$.

If the function g is the inverse function of the function f, it must also be true that the function f is the inverse function of the function g. For this reason, you can say that the functions f and g are *inverse functions of each other*.

Example 2   Verifying Inverse Functions

Which of the functions is the inverse function of $f(x) = \dfrac{5}{x-2}$?

$$g(x) = \frac{x-2}{5} \qquad h(x) = \frac{5}{x} + 2$$

Solution

By forming the composition of f with g, you have

$$f(g(x)) = f\left(\frac{x-2}{5}\right) = \frac{5}{\left(\dfrac{x-2}{5}\right) - 2} = \frac{25}{x-12} \neq x.$$

Because this composition is not equal to the identity function x, it follows that g *is not* the inverse function of f. By forming the composition of f with h, you have

$$f(h(x)) = f\left(\frac{5}{x} + 2\right) = \frac{5}{\left(\dfrac{5}{x} + 2\right) - 2} = \frac{5}{\left(\dfrac{5}{x}\right)} = x.$$

So, it appears that h *is* the inverse function of f. You can confirm this by showing that the composition of h with f is also equal to the identity function, as shown below.

$$h(f(x)) = h\left(\frac{5}{x-2}\right) = \frac{5}{\left(\dfrac{5}{x-2}\right)} + 2 = x - 2 + 2 = x$$

CHECK*Point*   Now try Exercise 19.

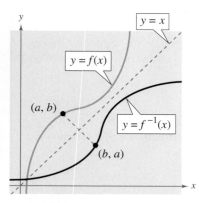

FIGURE **1.94**

The Graph of an Inverse Function

The graphs of a function f and its inverse function f^{-1} are related to each other in the following way. If the point (a, b) lies on the graph of f, then the point (b, a) must lie on the graph of f^{-1}, and vice versa. This means that the graph of f^{-1} is a *reflection* of the graph of f in the line $y = x$, as shown in Figure 1.94.

| **Example 3**   **Finding Inverse Functions Graphically**

Sketch the graphs of the inverse functions $f(x) = 2x - 3$ and $f^{-1}(x) = \frac{1}{2}(x + 3)$ on the same rectangular coordinate system and show that the graphs are reflections of each other in the line $y = x$.

Solution

The graphs of f and f^{-1} are shown in Figure 1.95. It appears that the graphs are reflections of each other in the line $y = x$. You can further verify this reflective property by testing a few points on each graph. Note in the following list that if the point (a, b) is on the graph of f, the point (b, a) is on the graph of f^{-1}.

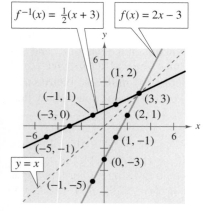

FIGURE **1.95**

Graph of $f(x) = 2x - 3$	*Graph of* $f^{-1}(x) = \frac{1}{2}(x + 3)$
$(-1, -5)$	$(-5, -1)$
$(0, -3)$	$(-3, 0)$
$(1, -1)$	$(-1, 1)$
$(2, 1)$	$(1, 2)$
$(3, 3)$	$(3, 3)$

CHECK**Point**   Now try Exercise 25.

| **Example 4**   **Finding Inverse Functions Graphically**

Sketch the graphs of the inverse functions $f(x) = x^2$ $(x \geq 0)$ and $f^{-1}(x) = \sqrt{x}$ on the same rectangular coordinate system and show that the graphs are reflections of each other in the line $y = x$.

Solution

The graphs of f and f^{-1} are shown in Figure 1.96. It appears that the graphs are reflections of each other in the line $y = x$. You can further verify this reflective property by testing a few points on each graph. Note in the following list that if the point (a, b) is on the graph of f, the point (b, a) is on the graph of f^{-1}.

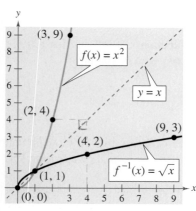

FIGURE **1.96**

Graph of $f(x) = x^2$, $x \geq 0$	*Graph of* $f^{-1}(x) = \sqrt{x}$
$(0, 0)$	$(0, 0)$
$(1, 1)$	$(1, 1)$
$(2, 4)$	$(4, 2)$
$(3, 9)$	$(9, 3)$

Try showing that $f(f^{-1}(x)) = x$ and $f^{-1}(f(x)) = x$.

CHECK**Point**   Now try Exercise 27.

One-to-One Functions

The reflective property of the graphs of inverse functions gives you a nice *geometric* test for determining whether a function has an inverse function. This test is called the **Horizontal Line Test** for inverse functions.

Horizontal Line Test for Inverse Functions

A function f has an inverse function if and only if no *horizontal* line intersects the graph of f at more than one point.

If no horizontal line intersects the graph of f at more than one point, then no y-value is matched with more than one x-value. This is the essential characteristic of what are called **one-to-one functions.**

One-to-One Functions

A function f is **one-to-one** if each value of the dependent variable corresponds to exactly one value of the independent variable. A function f has an inverse function if and only if f is one-to-one.

Consider the function given by $f(x) = x^2$. The table on the left is a table of values for $f(x) = x^2$. The table of values on the right is made up by interchanging the columns of the first table. The table on the right does not represent a function because the input $x = 4$ is matched with two different outputs: $y = -2$ and $y = 2$. So, $f(x) = x^2$ is not one-to-one and does not have an inverse function.

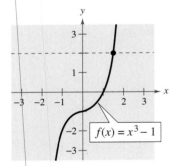

FIGURE 1.97

x	$f(x) = x^2$
-2	4
-1	1
0	0
1	1
2	4
3	9

x	y
4	-2
1	-1
0	0
1	1
4	2
9	3

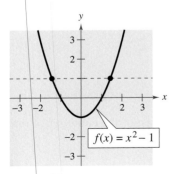

FIGURE 1.98

Example 5   Applying the Horizontal Line Test

a. The graph of the function given by $f(x) = x^3 - 1$ is shown in Figure 1.97. Because no horizontal line intersects the graph of f at more than one point, you can conclude that f *is* a one-to-one function and *does* have an inverse function.

b. The graph of the function given by $f(x) = x^2 - 1$ is shown in Figure 1.98. Because it is possible to find a horizontal line that intersects the graph of f at more than one point, you can conclude that f *is not* a one-to-one function and *does not* have an inverse function.

CHECK*Point*   Now try Exercise 39.

Finding Inverse Functions Algebraically

⚠️ **WARNING / CAUTION**

Note what happens when you try to find the inverse function of a function that is not one-to-one.

$f(x) = x^2 + 1$ Original function

$y = x^2 + 1$ Replace $f(x)$ by y.

$x = y^2 + 1$ Interchange x and y.

$x - 1 = y^2$ Isolate y-term.

$y = \pm\sqrt{x - 1}$ Solve for y.

You obtain two y-values for each x.

For simple functions (such as the one in Example 1), you can find inverse functions by inspection. For more complicated functions, however, it is best to use the following guidelines. The key step in these guidelines is Step 3—interchanging the roles of x and y. This step corresponds to the fact that inverse functions have ordered pairs with the coordinates reversed.

Finding an Inverse Function

1. Use the Horizontal Line Test to decide whether f has an inverse function.

2. In the equation for $f(x)$, replace $f(x)$ by y.

3. Interchange the roles of x and y, and solve for y.

4. Replace y by $f^{-1}(x)$ in the new equation.

5. Verify that f and f^{-1} are inverse functions of each other by showing that the domain of f is equal to the range of f^{-1}, the range of f is equal to the domain of f^{-1}, and $f(f^{-1}(x)) = x$ and $f^{-1}(f(x)) = x$.

Example 6   Finding an Inverse Function Algebraically

Find the inverse function of

$$f(x) = \frac{5 - 3x}{2}.$$

Solution

The graph of f is a line, as shown in Figure 1.99. This graph passes the Horizontal Line Test. So, you know that f is one-to-one and has an inverse function.

$f(x) = \dfrac{5 - 3x}{2}$ Write original function.

$y = \dfrac{5 - 3x}{2}$ Replace $f(x)$ by y.

$x = \dfrac{5 - 3y}{2}$ Interchange x and y.

$2x = 5 - 3y$ Multiply each side by 2.

$3y = 5 - 2x$ Isolate the y-term.

$y = \dfrac{5 - 2x}{3}$ Solve for y.

$f^{-1}(x) = \dfrac{5 - 2x}{3}$ Replace y by $f^{-1}(x)$.

Note that both f and f^{-1} have domains and ranges that consist of the entire set of real numbers. Check that $f(f^{-1}(x)) = x$ and $f^{-1}(f(x)) = x$.

CHECK*Point*   Now try Exercise 63.

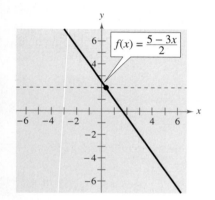

FIGURE **1.99**

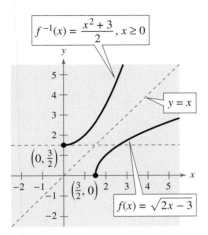

$$f^{-1}(x) = \frac{x^2 + 3}{2}, \; x \geq 0$$

$y = x$

$\left(0, \frac{3}{2}\right)$

$\left(\frac{3}{2}, 0\right)$

$f(x) = \sqrt{2x - 3}$

FIGURE **1.100**

Example 7   Finding an Inverse Function

Find the inverse function of

$$f(x) = \sqrt{2x - 3}.$$

Solution

The graph of f is a curve, as shown in Figure 1.100. Because this graph passes the Horizontal Line Test, you know that f is one-to-one and has an inverse function.

$$f(x) = \sqrt{2x - 3} \qquad \text{Write original function.}$$

$$y = \sqrt{2x - 3} \qquad \text{Replace } f(x) \text{ by } y.$$

$$x = \sqrt{2y - 3} \qquad \text{Interchange } x \text{ and } y.$$

$$x^2 = 2y - 3 \qquad \text{Square each side.}$$

$$2y = x^2 + 3 \qquad \text{Isolate } y.$$

$$y = \frac{x^2 + 3}{2} \qquad \text{Solve for } y.$$

$$f^{-1}(x) = \frac{x^2 + 3}{2}, \quad x \geq 0 \qquad \text{Replace } y \text{ by } f^{-1}(x).$$

The graph of f^{-1} in Figure 1.100 is the reflection of the graph of f in the line $y = x$. Note that the range of f is the interval $[0, \infty)$, which implies that the domain of f^{-1} is the interval $[0, \infty)$. Moreover, the domain of f is the interval $\left[\frac{3}{2}, \infty\right)$, which implies that the range of f^{-1} is the interval $\left[\frac{3}{2}, \infty\right)$. Verify that $f(f^{-1}(x)) = x$ and $f^{-1}(f(x)) = x$.

CHECK*Point*   Now try Exercise 69.

CLASSROOM DISCUSSION

The Existence of an Inverse Function   Write a short paragraph describing why the following functions do or do not have inverse functions.

a. Let x represent the retail price of an item (in dollars), and let $f(x)$ represent the sales tax on the item. Assume that the sales tax is 6% of the retail price *and* that the sales tax is rounded to the nearest cent. Does this function have an inverse function? (*Hint:* Can you undo this function? For instance, if you know that the sales tax is $0.12, can you determine exactly what the retail price is?)

b. Let x represent the temperature in degrees Celsius, and let $f(x)$ represent the temperature in degrees Fahrenheit. Does this function have an inverse function? (*Hint:* The formula for converting from degrees Celsius to degrees Fahrenheit is $F = \frac{9}{5}C + 32$.)

1.9   EXERCISES

VOCABULARY: Fill in the blanks.

1. If the composite functions $f(g(x))$ and $g(f(x))$ both equal x, then the function g is the _____ function of f.
2. The inverse function of f is denoted by _____.
3. The domain of f is the _____ of f^{-1}, and the _____ of f^{-1} is the range of f.
4. The graphs of f and f^{-1} are reflections of each other in the line _____.
5. A function f is _____ if each value of the dependent variable corresponds to exactly one value of the independent variable.
6. A graphical test for the existence of an inverse function of f is called the _____ Line Test.

SKILLS AND APPLICATIONS

In Exercises 7–14, find the inverse function of f informally. Verify that $f(f^{-1}(x)) = x$ and $f^{-1}(f(x)) = x$.

7. $f(x) = 6x$

8. $f(x) = \frac{1}{3}x$

9. $f(x) = x + 9$

10. $f(x) = x - 4$

11. $f(x) = 3x + 1$

12. $f(x) = \dfrac{x - 1}{5}$

13. $f(x) = \sqrt[3]{x}$

14. $f(x) = x^5$

In Exercises 15–18, match the graph of the function with the graph of its inverse function. [The graphs of the inverse functions are labeled (a), (b), (c), and (d).]

(a)
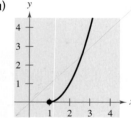

(b)
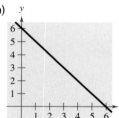

(c)
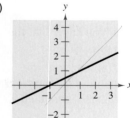

(d)
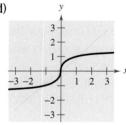

15.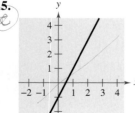

16.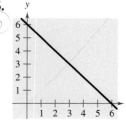

17.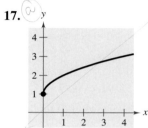

18.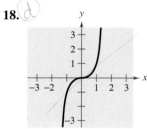

In Exercises 19–22, verify that f and g are inverse functions.

19. $f(x) = -\dfrac{7}{2}x - 3$, $g(x) = -\dfrac{2x + 6}{7}$

20. $f(x) = \dfrac{x - 9}{4}$, $g(x) = 4x + 9$

21. $f(x) = x^3 + 5$, $g(x) = \sqrt[3]{x - 5}$

22. $f(x) = \dfrac{x^3}{2}$, $g(x) = \sqrt[3]{2x}$

In Exercises 23–34, show that f and g are inverse functions (a) algebraically and (b) graphically.

23. $f(x) = 2x$, $g(x) = \dfrac{x}{2}$

24. $f(x) = x - 5$, $g(x) = x + 5$

25. $f(x) = 7x + 1$, $g(x) = \dfrac{x - 1}{7}$

26. $f(x) = 3 - 4x$, $g(x) = \dfrac{3 - x}{4}$

27. $f(x) = \dfrac{x^3}{8}$, $g(x) = \sqrt[3]{8x}$

28. $f(x) = \dfrac{1}{x}$, $g(x) = \dfrac{1}{x}$

29. $f(x) = \sqrt{x - 4}$, $g(x) = x^2 + 4$, $x \geq 0$

30. $f(x) = 1 - x^3$, $g(x) = \sqrt[3]{1 - x}$

31. $f(x) = 9 - x^2$, $x \geq 0$, $g(x) = \sqrt{9 - x}$, $x \leq 9$

32. $f(x) = \dfrac{1}{1+x}$, $x \geq 0$, $g(x) = \dfrac{1-x}{x}$, $0 < x \leq 1$

33. $f(x) = \dfrac{x-1}{x+5}$, $g(x) = -\dfrac{5x+1}{x-1}$

34. $f(x) = \dfrac{x+3}{x-2}$, $g(x) = \dfrac{2x+3}{x-1}$

In Exercises 35 and 36, does the function have an inverse function?

35.

x	−1	0	1	2	3	4
f(x)	−2	1	2	1	−2	−6

No

36.

x	−3	−2	−1	0	2	3
f(x)	10	6	4	1	−3	−10

yes

In Exercises 37 and 38, use the table of values for $y = f(x)$ to complete a table for $y = f^{-1}(x)$.

37.

x	−2	−1	0	1	2	3
f(x)	−2	0	2	4	6	8

38.

x	−3	−2	−1	0	1	2
f(x)	−10	−7	−4	−1	2	5

In Exercises 39–42, does the function have an inverse function?

39.
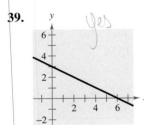
yes

40.
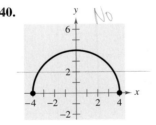
No

41.
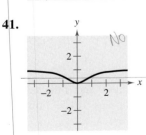
No

42.
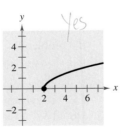
yes

In Exercises 43–48, use a graphing utility to graph the function, and use the Horizontal Line Test to determine whether the function is one-to-one and so has an inverse function.

43. $g(x) = \dfrac{4-x}{6}$ yes.

44. $f(x) = 10$

45. $h(x) = |x+4| - |x-4|$ No.

46. $g(x) = (x+5)^3$

47. $f(x) = -2x\sqrt{16-x^2}$ No

48. $f(x) = \frac{1}{8}(x+2)^2 - 1$

In Exercises 49–62, (a) find the inverse function of f, (b) graph both f and f^{-1} on the same set of coordinate axes, (c) describe the relationship between the graphs of f and f^{-1}, and (d) state the domain and range of f and f^{-1}.

49. $f(x) = 2x - 3$ **50.** $f(x) = 3x + 1$

51. $f(x) = x^5 - 2$ **52.** $f(x) = x^3 + 1$

53. $f(x) = \sqrt{4-x^2}$, $0 \leq x \leq 2$

54. $f(x) = x^2 - 2$, $x \leq 0$

55. $f(x) = \dfrac{4}{x}$ **56.** $f(x) = -\dfrac{2}{x}$

57. $f(x) = \dfrac{x+1}{x-2}$ **58.** $f(x) = \dfrac{x-3}{x+2}$

59. $f(x) = \sqrt[3]{x-1}$ **60.** $f(x) = x^{3/5}$

61. $f(x) = \dfrac{6x+4}{4x+5}$ **62.** $f(x) = \dfrac{8x-4}{2x+6}$

In Exercises 63–76, determine whether the function has an inverse function. If it does, find the inverse function.

63. $f(x) = x^4$ **64.** $f(x) = \dfrac{1}{x^2}$

65. $g(x) = \dfrac{x}{8}$ **66.** $f(x) = 3x + 5$

67. $p(x) = -4$ **68.** $f(x) = \dfrac{3x+4}{5}$

69. $f(x) = (x+3)^2$, $x \geq -3$

70. $q(x) = (x-5)^2$

71. $f(x) = \begin{cases} x+3, & x < 0 \\ 6-x, & x \geq 0 \end{cases}$

72. $f(x) = \begin{cases} -x, & x \leq 0 \\ x^2 - 3x, & x > 0 \end{cases}$

73. $h(x) = -\dfrac{4}{x^2}$ **74.** $f(x) = |x-2|$, $x \leq 2$

75. $f(x) = \sqrt{2x+3}$ **76.** $f(x) = \sqrt{x-2}$

THINK ABOUT IT In Exercises 77–86, restrict the domain of the function f so that the function is one-to-one and has an inverse function. Then find the inverse function f^{-1}. State the domains and ranges of f and f^{-1}. Explain your results. (There are many correct answers.)

77. $f(x) = (x - 2)^2$

78. $f(x) = 1 - x^4$

79. $f(x) = |x + 2|$

80. $f(x) = |x - 5|$

81. $f(x) = (x + 6)^2$

82. $f(x) = (x - 4)^2$

83. $f(x) = -2x^2 + 5$

84. $f(x) = \frac{1}{2}x^2 - 1$

85. $f(x) = |x - 4| + 1$

86. $f(x) = -|x - 1| - 2$

In Exercises 87–92, use the functions given by $f(x) = \frac{1}{8}x - 3$ and $g(x) = x^3$ to find the indicated value or function.

87. $(f^{-1} \circ g^{-1})(1)$

88. $(g^{-1} \circ f^{-1})(-3)$

89. $(f^{-1} \circ f^{-1})(6)$

90. $(g^{-1} \circ g^{-1})(-4)$

91. $(f \circ g)^{-1}$

92. $g^{-1} \circ f^{-1}$

In Exercises 93–96, use the functions given by $f(x) = x + 4$ and $g(x) = 2x - 5$ to find the specified function.

93. $g^{-1} \circ f^{-1}$

94. $f^{-1} \circ g^{-1}$

95. $(f \circ g)^{-1}$

96. $(g \circ f)^{-1}$

97. SHOE SIZES The table shows men's shoe sizes in the United States and the corresponding European shoe sizes. Let $y = f(x)$ represent the function that gives the men's European shoe size in terms of x, the men's U.S. size.

Men's U.S. shoe size	Men's European shoe size
8	41
9	42
10	43
11	45
12	46
13	47

(a) Is f one-to-one? Explain.

(b) Find $f(11)$.

(c) Find $f^{-1}(43)$, if possible.

(d) Find $f(f^{-1}(41))$.

(e) Find $f^{-1}(f(13))$.

98. SHOE SIZES The table shows women's shoe sizes in the United States and the corresponding European shoe sizes. Let $y = g(x)$ represent the function that gives the women's European shoe size in terms of x, the women's U.S. size.

Women's U.S. shoe size	Women's European shoe size
4	35
5	37
6	38
7	39
8	40
9	42

(a) Is g one-to-one? Explain.

(b) Find $g(6)$.

(c) Find $g^{-1}(42)$.

(d) Find $g(g^{-1}(39))$.

(e) Find $g^{-1}(g(5))$.

99. LCD TVS The sales S (in millions of dollars) of LCD televisions in the United States from 2001 through 2007 are shown in the table. The time (in years) is given by t, with $t = 1$ corresponding to 2001. (Source: Consumer Electronics Association)

Year, t	Sales, $S(t)$
1	62
2	246
3	664
4	1579
5	3258
6	8430
7	14,532

(a) Does S^{-1} exist?

(b) If S^{-1} exists, what does it represent in the context of the problem?

(c) If S^{-1} exists, find $S^{-1}(8430)$.

(d) If the table was extended to 2009 and if the sales of LCD televisions for that year was $14,532 million, would S^{-1} exist? Explain.

100. POPULATION The projected populations P (in millions of people) in the United States for 2015 through 2040 are shown in the table. The time (in years) is given by t, with $t = 15$ corresponding to 2015. (Source: U.S. Census Bureau)

Year, t	Population, $P(t)$
15	325.5
20	341.4
25	357.5
30	373.5
35	389.5
40	405.7

(a) Does P^{-1} exist?

(b) If P^{-1} exists, what does it represent in the context of the problem?

(c) If P^{-1} exists, find $P^{-1}(357.5)$.

(d) If the table was extended to 2050 and if the projected population of the U.S. for that year was 373.5 million, would P^{-1} exist? Explain.

101. HOURLY WAGE Your wage is $10.00 per hour plus $0.75 for each unit produced per hour. So, your hourly wage y in terms of the number of units produced x is $y = 10 + 0.75x$.

(a) Find the inverse function. What does each variable represent in the inverse function?

(b) Determine the number of units produced when your hourly wage is $24.25.

102. DIESEL MECHANICS The function given by

$$y = 0.03x^2 + 245.50, \quad 0 < x < 100$$

approximates the exhaust temperature y in degrees Fahrenheit, where x is the percent load for a diesel engine.

(a) Find the inverse function. What does each variable represent in the inverse function?

(b) Use a graphing utility to graph the inverse function.

(c) The exhaust temperature of the engine must not exceed 500 degrees Fahrenheit. What is the percent load interval?

EXPLORATION

TRUE OR FALSE? In Exercises 103 and 104, determine whether the statement is true or false. Justify your answer.

103. If f is an even function, then f^{-1} exists.

104. If the inverse function of f exists and the graph of f has a y-intercept, then the y-intercept of f is an x-intercept of f^{-1}.

105. PROOF Prove that if f and g are one-to-one functions, then $(f \circ g)^{-1}(x) = (g^{-1} \circ f^{-1})(x)$.

106. PROOF Prove that if f is a one-to-one odd function, then f^{-1} is an odd function.

In Exercises 107 and 108, use the graph of the function f to create a table of values for the given points. Then create a second table that can be used to find f^{-1}, and sketch the graph of f^{-1} if possible.

107.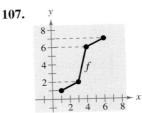

108.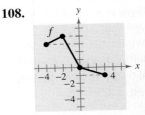

In Exercises 109–112, determine if the situation could be represented by a one-to-one function. If so, write a statement that describes the inverse function.

109. The number of miles n a marathon runner has completed in terms of the time t in hours

110. The population p of South Carolina in terms of the year t from 1960 through 2008

111. The depth of the tide d at a beach in terms of the time t over a 24-hour period

112. The height h in inches of a human born in the year 2000 in terms of his or her age n in years.

113. THINK ABOUT IT The function given by $f(x) = k(2 - x - x^3)$ has an inverse function, and $f^{-1}(3) = -2$. Find k.

114. THINK ABOUT IT Consider the functions given by $f(x) = x + 2$ and $f^{-1}(x) = x - 2$. Evaluate $f(f^{-1}(x))$ and $f^{-1}(f(x))$ for the indicated values of x. What can you conclude about the functions?

x	-10	0	7	45
$f(f^{-1}(x))$				
$f^{-1}(f(x))$				

115. THINK ABOUT IT Restrict the domain of $f(x) = x^2 + 1$ to $x \geq 0$. Use a graphing utility to graph the function. Does the restricted function have an inverse function? Explain.

116. CAPSTONE Describe and correct the error.

Given $f(x) = \sqrt{x} - 6$, then $f^{-1}(x) = \dfrac{1}{\sqrt{x} - 6}$.

1.10 MATHEMATICAL MODELING AND VARIATION

Introduction

You have already studied some techniques for fitting models to data. For instance, in Section 1.3, you learned how to find the equation of a line that passes through two points. In this section, you will study other techniques for fitting models to data: *least squares regression* and *direct and inverse variation*. The resulting models are either polynomial functions or rational functions. (Rational functions will be studied in Chapter 2.)

Example 1 A Mathematical Model

The populations y (in millions) of the United States from 2000 through 2007 are shown in the table. (Source: U.S. Census Bureau)

Year	Population, y
2000	282.4
2001	285.3
2002	288.2
2003	290.9
2004	293.6
2005	296.3
2006	299.2
2007	302.0

A linear model that approximates the data is $y = 2.78t + 282.5$ for $0 \le t \le 7$, where t is the year, with $t = 0$ corresponding to 2000. Plot the actual data *and* the model on the same graph. How closely does the model represent the data?

Solution

The actual data are plotted in Figure 1.101, along with the graph of the linear model. From the graph, it appears that the model is a "good fit" for the actual data. You can see how well the model fits by comparing the actual values of y with the values of y given by the model. The values given by the model are labeled $y*$ in the table below.

t	0	1	2	3	4	5	6	7
y	282.4	285.3	288.2	290.9	293.6	296.3	299.2	302.0
$y*$	282.5	285.3	288.1	290.8	293.6	296.4	299.2	302.0

CHECK**Point** Now try Exercise 11.

Note in Example 1 that you could have chosen any two points to find a line that fits the data. However, the given linear model was found using the *regression* feature of a graphing utility and is the line that *best* fits the data. This concept of a "best-fitting" line is discussed on the next page.

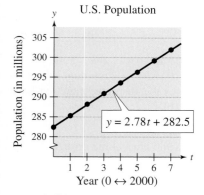

U.S. Population

FIGURE **1.101**

$y = 2.78t + 282.5$

Year (0 ↔ 2000)

Population (in millions)

Least Squares Regression and Graphing Utilities

So far in this text, you have worked with many different types of mathematical models that approximate real-life data. In some instances the model was given (as in Example 1), whereas in other instances you were asked to find the model using simple algebraic techniques or a graphing utility.

To find a model that approximates the data most accurately, statisticians use a measure called the **sum of square differences,** which is the sum of the squares of the differences between actual data values and model values. The "best-fitting" linear model, called the **least squares regression line,** is the one with the least sum of square differences. Recall that you can approximate this line visually by plotting the data points and drawing the line that appears to fit best—or you can enter the data points into a calculator or computer and use the *linear regression* feature of the calculator or computer. When you use the *regression* feature of a graphing calculator or computer program, you will notice that the program may also output an "*r*-value." This *r*-value is the **correlation coefficient** of the data and gives a measure of how well the model fits the data. The closer the value of $|r|$ is to 1, the better the fit.

Example 2   Finding a Least Squares Regression Line

The data in the table show the outstanding household credit market debt D (in trillions of dollars) from 2000 through 2007. Construct a scatter plot that represents the data and find the least squares regression line for the data.   (Source: Board of Governors of the Federal Reserve System)

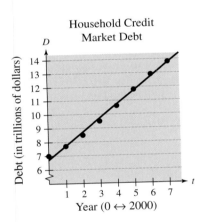

Household Credit
Market Debt

FIGURE **1.102**

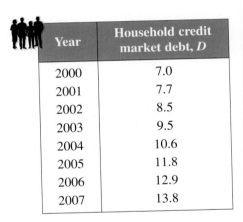

Year	Household credit market debt, D
2000	7.0
2001	7.7
2002	8.5
2003	9.5
2004	10.6
2005	11.8
2006	12.9
2007	13.8

t	D	$D*$
0	7.0	6.7
1	7.7	7.7
2	8.5	8.7
3	9.5	9.7
4	10.6	10.7
5	11.8	11.8
6	12.9	12.8
7	13.8	13.8

Solution

Let $t = 0$ represent 2000. The scatter plot for the points is shown in Figure 1.102. Using the *regression* feature of a graphing utility, you can determine that the equation of the least squares regression line is

$$D = 1.01t + 6.7.$$

To check this model, compare the actual D-values with the D-values given by the model, which are labeled $D*$ in the table at the left. The correlation coefficient for this model is $r \approx 0.997$, which implies that the model is a good fit.

CHECK *Point*   Now try Exercise 17.

Direct Variation

There are two basic types of linear models. The more general model has a y-intercept that is nonzero.

$$y = mx + b, \quad b \neq 0$$

The simpler model

$$y = kx$$

has a y-intercept that is zero. In the simpler model, y is said to **vary directly** as x, or to be **directly proportional** to x.

Direct Variation

The following statements are equivalent.

1. y **varies directly** as x.

2. y is **directly proportional** to x.

3. $y = kx$ for some nonzero constant k.

k is the **constant of variation** or the **constant of proportionality.**

Example 3   Direct Variation

In Pennsylvania, the state income tax is directly proportional to *gross income*. You are working in Pennsylvania and your state income tax deduction is $46.05 for a gross monthly income of $1500. Find a mathematical model that gives the Pennsylvania state income tax in terms of gross income.

Solution

Verbal Model: $\boxed{\text{State income tax}} = \boxed{k} \cdot \boxed{\text{Gross income}}$

Labels:
State income tax $= y$	(dollars)
Gross income $= x$	(dollars)
Income tax rate $= k$	(percent in decimal form)

Equation: $y = kx$

To solve for k, substitute the given information into the equation $y = kx$, and then solve for k.

$$y = kx \qquad \text{Write direct variation model.}$$
$$46.05 = k(1500) \qquad \text{Substitute } y = 46.05 \text{ and } x = 1500.$$
$$0.0307 = k \qquad \text{Simplify.}$$

So, the equation (or model) for state income tax in Pennsylvania is

$$y = 0.0307x.$$

In other words, Pennsylvania has a state income tax rate of 3.07% of gross income. The graph of this equation is shown in Figure 1.103.

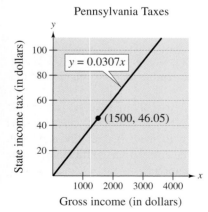

Pennsylvania Taxes

$y = 0.0307x$

(1500, 46.05)

State income tax (in dollars)

Gross income (in dollars)

FIGURE **1.103**

CHECK*Point*▶   Now try Exercise 43.

Direct Variation as an *n*th Power

Another type of direct variation relates one variable to a *power* of another variable. For example, in the formula for the area of a circle

$$A = \pi r^2$$

the area A is directly proportional to the square of the radius r. Note that for this formula, π is the constant of proportionality.

> **Direct Variation as an *n*th Power**
>
> The following statements are equivalent.
>
> **1.** y **varies directly as the *n*th power** of x.
>
> **2.** y is **directly proportional to the *n*th power** of x.
>
> **3.** $y = kx^n$ for some constant k.

Example 4   Direct Variation as *n*th Power

The distance a ball rolls down an inclined plane is directly proportional to the square of the time it rolls. During the first second, the ball rolls 8 feet. (See Figure 1.104.)

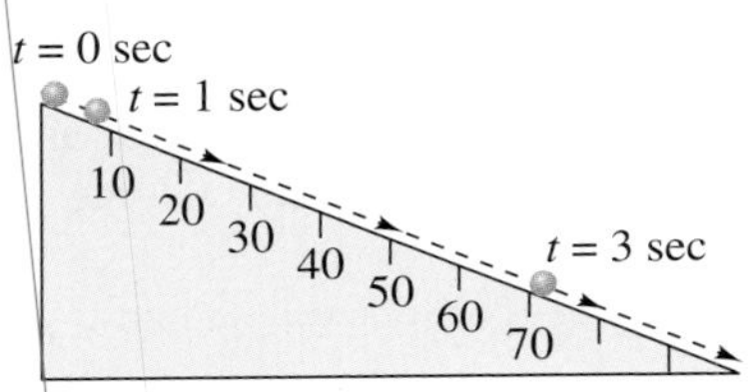

a. Write an equation relating the distance traveled to the time.

b. How far will the ball roll during the first 3 seconds?

Solution

a. Letting d be the distance (in feet) the ball rolls and letting t be the time (in seconds), you have

$$d = kt^2.$$

Now, because $d = 8$ when $t = 1$, you can see that $k = 8$, as follows.

$$d = kt^2$$
$$8 = k(1)^2$$
$$8 = k$$

So, the equation relating distance to time is

$$d = 8t^2.$$

b. When $t = 3$, the distance traveled is $d = 8(3)^2 = 8(9) = 72$ feet.

CHECK Point Now try Exercise 75.

t = 0 sec, t = 1 sec, t = 3 sec; 10 20 30 40 50 60 70

FIGURE 1.104

In Examples 3 and 4, the direct variations are such that an *increase* in one variable corresponds to an *increase* in the other variable. This is also true in the model $d = \frac{1}{5}F, F > 0$, where an increase in F results in an increase in d. You should not, however, assume that this always occurs with direct variation. For example, in the model $y = -3x$, an increase in x results in a *decrease* in y, and yet y is said to vary directly as x.

Inverse Variation

> ### Inverse Variation
>
> The following statements are equivalent.
>
> **1.** y **varies inversely** as x.   **2.** y is **inversely proportional** to x.
>
> **3.** $y = \dfrac{k}{x}$ for some constant k.

If x and y are related by an equation of the form $y = k/x^n$, then y varies inversely as the nth power of x (or y is inversely proportional to the nth power of x).

Some applications of variation involve problems with *both* direct and inverse variation in the same model. These types of models are said to have **combined variation.**

| Example 5   Direct and Inverse Variation

A gas law states that the volume of an enclosed gas varies directly as the temperature *and* inversely as the pressure, as shown in Figure 1.105. The pressure of a gas is 0.75 kilogram per square centimeter when the temperature is 294 K and the volume is 8000 cubic centimeters. (a) Write an equation relating pressure, temperature, and volume. (b) Find the pressure when the temperature is 300 K and the volume is 7000 cubic centimeters.

$P_2 > P_1$ then $V_2 < V_1$

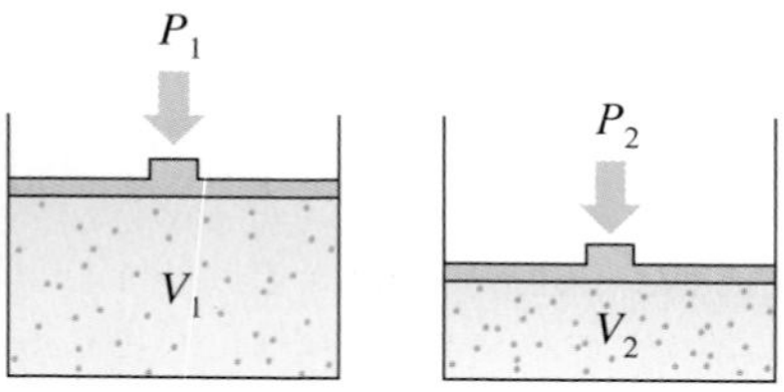

FIGURE **1.105**   If the temperature is held constant and pressure increases, volume decreases.

Solution

a. Let V be volume (in cubic centimeters), let P be pressure (in kilograms per square centimeter), and let T be temperature (in Kelvin). Because V varies directly as T and inversely as P, you have

$$V = \frac{kT}{P}.$$

Now, because $P = 0.75$ when $T = 294$ and $V = 8000$, you have

$$8000 = \frac{k(294)}{0.75}$$

$$k = \frac{6000}{294} = \frac{1000}{49}.$$

So, the equation relating pressure, temperature, and volume is

$$V = \frac{1000}{49}\left(\frac{T}{P}\right).$$

b. When $T = 300$ and $V = 7000$, the pressure is

$$P = \frac{1000}{49}\left(\frac{300}{7000}\right) = \frac{300}{343} \approx 0.87 \text{ kilogram per square centimeter.}$$

CHECK*Point*   Now try Exercise 77.

Joint Variation

In Example 5, note that when a direct variation and an inverse variation occur in the same statement, they are coupled with the word "and." To describe two different *direct* variations in the same statement, the word **jointly** is used.

Joint Variation

The following statements are equivalent.

1. z **varies jointly** as x and y.

2. z is **jointly proportional** to x and y.

3. $z = kxy$ for some constant k.

If x, y, and z are related by an equation of the form

$$z = kx^n y^m$$

then z varies jointly as the nth power of x and the mth power of y.

Example 6   Joint Variation

The *simple* interest for a certain savings account is jointly proportional to the time and the principal. After one quarter (3 months), the interest on a principal of $5000 is $43.75.

a. Write an equation relating the interest, principal, and time.

b. Find the interest after three quarters.

Solution

a. Let $I =$ interest (in dollars), $P =$ principal (in dollars), and $t =$ time (in years). Because I is jointly proportional to P and t, you have

$$I = kPt.$$

For $I = 43.75$, $P = 5000$, and $t = \frac{1}{4}$, you have

$$43.75 = k(5000)\left(\frac{1}{4}\right)$$

which implies that $k = 4(43.75)/5000 = 0.035$. So, the equation relating interest, principal, and time is

$$I = 0.035Pt$$

which is the familiar equation for simple interest where the constant of proportionality, 0.035, represents an annual interest rate of 3.5%.

b. When $P = \$5000$ and $t = \frac{3}{4}$, the interest is

$$I = (0.035)(5000)\left(\frac{3}{4}\right)$$

$$= \$131.25.$$

CHECK Point Now try Exercise 79.

1.10 EXERCISES

See www.CalcChat.com for worked-out solutions to odd-numbered exercises.

VOCABULARY: Fill in the blanks.

1. Two techniques for fitting models to data are called direct _____ and least squares _____.
2. Statisticians use a measure called _____ of_____ _____ to find a model that approximates a set of data most accurately.
3. The linear model with the least sum of square differences is called the _____ _____ _____ line.
4. An *r*-value of a set of data, also called a _____ _____, gives a measure of how well a model fits a set of data.
5. Direct variation models can be described as "*y* varies directly as *x*," or "*y* is _____ _____ to *x*."
6. In direct variation models of the form $y = kx$, k is called the _____ of _____.
7. The direct variation model $y = kx^n$ can be described as "*y* varies directly as the *n*th power of *x*," or "*y* is _____ _____ to the *n*th power of *x*."
8. The mathematical model $y = \dfrac{k}{x}$ is an example of _____ variation.
9. Mathematical models that involve both direct and inverse variation are said to have _____ variation.
10. The joint variation model $z = kxy$ can be described as "*z* varies jointly as *x* and *y*," or "*z* is _____ _____ to *x* and *y*."

SKILLS AND APPLICATIONS

11. **EMPLOYMENT** The total numbers of people (in thousands) in the U.S. civilian labor force from 1992 through 2007 are given by the following ordered pairs.

(1992, 128,105)	(2000, 142,583)
(1993, 129,200)	(2001, 143,734)
(1994, 131,056)	(2002, 144,863)
(1995, 132,304)	(2003, 146,510)
(1996, 133,943)	(2004, 147,401)
(1997, 136,297)	(2005, 149,320)
(1998, 137,673)	(2006, 151,428)
(1999, 139,368)	(2007, 153,124)

A linear model that approximates the data is $y = 1695.9t + 124{,}320$, where *y* represents the number of employees (in thousands) and $t = 2$ represents 1992. Plot the actual data and the model on the same set of coordinate axes. How closely does the model represent the data? (Source: U.S. Bureau of Labor Statistics)

12. **SPORTS** The winning times (in minutes) in the women's 400-meter freestyle swimming event in the Olympics from 1948 through 2008 are given by the following ordered pairs.

(1948, 5.30)	(1972, 4.32)	(1996, 4.12)
(1952, 5.20)	(1976, 4.16)	(2000, 4.10)
(1956, 4.91)	(1980, 4.15)	(2004, 4.09)
(1960, 4.84)	(1984, 4.12)	(2008, 4.05)
(1964, 4.72)	(1988, 4.06)	
(1968, 4.53)	(1992, 4.12)	

A linear model that approximates the data is $y = -0.020t + 5.00$, where *y* represents the winning time (in minutes) and $t = 0$ represents 1950. Plot the actual data and the model on the same set of coordinate axes. How closely does the model represent the data? Does it appear that another type of model may be a better fit? Explain. (Source: International Olympic Committee)

In Exercises 13–16, sketch the line that you think best approximates the data in the scatter plot. Then find an equation of the line. To print an enlarged copy of the graph, go to the website *www.mathgraphs.com*.

13.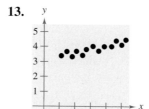

14.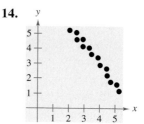

15.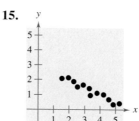

16.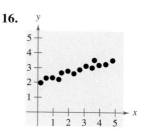

17. SPORTS The lengths (in feet) of the winning men's discus throws in the Olympics from 1920 through 2008 are listed below. (Source: International Olympic Committee)

1920	146.6	1956	184.9	1984	218.5
1924	151.3	1960	194.2	1988	225.8
1928	155.3	1964	200.1	1992	213.7
1932	162.3	1968	212.5	1996	227.7
1936	165.6	1972	211.3	2000	227.3
1948	173.2	1976	221.5	2004	229.3
1952	180.5	1980	218.7	2008	225.8

(a) Sketch a scatter plot of the data. Let y represent the length of the winning discus throw (in feet) and let $t = 20$ represent 1920.

(b) Use a straightedge to sketch the best-fitting line through the points and find an equation of the line.

(c) Use the *regression* feature of a graphing utility to find the least squares regression line that fits the data.

(d) Compare the linear model you found in part (b) with the linear model given by the graphing utility in part (c).

(e) Use the models from parts (b) and (c) to estimate the winning men's discus throw in the year 2012.

 18. SALES The total sales (in billions of dollars) for Coca-Cola Enterprises from 2000 through 2007 are listed below. (Source: Coca-Cola Enterprises, Inc.)

2000	14.750	2004	18.185
2001	15.700	2005	18.706
2002	16.899	2006	19.804
2003	17.330	2007	20.936

(a) Sketch a scatter plot of the data. Let y represent the total revenue (in billions of dollars) and let $t = 0$ represent 2000.

(b) Use a straightedge to sketch the best-fitting line through the points and find an equation of the line.

(c) Use the *regression* feature of a graphing utility to find the least squares regression line that fits the data.

(d) Compare the linear model you found in part (b) with the linear model given by the graphing utility in part (c).

(e) Use the models from parts (b) and (c) to estimate the sales of Coca-Cola Enterprises in 2008.

(f) Use your school's library, the Internet, or some other reference source to analyze the accuracy of the estimate in part (e).

 19. DATA ANALYSIS: BROADWAY SHOWS The table shows the annual gross ticket sales S (in millions of dollars) for Broadway shows in New York City from 1995 through 2006. (Source: The League of American Theatres and Producers, Inc.)

Year	Sales, S
1995	406
1996	436
1997	499
1998	558
1999	588
2000	603
2001	666
2002	643
2003	721
2004	771
2005	769
2006	862

(a) Use a graphing utility to create a scatter plot of the data. Let $t = 5$ represent 1995.

(b) Use the *regression* feature of a graphing utility to find the equation of the least squares regression line that fits the data.

(c) Use the graphing utility to graph the scatter plot you created in part (a) and the model you found in part (b) in the same viewing window. How closely does the model represent the data?

(d) Use the model to estimate the annual gross ticket sales in 2007 and 2009.

(e) Interpret the meaning of the slope of the linear model in the context of the problem.

 20. DATA ANALYSIS: TELEVISION SETS The table shows the numbers N (in millions) of television sets in U.S. households from 2000 through 2006. (Source: Television Bureau of Advertising, Inc.)

Year	Television sets, N
2000	245
2001	248
2002	254
2003	260
2004	268
2005	287
2006	301

(a) Use the *regression* feature of a graphing utility to find the equation of the least squares regression line that fits the data. Let $t = 0$ represent 2000.

(b) Use the graphing utility to create a scatter plot of the data. Then graph the model you found in part (a) and the scatter plot in the same viewing window. How closely does the model represent the data?

(c) Use the model to estimate the number of television sets in U.S. households in 2008.

(d) Use your school's library, the Internet, or some other reference source to analyze the accuracy of the estimate in part (c).

THINK ABOUT IT In Exercises 21 and 22, use the graph to determine whether y varies directly as some power of x or inversely as some power of x. Explain.

21.

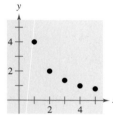

22.

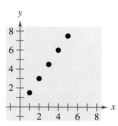

In Exercises 23–26, use the given value of k to complete the table for the direct variation model

$y = kx^2$.

Plot the points on a rectangular coordinate system.

x	2	4	6	8	10
$y = kx^2$					

23. $k = 1$ 　　　　24. $k = 2$
25. $k = \frac{1}{2}$ 　　　　26. $k = \frac{1}{4}$

In Exercises 27–30, use the given value of k to complete the table for the inverse variation model

$y = \dfrac{k}{x^2}$.

Plot the points on a rectangular coordinate system.

x	2	4	6	8	10
$y = \dfrac{k}{x^2}$					

27. $k = 2$ 　　　　28. $k = 5$
29. $k = 10$ 　　　　30. $k = 20$

In Exercises 31–34, determine whether the variation model is of the form $y = kx$ or $y = k/x$, and find k. Then write a model that relates y and x.

31.

x	5	10	15	20	25
y	1	$\frac{1}{2}$	$\frac{1}{3}$	$\frac{1}{4}$	$\frac{1}{5}$

32.

x	5	10	15	20	25
y	2	4	6	8	10

33.

x	5	10	15	20	25
y	-3.5	-7	-10.5	-14	-17.5

34.

x	5	10	15	20	25
y	24	12	8	6	$\frac{24}{5}$

DIRECT VARIATION In Exercises 35–38, assume that y is directly proportional to x. Use the given x-value and y-value to find a linear model that relates y and x.

35. $x = 5, y = 12$ 　　　　36. $x = 2, y = 14$
37. $x = 10, y = 2050$ 　　　　38. $x = 6, y = 580$

39. **SIMPLE INTEREST** The simple interest on an investment is directly proportional to the amount of the investment. By investing \$3250 in a certain bond issue, you obtained an interest payment of \$113.75 after 1 year. Find a mathematical model that gives the interest I for this bond issue after 1 year in terms of the amount invested P.

40. **SIMPLE INTEREST** The simple interest on an investment is directly proportional to the amount of the investment. By investing \$6500 in a municipal bond, you obtained an interest payment of \$211.25 after 1 year. Find a mathematical model that gives the interest I for this municipal bond after 1 year in terms of the amount invested P.

41. **MEASUREMENT** On a yardstick with scales in inches and centimeters, you notice that 13 inches is approximately the same length as 33 centimeters. Use this information to find a mathematical model that relates centimeters y to inches x. Then use the model to find the numbers of centimeters in 10 inches and 20 inches.

42. **MEASUREMENT** When buying gasoline, you notice that 14 gallons of gasoline is approximately the same amount of gasoline as 53 liters. Use this information to find a linear model that relates liters y to gallons x. Then use the model to find the numbers of liters in 5 gallons and 25 gallons.

43. TAXES Property tax is based on the assessed value of a property. A house that has an assessed value of $150,000 has a property tax of $5520. Find a mathematical model that gives the amount of property tax y in terms of the assessed value x of the property. Use the model to find the property tax on a house that has an assessed value of $225,000.

44. TAXES State sales tax is based on retail price. An item that sells for $189.99 has a sales tax of $11.40. Find a mathematical model that gives the amount of sales tax y in terms of the retail price x. Use the model to find the sales tax on a $639.99 purchase.

HOOKE'S LAW In Exercises 45–48, use Hooke's Law for springs, which states that the distance a spring is stretched (or compressed) varies directly as the force on the spring.

45. A force of 265 newtons stretches a spring 0.15 meter (see figure).

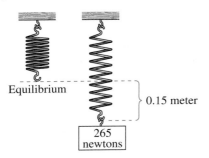

(a) How far will a force of 90 newtons stretch the spring?

(b) What force is required to stretch the spring 0.1 meter?

46. A force of 220 newtons stretches a spring 0.12 meter. What force is required to stretch the spring 0.16 meter?

47. The coiled spring of a toy supports the weight of a child. The spring is compressed a distance of 1.9 inches by the weight of a 25-pound child. The toy will not work properly if its spring is compressed more than 3 inches. What is the weight of the heaviest child who should be allowed to use the toy?

48. An overhead garage door has two springs, one on each side of the door (see figure). A force of 15 pounds is required to stretch each spring 1 foot. Because of a pulley system, the springs stretch only one-half the distance the door travels. The door moves a total of 8 feet, and the springs are at their natural length when the door is open. Find the combined lifting force applied to the door by the springs when the door is closed.

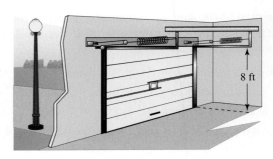

FIGURE FOR **48**

In Exercises 49–58, find a mathematical model for the verbal statement.

49. A varies directly as the square of r.

50. V varies directly as the cube of e.

51. y varies inversely as the square of x.

52. h varies inversely as the square root of s.

53. F varies directly as g and inversely as r^2.

54. z is jointly proportional to the square of x and the cube of y.

55. BOYLE'S LAW: For a constant temperature, the pressure P of a gas is inversely proportional to the volume V of the gas.

56. NEWTON'S LAW OF COOLING: The rate of change R of the temperature of an object is proportional to the difference between the temperature T of the object and the temperature T_e of the environment in which the object is placed.

57. NEWTON'S LAW OF UNIVERSAL GRAVITATION: The gravitational attraction F between two objects of masses m_1 and m_2 is proportional to the product of the masses and inversely proportional to the square of the distance r between the objects.

58. LOGISTIC GROWTH: The rate of growth R of a population is jointly proportional to the size S of the population and the difference between S and the maximum population size L that the environment can support.

In Exercises 59–66, write a sentence using the variation terminology of this section to describe the formula.

59. *Area of a triangle:* $A = \frac{1}{2}bh$

60. *Area of a rectangle:* $A = lw$

61. *Area of an equilateral triangle:* $A = \left(\sqrt{3}s^2\right)/4$

62. *Surface area of a sphere:* $S = 4\pi r^2$

63. *Volume of a sphere:* $V = \frac{4}{3}\pi r^3$

64. *Volume of a right circular cylinder:* $V = \pi r^2 h$

65. *Average speed:* $r = d/t$

66. *Free vibrations:* $\omega = \sqrt{(kg)/W}$

In Exercises 67–74, find a mathematical model representing the statement. (In each case, determine the constant of proportionality.)

67. A varies directly as r^2. ($A = 9\pi$ when $r = 3$.)

68. y varies inversely as x. ($y = 3$ when $x = 25$.)

69. y is inversely proportional to x. ($y = 7$ when $x = 4$.)

70. z varies jointly as x and y. ($z = 64$ when $x = 4$ and $y = 8$.)

71. F is jointly proportional to r and the third power of s. ($F = 4158$ when $r = 11$ and $s = 3$.)

72. P varies directly as x and inversely as the square of y. $\left(P = \frac{28}{3}\text{ when }x = 42\text{ and }y = 9.\right)$

73. z varies directly as the square of x and inversely as y. ($z = 6$ when $x = 6$ and $y = 4$.)

74. v varies jointly as p and q and inversely as the square of s. ($v = 1.5$ when $p = 4.1$, $q = 6.3$, and $s = 1.2$.)

ECOLOGY   In Exercises 75 and 76, use the fact that the diameter of the largest particle that can be moved by a stream varies approximately directly as the square of the velocity of the stream.

75. A stream with a velocity of $\frac{1}{4}$ mile per hour can move coarse sand particles about 0.02 inch in diameter. Approximate the velocity required to carry particles 0.12 inch in diameter.

76. A stream of velocity v can move particles of diameter d or less. By what factor does d increase when the velocity is doubled?

RESISTANCE   In Exercises 77 and 78, use the fact that the resistance of a wire carrying an electrical current is directly proportional to its length and inversely proportional to its cross-sectional area.

77. If #28 copper wire (which has a diameter of 0.0126 inch) has a resistance of 66.17 ohms per thousand feet, what length of #28 copper wire will produce a resistance of 33.5 ohms?

78. A 14-foot piece of copper wire produces a resistance of 0.05 ohm. Use the constant of proportionality from Exercise 77 to find the diameter of the wire.

79. WORK   The work W (in joules) done when lifting an object varies jointly with the mass m (in kilograms) of the object and the height h (in meters) that the object is lifted. The work done when a 120-kilogram object is lifted 1.8 meters is 2116.8 joules. How much work is done when lifting a 100-kilogram object 1.5 meters?

80. MUSIC   The frequency of vibrations of a piano string varies directly as the square root of the tension on the string and inversely as the length of the string. The middle A string has a frequency of 440 vibrations per second. Find the frequency of a string that has 1.25 times as much tension and is 1.2 times as long.

81. FLUID FLOW   The velocity v of a fluid flowing in a conduit is inversely proportional to the cross-sectional area of the conduit. (Assume that the volume of the flow per unit of time is held constant.) Determine the change in the velocity of water flowing from a hose when a person places a finger over the end of the hose to decrease its cross-sectional area by 25%.

82. BEAM LOAD   The maximum load that can be safely supported by a horizontal beam varies jointly as the width of the beam and the square of its depth, and inversely as the length of the beam. Determine the changes in the maximum safe load under the following conditions.

(a) The width and length of the beam are doubled.

(b) The width and depth of the beam are doubled.

(c) All three of the dimensions are doubled.

(d) The depth of the beam is halved.

83. DATA ANALYSIS: OCEAN TEMPERATURES   An oceanographer took readings of the water temperatures C (in degrees Celsius) at several depths d (in meters). The data collected are shown in the table.

Depth, d	Temperature, C
1000	4.2°
2000	1.9°
3000	1.4°
4000	1.2°
5000	0.9°

(a) Sketch a scatter plot of the data.

(b) Does it appear that the data can be modeled by the inverse variation model $C = k/d$? If so, find k for each pair of coordinates.

(c) Determine the mean value of k from part (b) to find the inverse variation model $C = k/d$.

(d) Use a graphing utility to plot the data points and the inverse model from part (c).

(e) Use the model to approximate the depth at which the water temperature is 3°C.

84. DATA ANALYSIS: PHYSICS EXPERIMENT An experiment in a physics lab requires a student to measure the compressed lengths y (in centimeters) of a spring when various forces of F pounds are applied. The data are shown in the table.

Force, F	Length, y
0	0
2	1.15
4	2.3
6	3.45
8	4.6
10	5.75
12	6.9

(a) Sketch a scatter plot of the data.

(b) Does it appear that the data can be modeled by Hooke's Law? If so, estimate k. (See Exercises 45–48.)

(c) Use the model in part (b) to approximate the force required to compress the spring 9 centimeters.

85. DATA ANALYSIS: LIGHT INTENSITY A light probe is located x centimeters from a light source, and the intensity y (in microwatts per square centimeter) of the light is measured. The results are shown as ordered pairs (x, y).

(30, 0.1881)   (34, 0.1543)   (38, 0.1172)
(42, 0.0998)   (46, 0.0775)   (50, 0.0645)

A model for the data is $y = 262.76/x^{2.12}$.

(a) Use a graphing utility to plot the data points and the model in the same viewing window.

(b) Use the model to approximate the light intensity 25 centimeters from the light source.

86. ILLUMINATION The illumination from a light source varies inversely as the square of the distance from the light source. When the distance from a light source is doubled, how does the illumination change? Discuss this model in terms of the data given in Exercise 85. Give a possible explanation of the difference.

EXPLORATION

TRUE OR FALSE? In Exercises 87 and 88, decide whether the statement is true or false. Justify your answer.

87. In the equation for kinetic energy, $E = \frac{1}{2}mv^2$, the amount of kinetic energy E is directly proportional to the mass m of an object and the square of its velocity v.

88. If the correlation coefficient for a least squares regression line is close to -1, the regression line cannot be used to describe the data.

89. Discuss how well the data shown in each scatter plot can be approximated by a linear model.

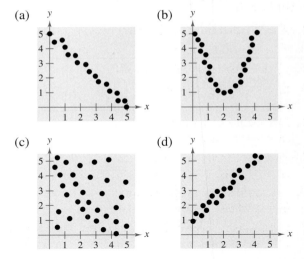

90. WRITING A linear model for predicting prize winnings at a race is based on data for 3 years. Write a paragraph discussing the potential accuracy or inaccuracy of such a model.

91. WRITING Suppose the constant of proportionality is positive and y varies directly as x. When one of the variables increases, how will the other change? Explain your reasoning.

92. WRITING Suppose the constant of proportionality is positive and y varies inversely as x. When one of the variables increases, how will the other change? Explain your reasoning.

93. WRITING

(a) Given that y varies inversely as the square of x and x is doubled, how will y change? Explain.

(b) Given that y varies directly as the square of x and x is doubled, how will y change? Explain.

94. CAPSTONE The prices of three sizes of pizza at a pizza shop are as follows.

9-inch: $8.78, 12-inch: $11.78, 15-inch: $14.18

You would expect that the price of a certain size of pizza would be directly proportional to its surface area. Is that the case for this pizza shop? If not, which size of pizza is the best buy?

PROJECT: FRAUD AND IDENTITY THEFT To work an extended application analyzing the numbers of fraud complaints and identity theft victims in the United States in 2007, visit this text's website at *academic.cengage.com*. (Data Source: U.S. Census Bureau)

1 CHAPTER SUMMARY

What Did You Learn?	Explanation/Examples	Review Exercises
Section 1.1 Plot points in the Cartesian plane (*p. 2*).	For an ordered pair (x, y), the x-coordinate is the directed distance from the y-axis to the point, and the y-coordinate is the directed distance from the x-axis to the point.	1–4
Use the Distance Formula (*p. 4*) and the Midpoint Formula (*p. 5*).	**Distance Formula:** $d = \sqrt{(x_2 - x_1)^2 + (y_2 - y_1)^2}$ **Midpoint Formula:** Midpoint $= \left(\dfrac{x_1 + x_2}{2}, \dfrac{y_1 + y_2}{2} \right)$	5–8
Use a coordinate plane to model and solve real-life problems (*p. 6*).	The coordinate plane can be used to find the length of a football pass (See Example 6).	9–12
Section 1.2 Sketch graphs of equations (*p. 13*), find x- and y-intercepts of graphs (*p. 16*), and use symmetry to sketch graphs of equations (*p. 17*).	To graph an equation, make a table of values, plot the points, and connect the points with a smooth curve or line. To find x-intercepts, let y be zero and solve for x. To find y-intercepts, let x be zero and solve for y. Graphs can have symmetry with respect to one of the coordinate axes or with respect to the origin.	13–34
Find equations of and sketch graphs of circles (*p. 19*).	The point (x, y) lies on the circle of radius r and center (h, k) if and only if $(x - h)^2 + (y - k)^2 = r^2$.	35–42
Use graphs of equations in solving real-life problems (*p. 20*).	The graph of an equation can be used to estimate the recommended weight for a man. (See Example 9.)	43, 44
Section 1.3 Use slope to graph linear equations in two variables (*p. 24*).	The graph of the equation $y = mx + b$ is a line whose slope is m and whose y-intercept is $(0, b)$.	45–48
Find the slope of a line given two points on the line (*p. 26*).	The slope m of the nonvertical line through (x_1, y_1) and (x_2, y_2) is $m = (y_2 - y_1)/(x_2 - x_1)$, where $x_1 \neq x_2$.	49–52
Write linear equations in two variables (*p. 28*).	The equation of the line with slope m passing through the point (x_1, y_1) is $y - y_1 = m(x - x_1)$.	53–60
Use slope to identify parallel and perpendicular lines (*p. 29*).	**Parallel lines:** Slopes are equal. **Perpendicular lines:** Slopes are negative reciprocals of each other.	61, 62
Use slope and linear equations in two variables to model and solve real-life problems (*p. 30*).	A linear equation in two variables can be used to describe the book value of exercise equipment in a given year. (See Example 7.)	63, 64
Section 1.4 Determine whether relations between two variables are functions (*p. 39*).	A function f from a set A (domain) to a set B (range) is a relation that assigns to each element x in the set A exactly one element y in the set B.	65–68
Use function notation, evaluate functions, and find domains (*p. 41*).	**Equation:** $f(x) = 5 - x^2$ $f(2)$: $f(2) = 5 - 2^2 = 1$ **Domain of** $f(x) = 5 - x^2$: All real numbers	69–74
Use functions to model and solve real-life problems (*p. 45*).	A function can be used to model the number of alternative-fueled vehicles in the United States (See Example 10.)	75, 76
Evaluate difference quotients (*p. 46*).	**Difference quotient:** $[f(x + h) - f(x)]/h,\ h \neq 0$	77, 78
Section 1.5 Use the Vertical Line Test for functions (*p. 55*).	A graph represents a function if and only if no *vertical* line intersects the graph at more than one point.	79–82
Find the zeros of functions (*p. 56*).	**Zeros of** $f(x)$: x-values for which $f(x) = 0$	83–86

	What Did You Learn?	Explanation/Examples	Review Exercises
Section 1.5	Determine intervals on which functions are increasing or decreasing *(p. 57)*, find relative minimum and maximum values *(p. 58)*, and find the average rate of change of a function *(p. 59)*.	To determine whether a function is increasing, decreasing, or constant on an interval, evaluate the function for several values of x. The points at which the behavior of a function changes can help determine the relative minimum or relative maximum. The average rate of change between any two points is the slope of the line (secant line) through the two points.	87–96
	Identify even and odd functions *(p. 60)*.	**Even:** For each x in the domain of f, $f(-x) = f(x)$. **Odd:** For each x in the domain of f, $f(-x) = -f(x)$.	97–100
Section 1.6	Identify and graph different types of functions *(p. 66)*, and recognize graphs of parent function *(p. 70)*.	**Linear:** $f(x) = ax + b$; **Squaring:** $f(x) = x^2$; **Cubic:** $f(x) = x^3$; **Square Root:** $f(x) = \sqrt{x}$; **Reciprocal:** $f(x) = 1/x$. Eight of the most commonly used functions in algebra are shown in Figure 1.75.	101–114
Section 1.7	Use vertical and horizontal shifts *(p. 73)*, reflections *(p. 75)*, and nonrigid transformations *(p. 77)* to sketch graphs of functions.	**Vertical shifts:** $h(x) = f(x) + c$ or $h(x) = f(x) - c$ **Horizontal shifts:** $h(x) = f(x - c)$ or $h(x) = f(x + c)$ **Reflection in x-axis:** $h(x) = -f(x)$ **Reflection in y-axis:** $h(x) = f(-x)$ **Nonrigid transformations:** $h(x) = cf(x)$ or $h(x) = f(cx)$	115–128
Section 1.8	Add, subtract, multiply, and divide functions *(p. 83)*, and find the compositions of functions *(p. 85)*.	$(f + g)(x) = f(x) + g(x)$ $(f - g)(x) = f(x) - g(x)$ $(fg)(x) = f(x) \cdot g(x)$ $(f/g)(x) = f(x)/g(x), g(x) \neq 0$ **Composition of Functions:** $(f \circ g)(x) = f(g(x))$	129–134
	Use combinations and compositions of functions to model and solve real-life problems *(p. 87)*.	A composite function can be used to represent the number of bacteria in food as a function of the amount of time the food has been out of refrigeration. (See Example 8.)	135, 136
Section 1.9	Find inverse functions informally and verify that two functions are inverse functions of each other *(p. 92)*.	Let f and g be two functions such that $f(g(x)) = x$ for every x in the domain of g and $g(f(x)) = x$ for every x in the domain of f. Under these conditions, the function g is the inverse function of the function f.	137, 138
	Use graphs of functions to determine whether functions have inverse functions *(p. 94)*.	If the point (a, b) lies on the graph of f, then the point (b, a) must lie on the graph of f^{-1}, and vice versa. In short, f^{-1} is a reflection of f in the line $y = x$.	139, 140
	Use the Horizontal Line Test to determine if functions are one-to-one *(p. 95)*.	**Horizontal Line Test for Inverse Functions** A function f has an inverse function if and only if no *horizontal* line intersects f at more than one point.	141–144
	Find inverse functions algebraically *(p.96)*.	To find inverse functions, replace $f(x)$ by y, interchange the roles of x and y, and solve for y. Replace y by $f^{-1}(x)$.	145–150
Section 1.10	Use mathematical models to approximate sets of data points *(p. 102)*, and use the *regression* feature of a graphing utility to find the equation of a least squares regression line *(p. 103)*.	To see how well a model fits a set of data, compare the actual values and model values of y. The sum of square differences is the sum of the squares of the differences between actual data values and model values. The least squares regression line is the linear model with the least sum of square differences.	151, 152
	Write mathematical models for direct variation, direct variation as an *n*th power, inverse variation, and joint variation *(pp. 104–107)*.	**Direct variation:** $y = kx$ for some nonzero constant k **Direct variation as an nth power:** $y = kx^n$ for some constant k **Inverse variation:** $y = k/x$ for some constant k **Joint variation:** $z = kxy$ for some constant k	153–158

1 REVIEW EXERCISES

See www.CalcChat.com for worked-out solutions to odd-numbered exercises.

1.1 In Exercises 1 and 2, plot the points in the Cartesian plane.

1. $(5, 5), (-2, 0), (-3, 6), (-1, -7)$

2. $(0, 6), (8, 1), (4, -2), (-3, -3)$

In Exercises 3 and 4, determine the quadrant(s) in which (x, y) is located so that the condition(s) is (are) satisfied.

3. $x > 0$ and $y = -2$ **4.** $xy = 4$

In Exercises 5–8, (a) plot the points, (b) find the distance between the points, and (c) find the midpoint of the line segment joining the points.

5. $(-3, 8), (1, 5)$

6. $(-2, 6), (4, -3)$

7. $(5.6, 0), (0, 8.2)$

8. $(1.8, 7.4), (-0.6, -14.5)$

In Exercises 9 and 10, the polygon is shifted to a new position in the plane. Find the coordinates of the vertices of the polygon in its new position.

9. Original coordinates of vertices:

$(4, 8), (6, 8), (4, 3), (6, 3)$

Shift: eight units downward, four units to the left

10. Original coordinates of vertices:

$(0, 1), (3, 3), (0, 5), (-3, 3)$

Shift: three units upward, two units to the left

11. SALES Starbucks had annual sales of $2.17 billion in 2000 and $10.38 billion in 2008. Use the Midpoint Formula to estimate the sales in 2004. (Source: Starbucks Corp.)

12. METEOROLOGY The apparent temperature is a measure of relative discomfort to a person from heat and high humidity. The table shows the actual temperatures x (in degrees Fahrenheit) versus the apparent temperatures y (in degrees Fahrenheit) for a relative humidity of 75%.

x	70	75	80	85	90	95	100
y	70	77	85	95	109	130	150

(a) Sketch a scatter plot of the data shown in the table.

(b) Find the change in the apparent temperature when the actual temperature changes from 70°F to 100°F.

1.2 In Exercises 13–16, complete a table of values. Use the solution points to sketch the graph of the equation.

13. $y = 3x - 5$ **14.** $y = -\frac{1}{2}x + 2$

15. $y = x^2 - 3x$ **16.** $y = 2x^2 - x - 9$

In Exercises 17–22, sketch the graph *by hand*.

17. $y - 2x - 3 = 0$ **18.** $3x + 2y + 6 = 0$

19. $y = \sqrt{5 - x}$ **20.** $y = \sqrt{x + 2}$

21. $y + 2x^2 = 0$ **22.** $y = x^2 - 4x$

In Exercises 23–26, find the x- and y-intercepts of the graph of the equation.

23. $y = 2x + 7$ **24.** $y = |x + 1| - 3$

25. $y = (x - 3)^2 - 4$ **26.** $y = x\sqrt{4 - x^2}$

In Exercises 27–34, identify any intercepts and test for symmetry. Then sketch the graph of the equation.

27. $y = -4x + 1$ **28.** $y = 5x - 6$

29. $y = 5 - x^2$ **30.** $y = x^2 - 10$

31. $y = x^3 + 3$ **32.** $y = -6 - x^3$

33. $y = \sqrt{x + 5}$ **34.** $y = |x| + 9$

In Exercises 35–40, find the center and radius of the circle and sketch its graph.

35. $x^2 + y^2 = 9$ **36.** $x^2 + y^2 = 4$

37. $(x + 2)^2 + y^2 = 16$

38. $x^2 + (y - 8)^2 = 81$

39. $\left(x - \frac{1}{2}\right)^2 + (y + 1)^2 = 36$

40. $(x + 4)^2 + \left(y - \frac{3}{2}\right)^2 = 100$

41. Find the standard form of the equation of the circle for which the endpoints of a diameter are $(0, 0)$ and $(4, -6)$.

42. Find the standard form of the equation of the circle for which the endpoints of a diameter are $(-2, -3)$ and $(4, -10)$.

43. NUMBER OF STORES The numbers N of Walgreen stores for the years 2000 through 2008 can be approximated by the model

$$N = 439.9t + 2987, \quad 0 \le t \le 8$$

where t represents the year, with $t = 0$ corresponding to 2000. (Source: Walgreen Co.)

(a) Sketch a graph of the model.

(b) Use the graph to estimate the year in which the number of stores was 6500.

44. PHYSICS The force F (in pounds) required to stretch a spring x inches from its natural length (see figure) is

$$F = \frac{5}{4}x, \quad 0 \le x \le 20.$$

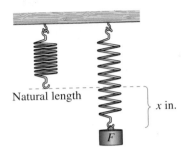

Natural length

x in.

F

(a) Use the model to complete the table.

x	0	4	8	12	16	20
Force, F						

(b) Sketch a graph of the model.

(c) Use the graph to estimate the force necessary to stretch the spring 10 inches.

1.3 In Exercises 45–48, find the slope and y-intercept (if possible) of the equation of the line. Sketch the line.

45. $y = 6$

46. $x = -3$

47. $y = 3x + 13$

48. $y = -10x + 9$

In Exercises 49–52, plot the points and find the slope of the line passing through the pair of points.

49. $(6, 4), (-3, -4)$

50. $\left(\frac{3}{2}, 1\right), \left(5, \frac{5}{2}\right)$

51. $(-4.5, 6), (2.1, 3)$

52. $(-3, 2), (8, 2)$

In Exercises 53–56, find the slope-intercept form of the equation of the line that passes through the given point and has the indicated slope. Sketch the line.

Point	Slope
53. $(3, 0)$	$m = \frac{2}{3}$
54. $(-8, 5)$	$m = 0$
55. $(10, -3)$	$m = -\frac{1}{2}$
56. $(12, -6)$	m is undefined.

In Exercises 57–60, find the slope-intercept form of the equation of the line passing through the points.

57. $(0, 0), (0, 10)$

58. $(2, -1), (4, -1)$

59. $(-1, 0), (6, 2)$

60. $(11, -2), (6, -1)$

In Exercises 61 and 62, write the slope-intercept forms of the equations of the lines through the given point (a) parallel to the given line and (b) perpendicular to the given line.

Point	Line
61. $(3, -2)$	$5x - 4y = 8$
62. $(-8, 3)$	$2x + 3y = 5$

RATE OF CHANGE In Exercises 63 and 64, you are given the dollar value of a product in 2010 and the rate at which the value of the product is expected to change during the next 5 years. Use this information to write a linear equation that gives the dollar value V of the product in terms of the year t. (Let $t = 10$ represent 2010.)

2010 Value	Rate
63. $12,500	$850 decrease per year
64. $72.95	$5.15 increase per year

1.4 In Exercises 65–68, determine whether the equation represents y as a function of x.

65. $16x - y^4 = 0$

66. $2x - y - 3 = 0$

67. $y = \sqrt{1 - x}$

68. $|y| = x + 2$

In Exercises 69 and 70, evaluate the function at each specified value of the independent variable and simplify.

69. $f(x) = x^2 + 1$
   (a) $f(2)$ (b) $f(-4)$ (c) $f(t^2)$ (d) $f(t + 1)$

70. $h(x) = \begin{cases} 2x + 1, & x \le -1 \\ x^2 + 2, & x > -1 \end{cases}$
   (a) $h(-2)$ (b) $h(-1)$ (c) $h(0)$ (d) $h(2)$

In Exercises 71–74, find the domain of the function. Verify your result with a graph.

71. $f(x) = \sqrt{25 - x^2}$

72. $g(s) = \dfrac{5s + 5}{3s - 9}$

73. $h(x) = \dfrac{x}{x^2 - x - 6}$

74. $h(t) = |t + 1|$

75. PHYSICS The velocity of a ball projected upward from ground level is given by $v(t) = -32t + 48$, where t is the time in seconds and v is the velocity in feet per second.
   (a) Find the velocity when $t = 1$.
   (b) Find the time when the ball reaches its maximum height. [*Hint:* Find the time when $v(t) = 0$.]
   (c) Find the velocity when $t = 2$.

76. MIXTURE PROBLEM   From a full 50-liter container of a 40% concentration of acid, x liters is removed and replaced with 100% acid.

(a) Write the amount of acid in the final mixture as a function of x.

(b) Determine the domain and range of the function.

(c) Determine x if the final mixture is 50% acid.

In Exercises 77 and 78, find the difference quotient and simplify your answer.

77. $f(x) = 2x^2 + 3x - 1$, $\dfrac{f(x + h) - f(x)}{h}$, $h \neq 0$

78. $f(x) = x^3 - 5x^2 + x$, $\dfrac{f(x + h) - f(x)}{h}$, $h \neq 0$

1.5   In Exercises 79–82, use the Vertical Line Test to determine whether y is a function of x. To print an enlarged copy of the graph, go to the website *www.mathgraphs.com*.

79. $y = (x - 3)^2$

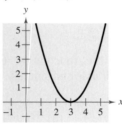

80. $y = -\frac{3}{5}x^3 - 2x + 1$

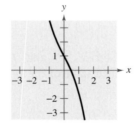

81. $x - 4 = y^2$

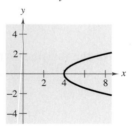

82. $x = -|4 - y|$

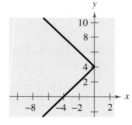

In Exercises 83–86, find the zeros of the function algebraically.

83. $f(x) = 3x^2 - 16x + 21$

84. $f(x) = 5x^2 + 4x - 1$

85. $f(x) = \dfrac{8x + 3}{11 - x}$

86. $f(x) = x^3 - x^2 - 25x + 25$

In Exercises 87 and 88, use a graphing utility to graph the function and visually determine the intervals over which the function is increasing, decreasing, or constant.

87. $f(x) = |x| + |x + 1|$ **88.** $f(x) = (x^2 - 4)^2$

In Exercises 89–92, use a graphing utility to graph the function and approximate any relative minimum or relative maximum values.

89. $f(x) = -x^2 + 2x + 1$

90. $f(x) = x^4 - 4x^2 - 2$

91. $f(x) = x^3 - 6x^4$

92. $f(x) = x^3 - 4x^2 - 1$

In Exercises 93–96, find the average rate of change of the function from x_1 to x_2.

Function	*x-Values*
93. $f(x) = -x^2 + 8x - 4$	$x_1 = 0, x_2 = 4$
94. $f(x) = x^3 + 12x - 2$	$x_1 = 0, x_2 = 4$
95. $f(x) = 2 - \sqrt{x + 1}$	$x_1 = 3, x_2 = 7$
96. $f(x) = 1 - \sqrt{x + 3}$	$x_1 = 1, x_2 = 6$

In Exercises 97–100, determine whether the function is even, odd, or neither.

97. $f(x) = x^5 + 4x - 7$

98. $f(x) = x^4 - 20x^2$

99. $f(x) = 2x\sqrt{x^2 + 3}$

100. $f(x) = \sqrt[5]{6x^2}$

1.6   In Exercises 101 and 102, write the linear function f such that it has the indicated function values. Then sketch the graph of the function.

101. $f(2) = -6, f(-1) = 3$

102. $f(0) = -5, f(4) = -8$

In Exercises 103–112, graph the function.

103. $f(x) = 3 - x^2$ **104.** $h(x) = x^3 - 2$

105. $f(x) = -\sqrt{x}$ **106.** $f(x) = \sqrt{x + 1}$

107. $g(x) = \dfrac{3}{x}$ **108.** $g(x) = \dfrac{1}{x + 5}$

109. $f(x) = [\![x]\!] + 2$

110. $g(x) = [\![x + 4]\!]$

111. $f(x) = \begin{cases} 5x - 3, & x \geq -1 \\ -4x + 5, & x < -1 \end{cases}$

112. $f(x) = \begin{cases} x^2 - 2, & x < -2 \\ 5, & -2 \leq x \leq 0 \\ 8x - 5, & x > 0 \end{cases}$

In Exercises 113 and 114, the figure shows the graph of a transformed parent function. Identify the parent function.

113.

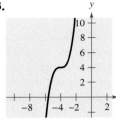

114.

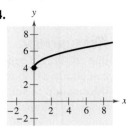

1.7 In Exercises 115–128, h is related to one of the parent functions described in this chapter. (a) Identify the parent function f. (b) Describe the sequence of transformations from f to h. (c) Sketch the graph of h. (d) Use function notation to write h in terms of f.

115. $h(x) = x^2 - 9$
116. $h(x) = (x - 2)^3 + 2$
117. $h(x) = -\sqrt{x} + 4$
118. $h(x) = |x + 3| - 5$
119. $h(x) = -(x + 2)^2 + 3$
120. $h(x) = \frac{1}{2}(x - 1)^2 - 2$
121. $h(x) = -[\![x]\!] + 6$
122. $h(x) = -\sqrt{x + 1} + 9$
123. $h(x) = -|-x + 4| + 6$
124. $h(x) = -(x + 1)^2 - 3$
125. $h(x) = 5[\![x - 9]\!]$
126. $h(x) = -\frac{1}{3}x^3$
127. $h(x) = -2\sqrt{x - 4}$
128. $h(x) = \frac{1}{2}|x| - 1$

1.8 In Exercises 129 and 130, find (a) $(f + g)(x)$, (b) $(f - g)(x)$, (c) $(fg)(x)$, and (d) $(f/g)(x)$. What is the domain of f/g?

129. $f(x) = x^2 + 3$, $g(x) = 2x - 1$
130. $f(x) = x^2 - 4$, $g(x) = \sqrt{3 - x}$

In Exercises 131 and 132, find (a) $f \circ g$ and (b) $g \circ f$. Find the domain of each function and each composite function.

131. $f(x) = \frac{1}{3}x - 3$, $g(x) = 3x + 1$
132. $f(x) = x^3 - 4$, $g(x) = \sqrt[3]{x + 7}$

In Exercises 133 and 134, find two functions f and g such that $(f \circ g)(x) = h(x)$. (There are many correct answers.)

133. $h(x) = (1 - 2x)^3$
134. $h(x) = \sqrt[3]{x + 2}$

135. PHONE EXPENDITURES The average annual expenditures (in dollars) for residential $r(t)$ and cellular $c(t)$ phone services from 2001 through 2006 can be approximated by the functions $r(t) = 27.5t + 705$ and $c(t) = 151.3t + 151$, where t represents the year, with $t = 1$ corresponding to 2001. (Source: Bureau of Labor Statistics)

(a) Find and interpret $(r + c)(t)$.

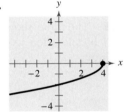
(b) Use a graphing utility to graph $r(t)$, $c(t)$, and $(r + c)(t)$ in the same viewing window.

(c) Find $(r + c)(13)$. Use the graph in part (b) to verify your result.

136. BACTERIA COUNT The number N of bacteria in a refrigerated food is given by

$$N(T) = 25T^2 - 50T + 300, \quad 2 \le T \le 20$$

where T is the temperature of the food in degrees Celsius. When the food is removed from refrigeration, the temperature of the food is given by

$$T(t) = 2t + 1, \quad 0 \le t \le 9$$

where t is the time in hours. (a) Find the composition $N(T(t))$, and interpret its meaning in context, and (b) find the time when the bacteria count reaches 750.

1.9 In Exercises 137 and 138, find the inverse function of f informally. Verify that $f(f^{-1}(x)) = x$ and $f^{-1}(f(x)) = x$.

137. $f(x) = 3x + 8$
138. $f(x) = \dfrac{x - 4}{5}$

In Exercises 139 and 140, determine whether the function has an inverse function.

139.

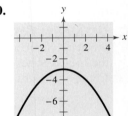

140.

In Exercises 141–144, use a graphing utility to graph the function, and use the Horizontal Line Test to determine whether the function is one-to-one and so has an inverse function.

141. $f(x) = 4 - \frac{1}{3}x$
142. $f(x) = (x - 1)^2$
143. $h(t) = \dfrac{2}{t - 3}$
144. $g(x) = \sqrt{x + 6}$

In Exercises 145–148, (a) find the inverse function of f, (b) graph both f and f^{-1} on the same set of coordinate axes, (c) describe the relationship between the graphs of f and f^{-1}, and (d) state the domains and ranges of f and f^{-1}.

145. $f(x) = \frac{1}{2}x - 3$
146. $f(x) = 5x - 7$
147. $f(x) = \sqrt{x + 1}$
148. $f(x) = x^3 + 2$

In Exercises 149 and 150, restrict the domain of the function f to an interval over which the function is increasing and determine f^{-1} over that interval.

149. $f(x) = 2(x - 4)^2$
150. $f(x) = |x - 2|$

151. COMPACT DISCS   The values V (in billions of dollars) of shipments of compact discs in the United States from 2000 through 2007 are shown in the table. A linear model that approximates these data is

$$V = -0.742t + 13.62$$

where t represents the year, with $t = 0$ corresponding to 2000. (Source: Recording Industry Association of America)

Year	Value, V
2000	13.21
2001	12.91
2002	12.04
2003	11.23
2004	11.45
2005	10.52
2006	9.37
2007	7.45

(a) Plot the actual data and the model on the same set of coordinate axes.

(b) How closely does the model represent the data?

152. DATA ANALYSIS: TV USAGE   The table shows the projected numbers of hours H of television usage in the United States from 2003 through 2011. (Source: Communications Industry Forecast and Report)

Year	Hours, H
2003	1615
2004	1620
2005	1659
2006	1673
2007	1686
2008	1704
2009	1714
2010	1728
2011	1742

(a) Use a graphing utility to create a scatter plot of the data. Let t represent the year, with $t = 3$ corresponding to 2003.

(b) Use the *regression* feature of the graphing utility to find the equation of the least squares regression line that fits the data. Then graph the model and the scatter plot you found in part (a) in the same viewing window. How closely does the model represent the data?

(c) Use the model to estimate the projected number of hours of television usage in 2020.

(d) Interpret the meaning of the slope of the linear model in the context of the problem.

153. MEASUREMENT   You notice a billboard indicating that it is 2.5 miles or 4 kilometers to the next restaurant of a national fast-food chain. Use this information to find a mathematical model that relates miles to kilometers. Then use the model to find the numbers of kilometers in 2 miles and 10 miles.

154. ENERGY   The power P produced by a wind turbine is proportional to the cube of the wind speed S. A wind speed of 27 miles per hour produces a power output of 750 kilowatts. Find the output for a wind speed of 40 miles per hour.

155. FRICTIONAL FORCE   The frictional force F between the tires and the road required to keep a car on a curved section of a highway is directly proportional to the square of the speed s of the car. If the speed of the car is doubled, the force will change by what factor?

156. DEMAND   A company has found that the daily demand x for its boxes of chocolates is inversely proportional to the price p. When the price is \$5, the demand is 800 boxes. Approximate the demand when the price is increased to \$6.

157. TRAVEL TIME   The travel time between two cities is inversely proportional to the average speed. A train travels between the cities in 3 hours at an average speed of 65 miles per hour. How long would it take to travel between the cities at an average speed of 80 miles per hour?

158. COST   The cost of constructing a wooden box with a square base varies jointly as the height of the box and the square of the width of the box. A box of height 16 inches and width 6 inches costs \$28.80. How much would a box of height 14 inches and width 8 inches cost?

EXPLORATION

TRUE OR FALSE? In Exercises 159 and 160, determine whether the statement is true or false. Justify your answer.

159. Relative to the graph of $f(x) = \sqrt{x}$, the function given by $h(x) = -\sqrt{x + 9} - 13$ is shifted 9 units to the left and 13 units downward, then reflected in the x-axis.

160. If f and g are two inverse functions, then the domain of g is equal to the range of f.

161. WRITING   Explain the difference between the Vertical Line Test and the Horizontal Line Test.

162. WRITING   Explain how to tell whether a relation between two variables is a function.

1 CHAPTER TEST

See www.CalcChat.com for worked-out solutions to odd-numbered exercises.

Take this test as you would take a test in class. When you are finished, check your work against the answers given in the back of the book.

1. Plot the points $(-2, 5)$ and $(6, 0)$. Find the coordinates of the midpoint of the line segment joining the points and the distance between the points.

2. A cylindrical can has a volume of 600 cubic centimeters and a radius of 4 centimeters. Find the height of the can.

In Exercises 3–5, use intercepts and symmetry to sketch the graph of the equation.

3. $y = 3 - 5x$

4. $y = 4 - |x|$

5. $y = x^2 - 1$

6. Write the standard form of the equation of the circle shown at the left.

In Exercises 7 and 8, find the slope-intercept form of the equation of the line passing through the points.

7. $(2, -3), (-4, 9)$

8. $(3, 0.8), (7, -6)$

9. Find equations of the lines that pass through the point $(0, 4)$ and are (a) parallel to and (b) perpendicular to the line $5x + 2y = 3$.

10. Evaluate $f(x) = \dfrac{\sqrt{x+9}}{x^2 - 81}$ at each value: (a) $f(7)$ (b) $f(-5)$ (c) $f(x - 9)$.

11. Find the domain of $f(x) = 10 - \sqrt{3 - x}$.

In Exercises 12–14, (a) find the zeros of the function, (b) use a graphing utility to graph the function, (c) approximate the intervals over which the function is increasing, decreasing, or constant, and (d) determine whether the function is even, odd, or neither.

12. $f(x) = 2x^6 + 5x^4 - x^2$

13. $f(x) = 4x\sqrt{3 - x}$

14. $f(x) = |x + 5|$

15. Sketch the graph of $f(x) = \begin{cases} 3x + 7, & x \le -3 \\ 4x^2 - 1, & x > -3 \end{cases}$.

In Exercises 16–18, identify the parent function in the transformation. Then sketch a graph of the function.

16. $h(x) = -[\![x]\!]$

17. $h(x) = -\sqrt{x + 5} + 8$

18. $h(x) = -2(x - 5)^3 + 3$

In Exercises 19 and 20, find (a) $(f + g)(x)$, (b) $(f - g)(x)$, (c) $(fg)(x)$, (d) $(f/g)(x)$, (e) $(f \circ g)(x)$, and (f) $(g \circ f)(x)$.

19. $f(x) = 3x^2 - 7, \quad g(x) = -x^2 - 4x + 5$

20. $f(x) = 1/x, \quad g(x) = 2\sqrt{x}$

In Exercises 21–23, determine whether or not the function has an inverse function, and if so, find the inverse function.

21. $f(x) = x^3 + 8$

22. $f(x) = |x^2 - 3| + 6$

23. $f(x) = 3x\sqrt{x}$

In Exercises 24–26, find a mathematical model representing the statement. (In each case, determine the constant of proportionality.)

24. v varies directly as the square root of s. ($v = 24$ when $s = 16$.)

25. A varies jointly as x and y. ($A = 500$ when $x = 15$ and $y = 8$.)

26. b varies inversely as a. ($b = 32$ when $a = 1.5$.)

FIGURE FOR 6

$V = \pi \cdot R^2 \cdot H$

$600 = \pi (4)^2 H$

$H = \dfrac{600}{16\pi} = \dfrac{25}{2\pi}$ cm

PROOFS IN MATHEMATICS

What does the word *proof* mean to you? In mathematics, the word *proof* is used to mean simply a valid argument. When you are proving a statement or theorem, you must use facts, definitions, and accepted properties in a logical order. You can also use previously proved theorems in your proof. For instance, the Distance Formula is used in the proof of the Midpoint Formula below. There are several different proof methods, which you will see in later chapters.

The Midpoint Formula   (p. 5)

The midpoint of the line segment joining the points (x_1, y_1) and (x_2, y_2) is given by the Midpoint Formula

$$\text{Midpoint} = \left(\frac{x_1 + x_2}{2}, \frac{y_1 + y_2}{2} \right).$$

The Cartesian Plane

The Cartesian plane was named after the French mathematician René Descartes (1596–1650). While Descartes was lying in bed, he noticed a fly buzzing around on the square ceiling tiles. He discovered that the position of the fly could be described by which ceiling tile the fly landed on. This led to the development of the Cartesian plane. Descartes felt that a coordinate plane could be used to facilitate description of the positions of objects.

Proof

Using the figure, you must show that $d_1 = d_2$ and $d_1 + d_2 = d_3$.

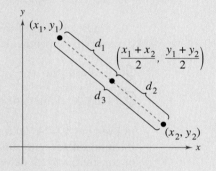

By the Distance Formula, you obtain

$$d_1 = \sqrt{\left(\frac{x_1 + x_2}{2} - x_1 \right)^2 + \left(\frac{y_1 + y_2}{2} - y_1 \right)^2}$$

$$= \frac{1}{2}\sqrt{(x_2 - x_1)^2 + (y_2 - y_1)^2}$$

$$d_2 = \sqrt{\left(x_2 - \frac{x_1 + x_2}{2} \right)^2 + \left(y_2 - \frac{y_1 + y_2}{2} \right)^2}$$

$$= \frac{1}{2}\sqrt{(x_2 - x_1)^2 + (y_2 - y_1)^2}$$

$$d_3 = \sqrt{(x_2 - x_1)^2 + (y_2 - y_1)^2}$$

So, it follows that $d_1 = d_2$ and $d_1 + d_2 = d_3$.

PROBLEM SOLVING

This collection of thought-provoking and challenging exercises further explores and expands upon concepts learned in this chapter.

1. As a salesperson, you receive a monthly salary of $2000, plus a commission of 7% of sales. You are offered a new job at $2300 per month, plus a commission of 5% of sales.

   (a) Write a linear equation for your current monthly wage W_1 in terms of your monthly sales S.

   (b) Write a linear equation for the monthly wage W_2 of your new job offer in terms of the monthly sales S.

   (c) Use a graphing utility to graph both equations in the same viewing window. Find the point of intersection. What does it signify?

   (d) You think you can sell $20,000 per month. Should you change jobs? Explain.

2. For the numbers 2 through 9 on a telephone keypad (see figure), create two relations: one mapping numbers onto letters, and the other mapping letters onto numbers. Are both relations functions? Explain.

3. What can be said about the sum and difference of each of the following?

   (a) Two even functions   (b) Two odd functions

   (c) An odd function and an even function

4. The two functions given by

   $$f(x) = x \quad \text{and} \quad g(x) = -x$$

   are their own inverse functions. Graph each function and explain why this is true. Graph other linear functions that are their own inverse functions. Find a general formula for a family of linear functions that are their own inverse functions.

5. Prove that a function of the following form is even.

   $$y = a_{2n}x^{2n} + a_{2n-2}x^{2n-2} + \cdots + a_2x^2 + a_0$$

6. A miniature golf professional is trying to make a hole-in-one on the miniature golf green shown. A coordinate plane is placed over the golf green. The golf ball is at the point $(2.5, 2)$ and the hole is at the point $(9.5, 2)$. The professional wants to bank the ball off the side wall of the green at the point (x, y). Find the coordinates of the point (x, y). Then write an equation for the path of the ball.

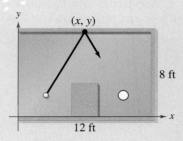

FIGURE FOR 6

7. At 2:00 P.M. on April 11, 1912, the *Titanic* left Cobh, Ireland, on her voyage to New York City. At 11:40 P.M. on April 14, the *Titanic* struck an iceberg and sank, having covered only about 2100 miles of the approximately 3400-mile trip.

   (a) What was the total duration of the voyage in hours?

   (b) What was the average speed in miles per hour?

   (c) Write a function relating the distance of the *Titanic* from New York City and the number of hours traveled. Find the domain and range of the function.

   (d) Graph the function from part (c).

8. Consider the function given by $f(x) = -x^2 + 4x - 3$. Find the average rate of change of the function from x_1 to x_2.

   (a) $x_1 = 1, x_2 = 2$ (b) $x_1 = 1, x_2 = 1.5$

   (c) $x_1 = 1, x_2 = 1.25$ (d) $x_1 = 1, x_2 = 1.125$

   (e) $x_1 = 1, x_2 = 1.0625$

   (f) Does the average rate of change seem to be approaching one value? If so, what value?

   (g) Find the equations of the secant lines through the points $(x_1, f(x_1))$ and $(x_2, f(x_2))$ for parts (a)–(e).

   (h) Find the equation of the line through the point $(1, f(1))$ using your answer from part (f) as the slope of the line.

9. Consider the functions given by $f(x) = 4x$ and $g(x) = x + 6$.

   (a) Find $(f \circ g)(x)$.   (b) Find $(f \circ g)^{-1}(x)$.

   (c) Find $f^{-1}(x)$ and $g^{-1}(x)$.

   (d) Find $(g^{-1} \circ f^{-1})(x)$ and compare the result with that of part (b).

   (e) Repeat parts (a) through (d) for $f(x) = x^3 + 1$ and $g(x) = 2x$.

   (f) Write two one-to-one functions f and g, and repeat parts (a) through (d) for these functions.

   (g) Make a conjecture about $(f \circ g)^{-1}(x)$ and $(g^{-1} \circ f^{-1})(x)$.

123

10. You are in a boat 2 miles from the nearest point on the coast. You are to travel to a point Q, 3 miles down the coast and 1 mile inland (see figure). You can row at 2 miles per hour and you can walk at 4 miles per hour.

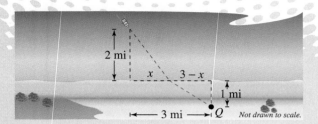

Not drawn to scale.

(a) Write the total time T of the trip as a function of x.

(b) Determine the domain of the function.

(c) Use a graphing utility to graph the function. Be sure to choose an appropriate viewing window.

(d) Use the *zoom* and *trace* features to find the value of x that minimizes T.

(e) Write a brief paragraph interpreting these values.

11. The **Heaviside function** $H(x)$ is widely used in engineering applications. (See figure.) To print an enlarged copy of the graph, go to the website *www.mathgraphs.com*.

$$H(x) = \begin{cases} 1, & x \ge 0 \\ 0, & x < 0 \end{cases}$$

Sketch the graph of each function by hand.

(a) $H(x) - 2$ (b) $H(x - 2)$ (c) $-H(x)$

(d) $H(-x)$ (e) $\frac{1}{2}H(x)$ (f) $-H(x - 2) + 2$

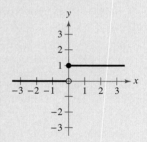

12. Let $f(x) = \dfrac{1}{1 - x}$.

(a) What are the domain and range of f?

(b) Find $f(f(x))$. What is the domain of this function?

(c) Find $f(f(f(x)))$. Is the graph a line? Why or why not?

13. Show that the Associative Property holds for compositions of functions—that is,

$$(f \circ (g \circ h))(x) = ((f \circ g) \circ h)(x).$$

14. Consider the graph of the function f shown in the figure. Use this graph to sketch the graph of each function. To print an enlarged copy of the graph, go to the website *www.mathgraphs.com*.

(a) $f(x + 1)$ (b) $f(x) + 1$ (c) $2f(x)$ (d) $f(-x)$

(e) $-f(x)$ (f) $|f(x)|$ (g) $f(|x|)$

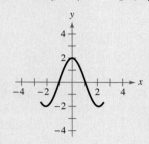

15. Use the graphs of f and f^{-1} to complete each table of function values.

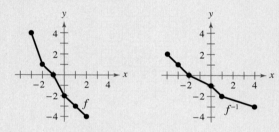

(a)

x	-4	-2	0	4
$(f(f^{-1}(x))$				

(b)

x	-3	-2	0	1
$(f + f^{-1})(x)$				

(c)

x	-3	-2	0	1
$(f \cdot f^{-1})(x)$				

(d)

x	-4	-3	0	4
$|f^{-1}(x)|$				

Polynomial and Rational Functions

2

In Mathematics

Functions defined by polynomial expressions are called polynomial functions, and functions defined by rational expressions are called rational functions.

In Real Life

Polynomial and rational functions are often used to model real-life phenomena. For instance, you can model the per capita cigarette consumption in the United States with a polynomial function. You can use the model to determine whether the addition of cigarette warnings affected consumption. (See Exercise 85, page 134.)

Michael Newman/PhotoEdit

IN CAREERS

There are many careers that use polynomial and rational functions. Several are listed below.

- Architect
  Exercise 82, page 134

- Forester
  Exercise 103, page 148

- Chemist
  Example 80, page 192

- Safety Engineer
  Exercise 78, page 203

2.1   QUADRATIC FUNCTIONS AND MODELS

What you should learn

- Analyze graphs of quadratic functions.
- Write quadratic functions in standard form and use the results to sketch graphs of functions.
- Find minimum and maximum values of quadratic functions in real-life applications.

Why you should learn it

Quadratic functions can be used to model data to analyze consumer behavior. For instance, in Exercise 79 on page 134, you will use a quadratic function to model the revenue earned from manufacturing handheld video games.

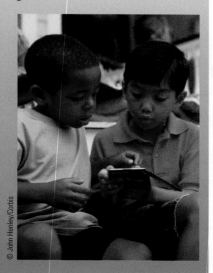

The Graph of a Quadratic Function

In this and the next section, you will study the graphs of polynomial functions. In Section 1.6, you were introduced to the following basic functions.

$$f(x) = ax + b \qquad \text{Linear function}$$

$$f(x) = c \qquad \text{Constant function}$$

$$f(x) = x^2 \qquad \text{Squaring function}$$

These functions are examples of **polynomial functions.**

Definition of Polynomial Function

Let n be a nonnegative integer and let $a_n, a_{n-1}, \ldots, a_2, a_1, a_0$ be real numbers with $a_n \neq 0$. The function given by

$$f(x) = a_n x^n + a_{n-1} x^{n-1} + \cdots + a_2 x^2 + a_1 x + a_0$$

is called a **polynomial function of x with degree n.**

Polynomial functions are classified by degree. For instance, a constant function $f(x) = c$ with $c \neq 0$ has degree 0, and a linear function $f(x) = ax + b$ with $a \neq 0$ has degree 1. In this section, you will study second-degree polynomial functions, which are called **quadratic functions.**

For instance, each of the following functions is a quadratic function.

$$f(x) = x^2 + 6x + 2$$

$$g(x) = 2(x + 1)^2 - 3$$

$$h(x) = 9 + \tfrac{1}{4}x^2$$

$$k(x) = -3x^2 + 4$$

$$m(x) = (x - 2)(x + 1)$$

Note that the squaring function is a simple quadratic function that has degree 2.

Definition of Quadratic Function

Let a, b, and c be real numbers with $a \neq 0$. The function given by

$$f(x) = ax^2 + bx + c \qquad \text{Quadratic function}$$

is called a **quadratic function.**

The graph of a quadratic function is a special type of "U"-shaped curve called a **parabola.** Parabolas occur in many real-life applications—especially those involving reflective properties of satellite dishes and flashlight reflectors. You will study these properties in Section 10.2.

All parabolas are symmetric with respect to a line called the **axis of symmetry,** or simply the **axis** of the parabola. The point where the axis intersects the parabola is the **vertex** of the parabola, as shown in Figure 2.1. If the leading coefficient is positive, the graph of

$$f(x) = ax^2 + bx + c$$

is a parabola that opens upward. If the leading coefficient is negative, the graph of

$$f(x) = ax^2 + bx + c$$

is a parabola that opens downward.

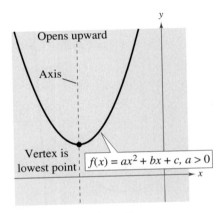

Leading coefficient is positive.

FIGURE 2.1

Leading coefficient is negative.

The simplest type of quadratic function is

$$f(x) = ax^2.$$

Its graph is a parabola whose vertex is $(0, 0)$. If $a > 0$, the vertex is the point with the *minimum* y-value on the graph, and if $a < 0$, the vertex is the point with the *maximum* y-value on the graph, as shown in Figure 2.2.

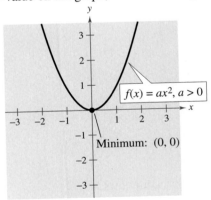

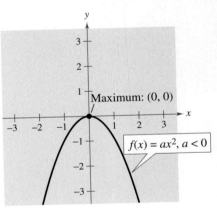

Leading coefficient is positive.

FIGURE 2.2

Leading coefficient is negative.

When sketching the graph of $f(x) = ax^2$, it is helpful to use the graph of $y = x^2$ as a reference, as discussed in Section 1.7.

Example 1   Sketching Graphs of Quadratic Functions

a. Compare the graphs of $y = x^2$ and $f(x) = \frac{1}{3}x^2$.

b. Compare the graphs of $y = x^2$ and $g(x) = 2x^2$.

Solution

a. Compared with $y = x^2$, each output of $f(x) = \frac{1}{3}x^2$ "shrinks" by a factor of $\frac{1}{3}$, creating the broader parabola shown in Figure 2.3.

b. Compared with $y = x^2$, each output of $g(x) = 2x^2$ "stretches" by a factor of 2, creating the narrower parabola shown in Figure 2.4.

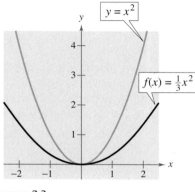

FIGURE 2.3

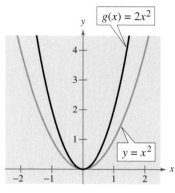

FIGURE 2.4

CHECK**Point**   Now try Exercise 13.

In Example 1, note that the coefficient a determines how widely the parabola given by $f(x) = ax^2$ opens. If $|a|$ is small, the parabola opens more widely than if $|a|$ is large.

Recall from Section 1.7 that the graphs of $y = f(x \pm c)$, $y = f(x) \pm c$, $y = f(-x)$, and $y = -f(x)$ are rigid transformations of the graph of $y = f(x)$. For instance, in Figure 2.5, notice how the graph of $y = x^2$ can be transformed to produce the graphs of $f(x) = -x^2 + 1$ and $g(x) = (x + 2)^2 - 3$.

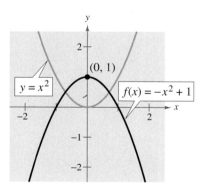

Reflection in x-axis followed by
an upward shift of one unit
FIGURE 2.5

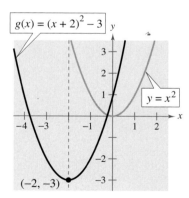

Left shift of two units followed by
a downward shift of three units

Algebra Help

You can review the techniques for shifting, reflecting, and stretching graphs in Section 1.7.

The Standard Form of a Quadratic Function

The **standard form** of a quadratic function is $f(x) = a(x - h)^2 + k$. This form is especially convenient for sketching a parabola because it identifies the vertex of the parabola as (h, k).

> ### Standard Form of a Quadratic Function
>
> The quadratic function given by
>
> $$f(x) = a(x - h)^2 + k, \quad a \neq 0$$
>
> is in **standard form.** The graph of f is a parabola whose axis is the vertical line $x = h$ and whose vertex is the point (h, k). If $a > 0$, the parabola opens upward, and if $a < 0$, the parabola opens downward.

To graph a parabola, it is helpful to begin by writing the quadratic function in standard form using the process of completing the square, as illustrated in Example 2. In this example, notice that when completing the square, you *add and subtract* the square of half the coefficient of x within the parentheses instead of adding the value to each side of the equation as is done in Appendix A.5.

Example 2   Graphing a Parabola in Standard Form

Sketch the graph of $f(x) = 2x^2 + 8x + 7$ and identify the vertex and the axis of the parabola.

Solution

Begin by writing the quadratic function in standard form. Notice that the first step in completing the square is to factor out any coefficient of x^2 that is not 1.

$$f(x) = 2x^2 + 8x + 7 \qquad \text{Write original function.}$$
$$= 2(x^2 + 4x) + 7 \qquad \text{Factor 2 out of } x\text{-terms.}$$
$$= 2(x^2 + 4x + 4 - 4) + 7 \qquad \text{Add and subtract 4 within parentheses.}$$

$$(4/2)^2$$

After adding and subtracting 4 within the parentheses, you must now regroup the terms to form a perfect square trinomial. The -4 can be removed from inside the parentheses; however, because of the 2 outside of the parentheses, you must multiply -4 by 2, as shown below.

$$f(x) = 2(x^2 + 4x + 4) - 2(4) + 7 \qquad \text{Regroup terms.}$$
$$= 2(x^2 + 4x + 4) - 8 + 7 \qquad \text{Simplify.}$$
$$= 2(x + 2)^2 - 1 \qquad \text{Write in standard form.}$$

From this form, you can see that the graph of f is a parabola that opens upward and has its vertex at $(-2, -1)$. This corresponds to a left shift of two units and a downward shift of one unit relative to the graph of $y = 2x^2$, as shown in Figure 2.6. In the figure, you can see that the axis of the parabola is the vertical line through the vertex, $x = -2$.

CHECK *Point*   Now try Exercise 19.

Study Tip

The standard form of a quadratic function identifies four basic transformations of the graph of $y = x^2$.

a. The factor $|a|$ produces a vertical stretch or shrink.

b. If $a < 0$, the graph is reflected in the x-axis.

c. The factor $(x - h)^2$ represents a horizontal shift of h units.

d. The term k represents a vertical shift of k units.

Algebra Help

You can review the techniques for completing the square in Appendix A.5.

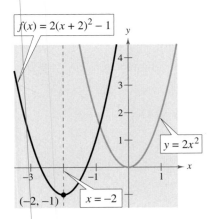

$f(x) = 2(x + 2)^2 - 1$

$y = 2x^2$

$(-2, -1)$ $x = -2$

FIGURE **2.6**

Algebra Help

You can review the techniques for using the Quadratic Formula in Appendix A.5.

To find the *x*-intercepts of the graph of $f(x) = ax^2 + bx + c$, you must solve the equation $ax^2 + bx + c = 0$. If $ax^2 + bx + c$ does not factor, you can use the Quadratic Formula to find the *x*-intercepts. Remember, however, that a parabola may not have *x*-intercepts.

Example 3   Finding the Vertex and *x*-Intercepts of a Parabola

Sketch the graph of $f(x) = -x^2 + 6x - 8$ and identify the vertex and *x*-intercepts.

Solution

$$f(x) = -x^2 + 6x - 8 \qquad \text{Write original function.}$$
$$= -(x^2 - 6x) - 8 \qquad \text{Factor } -1 \text{ out of } x\text{-terms.}$$
$$= -(x^2 - 6x + 9 - 9) - 8 \qquad \text{Add and subtract 9 within parentheses.}$$

$$(-6/2)^2$$

$$= -(x^2 - 6x + 9) - (-9) - 8 \qquad \text{Regroup terms.}$$
$$= -(x - 3)^2 + 1 \qquad \text{Write in standard form.}$$

From this form, you can see that *f* is a parabola that opens downward with vertex (3, 1). The *x*-intercepts of the graph are determined as follows.

$$-(x^2 - 6x + 8) = 0 \qquad \text{Factor out } -1.$$
$$-(x - 2)(x - 4) = 0 \qquad \text{Factor.}$$
$$x - 2 = 0 \quad \Longrightarrow \quad x = 2 \qquad \text{Set 1st factor equal to 0.}$$
$$x - 4 = 0 \quad \Longrightarrow \quad x = 4 \qquad \text{Set 2nd factor equal to 0.}$$

So, the *x*-intercepts are (2, 0) and (4, 0), as shown in Figure 2.7.

CHECK *Point*   Now try Exercise 25.

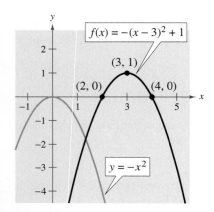

FIGURE **2.7**

Example 4   Writing the Equation of a Parabola

Write the standard form of the equation of the parabola whose vertex is (1, 2) and that passes through the point (3, −6).

Solution

Because the vertex of the parabola is at $(h, k) = (1, 2)$, the equation has the form

$$f(x) = a(x - 1)^2 + 2. \qquad \text{Substitute for } h \text{ and } k \text{ in standard form.}$$

Because the parabola passes through the point (3, −6), it follows that $f(3) = -6$. So,

$$f(x) = a(x - 1)^2 + 2 \qquad \text{Write in standard form.}$$
$$-6 = a(3 - 1)^2 + 2 \qquad \text{Substitute 3 for } x \text{ and } -6 \text{ for } f(x).$$
$$-6 = 4a + 2 \qquad \text{Simplify.}$$
$$-8 = 4a \qquad \text{Subtract 2 from each side.}$$
$$-2 = a. \qquad \text{Divide each side by 4.}$$

The equation in standard form is $f(x) = -2(x - 1)^2 + 2$. The graph of *f* is shown in Figure 2.8.

CHECK *Point*   Now try Exercise 47.

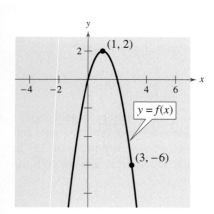

FIGURE **2.8**

Finding Minimum and Maximum Values

Many applications involve finding the maximum or minimum value of a quadratic function. By completing the square of the quadratic function $f(x) = ax^2 + bx + c$, you can rewrite the function in standard form (see Exercise 95).

$$f(x) = a\left(x + \frac{b}{2a}\right)^2 + \left(c - \frac{b^2}{4a}\right) \qquad \text{Standard form}$$

So, the vertex of the graph of f is $\left(-\frac{b}{2a}, f\left(-\frac{b}{2a}\right)\right)$, which implies the following.

Minimum and Maximum Values of Quadratic Functions

Consider the function $f(x) = ax^2 + bx + c$ with vertex $\left(-\frac{b}{2a},\ f\left(-\frac{b}{2a}\right)\right)$.

1. If $a > 0$, f has a *minimum* at $x = -\frac{b}{2a}$. The minimum value is $f\left(-\frac{b}{2a}\right)$.

2. If $a < 0$, f has a *maximum* at $x = -\frac{b}{2a}$. The maximum value is $f\left(-\frac{b}{2a}\right)$.

Example 5   The Maximum Height of a Baseball

A baseball is hit at a point 3 feet above the ground at a velocity of 100 feet per second and at an angle of $45°$ with respect to the ground. The path of the baseball is given by the function $f(x) = -0.0032x^2 + x + 3$, where $f(x)$ is the height of the baseball (in feet) and x is the horizontal distance from home plate (in feet). What is the maximum height reached by the baseball?

Algebraic Solution

For this quadratic function, you have

$$f(x) = ax^2 + bx + c$$
$$= -0.0032x^2 + x + 3$$

which implies that $a = -0.0032$ and $b = 1$. Because $a < 0$, the function has a maximum when $x = -b/(2a)$. So, you can conclude that the baseball reaches its maximum height when it is x feet from home plate, where x is

$$x = -\frac{b}{2a}$$
$$= -\frac{1}{2(-0.0032)}$$
$$= 156.25 \text{ feet.}$$

At this distance, the maximum height is

$$f(156.25) = -0.0032(156.25)^2 + 156.25 + 3$$
$$= 81.125 \text{ feet.}$$

CHECK **Point** Now try Exercise 75.

Graphical Solution

Use a graphing utility to graph

$$y = -0.0032x^2 + x + 3$$

so that you can see the important features of the parabola. Use the *maximum* feature (see Figure 2.9) or the *zoom* and *trace* features (see Figure 2.10) of the graphing utility to approximate the maximum height on the graph to be $y \approx 81.125$ feet at $x \approx 156.25$.

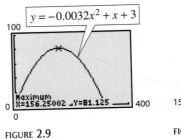

FIGURE **2.9**

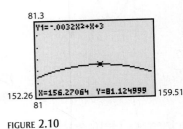

FIGURE **2.10**

2.1   EXERCISES

See www.CalcChat.com for worked-out solutions to odd-numbered exercises.

VOCABULARY: Fill in the blanks.

1. Linear, constant, and squaring functions are examples of _____ functions.
2. A polynomial function of degree n and leading coefficient a_n is a function of the form
   $f(x) = a_n x^n + a_{n-1}x^{n-1} + \cdots + a_1 x + a_0 \ (a_n \neq 0)$ where n is a _____ _____ and $a_n, a_{n-1}, \ldots, a_1, a_0$ are _____ numbers.
3. A _____ function is a second-degree polynomial function, and its graph is called a _____.
4. The graph of a quadratic function is symmetric about its _____.
5. If the graph of a quadratic function opens upward, then its leading coefficient is _____ and the vertex of the graph is a _____.
6. If the graph of a quadratic function opens downward, then its leading coefficient is _____ and the vertex of the graph is a _____.

SKILLS AND APPLICATIONS

In Exercises 7–12, match the quadratic function with its graph. [The graphs are labeled (a), (b), (c), (d), (e), and (f).]

(a)
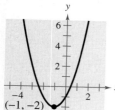
(b)
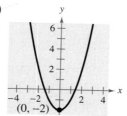

(c)
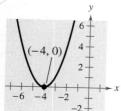
(d)
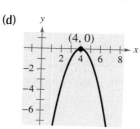

(e)
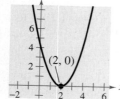
(f)
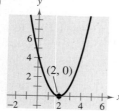

7. $f(x) = (x-2)^2$
8. $f(x) = (x+4)^2$
9. $f(x) = x^2 - 2$
10. $f(x) = (x+1)^2 - 2$
11. $f(x) = 4 - (x-2)^2$
12. $f(x) = -(x-4)^2$

In Exercises 13–16, graph each function. Compare the graph of each function with the graph of $y = x^2$.

13. (a) $f(x) = \frac{1}{2}x^2$ (b) $g(x) = -\frac{1}{8}x^2$
    (c) $h(x) = \frac{3}{2}x^2$ (d) $k(x) = -3x^2$

14. (a) $f(x) = x^2 + 1$ (b) $g(x) = x^2 - 1$
    (c) $h(x) = x^2 + 3$ (d) $k(x) = x^2 - 3$
15. (a) $f(x) = (x-1)^2$ (b) $g(x) = (3x)^2 + 1$
    (c) $h(x) = \left(\frac{1}{3}x\right)^2 - 3$ (d) $k(x) = (x+3)^2$
16. (a) $f(x) = -\frac{1}{2}(x-2)^2 + 1$
    (b) $g(x) = \left[\frac{1}{2}(x-1)\right]^2 - 3$
    (c) $h(x) = -\frac{1}{2}(x+2)^2 - 1$
    (d) $k(x) = [2(x+1)]^2 + 4$

In Exercises 17–34, sketch the graph of the quadratic function without using a graphing utility. Identify the vertex, axis of symmetry, and x-intercept(s).

17. $f(x) = 1 - x^2$ 18. $g(x) = x^2 - 8$
19. $f(x) = x^2 + 7$ 20. $h(x) = 12 - x^2$
21. $f(x) = \frac{1}{2}x^2 - 4$ 22. $f(x) = 16 - \frac{1}{4}x^2$
23. $f(x) = (x+4)^2 - 3$ 24. $f(x) = (x-6)^2 + 8$
25. $h(x) = x^2 - 8x + 16$ 26. $g(x) = x^2 + 2x + 1$
27. $f(x) = x^2 - x + \frac{5}{4}$ 28. $f(x) = x^2 + 3x + \frac{1}{4}$
29. $f(x) = -x^2 + 2x + 5$ 30. $f(x) = -x^2 - 4x + 1$
31. $h(x) = 4x^2 - 4x + 21$ 32. $f(x) = 2x^2 - x + 1$
33. $f(x) = \frac{1}{4}x^2 - 2x - 12$ 34. $f(x) = -\frac{1}{3}x^2 + 3x - 6$

In Exercises 35–42, use a graphing utility to graph the quadratic function. Identify the vertex, axis of symmetry, and x-intercepts. Then check your results algebraically by writing the quadratic function in standard form.

35. $f(x) = -(x^2 + 2x - 3)$ 36. $f(x) = -(x^2 + x - 30)$
37. $g(x) = x^2 + 8x + 11$ 38. $f(x) = x^2 + 10x + 14$
39. $f(x) = 2x^2 - 16x + 31$
40. $f(x) = -4x^2 + 24x - 41$
41. $g(x) = \frac{1}{2}(x^2 + 4x - 2)$ 42. $f(x) = \frac{3}{5}(x^2 + 6x - 5)$

In Exercises 43–46, write an equation for the parabola in standard form.

43.

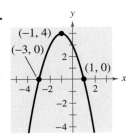

44.

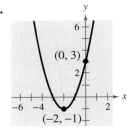

45.

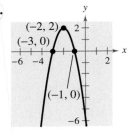

46.

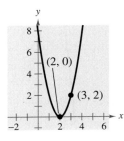

In Exercises 47–56, write the standard form of the equation of the parabola that has the indicated vertex and whose graph passes through the given point.

47. Vertex: $(-2, 5)$; point: $(0, 9)$

48. Vertex: $(4, -1)$; point: $(2, 3)$

49. Vertex: $(1, -2)$; point: $(-1, 14)$

50. Vertex: $(2, 3)$; point: $(0, 2)$

51. Vertex: $(5, 12)$; point: $(7, 15)$

52. Vertex: $(-2, -2)$; point: $(-1, 0)$

53. Vertex: $\left(-\frac{1}{4}, \frac{3}{2}\right)$; point: $(-2, 0)$

54. Vertex: $\left(\frac{5}{2}, -\frac{3}{4}\right)$; point: $(-2, 4)$

55. Vertex: $\left(-\frac{5}{2}, 0\right)$; point: $\left(-\frac{7}{2}, -\frac{16}{3}\right)$

56. Vertex: $(6, 6)$; point: $\left(\frac{61}{10}, \frac{3}{2}\right)$

GRAPHICAL REASONING In Exercises 57 and 58, determine the *x*-intercept(s) of the graph visually. Then find the *x*-intercept(s) algebraically to confirm your results.

57. $y = x^2 - 4x - 5$

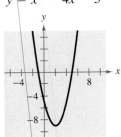

58. $y = 2x^2 + 5x - 3$

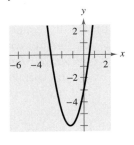

In Exercises 59–64, use a graphing utility to graph the quadratic function. Find the *x*-intercepts of the graph and compare them with the solutions of the corresponding quadratic equation when $f(x) = 0$.

59. $f(x) = x^2 - 4x$

60. $f(x) = -2x^2 + 10x$

61. $f(x) = x^2 - 9x + 18$

62. $f(x) = x^2 - 8x - 20$

63. $f(x) = 2x^2 - 7x - 30$

64. $f(x) = \frac{7}{10}(x^2 + 12x - 45)$

In Exercises 65–70, find two quadratic functions, one that opens upward and one that opens downward, whose graphs have the given *x*-intercepts. (There are many correct answers.)

65. $(-1, 0), (3, 0)$

66. $(-5, 0), (5, 0)$

67. $(0, 0), (10, 0)$

68. $(4, 0), (8, 0)$

69. $(-3, 0), \left(-\frac{1}{2}, 0\right)$

70. $\left(-\frac{5}{2}, 0\right), (2, 0)$

In Exercises 71–74, find two positive real numbers whose product is a maximum.

71. The sum is 110.

72. The sum is *S*.

73. The sum of the first and twice the second is 24.

74. The sum of the first and three times the second is 42.

75. PATH OF A DIVER   The path of a diver is given by

$$y = -\frac{4}{9}x^2 + \frac{24}{9}x + 12$$

where *y* is the height (in feet) and *x* is the horizontal distance from the end of the diving board (in feet). What is the maximum height of the diver?

76. HEIGHT OF A BALL   The height *y* (in feet) of a punted football is given by

$$y = -\frac{16}{2025}x^2 + \frac{9}{5}x + 1.5$$

where *x* is the horizontal distance (in feet) from the point at which the ball is punted.

(a) How high is the ball when it is punted?

(b) What is the maximum height of the punt?

(c) How long is the punt?

77. MINIMUM COST   A manufacturer of lighting fixtures has daily production costs of $C = 800 - 10x + 0.25x^2$, where *C* is the total cost (in dollars) and *x* is the number of units produced. How many fixtures should be produced each day to yield a minimum cost?

78. MAXIMUM PROFIT   The profit *P* (in hundreds of dollars) that a company makes depends on the amount *x* (in hundreds of dollars) the company spends on advertising according to the model $P = 230 + 20x - 0.5x^2$. What expenditure for advertising will yield a maximum profit?

79. MAXIMUM REVENUE   The total revenue R earned (in thousands of dollars) from manufacturing handheld video games is given by

$$R(p) = -25p^2 + 1200p$$

where p is the price per unit (in dollars).

(a) Find the revenues when the price per unit is $20, $25, and $30.

(b) Find the unit price that will yield a maximum revenue. What is the maximum revenue? Explain your results.

80. MAXIMUM REVENUE   The total revenue R earned per day (in dollars) from a pet-sitting service is given by $R(p) = -12p^2 + 150p$, where p is the price charged per pet (in dollars).

(a) Find the revenues when the price per pet is $4, $6, and $8.

(b) Find the price that will yield a maximum revenue. What is the maximum revenue? Explain your results.

81. NUMERICAL, GRAPHICAL, AND ANALYTICAL ANALYSIS   A rancher has 200 feet of fencing to enclose two adjacent rectangular corrals (see figure).

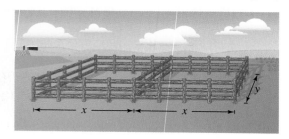

(a) Write the area A of the corrals as a function of x.

(b) Create a table showing possible values of x and the corresponding areas of the corral. Use the table to estimate the dimensions that will produce the maximum enclosed area.

 (c) Use a graphing utility to graph the area function. Use the graph to approximate the dimensions that will produce the maximum enclosed area.

(d) Write the area function in standard form to find analytically the dimensions that will produce the maximum area.

(e) Compare your results from parts (b), (c), and (d).

82. GEOMETRY   An indoor physical fitness room consists of a rectangular region with a semicircle on each end. The perimeter of the room is to be a 200-meter single-lane running track.

(a) Draw a diagram that illustrates the problem. Let x and y represent the length and width of the rectangular region, respectively.

(b) Determine the radius of each semicircular end of the room. Determine the distance, in terms of y, around the inside edge of each semicircular part of the track.

(c) Use the result of part (b) to write an equation, in terms of x and y, for the distance traveled in one lap around the track. Solve for y.

(d) Use the result of part (c) to write the area A of the rectangular region as a function of x. What dimensions will produce a rectangle of maximum area?

83. MAXIMUM REVENUE   A small theater has a seating capacity of 2000. When the ticket price is $20, attendance is 1500. For each $1 decrease in price, attendance increases by 100.

(a) Write the revenue R of the theater as a function of ticket price x.

(b) What ticket price will yield a maximum revenue? What is the maximum revenue?

84. MAXIMUM AREA   A Norman window is constructed by adjoining a semicircle to the top of an ordinary rectangular window (see figure). The perimeter of the window is 16 feet.

(a) Write the area A of the window as a function of x.

(b) What dimensions will produce a window of maximum area?

85. GRAPHICAL ANALYSIS   From 1950 through 2005, the per capita consumption C of cigarettes by Americans (age 18 and older) can be modeled by $C = 3565.0 + 60.30t - 1.783t^2$, $0 \leq t \leq 55$, where t is the year, with $t = 0$ corresponding to 1950. (Source: *Tobacco Outlook Report*)

(a) Use a graphing utility to graph the model.

(b) Use the graph of the model to approximate the maximum average annual consumption. Beginning in 1966, all cigarette packages were required by law to carry a health warning. Do you think the warning had any effect? Explain.

(c) In 2005, the U.S. population (age 18 and over) was 296,329,000. Of those, about 59,858,458 were smokers. What was the average annual cigarette consumption *per smoker* in 2005? What was the average daily cigarette consumption *per smoker*?

86. DATA ANALYSIS: SALES The sales y (in billions of dollars) for Harley-Davidson from 2000 through 2007 are shown in the table. (Source: U.S. Harley-Davidson, Inc.)

Year	Sales, y
2000	2.91
2001	3.36
2002	4.09
2003	4.62
2004	5.02
2005	5.34
2006	5.80
2007	5.73

(a) Use a graphing utility to create a scatter plot of the data. Let x represent the year, with $x = 0$ corresponding to 2000.

(b) Use the *regression* feature of the graphing utility to find a quadratic model for the data.

(c) Use the graphing utility to graph the model in the same viewing window as the scatter plot. How well does the model fit the data?

(d) Use the *trace* feature of the graphing utility to approximate the year in which the sales for Harley-Davidson were the greatest.

(e) Verify your answer to part (d) algebraically.

(f) Use the model to predict the sales for Harley-Davidson in 2010.

EXPLORATION

TRUE OR FALSE? In Exercises 87–90, determine whether the statement is true or false. Justify your answer.

87. The function given by $f(x) = -12x^2 - 1$ has no x-intercepts.

88. The graphs of $f(x) = -4x^2 - 10x + 7$ and $g(x) = 12x^2 + 30x + 1$ have the same axis of symmetry.

89. The graph of a quadratic function with a negative leading coefficient will have a maximum value at its vertex.

90. The graph of a quadratic function with a positive leading coefficient will have a minimum value at its vertex.

THINK ABOUT IT In Exercises 91–94, find the values of b such that the function has the given maximum or minimum value.

91. $f(x) = -x^2 + bx - 75$; Maximum value: 25

92. $f(x) = -x^2 + bx - 16$; Maximum value: 48

93. $f(x) = x^2 + bx + 26$; Minimum value: 10

94. $f(x) = x^2 + bx - 25$; Minimum value: -50

95. Write the quadratic function

$$f(x) = ax^2 + bx + c$$

in standard form to verify that the vertex occurs at

$$\left(-\frac{b}{2a},\ f\left(-\frac{b}{2a} \right) \right).$$

96. CAPSTONE The profit P (in millions of dollars) for a recreational vehicle retailer is modeled by a quadratic function of the form

$$P = at^2 + bt + c$$

where t represents the year. If you were president of the company, which of the models below would you prefer? Explain your reasoning.

(a) a is positive and $-b/(2a) \le t$.

(b) a is positive and $t \le -b/(2a)$.

(c) a is negative and $-b/(2a) \le t$.

(d) a is negative and $t \le -b/(2a)$.

97. GRAPHICAL ANALYSIS

(a) Graph $y = ax^2$ for $a = -2, -1, -0.5, 0.5, 1$ and 2. How does changing the value of a affect the graph?

(b) Graph $y = (x - h)^2$ for $h = -4, -2, 2,$ and 4. How does changing the value of h affect the graph?

(c) Graph $y = x^2 + k$ for $k = -4, -2, 2,$ and 4. How does changing the value of k affect the graph?

98. Describe the sequence of transformation from f to g given that $f(x) = x^2$ and $g(x) = a(x - h)^2 + k$. (Assume a, h, and k are positive.)

99. Is it possible for a quadratic equation to have only one x-intercept? Explain.

100. Assume that the function given by

$$f(x) = ax^2 + bx + c, \quad a \neq 0$$

has two real zeros. Show that the x-coordinate of the vertex of the graph is the average of the zeros of f. (*Hint:* Use the Quadratic Formula.)

PROJECT: HEIGHT OF A BASKETBALL To work an extended application analyzing the height of a basketball after it has been dropped, visit this text's website at *academic.cengage.com*.

2.2 POLYNOMIAL FUNCTIONS OF HIGHER DEGREE

Graphs of Polynomial Functions

In this section, you will study basic features of the graphs of polynomial functions. The first feature is that the graph of a polynomial function is **continuous.** Essentially, this means that the graph of a polynomial function has no breaks, holes, or gaps, as shown in Figure 2.11(a). The graph shown in Figure 2.11(b) is an example of a piecewise-defined function that is not continuous.

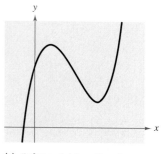

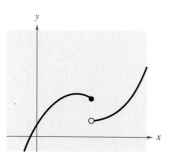

(a) Polynomial functions have continuous graphs.

(b) Functions with graphs that are not continuous are not polynomial functions.

FIGURE 2.11

The second feature is that the graph of a polynomial function has only smooth, rounded turns, as shown in Figure 2.12. A polynomial function cannot have a sharp turn. For instance, the function given by $f(x) = |x|$, which has a sharp turn at the point $(0, 0)$, as shown in Figure 2.13, is not a polynomial function.

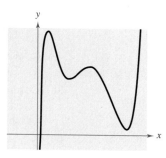

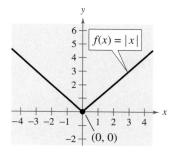

Polynomial functions have graphs with smooth, rounded turns.

FIGURE 2.12

Graphs of polynomial functions cannot have sharp turns.

FIGURE 2.13

The graphs of polynomial functions of degree greater than 2 are more difficult to analyze than the graphs of polynomials of degree 0, 1, or 2. However, using the features presented in this section, coupled with your knowledge of point plotting, intercepts, and symmetry, you should be able to make reasonably accurate sketches *by hand.*

The polynomial functions that have the simplest graphs are monomials of the form $f(x) = x^n$, where n is an integer greater than zero. From Figure 2.14, you can see that when n is *even*, the graph is similar to the graph of $f(x) = x^2$, and when n is *odd*, the graph is similar to the graph of $f(x) = x^3$. Moreover, the greater the value of n, the flatter the graph near the origin. Polynomial functions of the form $f(x) = x^n$ are often referred to as **power functions.**

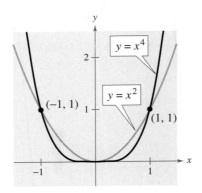   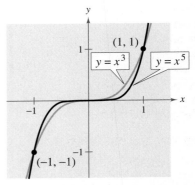

(a) If n is even, the graph of $y = x^n$ touches the axis at the x-intercept.

(b) If n is odd, the graph of $y = x^n$ crosses the axis at the x-intercept.

FIGURE 2.14

Example 1   Sketching Transformations of Polynomial Functions

Sketch the graph of each function.

a. $f(x) = -x^5$ **b.** $h(x) = (x + 1)^4$

Solution

a. Because the degree of $f(x) = -x^5$ is odd, its graph is similar to the graph of $y = x^3$. In Figure 2.15, note that the negative coefficient has the effect of reflecting the graph in the x-axis.

b. The graph of $h(x) = (x + 1)^4$, as shown in Figure 2.16, is a left shift by one unit of the graph of $y = x^4$.

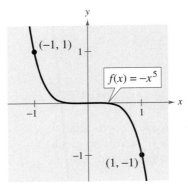   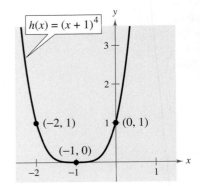

FIGURE 2.15

FIGURE 2.16

CHECK*Point*   Now try Exercise 17.

The Leading Coefficient Test

In Example 1, note that both graphs eventually rise or fall without bound as x moves to the right. Whether the graph of a polynomial function eventually rises or falls can be determined by the function's degree (even or odd) and by its leading coefficient, as indicated in the **Leading Coefficient Test.**

Leading Coefficient Test

As x moves without bound to the left or to the right, the graph of the polynomial function $f(x) = a_n x^n + \cdots + a_1 x + a_0$ eventually rises or falls in the following manner.

1. When n is *odd:*

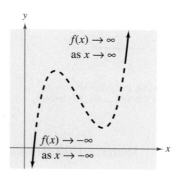

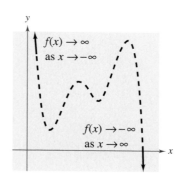

If the leading coefficient is positive $(a_n > 0)$, the graph falls to the left and rises to the right.

If the leading coefficient is negative $(a_n < 0)$, the graph rises to the left and falls to the right.

2. When n is *even:*

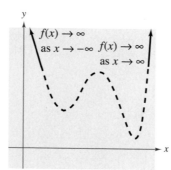

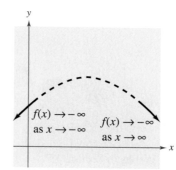

If the leading coefficient is positive $(a_n > 0)$, the graph rises to the left and right.

If the leading coefficient is negative $(a_n < 0)$, the graph falls to the left and right.

The dashed portions of the graphs indicate that the test determines *only* the right-hand and left-hand behavior of the graph.

| **Example 2** | **Applying the Leading Coefficient Test** |

Describe the right-hand and left-hand behavior of the graph of each function.

a. $f(x) = -x^3 + 4x$ **b.** $f(x) = x^4 - 5x^2 + 4$ **c.** $f(x) = x^5 - x$

Solution

a. Because the degree is odd and the leading coefficient is negative, the graph rises to the left and falls to the right, as shown in Figure 2.17.

b. Because the degree is even and the leading coefficient is positive, the graph rises to the left and right, as shown in Figure 2.18.

c. Because the degree is odd and the leading coefficient is positive, the graph falls to the left and rises to the right, as shown in Figure 2.19.

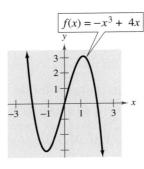

FIGURE **2.17**

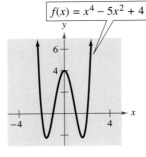

FIGURE **2.18**

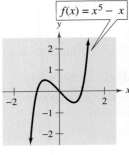

FIGURE **2.19**

CHECK *Point*   Now try Exercise 23.

In Example 2, note that the Leading Coefficient Test tells you only whether the graph *eventually* rises or falls to the right or left. Other characteristics of the graph, such as intercepts and minimum and maximum points, must be determined by other tests.

Zeros of Polynomial Functions

It can be shown that for a polynomial function f of degree n, the following statements are true.

1. The function f has, at most, n real zeros. (You will study this result in detail in the discussion of the Fundamental Theorem of Algebra in Section 2.5.)

2. The graph of f has, at most, $n - 1$ turning points. (Turning points, also called relative minima or relative maxima, are points at which the graph changes from increasing to decreasing or vice versa.)

Finding the zeros of polynomial functions is one of the most important problems in algebra. There is a strong interplay between graphical and algebraic approaches to this problem. Sometimes you can use information about the graph of a function to help find its zeros, and in other cases you can use information about the zeros of a function to help sketch its graph. Finding zeros of polynomial functions is closely related to factoring and finding x-intercepts.

Study Tip

Remember that the *zeros* of a function of x are the x-values for which the function is zero.

Algebra Help

To do Example 3 algebraically, you need to be able to completely factor polynomials. You can review the techniques for factoring in Appendix A.3.

Real Zeros of Polynomial Functions

If f is a polynomial function and a is a real number, the following statements are equivalent.

1. $x = a$ is a *zero* of the function f.

2. $x = a$ is a *solution* of the polynomial equation $f(x) = 0$.

3. $(x - a)$ is a *factor* of the polynomial $f(x)$.

4. $(a, 0)$ is an *x-intercept* of the graph of f.

Example 3   Finding the Zeros of a Polynomial Function

Find all real zeros of

$$f(x) = -2x^4 + 2x^2.$$

Then determine the number of turning points of the graph of the function.

Algebraic Solution

To find the real zeros of the function, set $f(x)$ equal to zero and solve for x.

$$-2x^4 + 2x^2 = 0 \qquad \text{Set } f(x) \text{ equal to 0.}$$

$$-2x^2(x^2 - 1) = 0 \qquad \text{Remove common monomial factor.}$$

$$-2x^2(x - 1)(x + 1) = 0 \qquad \text{Factor completely.}$$

So, the real zeros are $x = 0$, $x = 1$, and $x = -1$. Because the function is a fourth-degree polynomial, the graph of f can have at most $4 - 1 = 3$ turning points.

Graphical Solution

Use a graphing utility to graph $y = -2x^4 + 2x^2$. In Figure 2.20, the graph appears to have zeros at $(0, 0)$, $(1, 0)$, and $(-1, 0)$. Use the *zero* or *root* feature, or the *zoom* and *trace* features, of the graphing utility to verify these zeros. So, the real zeros are $x = 0$, $x = 1$, and $x = -1$. From the figure, you can see that the graph has three turning points. This is consistent with the fact that a fourth-degree polynomial can have at most three turning points.

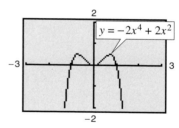

FIGURE 2.20

CHECK Point   Now try Exercise 35.

In Example 3, note that because the exponent is greater than 1, the factor $-2x^2$ yields the *repeated* zero $x = 0$. Because the exponent is even, the graph touches the x-axis at $x = 0$, as shown in Figure 2.20.

Repeated Zeros

A factor $(x - a)^k$, $k > 1$, yields a **repeated zero** $x = a$ of **multiplicity** k.

1. If k is odd, the graph *crosses* the x-axis at $x = a$.

2. If k is even, the graph *touches* the x-axis (but does not cross the x-axis) at $x = a$.

To graph polynomial functions, you can use the fact that a polynomial function can change signs only at its zeros. Between two consecutive zeros, a polynomial must be entirely positive or entirely negative. (This follows from the Intermediate Value Theorem, which you will study later in this section.) This means that when the real zeros of a polynomial function are put in order, they divide the real number line into intervals in which the function has no sign changes. These resulting intervals are **test intervals** in which a representative x-value in the interval is chosen to determine if the value of the polynomial function is positive (the graph lies above the x-axis) or negative (the graph lies below the x-axis).

TECHNOLOGY

Example 4 uses an *algebraic approach* to describe the graph of the function. A graphing utility is a complement to this approach. Remember that an important aspect of using a graphing utility is to find a viewing window that shows all significant features of the graph. For instance, the viewing window in part (a) illustrates all of the significant features of the function in Example 4 while the viewing window in part (b) does not.

a.
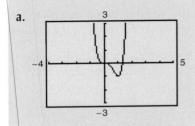

b.
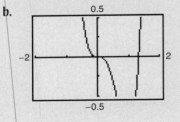

| Example 4 | **Sketching the Graph of a Polynomial Function**

Sketch the graph of $f(x) = 3x^4 - 4x^3$.

Solution

1. *Apply the Leading Coefficient Test.*   Because the leading coefficient is positive and the degree is even, you know that the graph eventually rises to the left and to the right (see Figure 2.21).

2. *Find the Zeros of the Polynomial.*   By factoring $f(x) = 3x^4 - 4x^3$ as $f(x) = x^3(3x - 4)$, you can see that the zeros of f are $x = 0$ and $x = \frac{4}{3}$ (both of odd multiplicity). So, the x-intercepts occur at $(0, 0)$ and $\left(\frac{4}{3}, 0\right)$. Add these points to your graph, as shown in Figure 2.21.

3. *Plot a Few Additional Points.*   Use the zeros of the polynomial to find the test intervals. In each test interval, choose a representative x-value and evaluate the polynomial function, as shown in the table.

Test interval	Representative x-value	Value of f	Sign	Point on graph
$(-\infty, 0)$	-1	$f(-1) = 7$	Positive	$(-1, 7)$
$\left(0, \frac{4}{3}\right)$	1	$f(1) = -1$	Negative	$(1, -1)$
$\left(\frac{4}{3}, \infty\right)$	1.5	$f(1.5) = 1.6875$	Positive	$(1.5, 1.6875)$

4. *Draw the Graph.*   Draw a continuous curve through the points, as shown in Figure 2.22. Because both zeros are of odd multiplicity, you know that the graph should cross the x-axis at $x = 0$ and $x = \frac{4}{3}$.

⚠ WARNING / CAUTION

If you are unsure of the shape of a portion of the graph of a polynomial function, plot some additional points, such as the point $(0.5, -0.3125)$, as shown in Figure 2.22.

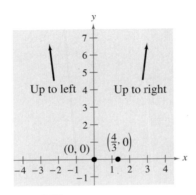

FIGURE 2.21

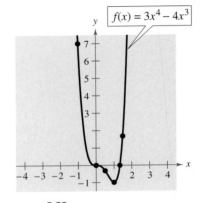

FIGURE 2.22

CHECK*Point*   Now try Exercise 75.

Example 5   Sketching the Graph of a Polynomial Function

Sketch the graph of $f(x) = -2x^3 + 6x^2 - \frac{9}{2}x$.

Solution

1. *Apply the Leading Coefficient Test.*   Because the leading coefficient is negative and the degree is odd, you know that the graph eventually rises to the left and falls to the right (see Figure 2.23).

2. *Find the Zeros of the Polynomial.*   By factoring

$$f(x) = -2x^3 + 6x^2 - \tfrac{9}{2}x$$
$$= -\tfrac{1}{2}x(4x^2 - 12x + 9)$$
$$= -\tfrac{1}{2}x(2x - 3)^2$$

you can see that the zeros of f are $x = 0$ (odd multiplicity) and $x = \frac{3}{2}$ (even multiplicity). So, the x-intercepts occur at $(0, 0)$ and $\left(\frac{3}{2}, 0\right)$. Add these points to your graph, as shown in Figure 2.23.

3. *Plot a Few Additional Points.*   Use the zeros of the polynomial to find the test intervals. In each test interval, choose a representative x-value and evaluate the polynomial function, as shown in the table.

Test interval	Representative x-value	Value of f	Sign	Point on graph
$(-\infty, 0)$	-0.5	$f(-0.5) = 4$	Positive	$(-0.5, 4)$
$\left(0, \frac{3}{2}\right)$	0.5	$f(0.5) = -1$	Negative	$(0.5, -1)$
$\left(\frac{3}{2}, \infty\right)$	2	$f(2) = -1$	Negative	$(2, -1)$

4. *Draw the Graph.*   Draw a continuous curve through the points, as shown in Figure 2.24. As indicated by the multiplicities of the zeros, the graph crosses the x-axis at $(0, 0)$ but does not cross the x-axis at $\left(\frac{3}{2}, 0\right)$.

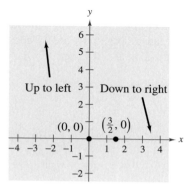

FIGURE 2.23

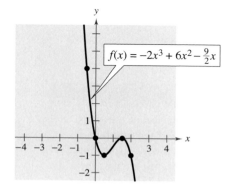

$f(x) = -2x^3 + 6x^2 - \frac{9}{2}x$

FIGURE 2.24

CHECK*Point*   Now try Exercise 77.

The Intermediate Value Theorem

The next theorem, called the **Intermediate Value Theorem,** illustrates the existence of real zeros of polynomial functions. This theorem implies that if $(a, f(a))$ and $(b, f(b))$ are two points on the graph of a polynomial function such that $f(a) \neq f(b)$, then for any number d between $f(a)$ and $f(b)$ there must be a number c between a and b such that $f(c) = d$. (See Figure 2.25.)

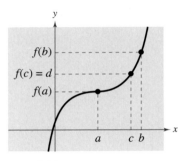

FIGURE 2.25

Intermediate Value Theorem

Let a and b be real numbers such that $a < b$. If f is a polynomial function such that $f(a) \neq f(b)$, then, in the interval $[a, b]$, f takes on every value between $f(a)$ and $f(b)$.

The Intermediate Value Theorem helps you locate the real zeros of a polynomial function in the following way. If you can find a value $x = a$ at which a polynomial function is positive, and another value $x = b$ at which it is negative, you can conclude that the function has at least one real zero between these two values. For example, the function given by $f(x) = x^3 + x^2 + 1$ is negative when $x = -2$ and positive when $x = -1$. Therefore, it follows from the Intermediate Value Theorem that f must have a real zero somewhere between -2 and -1, as shown in Figure 2.26.

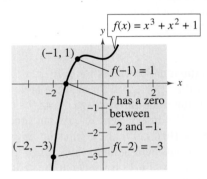

FIGURE 2.26

By continuing this line of reasoning, you can approximate any real zeros of a polynomial function to any desired accuracy. This concept is further demonstrated in Example 6.

Example 6 Approximating a Zero of a Polynomial Function

Use the Intermediate Value Theorem to approximate the real zero of

$$f(x) = x^3 - x^2 + 1.$$

Solution

Begin by computing a few function values, as follows.

x	$f(x)$
-2	-11
-1	-1
0	1
1	1

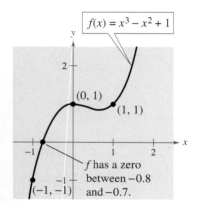

$f(x) = x^3 - x^2 + 1$

(0, 1)

(1, 1)

f has a zero between −0.8 and −0.7.

(−1, −1)

FIGURE 2.27

Because $f(-1)$ is negative and $f(0)$ is positive, you can apply the Intermediate Value Theorem to conclude that the function has a zero between -1 and 0. To pinpoint this zero more closely, divide the interval $[-1, 0]$ into tenths and evaluate the function at each point. When you do this, you will find that

$$f(-0.8) = -0.152 \quad \text{and} \quad f(-0.7) = 0.167.$$

So, f must have a zero between -0.8 and -0.7, as shown in Figure 2.27. For a more accurate approximation, compute function values between $f(-0.8)$ and $f(-0.7)$ and apply the Intermediate Value Theorem again. By continuing this process, you can approximate this zero to any desired accuracy.

CHECK*Point* Now try Exercise 93.

TECHNOLOGY

You can use the *table* feature of a graphing utility to approximate the zeros of a polynomial function. For instance, for the function given by

$$f(x) = -2x^3 - 3x^2 + 3$$

create a table that shows the function values for $-20 \le x \le 20$, as shown in the first table at the right. Scroll through the table looking for consecutive function values that differ in sign. From the table, you can see that $f(0)$ and $f(1)$ differ in sign. So, you can conclude from the Intermediate Value Theorem that the function has a zero between 0 and 1. You can adjust your table to show function values for $0 \le x \le 1$ using increments of 0.1, as shown in the second table at the right. By scrolling through the table you can see that $f(0.8)$ and $f(0.9)$ differ in sign. So, the function has a zero between 0.8 and 0.9. If you repeat this process several times, you should obtain $x \approx 0.806$ as the zero of the function. Use the *zero* or *root* feature of a graphing utility to confirm this result.

X	Y1		
-2	7		
-1	2		
0	3		
1			
2	-25		
3	-78		
4	-173		

X=1

X	Y1		
.4	2.392		
.5	2		
.6	1.488		
.7	.844		
.8	.056		
.9	-.888		
1	-2		

X=.9

2.2   EXERCISES

See www.CalcChat.com for worked-out solutions to odd-numbered exercises.

VOCABULARY: Fill in the blanks.

1. The graphs of all polynomial functions are _____, which means that the graphs have no breaks, holes, or gaps.

2. The _____ _____ _____ is used to determine the left-hand and right-hand behavior of the graph of a polynomial function.

3. Polynomial functions of the form $f(x) =$ _____ are often referred to as power functions.

4. A polynomial function of degree n has at most _____ real zeros and at most _____ turning points.

5. If $x = a$ is a zero of a polynomial function f, then the following three statements are true.
    (a) $x = a$ is a _____ of the polynomial equation $f(x) = 0$.
    (b) _____ is a factor of the polynomial $f(x)$.
    (c) $(a, 0)$ is an _____ of the graph of f.

6. If a real zero of a polynomial function is of even multiplicity, then the graph of f _____ the x-axis at $x = a$, and if it is of odd multiplicity, then the graph of f _____ the x-axis at $x = a$.

7. A polynomial function is written in _____ form if its terms are written in descending order of exponents from left to right.

8. The _____ _____ Theorem states that if f is a polynomial function such that $f(a) \neq f(b)$, then, in the interval $[a, b]$, f takes on every value between $f(a)$ and $f(b)$.

SKILLS AND APPLICATIONS

In Exercises 9–16, match the polynomial function with its graph. [The graphs are labeled (a), (b), (c), (d), (e), (f), (g), and (h).]

(a)
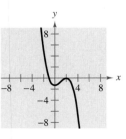

(b)
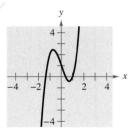

(g)
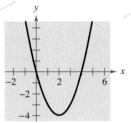

(h)
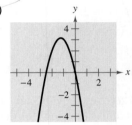

9. $f(x) = -2x + 3$ (C)   10. $f(x) = x^2 - 4x$ (g)
11. $f(x) = -2x^2 - 5x$ (h)   12. $f(x) = 2x^3 - 3x + 1$ (f)
13. $f(x) = -\frac{1}{4}x^4 + 3x^2$ (a)   14. $f(x) = -\frac{1}{3}x^3 + x^2 - \frac{4}{3}$ (e)
15. $f(x) = x^4 + 2x^3$ (d)   16. $f(x) = \frac{1}{5}x^5 - 2x^3 + \frac{9}{5}x$ (b)

(c)
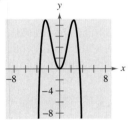

(d)
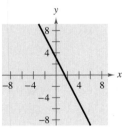

In Exercises 17–20, sketch the graph of $y = x^n$ and each transformation.

17. $y = x^3$
    (a) $f(x) = (x - 4)^3$ (b) $f(x) = x^3 - 4$
    (c) $f(x) = -\frac{1}{4}x^3$ (d) $f(x) = (x - 4)^3 - 4$

18. $y = x^5$
    (a) $f(x) = (x + 1)^5$ (b) $f(x) = x^5 + 1$
    (c) $f(x) = 1 - \frac{1}{2}x^5$ (d) $f(x) = -\frac{1}{2}(x + 1)^5$

(e)
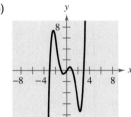

(f)

19. $y = x^4$
    (a) $f(x) = (x + 3)^4$ (b) $f(x) = x^4 - 3$
    (c) $f(x) = 4 - x^4$ (d) $f(x) = \frac{1}{2}(x - 1)^4$
    (e) $f(x) = (2x)^4 + 1$ (f) $f(x) = \left(\frac{1}{2}x\right)^4 - 2$

20. $y = x^6$

  (a) $f(x) = -\frac{1}{8}x^6$ (b) $f(x) = (x + 2)^6 - 4$

  (c) $f(x) = x^6 - 5$ (d) $f(x) = -\frac{1}{4}x^6 + 1$

  (e) $f(x) = \left(\frac{1}{4}x\right)^6 - 2$ (f) $f(x) = (2x)^6 - 1$

In Exercises 21–30, describe the right-hand and left-hand behavior of the graph of the polynomial function.

21. $f(x) = \frac{1}{5}x^3 + 4x$ **22.** $f(x) = 2x^2 - 3x + 1$

23. $g(x) = 5 - \frac{7}{2}x - 3x^2$ **24.** $h(x) = 1 - x^6$

25. $f(x) = -2.1x^5 + 4x^3 - 2$

26. $f(x) = 4x^5 - 7x + 6.5$

27. $f(x) = 6 - 2x + 4x^2 - 5x^3$

28. $f(x) = (3x^4 - 2x + 5)/4$

29. $h(t) = -\frac{3}{4}(t^2 - 3t + 6)$

30. $f(s) = -\frac{7}{8}(s^3 + 5s^2 - 7s + 1)$

 GRAPHICAL ANALYSIS In Exercises 31–34, use a graphing utility to graph the functions f and g in the same viewing window. Zoom out sufficiently far to show that the right-hand and left-hand behaviors of f and g appear identical.

31. $f(x) = 3x^3 - 9x + 1, \quad g(x) = 3x^3$

32. $f(x) = -\frac{1}{3}(x^3 - 3x + 2), \quad g(x) = -\frac{1}{3}x^3$

33. $f(x) = -(x^4 - 4x^3 + 16x), \quad g(x) = -x^4$

34. $f(x) = 3x^4 - 6x^2, \quad g(x) = 3x^4$

In Exercises 35–50, (a) find all the real zeros of the polynomial function, (b) determine the multiplicity of each zero and the number of turning points of the graph of the function, and (c) use a graphing utility to graph the function and verify your answers.

35. $f(x) = x^2 - 36$ **36.** $f(x) = 81 - x^2$

37. $h(t) = t^2 - 6t + 9$ **38.** $f(x) = x^2 + 10x + 25$

39. $f(x) = \frac{1}{3}x^2 + \frac{1}{3}x - \frac{2}{3}$ **40.** $f(x) = \frac{1}{2}x^2 + \frac{5}{2}x - \frac{3}{2}$

41. $f(x) = 3x^3 - 12x^2 + 3x$ **42.** $g(x) = 5x(x^2 - 2x - 1)$

43. $f(t) = t^3 - 8t^2 + 16t$ **44.** $f(x) = x^4 - x^3 - 30x^2$

45. $g(t) = t^5 - 6t^3 + 9t$ **46.** $f(x) = x^5 + x^3 - 6x$

47. $f(x) = 3x^4 + 9x^2 + 6$ **48.** $f(x) = 2x^4 - 2x^2 - 40$

49. $g(x) = x^3 + 3x^2 - 4x - 12$

50. $f(x) = x^3 - 4x^2 - 25x + 100$

 GRAPHICAL ANALYSIS In Exercises 51–54, (a) use a graphing utility to graph the function, (b) use the graph to approximate any x-intercepts of the graph, (c) set $y = 0$ and solve the resulting equation, and (d) compare the results of part (c) with any x-intercepts of the graph.

51. $y = 4x^3 - 20x^2 + 25x$

52. $y = 4x^3 + 4x^2 - 8x - 8$

53. $y = x^5 - 5x^3 + 4x$ **54.** $y = \frac{1}{4}x^3(x^2 - 9)$

In Exercises 55–64, find a polynomial function that has the given zeros. (There are many correct answers.)

55. $0, 8$ **56.** $0, -7$

57. $2, -6$ **58.** $-4, 5$

59. $0, -4, -5$ **60.** $0, 1, 10$

61. $4, -3, 3, 0$ **62.** $-2, -1, 0, 1, 2$

63. $1 + \sqrt{3}, 1 - \sqrt{3}$ **64.** $2, 4 + \sqrt{5}, 4 - \sqrt{5}$

In Exercises 65–74, find a polynomial of degree n that has the given zero(s). (There are many correct answers.)

	Zero(s)	*Degree*
65.	$x = -3$	$n = 2$
66.	$x = -12, -6$	$n = 2$
67.	$x = -5, 0, 1$	$n = 3$
68.	$x = -2, 4, 7$	$n = 3$
69.	$x = 0, \sqrt{3}, -\sqrt{3}$	$n = 3$
70.	$x = 9$	$n = 3$
71.	$x = -5, 1, 2$	$n = 4$
72.	$x = -4, -1, 3, 6$	$n = 4$
73.	$x = 0, -4$	$n = 5$
74.	$x = -1, 4, 7, 8$	$n = 5$

In Exercises 75–88, sketch the graph of the function by (a) applying the Leading Coefficient Test, (b) finding the zeros of the polynomial, (c) plotting sufficient solution points, and (d) drawing a continuous curve through the points.

75. $f(x) = x^3 - 25x$ **76.** $g(x) = x^4 - 9x^2$

77. $f(t) = \frac{1}{4}(t^2 - 2t + 15)$

78. $g(x) = -x^2 + 10x - 16$

79. $f(x) = x^3 - 2x^2$ **80.** $f(x) = 8 - x^3$

81. $f(x) = 3x^3 - 15x^2 + 18x$

82. $f(x) = -4x^3 + 4x^2 + 15x$

83. $f(x) = -5x^2 - x^3$ **84.** $f(x) = -48x^2 + 3x^4$

85. $f(x) = x^2(x - 4)$ **86.** $h(x) = \frac{1}{3}x^3(x - 4)^2$

87. $g(t) = -\frac{1}{4}(t - 2)^2(t + 2)^2$

88. $g(x) = \frac{1}{10}(x + 1)^2(x - 3)^3$

In Exercises 89–92, use a graphing utility to graph the function. Use the *zero* or *root* feature to approximate the real zeros of the function. Then determine the multiplicity of each zero.

89. $f(x) = x^3 - 16x$ **90.** $f(x) = \frac{1}{4}x^4 - 2x^2$

91. $g(x) = \frac{1}{5}(x + 1)^2(x - 3)(2x - 9)$

92. $h(x) = \frac{1}{5}(x + 2)^2(3x - 5)^2$

In Exercises 93–96, use the Intermediate Value Theorem and the *table* feature of a graphing utility to find intervals one unit in length in which the polynomial function is guaranteed to have a zero. Adjust the table to approximate the zeros of the function. Use the *zero* or *root* feature of the graphing utility to verify your results.

93. $f(x) = x^3 - 3x^2 + 3$

94. $f(x) = 0.11x^3 - 2.07x^2 + 9.81x - 6.88$

95. $g(x) = 3x^4 + 4x^3 - 3$

96. $h(x) = x^4 - 10x^2 + 3$

97. NUMERICAL AND GRAPHICAL ANALYSIS   An open box is to be made from a square piece of material, 36 inches on a side, by cutting equal squares with sides of length x from the corners and turning up the sides (see figure).

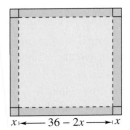
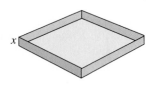

(a) Write a function $V(x)$ that represents the volume of the box.

(b) Determine the domain of the function.

(c) Use a graphing utility to create a table that shows box heights x and the corresponding volumes V. Use the table to estimate the dimensions that will produce a maximum volume.

(d) Use a graphing utility to graph V and use the graph to estimate the value of x for which $V(x)$ is maximum. Compare your result with that of part (c).

98. MAXIMUM VOLUME   An open box with locking tabs is to be made from a square piece of material 24 inches on a side. This is to be done by cutting equal squares from the corners and folding along the dashed lines shown in the figure.

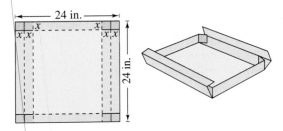

(a) Write a function $V(x)$ that represents the volume of the box.

(b) Determine the domain of the function V.

(c) Sketch a graph of the function and estimate the value of x for which $V(x)$ is maximum.

99. CONSTRUCTION   A roofing contractor is fabricating gutters from 12-inch aluminum sheeting. The contractor plans to use an aluminum siding folding press to create the gutter by creasing equal lengths for the sidewalls (see figure).

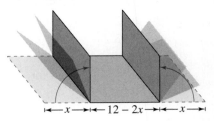

(a) Let x represent the height of the sidewall of the gutter. Write a function A that represents the cross-sectional area of the gutter.

(b) The length of the aluminum sheeting is 16 feet. Write a function V that represents the volume of one run of gutter in terms of x.

(c) Determine the domain of the function in part (b).

(d) Use a graphing utility to create a table that shows sidewall heights x and the corresponding volumes V. Use the table to estimate the dimensions that will produce a maximum volume.

(e) Use a graphing utility to graph V. Use the graph to estimate the value of x for which $V(x)$ is a maximum. Compare your result with that of part (d).

(f) Would the value of x change if the aluminum sheeting were of different lengths? Explain.

100. CONSTRUCTION   An industrial propane tank is formed by adjoining two hemispheres to the ends of a right circular cylinder. The length of the cylindrical portion of the tank is four times the radius of the hemispherical components (see figure).

(a) Write a function that represents the total volume V of the tank in terms of r.

(b) Find the domain of the function.

(c) Use a graphing utility to graph the function.

(d) The total volume of the tank is to be 120 cubic feet. Use the graph from part (c) to estimate the radius and length of the cylindrical portion of the tank.

101. REVENUE The total revenues R (in millions of dollars) for Krispy Kreme from 2000 through 2007 are shown in the table.

Year	Revenue, R
2000	300.7
2001	394.4
2002	491.5
2003	665.6
2004	707.8
2005	543.4
2006	461.2
2007	429.3

A model that represents these data is given by $R = 3.0711t^4 - 42.803t^3 + 160.59t^2 - 62.6t + 307$, $0 \le t \le 7$, where t represents the year, with $t = 0$ corresponding to 2000. (Source: Krispy Kreme)

(a) Use a graphing utility to create a scatter plot of the data. Then graph the model in the same viewing window.

(b) How well does the model fit the data?

(c) Use a graphing utility to approximate any relative extrema of the model over its domain.

(d) Use a graphing utility to approximate the intervals over which the revenue for Krispy Kreme was increasing and decreasing over its domain.

(e) Use the results of parts (c) and (d) to write a short paragraph about Krispy Kreme's revenue during this time period.

102. REVENUE The total revenues R (in millions of dollars) for Papa John's International from 2000 through 2007 are shown in the table.

Year	Revenue, R
2000	944.7
2001	971.2
2002	946.2
2003	917.4
2004	942.4
2005	968.8
2006	1001.6
2007	1063.6

A model that represents these data is given by $R = -0.5635t^4 + 9.019t^3 - 40.20t^2 + 49.0t + 947$, $0 \le t \le 7$, where t represents the year, with $t = 0$ corresponding to 2000. (Source: Papa John's International)

(a) Use a graphing utility to create a scatter plot of the data. Then graph the model in the same viewing window.

(b) How well does the model fit the data?

(c) Use a graphing utility to approximate any relative extrema of the model over its domain.

(d) Use a graphing utility to approximate the intervals over which the revenue for Papa John's International was increasing and decreasing over its domain.

(e) Use the results of parts (c) and (d) to write a short paragraph about the revenue for Papa John's International during this time period.

103. TREE GROWTH The growth of a red oak tree is approximated by the function

$$G = -0.003t^3 + 0.137t^2 + 0.458t - 0.839$$

where G is the height of the tree (in feet) and t ($2 \le t \le 34$) is its age (in years).

(a) Use a graphing utility to graph the function. (*Hint:* Use a viewing window in which $-10 \le x \le 45$ and $-5 \le y \le 60$.)

(b) Estimate the age of the tree when it is growing most rapidly. This point is called the *point of diminishing returns* because the increase in size will be less with each additional year.

(c) Using calculus, the point of diminishing returns can also be found by finding the vertex of the parabola given by

$$y = -0.009t^2 + 0.274t + 0.458.$$

Find the vertex of this parabola.

(d) Compare your results from parts (b) and (c).

104. REVENUE The total revenue R (in millions of dollars) for a company is related to its advertising expense by the function

$$R = \frac{1}{100,000}(-x^3 + 600x^2), \quad 0 \le x \le 400$$

where x is the amount spent on advertising (in tens of thousands of dollars). Use the graph of this function, shown in the figure on the next page, to estimate the point on the graph at which the function is increasing most rapidly. This point is called the *point of diminishing returns* because any expense above this amount will yield less return per dollar invested in advertising.

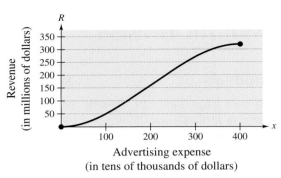

FIGURE FOR 104

EXPLORATION

TRUE OR FALSE? In Exercises 105–107, determine whether the statement is true or false. Justify your answer.

105. A fifth-degree polynomial can have five turning points in its graph.

106. It is possible for a sixth-degree polynomial to have only one solution.

107. The graph of the function given by

$$f(x) = 2 + x - x^2 + x^3 - x^4 + x^5 + x^6 - x^7$$

rises to the left and falls to the right.

108. **CAPSTONE**   For each graph, describe a polynomial function that could represent the graph. (Indicate the degree of the function and the sign of its leading coefficient.)

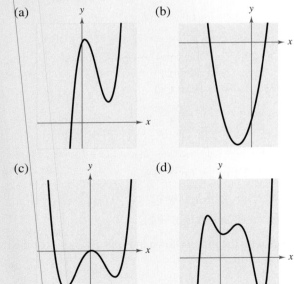

109. **GRAPHICAL REASONING** Sketch a graph of the function given by $f(x) = x^4$. Explain how the graph of each function g differs (if it does) from the graph of each function f. Determine whether g is odd, even, or neither.

(a) $g(x) = f(x) + 2$ (b) $g(x) = f(x + 2)$

(c) $g(x) = f(-x)$ (d) $g(x) = -f(x)$

(e) $g(x) = f\left(\frac{1}{2}x\right)$ (f) $g(x) = \frac{1}{2}f(x)$

(g) $g(x) = f\left(x^{3/4}\right)$ (h) $g(x) = (f \circ f)(x)$

110. **THINK ABOUT IT**   For each function, identify the degree of the function and whether the degree of the function is even or odd. Identify the leading coefficient and whether the leading coefficient is positive or negative. Use a graphing utility to graph each function. Describe the relationship between the degree of the function and the sign of the leading coefficient of the function and the right-hand and left-hand behavior of the graph of the function.

(a) $f(x) = x^3 - 2x^2 - x + 1$

(b) $f(x) = 2x^5 + 2x^2 - 5x + 1$

(c) $f(x) = -2x^5 - x^2 + 5x + 3$

(d) $f(x) = -x^3 + 5x - 2$

(e) $f(x) = 2x^2 + 3x - 4$

(f) $f(x) = x^4 - 3x^2 + 2x - 1$

(g) $f(x) = x^2 + 3x + 2$

111. **THINK ABOUT IT**   Sketch the graph of each polynomial function. Then count the number of zeros of the function and the numbers of relative minima and relative maxima. Compare these numbers with the degree of the polynomial. What do you observe?

(a) $f(x) = -x^3 + 9x$ (b) $f(x) = x^4 - 10x^2 + 9$

(c) $f(x) = x^5 - 16x$

112. Explore the transformations of the form

$$g(x) = a(x - h)^5 + k.$$

(a) Use a graphing utility to graph the functions $y_1 = -\frac{1}{3}(x - 2)^5 + 1$ and $y_2 = \frac{3}{5}(x + 2)^5 - 3$. Determine whether the graphs are increasing or decreasing. Explain.

(b) Will the graph of g always be increasing or decreasing? If so, is this behavior determined by a, h, or k? Explain.

(c) Use a graphing utility to graph the function given by $H(x) = x^5 - 3x^3 + 2x + 1$. Use the graph and the result of part (b) to determine whether H can be written in the form $H(x) = a(x - h)^5 + k$. Explain.

What you should learn

- Use long division to divide polynomials by other polynomials.
- Use synthetic division to divide polynomials by binomials of the form $(x - k)$.
- Use the Remainder Theorem and the Factor Theorem.

Why you should learn it

Synthetic division can help you evaluate polynomial functions. For instance, in Exercise 85 on page 157, you will use synthetic division to determine the amount donated to support higher education in the United States in 2010.

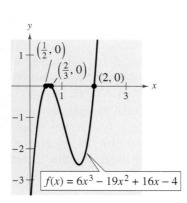

2.3   POLYNOMIAL AND SYNTHETIC DIVISION

Long Division of Polynomials

In this section, you will study two procedures for *dividing* polynomials. These procedures are especially valuable in factoring and finding the zeros of polynomial functions. To begin, suppose you are given the graph of

$$f(x) = 6x^3 - 19x^2 + 16x - 4.$$

Notice that a zero of f occurs at $x = 2$, as shown in Figure 2.28. Because $x = 2$ is a zero of f, you know that $(x - 2)$ is a factor of $f(x)$. This means that there exists a second-degree polynomial $q(x)$ such that

$$f(x) = (x - 2) \cdot q(x).$$

To find $q(x)$, you can use **long division,** as illustrated in Example 1.

Example 1   Long Division of Polynomials

Divide $6x^3 - 19x^2 + 16x - 4$ by $x - 2$, and use the result to factor the polynomial completely.

Solution

Think $\dfrac{6x^3}{x} = 6x^2$.

Think $\dfrac{-7x^2}{x} = -7x$.

Think $\dfrac{2x}{x} = 2$.

$$
\begin{array}{r}
6x^2 - 7x + 2 \\
x - 2 \overline{)\; 6x^3 - 19x^2 + 16x - 4} \\
\end{array}
$$

$6x^3 - 12x^2$	Multiply: $6x^2(x - 2)$.
$-7x^2 + 16x$	Subtract.
$-7x^2 + 14x$	Multiply: $-7x(x - 2)$.
$2x - 4$	Subtract.
$2x - 4$	Multiply: $2(x - 2)$.
0	Subtract.

From this division, you can conclude that

$$6x^3 - 19x^2 + 16x - 4 = (x - 2)(6x^2 - 7x + 2)$$

and by factoring the quadratic $6x^2 - 7x + 2$, you have

$$6x^3 - 19x^2 + 16x - 4 = (x - 2)(2x - 1)(3x - 2).$$

Note that this factorization agrees with the graph shown in Figure 2.28 in that the three x-intercepts occur at $x = 2$, $x = \frac{1}{2}$, and $x = \frac{2}{3}$.

CHECK*Point* ▶   Now try Exercise 11.

FIGURE 2.28

In Example 1, $x - 2$ is a factor of the polynomial $6x^3 - 19x^2 + 16x - 4$, and the long division process produces a remainder of zero. Often, long division will produce a nonzero remainder. For instance, if you divide $x^2 + 3x + 5$ by $x + 1$, you obtain the following.

$$
\begin{array}{r}
x + 2 \quad \longleftarrow \text{ Quotient} \\
\text{Divisor} \longrightarrow \quad x + 1 \overline{) x^2 + 3x + 5} \quad \longleftarrow \text{ Dividend} \\
\underline{x^2 + x} \\
2x + 5 \\
\underline{2x + 2} \\
3 \quad \longleftarrow \text{ Remainder}
\end{array}
$$

In fractional form, you can write this result as follows.

$$
\underbrace{\frac{x^2 + 3x + 5}{x + 1}}_{\text{Dividend} / \text{Divisor}} = \underbrace{x + 2}_{\text{Quotient}} + \underbrace{\frac{3}{x + 1}}_{\text{Remainder} / \text{Divisor}}
$$

This implies that

$$x^2 + 3x + 5 = (x + 1)(x + 2) + 3 \qquad \text{Multiply each side by } (x + 1).$$

which illustrates the following theorem, called the **Division Algorithm.**

The Division Algorithm

If $f(x)$ and $d(x)$ are polynomials such that $d(x) \neq 0$, and the degree of $d(x)$ is less than or equal to the degree of $f(x)$, there exist unique polynomials $q(x)$ and $r(x)$ such that

$$
\underbrace{f(x)}_{\text{Dividend}} = \underbrace{d(x)}_{\text{Divisor}} \underbrace{q(x)}_{\text{Quotient}} + \underbrace{r(x)}_{\text{Remainder}}
$$

where $r(x) = 0$ *or* the degree of $r(x)$ is less than the degree of $d(x)$. If the remainder $r(x)$ is zero, $d(x)$ *divides evenly* into $f(x)$.

The Division Algorithm can also be written as

$$\frac{f(x)}{d(x)} = q(x) + \frac{r(x)}{d(x)}.$$

In the Division Algorithm, the rational expression $f(x)/d(x)$ is **improper** because the degree of $f(x)$ is greater than or equal to the degree of $d(x)$. On the other hand, the rational expression $r(x)/d(x)$ is **proper** because the degree of $r(x)$ is less than the degree of $d(x)$.

Before you apply the Division Algorithm, follow these steps.

1. Write the dividend and divisor in descending powers of the variable.

2. Insert placeholders with zero coefficients for missing powers of the variable.

Example 2   Long Division of Polynomials

Divide $x^3 - 1$ by $x - 1$.

Solution

Because there is no x^2-term or x-term in the dividend, you need to line up the subtraction by using zero coefficients (or leaving spaces) for the missing terms.

$$
\begin{array}{r}
x^2 + x + 1 \\
x - 1 \overline{\smash{)}\, x^3 + 0x^2 + 0x - 1} \\
\underline{x^3 - x^2} \\
x^2 + 0x \\
\underline{x^2 - x} \\
x - 1 \\
\underline{x - 1} \\
0
\end{array}
$$

So, $x - 1$ divides evenly into $x^3 - 1$, and you can write

$$
\frac{x^3 - 1}{x - 1} = x^2 + x + 1, \quad x \neq 1.
$$

CHECK**Point** Now try Exercise 17.

You can check the result of Example 2 by multiplying.

$$(x - 1)(x^2 + x + 1) = x^3 + x^2 + x - x^2 - x - 1 = x^3 - 1$$

Example 3   Long Division of Polynomials

Divide $-5x^2 - 2 + 3x + 2x^4 + 4x^3$ by $2x - 3 + x^2$.

Solution

Begin by writing the dividend and divisor in descending powers of x.

$$
\begin{array}{r}
2x^2 + 1 \\
x^2 + 2x - 3 \overline{\smash{)}\, 2x^4 + 4x^3 - 5x^2 + 3x - 2} \\
\underline{2x^4 + 4x^3 - 6x^2} \\
x^2 + 3x - 2 \\
\underline{x^2 + 2x - 3} \\
x + 1
\end{array}
$$

Note that the first subtraction eliminated two terms from the dividend. When this happens, the quotient skips a term. You can write the result as

$$
\frac{2x^4 + 4x^3 - 5x^2 + 3x - 2}{x^2 + 2x - 3} = 2x^2 + 1 + \frac{x + 1}{x^2 + 2x - 3}.
$$

CHECK**Point** Now try Exercise 23.

Algebra Help

You can check a long division problem by multiplying. You can review the techniques for multiplying polynomials in Appendix A.3.

Synthetic Division

There is a nice shortcut for long division of polynomials by divisors of the form $x - k$. This shortcut is called **synthetic division.** The pattern for synthetic division of a cubic polynomial is summarized as follows. (The pattern for higher-degree polynomials is similar.)

Synthetic Division (for a Cubic Polynomial)

To divide $ax^3 + bx^2 + cx + d$ by $x - k$, use the following pattern.

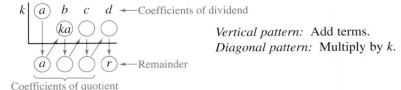

Vertical pattern: Add terms.
Diagonal pattern: Multiply by k.

This algorithm for synthetic division works only for divisors of the form $x - k$. Remember that $x + k = x - (-k)$.

▎Example 4   Using Synthetic Division

Use synthetic division to divide $x^4 - 10x^2 - 2x + 4$ by $x + 3$.

Solution

You should set up the array as follows. Note that a zero is included for the missing x^3-term in the dividend.

$$-3 \;\big|\; 1 \quad 0 \quad -10 \quad -2 \quad 4$$

Then, use the synthetic division pattern by adding terms in columns and multiplying the results by -3.

Divisor: $x + 3$ Dividend: $x^4 - 10x^2 - 2x + 4$

$$
\begin{array}{r|rrrrr}
-3 & 1 & 0 & -10 & -2 & 4 \\
   &   & -3 & 9 & 3 & -3 \\
\hline
   & 1 & -3 & -1 & 1 & (1) \quad \longleftarrow \text{Remainder: } 1
\end{array}
$$

Quotient: $x^3 - 3x^2 - x + 1$

So, you have

$$\frac{x^4 - 10x^2 - 2x + 4}{x + 3} = x^3 - 3x^2 - x + 1 + \frac{1}{x + 3}.$$

CHECK *Point*   Now try Exercise 27.

The Remainder and Factor Theorems

The remainder obtained in the synthetic division process has an important interpretation, as described in the **Remainder Theorem.**

> ### The Remainder Theorem
> If a polynomial $f(x)$ is divided by $x - k$, the remainder is
> $$r = f(k).$$

For a proof of the Remainder Theorem, see Proofs in Mathematics on page 211.

The Remainder Theorem tells you that synthetic division can be used to evaluate a polynomial function. That is, to evaluate a polynomial function $f(x)$ when $x = k$, divide $f(x)$ by $x - k$. The remainder will be $f(k)$, as illustrated in Example 5.

Example 5   Using the Remainder Theorem

Use the Remainder Theorem to evaluate the following function at $x = -2$.
$$f(x) = 3x^3 + 8x^2 + 5x - 7$$

Solution

Using synthetic division, you obtain the following.

```
-2 | 3    8    5   -7
   |     -6   -4   -2
     ----------------
     3    2    1   -9
```

Because the remainder is $r = -9$, you can conclude that

$$f(-2) = -9. \qquad {\scriptstyle r = f(k)}$$

This means that $(-2, -9)$ is a point on the graph of f. You can check this by substituting $x = -2$ in the original function.

Check

$$f(-2) = 3(-2)^3 + 8(-2)^2 + 5(-2) - 7$$
$$= 3(-8) + 8(4) - 10 - 7 = -9$$

CHECK*Point*   Now try Exercise 55.

Another important theorem is the **Factor Theorem,** stated below. This theorem states that you can test to see whether a polynomial has $(x - k)$ as a factor by evaluating the polynomial at $x = k$. If the result is 0, $(x - k)$ is a factor.

> ### The Factor Theorem
> A polynomial $f(x)$ has a factor $(x - k)$ if and only if $f(k) = 0$.

For a proof of the Factor Theorem, see Proofs in Mathematics on page 211.

Example 6   Factoring a Polynomial: Repeated Division

Show that $(x - 2)$ and $(x + 3)$ are factors of $f(x) = 2x^4 + 7x^3 - 4x^2 - 27x - 18$. Then find the remaining factors of $f(x)$.

Algebraic Solution

Using synthetic division with the factor $(x - 2)$, you obtain the following.

$$
\begin{array}{r|rrrrr}
2 & 2 & 7 & -4 & -27 & -18 \\
  &   & 4 & 22 & 36 & 18 \\
\hline
  & 2 & 11 & 18 & 9 & 0
\end{array}
$$

0 remainder, so $f(2) = 0$ and $(x - 2)$ is a factor.

Take the result of this division and perform synthetic division again using the factor $(x + 3)$.

$$
\begin{array}{r|rrrr}
-3 & 2 & 11 & 18 & 9 \\
   &   & -6 & -15 & -9 \\
\hline
   & 2 & 5 & 3 & 0
\end{array}
$$

0 remainder, so $f(-3) = 0$ and $(x + 3)$ is a factor.

$2x^2 + 5x + 3$

Because the resulting quadratic expression factors as

$$2x^2 + 5x + 3 = (2x + 3)(x + 1)$$

the complete factorization of $f(x)$ is

$$f(x) = (x - 2)(x + 3)(2x + 3)(x + 1).$$

CHECK *Point*   Now try Exercise 67.

Graphical Solution

From the graph of $f(x) = 2x^4 + 7x^3 - 4x^2 - 27x - 18$, you can see that there are four x-intercepts (see Figure 2.29). These occur at $x = -3$, $x = -\frac{3}{2}$, $x = -1$, and $x = 2$. (Check this algebraically.) This implies that $(x + 3)$, $\left(x + \frac{3}{2}\right)$, $(x + 1)$, and $(x - 2)$ are factors of $f(x)$. [Note that $\left(x + \frac{3}{2}\right)$ and $(2x + 3)$ are equivalent factors because they both yield the same zero, $x = -\frac{3}{2}$.]

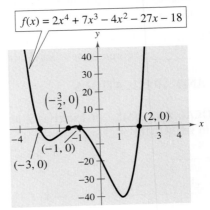

$f(x) = 2x^4 + 7x^3 - 4x^2 - 27x - 18$

FIGURE 2.29

Study Tip

Note in Example 6 that the complete factorization of $f(x)$ implies that f has four real zeros: $x = 2$, $x = -3$, $x = -\frac{3}{2}$, and $x = -1$. This is confirmed by the graph of f, which is shown in the Figure 2.29.

Uses of the Remainder in Synthetic Division

The remainder r, obtained in the synthetic division of $f(x)$ by $x - k$, provides the following information.

1. The remainder r gives the value of f at $x = k$. That is, $r = f(k)$.

2. If $r = 0$, $(x - k)$ is a factor of $f(x)$.

3. If $r = 0$, $(k, 0)$ is an x-intercept of the graph of f.

Throughout this text, the importance of developing several problem-solving strategies is emphasized. In the exercises for this section, try using more than one strategy to solve several of the exercises. For instance, if you find that $x - k$ divides evenly into $f(x)$ (with no remainder), try sketching the graph of f. You should find that $(k, 0)$ is an x-intercept of the graph.

2.3   EXERCISES

See www.CalcChat.com for worked-out solutions to odd-numbered exercises.

VOCABULARY

1. Two forms of the Division Algorithm are shown below. Identify and label each term or function.

$$f(x) = d(x)q(x) + r(x) \qquad \frac{f(x)}{d(x)} = q(x) + \frac{r(x)}{d(x)}$$

In Exercises 2–6, fill in the blanks.

2. The rational expression $p(x)/q(x)$ is called _____ if the degree of the numerator is greater than or equal to that of the denominator, and is called _____ if the degree of the numerator is less than that of the denominator.

3. In the Division Algorithm, the rational expression $f(x)/d(x)$ is _____ because the degree of $f(x)$ is greater than or equal to the degree of $d(x)$.

4. An alternative method to long division of polynomials is called _____ _____, in which the divisor must be of the form $x - k$.

5. The _____ Theorem states that a polynomial $f(x)$ has a factor $(x - k)$ if and only if $f(k) = 0$.

6. The _____ Theorem states that if a polynomial $f(x)$ is divided by $x - k$, the remainder is $r = f(k)$.

SKILLS AND APPLICATIONS

ANALYTICAL ANALYSIS In Exercises 7 and 8, use long division to verify that $y_1 = y_2$.

7. $y_1 = \dfrac{x^2}{x + 2}$, $y_2 = x - 2 + \dfrac{4}{x + 2}$

8. $y_1 = \dfrac{x^4 - 3x^2 - 1}{x^2 + 5}$, $y_2 = x^2 - 8 + \dfrac{39}{x^2 + 5}$

GRAPHICAL ANALYSIS In Exercises 9 and 10, (a) use a graphing utility to graph the two equations in the same viewing window, (b) use the graphs to verify that the expressions are equivalent, and (c) use long division to verify the results algebraically.

9. $y_1 = \dfrac{x^2 + 2x - 1}{x + 3}$, $y_2 = x - 1 + \dfrac{2}{x + 3}$

10. $y_1 = \dfrac{x^4 + x^2 - 1}{x^2 + 1}$, $y_2 = x^2 - \dfrac{1}{x^2 + 1}$

In Exercises 11–26, use long division to divide.

11. $(2x^2 + 10x + 12) \div (x + 3)$

12. $(5x^2 - 17x - 12) \div (x - 4)$

13. $(4x^3 - 7x^2 - 11x + 5) \div (4x + 5)$

14. $(6x^3 - 16x^2 + 17x - 6) \div (3x - 2)$

15. $(x^4 + 5x^3 + 6x^2 - x - 2) \div (x + 2)$

16. $(x^3 + 4x^2 - 3x - 12) \div (x - 3)$

17. $(x^3 - 27) \div (x - 3)$ **18.** $(x^3 + 125) \div (x + 5)$

19. $(7x + 3) \div (x + 2)$ **20.** $(8x - 5) \div (2x + 1)$

21. $(x^3 - 9) \div (x^2 + 1)$ **22.** $(x^5 + 7) \div (x^3 - 1)$

23. $(3x + 2x^3 - 9 - 8x^2) \div (x^2 + 1)$

24. $(5x^3 - 16 - 20x + x^4) \div (x^2 - x - 3)$

25. $\dfrac{x^4}{(x - 1)^3}$ **26.** $\dfrac{2x^3 - 4x^2 - 15x + 5}{(x - 1)^2}$

In Exercises 27–46, use synthetic division to divide.

27. $(3x^3 - 17x^2 + 15x - 25) \div (x - 5)$

28. $(5x^3 + 18x^2 + 7x - 6) \div (x + 3)$

29. $(6x^3 + 7x^2 - x + 26) \div (x - 3)$

30. $(2x^3 + 14x^2 - 20x + 7) \div (x + 6)$

31. $(4x^3 - 9x + 8x^2 - 18) \div (x + 2)$

32. $(9x^3 - 16x - 18x^2 + 32) \div (x - 2)$

33. $(-x^3 + 75x - 250) \div (x + 10)$

34. $(3x^3 - 16x^2 - 72) \div (x - 6)$

35. $(5x^3 - 6x^2 + 8) \div (x - 4)$

36. $(5x^3 + 6x + 8) \div (x + 2)$

37. $\dfrac{10x^4 - 50x^3 - 800}{x - 6}$ **38.** $\dfrac{x^5 - 13x^4 - 120x + 80}{x + 3}$

39. $\dfrac{x^3 + 512}{x + 8}$ **40.** $\dfrac{x^3 - 729}{x - 9}$

41. $\dfrac{-3x^4}{x - 2}$ **42.** $\dfrac{-3x^4}{x + 2}$

43. $\dfrac{180x - x^4}{x - 6}$ **44.** $\dfrac{5 - 3x + 2x^2 - x^3}{x + 1}$

45. $\dfrac{4x^3 + 16x^2 - 23x - 15}{x + \frac{1}{2}}$

46. $\dfrac{3x^3 - 4x^2 + 5}{x - \frac{3}{2}}$

In Exercises 47–54, write the function in the form $f(x) = (x - k)q(x) + r$ for the given value of k, and demonstrate that $f(k) = r$.

47. $f(x) = x^3 - x^2 - 14x + 11$, $k = 4$
48. $f(x) = x^3 - 5x^2 - 11x + 8$, $k = -2$
49. $f(x) = 15x^4 + 10x^3 - 6x^2 + 14$, $k = -\frac{2}{3}$
50. $f(x) = 10x^3 - 22x^2 - 3x + 4$, $k = \frac{1}{5}$
51. $f(x) = x^3 + 3x^2 - 2x - 14$, $k = \sqrt{2}$
52. $f(x) = x^3 + 2x^2 - 5x - 4$, $k = -\sqrt{5}$
53. $f(x) = -4x^3 + 6x^2 + 12x + 4$, $k = 1 - \sqrt{3}$
54. $f(x) = -3x^3 + 8x^2 + 10x - 8$, $k = 2 + \sqrt{2}$

In Exercises 55–58, use the Remainder Theorem and synthetic division to find each function value. Verify your answers using another method.

55. $f(x) = 2x^3 - 7x + 3$
   (a) $f(1)$ (b) $f(-2)$ (c) $f(\frac{1}{2})$ (d) $f(2)$
56. $g(x) = 2x^6 + 3x^4 - x^2 + 3$
   (a) $g(2)$ (b) $g(1)$ (c) $g(3)$ (d) $g(-1)$
57. $h(x) = x^3 - 5x^2 - 7x + 4$
   (a) $h(3)$ (b) $h(2)$ (c) $h(-2)$ (d) $h(-5)$
58. $f(x) = 4x^4 - 16x^3 + 7x^2 + 20$
   (a) $f(1)$ (b) $f(-2)$ (c) $f(5)$ (d) $f(-10)$

In Exercises 59–66, use synthetic division to show that x is a solution of the third-degree polynomial equation, and use the result to factor the polynomial completely. List all real solutions of the equation.

59. $x^3 - 7x + 6 = 0$, $x = 2$
60. $x^3 - 28x - 48 = 0$, $x = -4$
61. $2x^3 - 15x^2 + 27x - 10 = 0$, $x = \frac{1}{2}$
62. $48x^3 - 80x^2 + 41x - 6 = 0$, $x = \frac{2}{3}$
63. $x^3 + 2x^2 - 3x - 6 = 0$, $x = \sqrt{3}$
64. $x^3 + 2x^2 - 2x - 4 = 0$, $x = \sqrt{2}$
65. $x^3 - 3x^2 + 2 = 0$, $x = 1 + \sqrt{3}$
66. $x^3 - x^2 - 13x - 3 = 0$, $x = 2 - \sqrt{5}$

In Exercises 67–74, (a) verify the given factors of the function f, (b) find the remaining factor(s) of f, (c) use your results to write the complete factorization of f, (d) list all real zeros of f, and (e) confirm your results by using a graphing utility to graph the function.

Function	Factors
67. $f(x) = 2x^3 + x^2 - 5x + 2$	$(x + 2), (x - 1)$
68. $f(x) = 3x^3 + 2x^2 - 19x + 6$	$(x + 3), (x - 2)$
69. $f(x) = x^4 - 4x^3 - 15x^2$ $+ 58x - 40$	$(x - 5), (x + 4)$
70. $f(x) = 8x^4 - 14x^3 - 71x^2$ $- 10x + 24$	$(x + 2), (x - 4)$
71. $f(x) = 6x^3 + 41x^2 - 9x - 14$	$(2x + 1), (3x - 2)$
72. $f(x) = 10x^3 - 11x^2 - 72x + 45$	$(2x + 5), (5x - 3)$
73. $f(x) = 2x^3 - x^2 - 10x + 5$	$(2x - 1), (x + \sqrt{5})$
74. $f(x) = x^3 + 3x^2 - 48x - 144$	$(x + 4\sqrt{3}), (x + 3)$

 GRAPHICAL ANALYSIS   In Exercises 75–80, (a) use the *zero* or *root* feature of a graphing utility to approximate the zeros of the function accurate to three decimal places, (b) determine one of the exact zeros, and (c) use synthetic division to verify your result from part (b), and then factor the polynomial completely.

75. $f(x) = x^3 - 2x^2 - 5x + 10$
76. $g(x) = x^3 - 4x^2 - 2x + 8$
77. $h(t) = t^3 - 2t^2 - 7t + 2$
78. $f(s) = s^3 - 12s^2 + 40s - 24$
79. $h(x) = x^5 - 7x^4 + 10x^3 + 14x^2 - 24x$
80. $g(x) = 6x^4 - 11x^3 - 51x^2 + 99x - 27$

In Exercises 81–84, simplify the rational expression by using long division or synthetic division.

81. $\dfrac{4x^3 - 8x^2 + x + 3}{2x - 3}$ 82. $\dfrac{x^3 + x^2 - 64x - 64}{x + 8}$

83. $\dfrac{x^4 + 6x^3 + 11x^2 + 6x}{x^2 + 3x + 2}$

84. $\dfrac{x^4 + 9x^3 - 5x^2 - 36x + 4}{x^2 - 4}$

85. **DATA ANALYSIS: HIGHER EDUCATION**   The amounts A (in billions of dollars) donated to support higher education in the United States from 2000 through 2007 are shown in the table, where t represents the year, with $t = 0$ corresponding to 2000.

Year, t	Amount, A
0	23.2
1	24.2
2	23.9
3	23.9
4	24.4
5	25.6
6	28.0
7	29.8

(a) Use a graphing utility to create a scatter plot of the data.

(b) Use the *regression* feature of the graphing utility to find a cubic model for the data. Graph the model in the same viewing window as the scatter plot.

(c) Use the model to create a table of estimated values of A. Compare the model with the original data.

(d) Use synthetic division to evaluate the model for the year 2010. Even though the model is relatively accurate for estimating the given data, would you use this model to predict the amount donated to higher education in the future? Explain.

86. DATA ANALYSIS: HEALTH CARE The amounts A (in billions of dollars) of national health care expenditures in the United States from 2000 through 2007 are shown in the table, where t represents the year, with $t = 0$ corresponding to 2000.

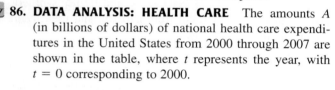

Year, t	Amount, A
0	30.5
1	32.2
2	34.2
3	38.0
4	42.7
5	47.9
6	52.7
7	57.6

(a) Use a graphing utility to create a scatter plot of the data.

(b) Use the *regression* feature of the graphing utility to find a cubic model for the data. Graph the model in the same viewing window as the scatter plot.

(c) Use the model to create a table of estimated values of A. Compare the model with the original data.

(d) Use synthetic division to evaluate the model for the year 2010.

EXPLORATION

TRUE OR FALSE? In Exercises 87–89, determine whether the statement is true or false. Justify your answer.

87. If $(7x + 4)$ is a factor of some polynomial function f, then $\frac{4}{7}$ is a zero of f.

88. $(2x - 1)$ is a factor of the polynomial
$$6x^6 + x^5 - 92x^4 + 45x^3 + 184x^2 + 4x - 48.$$

89. The rational expression
$$\frac{x^3 + 2x^2 - 13x + 10}{x^2 - 4x - 12}$$
is improper.

90. Use the form $f(x) = (x - k)q(x) + r$ to create a cubic function that (a) passes through the point $(2, 5)$ and rises to the right, and (b) passes through the point $(-3, 1)$ and falls to the right. (There are many correct answers.)

THINK ABOUT IT In Exercises 91 and 92, perform the division by assuming that n is a positive integer.

91. $\dfrac{x^{3n} + 9x^{2n} + 27x^n + 27}{x^n + 3}$ **92.** $\dfrac{x^{3n} - 3x^{2n} + 5x^n - 6}{x^n - 2}$

93. WRITING Briefly explain what it means for a divisor to divide evenly into a dividend.

94. WRITING Briefly explain how to check polynomial division, and justify your reasoning. Give an example.

EXPLORATION In Exercises 95 and 96, find the constant c such that the denominator will divide evenly into the numerator.

95. $\dfrac{x^3 + 4x^2 - 3x + c}{x - 5}$ **96.** $\dfrac{x^5 - 2x^2 + x + c}{x + 2}$

97. THINK ABOUT IT Find the value of k such that $x - 4$ is a factor of $x^3 - kx^2 + 2kx - 8$.

98. THINK ABOUT IT Find the value of k such that $x - 3$ is a factor of $x^3 - kx^2 + 2kx - 12$.

99. WRITING Complete each polynomial division. Write a brief description of the pattern that you obtain, and use your result to find a formula for the polynomial division $(x^n - 1)/(x - 1)$. Create a numerical example to test your formula.

(a) $\dfrac{x^2 - 1}{x - 1} =$ (b) $\dfrac{x^3 - 1}{x - 1} =$

(c) $\dfrac{x^4 - 1}{x - 1} =$

100. CAPSTONE Consider the division
$$f(x) \div (x - k)$$
where
$$f(x) = (x + 3)^2(x - 3)(x + 1)^3.$$

(a) What is the remainder when $k = -3$? Explain.

(b) If it is necessary to find $f(2)$, is it easier to evaluate the function directly or to use synthetic division? Explain.

2.4 COMPLEX NUMBERS

What you should learn

- Use the imaginary unit *i* to write complex numbers.
- Add, subtract, and multiply complex numbers.
- Use complex conjugates to write the quotient of two complex numbers in standard form.
- Find complex solutions of quadratic equations.

Why you should learn it

You can use complex numbers to model and solve real-life problems in electronics. For instance, in Exercise 89 on page 165, you will learn how to use complex numbers to find the impedance of an electrical circuit.

The Imaginary Unit *i*

You have learned that some quadratic equations have no real solutions. For instance, the quadratic equation $x^2 + 1 = 0$ has no real solution because there is no real number x that can be squared to produce -1. To overcome this deficiency, mathematicians created an expanded system of numbers using the **imaginary unit *i*,** defined as

$$i = \sqrt{-1} \qquad \text{Imaginary unit}$$

where $i^2 = -1$. By adding real numbers to real multiples of this imaginary unit, the set of **complex numbers** is obtained. Each complex number can be written in the **standard form $a + bi$.** For instance, the standard form of the complex number $-5 + \sqrt{-9}$ is $-5 + 3i$ because

$$-5 + \sqrt{-9} = -5 + \sqrt{3^2(-1)} = -5 + 3\sqrt{-1} = -5 + 3i.$$

In the standard form $a + bi$, the real number a is called the **real part** of the **complex number $a + bi$,** and the number bi (where b is a real number) is called the **imaginary part** of the complex number.

Definition of a Complex Number

If a and b are real numbers, the number $a + bi$ is a **complex number,** and it is said to be written in **standard form.** If $b = 0$, the number $a + bi = a$ is a real number. If $b \neq 0$, the number $a + bi$ is called an **imaginary number.** A number of the form bi, where $b \neq 0$, is called a **pure imaginary number.**

The set of real numbers is a subset of the set of complex numbers, as shown in Figure 2.30. This is true because every real number a can be written as a complex number using $b = 0$. That is, for every real number a, you can write $a = a + 0i$.

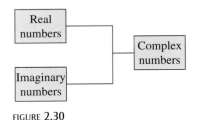

FIGURE **2.30**

Equality of Complex Numbers

Two complex numbers $a + bi$ and $c + di$, written in standard form, are equal to each other

$$a + bi = c + di \qquad \text{Equality of two complex numbers}$$

if and only if $a = c$ and $b = d$.

Operations with Complex Numbers

To add (or subtract) two complex numbers, you add (or subtract) the real and imaginary parts of the numbers separately.

Addition and Subtraction of Complex Numbers

If $a + bi$ and $c + di$ are two complex numbers written in standard form, their sum and difference are defined as follows.

Sum: $(a + bi) + (c + di) = (a + c) + (b + d)i$

Difference: $(a + bi) - (c + di) = (a - c) + (b - d)i$

The **additive identity** in the complex number system is zero (the same as in the real number system). Furthermore, the **additive inverse** of the complex number $a + bi$ is

$$-(a + bi) = -a - bi. \qquad \text{Additive inverse}$$

So, you have

$$(a + bi) + (-a - bi) = 0 + 0i = 0.$$

Example 1   Adding and Subtracting Complex Numbers

a. $(4 + 7i) + (1 - 6i) = 4 + 7i + 1 - 6i$ Remove parentheses.

$\qquad\qquad\qquad\qquad = (4 + 1) + (7i - 6i)$ Group like terms.

$\qquad\qquad\qquad\qquad = 5 + i$ Write in standard form.

b. $(1 + 2i) - (4 + 2i) = 1 + 2i - 4 - 2i$ Remove parentheses.

$\qquad\qquad\qquad\qquad = (1 - 4) + (2i - 2i)$ Group like terms.

$\qquad\qquad\qquad\qquad = -3 + 0$ Simplify.

$\qquad\qquad\qquad\qquad = -3$ Write in standard form.

c. $3i - (-2 + 3i) - (2 + 5i) = 3i + 2 - 3i - 2 - 5i$

$\qquad\qquad\qquad\qquad\qquad = (2 - 2) + (3i - 3i - 5i)$

$\qquad\qquad\qquad\qquad\qquad = 0 - 5i$

$\qquad\qquad\qquad\qquad\qquad = -5i$

d. $(3 + 2i) + (4 - i) - (7 + i) = 3 + 2i + 4 - i - 7 - i$

$\qquad\qquad\qquad\qquad\qquad = (3 + 4 - 7) + (2i - i - i)$

$\qquad\qquad\qquad\qquad\qquad = 0 + 0i$

$\qquad\qquad\qquad\qquad\qquad = 0$

CHECK*Point* Now try Exercise 21.

Note in Examples 1(b) and 1(d) that the sum of two complex numbers can be a real number.

Many of the properties of real numbers are valid for complex numbers as well. Here are some examples.

Associative Properties of Addition and Multiplication

Commutative Properties of Addition and Multiplication

Distributive Property of Multiplication Over Addition

Notice below how these properties are used when two complex numbers are multiplied.

$(a + bi)(c + di) = a(c + di) + bi(c + di)$	Distributive Property
$= ac + (ad)i + (bc)i + (bd)i^2$	Distributive Property
$= ac + (ad)i + (bc)i + (bd)(-1)$	$i^2 = -1$
$= ac - bd + (ad)i + (bc)i$	Commutative Property
$= (ac - bd) + (ad + bc)i$	Associative Property

Rather than trying to memorize this multiplication rule, you should simply remember how the Distributive Property is used to multiply two complex numbers.

Example 2   Multiplying Complex Numbers

a. $4(-2 + 3i) = 4(-2) + 4(3i)$ — Distributive Property

$= -8 + 12i$ — Simplify.

b. $(2 - i)(4 + 3i) = 2(4 + 3i) - i(4 + 3i)$ — Distributive Property

$= 8 + 6i - 4i - 3i^2$ — Distributive Property

$= 8 + 6i - 4i - 3(-1)$ — $i^2 = -1$

$= (8 + 3) + (6i - 4i)$ — Group like terms.

$= 11 + 2i$ — Write in standard form.

c. $(3 + 2i)(3 - 2i) = 3(3 - 2i) + 2i(3 - 2i)$ — Distributive Property

$= 9 - 6i + 6i - 4i^2$ — Distributive Property

$= 9 - 6i + 6i - 4(-1)$ — $i^2 = -1$

$= 9 + 4$ — Simplify.

$= 13$ — Write in standard form.

d. $(3 + 2i)^2 = (3 + 2i)(3 + 2i)$ — Square of a binomial

$= 3(3 + 2i) + 2i(3 + 2i)$ — Distributive Property

$= 9 + 6i + 6i + 4i^2$ — Distributive Property

$= 9 + 6i + 6i + 4(-1)$ — $i^2 = -1$

$= 9 + 12i - 4$ — Simplify.

$= 5 + 12i$ — Write in standard form.

CHECK*Point*   Now try Exercise 31.

Study Tip

The procedure described above is similar to multiplying two polynomials and combining like terms, as in the FOIL Method shown in Appendix A.3. For instance, you can use the FOIL Method to multiply the two complex numbers from Example 2(b).

$$\underset{\text{F} \quad \text{O} \quad \text{I} \quad \text{L}}{(2 - i)(4 + 3i) = 8 + 6i - 4i - 3i^2}$$

Complex Conjugates

Notice in Example 2(c) that the product of two complex numbers can be a real number. This occurs with pairs of complex numbers of the form $a + bi$ and $a - bi$, called **complex conjugates.**

$$(a + bi)(a - bi) = a^2 - abi + abi - b^2i^2$$
$$= a^2 - b^2(-1)$$
$$= a^2 + b^2$$

Example 3 Multiplying Conjugates

Multiply each complex number by its complex conjugate.

a. $1 + i$ **b.** $4 - 3i$

Solution

a. The complex conjugate of $1 + i$ is $1 - i$.
$$(1 + i)(1 - i) = 1^2 - i^2 = 1 - (-1) = 2$$

b. The complex conjugate of $4 - 3i$ is $4 + 3i$.
$$(4 - 3i)(4 + 3i) = 4^2 - (3i)^2 = 16 - 9i^2 = 16 - 9(-1) = 25$$

CHECK*Point* Now try Exercise 41.

To write the quotient of $a + bi$ and $c + di$ in standard form, where c and d are not both zero, multiply the numerator and denominator by the complex conjugate of the *denominator* to obtain

$$\frac{a + bi}{c + di} = \frac{a + bi}{c + di}\left(\frac{c - di}{c - di}\right)$$

$$= \frac{(ac + bd) + (bc - ad)i}{c^2 + d^2}.$$ Standard form

Example 4 Writing a Quotient of Complex Numbers in Standard Form

$$\frac{2 + 3i}{4 - 2i} = \frac{2 + 3i}{4 - 2i}\left(\frac{4 + 2i}{4 + 2i}\right)$$ Multiply numerator and denominator by complex conjugate of denominator.

$$= \frac{8 + 4i + 12i + 6i^2}{16 - 4i^2}$$ Expand.

$$= \frac{8 - 6 + 16i}{16 + 4}$$ $i^2 = -1$

$$= \frac{2 + 16i}{20}$$ Simplify.

$$= \frac{1}{10} + \frac{4}{5}i$$ Write in standard form.

CHECK*Point* Now try Exercise 53.

Complex Solutions of Quadratic Equations

When using the Quadratic Formula to solve a quadratic equation, you often obtain a result such as $\sqrt{-3}$, which you know is not a real number. By factoring out $i = \sqrt{-1}$, you can write this number in standard form.

$$\sqrt{-3} = \sqrt{3(-1)} = \sqrt{3}\sqrt{-1} = \sqrt{3}i$$

The number $\sqrt{3}i$ is called the *principal square root* of -3.

> ### Principal Square Root of a Negative Number
>
> If a is a positive number, the **principal square root** of the negative number $-a$ is defined as
>
> $$\sqrt{-a} = \sqrt{a}i.$$

Example 5   Writing Complex Numbers in Standard Form

a. $\sqrt{-3}\sqrt{-12} = \sqrt{3}i\sqrt{12}i = \sqrt{36}i^2 = 6(-1) = -6$

b. $\sqrt{-48} - \sqrt{-27} = \sqrt{48}i - \sqrt{27}i = 4\sqrt{3}i - 3\sqrt{3}i = \sqrt{3}i$

c. $\left(-1 + \sqrt{-3}\right)^2 = \left(-1 + \sqrt{3}i\right)^2$

$$= (-1)^2 - 2\sqrt{3}i + \left(\sqrt{3}\right)^2(i^2)$$
$$= 1 - 2\sqrt{3}i + 3(-1)$$
$$= -2 - 2\sqrt{3}i$$

CHECK*Point*   Now try Exercise 63.

Example 6   Complex Solutions of a Quadratic Equation

Solve (a) $x^2 + 4 = 0$ and (b) $3x^2 - 2x + 5 = 0$.

Solution

a. $x^2 + 4 = 0$ Write original equation.

$\quad x^2 = -4$ Subtract 4 from each side.

$\quad\quad x = \pm 2i$ Extract square roots.

b. $3x^2 - 2x + 5 = 0$ Write original equation.

$$x = \frac{-(-2) \pm \sqrt{(-2)^2 - 4(3)(5)}}{2(3)}$$
     Quadratic Formula

$$= \frac{2 \pm \sqrt{-56}}{6}$$
     Simplify.

$$= \frac{2 \pm 2\sqrt{14}i}{6}$$
     Write $\sqrt{-56}$ in standard form.

$$= \frac{1}{3} \pm \frac{\sqrt{14}}{3}i$$
     Write in standard form.

CHECK*Point*   Now try Exercise 69.

2.4 EXERCISES

See www.CalcChat.com for worked-out solutions to odd-numbered exercises.

VOCABULARY

1. Match the type of complex number with its definition.

(a) Real number   (i) $a + bi$, $a \neq 0$, $b \neq 0$

(b) Imaginary number   (ii) $a + bi$, $a = 0$, $b \neq 0$

(c) Pure imaginary number   (iii) $a + bi$, $b = 0$

In Exercises 2–4, fill in the blanks.

2. The imaginary unit i is defined as $i =$ _____, where $i^2 =$ _____.

3. If a is a positive number, the _____ _____ root of the negative number $-a$ is defined as $\sqrt{-a} = \sqrt{a}\,i$.

4. The numbers $a + bi$ and $a - bi$ are called _____ _____, and their product is a real number $a^2 + b^2$.

SKILLS AND APPLICATIONS

In Exercises 5–8, find real numbers a and b such that the equation is true.

5. $a + bi = -12 + 7i$ **6.** $a + bi = 13 + 4i$

7. $(a - 1) + (b + 3)i = 5 + 8i$

8. $(a + 6) + 2bi = 6 - 5i$

In Exercises 9–20, write the complex number in standard form.

9. $8 + \sqrt{-25}$ **10.** $5 + \sqrt{-36}$

11. $2 - \sqrt{-27}$ **12.** $1 + \sqrt{-8}$

13. $\sqrt{-80}$ **14.** $\sqrt{-4}$

15. 14 **16.** 75

17. $-10i + i^2$ **18.** $-4i^2 + 2i$

19. $\sqrt{-0.09}$ **20.** $\sqrt{-0.0049}$

In Exercises 21–30, perform the addition or subtraction and write the result in standard form.

21. $(7 + i) + (3 - 4i)$ **22.** $(13 - 2i) + (-5 + 6i)$

23. $(9 - i) - (8 - i)$ **24.** $(3 + 2i) - (6 + 13i)$

25. $\left(-2 + \sqrt{-8}\right) + \left(5 - \sqrt{-50}\right)$

26. $\left(8 + \sqrt{-18}\right) - \left(4 + 3\sqrt{2}\,i\right)$

27. $13i - (14 - 7i)$

28. $25 + (-10 + 11i) + 15i$

29. $-\left(\frac{3}{2} + \frac{5}{2}i\right) + \left(\frac{5}{3} + \frac{11}{3}i\right)$

30. $(1.6 + 3.2i) + (-5.8 + 4.3i)$

In Exercises 31–40, perform the operation and write the result in standard form.

31. $(1 + i)(3 - 2i)$ **32.** $(7 - 2i)(3 - 5i)$

33. $12i(1 - 9i)$ **34.** $-8i(9 + 4i)$

35. $\left(\sqrt{14} + \sqrt{10}\,i\right)\left(\sqrt{14} - \sqrt{10}\,i\right)$

36. $\left(\sqrt{3} + \sqrt{15}\,i\right)\left(\sqrt{3} - \sqrt{15}\,i\right)$

37. $(6 + 7i)^2$ **38.** $(5 - 4i)^2$

39. $(2 + 3i)^2 + (2 - 3i)^2$ **40.** $(1 - 2i)^2 - (1 + 2i)^2$

In Exercises 41–48, write the complex conjugate of the complex number. Then multiply the number by its complex conjugate.

41. $9 + 2i$ **42.** $8 - 10i$

43. $-1 - \sqrt{5}\,i$ **44.** $-3 + \sqrt{2}\,i$

45. $\sqrt{-20}$ **46.** $\sqrt{-15}$

47. $\sqrt{6}$ **48.** $1 + \sqrt{8}$

In Exercises 49–58, write the quotient in standard form.

49. $\dfrac{3}{i}$ **50.** $-\dfrac{14}{2i}$

51. $\dfrac{2}{4 - 5i}$ **52.** $\dfrac{13}{1 - i}$

53. $\dfrac{5 + i}{5 - i}$ **54.** $\dfrac{6 - 7i}{1 - 2i}$

55. $\dfrac{9 - 4i}{i}$ **56.** $\dfrac{8 + 16i}{2i}$

57. $\dfrac{3i}{(4 - 5i)^2}$ **58.** $\dfrac{5i}{(2 + 3i)^2}$

In Exercises 59–62, perform the operation and write the result in standard form.

59. $\dfrac{2}{1 + i} - \dfrac{3}{1 - i}$

60. $\dfrac{2i}{2 + i} + \dfrac{5}{2 - i}$

61. $\dfrac{i}{3 - 2i} + \dfrac{2i}{3 + 8i}$

62. $\dfrac{1 + i}{i} - \dfrac{3}{4 - i}$

In Exercises 63–68, write the complex number in standard form.

63. $\sqrt{-6} \cdot \sqrt{-2}$

64. $\sqrt{-5} \cdot \sqrt{-10}$

65. $\left(\sqrt{-15}\right)^2$

66. $\left(\sqrt{-75}\right)^2$

67. $\left(3 + \sqrt{-5}\right)\left(7 - \sqrt{-10}\right)$

68. $\left(2 - \sqrt{-6}\right)^2$

In Exercises 69–78, use the Quadratic Formula to solve the quadratic equation.

69. $x^2 - 2x + 2 = 0$

70. $x^2 + 6x + 10 = 0$

71. $4x^2 + 16x + 17 = 0$

72. $9x^2 - 6x + 37 = 0$

73. $4x^2 + 16x + 15 = 0$

74. $16t^2 - 4t + 3 = 0$

75. $\frac{3}{2}x^2 - 6x + 9 = 0$

76. $\frac{7}{8}x^2 - \frac{3}{4}x + \frac{5}{16} = 0$

77. $1.4x^2 - 2x - 10 = 0$

78. $4.5x^2 - 3x + 12 = 0$

In Exercises 79–88, simplify the complex number and write it in standard form.

79. $-6i^3 + i^2$

80. $4i^2 - 2i^3$

81. $-14i^5$

82. $(-i)^3$

83. $\left(\sqrt{-72}\right)^3$

84. $\left(\sqrt{-2}\right)^6$

85. $\dfrac{1}{i^3}$

86. $\dfrac{1}{(2i)^3}$

87. $(3i)^4$

88. $(-i)^6$

89. IMPEDANCE   The opposition to current in an electrical circuit is called its impedance. The impedance z in a parallel circuit with two pathways satisfies the equation

$$\frac{1}{z} = \frac{1}{z_1} + \frac{1}{z_2}$$

where z_1 is the impedance (in ohms) of pathway 1 and z_2 is the impedance of pathway 2.

(a) The impedance of each pathway in a parallel circuit is found by adding the impedances of all components in the pathway. Use the table to find z_1 and z_2.

(b) Find the impedance z.

	Resistor	Inductor	Capacitor
Symbol	$-\!\!\wedge\!\!\wedge\!\!\wedge\!\!-$ $a\Omega$	$-\!\!\text{000}\!\!-$ $b\Omega$	$-\!\!\vert\vert\!\!-$ $c\Omega$
Impedance	a	bi	$-ci$

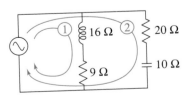

90. Cube each complex number.

(a) 2   (b) $-1 + \sqrt{3}i$ (c) $-1 - \sqrt{3}i$

91. Raise each complex number to the fourth power.

(a) 2   (b) -2 (c) $2i$ (d) $-2i$

92. Write each of the powers of i as i, $-i$, 1, or -1.

(a) i^{40} (b) i^{25} (c) i^{50} (d) i^{67}

EXPLORATION

TRUE OR FALSE?   In Exercises 93–96, determine whether the statement is true or false. Justify your answer.

93. There is no complex number that is equal to its complex conjugate.

94. $-i\sqrt{6}$ is a solution of $x^4 - x^2 + 14 = 56$.

95. $i^{44} + i^{150} - i^{74} - i^{109} + i^{61} = -1$

96. The sum of two complex numbers is always a real number.

97. PATTERN RECOGNITION   Complete the following.

$i^1 = i$ $i^2 = -1$ $i^3 = -i$ $i^4 = 1$

$i^5 = \rule{1cm}{0.15mm}$ $i^6 = \rule{1cm}{0.15mm}$ $i^7 = \rule{1cm}{0.15mm}$ $i^8 = \rule{1cm}{0.15mm}$

$i^9 = \rule{1cm}{0.15mm}$ $i^{10} = \rule{1cm}{0.15mm}$ $i^{11} = \rule{1cm}{0.15mm}$ $i^{12} = \rule{1cm}{0.15mm}$

What pattern do you see? Write a brief description of how you would find i raised to any positive integer power.

98. CAPSTONE   Consider the functions

$$f(x) = 2(x - 3)^2 - 4 \text{ and } g(x) = -2(x - 3)^2 - 4.$$

(a) Without graphing either function, determine whether the graph of f and the graph of g have x-intercepts. Explain your reasoning.

(b) Solve $f(x) = 0$ and $g(x) = 0$.

(c) Explain how the zeros of f and g are related to whether their graphs have x-intercepts.

(d) For the function $f(x) = a(x - h)^2 + k$, make a general statement about how a, h, and k affect whether the graph of f has x-intercepts, and whether the zeros of f are real or complex.

99. ERROR ANALYSIS   Describe the error.

$$\sqrt{-6}\sqrt{-6} = \sqrt{(-6)(-6)} = \sqrt{36} = 6$$

100. PROOF   Prove that the complex conjugate of the product of two complex numbers $a_1 + b_1i$ and $a_2 + b_2i$ is the product of their complex conjugates.

101. PROOF   Prove that the complex conjugate of the sum of two complex numbers $a_1 + b_1i$ and $a_2 + b_2i$ is the sum of their complex conjugates.

2.5 ZEROS OF POLYNOMIAL FUNCTIONS

What you should learn

- Use the Fundamental Theorem of Algebra to determine the number of zeros of polynomial functions.
- Find rational zeros of polynomial functions.
- Find conjugate pairs of complex zeros.
- Find zeros of polynomials by factoring.
- Use Descartes's Rule of Signs and the Upper and Lower Bound Rules to find zeros of polynomials.

Why you should learn it

Finding zeros of polynomial functions is an important part of solving real-life problems. For instance, in Exercise 120 on page 179, the zeros of a polynomial function can help you analyze the attendance at women's college basketball games.

The Fundamental Theorem of Algebra

You know that an nth-degree polynomial can have at most n real zeros. In the complex number system, this statement can be improved. That is, in the complex number system, every nth-degree polynomial function has *precisely* n zeros. This important result is derived from the **Fundamental Theorem of Algebra,** first proved by the German mathematician Carl Friedrich Gauss (1777–1855).

The Fundamental Theorem of Algebra

If $f(x)$ is a polynomial of degree n, where $n > 0$, then f has at least one zero in the complex number system.

Using the Fundamental Theorem of Algebra and the equivalence of zeros and factors, you obtain the **Linear Factorization Theorem.**

Linear Factorization Theorem

If $f(x)$ is a polynomial of degree n, where $n > 0$, then f has precisely n linear factors

$$f(x) = a_n(x - c_1)(x - c_2) \cdots (x - c_n)$$

where $c_1, c_2, \ldots, c_n$ are complex numbers.

For a proof of the Linear Factorization Theorem, see Proofs in Mathematics on page 212.

Note that the Fundamental Theorem of Algebra and the Linear Factorization Theorem tell you only that the zeros or factors of a polynomial exist, not how to find them. Such theorems are called *existence theorems.* Remember that the n zeros of a polynomial function can be real or complex, and they may be repeated.

Study Tip

Recall that in order to find the zeros of a function $f(x)$, set $f(x)$ equal to 0 and solve the resulting equation for x. For instance, the function in Example 1(a) has a zero at $x = 2$ because

$$x - 2 = 0$$
$$x = 2.$$

Algebra Help

Examples 1(b), 1(c), and 1(d) involve factoring polynomials. You can review the techniques for factoring polynomials in Appendix A.3.

Example 1   Zeros of Polynomial Functions

a. The first-degree polynomial $f(x) = x - 2$ has exactly *one* zero: $x = 2$.

b. Counting multiplicity, the second-degree polynomial function

$$f(x) = x^2 - 6x + 9 = (x - 3)(x - 3)$$

has exactly *two* zeros: $x = 3$ and $x = 3$. (This is called a *repeated zero.*)

c. The third-degree polynomial function

$$f(x) = x^3 + 4x = x(x^2 + 4) = x(x - 2i)(x + 2i)$$

has exactly *three* zeros: $x = 0$, $x = 2i$, and $x = -2i$.

d. The fourth-degree polynomial function

$$f(x) = x^4 - 1 = (x - 1)(x + 1)(x - i)(x + i)$$

has exactly *four* zeros: $x = 1$, $x = -1$, $x = i$, and $x = -i$.

CHECK*Point*   Now try Exercise 9.

The Rational Zero Test

The **Rational Zero Test** relates the possible rational zeros of a polynomial (having integer coefficients) to the leading coefficient and to the constant term of the polynomial.

The Rational Zero Test

If the polynomial $f(x) = a_n x^n + a_{n-1} x^{n-1} + \cdots + a_2 x^2 + a_1 x + a_0$ has *integer* coefficients, every rational zero of f has the form

$$\text{Rational zero} = \frac{p}{q}$$

where p and q have no common factors other than 1, and

p = a factor of the constant term a_0

q = a factor of the leading coefficient a_n.

To use the Rational Zero Test, you should first list all rational numbers whose numerators are factors of the constant term and whose denominators are factors of the leading coefficient.

$$\text{Possible rational zeros} = \frac{\text{factors of constant term}}{\text{factors of leading coefficient}}$$

Having formed this list of *possible rational zeros*, use a trial-and-error method to determine which, if any, are actual zeros of the polynomial. Note that when the leading coefficient is 1, the possible rational zeros are simply the factors of the constant term.

Example 2   Rational Zero Test with Leading Coefficient of 1

Find the rational zeros of

$$f(x) = x^3 + x + 1.$$

Solution

Because the leading coefficient is 1, the possible rational zeros are ± 1, the factors of the constant term. By testing these possible zeros, you can see that neither works.

$$f(1) = (1)^3 + 1 + 1$$

$$= 3$$

$$f(-1) = (-1)^3 + (-1) + 1$$

$$= -1$$

So, you can conclude that the given polynomial has *no* rational zeros. Note from the graph of f in Figure 2.31 that f does have one real zero between -1 and 0. However, by the Rational Zero Test, you know that this real zero is *not* a rational number.

CHECK **Point** Now try Exercise 15.

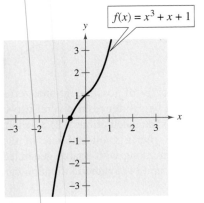

$f(x) = x^3 + x + 1$

FIGURE 2.31

Example 3   Rational Zero Test with Leading Coefficient of 1

Find the rational zeros of $f(x) = x^4 - x^3 + x^2 - 3x - 6$.

Solution

Because the leading coefficient is 1, the possible rational zeros are the factors of the constant term.

Possible rational zeros: $\pm 1, \pm 2, \pm 3, \pm 6$

By applying synthetic division successively, you can determine that $x = -1$ and $x = 2$ are the only two rational zeros.

```
-1 |  1   -1    1   -3   -6
   |      -1    2   -3    6
   ----------------------------
      1   -2    3   -6    0   ───→  0 remainder, so x = -1 is a zero.

 2 |  1   -2    3   -6
   |       2    0    6
   -----------------------
      1    0    3    0        ───→  0 remainder, so x = 2 is a zero.
```

So, $f(x)$ factors as

$$f(x) = (x + 1)(x - 2)(x^2 + 3).$$

Because the factor $(x^2 + 3)$ produces no real zeros, you can conclude that $x = -1$ and $x = 2$ are the only *real* zeros of f, which is verified in Figure 2.32.

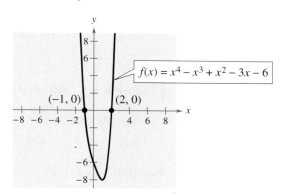

FIGURE 2.32

CHECK Point   Now try Exercise 19.

    If the leading coefficient of a polynomial is not 1, the list of possible rational zeros can increase dramatically. In such cases, the search can be shortened in several ways: (1) a programmable calculator can be used to speed up the calculations; (2) a graph, drawn either by hand or with a graphing utility, can give a good estimate of the locations of the zeros; (3) the Intermediate Value Theorem along with a table generated by a graphing utility can give approximations of zeros; and (4) synthetic division can be used to test the possible rational zeros.

    Finding the first zero is often the most difficult part. After that, the search is simplified by working with the lower-degree polynomial obtained in synthetic division, as shown in Example 3.

Example 4   Using the Rational Zero Test

Find the rational zeros of $f(x) = 2x^3 + 3x^2 - 8x + 3$.

Solution

The leading coefficient is 2 and the constant term is 3.

$$\text{Possible rational zeros: } \frac{\text{Factors of } 3}{\text{Factors of } 2} = \frac{\pm 1, \pm 3}{\pm 1, \pm 2} = \pm 1, \pm 3, \pm \frac{1}{2}, \pm \frac{3}{2}$$

By synthetic division, you can determine that $x = 1$ is a rational zero.

$$
\begin{array}{r|rrrr}
1 & 2 & 3 & -8 & 3 \\
  &   & 2 & 5 & -3 \\
\hline
  & 2 & 5 & -3 & 0
\end{array}
$$

So, $f(x)$ factors as

$$f(x) = (x - 1)(2x^2 + 5x - 3)$$
$$= (x - 1)(2x - 1)(x + 3)$$

and you can conclude that the rational zeros of f are $x = 1$, $x = \frac{1}{2}$, and $x = -3$.

CHECK Point   Now try Exercise 25.

Recall from Section 2.2 that if $x = a$ is a zero of the polynomial function f, then $x = a$ is a solution of the polynomial equation $f(x) = 0$.

Example 5   Solving a Polynomial Equation

Find all the real solutions of $-10x^3 + 15x^2 + 16x - 12 = 0$.

Solution

The leading coefficient is -10 and the constant term is -12.

$$\text{Possible rational solutions: } \frac{\text{Factors of } -12}{\text{Factors of } -10} = \frac{\pm 1, \pm 2, \pm 3, \pm 4, \pm 6, \pm 12}{\pm 1, \pm 2, \pm 5, \pm 10}$$

With so many possibilities (32, in fact), it is worth your time to stop and sketch a graph. From Figure 2.33, it looks like three reasonable solutions would be $x = -\frac{6}{5}$, $x = \frac{1}{2}$, and $x = 2$. Testing these by synthetic division shows that $x = 2$ is the only rational solution. So, you have

$$(x - 2)(-10x^2 - 5x + 6) = 0.$$

Using the Quadratic Formula for the second factor, you find that the two additional solutions are irrational numbers.

$$x = \frac{-5 - \sqrt{265}}{20} \approx -1.0639$$

and

$$x = \frac{-5 + \sqrt{265}}{20} \approx 0.5639$$

CHECK Point   Now try Exercise 31.

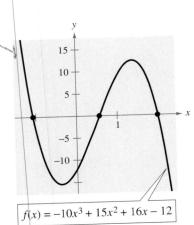

$$f(x) = -10x^3 + 15x^2 + 16x - 12$$

FIGURE 2.33

Conjugate Pairs

In Examples 1(c) and 1(d), note that the pairs of complex zeros are **conjugates.** That is, they are of the form $a + bi$ and $a - bi$.

Complex Zeros Occur in Conjugate Pairs

Let $f(x)$ be a polynomial function that has *real coefficients*. If $a + bi$, where $b \neq 0$, is a zero of the function, the conjugate $a - bi$ is also a zero of the function.

Be sure you see that this result is true only if the polynomial function has *real coefficients*. For instance, the result applies to the function given by $f(x) = x^2 + 1$ but not to the function given by $g(x) = x - i$.

Example 6   Finding a Polynomial with Given Zeros

Find a fourth-degree polynomial function with real coefficients that has -1, -1, and $3i$ as zeros.

Solution

Because $3i$ is a zero *and* the polynomial is stated to have real coefficients, you know that the conjugate $-3i$ must also be a zero. So, from the Linear Factorization Theorem, $f(x)$ can be written as

$$f(x) = a(x + 1)(x + 1)(x - 3i)(x + 3i).$$

For simplicity, let $a = 1$ to obtain

$$f(x) = (x^2 + 2x + 1)(x^2 + 9)$$

$$= x^4 + 2x^3 + 10x^2 + 18x + 9.$$

CHECK **Point**   Now try Exercise 45.

Factoring a Polynomial

The Linear Factorization Theorem shows that you can write any nth-degree polynomial as the product of n linear factors.

$$f(x) = a_n(x - c_1)(x - c_2)(x - c_3) \cdots (x - c_n)$$

However, this result includes the possibility that some of the values of c_i are complex. The following theorem says that even if you do not want to get involved with "complex factors," you can still write $f(x)$ as the product of linear and/or quadratic factors. For a proof of this theorem, see Proofs in Mathematics on page 212.

Factors of a Polynomial

Every polynomial of degree $n > 0$ with real coefficients can be written as the product of linear and quadratic factors with real coefficients, where the quadratic factors have no real zeros.

A quadratic factor with no real zeros is said to be *prime* or **irreducible over the reals.** Be sure you see that this is not the same as being *irreducible over the rationals.* For example, the quadratic $x^2 + 1 = (x - i)(x + i)$ is irreducible over the reals (and therefore over the rationals). On the other hand, the quadratic $x^2 - 2 = \left(x - \sqrt{2}\right)\left(x + \sqrt{2}\right)$ is irreducible over the rationals but *reducible* over the reals.

Example 7   Finding the Zeros of a Polynomial Function

Find all the zeros of $f(x) = x^4 - 3x^3 + 6x^2 + 2x - 60$ given that $1 + 3i$ is a zero of f.

Algebraic Solution

Because complex zeros occur in conjugate pairs, you know that $1 - 3i$ is also a zero of f. This means that both

$$[x - (1 + 3i)] \quad \text{and} \quad [x - (1 - 3i)]$$

are factors of f. Multiplying these two factors produces

$$[x - (1 + 3i)][x - (1 - 3i)] = [(x - 1) - 3i][(x - 1) + 3i]$$
$$= (x - 1)^2 - 9i^2$$
$$= x^2 - 2x + 10.$$

Using long division, you can divide $x^2 - 2x + 10$ into f to obtain the following.

$$
\begin{array}{r}
x^2 - x - 6 \\
x^2 - 2x + 10 \overline{)\,x^4 - 3x^3 + 6x^2 + 2x - 60} \\
\underline{x^4 - 2x^3 + 10x^2} \\
-x^3 - 4x^2 + 2x \\
\underline{-x^3 + 2x^2 - 10x} \\
-6x^2 + 12x - 60 \\
\underline{-6x^2 + 12x - 60} \\
0
\end{array}
$$

So, you have

$$f(x) = (x^2 - 2x + 10)(x^2 - x - 6)$$
$$= (x^2 - 2x + 10)(x - 3)(x + 2)$$

and you can conclude that the zeros of f are $x = 1 + 3i$, $x = 1 - 3i$, $x = 3$, and $x = -2$.

CHECK**Point** Now try Exercise 55.

Graphical Solution

Because complex zeros always occur in conjugate pairs, you know that $1 - 3i$ is also a zero of f. Because the polynomial is a fourth-degree polynomial, you know that there are at most two other zeros of the function. Use a graphing utility to graph

$$y = x^4 - 3x^3 + 6x^2 + 2x - 60$$

as shown in Figure 2.34.

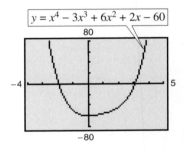

FIGURE 2.34

You can see that -2 and 3 appear to be zeros of the graph of the function. Use the *zero* or *root* feature or the *zoom* and *trace* features of the graphing utility to confirm that $x = -2$ and $x = 3$ are zeros of the graph. So, you can conclude that the zeros of f are $x = 1 + 3i$, $x = 1 - 3i$, $x = 3$, and $x = -2$.

Algebra Help

You can review the techniques for polynomial long division in Section 2.3.

In Example 7, if you were not told that $1 + 3i$ is a zero of f, you could still find all zeros of the function by using synthetic division to find the real zeros -2 and 3. Then you could factor the polynomial as $(x + 2)(x - 3)(x^2 - 2x + 10)$. Finally, by using the Quadratic Formula, you could determine that the zeros are $x = -2$, $x = 3$, $x = 1 + 3i$, and $x = 1 - 3i$.

Example 8 shows how to find all the zeros of a polynomial function, including complex zeros.

Example 8   Finding the Zeros of a Polynomial Function

Write $f(x) = x^5 + x^3 + 2x^2 - 12x + 8$ as the product of linear factors, and list all of its zeros.

Solution

The possible rational zeros are $\pm 1, \pm 2, \pm 4$, and ± 8. Synthetic division produces the following.

$$
\begin{array}{r|rrrrrr}
1 & 1 & 0 & 1 & 2 & -12 & 8 \\
  &   & 1 & 1 & 2 & 4   & -8 \\
\hline
  & 1 & 1 & 2 & 4 & -8 & 0 \\
\end{array} \longrightarrow \text{1 is a zero.}
$$

$$
\begin{array}{r|rrrrr}
-2 & 1 & 1 & 2 & 4 & -8 \\
   &   & -2 & 2 & -8 & 8 \\
\hline
   & 1 & -1 & 4 & -4 & 0 \\
\end{array} \longrightarrow \text{-2 is a zero.}
$$

So, you have

$$f(x) = x^5 + x^3 + 2x^2 - 12x + 8$$
$$= (x - 1)(x + 2)(x^3 - x^2 + 4x - 4).$$

You can factor $x^3 - x^2 + 4x - 4$ as $(x - 1)(x^2 + 4)$, and by factoring $x^2 + 4$ as

$$x^2 - (-4) = \left(x - \sqrt{-4}\right)\left(x + \sqrt{-4}\right)$$
$$= (x - 2i)(x + 2i)$$

you obtain

$$f(x) = (x - 1)(x - 1)(x + 2)(x - 2i)(x + 2i)$$

which gives the following five zeros of f.

$$x = 1, \; x = 1, \; x = -2, \; x = 2i, \quad \text{and} \quad x = -2i$$

From the graph of f shown in Figure 2.35, you can see that the *real* zeros are the only ones that appear as x-intercepts. Note that $x = 1$ is a repeated zero.

CHECK *Point*   Now try Exercise 77.

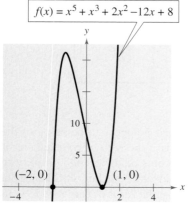

$f(x) = x^5 + x^3 + 2x^2 - 12x + 8$

FIGURE 2.35

TECHNOLOGY

You can use the *table* feature of a graphing utility to help you determine which of the possible rational zeros are zeros of the polynomial in Example 8. The table should be set to *ask* mode. Then enter each of the possible rational zeros in the table. When you do this, you will see that there are two rational zeros, -2 and 1, as shown at the right.

X	Y1
-8	-33048
-4	-1000
-2	0
-1	20
1	0
2	32
4	1080

X=4

Other Tests for Zeros of Polynomials

You know that an *n*th-degree polynomial function can have *at most n* real zeros. Of course, many *n*th-degree polynomials do not have that many real zeros. For instance, $f(x) = x^2 + 1$ has no real zeros, and $f(x) = x^3 + 1$ has only one real zero. The following theorem, called **Descartes's Rule of Signs,** sheds more light on the number of real zeros of a polynomial.

Descartes's Rule of Signs

Let $f(x) = a_n x^n + a_{n-1}x^{n-1} + \cdots + a_2 x^2 + a_1 x + a_0$ be a polynomial with real coefficients and $a_0 \neq 0$.

1. The number of *positive real zeros* of f is either equal to the number of variations in sign of $f(x)$ or less than that number by an even integer.

2. The number of *negative real zeros* of f is either equal to the number of variations in sign of $f(-x)$ or less than that number by an even integer.

A **variation in sign** means that two consecutive coefficients have opposite signs. When using Descartes's Rule of Signs, a zero of multiplicity k should be counted as k zeros. For instance, the polynomial $x^3 - 3x + 2$ has two variations in sign, and so has either two positive or no positive real zeros. Because

$$x^3 - 3x + 2 = (x - 1)(x - 1)(x + 2)$$

you can see that the two positive real zeros are $x = 1$ of multiplicity 2.

Example 9   Using Descartes's Rule of Signs

Describe the possible real zeros of

$$f(x) = 3x^3 - 5x^2 + 6x - 4.$$

Solution

The original polynomial has *three* variations in sign.

$$
\begin{array}{cc}
\text{+ to −} & \text{+ to −} \\
\downarrow \quad \downarrow & \downarrow \quad \downarrow \\
\end{array}
$$

$$f(x) = 3x^3 - 5x^2 + 6x - 4$$

$$
\begin{array}{c}
\uparrow \qquad \uparrow \\
\text{− to +}
\end{array}
$$

The polynomial

$$f(-x) = 3(-x)^3 - 5(-x)^2 + 6(-x) - 4$$

$$= -3x^3 - 5x^2 - 6x - 4$$

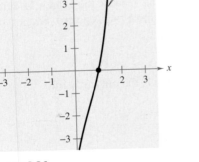

$$f(x) = 3x^3 - 5x^2 + 6x - 4$$

has no variations in sign. So, from Descartes's Rule of Signs, the polynomial $f(x) = 3x^3 - 5x^2 + 6x - 4$ has either three positive real zeros or one positive real zero, and has no negative real zeros. From the graph in Figure 2.36, you can see that the function has only one real zero, at $x = 1$.

CHECK Point   Now try Exercise 87.

FIGURE 2.36

Another test for zeros of a polynomial function is related to the sign pattern in the last row of the synthetic division array. This test can give you an upper or lower bound of the real zeros of f. A real number b is an **upper bound** for the real zeros of f if no zeros are greater than b. Similarly, b is a **lower bound** if no real zeros of f are less than b.

Upper and Lower Bound Rules

Let $f(x)$ be a polynomial with real coefficients and a positive leading coefficient. Suppose $f(x)$ is divided by $x - c$, using synthetic division.

1. If $c > 0$ and each number in the last row is either positive or zero, c is an **upper bound** for the real zeros of f.

2. If $c < 0$ and the numbers in the last row are alternately positive and negative (zero entries count as positive or negative), c is a **lower bound** for the real zeros of f.

Example 10   Finding the Zeros of a Polynomial Function

Find the real zeros of $f(x) = 6x^3 - 4x^2 + 3x - 2$.

Solution

The possible real zeros are as follows.

$$\frac{\text{Factors of 2}}{\text{Factors of 6}} = \frac{\pm 1, \pm 2}{\pm 1, \pm 2, \pm 3, \pm 6} = \pm 1, \pm\frac{1}{2}, \pm\frac{1}{3}, \pm\frac{1}{6}, \pm\frac{2}{3}, \pm 2$$

The original polynomial $f(x)$ has three variations in sign. The polynomial

$$f(-x) = 6(-x)^3 - 4(-x)^2 + 3(-x) - 2$$
$$= -6x^3 - 4x^2 - 3x - 2$$

has no variations in sign. As a result of these two findings, you can apply Descartes's Rule of Signs to conclude that there are three positive real zeros or one positive real zero, and no negative zeros. Trying $x = 1$ produces the following.

```
1 | 6   -4    3   -2
  |      6    2    5
  ----------------------
    6    2    5    3
```

So, $x = 1$ is not a zero, but because the last row has all positive entries, you know that $x = 1$ is an upper bound for the real zeros. So, you can restrict the search to zeros between 0 and 1. By trial and error, you can determine that $x = \frac{2}{3}$ is a zero. So,

$$f(x) = \left(x - \frac{2}{3}\right)(6x^2 + 3).$$

Because $6x^2 + 3$ has no real zeros, it follows that $x = \frac{2}{3}$ is the only real zero.

CHECK**Point**   Now try Exercise 95.

Before concluding this section, here are two additional hints that can help you find the real zeros of a polynomial.

1. If the terms of $f(x)$ have a common monomial factor, it should be factored out before applying the tests in this section. For instance, by writing

$$f(x) = x^4 - 5x^3 + 3x^2 + x$$
$$= x(x^3 - 5x^2 + 3x + 1)$$

you can see that $x = 0$ is a zero of f and that the remaining zeros can be obtained by analyzing the cubic factor.

2. If you are able to find all but two zeros of $f(x)$, you can always use the Quadratic Formula on the remaining quadratic factor. For instance, if you succeeded in writing

$$f(x) = x^4 - 5x^3 + 3x^2 + x$$
$$= x(x - 1)(x^2 - 4x - 1)$$

you can apply the Quadratic Formula to $x^2 - 4x - 1$ to conclude that the two remaining zeros are $x = 2 + \sqrt{5}$ and $x = 2 - \sqrt{5}$.

Example 11   Using a Polynomial Model

You are designing candle-making kits. Each kit contains 25 cubic inches of candle wax and a mold for making a pyramid-shaped candle. You want the height of the candle to be 2 inches less than the length of each side of the candle's square base. What should the dimensions of your candle mold be?

Solution

The volume of a pyramid is $V = \frac{1}{3}Bh$, where B is the area of the base and h is the height. The area of the base is x^2 and the height is $(x - 2)$. So, the volume of the pyramid is $V = \frac{1}{3}x^2(x - 2)$. Substituting 25 for the volume yields the following.

$$25 = \frac{1}{3}x^2(x - 2) \qquad \text{Substitute 25 for } V.$$

$$75 = x^3 - 2x^2 \qquad \text{Multiply each side by 3.}$$

$$0 = x^3 - 2x^2 - 75 \qquad \text{Write in general form.}$$

The possible rational solutions are $x = \pm1, \pm3, \pm5, \pm15, \pm25, \pm75$. Use synthetic division to test some of the possible solutions. Note that in this case, it makes sense to test only positive x-values. Using synthetic division, you can determine that $x = 5$ is a solution.

$$
\begin{array}{r|rrrr}
5 & 1 & -2 & 0 & -75 \\
  &   & 5 & 15 & 75 \\
\hline
  & 1 & 3 & 15 & 0
\end{array}
$$

The other two solutions, which satisfy $x^2 + 3x + 15 = 0$, are imaginary and can be discarded. You can conclude that the base of the candle mold should be 5 inches by 5 inches and the height of the mold should be $5 - 2 = 3$ inches.

CHECK**Point** ▶ Now try Exercise 115.

2.5 EXERCISES

See www.CalcChat.com for worked-out solutions to odd-numbered exercises.

VOCABULARY: Fill in the blanks.

1. The _____ _____ of _____ states that if $f(x)$ is a polynomial of degree n ($n > 0$), then f has at least one zero in the complex number system.

2. The _____ _____ _____ states that if $f(x)$ is a polynomial of degree n ($n > 0$), then f has precisely n linear factors, $f(x) = a_n(x - c_1)(x - c_2) \cdots (x - c_n)$, where $c_1, c_2, \ldots, c_n$ are complex numbers.

3. The test that gives a list of the possible rational zeros of a polynomial function is called the _____ _____ Test.

4. If $a + bi$ is a complex zero of a polynomial with real coefficients, then so is its _____, $a - bi$.

5. Every polynomial of degree $n > 0$ with real coefficients can be written as the product of _____ and _____ factors with real coefficients, where the _____ factors have no real zeros.

6. A quadratic factor that cannot be factored further as a product of linear factors containing real numbers is said to be _____ over the _____.

7. The theorem that can be used to determine the possible numbers of positive real zeros and negative real zeros of a function is called _____ _____ of _____.

8. A real number b is a(n) _____ bound for the real zeros of f if no real zeros are less than b, and is a(n) _____ bound if no real zeros are greater than b.

SKILLS AND APPLICATIONS

In Exercises 9–14, find all the zeros of the function.

9. $f(x) = x(x - 6)^2$

10. $f(x) = x^2(x + 3)(x^2 - 1)$

11. $g(x) = (x - 2)(x + 4)^3$

12. $f(x) = (x + 5)(x - 8)^2$

13. $f(x) = (x + 6)(x + i)(x - i)$

14. $h(t) = (t - 3)(t - 2)(t - 3i)(t + 3i)$

In Exercises 15–18, use the Rational Zero Test to list all possible rational zeros of f. Verify that the zeros of f shown on the graph are contained in the list.

15. $f(x) = x^3 + 2x^2 - x - 2$

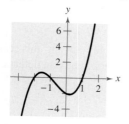

16. $f(x) = x^3 - 4x^2 - 4x + 16$

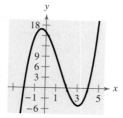

17. $f(x) = 2x^4 - 17x^3 + 35x^2 + 9x - 45$

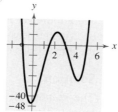

18. $f(x) = 4x^5 - 8x^4 - 5x^3 + 10x^2 + x - 2$

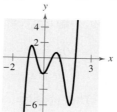

In Exercises 19–28, find all the rational zeros of the function.

19. $f(x) = x^3 - 6x^2 + 11x - 6$

20. $f(x) = x^3 - 7x - 6$

21. $g(x) = x^3 - 4x^2 - x + 4$

22. $h(x) = x^3 - 9x^2 + 20x - 12$

23. $h(t) = t^3 + 8t^2 + 13t + 6$

24. $p(x) = x^3 - 9x^2 + 27x - 27$

25. $C(x) = 2x^3 + 3x^2 - 1$

26. $f(x) = 3x^3 - 19x^2 + 33x - 9$

27. $f(x) = 9x^4 - 9x^3 - 58x^2 + 4x + 24$

28. $f(x) = 2x^4 - 15x^3 + 23x^2 + 15x - 25$

In Exercises 29–32, find all real solutions of the polynomial equation.

29. $z^4 + z^3 + z^2 + 3z - 6 = 0$

30. $x^4 - 13x^2 - 12x = 0$

31. $2y^4 + 3y^3 - 16y^2 + 15y - 4 = 0$

32. $x^5 - x^4 - 3x^3 + 5x^2 - 2x = 0$

In Exercises 33–36, (a) list the possible rational zeros of f, (b) sketch the graph of f so that some of the possible zeros in part (a) can be disregarded, and then (c) determine all real zeros of f.

33. $f(x) = x^3 + x^2 - 4x - 4$

34. $f(x) = -3x^3 + 20x^2 - 36x + 16$

35. $f(x) = -4x^3 + 15x^2 - 8x - 3$

36. $f(x) = 4x^3 - 12x^2 - x + 15$

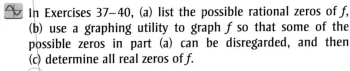 In Exercises 37–40, (a) list the possible rational zeros of f, (b) use a graphing utility to graph f so that some of the possible zeros in part (a) can be disregarded, and then (c) determine all real zeros of f.

37. $f(x) = -2x^4 + 13x^3 - 21x^2 + 2x + 8$

38. $f(x) = 4x^4 - 17x^2 + 4$

39. $f(x) = 32x^3 - 52x^2 + 17x + 3$

40. $f(x) = 4x^3 + 7x^2 - 11x - 18$

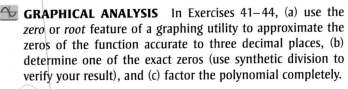 **GRAPHICAL ANALYSIS** In Exercises 41–44, (a) use the *zero* or *root* feature of a graphing utility to approximate the zeros of the function accurate to three decimal places, (b) determine one of the exact zeros (use synthetic division to verify your result), and (c) factor the polynomial completely.

41. $f(x) = x^4 - 3x^2 + 2$ **42.** $P(t) = t^4 - 7t^2 + 12$

43. $h(x) = x^5 - 7x^4 + 10x^3 + 14x^2 - 24x$

44. $g(x) = 6x^4 - 11x^3 - 51x^2 + 99x - 27$

In Exercises 45–50, find a polynomial function with real coefficients that has the given zeros. (There are many correct answers.)

45. $1, 5i$

46. $4, -3i$

47. $2, 5 + i$

48. $5, 3 - 2i$

49. $\frac{2}{3}, -1, 3 + \sqrt{2}i$

50. $-5, -5, 1 + \sqrt{3}i$

In Exercises 51–54, write the polynomial (a) as the product of factors that are irreducible over the *rationals*, (b) as the product of linear and quadratic factors that are irreducible over the *reals*, and (c) in completely factored form.

51. $f(x) = x^4 + 6x^2 - 27$

52. $f(x) = x^4 - 2x^3 - 3x^2 + 12x - 18$
    (*Hint:* One factor is $x^2 - 6$.)

53. $f(x) = x^4 - 4x^3 + 5x^2 - 2x - 6$
    (*Hint:* One factor is $x^2 - 2x - 2$.)

54. $f(x) = x^4 - 3x^3 - x^2 - 12x - 20$
    (*Hint:* One factor is $x^2 + 4$.)

In Exercises 55–62, use the given zero to find all the zeros of the function.

	Function	*Zero*
55.	$f(x) = x^3 - x^2 + 4x - 4$	$2i$
56.	$f(x) = 2x^3 + 3x^2 + 18x + 27$	$3i$
57.	$f(x) = 2x^4 - x^3 + 49x^2 - 25x - 25$	$5i$
58.	$g(x) = x^3 - 7x^2 - x + 87$	$5 + 2i$
59.	$g(x) = 4x^3 + 23x^2 + 34x - 10$	$-3 + i$
60.	$h(x) = 3x^3 - 4x^2 + 8x + 8$	$1 - \sqrt{3}i$
61.	$f(x) = x^4 + 3x^3 - 5x^2 - 21x + 22$	$-3 + \sqrt{2}i$
62.	$f(x) = x^3 + 4x^2 + 14x + 20$	$-1 - 3i$

In Exercises 63–80, find all the zeros of the function and write the polynomial as a product of linear factors.

63. $f(x) = x^2 + 36$ **64.** $f(x) = x^2 - x + 56$

65. $h(x) = x^2 - 2x + 17$ **66.** $g(x) = x^2 + 10x + 17$

67. $f(x) = x^4 - 16$ **68.** $f(y) = y^4 - 256$

69. $f(z) = z^2 - 2z + 2$

70. $h(x) = x^3 - 3x^2 + 4x - 2$

71. $g(x) = x^3 - 3x^2 + x + 5$

72. $f(x) = x^3 - x^2 + x + 39$

73. $h(x) = x^3 - x + 6$

74. $h(x) = x^3 + 9x^2 + 27x + 35$

75. $f(x) = 5x^3 - 9x^2 + 28x + 6$

76. $g(x) = 2x^3 - x^2 + 8x + 21$

77. $g(x) = x^4 - 4x^3 + 8x^2 - 16x + 16$

78. $h(x) = x^4 + 6x^3 + 10x^2 + 6x + 9$

79. $f(x) = x^4 + 10x^2 + 9$

80. $f(x) = x^4 + 29x^2 + 100$

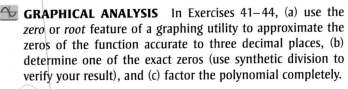 In Exercises 81–86, find all the zeros of the function. When there is an extended list of possible rational zeros, use a graphing utility to graph the function in order to discard any rational zeros that are obviously not zeros of the function.

81. $f(x) = x^3 + 24x^2 + 214x + 740$

82. $f(s) = 2s^3 - 5s^2 + 12s - 5$

83. $f(x) = 16x^3 - 20x^2 - 4x + 15$

84. $f(x) = 9x^3 - 15x^2 + 11x - 5$

85. $f(x) = 2x^4 + 5x^3 + 4x^2 + 5x + 2$

86. $g(x) = x^5 - 8x^4 + 28x^3 - 56x^2 + 64x - 32$

In Exercises 87–94, use Descartes's Rule of Signs to determine the possible numbers of positive and negative zeros of the function.

87. $g(x) = 2x^3 - 3x^2 - 3$ **88.** $h(x) = 4x^2 - 8x + 3$

89. $h(x) = 2x^3 + 3x^2 + 1$ **90.** $h(x) = 2x^4 - 3x + 2$

91. $g(x) = 5x^5 - 10x$

92. $f(x) = 4x^3 - 3x^2 + 2x - 1$

93. $f(x) = -5x^3 + x^2 - x + 5$

94. $f(x) = 3x^3 + 2x^2 + x + 3$

In Exercises 95–98, use synthetic division to verify the upper and lower bounds of the real zeros of f.

95. $f(x) = x^3 + 3x^2 - 2x + 1$

(a) Upper: $x = 1$ (b) Lower: $x = -4$

96. $f(x) = x^3 - 4x^2 + 1$

(a) Upper: $x = 4$ (b) Lower: $x = -1$

97. $f(x) = x^4 - 4x^3 + 16x - 16$

(a) Upper: $x = 5$ (b) Lower: $x = -3$

98. $f(x) = 2x^4 - 8x + 3$

(a) Upper: $x = 3$ (b) Lower: $x = -4$

In Exercises 99–102, find all the real zeros of the function.

99. $f(x) = 4x^3 - 3x - 1$

100. $f(z) = 12z^3 - 4z^2 - 27z + 9$

101. $f(y) = 4y^3 + 3y^2 + 8y + 6$

102. $g(x) = 3x^3 - 2x^2 + 15x - 10$

In Exercises 103–106, find all the rational zeros of the polynomial function.

103. $P(x) = x^4 - \frac{25}{4}x^2 + 9 = \frac{1}{4}(4x^4 - 25x^2 + 36)$

104. $f(x) = x^3 - \frac{3}{2}x^2 - \frac{23}{2}x + 6 = \frac{1}{2}(2x^3 - 3x^2 - 23x + 12)$

105. $f(x) = x^3 - \frac{1}{4}x^2 - x + \frac{1}{4} = \frac{1}{4}(4x^3 - x^2 - 4x + 1)$

106. $f(z) = z^3 + \frac{11}{6}z^2 - \frac{1}{2}z - \frac{1}{3} = \frac{1}{6}(6z^3 + 11z^2 - 3z - 2)$

In Exercises 107–110, match the cubic function with the numbers of rational and irrational zeros.

(a) Rational zeros: 0; irrational zeros: 1

(b) Rational zeros: 3; irrational zeros: 0

(c) Rational zeros: 1; irrational zeros: 2

(d) Rational zeros: 1; irrational zeros: 0

107. $f(x) = x^3 - 1$ **108.** $f(x) = x^3 - 2$

109. $f(x) = x^3 - x$ **110.** $f(x) = x^3 - 2x$

111. GEOMETRY An open box is to be made from a rectangular piece of material, 15 centimeters by 9 centimeters, by cutting equal squares from the corners and turning up the sides.

(a) Let x represent the length of the sides of the squares removed. Draw a diagram showing the squares removed from the original piece of material and the resulting dimensions of the open box.

(b) Use the diagram to write the volume V of the box as a function of x. Determine the domain of the function.

(c) Sketch the graph of the function and approximate the dimensions of the box that will yield a maximum volume.

(d) Find values of x such that $V = 56$. Which of these values is a physical impossibility in the construction of the box? Explain.

112. GEOMETRY A rectangular package to be sent by a delivery service (see figure) can have a maximum combined length and girth (perimeter of a cross section) of 120 inches.

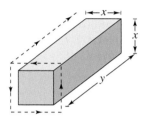

(a) Write a function $V(x)$ that represents the volume of the package.

 (b) Use a graphing utility to graph the function and approximate the dimensions of the package that will yield a maximum volume.

(c) Find values of x such that $V = 13,500$. Which of these values is a physical impossibility in the construction of the package? Explain.

113. ADVERTISING COST A company that produces MP3 players estimates that the profit P (in dollars) for selling a particular model is given by

$$P = -76x^3 + 4830x^2 - 320,000, \quad 0 \le x \le 60$$

where x is the advertising expense (in tens of thousands of dollars). Using this model, find the smaller of two advertising amounts that will yield a profit of $2,500,000.

114. ADVERTISING COST A company that manufactures bicycles estimates that the profit P (in dollars) for selling a particular model is given by

$$P = -45x^3 + 2500x^2 - 275,000, \quad 0 \le x \le 50$$

where x is the advertising expense (in tens of thousands of dollars). Using this model, find the smaller of two advertising amounts that will yield a profit of $800,000.

115. GEOMETRY   A bulk food storage bin with dimensions 2 feet by 3 feet by 4 feet needs to be increased in size to hold five times as much food as the current bin. (Assume each dimension is increased by the same amount.)

(a) Write a function that represents the volume V of the new bin.

(b) Find the dimensions of the new bin.

116. GEOMETRY   A manufacturer wants to enlarge an existing manufacturing facility such that the total floor area is 1.5 times that of the current facility. The floor area of the current facility is rectangular and measures 250 feet by 160 feet. The manufacturer wants to increase each dimension by the same amount.

(a) Write a function that represents the new floor area A.

(b) Find the dimensions of the new floor.

(c) Another alternative is to increase the current floor's length by an amount that is twice an increase in the floor's width. The total floor area is 1.5 times that of the current facility. Repeat parts (a) and (b) using these criteria.

117. COST   The ordering and transportation cost C (in thousands of dollars) for the components used in manufacturing a product is given by

$$C = 100\left(\frac{200}{x^2} + \frac{x}{x + 30}\right), \quad x \geq 1$$

where x is the order size (in hundreds). In calculus, it can be shown that the cost is a minimum when

$$3x^3 - 40x^2 - 2400x - 36{,}000 = 0.$$

Use a calculator to approximate the optimal order size to the nearest hundred units.

118. HEIGHT OF A BASEBALL   A baseball is thrown upward from a height of 6 feet with an initial velocity of 48 feet per second, and its height h (in feet) is

$$h(t) = -16t^2 + 48t + 6, \quad 0 \leq t \leq 3$$

where t is the time (in seconds). You are told the ball reaches a height of 64 feet. Is this possible?

119. PROFIT   The demand equation for a certain product is $p = 140 - 0.0001x$, where p is the unit price (in dollars) of the product and x is the number of units produced and sold. The cost equation for the product is $C = 80x + 150{,}000$, where C is the total cost (in dollars) and x is the number of units produced. The total profit obtained by producing and selling x units is $P = R - C = xp - C$. You are working in the marketing department of the company that produces this product, and you are asked to determine a price p that will yield a profit of 9 million dollars. Is this possible? Explain.

120. ATHLETICS   The attendance A (in millions) at NCAA women's college basketball games for the years 2000 through 2007 is shown in the table. (Source: National Collegiate Athletic Association, Indianapolis, IN)

Year	Attendance, A
2000	8.7
2001	8.8
2002	9.5
2003	10.2
2004	10.0
2005	9.9
2006	9.9
2007	10.9

(a) Use a graphing utility to create a scatter plot of the data. Let t represent the year, with $t = 0$ corresponding to 2000.

(b) Use the *regression* feature of the graphing utility to find a quartic model for the data.

(c) Graph the model and the scatter plot in the same viewing window. How well does the model fit the data?

(d) According to the model in part (b), in what year(s) was the attendance at least 10 million?

(e) According to the model, will the attendance continue to increase? Explain.

EXPLORATION

TRUE OR FALSE?   In Exercises 121 and 122, decide whether the statement is true or false. Justify your answer.

121. It is possible for a third-degree polynomial function with integer coefficients to have no real zeros.

122. If $x = -i$ is a zero of the function given by

$$f(x) = x^3 + ix^2 + ix - 1$$

then $x = i$ must also be a zero of f.

THINK ABOUT IT   In Exercises 123–128, determine (if possible) the zeros of the function g if the function f has zeros at $x = r_1$, $x = r_2$, and $x = r_3$.

123. $g(x) = -f(x)$ **124.** $g(x) = 3f(x)$

125. $g(x) = f(x - 5)$ **126.** $g(x) = f(2x)$

127. $g(x) = 3 + f(x)$ **128.** $g(x) = f(-x)$

129. THINK ABOUT IT A third-degree polynomial function f has real zeros -2, $\frac{1}{2}$, and 3, and its leading coefficient is negative. Write an equation for f. Sketch the graph of f. How many different polynomial functions are possible for f?

130. CAPSTONE Use a graphing utility to graph the function given by $f(x) = x^4 - 4x^2 + k$ for different values of k. Find values of k such that the zeros of f satisfy the specified characteristics. (Some parts do not have unique answers.)

(a) Four real zeros

(b) Two real zeros, each of multiplicity 2

(c) Two real zeros and two complex zeros

(d) Four complex zeros

(e) Will the answers to parts (a) through (d) change for the function g, where $g(x) = f(x - 2)$?

(f) Will the answers to parts (a) through (d) change for the function g, where $g(x) = f(2x)$?

131. THINK ABOUT IT Sketch the graph of a fifth-degree polynomial function whose leading coefficient is positive and that has a zero at $x = 3$ of multiplicity 2.

132. WRITING Compile a list of all the various techniques for factoring a polynomial that have been covered so far in the text. Give an example illustrating each technique, and write a paragraph discussing when the use of each technique is appropriate.

133. THINK ABOUT IT Let $y = f(x)$ be a quartic polynomial with leading coefficient $a = 1$ and $f(i) = f(2i) = 0$. Write an equation for f.

134. THINK ABOUT IT Let $y = f(x)$ be a cubic polynomial with leading coefficient $a = -1$ and $f(2) = f(i) = 0$. Write an equation for f.

In Exercises 135 and 136, the graph of a cubic polynomial function $y = f(x)$ is shown. It is known that one of the zeros is $1 + i$. Write an equation for f.

135.

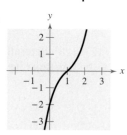

136.

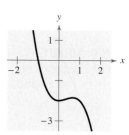

137. Use the information in the table to answer each question.

Interval	Value of $f(x)$
$(-\infty, -2)$	Positive
$(-2, 1)$	Negative
$(1, 4)$	Negative
$(4, \infty)$	Positive

(a) What are the three real zeros of the polynomial function f?

(b) What can be said about the behavior of the graph of f at $x = 1$?

(c) What is the least possible degree of f? Explain. Can the degree of f ever be odd? Explain.

(d) Is the leading coefficient of f positive or negative? Explain.

(e) Write an equation for f. (There are many correct answers.)

(f) Sketch a graph of the equation you wrote in part (e).

138. (a) Find a quadratic function f (with integer coefficients) that has $\pm \sqrt{b}\,i$ as zeros. Assume that b is a positive integer.

(b) Find a quadratic function f (with integer coefficients) that has $a \pm bi$ as zeros. Assume that b is a positive integer.

139. GRAPHICAL REASONING The graph of one of the following functions is shown below. Identify the function shown in the graph. Explain why each of the others is not the correct function. Use a graphing utility to verify your result.

(a) $f(x) = x^2(x + 2)(x - 3.5)$

(b) $g(x) = (x + 2)(x - 3.5)$

(c) $h(x) = (x + 2)(x - 3.5)(x^2 + 1)$

(d) $k(x) = (x + 1)(x + 2)(x - 3.5)$

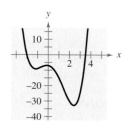

2.6 RATIONAL FUNCTIONS

What you should learn

- Find the domains of rational functions.
- Find the vertical and horizontal asymptotes of graphs of rational functions.
- Analyze and sketch graphs of rational functions.
- Sketch graphs of rational functions that have slant asymptotes.
- Use rational functions to model and solve real-life problems.

Why you should learn it

Rational functions can be used to model and solve real-life problems relating to business. For instance, in Exercise 83 on page 193, a rational function is used to model average speed over a distance.

Mike Powell/Getty Images

Introduction

A **rational function** is a quotient of polynomial functions. It can be written in the form

$$f(x) = \frac{N(x)}{D(x)}$$

where $N(x)$ and $D(x)$ are polynomials and $D(x)$ is not the zero polynomial.

In general, the *domain* of a rational function of x includes all real numbers except x-values that make the denominator zero. Much of the discussion of rational functions will focus on their graphical behavior near the x-values excluded from the domain.

Example 1   Finding the Domain of a Rational Function

Find the domain of the reciprocal function $f(x) = \dfrac{1}{x}$ and discuss the behavior of f near any excluded x-values.

Solution

Because the denominator is zero when $x = 0$, the domain of f is all real numbers except $x = 0$. To determine the behavior of f near this excluded value, evaluate $f(x)$ to the left and right of $x = 0$, as indicated in the following tables.

x	-1	-0.5	-0.1	-0.01	-0.001	$\rightarrow 0$
$f(x)$	-1	-2	-10	-100	-1000	$\rightarrow -\infty$

x	$0 \leftarrow$	0.001	0.01	0.1	0.5	1
$f(x)$	$\infty \leftarrow$	1000	100	10	2	1

Note that as x approaches 0 *from the left*, $f(x)$ decreases without bound. In contrast, as x approaches 0 *from the right*, $f(x)$ increases without bound. The graph of f is shown in Figure 2.37.

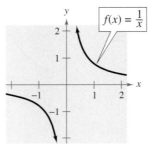

FIGURE 2.37

CHECK**Point**   Now try Exercise 5.

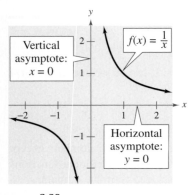

FIGURE **2.38**

Vertical and Horizontal Asymptotes

In Example 1, the behavior of f near $x = 0$ is denoted as follows.

$$f(x) \longrightarrow -\infty \text{ as } x \longrightarrow 0^- \qquad f(x) \longrightarrow \infty \text{ as } x \longrightarrow 0^+$$

$f(x)$ decreases without bound as x approaches 0 from the left. $f(x)$ increases without bound as x approaches 0 from the right.

The line $x = 0$ is a **vertical asymptote** of the graph of f, as shown in Figure 2.38. From this figure, you can see that the graph of f also has a **horizontal asymptote**—the line $y = 0$. This means that the values of $f(x) = \dfrac{1}{x}$ approach zero as x increases or decreases without bound.

$$f(x) \longrightarrow 0 \text{ as } x \longrightarrow -\infty \qquad f(x) \longrightarrow 0 \text{ as } x \longrightarrow \infty$$

$f(x)$ approaches 0 as x decreases without bound. $f(x)$ approaches 0 as x increases without bound.

Definitions of Vertical and Horizontal Asymptotes

1. The line $x = a$ is a **vertical asymptote** of the graph of f if

$$f(x) \longrightarrow \infty \quad \text{or} \quad f(x) \longrightarrow -\infty$$

as $x \longrightarrow a$, either from the right or from the left.

2. The line $y = b$ is a **horizontal asymptote** of the graph of f if

$$f(x) \longrightarrow b$$

as $x \longrightarrow \infty$ or $x \longrightarrow -\infty$.

Eventually (as $x \longrightarrow \infty$ or $x \longrightarrow -\infty$), the distance between the horizontal asymptote and the points on the graph must approach zero. Figure 2.39 shows the vertical and horizontal asymptotes of the graphs of three rational functions.

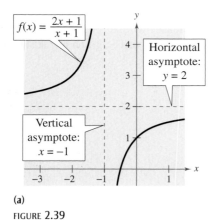

(a)

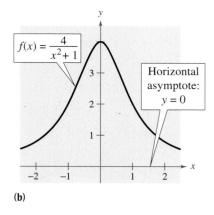

(b)

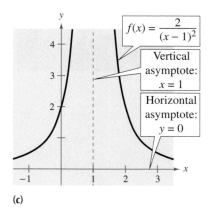

(c)

FIGURE **2.39**

The graphs of $f(x) = \dfrac{1}{x}$ in Figure 2.38 and $f(x) = \dfrac{2x + 1}{x + 1}$ in Figure 2.39(a) are **hyperbolas.** You will study hyperbolas in Section 10.4.

Vertical and Horizontal Asymptotes of a Rational Function

Let f be the rational function given by

$$f(x) = \frac{N(x)}{D(x)} = \frac{a_n x^n + a_{n-1} x^{n-1} + \cdots + a_1 x + a_0}{b_m x^m + b_{m-1} x^{m-1} + \cdots + b_1 x + b_0}$$

where $N(x)$ and $D(x)$ have no common factors.

1. The graph of f has *vertical* asymptotes at the zeros of $D(x)$.

2. The graph of f has one or no *horizontal* asymptote determined by comparing the degrees of $N(x)$ and $D(x)$.

    a. If $n < m$, the graph of f has the line $y = 0$ (the x-axis) as a horizontal asymptote.

    b. If $n = m$, the graph of f has the line $y = \dfrac{a_n}{b_m}$ (ratio of the leading coefficients) as a horizontal asymptote.

    c. If $n > m$, the graph of f has no horizontal asymptote.

Example 2   Finding Vertical and Horizontal Asymptotes

Find all vertical and horizontal asymptotes of the graph of each rational function.

a. $f(x) = \dfrac{2x^2}{x^2 - 1}$ **b.** $f(x) = \dfrac{x^2 + x - 2}{x^2 - x - 6}$

Solution

a. For this rational function, the degree of the numerator is *equal to* the degree of the denominator. The leading coefficient of the numerator is 2 and the leading coefficient of the denominator is 1, so the graph has the line $y = 2$ as a horizontal asymptote. To find any vertical asymptotes, set the denominator equal to zero and solve the resulting equation for x.

$$x^2 - 1 = 0 \qquad\qquad \text{Set denominator equal to zero.}$$

$$(x + 1)(x - 1) = 0 \qquad\qquad \text{Factor.}$$

$$x + 1 = 0 \implies x = -1 \qquad \text{Set 1st factor equal to 0.}$$

$$x - 1 = 0 \implies x = 1 \qquad \text{Set 2nd factor equal to 0.}$$

This equation has two real solutions, $x = -1$ and $x = 1$, so the graph has the lines $x = -1$ and $x = 1$ as vertical asymptotes. The graph of the function is shown in Figure 2.40.

b. For this rational function, the degree of the numerator is *equal to* the degree of the denominator. The leading coefficient of both the numerator and denominator is 1, so the graph has the line $y = 1$ as a horizontal asymptote. To find any vertical asymptotes, first factor the numerator and denominator as follows.

$$f(x) = \frac{x^2 + x - 2}{x^2 - x - 6} = \frac{(x - 1)(x + 2)}{(x + 2)(x - 3)} = \frac{x - 1}{x - 3}, \quad x \neq -2$$

By setting the denominator $x - 3$ (of the simplified function) equal to zero, you can determine that the graph has the line $x = 3$ as a vertical asymptote.

CHECK*Point* ▶ Now try Exercise 13.

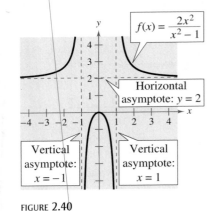

$f(x) = \dfrac{2x^2}{x^2 - 1}$

Horizontal asymptote: $y = 2$

Vertical asymptote: $x = -1$

Vertical asymptote: $x = 1$

FIGURE **2.40**

Algebra Help

You can review the techniques for factoring in Appendix A.3.

Analyzing Graphs of Rational Functions

To sketch the graph of a rational function, use the following guidelines.

Guidelines for Analyzing Graphs of Rational Functions

Let $f(x) = \dfrac{N(x)}{D(x)}$, where $N(x)$ and $D(x)$ are polynomials.

1. Simplify f, if possible.

2. Find and plot the y-intercept (if any) by evaluating $f(0)$.

3. Find the zeros of the numerator (if any) by solving the equation $N(x) = 0$. Then plot the corresponding x-intercepts.

4. Find the zeros of the denominator (if any) by solving the equation $D(x) = 0$. Then sketch the corresponding vertical asymptotes.

5. Find and sketch the horizontal asymptote (if any) by using the rule for finding the horizontal asymptote of a rational function.

6. Plot at least one point *between* and one point *beyond* each x-intercept and vertical asymptote.

7. Use smooth curves to complete the graph between and beyond the vertical asymptotes.

TECHNOLOGY

Some graphing utilities have difficulty graphing rational functions that have vertical asymptotes. Often, the utility will connect parts of the graph that are not supposed to be connected. For instance, the top screen on the right shows the graph of

$$f(x) = \frac{1}{x - 2}.$$

Notice that the graph should consist of two unconnected portions—one to the left of $x = 2$ and the other to the right of $x = 2$. To eliminate this problem, you can try changing the mode of the graphing utility to *dot mode*. The problem with this is that the graph is then represented as a collection of dots (as shown in the bottom screen on the right) rather than as a smooth curve.

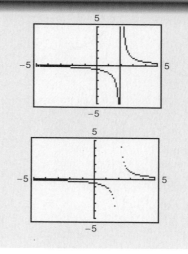

The concept of *test intervals* from Section 2.2 can be extended to graphing of rational functions. To do this, use the fact that a rational function can change signs only at its zeros and its undefined values (the x-values for which its denominator is zero). Between two consecutive zeros of the numerator and the denominator, a rational function must be entirely positive or entirely negative. This means that when the zeros of the numerator and the denominator of a rational function are put in order, they divide the real number line into test intervals in which the function has no sign changes. A representative x-value is chosen to determine if the value of the rational function is positive (the graph lies above the x-axis) or negative (the graph lies below the x-axis).

Study Tip

You can use transformations to help you sketch graphs of rational functions. For instance, the graph of g in Example 3 is a vertical stretch and a right shift of the graph of $f(x) = 1/x$ because

$$g(x) = \frac{3}{x-2}$$

$$= 3\left(\frac{1}{x-2}\right)$$

$$= 3f(x-2).$$

Example 3   Sketching the Graph of a Rational Function

Sketch the graph of $g(x) = \dfrac{3}{x-2}$ and state its domain.

Solution

y-intercept:	$\left(0, -\frac{3}{2}\right)$, because $g(0) = -\frac{3}{2}$
x-intercept:	None, because $3 \neq 0$
Vertical asymptote:	$x = 2$, zero of denominator
Horizontal asymptote:	$y = 0$, because degree of $N(x) <$ degree of $D(x)$
Additional points:	

Test interval	Representative x-value	Value of g	Sign	Point on graph
$(-\infty, 2)$	-4	$g(-4) = -0.5$	Negative	$(-4, -0.5)$
$(2, \infty)$	3	$g(3) = 3$	Positive	$(3, 3)$

By plotting the intercepts, asymptotes, and a few additional points, you can obtain the graph shown in Figure 2.41. The domain of g is all real numbers x except $x = 2$.

CHECK *Point*   Now try Exercise 31.

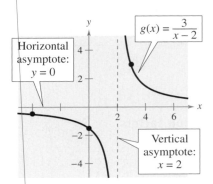

Horizontal asymptote: $y = 0$

$g(x) = \dfrac{3}{x-2}$

Vertical asymptote: $x = 2$

FIGURE **2.41**

Example 4   Sketching the Graph of a Rational Function

Sketch the graph of

$$f(x) = \frac{2x-1}{x}$$

and state its domain.

Solution

y-intercept:	None, because $x = 0$ is not in the domain
x-intercept:	$\left(\frac{1}{2}, 0\right)$, because $2x - 1 = 0$
Vertical asymptote:	$x = 0$, zero of denominator
Horizontal asymptote:	$y = 2$, because degree of $N(x) =$ degree of $D(x)$
Additional points:	

Test interval	Representative x-value	Value of f	Sign	Point on graph
$(-\infty, 0)$	-1	$f(-1) = 3$	Positive	$(-1, 3)$
$\left(0, \frac{1}{2}\right)$	$\frac{1}{4}$	$f\left(\frac{1}{4}\right) = -2$	Negative	$\left(\frac{1}{4}, -2\right)$
$\left(\frac{1}{2}, \infty\right)$	4	$f(4) = 1.75$	Positive	$(4, 1.75)$

By plotting the intercepts, asymptotes, and a few additional points, you can obtain the graph shown in Figure 2.42. The domain of f is all real numbers x except $x = 0$.

CHECK *Point*   Now try Exercise 35.

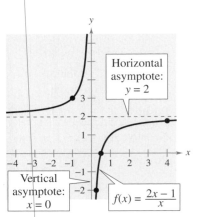

Horizontal asymptote: $y = 2$

Vertical asymptote: $x = 0$

$f(x) = \dfrac{2x-1}{x}$

FIGURE **2.42**

Example 5   Sketching the Graph of a Rational Function

Sketch the graph of $f(x) = x/(x^2 - x - 2)$.

Solution

Factoring the denominator, you have $f(x) = \dfrac{x}{(x + 1)(x - 2)}$.

y-intercept: $(0, 0)$, because $f(0) = 0$

x-intercept: $(0, 0)$

Vertical asymptotes: $x = -1$, $x = 2$, zeros of denominator

Horizontal asymptote: $y = 0$, because degree of $N(x) <$ degree of $D(x)$

Additional points:

Test interval	Representative x-value	Value of f	Sign	Point on graph
$(-\infty, -1)$	-3	$f(-3) = -0.3$	Negative	$(-3, -0.3)$
$(-1, 0)$	-0.5	$f(-0.5) = 0.4$	Positive	$(-0.5, 0.4)$
$(0, 2)$	1	$f(1) = -0.5$	Negative	$(1, -0.5)$
$(2, \infty)$	3	$f(3) = 0.75$	Positive	$(3, 0.75)$

The graph is shown in Figure 2.43.

CHECK*Point*   Now try Exercise 39.

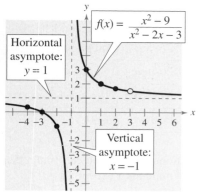

Vertical asymptote: $x = -1$

Vertical asymptote: $x = 2$

Horizontal asymptote: $y = 0$

$f(x) = \dfrac{x}{x^2 - x - 2}$

FIGURE 2.43

⚠ **WARNING / CAUTION**

If you are unsure of the shape of a portion of the graph of a rational function, plot some additional points. Also note that when the numerator and the denominator of a rational function have a common factor, the graph of the function has a *hole* at the zero of the common factor (see Example 6).

Example 6   A Rational Function with Common Factors

Sketch the graph of $f(x) = (x^2 - 9)/(x^2 - 2x - 3)$.

Solution

By factoring the numerator and denominator, you have

$$f(x) = \frac{x^2 - 9}{x^2 - 2x - 3} = \frac{(x - 3)(x + 3)}{(x - 3)(x + 1)} = \frac{x + 3}{x + 1}, \quad x \neq 3.$$

y-intercept: $(0, 3)$, because $f(0) = 3$

x-intercept: $(-3, 0)$, because $f(-3) = 0$

Vertical asymptote: $x = -1$, zero of (simplified) denominator

Horizontal asymptote: $y = 1$, because degree of $N(x) =$ degree of $D(x)$

Additional points:

Test interval	Representative x-value	Value of f	Sign	Point on graph
$(-\infty, -3)$	-4	$f(-4) = 0.33$	Positive	$(-4, 0.33)$
$(-3, -1)$	-2	$f(-2) = -1$	Negative	$(-2, -1)$
$(-1, \infty)$	2	$f(2) = 1.67$	Positive	$(2, 1.67)$

The graph is shown in Figure 2.44. Notice that there is a hole in the graph at $x = 3$, because the function is not defined when $x = 3$.

CHECK*Point*   Now try Exercise 45.

Horizontal asymptote: $y = 1$

$f(x) = \dfrac{x^2 - 9}{x^2 - 2x - 3}$

Vertical asymptote: $x = -1$

FIGURE 2.44   Hole at $x = 3$

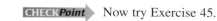

Slant Asymptotes

Consider a rational function whose denominator is of degree 1 or greater. If the degree of the numerator is exactly *one more* than the degree of the denominator, the graph of the function has a **slant** (or **oblique**) **asymptote.** For example, the graph of

$$f(x) = \frac{x^2 - x}{x + 1}$$

has a slant asymptote, as shown in Figure 2.45. To find the equation of a slant asymptote, use long division. For instance, by dividing $x + 1$ into $x^2 - x$, you obtain

$$f(x) = \frac{x^2 - x}{x + 1} = \underbrace{x - 2}_{\substack{\text{Slant asymptote} \\ (y = x - 2)}} + \frac{2}{x + 1}.$$

As x increases or decreases without bound, the remainder term $2/(x + 1)$ approaches 0, so the graph of f approaches the line $y = x - 2$, as shown in Figure 2.45.

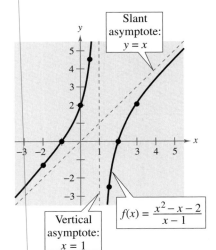

Vertical asymptote: $x = -1$

$f(x) = \dfrac{x^2 - x}{x + 1}$

Slant asymptote: $y = x - 2$

FIGURE **2.45**

Example 7   A Rational Function with a Slant Asymptote

Sketch the graph of $f(x) = \dfrac{x^2 - x - 2}{x - 1}$.

Solution

Factoring the numerator as $(x - 2)(x + 1)$ allows you to recognize the *x*-intercepts. Using long division

$$f(x) = \frac{x^2 - x - 2}{x - 1} = x - \frac{2}{x - 1}$$

allows you to recognize that the line $y = x$ is a slant asymptote of the graph.

y-intercept:	$(0, 2)$, because $f(0) = 2$
x-intercepts:	$(-1, 0)$ and $(2, 0)$
Vertical asymptote:	$x = 1$, zero of denominator
Slant asymptote:	$y = x$

Additional points:

Test interval	Representative *x*-value	Value of f	Sign	Point on graph
$(-\infty, -1)$	-2	$f(-2) = -1.33$	Negative	$(-2, -1.33)$
$(-1, 1)$	0.5	$f(0.5) = 4.5$	Positive	$(0.5, 4.5)$
$(1, 2)$	1.5	$f(1.5) = -2.5$	Negative	$(1.5, -2.5)$
$(2, \infty)$	3	$f(3) = 2$	Positive	$(3, 2)$

The graph is shown in Figure 2.46.

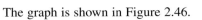 **CHECK** *Point* Now try Exercise 65.

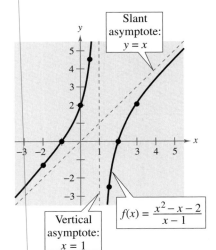

Slant asymptote: $y = x$

Vertical asymptote: $x = 1$

$f(x) = \dfrac{x^2 - x - 2}{x - 1}$

FIGURE **2.46**

Applications

There are many examples of asymptotic behavior in real life. For instance, Example 8 shows how a vertical asymptote can be used to analyze the cost of removing pollutants from smokestack emissions.

Example 8   Cost-Benefit Model

A utility company burns coal to generate electricity. The cost C (in dollars) of removing $p\%$ of the smokestack pollutants is given by

$$C = \frac{80{,}000p}{100 - p}$$

for $0 \leq p < 100$. You are a member of a state legislature considering a law that would require utility companies to remove 90% of the pollutants from their smokestack emissions. The current law requires 85% removal. How much additional cost would the utility company incur as a result of the new law?

Algebraic Solution

Because the current law requires 85% removal, the current cost to the utility company is

$$C = \frac{80{,}000(85)}{100 - 85} \approx \$453{,}333. \qquad \text{Evaluate } C \text{ when } p = 85.$$

If the new law increases the percent removal to 90%, the cost will be

$$C = \frac{80{,}000(90)}{100 - 90} = \$720{,}000. \qquad \text{Evaluate } C \text{ when } p = 90.$$

So, the new law would require the utility company to spend an additional

$$720{,}000 - 453{,}333 = \$266{,}667. \qquad \begin{array}{l}\text{Subtract 85\% removal cost}\\\text{from 90\% removal cost.}\end{array}$$

Graphical Solution

Use a graphing utility to graph the function

$$y_1 = \frac{80{,}000}{100 - x}$$

using a viewing window similar to that shown in Figure 2.47. Note that the graph has a vertical asymptote at $x = 100$. Then use the *trace* or *value* feature to approximate the values of y_1 when $x = 85$ and $x = 90$. You should obtain the following values.

When $x = 85$, $y_1 \approx 453{,}333$.

When $x = 90$, $y_1 = 720{,}000$.

So, the new law would require the utility company to spend an additional

$$720{,}000 - 453{,}333 = \$266{,}667.$$

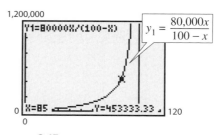

FIGURE 2.47

CHECK *Point*   Now try Exercise 77.

Example 9   Finding a Minimum Area

A rectangular page is designed to contain 48 square inches of print. The margins at the top and bottom of the page are each 1 inch deep. The margins on each side are $1\frac{1}{2}$ inches wide. What should the dimensions of the page be so that the least amount of paper is used?

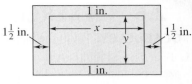

FIGURE **2.48**

Graphical Solution

Let A be the area to be minimized. From Figure 2.48, you can write

$$A = (x + 3)(y + 2).$$

The printed area inside the margins is modeled by $48 = xy$ or $y = 48/x$. To find the minimum area, rewrite the equation for A in terms of just one variable by substituting $48/x$ for y.

$$A = (x + 3)\left(\frac{48}{x} + 2\right)$$

$$= \frac{(x + 3)(48 + 2x)}{x}, \quad x > 0$$

The graph of this rational function is shown in Figure 2.49. Because x represents the width of the printed area, you need consider only the portion of the graph for which x is positive. Using a graphing utility, you can approximate the minimum value of A to occur when $x \approx 8.5$ inches. The corresponding value of y is $48/8.5 \approx 5.6$ inches. So, the dimensions should be

$$x + 3 \approx 11.5 \text{ inches} \quad \text{by} \quad y + 2 \approx 7.6 \text{ inches.}$$

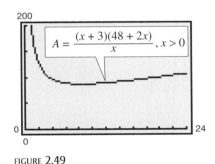

FIGURE **2.49**

Numerical Solution

Let A be the area to be minimized. From Figure 2.48, you can write

$$A = (x + 3)(y + 2).$$

The printed area inside the margins is modeled by $48 = xy$ or $y = 48/x$. To find the minimum area, rewrite the equation for A in terms of just one variable by substituting $48/x$ for y.

$$A = (x + 3)\left(\frac{48}{x} + 2\right)$$

$$= \frac{(x + 3)(48 + 2x)}{x}, \quad x > 0$$

Use the *table* feature of a graphing utility to create a table of values for the function

$$y_1 = \frac{(x + 3)(48 + 2x)}{x}$$

beginning at $x = 1$. From the table, you can see that the minimum value of y_1 occurs when x is somewhere between 8 and 9, as shown in Figure 2.50. To approximate the minimum value of y_1 to one decimal place, change the table so that it starts at $x = 8$ and increases by 0.1. The minimum value of y_1 occurs when $x \approx 8.5$, as shown in Figure 2.51. The corresponding value of y is $48/8.5 \approx 5.6$ inches. So, the dimensions should be $x + 3 \approx 11.5$ inches by $y + 2 \approx 7.6$ inches.

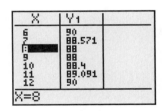   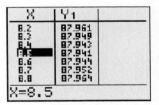

FIGURE **2.50**   FIGURE **2.51**

CHECK**Point**   Now try Exercise 81.

If you go on to take a course in calculus, you will learn an analytic technique for finding the exact value of x that produces a minimum area. In this case, that value is $x = 6\sqrt{2} \approx 8.485$.

2.6   EXERCISES

See www.CalcChat.com for worked-out solutions to odd-numbered exercises.

VOCABULARY: Fill in the blanks.

1. Functions of the form $f(x) = N(x)/D(x)$, where $N(x)$ and $D(x)$ are polynomials and $D(x)$ is not the zero polynomial, are called _____ _____.

2. If $f(x) \to \pm\infty$ as $x \to a$ from the left or the right, then $x = a$ is a _____ _____ of the graph of f.

3. If $f(x) \to b$ as $x \to \pm\infty$, then $y = b$ is a _____ _____ of the graph of f.

4. For the rational function given by $f(x) = N(x)/D(x)$, if the degree of $N(x)$ is exactly one more than the degree of $D(x)$, then the graph of f has a _____ (or oblique) _____.

SKILLS AND APPLICATIONS

In Exercises 5–8, (a) complete each table for the function, (b) determine the vertical and horizontal asymptotes of the graph of the function, and (c) find the domain of the function.

x	$f(x)$	x	$f(x)$	x	$f(x)$
0.5		1.5		5	
0.9		1.1		10	
0.99		1.01		100	
0.999		1.001		1000	

5. $f(x) = \dfrac{1}{x-1}$

6. $f(x) = \dfrac{5x}{x-1}$

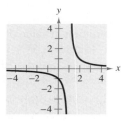

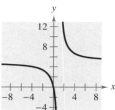

7. $f(x) = \dfrac{3x^2}{x^2-1}$

8. $f(x) = \dfrac{4x}{x^2-1}$

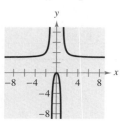

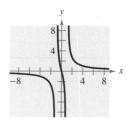

In Exercises 9–16, find the domain of the function and identify any vertical and horizontal asymptotes.

9. $f(x) = \dfrac{4}{x^2}$

10. $f(x) = \dfrac{4}{(x-2)^3}$

11. $f(x) = \dfrac{5+x}{5-x}$

12. $f(x) = \dfrac{3-7x}{3+2x}$

13. $f(x) = \dfrac{x^3}{x^2-1}$

14. $f(x) = \dfrac{4x^2}{x+2}$

15. $f(x) = \dfrac{3x^2+1}{x^2+x+9}$

16. $f(x) = \dfrac{3x^2+x-5}{x^2+1}$

In Exercises 17–20, match the rational function with its graph. [The graphs are labeled (a), (b), (c), and (d).]

(a)
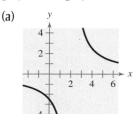

(b)
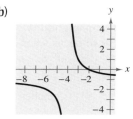

(c)
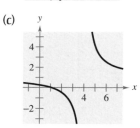

(d)
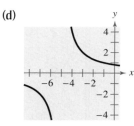

17. $f(x) = \dfrac{4}{x+5}$

18. $f(x) = \dfrac{5}{x-2}$

19. $f(x) = \dfrac{x-1}{x-4}$

20. $f(x) = -\dfrac{x+2}{x+4}$

In Exercises 21–24, find the zeros (if any) of the rational function.

21. $g(x) = \dfrac{x^2-9}{x+3}$

22. $h(x) = 4 + \dfrac{10}{x^2+5}$

23. $f(x) = 1 - \dfrac{2}{x-7}$

24. $g(x) = \dfrac{x^3-8}{x^2+1}$

In Exercises 25–30, find the domain of the function and identify any vertical and horizontal asymptotes.

25. $f(x) = \dfrac{x - 4}{x^2 - 16}$ **26.** $f(x) = \dfrac{x + 1}{x^2 - 1}$

27. $f(x) = \dfrac{x^2 - 25}{x^2 - 4x - 5}$ **28.** $f(x) = \dfrac{x^2 - 4}{x^2 - 3x + 2}$

29. $f(x) = \dfrac{x^2 - 3x - 4}{2x^2 + x - 1}$ **30.** $f(x) = \dfrac{6x^2 - 11x + 3}{6x^2 - 7x - 3}$

In Exercises 31–50, (a) state the domain of the function, (b) identify all intercepts, (c) find any vertical and horizontal asymptotes, and (d) plot additional solution points as needed to sketch the graph of the rational function.

31. $f(x) = \dfrac{1}{x + 2}$ **32.** $f(x) = \dfrac{1}{x - 3}$

33. $h(x) = \dfrac{-1}{x + 4}$ **34.** $g(x) = \dfrac{1}{6 - x}$

35. $C(x) = \dfrac{7 + 2x}{2 + x}$ **36.** $P(x) = \dfrac{1 - 3x}{1 - x}$

37. $f(x) = \dfrac{x^2}{x^2 + 9}$ **38.** $f(t) = \dfrac{1 - 2t}{t}$

39. $g(s) = \dfrac{4s}{s^2 + 4}$ **40.** $f(x) = -\dfrac{1}{(x - 2)^2}$

41. $h(x) = \dfrac{x^2 - 5x + 4}{x^2 - 4}$ **42.** $g(x) = \dfrac{x^2 - 2x - 8}{x^2 - 9}$

43. $f(x) = \dfrac{2x^2 - 5x - 3}{x^3 - 2x^2 - x + 2}$

44. $f(x) = \dfrac{x^2 - x - 2}{x^3 - 2x^2 - 5x + 6}$

45. $f(x) = \dfrac{x^2 + 3x}{x^2 + x - 6}$ **46.** $f(x) = \dfrac{5(x + 4)}{x^2 + x - 12}$

47. $f(x) = \dfrac{2x^2 - 5x + 2}{2x^2 - x - 6}$ **48.** $f(x) = \dfrac{3x^2 - 8x + 4}{2x^2 - 3x - 2}$

49. $f(t) = \dfrac{t^2 - 1}{t - 1}$ **50.** $f(x) = \dfrac{x^2 - 36}{x + 6}$

〰 ANALYTICAL, NUMERICAL, AND GRAPHICAL ANALYSIS

In Exercises 51–54, do the following.

(a) Determine the domains of f and g.

(b) Simplify f and find any vertical asymptotes of the graph of f.

(c) Compare the functions by completing the table.

(d) Use a graphing utility to graph f and g in the same viewing window.

(e) Explain why the graphing utility may not show the difference in the domains of f and g.

51. $f(x) = \dfrac{x^2 - 1}{x + 1}$, $g(x) = x - 1$

x	-3	-2	-1.5	-1	-0.5	0	1
$f(x)$							
$g(x)$							

52. $f(x) = \dfrac{x^2(x - 2)}{x^2 - 2x}$, $g(x) = x$

x	-1	0	1	1.5	2	2.5	3
$f(x)$							
$g(x)$							

53. $f(x) = \dfrac{x - 2}{x^2 - 2x}$, $g(x) = \dfrac{1}{x}$

x	-0.5	0	0.5	1	1.5	2	3
$f(x)$							
$g(x)$							

54. $f(x) = \dfrac{2x - 6}{x^2 - 7x + 12}$, $g(x) = \dfrac{2}{x - 4}$

x	0	1	2	3	4	5	6
$f(x)$							
$g(x)$							

In Exercises 55–68, (a) state the domain of the function, (b) identify all intercepts, (c) identify any vertical and slant asymptotes, and (d) plot additional solution points as needed to sketch the graph of the rational function.

55. $h(x) = \dfrac{x^2 - 9}{x}$ **56.** $g(x) = \dfrac{x^2 + 5}{x}$

57. $f(x) = \dfrac{2x^2 + 1}{x}$ **58.** $f(x) = \dfrac{1 - x^2}{x}$

59. $g(x) = \dfrac{x^2 + 1}{x}$ **60.** $h(x) = \dfrac{x^2}{x - 1}$

61. $f(t) = -\dfrac{t^2 + 1}{t + 5}$ **62.** $f(x) = \dfrac{x^2}{3x + 1}$

63. $f(x) = \dfrac{x^3}{x^2 - 4}$ **64.** $g(x) = \dfrac{x^3}{2x^2 - 8}$

65. $f(x) = \dfrac{x^2 - x + 1}{x - 1}$ **66.** $f(x) = \dfrac{2x^2 - 5x + 5}{x - 2}$

67. $f(x) = \dfrac{2x^3 - x^2 - 2x + 1}{x^2 + 3x + 2}$

68. $f(x) = \dfrac{2x^3 + x^2 - 8x - 4}{x^2 - 3x + 2}$

In Exercises 69–72, use a graphing utility to graph the rational function. Give the domain of the function and identify any asymptotes. Then zoom out sufficiently far so that the graph appears as a line. Identify the line.

69. $f(x) = \dfrac{x^2 + 5x + 8}{x + 3}$ **70.** $f(x) = \dfrac{2x^2 + x}{x + 1}$

71. $g(x) = \dfrac{1 + 3x^2 - x^3}{x^2}$ **72.** $h(x) = \dfrac{12 - 2x - x^2}{2(4 + x)}$

GRAPHICAL REASONING   In Exercises 73–76, (a) use the graph to determine any x-intercepts of the graph of the rational function and (b) set $y = 0$ and solve the resulting equation to confirm your result in part (a).

73. $y = \dfrac{x + 1}{x - 3}$ **74.** $y = \dfrac{2x}{x - 3}$

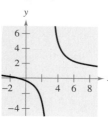   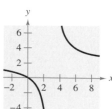

75. $y = \dfrac{1}{x} - x$ **76.** $y = x - 3 + \dfrac{2}{x}$

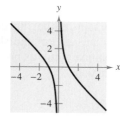   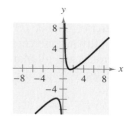

77. POLLUTION   The cost C (in millions of dollars) of removing $p\%$ of the industrial and municipal pollutants discharged into a river is given by

$$C = \dfrac{255p}{100 - p}, \quad 0 \le p < 100.$$

(a) Use a graphing utility to graph the cost function.

(b) Find the costs of removing 10%, 40%, and 75% of the pollutants.

(c) According to this model, would it be possible to remove 100% of the pollutants? Explain.

78. RECYCLING   In a pilot project, a rural township is given recycling bins for separating and storing recyclable products. The cost C (in dollars) of supplying bins to $p\%$ of the population is given by

$$C = \dfrac{25,000p}{100 - p}, \quad 0 \le p < 100.$$

(a) Use a graphing utility to graph the cost function.

(b) Find the costs of supplying bins to 15%, 50%, and 90% of the population.

(c) According to this model, would it be possible to supply bins to 100% of the residents? Explain.

79. POPULATION GROWTH   The game commission introduces 100 deer into newly acquired state game lands. The population N of the herd is modeled by

$$N = \dfrac{20(5 + 3t)}{1 + 0.04t}, \quad t \ge 0$$

where t is the time in years (see figure).

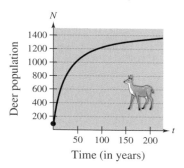

(a) Find the populations when $t = 5$, $t = 10$, and $t = 25$.

(b) What is the limiting size of the herd as time increases?

80. CONCENTRATION OF A MIXTURE   A 1000-liter tank contains 50 liters of a 25% brine solution. You add x liters of a 75% brine solution to the tank.

(a) Show that the concentration C, the proportion of brine to total solution, in the final mixture is

$$C = \dfrac{3x + 50}{4(x + 50)}.$$

(b) Determine the domain of the function based on the physical constraints of the problem.

(c) Sketch a graph of the concentration function.

(d) As the tank is filled, what happens to the rate at which the concentration of brine is increasing? What percent does the concentration of brine appear to approach?

81. PAGE DESIGN A page that is x inches wide and y inches high contains 30 square inches of print. The top and bottom margins are 1 inch deep, and the margins on each side are 2 inches wide (see figure).

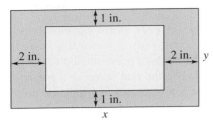

(a) Write a function for the total area A of the page in terms of x.

(b) Determine the domain of the function based on the physical constraints of the problem.

(c) Use a graphing utility to graph the area function and approximate the page size for which the least amount of paper will be used. Verify your answer numerically using the *table* feature of the graphing utility.

82. PAGE DESIGN A rectangular page is designed to contain 64 square inches of print. The margins at the top and bottom of the page are each 1 inch deep. The margins on each side are $1\frac{1}{2}$ inches wide. What should the dimensions of the page be so that the least amount of paper is used?

83. AVERAGE SPEED A driver averaged 50 miles per hour on the round trip between Akron, Ohio, and Columbus, Ohio, 100 miles away. The average speeds for going and returning were x and y miles per hour, respectively.

(a) Show that $y = \dfrac{25x}{x - 25}$.

(b) Determine the vertical and horizontal asymptotes of the graph of the function.

(c) Use a graphing utility to graph the function.

(d) Complete the table.

x	30	35	40	45	50	55	60
y							

(e) Are the results in the table what you expected? Explain.

(f) Is it possible to average 20 miles per hour in one direction and still average 50 miles per hour on the round trip? Explain.

EXPLORATION

84. WRITING Is every rational function a polynomial function? Is every polynomial function a rational function? Explain.

TRUE OR FALSE? In Exercises 85–87, determine whether the statement is true or false. Justify your answer.

85. A polynomial can have infinitely many vertical asymptotes.

86. The graph of a rational function can never cross one of its asymptotes.

87. The graph of a function can have a vertical asymptote, a horizontal asymptote, and a slant asymptote.

LIBRARY OF PARENT FUNCTIONS In Exercises 88 and 89, identify the rational function represented by the graph.

88.

(a) $f(x) = \dfrac{x^2 - 9}{x^2 - 4}$

(b) $f(x) = \dfrac{x^2 - 4}{x^2 - 9}$

(c) $f(x) = \dfrac{x - 4}{x^2 - 9}$

(d) $f(x) = \dfrac{x - 9}{x^2 - 4}$

89.
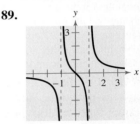

(a) $f(x) = \dfrac{x^2 - 1}{x^2 + 1}$

(b) $f(x) = \dfrac{x^2 + 1}{x^2 - 1}$

(c) $f(x) = \dfrac{x}{x^2 - 1}$

(d) $f(x) = \dfrac{x}{x^2 + 1}$

90. CAPSTONE Write a rational function f that has the specified characteristics. (There are many correct answers.)

(a) Vertical asymptote: $x = 2$
    Horizontal asymptote: $y = 0$
    Zero: $x = 1$

(b) Vertical asymptote: $x = -1$
    Horizontal asymptote: $y = 0$
    Zero: $x = 2$

(c) Vertical asymptotes: $x = -2, x = 1$
    Horizontal asymptote: $y = 2$
    Zeros: $x = 3, x = -3$,

(d) Vertical asymptotes: $x = -1, x = 2$
    Horizontal asymptote: $y = -2$
    Zeros: $x = -2, x = 3$

PROJECT: DEPARTMENT OF DEFENSE To work an extended application analyzing the total numbers of the Department of Defense personnel from 1980 through 2007, visit this text's website at *academic.cengage.com*. (Data Source: U.S. Department of Defense)

2.7   NONLINEAR INEQUALITIES

What you should learn

- Solve polynomial inequalities.
- Solve rational inequalities.
- Use inequalities to model and solve real-life problems.

Why you should learn it

Inequalities can be used to model and solve real-life problems. For instance, in Exercise 77 on page 202, a polynomial inequality is used to model school enrollment in the United States.

Ellen Senisi/The Image Works

Polynomial Inequalities

To solve a polynomial inequality such as $x^2 - 2x - 3 < 0$, you can use the fact that a polynomial can change signs only at its zeros (the x-values that make the polynomial equal to zero). Between two consecutive zeros, a polynomial must be entirely positive or entirely negative. This means that when the real zeros of a polynomial are put in order, they divide the real number line into intervals in which the polynomial has no sign changes. These zeros are the **key numbers** of the inequality, and the resulting intervals are the **test intervals** for the inequality. For instance, the polynomial above factors as

$$x^2 - 2x - 3 = (x + 1)(x - 3)$$

and has two zeros, $x = -1$ and $x = 3$. These zeros divide the real number line into three test intervals:

$$(-\infty, -1), \quad (-1, 3), \quad \text{and} \quad (3, \infty). \qquad \text{(See Figure 2.52.)}$$

So, to solve the inequality $x^2 - 2x - 3 < 0$, you need only test one value from each of these test intervals to determine whether the value satisfies the original inequality. If so, you can conclude that the interval is a solution of the inequality.

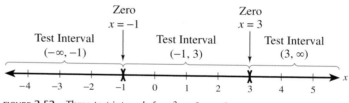

FIGURE 2.52   Three test intervals for $x^2 - 2x - 3$

You can use the same basic approach to determine the test intervals for any polynomial.

Finding Test Intervals for a Polynomial

To determine the intervals on which the values of a polynomial are entirely negative or entirely positive, use the following steps.

1. Find all real zeros of the polynomial, and arrange the zeros in increasing order (from smallest to largest). These zeros are the key numbers of the polynomial.

2. Use the key numbers of the polynomial to determine its test intervals.

3. Choose one representative x-value in each test interval and evaluate the polynomial at that value. If the value of the polynomial is negative, the polynomial will have negative values for every x-value in the interval. If the value of the polynomial is positive, the polynomial will have positive values for every x-value in the interval.

Example 1   Solving a Polynomial Inequality

Solve $x^2 - x - 6 < 0$.

Solution

By factoring the polynomial as

$$x^2 - x - 6 = (x + 2)(x - 3)$$

you can see that the key numbers are $x = -2$ and $x = 3$. So, the polynomial's test intervals are

$(-\infty, -2)$, $(-2, 3)$,   and $(3, \infty)$.   Test intervals

In each test interval, choose a representative x-value and evaluate the polynomial.

Test Interval	x-Value	Polynomial Value	Conclusion
$(-\infty, -2)$	$x = -3$	$(-3)^2 - (-3) - 6 = 6$	Positive
$(-2, 3)$	$x = 0$	$(0)^2 - (0) - 6 = -6$	Negative
$(3, \infty)$	$x = 4$	$(4)^2 - (4) - 6 = 6$	Positive

From this you can conclude that the inequality is satisfied for all x-values in $(-2, 3)$. This implies that the solution of the inequality $x^2 - x - 6 < 0$ is the interval $(-2, 3)$, as shown in Figure 2.53. Note that the original inequality contains a "less than" symbol. This means that the solution set does not contain the endpoints of the test interval $(-2, 3)$.

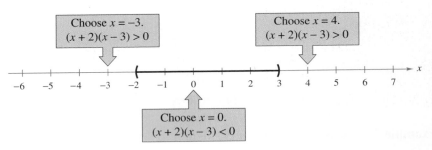

Choose $x = -3$.
$(x + 2)(x - 3) > 0$

Choose $x = 4$.
$(x + 2)(x - 3) > 0$

Choose $x = 0$.
$(x + 2)(x - 3) < 0$

FIGURE 2.53

CHECK Point   Now try Exercise 21.

As with linear inequalities, you can check the reasonableness of a solution by substituting x-values into the original inequality. For instance, to check the solution found in Example 1, try substituting several x-values from the interval $(-2, 3)$ into the inequality

$$x^2 - x - 6 < 0.$$

Regardless of which x-values you choose, the inequality should be satisfied.

You can also use a graph to check the result of Example 1. Sketch the graph of $y = x^2 - x - 6$, as shown in Figure 2.54. Notice that the graph is below the x-axis on the interval $(-2, 3)$.

In Example 1, the polynomial inequality was given in general form (with the polynomial on one side and zero on the other). Whenever this is not the case, you should begin the solution process by writing the inequality in general form.

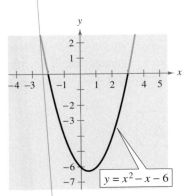

$y = x^2 - x - 6$

FIGURE 2.54

Example 2   Solving a Polynomial Inequality

Solve $2x^3 - 3x^2 - 32x > -48$.

Solution

$$2x^3 - 3x^2 - 32x + 48 > 0 \qquad \text{Write in general form.}$$

$$(x - 4)(x + 4)(2x - 3) > 0 \qquad \text{Factor.}$$

The key numbers are $x = -4$, $x = \frac{3}{2}$, and $x = 4$, and the test intervals are $(-\infty, -4)$, $\left(-4, \frac{3}{2}\right)$, $\left(\frac{3}{2}, 4\right)$, and $(4, \infty)$.

Test Interval	x-Value	Polynomial Value	Conclusion
$(-\infty, -4)$	$x = -5$	$2(-5)^3 - 3(-5)^2 - 32(-5) + 48$	Negative
$\left(-4, \frac{3}{2}\right)$	$x = 0$	$2(0)^3 - 3(0)^2 - 32(0) + 48$	Positive
$\left(\frac{3}{2}, 4\right)$	$x = 2$	$2(2)^3 - 3(2)^2 - 32(2) + 48$	Negative
$(4, \infty)$	$x = 5$	$2(5)^3 - 3(5)^2 - 32(5) + 48$	Positive

From this you can conclude that the inequality is satisfied on the open intervals $\left(-4, \frac{3}{2}\right)$ and $(4, \infty)$. So, the solution set is $\left(-4, \frac{3}{2}\right) \cup (4, \infty)$, as shown in Figure 2.55.

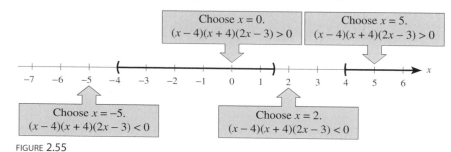

FIGURE 2.55

CHECK *Point*   Now try Exercise 27.

Example 3   Solving a Polynomial Inequality

Solve $4x^2 - 5x > 6$.

Algebraic Solution

$$4x^2 - 5x - 6 > 0 \qquad \text{Write in general form.}$$

$$(x - 2)(4x + 3) > 0 \qquad \text{Factor.}$$

Key Numbers: $x = -\frac{3}{4}$, $x = 2$

Test Intervals: $\left(-\infty, -\frac{3}{4}\right)$, $\left(-\frac{3}{4}, 2\right)$, $(2, \infty)$

Test: Is $(x - 2)(4x + 3) > 0$?

After testing these intervals, you can see that the polynomial $4x^2 - 5x - 6$ is positive on the open intervals $\left(-\infty, -\frac{3}{4}\right)$ and $(2, \infty)$. So, the solution set of the inequality is $\left(-\infty, -\frac{3}{4}\right) \cup (2, \infty)$.

CHECK *Point*   Now try Exercise 23.

Graphical Solution

First write the polynomial inequality $4x^2 - 5x > 6$ as $4x^2 - 5x - 6 > 0$. Then use a graphing utility to graph $y = 4x^2 - 5x - 6$. In Figure 2.56, you can see that the graph is *above* the x-axis when x is less than $-\frac{3}{4}$ or when x is greater than 2. So, you can graphically approximate the solution set to be $\left(-\infty, -\frac{3}{4}\right) \cup (2, \infty)$.

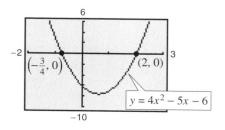

FIGURE 2.56

Study Tip

You may find it easier to determine the sign of a polynomial from its *factored* form. For instance, in Example 3, if the test value $x = 1$ is substituted into the factored form

$$(x - 2)(4x + 3)$$

you can see that the sign pattern of the factors is

$$(-)(+)$$

which yields a negative result. Try using the factored forms of the polynomials to determine the signs of the polynomials in the test intervals of the other examples in this section.

When solving a polynomial inequality, be sure you have accounted for the particular type of inequality symbol given in the inequality. For instance, in Example 3, note that the original inequality contained a "greater than" symbol and the solution consisted of two open intervals. If the original inequality had been

$$4x^2 - 5x \geq 6$$

the solution would have consisted of the intervals $\left(-\infty, -\frac{3}{4}\right]$ and $[2, \infty)$.

Each of the polynomial inequalities in Examples 1, 2, and 3 has a solution set that consists of a single interval or the union of two intervals. When solving the exercises for this section, watch for unusual solution sets, as illustrated in Example 4.

Example 4   Unusual Solution Sets

a. The solution set of the following inequality consists of the entire set of real numbers, $(-\infty, \infty)$. In other words, the value of the quadratic $x^2 + 2x + 4$ is positive for every real value of x.

$$x^2 + 2x + 4 > 0$$

b. The solution set of the following inequality consists of the single real number $\{-1\}$, because the quadratic $x^2 + 2x + 1$ has only one key number, $x = -1$, and it is the only value that satisfies the inequality.

$$x^2 + 2x + 1 \leq 0$$

c. The solution set of the following inequality is empty. In other words, the quadratic $x^2 + 3x + 5$ is not less than zero for any value of x.

$$x^2 + 3x + 5 < 0$$

d. The solution set of the following inequality consists of all real numbers except $x = 2$. In interval notation, this solution set can be written as $(-\infty, 2) \cup (2, \infty)$.

$$x^2 - 4x + 4 > 0$$

CHECK*Point*   Now try Exercise 29.

Rational Inequalities

The concepts of key numbers and test intervals can be extended to rational inequalities. To do this, use the fact that the value of a rational expression can change sign only at its *zeros* (the x-values for which its numerator is zero) and its *undefined values* (the x-values for which its denominator is zero). These two types of numbers make up the *key numbers* of a rational inequality. When solving a rational inequality, begin by writing the inequality in general form with the rational expression on the left and zero on the right.

Example 5   Solving a Rational Inequality

Solve $\dfrac{2x - 7}{x - 5} \le 3$.

Solution

$$\frac{2x - 7}{x - 5} \le 3 \qquad \text{Write original inequality.}$$

$$\frac{2x - 7}{x - 5} - 3 \le 0 \qquad \text{Write in general form.}$$

$$\frac{2x - 7 - 3x + 15}{x - 5} \le 0 \qquad \text{Find the LCD and subtract fractions.}$$

$$\frac{-x + 8}{x - 5} \le 0 \qquad \text{Simplify.}$$

Key numbers: $x = 5, x = 8$ Zeros and undefined values of rational expression

Test intervals: $(-\infty, 5), (5, 8), (8, \infty)$

Test: Is $\dfrac{-x + 8}{x - 5} \le 0$?

After testing these intervals, as shown in Figure 2.57, you can see that the inequality is satisfied on the open intervals $(-\infty, 5)$ and $(8, \infty)$. Moreover, because $\dfrac{-x + 8}{x - 5} = 0$ when $x = 8$, you can conclude that the solution set consists of all real numbers in the intervals $(-\infty, 5) \cup [8, \infty)$. (Be sure to use a closed interval to indicate that x can equal 8.)

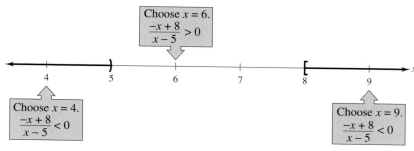

FIGURE 2.57

CHECKPoint   Now try Exercise 45.

Applications

One common application of inequalities comes from business and involves profit, revenue, and cost. The formula that relates these three quantities is

$$\boxed{\text{Profit}} = \boxed{\text{Revenue}} - \boxed{\text{Cost}}$$

$$P = R - C.$$

| Example 6   Increasing the Profit for a Product

The marketing department of a calculator manufacturer has determined that the demand for a new model of calculator is

$$p = 100 - 0.00001x, \quad 0 \le x \le 10,000,000 \qquad \text{Demand equation}$$

where p is the price per calculator (in dollars) and x represents the number of calculators sold. (If this model is accurate, no one would be willing to pay $100 for the calculator. At the other extreme, the company couldn't sell more than 10 million calculators.) The revenue for selling x calculators is

$$R = xp = x(100 - 0.00001x) \qquad \text{Revenue equation}$$

as shown in Figure 2.58. The total cost of producing x calculators is $10 per calculator plus a development cost of $2,500,000. So, the total cost is

$$C = 10x + 2,500,000. \qquad \text{Cost equation}$$

What price should the company charge per calculator to obtain a profit of at least $190,000,000?

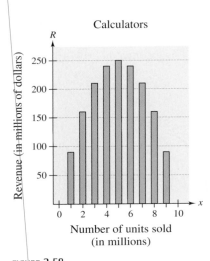

Calculators

Number of units sold
(in millions)

FIGURE 2.58

Solution

Verbal Model: $\boxed{\text{Profit}} = \boxed{\text{Revenue}} - \boxed{\text{Cost}}$

Equation: $P = R - C$

$$P = 100x - 0.00001x^2 - (10x + 2,500,000)$$

$$P = -0.00001x^2 + 90x - 2,500,000$$

To answer the question, solve the inequality

$$P \ge 190,000,000$$

$$-0.00001x^2 + 90x - 2,500,000 \ge 190,000,000.$$

When you write the inequality in general form, find the key numbers and the test intervals, and then test a value in each test interval, you can find the solution to be

$$3,500,000 \le x \le 5,500,000$$

as shown in Figure 2.59. Substituting the x-values in the original price equation shows that prices of

$$\$45.00 \le p \le \$65.00$$

will yield a profit of at least $190,000,000.

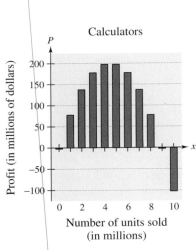

Calculators

Number of units sold
(in millions)

FIGURE 2.59

CHECK *Point*   Now try Exercise 75.

Another common application of inequalities is finding the domain of an expression that involves a square root, as shown in Example 7.

Example 7   Finding the Domain of an Expression

Find the domain of $\sqrt{64 - 4x^2}$.

Algebraic Solution

Remember that the domain of an expression is the set of all x-values for which the expression is defined. Because $\sqrt{64 - 4x^2}$ is defined (has real values) only if $64 - 4x^2$ is nonnegative, the domain is given by $64 - 4x^2 \geq 0$.

$$64 - 4x^2 \geq 0 \qquad \text{Write in general form.}$$

$$16 - x^2 \geq 0 \qquad \text{Divide each side by 4.}$$

$$(4 - x)(4 + x) \geq 0 \qquad \text{Write in factored form.}$$

So, the inequality has two key numbers: $x = -4$ and $x = 4$. You can use these two numbers to test the inequality as follows.

Key numbers: $\qquad x = -4, x = 4$

Test intervals: $\qquad (-\infty, -4), (-4, 4), (4, \infty)$

Test: $\qquad$ For what values of x is $\sqrt{64 - 4x^2} \geq 0$?

A test shows that the inequality is satisfied in the *closed interval* $[-4, 4]$. So, the domain of the expression $\sqrt{64 - 4x^2}$ is the interval $[-4, 4]$.

CHECK**Point** Now try Exercise 59.

Graphical Solution

Begin by sketching the graph of the equation $y = \sqrt{64 - 4x^2}$, as shown in Figure 2.60. From the graph, you can determine that the x-values extend from -4 to 4 (including -4 and 4). So, the domain of the expression $\sqrt{64 - 4x^2}$ is the interval $[-4, 4]$.

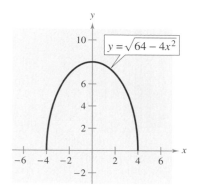

FIGURE **2.60**

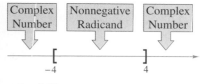

FIGURE **2.61**

To analyze a test interval, choose a representative x-value in the interval and evaluate the expression at that value. For instance, in Example 7, if you substitute any number from the interval $[-4, 4]$ into the expression $\sqrt{64 - 4x^2}$, you will obtain a nonnegative number under the radical symbol that simplifies to a real number. If you substitute any number from the intervals $(-\infty, -4)$ and $(4, \infty)$, you will obtain a complex number. It might be helpful to draw a visual representation of the intervals, as shown in Figure 2.61.

CLASSROOM DISCUSSION

Profit Analysis Consider the relationship

$$P = R - C$$

described on page 199. Write a paragraph discussing why it might be beneficial to solve $P < 0$ if you owned a business. Use the situation described in Example 6 to illustrate your reasoning.

2.7   EXERCISES

See www.CalcChat.com for worked-out solutions to odd-numbered exercises.

VOCABULARY: Fill in the blanks.

1. Between two consecutive zeros, a polynomial must be entirely _____ or entirely _____.

2. To solve a polynomial inequality, find the _____ numbers of the polynomial, and use these numbers to create _____ _____ for the inequality.

3. The key numbers of a rational expression are its _____ and its _____ _____.

4. The formula that relates cost, revenue, and profit is _____.

SKILLS AND APPLICATIONS

In Exercises 5–8, determine whether each value of x is a solution of the inequality.

Inequality	*Values*
5. $x^2 - 3 < 0$	(a) $x = 3$ (b) $x = 0$
	(c) $x = \frac{3}{2}$ (d) $x = -5$
6. $x^2 - x - 12 \geq 0$	(a) $x = 5$ (b) $x = 0$
	(c) $x = -4$ (d) $x = -3$
7. $\dfrac{x+2}{x-4} \geq 3$	(a) $x = 5$ (b) $x = 4$
	(c) $x = -\frac{9}{2}$ (d) $x = \frac{9}{2}$
8. $\dfrac{3x^2}{x^2+4} < 1$	(a) $x = -2$ (b) $x = -1$
	(c) $x = 0$ (d) $x = 3$

In Exercises 9–12, find the key numbers of the expression.

9. $3x^2 - x - 2$

10. $9x^3 - 25x^2$

11. $\dfrac{1}{x-5} + 1$

12. $\dfrac{x}{x+2} - \dfrac{2}{x-1}$

In Exercises 13–30, solve the inequality and graph the solution on the real number line.

13. $x^2 < 9$

14. $x^2 \leq 16$

15. $(x+2)^2 \leq 25$

16. $(x-3)^2 \geq 1$

17. $x^2 + 4x + 4 \geq 9$

18. $x^2 - 6x + 9 < 16$

19. $x^2 + x < 6$

20. $x^2 + 2x > 3$

21. $x^2 + 2x - 3 < 0$

22. $x^2 > 2x + 8$

23. $3x^2 - 11x > 20$

24. $-2x^2 + 6x + 15 \leq 0$

25. $x^2 + 3x - 18 > 0$

26. $x^3 + 2x^2 - 4x - 8 \leq 0$

27. $x^3 - 3x^2 - x > -3$

28. $2x^3 + 13x^2 - 8x - 46 \geq 6$

29. $4x^2 + 4x + 1 \leq 0$

30. $x^2 + 3x + 8 > 0$

In Exercises 31–36, solve the inequality and write the solution set in interval notation.

31. $4x^3 - 6x^2 < 0$

32. $4x^3 - 12x^2 > 0$

33. $x^3 - 4x \geq 0$

34. $2x^3 - x^4 \leq 0$

35. $(x-1)^2(x+2)^3 \geq 0$

36. $x^4(x-3) \leq 0$

GRAPHICAL ANALYSIS   In Exercises 37–40, use a graphing utility to graph the equation. Use the graph to approximate the values of x that satisfy each inequality.

Equation	*Inequalities*
37. $y = -x^2 + 2x + 3$	(a) $y \leq 0$ (b) $y \geq 3$
38. $y = \frac{1}{2}x^2 - 2x + 1$	(a) $y \leq 0$ (b) $y \geq 7$
39. $y = \frac{1}{8}x^3 - \frac{1}{2}x$	(a) $y \geq 0$ (b) $y \leq 6$
40. $y = x^3 - x^2 - 16x + 16$	(a) $y \leq 0$ (b) $y \geq 36$

In Exercises 41–54, solve the inequality and graph the solution on the real number line.

41. $\dfrac{4x-1}{x} > 0$

42. $\dfrac{x^2-1}{x} < 0$

43. $\dfrac{3x-5}{x-5} \geq 0$

44. $\dfrac{5+7x}{1+2x} \leq 4$

45. $\dfrac{x+6}{x+1} - 2 < 0$

46. $\dfrac{x+12}{x+2} - 3 \geq 0$

47. $\dfrac{2}{x+5} > \dfrac{1}{x-3}$

48. $\dfrac{5}{x-6} > \dfrac{3}{x+2}$

49. $\dfrac{1}{x-3} \leq \dfrac{9}{4x+3}$

50. $\dfrac{1}{x} \geq \dfrac{1}{x+3}$

51. $\dfrac{x^2+2x}{x^2-9} \leq 0$

52. $\dfrac{x^2+x-6}{x} \geq 0$

53. $\dfrac{3}{x-1} + \dfrac{2x}{x+1} > -1$

54. $\dfrac{3x}{x-1} \leq \dfrac{x}{x+4} + 3$

GRAPHICAL ANALYSIS In Exercises 55–58, use a graphing utility to graph the equation. Use the graph to approximate the values of x that satisfy each inequality.

Equation	Inequalities	
55. $y = \dfrac{3x}{x-2}$	(a) $y \le 0$	(b) $y \ge 6$
56. $y = \dfrac{2(x-2)}{x+1}$	(a) $y \le 0$	(b) $y \ge 8$
57. $y = \dfrac{2x^2}{x^2+4}$	(a) $y \ge 1$	(b) $y \le 2$
58. $y = \dfrac{5x}{x^2+4}$	(a) $y \ge 1$	(b) $y \le 0$

In Exercises 59–64, find the domain of x in the expression. Use a graphing utility to verify your result.

59. $\sqrt{4 - x^2}$ **60.** $\sqrt{x^2 - 4}$

61. $\sqrt{x^2 - 9x + 20}$ **62.** $\sqrt{81 - 4x^2}$

63. $\sqrt{\dfrac{x}{x^2 - 2x - 35}}$ **64.** $\sqrt{\dfrac{x}{x^2 - 9}}$

In Exercises 65–70, solve the inequality. (Round your answers to two decimal places.)

65. $0.4x^2 + 5.26 < 10.2$

66. $-1.3x^2 + 3.78 > 2.12$

67. $-0.5x^2 + 12.5x + 1.6 > 0$

68. $1.2x^2 + 4.8x + 3.1 < 5.3$

69. $\dfrac{1}{2.3x - 5.2} > 3.4$ **70.** $\dfrac{2}{3.1x - 3.7} > 5.8$

HEIGHT OF A PROJECTILE In Exercises 71 and 72, use the position equation $s = -16t^2 + v_0t + s_0$, where s represents the height of an object (in feet), v_0 represents the initial velocity of the object (in feet per second), s_0 represents the initial height of the object (in feet), and t represents the time (in seconds).

71. A projectile is fired straight upward from ground level ($s_0 = 0$) with an initial velocity of 160 feet per second.

(a) At what instant will it be back at ground level?

(b) When will the height exceed 384 feet?

72. A projectile is fired straight upward from ground level ($s_0 = 0$) with an initial velocity of 128 feet per second.

(a) At what instant will it be back at ground level?

(b) When will the height be less than 128 feet?

73. GEOMETRY A rectangular playing field with a perimeter of 100 meters is to have an area of at least 500 square meters. Within what bounds must the length of the rectangle lie?

74. GEOMETRY A rectangular parking lot with a perimeter of 440 feet is to have an area of at least 8000 square feet. Within what bounds must the length of the rectangle lie?

75. COST, REVENUE, AND PROFIT The revenue and cost equations for a product are $R = x(75 - 0.0005x)$ and $C = 30x + 250,000$, where R and C are measured in dollars and x represents the number of units sold. How many units must be sold to obtain a profit of at least \$750,000? What is the price per unit?

76. COST, REVENUE, AND PROFIT The revenue and cost equations for a product are

$$R = x(50 - 0.0002x) \quad \text{and} \quad C = 12x + 150,000$$

where R and C are measured in dollars and x represents the number of units sold. How many units must be sold to obtain a profit of at least \$1,650,000? What is the price per unit?

 77. SCHOOL ENROLLMENT The numbers N (in millions) of students enrolled in schools in the United States from 1995 through 2006 are shown in the table. (Source: U.S. Census Bureau)

Year	Number, N
1995	69.8
1996	70.3
1997	72.0
1998	72.1
1999	72.4
2000	72.2
2001	73.1
2002	74.0
2003	74.9
2004	75.5
2005	75.8
2006	75.2

(a) Use a graphing utility to create a scatter plot of the data. Let t represent the year, with $t = 5$ corresponding to 1995.

(b) Use the *regression* feature of a graphing utility to find a quartic model for the data.

(c) Graph the model and the scatter plot in the same viewing window. How well does the model fit the data?

(d) According to the model, during what range of years will the number of students enrolled in schools exceed 74 million?

(e) Is the model valid for long-term predictions of student enrollment in schools? Explain.

78. SAFE LOAD The maximum safe load uniformly distributed over a one-foot section of a two-inch-wide wooden beam is approximated by the model Load $= 168.5d^2 - 472.1$, where d is the depth of the beam.

(a) Evaluate the model for $d = 4$, $d = 6$, $d = 8$, $d = 10$, and $d = 12$. Use the results to create a bar graph.

(b) Determine the minimum depth of the beam that will safely support a load of 2000 pounds.

79. RESISTORS When two resistors of resistances R_1 and R_2 are connected in parallel (see figure), the total resistance R satisfies the equation

$$\frac{1}{R} = \frac{1}{R_1} + \frac{1}{R_2}.$$

Find R_1 for a parallel circuit in which $R_2 = 2$ ohms and R must be at least 1 ohm.

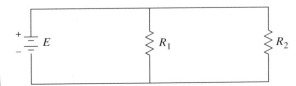

80. TEACHER SALARIES The mean salaries S (in thousands of dollars) of classroom teachers in the United States from 2000 through 2007 are shown in the table.

Year	Salary, S
2000	42.2
2001	43.7
2002	43.8
2003	45.0
2004	45.6
2005	45.9
2006	48.2
2007	49.3

A model that approximates these data is given by

$$S = \frac{42.6 - 1.95t}{1 - 0.06t}$$

where t represents the year, with $t = 0$ corresponding to 2000. (Source: Educational Research Service, Arlington, VA)

(a) Use a graphing utility to create a scatter plot of the data. Then graph the model in the same viewing window.

(b) How well does the model fit the data? Explain.

(c) According to the model, in what year will the salary for classroom teachers exceed $60,000?

(d) Is the model valid for long-term predictions of classroom teacher salaries? Explain.

EXPLORATION

TRUE OR FALSE? In Exercises 81 and 82, determine whether the statement is true or false. Justify your answer.

81. The zeros of the polynomial $x^3 - 2x^2 - 11x + 12 \geq 0$ divide the real number line into four test intervals.

82. The solution set of the inequality $\frac{3}{2}x^2 + 3x + 6 \geq 0$ is the entire set of real numbers.

In Exercises 83–86, (a) find the interval(s) for b such that the equation has at least one real solution and (b) write a conjecture about the interval(s) based on the values of the coefficients.

83. $x^2 + bx + 4 = 0$ **84.** $x^2 + bx - 4 = 0$

85. $3x^2 + bx + 10 = 0$ **86.** $2x^2 + bx + 5 = 0$

87. GRAPHICAL ANALYSIS You can use a graphing utility to verify the results in Example 4. For instance, the graph of $y = x^2 + 2x + 4$ is shown below. Notice that the y-values are greater than 0 for all values of x, as stated in Example 4(a). Use the graphing utility to graph $y = x^2 + 2x + 1$, $y = x^2 + 3x + 5$, and $y = x^2 - 4x + 4$. Explain how you can use the graphs to verify the results of parts (b), (c), and (d) of Example 4.

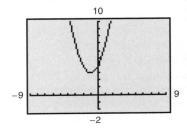

88. CAPSTONE Consider the polynomial

$$(x - a)(x - b)$$

and the real number line shown below.

(a) Identify the points on the line at which the polynomial is zero.

(b) In each of the three subintervals of the line, write the sign of each factor and the sign of the product.

(c) At what x-values does the polynomial change signs?

2   CHAPTER SUMMARY

	What Did You Learn?	**Explanation/Examples**	**Review Exercises**
Section 2.1	Analyze graphs of quadratic functions (p. 126).	Let a, b, and c be real numbers with $a \neq 0$. The function given by $f(x) = ax^2 + bx + c$ is called a quadratic function. Its graph is a "U-shaped" curve called a parabola.	1, 2
	Write quadratic functions in standard form and use the results to sketch graphs of functions (p. 129).	The quadratic function $f(x) = a(x - h)^2 + k, a \neq 0$, is in standard form. The graph of f is a parabola whose axis is the vertical line $x = h$ and whose vertex is (h, k). If $a > 0$, the parabola opens upward. If $a < 0$, the parabola opens downward.	3–20
	Find minimum and maximum values of quadratic functions in real-life applications (p. 131).	Consider $f(x) = ax^2 + bx + c$ with vertex $\left(-\dfrac{b}{2a}, f\left(\dfrac{b}{2a}\right)\right)$. If $a > 0$, f has a *minimum* at $x = -b/(2a)$. If $a < 0$, f has a *maximum* at $x = -b/(2a)$.	21–24
Section 2.2	Use transformations to sketch graphs of polynomial functions (p. 136).	The graph of a polynomial function is continuous (no breaks, holes, or gaps) and has only smooth, rounded turns.	25–30
	Use the Leading Coefficient Test to determine the end behavior of graphs of polynomial functions (p. 138).	Consider the graph of $f(x) = a_n x^n + \cdots + a_1 x + a_0$. **When n is odd:** If $a_n > 0$, the graph falls to the left and rises to the right. If $a_n < 0$, the graph rises to the left and falls to the right. **When n is even:** If $a_n > 0$, the graph rises to the left and right. If $a_n < 0$, the graph falls to the left and right.	31–34
	Find and use zeros of polynomial functions as sketching aids (p. 139).	If f is a polynomial function and a is a real number, the following are equivalent: (1) $x = a$ is a *zero* of f, (2) $x = a$ is a *solution* of the equation $f(x) = 0$, (3) $(x - a)$ is a *factor* of $f(x)$, and (4) $(a, 0)$ is an *x-intercept* of the graph of f.	35–44
	Use the Intermediate Value Theorem to help locate zeros of polynomial functions (p. 143).	Let a and b be real numbers such that $a < b$. If f is a polynomial function such that $f(a) \neq f(b)$, then, in $[a, b]$, f takes on every value between $f(a)$ and $f(b)$.	45–48
Section 2.3	Use long division to divide polynomials by other polynomials (p. 150).	Dividend ... Quotient ... Remainder $$\frac{x^2 + 3x + 5}{x + 1} = x + 2 + \frac{3}{x + 1}$$ Divisor $\longrightarrow$ ... $\longleftarrow$ Divisor	49–54
	Use synthetic division to divide polynomials by binomials of the form $(x - k)$ (p. 153).	Divisor: $x + 3$ Dividend: $x^4 - 10x^2 - 2x + 4$ $$\begin{array}{r|rrrrr} -3 & 1 & 0 & -10 & -2 & 4 \\ & & -3 & 9 & 3 & -3 \\ \hline & 1 & -3 & -1 & 1 & (1) \end{array}$$ $\longleftarrow$ Remainder: 1 Quotient: $x^3 - 3x^2 - x + 1$	55–60
	Use the Remainder Theorem and the Factor Theorem (p. 154).	**The Remainder Theorem:** If a polynomial $f(x)$ is divided by $x - k$, the remainder is $r = f(k)$. **The Factor Theorem:** A polynomial $f(x)$ has a factor $(x - k)$ if and only if $f(k) = 0$.	61–66
Section 2.4	Use the imaginary unit i to write complex numbers (p. 159).	If a and b are real numbers, $a + bi$ is a complex number. Two complex numbers $a + bi$ and $c + di$, written in standard form, are equal to each other if and only if $a = c$ and $b = d$.	67–70

	What Did You Learn?	**Explanation/Examples**	**Review Exercises**
Section 2.4	Add, subtract, and multiply complex numbers (*p. 160*).	**Sum:** $(a + bi) + (c + di) = (a + c) + (b + d)i$ **Difference:** $(a + bi) - (c + di) = (a - c) + (b - d)i$	71–78
	Use complex conjugates to write the quotient of two complex numbers in standard form (*p. 162*).	The numbers $a + bi$ and $a - bi$ are complex conjugates. To write $(a + bi)/(c + di)$ in standard form, multiply the numerator and denominator by $c - di$.	79–82
	Find complex solutions of quadratic equations (*p. 163*).	If a is a positive number, the principal square root of the negative number $-a$ is defined as $\sqrt{-a} = \sqrt{a}i$.	83–86
Section 2.5	Use the Fundamental Theorem of Algebra to find the number of zeros of polynomial functions (*p. 166*).	**The Fundamental Theorem of Algebra** If $f(x)$ is a polynomial of degree n, where $n > 0$, then f has at least one zero in the complex number system.	87–92
	Find rational zeros of polynomial functions (*p. 167*), and conjugate pairs of complex zeros (*p. 170*).	The Rational Zero Test relates the possible rational zeros of a polynomial to the leading coefficient and to the constant term of the polynomial. Let $f(x)$ be a polynomial function that has real coefficients. If $a + bi$ $(b \neq 0)$ is a zero of the function, the conjugate $a - bi$ is also a zero of the function.	93–102
	Find zeros of polynomials by factoring (*p. 170*).	Every polynomial of degree $n > 0$ with real coefficients can be written as the product of linear and quadratic factors with real coefficients, where the quadratic factors have no real zeros.	103–110
	Use Descartes's Rule of Signs (*p. 173*) and the Upper and Lower Bound Rules (*p. 174*) to find zeros of polynomials.	**Descartes's Rule of Signs** Let $f(x) = a_n x^n + a_{n-1} x^{n-1} + \cdots + a_2 x^2 + a_1 x + a_0$ be a polynomial with real coefficients and $a_0 \neq 0$. **1.** The number of *positive real zeros* of f is either equal to the number of variations in sign of $f(x)$ or less than that number by an even integer. **2.** The number of *negative real zeros* of f is either equal to the number of variations in sign of $f(-x)$ or less than that number by an even integer.	111–114
Section 2.6	Find the domains (*p. 181*), and vertical and horizontal asymptotes (*p. 182*) of rational functions.	The domain of a rational function of x includes all real numbers except x-values that make the denominator zero. The line $x = a$ is a vertical asymptote of the graph of f if $f(x) \to \infty$ or $f(x) \to -\infty$ as $x \to a$, either from the right or from the left. The line $y = b$ is a horizontal asymptote of the graph of f if $f(x) \to b$ as $x \to \infty$. or $x \to -\infty$.	115–122
	Analyze and sketch graphs of rational functions (*p. 184*) including functions with slant asymptotes (*p. 187*).	Consider a rational function whose denominator is of degree 1 or greater. If the degree of the numerator is exactly *one more* than the degree of the denominator, the graph of the function has a slant asymptote.	123–138
	Use rational functions to model and solve real-life problems (*p. 188*).	A rational function can be used to model the cost of removing a given percent of smokestack pollutants at a utility company that burns coal. (See Example 8.)	139–142
Section 2.7	Solve polynomial (*p. 194*) and rational inequalities (*p. 198*).	Use the concepts of key numbers and test intervals to solve both polynomial and rational inequalities.	143–150
	Use inequalities to model and solve real-life problems (*p. 199*).	A common application of inequalities involves profit P, revenue R, and cost C. (See Example 6.)	151, 152

2 REVIEW EXERCISES

See www.CalcChat.com for worked-out solutions to odd-numbered exercises.

2.1 In Exercises 1 and 2, graph each function. Compare the graph of each function with the graph of $y = x^2$.

1. (a) $f(x) = 2x^2$
   (b) $g(x) = -2x^2$
   (c) $h(x) = x^2 + 2$
   (d) $k(x) = (x + 2)^2$

2. (a) $f(x) = x^2 - 4$
   (b) $g(x) = 4 - x^2$
   (c) $h(x) = (x - 3)^2$
   (d) $k(x) = \frac{1}{2}x^2 - 1$

In Exercises 3–14, write the quadratic function in standard form and sketch its graph. Identify the vertex, axis of symmetry, and x-intercept(s).

3. $g(x) = x^2 - 2x$
4. $f(x) = 6x - x^2$
5. $f(x) = x^2 + 8x + 10$
6. $h(x) = 3 + 4x - x^2$
7. $f(t) = -2t^2 + 4t + 1$
8. $f(x) = x^2 - 8x + 12$
9. $h(x) = 4x^2 + 4x + 13$
10. $f(x) = x^2 - 6x + 1$
11. $h(x) = x^2 + 5x - 4$
12. $f(x) = 4x^2 + 4x + 5$
13. $f(x) = \frac{1}{3}(x^2 + 5x - 4)$
14. $f(x) = \frac{1}{2}(6x^2 - 24x + 22)$

In Exercises 15–20, write the standard form of the equation of the parabola that has the indicated vertex and whose graph passes through the given point.

15.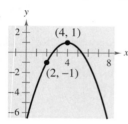

16.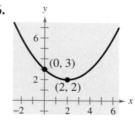

17. Vertex: $(1, -4)$; point: $(2, -3)$
18. Vertex: $(2, 3)$; point: $(-1, 6)$
19. Vertex: $\left(-\frac{3}{2}, 0\right)$; point: $\left(-\frac{9}{2}, -\frac{11}{4}\right)$
20. Vertex: $(3, 3)$; point: $\left(\frac{1}{4}, \frac{4}{5}\right)$

21. **GEOMETRY** The perimeter of a rectangle is 1000 meters.
   (a) Draw a diagram that gives a visual representation of the problem. Label the length and width as x and y, respectively.
   (b) Write y as a function of x. Use the result to write the area as a function of x.
   (c) Of all possible rectangles with perimeters of 1000 meters, find the dimensions of the one with the maximum area.

22. **MAXIMUM REVENUE** The total revenue R earned (in dollars) from producing a gift box of candles is given by

$$R(p) = -10p^2 + 800p$$

   where p is the price per unit (in dollars).
   (a) Find the revenues when the prices per box are $20, $25, and $30.
   (b) Find the unit price that will yield a maximum revenue. What is the maximum revenue? Explain your results.

23. **MINIMUM COST** A soft-drink manufacturer has daily production costs of

$$C = 70{,}000 - 120x + 0.055x^2$$

   where C is the total cost (in dollars) and x is the number of units produced. How many units should be produced each day to yield a minimum cost?

24. **SOCIOLOGY** The average age of the groom at a first marriage for a given age of the bride can be approximated by the model

$$y = -0.107x^2 + 5.68x - 48.5, \quad 20 \le x \le 25$$

   where y is the age of the groom and x is the age of the bride. Sketch a graph of the model. For what age of the bride is the average age of the groom 26? (Source: U.S. Census Bureau)

2.2 In Exercises 25–30, sketch the graphs of $y = x^n$ and the transformation.

25. $y = x^3$, $f(x) = -(x - 2)^3$
26. $y = x^3$, $f(x) = -4x^3$
27. $y = x^4$, $f(x) = 6 - x^4$
28. $y = x^4$, $f(x) = 2(x - 8)^4$
29. $y = x^5$, $f(x) = (x - 5)^5$
30. $y = x^5$, $f(x) = \frac{1}{2}x^5 + 3$

In Exercises 31–34, describe the right-hand and left-hand behavior of the graph of the polynomial function.

31. $f(x) = -2x^2 - 5x + 12$

32. $f(x) = \frac{1}{2}x^3 + 2x$

33. $g(x) = \frac{3}{4}(x^4 + 3x^2 + 2)$

34. $h(x) = -x^7 + 8x^2 - 8x$

In Exercises 35–40, find all the real zeros of the polynomial function. Determine the multiplicity of each zero and the number of turning points of the graph of the function. Use a graphing utility to verify your answers.

35. $f(x) = 3x^2 + 20x - 32$ **36.** $f(x) = x(x + 3)^2$

37. $f(t) = t^3 - 3t$ **38.** $f(x) = x^3 - 8x^2$

39. $f(x) = -18x^3 + 12x^2$ **40.** $g(x) = x^4 + x^3 - 12x^2$

In Exercises 41–44, sketch the graph of the function by (a) applying the Leading Coefficient Test, (b) finding the zeros of the polynomial, (c) plotting sufficient solution points, and (d) drawing a continuous curve through the points.

41. $f(x) = -x^3 + x^2 - 2$

42. $g(x) = 2x^3 + 4x^2$

43. $f(x) = x(x^3 + x^2 - 5x + 3)$

44. $h(x) = 3x^2 - x^4$

In Exercises 45–48, (a) use the Intermediate Value Theorem and the *table* feature of a graphing utility to find intervals one unit in length in which the polynomial function is guaranteed to have a zero. (b) Adjust the table to approximate the zeros of the function. Use the *zero* or *root* feature of the graphing utility to verify your results.

45. $f(x) = 3x^3 - x^2 + 3$

46. $f(x) = 0.25x^3 - 3.65x + 6.12$

47. $f(x) = x^4 - 5x - 1$

48. $f(x) = 7x^4 + 3x^3 - 8x^2 + 2$

2.3 In Exercises 49–54, use long division to divide.

49. $\dfrac{30x^2 - 3x + 8}{5x - 3}$ **50.** $\dfrac{4x + 7}{3x - 2}$

51. $\dfrac{5x^3 - 21x^2 - 25x - 4}{x^2 - 5x - 1}$

52. $\dfrac{3x^4}{x^2 - 1}$

53. $\dfrac{x^4 + 3x^3 + 4x^2 - 6x + 3}{x^2 + 2}$

54. $\dfrac{6x^4 + 10x^3 + 13x^2 - 5x + 2}{2x^2 - 1}$

In Exercises 55–58, use synthetic division to divide.

55. $\dfrac{6x^4 - 4x^3 - 27x^2 + 18x}{x - 2}$ **56.** $\dfrac{0.1x^3 + 0.3x^2 - 0.5}{x - 5}$

57. $\dfrac{2x^3 - 25x^2 + 66x + 48}{x - 8}$

58. $\dfrac{5x^3 + 33x^2 + 50x - 8}{x + 4}$

In Exercises 59 and 60, use synthetic division to determine whether the given values of x are zeros of the function.

59. $f(x) = 20x^4 + 9x^3 - 14x^2 - 3x$

 (a) $x = -1$ (b) $x = \frac{3}{4}$ (c) $x = 0$ (d) $x = 1$

60. $f(x) = 3x^3 - 8x^2 - 20x + 16$

 (a) $x = 4$ (b) $x = -4$ (c) $x = \frac{2}{3}$ (d) $x = -1$

In Exercises 61 and 62, use the Remainder Theorem and synthetic division to find each function value.

61. $f(x) = x^4 + 10x^3 - 24x^2 + 20x + 44$

 (a) $f(-3)$ (b) $f(-1)$

62. $g(t) = 2t^5 - 5t^4 - 8t + 20$

 (a) $g(-4)$ (b) $g(\sqrt{2})$

In Exercises 63–66, (a) verify the given factor(s) of the function f, (b) find the remaining factors of f, (c) use your results to write the complete factorization of f, (d) list all real zeros of f, and (e) confirm your results by using a graphing utility to graph the function.

	Function	Factor(s)
63.	$f(x) = x^3 + 4x^2 - 25x - 28$	$(x - 4)$
64.	$f(x) = 2x^3 + 11x^2 - 21x - 90$	$(x + 6)$
65.	$f(x) = x^4 - 4x^3 - 7x^2 + 22x + 24$	$(x + 2)(x - 3)$
66.	$f(x) = x^4 - 11x^3 + 41x^2 - 61x + 30$	$(x - 2)(x - 5)$

2.4 In Exercises 67–70, write the complex number in standard form.

67. $8 + \sqrt{-100}$ **68.** $5 - \sqrt{-49}$

69. $i^2 + 3i$ **70.** $-5i + i^2$

In Exercises 71–78, perform the operation and write the result in standard form.

71. $(7 + 5i) + (-4 + 2i)$

72. $\left(\dfrac{\sqrt{2}}{2} - \dfrac{\sqrt{2}}{2}i\right) - \left(\dfrac{\sqrt{2}}{2} + \dfrac{\sqrt{2}}{2}i\right)$

73. $7i(11 - 9i)$ **74.** $(1 + 6i)(5 - 2i)$

75. $(10 - 8i)(2 - 3i)$ **76.** $i(6 + i)(3 - 2i)$

77. $(8 - 5i)^2$

78. $(4 + 7i)^2 + (4 - 7i)^2$

In Exercises 79 and 80, write the quotient in standard form.

79. $\dfrac{6 + i}{4 - i}$

80. $\dfrac{8 - 5i}{i}$

In Exercises 81 and 82, perform the operation and write the result in standard form.

81. $\dfrac{4}{2 - 3i} + \dfrac{2}{1 + i}$

82. $\dfrac{1}{2 + i} - \dfrac{5}{1 + 4i}$

In Exercises 83–86, find all solutions of the equation.

83. $5x^2 + 2 = 0$

84. $2 + 8x^2 = 0$

85. $x^2 - 2x + 10 = 0$

86. $6x^2 + 3x + 27 = 0$

2.5 In Exercises 87–92, find all the zeros of the function.

87. $f(x) = 4x(x - 3)^2$

88. $f(x) = (x - 4)(x + 9)^2$

89. $f(x) = x^2 - 11x + 18$

90. $f(x) = x^3 + 10x$

91. $f(x) = (x + 4)(x - 6)(x - 2i)(x + 2i)$

92. $f(x) = (x - 8)(x - 5)^2(x - 3 + i)(x - 3 - i)$

In Exercises 93 and 94, use the Rational Zero Test to list all possible rational zeros of f.

93. $f(x) = -4x^3 + 8x^2 - 3x + 15$

94. $f(x) = 3x^4 + 4x^3 - 5x^2 - 8$

In Exercises 95–100, find all the rational zeros of the function.

95. $f(x) = x^3 + 3x^2 - 28x - 60$

96. $f(x) = 4x^3 - 27x^2 + 11x + 42$

97. $f(x) = x^3 - 10x^2 + 17x - 8$

98. $f(x) = x^3 + 9x^2 + 24x + 20$

99. $f(x) = x^4 + x^3 - 11x^2 + x - 12$

100. $f(x) = 25x^4 + 25x^3 - 154x^2 - 4x + 24$

In Exercises 101 and 102, find a polynomial function with real coefficients that has the given zeros. (There are many correct answers.)

101. $\frac{2}{3}, 4, \sqrt{3}\,i$

102. $2, -3, 1 - 2i$

In Exercises 103–106, use the given zero to find all the zeros of the function.

	Function	Zero
103.	$f(x) = x^3 - 4x^2 + x - 4$	i
104.	$h(x) = -x^3 + 2x^2 - 16x + 32$	$-4i$
105.	$g(x) = 2x^4 - 3x^3 - 13x^2 + 37x - 15$	$2 + i$
106.	$f(x) = 4x^4 - 11x^3 + 14x^2 - 6x$	$1 - i$

In Exercises 107–110, find all the zeros of the function and write the polynomial as a product of linear factors.

107. $f(x) = x^3 + 4x^2 - 5x$

108. $g(x) = x^3 - 7x^2 + 36$

109. $g(x) = x^4 + 4x^3 - 3x^2 + 40x + 208$

110. $f(x) = x^4 + 8x^3 + 8x^2 - 72x - 153$

In Exercises 111 and 112, use Descartes's Rule of Signs to determine the possible numbers of positive and negative zeros of the function.

111. $g(x) = 5x^3 + 3x^2 - 6x + 9$

112. $h(x) = -2x^5 + 4x^3 - 2x^2 + 5$

In Exercises 113 and 114, use synthetic division to verify the upper and lower bounds of the real zeros of f.

113. $f(x) = 4x^3 - 3x^2 + 4x - 3$
    (a) Upper: $x = 1$ (b) Lower: $x = -\frac{1}{4}$

114. $f(x) = 2x^3 - 5x^2 - 14x + 8$
    (a) Upper: $x = 8$ (b) Lower: $x = -4$

2.6 In Exercises 115–118, find the domain of the rational function.

115. $f(x) = \dfrac{3x}{x + 10}$

116. $f(x) = \dfrac{4x^3}{2 + 5x}$

117. $f(x) = \dfrac{8}{x^2 - 10x + 24}$

118. $f(x) = \dfrac{x^2 + x - 2}{x^2 + 4}$

In Exercises 119–122, identify any vertical or horizontal asymptotes.

119. $f(x) = \dfrac{4}{x + 3}$

120. $f(x) = \dfrac{2x^2 + 5x - 3}{x^2 + 2}$

121. $h(x) = \dfrac{5x + 20}{x^2 - 2x - 24}$

122. $h(x) = \dfrac{x^3 - 4x^2}{x^2 + 3x + 2}$

In Exercises 123–134, (a) state the domain of the function, (b) identify all intercepts, (c) find any vertical and horizontal asymptotes, and (d) plot additional solution points as needed to sketch the graph of the rational function.

123. $f(x) = \dfrac{-3}{2x^2}$

124. $f(x) = \dfrac{4}{x}$

125. $g(x) = \dfrac{2 + x}{1 - x}$

126. $h(x) = \dfrac{x - 4}{x - 7}$

127. $p(x) = \dfrac{5x^2}{4x^2 + 1}$

128. $f(x) = \dfrac{2x}{x^2 + 4}$

129. $f(x) = \dfrac{x}{x^2 + 1}$

130. $h(x) = \dfrac{9}{(x - 3)^2}$

131. $f(x) = \dfrac{-6x^2}{x^2 + 1}$ **132.** $f(x) = \dfrac{2x^2}{x^2 - 4}$

133. $f(x) = \dfrac{6x^2 - 11x + 3}{3x^2 - x}$ **134.** $f(x) = \dfrac{6x^2 - 7x + 2}{4x^2 - 1}$

In Exercises 135–138, (a) state the domain of the function, (b) identify all intercepts, (c) identify any vertical and slant asymptotes, and (d) plot additional solution points as needed to sketch the graph of the rational function.

135. $f(x) = \dfrac{2x^3}{x^2 + 1}$ **136.** $f(x) = \dfrac{x^2 + 1}{x + 1}$

137. $f(x) = \dfrac{3x^3 - 2x^2 - 3x + 2}{3x^2 - x - 4}$

138. $f(x) = \dfrac{3x^3 - 4x^2 - 12x + 16}{3x^2 + 5x - 2}$

139. AVERAGE COST A business has a production cost of $C = 0.5x + 500$ for producing x units of a product. The average cost per unit, $\overline{C}$, is given by

$$\overline{C} = \frac{C}{x} = \frac{0.5x + 500}{x}, \quad x > 0.$$

Determine the average cost per unit as x increases without bound. (Find the horizontal asymptote.)

140. SEIZURE OF ILLEGAL DRUGS The cost C (in millions of dollars) for the federal government to seize $p\%$ of an illegal drug as it enters the country is given by

$$C = \frac{528p}{100 - p}, \quad 0 \le p < 100.$$

 (a) Use a graphing utility to graph the cost function.

(b) Find the costs of seizing 25%, 50%, and 75% of the drug.

(c) According to this model, would it be possible to seize 100% of the drug?

141. PAGE DESIGN A page that is x inches wide and y inches high contains 30 square inches of print. The top and bottom margins are 2 inches deep and the margins on each side are 2 inches wide.

(a) Draw a diagram that gives a visual representation of the problem.

(b) Write a function for the total area A of the page in terms of x.

(c) Determine the domain of the function based on the physical constraints of the problem.

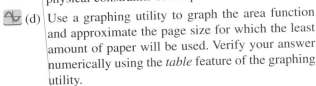 (d) Use a graphing utility to graph the area function and approximate the page size for which the least amount of paper will be used. Verify your answer numerically using the *table* feature of the graphing utility.

 142. PHOTOSYNTHESIS The amount y of CO_2 uptake (in milligrams per square decimeter per hour) at optimal temperatures and with the natural supply of CO_2 is approximated by the model

$$y = \frac{18.47x - 2.96}{0.23x + 1}, \quad x > 0$$

where x is the light intensity (in watts per square meter). Use a graphing utility to graph the function and determine the limiting amount of CO_2 uptake.

2.7 In Exercises 143–150, solve the inequality.

143. $12x^2 + 5x < 2$ **144.** $3x^2 + x \ge 24$

145. $x^3 - 16x \ge 0$ **146.** $12x^3 - 20x^2 < 0$

147. $\dfrac{2}{x + 1} \le \dfrac{3}{x - 1}$ **148.** $\dfrac{x - 5}{3 - x} < 0$

149. $\dfrac{x^2 - 9x + 20}{x} \le 0$ **150.** $\dfrac{1}{x - 2} > \dfrac{1}{x}$

151. INVESTMENT P dollars invested at interest rate r compounded annually increases to an amount

$$A = P(1 + r)^2$$

in 2 years. An investment of $5000 is to increase to an amount greater than $5500 in 2 years. The interest rate must be greater than what percent?

152. POPULATION OF A SPECIES A biologist introduces 200 ladybugs into a crop field. The population P of the ladybugs is approximated by the model

$$P = \frac{1000(1 + 3t)}{5 + t}$$

where t is the time in days. Find the time required for the population to increase to at least 2000 ladybugs.

EXPLORATION

TRUE OR FALSE? In Exercises 153 and 154, determine whether the statement is true or false. Justify your answer.

153. A fourth-degree polynomial with real coefficients can have -5, $-8i$, $4i$, and 5 as its zeros.

154. The domain of a rational function can never be the set of all real numbers.

155. WRITING Explain how to determine the maximum or minimum value of a quadratic function.

156. WRITING Explain the connections among factors of a polynomial, zeros of a polynomial function, and solutions of a polynomial equation.

157. WRITING Describe what is meant by an asymptote of a graph.

2 CHAPTER TEST

See www.CalcChat.com for worked-out solutions to odd-numbered exercises.

Take this test as you would take a test in class. When you are finished, check your work against the answers given in the back of the book.

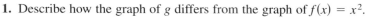

1. Describe how the graph of g differs from the graph of $f(x) = x^2$.

    (a) $g(x) = 2 - x^2$ (b) $g(x) = \left(x - \frac{3}{2}\right)^2$

2. Find an equation of the parabola shown in the figure at the left.

3. The path of a ball is given by $y = -\frac{1}{20}x^2 + 3x + 5$, where y is the height (in feet) of the ball and x is the horizontal distance (in feet) from where the ball was thrown.

    (a) Find the maximum height of the ball.

    (b) Which number determines the height at which the ball was thrown? Does changing this value change the coordinates of the maximum height of the ball? Explain.

4. Determine the right-hand and left-hand behavior of the graph of the function $h(t) = -\frac{3}{4}t^5 + 2t^2$. Then sketch its graph.

5. Divide using long division.

$$\frac{3x^3 + 4x - 1}{x^2 + 1}$$

6. Divide using synthetic division.

$$\frac{2x^4 - 5x^2 - 3}{x - 2}$$

7. Use synthetic division to show that $x = \frac{5}{2}$ is a zero of the function given by

$$f(x) = 2x^3 - 5x^2 - 6x + 15.$$

Use the result to factor the polynomial function completely and list all the real zeros of the function.

8. Perform each operation and write the result in standard form.

    (a) $10i - \left(3 + \sqrt{-25}\right)$ (b) $\left(2 + \sqrt{3}i\right)\left(2 - \sqrt{3}i\right)$

9. Write the quotient in standard form: $\dfrac{5}{2 + i}$.

In Exercises 10 and 11, find a polynomial function with real coefficients that has the given zeros. (There are many correct answers.)

10. $0, 3, 2 + i$ **11.** $1 - \sqrt{3}i, 2, 2$

In Exercises 12 and 13, find all the zeros of the function.

12. $f(x) = 3x^3 + 14x^2 - 7x - 10$ **13.** $f(x) = x^4 - 9x^2 - 22x - 24$

In Exercises 14–16, identify any intercepts and asymptotes of the graph of the function. Then sketch a graph of the function.

14. $h(x) = \dfrac{4}{x^2} - 1$ **15.** $f(x) = \dfrac{2x^2 - 5x - 12}{x^2 - 16}$ **16.** $g(x) = \dfrac{x^2 + 2}{x - 1}$

In Exercises 17 and 18, solve the inequality. Sketch the solution set on the real number line.

17. $2x^2 + 5x > 12$ **18.** $\dfrac{2}{x} \leq \dfrac{1}{x + 6}$

FIGURE FOR **2**

PROOFS IN MATHEMATICS

These two pages contain proofs of four important theorems about polynomial functions. The first two theorems are from Section 2.3, and the second two theorems are from Section 2.5.

The Remainder Theorem *(p. 154)*

If a polynomial $f(x)$ is divided by $x - k$, the remainder is

$$r = f(k).$$

Proof

From the Division Algorithm, you have

$$f(x) = (x - k)q(x) + r(x)$$

and because either $r(x) = 0$ or the degree of $r(x)$ is less than the degree of $x - k$, you know that $r(x)$ must be a constant. That is, $r(x) = r$. Now, by evaluating $f(x)$ at $x = k$, you have

$$f(k) = (k - k)q(k) + r$$
$$= (0)q(k) + r = r.$$

To be successful in algebra, it is important that you understand the connection among *factors* of a polynomial, *zeros* of a polynomial function, and *solutions* or *roots* of a polynomial equation. The Factor Theorem is the basis for this connection.

The Factor Theorem *(p. 154)*

A polynomial $f(x)$ has a factor $(x - k)$ if and only if $f(k) = 0$.

Proof

Using the Division Algorithm with the factor $(x - k)$, you have

$$f(x) = (x - k)q(x) + r(x).$$

By the Remainder Theorem, $r(x) = r = f(k)$, and you have

$$f(x) = (x - k)q(x) + f(k)$$

where $q(x)$ is a polynomial of lesser degree than $f(x)$. If $f(k) = 0$, then

$$f(x) = (x - k)q(x)$$

and you see that $(x - k)$ is a factor of $f(x)$. Conversely, if $(x - k)$ is a factor of $f(x)$, division of $f(x)$ by $(x - k)$ yields a remainder of 0. So, by the Remainder Theorem, you have $f(k) = 0$.

The Linear Factorization Theorem is closely related to the Fundamental Theorem of Algebra. The Fundamental Theorem of Algebra has a long and interesting history. In the early work with polynomial equations, The Fundamental Theorem of Algebra was thought to have been not true, because imaginary solutions were not considered. In fact, in the very early work by mathematicians such as Abu al-Khwarizmi (c. 800 A.D.), negative solutions were also not considered.

Once imaginary numbers were accepted, several mathematicians attempted to give a general proof of the Fundamental Theorem of Algebra. These included Gottfried von Leibniz (1702), Jean d'Alembert (1746), Leonhard Euler (1749), Joseph-Louis Lagrange (1772), and Pierre Simon Laplace (1795). The mathematician usually credited with the first correct proof of the Fundamental Theorem of Algebra is Carl Friedrich Gauss, who published the proof in his doctoral thesis in 1799.

Linear Factorization Theorem   (p. 166)

If $f(x)$ is a polynomial of degree n, where $n > 0$, then f has precisely n linear factors

$$f(x) = a_n(x - c_1)(x - c_2) \cdots (x - c_n)$$

where $c_1, c_2, \ldots, c_n$ are complex numbers.

Proof

Using the Fundamental Theorem of Algebra, you know that f must have at least one zero, c_1. Consequently, $(x - c_1)$ is a factor of $f(x)$, and you have

$$f(x) = (x - c_1)f_1(x).$$

If the degree of $f_1(x)$ is greater than zero, you again apply the Fundamental Theorem to conclude that f_1 must have a zero c_2, which implies that

$$f(x) = (x - c_1)(x - c_2)f_2(x).$$

It is clear that the degree of $f_1(x)$ is $n - 1$, that the degree of $f_2(x)$ is $n - 2$, and that you can repeatedly apply the Fundamental Theorem n times until you obtain

$$f(x) = a_n(x - c_1)(x - c_2) \cdots (x - c_n)$$

where a_n is the leading coefficient of the polynomial $f(x)$.

Factors of a Polynomial   (p. 170)

Every polynomial of degree $n > 0$ with real coefficients can be written as the product of linear and quadratic factors with real coefficients, where the quadratic factors have no real zeros.

Proof

To begin, you use the Linear Factorization Theorem to conclude that $f(x)$ can be *completely* factored in the form

$$f(x) = d(x - c_1)(x - c_2)(x - c_3) \cdots (x - c_n).$$

If each c_i is real, there is nothing more to prove. If any c_i is complex ($c_i = a + bi$, $b \neq 0$), then, because the coefficients of $f(x)$ are real, you know that the conjugate $c_j = a - bi$ is also a zero. By multiplying the corresponding factors, you obtain

$$(x - c_i)(x - c_j) = [x - (a + bi)][x - (a - bi)]$$
$$= x^2 - 2ax + (a^2 + b^2)$$

where each coefficient is real.

PROBLEM SOLVING

This collection of thought-provoking and challenging exercises further explores and expands upon concepts learned in this chapter.

1. Show that if $f(x) = ax^3 + bx^2 + cx + d$, then $f(k) = r$, where $r = ak^3 + bk^2 + ck + d$, using long division. In other words, verify the Remainder Theorem for a third-degree polynomial function.

2. In 2000 B.C., the Babylonians solved polynomial equations by referring to tables of values. One such table gave the values of $y^3 + y^2$. To be able to use this table, the Babylonians sometimes had to manipulate the equation, as shown below.

$$ax^3 + bx^2 = c \qquad \text{Original equation}$$

$$\frac{a^3 x^3}{b^3} + \frac{a^2 x^2}{b^2} = \frac{a^2 c}{b^3} \qquad \text{Multiply each side by } \frac{a^2}{b^3}.$$

$$\left(\frac{ax}{b}\right)^3 + \left(\frac{ax}{b}\right)^2 = \frac{a^2 c}{b^3} \qquad \text{Rewrite.}$$

Then they would find $(a^2c)/b^3$ in the $y^3 + y^2$ column of the table. Because they knew that the corresponding y-value was equal to $(ax)/b$, they could conclude that $x = (by)/a$.

(a) Calculate $y^3 + y^2$ for $y = 1, 2, 3, \ldots, 10$. Record the values in a table.

Use the table from part (a) and the method above to solve each equation.

(b) $x^3 + x^2 = 252$

(c) $x^3 + 2x^2 = 288$

(d) $3x^3 + x^2 = 90$

(e) $2x^3 + 5x^2 = 2500$

(f) $7x^3 + 6x^2 = 1728$

(g) $10x^3 + 3x^2 = 297$

Using the methods from this chapter, verify your solution to each equation.

3. At a glassware factory, molten cobalt glass is poured into molds to make paperweights. Each mold is a rectangular prism whose height is 3 inches greater than the length of each side of the square base. A machine pours 20 cubic inches of liquid glass into each mold. What are the dimensions of the mold?

4. Determine whether the statement is true or false. If false, provide one or more reasons why the statement is false and correct the statement. Let $f(x) = ax^3 + bx^2 + cx + d$, $a \neq 0$, and let $f(2) = -1$. Then

$$\frac{f(x)}{x+1} = q(x) + \frac{2}{x+1}$$

where $q(x)$ is a second-degree polynomial.

5. The parabola shown in the figure has an equation of the form $y = ax^2 + bx + c$. Find the equation of this parabola by the following methods. (a) Find the equation analytically. (b) Use the *regression* feature of a graphing utility to find the equation.

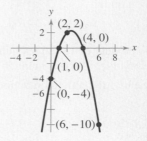

6. One of the fundamental themes of calculus is to find the slope of the tangent line to a curve at a point. To see how this can be done, consider the point $(2, 4)$ on the graph of the quadratic function $f(x) = x^2$, which is shown in the figure.

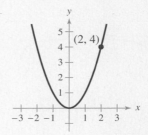

(a) Find the slope m_1 of the line joining $(2, 4)$ and $(3, 9)$. Is the slope of the tangent line at $(2, 4)$ greater than or less than the slope of the line through $(2, 4)$ and $(3, 9)$?

(b) Find the slope m_2 of the line joining $(2, 4)$ and $(1, 1)$. Is the slope of the tangent line at $(2, 4)$ greater than or less than the slope of the line through $(2, 4)$ and $(1, 1)$?

(c) Find the slope m_3 of the line joining $(2, 4)$ and $(2.1, 4.41)$. Is the slope of the tangent line at $(2, 4)$ greater than or less than the slope of the line through $(2, 4)$ and $(2.1, 4.41)$?

(d) Find the slope m_h of the line joining $(2, 4)$ and $(2 + h, f(2 + h))$ in terms of the nonzero number h.

(e) Evaluate the slope formula from part (d) for $h = -1$, 1, and 0.1. Compare these values with those in parts (a)–(c).

(f) What can you conclude the slope m_{tan} of the tangent line at $(2, 4)$ to be? Explain your answer.

7. Use the form $f(x) = (x - k)q(x) + r$ to create a cubic function that (a) passes through the point $(2, 5)$ and rises to the right and (b) passes through the point $(-3, 1)$ and falls to the right. (There are many correct answers.)

8. The multiplicative inverse of z is a complex number z_m such that $z \cdot z_m = 1$. Find the multiplicative inverse of each complex number.

(a) $z = 1 + i$ (b) $z = 3 - i$ (c) $z = -2 + 8i$

9. Prove that the product of a complex number $a + bi$ and its complex conjugate is a real number.

10. Match the graph of the rational function given by

$$f(x) = \frac{ax + b}{cx + d}$$

with the given conditions.

(a) 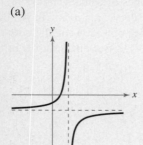 (b)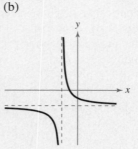

(c) 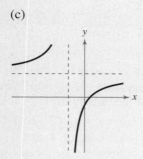 (d)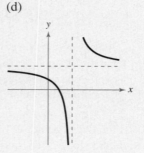

(i) $a > 0$ (ii) $a > 0$ (iii) $a < 0$ (iv) $a > 0$
$b < 0$ $b > 0$ $b > 0$ $b < 0$
$c > 0$ $c < 0$ $c > 0$ $c > 0$
$d < 0$ $d < 0$ $d < 0$ $d > 0$

11. Consider the function given by

$$f(x) = \frac{ax}{(x - b)^2}.$$

(a) Determine the effect on the graph of f if $b \neq 0$ and a is varied. Consider cases in which a is positive and a is negative.

(b) Determine the effect on the graph of f if $a \neq 0$ and b is varied.

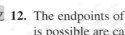 **12.** The endpoints of the interval over which distinct vision is possible are called the *near point* and *far point* of the eye (see figure). With increasing age, these points normally change. The table shows the approximate near points y (in inches) for various ages x (in years).

Age, x	Near point, y
16	3.0
32	4.7
44	9.8
50	19.7
60	39.4

(a) Use the *regression* feature of a graphing utility to find a quadratic model y_1 for the data. Use a graphing utility to plot the data and graph the model in the same viewing window.

(b) Find a rational model y_2 for the data. Take the reciprocals of the near points to generate the points $(x, 1/y)$. Use the *regression* feature of a graphing utility to find a linear model for the data. The resulting line has the form

$$\frac{1}{y} = ax + b.$$

Solve for y. Use a graphing utility to plot the data and graph the model in the same viewing window.

(c) Use the *table* feature of a graphing utility to create a table showing the predicted near point based on each model for each of the ages in the original table. How well do the models fit the original data?

(d) Use both models to estimate the near point for a person who is 25 years old. Which model is a better fit?

(e) Do you think either model can be used to predict the near point for a person who is 70 years old? Explain.

Exponential and Logarithmic Functions

3

In Mathematics

Exponential functions involve a constant base and a variable exponent. The inverse of an exponential function is a logarithmic function.

In Real Life

Exponential and logarithmic functions are widely used in describing economic and physical phenomena such as compound interest, population growth, memory retention, and decay of radioactive material. For instance, a logarithmic function can be used to relate an animal's weight and its lowest galloping speed. (See Exercise 95, page 242.)

Juniors Bildarchiv / Alamy

IN CAREERS

There are many careers that use exponential and logarithmic functions. Several are listed below.

- Astronomer
  Example 7, page 240

- Psychologist
  Exercise 136, page 253

- Archeologist
  Example 3, page 258

- Forensic Scientist
  Exercise 75, page 266

3.1   EXPONENTIAL FUNCTIONS AND THEIR GRAPHS

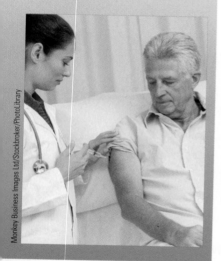

Monkey Business Images Ltd/Stockbroker/PhotoLibrary

Exponential Functions

So far, this text has dealt mainly with **algebraic functions,** which include polynomial functions and rational functions. In this chapter, you will study two types of nonalgebraic functions—*exponential functions* and *logarithmic functions*. These functions are examples of **transcendental functions.**

> **Definition of Exponential Function**
>
> The **exponential function** f with base a is denoted by
>
> $$f(x) = a^x$$
>
> where $a > 0$, $a \neq 1$, and x is any real number.

The base $a = 1$ is excluded because it yields $f(x) = 1^x = 1$. This is a constant function, not an exponential function.

You have evaluated a^x for integer and rational values of x. For example, you know that $4^3 = 64$ and $4^{1/2} = 2$. However, to evaluate 4^x for any real number x, you need to interpret forms with *irrational* exponents. For the purposes of this text, it is sufficient to think of

$$a^{\sqrt{2}} \quad (\text{where } \sqrt{2} \approx 1.41421356)$$

as the number that has the successively closer approximations

$$a^{1.4}, a^{1.41}, a^{1.414}, a^{1.4142}, a^{1.41421}, \dots$$

Example 1   Evaluating Exponential Functions

Use a calculator to evaluate each function at the indicated value of x.

Function	Value
a. $f(x) = 2^x$	$x = -3.1$
b. $f(x) = 2^{-x}$	$x = \pi$
c. $f(x) = 0.6^x$	$x = \frac{3}{2}$

Solution

Function Value	Graphing Calculator Keystrokes	Display
a. $f(-3.1) = 2^{-3.1}$	2 $\boxed{\wedge}$ $\boxed{(-)}$ 3.1 $\boxed{\text{ENTER}}$	0.1166291
b. $f(\pi) = 2^{-\pi}$	2 $\boxed{\wedge}$ $\boxed{(-)}$ π $\boxed{\text{ENTER}}$	0.1133147
c. $f\left(\frac{3}{2}\right) = (0.6)^{3/2}$	.6 $\boxed{\wedge}$ $\boxed{(}$ 3 $\boxed{\div}$ 2 $\boxed{)}$ $\boxed{\text{ENTER}}$	0.4647580

CHECK**Point** Now try Exercise 7.

When evaluating exponential functions with a calculator, remember to enclose fractional exponents in parentheses. Because the calculator follows the order of operations, parentheses are crucial in order to obtain the correct result.

Graphs of Exponential Functions

The graphs of all exponential functions have similar characteristics, as shown in Examples 2, 3, and 5.

Example 2   Graphs of $y = a^x$

In the same coordinate plane, sketch the graph of each function.

a. $f(x) = 2^x$ **b.** $g(x) = 4^x$

Solution

The table below lists some values for each function, and Figure 3.1 shows the graphs of the two functions. Note that both graphs are increasing. Moreover, the graph of $g(x) = 4^x$ is increasing more rapidly than the graph of $f(x) = 2^x$.

x	-3	-2	-1	0	1	2
2^x	$\frac{1}{8}$	$\frac{1}{4}$	$\frac{1}{2}$	1	2	4
4^x	$\frac{1}{64}$	$\frac{1}{16}$	$\frac{1}{4}$	1	4	16

CHECK**Point** Now try Exercise 17.

The table in Example 2 was evaluated by hand. You could, of course, use a graphing utility to construct tables with even more values.

Example 3   Graphs of $y = a^{-x}$

In the same coordinate plane, sketch the graph of each function.

a. $F(x) = 2^{-x}$ **b.** $G(x) = 4^{-x}$

Solution

The table below lists some values for each function, and Figure 3.2 shows the graphs of the two functions. Note that both graphs are decreasing. Moreover, the graph of $G(x) = 4^{-x}$ is decreasing more rapidly than the graph of $F(x) = 2^{-x}$.

x	-2	-1	0	1	2	3
2^{-x}	4	2	1	$\frac{1}{2}$	$\frac{1}{4}$	$\frac{1}{8}$
4^{-x}	16	4	1	$\frac{1}{4}$	$\frac{1}{16}$	$\frac{1}{64}$

CHECK**Point** Now try Exercise 19.

In Example 3, note that by using one of the properties of exponents, the functions $F(x) = 2^{-x}$ and $G(x) = 4^{-x}$ can be rewritten with positive exponents.

$$F(x) = 2^{-x} = \frac{1}{2^x} = \left(\frac{1}{2}\right)^x \quad \text{and} \quad G(x) = 4^{-x} = \frac{1}{4^x} = \left(\frac{1}{4}\right)^x$$

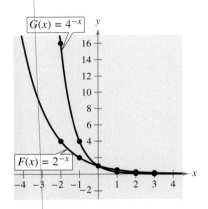

Algebra Help

You can review the techniques for sketching the graph of an equation in Section 1.2.

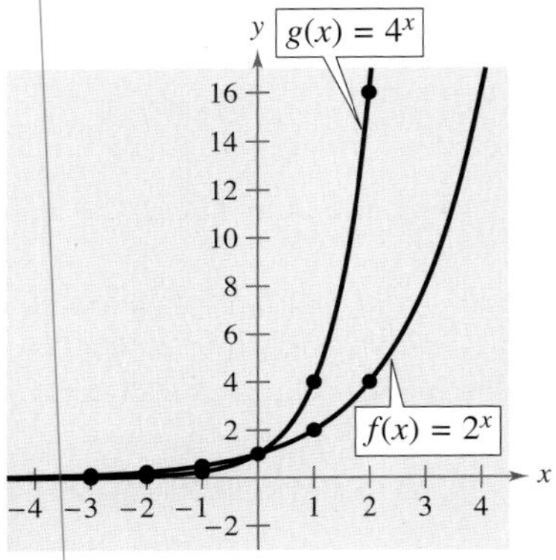

FIGURE 3.1

FIGURE 3.2

Comparing the functions in Examples 2 and 3, observe that

$$F(x) = 2^{-x} = f(-x) \qquad \text{and} \qquad G(x) = 4^{-x} = g(-x).$$

Consequently, the graph of F is a reflection (in the y-axis) of the graph of f. The graphs of G and g have the same relationship. The graphs in Figures 3.1 and 3.2 are typical of the exponential functions $y = a^x$ and $y = a^{-x}$. They have one y-intercept and one horizontal asymptote (the x-axis), and they are continuous. The basic characteristics of these exponential functions are summarized in Figures 3.3 and 3.4.

Study Tip

Notice that the range of an exponential function is $(0, \infty)$, which means that $a^x > 0$ for all values of x.

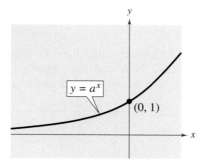

FIGURE **3.3**

Graph of $y = a^x$, $a > 1$

- Domain: $(-\infty, \infty)$
- Range: $(0, \infty)$
- y-intercept: $(0, 1)$
- Increasing
- x-axis is a horizontal asymptote $(a^x \to 0$ as $x \to -\infty)$.
- Continuous

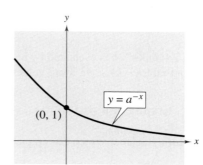

FIGURE **3.4**

Graph of $y = a^{-x}$, $a > 1$

- Domain: $(-\infty, \infty)$
- Range: $(0, \infty)$
- y-intercept: $(0, 1)$
- Decreasing
- x-axis is a horizontal asymptote $(a^{-x} \to 0$ as $x \to \infty)$.
- Continuous

From Figures 3.3 and 3.4, you can see that the graph of an exponential function is always increasing or always decreasing. As a result, the graphs pass the Horizontal Line Test, and therefore the functions are one-to-one functions. You can use the following **One-to-One Property** to solve simple exponential equations.

For $a > 0$ and $a \neq 1$, $a^x = a^y$ if and only if $x = y$.   One-to-One Property

Example 4   Using the One-to-One Property

a. $9 = 3^{x+1}$ Original equation

   $3^2 = 3^{x+1}$ $9 = 3^2$

   $2 = x + 1$ One-to-One Property

   $1 = x$ Solve for x.

b. $\left(\tfrac{1}{2}\right)^x = 8 \Longrightarrow 2^{-x} = 2^3 \Longrightarrow x = -3$

CHECK**Point**   Now try Exercise 51.

In the following example, notice how the graph of $y = a^x$ can be used to sketch the graphs of functions of the form $f(x) = b \pm a^{x+c}$.

Algebra Help

You can review the techniques for transforming the graph of a function in Section 1.7.

Example 5   Transformations of Graphs of Exponential Functions

Each of the following graphs is a transformation of the graph of $f(x) = 3^x$.

a. Because $g(x) = 3^{x+1} = f(x + 1)$, the graph of g can be obtained by shifting the graph of f one unit to the *left*, as shown in Figure 3.5.

b. Because $h(x) = 3^x - 2 = f(x) - 2$, the graph of h can be obtained by shifting the graph of f *downward* two units, as shown in Figure 3.6.

c. Because $k(x) = -3^x = -f(x)$, the graph of k can be obtained by *reflecting* the graph of f in the x-axis, as shown in Figure 3.7.

d. Because $j(x) = 3^{-x} = f(-x)$, the graph of j can be obtained by *reflecting* the graph of f in the y-axis, as shown in Figure 3.8.

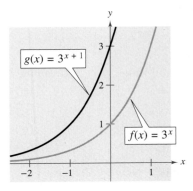

FIGURE **3.5**   Horizontal shift

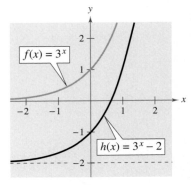

FIGURE **3.6**   Vertical shift

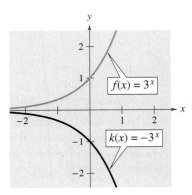

FIGURE **3.7**   Reflection in x-axis

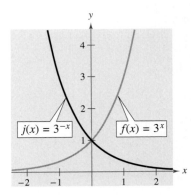

FIGURE **3.8**   Reflection in y-axis

CHECK *Point*   Now try Exercise 23.

Notice that the transformations in Figures 3.5, 3.7, and 3.8 keep the x-axis as a horizontal asymptote, but the transformation in Figure 3.6 yields a new horizontal asymptote of $y = -2$. Also, be sure to note how the y-intercept is affected by each transformation.

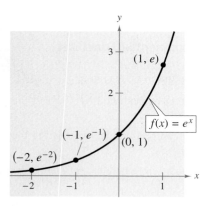

FIGURE **3.9**

The Natural Base *e*

In many applications, the most convenient choice for a base is the irrational number

$$e \approx 2.718281828 \ldots .$$

This number is called the **natural base.** The function given by $f(x) = e^x$ is called the **natural exponential function.** Its graph is shown in Figure 3.9. Be sure you see that for the exponential function $f(x) = e^x$, e is the constant $2.718281828 \ldots$, whereas x is the variable.

Example 6   Evaluating the Natural Exponential Function

Use a calculator to evaluate the function given by $f(x) = e^x$ at each indicated value of x.

a. $x = -2$

b. $x = -1$

c. $x = 0.25$

d. $x = -0.3$

Solution

	Function Value	*Graphing Calculator Keystrokes*	*Display*
a.	$f(-2) = e^{-2}$	e^x $(-)$ 2 ENTER	0.1353353
b.	$f(-1) = e^{-1}$	e^x $(-)$ 1 ENTER	0.3678794
c.	$f(0.25) = e^{0.25}$	e^x 0.25 ENTER	1.2840254
d.	$f(-0.3) = e^{-0.3}$	e^x $(-)$ 0.3 ENTER	0.7408182

CHECK*Point*   Now try Exercise 33.

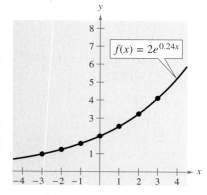

FIGURE **3.10**

Example 7   Graphing Natural Exponential Functions

Sketch the graph of each natural exponential function.

a. $f(x) = 2e^{0.24x}$

b. $g(x) = \frac{1}{2}e^{-0.58x}$

Solution

To sketch these two graphs, you can use a graphing utility to construct a table of values, as shown below. After constructing the table, plot the points and connect them with smooth curves, as shown in Figures 3.10 and 3.11. Note that the graph in Figure 3.10 is increasing, whereas the graph in Figure 3.11 is decreasing.

x	-3	-2	-1	0	1	2	3
$f(x)$	0.974	1.238	1.573	2.000	2.542	3.232	4.109
$g(x)$	2.849	1.595	0.893	0.500	0.280	0.157	0.088

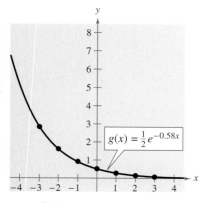

FIGURE **3.11**

CHECK*Point*   Now try Exercise 41.

✳ Applications ✳

One of the most familiar examples of exponential growth is an investment earning *continuously compounded interest*. Using exponential functions, you can *develop* a formula for interest compounded n times per year and show how it leads to continuous compounding.

Suppose a principal P is invested at an annual interest rate r, compounded once per year. If the interest is added to the principal at the end of the year, the new balance P_1 is

$$P_1 = P + Pr$$
$$= P(1 + r).$$

This pattern of multiplying the previous principal by $1 + r$ is then repeated each successive year, as shown below.

Year	Balance After Each Compounding
0	$P = P$
1	$P_1 = P(1 + r)$
2	$P_2 = P_1(1 + r) = P(1 + r)(1 + r) = P(1 + r)^2$
3	$P_3 = P_2(1 + r) = P(1 + r)^2(1 + r) = P(1 + r)^3$
$\vdots$	$\vdots$
t	$P_t = P(1 + r)^t$

To accommodate more frequent (quarterly, monthly, or daily) compounding of interest, let n be the number of compoundings per year and let t be the number of years. Then the rate per compounding is r/n and the account balance after t years is

$$A = P\left(1 + \frac{r}{n}\right)^{nt}. \qquad \text{Amount (balance) with } n \text{ compoundings per year}$$

If you let the number of compoundings n increase without bound, the process approaches what is called **continuous compounding.** In the formula for n compoundings per year, let $m = n/r$. This produces

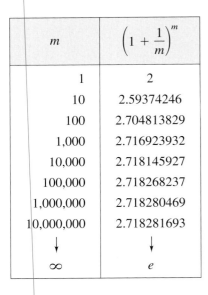

m	$\left(1 + \dfrac{1}{m}\right)^m$
1	2
10	2.59374246
100	2.704813829
1,000	2.716923932
10,000	2.718145927
100,000	2.718268237
1,000,000	2.718280469
10,000,000	2.718281693
$\downarrow$	$\downarrow$
∞	e

$$A = P\left(1 + \frac{r}{n}\right)^{nt} \qquad \text{Amount with } n \text{ compoundings per year}$$

$$= P\left(1 + \frac{r}{mr}\right)^{mrt} \qquad \text{Substitute } mr \text{ for } n.$$

$$= P\left(1 + \frac{1}{m}\right)^{mrt} \qquad \text{Simplify.}$$

$$= P\left[\left(1 + \frac{1}{m}\right)^m\right]^{rt}. \qquad \text{Property of exponents}$$

As m increases without bound, the table at the left shows that $[1 + (1/m)]^m \to e$ as $m \to \infty$. From this, you can conclude that the formula for continuous compounding is

$$A = Pe^{rt}. \qquad \text{Substitute } e \text{ for } (1 + 1/m)^m.$$

Formulas for Compound Interest

After t years, the balance A in an account with principal P and annual interest rate r (in decimal form) is given by the following formulas.

1. For n compoundings per year: $A = P\left(1 + \dfrac{r}{n}\right)^{nt}$

2. For continuous compounding: $A = Pe^{rt}$

Example 8   Compound Interest

A total of $12,000 is invested at an annual interest rate of 9%. Find the balance after 5 years if it is compounded

a. quarterly.

b. monthly.

c. continuously.

Solution

a. For quarterly compounding, you have $n = 4$. So, in 5 years at 9%, the balance is

$$A = P\left(1 + \frac{r}{n}\right)^{nt} \qquad \text{Formula for compound interest}$$

$$= 12{,}000\left(1 + \frac{0.09}{4}\right)^{4(5)} \qquad \text{Substitute for } P, r, n, \text{ and } t.$$

$$\approx \$18{,}726.11. \qquad \text{Use a calculator.}$$

b. For monthly compounding, you have $n = 12$. So, in 5 years at 9%, the balance is

$$A = P\left(1 + \frac{r}{n}\right)^{nt} \qquad \text{Formula for compound interest}$$

$$= 12{,}000\left(1 + \frac{0.09}{12}\right)^{12(5)} \qquad \text{Substitute for } P, r, n, \text{ and } t.$$

$$\approx \$18{,}788.17. \qquad \text{Use a calculator.}$$

c. For continuous compounding, the balance is

$$A = Pe^{rt} \qquad \text{Formula for continuous compounding}$$

$$= 12{,}000e^{0.09(5)} \qquad \text{Substitute for } P, r, \text{ and } t.$$

$$\approx \$18{,}819.75. \qquad \text{Use a calculator.}$$

CHECK**Point**   Now try Exercise 59.

In Example 8, note that continuous compounding yields more than quarterly or monthly compounding. This is typical of the two types of compounding. That is, for a given principal, interest rate, and time, continuous compounding will always yield a larger balance than compounding n times per year.

Example 9   Radioactive Decay

The *half-life* of radioactive radium (^{226}Ra) is about 1599 years. That is, for a given amount of radium, *half* of the original amount will remain after 1599 years. After another 1599 years, one-quarter of the original amount will remain, and so on. Let y represent the mass, in grams, of a quantity of radium. The quantity present after t years, then, is $y = 25\left(\frac{1}{2}\right)^{t/1599}$.

a. What is the initial mass (when $t = 0$)?   25

b. How much of the initial mass is present after 2500 years? $y = 25\left(\frac{1}{2}\right)^{\frac{2500}{1599}} = 8.45$

Algebraic Solution

a. $y = 25\left(\frac{1}{2}\right)^{t/1599}$ Write original equation.

$= 25\left(\frac{1}{2}\right)^{0/1599}$ Substitute 0 for t.

$= 25$ Simplify.

So, the initial mass is 25 grams.

b. $y = 25\left(\frac{1}{2}\right)^{t/1599}$ Write original equation.

$= 25\left(\frac{1}{2}\right)^{2500/1599}$ Substitute 2500 for t.

$\approx 25\left(\frac{1}{2}\right)^{1.563}$ Simplify.

≈ 8.46 Use a calculator.

So, about 8.46 grams is present after 2500 years.

CHECK *Point*   Now try Exercise 73.

Graphical Solution

Use a graphing utility to graph $y = 25\left(\frac{1}{2}\right)^{t/1599}$.

a. Use the *value* feature or the *zoom* and *trace* features of the graphing utility to determine that when $x = 0$, the value of y is 25, as shown in Figure 3.12. So, the initial mass is 25 grams.

b. Use the *value* feature or the *zoom* and *trace* features of the graphing utility to determine that when $x = 2500$, the value of y is about 8.46, as shown in Figure 3.13. So, about 8.46 grams is present after 2500 years.

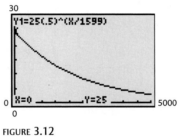

FIGURE **3.12**

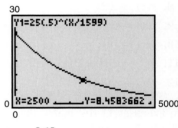

FIGURE **3.13**

CLASSROOM DISCUSSION

Identifying Exponential Functions   Which of the following functions generated the two tables below? Discuss how you were able to decide. What do these functions have in common? Are any of them the same? If so, explain why.

a. $f_1(x) = 2^{(x+3)}$ **b.** $f_2(x) = 8\left(\frac{1}{2}\right)^x$ **c.** $f_3(x) = \left(\frac{1}{2}\right)^{(x-3)}$

d. $f_4(x) = \left(\frac{1}{2}\right)^x + 7$ **e.** $f_5(x) = 7 + 2^x$ **f.** $f_6(x) = 8(2^x)$

x	-1	0	1	2	3
$g(x)$	7.5	8	9	11	15

x	-2	-1	0	1	2
$h(x)$	32	16	8	4	2

Create two different exponential functions of the forms $y = a(b)^x$ and $y = c^x + d$ with y-intercepts of $(0, -3)$.

3.1 EXERCISES

See www.CalcChat.com for worked-out solutions to odd-numbered exercises.

VOCABULARY: Fill in the blanks.

1. Polynomial and rational functions are examples of _____ functions.

2. Exponential and logarithmic functions are examples of nonalgebraic functions, also called _____ functions.

3. You can use the _____ Property to solve simple exponential equations.

4. The exponential function given by $f(x) = e^x$ is called the _____ _____ function, and the base e is called the _____ base.

5. To find the amount A in an account after t years with principal P and an annual interest rate r compounded n times per year, you can use the formula _____.

6. To find the amount A in an account after t years with principal P and an annual interest rate r compounded continuously, you can use the formula _____.

SKILLS AND APPLICATIONS

In Exercises 7–12, evaluate the function at the indicated value of x. Round your result to three decimal places.

Function	Value
7. $f(x) = 0.9^x$	$x = 1.4$
8. $f(x) = 2.3^x$	$x = \frac{3}{2}$
9. $f(x) = 5^x$	$x = -\pi$
10. $f(x) = \left(\frac{2}{3}\right)^{5x}$	$x = \frac{3}{10}$
11. $g(x) = 5000(2^x)$	$x = -1.5$
12. $f(x) = 200(1.2)^{12x}$	$x = 24$

In Exercises 13–16, match the exponential function with its graph. [The graphs are labeled (a), (b), (c), and (d).]

(a)
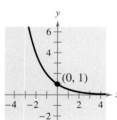

(b)
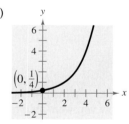

(c)
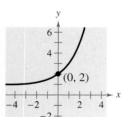

(d)
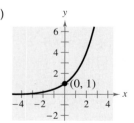

13. $f(x) = 2^x$

14. $f(x) = 2^x + 1$

15. $f(x) = 2^{-x}$

16. $f(x) = 2^{x-2}$

In Exercises 17–22, use a graphing utility to construct a table of values for the function. Then sketch the graph of the function.

17. $f(x) = \left(\frac{1}{2}\right)^x$

18. $f(x) = \left(\frac{1}{2}\right)^{-x}$

19. $f(x) = 6^{-x}$

20. $f(x) = 6^x$

21. $f(x) = 2^{x-1}$

22. $f(x) = 4^{x-3} + 3$

In Exercises 23–28, use the graph of f to describe the transformation that yields the graph of g.

23. $f(x) = 3^x$, $g(x) = 3^x + 1$

24. $f(x) = 4^x$, $g(x) = 4^{x-3}$

25. $f(x) = 2^x$, $g(x) = 3 - 2^x$

26. $f(x) = 10^x$, $g(x) = 10^{-x+3}$

27. $f(x) = \left(\frac{7}{2}\right)^x$, $g(x) = -\left(\frac{7}{2}\right)^{-x}$

28. $f(x) = 0.3^x$, $g(x) = -0.3^x + 5$

In Exercises 29–32, use a graphing utility to graph the exponential function.

29. $y = 2^{-x^2}$

30. $y = 3^{-|x|}$

31. $y = 3^{x-2} + 1$

32. $y = 4^{x+1} - 2$

In Exercises 33–38, evaluate the function at the indicated value of x. Round your result to three decimal places.

Function	Value
33. $h(x) = e^{-x}$	$x = \frac{3}{4}$
34. $f(x) = e^x$	$x = 3.2$
35. $f(x) = 2e^{-5x}$	$x = 10$
36. $f(x) = 1.5e^{x/2}$	$x = 240$
37. $f(x) = 5000e^{0.06x}$	$x = 6$
38. $f(x) = 250e^{0.05x}$	$x = 20$

In Exercises 39–44, use a graphing utility to construct a table of values for the function. Then sketch the graph of the function.

39. $f(x) = e^x$

40. $f(x) = e^{-x}$

41. $f(x) = 3e^{x+4}$

42. $f(x) = 2e^{-0.5x}$

43. $f(x) = 2e^{x-2} + 4$

44. $f(x) = 2 + e^{x-5}$

In Exercises 45–50, use a graphing utility to graph the exponential function.

45. $y = 1.08^{-5x}$

46. $y = 1.08^{5x}$

47. $s(t) = 2e^{0.12t}$

48. $s(t) = 3e^{-0.2t}$

49. $g(x) = 1 + e^{-x}$

50. $h(x) = e^{x-2}$

In Exercises 51–58, use the One-to-One Property to solve the equation for x.

51. $3^{x+1} = 27$

52. $2^{x-3} = 16$

53. $\left(\frac{1}{2}\right)^x = 32$

54. $5^{x-2} = \frac{1}{125}$

55. $e^{3x+2} = e^3$

56. $e^{2x-1} = e^4$

57. $e^{x^2-3} = e^{2x}$

58. $e^{x^2+6} = e^{5x}$

COMPOUND INTEREST In Exercises 59–62, complete the table to determine the balance A for P dollars invested at rate r for t years and compounded n times per year.

n	1	2	4	12	365	Continuous
A						

59. $P = \$1500$, $r = 2\%$, $t = 10$ years

60. $P = \$2500$, $r = 3.5\%$, $t = 10$ years

61. $P = \$2500$, $r = 4\%$, $t = 20$ years

62. $P = \$1000$, $r = 6\%$, $t = 40$ years

COMPOUND INTEREST In Exercises 63–66, complete the table to determine the balance A for $\$12,000$ invested at rate r for t years, compounded continuously.

t	10	20	30	40	50
A					

63. $r = 4\%$

64. $r = 6\%$

65. $r = 6.5\%$

66. $r = 3.5\%$

67. TRUST FUND On the day of a child's birth, a deposit of $30,000 is made in a trust fund that pays 5% interest, compounded continuously. Determine the balance in this account on the child's 25th birthday.

68. TRUST FUND A deposit of $5000 is made in a trust fund that pays 7.5% interest, compounded continuously. It is specified that the balance will be given to the college from which the donor graduated after the money has earned interest for 50 years. How much will the college receive?

69. INFLATION If the annual rate of inflation averages 4% over the next 10 years, the approximate costs C of goods or services during any year in that decade will be modeled by $C(t) = P(1.04)^t$, where t is the time in years and P is the present cost. The price of an oil change for your car is presently $23.95. Estimate the price 10 years from now.

70. COMPUTER VIRUS The number V of computers infected by a computer virus increases according to the model $V(t) = 100e^{4.6052t}$, where t is the time in hours. Find the number of computers infected after (a) 1 hour, (b) 1.5 hours, and (c) 2 hours.

71. POPULATION GROWTH The projected populations of California for the years 2015 through 2030 can be modeled by $P = 34.696e^{0.0098t}$, where P is the population (in millions) and t is the time (in years), with $t = 15$ corresponding to 2015. (Source: U.S. Census Bureau)

(a) Use a graphing utility to graph the function for the years 2015 through 2030.

(b) Use the *table* feature of a graphing utility to create a table of values for the same time period as in part (a).

(c) According to the model, when will the population of California exceed 50 million?

72. POPULATION The populations P (in millions) of Italy from 1990 through 2008 can be approximated by the model $P = 56.8e^{0.0015t}$, where t represents the year, with $t = 0$ corresponding to 1990. (Source: U.S. Census Bureau, International Data Base)

(a) According to the model, is the population of Italy increasing or decreasing? Explain.

(b) Find the populations of Italy in 2000 and 2008.

(c) Use the model to predict the populations of Italy in 2015 and 2020.

73. RADIOACTIVE DECAY Let Q represent a mass of radioactive plutonium (^{239}Pu) (in grams), whose half-life is 24,100 years. The quantity of plutonium present after t years is $Q = 16\left(\frac{1}{2}\right)^{t/24,100}$.

(a) Determine the initial quantity (when $t = 0$).

(b) Determine the quantity present after 75,000 years.

(c) Use a graphing utility to graph the function over the interval $t = 0$ to $t = 150,000$.

74. RADIOACTIVE DECAY Let Q represent a mass of carbon 14 (^{14}C) (in grams), whose half-life is 5715 years. The quantity of carbon 14 present after t years is $Q = 10\left(\frac{1}{2}\right)^{t/5715}$.

(a) Determine the initial quantity (when $t = 0$).

(b) Determine the quantity present after 2000 years.

(c) Sketch the graph of this function over the interval $t = 0$ to $t = 10,000$.

75. DEPRECIATION After t years, the value of a wheelchair conversion van that originally cost $30,500 depreciates so that each year it is worth $\frac{7}{8}$ of its value for the previous year.

(a) Find a model for $V(t)$, the value of the van after t years.

(b) Determine the value of the van 4 years after it was purchased.

76. DRUG CONCENTRATION Immediately following an injection, the concentration of a drug in the bloodstream is 300 milligrams per milliliter. After t hours, the concentration is 75% of the level of the previous hour.

(a) Find a model for $C(t)$, the concentration of the drug after t hours.

(b) Determine the concentration of the drug after 8 hours.

EXPLORATION

TRUE OR FALSE? In Exercises 77 and 78, determine whether the statement is true or false. Justify your answer.

77. The line $y = -2$ is an asymptote for the graph of $f(x) = 10^x - 2$.

78. $e = \dfrac{271,801}{99,990}$

THINK ABOUT IT In Exercises 79–82, use properties of exponents to determine which functions (if any) are the same.

79. $f(x) = 3^{x-2}$
$g(x) = 3^x - 9$
$h(x) = \frac{1}{9}(3^x)$

80. $f(x) = 4^x + 12$
$g(x) = 2^{2x+6}$
$h(x) = 64(4^x)$

81. $f(x) = 16(4^{-x})$
$g(x) = \left(\frac{1}{4}\right)^{x-2}$
$h(x) = 16(2^{-2x})$

82. $f(x) = e^{-x} + 3$
$g(x) = e^{3-x}$
$h(x) = -e^{x-3}$

83. Graph the functions given by $y = 3^x$ and $y = 4^x$ and use the graphs to solve each inequality.

(a) $4^x < 3^x$ (b) $4^x > 3^x$

 84. Use a graphing utility to graph each function. Use the graph to find where the function is increasing and decreasing, and approximate any relative maximum or minimum values.

(a) $f(x) = x^2 e^{-x}$ (b) $g(x) = x2^{3-x}$

 85. GRAPHICAL ANALYSIS Use a graphing utility to graph $y_1 = (1 + 1/x)^x$ and $y_2 = e$ in the same viewing window. Using the *trace* feature, explain what happens to the graph of y_1 as x increases.

 86. GRAPHICAL ANALYSIS Use a graphing utility to graph

$$f(x) = \left(1 + \frac{0.5}{x}\right)^x \quad \text{and} \quad g(x) = e^{0.5}$$

in the same viewing window. What is the relationship between f and g as x increases and decreases without bound?

 87. GRAPHICAL ANALYSIS Use a graphing utility to graph each pair of functions in the same viewing window. Describe any similarities and differences in the graphs.

(a) $y_1 = 2^x, y_2 = x^2$ (b) $y_1 = 3^x, y_2 = x^3$

88. THINK ABOUT IT Which functions are exponential?

(a) $3x$ (b) $3x^2$ (c) 3^x (d) 2^{-x}

89. COMPOUND INTEREST Use the formula

$$A = P\left(1 + \frac{r}{n}\right)^{nt}$$

to calculate the balance of an account when $P = \$3000$, $r = 6\%$, and $t = 10$ years, and compounding is done (a) by the day, (b) by the hour, (c) by the minute, and (d) by the second. Does increasing the number of compoundings per year result in unlimited growth of the balance of the account? Explain.

90. CAPSTONE The figure shows the graphs of $y = 2^x$, $y = e^x$, $y = 10^x$, $y = 2^{-x}$, $y = e^{-x}$, and $y = 10^{-x}$. Match each function with its graph. [The graphs are labeled (a) through (f).] Explain your reasoning.

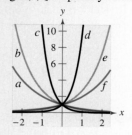

PROJECT: POPULATION PER SQUARE MILE To work an extended application analyzing the population per square mile of the United States, visit this text's website at *academic.cengage.com*. (Data Source: U.S. Census Bureau)

3.2 LOGARITHMIC FUNCTIONS AND THEIR GRAPHS

What you should learn

- Recognize and evaluate logarithmic functions with base a.
- Graph logarithmic functions.
- Recognize, evaluate, and graph natural logarithmic functions.
- Use logarithmic functions to model and solve real-life problems.

Why you should learn it

Logarithmic functions are often used to model scientific observations. For instance, in Exercise 97 on page 236, a logarithmic function is used to model human memory.

Logarithmic Functions

In Section 1.9, you studied the concept of an inverse function. There, you learned that if a function is one-to-one—that is, if the function has the property that no horizontal line intersects the graph of the function more than once—the function must have an inverse function. By looking back at the graphs of the exponential functions introduced in Section 3.1, you will see that every function of the form $f(x) = a^x$ passes the Horizontal Line Test and therefore must have an inverse function. This inverse function is called the **logarithmic function with base a.**

Definition of Logarithmic Function with Base a

For $x > 0$, $a > 0$, and $a \neq 1$,

$$y = \log_a x \text{ if and only if } x = a^y.$$

The function given by

$$f(x) = \log_a x \qquad \text{Read as "log base } a \text{ of } x\text{."}$$

is called the **logarithmic function with base a.**

The equations

$$y = \log_a x \qquad \text{and} \qquad x = a^y$$

are equivalent. The first equation is in logarithmic form and the second is in exponential form. For example, the logarithmic equation $2 = \log_3 9$ can be rewritten in exponential form as $9 = 3^2$. The exponential equation $5^3 = 125$ can be rewritten in logarithmic form as $\log_5 125 = 3$.

When evaluating logarithms, remember that *a logarithm is an exponent.* This means that $\log_a x$ is the exponent to which a must be raised to obtain x. For instance, $\log_2 8 = 3$ because 2 must be raised to the third power to get 8.

Example 1   Evaluating Logarithms

Use the definition of logarithmic function to evaluate each logarithm at the indicated value of x.

a. $f(x) = \log_2 x, \quad x = 32$ **b.** $f(x) = \log_3 x, \quad x = 1$

c. $f(x) = \log_4 x, \quad x = 2$ **d.** $f(x) = \log_{10} x, \quad x = \frac{1}{100}$

Solution

a. $f(32) = \log_2 32 = 5$ because $2^5 = 32$.

b. $f(1) = \log_3 1 = 0$ because $3^0 = 1$.

c. $f(2) = \log_4 2 = \frac{1}{2}$ because $4^{1/2} = \sqrt{4} = 2$.

d. $f\left(\frac{1}{100}\right) = \log_{10} \frac{1}{100} = -2$ because $10^{-2} = \frac{1}{10^2} = \frac{1}{100}$.

CHECK **Point**   Now try Exercise 23.

The logarithmic function with base 10 is called the **common logarithmic function.** It is denoted by $\log_{10}$ or simply by log. On most calculators, this function is denoted by (LOG). Example 2 shows how to use a calculator to evaluate common logarithmic functions. You will learn how to use a calculator to calculate logarithms to any base in the next section.

Example 2 Evaluating Common Logarithms on a Calculator

Use a calculator to evaluate the function given by $f(x) = \log x$ at each value of x.

a. $x = 10$ **b.** $x = \frac{1}{3}$ **c.** $x = 2.5$ **d.** $x = -2$

Solution

Function Value	Graphing Calculator Keystrokes	Display
a. $f(10) = \log 10$	(LOG) 10 (ENTER)	1
b. $f\left(\frac{1}{3}\right) = \log \frac{1}{3}$	(LOG) (1 ÷ 3) (ENTER)	-0.4771213
c. $f(2.5) = \log 2.5$	(LOG) 2.5 (ENTER)	0.3979400
d. $f(-2) = \log(-2)$	(LOG) (−) 2 (ENTER)	ERROR

Note that the calculator displays an error message (or a complex number) when you try to evaluate $\log(-2)$. The reason for this is that there is no real number power to which 10 can be raised to obtain -2.

CHECK *Point* Now try Exercise 29.

The following properties follow directly from the definition of the logarithmic function with base a.

Properties of Logarithms

1. $\log_a 1 = 0$ because $a^0 = 1$.

2. $\log_a a = 1$ because $a^1 = a$.

3. $\log_a a^x = x$ and $a^{\log_a x} = x$ Inverse Properties

4. If $\log_a x = \log_a y$, then $x = y$.   One-to-One Property

---

Example 3 Using Properties of Logarithms

a. Simplify: $\log_4 1$ **b.** Simplify: $\log_{\sqrt{7}} \sqrt{7}$ **c.** Simplify: $6^{\log_6 20}$

Solution

a. Using Property 1, it follows that $\log_4 1 = 0$.

b. Using Property 2, you can conclude that $\log_{\sqrt{7}} \sqrt{7} = 1$.

c. Using the Inverse Property (Property 3), it follows that $6^{\log_6 20} = 20$.

CHECK *Point* Now try Exercise 33.

You can use the One-to-One Property (Property 4) to solve simple logarithmic equations, as shown in Example 4.

Example 4   Using the One-to-One Property

a. $\log_3 x = \log_3 12$ Original equation

   $x = 12$ One-to-One Property

b. $\log(2x + 1) = \log 3x \implies 2x + 1 = 3x \implies 1 = x$

c. $\log_4(x^2 - 6) = \log_4 10 \implies x^2 - 6 = 10 \implies x^2 = 16 \implies x = \pm 4$

CHECK Point ▶ Now try Exercise 85.

Graphs of Logarithmic Functions

To sketch the graph of $y = \log_a x$, you can use the fact that the graphs of inverse functions are reflections of each other in the line $y = x$.

Example 5   Graphs of Exponential and Logarithmic Functions

In the same coordinate plane, sketch the graph of each function.

a. $f(x) = 2^x$ **b.** $g(x) = \log_2 x$

Solution

a. For $f(x) = 2^x$, construct a table of values. By plotting these points and connecting them with a smooth curve, you obtain the graph shown in Figure 3.14.

x	-2	-1	0	1	2	3
$f(x) = 2^x$	$\frac{1}{4}$	$\frac{1}{2}$	1	2	4	8

b. Because $g(x) = \log_2 x$ is the inverse function of $f(x) = 2^x$, the graph of g is obtained by plotting the points $(f(x), x)$ and connecting them with a smooth curve. The graph of g is a reflection of the graph of f in the line $y = x$, as shown in Figure 3.14.

CHECK Point ▶ Now try Exercise 37.

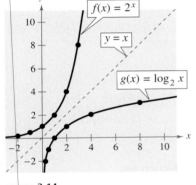

FIGURE 3.14

Example 6   Sketching the Graph of a Logarithmic Function

Sketch the graph of the common logarithmic function $f(x) = \log x$. Identify the vertical asymptote.

Solution

Begin by constructing a table of values. Note that some of the values can be obtained without a calculator by using the Inverse Property of Logarithms. Others require a calculator. Next, plot the points and connect them with a smooth curve, as shown in Figure 3.15. The vertical asymptote is $x = 0$ (y-axis).

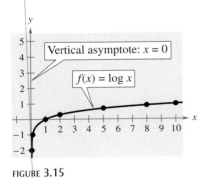

FIGURE 3.15

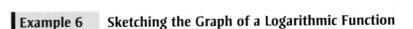

	Without calculator				With calculator		
x	$\frac{1}{100}$	$\frac{1}{10}$	1	10	2	5	8
$f(x) = \log x$	-2	-1	0	1	0.301	0.699	0.903

CHECK Point ▶ Now try Exercise 43.

The nature of the graph in Figure 3.15 is typical of functions of the form $f(x) = \log_a x, a > 1$. They have one x-intercept and one vertical asymptote. Notice how slowly the graph rises for $x > 1$. The basic characteristics of logarithmic graphs are summarized in Figure 3.16.

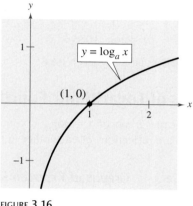

FIGURE **3.16**

Graph of $y = \log_a x, a > 1$

- Domain: $(0, \infty)$
- Range: $(-\infty, \infty)$
- x-intercept: $(1, 0)$
- Increasing
- One-to-one, therefore has an inverse function
- y-axis is a vertical asymptote ($\log_a x \to -\infty$ as $x \to 0^+$).
- Continuous
- Reflection of graph of $y = a^x$ about the line $y = x$

The basic characteristics of the graph of $f(x) = a^x$ are shown below to illustrate the inverse relation between $f(x) = a^x$ and $g(x) = \log_a x$.

- Domain: $(-\infty, \infty)$ • Range: $(0, \infty)$
- y-intercept: $(0, 1)$ • x-axis is a horizontal asymptote ($a^x \to 0$ as $x \to -\infty$).

In the next example, the graph of $y = \log_a x$ is used to sketch the graphs of functions of the form $f(x) = b \pm \log_a(x + c)$. Notice how a horizontal shift of the graph results in a horizontal shift of the vertical asymptote.

Example 7   Shifting Graphs of Logarithmic Functions

The graph of each of the functions is similar to the graph of $f(x) = \log x$.

a. Because $g(x) = \log(x - 1) = f(x - 1)$, the graph of g can be obtained by shifting the graph of f one unit to the right, as shown in Figure 3.17.

b. Because $h(x) = 2 + \log x = 2 + f(x)$, the graph of h can be obtained by shifting the graph of f two units upward, as shown in Figure 3.18.

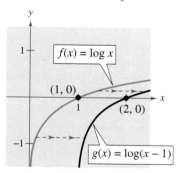

FIGURE **3.17**

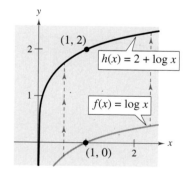

FIGURE **3.18**

CHECK*Point*   Now try Exercise 45.

Properties of natural log

1) $\ln 1 = 0$

2) $\ln e = 1$

3) $\ln e^x = x$

4) $\ln x = \ln y$, then
   $x = y$

The Natural Logarithmic Function

By looking back at the graph of the natural exponential function introduced on page 220 in Section 3.1, you will see that $f(x) = e^x$ is one-to-one and so has an inverse function. This inverse function is called the **natural logarithmic function** and is denoted by the special symbol ln x, read as "the natural log of x" or "el en of x." Note that the natural logarithm is written without a base. The base is understood to be e.

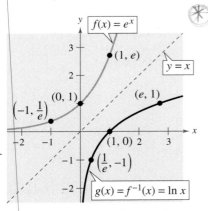

Reflection of graph of $f(x) = e^x$ about the line $y = x$

FIGURE 3.19

The Natural Logarithmic Function

The function defined by

$$f(x) = \log_e x = \ln x, \quad x > 0$$

is called the **natural logarithmic function.**

The definition above implies that the natural logarithmic function and the natural exponential function are inverse functions of each other. So, every logarithmic equation can be written in an equivalent exponential form, and every exponential equation can be written in logarithmic form. That is, $y = \ln x$ and $x = e^y$ are equivalent equations.

Because the functions given by $f(x) = e^x$ and $g(x) = \ln x$ are inverse functions of each other, their graphs are reflections of each other in the line $y = x$. This reflective property is illustrated in Figure 3.19.

On most calculators, the natural logarithm is denoted by $\boxed{\text{LN}}$, as illustrated in Example 8.

Example 8   Evaluating the Natural Logarithmic Function

Use a calculator to evaluate the function given by $f(x) = \ln x$ for each value of x.

a. $x = 2$

b. $x = 0.3$

c. $x = -1$

d. $x = 1 + \sqrt{2}$

Solution

Function Value	Graphing Calculator Keystrokes	Display
a. $f(2) = \ln 2$	$\boxed{\text{LN}}$ 2 $\boxed{\text{ENTER}}$	0.6931472
b. $f(0.3) = \ln 0.3$	$\boxed{\text{LN}}$ .3 $\boxed{\text{ENTER}}$	−1.2039728
c. $f(-1) = \ln(-1)$	$\boxed{\text{LN}}$ $\boxed{(-)}$ 1 $\boxed{\text{ENTER}}$	ERROR
d. $f\left(1 + \sqrt{2}\right) = \ln\left(1 + \sqrt{2}\right)$	$\boxed{\text{LN}}$ $\boxed{(}$ 1 $\boxed{+}$ $\boxed{\sqrt{}}$ 2 $\boxed{)}$ $\boxed{\text{ENTER}}$	0.8813736

CHECK **Point**   Now try Exercise 67.

In Example 8, be sure you see that $\ln(-1)$ gives an error message on most calculators. (Some calculators may display a complex number.) This occurs because the domain of $\ln x$ is the set of positive real numbers (see Figure 3.19). So, $\ln(-1)$ is undefined.

The four properties of logarithms listed on page 228 are also valid for natural logarithms.

⚠ **WARNING / CAUTION**

Notice that as with every other logarithmic function, the domain of the natural logarithmic function is the set of *positive real numbers*—be sure you see that ln x is not defined for zero or for negative numbers.

> ## Properties of Natural Logarithms
> 1. $\ln 1 = 0$ because $e^0 = 1$.
> 2. $\ln e = 1$ because $e^1 = e$.
> 3. $\ln e^x = x$ and $e^{\ln x} = x$ Inverse Properties
> 4. If $\ln x = \ln y$, then $x = y$.   One-to-One Property

Example 9   Using Properties of Natural Logarithms

Use the properties of natural logarithms to simplify each expression.

a. $\ln \dfrac{1}{e}$ **b.** $e^{\ln 5}$ **c.** $\dfrac{\ln 1}{3}$ **d.** $2 \ln e$

Solution

a. $\ln \dfrac{1}{e} = \ln e^{-1} = -1$ Inverse Property   **b.** $e^{\ln 5} = 5$ Inverse Property

c. $\dfrac{\ln 1}{3} = \dfrac{0}{3} = 0$ Property 1   **d.** $2 \ln e = 2(1) = 2$ Property 2

CHECK Point ⟩ Now try Exercise 71.

Example 10   Finding the Domains of Logarithmic Functions

Find the domain of each function.

a. $f(x) = \ln(x - 2)$ **b.** $g(x) = \ln(2 - x)$ **c.** $h(x) = \ln x^2$

Solution

a. Because $\ln(x - 2)$ is defined only if $x - 2 > 0$, it follows that the domain of f is $(2, \infty)$. The graph of f is shown in Figure 3.20.

b. Because $\ln(2 - x)$ is defined only if $2 - x > 0$, it follows that the domain of g is $(-\infty, 2)$. The graph of g is shown in Figure 3.21.

c. Because $\ln x^2$ is defined only if $x^2 > 0$, it follows that the domain of h is all real numbers except $x = 0$. The graph of h is shown in Figure 3.22.

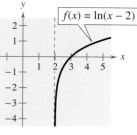

FIGURE 3.20

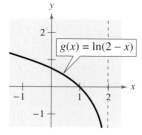

FIGURE 3.21

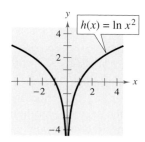

FIGURE 3.22

CHECK Point ⟩ Now try Exercise 75.

Application

Example 11 Human Memory Model

Students participating in a psychology experiment attended several lectures on a subject and were given an exam. Every month for a year after the exam, the students were retested to see how much of the material they remembered. The average scores for the group are given by the *human memory model* $f(t) = 75 - 6\ln(t+1)$, $0 \le t \le 12$, where t is the time in months.

a. What was the average score on the original $(t = 0)$ exam?

b. What was the average score at the end of $t = 2$ months?

c. What was the average score at the end of $t = 6$ months?

Algebraic Solution

a. The original average score was

$$\begin{aligned} f(0) &= 75 - 6\ln(0+1) && \text{Substitute 0 for } t. \\ &= 75 - 6\ln 1 && \text{Simplify.} \\ &= 75 - 6(0) && \text{Property of natural logarithms} \\ &= 75. && \text{Solution} \end{aligned}$$

b. After 2 months, the average score was

$$\begin{aligned} f(2) &= 75 - 6\ln(2+1) && \text{Substitute 2 for } t. \\ &= 75 - 6\ln 3 && \text{Simplify.} \\ &\approx 75 - 6(1.0986) && \text{Use a calculator.} \\ &\approx 68.4. && \text{Solution} \end{aligned}$$

c. After 6 months, the average score was

$$\begin{aligned} f(6) &= 75 - 6\ln(6+1) && \text{Substitute 6 for } t. \\ &= 75 - 6\ln 7 && \text{Simplify.} \\ &\approx 75 - 6(1.9459) && \text{Use a calculator.} \\ &\approx 63.3. && \text{Solution} \end{aligned}$$

Graphical Solution

Use a graphing utility to graph the model $y = 75 - 6\ln(x+1)$. Then use the *value* or *trace* feature to approximate the following.

a. When $x = 0$, $y = 75$ (see Figure 3.23). So, the original average score was 75.

b. When $x = 2$, $y \approx 68.4$ (see Figure 3.24). So, the average score after 2 months was about 68.4.

c. When $x = 6$, $y \approx 63.3$ (see Figure 3.25). So, the average score after 6 months was about 63.3.

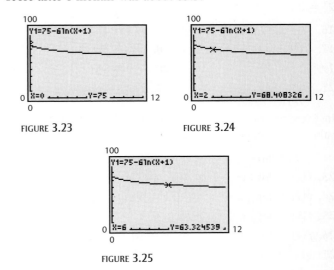

FIGURE 3.23

FIGURE 3.24

FIGURE 3.25

CHECK Point Now try Exercise 97.

Classroom Discussion

Analyzing a Human Memory Model Use a graphing utility to determine the time in months when the average score in Example 11 was 60. Explain your method of solving the problem. Describe another way that you can use a graphing utility to determine the answer.

3.2　EXERCISES

See www.CalcChat.com for worked-out solutions to odd-numbered exercises.

VOCABULARY: Fill in the blanks.

1. The inverse function of the exponential function given by $f(x) = a^x$ is called the _____ function with base a.

2. The common logarithmic function has base _____ .

3. The logarithmic function given by $f(x) = \ln x$ is called the _____ logarithmic function and has base _____.

4. The Inverse Properties of logarithms and exponentials state that $\log_a a^x = x$ and _____.

5. The One-to-One Property of natural logarithms states that if $\ln x = \ln y$, then _____.

6. The domain of the natural logarithmic function is the set of _____ _____ _____ .

SKILLS AND APPLICATIONS

In Exercises 7–14, write the logarithmic equation in exponential form. For example, the exponential form of $\log_5 25 = 2$ is $5^2 = 25$.

7. $\log_4 16 = 2$

8. $\log_7 343 = 3$

9. $\log_9 \frac{1}{81} = -2$

10. $\log \frac{1}{1000} = -3$

11. $\log_{32} 4 = \frac{2}{5}$

12. $\log_{16} 8 = \frac{3}{4}$

13. $\log_{64} 8 = \frac{1}{2}$

14. $\log_8 4 = \frac{2}{3}$

In Exercises 15–22, write the exponential equation in logarithmic form. For example, the logarithmic form of $2^3 = 8$ is $\log_2 8 = 3$.

15. $5^3 = 125$

16. $13^2 = 169$

17. $81^{1/4} = 3$

18. $9^{3/2} = 27$

19. $6^{-2} = \frac{1}{36}$

20. $4^{-3} = \frac{1}{64}$

21. $24^0 = 1$

22. $10^{-3} = 0.001$

In Exercises 23–28, evaluate the function at the indicated value of x without using a calculator.

Function	Value
23. $f(x) = \log_2 x$	$x = 64$
24. $f(x) = \log_{25} x$	$x = 5$
25. $f(x) = \log_8 x$	$x = 1$
26. $f(x) = \log x$	$x = 10$
27. $g(x) = \log_a x$	$x = a^2$
28. $g(x) = \log_b x$	$x = b^{-3}$

In Exercises 29–32, use a calculator to evaluate $f(x) = \log x$ at the indicated value of x. Round your result to three decimal places.

29. $x = \frac{7}{8}$

30. $x = \frac{1}{500}$

31. $x = 12.5$

32. $x = 96.75$

In Exercises 33–36, use the properties of logarithms to simplify the expression.

33. $\log_{11} 11^7 = 7$

34. $\log_{3.2} 1 = 0$

35. $\log_\pi \pi = 1$

36. $9^{\log_9 15} = 15$

In Exercises 37–44, find the domain, x-intercept, and vertical asymptote of the logarithmic function and sketch its graph.

37. $f(x) = \log_4 x$

38. $g(x) = \log_6 x$

39. $y = -\log_3 x + 2$

40. $h(x) = \log_4(x - 3)$

41. $f(x) = -\log_6(x + 2)$

42. $y = \log_5(x - 1) + 4$

43. $y = \log\left(\frac{x}{7}\right)$

44. $y = \log(-x)$

In Exercises 45–50, use the graph of $g(x) = \log_3 x$ to match the given function with its graph. Then describe the relationship between the graphs of f and g. [The graphs are labeled (a), (b), (c), (d), (e), and (f).]

(a)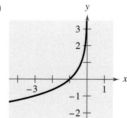

(b)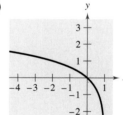

(c)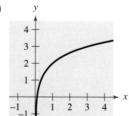

(d)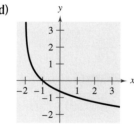

(e)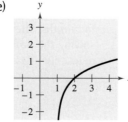

(f)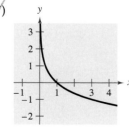

45. $f(x) = \log_3 x + 2$

46. $f(x) = -\log_3 x$

47. $f(x) = -\log_3(x + 2)$

48. $f(x) = \log_3(x - 1)$

49. $f(x) = \log_3(1 - x)$

50. $f(x) = -\log_3(-x)$

In Exercises 51–58, write the logarithmic equation in exponential form.

51. $\ln \frac{1}{2} = -0.693 \ldots$

52. $\ln \frac{2}{5} = -0.916 \ldots$

53. $\ln 7 = 1.945 \ldots$

54. $\ln 10 = 2.302 \ldots$

55. $\ln 250 = 5.521 \ldots$

56. $\ln 1084 = 6.988 \ldots$

57. $\ln 1 = 0$

58. $\ln e = 1$

In Exercises 59–66, write the exponential equation in logarithmic form.

59. $e^4 = 54.598 \ldots$

60. $e^2 = 7.3890 \ldots$

61. $e^{1/2} = 1.6487 \ldots$

62. $e^{1/3} = 1.3956 \ldots$

63. $e^{-0.9} = 0.406 \ldots$

64. $e^{-4.1} = 0.0165 \ldots$

65. $e^x = 4$

66. $e^{2x} = 3$

 In Exercises 67–70, use a calculator to evaluate the function at the indicated value of x. Round your result to three decimal places.

Function	Value
67. $f(x) = \ln x$	$x = 18.42$
68. $f(x) = 3 \ln x$	$x = 0.74$
69. $g(x) = 8 \ln x$	$x = 0.05$
70. $g(x) = -\ln x$	$x = \frac{1}{2}$

In Exercises 71–74, evaluate $g(x) = \ln x$ at the indicated value of x without using a calculator.

71. $x = e^5$

72. $x = e^{-4}$

73. $x = e^{-5/6}$

74. $x = e^{-5/2}$

In Exercises 75–78, find the domain, x-intercept, and vertical asymptote of the logarithmic function and sketch its graph.

75. $f(x) = \ln(x - 4)$

76. $h(x) = \ln(x + 5)$

77. $g(x) = \ln(-x)$

78. $f(x) = \ln(3 - x)$

 In Exercises 79–84, use a graphing utility to graph the function. Be sure to use an appropriate viewing window.

79. $f(x) = \log(x + 9)$

80. $f(x) = \log(x - 6)$

81. $f(x) = \ln(x - 1)$

82. $f(x) = \ln(x + 2)$

83. $f(x) = \ln x + 8$

84. $f(x) = 3 \ln x - 1$

In Exercises 85–92, use the One-to-One Property to solve the equation for x.

85. $\log_5(x + 1) = \log_5 6$

86. $\log_2(x - 3) = \log_2 9$

87. $\log(2x + 1) = \log 15$

88. $\log(5x + 3) = \log 12$

89. $\ln(x + 4) = \ln 12$

90. $\ln(x - 7) = \ln 7$

91. $\ln(x^2 - 2) = \ln 23$

92. $\ln(x^2 - x) = \ln 6$

93. MONTHLY PAYMENT   The model

$$t = 16.625 \ln\left(\frac{x}{x - 750}\right), \quad x > 750$$

approximates the length of a home mortgage of $150,000 at 6% in terms of the monthly payment. In the model, t is the length of the mortgage in years and x is the monthly payment in dollars.

(a) Use the model to approximate the lengths of a $150,000 mortgage at 6% when the monthly payment is $897.72 and when the monthly payment is $1659.24.

(b) Approximate the total amounts paid over the term of the mortgage with a monthly payment of $897.72 and with a monthly payment of $1659.24.

(c) Approximate the total interest charges for a monthly payment of $897.72 and for a monthly payment of $1659.24.

(d) What is the vertical asymptote for the model? Interpret its meaning in the context of the problem.

94. COMPOUND INTEREST   A principal P, invested at $5\frac{1}{2}\%$ and compounded continuously, increases to an amount K times the original principal after t years, where t is given by $t = (\ln K)/0.055$.

(a) Complete the table and interpret your results.

K	1	2	4	6	8	10	12
t							

(b) Sketch a graph of the function.

95. CABLE TELEVISION   The numbers of cable television systems C (in thousands) in the United States from 2001 through 2006 can be approximated by the model

$$C = 10.355 - 0.298t \ln t, \quad 1 \le t \le 6$$

where t represents the year, with $t = 1$ corresponding to 2001.   (Source: Warren Communication News)

(a) Complete the table.

t	1	2	3	4	5	6
C						

 (b) Use a graphing utility to graph the function.

(c) Can the model be used to predict the numbers of cable television systems beyond 2006? Explain.

96. POPULATION   The time t in years for the world population to double if it is increasing at a continuous rate of r is given by $t = (\ln 2)/r$.

(a) Complete the table and interpret your results.

r	0.005	0.010	0.015	0.020	0.025	0.030
t						

(b) Use a graphing utility to graph the function.

97. HUMAN MEMORY MODEL   Students in a mathematics class were given an exam and then retested monthly with an equivalent exam. The average scores for the class are given by the human memory model $f(t) = 80 - 17 \log(t + 1), 0 \le t \le 12$, where t is the time in months.

 (a) Use a graphing utility to graph the model over the specified domain.

(b) What was the average score on the original exam $(t = 0)$?

(c) What was the average score after 4 months?

(d) What was the average score after 10 months?

98. SOUND INTENSITY   The relationship between the number of decibels β and the intensity of a sound I in watts per square meter is

$$\beta = 10 \log\left(\frac{I}{10^{-12}}\right).$$

(a) Determine the number of decibels of a sound with an intensity of 1 watt per square meter.

(b) Determine the number of decibels of a sound with an intensity of 10^{-2} watt per square meter.

(c) The intensity of the sound in part (a) is 100 times as great as that in part (b). Is the number of decibels 100 times as great? Explain.

EXPLORATION

TRUE OR FALSE?   In Exercises 99 and 100, determine whether the statement is true or false. Justify your answer.

99. You can determine the graph of $f(x) = \log_6 x$ by graphing $g(x) = 6^x$ and reflecting it about the x-axis.

100. The graph of $f(x) = \log_3 x$ contains the point $(27, 3)$.

In Exercises 101–104, sketch the graphs of f and g and describe the relationship between the graphs of f and g. What is the relationship between the functions f and g?

101. $f(x) = 3^x$, $g(x) = \log_3 x$

102. $f(x) = 5^x$, $g(x) = \log_5 x$

103. $f(x) = e^x$, $g(x) = \ln x$

104. $f(x) = 8^x$, $g(x) = \log_8 x$

105. THINK ABOUT IT   Complete the table for $f(x) = 10^x$.

x	-2	-1	0	1	2
$f(x)$					

Complete the table for $f(x) = \log x$.

x	$\frac{1}{100}$	$\frac{1}{10}$	1	10	100
$f(x)$					

Compare the two tables. What is the relationship between $f(x) = 10^x$ and $f(x) = \log x$?

106. GRAPHICAL ANALYSIS   Use a graphing utility to graph f and g in the same viewing window and determine which is increasing at the greater rate as x approaches $+\infty$. What can you conclude about the rate of growth of the natural logarithmic function?

(a) $f(x) = \ln x$, $g(x) = \sqrt{x}$

(b) $f(x) = \ln x$, $g(x) = \sqrt[4]{x}$

107. (a) Complete the table for the function given by $f(x) = (\ln x)/x$.

x	1	5	10	10^2	10^4	10^6
$f(x)$						

(b) Use the table in part (a) to determine what value $f(x)$ approaches as x increases without bound.

(c) Use a graphing utility to confirm the result of part (b).

108. CAPSTONE   The table of values was obtained by evaluating a function. Determine which of the statements may be true and which must be false.

x	y
1	0
2	1
8	3

(a) y is an exponential function of x.

(b) y is a logarithmic function of x.

(c) x is an exponential function of y.

(d) y is a linear function of x.

109. WRITING   Explain why $\log_a x$ is defined only for $0 < a < 1$ and $a > 1$.

In Exercises 110 and 111, (a) use a graphing utility to graph the function, (b) use the graph to determine the intervals in which the function is increasing and decreasing, and (c) approximate any relative maximum or minimum values of the function.

110. $f(x) = |\ln x|$ **111.** $h(x) = \ln(x^2 + 1)$

3.3   PROPERTIES OF LOGARITHMS

What you should learn

• Use the change-of-base formula to rewrite and evaluate logarithmic expressions.

• Use properties of logarithms to evaluate or rewrite logarithmic expressions.

• Use properties of logarithms to expand or condense logarithmic expressions.

• Use logarithmic functions to model and solve real-life problems.

Why you should learn it

Logarithmic functions can be used to model and solve real-life problems. For instance, in Exercises 87–90 on page 242, a logarithmic function is used to model the relationship between the number of decibels and the intensity of a sound.

Dynamic Graphics / Jupiter Images

Change of Base

Most calculators have only two types of log keys, one for common logarithms (base 10) and one for natural logarithms (base e). Although common logarithms and natural logarithms are the most frequently used, you may occasionally need to evaluate logarithms with other bases. To do this, you can use the following **change-of-base formula.**

Change-of-Base Formula

Let a, b, and x be positive real numbers such that $a \neq 1$ and $b \neq 1$. Then $\log_a x$ can be converted to a different base as follows.

Base b	*Base 10*	*Base e*
$\log_a x = \dfrac{\log_b x}{\log_b a}$	$\log_a x = \dfrac{\log x}{\log a}$	$\log_a x = \dfrac{\ln x}{\ln a}$

One way to look at the change-of-base formula is that logarithms with base a are simply *constant multiples* of logarithms with base b. The constant multiplier is $1/(\log_b a)$.

Example 1   Changing Bases Using Common Logarithms

a. $\log_4 25 = \dfrac{\log 25}{\log 4}$ $\log_a x = \dfrac{\log x}{\log a}$

$\approx \dfrac{1.39794}{0.60206}$ Use a calculator.

≈ 2.3219 Simplify.

b. $\log_2 12 = \dfrac{\log 12}{\log 2} \approx \dfrac{1.07918}{0.30103} \approx 3.5850$

CHECK *Point*   Now try Exercise 7(a).

Example 2   Changing Bases Using Natural Logarithms

a. $\log_4 25 = \dfrac{\ln 25}{\ln 4}$ $\log_a x = \dfrac{\ln x}{\ln a}$

$\approx \dfrac{3.21888}{1.38629}$ Use a calculator.

≈ 2.3219 Simplify.

b. $\log_2 12 = \dfrac{\ln 12}{\ln 2} \approx \dfrac{2.48491}{0.69315} \approx 3.5850$

CHECK *Point*   Now try Exercise 7(b).

Properties of Logarithms

You know from the preceding section that the logarithmic function with base a is the *inverse function* of the exponential function with base a. So, it makes sense that the properties of exponents should have corresponding properties involving logarithms. For instance, the exponential property $a^0 = 1$ has the corresponding logarithmic property $\log_a 1 = 0$.

Properties of Logarithms

Let a be a positive number such that $a \neq 1$, and let n be a real number. If u and v are positive real numbers, the following properties are true.

	Logarithm with Base a	*Natural Logarithm*
1. Product Property:	$\log_a(uv) = \log_a u + \log_a v$	$\ln(uv) = \ln u + \ln v$
2. Quotient Property:	$\log_a \dfrac{u}{v} = \log_a u - \log_a v$	$\ln \dfrac{u}{v} = \ln u - \ln v$
3. Power Property:	$\log_a u^n = n \log_a u$	$\ln u^n = n \ln u$

For proofs of the properties listed above, see Proofs in Mathematics on page 276.

Example 3   Using Properties of Logarithms

Write each logarithm in terms of $\ln 2$ and $\ln 3$.

a. $\ln 6$ **b.** $\ln \dfrac{2}{27}$

Solution

a. $\ln 6 = \ln(2 \cdot 3)$ Rewrite 6 as $2 \cdot 3$.

        $= \ln 2 + \ln 3$ Product Property

b. $\ln \dfrac{2}{27} = \ln 2 - \ln 27$ Quotient Property

          $= \ln 2 - \ln 3^3$ Rewrite 27 as 3^3.

          $= \ln 2 - 3 \ln 3$ Power Property

CHECK**Point** Now try Exercise 27.

Example 4   Using Properties of Logarithms

Find the exact value of each expression without using a calculator.

a. $\log_5 \sqrt[3]{5}$ **b.** $\ln e^6 - \ln e^2$

Solution

a. $\log_5 \sqrt[3]{5} = \log_5 5^{1/3} = \frac{1}{3} \log_5 5 = \frac{1}{3}(1) = \frac{1}{3}$

b. $\ln e^6 - \ln e^2 = \ln \dfrac{e^6}{e^2} = \ln e^4 = 4 \ln e = 4(1) = 4$

CHECK**Point** Now try Exercise 29.

Rewriting Logarithmic Expressions

The properties of logarithms are useful for rewriting logarithmic expressions in forms that simplify the operations of algebra. This is true because these properties convert complicated products, quotients, and exponential forms into simpler sums, differences, and products, respectively.

Example 5   Expanding Logarithmic Expressions

Expand each logarithmic expression.

a. $\log_4 5x^3y$ **b.** $\ln \dfrac{\sqrt{3x-5}}{7}$

Solution

a.
$$\log_4 5x^3y = \log_4 5 + \log_4 x^3 + \log_4 y \qquad \text{Product Property}$$
$$= \log_4 5 + 3\log_4 x + \log_4 y \qquad \text{Power Property}$$

b.
$$\ln \frac{\sqrt{3x-5}}{7} = \ln \frac{(3x-5)^{1/2}}{7} \qquad \text{Rewrite using rational exponent.}$$
$$= \ln(3x-5)^{1/2} - \ln 7 \qquad \text{Quotient Property}$$
$$= \frac{1}{2}\ln(3x-5) - \ln 7 \qquad \text{Power Property}$$

CHECK *Point*   Now try Exercise 53.

In Example 5, the properties of logarithms were used to *expand* logarithmic expressions. In Example 6, this procedure is reversed and the properties of logarithms are used to *condense* logarithmic expressions.

Example 6   Condensing Logarithmic Expressions

Condense each logarithmic expression.

a. $\frac{1}{2}\log x + 3\log(x+1)$ **b.** $2\ln(x+2) - \ln x$
c. $\frac{1}{3}[\log_2 x + \log_2(x+1)]$

Solution

a.
$$\frac{1}{2}\log x + 3\log(x+1) = \log x^{1/2} + \log(x+1)^3 \qquad \text{Power Property}$$
$$= \log\left[\sqrt{x}(x+1)^3\right] \qquad \text{Product Property}$$

b.
$$2\ln(x+2) - \ln x = \ln(x+2)^2 - \ln x \qquad \text{Power Property}$$
$$= \ln \frac{(x+2)^2}{x} \qquad \text{Quotient Property}$$

c.
$$\frac{1}{3}[\log_2 x + \log_2(x+1)] = \frac{1}{3}\{\log_2[x(x+1)]\} \qquad \text{Product Property}$$
$$= \log_2[x(x+1)]^{1/3} \qquad \text{Power Property}$$
$$= \log_2 \sqrt[3]{x(x+1)} \qquad \text{Rewrite with a radical.}$$

CHECK *Point*   Now try Exercise 75.

Algebra Help

You can review rewriting radicals and rational exponents in Appendix A.2.

Application

One method of determining how the x- and y-values for a set of nonlinear data are related is to take the natural logarithm of each of the x- and y-values. If the points are graphed and fall on a line, then you can determine that the x- and y-values are related by the equation

$$\ln y = m \ln x$$

where m is the slope of the line.

Example 7   Finding a Mathematical Model

The table shows the mean distance from the sun x and the period y (the time it takes a planet to orbit the sun) for each of the six planets that are closest to the sun. In the table, the mean distance is given in terms of astronomical units (where Earth's mean distance is defined as 1.0), and the period is given in years. Find an equation that relates y and x.

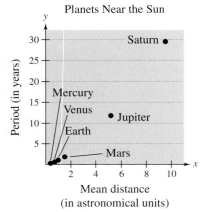

Planets Near the Sun

Period (in years)

Mean distance
(in astronomical units)

FIGURE 3.26

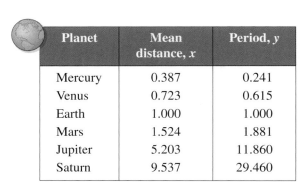

Planet	Mean distance, x	Period, y
Mercury	0.387	0.241
Venus	0.723	0.615
Earth	1.000	1.000
Mars	1.524	1.881
Jupiter	5.203	11.860
Saturn	9.537	29.460

Solution

The points in the table above are plotted in Figure 3.26. From this figure it is not clear how to find an equation that relates y and x. To solve this problem, take the natural logarithm of each of the x- and y-values in the table. This produces the following results.

Planet	Mercury	Venus	Earth	Mars	Jupiter	Saturn
$\ln x$	-0.949	-0.324	0.000	0.421	1.649	2.255
$\ln y$	-1.423	-0.486	0.000	0.632	2.473	3.383

Now, by plotting the points in the second table, you can see that all six of the points appear to lie in a line (see Figure 3.27). Choose any two points to determine the slope of the line. Using the two points $(0.421, 0.632)$ and $(0, 0)$, you can determine that the slope of the line is

$$m = \frac{0.632 - 0}{0.421 - 0} \approx 1.5 = \frac{3}{2}.$$

By the point-slope form, the equation of the line is $Y = \frac{3}{2}X$, where $Y = \ln y$ and $X = \ln x$. You can therefore conclude that $\ln y = \frac{3}{2}\ln x$.

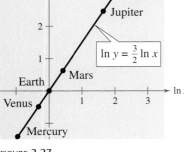

FIGURE 3.27

CHECK Point Now try Exercise 91.

3.3 EXERCISES

See www.CalcChat.com for worked-out solutions to odd-numbered exercises.

VOCABULARY

In Exercises 1–3, fill in the blanks.

1. To evaluate a logarithm to any base, you can use the _____ formula.

2. The change-of-base formula for base e is given by $\log_a x = $ _____.

3. You can consider $\log_a x$ to be a constant multiple of $\log_b x$; the constant multiplier is _____.

In Exercises 4–6, match the property of logarithms with its name.

4. $\log_a(uv) = \log_a u + \log_a v$ (a) Power Property

5. $\ln u^n = n \ln u$ (b) Quotient Property

6. $\log_a \dfrac{u}{v} = \log_a u - \log_a v$ (c) Product Property

SKILLS AND APPLICATIONS

In Exercises 7–14, rewrite the logarithm as a ratio of (a) common logarithms and (b) natural logarithms.

7. $\log_5 16$ **8.** $\log_3 47$

9. $\log_{1/5} x$ **10.** $\log_{1/3} x$

11. $\log_x \frac{3}{10}$ **12.** $\log_x \frac{3}{4}$

13. $\log_{2.6} x$ **14.** $\log_{7.1} x$

In Exercises 15–22, evaluate the logarithm using the change-of-base formula. Round your result to three decimal places.

15. $\log_3 7$ **16.** $\log_7 4$

17. $\log_{1/2} 4$ **18.** $\log_{1/4} 5$

19. $\log_9 0.1$ **20.** $\log_{20} 0.25$

21. $\log_{15} 1250$ **22.** $\log_3 0.015$

In Exercises 23–28, use the properties of logarithms to rewrite and simplify the logarithmic expression.

23. $\log_4 8$ **24.** $\log_2(4^2 \cdot 3^4)$

25. $\log_5 \frac{1}{250}$ **26.** $\log \frac{9}{300}$

27. $\ln(5e^6)$ **28.** $\ln \frac{6}{e^2}$

In Exercises 29–44, find the exact value of the logarithmic expression without using a calculator. (If this is not possible, state the reason.)

29. $\log_3 9$ **30.** $\log_5 \frac{1}{125}$

31. $\log_2 \sqrt[4]{8}$ **32.** $\log_6 \sqrt[3]{6}$

33. $\log_4 16^2$ **34.** $\log_3 81^{-3}$

35. $\log_2(-2)$ **36.** $\log_3(-27)$

37. $\ln e^{4.5}$ **38.** $3 \ln e^4$

39. $\ln \dfrac{1}{\sqrt{e}}$ **40.** $\ln \sqrt[4]{e^3}$

41. $\ln e^2 + \ln e^5$ **42.** $2 \ln e^6 - \ln e^5$

43. $\log_5 75 - \log_5 3$ **44.** $\log_4 2 + \log_4 32$

In Exercises 45–66, use the properties of logarithms to expand the expression as a sum, difference, and/or constant multiple of logarithms. (Assume all variables are positive.)

45. $\ln 4x$ **46.** $\log_3 10z$

47. $\log_8 x^4$ **48.** $\log_{10} \dfrac{y}{2}$

49. $\log_5 \dfrac{5}{x}$ **50.** $\log_6 \dfrac{1}{z^3}$

51. $\ln \sqrt{z}$ **52.** $\ln \sqrt[3]{t}$

53. $\ln xyz^2$ **54.** $\log 4x^2 y$

55. $\ln z(z-1)^2,\ z > 1$ **56.** $\ln\left(\dfrac{x^2-1}{x^3}\right),\ x > 1$

57. $\log_2 \dfrac{\sqrt{a-1}}{9},\ a > 1$ **58.** $\ln \dfrac{6}{\sqrt{x^2+1}}$

59. $\ln \sqrt[3]{\dfrac{x}{y}}$ **60.** $\ln \sqrt{\dfrac{x^2}{y^3}}$

61. $\ln x^2 \sqrt{\dfrac{y}{z}}$ **62.** $\log_2 x^4 \sqrt{\dfrac{y}{z^3}}$

63. $\log_5 \dfrac{x^2}{y^2 z^3}$ **64.** $\log_{10} \dfrac{xy^4}{z^5}$

65. $\ln \sqrt[4]{x^3(x^2+3)}$ **66.** $\ln \sqrt{x^2(x+2)}$

In Exercises 67–84, condense the expression to the logarithm of a single quantity.

67. $\ln 2 + \ln x$

68. $\ln y + \ln t$

69. $\log_4 z - \log_4 y$

70. $\log_5 8 - \log_5 t$

71. $2 \log_2 x + 4 \log_2 y$

72. $\frac{2}{3} \log_7(z - 2)$

73. $\frac{1}{4} \log_3 5x$

74. $-4 \log_6 2x$

75. $\log x - 2 \log(x + 1)$

76. $2 \ln 8 + 5 \ln(z - 4)$

77. $\log x - 2 \log y + 3 \log z$

78. $3 \log_3 x + 4 \log_3 y - 4 \log_3 z$

79. $\ln x - [\ln(x + 1) + \ln(x - 1)]$

80. $4[\ln z + \ln(z + 5)] - 2 \ln(z - 5)$

81. $\frac{1}{3}[2 \ln(x + 3) + \ln x - \ln(x^2 - 1)]$

82. $2[3 \ln x - \ln(x + 1) - \ln(x - 1)]$

83. $\frac{1}{3}[\log_8 y + 2 \log_8(y + 4)] - \log_8(y - 1)$

84. $\frac{1}{2}[\log_4(x + 1) + 2 \log_4(x - 1)] + 6 \log_4 x$

In Exercises 85 and 86, compare the logarithmic quantities. If two are equal, explain why.

85. $\dfrac{\log_2 32}{\log_2 4}$, $\log_2 \dfrac{32}{4}$, $\log_2 32 - \log_2 4$

86. $\log_7 \sqrt{70}$, $\log_7 35$, $\frac{1}{2} + \log_7 \sqrt{10}$

SOUND INTENSITY In Exercises 87–90, use the following information. The relationship between the number of decibels β and the intensity of a sound I in watts per square meter is given by

$$\beta = 10 \log\left(\frac{I}{10^{-12}}\right).$$

87. Use the properties of logarithms to write the formula in simpler form, and determine the number of decibels of a sound with an intensity of 10^{-6} watt per square meter.

88. Find the difference in loudness between an average office with an intensity of 1.26×10^{-7} watt per square meter and a broadcast studio with an intensity of 3.16×10^{-10} watt per square meter.

89. Find the difference in loudness between a vacuum cleaner with an intensity of 10^{-4} watt per square meter and rustling leaves with an intensity of 10^{-11} watt per square meter.

90. You and your roommate are playing your stereos at the same time and at the same intensity. How much louder is the music when both stereos are playing compared with just one stereo playing?

CURVE FITTING In Exercises 91–94, find a logarithmic equation that relates y and x. Explain the steps used to find the equation.

91.

x	1	2	3	4	5	6
y	1	1.189	1.316	1.414	1.495	1.565

92.

x	1	2	3	4	5	6
y	1	1.587	2.080	2.520	2.924	3.302

93.

x	1	2	3	4	5	6
y	2.5	2.102	1.9	1.768	1.672	1.597

94.

x	1	2	3	4	5	6
y	0.5	2.828	7.794	16	27.951	44.091

95. GALLOPING SPEEDS OF ANIMALS Four-legged animals run with two different types of motion: trotting and galloping. An animal that is trotting has at least one foot on the ground at all times, whereas an animal that is galloping has all four feet off the ground at some point in its stride. The number of strides per minute at which an animal breaks from a trot to a gallop depends on the weight of the animal. Use the table to find a logarithmic equation that relates an animal's weight x (in pounds) and its lowest galloping speed y (in strides per minute).

Weight, x	Galloping speed, y
25	191.5
35	182.7
50	173.8
75	164.2
500	125.9
1000	114.2

96. NAIL LENGTH The approximate lengths and diameters (in inches) of common nails are shown in the table. Find a logarithmic equation that relates the diameter y of a common nail to its length x.

Length, x	Diameter, y	Length, x	Diameter, y
1	0.072	4	0.203
2	0.120	5	0.238
3	0.148	6	0.284

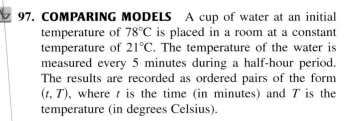

97. COMPARING MODELS A cup of water at an initial temperature of 78°C is placed in a room at a constant temperature of 21°C. The temperature of the water is measured every 5 minutes during a half-hour period. The results are recorded as ordered pairs of the form (t, T), where t is the time (in minutes) and T is the temperature (in degrees Celsius).

(0, 78.0°), (5, 66.0°), (10, 57.5°), (15, 51.2°), (20, 46.3°), (25, 42.4°), (30, 39.6°)

(a) The graph of the model for the data should be asymptotic with the graph of the temperature of the room. Subtract the room temperature from each of the temperatures in the ordered pairs. Use a graphing utility to plot the data points (t, T) and $(t, T - 21)$.

(b) An exponential model for the data $(t, T - 21)$ is given by $T - 21 = 54.4(0.964)^t$. Solve for T and graph the model. Compare the result with the plot of the original data.

(c) Take the natural logarithms of the revised temperatures. Use a graphing utility to plot the points $(t, \ln(T - 21))$ and observe that the points appear to be linear. Use the *regression* feature of the graphing utility to fit a line to these data. This resulting line has the form $\ln(T - 21) = at + b$. Solve for T, and verify that the result is equivalent to the model in part (b).

(d) Fit a rational model to the data. Take the reciprocals of the y-coordinates of the revised data points to generate the points

$$\left(t, \frac{1}{T - 21}\right).$$

Use a graphing utility to graph these points and observe that they appear to be linear. Use the *regression* feature of a graphing utility to fit a line to these data. The resulting line has the form

$$\frac{1}{T - 21} = at + b.$$

Solve for T, and use a graphing utility to graph the rational function and the original data points.

(e) Why did taking the logarithms of the temperatures lead to a linear scatter plot? Why did taking the reciprocals of the temperatures lead to a linear scatter plot?

EXPLORATION

98. PROOF Prove that $\log_b \dfrac{u}{v} = \log_b u - \log_b v$.

99. PROOF Prove that $\log_b u^n = n \log_b u$.

100. CAPSTONE A classmate claims that the following are true.

(a) $\ln(u + v) = \ln u + \ln v = \ln(uv)$

(b) $\ln(u - v) = \ln u - \ln v = \ln \dfrac{u}{v}$

(c) $(\ln u)^n = n(\ln u) = \ln u^n$

Discuss how you would demonstrate that these claims are not true.

TRUE OR FALSE? In Exercises 101–106, determine whether the statement is true or false given that $f(x) = \ln x$. Justify your answer.

101. $f(0) = 0$

102. $f(ax) = f(a) + f(x), \quad a > 0, x > 0$

103. $f(x - 2) = f(x) - f(2), \quad x > 2$

104. $\sqrt{f(x)} = \frac{1}{2}f(x)$

105. If $f(u) = 2f(v)$, then $v = u^2$.

106. If $f(x) < 0$, then $0 < x < 1$.

In Exercises 107–112, use the change-of-base formula to rewrite the logarithm as a ratio of logarithms. Then use a graphing utility to graph the ratio.

107. $f(x) = \log_2 x$

108. $f(x) = \log_4 x$

109. $f(x) = \log_{1/2} x$

110. $f(x) = \log_{1/4} x$

111. $f(x) = \log_{11.8} x$

112. $f(x) = \log_{12.4} x$

113. THINK ABOUT IT Consider the functions below.

$$f(x) = \ln \frac{x}{2}, \quad g(x) = \frac{\ln x}{\ln 2}, \quad h(x) = \ln x - \ln 2$$

Which two functions should have identical graphs? Verify your answer by sketching the graphs of all three functions on the same set of coordinate axes.

114. GRAPHICAL ANALYSIS Use a graphing utility to graph the functions given by $y_1 = \ln x - \ln(x - 3)$ and $y_2 = \ln \dfrac{x}{x - 3}$ in the same viewing window. Does the graphing utility show the functions with the same domain? If so, should it? Explain your reasoning.

115. THINK ABOUT IT For how many integers between 1 and 20 can the natural logarithms be approximated given the values $\ln 2 \approx 0.6931$, $\ln 3 \approx 1.0986$, and $\ln 5 \approx 1.6094$? Approximate these logarithms (do not use a calculator).

3.4 EXPONENTIAL AND LOGARITHMIC EQUATIONS

What you should learn

- Solve simple exponential and logarithmic equations.
- Solve more complicated exponential equations.
- Solve more complicated logarithmic equations.
- Use exponential and logarithmic equations to model and solve real-life problems.

Why you should learn it

Exponential and logarithmic equations are used to model and solve life science applications. For instance, in Exercise 132 on page 253, an exponential function is used to model the number of trees per acre given the average diameter of the trees.

Introduction

So far in this chapter, you have studied the definitions, graphs, and properties of exponential and logarithmic functions. In this section, you will study procedures for *solving equations* involving these exponential and logarithmic functions.

There are two basic strategies for solving exponential or logarithmic equations. The first is based on the One-to-One Properties and was used to solve simple exponential and logarithmic equations in Sections 3.1 and 3.2. The second is based on the Inverse Properties. For $a > 0$ and $a \neq 1$, the following properties are true for all x and y for which $\log_a x$ and $\log_a y$ are defined.

One-to-One Properties

$$a^x = a^y \text{ if and only if } x = y.$$
$$\log_a x = \log_a y \text{ if and only if } x = y.$$

Inverse Properties

$$a^{\log_a x} = x$$
$$\log_a a^x = x$$

Example 1 Solving Simple Equations

	Original Equation	Rewritten Equation	Solution	Property
a.	$2^x = 32$	$2^x = 2^5$	$x = 5$	One-to-One
b.	$\ln x - \ln 3 = 0$	$\ln x = \ln 3$	$x = 3$	One-to-One
c.	$\left(\frac{1}{3}\right)^x = 9$	$3^{-x} = 3^2$	$x = -2$	One-to-One
d.	$e^x = 7$	$\ln e^x = \ln 7$	$x = \ln 7$	Inverse
e.	$\ln x = -3$	$e^{\ln x} = e^{-3}$	$x = e^{-3}$	Inverse
f.	$\log x = -1$	$10^{\log x} = 10^{-1}$	$x = 10^{-1} = \frac{1}{10}$	Inverse
g.	$\log_3 x = 4$	$3^{\log_3 x} = 3^4$	$x = 81$	Inverse

CHECK*Point* Now try Exercise 17.

The strategies used in Example 1 are summarized as follows.

Strategies for Solving Exponential and Logarithmic Equations

1. Rewrite the original equation in a form that allows the use of the One-to-One Properties of exponential or logarithmic functions.

2. Rewrite an *exponential* equation in logarithmic form and apply the Inverse Property of logarithmic functions.

3. Rewrite a *logarithmic* equation in exponential form and apply the Inverse Property of exponential functions.

Solving Exponential Equations

Example 2   Solving Exponential Equations

Solve each equation and approximate the result to three decimal places, if necessary.

a. $e^{-x^2} = e^{-3x-4}$

b. $3(2^x) = 42$

Solution

a.

$e^{-x^2} = e^{-3x-4}$	Write original equation.
$-x^2 = -3x - 4$	One-to-One Property
$x^2 - 3x - 4 = 0$	Write in general form.
$(x + 1)(x - 4) = 0$	Factor.
$(x + 1) = 0 \Longrightarrow x = -1$	Set 1st factor equal to 0.
$(x - 4) = 0 \Longrightarrow x = 4$	Set 2nd factor equal to 0.

The solutions are $x = -1$ and $x = 4$. Check these in the original equation.

b.

$3(2^x) = 42$	Write original equation.
$2^x = 14$	Divide each side by 3.
$\log_2 2^x = \log_2 14$	Take log (base 2) of each side.
$x = \log_2 14$	Inverse Property
$x = \dfrac{\ln 14}{\ln 2} \approx 3.807$	Change-of-base formula

The solution is $x = \log_2 14 \approx 3.807$. Check this in the original equation.

CHECK *Point*   Now try Exercise 29.

In Example 2(b), the exact solution is $x = \log_2 14$ and the approximate solution is $x \approx 3.807$. An exact answer is preferred when the solution is an intermediate step in a larger problem. For a final answer, an approximate solution is easier to comprehend.

Example 3   Solving an Exponential Equation

Solve $e^x + 5 = 60$ and approximate the result to three decimal places.

Solution

$e^x + 5 = 60$	Write original equation.
$e^x = 55$	Subtract 5 from each side.
$\ln e^x = \ln 55$	Take natural log of each side.
$x = \ln 55 \approx 4.007$	Inverse Property

The solution is $x = \ln 55 \approx 4.007$. Check this in the original equation.

CHECK *Point*   Now try Exercise 55.

Study Tip

Another way to solve Example 2(b) is by taking the natural log of each side and then applying the Power Property, as follows.

$$3(2^x) = 42$$

$$2^x = 14$$

$$\ln 2^x = \ln 14$$

$$x \ln 2 = \ln 14$$

$$x = \frac{\ln 14}{\ln 2} \approx 3.807$$

As you can see, you obtain the same result as in Example 2(b).

Study Tip

Remember that the natural logarithmic function has a base of e.

Example 4 Solving an Exponential Equation

Solve $2(3^{2t-5}) - 4 = 11$ and approximate the result to three decimal places.

Solution

$$2(3^{2t-5}) - 4 = 11 \qquad \text{Write original equation.}$$

$$2(3^{2t-5}) = 15 \qquad \text{Add 4 to each side.}$$

$$3^{2t-5} = \frac{15}{2} \qquad \text{Divide each side by 2.}$$

$$\log_3 3^{2t-5} = \log_3 \frac{15}{2} \qquad \text{Take log (base 3) of each side.}$$

$$2t - 5 = \log_3 \frac{15}{2} \qquad \text{Inverse Property}$$

$$2t = 5 + \log_3 7.5 \qquad \text{Add 5 to each side.}$$

$$t = \frac{5}{2} + \frac{1}{2}\log_3 7.5 \qquad \text{Divide each side by 2.}$$

$$t \approx 3.417 \qquad \text{Use a calculator.}$$

The solution is $t = \frac{5}{2} + \frac{1}{2}\log_3 7.5 \approx 3.417$. Check this in the original equation.

CHECK*Point* Now try Exercise 57.

Study Tip

Remember that to evaluate a logarithm such as $\log_3 7.5$, you need to use the change-of-base formula.

$$\log_3 7.5 = \frac{\ln 7.5}{\ln 3} \approx 1.834$$

When an equation involves two or more exponential expressions, you can still use a procedure similar to that demonstrated in Examples 2, 3, and 4. However, the algebra is a bit more complicated.

Example 5 Solving an Exponential Equation of Quadratic Type

Solve $e^{2x} - 3e^x + 2 = 0$.

Algebraic Solution

$$e^{2x} - 3e^x + 2 = 0 \qquad \text{Write original equation.}$$

$$(e^x)^2 - 3e^x + 2 = 0 \qquad \text{Write in quadratic form.}$$

$$(e^x - 2)(e^x - 1) = 0 \qquad \text{Factor.}$$

$$e^x - 2 = 0 \qquad \text{Set 1st factor equal to 0.}$$

$$x = \ln 2 \qquad \text{Solution}$$

$$e^x - 1 = 0 \qquad \text{Set 2nd factor equal to 0.}$$

$$x = 0 \qquad \text{Solution}$$

The solutions are $x = \ln 2 \approx 0.693$ and $x = 0$. Check these in the original equation.

CHECK*Point* Now try Exercise 59.

Graphical Solution

Use a graphing utility to graph $y = e^{2x} - 3e^x + 2$. Use the *zero* or *root* feature or the *zoom* and *trace* features of the graphing utility to approximate the values of x for which $y = 0$. In Figure 3.28, you can see that the zeros occur at $x = 0$ and at $x \approx 0.693$. So, the solutions are $x = 0$ and $x \approx 0.693$.

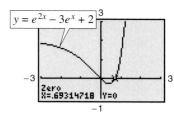

$y = e^{2x} - 3e^x + 2$

Zero
X=.69314718 Y=0

FIGURE **3.28**

Solving Logarithmic Equations

To solve a logarithmic equation, you can write it in exponential form.

$\ln x = 3$	Logarithmic form
$e^{\ln x} = e^3$	Exponentiate each side.
$x = e^3$	Exponential form

This procedure is called *exponentiating* each side of an equation.

Example 6   Solving Logarithmic Equations

a.

$\ln x = 2$	Original equation
$e^{\ln x} = e^2$	Exponentiate each side.
$x = e^2$	Inverse Property

b.

$\log_3(5x - 1) = \log_3(x + 7)$	Original equation
$5x - 1 = x + 7$	One-to-One Property
$4x = 8$	Add $-x$ and 1 to each side.
$x = 2$	Divide each side by 4.

c.

$\log_6(3x + 14) - \log_6 5 = \log_6 2x$	Original equation
$\log_6\left(\dfrac{3x + 14}{5}\right) = \log_6 2x$	Quotient Property of Logarithms
$\dfrac{3x + 14}{5} = 2x$	One-to-One Property
$3x + 14 = 10x$	Cross multiply.
$-7x = -14$	Isolate x.
$x = 2$	Divide each side by -7.

CHECK*Point* Now try Exercise 83.

> **! WARNING / CAUTION**
>
> Remember to check your solutions in the original equation when solving equations to verify that the answer is correct and to make sure that the answer lies in the domain of the original equation.

Example 7   Solving a Logarithmic Equation

Solve $5 + 2 \ln x = 4$ and approximate the result to three decimal places.

Algebraic Solution

$5 + 2 \ln x = 4$	Write original equation.
$2 \ln x = -1$	Subtract 5 from each side.
$\ln x = -\dfrac{1}{2}$	Divide each side by 2.
$e^{\ln x} = e^{-1/2}$	Exponentiate each side.
$x = e^{-1/2}$	Inverse Property
$x \approx 0.607$	Use a calculator.

Graphical Solution

Use a graphing utility to graph $y_1 = 5 + 2 \ln x$ and $y_2 = 4$ in the same viewing window. Use the *intersect* feature or the *zoom* and *trace* features to approximate the intersection point, as shown in Figure 3.29. So, the solution is $x \approx 0.607$.

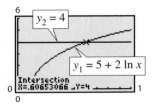

FIGURE **3.29**

CHECK*Point* Now try Exercise 93.

Example 8   Solving a Logarithmic Equation

Solve $2 \log_5 3x = 4$.

Solution

$2 \log_5 3x = 4$	Write original equation.
$\log_5 3x = 2$	Divide each side by 2.
$5^{\log_5 3x} = 5^2$	Exponentiate each side (base 5).
$3x = 25$	Inverse Property
$x = \dfrac{25}{3}$	Divide each side by 3.

The solution is $x = \frac{25}{3}$. Check this in the original equation.

CHECK Point   Now try Exercise 97.

> *Study Tip*
>
> Notice in Example 9 that the logarithmic part of the equation is condensed into a single logarithm before exponentiating each side of the equation.

Because the domain of a logarithmic function generally does not include all real numbers, you should be sure to check for extraneous solutions of logarithmic equations.

Example 9   Checking for Extraneous Solutions

Solve $\log 5x + \log(x - 1) = 2$.

Algebraic Solution

$\log 5x + \log(x - 1) = 2$	Write original equation.
$\log[5x(x - 1)] = 2$	Product Property of Logarithms
$10^{\log(5x^2 - 5x)} = 10^2$	Exponentiate each side (base 10).
$5x^2 - 5x = 100$	Inverse Property
$x^2 - x - 20 = 0$	Write in general form.
$(x - 5)(x + 4) = 0$	Factor.
$x - 5 = 0$	Set 1st factor equal to 0.
$x = 5$	Solution
$x + 4 = 0$	Set 2nd factor equal to 0.
$x = -4$	Solution

The solutions appear to be $x = 5$ and $x = -4$. However, when you check these in the original equation, you can see that $x = 5$ is the only solution.

CHECK Point   Now try Exercise 109.

Graphical Solution

Use a graphing utility to graph $y_1 = \log 5x + \log(x - 1)$ and $y_2 = 2$ in the same viewing window. From the graph shown in Figure 3.30, it appears that the graphs intersect at one point. Use the *intersect* feature or the *zoom* and *trace* features to determine that the graphs intersect at approximately $(5, 2)$. So, the solution is $x = 5$. Verify that 5 is an exact solution algebraically.

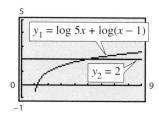

FIGURE **3.30**

In Example 9, the domain of $\log 5x$ is $x > 0$ and the domain of $\log(x - 1)$ is $x > 1$, so the domain of the original equation is $x > 1$. Because the domain is all real numbers greater than 1, the solution $x = -4$ is extraneous. The graph in Figure 3.30 verifies this conclusion.

Applications

Example 10   Doubling an Investment

You have deposited $500 in an account that pays 6.75% interest, compounded continuously. How long will it take your money to double?

Solution

Using the formula for continuous compounding, you can find that the balance in the account is

$$A = Pe^{rt}$$

$$A = 500e^{0.0675t}.$$

To find the time required for the balance to double, let $A = 1000$ and solve the resulting equation for t.

$500e^{0.0675t} = 1000$	Let $A = 1000$.
$e^{0.0675t} = 2$	Divide each side by 500.
$\ln e^{0.0675t} = \ln 2$	Take natural log of each side.
$0.0675t = \ln 2$	Inverse Property
$t = \dfrac{\ln 2}{0.0675}$	Divide each side by 0.0675.
$t \approx 10.27$	Use a calculator.

The balance in the account will double after approximately 10.27 years. This result is demonstrated graphically in Figure 3.31.

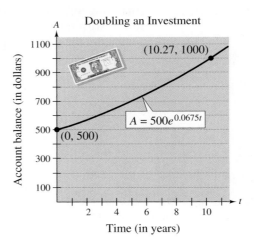

Doubling an Investment

(10.27, 1000)

$A = 500e^{0.0675t}$

(0, 500)

Account balance (in dollars)

Time (in years)

FIGURE **3.31**

CHECK*Point*   Now try Exercise 117.

In Example 10, an approximate answer of 10.27 years is given. Within the context of the problem, the exact solution, $(\ln 2)/0.0675$ years, does not make sense as an answer.

Retail Sales of e-Commerce
Companies

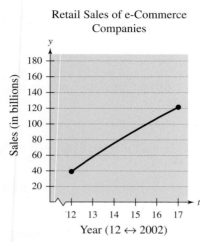

FIGURE 3.32

Example 11 Retail Sales

The retail sales y (in billions) of e-commerce companies in the United States from 2002 through 2007 can be modeled by

$$y = -549 + 236.7 \ln t, \quad 12 \le t \le 17$$

where t represents the year, with $t = 12$ corresponding to 2002 (see Figure 3.32). During which year did the sales reach \$108 billion? (Source: U.S. Census Bureau)

Solution

$-549 + 236.7 \ln t = y$	Write original equation.
$-549 + 236.7 \ln t = 108$	Substitute 108 for y.
$236.7 \ln t = 657$	Add 549 to each side.
$\ln t = \dfrac{657}{236.7}$	Divide each side by 236.7.
$e^{\ln t} = e^{657/236.7}$	Exponentiate each side.
$t = e^{657/236.7}$	Inverse Property
$t \approx 16$	Use a calculator.

The solution is $t \approx 16$. Because $t = 12$ represents 2002, it follows that the sales reached \$108 billion in 2006.

CHECK Point Now try Exercise 133.

CLASSROOM DISCUSSION

Analyzing Relationships Numerically Use a calculator to fill in the table row-by-row. Discuss the resulting pattern. What can you conclude? Find two equations that summarize the relationships you discovered.

x	$\dfrac{1}{2}$	1	2	10	25	50
e^x						
$\ln(e^x)$						
$\ln x$						
$e^{\ln x}$						

3.4   EXERCISES

See www.CalcChat.com for worked-out solutions to odd-numbered exercises.

VOCABULARY: Fill in the blanks.

1. To _____ an equation in x means to find all values of x for which the equation is true.

2. To solve exponential and logarithmic equations, you can use the following One-to-One and Inverse Properties.
   (a) $a^x = a^y$ if and only if _____.
   (b) $\log_a x = \log_a y$ if and only if _____.
   (c) $a^{\log_a x} = $ _____
   (d) $\log_a a^x = $ _____

3. To solve exponential and logarithmic equations, you can use the following strategies.
   (a) Rewrite the original equation in a form that allows the use of the _____ Properties of exponential or logarithmic functions.
   (b) Rewrite an exponential equation in _____ form and apply the Inverse Property of _____ functions.
   (c) Rewrite a logarithmic equation in _____ form and apply the Inverse Property of _____ functions.

4. An _____ solution does not satisfy the original equation.

SKILLS AND APPLICATIONS

In Exercises 5–12, determine whether each x-value is a solution (or an approximate solution) of the equation.

5. $4^{2x-7} = 64$
   (a) $x = 5$
   (b) $x = 2$

6. $2^{3x+1} = 32$
   (a) $x = -1$
   (b) $x = 2$

7. $3e^{x+2} = 75$
   (a) $x = -2 + e^{25}$
   (b) $x = -2 + \ln 25$
   (c) $x \approx 1.219$

8. $4e^{x-1} = 60$
   (a) $x = 1 + \ln 15$
   (b) $x \approx 3.7081$
   (c) $x = \ln 16$

9. $\log_4(3x) = 3$
   (a) $x \approx 21.333$
   (b) $x = -4$
   (c) $x = \frac{64}{3}$

10. $\log_2(x + 3) = 10$
    (a) $x = 1021$
    (b) $x = 17$
    (c) $x = 10^2 - 3$

11. $\ln(2x + 3) = 5.8$
    (a) $x = \frac{1}{2}(-3 + \ln 5.8)$
    (b) $x = \frac{1}{2}(-3 + e^{5.8})$
    (c) $x \approx 163.650$

12. $\ln(x - 1) = 3.8$
    (a) $x = 1 + e^{3.8}$
    (b) $x \approx 45.701$
    (c) $x = 1 + \ln 3.8$

In Exercises 13–24, solve for x.

13. $4^x = 16$
14. $3^x = 243$
15. $\left(\frac{1}{2}\right)^x = 32$
16. $\left(\frac{1}{4}\right)^x = 64$
17. $\ln x - \ln 2 = 0$
18. $\ln x - \ln 5 = 0$
19. $e^x = 2$
20. $e^x = 4$
21. $\ln x = -1$
22. $\log x = -2$
23. $\log_4 x = 3$
24. $\log_5 x = \frac{1}{2}$

In Exercises 25–28, approximate the point of intersection of the graphs of f and g. Then solve the equation $f(x) = g(x)$ algebraically to verify your approximation.

25. $f(x) = 2^x$
    $g(x) = 8$
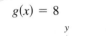
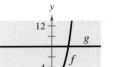

26. $f(x) = 27^x$
    $g(x) = 9$
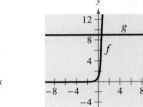

27. $f(x) = \log_3 x$
    $g(x) = 2$
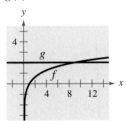

28. $f(x) = \ln(x - 4)$
    $g(x) = 0$
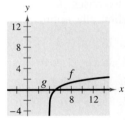

In Exercises 29–70, solve the exponential equation algebraically. Approximate the result to three decimal places.

29. $e^x = e^{x^2-2}$
30. $e^{2x} = e^{x^2-8}$
31. $e^{x^2-3} = e^{x-2}$
32. $e^{-x^2} = e^{x^2-2x}$
33. $4(3^x) = 20$
34. $2(5^x) = 32$
35. $2e^x = 10$
36. $4e^x = 91$
37. $e^x - 9 = 19$
38. $6^x + 10 = 47$
39. $3^{2x} = 80$
40. $6^{5x} = 3000$
41. $5^{-t/2} = 0.20$
42. $4^{-3t} = 0.10$
43. $3^{x-1} = 27$
44. $2^{x-3} = 32$
45. $2^{3-x} = 565$
46. $8^{-2-x} = 431$

47. $8(10^{3x}) = 12$

48. $5(10^{x-6}) = 7$

49. $3(5^{x-1}) = 21$

50. $8(3^{6-x}) = 40$

51. $e^{3x} = 12$

52. $e^{2x} = 50$

53. $500e^{-x} = 300$

54. $1000e^{-4x} = 75$

55. $7 - 2e^x = 5$

56. $-14 + 3e^x = 11$

57. $6(2^{3x-1}) - 7 = 9$

58. $8(4^{6-2x}) + 13 = 41$

59. $e^{2x} - 4e^x - 5 = 0$

60. $e^{2x} - 5e^x + 6 = 0$

61. $e^{2x} - 3e^x - 4 = 0$

62. $e^{2x} + 9e^x + 36 = 0$

63. $\dfrac{500}{100 - e^{x/2}} = 20$

64. $\dfrac{400}{1 + e^{-x}} = 350$

65. $\dfrac{3000}{2 + e^{2x}} = 2$

66. $\dfrac{119}{e^{6x} - 14} = 7$

67. $\left(1 + \dfrac{0.065}{365}\right)^{365t} = 4$

68. $\left(4 - \dfrac{2.471}{40}\right)^{9t} = 21$

69. $\left(1 + \dfrac{0.10}{12}\right)^{12t} = 2$

70. $\left(16 - \dfrac{0.878}{26}\right)^{3t} = 30$

In Exercises 71–80, use a graphing utility to graph and solve the equation. Approximate the result to three decimal places. Verify your result algebraically.

71. $7 = 2^x$

72. $5^x = 212$

73. $6e^{1-x} = 25$

74. $-4e^{-x-1} + 15 = 0$

75. $3e^{3x/2} = 962$

76. $8e^{-2x/3} = 11$

77. $e^{0.09t} = 3$

78. $-e^{1.8x} + 7 = 0$

79. $e^{0.125t} - 8 = 0$

80. $e^{2.724x} = 29$

In Exercises 81–112, solve the logarithmic equation algebraically. Approximate the result to three decimal places.

81. $\ln x = -3$

82. $\ln x = 1.6$

83. $\ln x - 7 = 0$

84. $\ln x + 1 = 0$

85. $\ln 2x = 2.4$

86. $2.1 = \ln 6x$

87. $\log x = 6$

88. $\log 3z = 2$

89. $3\ln 5x = 10$

90. $2\ln x = 7$

91. $\ln\sqrt{x + 2} = 1$

92. $\ln\sqrt{x - 8} = 5$

93. $7 + 3\ln x = 5$

94. $2 - 6\ln x = 10$

95. $-2 + 2\ln 3x = 17$

96. $2 + 3\ln x = 12$

97. $6\log_3(0.5x) = 11$

98. $4\log(x - 6) = 11$

99. $\ln x - \ln(x + 1) = 2$

100. $\ln x + \ln(x + 1) = 1$

101. $\ln x + \ln(x - 2) = 1$

102. $\ln x + \ln(x + 3) = 1$

103. $\ln(x + 5) = \ln(x - 1) - \ln(x + 1)$

104. $\ln(x + 1) - \ln(x - 2) = \ln x$

105. $\log_2(2x - 3) = \log_2(x + 4)$

106. $\log(3x + 4) = \log(x - 10)$

107. $\log(x + 4) - \log x = \log(x + 2)$

108. $\log_2 x + \log_2(x + 2) = \log_2(x + 6)$

109. $\log_4 x - \log_4(x - 1) = \frac{1}{2}$

110. $\log_3 x + \log_3(x - 8) = 2$

111. $\log 8x - \log\left(1 + \sqrt{x}\right) = 2$

112. $\log 4x - \log\left(12 + \sqrt{x}\right) = 2$

In Exercises 113–116, use a graphing utility to graph and solve the equation. Approximate the result to three decimal places. Verify your result algebraically.

113. $3 - \ln x = 0$

114. $10 - 4\ln(x - 2) = 0$

115. $2\ln(x + 3) = 3$

116. $\ln(x + 1) = 2 - \ln x$

COMPOUND INTEREST In Exercises 117–120, $2500 is invested in an account at interest rate r, compounded continuously. Find the time required for the amount to (a) double and (b) triple.

117. $r = 0.05$

118. $r = 0.045$

119. $r = 0.025$

120. $r = 0.0375$

In Exercises 121–128, solve the equation algebraically. Round the result to three decimal places. Verify your answer using a graphing utility.

121. $2x^2e^{2x} + 2xe^{2x} = 0$

122. $-x^2e^{-x} + 2xe^{-x} = 0$

123. $-xe^{-x} + e^{-x} = 0$

124. $e^{-2x} - 2xe^{-2x} = 0$

125. $2x\ln x + x = 0$

126. $\dfrac{1 - \ln x}{x^2} = 0$

127. $\dfrac{1 + \ln x}{2} = 0$

128. $2x\ln\left(\dfrac{1}{x}\right) - x = 0$

129. DEMAND The demand equation for a limited edition coin set is

$$p = 1000\left(1 - \frac{5}{5 + e^{-0.001x}}\right).$$

Find the demand x for a price of (a) $p = \$139.50$ and (b) $p = \$99.99$.

130. DEMAND The demand equation for a hand-held electronic organizer is

$$p = 5000\left(1 - \frac{4}{4 + e^{-0.002x}}\right).$$

Find the demand x for a price of (a) $p = \$600$ and (b) $p = \$400$.

131. FOREST YIELD The yield V (in millions of cubic feet per acre) for a forest at age t years is given by $V = 6.7e^{-48.1/t}$.

 (a) Use a graphing utility to graph the function.

(b) Determine the horizontal asymptote of the function. Interpret its meaning in the context of the problem.

(c) Find the time necessary to obtain a yield of 1.3 million cubic feet.

132. TREES PER ACRE The number N of trees of a given species per acre is approximated by the model $N = 68(10^{-0.04x})$, $5 \le x \le 40$, where x is the average diameter of the trees (in inches) 3 feet above the ground. Use the model to approximate the average diameter of the trees in a test plot when $N = 21$.

133. U.S. CURRENCY The values y (in billions of dollars) of U.S. currency in circulation in the years 2000 through 2007 can be modeled by $y = -451 + 444 \ln t$, $10 \le t \le 17$, where t represents the year, with $t = 10$ corresponding to 2000. During which year did the value of U.S. currency in circulation exceed $690 billion? (Source: Board of Governors of the Federal Reserve System)

 134. MEDICINE The numbers y of freestanding ambulatory care surgery centers in the United States from 2000 through 2007 can be modeled by

$$y = 2875 + \frac{2635.11}{1 + 14.215e^{-0.8038t}}, \quad 0 \le t \le 7$$

where t represents the year, with $t = 0$ corresponding to 2000. (Source: Verispan)

(a) Use a graphing utility to graph the model.

(b) Use the *trace* feature of the graphing utility to estimate the year in which the number of surgery centers exceeded 3600.

135. AVERAGE HEIGHTS The percent m of American males between the ages of 18 and 24 who are no more than x inches tall is modeled by

$$m(x) = \frac{100}{1 + e^{-0.6114(x - 69.71)}}$$

and the percent f of American females between the ages of 18 and 24 who are no more than x inches tall is modeled by

$$f(x) = \frac{100}{1 + e^{-0.66607(x - 64.51)}}.$$

(Source: U.S. National Center for Health Statistics)

(a) Use the graph to determine any horizontal asymptotes of the graphs of the functions. Interpret the meaning in the context of the problem.

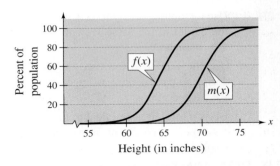

(b) What is the average height of each sex?

 136. LEARNING CURVE In a group project in learning theory, a mathematical model for the proportion P of correct responses after n trials was found to be $P = 0.83/(1 + e^{-0.2n})$.

(a) Use a graphing utility to graph the function.

(b) Use the graph to determine any horizontal asymptotes of the graph of the function. Interpret the meaning of the upper asymptote in the context of this problem.

(c) After how many trials will 60% of the responses be correct?

137. AUTOMOBILES Automobiles are designed with crumple zones that help protect their occupants in crashes. The crumple zones allow the occupants to move short distances when the automobiles come to abrupt stops. The greater the distance moved, the fewer g's the crash victims experience. (One g is equal to the acceleration due to gravity. For very short periods of time, humans have withstood as much as 40 g's.) In crash tests with vehicles moving at 90 kilometers per hour, analysts measured the numbers of g's experienced during deceleration by crash dummies that were permitted to move x meters during impact. The data are shown in the table. A model for the data is given by $y = -3.00 + 11.88 \ln x + (36.94/x)$, where y is the number of g's.

x	g's
0.2	158
0.4	80
0.6	53
0.8	40
1.0	32

(a) Complete the table using the model.

x	0.2	0.4	0.6	0.8	1.0
y					

(b) Use a graphing utility to graph the data points and the model in the same viewing window. How do they compare?

(c) Use the model to estimate the distance traveled during impact if the passenger deceleration must not exceed 30 g's.

(d) Do you think it is practical to lower the number of g's experienced during impact to fewer than 23? Explain your reasoning.

138. DATA ANALYSIS An object at a temperature of 160°C was removed from a furnace and placed in a room at 20°C. The temperature T of the object was measured each hour h and recorded in the table. A model for the data is given by $T = 20[1 + 7(2^{-h})]$. The graph of this model is shown in the figure.

Hour, h	Temperature, T
0	160°
1	90°
2	56°
3	38°
4	29°
5	24°

(a) Use the graph to identify the horizontal asymptote of the model and interpret the asymptote in the context of the problem.

(b) Use the model to approximate the time when the temperature of the object was 100°C.

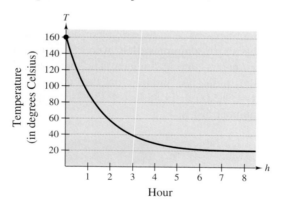

Hour

EXPLORATION

TRUE OR FALSE? In Exercises 139–142, rewrite each verbal statement as an equation. Then decide whether the statement is true or false. Justify your answer.

139. The logarithm of the product of two numbers is equal to the sum of the logarithms of the numbers.

140. The logarithm of the sum of two numbers is equal to the product of the logarithms of the numbers.

141. The logarithm of the difference of two numbers is equal to the difference of the logarithms of the numbers.

142. The logarithm of the quotient of two numbers is equal to the difference of the logarithms of the numbers.

143. THINK ABOUT IT Is it possible for a logarithmic equation to have more than one extraneous solution? Explain.

144. FINANCE You are investing P dollars at an annual interest rate of r, compounded continuously, for t years. Which of the following would result in the highest value of the investment? Explain your reasoning.

(a) Double the amount you invest.

(b) Double your interest rate.

(c) Double the number of years.

145. THINK ABOUT IT Are the times required for the investments in Exercises 117–120 to quadruple twice as long as the times for them to double? Give a reason for your answer and verify your answer algebraically.

146. The *effective yield* of a savings plan is the percent increase in the balance after 1 year. Find the effective yield for each savings plan when $1000 is deposited in a savings account. Which savings plan has the greatest effective yield? Which savings plan will have the highest balance after 5 years?

(a) 7% annual interest rate, compounded annually

(b) 7% annual interest rate, compounded continuously

(c) 7% annual interest rate, compounded quarterly

(d) 7.25% annual interest rate, compounded quarterly

 147. GRAPHICAL ANALYSIS Let $f(x) = \log_a x$ and $g(x) = a^x$, where $a > 1$.

(a) Let $a = 1.2$ and use a graphing utility to graph the two functions in the same viewing window. What do you observe? Approximate any points of intersection of the two graphs.

(b) Determine the value(s) of a for which the two graphs have one point of intersection.

(c) Determine the value(s) of a for which the two graphs have two points of intersection.

148. CAPSTONE Write two or three sentences stating the general guidelines that you follow when solving (a) exponential equations and (b) logarithmic equations.

3.5   EXPONENTIAL AND LOGARITHMIC MODELS

What you should learn

- Recognize the five most common types of models involving exponential and logarithmic functions.
- Use exponential growth and decay functions to model and solve real-life problems.
- Use Gaussian functions to model and solve real-life problems.
- Use logistic growth functions to model and solve real-life problems.
- Use logarithmic functions to model and solve real-life problems.

Why you should learn it

Exponential growth and decay models are often used to model the populations of countries. For instance, in Exercise 44 on page 263, you will use exponential growth and decay models to compare the populations of several countries.

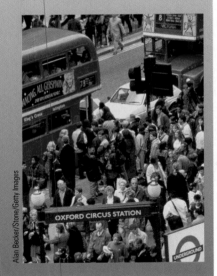

Introduction

The five most common types of mathematical models involving exponential functions and logarithmic functions are as follows.

1. **Exponential growth model:** $\quad y = ae^{bx}, \quad b > 0$

2. **Exponential decay model:** $\quad y = ae^{-bx}, \quad b > 0$

3. **Gaussian model:** $\quad y = ae^{-(x-b)^2/c}$

4. **Logistic growth model:** $\quad y = \dfrac{a}{1 + be^{-rx}}$

5. **Logarithmic models:** $\quad y = a + b \ln x, \quad y = a + b \log x$

The basic shapes of the graphs of these functions are shown in Figure 3.33.

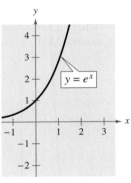

Exponential growth model

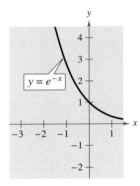

Exponential decay model

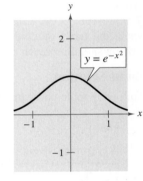

Gaussian model

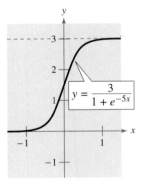

Logistic growth model

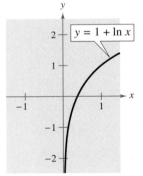

Natural logarithmic model

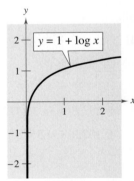

Common logarithmic model

FIGURE **3.33**

You can often gain quite a bit of insight into a situation modeled by an exponential or logarithmic function by identifying and interpreting the function's asymptotes. Use the graphs in Figure 3.33 to identify the asymptotes of the graph of each function.

Exponential Growth and Decay

Example 1   Online Advertising

Estimates of the amounts (in billions of dollars) of U.S. online advertising spending from 2007 through 2011 are shown in the table. A scatter plot of the data is shown in Figure 3.34.   (Source: eMarketer)

Year	Advertising spending
2007	21.1
2008	23.6
2009	25.7
2010	28.5
2011	32.0

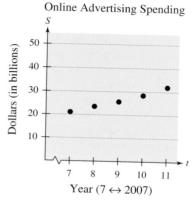

Online Advertising Spending

FIGURE 3.34

An exponential growth model that approximates these data is given by $S = 10.33e^{0.1022t}$, $7 \le t \le 11$, where S is the amount of spending (in billions) and $t = 7$ represents 2007. Compare the values given by the model with the estimates shown in the table. According to this model, when will the amount of U.S. online advertising spending reach $40 billion?

Algebraic Solution

The following table compares the two sets of advertising spending figures.

Year	2007	2008	2009	2010	2011
Advertising spending	21.1	23.6	25.7	28.5	32.0
Model	21.1	23.4	25.9	28.7	31.8

To find when the amount of U.S. online advertising spending will reach $40 billion, let $S = 40$ in the model and solve for t.

$10.33e^{0.1022t} = S$	Write original model.
$10.33e^{0.1022t} = 40$	Substitute 40 for S.
$e^{0.1022t} \approx 3.8722$	Divide each side by 10.33.
$\ln e^{0.1022t} \approx \ln 3.8722$	Take natural log of each side.
$0.1022t \approx 1.3538$	Inverse Property
$t \approx 13.2$	Divide each side by 0.1022.

According to the model, the amount of U.S. online advertising spending will reach $40 billion in 2013.

CHECK**Point**   Now try Exercise 43.

Graphical Solution

Use a graphing utility to graph the model $y = 10.33e^{0.1022x}$ and the data in the same viewing window. You can see in Figure 3.35 that the model appears to fit the data closely.

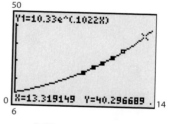

FIGURE 3.35

Use the *zoom* and *trace* features of the graphing utility to find that the approximate value of x for $y = 40$ is $x \approx 13.2$. So, according to the model, the amount of U.S. online advertising spending will reach $40 billion in 2013.

TECHNOLOGY

Some graphing utilities have an *exponential regression* feature that can be used to find exponential models that represent data. If you have such a graphing utility, try using it to find an exponential model for the data given in Example 1. How does your model compare with the model given in Example 1?

In Example 1, you were given the exponential growth model. But suppose this model were not given; how could you find such a model? One technique for doing this is demonstrated in Example 2.

Example 2   Modeling Population Growth

In a research experiment, a population of fruit flies is increasing according to the law of exponential growth. After 2 days there are 100 flies, and after 4 days there are 300 flies. How many flies will there be after 5 days?

Solution

Let y be the number of flies at time t. From the given information, you know that $y = 100$ when $t = 2$ and $y = 300$ when $t = 4$. Substituting this information into the model $y = ae^{bt}$ produces

$$100 = ae^{2b} \qquad \text{and} \qquad 300 = ae^{4b}.$$

To solve for b, solve for a in the first equation.

$$100 = ae^{2b} \quad \Longrightarrow \quad a = \frac{100}{e^{2b}} \qquad \text{Solve for } a \text{ in the first equation.}$$

Then substitute the result into the second equation.

$$300 = ae^{4b} \qquad\qquad \text{Write second equation.}$$

$$300 = \left(\frac{100}{e^{2b}}\right)e^{4b} \qquad \text{Substitute } \frac{100}{e^{2b}} \text{ for } a.$$

$$\frac{300}{100} = e^{2b} \qquad\qquad \text{Divide each side by 100.}$$

$$\ln 3 = 2b \qquad\qquad \text{Take natural log of each side.}$$

$$\frac{1}{2}\ln 3 = b \qquad\qquad \text{Solve for } b.$$

Using $b = \frac{1}{2}\ln 3$ and the equation you found for a, you can determine that

$$a = \frac{100}{e^{2[(1/2)\ln 3]}} \qquad \text{Substitute } \tfrac{1}{2}\ln 3 \text{ for } b.$$

$$= \frac{100}{e^{\ln 3}} \qquad\qquad \text{Simplify.}$$

$$= \frac{100}{3} \qquad\qquad \text{Inverse Property}$$

$$\approx 33.33. \qquad\qquad \text{Simplify.}$$

So, with $a \approx 33.33$ and $b = \frac{1}{2}\ln 3 \approx 0.5493$, the exponential growth model is

$$y = 33.33e^{0.5493t}$$

as shown in Figure 3.36. This implies that, after 5 days, the population will be

$$y = 33.33e^{0.5493(5)} \approx 520 \text{ flies.}$$

CHECK Point Now try Exercise 49.

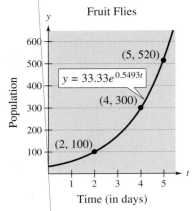

FIGURE **3.36**

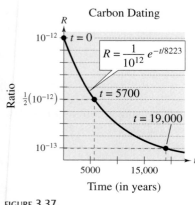

Carbon Dating

$$R = \frac{1}{10^{12}}e^{-t/8223}$$

$t = 0$

$\frac{1}{2}(10^{-12})$

$t = 5700$

$t = 19{,}000$

Ratio

5000   15,000

Time (in years)

FIGURE **3.37**

In living organic material, the ratio of the number of radioactive carbon isotopes (carbon 14) to the number of nonradioactive carbon isotopes (carbon 12) is about 1 to 10^{12}. When organic material dies, its carbon 12 content remains fixed, whereas its radioactive carbon 14 begins to decay with a half-life of about 5700 years. To estimate the age of dead organic material, scientists use the following formula, which denotes the ratio of carbon 14 to carbon 12 present at any time t (in years).

$$R = \frac{1}{10^{12}}e^{-t/8223}$$ Carbon dating model

The graph of R is shown in Figure 3.37. Note that R decreases as t increases.

Example 3   Carbon Dating

Estimate the age of a newly discovered fossil in which the ratio of carbon 14 to carbon 12 is

$R = 1/10^{13}$.

Algebraic Solution

In the carbon dating model, substitute the given value of R to obtain the following.

$\dfrac{1}{10^{12}}e^{-t/8223} = R$	Write original model.
$\dfrac{e^{-t/8223}}{10^{12}} = \dfrac{1}{10^{13}}$	Let $R = \dfrac{1}{10^{13}}$.
$e^{-t/8223} = \dfrac{1}{10}$	Multiply each side by 10^{12}.
$\ln e^{-t/8223} = \ln \dfrac{1}{10}$	Take natural log of each side.
$-\dfrac{t}{8223} \approx -2.3026$	Inverse Property
$t \approx 18{,}934$	Multiply each side by -8223.

So, to the nearest thousand years, the age of the fossil is about 19,000 years.

CHECK *Point*   Now try Exercise 51.

Graphical Solution

Use a graphing utility to graph the formula for the ratio of carbon 14 to carbon 12 at any time t as

$$y_1 = \frac{1}{10^{12}}e^{-x/8223}.$$

In the same viewing window, graph $y_2 = 1/(10^{13})$. Use the *intersect* feature or the *zoom* and *trace* features of the graphing utility to estimate that $x \approx 18{,}934$ when $y = 1/(10^{13})$, as shown in Figure 3.38.

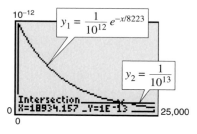

10^{-12}

$y_1 = \dfrac{1}{10^{12}}e^{-x/8223}$

$y_2 = \dfrac{1}{10^{13}}$

Intersection
X=18934.157 Y=1E-13   25,000

FIGURE **3.38**

So, to the nearest thousand years, the age of the fossil is about 19,000 years.

The value of b in the exponential decay model $y = ae^{-bt}$ determines the *decay* of radioactive isotopes. For instance, to find how much of an initial 10 grams of ^{226}Ra isotope with a half-life of 1599 years is left after 500 years, substitute this information into the model $y = ae^{-bt}$.

$$\frac{1}{2}(10) = 10e^{-b(1599)} \implies \ln\frac{1}{2} = -1599b \implies b = -\frac{\ln\frac{1}{2}}{1599}$$

Using the value of b found above and $a = 10$, the amount left is

$$y = 10e^{-[-\ln(1/2)/1599](500)} \approx 8.05 \text{ grams.}$$

Gaussian Models

As mentioned at the beginning of this section, Gaussian models are of the form

$$y = ae^{-(x-b)^2/c}.$$

This type of model is commonly used in probability and statistics to represent populations that are **normally distributed.** The graph of a Gaussian model is called a **bell-shaped curve.** Try graphing the normal distribution with a graphing utility. Can you see why it is called a bell-shaped curve?

For *standard* normal distributions, the model takes the form

$$y = \frac{1}{\sqrt{2\pi}}e^{-x^2/2}.$$

The **average value** of a population can be found from the bell-shaped curve by observing where the maximum *y*-value of the function occurs. The *x*-value corresponding to the maximum *y*-value of the function represents the average value of the independent variable—in this case, *x*.

Example 4   SAT Scores

In 2008, the Scholastic Aptitude Test (SAT) math scores for college-bound seniors roughly followed the normal distribution given by

$$y = 0.0034e^{-(x-515)^2/26,912}, \quad 200 \le x \le 800$$

where *x* is the SAT score for mathematics. Sketch the graph of this function. From the graph, estimate the average SAT score.   (Source: College Board)

Solution

The graph of the function is shown in Figure 3.39. On this bell-shaped curve, the maximum value of the curve represents the average score. From the graph, you can estimate that the average mathematics score for college-bound seniors in 2008 was 515.

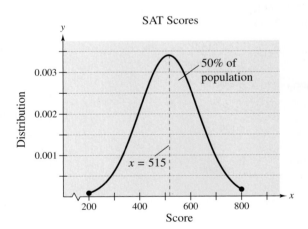

FIGURE **3.39**

CHECK *Point*   Now try Exercise 57.

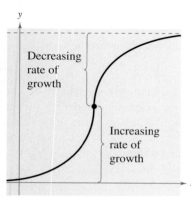

FIGURE **3.40**

Logistic Growth Models

Some populations initially have rapid growth, followed by a declining rate of growth, as indicated by the graph in Figure 3.40. One model for describing this type of growth pattern is the **logistic curve** given by the function

$$y = \frac{a}{1 + be^{-rx}}$$

where y is the population size and x is the time. An example is a bacteria culture that is initially allowed to grow under ideal conditions, and then under less favorable conditions that inhibit growth. A logistic growth curve is also called a **sigmoidal curve.**

| **Example 5** | **Spread of a Virus** |

On a college campus of 5000 students, one student returns from vacation with a contagious and long-lasting flu virus. The spread of the virus is modeled by

$$y = \frac{5000}{1 + 4999e^{-0.8t}}, \quad t \geq 0$$

where y is the total number of students infected after t days. The college will cancel classes when 40% or more of the students are infected.

a. How many students are infected after 5 days?

b. After how many days will the college cancel classes?

Algebraic Solution

a. After 5 days, the number of students infected is

$$y = \frac{5000}{1 + 4999e^{-0.8(5)}} = \frac{5000}{1 + 4999e^{-4}} \approx 54.$$

b. Classes are canceled when the number infected is $(0.40)(5000) = 2000$.

$$2000 = \frac{5000}{1 + 4999e^{-0.8t}}$$

$$1 + 4999e^{-0.8t} = 2.5$$

$$e^{-0.8t} = \frac{1.5}{4999}$$

$$\ln e^{-0.8t} = \ln \frac{1.5}{4999}$$

$$-0.8t = \ln \frac{1.5}{4999}$$

$$t = -\frac{1}{0.8} \ln \frac{1.5}{4999}$$

$$t \approx 10.1$$

So, after about 10 days, at least 40% of the students will be infected, and the college will cancel classes.

CHECK *Point*   Now try Exercise 59.

Graphical Solution

a. Use a graphing utility to graph $y = \dfrac{5000}{1 + 4999e^{-0.8x}}$. Use the *value* feature or the *zoom* and *trace* features of the graphing utility to estimate that $y \approx 54$ when $x = 5$. So, after 5 days, about 54 students will be infected.

b. Classes are canceled when the number of infected students is $(0.40)(5000) = 2000$. Use a graphing utility to graph

$$y_1 = \frac{5000}{1 + 4999e^{-0.8x}} \quad \text{and} \quad y_2 = 2000$$

in the same viewing window. Use the *intersect* feature or the *zoom* and *trace* features of the graphing utility to find the point of intersection of the graphs. In Figure 3.41, you can see that the point of intersection occurs near $x \approx 10.1$. So, after about 10 days, at least 40% of the students will be infected, and the college will cancel classes.

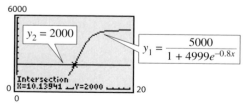

FIGURE **3.41**

Claro Cortes IV/Reuters /Landow

On May 12, 2008, an earthquake of magnitude 7.9 struck Eastern Sichuan Province, China. The total economic loss was estimated at 86 billion U.S. dollars.

Logarithmic Models

Example 6   Magnitudes of Earthquakes

On the Richter scale, the magnitude R of an earthquake of intensity I is given by

$$R = \log \frac{I}{I_0}$$

where $I_0 = 1$ is the minimum intensity used for comparison. Find the intensity of each earthquake. (Intensity is a measure of the wave energy of an earthquake.)

a. Nevada in 2008: $R = 6.0$

b. Eastern Sichuan, China in 2008: $R = 7.9$

Solution

a. Because $I_0 = 1$ and $R = 6.0$, you have

$$6.0 = \log \frac{I}{1}$$ Substitute 1 for I_0 and 6.0 for R.

$$10^{6.0} = 10^{\log I}$$ Exponentiate each side.

$$I = 10^{6.0} = 1{,}000{,}000.$$ Inverse Property

b. For $R = 7.9$, you have

$$7.9 = \log \frac{I}{1}$$ Substitute 1 for I_0 and 7.9 for R.

$$10^{7.9} = 10^{\log I}$$ Exponentiate each side.

$$I = 10^{7.9} \approx 79{,}400{,}000.$$ Inverse Property

Note that an increase of 1.9 units on the Richter scale (from 6.0 to 7.9) represents an increase in intensity by a factor of

$$\frac{79{,}400{,}000}{1{,}000{,}000} = 79.4.$$

In other words, the intensity of the earthquake in Eastern Sichuan was about 79 times as great as that of the earthquake in Nevada.

CHECK**Point** Now try Exercise 63.

CLASSROOM DISCUSSION

Comparing Population Models The populations P (in millions) of the United States for the census years from 1910 to 2000 are shown in the table at the left. Least squares regression analysis gives the best quadratic model for these data as $P = 1.0328t^2 + 9.607t + 81.82$, and the best exponential model for these data as $P = 82.677e^{0.124t}$. Which model better fits the data? Describe how you reached your conclusion.   (Source: U.S. Census Bureau)

t	Year	Population, P
1	1910	92.23
2	1920	106.02
3	1930	123.20
4	1940	132.16
5	1950	151.33
6	1960	179.32
7	1970	203.30
8	1980	226.54
9	1990	248.72
10	2000	281.42

3.5   EXERCISES

See www.CalcChat.com for worked-out solutions to odd-numbered exercises.

VOCABULARY: Fill in the blanks.

1. An exponential growth model has the form _____ and an exponential decay model has the form _____.
2. A logarithmic model has the form _____ or _____.
3. Gaussian models are commonly used in probability and statistics to represent populations that are _____ _____.
4. The graph of a Gaussian model is _____ shaped, where the _____ _____ is the maximum y-value of the graph.
5. A logistic growth model has the form _____.
6. A logistic curve is also called a _____ curve.

SKILLS AND APPLICATIONS

In Exercises 7–12, match the function with its graph. [The graphs are labeled (a), (b), (c), (d), (e), and (f).]

(a)
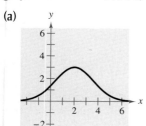

(b)
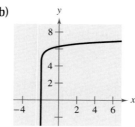

(c)
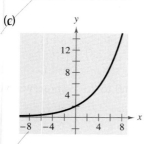

(d)
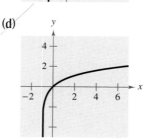

(e)
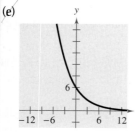

(f)
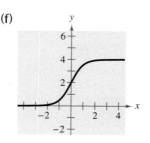

7. $y = 2e^{x/4}$

8. $y = 6e^{-x/4}$

9. $y = 6 + \log(x + 2)$

10. $y = 3e^{-(x-2)^2/5}$

11. $y = \ln(x + 1)$

12. $y = \dfrac{4}{1 + e^{-2x}}$

In Exercises 13 and 14, (a) solve for P and (b) solve for t.

13. $A = Pe^{rt}$

14. $A = P\left(1 + \dfrac{r}{n}\right)^{nt}$

COMPOUND INTEREST   In Exercises 15–22, complete the table for a savings account in which interest is compounded continuously.

	Initial Investment	Annual % Rate	Time to Double	Amount After 10 Years
15.	$1000	3.5%		
16.	$750	$10\frac{1}{2}\%$		
17.	$750		$7\frac{3}{4}$ yr	
18.	$10,000		12 yr	
19.	$500			$1505.00
20.	$600			$19,205.00
21.		4.5%		$10,000.00
22.		2%		$2000.00

COMPOUND INTEREST   In Exercises 23 and 24, determine the principal P that must be invested at rate r, compounded monthly, so that $500,000 will be available for retirement in t years.

23. $r = 5\%, t = 10$

24. $r = 3\frac{1}{2}\%, t = 15$

COMPOUND INTEREST   In Exercises 25 and 26, determine the time necessary for $1000 to double if it is invested at interest rate r compounded (a) annually, (b) monthly, (c) daily, and (d) continuously.

25. $r = 10\%$

26. $r = 6.5\%$

27. **COMPOUND INTEREST**   Complete the table for the time t (in years) necessary for P dollars to triple if interest is compounded continuously at rate r.

r	2%	4%	6%	8%	10%	12%
t						

28. **MODELING DATA**   Draw a scatter plot of the data in Exercise 27. Use the *regression* feature of a graphing utility to find a model for the data.

29. COMPOUND INTEREST   Complete the table for the time t (in years) necessary for P dollars to triple if interest is compounded annually at rate r.

r	2%	4%	6%	8%	10%	12%
t						

30. MODELING DATA   Draw a scatter plot of the data in Exercise 29. Use the *regression* feature of a graphing utility to find a model for the data.

31. COMPARING MODELS   If $1 is invested in an account over a 10-year period, the amount in the account, where t represents the time in years, is given by $A = 1 + 0.075[\![t]\!]$ or $A = e^{0.07t}$ depending on whether the account pays simple interest at $7\frac{1}{2}$% or continuous compound interest at 7%. Graph each function on the same set of axes. Which grows at a higher rate? (Remember that $[\![t]\!]$ is the greatest integer function discussed in Section 1.6.)

32. COMPARING MODELS   If $1 is invested in an account over a 10-year period, the amount in the account, where t represents the time in years, is given by $A = 1 + 0.06[\![t]\!]$ or $A = [1 + (0.055/365)]^{[\![365t]\!]}$ depending on whether the account pays simple interest at 6% or compound interest at $5\frac{1}{2}$% compounded daily. Use a graphing utility to graph each function in the same viewing window. Which grows at a higher rate?

RADIOACTIVE DECAY   In Exercises 33–38, complete the table for the radioactive isotope.

Isotope	Half-life (years)	Initial Quantity	Amount After 1000 Years
33. ^{226}Ra	1599	10 g	
34. ^{14}C	5715	6.5 g	
35. ^{239}Pu	24,100	2.1g	
36. ^{226}Ra	1599		2 g
37. ^{14}C	5715		2 g
38. ^{239}Pu	24,100		0.4 g

In Exercises 39–42, find the exponential model $y = ae^{bx}$ that fits the points shown in the graph or table.

39.
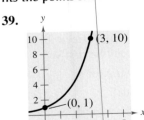
(3, 10)
(0, 1)

40.
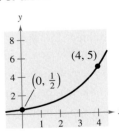
(4, 5)
$(0, \frac{1}{2})$

41.

x	0	4
y	5	1

42.

x	0	3
y	1	$\frac{1}{4}$

43. POPULATION   The populations P (in thousands) of Horry County, South Carolina from 1970 through 2007 can be modeled by

$$P = -18.5 + 92.2e^{0.0282t}$$

where t represents the year, with $t = 0$ corresponding to 1970.   (Source: U.S. Census Bureau)

(a) Use the model to complete the table.

Year	1970	1980	1990	2000	2007
Population					

(b) According to the model, when will the population of Horry County reach 300,000?

(c) Do you think the model is valid for long-term predictions of the population? Explain.

44. POPULATION   The table shows the populations (in millions) of five countries in 2000 and the projected populations (in millions) for the year 2015.   (Source: U.S. Census Bureau)

Country	2000	2015
Bulgaria	7.8	6.9
Canada	31.1	35.1
China	1268.9	1393.4
United Kingdom	59.5	62.2
United States	282.2	325.5

(a) Find the exponential growth or decay model $y = ae^{bt}$ or $y = ae^{-bt}$ for the population of each country by letting $t = 0$ correspond to 2000. Use the model to predict the population of each country in 2030.

(b) You can see that the populations of the United States and the United Kingdom are growing at different rates. What constant in the equation $y = ae^{bt}$ is determined by these different growth rates? Discuss the relationship between the different growth rates and the magnitude of the constant.

(c) You can see that the population of China is increasing while the population of Bulgaria is decreasing. What constant in the equation $y = ae^{bt}$ reflects this difference? Explain.

45. WEBSITE GROWTH The number y of hits a new search-engine website receives each month can be modeled by $y = 4080e^{kt}$, where t represents the number of months the website has been operating. In the website's third month, there were 10,000 hits. Find the value of k, and use this value to predict the number of hits the website will receive after 24 months.

46. VALUE OF A PAINTING The value V (in millions of dollars) of a famous painting can be modeled by $V = 10e^{kt}$, where t represents the year, with $t = 0$ corresponding to 2000. In 2008, the same painting was sold for $65 million. Find the value of k, and use this value to predict the value of the painting in 2014.

47. POPULATION The populations P (in thousands) of Reno, Nevada from 2000 through 2007 can be modeled by $P = 346.8e^{kt}$, where t represents the year, with $t = 0$ corresponding to 2000. In 2005, the population of Reno was about 395,000.   (Source: U.S. Census Bureau)

(a) Find the value of k. Is the population increasing or decreasing? Explain.

(b) Use the model to find the populations of Reno in 2010 and 2015. Are the results reasonable? Explain.

(c) According to the model, during what year will the population reach 500,000?

48. POPULATION The populations P (in thousands) of Orlando, Florida from 2000 through 2007 can be modeled by $P = 1656.2e^{kt}$, where t represents the year, with $t = 0$ corresponding to 2000. In 2005, the population of Orlando was about 1,940,000.   (Source: U.S. Census Bureau)

(a) Find the value of k. Is the population increasing or decreasing? Explain.

(b) Use the model to find the populations of Orlando in 2010 and 2015. Are the results reasonable? Explain.

(c) According to the model, during what year will the population reach 2.2 million?

49. BACTERIA GROWTH The number of bacteria in a culture is increasing according to the law of exponential growth. After 3 hours, there are 100 bacteria, and after 5 hours, there are 400 bacteria. How many bacteria will there be after 6 hours?

50. BACTERIA GROWTH The number of bacteria in a culture is increasing according to the law of exponential growth. The initial population is 250 bacteria, and the population after 10 hours is double the population after 1 hour. How many bacteria will there be after 6 hours?

51. CARBON DATING

(a) The ratio of carbon 14 to carbon 12 in a piece of wood discovered in a cave is $R = 1/8^{14}$. Estimate the age of the piece of wood.

(b) The ratio of carbon 14 to carbon 12 in a piece of paper buried in a tomb is $R = 1/13^{11}$. Estimate the age of the piece of paper.

52. RADIOACTIVE DECAY Carbon 14 dating assumes that the carbon dioxide on Earth today has the same radioactive content as it did centuries ago. If this is true, the amount of ^{14}C absorbed by a tree that grew several centuries ago should be the same as the amount of ^{14}C absorbed by a tree growing today. A piece of ancient charcoal contains only 15% as much radioactive carbon as a piece of modern charcoal. How long ago was the tree burned to make the ancient charcoal if the half-life of ^{14}C is 5715 years?

53. DEPRECIATION A sport utility vehicle that costs $23,300 new has a book value of $12,500 after 2 years.

(a) Find the linear model $V = mt + b$.

(b) Find the exponential model $V = ae^{kt}$.

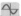 (c) Use a graphing utility to graph the two models in the same viewing window. Which model depreciates faster in the first 2 years?

(d) Find the book values of the vehicle after 1 year and after 3 years using each model.

(e) Explain the advantages and disadvantages of using each model to a buyer and a seller.

54. DEPRECIATION A laptop computer that costs $1150 new has a book value of $550 after 2 years.

(a) Find the linear model $V = mt + b$.

(b) Find the exponential model $V = ae^{kt}$.

 (c) Use a graphing utility to graph the two models in the same viewing window. Which model depreciates faster in the first 2 years?

(d) Find the book values of the computer after 1 year and after 3 years using each model.

(e) Explain the advantages and disadvantages of using each model to a buyer and a seller.

55. SALES The sales S (in thousands of units) of a new CD burner after it has been on the market for t years are modeled by $S(t) = 100(1 - e^{kt})$. Fifteen thousand units of the new product were sold the first year.

(a) Complete the model by solving for k.

(b) Sketch the graph of the model.

(c) Use the model to estimate the number of units sold after 5 years.

56. LEARNING CURVE The management at a plastics factory has found that the maximum number of units a worker can produce in a day is 30. The learning curve for the number N of units produced per day after a new employee has worked t days is modeled by $N = 30(1 - e^{kt})$. After 20 days on the job, a new employee produces 19 units.

(a) Find the learning curve for this employee (first, find the value of k).

(b) How many days should pass before this employee is producing 25 units per day?

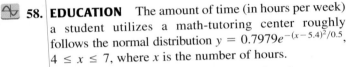

57. IQ SCORES The IQ scores for a sample of a class of returning adult students at a small northeastern college roughly follow the normal distribution $y = 0.0266e^{-(x-100)^2/450}$, $70 \le x \le 115$, where x is the IQ score.

(a) Use a graphing utility to graph the function.

(b) From the graph in part (a), estimate the average IQ score of an adult student.

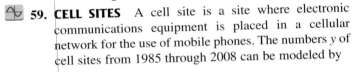

58. EDUCATION The amount of time (in hours per week) a student utilizes a math-tutoring center roughly follows the normal distribution $y = 0.7979e^{-(x-5.4)^2/0.5}$, $4 \le x \le 7$, where x is the number of hours.

(a) Use a graphing utility to graph the function.

(b) From the graph in part (a), estimate the average number of hours per week a student uses the tutoring center.

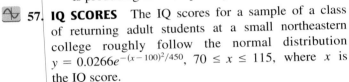

59. CELL SITES A cell site is a site where electronic communications equipment is placed in a cellular network for the use of mobile phones. The numbers y of cell sites from 1985 through 2008 can be modeled by

$$y = \frac{237,101}{1 + 1950e^{-0.355t}}$$

where t represents the year, with $t = 5$ corresponding to 1985. (Source: CTIA-The Wireless Association)

(a) Use the model to find the numbers of cell sites in the years 1985, 2000, and 2006.

(b) Use a graphing utility to graph the function.

(c) Use the graph to determine the year in which the number of cell sites will reach 235,000.

(d) Confirm your answer to part (c) algebraically.

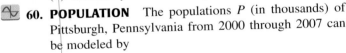

60. POPULATION The populations P (in thousands) of Pittsburgh, Pennsylvania from 2000 through 2007 can be modeled by

$$P = \frac{2632}{1 + 0.083e^{0.0500t}}$$

where t represents the year, with $t = 0$ corresponding to 2000. (Source: U.S. Census Bureau)

(a) Use the model to find the populations of Pittsburgh in the years 2000, 2005, and 2007.

(b) Use a graphing utility to graph the function.

(c) Use the graph to determine the year in which the population will reach 2.2 million.

(d) Confirm your answer to part (c) algebraically.

61. POPULATION GROWTH A conservation organization releases 100 animals of an endangered species into a game preserve. The organization believes that the preserve has a carrying capacity of 1000 animals and that the growth of the pack will be modeled by the logistic curve

$$p(t) = \frac{1000}{1 + 9e^{-0.1656t}}$$

where t is measured in months (see figure).

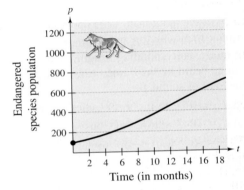

Time (in months)

(a) Estimate the population after 5 months.

(b) After how many months will the population be 500?

(c) Use a graphing utility to graph the function. Use the graph to determine the horizontal asymptotes, and interpret the meaning of the asymptotes in the context of the problem.

62. SALES After discontinuing all advertising for a tool kit in 2004, the manufacturer noted that sales began to drop according to the model

$$S = \frac{500,000}{1 + 0.4e^{kt}}$$

where S represents the number of units sold and $t = 4$ represents 2004. In 2008, the company sold 300,000 units.

(a) Complete the model by solving for k.

(b) Estimate sales in 2012.

GEOLOGY   In Exercises 63 and 64, use the Richter scale

$$R = \log \frac{I}{I_0}$$

for measuring the magnitudes of earthquakes.

63. Find the intensity I of an earthquake measuring R on the Richter scale (let $I_0 = 1$).

   (a) Southern Sumatra, Indonesia in 2007, $R = 8.5$

   (b) Illinois in 2008, $R = 5.4$

   (c) Costa Rica in 2009, $R = 6.1$

64. Find the magnitude R of each earthquake of intensity I (let $I_0 = 1$).

   (a) $I = 199,500,000$ (b) $I = 48,275,000$

   (c) $I = 17,000$

INTENSITY OF SOUND   In Exercises 65–68, use the following information for determining sound intensity. The level of sound β, in decibels, with an intensity of I, is given by $\beta = 10 \log(I/I_0)$, where I_0 is an intensity of 10^{-12} watt per square meter, corresponding roughly to the faintest sound that can be heard by the human ear. In Exercises 65 and 66, find the level of sound β.

65. (a) $I = 10^{-10}$ watt per m^2 (quiet room)

   (b) $I = 10^{-5}$ watt per m^2 (busy street corner)

   (c) $I = 10^{-8}$ watt per m^2 (quiet radio)

   (d) $I = 10^0$ watt per m^2 (threshold of pain)

66. (a) $I = 10^{-11}$ watt per m^2 (rustle of leaves)

   (b) $I = 10^2$ watt per m^2 (jet at 30 meters)

   (c) $I = 10^{-4}$ watt per m^2 (door slamming)

   (d) $I = 10^{-2}$ watt per m^2 (siren at 30 meters)

67. Due to the installation of noise suppression materials, the noise level in an auditorium was reduced from 93 to 80 decibels. Find the percent decrease in the intensity level of the noise as a result of the installation of these materials.

68. Due to the installation of a muffler, the noise level of an engine was reduced from 88 to 72 decibels. Find the percent decrease in the intensity level of the noise as a result of the installation of the muffler.

pH LEVELS   In Exercises 69–74, use the acidity model given by pH $= -\log[\text{H}^+]$, where acidity (pH) is a measure of the hydrogen ion concentration $[\text{H}^+]$ (measured in moles of hydrogen per liter) of a solution.

69. Find the pH if $[\text{H}^+] = 2.3 \times 10^{-5}$.

70. Find the pH if $[\text{H}^+] = 1.13 \times 10^{-5}$.

71. Compute $[\text{H}^+]$ for a solution in which pH $= 5.8$.

72. Compute $[\text{H}^+]$ for a solution in which pH $= 3.2$.

73. Apple juice has a pH of 2.9 and drinking water has a pH of 8.0. The hydrogen ion concentration of the apple juice is how many times the concentration of drinking water?

74. The pH of a solution is decreased by one unit. The hydrogen ion concentration is increased by what factor?

75. **FORENSICS**   At 8:30 A.M., a coroner was called to the home of a person who had died during the night. In order to estimate the time of death, the coroner took the person's temperature twice. At 9:00 A.M. the temperature was 85.7°F, and at 11:00 A.M. the temperature was 82.8°F. From these two temperatures, the coroner was able to determine that the time elapsed since death and the body temperature were related by the formula

$$t = -10 \ln \frac{T - 70}{98.6 - 70}$$

where t is the time in hours elapsed since the person died and T is the temperature (in degrees Fahrenheit) of the person's body. (This formula is derived from a general cooling principle called *Newton's Law of Cooling*. It uses the assumptions that the person had a normal body temperature of 98.6°F at death, and that the room temperature was a constant 70°F.) Use the formula to estimate the time of death of the person.

 76. **HOME MORTGAGE**   A $120,000 home mortgage for 30 years at $7\frac{1}{2}\%$ has a monthly payment of $839.06. Part of the monthly payment is paid toward the interest charge on the unpaid balance, and the remainder of the payment is used to reduce the principal. The amount that is paid toward the interest is

$$u = M - \left(M - \frac{Pr}{12}\right)\left(1 + \frac{r}{12}\right)^{12t}$$

and the amount that is paid toward the reduction of the principal is

$$v = \left(M - \frac{Pr}{12}\right)\left(1 + \frac{r}{12}\right)^{12t}.$$

In these formulas, P is the size of the mortgage, r is the interest rate, M is the monthly payment, and t is the time (in years).

   (a) Use a graphing utility to graph each function in the same viewing window. (The viewing window should show all 30 years of mortgage payments.)

   (b) In the early years of the mortgage, is the larger part of the monthly payment paid toward the interest or the principal? Approximate the time when the monthly payment is evenly divided between interest and principal reduction.

   (c) Repeat parts (a) and (b) for a repayment period of 20 years ($M = \$966.71$). What can you conclude?

77. HOME MORTGAGE The total interest u paid on a home mortgage of P dollars at interest rate r for t years is

$$u = P\left[\frac{rt}{1 - \left(\dfrac{1}{1 + r/12}\right)^{12t}} - 1\right].$$

Consider a \$120,000 home mortgage at $7\frac{1}{2}\%$.

(a) Use a graphing utility to graph the total interest function.

(b) Approximate the length of the mortgage for which the total interest paid is the same as the size of the mortgage. Is it possible that some people are paying twice as much in interest charges as the size of the mortgage?

78. DATA ANALYSIS The table shows the time t (in seconds) required for a car to attain a speed of s miles per hour from a standing start.

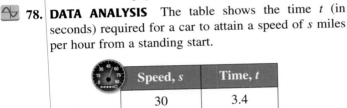

Speed, s	Time, t
30	3.4
40	5.0
50	7.0
60	9.3
70	12.0
80	15.8
90	20.0

Two models for these data are as follows.

$$t_1 = 40.757 + 0.556s - 15.817 \ln s$$

$$t_2 = 1.2259 + 0.0023s^2$$

(a) Use the *regression* feature of a graphing utility to find a linear model t_3 and an exponential model t_4 for the data.

(b) Use a graphing utility to graph the data and each model in the same viewing window.

(c) Create a table comparing the data with estimates obtained from each model.

(d) Use the results of part (c) to find the sum of the absolute values of the differences between the data and the estimated values given by each model. Based on the four sums, which model do you think best fits the data? Explain.

EXPLORATION

TRUE OR FALSE? In Exercises 79–82, determine whether the statement is true or false. Justify your answer.

79. The domain of a logistic growth function cannot be the set of real numbers.

80. A logistic growth function will always have an x-intercept.

81. The graph of $f(x) = \dfrac{4}{1 + 6e^{-2x}} + 5$ is the graph of $g(x) = \dfrac{4}{1 + 6e^{-2x}}$ shifted to the right five units.

82. The graph of a Gaussian model will never have an x-intercept.

83. WRITING Use your school's library, the Internet, or some other reference source to write a paper describing John Napier's work with logarithms.

84. CAPSTONE Identify each model as exponential, Gaussian, linear, logarithmic, logistic, quadratic, or none of the above. Explain your reasoning.

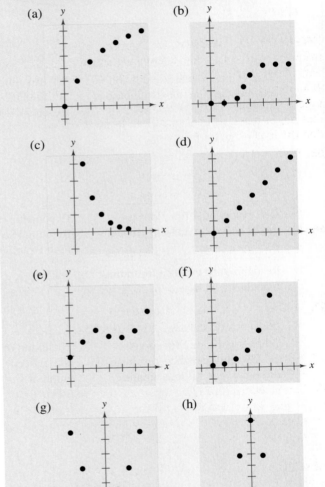

PROJECT: SALES PER SHARE To work an extended application analyzing the sales per share for Kohl's Corporation from 1992 through 2007, visit this text's website at *academic.cengage.com*. (Data Source: Kohl's Corporation)

3 CHAPTER SUMMARY

What Did You Learn?	Explanation/Examples	Review Exercises

Section 3.1

What Did You Learn?	Explanation/Examples	Review Exercises
Recognize and evaluate exponential functions with base a (p. 216).	The exponential function f with base a is denoted by $f(x) = a^x$ where $a > 0$, $a \neq 1$, and x is any real number.	1–6
Graph exponential functions and use the One-to-One Property (p. 217).	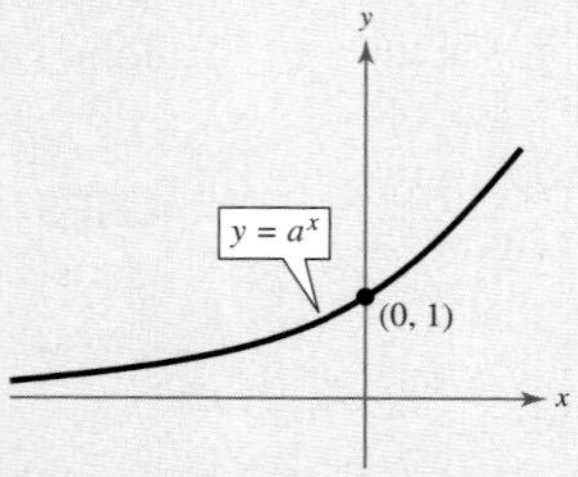 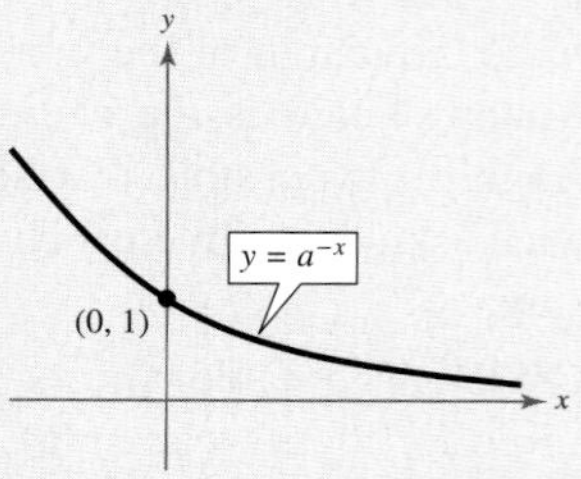 **One-to-One Property:** For $a > 0$ and $a \neq 1$, $a^x = a^y$ if and only if $x = y$.	7–24
Recognize, evaluate, and graph exponential functions with base e (p. 220).	The function $f(x) = e^x$ is called the natural exponential function. 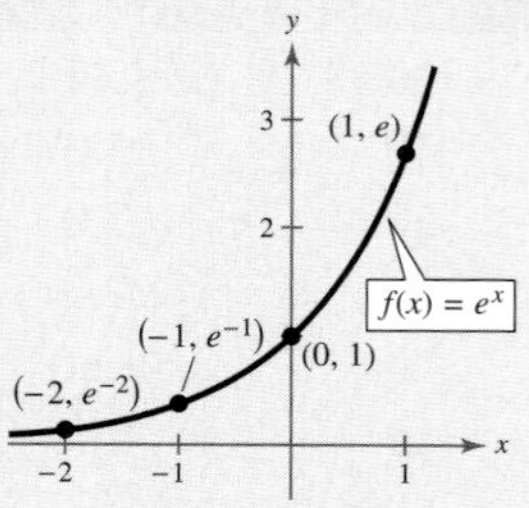	25–32
Use exponential functions to model and solve real-life problems (p. 221).	Exponential functions are used in compound interest formulas (See Example 8.) and in radioactive decay models. (See Example 9.)	33–36

Section 3.2

What Did You Learn?	Explanation/Examples	Review Exercises
Recognize and evaluate logarithmic functions with base a (p. 227).	For $x > 0$, $a > 0$, and $a \neq 1$, $y = \log_a x$ if and only if $x = a^y$. The function $f(x) = \log_a x$ is called the logarithmic function with base a. The logarithmic function with base 10 is the common logarithmic function. It is denoted by $\log_{10}$ or $\log$.	37–48
Graph logarithmic functions (p. 229) and recognize, evaluate, and graph natural logarithmic functions (p. 231).	The graph of $y = \log_a x$ is a reflection of the graph of $y = a^x$ about the line $y = x$. The function defined by $f(x) = \ln x$, $x > 0$, is called the natural logarithmic function. Its graph is a reflection of the graph of $f(x) = e^x$ about the line $y = x$.	49–52 53–58
Use logarithmic functions to model and solve real-life problems (p. 233).	A logarithmic function is used in the human memory model. (See Example 11.)	59, 60

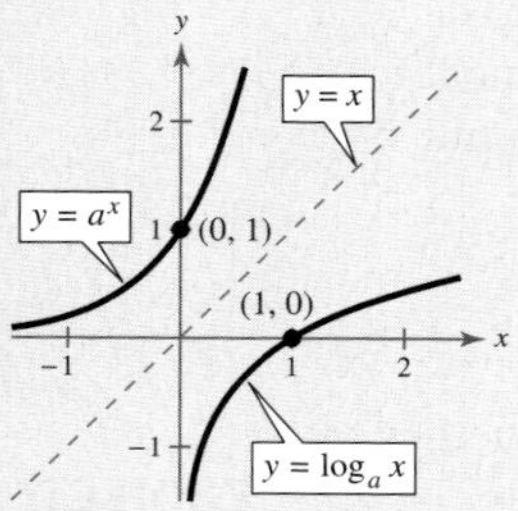

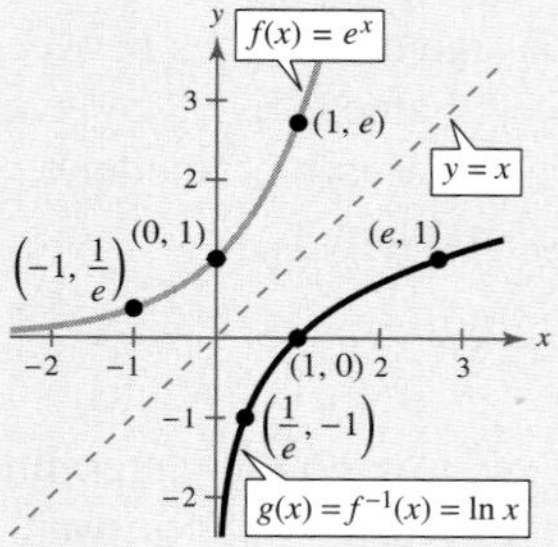

	What Did You Learn?	Explanation/Examples	Review Exercises
Section 3.3	Use the change-of-base formula to rewrite and evaluate logarithmic expressions *(p. 237)*.	Let a, b, and x be positive real numbers such that $a \neq 1$ and $b \neq 1$. Then $\log_a x$ can be converted to a different base as follows. *Base b* $\qquad$ *Base 10* $\qquad$ *Base e* $\log_a x = \dfrac{\log_b x}{\log_b a} \qquad \log_a x = \dfrac{\log x}{\log a} \qquad \log_a x = \dfrac{\ln x}{\ln a}$	61–64
	Use properties of logarithms to evaluate, rewrite, expand, or condense logarithmic expressions *(p. 238)*.	Let a be a positive number $(a \neq 1)$, n be a real number, and u and v be positive real numbers. **1. Product Property:** $\log_a(uv) = \log_a u + \log_a v$ $\qquad\qquad\qquad\quad \ln(uv) = \ln u + \ln v$ **2. Quotient Property:** $\log_a(u/v) = \log_a u - \log_a v$ $\qquad\qquad\qquad\quad\; \ln(u/v) = \ln u - \ln v$ **3. Power Property:** $\log_a u^n = n \log_a u$, $\ln u^n = n \ln u$	65–80
	Use logarithmic functions to model and solve real-life problems *(p. 240)*.	Logarithmic functions can be used to find an equation that relates the periods of several planets and their distances from the sun. (See Example 7.)	81, 82
Section 3.4	Solve simple exponential and logarithmic equations *(p. 244)*.	One-to-One Properties and Inverse Properties of exponential or logarithmic functions can be used to help solve exponential or logarithmic equations.	83–88
	Solve more complicated exponential equations *(p. 245)* and logarithmic equations *(p. 247)*.	To solve more complicated equations, rewrite the equations so that the One-to-One Properties and Inverse Properties of exponential or logarithmic functions can be used. (See Examples 2–8.)	89–108
	Use exponential and logarithmic equations to model and solve real-life problems *(p. 249)*.	Exponential and logarithmic equations can be used to find how long it will take to double an investment (see Example 10) and to find the year in which companies reached a given amount of sales. (See Example 11.)	109, 110
Section 3.5	Recognize the five most common types of models involving exponential and logarithmic functions *(p. 255)*.	**1. Exponential growth model:** $y = ae^{bx}$, $b > 0$ **2. Exponential decay model:** $y = ae^{-bx}$, $b > 0$ **3. Gaussian model:** $y = ae^{-(x-b)^2/c}$ **4. Logistic growth model:** $y = \dfrac{a}{1 + be^{-rx}}$ **5. Logarithmic models:** $y = a + b \ln x$, $y = a + b \log x$	111–116
	Use exponential growth and decay functions to model and solve real-life problems *(p. 256)*.	An exponential growth function can be used to model a population of fruit flies (see Example 2) and an exponential decay function can be used to find the age of a fossil (see Example 3).	117–120
	Use Gaussian functions *(p. 259)*, logistic growth functions *(p. 260)*, and logarithmic functions *(p. 261)* to model and solve real-life problems.	A Gaussian function can be used to model SAT math scores for college-bound seniors. (See Example 4.) A logistic growth function can be used to model the spread of a flu virus. (See Example 5.) A logarithmic function can be used to find the intensity of an earthquake using its magnitude. (See Example 6.)	121–123

3 REVIEW EXERCISES

See www.CalcChat.com for worked-out solutions to odd-numbered exercises.

3.1 In Exercises 1–6, evaluate the function at the indicated value of x. Round your result to three decimal places.

1. $f(x) = 0.3^x$, $x = 1.5$
2. $f(x) = 30^x$, $x = \sqrt{3}$
3. $f(x) = 2^{-0.5x}$, $x = \pi$
4. $f(x) = 1278^{x/5}$, $x = 1$
5. $f(x) = 7(0.2^x)$, $x = -\sqrt{11}$
6. $f(x) = -14(5^x)$, $x = -0.8$

In Exercises 7–14, use the graph of f to describe the transformation that yields the graph of g.

7. $f(x) = 2^x$, $g(x) = 2^x - 2$
8. $f(x) = 5^x$, $g(x) = 5^x + 1$
9. $f(x) = 4^x$, $g(x) = 4^{-x+2}$
10. $f(x) = 6^x$, $g(x) = 6^{x+1}$
11. $f(x) = 3^x$, $g(x) = 1 - 3^x$
12. $f(x) = 0.1^x$, $g(x) = -0.1^x$
13. $f(x) = \left(\frac{1}{2}\right)^x$, $g(x) = -\left(\frac{1}{2}\right)^{x+2}$
14. $f(x) = \left(\frac{2}{3}\right)^x$, $g(x) = 8 - \left(\frac{2}{3}\right)^x$

In Exercises 15–20, use a graphing utility to construct a table of values for the function. Then sketch the graph of the function.

15. $f(x) = 4^{-x} + 4$
16. $f(x) = 2.65^{x-1}$
17. $f(x) = 5^{x-2} + 4$
18. $f(x) = 2^{x-6} - 5$
19. $f(x) = \left(\frac{1}{2}\right)^{-x} + 3$
20. $f(x) = \left(\frac{1}{8}\right)^{x+2} - 5$

In Exercises 21–24, use the One-to-One Property to solve the equation for x.

21. $\left(\frac{1}{3}\right)^{x-3} = 9$
22. $3^{x+3} = \frac{1}{81}$
23. $e^{3x-5} = e^7$
24. $e^{8-2x} = e^{-3}$

In Exercises 25–28, evaluate $f(x) = e^x$ at the indicated value of x. Round your result to three decimal places.

25. $x = 8$
26. $x = \frac{5}{8}$
27. $x = -1.7$
28. $x = 0.278$

In Exercises 29–32, use a graphing utility to construct a table of values for the function. Then sketch the graph of the function.

29. $h(x) = e^{-x/2}$
30. $h(x) = 2 - e^{-x/2}$
31. $f(x) = e^{x+2}$
32. $s(t) = 4e^{-2/t}$, $t > 0$

COMPOUND INTEREST In Exercises 33 and 34, complete the table to determine the balance A for P dollars invested at rate r for t years and compounded n times per year.

n	1	2	4	12	365	Continuous
A						

TABLE FOR 33 AND 34

33. $P = \$5000$, $r = 3\%$, $t = 10$ years
34. $P = \$4500$, $r = 2.5\%$, $t = 30$ years

35. WAITING TIMES The average time between incoming calls at a switchboard is 3 minutes. The probability F of waiting less than t minutes until the next incoming call is approximated by the model $F(t) = 1 - e^{-t/3}$. A call has just come in. Find the probability that the next call will be within

(a) $\frac{1}{2}$ minute.   (b) 2 minutes.   (c) 5 minutes.

36. DEPRECIATION After t years, the value V of a car that originally cost $23,970 is given by $V(t) = 23,970\left(\frac{3}{4}\right)^t$.

(a) Use a graphing utility to graph the function.
(b) Find the value of the car 2 years after it was purchased.
(c) According to the model, when does the car depreciate most rapidly? Is this realistic? Explain.
(d) According to the model, when will the car have no value?

3.2 In Exercises 37–40, write the exponential equation in logarithmic form. For example, the logarithmic form of $2^3 = 8$ is $\log_2 8 = 3$.

37. $3^3 = 27$
38. $25^{3/2} = 125$
39. $e^{0.8} = 2.2255\ldots$
40. $e^0 = 1$

In Exercises 41–44, evaluate the function at the indicated value of x without using a calculator.

41. $f(x) = \log x$, $x = 1000$
42. $g(x) = \log_9 x$, $x = 3$
43. $g(x) = \log_2 x$, $x = \frac{1}{4}$
44. $f(x) = \log_3 x$, $x = \frac{1}{81}$

In Exercises 45–48, use the One-to-One Property to solve the equation for x.

45. $\log_4(x + 7) = \log_4 14$
46. $\log_8(3x - 10) = \log_8 5$
47. $\ln(x + 9) = \ln 4$
48. $\ln(2x - 1) = \ln 11$

In Exercises 49–52, find the domain, x-intercept, and vertical asymptote of the logarithmic function and sketch its graph.

49. $g(x) = \log_7 x$
50. $f(x) = \log\left(\frac{x}{3}\right)$
51. $f(x) = 4 - \log(x + 5)$
52. $f(x) = \log(x - 3) + 1$

53. Use a calculator to evaluate $f(x) = \ln x$ at (a) $x = 22.6$ and (b) $x = 0.98$. Round your results to three decimal places if necessary.

 54. Use a calculator to evaluate $f(x) = 5 \ln x$ at (a) $x = e^{-12}$ and (b) $x = \sqrt{3}$. Round your results to three decimal places if necessary.

In Exercises 55–58, find the domain, x-intercept, and vertical asymptote of the logarithmic function and sketch its graph.

55. $f(x) = \ln x + 3$ **56.** $f(x) = \ln(x - 3)$

57. $h(x) = \ln(x^2)$ **58.** $f(x) = \frac{1}{4} \ln x$

59. ANTLER SPREAD The antler spread a (in inches) and shoulder height h (in inches) of an adult male American elk are related by the model $h = 116 \log(a + 40) - 176$. Approximate the shoulder height of a male American elk with an antler spread of 55 inches.

60. SNOW REMOVAL The number of miles s of roads cleared of snow is approximated by the model

$$s = 25 - \frac{13 \ln(h/12)}{\ln 3}, \quad 2 \le h \le 15$$

where h is the depth of the snow in inches. Use this model to find s when $h = 10$ inches.

3.3 In Exercises 61–64, evaluate the logarithm using the change-of-base formula. Do each exercise twice, once with common logarithms and once with natural logarithms. Round the results to three decimal places.

61. $\log_2 6$ **62.** $\log_{12} 200$

63. $\log_{1/2} 5$ **64.** $\log_3 0.28$

In Exercises 65–68, use the properties of logarithms to rewrite and simplify the logarithmic expression.

65. $\log 18$ **66.** $\log_2\left(\frac{1}{12}\right)$

67. $\ln 20$ **68.** $\ln(3e^{-4})$

In Exercises 69–74, use the properties of logarithms to expand the expression as a sum, difference, and/or constant multiple of logarithms. (Assume all variables are positive.)

69. $\log_5 5x^2$ **70.** $\log 7x^4$

71. $\log_3 \dfrac{9}{\sqrt{x}}$ **72.** $\log_7 \dfrac{\sqrt[3]{x}}{14}$

73. $\ln x^2 y^2 z$ **74.** $\ln\left(\dfrac{y-1}{4}\right)^2, \quad y > 1$

In Exercises 75–80, condense the expression to the logarithm of a single quantity.

75. $\log_2 5 + \log_2 x$ **76.** $\log_6 y - 2 \log_6 z$

77. $\ln x - \frac{1}{4} \ln y$ **78.** $3 \ln x + 2 \ln(x + 1)$

79. $\frac{1}{2} \log_3 x - 2 \log_3(y + 8)$

80. $5 \ln(x - 2) - \ln(x + 2) - 3 \ln x$

81. CLIMB RATE The time t (in minutes) for a small plane to climb to an altitude of h feet is modeled by $t = 50 \log[18{,}000/(18{,}000 - h)]$, where 18,000 feet is the plane's absolute ceiling.

(a) Determine the domain of the function in the context of the problem.

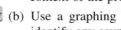 (b) Use a graphing utility to graph the function and identify any asymptotes.

(c) As the plane approaches its absolute ceiling, what can be said about the time required to increase its altitude?

(d) Find the time for the plane to climb to an altitude of 4000 feet.

82. HUMAN MEMORY MODEL Students in a learning theory study were given an exam and then retested monthly for 6 months with an equivalent exam. The data obtained in the study are given as the ordered pairs (t, s), where t is the time in months after the initial exam and s is the average score for the class. Use these data to find a logarithmic equation that relates t and s.

(1, 84.2), (2, 78.4), (3, 72.1),
(4, 68.5), (5, 67.1), (6, 65.3)

3.4 In Exercises 83–88, solve for x.

83. $5^x = 125$ **84.** $6^x = \frac{1}{216}$

85. $e^x = 3$ **86.** $\log_6 x = -1$

87. $\ln x = 4$ **88.** $\ln x = -1.6$

In Exercises 89–92, solve the exponential equation algebraically. Approximate your result to three decimal places.

89. $e^{4x} = e^{x^2+3}$ **90.** $e^{3x} = 25$

91. $2^x - 3 = 29$ **92.** $e^{2x} - 6e^x + 8 = 0$

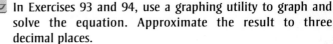 In Exercises 93 and 94, use a graphing utility to graph and solve the equation. Approximate the result to three decimal places.

93. $25e^{-0.3x} = 12$ **94.** $2^x = 3 + x - e^x$

In Exercises 95–104, solve the logarithmic equation algebraically. Approximate the result to three decimal places.

95. $\ln 3x = 8.2$ **96.** $4 \ln 3x = 15$

97. $\ln x - \ln 3 = 2$ **98.** $\ln x - \ln 5 = 4$

99. $\ln \sqrt{x} = 4$ **100.** $\ln \sqrt{x + 8} = 3$

101. $\log_8(x - 1) = \log_8(x - 2) - \log_8(x + 2)$

102. $\log_6(x + 2) - \log_6 x = \log_6(x + 5)$

103. $\log(1 - x) = -1$ **104.** $\log(-x - 4) = 2$

 In Exercises 105–108, use a graphing utility to graph and solve the equation. Approximate the result to three decimal places.

105. $2 \ln(x + 3) - 3 = 0$ **106.** $x - 2 \log(x + 4) = 0$

107. $6 \log(x^2 + 1) - x = 0$

108. $3 \ln x + 2 \log x = e^x - 25$

109. COMPOUND INTEREST   You deposit $8500 in an account that pays 3.5% interest, compounded continuously. How long will it take for the money to triple?

110. METEOROLOGY   The speed of the wind S (in miles per hour) near the center of a tornado and the distance d (in miles) the tornado travels are related by the model $S = 93 \log d + 65$. On March 18, 1925, a large tornado struck portions of Missouri, Illinois, and Indiana with a wind speed at the center of about 283 miles per hour. Approximate the distance traveled by this tornado.

3.5   In Exercises 111–116, match the function with its graph. [The graphs are labeled (a), (b), (c), (d), (e), and (f).]

(a)

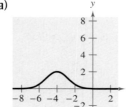

(b)

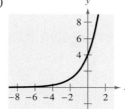

(c)

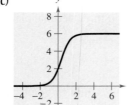

(d)

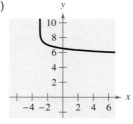

(e)

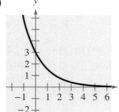

(f)

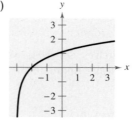

111. $y = 3e^{-2x/3}$ **112.** $y = 4e^{2x/3}$

113. $y = \ln(x + 3)$ **114.** $y = 7 - \log(x + 3)$

115. $y = 2e^{-(x+4)^2/3}$ **116.** $y = \dfrac{6}{1 + 2e^{-2x}}$

In Exercises 117 and 118, find the exponential model $y = ae^{bx}$ that passes through the points.

117. $(0, 2), (4, 3)$ **118.** $\left(0, \frac{1}{2}\right), (5, 5)$

119. POPULATION   In 2007, the population of Florida residents aged 65 and over was about 3.10 million. In 2015 and 2020, the populations of Florida residents aged 65 and over are projected to be about 4.13 million and 5.11 million, respectively. An exponential growth model that approximates these data is given by $P = 2.36e^{0.0382t}$, $7 \le t \le 20$, where P is the population (in millions) and $t = 7$ represents 2007. (Source: U.S. Census Bureau)

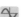 (a) Use a graphing utility to graph the model and the data in the same viewing window. Is the model a good fit for the data? Explain.

(b) According to the model, when will the population of Florida residents aged 65 and over reach 5.5 million? Does your answer seem reasonable? Explain.

120. WILDLIFE POPULATION   A species of bat is in danger of becoming extinct. Five years ago, the total population of the species was 2000. Two years ago, the total population of the species was 1400. What was the total population of the species one year ago?

121. TEST SCORES   The test scores for a biology test follow a normal distribution modeled by $y = 0.0499e^{-(x-71)^2/128}$, $40 \le x \le 100$, where x is the test score. Use a graphing utility to graph the equation and estimate the average test score.

122. TYPING SPEED   In a typing class, the average number N of words per minute typed after t weeks of lessons was found to be $N = 157/(1 + 5.4e^{-0.12t})$. Find the time necessary to type (a) 50 words per minute and (b) 75 words per minute.

123. SOUND INTENSITY   The relationship between the number of decibels β and the intensity of a sound I in watts per square meter is $\beta = 10 \log(I/10^{-12})$. Find I for each decibel level β.

(a) $\beta = 60$ (b) $\beta = 135$ (c) $\beta = 1$

EXPLORATION

124. Consider the graph of $y = e^{kt}$. Describe the characteristics of the graph when k is positive and when k is negative.

TRUE OR FALSE? In Exercises 125 and 126, determine whether the equation is true or false. Justify your answer.

125. $\log_b b^{2x} = 2x$ **126.** $\ln(x + y) = \ln x + \ln y$

3 CHAPTER TEST

See www.CalcChat.com for worked-out solutions to odd-numbered exercises.

Take this test as you would take a test in class. When you are finished, check your work against the answers given in the back of the book.

In Exercises 1–4, evaluate the expression. Approximate your result to three decimal places.

1. $4.2^{0.6}$ **2.** $4^{3\pi/2}$ **3.** $e^{-7/10}$ **4.** $e^{3.1}$

In Exercises 5–7, construct a table of values. Then sketch the graph of the function.

5. $f(x) = 10^{-x}$ **6.** $f(x) = -6^{x-2}$ **7.** $f(x) = 1 - e^{2x}$

8. Evaluate (a) $\log_7 7^{-0.89}$ and (b) $4.6 \ln e^2$.

In Exercises 9–11, construct a table of values. Then sketch the graph of the function. Identify any asymptotes.

9. $f(x) = -\log x - 6$ **10.** $f(x) = \ln(x - 4)$ **11.** $f(x) = 1 + \ln(x + 6)$

In Exercises 12–14, evaluate the logarithm using the change-of-base formula. Round your result to three decimal places.

12. $\log_7 44$ **13.** $\log_{16} 0.63$ **14.** $\log_{3/4} 24$

In Exercises 15–17, use the properties of logarithms to expand the expression as a sum, difference, and/or constant multiple of logarithms.

15. $\log_2 3a^4$ **16.** $\ln \dfrac{5\sqrt{x}}{6}$ **17.** $\log \dfrac{(x-1)^3}{y^2 z}$

In Exercises 18–20, condense the expression to the logarithm of a single quantity.

18. $\log_3 13 + \log_3 y$ **19.** $4 \ln x - 4 \ln y$
20. $3 \ln x - \ln(x + 3) + 2 \ln y$

In Exercises 21–26, solve the equation algebraically. Approximate your result to three decimal places.

21. $5^x = \dfrac{1}{25}$ **22.** $3e^{-5x} = 132$

23. $\dfrac{1025}{8 + e^{4x}} = 5$ **24.** $\ln x = \dfrac{1}{2}$

25. $18 + 4 \ln x = 7$ **26.** $\log x + \log(x - 15) = 2$

27. Find an exponential growth model for the graph shown in the figure.

28. The half-life of radioactive actinium (^{227}Ac) is 21.77 years. What percent of a present amount of radioactive actinium will remain after 19 years?

29. A model that can be used for predicting the height H (in centimeters) of a child based on his or her age is $H = 70.228 + 5.104x + 9.222 \ln x$, $\frac{1}{4} \le x \le 6$, where x is the age of the child in years. (Source: Snapshots of Applications in Mathematics)

(a) Construct a table of values. Then sketch the graph of the model.

(b) Use the graph from part (a) to estimate the height of a four-year-old child. Then calculate the actual height using the model.

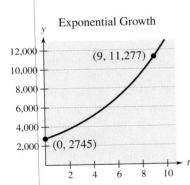

Exponential Growth

FIGURE FOR 27

3 CUMULATIVE TEST FOR CHAPTERS 1–3   See www.CalcChat.com for worked-out solutions to odd-numbered exercises.

Take this test as you would take a test in class. When you are finished, check your work against the answers given in the back of the book.

1. Plot the points $(-2, 5)$ and $(3, -1)$. Find the coordinates of the midpoint of the line segment joining the points and the distance between the points.

In Exercises 2–4, graph the equation without using a graphing utility.

2. $x - 3y + 12 = 0$ 3. $y = x^2 - 9$ 4. $y = \sqrt{4 - x}$

5. Find an equation of the line passing through $\left(-\frac{1}{2}, 1\right)$ and $(3, 8)$.

6. Explain why the graph at the left does not represent y as a function of x.

7. Evaluate (if possible) the function given by $f(x) = \dfrac{x}{x - 2}$ for each value.

(a) $f(6)$ (b) $f(2)$ (c) $f(s + 2)$

8. Compare the graph of each function with the graph of $y = \sqrt[3]{x}$. (*Note:* It is not necessary to sketch the graphs.)

(a) $r(x) = \frac{1}{2}\sqrt[3]{x}$ (b) $h(x) = \sqrt[3]{x} + 2$ (c) $g(x) = \sqrt[3]{x + 2}$

In Exercises 9 and 10, find (a) $(f + g)(x)$, (b) $(f - g)(x)$, (c) $(fg)(x)$, and (d) $(f/g)(x)$. What is the domain of f/g?

9. $f(x) = x - 3,\quad g(x) = 4x + 1$ 10. $f(x) = \sqrt{x - 1},\quad g(x) = x^2 + 1$

In Exercises 11 and 12, find (a) $f \circ g$ and (b) $g \circ f$. Find the domain of each composite function.

11. $f(x) = 2x^2,\quad g(x) = \sqrt{x + 6}$
12. $f(x) = x - 2,\quad g(x) = |x|$

13. Determine whether $h(x) = -5x + 3$ has an inverse function. If so, find the inverse function.

14. The power P produced by a wind turbine is proportional to the cube of the wind speed S. A wind speed of 27 miles per hour produces a power output of 750 kilowatts. Find the output for a wind speed of 40 miles per hour.

15. Find the quadratic function whose graph has a vertex at $(-8, 5)$ and passes through the point $(-4, -7)$.

In Exercises 16–18, sketch the graph of the function without the aid of a graphing utility.

16. $h(x) = -(x^2 + 4x)$ 17. $f(t) = \frac{1}{4}t(t - 2)^2$
18. $g(s) = s^2 + 4s + 10$

In Exercises 19–21, find all the zeros of the function and write the function as a product of linear factors.

19. $f(x) = x^3 + 2x^2 + 4x + 8$
20. $f(x) = x^4 + 4x^3 - 21x^2$
21. $f(x) = 2x^4 - 11x^3 + 30x^2 - 62x - 40$

22. Use long division to divide $6x^3 - 4x^2$ by $2x^2 + 1$.

23. Use synthetic division to divide $3x^4 + 2x^2 - 5x + 3$ by $x - 2$.

24. Use the Intermediate Value Theorem and a graphing utility to find intervals one unit in length in which the function $g(x) = x^3 + 3x^2 - 6$ is guaranteed to have a zero. Approximate the real zeros of the function.

In Exercises 25–27, sketch the graph of the rational function by hand. Be sure to identify all intercepts and asymptotes.

25. $f(x) = \dfrac{2x}{x^2 + 2x - 3}$

26. $f(x) = \dfrac{x^2 - 4}{x^2 + x - 2}$

27. $f(x) = \dfrac{x^3 - 2x^2 - 9x + 18}{x^2 + 4x + 3}$

In Exercises 28 and 29, solve the inequality. Sketch the solution set on the real number line.

28. $2x^3 - 18x \le 0$

29. $\dfrac{1}{x + 1} \ge \dfrac{1}{x + 5}$

In Exercises 30 and 31, use the graph of f to describe the transformation that yields the graph of g.

30. $f(x) = \left(\frac{2}{5}\right)^x, \quad g(x) = -\left(\frac{2}{5}\right)^{-x+3}$

31. $f(x) = 2.2^x, \quad g(x) = -2.2^x + 4$

In Exercises 32–35, use a calculator to evaluate the expression. Round your result to three decimal places.

32. $\log 98$

33. $\log\left(\frac{6}{7}\right)$

34. $\ln\sqrt{31}$

35. $\ln\left(\sqrt{40} - 5\right)$

36. Use the properties of logarithms to expand $\ln\left(\dfrac{x^2 - 16}{x^4}\right)$, where $x > 4$.

37. Write $2\ln x - \frac{1}{2}\ln(x + 5)$ as a logarithm of a single quantity.

In Exercises 38–40, solve the equation algebraically. Approximate the result to three decimal places.

38. $6e^{2x} = 72$

39. $e^{2x} - 13e^x + 42 = 0$

40. $\ln\sqrt{x + 2} = 3$

41. The sales S (in billions of dollars) of lottery tickets in the United States from 1997 through 2007 are shown in the table. (Source: TLF Publications, Inc.)

(a) Use a graphing utility to create a scatter plot of the data. Let t represent the year, with $t = 7$ corresponding to 1997.

(b) Use the *regression* feature of the graphing utility to find a cubic model for the data.

(c) Use the graphing utility to graph the model in the same viewing window used for the scatter plot. How well does the model fit the data?

(d) Use the model to predict the sales of lottery tickets in 2015. Does your answer seem reasonable? Explain.

42. The number N of bacteria in a culture is given by the model $N = 175e^{kt}$, where t is the time in hours. If $N = 420$ when $t = 8$, estimate the time required for the population to double in size.

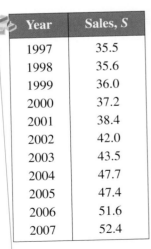

Year	Sales, S
1997	35.5
1998	35.6
1999	36.0
2000	37.2
2001	38.4
2002	42.0
2003	43.5
2004	47.7
2005	47.4
2006	51.6
2007	52.4

TABLE FOR 41

PROOFS IN MATHEMATICS

Each of the following three properties of logarithms can be proved by using properties of exponential functions.

Slide Rules

The slide rule was invented by William Oughtred (1574–1660) in 1625. The slide rule is a computational device with a sliding portion and a fixed portion. A slide rule enables you to perform multiplication by using the Product Property of Logarithms. There are other slide rules that allow for the calculation of roots and trigonometric functions. Slide rules were used by mathematicians and engineers until the invention of the hand-held calculator in 1972.

Properties of Logarithms (p. 238)

Let a be a positive number such that $a \neq 1$, and let n be a real number. If u and v are positive real numbers, the following properties are true.

	Logarithm with Base a	Natural Logarithm
1. **Product Property:**	$\log_a(uv) = \log_a u + \log_a v$	$\ln(uv) = \ln u + \ln v$
2. **Quotient Property:**	$\log_a \dfrac{u}{v} = \log_a u - \log_a v$	$\ln \dfrac{u}{v} = \ln u - \ln v$
3. **Power Property:**	$\log_a u^n = n \log_a u$	$\ln u^n = n \ln u$

Proof

Let

$$x = \log_a u \quad \text{and} \quad y = \log_a v.$$

The corresponding exponential forms of these two equations are

$$a^x = u \quad \text{and} \quad a^y = v.$$

To prove the Product Property, multiply u and v to obtain

$$uv = a^x a^y = a^{x+y}.$$

The corresponding logarithmic form of $uv = a^{x+y}$ is $\log_a(uv) = x + y$. So,

$$\log_a(uv) = \log_a u + \log_a v.$$

To prove the Quotient Property, divide u by v to obtain

$$\frac{u}{v} = \frac{a^x}{a^y} = a^{x-y}.$$

The corresponding logarithmic form of $\dfrac{u}{v} = a^{x-y}$ is $\log_a \dfrac{u}{v} = x - y$. So,

$$\log_a \frac{u}{v} = \log_a u - \log_a v.$$

To prove the Power Property, substitute a^x for u in the expression $\log_a u^n$, as follows.

$\log_a u^n = \log_a (a^x)^n$	Substitute a^x for u.
$= \log_a a^{nx}$	Property of Exponents
$= nx$	Inverse Property of Logarithms
$= n \log_a u$	Substitute $\log_a u$ for x.

So, $\log_a u^n = n \log_a u.$

PROBLEM SOLVING

This collection of thought-provoking and challenging exercises further explores and expands upon concepts learned in this chapter.

1. Graph the exponential function given by $y = a^x$ for $a = 0.5$, 1.2, and 2.0. Which of these curves intersects the line $y = x$? Determine all positive numbers a for which the curve $y = a^x$ intersects the line $y = x$.

2. Use a graphing utility to graph $y_1 = e^x$ and each of the functions $y_2 = x^2$, $y_3 = x^3$, $y_4 = \sqrt{x}$, and $y_5 = |x|$. Which function increases at the greatest rate as x approaches $+\infty$?

3. Use the result of Exercise 2 to make a conjecture about the rate of growth of $y_1 = e^x$ and $y = x^n$, where n is a natural number and x approaches $+\infty$.

4. Use the results of Exercises 2 and 3 to describe what is implied when it is stated that a quantity is growing exponentially.

5. Given the exponential function

$$f(x) = a^x$$

show that

(a) $f(u + v) = f(u) \cdot f(v)$.   (b) $f(2x) = [f(x)]^2$.

6. Given that

$$f(x) = \frac{e^x + e^{-x}}{2} \text{ and } g(x) = \frac{e^x - e^{-x}}{2}$$

show that

$$[f(x)]^2 - [g(x)]^2 = 1.$$

7. Use a graphing utility to compare the graph of the function given by $y = e^x$ with the graph of each given function. [$n!$ (read "n factorial") is defined as $n! = 1 \cdot 2 \cdot 3 \cdots (n - 1) \cdot n$.]

(a) $y_1 = 1 + \dfrac{x}{1!}$

(b) $y_2 = 1 + \dfrac{x}{1!} + \dfrac{x^2}{2!}$

(c) $y_3 = 1 + \dfrac{x}{1!} + \dfrac{x^2}{2!} + \dfrac{x^3}{3!}$

8. Identify the pattern of successive polynomials given in Exercise 7. Extend the pattern one more term and compare the graph of the resulting polynomial function with the graph of $y = e^x$. What do you think this pattern implies?

9. Graph the function given by

$$f(x) = e^x - e^{-x}.$$

From the graph, the function appears to be one-to-one. Assuming that the function has an inverse function, find $f^{-1}(x)$.

10. Find a pattern for $f^{-1}(x)$ if

$$f(x) = \frac{a^x + 1}{a^x - 1}$$

where $a > 0$, $a \neq 1$.

11. By observation, identify the equation that corresponds to the graph. Explain your reasoning.

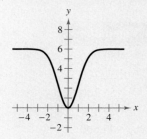

(a) $y = 6e^{-x^2/2}$

(b) $y = \dfrac{6}{1 + e^{-x/2}}$

(c) $y = 6(1 - e^{-x^2/2})$

12. You have two options for investing $500. The first earns 7% compounded annually and the second earns 7% simple interest. The figure shows the growth of each investment over a 30-year period.

(a) Identify which graph represents each type of investment. Explain your reasoning.

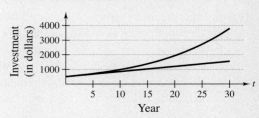

(b) Verify your answer in part (a) by finding the equations that model the investment growth and graphing the models.

(c) Which option would you choose? Explain your reasoning.

13. Two different samples of radioactive isotopes are decaying. The isotopes have initial amounts of c_1 and c_2, as well as half-lives of k_1 and k_2, respectively. Find the time t required for the samples to decay to equal amounts.

277

14. A lab culture initially contains 500 bacteria. Two hours later, the number of bacteria has decreased to 200. Find the exponential decay model of the form

$$B = B_0 a^{kt}$$

that can be used to approximate the number of bacteria after t hours.

15. The table shows the colonial population estimates of the American colonies from 1700 to 1780. (Source: U.S. Census Bureau)

Year	Population
1700	250,900
1710	331,700
1720	466,200
1730	629,400
1740	905,600
1750	1,170,800
1760	1,593,600
1770	2,148,100
1780	2,780,400

In each of the following, let y represent the population in the year t, with $t = 0$ corresponding to 1700.

(a) Use the *regression* feature of a graphing utility to find an exponential model for the data.

(b) Use the *regression* feature of the graphing utility to find a quadratic model for the data.

(c) Use the graphing utility to plot the data and the models from parts (a) and (b) in the same viewing window.

(d) Which model is a better fit for the data? Would you use this model to predict the population of the United States in 2015? Explain your reasoning.

16. Show that $\dfrac{\log_a x}{\log_{a/b} x} = 1 + \log_a \dfrac{1}{b}$.

17. Solve $(\ln x)^2 = \ln x^2$.

18. Use a graphing utility to compare the graph of the function $y = \ln x$ with the graph of each given function.

(a) $y_1 = x - 1$

(b) $y_2 = (x - 1) - \frac{1}{2}(x - 1)^2$

(c) $y_3 = (x - 1) - \frac{1}{2}(x - 1)^2 + \frac{1}{3}(x - 1)^3$

19. Identify the pattern of successive polynomials given in Exercise 18. Extend the pattern one more term and compare the graph of the resulting polynomial function with the graph of $y = \ln x$. What do you think the pattern implies?

20. Using

$$y = ab^x \qquad \text{and} \qquad y = ax^b$$

take the natural logarithm of each side of each equation. What are the slope and y-intercept of the line relating x and $\ln y$ for $y = ab^x$? What are the slope and y-intercept of the line relating $\ln x$ and $\ln y$ for $y = ax^b$?

In Exercises 21 and 22, use the model

$$y = 80.4 - 11 \ln x, \quad 100 \le x \le 1500$$

which approximates the minimum required ventilation rate in terms of the air space per child in a public school classroom. In the model, x is the air space per child in cubic feet and y is the ventilation rate per child in cubic feet per minute.

21. Use a graphing utility to graph the model and approximate the required ventilation rate if there is 300 cubic feet of air space per child.

22. A classroom is designed for 30 students. The air conditioning system in the room has the capacity of moving 450 cubic feet of air per minute.

(a) Determine the ventilation rate per child, assuming that the room is filled to capacity.

(b) Estimate the air space required per child.

(c) Determine the minimum number of square feet of floor space required for the room if the ceiling height is 30 feet.

In Exercises 23–26, (a) use a graphing utility to create a scatter plot of the data, (b) decide whether the data could best be modeled by a linear model, an exponential model, or a logarithmic model, (c) explain why you chose the model you did in part (b), (d) use the *regression* feature of a graphing utility to find the model you chose in part (b) for the data and graph the model with the scatter plot, and (e) determine how well the model you chose fits the data.

23. $(1, 2.0), (1.5, 3.5), (2, 4.0), (4, 5.8), (6, 7.0), (8, 7.8)$

24. $(1, 4.4), (1.5, 4.7), (2, 5.5), (4, 9.9), (6, 18.1), (8, 33.0)$

25. $(1, 7.5), (1.5, 7.0), (2, 6.8), (4, 5.0), (6, 3.5), (8, 2.0)$

26. $(1, 5.0), (1.5, 6.0), (2, 6.4), (4, 7.8), (6, 8.6), (8, 9.0)$

Trigonometry

4

In Mathematics

Trigonometry is used to find relationships between the sides and angles of triangles, and to write trigonometric functions as models of real-life quantities.

In Real Life

Trigonometric functions are used to model quantities that are periodic. For instance, throughout the day, the depth of water at the end of a dock in Bar Harbor, Maine varies with the tides. The depth can be modeled by a trigonometric function. (See Example 7, page 325.)

Andre Jenny/Alamy

IN CAREERS

There are many careers that use trigonometry. Several are listed below.

- Biologist
  Exercise 70, page 308

- Meteorologist
  Exercise 99, page 318

- Mechanical Engineer
  Exercise 95, page 339

- Surveyor
  Exercise 41, page 359

4.1 RADIAN AND DEGREE MEASURE

What you should learn

- Describe angles.
- Use radian measure.
- Use degree measure.
- Use angles to model and solve real-life problems.

Why you should learn it

You can use angles to model and solve real-life problems. For instance, in Exercise 119 on page 291, you are asked to use angles to find the speed of a bicycle.

Angles

As derived from the Greek language, the word **trigonometry** means "measurement of triangles." Initially, trigonometry dealt with relationships among the sides and angles of triangles and was used in the development of astronomy, navigation, and surveying. With the development of calculus and the physical sciences in the 17th century, a different perspective arose—one that viewed the classic trigonometric relationships as *functions* with the set of real numbers as their domains. Consequently, the applications of trigonometry expanded to include a vast number of physical phenomena involving rotations and vibrations. These phenomena include sound waves, light rays, planetary orbits, vibrating strings, pendulums, and orbits of atomic particles.

The approach in this text incorporates *both* perspectives, starting with angles and their measure.

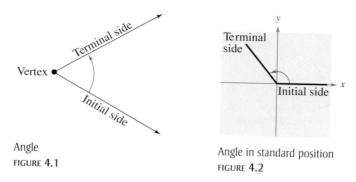

Angle
FIGURE 4.1

Angle in standard position
FIGURE 4.2

An **angle** is determined by rotating a ray (half-line) about its endpoint. The starting position of the ray is the **initial side** of the angle, and the position after rotation is the **terminal side,** as shown in Figure 4.1. The endpoint of the ray is the **vertex** of the angle. This perception of an angle fits a coordinate system in which the origin is the vertex and the initial side coincides with the positive x-axis. Such an angle is in **standard position,** as shown in Figure 4.2. **Positive angles** are generated by counterclockwise rotation, and **negative angles** by clockwise rotation, as shown in Figure 4.3. Angles are labeled with Greek letters α (alpha), β (beta), and θ (theta), as well as uppercase letters $A, B,$ and C. In Figure 4.4, note that angles α and β have the same initial and terminal sides. Such angles are **coterminal.**

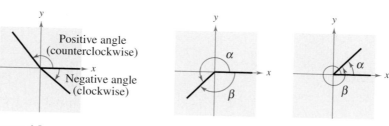

FIGURE 4.3

FIGURE 4.4 Coterminal angles

Radian Measure

The **measure of an angle** is determined by the amount of rotation from the initial side to the terminal side. One way to measure angles is in *radians*. This type of measure is especially useful in calculus. To define a radian, you can use a **central angle** of a circle, one whose vertex is the center of the circle, as shown in Figure 4.5.

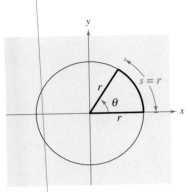

Arc length = radius when θ = 1 radian

FIGURE **4.5**

> ## Definition of Radian
>
> One **radian** is the measure of a central angle θ that intercepts an arc s equal in length to the radius r of the circle. See Figure 4.5. Algebraically, this means that
>
> $$\theta = \frac{s}{r}$$
>
> where θ is measured in radians.

Because the circumference of a circle is $2\pi r$ units, it follows that a central angle of one full revolution (counterclockwise) corresponds to an arc length of

$$s = 2\pi r.$$

Moreover, because $2\pi \approx 6.28$, there are just over six radius lengths in a full circle, as shown in Figure 4.6. Because the units of measure for s and r are the same, the ratio s/r has no units—it is simply a real number.

Because the radian measure of an angle of one full revolution is 2π, you can obtain the following.

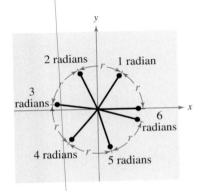

FIGURE **4.6**

$$\frac{1}{2} \text{ revolution} = \frac{2\pi}{2} = \pi \text{ radians}$$

$$\frac{1}{4} \text{ revolution} = \frac{2\pi}{4} = \frac{\pi}{2} \text{ radians}$$

$$\frac{1}{6} \text{ revolution} = \frac{2\pi}{6} = \frac{\pi}{3} \text{ radians}$$

These and other common angles are shown in Figure 4.7.

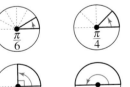

FIGURE **4.7**

Study Tip

One revolution around a circle of radius r corresponds to an angle of 2π radians because

$$\theta = \frac{s}{r} = \frac{2\pi r}{r} = 2\pi \text{ radians.}$$

Recall that the four quadrants in a coordinate system are numbered I, II, III, and IV. Figure 4.8 on page 282 shows which angles between 0 and 2π lie in each of the four quadrants. Note that angles between 0 and $\pi/2$ are **acute** angles and angles between $\pi/2$ and π are **obtuse** angles.

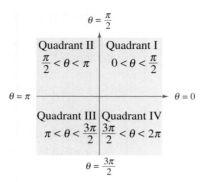

FIGURE **4.8**

Study Tip

The phrase "the terminal side of θ lies in a quadrant" is often abbreviated by simply saying that "θ lies in a quadrant." The terminal sides of the "quadrant angles" 0, π/2, π, and 3π/2 do not lie within quadrants.

Algebra Help

You can review operations involving fractions in Appendix A.1.

Two angles are coterminal if they have the same initial and terminal sides. For instance, the angles 0 and 2π are coterminal, as are the angles $\pi/6$ and $13\pi/6$. You can find an angle that is coterminal to a given angle θ by adding or subtracting 2π (one revolution), as demonstrated in Example 1. A given angle θ has infinitely many coterminal angles. For instance, $\theta = \pi/6$ is coterminal with

$$\frac{\pi}{6} + 2n\pi$$

where n is an integer.

Example 1 Sketching and Finding Coterminal Angles

a. For the positive angle $13\pi/6$, subtract 2π to obtain a coterminal angle

$$\frac{13\pi}{6} - 2\pi = \frac{\pi}{6}.$$ See Figure 4.9.

b. For the positive angle $3\pi/4$, subtract 2π to obtain a coterminal angle

$$\frac{3\pi}{4} - 2\pi = -\frac{5\pi}{4}.$$ See Figure 4.10.

c. For the negative angle $-2\pi/3$, add 2π to obtain a coterminal angle

$$-\frac{2\pi}{3} + 2\pi = \frac{4\pi}{3}.$$ See Figure 4.11.

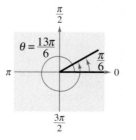

FIGURE **4.9**

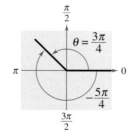

FIGURE **4.10**

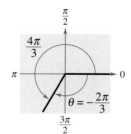

FIGURE **4.11**

CHECK*Point* Now try Exercise 27.

Two positive angles α and β are **complementary** (complements of each other) if their sum is $\pi/2$. Two positive angles are **supplementary** (supplements of each other) if their sum is π. See Figure 4.12.

Complementary angles   Supplementary angles

FIGURE **4.12**

| ▌**Example 2** | **Complementary and Supplementary Angles** |

If possible, find the complement and the supplement of (a) $2\pi/5$ and (b) $4\pi/5$.

Solution

a. The complement of $2\pi/5$ is

$$\frac{\pi}{2} - \frac{2\pi}{5} = \frac{5\pi}{10} - \frac{4\pi}{10} = \frac{\pi}{10}.$$

The supplement of $2\pi/5$ is

$$\pi - \frac{2\pi}{5} = \frac{5\pi}{5} - \frac{2\pi}{5} = \frac{3\pi}{5}.$$

b. Because $4\pi/5$ is greater than $\pi/2$, it has no complement. (Remember that complements are *positive* angles.) The supplement is

$$\pi - \frac{4\pi}{5} = \frac{5\pi}{5} - \frac{4\pi}{5} = \frac{\pi}{5}.$$

CHECK*Point*   Now try Exercise 31.   ▮

Degree Measure

A second way to measure angles is in terms of **degrees,** denoted by the symbol $^{\circ}$. A measure of one degree (1°) is equivalent to a rotation of $\frac{1}{360}$ of a complete revolution about the vertex. To measure angles, it is convenient to mark degrees on the circumference of a circle, as shown in Figure 4.13. So, a full revolution (counterclockwise) corresponds to 360°, a half revolution to 180°, a quarter revolution to 90°, and so on.

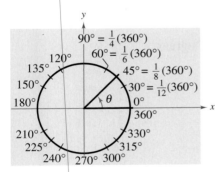

FIGURE **4.13**

Because 2π radians corresponds to one complete revolution, degrees and radians are related by the equations

$$360^{\circ} = 2\pi \text{ rad} \qquad \text{and} \qquad 180^{\circ} = \pi \text{ rad}.$$

From the latter equation, you obtain

$$1^{\circ} = \frac{\pi}{180} \text{ rad} \qquad \text{and} \qquad 1 \text{ rad} = \left(\frac{180^{\circ}}{\pi}\right)$$

which lead to the conversion rules at the top of the next page.

Conversions Between Degrees and Radians

1. To convert degrees to radians, multiply degrees by $\dfrac{\pi \text{ rad}}{180°}$.

2. To convert radians to degrees, multiply radians by $\dfrac{180°}{\pi \text{ rad}}$.

To apply these two conversion rules, use the basic relationship $\pi \text{ rad} = 180°$. (See Figure 4.14.)

$\dfrac{\pi}{6}$
30°

$\dfrac{\pi}{4}$
45°

$\dfrac{\pi}{3}$
60°

$\dfrac{\pi}{2}$
90°

π
180°

2π
360°

FIGURE **4.14**

When no units of angle measure are specified, *radian measure is implied.* For instance, if you write $\theta = 2$, you imply that $\theta = 2$ radians.

Example 3   Converting from Degrees to Radians

a. $135° = (135 \text{ deg})\left(\dfrac{\pi \text{ rad}}{180 \text{ deg}}\right) = \dfrac{3\pi}{4}$ radians   Multiply by $\pi/180$.

b. $540° = (540 \text{ deg})\left(\dfrac{\pi \text{ rad}}{180 \text{ deg}}\right) = 3\pi$ radians   Multiply by $\pi/180$.

c. $-270° = (-270 \text{ deg})\left(\dfrac{\pi \text{ rad}}{180 \text{ deg}}\right) = -\dfrac{3\pi}{2}$ radians   Multiply by $\pi/180$.

CHECK Point ▶ Now try Exercise 57.

Example 4   Converting from Radians to Degrees

a. $-\dfrac{\pi}{2} \text{ rad} = \left(-\dfrac{\pi}{2} \text{ rad}\right)\left(\dfrac{180 \text{ deg}}{\pi \text{ rad}}\right) = -90°$ Multiply by $180/\pi$.

b. $\dfrac{9\pi}{2} \text{ rad} = \left(\dfrac{9\pi}{2} \text{ rad}\right)\left(\dfrac{180 \text{ deg}}{\pi \text{ rad}}\right) = 810°$ Multiply by $180/\pi$.

c. $2 \text{ rad} = (2 \text{ rad})\left(\dfrac{180 \text{ deg}}{\pi \text{ rad}}\right) = \dfrac{360°}{\pi} \approx 114.59°$ Multiply by $180/\pi$.

CHECK Point ▶ Now try Exercise 61.

If you have a calculator with a "radian-to-degree" conversion key, try using it to verify the result shown in part (b) of Example 4.

TECHNOLOGY

With calculators it is convenient to use *decimal* degrees to denote fractional parts of degrees. Historically, however, fractional parts of degrees were expressed in *minutes* and *seconds*, using the prime ($'$) and double prime ($''$) notations, respectively. That is,

$1' = $ one minute $= \dfrac{1}{60}(1°)$

$1'' = $ one second $= \dfrac{1}{3600}(1°)$.

Consequently, an angle of 64 degrees, 32 minutes, and 47 seconds is represented by $\theta = 64° \, 32' \, 47''$. Many calculators have special keys for converting an angle in degrees, minutes, and seconds ($D° \, M' \, S''$) to decimal degree form, and vice versa.

Applications

The *radian measure* formula, $\theta = s/r$, can be used to measure arc length along a circle.

Arc Length

For a circle of radius r, a central angle θ intercepts an arc of length s given by

$$s = r\theta \qquad \text{Length of circular arc}$$

where θ is measured in radians. Note that if $r = 1$, then $s = \theta$, and the radian measure of θ equals the arc length.

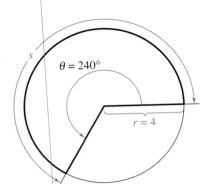

$\theta = 240°$

$r = 4$

s

FIGURE 4.15

| Example 5   Finding Arc Length

A circle has a radius of 4 inches. Find the length of the arc intercepted by a central angle of 240°, as shown in Figure 4.15.

Solution

To use the formula $s = r\theta$, first convert 240° to radian measure.

$$240° = (240 \text{ deg})\left(\frac{\pi \text{ rad}}{180 \text{ deg}}\right)$$

$$= \frac{4\pi}{3} \text{ radians}$$

Then, using a radius of $r = 4$ inches, you can find the arc length to be

$$s = r\theta$$

$$= 4\left(\frac{4\pi}{3}\right)$$

$$= \frac{16\pi}{3} \approx 16.76 \text{ inches.}$$

Note that the units for $r\theta$ are determined by the units for r because θ is given in radian measure, which has no units.

CHECK**Point** Now try Exercise 89.

The formula for the length of a circular arc can be used to analyze the motion of a particle moving at a *constant speed* along a circular path.

Linear and Angular Speeds

Consider a particle moving at a constant speed along a circular arc of radius r. If s is the length of the arc traveled in time t, then the **linear speed** v of the particle is

$$\text{Linear speed } v = \frac{\text{arc length}}{\text{time}} = \frac{s}{t}.$$

Moreover, if θ is the angle (in radian measure) corresponding to the arc length s, then the **angular speed** ω (the lowercase Greek letter omega) of the particle is

$$\text{Angular speed } \omega = \frac{\text{central angle}}{\text{time}} = \frac{\theta}{t}.$$

Study Tip

Linear speed measures how fast the particle moves, and angular speed measures how fast the angle changes. By dividing the formula for arc length by t, you can establish a relationship between linear speed v and angular speed ω, as shown.

$$s = r\theta$$

$$\frac{s}{t} = \frac{r\theta}{t}$$

$$v = r\omega$$

FIGURE **4.16**

FIGURE **4.17**

Example 6   Finding Linear Speed

The second hand of a clock is 10.2 centimeters long, as shown in Figure 4.16. Find the linear speed of the tip of this second hand as it passes around the clock face.

Solution

In one revolution, the arc length traveled is

$$s = 2\pi r$$

$$= 2\pi(10.2) \qquad \text{Substitute for } r.$$

$$= 20.4\pi \text{ centimeters.}$$

The time required for the second hand to travel this distance is

$$t = 1 \text{ minute} = 60 \text{ seconds.}$$

So, the linear speed of the tip of the second hand is

$$\text{Linear speed} = \frac{s}{t}$$

$$= \frac{20.4\pi \text{ centimeters}}{60 \text{ seconds}}$$

$$\approx 1.068 \text{ centimeters per second.}$$

CHECK *Point*   Now try Exercise 111.

Example 7   Finding Angular and Linear Speeds

The blades of a wind turbine are 116 feet long (see Figure 4.17). The propeller rotates at 15 revolutions per minute.

a. Find the angular speed of the propeller in radians per minute.

b. Find the linear speed of the tips of the blades.

Solution

a. Because each revolution generates 2π radians, it follows that the propeller turns $(15)(2\pi) = 30\pi$ radians per minute. In other words, the angular speed is

$$\text{Angular speed} = \frac{\theta}{t}$$

$$= \frac{30\pi \text{ radians}}{1 \text{ minute}} = 30\pi \text{ radians per minute.}$$

b. The linear speed is

$$\text{Linear speed} = \frac{s}{t}$$

$$= \frac{r\theta}{t}$$

$$= \frac{(116)(30\pi) \text{ feet}}{1 \text{ minute}} \approx 10{,}933 \text{ feet per minute.}$$

CHECK *Point*   Now try Exercise 113.

A **sector** of a circle is the region bounded by two radii of the circle and their intercepted arc (see Figure 4.18).

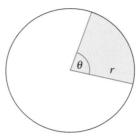

FIGURE **4.18**

Area of a Sector of a Circle

For a circle of radius r, the area A of a sector of the circle with central angle θ is given by

$$A = \frac{1}{2}r^2\theta$$

where θ is measured in radians.

Example 8   Area of a Sector of a Circle

A sprinkler on a golf course fairway sprays water over a distance of 70 feet and rotates through an angle of 120° (see Figure 4.19). Find the area of the fairway watered by the sprinkler.

Solution

First convert 120° to radian measure as follows.

$$\theta = 120°$$

$$= (120 \text{ deg})\left(\frac{\pi \text{ rad}}{180 \text{ deg}}\right) \qquad \text{Multiply by } \pi/180.$$

$$= \frac{2\pi}{3} \text{ radians}$$

Then, using $\theta = 2\pi/3$ and $r = 70$, the area is

$$A = \frac{1}{2}r^2\theta \qquad \text{Formula for the area of a sector of a circle}$$

$$= \frac{1}{2}(70)^2\left(\frac{2\pi}{3}\right) \qquad \text{Substitute for } r \text{ and } \theta.$$

$$= \frac{4900\pi}{3} \qquad \text{Simplify.}$$

$$\approx 5131 \text{ square feet.} \qquad \text{Simplify.}$$

CHECK *Point*   Now try Exercise 117.

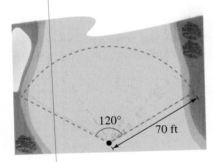

FIGURE **4.19**

4.1   EXERCISES

See www.CalcChat.com for worked-out solutions to odd-numbered exercises.

VOCABULARY: Fill in the blanks.

1. _____ means "measurement of triangles."

2. An _____ is determined by rotating a ray about its endpoint.

3. Two angles that have the same initial and terminal sides are _____.

4. One _____ is the measure of a central angle that intercepts an arc equal to the radius of the circle.

5. Angles that measure between 0 and $\pi/2$ are _____ angles, and angles that measure between $\pi/2$ and π are _____ angles.

6. Two positive angles that have a sum of $\pi/2$ are _____ angles, whereas two positive angles that have a sum of π are _____ angles.

7. The angle measure that is equivalent to a rotation of $\frac{1}{360}$ of a complete revolution about an angle's vertex is one _____.

8. 180 degrees = _____ radians.

9. The _____ speed of a particle is the ratio of arc length to time traveled, and the _____ speed of a particle is the ratio of central angle to time traveled.

10. The area A of a sector of a circle with radius r and central angle θ, where θ is measured in radians, is given by the formula _____.

SKILLS AND APPLICATIONS

In Exercises 11–16, estimate the angle to the nearest one-half radian.

11.

12.

13.

14.

15.

16.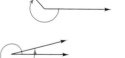

In Exercises 17–22, determine the quadrant in which each angle lies. (The angle measure is given in radians.)

17. (a) $\dfrac{\pi}{4}$ (b) $\dfrac{5\pi}{4}$ **18.** (a) $\dfrac{11\pi}{8}$ (b) $\dfrac{9\pi}{8}$

19. (a) $-\dfrac{\pi}{6}$ (b) $-\dfrac{\pi}{3}$ **20.** (a) $-\dfrac{5\pi}{6}$ (b) $-\dfrac{11\pi}{9}$

21. (a) 3.5   (b) 2.25   **22.** (a) 6.02   (b) -4.25

In Exercises 23–26, sketch each angle in standard position.

23. (a) $\dfrac{\pi}{3}$ (b) $-\dfrac{2\pi}{3}$ **24.** (a) $-\dfrac{7\pi}{4}$ (b) $\dfrac{5\pi}{2}$

25. (a) $\dfrac{11\pi}{6}$ (b) -3

26. (a) 4   (b) 7π

In Exercises 27–30, determine two coterminal angles (one positive and one negative) for each angle. Give your answers in radians.

27. (a) 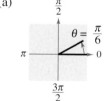   (b)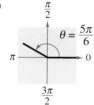

28. (a) 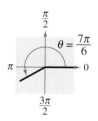   (b)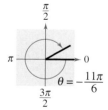

29. (a) $\theta = \dfrac{2\pi}{3}$ (b) $\theta = \dfrac{\pi}{12}$

30. (a) $\theta = -\dfrac{9\pi}{4}$ (b) $\theta = -\dfrac{2\pi}{15}$

In Exercises 31–34, find (if possible) the complement and supplement of each angle.

31. (a) $\pi/3$ (b) $\pi/4$ **32.** (a) $\pi/12$ (b) $11\pi/12$
33. (a) 1   (b) 2   **34.** (a) 3   (b) 1.5

In Exercises 35–40, estimate the number of degrees in the angle. Use a protractor to check your answer.

35.    **36.**

37.    **38.**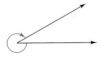

39.    **40.**

In Exercises 41–44, determine the quadrant in which each angle lies.

41. (a) $130°$ (b) $285°$
42. (a) $8.3°$ (b) $257°\ 30'$
43. (a) $-132°\ 50'$ (b) $-336°$
44. (a) $-260°$ (b) $-3.4°$

In Exercises 45–48, sketch each angle in standard position.

45. (a) $90°$ (b) $180°$ **46.** (a) $270°$ (b) $120°$
47. (a) $-30°$ (b) $-135°$
48. (a) $-750°$ (b) $-600°$

In Exercises 49–52, determine two coterminal angles (one positive and one negative) for each angle. Give your answers in degrees.

49. (a)   (b)

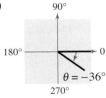

50. (a)   (b)

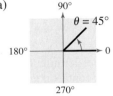   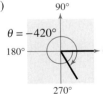

51. (a) $\theta = 240°$ (b) $\theta = -180°$

52. (a) $\theta = -390°$ (b) $\theta = 230°$

In Exercises 53–56, find (if possible) the complement and supplement of each angle.

53. (a) $18°$ (b) $85°$ **54.** (a) $46°$ (b) $93°$
55. (a) $150°$ (b) $79°$ **56.** (a) $130°$ (b) $170°$

In Exercises 57–60, rewrite each angle in radian measure as a multiple of π. (Do not use a calculator.)

57. (a) $30°$ (b) $45°$ **58.** (a) $315°$ (b) $120°$
59. (a) $-20°$ (b) $-60°$ **60.** (a) $-270°$ (b) $144°$

In Exercises 61–64, rewrite each angle in degree measure. (Do not use a calculator.)

61. (a) $\dfrac{3\pi}{2}$ (b) $\dfrac{7\pi}{6}$ **62.** (a) $-\dfrac{7\pi}{12}$ (b) $\dfrac{\pi}{9}$

63. (a) $\dfrac{5\pi}{4}$ (b) $-\dfrac{7\pi}{3}$ **64.** (a) $\dfrac{11\pi}{6}$ (b) $\dfrac{34\pi}{15}$

In Exercises 65–72, convert the angle measure from degrees to radians. Round to three decimal places.

65. $45°$ **66.** $87.4°$
67. $-216.35°$ **68.** $-48.27°$
69. $532°$ **70.** $345°$
71. $-0.83°$ **72.** $0.54°$

In Exercises 73–80, convert the angle measure from radians to degrees. Round to three decimal places.

73. $\pi/7$ **74.** $5\pi/11$
75. $15\pi/8$ **76.** $13\pi/2$
77. -4.2π **78.** 4.8π
79. -2 **80.** -0.57

In Exercises 81–84, convert each angle measure to decimal degree form without using a calculator. Then check your answers using a calculator.

81. (a) $54°\ 45'$ (b) $-128°\ 30'$
82. (a) $245°\ 10'$ (b) $2°\ 12'$
83. (a) $85°\ 18'\ 30''$ (b) $330°\ 25''$
84. (a) $-135°\ 36''$ (b) $-408°\ 16'20''$

In Exercises 85–88, convert each angle measure to degrees, minutes, and seconds without using a calculator. Then check your answers using a calculator.

85. (a) $240.6°$ (b) $-145.8°$
86. (a) $-345.12°$ (b) $0.45°$
87. (a) $2.5°$ (b) $-3.58°$
88. (a) $-0.36°$ (b) $0.79°$

In Exercises 89–92, find the length of the arc on a circle of radius *r* intercepted by a central angle *θ*.

	Radius r	Central Angle θ
89.	15 inches	120°
90.	9 feet	60°
91.	3 meters	150°
92.	20 centimeters	45°

In Exercises 93–96, find the radian measure of the central angle of a circle of radius *r* that intercepts an arc of length *s*.

	Radius r	Arc Length s
93.	4 inches	18 inches
94.	14 feet	8 feet
95.	25 centimeters	10.5 centimeters
96.	80 kilometers	150 kilometers

In Exercises 97–100, use the given arc length and radius to find the angle *θ* (in radians).

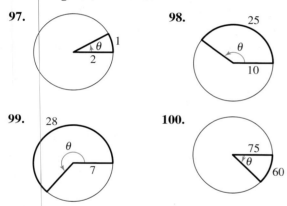

97.   98.

99.   100.

In Exercises 101–104, find the area of the sector of the circle with radius *r* and central angle *θ*.

	Radius r	Central Angle θ
101.	6 inches	$\pi/3$
102.	12 millimeters	$\pi/4$
103.	2.5 feet	225°
104.	1.4 miles	330°

DISTANCE BETWEEN CITIES In Exercises 105 and 106, find the distance between the cities. Assume that Earth is a sphere of radius 4000 miles and that the cities are on the same longitude (one city is due north of the other).

	City	Latitude
105.	Dallas, Texas	32° 47′ 39″ N
	Omaha, Nebraska	41° 15′ 50″ N

	City	Latitude
106.	San Francisco, California	37° 47′ 36″ N
	Seattle, Washington	47° 37′ 18″ N

107. DIFFERENCE IN LATITUDES Assuming that Earth is a sphere of radius 6378 kilometers, what is the difference in the latitudes of Syracuse, New York and Annapolis, Maryland, where Syracuse is about 450 kilometers due north of Annapolis?

108. DIFFERENCE IN LATITUDES Assuming that Earth is a sphere of radius 6378 kilometers, what is the difference in the latitudes of Lynchburg, Virginia and Myrtle Beach, South Carolina, where Lynchburg is about 400 kilometers due north of Myrtle Beach?

109. INSTRUMENTATION The pointer on a voltmeter is 6 centimeters in length (see figure). Find the angle through which the pointer rotates when it moves 2.5 centimeters on the scale.

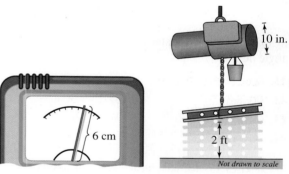

FIGURE FOR **109**   FIGURE FOR **110**

110. ELECTRIC HOIST An electric hoist is being used to lift a beam (see figure). The diameter of the drum on the hoist is 10 inches, and the beam must be raised 2 feet. Find the number of degrees through which the drum must rotate.

111. LINEAR AND ANGULAR SPEEDS A circular power saw has a $7\frac{1}{4}$-inch-diameter blade that rotates at 5000 revolutions per minute.

(a) Find the angular speed of the saw blade in radians per minute.

(b) Find the linear speed (in feet per minute) of one of the 24 cutting teeth as they contact the wood being cut.

112. LINEAR AND ANGULAR SPEEDS A carousel with a 50-foot diameter makes 4 revolutions per minute.

(a) Find the angular speed of the carousel in radians per minute.

(b) Find the linear speed (in feet per minute) of the platform rim of the carousel.

113. LINEAR AND ANGULAR SPEEDS   The diameter of a DVD is approximately 12 centimeters. The drive motor of the DVD player is controlled to rotate precisely between 200 and 500 revolutions per minute, depending on what track is being read.

(a) Find an interval for the angular speed of a DVD as it rotates.

(b) Find an interval for the linear speed of a point on the outermost track as the DVD rotates.

114. ANGULAR SPEED   A two-inch-diameter pulley on an electric motor that runs at 1700 revolutions per minute is connected by a belt to a four-inch-diameter pulley on a saw arbor.

(a) Find the angular speed (in radians per minute) of each pulley.

(b) Find the revolutions per minute of the saw.

115. ANGULAR SPEED   A car is moving at a rate of 65 miles per hour, and the diameter of its wheels is 2 feet.

(a) Find the number of revolutions per minute the wheels are rotating.

(b) Find the angular speed of the wheels in radians per minute.

116. ANGULAR SPEED   A computerized spin balance machine rotates a 25-inch-diameter tire at 480 revolutions per minute.

(a) Find the road speed (in miles per hour) at which the tire is being balanced.

(b) At what rate should the spin balance machine be set so that the tire is being tested for 55 miles per hour?

117. AREA   A sprinkler on a golf green is set to spray water over a distance of 15 meters and to rotate through an angle of 140°. Draw a diagram that shows the region that can be irrigated with the sprinkler. Find the area of the region.

118. AREA   A car's rear windshield wiper rotates 125°. The total length of the wiper mechanism is 25 inches and wipes the windshield over a distance of 14 inches. Find the area covered by the wiper.

119. SPEED OF A BICYCLE   The radii of the pedal sprocket, the wheel sprocket, and the wheel of the bicycle in the figure are 4 inches, 2 inches, and 14 inches, respectively. A cyclist is pedaling at a rate of 1 revolution per second.

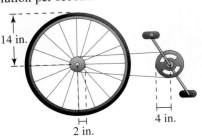

14 in.

4 in.

2 in.

(a) Find the speed of the bicycle in feet per second and miles per hour.

(b) Use your result from part (a) to write a function for the distance d (in miles) a cyclist travels in terms of the number n of revolutions of the pedal sprocket.

(c) Write a function for the distance d (in miles) a cyclist travels in terms of the time t (in seconds). Compare this function with the function from part (b).

(d) Classify the types of functions you found in parts (b) and (c). Explain your reasoning.

120. CAPSTONE   Write a short paper in your own words explaining the meaning of each of the following concepts to a classmate.

(a) an angle in standard position

(b) positive and negative angles

(c) coterminal angles

(d) angle measure in degrees and radians

(e) obtuse and acute angles

(f) complementary and supplementary angles

EXPLORATION

TRUE OR FALSE?   In Exercises 121–123, determine whether the statement is true or false. Justify your answer.

121. A measurement of 4 radians corresponds to two complete revolutions from the initial side to the terminal side of an angle.

122. The difference between the measures of two coterminal angles is always a multiple of 360° if expressed in degrees and is always a multiple of 2π radians if expressed in radians.

123. An angle that measures $-1260°$ lies in Quadrant III.

124. THINK ABOUT IT   A fan motor turns at a given angular speed. How does the speed of the tips of the blades change if a fan of greater diameter is installed on the motor? Explain.

125. THINK ABOUT IT   Is a degree or a radian the larger unit of measure? Explain.

126. WRITING   If the radius of a circle is increasing and the magnitude of a central angle is held constant, how is the length of the intercepted arc changing? Explain your reasoning.

127. PROOF   Prove that the area of a circular sector of radius r with central angle θ is $A = \frac{1}{2}\theta r^2$, where θ is measured in radians.

4.2 TRIGONOMETRIC FUNCTIONS: THE UNIT CIRCLE

What you should learn

- Identify a unit circle and describe its relationship to real numbers.
- Evaluate trigonometric functions using the unit circle.
- Use the domain and period to evaluate sine and cosine functions.
- Use a calculator to evaluate trigonometric functions.

Why you should learn it

Trigonometric functions are used to model the movement of an oscillating weight. For instance, in Exercise 60 on page 298, the displacement from equilibrium of an oscillating weight suspended by a spring is modeled as a function of time.

The Unit Circle

The two historical perspectives of trigonometry incorporate different methods for introducing the trigonometric functions. Our first introduction to these functions is based on the unit circle.

Consider the **unit circle** given by

$$x^2 + y^2 = 1 \qquad \text{Unit circle}$$

as shown in Figure 4.20.

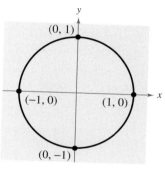

FIGURE **4.20**

Imagine that the real number line is wrapped around this circle, with positive numbers corresponding to a counterclockwise wrapping and negative numbers corresponding to a clockwise wrapping, as shown in Figure 4.21.

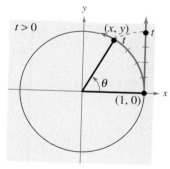
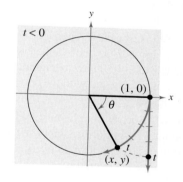

FIGURE **4.21**

As the real number line is wrapped around the unit circle, each real number t corresponds to a point (x, y) on the circle. For example, the real number 0 corresponds to the point $(1, 0)$. Moreover, because the unit circle has a circumference of 2π, the real number 2π also corresponds to the point $(1, 0)$.

In general, each real number t also corresponds to a central angle θ (in standard position) whose radian measure is t. With this interpretation of t, the arc length formula $s = r\theta$ (with $r = 1$) indicates that the real number t is the (directional) length of the arc intercepted by the angle θ, given in radians.

The Trigonometric Functions

From the preceding discussion, it follows that the coordinates x and y are two functions of the real variable t. You can use these coordinates to define the six trigonometric functions of t.

sine   cosecant   cosine   secant   tangent   cotangent

These six functions are normally abbreviated sin, csc, cos, sec, tan, and cot, respectively.

Definitions of Trigonometric Functions

Let t be a real number and let (x, y) be the point on the unit circle corresponding to t.

$$\sin t = y \qquad\qquad \cos t = x \qquad\qquad \tan t = \frac{y}{x}, \quad x \neq 0$$

$$\csc t = \frac{1}{y}, \quad y \neq 0 \qquad \sec t = \frac{1}{x}, \quad x \neq 0 \qquad \cot t = \frac{x}{y}, \quad y \neq 0$$

In the definitions of the trigonometric functions, note that the tangent and secant are not defined when $x = 0$. For instance, because $t = \pi/2$ corresponds to $(x, y) = (0, 1)$, it follows that $\tan(\pi/2)$ and $\sec(\pi/2)$ are *undefined*. Similarly, the cotangent and cosecant are not defined when $y = 0$. For instance, because $t = 0$ corresponds to $(x, y) = (1, 0)$, $\cot 0$ and $\csc 0$ are *undefined*.

In Figure 4.22, the unit circle has been divided into eight equal arcs, corresponding to t-values of

$$0, \frac{\pi}{4}, \frac{\pi}{2}, \frac{3\pi}{4}, \pi, \frac{5\pi}{4}, \frac{3\pi}{2}, \frac{7\pi}{4}, \text{ and } 2\pi.$$

Similarly, in Figure 4.23, the unit circle has been divided into 12 equal arcs, corresponding to t-values of

$$0, \frac{\pi}{6}, \frac{\pi}{3}, \frac{\pi}{2}, \frac{2\pi}{3}, \frac{5\pi}{6}, \pi, \frac{7\pi}{6}, \frac{4\pi}{3}, \frac{3\pi}{2}, \frac{5\pi}{3}, \frac{11\pi}{6}, \text{ and } 2\pi.$$

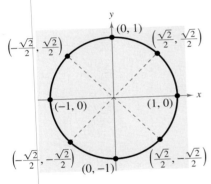

FIGURE 4.22

To verify the points on the unit circle in Figure 4.22, note that $\left(\frac{\sqrt{2}}{2}, \frac{\sqrt{2}}{2}\right)$ also lies on the line $y = x$. So, substituting x for y in the equation of the unit circle produces the following.

$$x^2 + x^2 = 1 \implies 2x^2 = 1 \implies x^2 = \frac{1}{2} \implies x = \pm\frac{\sqrt{2}}{2}$$

Because the point is in the first quadrant, $x = \frac{\sqrt{2}}{2}$ and because $y = x$, you also have $y = \frac{\sqrt{2}}{2}$. You can use similar reasoning to verify the rest of the points in Figure 4.22 and the points in Figure 4.23.

Using the (x, y) coordinates in Figures 4.22 and 4.23, you can evaluate the trigonometric functions for common t-values. This procedure is demonstrated in Examples 1, 2, and 3. You should study and learn these exact function values for common t-values because they will help you in later sections to perform calculations.

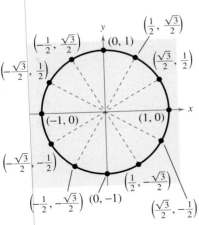

FIGURE 4.23

You can review dividing fractions and rationalizing denominators in Appendix A.1 and Appendix A.2, respectively.

Algebra Help

Example 1   Evaluating Trigonometric Functions

Evaluate the six trigonometric functions at each real number.

a. $t = \dfrac{\pi}{6}$ **b.** $t = \dfrac{5\pi}{4}$ **c.** $t = 0$ **d.** $t = \pi$

Solution

For each t-value, begin by finding the corresponding point (x, y) on the unit circle. Then use the definitions of trigonometric functions listed on page 293.

a. $t = \dfrac{\pi}{6}$ corresponds to the point $(x, y) = \left(\dfrac{\sqrt{3}}{2}, \dfrac{1}{2}\right)$.

$$\sin \frac{\pi}{6} = y = \frac{1}{2} \qquad\qquad \csc \frac{\pi}{6} = \frac{1}{y} = \frac{1}{1/2} = 2$$

$$\cos \frac{\pi}{6} = x = \frac{\sqrt{3}}{2} \qquad\qquad \sec \frac{\pi}{6} = \frac{1}{x} = \frac{2}{\sqrt{3}} = \frac{2\sqrt{3}}{3}$$

$$\tan \frac{\pi}{6} = \frac{y}{x} = \frac{1/2}{\sqrt{3}/2} = \frac{1}{\sqrt{3}} = \frac{\sqrt{3}}{3} \qquad\qquad \cot \frac{\pi}{6} = \frac{x}{y} = \frac{\sqrt{3}/2}{1/2} = \sqrt{3}$$

b. $t = \dfrac{5\pi}{4}$ corresponds to the point $(x, y) = \left(-\dfrac{\sqrt{2}}{2}, -\dfrac{\sqrt{2}}{2}\right)$.

$$\sin \frac{5\pi}{4} = y = -\frac{\sqrt{2}}{2} \qquad\qquad \csc \frac{5\pi}{4} = \frac{1}{y} = -\frac{2}{\sqrt{2}} = -\sqrt{2}$$

$$\cos \frac{5\pi}{4} = x = -\frac{\sqrt{2}}{2} \qquad\qquad \sec \frac{5\pi}{4} = \frac{1}{x} = -\frac{2}{\sqrt{2}} = -\sqrt{2}$$

$$\tan \frac{5\pi}{4} = \frac{y}{x} = \frac{-\sqrt{2}/2}{-\sqrt{2}/2} = 1 \qquad\qquad \cot \frac{5\pi}{4} = \frac{x}{y} = \frac{-\sqrt{2}/2}{-\sqrt{2}/2} = 1$$

c. $t = 0$ corresponds to the point $(x, y) = (1, 0)$.

$$\sin 0 = y = 0 \qquad\qquad \csc 0 = \frac{1}{y} \text{ is undefined.}$$

$$\cos 0 = x = 1 \qquad\qquad \sec 0 = \frac{1}{x} = \frac{1}{1} = 1$$

$$\tan 0 = \frac{y}{x} = \frac{0}{1} = 0 \qquad\qquad \cot 0 = \frac{x}{y} \text{ is undefined.}$$

d. $t = \pi$ corresponds to the point $(x, y) = (-1, 0)$.

$$\sin \pi = y = 0 \qquad\qquad \csc \pi = \frac{1}{y} \text{ is undefined.}$$

$$\cos \pi = x = -1 \qquad\qquad \sec \pi = \frac{1}{x} = \frac{1}{-1} = -1$$

$$\tan \pi = \frac{y}{x} = \frac{0}{-1} = 0 \qquad\qquad \cot \pi = \frac{x}{y} \text{ is undefined.}$$

CHECK *Point*   Now try Exercise 23.

| **Example 2** | **Evaluating Trigonometric Functions** |

Evaluate the six trigonometric functions at $t = -\dfrac{\pi}{3}$.

Solution

Moving *clockwise* around the unit circle, it follows that $t = -\pi/3$ corresponds to the point $(x, y) = \left(1/2, -\sqrt{3}/2\right)$.

$$\sin\left(-\frac{\pi}{3}\right) = -\frac{\sqrt{3}}{2} \qquad\qquad \csc\left(-\frac{\pi}{3}\right) = -\frac{2}{\sqrt{3}} = -\frac{2\sqrt{3}}{3}$$

$$\cos\left(-\frac{\pi}{3}\right) = \frac{1}{2} \qquad\qquad \sec\left(-\frac{\pi}{3}\right) = 2$$

$$\tan\left(-\frac{\pi}{3}\right) = \frac{-\sqrt{3}/2}{1/2} = -\sqrt{3} \qquad \cot\left(-\frac{\pi}{3}\right) = \frac{1/2}{-\sqrt{3}/2} = -\frac{1}{\sqrt{3}} = -\frac{\sqrt{3}}{3}$$

CHECK*Point* ▶ Now try Exercise 33.

Domain and Period of Sine and Cosine

The *domain* of the sine and cosine functions is the set of all real numbers. To determine the *range* of these two functions, consider the unit circle shown in Figure 4.24. By definition, $\sin t = y$ and $\cos t = x$. Because (x, y) is on the unit circle, you know that $-1 \le y \le 1$ and $-1 \le x \le 1$. So, the values of sine and cosine also range between -1 and 1.

$$\begin{array}{ccc} -1 \le & y & \le 1 \\ -1 \le & \sin t \le 1 \end{array} \quad \text{and} \quad \begin{array}{ccc} -1 \le & x & \le 1 \\ -1 \le & \cos t \le 1 \end{array}$$

Adding 2π to each value of t in the interval $[0, 2\pi]$ completes a second revolution around the unit circle, as shown in Figure 4.25. The values of $\sin(t + 2\pi)$ and $\cos(t + 2\pi)$ correspond to those of $\sin t$ and $\cos t$. Similar results can be obtained for repeated revolutions (positive or negative) on the unit circle. This leads to the general result

$$\sin(t + 2\pi n) = \sin t$$

and

$$\cos(t + 2\pi n) = \cos t$$

for any integer n and real number t. Functions that behave in such a repetitive (or cyclic) manner are called **periodic**.

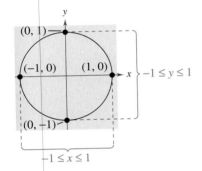

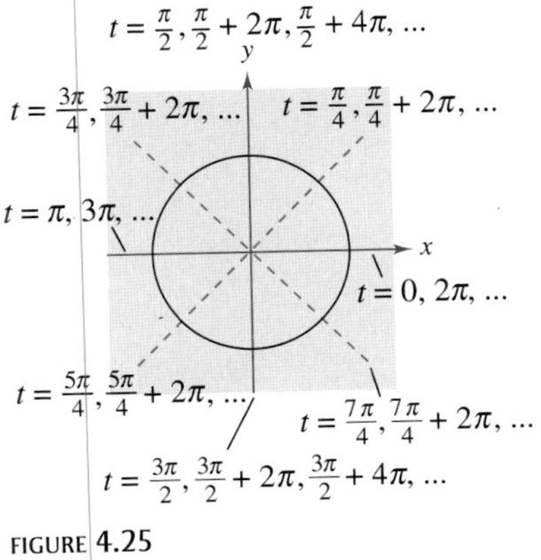

FIGURE **4.24**

$t = \frac{\pi}{2}, \frac{\pi}{2} + 2\pi, \frac{\pi}{2} + 4\pi, \dots$

$t = \frac{3\pi}{4}, \frac{3\pi}{4} + 2\pi, \dots$ $t = \frac{\pi}{4}, \frac{\pi}{4} + 2\pi, \dots$

$t = \pi, 3\pi, \dots$

$t = 0, 2\pi, \dots$

$t = \frac{5\pi}{4}, \frac{5\pi}{4} + 2\pi, \dots$ $t = \frac{7\pi}{4}, \frac{7\pi}{4} + 2\pi, \dots$

$t = \frac{3\pi}{2}, \frac{3\pi}{2} + 2\pi, \frac{3\pi}{2} + 4\pi, \dots$

FIGURE **4.25**

Definition of Periodic Function

A function f is **periodic** if there exists a positive real number c such that

$$f(t + c) = f(t)$$

for all t in the domain of f. The smallest number c for which f is periodic is called the **period** of f.

Recall from Section 1.5 that a function f is *even* if $f(-t) = f(t)$, and is *odd* if $f(-t) = -f(t)$.

Even and Odd Trigonometric Functions

The cosine and secant functions are *even*.

$$\cos(-t) = \cos t \qquad \sec(-t) = \sec t$$

The sine, cosecant, tangent, and cotangent functions are *odd*.

$$\sin(-t) = -\sin t \qquad \csc(-t) = -\csc t$$
$$\tan(-t) = -\tan t \qquad \cot(-t) = -\cot t$$

Example 3   Using the Period to Evaluate the Sine and Cosine

a. Because $\dfrac{13\pi}{6} = 2\pi + \dfrac{\pi}{6}$, you have $\sin\dfrac{13\pi}{6} = \sin\left(2\pi + \dfrac{\pi}{6}\right) = \sin\dfrac{\pi}{6} = \dfrac{1}{2}$.

b. Because $-\dfrac{7\pi}{2} = -4\pi + \dfrac{\pi}{2}$, you have

$$\cos\left(-\dfrac{7\pi}{2}\right) = \cos\left(-4\pi + \dfrac{\pi}{2}\right) = \cos\dfrac{\pi}{2} = 0.$$

c. For $\sin t = \dfrac{4}{5}$, $\sin(-t) = -\dfrac{4}{5}$ because the sine function is odd.

CHECK**Point**   Now try Exercise 37.

Evaluating Trigonometric Functions with a Calculator

When evaluating a trigonometric function with a calculator, you need to set the calculator to the desired *mode* of measurement (*degree* or *radian*).

Most calculators do not have keys for the cosecant, secant, and cotangent functions. To evaluate these functions, you can use the $\boxed{x^{-1}}$ key with their respective reciprocal functions sine, cosine, and tangent. For instance, to evaluate $\csc(\pi/8)$, use the fact that

$$\csc\dfrac{\pi}{8} = \dfrac{1}{\sin(\pi/8)}$$

and enter the following keystroke sequence in *radian* mode.

(SIN (π ÷ 8)) x⁻¹ ENTER   Display 2.6131259

Example 4   Using a Calculator

Function	Mode	Calculator Keystrokes	Display
a. $\sin\dfrac{2\pi}{3}$	Radian	SIN (2 π ÷ 3) ENTER	0.8660254
b. $\cot 1.5$	Radian	(TAN (1.5)) x⁻¹ ENTER	0.0709148

CHECK**Point**   Now try Exercise 55.

Study Tip

From the definition of periodic function, it follows that the sine and cosine functions are periodic and have a period of 2π. The other four trigonometric functions are also periodic, and will be discussed further in Section 4.6.

TECHNOLOGY

When evaluating trigonometric functions with a calculator, remember to enclose all fractional angle measures in parentheses. For instance, if you want to evaluate *sin t* for $t = \pi/6$, you should enter

SIN (π ÷ 6) ENTER.

These keystrokes yield the correct value of 0.5. Note that some calculators automatically place a left parenthesis after trigonometric functions. Check the user's guide for your calculator for specific keystrokes on how to evaluate trigonometric functions.

4.2   EXERCISES

See www.CalcChat.com for worked-out solutions to odd-numbered exercises.

VOCABULARY: Fill in the blanks.

1. Each real number t corresponds to a point (x, y) on the _____ _____.
2. A function f is _____ if there exists a positive real number c such that $f(t + c) = f(t)$ for all t in the domain of f.
3. The smallest number c for which a function f is periodic is called the _____ of f.
4. A function f is _____ if $f(-t) = -f(t)$ and _____ if $f(-t) = f(t)$.

SKILLS AND APPLICATIONS

In Exercises 5–8, determine the exact values of the six trigonometric functions of the real number t.

5.

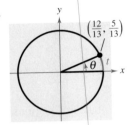

6. $\left(-\frac{8}{17}, \frac{15}{17}\right)$

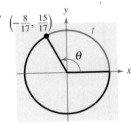

7.

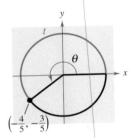

$\left(-\frac{4}{5}, -\frac{3}{5}\right)$

8.

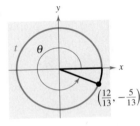

$\left(\frac{12}{13}, -\frac{5}{13}\right)$

In Exercises 9–16, find the point (x, y) on the unit circle that corresponds to the real number t.

9. $t = \dfrac{\pi}{2}$

10. $t = \pi$

11. $t = \dfrac{\pi}{4}$

12. $t = \dfrac{\pi}{3}$

13. $t = \dfrac{5\pi}{6}$

14. $t = \dfrac{3\pi}{4}$

15. $t = \dfrac{4\pi}{3}$

16. $t = \dfrac{5\pi}{3}$

In Exercises 17–26, evaluate (if possible) the sine, cosine, and tangent of the real number.

17. $t = \dfrac{\pi}{4}$

18. $t = \dfrac{\pi}{3}$

19. $t = -\dfrac{\pi}{6}$

20. $t = -\dfrac{\pi}{4}$

21. $t = -\dfrac{7\pi}{4}$

22. $t = -\dfrac{4\pi}{3}$

23. $t = \dfrac{11\pi}{6}$

24. $t = \dfrac{5\pi}{3}$

25. $t = -\dfrac{3\pi}{2}$

26. $t = -2\pi$

In Exercises 27–34, evaluate (if possible) the six trigonometric functions of the real number.

27. $t = \dfrac{2\pi}{3}$

28. $t = \dfrac{5\pi}{6}$

29. $t = \dfrac{4\pi}{3}$

30. $t = \dfrac{7\pi}{4}$

31. $t = \dfrac{3\pi}{4}$

32. $t = \dfrac{3\pi}{2}$

33. $t = -\dfrac{\pi}{2}$

34. $t = -\pi$

In Exercises 35–42, evaluate the trigonometric function using its period as an aid.

35. $\sin 4\pi$

36. $\cos 3\pi$

37. $\cos \dfrac{7\pi}{3}$

38. $\sin \dfrac{9\pi}{4}$

39. $\cos \dfrac{17\pi}{4}$

40. $\sin \dfrac{19\pi}{6}$

41. $\sin\left(-\dfrac{8\pi}{3}\right)$

42. $\cos\left(-\dfrac{9\pi}{4}\right)$

In Exercises 43–48, use the value of the trigonometric function to evaluate the indicated functions.

43. $\sin t = \frac{1}{2}$
   (a) $\sin(-t)$
   (b) $\csc(-t)$

44. $\sin(-t) = \frac{3}{8}$
   (a) $\sin t$
   (b) $\csc t$

45. $\cos(-t) = -\frac{1}{5}$
   (a) $\cos t$
   (b) $\sec(-t)$

46. $\cos t = -\frac{3}{4}$
   (a) $\cos(-t)$
   (b) $\sec(-t)$

47. $\sin t = \frac{4}{5}$
   (a) $\sin(\pi - t)$
   (b) $\sin(t + \pi)$

48. $\cos t = \frac{4}{5}$
   (a) $\cos(\pi - t)$
   (b) $\cos(t + \pi)$

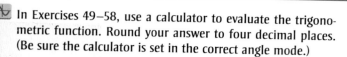

In Exercises 49–58, use a calculator to evaluate the trigono-metric function. Round your answer to four decimal places. (Be sure the calculator is set in the correct angle mode.)

49. $\sin \dfrac{\pi}{4}$

50. $\tan \dfrac{\pi}{3}$

51. $\cot \dfrac{\pi}{4}$

52. $\csc \dfrac{2\pi}{3}$

53. $\cos(-1.7)$

54. $\cos(-2.5)$

55. $\csc 0.8$

56. $\sec 1.8$

57. $\sec(-22.8)$

58. $\cot(-0.9)$

59. HARMONIC MOTION The displacement from equilibrium of an oscillating weight suspended by a spring is given by $y(t) = \frac{1}{4} \cos 6t$, where y is the displacement (in feet) and t is the time (in seconds). Find the displacements when (a) $t = 0$, (b) $t = \frac{1}{4}$, and (c) $t = \frac{1}{2}$.

60. HARMONIC MOTION The displacement from equilibrium of an oscillating weight suspended by a spring and subject to the damping effect of friction is given by $y(t) = \frac{1}{4}e^{-t} \cos 6t$, where y is the displacement (in feet) and t is the time (in seconds).

(a) Complete the table.

t	0	$\frac{1}{4}$	$\frac{1}{2}$	$\frac{3}{4}$	1
y					

(b) Use the *table* feature of a graphing utility to approximate the time when the weight reaches equilibrium.

(c) What appears to happen to the displacement as t increases?

EXPLORATION

TRUE OR FALSE? In Exercises 61–64, determine whether the statement is true or false. Justify your answer.

61. Because $\sin(-t) = -\sin t$, it can be said that the sine of a negative angle is a negative number.

62. $\tan a = \tan(a - 6\pi)$

63. The real number 0 corresponds to the point $(0, 1)$ on the unit circle.

64. $\cos\left(-\dfrac{7\pi}{2}\right) = \cos\left(\pi + \dfrac{\pi}{2}\right)$

65. Let (x_1, y_1) and (x_2, y_2) be points on the unit circle corresponding to $t = t_1$ and $t = \pi - t_1$, respectively.

(a) Identify the symmetry of the points (x_1, y_1) and (x_2, y_2).

(b) Make a conjecture about any relationship between $\sin t_1$ and $\sin(\pi - t_1)$.

(c) Make a conjecture about any relationship between $\cos t_1$ and $\cos(\pi - t_1)$.

66. Use the unit circle to verify that the cosine and secant functions are even and that the sine, cosecant, tangent, and cotangent functions are odd.

67. Verify that $\cos 2t \neq 2 \cos t$ by approximating $\cos 1.5$ and $2 \cos 0.75$.

68. Verify that $\sin(t_1 + t_2) \neq \sin t_1 + \sin t_2$ by approximating $\sin 0.25$, $\sin 0.75$, and $\sin 1$.

69. THINK ABOUT IT Because $f(t) = \sin t$ is an odd function and $g(t) = \cos t$ is an even function, what can be said about the function $h(t) = f(t)g(t)$?

70. THINK ABOUT IT Because $f(t) = \sin t$ and $g(t) = \tan t$ are odd functions, what can be said about the function $h(t) = f(t)g(t)$?

71. GRAPHICAL ANALYSIS With your graphing utility in *radian* and *parametric* modes, enter the equations

$$X_{1T} = \cos T \quad \text{and} \quad Y_{1T} = \sin T$$

and use the following settings.

Tmin = 0, Tmax = 6.3, Tstep = 0.1

Xmin = −1.5, Xmax = 1.5, Xscl = 1

Ymin = −1, Ymax = 1, Yscl = 1

(a) Graph the entered equations and describe the graph.

(b) Use the *trace* feature to move the cursor around the graph. What do the t-values represent? What do the x- and y-values represent?

(c) What are the least and greatest values of x and y?

72. CAPSTONE A student you are tutoring has used a unit circle divided into 8 equal parts to complete the table for selected values of t. What is wrong?

t	0	$\dfrac{\pi}{4}$	$\dfrac{\pi}{2}$	$\dfrac{3\pi}{4}$	π
x	1	$\dfrac{\sqrt{2}}{2}$	0	$-\dfrac{\sqrt{2}}{2}$	−1
y	0	$\dfrac{\sqrt{2}}{2}$	1	$\dfrac{\sqrt{2}}{2}$	0
$\sin t$	1	$\dfrac{\sqrt{2}}{2}$	0	$-\dfrac{\sqrt{2}}{2}$	−1
$\cos t$	0	$\dfrac{\sqrt{2}}{2}$	1	$\dfrac{\sqrt{2}}{2}$	0
$\tan t$	Undef.	1	0	−1	Undef.

4.3

RIGHT TRIANGLE TRIGONOMETRY

What you should learn

- Evaluate trigonometric functions of acute angles.
- Use fundamental trigonometric identities.
- Use a calculator to evaluate trigonometric functions.
- Use trigonometric functions to model and solve real-life problems.

Why you should learn it

Trigonometric functions are often used to analyze real-life situations. For instance, in Exercise 76 on page 309, you can use trigonometric functions to find the height of a helium-filled balloon.

The Six Trigonometric Functions

Our second look at the trigonometric functions is from a *right triangle* perspective. Consider a right triangle, with one acute angle labeled θ, as shown in Figure 4.26. Relative to the angle θ, the three sides of the triangle are the **hypotenuse,** the **opposite side** (the side opposite the angle θ), and the **adjacent side** (the side adjacent to the angle θ).

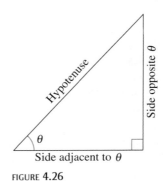

FIGURE **4.26**

Using the lengths of these three sides, you can form six ratios that define the six trigonometric functions of the acute angle θ.

sine   cosecant   cosine   secant   tangent   cotangent

In the following definitions, it is important to see that $0° < \theta < 90°$ (θ lies in the first quadrant) and that for such angles the value of each trigonometric function is *positive*.

Right Triangle Definitions of Trigonometric Functions

Let θ be an *acute* angle of a right triangle. The six trigonometric functions of the angle θ are defined as follows. (Note that the functions in the second row are the *reciprocals* of the corresponding functions in the first row.)

$$\sin \theta = \frac{\text{opp}}{\text{hyp}} \qquad \cos \theta = \frac{\text{adj}}{\text{hyp}} \qquad \tan \theta = \frac{\text{opp}}{\text{adj}}$$

$$\csc \theta = \frac{\text{hyp}}{\text{opp}} \qquad \sec \theta = \frac{\text{hyp}}{\text{adj}} \qquad \cot \theta = \frac{\text{adj}}{\text{opp}}$$

The abbreviations opp, adj, and hyp represent the lengths of the three sides of a right triangle.

opp = the length of the side *opposite* θ

adj = the length of the side *adjacent to* θ

hyp = the length of the *hypotenuse*

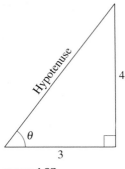

FIGURE **4.27**

Algebra Help

You can review the Pythagorean Theorem in Section 1.1.

HISTORICAL NOTE

Georg Joachim Rhaeticus (1514–1574) was the leading Teutonic mathematical astronomer of the 16th century. He was the first to define the trigonometric functions as ratios of the sides of a right triangle.

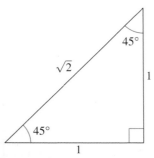

FIGURE **4.28**

Example 1   Evaluating Trigonometric Functions

Use the triangle in Figure 4.27 to find the values of the six trigonometric functions of θ.

Solution

By the Pythagorean Theorem, $(\text{hyp})^2 = (\text{opp})^2 + (\text{adj})^2$, it follows that

$$\text{hyp} = \sqrt{4^2 + 3^2}$$
$$= \sqrt{25}$$
$$= 5.$$

So, the six trigonometric functions of θ are

$$\sin \theta = \frac{\text{opp}}{\text{hyp}} = \frac{4}{5} \qquad \csc \theta = \frac{\text{hyp}}{\text{opp}} = \frac{5}{4}$$

$$\cos \theta = \frac{\text{adj}}{\text{hyp}} = \frac{3}{5} \qquad \sec \theta = \frac{\text{hyp}}{\text{adj}} = \frac{5}{3}$$

$$\tan \theta = \frac{\text{opp}}{\text{adj}} = \frac{4}{3} \qquad \cot \theta = \frac{\text{adj}}{\text{opp}} = \frac{3}{4}.$$

CHECK*Point*   Now try Exercise 7.

In Example 1, you were given the lengths of two sides of the right triangle, but not the angle θ. Often, you will be asked to find the trigonometric functions of a *given* acute angle θ. To do this, construct a right triangle having θ as one of its angles.

Example 2   Evaluating Trigonometric Functions of 45°

Find the values of sin 45°, cos 45°, and tan 45°.

Solution

Construct a right triangle having 45° as one of its acute angles, as shown in Figure 4.28. Choose the length of the adjacent side to be 1. From geometry, you know that the other acute angle is also 45°. So, the triangle is isosceles and the length of the opposite side is also 1. Using the Pythagorean Theorem, you find the length of the hypotenuse to be $\sqrt{2}$.

$$\sin 45° = \frac{\text{opp}}{\text{hyp}} = \frac{1}{\sqrt{2}} = \frac{\sqrt{2}}{2}$$

$$\cos 45° = \frac{\text{adj}}{\text{hyp}} = \frac{1}{\sqrt{2}} = \frac{\sqrt{2}}{2}$$

$$\tan 45° = \frac{\text{opp}}{\text{adj}} = \frac{1}{1} = 1$$

CHECK*Point*   Now try Exercise 23.

Example 3   Evaluating Trigonometric Functions of 30° and 60°

Use the equilateral triangle shown in Figure 4.29 to find the values of sin 60°, cos 60°, sin 30°, and cos 30°.

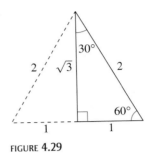

FIGURE **4.29**

Solution

Use the Pythagorean Theorem and the equilateral triangle in Figure 4.29 to verify the lengths of the sides shown in the figure. For $\theta = 60°$, you have adj $= 1$, opp $= \sqrt{3}$, and hyp $= 2$. So,

$$\sin 60° = \frac{\text{opp}}{\text{hyp}} = \frac{\sqrt{3}}{2} \qquad \text{and} \qquad \cos 60° = \frac{\text{adj}}{\text{hyp}} = \frac{1}{2}.$$

For $\theta = 30°$, adj $= \sqrt{3}$, opp $= 1$, and hyp $= 2$. So,

$$\sin 30° = \frac{\text{opp}}{\text{hyp}} = \frac{1}{2} \qquad \text{and} \qquad \cos 30° = \frac{\text{adj}}{\text{hyp}} = \frac{\sqrt{3}}{2}.$$

CHECK**Point**   Now try Exercise 27.

Sines, Cosines, and Tangents of Special Angles

$$\sin 30° = \sin \frac{\pi}{6} = \frac{1}{2} \qquad \cos 30° = \cos \frac{\pi}{6} = \frac{\sqrt{3}}{2} \qquad \tan 30° = \tan \frac{\pi}{6} = \frac{\sqrt{3}}{3}$$

$$\sin 45° = \sin \frac{\pi}{4} = \frac{\sqrt{2}}{2} \qquad \cos 45° = \cos \frac{\pi}{4} = \frac{\sqrt{2}}{2} \qquad \tan 45° = \tan \frac{\pi}{4} = 1$$

$$\sin 60° = \sin \frac{\pi}{3} = \frac{\sqrt{3}}{2} \qquad \cos 60° = \cos \frac{\pi}{3} = \frac{1}{2} \qquad \tan 60° = \tan \frac{\pi}{3} = \sqrt{3}$$

In the box, note that $\sin 30° = \frac{1}{2} = \cos 60°$. This occurs because 30° and 60° are complementary angles. In general, it can be shown from the right triangle definitions that *cofunctions of complementary angles are equal.* That is, if θ is an acute angle, the following relationships are true.

$$\sin(90° - \theta) = \cos \theta \qquad\qquad \cos(90° - \theta) = \sin \theta$$

$$\tan(90° - \theta) = \cot \theta \qquad\qquad \cot(90° - \theta) = \tan \theta$$

$$\sec(90° - \theta) = \csc \theta \qquad\qquad \csc(90° - \theta) = \sec \theta$$

Trigonometric Identities

In trigonometry, a great deal of time is spent studying relationships between trigono-metric functions (identities).

Fundamental Trigonometric Identities

Reciprocal Identities

$$\sin \theta = \frac{1}{\csc \theta} \qquad \cos \theta = \frac{1}{\sec \theta} \qquad \tan \theta = \frac{1}{\cot \theta}$$

$$\csc \theta = \frac{1}{\sin \theta} \qquad \sec \theta = \frac{1}{\cos \theta} \qquad \cot \theta = \frac{1}{\tan \theta}$$

Quotient Identities

$$\tan \theta = \frac{\sin \theta}{\cos \theta} \qquad \cot \theta = \frac{\cos \theta}{\sin \theta}$$

Pythagorean Identities

$$\sin^2 \theta + \cos^2 \theta = 1 \qquad\qquad 1 + \tan^2 \theta = \sec^2 \theta$$

$$1 + \cot^2 \theta = \csc^2 \theta$$

Note that $\sin^2 \theta$ represents $(\sin \theta)^2$, $\cos^2 \theta$ represents $(\cos \theta)^2$, and so on.

Example 4   Applying Trigonometric Identities

Let θ be an acute angle such that $\sin \theta = 0.6$. Find the values of (a) $\cos \theta$ and (b) $\tan \theta$ using trigonometric identities.

Solution

a. To find the value of $\cos \theta$, use the Pythagorean identity

$$\sin^2 \theta + \cos^2 \theta = 1.$$

So, you have

$$(0.6)^2 + \cos^2 \theta = 1 \qquad \text{Substitute 0.6 for } \sin \theta.$$

$$\cos^2 \theta = 1 - (0.6)^2 = 0.64 \qquad \text{Subtract } (0.6)^2 \text{ from each side.}$$

$$\cos \theta = \sqrt{0.64} = 0.8. \qquad \text{Extract the positive square root.}$$

b. Now, knowing the sine and cosine of θ, you can find the tangent of θ to be

$$\tan \theta = \frac{\sin \theta}{\cos \theta}$$

$$= \frac{0.6}{0.8}$$

$$= 0.75.$$

Use the definitions of $\cos \theta$ and $\tan \theta$, and the triangle shown in Figure 4.30, to check these results.

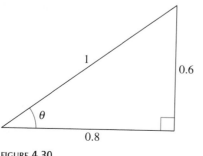

FIGURE 4.30

CHECK*Point*   Now try Exercise 33.

Example 5   Applying Trigonometric Identities

Let θ be an acute angle such that $\tan \theta = 3$. Find the values of (a) $\cot \theta$ and (b) $\sec \theta$ using trigonometric identities.

Solution

a. $\cot \theta = \dfrac{1}{\tan \theta}$ Reciprocal identity

$\cot \theta = \dfrac{1}{3}$

b. $\sec^2 \theta = 1 + \tan^2 \theta$ Pythagorean identity

$\sec^2 \theta = 1 + 3^2$

$\sec^2 \theta = 10$

$\sec \theta = \sqrt{10}$

Use the definitions of $\cot \theta$ and $\sec \theta$, and the triangle shown in Figure 4.31, to check these results.

CHECK**Point**   Now try Exercise 35.

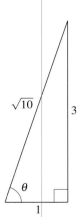

$\sqrt{10}$ 3

θ

1

FIGURE 4.31

You can also use the reciprocal identities for sine, cosine, and tangent to evaluate the cosecant, secant, and cotangent functions with a calculator. For instance, you could use the following keystroke sequence to evaluate sec 28°.

1 ÷ COS 28 ENTER

The calculator should display 1.1325701.

Evaluating Trigonometric Functions with a Calculator

To use a calculator to evaluate trigonometric functions of angles measured in degrees, first set the calculator to *degree* mode and then proceed as demonstrated in Section 4.2. For instance, you can find values of cos 28° and sec 28° as follows.

Function	*Mode*	*Calculator Keystrokes*	*Display*
a. cos 28°	Degree	COS 28 ENTER	0.8829476
b. sec 28°	Degree	(COS (28)) x⁻¹ ENTER	1.1325701

Throughout this text, angles are assumed to be measured in radians unless noted otherwise. For example, sin 1 means the sine of 1 radian and sin 1° means the sine of 1 degree.

Example 6   Using a Calculator

Use a calculator to evaluate $\sec(5° \, 40' \, 12'')$.

Solution

Begin by converting to decimal degree form. [Recall that $1' = \frac{1}{60}(1°)$ and $1'' = \frac{1}{3600}(1°)$].

$$5° \, 40' \, 12'' = 5° + \left(\frac{40}{60}\right)^{\circ} + \left(\frac{12}{3600}\right)^{\circ} = 5.67°$$

Then, use a calculator to evaluate sec 5.67°.

Function	*Calculator Keystrokes*	*Display*
$\sec(5° \, 40' \, 12'') = \sec 5.67°$	(COS (5.67)) x⁻¹ ENTER	1.0049166

CHECK**Point**   Now try Exercise 51.

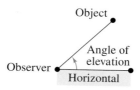

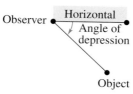

FIGURE 4.32

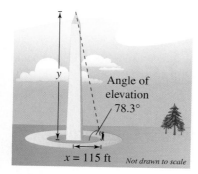

FIGURE 4.33

Applications Involving Right Triangles

Many applications of trigonometry involve a process called **solving right triangles.** In this type of application, you are usually given one side of a right triangle and one of the acute angles and are asked to find one of the other sides, *or* you are given two sides and are asked to find one of the acute angles.

In Example 7, the angle you are given is the **angle of elevation,** which represents the angle from the horizontal upward to an object. For objects that lie below the horizontal, it is common to use the term **angle of depression,** as shown in Figure 4.32.

Example 7   Using Trigonometry to Solve a Right Triangle

A surveyor is standing 115 feet from the base of the Washington Monument, as shown in Figure 4.33. The surveyor measures the angle of elevation to the top of the monument as 78.3°. How tall is the Washington Monument?

Solution

From Figure 4.33, you can see that

$$\tan 78.3° = \frac{\text{opp}}{\text{adj}} = \frac{y}{x}$$

where $x = 115$ and y is the height of the monument. So, the height of the Washington Monument is

$$y = x \tan 78.3° \approx 115(4.82882) \approx 555 \text{ feet.}$$

CHECK *Point*   Now try Exercise 67.

Example 8   Using Trigonometry to Solve a Right Triangle

A historic lighthouse is 200 yards from a bike path along the edge of a lake. A walkway to the lighthouse is 400 yards long. Find the acute angle θ between the bike path and the walkway, as illustrated in Figure 4.34.

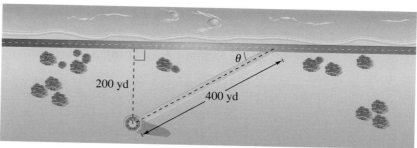

FIGURE 4.34

Solution

From Figure 4.34, you can see that the sine of the angle θ is

$$\sin \theta = \frac{\text{opp}}{\text{hyp}} = \frac{200}{400} = \frac{1}{2}.$$

Now you should recognize that $\theta = 30°$.

CHECK *Point*   Now try Exercise 69.

By now you are able to recognize that $\theta = 30°$ is the acute angle that satisfies the equation $\sin\theta = \frac{1}{2}$. Suppose, however, that you were given the equation $\sin\theta = 0.6$ and were asked to find the acute angle θ. Because

$$\sin 30° = \frac{1}{2}$$

$$= 0.5000$$

and

$$\sin 45° = \frac{1}{\sqrt{2}}$$

$$\approx 0.7071$$

you might guess that θ lies somewhere between $30°$ and $45°$. In a later section, you will study a method by which a more precise value of θ can be determined.

Example 9   Solving a Right Triangle

Find the length c of the skateboard ramp shown in Figure 4.35.

FIGURE 4.35

Solution

From Figure 4.35, you can see that

$$\sin 18.4° = \frac{\text{opp}}{\text{hyp}}$$

$$= \frac{4}{c}.$$

So, the length of the skateboard ramp is

$$c = \frac{4}{\sin 18.4°}$$

$$\approx \frac{4}{0.3156}$$

$$\approx 12.7 \text{ feet.}$$

CHECK*Point*   Now try Exercise 71.

4.3 EXERCISES

See www.CalcChat.com for worked-out solutions to odd-numbered exercises.

VOCABULARY

1. Match the trigonometric function with its right triangle definition.

(a) Sine   (b) Cosine   (c) Tangent   (d) Cosecant   (e) Secant   (f) Cotangent

(i) $\dfrac{\text{hypotenuse}}{\text{adjacent}}$ (ii) $\dfrac{\text{adjacent}}{\text{opposite}}$ (iii) $\dfrac{\text{hypotenuse}}{\text{opposite}}$ (iv) $\dfrac{\text{adjacent}}{\text{hypotenuse}}$ (v) $\dfrac{\text{opposite}}{\text{hypotenuse}}$ (vi) $\dfrac{\text{opposite}}{\text{adjacent}}$

In Exercises 2–4, fill in the blanks.

2. Relative to the angle θ, the three sides of a right triangle are the _____ side, the _____ side, and the _____.

3. Cofunctions of _____ angles are equal.

4. An angle that measures from the horizontal upward to an object is called the angle of _____, whereas an angle that measures from the horizontal downward to an object is called the angle of _____.

SKILLS AND APPLICATIONS

In Exercises 5–8, find the exact values of the six trigonometric functions of the angle θ shown in the figure. (Use the Pythagorean Theorem to find the third side of the triangle.)

5.

6.

7.

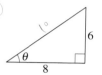

8.

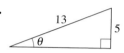

In Exercises 9–12, find the exact values of the six trigonometric functions of the angle θ for each of the two triangles. Explain why the function values are the same.

9.

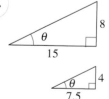

10.

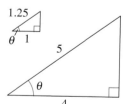

11.

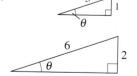

12.

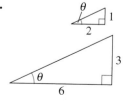

In Exercises 13–20, sketch a right triangle corresponding to the trigonometric function of the acute angle θ. Use the Pythagorean Theorem to determine the third side and then find the other five trigonometric functions of θ.

13. $\tan \theta = \frac{3}{4}$

14. $\cos \theta = \frac{5}{6}$

15. $\sec \theta = \frac{3}{2}$

16. $\tan \theta = \frac{4}{5}$

17. $\sin \theta = \frac{1}{5}$

18. $\sec \theta = \frac{17}{7}$

19. $\cot \theta = 3$

20. $\csc \theta = 9$

In Exercises 21–30, construct an appropriate triangle to complete the table. $(0° \le \theta \le 90°, 0 \le \theta \le \pi/2)$

Function	θ (deg)	θ (rad)	Function Value
21. sin	30°		
22. cos	45°		
23. sec		$\dfrac{\pi}{4}$	
24. tan		$\dfrac{\pi}{3}$	
25. cot			$\dfrac{\sqrt{3}}{3}$
26. csc			$\sqrt{2}$
27. csc		$\dfrac{\pi}{6}$	
28. sin		$\dfrac{\pi}{4}$	
29. cot			1
30. tan			$\dfrac{\sqrt{3}}{3}$

In Exercises 31–36, use the given function value(s), and trigonometric identities (including the cofunction identities), to find the indicated trigonometric functions.

31. $\sin 60° = \dfrac{\sqrt{3}}{2}$, $\cos 60° = \dfrac{1}{2}$

   (a) $\sin 30°$ (b) $\cos 30°$

   (c) $\tan 60°$ (d) $\cot 60°$

32. $\sin 30° = \dfrac{1}{2}$, $\tan 30° = \dfrac{\sqrt{3}}{3}$

   (a) $\csc 30°$ (b) $\cot 60°$

   (c) $\cos 30°$ (d) $\cot 30°$

33. $\cos \theta = \dfrac{1}{3}$

   (a) $\sin \theta$ (b) $\tan \theta$

   (c) $\sec \theta$ (d) $\csc(90° - \theta)$

34. $\sec \theta = 5$

   (a) $\cos \theta$ (b) $\cot \theta$

   (c) $\cot(90° - \theta)$ (d) $\sin \theta$

35. $\cot \alpha = 5$

   (a) $\tan \alpha$ (b) $\csc \alpha$

   (c) $\cot(90° - \alpha)$ (d) $\cos \alpha$

36. $\cos \beta = \dfrac{\sqrt{7}}{4}$

   (a) $\sec \beta$ (b) $\sin \beta$

   (c) $\cot \beta$ (d) $\sin(90° - \beta)$

In Exercises 37–46, use trigonometric identities to transform the left side of the equation into the right side $(0 < \theta < \pi/2)$.

37. $\tan \theta \cot \theta = 1$

38. $\cos \theta \sec \theta = 1$

39. $\tan \alpha \cos \alpha = \sin \alpha$

40. $\cot \alpha \sin \alpha = \cos \alpha$

41. $(1 + \sin \theta)(1 - \sin \theta) = \cos^2 \theta$

42. $(1 + \cos \theta)(1 - \cos \theta) = \sin^2 \theta$

43. $(\sec \theta + \tan \theta)(\sec \theta - \tan \theta) = 1$

44. $\sin^2 \theta - \cos^2 \theta = 2 \sin^2 \theta - 1$

45. $\dfrac{\sin \theta}{\cos \theta} + \dfrac{\cos \theta}{\sin \theta} = \csc \theta \sec \theta$

46. $\dfrac{\tan \beta + \cot \beta}{\tan \beta} = \csc^2 \beta$

In Exercises 47–56, use a calculator to evaluate each function. Round your answers to four decimal places. (Be sure the calculator is in the correct angle mode.)

47. (a) $\sin 10°$ (b) $\cos 80°$

48. (a) $\tan 23.5°$ (b) $\cot 66.5°$

49. (a) $\sin 16.35°$ (b) $\csc 16.35°$

50. (a) $\cot 79.56°$ (b) $\sec 79.56°$

51. (a) $\cos 4° 50' 15''$ (b) $\sec 4° 50' 15''$

52. (a) $\sec 42° 12'$ (b) $\csc 48° 7'$

53. (a) $\cot 11° 15'$ (b) $\tan 11° 15'$

54. (a) $\sec 56° 8' 10''$ (b) $\cos 56° 8' 10''$

55. (a) $\csc 32° 40' 3''$ (b) $\tan 44° 28' 16''$

56. (a) $\sec\!\left(\frac{9}{5} \cdot 20 + 32\right)°$ (b) $\cot\!\left(\frac{9}{5} \cdot 30 + 32\right)°$

In Exercises 57–62, find the values of θ in degrees $(0° < \theta < 90°)$ and radians $(0 < \theta < \pi/2)$ without the aid of a calculator.

57. (a) $\sin \theta = \dfrac{1}{2}$ (b) $\csc \theta = 2$

58. (a) $\cos \theta = \dfrac{\sqrt{2}}{2}$ (b) $\tan \theta = 1$

59. (a) $\sec \theta = 2$ (b) $\cot \theta = 1$

60. (a) $\tan \theta = \sqrt{3}$ (b) $\cos \theta = \dfrac{1}{2}$

61. (a) $\csc \theta = \dfrac{2\sqrt{3}}{3}$ (b) $\sin \theta = \dfrac{\sqrt{2}}{2}$

62. (a) $\cot \theta = \dfrac{\sqrt{3}}{3}$ (b) $\sec \theta = \sqrt{2}$

In Exercises 63–66, solve for x, y, or r as indicated.

63. Solve for y.

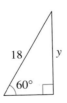

64. Solve for x.

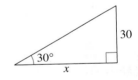

65. Solve for x.

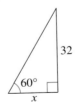

66. Solve for r.

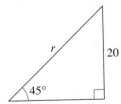

67. EMPIRE STATE BUILDING You are standing 45 meters from the base of the Empire State Building. You estimate that the angle of elevation to the top of the 86th floor (the observatory) is 82°. If the total height of the building is another 123 meters above the 86th floor, what is the approximate height of the building? One of your friends is on the 86th floor. What is the distance between you and your friend?

68. HEIGHT   A six-foot person walks from the base of a broadcasting tower directly toward the tip of the shadow cast by the tower. When the person is 132 feet from the tower and 3 feet from the tip of the shadow, the person's shadow starts to appear beyond the tower's shadow.

(a) Draw a right triangle that gives a visual representation of the problem. Show the known quantities of the triangle and use a variable to indicate the height of the tower.

(b) Use a trigonometric function to write an equation involving the unknown quantity.

(c) What is the height of the tower?

69. ANGLE OF ELEVATION   You are skiing down a mountain with a vertical height of 1500 feet. The distance from the top of the mountain to the base is 3000 feet. What is the angle of elevation from the base to the top of the mountain?

70. WIDTH OF A RIVER   A biologist wants to know the width w of a river so that instruments for studying the pollutants in the water can be set properly. From point A, the biologist walks downstream 100 feet and sights to point C (see figure). From this sighting, it is determined that $\theta = 54°$. How wide is the river?

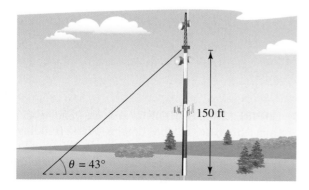

71. LENGTH   A guy wire runs from the ground to a cell tower. The wire is attached to the cell tower 150 feet above the ground. The angle formed between the wire and the ground is 43° (see figure).

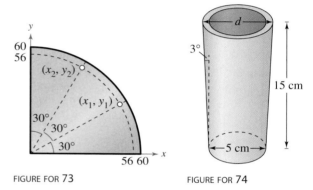

(a) How long is the guy wire?

(b) How far from the base of the tower is the guy wire anchored to the ground?

72. HEIGHT OF A MOUNTAIN   In traveling across flat land, you notice a mountain directly in front of you. Its angle of elevation (to the peak) is 3.5°. After you drive 13 miles closer to the mountain, the angle of elevation is 9°. Approximate the height of the mountain.

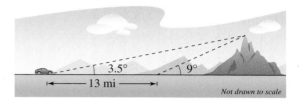

Not drawn to scale

73. MACHINE SHOP CALCULATIONS   A steel plate has the form of one-fourth of a circle with a radius of 60 centimeters. Two two-centimeter holes are to be drilled in the plate positioned as shown in the figure. Find the coordinates of the center of each hole.

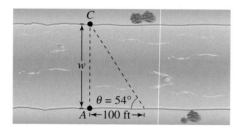

FIGURE FOR 73

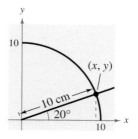

FIGURE FOR 74

74. MACHINE SHOP CALCULATIONS   A tapered shaft has a diameter of 5 centimeters at the small end and is 15 centimeters long (see figure). The taper is 3°. Find the diameter d of the large end of the shaft.

75. GEOMETRY   Use a compass to sketch a quarter of a circle of radius 10 centimeters. Using a protractor, construct an angle of 20° in standard position (see figure). Drop a perpendicular line from the point of intersection of the terminal side of the angle and the arc of the circle. By actual measurement, calculate the coordinates (x, y) of the point of intersection and use these measurements to approximate the six trigonometric functions of a 20° angle.

76. HEIGHT A 20-meter line is used to tether a helium-filled balloon. Because of a breeze, the line makes an angle of approximately 85° with the ground.

(a) Draw a right triangle that gives a visual representation of the problem. Show the known quantities of the triangle and use a variable to indicate the height of the balloon.

(b) Use a trigonometric function to write an equation involving the unknown quantity.

(c) What is the height of the balloon?

(d) The breeze becomes stronger and the angle the balloon makes with the ground decreases. How does this affect the triangle you drew in part (a)?

(e) Complete the table, which shows the heights (in meters) of the balloon for decreasing angle measures θ.

Angle, θ	80°	70°	60°	50°
Height				

Angle, θ	40°	30°	20°	10°
Height				

(f) As the angle the balloon makes with the ground approaches 0°, how does this affect the height of the balloon? Draw a right triangle to explain your reasoning.

EXPLORATION

TRUE OR FALSE? In Exercises 77–82, determine whether the statement is true or false. Justify your answer.

77. $\sin 60° \csc 60° = 1$ **78.** $\sec 30° = \csc 60°$

79. $\sin 45° + \cos 45° = 1$ **80.** $\cot^2 10° - \csc^2 10° = -1$

81. $\dfrac{\sin 60°}{\sin 30°} = \sin 2°$ **82.** $\tan[(5°)^2] = \tan^2 5°$

83. THINK ABOUT IT

(a) Complete the table.

θ	0.1	0.2	0.3	0.4	0.5
$\sin \theta$					

(b) Is θ or $\sin \theta$ greater for θ in the interval $(0, 0.5]$?

(c) As θ approaches 0, how do θ and $\sin \theta$ compare? Explain.

84. THINK ABOUT IT

(a) Complete the table.

θ	0°	18°	36°	54°	72°	90°
$\sin \theta$						
$\cos \theta$						

(b) Discuss the behavior of the sine function for θ in the range from 0° to 90°.

(c) Discuss the behavior of the cosine function for θ in the range from 0° to 90°.

(d) Use the definitions of the sine and cosine functions to explain the results of parts (b) and (c).

85. WRITING In right triangle trigonometry, explain why $\sin 30° = \frac{1}{2}$ regardless of the size of the triangle.

86. GEOMETRY Use the equilateral triangle shown in Figure 4.29 and similar triangles to verify the points in Figure 4.23 (in Section 4.2) that do not lie on the axes.

87. THINK ABOUT IT You are given only the value $\tan \theta$. Is it possible to find the value of $\sec \theta$ without finding the measure of θ? Explain.

88. CAPSTONE The Johnstown Inclined Plane in Pennsylvania is one of the longest and steepest hoists in the world. The railway cars travel a distance of 896.5 feet at an angle of approximately 35.4°, rising to a height of 1693.5 feet above sea level.

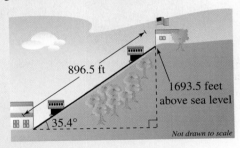

896.5 ft

1693.5 feet above sea level

35.4°

Not drawn to scale

(a) Find the vertical rise of the inclined plane.

(b) Find the elevation of the lower end of the inclined plane.

(c) The cars move up the mountain at a rate of 300 feet per minute. Find the rate at which they rise vertically.

4.4 TRIGONOMETRIC FUNCTIONS OF ANY ANGLE

What you should learn

• Evaluate trigonometric functions of any angle.
• Find reference angles.
• Evaluate trigonometric functions of real numbers.

Why you should learn it

You can use trigonometric functions to model and solve real-life problems. For instance, in Exercise 99 on page 318, you can use trigonometric functions to model the monthly normal temperatures in New York City and Fairbanks, Alaska.

Introduction

In Section 4.3, the definitions of trigonometric functions were restricted to acute angles. In this section, the definitions are extended to cover *any* angle. If θ is an *acute* angle, these definitions coincide with those given in the preceding section.

> **Definitions of Trigonometric Functions of Any Angle**
>
> Let θ be an angle in standard position with (x, y) a point on the terminal side of θ and $r = \sqrt{x^2 + y^2} \neq 0$.
>
> $\sin \theta = \dfrac{y}{r}$ $\cos \theta = \dfrac{x}{r}$
>
> $\tan \theta = \dfrac{y}{x}, \quad x \neq 0$ $\cot \theta = \dfrac{x}{y}, \quad y \neq 0$
>
> $\sec \theta = \dfrac{r}{x}, \quad x \neq 0$ $\csc \theta = \dfrac{r}{y}, \quad y \neq 0$
>
>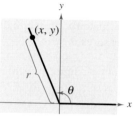

Because $r = \sqrt{x^2 + y^2}$ *cannot* be zero, it follows that the sine and cosine functions are defined for any real value of θ. However, if $x = 0$, the tangent and secant of θ are undefined. For example, the tangent of $90°$ is undefined. Similarly, if $y = 0$, the cotangent and cosecant of θ are undefined.

Example 1   Evaluating Trigonometric Functions

Let $(-3, 4)$ be a point on the terminal side of θ. Find the sine, cosine, and tangent of θ.

Solution

Referring to Figure 4.36, you can see that $x = -3$, $y = 4$, and

$$r = \sqrt{x^2 + y^2} = \sqrt{(-3)^2 + 4^2} = \sqrt{25} = 5.$$

So, you have the following.

$$\sin \theta = \frac{y}{r} = \frac{4}{5} \qquad \cos \theta = \frac{x}{r} = -\frac{3}{5} \qquad \tan \theta = \frac{y}{x} = -\frac{4}{3}$$

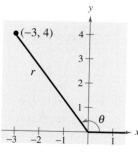

FIGURE **4.36**

Algebra Help

The formula $r = \sqrt{x^2 + y^2}$ is a result of the Distance Formula. You can review the Distance Formula in Section 1.1.

CHECK**Point**   Now try Exercise 9.

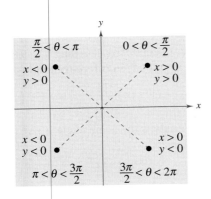

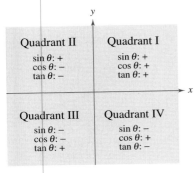

FIGURE 4.37

The *signs* of the trigonometric functions in the four quadrants can be determined from the definitions of the functions. For instance, because $\cos \theta = x/r$, it follows that $\cos \theta$ is positive wherever $x > 0$, which is in Quadrants I and IV. (Remember, r is always positive.) In a similar manner, you can verify the results shown in Figure 4.37.

Example 2   Evaluating Trigonometric Functions

Given $\tan \theta = -\frac{5}{4}$ and $\cos \theta > 0$, find $\sin \theta$ and $\sec \theta$.

Solution

Note that θ lies in Quadrant IV because that is the only quadrant in which the tangent is negative and the cosine is positive. Moreover, using

$$\tan \theta = \frac{y}{x}$$

$$= -\frac{5}{4}$$

and the fact that y is negative in Quadrant IV, you can let $y = -5$ and $x = 4$. So, $r = \sqrt{16 + 25} = \sqrt{41}$ and you have

$$\sin \theta = \frac{y}{r} = \frac{-5}{\sqrt{41}}$$

$$\approx -0.7809$$

$$\sec \theta = \frac{r}{x} = \frac{\sqrt{41}}{4}$$

$$\approx 1.6008.$$

CHECK *Point* ▸   Now try Exercise 23.

Example 3   Trigonometric Functions of Quadrant Angles

Evaluate the cosine and tangent functions at the four quadrant angles 0, $\frac{\pi}{2}$, π, and $\frac{3\pi}{2}$.

Solution

To begin, choose a point on the terminal side of each angle, as shown in Figure 4.38. For each of the four points, $r = 1$, and you have the following.

$$\cos 0 = \frac{x}{r} = \frac{1}{1} = 1 \qquad\qquad \tan 0 = \frac{y}{x} = \frac{0}{1} = 0 \qquad\qquad (x, y) = (1, 0)$$

$$\cos \frac{\pi}{2} = \frac{x}{r} = \frac{0}{1} = 0 \qquad\qquad \tan \frac{\pi}{2} = \frac{y}{x} = \frac{1}{0} \Rightarrow \text{undefined} \qquad (x, y) = (0, 1)$$

$$\cos \pi = \frac{x}{r} = \frac{-1}{1} = -1 \qquad\qquad \tan \pi = \frac{y}{x} = \frac{0}{-1} = 0 \qquad\qquad (x, y) = (-1, 0)$$

$$\cos \frac{3\pi}{2} = \frac{x}{r} = \frac{0}{1} = 0 \qquad\qquad \tan \frac{3\pi}{2} = \frac{y}{x} = \frac{-1}{0} \Rightarrow \text{undefined} \qquad (x, y) = (0, -1)$$

CHECK *Point* ▸   Now try Exercise 37.

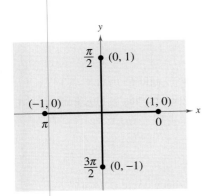

FIGURE 4.38

Reference Angles

The values of the trigonometric functions of angles greater than 90° (or less than 0°) can be determined from their values at corresponding acute angles called **reference angles.**

Definition of Reference Angle

Let θ be an angle in standard position. Its **reference angle** is the acute angle θ' formed by the terminal side of θ and the horizontal axis.

Figure 4.39 shows the reference angles for θ in Quadrants II, III, and IV.

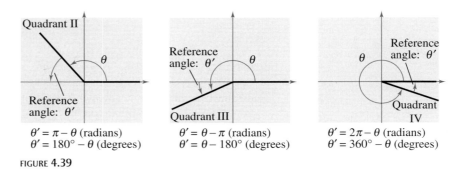

$\theta' = \pi - \theta$ (radians)
$\theta' = 180° - \theta$ (degrees)

$\theta' = \theta - \pi$ (radians)
$\theta' = \theta - 180°$ (degrees)

$\theta' = 2\pi - \theta$ (radians)
$\theta' = 360° - \theta$ (degrees)

FIGURE 4.39

| **Example 4** | **Finding Reference Angles** |

Find the reference angle θ'.

a. $\theta = 300°$ **b.** $\theta = 2.3$ **c.** $\theta = -135°$

Solution

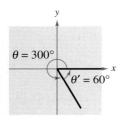

FIGURE 4.40

a. Because 300° lies in Quadrant IV, the angle it makes with the x-axis is

$$\theta' = 360° - 300°$$

$$= 60°. \qquad \text{Degrees}$$

Figure 4.40 shows the angle $\theta = 300°$ and its reference angle $\theta' = 60°$.

b. Because 2.3 lies between $\pi/2 \approx 1.5708$ and $\pi \approx 3.1416$, it follows that it is in Quadrant II and its reference angle is

$$\theta' = \pi - 2.3$$

$$\approx 0.8416. \qquad \text{Radians}$$

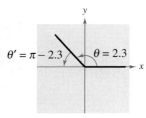

FIGURE 4.41

Figure 4.41 shows the angle $\theta = 2.3$ and its reference angle $\theta' = \pi - 2.3$.

c. First, determine that $-135°$ is coterminal with 225°, which lies in Quadrant III. So, the reference angle is

$$\theta' = 225° - 180°$$

$$= 45°. \qquad \text{Degrees}$$

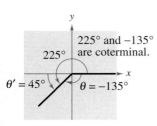

FIGURE 4.42

Figure 4.42 shows the angle $\theta = -135°$ and its reference angle $\theta' = 45°$.

CHECK*Point* Now try Exercise 45.

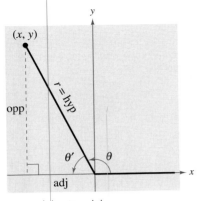

opp = $|y|$, adj = $|x|$

FIGURE 4.43

Trigonometric Functions of Real Numbers

To see how a reference angle is used to evaluate a trigonometric function, consider the point (x, y) on the terminal side of θ, as shown in Figure 4.43. By definition, you know that

$$\sin \theta = \frac{y}{r} \qquad \text{and} \qquad \tan \theta = \frac{y}{x}.$$

For the right triangle with acute angle θ' and sides of lengths $|x|$ and $|y|$, you have

$$\sin \theta' = \frac{\text{opp}}{\text{hyp}} = \frac{|y|}{r}$$

and

$$\tan \theta' = \frac{\text{opp}}{\text{adj}} = \frac{|y|}{|x|}.$$

So, it follows that $\sin \theta$ and $\sin \theta'$ are equal, *except possibly in sign*. The same is true for $\tan \theta$ and $\tan \theta'$ *and* for the other four trigonometric functions. In all cases, the sign of the function value can be determined by the quadrant in which θ lies.

Evaluating Trigonometric Functions of Any Angle

To find the value of a trigonometric function of any angle θ:

1. Determine the function value for the associated reference angle θ'.

2. Depending on the quadrant in which θ lies, affix the appropriate sign to the function value.

By using reference angles and the special angles discussed in the preceding section, you can greatly extend the scope of *exact* trigonometric values. For instance, knowing the function values of 30° means that you know the function values of all angles for which 30° is a reference angle. For convenience, the table below shows the exact values of the trigonometric functions of special angles and quadrant angles.

Trigonometric Values of Common Angles

θ (degrees)	0°	30°	45°	60°	90°	180°	270°
θ (radians)	0	$\dfrac{\pi}{6}$	$\dfrac{\pi}{4}$	$\dfrac{\pi}{3}$	$\dfrac{\pi}{2}$	π	$\dfrac{3\pi}{2}$
$\sin \theta$	0	$\dfrac{1}{2}$	$\dfrac{\sqrt{2}}{2}$	$\dfrac{\sqrt{3}}{2}$	1	0	-1
$\cos \theta$	1	$\dfrac{\sqrt{3}}{2}$	$\dfrac{\sqrt{2}}{2}$	$\dfrac{1}{2}$	0	-1	0
$\tan \theta$	0	$\dfrac{\sqrt{3}}{3}$	1	$\sqrt{3}$	Undef.	0	Undef.

Study Tip

Learning the table of values at the right is worth the effort because doing so will increase both your efficiency and your confidence. Here is a pattern for the sine function that may help you remember the values.

θ	0°	30°	45°	60°	90°
$\sin \theta$	$\dfrac{\sqrt{0}}{2}$	$\dfrac{\sqrt{1}}{2}$	$\dfrac{\sqrt{2}}{2}$	$\dfrac{\sqrt{3}}{2}$	$\dfrac{\sqrt{4}}{2}$

Reverse the order to get cosine values of the same angles.

Example 5 Using Reference Angles

Evaluate each trigonometric function.

a. $\cos\dfrac{4\pi}{3}$ **b.** $\tan(-210°)$ **c.** $\csc\dfrac{11\pi}{4}$

Solution

a. Because $\theta = 4\pi/3$ lies in Quadrant III, the reference angle is

$$\theta' = \frac{4\pi}{3} - \pi = \frac{\pi}{3}$$

as shown in Figure 6.41. Moreover, the cosine is negative in Quadrant III, so

$$\cos\frac{4\pi}{3} = (-)\cos\frac{\pi}{3}$$

$$= -\frac{1}{2}.$$

b. Because $-210° + 360° = 150°$, it follows that $-210°$ is coterminal with the second-quadrant angle 150°. So, the reference angle is $\theta' = 180° - 150° = 30°$, as shown in Figure 4.45. Finally, because the tangent is negative in Quadrant II, you have

$$\tan(-210°) = (-)\tan 30°$$

$$= -\frac{\sqrt{3}}{3}.$$

c. Because $(11\pi/4) - 2\pi = 3\pi/4$, it follows that $11\pi/4$ is coterminal with the second-quadrant angle $3\pi/4$. So, the reference angle is $\theta' = \pi - (3\pi/4) = \pi/4$, as shown in Figure 4.46. Because the cosecant is positive in Quadrant II, you have

$$\csc\frac{11\pi}{4} = (+)\csc\frac{\pi}{4}$$

$$= \frac{1}{\sin(\pi/4)}$$

$$= \sqrt{2}.$$

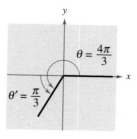

FIGURE 4.44

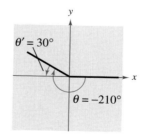

FIGURE 4.45

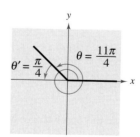

FIGURE 4.46

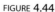 **CHECK***Point* Now try Exercise 59.

Example 6   Using Trigonometric Identities

Let θ be an angle in Quadrant II such that $\sin \theta = \frac{1}{3}$. Find (a) $\cos \theta$ and (b) $\tan \theta$ by using trigonometric identities.

Solution

a. Using the Pythagorean identity $\sin^2 \theta + \cos^2 \theta = 1$, you obtain

$$\left(\frac{1}{3}\right)^2 + \cos^2 \theta = 1 \qquad\qquad \text{Substitute } \tfrac{1}{3} \text{ for } \sin \theta.$$

$$\cos^2 \theta = 1 - \frac{1}{9} = \frac{8}{9}.$$

Because $\cos \theta < 0$ in Quadrant II, you can use the negative root to obtain

$$\cos \theta = -\frac{\sqrt{8}}{\sqrt{9}}$$

$$= -\frac{2\sqrt{2}}{3}.$$

b. Using the trigonometric identity $\tan \theta = \dfrac{\sin \theta}{\cos \theta}$, you obtain

$$\tan \theta = \frac{1/3}{-2\sqrt{2}/3} \qquad\qquad \text{Substitute for } \sin \theta \text{ and } \cos \theta.$$

$$= -\frac{1}{2\sqrt{2}}$$

$$= -\frac{\sqrt{2}}{4}.$$

CHECK **Point**   Now try Exercise 69.

You can use a calculator to evaluate trigonometric functions, as shown in the next example.

Example 7   Using a Calculator

Use a calculator to evaluate each trigonometric function.

a. $\cot 410°$ **b.** $\sin(-7)$ **c.** $\sec \dfrac{\pi}{9}$

Solution

Function	Mode	Calculator Keystrokes	Display
a. $\cot 410°$	Degree	$($ TAN $($ 410 $)$ $)$ x^{-1} ENTER	0.8390996
b. $\sin(-7)$	Radian	SIN $($ (−) 7 $)$ ENTER	−0.6569866
c. $\sec \dfrac{\pi}{9}$	Radian	$($ COS $($ π ÷ 9 $)$ $)$ x^{-1} ENTER	1.0641778

CHECK **Point**   Now try Exercise 79.

4.4 EXERCISES

See www.CalcChat.com for worked-out solutions to odd-numbered exercises.

VOCABULARY: Fill in the blanks.

In Exercises 1–6, let θ be an angle in standard position, with (x, y) a point on the terminal side of θ and $r = \sqrt{x^2 + y^2} \neq 0$.

1. $\sin \theta =$ _____
2. $\dfrac{r}{y} =$ _____
3. $\tan \theta =$ _____
4. $\sec \theta =$ _____
5. $\dfrac{x}{r} =$ _____
6. $\dfrac{x}{y} =$ _____
7. Because $r = \sqrt{x^2 + y^2}$ cannot be _____, the sine and cosine functions are _____ for any real value of θ.
8. The acute positive angle that is formed by the terminal side of the angle θ and the horizontal axis is called the _____ angle of θ and is denoted by θ'.

SKILLS AND APPLICATIONS

In Exercises 9–12, determine the exact values of the six trigonometric functions of the angle θ.

9. (a) 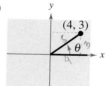 (b)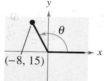

10. (a) 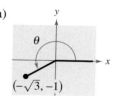 (b)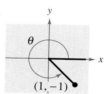

11. (a) 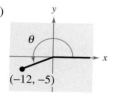 (b)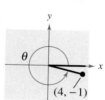

12. (a) 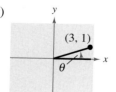 (b)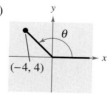

In Exercises 13–18, the point is on the terminal side of an angle in standard position. Determine the exact values of the six trigonometric functions of the angle.

13. $(5, 12)$
14. $(8, 15)$
15. $(-5, -2)$
16. $(-4, 10)$
17. $(-5.4, 7.2)$
18. $\left(3\frac{1}{2}, -7\frac{3}{4}\right)$

In Exercises 19–22, state the quadrant in which θ lies.

19. $\sin \theta > 0$ and $\cos \theta > 0$
20. $\sin \theta < 0$ and $\cos \theta < 0$
21. $\sin \theta > 0$ and $\cos \theta < 0$
22. $\sec \theta > 0$ and $\cot \theta < 0$

In Exercises 23–32, find the values of the six trigonometric functions of θ with the given constraint.

	Function Value	*Constraint*
23.	$\tan \theta = -\dfrac{15}{8}$	$\sin \theta > 0$
24.	$\cos \theta = \dfrac{8}{17}$	$\tan \theta < 0$
25.	$\sin \theta = \dfrac{3}{5}$	θ lies in Quadrant II.
26.	$\cos \theta = -\dfrac{4}{5}$	θ lies in Quadrant III.
27.	$\cot \theta = -3$	$\cos \theta > 0$
28.	$\csc \theta = 4$	$\cot \theta < 0$
29.	$\sec \theta = -2$	$\sin \theta < 0$
30.	$\sin \theta = 0$	$\sec \theta = -1$
31.	$\cot \theta$ is undefined.	$\pi/2 \leq \theta \leq 3\pi/2$
32.	$\tan \theta$ is undefined.	$\pi \leq \theta \leq 2\pi$

In Exercises 33–36, the terminal side of θ lies on the given line in the specified quadrant. Find the values of the six trigonometric functions of θ by finding a point on the line.

	Line	*Quadrant*
33.	$y = -x$	II
34.	$y = \frac{1}{3}x$	III
35.	$2x - y = 0$	III
36.	$4x + 3y = 0$	IV

In Exercises 37–44, evaluate the trigonometric function of the quadrant angle.

37. $\sin \pi$

38. $\csc \dfrac{3\pi}{2}$

39. $\sec \dfrac{3\pi}{2}$

40. $\sec \pi$

41. $\sin \dfrac{\pi}{2}$

42. $\cot \pi$

43. $\csc \pi$

44. $\cot \dfrac{\pi}{2}$

In Exercises 45–52, find the reference angle θ', and sketch θ and θ' in standard position.

45. $\theta = 160°$

46. $\theta = 309°$

47. $\theta = -125°$

48. $\theta = -215°$

49. $\theta = \dfrac{2\pi}{3}$

50. $\theta = \dfrac{7\pi}{6}$

51. $\theta = 4.8$

52. $\theta = 11.6$

In Exercises 53–68, evaluate the sine, cosine, and tangent of the angle without using a calculator.

53. $225°$

54. $300°$

55. $750°$

56. $-405°$

57. $-150°$

58. $-840°$

59. $\dfrac{2\pi}{3}$

60. $\dfrac{3\pi}{4}$

61. $\dfrac{5\pi}{4}$

62. $\dfrac{7\pi}{6}$

63. $-\dfrac{\pi}{6}$

64. $-\dfrac{\pi}{2}$

65. $\dfrac{9\pi}{4}$

66. $\dfrac{10\pi}{3}$

67. $-\dfrac{3\pi}{2}$

68. $-\dfrac{23\pi}{4}$

In Exercises 69–74, find the indicated trigonometric value in the specified quadrant.

Function	Quadrant	Trigonometric Value
69. $\sin \theta = -\dfrac{3}{5}$	IV	$\cos \theta$
70. $\cot \theta = -3$	II	$\sin \theta$
71. $\tan \theta = \dfrac{3}{2}$	III	$\sec \theta$
72. $\csc \theta = -2$	IV	$\cot \theta$
73. $\cos \theta = \dfrac{5}{8}$	I	$\sec \theta$
74. $\sec \theta = -\dfrac{9}{4}$	III	$\tan \theta$

In Exercises 75–90, use a calculator to evaluate the trigonometric function. Round your answer to four decimal places. (Be sure the calculator is set in the correct angle mode.)

75. $\sin 10°$

76. $\sec 225°$

77. $\cos(-110°)$

78. $\csc(-330°)$

79. $\tan 304°$

80. $\cot 178°$

81. $\sec 72°$

82. $\tan(-188°)$

83. $\tan 4.5$

84. $\cot 1.35$

85. $\tan \dfrac{\pi}{9}$

86. $\tan\left(-\dfrac{\pi}{9}\right)$

87. $\sin(-0.65)$

88. $\sec 0.29$

89. $\cot\left(-\dfrac{11\pi}{8}\right)$

90. $\csc\left(-\dfrac{15\pi}{14}\right)$

In Exercises 91–96, find two solutions of the equation. Give your answers in degrees ($0° \le \theta < 360°$) and in radians ($0 \le \theta < 2\pi$). Do not use a calculator.

91. (a) $\sin \theta = \dfrac{1}{2}$ (b) $\sin \theta = -\dfrac{1}{2}$

92. (a) $\cos \theta = \dfrac{\sqrt{2}}{2}$ (b) $\cos \theta = -\dfrac{\sqrt{2}}{2}$

93. (a) $\csc \theta = \dfrac{2\sqrt{3}}{3}$ (b) $\cot \theta = -1$

94. (a) $\sec \theta = 2$ (b) $\sec \theta = -2$

95. (a) $\tan \theta = 1$ (b) $\cot \theta = -\sqrt{3}$

96. (a) $\sin \theta = \dfrac{\sqrt{3}}{2}$ (b) $\sin \theta = -\dfrac{\sqrt{3}}{2}$

97. DISTANCE   An airplane, flying at an altitude of 6 miles, is on a flight path that passes directly over an observer (see figure). If θ is the angle of elevation from the observer to the plane, find the distance d from the observer to the plane when (a) $\theta = 30°$, (b) $\theta = 90°$, and (c) $\theta = 120°$.

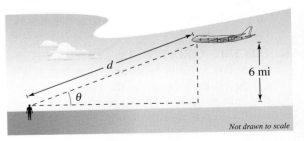

Not drawn to scale

98. HARMONIC MOTION   The displacement from equilibrium of an oscillating weight suspended by a spring is given by $y(t) = 2 \cos 6t$, where y is the displacement (in centimeters) and t is the time (in seconds). Find the displacement when (a) $t = 0$, (b) $t = \dfrac{1}{4}$, and (c) $t = \dfrac{1}{2}$.

99. DATA ANALYSIS: METEOROLOGY The table shows the monthly normal temperatures (in degrees Fahrenheit) for selected months in New York City (N) and Fairbanks, Alaska (F). (Source: National Climatic Data Center)

Month	New York City, N	Fairbanks, F
January	33	-10
April	52	32
July	77	62
October	58	24
December	38	-6

(a) Use the *regression* feature of a graphing utility to find a model of the form $y = a \sin(bt + c) + d$ for each city. Let t represent the month, with $t = 1$ corresponding to January.

(b) Use the models from part (a) to find the monthly normal temperatures for the two cities in February, March, May, June, August, September, and November.

(c) Compare the models for the two cities.

100. SALES A company that produces snowboards, which are seasonal products, forecasts monthly sales over the next 2 years to be $S = 23.1 + 0.442t + 4.3 \cos(\pi t/6)$, where S is measured in thousands of units and t is the time in months, with $t = 1$ representing January 2010. Predict sales for each of the following months.

(a) February 2010   (b) February 2011

(c) June 2010   (d) June 2011

101. HARMONIC MOTION The displacement from equilibrium of an oscillating weight suspended by a spring and subject to the damping effect of friction is given by $y(t) = 2e^{-t} \cos 6t$, where y is the displacement (in centimeters) and t is the time (in seconds). Find the displacement when (a) $t = 0$, (b) $t = \frac{1}{4}$, and (c) $t = \frac{1}{2}$.

102. ELECTRIC CIRCUITS The current I (in amperes) when 100 volts is applied to a circuit is given by $I = 5e^{-2t} \sin t$, where t is the time (in seconds) after the voltage is applied. Approximate the current at $t = 0.7$ second after the voltage is applied.

EXPLORATION

TRUE OR FALSE? In Exercises 103 and 104, determine whether the statement is true or false. Justify your answer.

103. In each of the four quadrants, the signs of the secant function and sine function will be the same.

104. To find the reference angle for an angle θ (given in degrees), find the integer n such that $0 \le 360°n - \theta \le 360°$. The difference $360°n - \theta$ is the reference angle.

105. WRITING Consider an angle in standard position with $r = 12$ centimeters, as shown in the figure. Write a short paragraph describing the changes in the values of x, y, $\sin \theta$, $\cos \theta$, and $\tan \theta$ as θ increases continuously from $0°$ to $90°$.

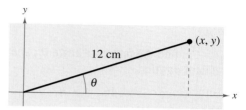

106. CAPSTONE Write a short paper in your own words explaining to a classmate how to evaluate the six trigonometric functions of any angle θ in standard position. Include an explanation of reference angles and how to use them, the signs of the functions in each of the four quadrants, and the trigonometric values of common angles. Be sure to include figures or diagrams in your paper.

107. THINK ABOUT IT The figure shows point $P(x, y)$ on a unit circle and right triangle OAP.

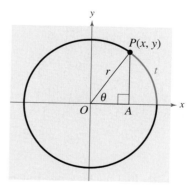

(a) Find $\sin t$ and $\cos t$ using the unit circle definitions of sine and cosine (from Section 4.2).

(b) What is the value of r? Explain.

(c) Use the definitions of sine and cosine given in this section to find $\sin \theta$ and $\cos \theta$. Write your answers in terms of x and y.

(d) Based on your answers to parts (a) and (c), what can you conclude?

4.5 GRAPHS OF SINE AND COSINE FUNCTIONS

What you should learn

- Sketch the graphs of basic sine and cosine functions.
- Use amplitude and period to help sketch the graphs of sine and cosine functions.
- Sketch translations of the graphs of sine and cosine functions.
- Use sine and cosine functions to model real-life data.

Why you should learn it

Sine and cosine functions are often used in scientific calculations. For instance, in Exercise 87 on page 328, you can use a trigonometric function to model the airflow of your respiratory cycle.

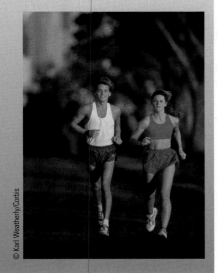

© Karl Weatherly/Corbis

Basic Sine and Cosine Curves

In this section, you will study techniques for sketching the graphs of the sine and cosine functions. The graph of the sine function is a **sine curve.** In Figure 4.47, the black portion of the graph represents one period of the function and is called **one cycle** of the sine curve. The gray portion of the graph indicates that the basic sine curve repeats indefinitely in the positive and negative directions. The graph of the cosine function is shown in Figure 4.48.

Recall from Section 4.2 that the domain of the sine and cosine functions is the set of all real numbers. Moreover, the range of each function is the interval $[-1, 1]$, and each function has a period of 2π. Do you see how this information is consistent with the basic graphs shown in Figures 4.47 and 4.48?

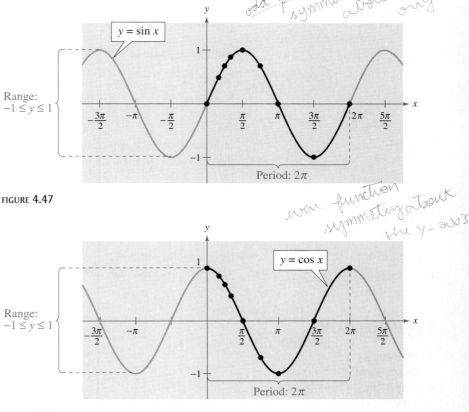

FIGURE **4.47**

FIGURE **4.48**

Note in Figures 4.47 and 4.48 that the sine curve is symmetric with respect to the *origin*, whereas the cosine curve is symmetric with respect to the *y-axis*. These properties of symmetry follow from the fact that the sine function is odd and the cosine function is even.

To sketch the graphs of the basic sine and cosine functions by hand, it helps to note five **key points** in one period of each graph: the *intercepts*, *maximum points*, and *minimum points* (see Figure 4.49).

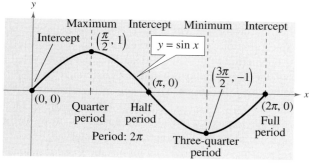
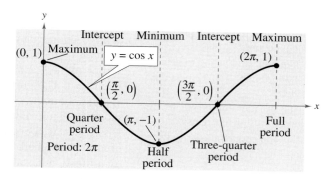

FIGURE **4.49**

Example 1   Using Key Points to Sketch a Sine Curve

Sketch the graph of $y = 2 \sin x$ on the interval $[-\pi, 4\pi]$.

Solution

Note that

$$y = 2 \sin x = 2(\sin x)$$

indicates that the y-values for the key points will have twice the magnitude of those on the graph of $y = \sin x$. Divide the period 2π into four equal parts to get the key points for $y = 2 \sin x$.

Intercept	*Maximum*	*Intercept*	*Minimum*	*Intercept*
$(0, 0)$,	$\left(\dfrac{\pi}{2}, 2\right)$,	$(\pi, 0)$;	$\left(\dfrac{3\pi}{2}, -2\right)$,	and $(2\pi, 0)$

By connecting these key points with a smooth curve and extending the curve in both directions over the interval $[-\pi, 4\pi]$, you obtain the graph shown in Figure 4.50.

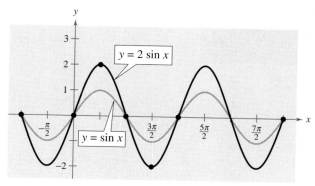

FIGURE **4.50**

CHECK Point   Now try Exercise 39.

TECHNOLOGY

When using a graphing utility to graph trigonometric functions, pay special attention to the viewing window you use. For instance, try graphing $y = [\sin(10x)]/10$ in the standard viewing window in *radian* mode. What do you observe? Use the *zoom* feature to find a viewing window that displays a good view of the graph.

amplitude

Amplitude and Period

In the remainder of this section you will study the graphic effect of each of the constants *a*, *b*, *c*, and *d* in equations of the forms

$$y = d + a \sin(bx - c)$$

and

$$y = d + a \cos(bx - c).$$

A quick review of the transformations you studied in Section 1.7 should help in this investigation.

The constant factor *a* in $y = a \sin x$ acts as a *scaling factor*—a *vertical stretch* or *vertical shrink* of the basic sine curve. If $|a| > 1$, the basic sine curve is stretched, and if $|a| < 1$, the basic sine curve is shrunk. The result is that the graph of $y = a \sin x$ ranges between $-a$ and a instead of between -1 and 1. The absolute value of *a* is the **amplitude** of the function $y = a \sin x$. The range of the function $y = a \sin x$ for $a > 0$ is $-a \le y \le a$.

Definition of Amplitude of Sine and Cosine Curves

The **amplitude** of $y = a \sin x$ and $y = a \cos x$ represents half the distance between the maximum and minimum values of the function and is given by

$$\text{Amplitude} = |a|.$$

Example 2   Scaling: Vertical Shrinking and Stretching

On the same coordinate axes, sketch the graph of each function.

a. $y = \dfrac{1}{2} \cos x$ **b.** $y = 3 \cos x$

Solution

a. Because the amplitude of $y = \frac{1}{2} \cos x$ is $\frac{1}{2}$, the maximum value is $\frac{1}{2}$ and the minimum value is $-\frac{1}{2}$. Divide one cycle, $0 \le x \le 2\pi$, into four equal parts to get the key points

Maximum	*Intercept*	*Minimum*	*Intercept*		*Maximum*
$\left(0, \dfrac{1}{2}\right)$,	$\left(\dfrac{\pi}{2}, 0\right)$,	$\left(\pi, -\dfrac{1}{2}\right)$,	$\left(\dfrac{3\pi}{2}, 0\right)$,	and	$\left(2\pi, \dfrac{1}{2}\right)$.

b. A similar analysis shows that the amplitude of $y = 3 \cos x$ is 3, and the key points are

Maximum	*Intercept*	*Minimum*	*Intercept*		*Maximum*
$(0, 3)$,	$\left(\dfrac{\pi}{2}, 0\right)$,	$(\pi, -3)$,	$\left(\dfrac{3\pi}{2}, 0\right)$,	and	$(2\pi, 3)$.

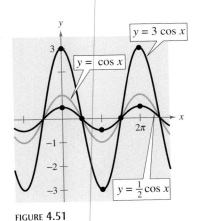

$y = 3 \cos x$

$y = \cos x$

$y = \frac{1}{2} \cos x$

FIGURE **4.51**

The graphs of these two functions are shown in Figure 4.51. Notice that the graph of $y = \frac{1}{2} \cos x$ is a vertical *shrink* of the graph of $y = \cos x$ and the graph of $y = 3 \cos x$ is a vertical *stretch* of the graph of $y = \cos x$.

CHECK *Point*   Now try Exercise 41.

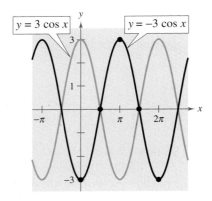

FIGURE **4.52**

You know from Section 1.7 that the graph of $y = -f(x)$ is a **reflection** in the x-axis of the graph of $y = f(x)$. For instance, the graph of $y = -3 \cos x$ is a reflection of the graph of $y = 3 \cos x$, as shown in Figure 4.52.

Because $y = a \sin x$ completes one cycle from $x = 0$ to $x = 2\pi$, it follows that $y = a \sin bx$ completes one cycle from $x = 0$ to $x = 2\pi/b$.

Period of Sine and Cosine Functions

Let b be a positive real number. The **period** of $y = a \sin bx$ and $y = a \cos bx$ is given by

$$\text{Period} = \frac{2\pi}{b}.$$

Note that if $0 < b < 1$, the period of $y = a \sin bx$ is greater than 2π and represents a *horizontal stretching* of the graph of $y = a \sin x$. Similarly, if $b > 1$, the period of $y = a \sin bx$ is less than 2π and represents a *horizontal shrinking* of the graph of $y = a \sin x$. If b is negative, the identities $\sin(-x) = -\sin x$ and $\cos(-x) = \cos x$ are used to rewrite the function.

Example 3   Scaling: Horizontal Stretching

Sketch the graph of $y = \sin \dfrac{x}{2}$.

Solution

The amplitude is 1. Moreover, because $b = \frac{1}{2}$, the period is

$$\frac{2\pi}{b} = \frac{2\pi}{\frac{1}{2}} = 4\pi. \qquad \text{Substitute for } b.$$

Now, divide the period-interval $[0, 4\pi]$ into four equal parts with the values π, 2π, and 3π to obtain the key points on the graph.

Intercept	Maximum	Intercept	Minimum	Intercept
$(0, 0)$,	$(\pi, 1)$,	$(2\pi, 0)$,	$(3\pi, -1)$,	and $(4\pi, 0)$

The graph is shown in Figure 4.53.

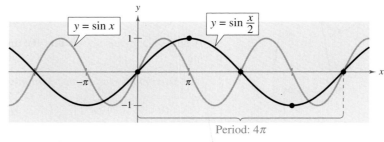

FIGURE **4.53**

CHECK*Point*   Now try Exercise 43.

Study Tip

In general, to divide a period-interval into four equal parts, successively add "period/4," starting with the left endpoint of the interval. For instance, for the period-interval $[-\pi/6, \pi/2]$ of length $2\pi/3$, you would successively add

$$\frac{2\pi/3}{4} = \frac{\pi}{6}$$

to get $-\pi/6, 0, \pi/6, \pi/3,$ and $\pi/2$ as the x-values for the key points on the graph.

Translations of Sine and Cosine Curves

The constant c in the general equations

$$y = a \sin(bx - c) \qquad \text{and} \qquad y = a \cos(bx - c)$$

creates a *horizontal translation* (shift) of the basic sine and cosine curves. Comparing $y = a \sin bx$ with $y = a \sin(bx - c)$, you find that the graph of $y = a \sin(bx - c)$ completes one cycle from $bx - c = 0$ to $bx - c = 2\pi$. By solving for x, you can find the interval for one cycle to be

Left endpoint   Right endpoint

$$\overbrace{\frac{c}{b}} \leq x \leq \overbrace{\frac{c}{b} + \frac{2\pi}{b}}.$$

Period

This implies that the period of $y = a \sin(bx - c)$ is $2\pi/b$, and the graph of $y = a \sin bx$ is shifted by an amount c/b. The number c/b is the **phase shift.**

Graphs of Sine and Cosine Functions

The graphs of $y = a \sin(bx - c)$ and $y = a \cos(bx - c)$ have the following characteristics. (Assume $b > 0$.)

$$\text{Amplitude} = |a| \qquad \text{Period} = \frac{2\pi}{b}$$

The left and right endpoints of a one-cycle interval can be determined by solving the equations $bx - c = 0$ and $bx - c = 2\pi$.

Example 4   Horizontal Translation

Analyze the graph of $y = \dfrac{1}{2} \sin\left(x - \dfrac{\pi}{3}\right)$.

Algebraic Solution

The amplitude is $\frac{1}{2}$ and the period is 2π. By solving the equations

$$x - \frac{\pi}{3} = 0 \quad \Longrightarrow \quad x = \frac{\pi}{3}$$

and

$$x - \frac{\pi}{3} = 2\pi \quad \Longrightarrow \quad x = \frac{7\pi}{3}$$

you see that the interval $[\pi/3, 7\pi/3]$ corresponds to one cycle of the graph. Dividing this interval into four equal parts produces the key points

Intercept Maximum Intercept   Minimum   Intercept

$$\left(\frac{\pi}{3}, 0\right), \quad \left(\frac{5\pi}{6}, \frac{1}{2}\right), \quad \left(\frac{4\pi}{3}, 0\right), \quad \left(\frac{11\pi}{6}, -\frac{1}{2}\right), \quad \text{and} \quad \left(\frac{7\pi}{3}, 0\right).$$

✓CHECK Point▶   Now try Exercise 49.

Graphical Solution

Use a graphing utility set in *radian* mode to graph $y = (1/2) \sin(x - \pi/3)$, as shown in Figure 4.54. Use the *minimum, maximum,* and *zero* or *root* features of the graphing utility to approximate the key points $(1.05, 0)$, $(2.62, 0.5)$, $(4.19, 0)$, $(5.76, -0.5)$, and $(7.33, 0)$.

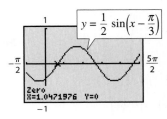

FIGURE **4.54**

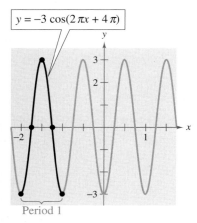

$y = -3 \cos(2\pi x + 4\pi)$

Period 1

FIGURE **4.55**

Example 5 Horizontal Translation

Sketch the graph of

$$y = -3 \cos(2\pi x + 4\pi).$$

Solution

The amplitude is 3 and the period is $2\pi/2\pi = 1$. By solving the equations

$$2\pi x + 4\pi = 0$$
$$2\pi x = -4\pi$$
$$x = -2$$

and

$$2\pi x + 4\pi = 2\pi$$
$$2\pi x = -2\pi$$
$$x = -1$$

you see that the interval $[-2, -1]$ corresponds to one cycle of the graph. Dividing this interval into four equal parts produces the key points

Minimum	Intercept	Maximum	Intercept		Minimum
$(-2, -3),$	$\left(-\dfrac{7}{4}, 0\right),$	$\left(-\dfrac{3}{2}, 3\right),$	$\left(-\dfrac{5}{4}, 0\right),$	and	$(-1, -3).$

The graph is shown in Figure 4.55.

CHECK *Point* Now try Exercise 51.

The final type of transformation is the *vertical translation* caused by the constant d in the equations

$$y = d + a \sin(bx - c)$$

and

$$y = d + a \cos(bx - c).$$

The shift is d units upward for $d > 0$ and d units downward for $d < 0$. In other words, the graph oscillates about the horizontal line $y = d$ instead of about the x-axis.

Example 6 Vertical Translation

Sketch the graph of

$$y = 2 + 3 \cos 2x.$$

Solution

The amplitude is 3 and the period is π. The key points over the interval $[0, \pi]$ are

$(0, 5),$	$\left(\dfrac{\pi}{4}, 2\right),$	$\left(\dfrac{\pi}{2}, -1\right),$	$\left(\dfrac{3\pi}{4}, 2\right),$	and	$(\pi, 5).$

The graph is shown in Figure 4.56. Compared with the graph of $f(x) = 3 \cos 2x$, the graph of $y = 2 + 3 \cos 2x$ is shifted upward two units.

CHECK *Point* Now try Exercise 57.

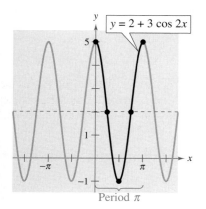

$y = 2 + 3 \cos 2x$

Period π

FIGURE **4.56**

Mathematical Modeling

Sine and cosine functions can be used to model many real-life situations, including electric currents, musical tones, radio waves, tides, and weather patterns.

| Example 7 | **Finding a Trigonometric Model**

Throughout the day, the depth of water at the end of a dock in Bar Harbor, Maine varies with the tides. The table shows the depths (in feet) at various times during the morning.   (Source: Nautical Software, Inc.)

Time, t	Depth, y
Midnight	3.4
2 A.M.	8.7
4 A.M.	11.3
6 A.M.	9.1
8 A.M.	3.8
10 A.M.	0.1
Noon	1.2

a. Use a trigonometric function to model the data.

b. Find the depths at 9 A.M. and 3 P.M.

c. A boat needs at least 10 feet of water to moor at the dock. During what times in the afternoon can it safely dock?

Solution

a. Begin by graphing the data, as shown in Figure 4.57. You can use either a sine or a cosine model. Suppose you use a cosine model of the form

$$y = a \cos(bt - c) + d.$$

The difference between the maximum height and the minimum height of the graph is twice the amplitude of the function. So, the amplitude is

$$a = \frac{1}{2}[(\text{maximum depth}) - (\text{minimum depth})] = \frac{1}{2}(11.3 - 0.1) = 5.6.$$

The cosine function completes one half of a cycle between the times at which the maximum and minimum depths occur. So, the period is

$$p = 2[(\text{time of min. depth}) - (\text{time of max. depth})] = 2(10 - 4) = 12$$

which implies that $b = 2\pi/p \approx 0.524$. Because high tide occurs 4 hours after midnight, consider the left endpoint to be $c/b = 4$, so $c \approx 2.094$. Moreover, because the average depth is $\frac{1}{2}(11.3 + 0.1) = 5.7$, it follows that $d = 5.7$. So, you can model the depth with the function given by

$$y = 5.6 \cos(0.524t - 2.094) + 5.7.$$

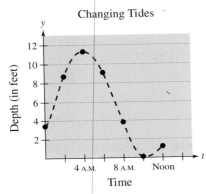

Changing Tides

FIGURE **4.57**

b. The depths at 9 A.M. and 3 P.M. are as follows.

$$y = 5.6 \cos(0.524 \cdot 9 - 2.094) + 5.7$$

$$\approx 0.84 \text{ foot} \qquad \qquad 9 \text{ A.M.}$$

$$y = 5.6 \cos(0.524 \cdot 15 - 2.094) + 5.7$$

$$\approx 10.57 \text{ feet} \qquad \qquad 3 \text{ P.M.}$$

c. To find out when the depth y is at least 10 feet, you can graph the model with the line $y = 10$ using a graphing utility, as shown in Figure 4.58. Using the *intersect* feature, you can determine that the depth is at least 10 feet between 2:42 P.M. ($t \approx 14.7$) and 5:18 P.M. ($t \approx 17.3$).

CHECK **Point**   Now try Exercise 91.

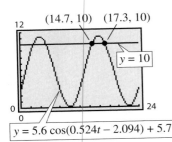

$y = 5.6 \cos(0.524t - 2.094) + 5.7$

FIGURE **4.58**

4.5   EXERCISES

See www.CalcChat.com for worked-out solutions to odd-numbered exercises.

VOCABULARY: Fill in the blanks.

1. One period of a sine or cosine function is called one _____ of the sine or cosine curve.

2. The _____ of a sine or cosine curve represents half the distance between the maximum and minimum values of the function.

3. For the function given by $y = a \sin(bx - c)$, $\frac{c}{b}$ represents the _____ _____ of the graph of the function.

4. For the function given by $y = d + a \cos(bx - c)$, d represents a _____ _____ of the graph of the function.

SKILLS AND APPLICATIONS

In Exercises 5–18, find the period and amplitude.

5. $y = 2 \sin 5x$

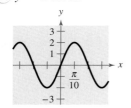

6. $y = 3 \cos 2x$

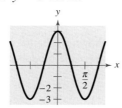

7. $y = \frac{3}{4} \cos \frac{x}{2}$

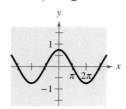

8. $y = -3 \sin \frac{x}{3}$

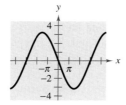

9. $y = \frac{1}{2} \sin \frac{\pi x}{3}$

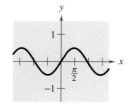

10. $y = \frac{3}{2} \cos \frac{\pi x}{2}$

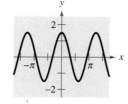

11. $y = -4 \sin x$

12. $y = -\cos \frac{2x}{3}$

13. $y = 3 \sin 10x$

14. $y = \frac{1}{5} \sin 6x$

15. $y = \frac{5}{3} \cos \frac{4x}{5}$

16. $y = \frac{5}{2} \cos \frac{x}{4}$

17. $y = \frac{1}{4} \sin 2\pi x$

18. $y = \frac{2}{3} \cos \frac{\pi x}{10}$

In Exercises 19–26, describe the relationship between the graphs of f and g. Consider amplitude, period, and shifts.

19. $f(x) = \sin x$
    $g(x) = \sin(x - \pi)$

20. $f(x) = \cos x$
    $g(x) = \cos(x + \pi)$

21. $f(x) = \cos 2x$
    $g(x) = -\cos 2x$

22. $f(x) = \sin 3x$
    $g(x) = \sin(-3x)$

23. $f(x) = \cos x$
    $g(x) = \cos 2x$

24. $f(x) = \sin x$
    $g(x) = \sin 3x$

25. $f(x) = \sin 2x$
    $g(x) = 3 + \sin 2x$

26. $f(x) = \cos 4x$
    $g(x) = -2 + \cos 4x$

In Exercises 27–30, describe the relationship between the graphs of f and g. Consider amplitude, period, and shifts.

27.

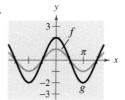

28.

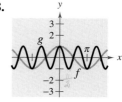

29.

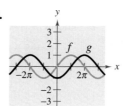

30.

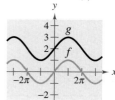

In Exercises 31–38, graph f and g on the same set of coordinate axes. (Include two full periods.)

31. $f(x) = -2 \sin x$
    $g(x) = 4 \sin x$

32. $f(x) = \sin x$
    $g(x) = \sin \frac{x}{3}$

33. $f(x) = \cos x$
    $g(x) = 2 + \cos x$

34. $f(x) = 2 \cos 2x$
    $g(x) = -\cos 4x$

35. $f(x) = -\dfrac{1}{2} \sin \dfrac{x}{2}$

$g(x) = 3 - \dfrac{1}{2} \sin \dfrac{x}{2}$

36. $f(x) = 4 \sin \pi x$

$g(x) = 4 \sin \pi x - 3$

37. $f(x) = 2 \cos x$

$g(x) = 2 \cos(x + \pi)$

38. $f(x) = -\cos x$

$g(x) = -\cos(x - \pi)$

In Exercises 39–60, sketch the graph of the function. (Include two full periods.)

39. $y = 5 \sin x$

40. $y = \frac{1}{4} \sin x$

41. $y = \frac{1}{3} \cos x$

42. $y = 4 \cos x$

43. $y = \cos \dfrac{x}{2}$

44. $y = \sin 4x$

45. $y = \cos 2\pi x$

46. $y = \sin \dfrac{\pi x}{4}$

47. $y = -\sin \dfrac{2\pi x}{3}$

48. $y = -10 \cos \dfrac{\pi x}{6}$

49. $y = \sin\left(x - \dfrac{\pi}{2}\right)$

50. $y = \sin(x - 2\pi)$

51. $y = 3 \cos(x + \pi)$

52. $y = 4 \cos\left(x + \dfrac{\pi}{4}\right)$

53. $y = 2 - \sin \dfrac{2\pi x}{3}$

54. $y = -3 + 5 \cos \dfrac{\pi t}{12}$

55. $y = 2 + \frac{1}{10} \cos 60\pi x$

56. $y = 2 \cos x - 3$

57. $y = 3 \cos(x + \pi) - 3$

58. $y = 4 \cos\left(x + \dfrac{\pi}{4}\right) + 4$

59. $y = \dfrac{2}{3} \cos\left(\dfrac{x}{2} - \dfrac{\pi}{4}\right)$

60. $y = -3 \cos(6x + \pi)$

In Exercises 61–66, g is related to a parent function $f(x) = \sin(x)$ or $f(x) = \cos(x)$. (a) Describe the sequence of transformations from f to g. (b) Sketch the graph of g. (c) Use function notation to write g in terms of f.

61. $g(x) = \sin(4x - \pi)$

62. $g(x) = \sin(2x + \pi)$

63. $g(x) = \cos(x - \pi) + 2$

64. $g(x) = 1 + \cos(x + \pi)$

65. $g(x) = 2 \sin(4x - \pi) - 3$

66. $g(x) = 4 - \sin(2x + \pi)$

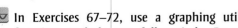 In Exercises 67–72, use a graphing utility to graph the function. Include two full periods. Be sure to choose an appropriate viewing window.

67. $y = -2 \sin(4x + \pi)$

68. $y = -4 \sin\left(\dfrac{2}{3}x - \dfrac{\pi}{3}\right)$

69. $y = \cos\left(2\pi x - \dfrac{\pi}{2}\right) + 1$

70. $y = 3 \cos\left(\dfrac{\pi x}{2} + \dfrac{\pi}{2}\right) - 2$

71. $y = -0.1 \sin\left(\dfrac{\pi x}{10} + \pi\right)$

72. $y = \dfrac{1}{100} \sin 120\pi t$

GRAPHICAL REASONING   In Exercises 73–76, find a and d for the function $f(x) = a \cos x + d$ such that the graph of f matches the figure.

73.

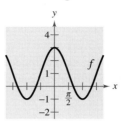

74.

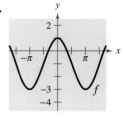

75.

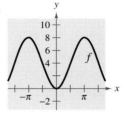

76.

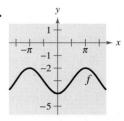

GRAPHICAL REASONING   In Exercises 77–80, find a, b, and c for the function $f(x) = a \sin(bx - c)$ such that the graph of f matches the figure.

77.

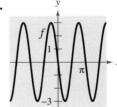

78.

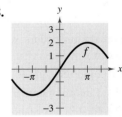

79.

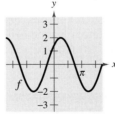

80.

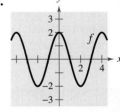

In Exercises 81 and 82, use a graphing utility to graph y_1 and y_2 in the interval $[-2\pi, 2\pi]$. Use the graphs to find real numbers x such that $y_1 = y_2$.

81. $y_1 = \sin x$

$y_2 = -\dfrac{1}{2}$

82. $y_1 = \cos x$

$y_2 = -1$

In Exercises 83–86, write an equation for the function that is described by the given characteristics.

83. A sine curve with a period of π, an amplitude of 2, a right phase shift of $\pi/2$, and a vertical translation up 1 unit

84. A sine curve with a period of 4π, an amplitude of 3, a left phase shift of $\pi/4$, and a vertical translation down 1 unit

85. A cosine curve with a period of π, an amplitude of 1, a left phase shift of π, and a vertical translation down $\frac{3}{2}$ units

86. A cosine curve with a period of 4π, an amplitude of 3, a right phase shift of $\pi/2$, and a vertical translation up 2 units

87. RESPIRATORY CYCLE For a person at rest, the velocity v (in liters per second) of airflow during a respiratory cycle (the time from the beginning of one breath to the beginning of the next) is given by $v = 0.85 \sin\dfrac{\pi t}{3}$, where t is the time (in seconds). (Inhalation occurs when $v > 0$, and exhalation occurs when $v < 0$.)

(a) Find the time for one full respiratory cycle.

(b) Find the number of cycles per minute.

(c) Sketch the graph of the velocity function.

88. RESPIRATORY CYCLE After exercising for a few minutes, a person has a respiratory cycle for which the velocity of airflow is approximated by $v = 1.75 \sin\dfrac{\pi t}{2}$, where t is the time (in seconds). (Inhalation occurs when $v > 0$, and exhalation occurs when $v < 0$.)

(a) Find the time for one full respiratory cycle.

(b) Find the number of cycles per minute.

(c) Sketch the graph of the velocity function.

89. DATA ANALYSIS: METEOROLOGY The table shows the maximum daily high temperatures in Las Vegas L and International Falls I (in degrees Fahrenheit) for month t, with $t = 1$ corresponding to January. (Source: National Climatic Data Center)

Month, t	Las Vegas, L	International Falls, I
1	57.1	13.8
2	63.0	22.4
3	69.5	34.9
4	78.1	51.5
5	87.8	66.6
6	98.9	74.2
7	104.1	78.6
8	101.8	76.3
9	93.8	64.7
10	80.8	51.7
11	66.0	32.5
12	57.3	18.1

(a) A model for the temperature in Las Vegas is given by
$$L(t) = 80.60 + 23.50 \cos\left(\frac{\pi t}{6} - 3.67\right).$$
Find a trigonometric model for International Falls.

(b) Use a graphing utility to graph the data points and the model for the temperatures in Las Vegas. How well does the model fit the data?

(c) Use a graphing utility to graph the data points and the model for the temperatures in International Falls. How well does the model fit the data?

(d) Use the models to estimate the average maximum temperature in each city. Which term of the models did you use? Explain.

(e) What is the period of each model? Are the periods what you expected? Explain.

(f) Which city has the greater variability in temperature throughout the year? Which factor of the models determines this variability? Explain.

90. HEALTH The function given by
$$P = 100 - 20\cos\frac{5\pi t}{3}$$
approximates the blood pressure P (in millimeters of mercury) at time t (in seconds) for a person at rest.

(a) Find the period of the function.

(b) Find the number of heartbeats per minute.

91. PIANO TUNING When tuning a piano, a technician strikes a tuning fork for the A above middle C and sets up a wave motion that can be approximated by $y = 0.001 \sin 880\pi t$, where t is the time (in seconds).

(a) What is the period of the function?

(b) The frequency f is given by $f = 1/p$. What is the frequency of the note?

92. DATA ANALYSIS: ASTRONOMY The percents y (in decimal form) of the moon's face that was illuminated on day x in the year 2009, where $x = 1$ represents January 1, are shown in the table. (Source: U.S. Naval Observatory)

x	y
4	0.5
11	1.0
18	0.5
26	0.0
33	0.5
40	1.0

(a) Create a scatter plot of the data.

(b) Find a trigonometric model that fits the data.

(c) Add the graph of your model in part (b) to the scatter plot. How well does the model fit the data?

(d) What is the period of the model?

(e) Estimate the moon's percent illumination for March 12, 2009.

93. FUEL CONSUMPTION   The daily consumption C (in gallons) of diesel fuel on a farm is modeled by

$$C = 30.3 + 21.6 \sin\left(\frac{2\pi t}{365} + 10.9\right)$$

where t is the time (in days), with $t = 1$ corresponding to January 1.

(a) What is the period of the model? Is it what you expected? Explain.

(b) What is the average daily fuel consumption? Which term of the model did you use? Explain.

(c) Use a graphing utility to graph the model. Use the graph to approximate the time of the year when consumption exceeds 40 gallons per day.

94. FERRIS WHEEL   A Ferris wheel is built such that the height h (in feet) above ground of a seat on the wheel at time t (in seconds) can be modeled by

$$h(t) = 53 + 50 \sin\left(\frac{\pi}{10}t - \frac{\pi}{2}\right).$$

(a) Find the period of the model. What does the period tell you about the ride?

(b) Find the amplitude of the model. What does the amplitude tell you about the ride?

(c) Use a graphing utility to graph one cycle of the model.

EXPLORATION

TRUE OR FALSE?   In Exercises 95–97, determine whether the statement is true or false. Justify your answer.

95. The graph of the function given by $f(x) = \sin(x + 2\pi)$ translates the graph of $f(x) = \sin x$ exactly one period to the right so that the two graphs look identical.

96. The function given by $y = \frac{1}{2}\cos 2x$ has an amplitude that is twice that of the function given by $y = \cos x$.

97. The graph of $y = -\cos x$ is a reflection of the graph of $y = \sin(x + \pi/2)$ in the x-axis.

98. WRITING   Sketch the graph of $y = \cos bx$ for $b = \frac{1}{2}$, 2, and 3. How does the value of b affect the graph? How many complete cycles occur between 0 and 2π for each value of b?

99. WRITING   Sketch the graph of $y = \sin(x - c)$ for $c = -\pi/4$, 0, and $\pi/4$. How does the value of c affect the graph?

100. CAPSTONE   Use a graphing utility to graph the function given by $y = d + a\sin(bx - c)$, for several different values of a, b, c, and d. Write a paragraph describing the changes in the graph corresponding to changes in each constant.

CONJECTURE   In Exercises 101 and 102, graph f and g on the same set of coordinate axes. Include two full periods. Make a conjecture about the functions.

101. $f(x) = \sin x$, $g(x) = \cos\left(x - \frac{\pi}{2}\right)$

102. $f(x) = \sin x$, $g(x) = -\cos\left(x + \frac{\pi}{2}\right)$

103. Using calculus, it can be shown that the sine and cosine functions can be approximated by the polynomials

$$\sin x \approx x - \frac{x^3}{3!} + \frac{x^5}{5!} \quad \text{and} \quad \cos x \approx 1 - \frac{x^2}{2!} + \frac{x^4}{4!}$$

where x is in radians.

(a) Use a graphing utility to graph the sine function and its polynomial approximation in the same viewing window. How do the graphs compare?

(b) Use a graphing utility to graph the cosine function and its polynomial approximation in the same viewing window. How do the graphs compare?

(c) Study the patterns in the polynomial approximations of the sine and cosine functions and predict the next term in each. Then repeat parts (a) and (b). How did the accuracy of the approximations change when an additional term was added?

104. Use the polynomial approximations of the sine and cosine functions in Exercise 103 to approximate the following function values. Compare the results with those given by a calculator. Is the error in the approximation the same in each case? Explain.

(a) $\sin\frac{1}{2}$ (b) $\sin 1$ (c) $\sin\frac{\pi}{6}$

(d) $\cos(-0.5)$ (e) $\cos 1$ (f) $\cos\frac{\pi}{4}$

PROJECT: METEOROLOGY   To work an extended application analyzing the mean monthly temperature and mean monthly precipitation in Honolulu, Hawaii, visit this text's website at *academic.cengage.com*. (Data Source: National Climatic Data Center)

4.6 GRAPHS OF OTHER TRIGONOMETRIC FUNCTIONS

What you should learn

- Sketch the graphs of tangent functions.
- Sketch the graphs of cotangent functions.
- Sketch the graphs of secant and cosecant functions.
- Sketch the graphs of damped trigonometric functions.

Why you should learn it

Graphs of trigonometric functions can be used to model real-life situations such as the distance from a television camera to a unit in a parade, as in Exercise 92 on page 339.

Alan Pappe/Photodisc/Getty Images

Algebra Help

- You can review odd and even functions in Section 1.5.
- You can review symmetry of a graph in Section 1.2.
- You can review trigonometric identities in Section 4.3.
- You can review asymptotes in Section 2.6.
- You can review domain and range of a function in Section 1.4.
- You can review intercepts of a graph in Section 1.2.

Graph of the Tangent Function

Recall that the tangent function is odd. That is, $\tan(-x) = -\tan x$. Consequently, the graph of $y = \tan x$ is symmetric with respect to the origin. You also know from the identity $\tan x = \sin x/\cos x$ that the tangent is undefined for values at which $\cos x = 0$. Two such values are $x = \pm\pi/2 \approx \pm 1.5708$.

x	$-\dfrac{\pi}{2}$	-1.57	-1.5	$-\dfrac{\pi}{4}$	0	$\dfrac{\pi}{4}$	1.5	1.57	$\dfrac{\pi}{2}$
$\tan x$	Undef.	-1255.8	-14.1	-1	0	1	14.1	1255.8	Undef.

As indicated in the table, $\tan x$ increases without bound as x approaches $\pi/2$ from the left, and decreases without bound as x approaches $-\pi/2$ from the right. So, the graph of $y = \tan x$ has *vertical asymptotes* at $x = \pi/2$ and $x = -\pi/2$, as shown in Figure 4.59. Moreover, because the period of the tangent function is π, vertical asymptotes also occur when $x = \pi/2 + n\pi$, where n is an integer. The domain of the tangent function is the set of all real numbers other than $x = \pi/2 + n\pi$, and the range is the set of all real numbers.

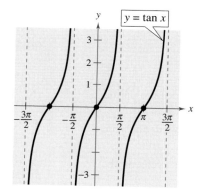

PERIOD: π
DOMAIN: ALL $x \neq \dfrac{\pi}{2} + n\pi$
RANGE: $(-\infty, \infty)$
VERTICAL ASYMPTOTES: $x = \dfrac{\pi}{2} + n\pi$
SYMMETRY: ORIGIN

FIGURE 4.59

Sketching the graph of $y = a\tan(bx - c)$ is similar to sketching the graph of $y = a\sin(bx - c)$ in that you locate key points that identify the intercepts and asymptotes. Two consecutive vertical asymptotes can be found by solving the equations

$$bx - c = -\frac{\pi}{2} \qquad \text{and} \qquad bx - c = \frac{\pi}{2}.$$

The midpoint between two consecutive vertical asymptotes is an x-intercept of the graph. The period of the function $y = a\tan(bx - c)$ is the distance between two consecutive vertical asymptotes. The amplitude of a tangent function is not defined. After plotting the asymptotes and the x-intercept, plot a few additional points between the two asymptotes and sketch one cycle. Finally, sketch one or two additional cycles to the left and right.

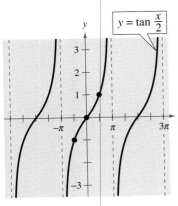

FIGURE **4.60**

Example 1   Sketching the Graph of a Tangent Function

Sketch the graph of $y = \tan(x/2)$.

Solution

By solving the equations

$$\frac{x}{2} = -\frac{\pi}{2} \quad \text{and} \quad \frac{x}{2} = \frac{\pi}{2}$$

$$x = -\pi \qquad\qquad x = \pi$$

you can see that two consecutive vertical asymptotes occur at $x = -\pi$ and $x = \pi$. Between these two asymptotes, plot a few points, including the x-intercept, as shown in the table. Three cycles of the graph are shown in Figure 4.60.

x	$-\pi$	$-\dfrac{\pi}{2}$	0	$\dfrac{\pi}{2}$	π
$\tan \dfrac{x}{2}$	Undef.	-1	0	1	Undef.

CHECK**Point**   Now try Exercise 15.

Example 2   Sketching the Graph of a Tangent Function

Sketch the graph of $y = -3 \tan 2x$.

Solution

By solving the equations

$$2x = -\frac{\pi}{2} \quad \text{and} \quad 2x = \frac{\pi}{2}$$

$$x = -\frac{\pi}{4} \qquad\qquad x = \frac{\pi}{4}$$

you can see that two consecutive vertical asymptotes occur at $x = -\pi/4$ and $x = \pi/4$. Between these two asymptotes, plot a few points, including the x-intercept, as shown in the table. Three cycles of the graph are shown in Figure 4.61.

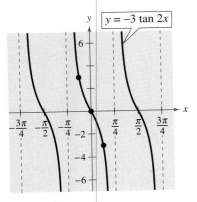

FIGURE **4.61**

x	$-\dfrac{\pi}{4}$	$-\dfrac{\pi}{8}$	0	$\dfrac{\pi}{8}$	$\dfrac{\pi}{4}$
$-3 \tan 2x$	Undef.	3	0	-3	Undef.

By comparing the graphs in Examples 1 and 2, you can see that the graph of $y = a \tan(bx - c)$ increases between consecutive vertical asymptotes when $a > 0$, and decreases between consecutive vertical asymptotes when $a < 0$. In other words, the graph for $a < 0$ is a reflection in the x-axis of the graph for $a > 0$.

CHECK**Point**   Now try Exercise 17.

Graph of the Cotangent Function

The graph of the cotangent function is similar to the graph of the tangent function. It also has a period of π. However, from the identity

$$y = \cot x = \frac{\cos x}{\sin x}$$

you can see that the cotangent function has vertical asymptotes when $\sin x$ is zero, which occurs at $x = n\pi$, where n is an integer. The graph of the cotangent function is shown in Figure 4.62. Note that two consecutive vertical asymptotes of the graph of $y = a \cot(bx - c)$ can be found by solving the equations $bx - c = 0$ and $bx - c = \pi$.

TECHNOLOGY

Some graphing utilities have difficulty graphing trigonometric functions that have vertical asymptotes. Your graphing utility may connect parts of the graphs of tangent, cotangent, secant, and cosecant functions that are not supposed to be connected. To eliminate this problem, change the mode of the graphing utility to *dot* mode.

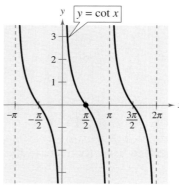

PERIOD: π
DOMAIN: ALL $x \ne n\pi$
RANGE: $(-\infty, \infty)$
VERTICAL ASYMPTOTES: $x = n\pi$
SYMMETRY: ORIGIN

FIGURE 4.62

FIGURE 4.63

| **Example 3** | **Sketching the Graph of a Cotangent Function**

Sketch the graph of $y = 2 \cot \dfrac{x}{3}$.

Solution

By solving the equations

$$\frac{x}{3} = 0 \quad \text{and} \quad \frac{x}{3} = \pi$$

$$x = 0 \qquad\qquad x = 3\pi$$

you can see that two consecutive vertical asymptotes occur at $x = 0$ and $x = 3\pi$. Between these two asymptotes, plot a few points, including the x-intercept, as shown in the table. Three cycles of the graph are shown in Figure 4.63. Note that the period is 3π, the distance between consecutive asymptotes.

x	0	$\dfrac{3\pi}{4}$	$\dfrac{3\pi}{2}$	$\dfrac{9\pi}{4}$	3π
$2 \cot \dfrac{x}{3}$	Undef.	2	0	-2	Undef.

CHECK **Point** Now try Exercise 27.

Graphs of the Reciprocal Functions

The graphs of the two remaining trigonometric functions can be obtained from the graphs of the sine and cosine functions using the reciprocal identities

$$\csc x = \frac{1}{\sin x} \qquad \text{and} \qquad \sec x = \frac{1}{\cos x}.$$

For instance, at a given value of x, the y-coordinate of $\sec x$ is the reciprocal of the y-coordinate of $\cos x$. Of course, when $\cos x = 0$, the reciprocal does not exist. Near such values of x, the behavior of the secant function is similar to that of the tangent function. In other words, the graphs of

$$\tan x = \frac{\sin x}{\cos x} \qquad \text{and} \qquad \sec x = \frac{1}{\cos x}$$

have vertical asymptotes at $x = \pi/2 + n\pi$, where n is an integer, and the cosine is zero at these x-values. Similarly,

$$\cot x = \frac{\cos x}{\sin x} \qquad \text{and} \qquad \csc x = \frac{1}{\sin x}$$

have vertical asymptotes where $\sin x = 0$—that is, at $x = n\pi$.

To sketch the graph of a secant or cosecant function, you should first make a sketch of its reciprocal function. For instance, to sketch the graph of $y = \csc x$, first sketch the graph of $y = \sin x$. Then take reciprocals of the y-coordinates to obtain points on the graph of $y = \csc x$. This procedure is used to obtain the graphs shown in Figure 4.64.

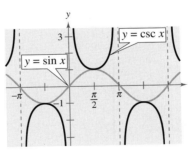

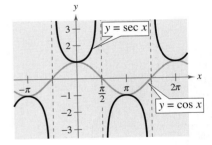

PERIOD: 2π

DOMAIN: ALL $x \neq n\pi$

RANGE: $(-\infty, -1] \cup [1, \infty)$

VERTICAL ASYMPTOTES: $x = n\pi$

SYMMETRY: ORIGIN

FIGURE **4.64**

PERIOD: 2π

DOMAIN: ALL $x \neq \frac{\pi}{2} + n\pi$

RANGE: $(-\infty, -1] \cup [1, \infty)$

VERTICAL ASYMPTOTES: $x = \frac{\pi}{2} + n\pi$

SYMMETRY: y-AXIS

In comparing the graphs of the cosecant and secant functions with those of the sine and cosine functions, note that the "hills" and "valleys" are interchanged. For example, a hill (or maximum point) on the sine curve corresponds to a valley (a relative minimum) on the cosecant curve, and a valley (or minimum point) on the sine curve corresponds to a hill (a relative maximum) on the cosecant curve, as shown in Figure 4.65. Additionally, x-intercepts of the sine and cosine functions become vertical asymptotes of the cosecant and secant functions, respectively (see Figure 4.65).

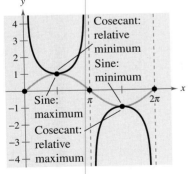

FIGURE **4.65**

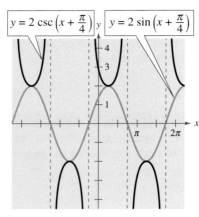

FIGURE **4.66**

Example 4   Sketching the Graph of a Cosecant Function

Sketch the graph of $y = 2 \csc\left(x + \dfrac{\pi}{4}\right)$.

Solution

Begin by sketching the graph of

$$y = 2 \sin\left(x + \frac{\pi}{4}\right).$$

For this function, the amplitude is 2 and the period is 2π. By solving the equations

$$x + \frac{\pi}{4} = 0 \qquad \text{and} \qquad x + \frac{\pi}{4} = 2\pi$$

$$x = -\frac{\pi}{4} \qquad\qquad\qquad x = \frac{7\pi}{4}$$

you can see that one cycle of the sine function corresponds to the interval from $x = -\pi/4$ to $x = 7\pi/4$. The graph of this sine function is represented by the gray curve in Figure 4.66. Because the sine function is zero at the midpoint and endpoints of this interval, the corresponding cosecant function

$$y = 2 \csc\left(x + \frac{\pi}{4}\right)$$

$$= 2\left(\frac{1}{\sin[x + (\pi/4)]}\right)$$

has vertical asymptotes at $x = -\pi/4, x = 3\pi/4, x = 7\pi/4$, etc. The graph of the cosecant function is represented by the black curve in Figure 4.66.

CHECK Point Now try Exercise 33.

Example 5   Sketching the Graph of a Secant Function

Sketch the graph of $y = \sec 2x$.

Solution

Begin by sketching the graph of $y = \cos 2x$, as indicated by the gray curve in Figure 4.67. Then, form the graph of $y = \sec 2x$ as the black curve in the figure. Note that the x-intercepts of $y = \cos 2x$

$$\left(-\frac{\pi}{4}, 0\right), \qquad \left(\frac{\pi}{4}, 0\right), \qquad \left(\frac{3\pi}{4}, 0\right), \ldots$$

correspond to the vertical asymptotes

$$x = -\frac{\pi}{4}, \qquad x = \frac{\pi}{4}, \qquad x = \frac{3\pi}{4}, \ldots$$

of the graph of $y = \sec 2x$. Moreover, notice that the period of $y = \cos 2x$ and $y = \sec 2x$ is π.

CHECK Point Now try Exercise 35.

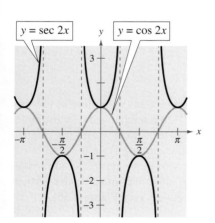

FIGURE **4.67**

Damped Trigonometric Graphs

A *product* of two functions can be graphed using properties of the individual functions. For instance, consider the function

$$f(x) = x \sin x$$

as the product of the functions $y = x$ and $y = \sin x$. Using properties of absolute value and the fact that $|\sin x| \leq 1$, you have $0 \leq |x||\sin x| \leq |x|$. Consequently,

$$-|x| \leq x \sin x \leq |x|$$

which means that the graph of $f(x) = x \sin x$ lies between the lines $y = -x$ and $y = x$. Furthermore, because

$$f(x) = x \sin x = \pm x \qquad \text{at} \qquad x = \frac{\pi}{2} + n\pi$$

and

$$f(x) = x \sin x = 0 \qquad \text{at} \qquad x = n\pi$$

the graph of f touches the line $y = -x$ or the line $y = x$ at $x = \pi/2 + n\pi$ and has x-intercepts at $x = n\pi$. A sketch of f is shown in Figure 4.68. In the function $f(x) = x \sin x$, the factor x is called the **damping factor.**

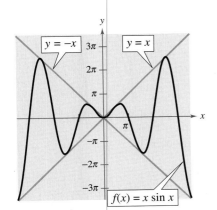

FIGURE **4.68**

Example 6   Damped Sine Wave

Sketch the graph of $f(x) = e^{-x} \sin 3x$.

Solution

Consider $f(x)$ as the product of the two functions

$$y = e^{-x} \qquad \text{and} \qquad y = \sin 3x$$

each of which has the set of real numbers as its domain. For any real number x, you know that $e^{-x} \geq 0$ and $|\sin 3x| \leq 1$. So, $e^{-x} |\sin 3x| \leq e^{-x}$, which means that

$$-e^{-x} \leq e^{-x} \sin 3x \leq e^{-x}.$$

Furthermore, because

$$f(x) = e^{-x} \sin 3x = \pm e^{-x} \quad \text{at} \quad x = \frac{\pi}{6} + \frac{n\pi}{3}$$

and

$$f(x) = e^{-x} \sin 3x = 0 \quad \text{at} \quad x = \frac{n\pi}{3}$$

the graph of f touches the curves $y = -e^{-x}$ and $y = e^{-x}$ at $x = \pi/6 + n\pi/3$ and has intercepts at $x = n\pi/3$. A sketch is shown in Figure 4.69.

CHECK *Point*   Now try Exercise 65.

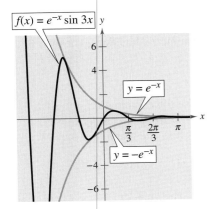

FIGURE **4.69**

Figure 4.70 summarizes the characteristics of the six basic trigonometric functions.

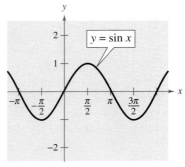

DOMAIN: $(-\infty, \infty)$
RANGE: $[-1, 1]$
PERIOD: 2π

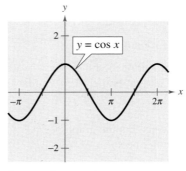

DOMAIN: $(-\infty, \infty)$
RANGE: $[-1, 1]$
PERIOD: 2π

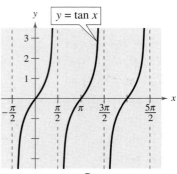

DOMAIN: ALL $x \neq \frac{\pi}{2} + n\pi$
RANGE: $(-\infty, \infty)$
PERIOD: π

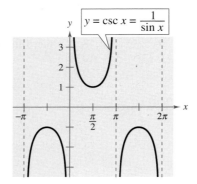

DOMAIN: ALL $x \neq n\pi$
RANGE: $(-\infty, -1] \cup [1, \infty)$
PERIOD: 2π
FIGURE 4.70

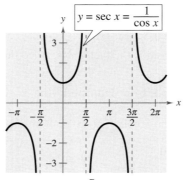

DOMAIN: ALL $x \neq \frac{\pi}{2} + n\pi$
RANGE: $(-\infty, -1] \cup [1, \infty)$
PERIOD: 2π

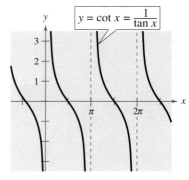

DOMAIN: ALL $x \neq n\pi$
RANGE: $(-\infty, \infty)$
PERIOD: π

CLASSROOM DISCUSSION

Combining Trigonometric Functions   Recall from Section 1.8 that functions can be combined arithmetically. This also applies to trigonometric functions. For each of the functions

$$h(x) = x + \sin x \quad \text{and} \quad h(x) = \cos x - \sin 3x$$

(a) identify two simpler functions f and g that comprise the combination, (b) use a table to show how to obtain the numerical values of $h(x)$ from the numerical values of $f(x)$ and $g(x)$, and (c) use graphs of f and g to show how the graph of h may be formed.

Can you find functions

$$f(x) = d + a \sin(bx + c) \quad \text{and} \quad g(x) = d + a \cos(bx + c)$$

such that $f(x) + g(x) = 0$ for all x?

4.6   EXERCISES

See www.CalcChat.com for worked-out solutions to odd-numbered exercises.

VOCABULARY: Fill in the blanks.

1. The tangent, cotangent, and cosecant functions are _____ , so the graphs of these functions have symmetry with respect to the _____ .

2. The graphs of the tangent, cotangent, secant, and cosecant functions all have _____ asymptotes.

3. To sketch the graph of a secant or cosecant function, first make a sketch of its corresponding _____ function.

4. For the functions given by $f(x) = g(x) \cdot \sin x$, $g(x)$ is called the _____ factor of the function $f(x)$.

5. The period of $y = \tan x$ is _____ .

6. The domain of $y = \cot x$ is all real numbers such that _____ .

7. The range of $y = \sec x$ is _____ .

8. The period of $y = \csc x$ is _____ .

SKILLS AND APPLICATIONS

In Exercises 9–14, match the function with its graph. State the period of the function. [The graphs are labeled (a), (b), (c), (d), (e), and (f).]

(a)
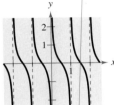

(b)
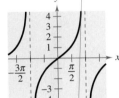

(c)
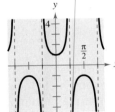

(d)
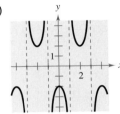

(e)
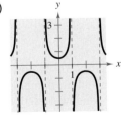

(f)
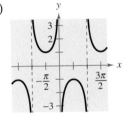

9. $y = \sec 2x$

10. $y = \tan \dfrac{x}{2}$

11. $y = \dfrac{1}{2} \cot \pi x$

12. $y = -\csc x$

13. $y = \dfrac{1}{2} \sec \dfrac{\pi x}{2}$

14. $y = -2 \sec \dfrac{\pi x}{2}$

In Exercises 15–38, sketch the graph of the function. Include two full periods.

15. $y = \dfrac{1}{3} \tan x$

16. $y = \tan 4x$

17. $y = -2 \tan 3x$

18. $y = -3 \tan \pi x$

19. $y = -\dfrac{1}{2} \sec x$

20. $y = \dfrac{1}{4} \sec x$

21. $y = \csc \pi x$

22. $y = 3 \csc 4x$

23. $y = \dfrac{1}{2} \sec \pi x$

24. $y = -2 \sec 4x + 2$

25. $y = \csc \dfrac{x}{2}$

26. $y = \csc \dfrac{x}{3}$

27. $y = 3 \cot 2x$

28. $y = 3 \cot \dfrac{\pi x}{2}$

29. $y = 2 \sec 3x$

30. $y = -\dfrac{1}{2} \tan x$

31. $y = \tan \dfrac{\pi x}{4}$

32. $y = \tan(x + \pi)$

33. $y = 2 \csc(x - \pi)$

34. $y = \csc(2x - \pi)$

35. $y = 2 \sec(x + \pi)$

36. $y = -\sec \pi x + 1$

37. $y = \dfrac{1}{4} \csc\left(x + \dfrac{\pi}{4}\right)$

38. $y = 2 \cot\left(x + \dfrac{\pi}{2}\right)$

 In Exercises 39–48, use a graphing utility to graph the function. Include two full periods.

39. $y = \tan \dfrac{x}{3}$

40. $y = -\tan 2x$

41. $y = -2 \sec 4x$

42. $y = \sec \pi x$

43. $y = \tan\left(x - \dfrac{\pi}{4}\right)$

44. $y = \dfrac{1}{4} \cot\left(x - \dfrac{\pi}{2}\right)$

45. $y = -\csc(4x - \pi)$

46. $y = 2 \sec(2x - \pi)$

47. $y = 0.1 \tan\left(\dfrac{\pi x}{4} + \dfrac{\pi}{4}\right)$

48. $y = \dfrac{1}{3} \sec\left(\dfrac{\pi x}{2} + \dfrac{\pi}{2}\right)$

In Exercises 49–56, use a graph to solve the equation on the interval $[-2\pi, 2\pi]$.

49. $\tan x = 1$

50. $\tan x = \sqrt{3}$

51. $\cot x = -\dfrac{\sqrt{3}}{3}$

52. $\cot x = 1$

53. $\sec x = -2$

54. $\sec x = 2$

55. $\csc x = \sqrt{2}$

56. $\csc x = -\dfrac{2\sqrt{3}}{3}$

In Exercises 57–64, use the graph of the function to determine whether the function is even, odd, or neither. Verify your answer algebraically.

57. $f(x) = \sec x$

58. $f(x) = \tan x$

59. $g(x) = \cot x$

60. $g(x) = \csc x$

61. $f(x) = x + \tan x$

62. $f(x) = x^2 - \sec x$

63. $g(x) = x \csc x$

64. $g(x) = x^2 \cot x$

65. GRAPHICAL REASONING   Consider the functions given by

$$f(x) = 2 \sin x \quad \text{and} \quad g(x) = \frac{1}{2} \csc x$$

on the interval $(0, \pi)$.

(a) Graph f and g in the same coordinate plane.

(b) Approximate the interval in which $f > g$.

(c) Describe the behavior of each of the functions as x approaches π. How is the behavior of g related to the behavior of f as x approaches π?

66. GRAPHICAL REASONING   Consider the functions given by

$$f(x) = \tan \frac{\pi x}{2} \quad \text{and} \quad g(x) = \frac{1}{2} \sec \frac{\pi x}{2}$$

on the interval $(-1, 1)$.

(a) Use a graphing utility to graph f and g in the same viewing window.

(b) Approximate the interval in which $f < g$.

(c) Approximate the interval in which $2f < 2g$. How does the result compare with that of part (b)? Explain.

In Exercises 67–72, use a graphing utility to graph the two equations in the same viewing window. Use the graphs to determine whether the expressions are equivalent. Verify the results algebraically.

67. $y_1 = \sin x \csc x, \quad y_2 = 1$

68. $y_1 = \sin x \sec x, \quad y_2 = \tan x$

69. $y_1 = \dfrac{\cos x}{\sin x}, \quad y_2 = \cot x$

70. $y_1 = \tan x \cot^2 x, \quad y_2 = \cot x$

71. $y_1 = 1 + \cot^2 x, \quad y_2 = \csc^2 x$

72. $y_1 = \sec^2 x - 1, \quad y_2 = \tan^2 x$

In Exercises 73–76, match the function with its graph. Describe the behavior of the function as x approaches zero. [The graphs are labeled (a), (b), (c), and (d).]

(a)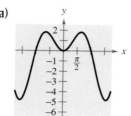

(b)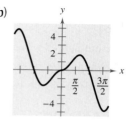

(c)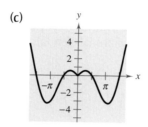

(d)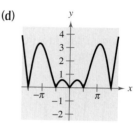

73. $f(x) = |x \cos x|$

74. $f(x) = x \sin x$

75. $g(x) = |x| \sin x$

76. $g(x) = |x| \cos x$

CONJECTURE   In Exercises 77–80, graph the functions f and g. Use the graphs to make a conjecture about the relationship between the functions.

77. $f(x) = \sin x + \cos\left(x + \dfrac{\pi}{2}\right), \quad g(x) = 0$

78. $f(x) = \sin x - \cos\left(x + \dfrac{\pi}{2}\right), \quad g(x) = 2 \sin x$

79. $f(x) = \sin^2 x, \quad g(x) = \frac{1}{2}(1 - \cos 2x)$

80. $f(x) = \cos^2 \dfrac{\pi x}{2}, \quad g(x) = \frac{1}{2}(1 + \cos \pi x)$

In Exercises 81–84, use a graphing utility to graph the function and the damping factor of the function in the same viewing window. Describe the behavior of the function as x increases without bound.

81. $g(x) = e^{-x^2/2} \sin x$

82. $f(x) = e^{-x} \cos x$

83. $f(x) = 2^{-x/4} \cos \pi x$

84. $h(x) = 2^{-x^2/4} \sin x$

In Exercises 85–90, use a graphing utility to graph the function. Describe the behavior of the function as x approaches zero.

85. $y = \dfrac{6}{x} + \cos x, \quad x > 0$

86. $y = \dfrac{4}{x} + \sin 2x, \quad x > 0$

87. $g(x) = \dfrac{\sin x}{x}$

88. $f(x) = \dfrac{1 - \cos x}{x}$

89. $f(x) = \sin \dfrac{1}{x}$

90. $h(x) = x \sin \dfrac{1}{x}$

91. DISTANCE A plane flying at an altitude of 7 miles above a radar antenna will pass directly over the radar antenna (see figure). Let d be the ground distance from the antenna to the point directly under the plane and let x be the angle of elevation to the plane from the antenna. (d is positive as the plane approaches the antenna.) Write d as a function of x and graph the function over the interval $0 < x < \pi$.

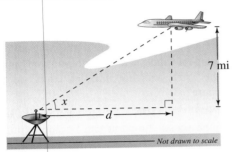

Not drawn to scale

92. TELEVISION COVERAGE A television camera is on a reviewing platform 27 meters from the street on which a parade will be passing from left to right (see figure). Write the distance d from the camera to a particular unit in the parade as a function of the angle x, and graph the function over the interval $-\pi/2 < x < \pi/2$. (Consider x as negative when a unit in the parade approaches from the left.)

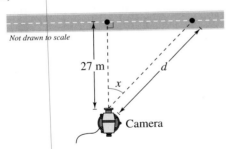

Not drawn to scale

93. METEOROLOGY The normal monthly high temperatures H (in degrees Fahrenheit) in Erie, Pennsylvania are approximated by

$$H(t) = 56.94 - 20.86 \cos(\pi t/6) - 11.58 \sin(\pi t/6)$$

and the normal monthly low temperatures L are approximated by

$$L(t) = 41.80 - 17.13 \cos(\pi t/6) - 13.39 \sin(\pi t/6)$$

where t is the time (in months), with $t = 1$ corresponding to January (see figure). (Source: National Climatic Data Center)

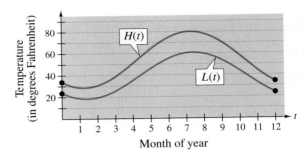

(a) What is the period of each function?

(b) During what part of the year is the difference between the normal high and normal low temperatures greatest? When is it smallest?

(c) The sun is northernmost in the sky around June 21, but the graph shows the warmest temperatures at a later date. Approximate the lag time of the temperatures relative to the position of the sun.

94. SALES The projected monthly sales S (in thousands of units) of lawn mowers (a seasonal product) are modeled by $S = 74 + 3t - 40 \cos(\pi t/6)$, where t is the time (in months), with $t = 1$ corresponding to January. Graph the sales function over 1 year.

95. HARMONIC MOTION An object weighing W pounds is suspended from the ceiling by a steel spring (see figure). The weight is pulled downward (positive direction) from its equilibrium position and released. The resulting motion of the weight is described by the function $y = \frac{1}{2}e^{-t/4} \cos 4t$, $t > 0$, where y is the distance (in feet) and t is the time (in seconds).

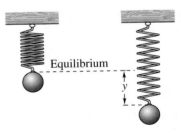

Equilibrium

(a) Use a graphing utility to graph the function.

(b) Describe the behavior of the displacement function for increasing values of time t.

EXPLORATION

TRUE OR FALSE? In Exercises 96 and 97, determine whether the statement is true or false. Justify your answer.

96. The graph of $y = \csc x$ can be obtained on a calculator by graphing the reciprocal of $y = \sin x$.

97. The graph of $y = \sec x$ can be obtained on a calculator by graphing a translation of the reciprocal of $y = \sin x$.

98. CAPSTONE Determine which function is represented by the graph. Do not use a calculator. Explain your reasoning.

(a) 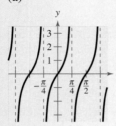 (b)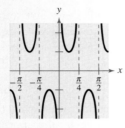

(i) $f(x) = \tan 2x$ (i) $f(x) = \sec 4x$

(ii) $f(x) = \tan(x/2)$ (ii) $f(x) = \csc 4x$

(iii) $f(x) = 2\tan x$ (iii) $f(x) = \csc(x/4)$

(iv) $f(x) = -\tan 2x$ (iv) $f(x) = \sec(x/4)$

(v) $f(x) = -\tan(x/2)$ (v) $f(x) = \csc(4x - \pi)$

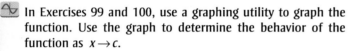 In Exercises 99 and 100, use a graphing utility to graph the function. Use the graph to determine the behavior of the function as $x \to c$.

(a) $x \to \dfrac{\pi^+}{2}$ $\left(\text{as } x \text{ approaches } \dfrac{\pi}{2} \text{ from the right}\right)$

(b) $x \to \dfrac{\pi^-}{2}$ $\left(\text{as } x \text{ approaches } \dfrac{\pi}{2} \text{ from the left}\right)$

(c) $x \to -\dfrac{\pi^+}{2}$ $\left(\text{as } x \text{ approaches } -\dfrac{\pi}{2} \text{ from the right}\right)$

(d) $x \to -\dfrac{\pi^-}{2}$ $\left(\text{as } x \text{ approaches } -\dfrac{\pi}{2} \text{ from the left}\right)$

99. $f(x) = \tan x$ **100.** $f(x) = \sec x$

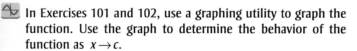 In Exercises 101 and 102, use a graphing utility to graph the function. Use the graph to determine the behavior of the function as $x \to c$.

(a) As $x \to 0^+$, the value of $f(x) \to$ ____ .

(b) As $x \to 0^-$, the value of $f(x) \to$ ____ .

(c) As $x \to \pi^+$, the value of $f(x) \to$ ____ .

(d) As $x \to \pi^-$, the value of $f(x) \to$ ____ .

101. $f(x) = \cot x$ **102.** $f(x) = \csc x$

103. THINK ABOUT IT Consider the function given by $f(x) = x - \cos x$.

 (a) Use a graphing utility to graph the function and verify that there exists a zero between 0 and 1. Use the graph to approximate the zero.

(b) Starting with $x_0 = 1$, generate a sequence $x_1, x_2, x_3, \ldots$, where $x_n = \cos(x_{n-1})$. For example,

$$x_0 = 1$$
$$x_1 = \cos(x_0)$$
$$x_2 = \cos(x_1)$$
$$x_3 = \cos(x_2)$$
$$\vdots$$

What value does the sequence approach?

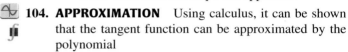 **104. APPROXIMATION** Using calculus, it can be shown that the tangent function can be approximated by the polynomial

$$\tan x \approx x + \frac{2x^3}{3!} + \frac{16x^5}{5!}$$

where x is in radians. Use a graphing utility to graph the tangent function and its polynomial approximation in the same viewing window. How do the graphs compare?

105. APPROXIMATION Using calculus, it can be shown that the secant function can be approximated by the polynomial

$$\sec x \approx 1 + \frac{x^2}{2!} + \frac{5x^4}{4!}$$

where x is in radians. Use a graphing utility to graph the secant function and its polynomial approximation in the same viewing window. How do the graphs compare?

106. PATTERN RECOGNITION

(a) Use a graphing utility to graph each function.

$$y_1 = \frac{4}{\pi}\left(\sin \pi x + \frac{1}{3}\sin 3\pi x\right)$$

$$y_2 = \frac{4}{\pi}\left(\sin \pi x + \frac{1}{3}\sin 3\pi x + \frac{1}{5}\sin 5\pi x\right)$$

(b) Identify the pattern started in part (a) and find a function y_3 that continues the pattern one more term. Use a graphing utility to graph y_3.

(c) The graphs in parts (a) and (b) approximate the periodic function in the figure. Find a function y_4 that is a better approximation.

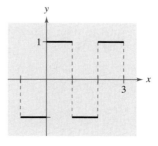

4.7

INVERSE TRIGONOMETRIC FUNCTIONS

What you should learn

- Evaluate and graph the inverse sine function.
- Evaluate and graph the other inverse trigonometric functions.
- Evaluate and graph the compositions of trigonometric functions.

Why you should learn it

You can use inverse trigonometric functions to model and solve real-life problems. For instance, in Exercise 106 on page 349, an inverse trigonometric function can be used to model the angle of elevation from a television camera to a space shuttle launch.

NASA

Inverse Sine Function

Recall from Section 1.9 that, for a function to have an inverse function, it must be one-to-one—that is, it must pass the Horizontal Line Test. From Figure 4.71, you can see that $y = \sin x$ does not pass the test because different values of x yield the same y-value.

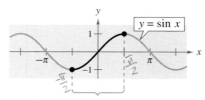

$\sin x$ has an inverse function on this interval.

FIGURE 4.71

However, if you restrict the domain to the interval $-\pi/2 \leq x \leq \pi/2$ (corresponding to the black portion of the graph in Figure 4.71), the following properties hold.

1. On the interval $[-\pi/2, \pi/2]$, the function $y = \sin x$ is increasing.

2. On the interval $[-\pi/2, \pi/2]$, $y = \sin x$ takes on its full range of values, $-1 \leq \sin x \leq 1$.

3. On the interval $[-\pi/2, \pi/2]$, $y = \sin x$ is one-to-one.

So, on the restricted domain $-\pi/2 \leq x \leq \pi/2$, $y = \sin x$ has a unique inverse function called the **inverse sine function.** It is denoted by

$$y = \arcsin x \qquad \text{or} \qquad y = \sin^{-1} x.$$

The notation $\sin^{-1} x$ is consistent with the inverse function notation $f^{-1}(x)$. The arcsin x notation (read as "the arcsine of x") comes from the association of a central angle with its intercepted *arc length* on a unit circle. So, arcsin x means the angle (or arc) whose sine is x. Both notations, arcsin x and $\sin^{-1} x$, are commonly used in mathematics, so remember that $\sin^{-1} x$ denotes the *inverse* sine function rather than $1/\sin x$. The values of arcsin x lie in the interval $-\pi/2 \leq \arcsin x \leq \pi/2$. The graph of $y = \arcsin x$ is shown in Example 2.

Study Tip

When evaluating the inverse sine function, it helps to remember the phrase "the arcsine of x is the angle (or number) whose sine is x."

Definition of Inverse Sine Function

The **inverse sine function** is defined by

$$y = \arcsin x \qquad \text{if and only if} \qquad \sin y = x$$

where $-1 \leq x \leq 1$ and $-\pi/2 \leq y \leq \pi/2$. The domain of $y = \arcsin x$ is $[-1, 1]$, and the range is $[-\pi/2, \pi/2]$.

Example 1   Evaluating the Inverse Sine Function

If possible, find the exact value.

a. $\arcsin\left(-\dfrac{1}{2}\right)$ **b.** $\sin^{-1}\dfrac{\sqrt{3}}{2}$ **c.** $\sin^{-1}2$

Solution

a. Because $\sin\left(-\dfrac{\pi}{6}\right) = -\dfrac{1}{2}$ for $-\dfrac{\pi}{2} \le y \le \dfrac{\pi}{2}$, it follows that

$$\arcsin\left(-\dfrac{1}{2}\right) = -\dfrac{\pi}{6}.$$ Angle whose sine is $-\frac{1}{2}$

b. Because $\sin\dfrac{\pi}{3} = \dfrac{\sqrt{3}}{2}$ for $-\dfrac{\pi}{2} \le y \le \dfrac{\pi}{2}$, it follows that

$$\sin^{-1}\dfrac{\sqrt{3}}{2} = \dfrac{\pi}{3}.$$ Angle whose sine is $\sqrt{3}/2$

c. It is not possible to evaluate $y = \sin^{-1}x$ when $x = 2$ because there is no angle whose sine is 2. Remember that the domain of the inverse sine function is $[-1, 1]$.

CHECK *Point*   Now try Exercise 5.

Example 2   Graphing the Arcsine Function

Sketch a graph of

$$y = \arcsin x.$$

Solution

By definition, the equations $y = \arcsin x$ and $\sin y = x$ are equivalent for $-\pi/2 \le y \le \pi/2$. So, their graphs are the same. From the interval $[-\pi/2, \pi/2]$, you can assign values to y in the second equation to make a table of values. Then plot the points and draw a smooth curve through the points.

y	$-\dfrac{\pi}{2}$	$-\dfrac{\pi}{4}$	$-\dfrac{\pi}{6}$	0	$\dfrac{\pi}{6}$	$\dfrac{\pi}{4}$	$\dfrac{\pi}{2}$
$x = \sin y$	-1	$-\dfrac{\sqrt{2}}{2}$	$-\dfrac{1}{2}$	0	$\dfrac{1}{2}$	$\dfrac{\sqrt{2}}{2}$	1

The resulting graph for $y = \arcsin x$ is shown in Figure 4.72. Note that it is the reflection (in the line $y = x$) of the black portion of the graph in Figure 4.71. Be sure you see that Figure 4.72 shows the *entire* graph of the inverse sine function. Remember that the domain of $y = \arcsin x$ is the closed interval $[-1, 1]$ and the range is the closed interval $[-\pi/2, \pi/2]$.

CHECK *Point*   Now try Exercise 21.

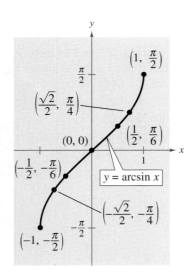

FIGURE 4.72

Other Inverse Trigonometric Functions

The cosine function is decreasing and one-to-one on the interval $0 \leq x \leq \pi$, as shown in Figure 4.73.

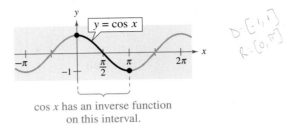

cos x has an inverse function
on this interval.

FIGURE **4.73**

Consequently, on this interval the cosine function has an inverse function—the **inverse cosine function**—denoted by

$$y = \arccos x \qquad \text{or} \qquad y = \cos^{-1} x.$$

Similarly, you can define an **inverse tangent function** by restricting the domain of $y = \tan x$ to the interval $(-\pi/2, \pi/2)$. The following list summarizes the definitions of the three most common inverse trigonometric functions. The remaining three are defined in Exercises 115–117.

Definitions of the Inverse Trigonometric Functions		
Function	*Domain*	*Range*
$y = \arcsin x$ if and only if $\sin y = x$	$-1 \leq x \leq 1$	$-\dfrac{\pi}{2} \leq y \leq \dfrac{\pi}{2}$
$y = \arccos x$ if and only if $\cos y = x$	$-1 \leq x \leq 1$	$0 \leq y \leq \pi$
$y = \arctan x$ if and only if $\tan y = x$	$-\infty < x < \infty$	$-\dfrac{\pi}{2} < y < \dfrac{\pi}{2}$

The graphs of these three inverse trigonometric functions are shown in Figure 4.74.

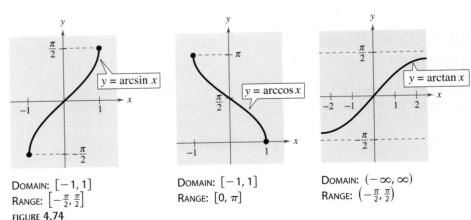

DOMAIN: $[-1, 1]$
RANGE: $\left[-\frac{\pi}{2}, \frac{\pi}{2}\right]$

DOMAIN: $[-1, 1]$
RANGE: $[0, \pi]$

DOMAIN: $(-\infty, \infty)$
RANGE: $\left(-\frac{\pi}{2}, \frac{\pi}{2}\right)$

FIGURE **4.74**

Example 3   Evaluating Inverse Trigonometric Functions

Find the exact value.

a. $\arccos \dfrac{\sqrt{2}}{2}$ **b.** $\cos^{-1}(-1)$

c. $\arctan 0$ **d.** $\tan^{-1}(-1)$

Solution

a. Because $\cos(\pi/4) = \sqrt{2}/2$, and $\pi/4$ lies in $[0, \pi]$, it follows that

$$\arccos \frac{\sqrt{2}}{2} = \frac{\pi}{4}. \qquad \text{Angle whose cosine is } \sqrt{2}/2$$

b. Because $\cos \pi = -1$, and π lies in $[0, \pi]$, it follows that

$$\cos^{-1}(-1) = \pi. \qquad \text{Angle whose cosine is } -1$$

c. Because $\tan 0 = 0$, and 0 lies in $(-\pi/2, \pi/2)$, it follows that

$$\arctan 0 = 0. \qquad \text{Angle whose tangent is } 0$$

d. Because $\tan(-\pi/4) = -1$, and $-\pi/4$ lies in $(-\pi/2, \pi/2)$, it follows that

$$\tan^{-1}(-1) = -\frac{\pi}{4}. \qquad \text{Angle whose tangent is } -1$$

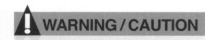 Now try Exercise 15.

Example 4   Calculators and Inverse Trigonometric Functions

Use a calculator to approximate the value (if possible).

a. $\arctan(-8.45)$

b. $\sin^{-1} 0.2447$

c. $\arccos 2$

Solution

Function	*Mode*	*Calculator Keystrokes*
a. $\arctan(-8.45)$	Radian	(TAN⁻¹) ( ) ((−)) 8.45 ( ) (ENTER)

From the display, it follows that $\arctan(-8.45) \approx -1.453001$.

b. $\sin^{-1} 0.2447$	Radian	(SIN⁻¹) (( ) 0.2447 ( ) (ENTER)

From the display, it follows that $\sin^{-1} 0.2447 \approx 0.2472103$.

c. $\arccos 2$	Radian	(COS⁻¹) (( ) 2 ( ) (ENTER)

In *real number* mode, the calculator should display an *error message* because the domain of the inverse cosine function is $[-1, 1]$.

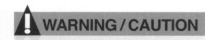 Now try Exercise 29.

WARNING / CAUTION

Remember that the domain of the inverse sine function and the inverse cosine function is $[-1, 1]$, as indicated in Example 4(c).

In Example 4, if you had set the calculator to *degree* mode, the displays would have been in degrees rather than radians. This convention is peculiar to calculators. By definition, the values of inverse trigonometric functions are *always in radians*.

Compositions of Functions

Algebra Help

You can review the composition of functions in Section 1.8.

Recall from Section 1.9 that for all x in the domains of f and f^{-1}, inverse functions have the properties

$$f(f^{-1}(x)) = x \qquad \text{and} \qquad f^{-1}(f(x)) = x.$$

Inverse Properties of Trigonometric Functions

If $-1 \le x \le 1$ and $-\pi/2 \le y \le \pi/2$, then

$$\sin(\arcsin x) = x \qquad \text{and} \qquad \arcsin(\sin y) = y.$$

If $-1 \le x \le 1$ and $0 \le y \le \pi$, then

$$\cos(\arccos x) = x \qquad \text{and} \qquad \arccos(\cos y) = y.$$

If x is a real number and $-\pi/2 < y < \pi/2$, then

$$\tan(\arctan x) = x \qquad \text{and} \qquad \arctan(\tan y) = y.$$

Keep in mind that these inverse properties do not apply for arbitrary values of x and y. For instance,

$$\arcsin\left(\sin \frac{3\pi}{2}\right) = \arcsin(-1) = -\frac{\pi}{2} \ne \frac{3\pi}{2}.$$

In other words, the property

$$\arcsin(\sin y) = y$$

is not valid for values of y outside the interval $[-\pi/2, \pi/2]$.

Example 5   Using Inverse Properties

If possible, find the exact value.

a. $\tan[\arctan(-5)]$ **b.** $\arcsin\left(\sin \dfrac{5\pi}{3}\right)$ **c.** $\cos(\cos^{-1} \pi)$

Solution

a. Because -5 lies in the domain of the arctan function, the inverse property applies, and you have

$$\tan[\arctan(-5)] = -5.$$

b. In this case, $5\pi/3$ does not lie within the range of the arcsine function, $-\pi/2 \le y \le \pi/2$. However, $5\pi/3$ is coterminal with

$$\frac{5\pi}{3} - 2\pi = -\frac{\pi}{3}$$

which does lie in the range of the arcsine function, and you have

$$\arcsin\left(\sin \frac{5\pi}{3}\right) = \arcsin\left[\sin\left(-\frac{\pi}{3}\right)\right] = -\frac{\pi}{3}.$$

c. The expression $\cos(\cos^{-1} \pi)$ is not defined because $\cos^{-1} \pi$ is not defined. Remember that the domain of the inverse cosine function is $[-1, 1]$.

CHECK *Point*   Now try Exercise 49.

Example 6 shows how to use right triangles to find exact values of compositions of inverse functions. Then, Example 7 shows how to use right triangles to convert a trigonometric expression into an algebraic expression. This conversion technique is used frequently in calculus.

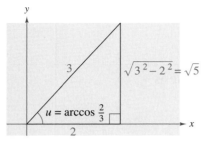

Angle whose cosine is $\frac{2}{3}$

FIGURE **4.75**

Example 6 Evaluating Compositions of Functions

Find the exact value.

a. $\tan\left(\arccos\dfrac{2}{3}\right)$ **b.** $\cos\left[\arcsin\left(-\dfrac{3}{5}\right)\right]$

Solution

a. If you let $u = \arccos\frac{2}{3}$, then $\cos u = \frac{2}{3}$. Because $\cos u$ is positive, u is a *first*-quadrant angle. You can sketch and label angle u as shown in Figure 4.75. Consequently,

$$\tan\left(\arccos\frac{2}{3}\right) = \tan u = \frac{\text{opp}}{\text{adj}} = \frac{\sqrt{5}}{2}.$$

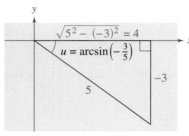

Angle whose sine is $-\frac{3}{5}$

FIGURE **4.76**

b. If you let $u = \arcsin\left(-\frac{3}{5}\right)$, then $\sin u = -\frac{3}{5}$. Because $\sin u$ is negative, u is a *fourth*-quadrant angle. You can sketch and label angle u as shown in Figure 4.76. Consequently,

$$\cos\left[\arcsin\left(-\frac{3}{5}\right)\right] = \cos u = \frac{\text{adj}}{\text{hyp}} = \frac{4}{5}.$$

CHECK *Point* Now try Exercise 57.

Example 7 Some Problems from Calculus

Write each of the following as an algebraic expression in x.

a. $\sin(\arccos 3x), \quad 0 \le x \le \dfrac{1}{3}$ **b.** $\cot(\arccos 3x), \quad 0 \le x < \dfrac{1}{3}$

Solution

If you let $u = \arccos 3x$, then $\cos u = 3x$, where $-1 \le 3x \le 1$. Because

$$\cos u = \frac{\text{adj}}{\text{hyp}} = \frac{3x}{1}$$

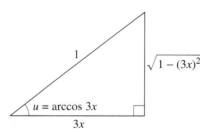

Angle whose cosine is $3x$

FIGURE **4.77**

you can sketch a right triangle with acute angle u, as shown in Figure 4.77. From this triangle, you can easily convert each expression to algebraic form.

a. $\sin(\arccos 3x) = \sin u = \dfrac{\text{opp}}{\text{hyp}} = \sqrt{1 - 9x^2}, \quad 0 \le x \le \dfrac{1}{3}$

b. $\cot(\arccos 3x) = \cot u = \dfrac{\text{adj}}{\text{opp}} = \dfrac{3x}{\sqrt{1 - 9x^2}}, \quad 0 \le x < \dfrac{1}{3}$

CHECK *Point* Now try Exercise 67.

In Example 7, similar arguments can be made for x-values lying in the interval $\left[-\frac{1}{3}, 0\right]$.

4.7   EXERCISES

See www.CalcChat.com for worked-out solutions to odd-numbered exercises.

VOCABULARY: Fill in the blanks.

Function	Alternative Notation	Domain	Range
1. $y = \arcsin x$	_____	_____	$-\dfrac{\pi}{2} \le y \le \dfrac{\pi}{2}$
2. _____	$y = \cos^{-1} x$	$-1 \le x \le 1$	_____
3. $y = \arctan x$	_____	_____	_____

4. Without restrictions, no trigonometric function has a(n) _____ function.

SKILLS AND APPLICATIONS

In Exercises 5–20, evaluate the expression without using a calculator.

5. $\arcsin \frac{1}{2}$

6. $\arcsin 0$

7. $\arccos \frac{1}{2}$

8. $\arccos 0$

9. $\arctan \dfrac{\sqrt{3}}{3}$

10. $\arctan(1)$

11. $\cos^{-1}\left(-\dfrac{\sqrt{3}}{2}\right)$

12. $\sin^{-1}\left(-\dfrac{\sqrt{2}}{2}\right)$

13. $\arctan\left(-\sqrt{3}\right)$

14. $\arctan \sqrt{3}$

15. $\arccos\left(-\dfrac{1}{2}\right)$

16. $\arcsin \dfrac{\sqrt{2}}{2}$

17. $\sin^{-1}\left(-\dfrac{\sqrt{3}}{2}\right)$

18. $\tan^{-1}\left(-\dfrac{\sqrt{3}}{3}\right)$

19. $\tan^{-1} 0$

20. $\cos^{-1} 1$

 In Exercises 21 and 22, use a graphing utility to graph f, g, and $y = x$ in the same viewing window to verify geometrically that g is the inverse function of f. (Be sure to restrict the domain of f properly.)

21. $f(x) = \sin x$, $g(x) = \arcsin x$

22. $f(x) = \tan x$, $g(x) = \arctan x$

 In Exercises 23–40, use a calculator to evaluate the expression. Round your result to two decimal places.

23. $\arccos 0.37$

24. $\arcsin 0.65$

25. $\arcsin(-0.75)$

26. $\arccos(-0.7)$

27. $\arctan(-3)$

28. $\arctan 25$

29. $\sin^{-1} 0.31$

30. $\cos^{-1} 0.26$

31. $\arccos(-0.41)$

32. $\arcsin(-0.125)$

33. $\arctan 0.92$

34. $\arctan 2.8$

35. $\arcsin \frac{7}{8}$

36. $\arccos\left(-\frac{1}{3}\right)$

37. $\tan^{-1} \frac{19}{4}$

38. $\tan^{-1}\left(-\frac{95}{7}\right)$

39. $\tan^{-1}\left(-\sqrt{372}\right)$

40. $\tan^{-1}\left(-\sqrt{2165}\right)$

In Exercises 41 and 42, determine the missing coordinates of the points on the graph of the function.

41.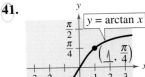

42.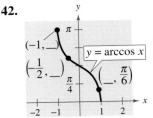

In Exercises 43–48, use an inverse trigonometric function to write θ as a function of x.

43.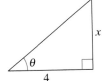

44.

45.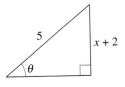

46.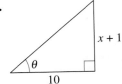

47.

48.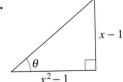

In Exercises 49–54, use the properties of inverse trigonometric functions to evaluate the expression.

49. $\sin(\arcsin 0.3)$

50. $\tan(\arctan 45)$

51. $\cos[\arccos(-0.1)]$

52. $\sin[\arcsin(-0.2)]$

53. $\arcsin(\sin 3\pi)$

54. $\arccos\left(\cos \dfrac{7\pi}{2}\right)$

In Exercises 55–66, find the exact value of the expression. (*Hint:* Sketch a right triangle.)

55. $\sin\left(\arctan\frac{3}{4}\right)$

56. $\sec\left(\arcsin\frac{4}{5}\right)$

57. $\cos(\tan^{-1} 2)$

58. $\sin\left(\cos^{-1}\frac{\sqrt{5}}{5}\right)$

59. $\cos\left(\arcsin\frac{5}{13}\right)$

60. $\csc\left[\arctan\left(-\frac{5}{12}\right)\right]$

61. $\sec\left[\arctan\left(-\frac{3}{5}\right)\right]$

62. $\tan\left[\arcsin\left(-\frac{3}{4}\right)\right]$

63. $\sin\left[\arccos\left(-\frac{2}{3}\right)\right]$

64. $\cot\left(\arctan\frac{5}{8}\right)$

65. $\csc\left[\cos^{-1}\left(\frac{\sqrt{3}}{2}\right)\right]$

66. $\sec\left[\sin^{-1}\left(-\frac{\sqrt{2}}{2}\right)\right]$

In Exercises 67–76, write an algebraic expression that is equivalent to the expression. (*Hint:* Sketch a right triangle, as demonstrated in Example 7.)

67. $\cot(\arctan x)$

68. $\sin(\arctan x)$

69. $\cos(\arcsin 2x)$

70. $\sec(\arctan 3x)$

71. $\sin(\arccos x)$

72. $\sec[\arcsin(x - 1)]$

73. $\tan\left(\arccos\frac{x}{3}\right)$

74. $\cot\left(\arctan\frac{1}{x}\right)$

75. $\csc\left(\arctan\frac{x}{\sqrt{2}}\right)$

76. $\cos\left(\arcsin\frac{x - h}{r}\right)$

In Exercises 77 and 78, use a graphing utility to graph f and g in the same viewing window to verify that the two functions are equal. Explain why they are equal. Identify any asymptotes of the graphs.

77. $f(x) = \sin(\arctan 2x)$, $g(x) = \dfrac{2x}{\sqrt{1 + 4x^2}}$

78. $f(x) = \tan\left(\arccos\dfrac{x}{2}\right)$, $g(x) = \dfrac{\sqrt{4 - x^2}}{x}$

In Exercises 79–82, fill in the blank.

79. $\arctan\dfrac{9}{x} = \arcsin\left(\boxed{}\right)$, $x \neq 0$

80. $\arcsin\dfrac{\sqrt{36 - x^2}}{6} = \arccos\left(\boxed{}\right)$, $0 \leq x \leq 6$

81. $\arccos\dfrac{3}{\sqrt{x^2 - 2x + 10}} = \arcsin\left(\boxed{}\right)$

82. $\arccos\dfrac{x - 2}{2} = \arctan\left(\,\boxed{}\,\right)$, $|x - 2| \leq 2$

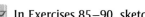

In Exercises 83 and 84, sketch a graph of the function and compare the graph of g with the graph of $f(x) = \arcsin x$.

83. $g(x) = \arcsin(x - 1)$

84. $g(x) = \arcsin\dfrac{x}{2}$

In Exercises 85–90, sketch a graph of the function.

85. $y = 2 \arccos x$

86. $g(t) = \arccos(t + 2)$

87. $f(x) = \arctan 2x$

88. $f(x) = \dfrac{\pi}{2} + \arctan x$

89. $h(v) = \tan(\arccos v)$

90. $f(x) = \arccos\dfrac{x}{4}$

In Exercises 91–96, use a graphing utility to graph the function.

91. $f(x) = 2 \arccos(2x)$

92. $f(x) = \pi \arcsin(4x)$

93. $f(x) = \arctan(2x - 3)$

94. $f(x) = -3 + \arctan(\pi x)$

95. $f(x) = \pi - \sin^{-1}\left(\dfrac{2}{3}\right)$

96. $f(x) = \dfrac{\pi}{2} + \cos^{-1}\left(\dfrac{1}{\pi}\right)$

In Exercises 97 and 98, write the function in terms of the sine function by using the identity

$$A \cos \omega t + B \sin \omega t = \sqrt{A^2 + B^2}\, \sin\left(\omega t + \arctan\frac{A}{B}\right).$$

Use a graphing utility to graph both forms of the function. What does the graph imply?

97. $f(t) = 3 \cos 2t + 3 \sin 2t$

98. $f(t) = 4 \cos \pi t + 3 \sin \pi t$

In Exercises 99–104, fill in the blank. If not possible, state the reason. (*Note:* The notation $x \to c^+$ indicates that x approaches c from the right and $x \to c^-$ indicates that x approaches c from the left.)

99. As $x \to 1^-$, the value of $\arcsin x \to$ $\boxed{}$.

100. As $x \to 1^-$, the value of $\arccos x \to$ $\boxed{}$.

101. As $x \to \infty$, the value of arctan $x \to$ ⬜.

102. As $x \to -1^+$, the value of arcsin $x \to$ ⬜.

103. As $x \to -1^+$, the value of arccos $x \to$ ⬜.

104. As $x \to -\infty$, the value of arctan $x \to$ ⬜.

105. DOCKING A BOAT A boat is pulled in by means of a winch located on a dock 5 feet above the deck of the boat (see figure). Let θ be the angle of elevation from the boat to the winch and let s be the length of the rope from the winch to the boat.

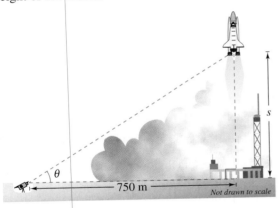

(a) Write θ as a function of s.

(b) Find θ when $s = 40$ feet and $s = 20$ feet.

106. PHOTOGRAPHY A television camera at ground level is filming the lift-off of a space shuttle at a point 750 meters from the launch pad (see figure). Let θ be the angle of elevation to the shuttle and let s be the height of the shuttle.

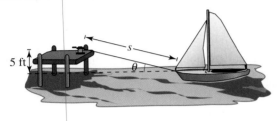

(a) Write θ as a function of s.

(b) Find θ when $s = 300$ meters and $s = 1200$ meters.

107. PHOTOGRAPHY A photographer is taking a picture of a three-foot-tall painting hung in an art gallery. The camera lens is 1 foot below the lower edge of the painting (see figure). The angle β subtended by the camera lens x feet from the painting is

$$\beta = \arctan \frac{3x}{x^2 + 4}, \quad x > 0.$$

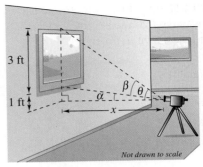

Not drawn to scale

(a) Use a graphing utility to graph β as a function of x.

(b) Move the cursor along the graph to approximate the distance from the picture when β is maximum.

(c) Identify the asymptote of the graph and discuss its meaning in the context of the problem.

108. GRANULAR ANGLE OF REPOSE Different types of granular substances naturally settle at different angles when stored in cone-shaped piles. This angle θ is called the *angle of repose* (see figure). When rock salt is stored in a cone-shaped pile 11 feet high, the diameter of the pile's base is about 34 feet. (Source: Bulk-Store Structures, Inc.)

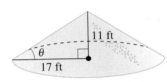

(a) Find the angle of repose for rock salt.

(b) How tall is a pile of rock salt that has a base diameter of 40 feet?

109. GRANULAR ANGLE OF REPOSE When whole corn is stored in a cone-shaped pile 20 feet high, the diameter of the pile's base is about 82 feet.

(a) Find the angle of repose for whole corn.

(b) How tall is a pile of corn that has a base diameter of 100 feet?

110. ANGLE OF ELEVATION An airplane flies at an altitude of 6 miles toward a point directly over an observer. Consider θ and x as shown in the figure.

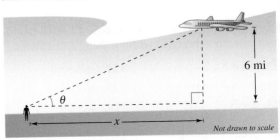

Not drawn to scale

(a) Write θ as a function of x.

(b) Find θ when $x = 7$ miles and $x = 1$ mile.

111. SECURITY PATROL   A security car with its spotlight on is parked 20 meters from a warehouse. Consider θ and x as shown in the figure.

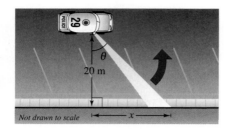

Not drawn to scale

(a) Write θ as a function of x.

(b) Find θ when $x = 5$ meters and $x = 12$ meters.

EXPLORATION

TRUE OR FALSE?   In Exercises 112–114, determine whether the statement is true or false. Justify your answer.

112. $\sin \dfrac{5\pi}{6} = \dfrac{1}{2}$ ⟹ $\arcsin \dfrac{1}{2} = \dfrac{5\pi}{6}$

113. $\tan \dfrac{5\pi}{4} = 1$ ⟹ $\arctan 1 = \dfrac{5\pi}{4}$

114. $\arctan x = \dfrac{\arcsin x}{\arccos x}$

115. Define the inverse cotangent function by restricting the domain of the cotangent function to the interval $(0, \pi)$, and sketch its graph.

116. Define the inverse secant function by restricting the domain of the secant function to the intervals $[0, \pi/2)$ and $(\pi/2, \pi]$, and sketch its graph.

117. Define the inverse cosecant function by restricting the domain of the cosecant function to the intervals $[-\pi/2, 0)$ and $(0, \pi/2]$, and sketch its graph.

118. CAPSTONE   Use the results of Exercises 115–117 to explain how to graph (a) the inverse cotangent function, (b) the inverse secant function, and (c) the inverse cosecant function on a graphing utility.

In Exercises 119–126, use the results of Exercises 115–117 to evaluate each expression without using a calculator.

119. $\operatorname{arcsec} \sqrt{2}$ **120.** $\operatorname{arcsec} 1$

121. $\operatorname{arccot}(-1)$ **122.** $\operatorname{arccot}\left(-\sqrt{3}\right)$

123. $\operatorname{arccsc} 2$ **124.** $\operatorname{arccsc}(-1)$

125. $\operatorname{arccsc}\left(\dfrac{2\sqrt{3}}{3}\right)$ **126.** $\operatorname{arcsec}\left(-\dfrac{2\sqrt{3}}{3}\right)$

In Exercises 127–134, use the results of Exercises 115–117 and a calculator to approximate the value of the expression. Round your result to two decimal places.

127. $\operatorname{arcsec} 2.54$ **128.** $\operatorname{arcsec}(-1.52)$

129. $\operatorname{arccot} 5.25$ **130.** $\operatorname{arccot}(-10)$

131. $\operatorname{arccot} \dfrac{5}{3}$ **132.** $\operatorname{arccot}\left(-\dfrac{16}{7}\right)$

133. $\operatorname{arccsc}\left(-\dfrac{25}{3}\right)$ **134.** $\operatorname{arccsc}(-12)$

135. AREA   In calculus, it is shown that the area of the region bounded by the graphs of $y = 0$, $y = 1/(x^2 + 1)$, $x = a$, and $x = b$ is given by

$$\text{Area} = \arctan b - \arctan a$$

(see figure). Find the area for the following values of a and b.

(a) $a = 0, b = 1$ (b) $a = -1, b = 1$

(c) $a = 0, b = 3$ (d) $a = -1, b = 3$

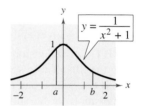

136. THINK ABOUT IT   Use a graphing utility to graph the functions

$$f(x) = \sqrt{x} \quad \text{and} \quad g(x) = 6 \arctan x.$$

For $x > 0$, it appears that $g > f$. Explain why you know that there exists a positive real number a such that $g < f$ for $x > a$. Approximate the number a.

137. THINK ABOUT IT   Consider the functions given by

$$f(x) = \sin x \quad \text{and} \quad f^{-1}(x) = \arcsin x.$$

(a) Use a graphing utility to graph the composite functions $f \circ f^{-1}$ and $f^{-1} \circ f$.

(b) Explain why the graphs in part (a) are not the graph of the line $y = x$. Why do the graphs of $f \circ f^{-1}$ and $f^{-1} \circ f$ differ?

138. PROOF   Prove each identity.

(a) $\arcsin(-x) = -\arcsin x$

(b) $\arctan(-x) = -\arctan x$

(c) $\arctan x + \arctan \dfrac{1}{x} = \dfrac{\pi}{2}, \quad x > 0$

(d) $\arcsin x + \arccos x = \dfrac{\pi}{2}$

(e) $\arcsin x = \arctan \dfrac{x}{\sqrt{1 - x^2}}$

4.8 APPLICATIONS AND MODELS

What you should learn
- Solve real-life problems involving right triangles.
- Solve real-life problems involving directional bearings.
- Solve real-life problems involving harmonic motion.

Why you should learn it
Right triangles often occur in real-life situations. For instance, in Exercise 65 on page 361, right triangles are used to determine the shortest grain elevator for a grain storage bin on a farm.

Applications Involving Right Triangles

In this section, the three angles of a right triangle are denoted by the letters A, B, and C (where C is the right angle), and the lengths of the sides opposite these angles by the letters a, b, and c (where c is the hypotenuse).

Example 1   Solving a Right Triangle

Solve the right triangle shown in Figure 4.78 for all unknown sides and angles.

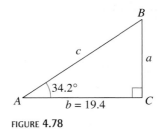

FIGURE **4.78**

Solution

Because $C = 90°$, it follows that $A + B = 90°$ and $B = 90° - 34.2° = 55.8°$. To solve for a, use the fact that

$$\tan A = \frac{\text{opp}}{\text{adj}} = \frac{a}{b} \implies a = b \tan A.$$

So, $a = 19.4 \tan 34.2° \approx 13.18$. Similarly, to solve for c, use the fact that

$$\cos A = \frac{\text{adj}}{\text{hyp}} = \frac{b}{c} \implies c = \frac{b}{\cos A}.$$

So, $c = \dfrac{19.4}{\cos 34.2°} \approx 23.46$.

CHECK *Point*   Now try Exercise 5.

Example 2   Finding a Side of a Right Triangle

A safety regulation states that the maximum angle of elevation for a rescue ladder is 72°. A fire department's longest ladder is 110 feet. What is the maximum safe rescue height?

Solution

A sketch is shown in Figure 4.79. From the equation $\sin A = a/c$, it follows that

$$a = c \sin A = 110 \sin 72° \approx 104.6.$$

So, the maximum safe rescue height is about 104.6 feet above the height of the fire truck.

CHECK *Point*   Now try Exercise 19.

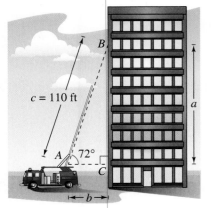

FIGURE **4.79**

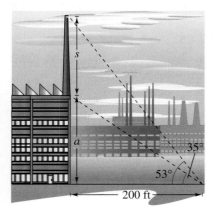

FIGURE **4.80**

Example 3 Finding a Side of a Right Triangle

At a point 200 feet from the base of a building, the angle of elevation to the *bottom* of a smokestack is 35°, whereas the angle of elevation to the *top* is 53°, as shown in Figure 4.80. Find the height s of the smokestack alone.

Solution

Note from Figure 4.80 that this problem involves two right triangles. For the smaller right triangle, use the fact that

$$\tan 35° = \frac{a}{200}$$

to conclude that the height of the building is

$$a = 200 \tan 35°.$$

For the larger right triangle, use the equation

$$\tan 53° = \frac{a + s}{200}$$

to conclude that $a + s = 200 \tan 53°$. So, the height of the smokestack is

$$s = 200 \tan 53° - a$$

$$= 200 \tan 53° - 200 \tan 35°$$

$$\approx 125.4 \text{ feet.}$$

CHECK*Point* Now try Exercise 23.

Example 4 Finding an Acute Angle of a Right Triangle

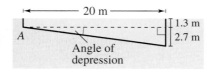

FIGURE **4.81**

A swimming pool is 20 meters long and 12 meters wide. The bottom of the pool is slanted so that the water depth is 1.3 meters at the shallow end and 4 meters at the deep end, as shown in Figure 4.81. Find the angle of depression of the bottom of the pool.

Solution

Using the tangent function, you can see that

$$\tan A = \frac{\text{opp}}{\text{adj}}$$

$$= \frac{2.7}{20}$$

$$= 0.135.$$

So, the angle of depression is

$$A = \arctan 0.135$$

$$\approx 0.13419 \text{ radian}$$

$$\approx 7.69°.$$

CHECK*Point* Now try Exercise 29.

Trigonometry and Bearings

In surveying and navigation, directions can be given in terms of **bearings.** A bearing measures the acute angle that a path or line of sight makes with a fixed north-south line, as shown in Figure 4.82. For instance, the bearing S 35° E in Figure 4.82 means 35 degrees east of south.

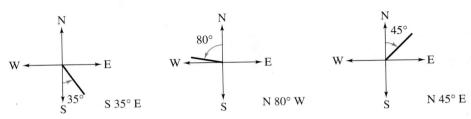

FIGURE **4.82**

Example 5   Finding Directions in Terms of Bearings

A ship leaves port at noon and heads due west at 20 knots, or 20 nautical miles (nm) per hour. At 2 P.M. the ship changes course to N 54° W, as shown in Figure 4.83. Find the ship's bearing and distance from the port of departure at 3 P.M.

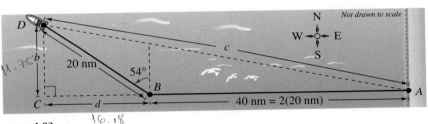

FIGURE **4.83**

Solution

For triangle BCD, you have $B = 90° - 54° = 36°$. The two sides of this triangle can be determined to be

$$b = 20 \sin 36° \qquad \text{and} \qquad d = 20 \cos 36°.$$

For triangle ACD, you can find angle A as follows.

$$\tan A = \frac{b}{d + 40} = \frac{20 \sin 36°}{20 \cos 36° + 40} \approx 0.2092494$$

$$A \approx \arctan 0.2092494 \approx 11.82°$$

The angle with the north-south line is $90° - 11.82° = 78.18°$. So, the bearing of the ship is N 78.18° W. Finally, from triangle ACD, you have $\sin A = b/c$, which yields

$$c = \frac{b}{\sin A} = \frac{20 \sin 36°}{\sin 11.82°}$$

$$\approx 57.4 \text{ nautical miles.} \qquad \text{Distance from port}$$

CHECK*Point*   Now try Exercise 37.

Study Tip

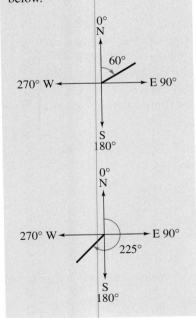

In *air navigation*, bearings are measured in degrees *clockwise* from north. Examples of air navigation bearings are shown below.

Harmonic Motion

The periodic nature of the trigonometric functions is useful for describing the motion of a point on an object that vibrates, oscillates, rotates, or is moved by wave motion.

For example, consider a ball that is bobbing up and down on the end of a spring, as shown in Figure 4.84. Suppose that 10 centimeters is the maximum distance the ball moves vertically upward or downward from its equilibrium (at rest) position. Suppose further that the time it takes for the ball to move from its maximum displacement above zero to its maximum displacement below zero and back again is $t = 4$ seconds. Assuming the ideal conditions of perfect elasticity and no friction or air resistance, the ball would continue to move up and down in a uniform and regular manner.

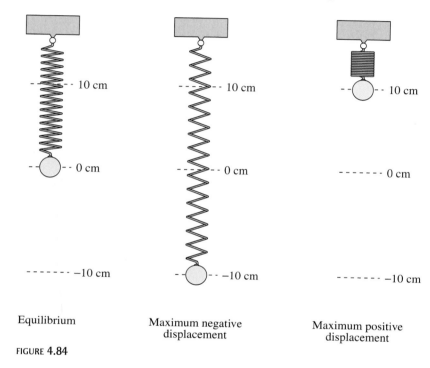

| Equilibrium | Maximum negative displacement | Maximum positive displacement |

FIGURE **4.84**

From this spring you can conclude that the period (time for one complete cycle) of the motion is

Period = 4 seconds

its amplitude (maximum displacement from equilibrium) is

Amplitude = 10 centimeters

and its **frequency** (number of cycles per second) is

Frequency = $\frac{1}{4}$ cycle per second.

Motion of this nature can be described by a sine or cosine function, and is called **simple harmonic motion.**

Definition of Simple Harmonic Motion

A point that moves on a coordinate line is said to be in **simple harmonic motion** if its distance d from the origin at time t is given by either

$$d = a \sin \omega t \qquad \text{or} \qquad d = a \cos \omega t$$

where a and ω are real numbers such that $\omega > 0$. The motion has amplitude $|a|$, period $\dfrac{2\pi}{\omega}$, and frequency $\dfrac{\omega}{2\pi}$.

$P = \dfrac{2\pi}{\omega}$ $\qquad P = \dfrac{\omega}{2\pi}$

Example 6   Simple Harmonic Motion

Write the equation for the simple harmonic motion of the ball described in Figure 4.84, where the period is 4 seconds. What is the frequency of this harmonic motion?

Solution

Because the spring is at equilibrium $(d = 0)$ when $t = 0$, you use the equation

$$d = a \sin \omega t.$$

Moreover, because the maximum displacement from zero is 10 and the period is 4, you have

$$\text{Amplitude} = |a| = 10$$

$$\text{Period} = \frac{2\pi}{\omega} = 4 \quad \Longrightarrow \quad \omega = \frac{\pi}{2}.$$

Consequently, the equation of motion is

$$d = 10 \sin \frac{\pi}{2} t.$$

Note that the choice of $a = 10$ or $a = -10$ depends on whether the ball initially moves up or down. The frequency is

$$\begin{aligned}
\text{Frequency} &= \frac{\omega}{2\pi} \\
&= \frac{\pi/2}{2\pi} \\
&= \frac{1}{4} \text{ cycle per second.}
\end{aligned}$$

CHECK*Point* Now try Exercise 53.

One illustration of the relationship between sine waves and harmonic motion can be seen in the wave motion resulting when a stone is dropped into a calm pool of water. The waves move outward in roughly the shape of sine (or cosine) waves, as shown in Figure 4.85. As an example, suppose you are fishing and your fishing bob is attached so that it does not move horizontally. As the waves move outward from the dropped stone, your fishing bob will move up and down in simple harmonic motion, as shown in Figure 4.86.

FIGURE **4.85**

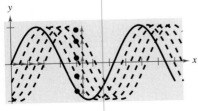

FIGURE **4.86**

Example 7   Simple Harmonic Motion

Given the equation for simple harmonic motion

$$d = 6 \cos \frac{3\pi}{4} t$$

find (a) the maximum displacement, (b) the frequency, (c) the value of d when $t = 4$, and (d) the least positive value of t for which $d = 0$.

Algebraic Solution

The given equation has the form $d = a \cos \omega t$, with $a = 6$ and $\omega = 3\pi/4$.

a. The maximum displacement (from the point of equilibrium) is given by the amplitude. So, the maximum displacement is 6.

b. Frequency $= \dfrac{\omega}{2\pi}$

$$= \frac{3\pi/4}{2\pi}$$

$$= \frac{3}{8} \text{ cycle per unit of time}$$

c. $d = 6 \cos \left[\dfrac{3\pi}{4} (4) \right]$

$$= 6 \cos 3\pi$$

$$= 6(-1)$$

$$= -6$$

d. To find the least positive value of t for which $d = 0$, solve the equation

$$d = 6 \cos \frac{3\pi}{4} t = 0.$$

First divide each side by 6 to obtain

$$\cos \frac{3\pi}{4} t = 0.$$

This equation is satisfied when

$$\frac{3\pi}{4} t = \frac{\pi}{2}, \frac{3\pi}{2}, \frac{5\pi}{2}, \ldots.$$

Multiply these values by $4/(3\pi)$ to obtain

$$t = \frac{2}{3}, 2, \frac{10}{3}, \ldots.$$

So, the least positive value of t is $t = \frac{2}{3}$.

CHECK *Point*   Now try Exercise 57.

Graphical Solution

Use a graphing utility set in *radian* mode to graph

$$y = 6 \cos \frac{3\pi}{4} x.$$

a. Use the *maximum* feature of the graphing utility to estimate that the maximum displacement from the point of equilibrium $y = 0$ is 6, as shown in Figure 4.87.

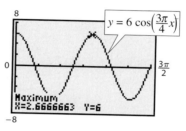

FIGURE 4.87

b. The period is the time for the graph to complete one cycle, which is $x \approx 2.667$. You can estimate the frequency as follows.

$$\text{Frequency} \approx \frac{1}{2.667} \approx 0.375 \text{ cycle per unit of time}$$

c. Use the *trace* or *value* feature to estimate that the value of y when $x = 4$ is $y = -6$, as shown in Figure 4.88.

d. Use the *zero* or *root* feature to estimate that the least positive value of x for which $y = 0$ is $x \approx 0.6667$, as shown in Figure 4.89.

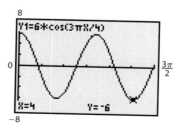

FIGURE 4.88

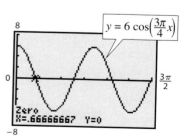

FIGURE 4.89

4.8   EXERCISES

See www.CalcChat.com for worked-out solutions to odd-numbered exercises.

VOCABULARY: Fill in the blanks.

1. A _____ measures the acute angle a path or line of sight makes with a fixed north-south line.

2. A point that moves on a coordinate line is said to be in simple _____ _____ if its distance d from the origin at time t is given by either $d = a \sin \omega t$ or $d = a \cos \omega t$.

3. The time for one complete cycle of a point in simple harmonic motion is its _____.

4. The number of cycles per second of a point in simple harmonic motion is its _____.

SKILLS AND APPLICATIONS

In Exercises 5–14, solve the right triangle shown in the figure for all unknown sides and angles. Round your answers to two decimal places.

5. $A = 30°$, $b = 3$
6. $B = 54°$, $c = 15$
7. $B = 71°$, $b = 24$
8. $A = 8.4°$, $a = 40.5$
9. $a = 3$, $b = 4$
10. $a = 25$, $c = 35$
11. $b = 16$, $c = 52$
12. $b = 1.32$, $c = 9.45$
13. $A = 12°\,15'$, $c = 430.5$
14. $B = 65°\,12'$, $a = 14.2$

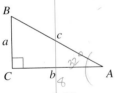

FIGURE FOR 5–14   FIGURE FOR 15–18

In Exercises 15–18, find the altitude of the isosceles triangle shown in the figure. Round your answers to two decimal places.

15. $\theta = 45°$, $b = 6$
16. $\theta = 18°$, $b = 10$
17. $\theta = 32°$, $b = 8$
18. $\theta = 27°$, $b = 11$

19. **LENGTH**   The sun is 25° above the horizon. Find the length of a shadow cast by a building that is 100 feet tall (see figure).

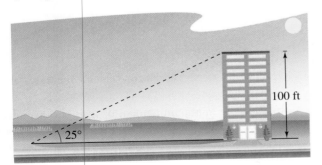

20. **LENGTH**   The sun is 20° above the horizon. Find the length of a shadow cast by a park statue that is 12 feet tall.

21. **HEIGHT**   A ladder 20 feet long leans against the side of a house. Find the height from the top of the ladder to the ground if the angle of elevation of the ladder is 80°.

22. **HEIGHT**   The length of a shadow of a tree is 125 feet when the angle of elevation of the sun is 33°. Approximate the height of the tree.

23. **HEIGHT**   From a point 50 feet in front of a church, the angles of elevation to the base of the steeple and the top of the steeple are 35° and 47° 40′, respectively. Find the height of the steeple.

24. **DISTANCE**   An observer in a lighthouse 350 feet above sea level observes two ships directly offshore. The angles of depression to the ships are 4° and 6.5° (see figure). How far apart are the ships?

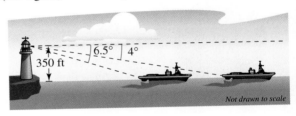

25. **DISTANCE**   A passenger in an airplane at an altitude of 10 kilometers sees two towns directly to the east of the plane. The angles of depression to the towns are 28° and 55° (see figure). How far apart are the towns?

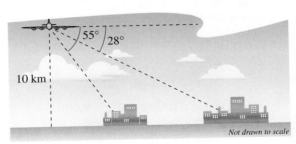

26. **ALTITUDE**   You observe a plane approaching overhead and assume that its speed is 550 miles per hour. The angle of elevation of the plane is 16° at one time and 57° one minute later. Approximate the altitude of the plane.

27. **ANGLE OF ELEVATION**   An engineer erects a 75-foot cellular telephone tower. Find the angle of elevation to the top of the tower at a point on level ground 50 feet from its base.

28. **ANGLE OF ELEVATION**   The height of an outdoor basketball backboard is $12\frac{1}{2}$ feet, and the backboard casts a shadow $17\frac{1}{3}$ feet long.

    (a) Draw a right triangle that gives a visual representation of the problem. Label the known and unknown quantities.

    (b) Use a trigonometric function to write an equation involving the unknown quantity.

    (c) Find the angle of elevation of the sun.

29. **ANGLE OF DEPRESSION**   A cellular telephone tower that is 150 feet tall is placed on top of a mountain that is 1200 feet above sea level. What is the angle of depression from the top of the tower to a cell phone user who is 5 horizontal miles away and 400 feet above sea level?

30. **ANGLE OF DEPRESSION**   A Global Positioning System satellite orbits 12,500 miles above Earth's surface (see figure). Find the angle of depression from the satellite to the horizon. Assume the radius of Earth is 4000 miles.

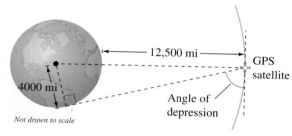

Not drawn to scale

31. **HEIGHT**   You are holding one of the tethers attached to the top of a giant character balloon in a parade. Before the start of the parade the balloon is upright and the bottom is floating approximately 20 feet above ground level. You are standing approximately 100 feet ahead of the balloon (see figure).

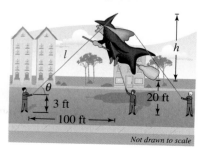

Not drawn to scale

(a) Find the length l of the tether you are holding in terms of h, the height of the balloon from top to bottom.

(b) Find an expression for the angle of elevation θ from you to the top of the balloon.

(c) Find the height h of the balloon if the angle of elevation to the top of the balloon is 35°.

32. **HEIGHT**   The designers of a water park are creating a new slide and have sketched some preliminary drawings. The length of the ladder is 30 feet, and its angle of elevation is 60° (see figure).

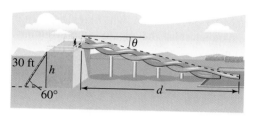

(a) Find the height h of the slide.

(b) Find the angle of depression θ from the top of the slide to the end of the slide at the ground in terms of the horizontal distance d the rider travels.

(c) The angle of depression of the ride is bounded by safety restrictions to be no less than 25° and not more than 30°. Find an interval for how far the rider travels horizontally.

33. **SPEED ENFORCEMENT**   A police department has set up a speed enforcement zone on a straight length of highway. A patrol car is parked parallel to the zone, 200 feet from one end and 150 feet from the other end (see figure).

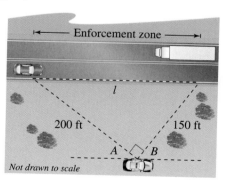

Not drawn to scale

(a) Find the length l of the zone and the measures of the angles A and B (in degrees).

(b) Find the minimum amount of time (in seconds) it takes for a vehicle to pass through the zone without exceeding the posted speed limit of 35 miles per hour.

34. AIRPLANE ASCENT During takeoff, an airplane's angle of ascent is 18° and its speed is 275 feet per second.

(a) Find the plane's altitude after 1 minute.

(b) How long will it take the plane to climb to an altitude of 10,000 feet?

35. NAVIGATION An airplane flying at 600 miles per hour has a bearing of 52°. After flying for 1.5 hours, how far north and how far east will the plane have traveled from its point of departure?

36. NAVIGATION A jet leaves Reno, Nevada and is headed toward Miami, Florida at a bearing of 100°. The distance between the two cities is approximately 2472 miles.

(a) How far north and how far west is Reno relative to Miami?

(b) If the jet is to return directly to Reno from Miami, at what bearing should it travel?

37. NAVIGATION A ship leaves port at noon and has a bearing of S 29° W. The ship sails at 20 knots.

(a) How many nautical miles south and how many nautical miles west will the ship have traveled by 6:00 P.M.?

(b) At 6:00 P.M., the ship changes course to due west. Find the ship's bearing and distance from the port of departure at 7:00 P.M.

38. NAVIGATION A privately owned yacht leaves a dock in Myrtle Beach, South Carolina and heads toward Freeport in the Bahamas at a bearing of S 1.4° E. The yacht averages a speed of 20 knots over the 428 nautical-mile trip.

(a) How long will it take the yacht to make the trip?

(b) How far east and south is the yacht after 12 hours?

(c) If a plane leaves Myrtle Beach to fly to Freeport, what bearing should be taken?

39. NAVIGATION A ship is 45 miles east and 30 miles south of port. The captain wants to sail directly to port. What bearing should be taken?

40. NAVIGATION An airplane is 160 miles north and 85 miles east of an airport. The pilot wants to fly directly to the airport. What bearing should be taken?

41. SURVEYING A surveyor wants to find the distance across a swamp (see figure). The bearing from A to B is N 32° W. The surveyor walks 50 meters from A, and at the point C the bearing to B is N 68° W. Find (a) the bearing from A to C and (b) the distance from A to B.

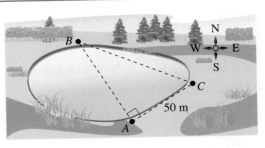

FIGURE FOR **41**

42. LOCATION OF A FIRE Two fire towers are 30 kilometers apart, where tower A is due west of tower B. A fire is spotted from the towers, and the bearings from A and B are N 76° E and N 56° W, respectively (see figure). Find the distance d of the fire from the line segment AB.

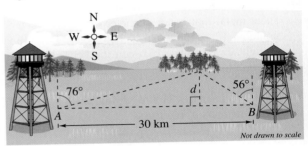

GEOMETRY In Exercises 43 and 44, find the angle α between two nonvertical lines L_1 and L_2. The angle α satisfies the equation

$$\tan \alpha = \left| \frac{m_2 - m_1}{1 + m_2 m_1} \right|$$

where m_1 and m_2 are the slopes of L_1 and L_2, respectively. (Assume that $m_1 m_2 \neq -1$.)

43. L_1: $3x - 2y = 5$
    L_2: $x + y = 1$

44. L_1: $2x - y = 8$
    L_2: $x - 5y = -4$

45. GEOMETRY Determine the angle between the diagonal of a cube and the diagonal of its base, as shown in the figure.

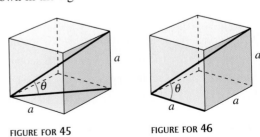

FIGURE FOR **45**   FIGURE FOR **46**

46. GEOMETRY Determine the angle between the diagonal of a cube and its edge, as shown in the figure.

47. GEOMETRY Find the length of the sides of a regular pentagon inscribed in a circle of radius 25 inches.

48. GEOMETRY Find the length of the sides of a regular hexagon inscribed in a circle of radius 25 inches.

49. HARDWARE Write the distance y across the flat sides of a hexagonal nut as a function of r (see figure).

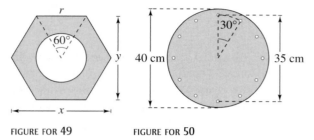

FIGURE FOR 49   FIGURE FOR 50

50. BOLT HOLES The figure shows a circular piece of sheet metal that has a diameter of 40 centimeters and contains 12 equally-spaced bolt holes. Determine the straight-line distance between the centers of consecutive bolt holes.

TRUSSES In Exercises 51 and 52, find the lengths of all the unknown members of the truss.

51.

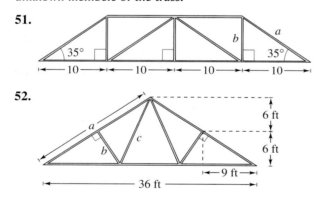

52.

HARMONIC MOTION In Exercises 53–56, find a model for simple harmonic motion satisfying the specified conditions.

Displacement ($t = 0$)	Amplitude	Period
53. 0	4 centimeters	2 seconds
54. 0	3 meters	6 seconds
55. 3 inches	3 inches	1.5 seconds
56. 2 feet	2 feet	10 seconds

HARMONIC MOTION In Exercises 57–60, for the simple harmonic motion described by the trigonometric function, find (a) the maximum displacement, (b) the frequency, (c) the value of d when $t = 5$, and (d) the least positive value of t for which $d = 0$. Use a graphing utility to verify your results.

57. $d = 9 \cos \frac{6\pi}{5} t$ **58.** $d = \frac{1}{2} \cos 20\pi t$

59. $d = \frac{1}{4} \sin 6\pi t$ **60.** $d = \frac{1}{64} \sin 792\pi t$

61. TUNING FORK A point on the end of a tuning fork moves in simple harmonic motion described by $d = a \sin \omega t$. Find ω given that the tuning fork for middle C has a frequency of 264 vibrations per second.

62. WAVE MOTION A buoy oscillates in simple harmonic motion as waves go past. It is noted that the buoy moves a total of 3.5 feet from its low point to its high point (see figure), and that it returns to its high point every 10 seconds. Write an equation that describes the motion of the buoy if its high point is at $t = 0$.

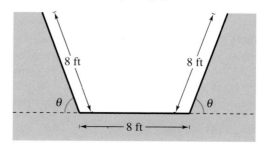

63. OSCILLATION OF A SPRING A ball that is bobbing up and down on the end of a spring has a maximum displacement of 3 inches. Its motion (in ideal conditions) is modeled by $y = \frac{1}{4} \cos 16t$ ($t > 0$), where y is measured in feet and t is the time in seconds.

(a) Graph the function.

(b) What is the period of the oscillations?

(c) Determine the first time the weight passes the point of equilibrium ($y = 0$).

64. NUMERICAL AND GRAPHICAL ANALYSIS The cross section of an irrigation canal is an isosceles trapezoid of which 3 of the sides are 8 feet long (see figure). The objective is to find the angle θ that maximizes the area of the cross section. [*Hint:* The area of a trapezoid is $(h/2)(b_1 + b_2)$.]

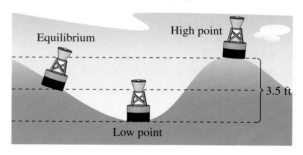

(a) Complete seven additional rows of the table.

Base 1	Base 2	Altitude	Area
8	$8 + 16 \cos 10°$	$8 \sin 10°$	22.1
8	$8 + 16 \cos 20°$	$8 \sin 20°$	42.5

(b) Use a graphing utility to generate additional rows of the table. Use the table to estimate the maximum cross-sectional area.

(c) Write the area A as a function of θ.

(d) Use a graphing utility to graph the function. Use the graph to estimate the maximum cross-sectional area. How does your estimate compare with that of part (b)?

65. NUMERICAL AND GRAPHICAL ANALYSIS A 2-meter-high fence is 3 meters from the side of a grain storage bin. A grain elevator must reach from ground level outside the fence to the storage bin (see figure). The objective is to determine the shortest elevator that meets the constraints.

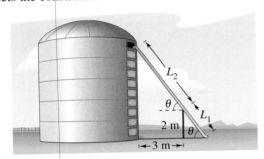

(a) Complete four rows of the table.

θ	L_1	L_2	$L_1 + L_2$
0.1	$\dfrac{2}{\sin 0.1}$	$\dfrac{3}{\cos 0.1}$	23.0
0.2	$\dfrac{2}{\sin 0.2}$	$\dfrac{3}{\cos 0.2}$	13.1

(b) Use a graphing utility to generate additional rows of the table. Use the table to estimate the minimum length of the elevator.

(c) Write the length $L_1 + L_2$ as a function of θ.

(d) Use a graphing utility to graph the function. Use the graph to estimate the minimum length. How does your estimate compare with that of part (b)?

66. DATA ANALYSIS The table shows the average sales S (in millions of dollars) of an outerwear manufacturer for each month t, where $t = 1$ represents January.

Time, t	1	2	3	4	5	6
Sales, S	13.46	11.15	8.00	4.85	2.54	1.70

Time, t	7	8	9	10	11	12
Sales, S	2.54	4.85	8.00	11.15	13.46	14.30

(a) Create a scatter plot of the data.

(b) Find a trigonometric model that fits the data. Graph the model with your scatter plot. How well does the model fit the data?

(c) What is the period of the model? Do you think it is reasonable given the context? Explain your reasoning.

(d) Interpret the meaning of the model's amplitude in the context of the problem.

67. DATA ANALYSIS The number of hours H of daylight in Denver, Colorado on the 15th of each month are: 1(9.67), 2(10.72), 3(11.92), 4(13.25), 5(14.37), 6(14.97), 7(14.72), 8(13.77), 9(12.48), 10(11.18), 11(10.00), 12(9.38). The month is represented by t, with $t = 1$ corresponding to January. A model for the data is given by

$$H(t) = 12.13 + 2.77 \sin[(\pi t/6) - 1.60].$$

(a) Use a graphing utility to graph the data points and the model in the same viewing window.

(b) What is the period of the model? Is it what you expected? Explain.

(c) What is the amplitude of the model? What does it represent in the context of the problem? Explain.

EXPLORATION

68. CAPSTONE While walking across flat land, you notice a wind turbine tower of height h feet directly in front of you. The angle of elevation to the top of the tower is A degrees. After you walk d feet closer to the tower, the angle of elevation increases to B degrees.

(a) Draw a diagram to represent the situation.

(b) Write an expression for the height h of the tower in terms of the angles A and B and the distance d.

TRUE OR FALSE? In Exercises 69 and 70, determine whether the statement is true or false. Justify your answer.

69. The Leaning Tower of Pisa is not vertical, but if you know the angle of elevation θ to the top of the tower when you stand d feet away from it, you can find its height h using the formula $h = d \tan \theta$.

70. N 24° E means 24 degrees north of east.

4 CHAPTER SUMMARY

What Did You Learn?	Explanation/Examples	Review Exercises
Section 4.1		
Describe angles (p. 280).		1–8
Convert between degrees and radians (p. 284).	To convert degrees to radians, multiply degrees by $(\pi \text{ rad})/180°$. To convert radians to degrees, multiply radians by $180°/(\pi \text{ rad})$.	9–20
Use angles to model and solve real-life problems (p. 285).	Angles can be used to find the length of a circular arc and the area of a sector of a circle. (See Examples 5 and 8.)	21–24
Section 4.2		
Identify a unit circle and describe its relationship to real numbers (p. 292).		25–28
Evaluate trigonometric functions using the unit circle (p. 293).	$t = \dfrac{2\pi}{3}$ corresponds to $(x, y) = \left(-\dfrac{1}{2}, \dfrac{\sqrt{3}}{2}\right)$. So $$\cos\frac{2\pi}{3} = -\frac{1}{2}, \ \sin\frac{2\pi}{3} = \frac{\sqrt{3}}{2}, \text{ and } \tan\frac{2\pi}{3} = -\sqrt{3}.$$	29–32
Use domain and period to evaluate sine and cosine functions (p. 295).	Because $\dfrac{9\pi}{4} = 2\pi + \dfrac{\pi}{4}$, $\sin\dfrac{9\pi}{4} = \sin\dfrac{\pi}{4} = \dfrac{\sqrt{2}}{2}$.	33–36
Use a calculator to evaluate trigonometric functions (p. 296).	$\sin\dfrac{3\pi}{8} \approx 0.9239, \ \cot(-1.2) \approx -0.3888$	37–40
Section 4.3		
Evaluate trigonometric functions of acute angles (p. 299).	$\sin\theta = \dfrac{\text{opp}}{\text{hyp}}, \quad \cos\theta = \dfrac{\text{adj}}{\text{hyp}}, \quad \tan\theta = \dfrac{\text{opp}}{\text{adj}}$ $\csc\theta = \dfrac{\text{hyp}}{\text{opp}}, \quad \sec\theta = \dfrac{\text{hyp}}{\text{adj}}, \quad \cot\theta = \dfrac{\text{adj}}{\text{opp}}$	41, 42
Use fundamental trigonometric identities (p. 302).	$\sin\theta = \dfrac{1}{\csc\theta}, \quad \tan\theta = \dfrac{\sin\theta}{\cos\theta}, \ \sin^2\theta + \cos^2\theta = 1$	43–46
Use a calculator to evaluate trigonometric functions (p. 303).	$\tan 34.7° \approx 0.6924, \ \csc 29° \ 15' \approx 2.0466$	47–54
Use trigonometric functions to model and solve real-life problems (p. 304).	Trigonometric functions can be used to find the height of a monument, the angle between two paths, and the length of a ramp. (See Examples 7–9.)	55, 56

What Did You Learn?	Explanation/Examples	Review Exercises
Section 4.4 Evaluate trigonometric functions of any angle (p. 310).	Let (3, 4) be a point on the terminal side of θ. Then $\sin\theta = \frac{4}{5}$, $\cos\theta = \frac{3}{5}$, and $\tan\theta = \frac{4}{3}$.	57–70
Find reference angles (p. 312).	Let θ be an angle in standard position. Its reference angle is the acute angle θ' formed by the terminal side of θ and the horizontal axis.	71–74
Evaluate trigonometric functions of real numbers (p. 313).	$\cos\dfrac{7\pi}{3} = \dfrac{1}{2}$ because $\theta' = \dfrac{7\pi}{3} - 2\pi = \dfrac{\pi}{3}$. So, $\cos\dfrac{7\pi}{3} = \cos\dfrac{\pi}{3} = \dfrac{1}{2}$.	75–84
Section 4.5 Sketch the graphs of sine and cosine functions using amplitude and period (p. 319).		85–88
Sketch translations of the graphs of sine and cosine functions (p. 323).	For $y = d + a\sin(bx - c)$ and $y = d + a\cos(bx - c)$, the constant c creates a horizontal translation. The constant d creates a vertical translation. (See Examples 4–6.)	89–92
Use sine and cosine functions to model real-life data (p. 325).	A cosine function can be used to model the depth of the water at the end of a dock at various times. (See Example 7.)	93, 94
Section 4.6 Sketch the graphs of tangent (p. 330), cotangent (p. 332), secant (p. 333), and cosecant (p. 333), functions.		95–102
Sketch the graphs of damped trigonometric functions (p. 335).	For $f(x) = x\cos 2x$ and $g(x) = \log x \sin 4x$, the factors x and $\log x$ are called damping factors.	103, 104
Section 4.7 Evaluate and graph inverse trigonometric functions (p. 341).	$\sin^{-1}\dfrac{1}{2} = \dfrac{\pi}{6}$, $\cos^{-1}\left(-\dfrac{\sqrt{2}}{2}\right) = \dfrac{3\pi}{4}$, $\tan^{-1}\sqrt{3} = \dfrac{\pi}{3}$	105–122, 131–138
Evaluate and graph the compositions of trigonometric functions (p. 345).	$\cos[\arctan(5/12)] = 12/13$, $\sin(\sin^{-1}0.4) = 0.4$	123–130
Section 4.8 Solve real-life problems involving right triangles (p. 351).	A trigonometric function can be used to find the height of a smokestack on top of a building. (See Example 3.)	139, 140
Solve real-life problems involving directional bearings (p. 353).	Trigonometric functions can be used to find a ship's bearing and distance from a port at a given time. (See Example 5.)	141
Solve real-life problems involving harmonic motion (p. 354).	Sine or cosine functions can be used to describe the motion of an object that vibrates, oscillates, rotates, or is moved by wave motion. (See Examples 6 and 7.)	142

4 REVIEW EXERCISES

See www.CalcChat.com for worked-out solutions to odd-numbered exercises.

4.1 In Exercises 1–8, (a) sketch the angle in standard position, (b) determine the quadrant in which the angle lies, and (c) determine one positive and one negative coterminal angle.

1. $15\pi/4$
2. $2\pi/9$
3. $-4\pi/3$
4. $-23\pi/3$
5. $70°$
6. $280°$
7. $-110°$
8. $-405°$

In Exercises 9–12, convert the angle measure from degrees to radians. Round your answer to three decimal places.

9. $450°$
10. $-112.5°$
11. $-33° \, 45'$
12. $197° \, 17'$

In Exercises 13–16, convert the angle measure from radians to degrees. Round your answer to three decimal places.

13. $3\pi/10$
14. $-11\pi/6$
15. -3.5
16. 5.7

In Exercises 17–20, convert each angle measure to degrees, minutes, and seconds without using a calculator.

17. $198.4°$
18. $-70.2°$
19. $0.65°$
20. $-5.96°$

21. **ARC LENGTH** Find the length of the arc on a circle with a radius of 20 inches intercepted by a central angle of $138°$.

22. **PHONOGRAPH** Phonograph records are vinyl discs that rotate on a turntable. A typical record album is 12 inches in diameter and plays at $33\frac{1}{3}$ revolutions per minute.
    (a) What is the angular speed of a record album?
    (b) What is the linear speed of the outer edge of a record album?

23. **CIRCULAR SECTOR** Find the area of the sector of a circle with a radius of 18 inches and central angle $\theta = 120°$.

24. **CIRCULAR SECTOR** Find the area of the sector of a circle with a radius of 6.5 millimeters and central angle $\theta = 5\pi/6$.

4.2 In Exercises 25–28, find the point (x, y) on the unit circle that corresponds to the real number t.

25. $t = 2\pi/3$
26. $t = 7\pi/4$
27. $t = 7\pi/6$
28. $t = -4\pi/3$

In Exercises 29–32, evaluate (if possible) the six trigonometric functions of the real number.

29. $t = 7\pi/6$
30. $t = 3\pi/4$
31. $t = -2\pi/3$
32. $t = 2\pi$

In Exercises 33–36, evaluate the trigonometric function using its period as an aid.

33. $\sin(11\pi/4)$
34. $\cos 4\pi$
35. $\sin(-17\pi/6)$
36. $\cos(-13\pi/3)$

 In Exercises 37–40, use a calculator to evaluate the trigonometric function. Round your answer to four decimal places.

37. $\tan 33$
38. $\csc 10.5$
39. $\sec(12\pi/5)$
40. $\sin(-\pi/9)$

4.3 In Exercises 41 and 42, find the exact values of the six trigonometric functions of the angle θ shown in the figure.

41.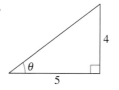
42.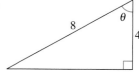

In Exercises 43–46, use the given function value and trigonometric identities (including the cofunction identities) to find the indicated trigonometric functions.

43. $\sin \theta = \frac{1}{3}$ (a) $\csc \theta$ (b) $\cos \theta$
               (c) $\sec \theta$ (d) $\tan \theta$
44. $\tan \theta = 4$ (a) $\cot \theta$ (b) $\sec \theta$
               (c) $\cos \theta$ (d) $\csc \theta$
45. $\csc \theta = 4$ (a) $\sin \theta$ (b) $\cos \theta$
               (c) $\sec \theta$ (d) $\tan \theta$
46. $\csc \theta = 5$ (a) $\sin \theta$ (b) $\cot \theta$
               (c) $\tan \theta$ (d) $\sec(90° - \theta)$

 In Exercises 47–54, use a calculator to evaluate the trigonometric function. Round your answer to four decimal places.

47. $\tan 33°$
48. $\csc 11°$
49. $\sin 34.2°$
50. $\sec 79.3°$
51. $\cot 15° \, 14'$
52. $\csc 44° \, 35'$
53. $\tan 31° \, 24' \, 5''$
54. $\cos 78° \, 11' \, 58''$

55. **RAILROAD GRADE** A train travels 3.5 kilometers on a straight track with a grade of $1° \, 10'$ (see figure on the next page). What is the vertical rise of the train in that distance?

3.5 km

1°10′

Not drawn to scale

FIGURE FOR 55

56. GUY WIRE A guy wire runs from the ground to the top of a 25-foot telephone pole. The angle formed between the wire and the ground is 52°. How far from the base of the pole is the wire attached to the ground?

4.4 In Exercises 57–64, the point is on the terminal side of an angle θ in standard position. Determine the exact values of the six trigonometric functions of the angle θ.

57. $(12, 16)$ **58.** $(3, -4)$

59. $\left(\frac{2}{3}, \frac{5}{2}\right)$ **60.** $\left(-\frac{10}{3}, -\frac{2}{3}\right)$

61. $(-0.5, 4.5)$ **62.** $(0.3, 0.4)$

63. $(x, 4x), \quad x > 0$ **64.** $(-2x, -3x), \quad x > 0$

In Exercises 65–70, find the values of the remaining five trigonometric functions of θ.

Function Value	Constraint
65. $\sec \theta = \frac{6}{5}$	$\tan \theta < 0$
66. $\csc \theta = \frac{3}{2}$	$\cos \theta < 0$
67. $\sin \theta = \frac{3}{8}$	$\cos \theta < 0$
68. $\tan \theta = \frac{5}{4}$	$\cos \theta < 0$
69. $\cos \theta = -\frac{2}{5}$	$\sin \theta > 0$
70. $\sin \theta = -\frac{1}{2}$	$\cos \theta > 0$

In Exercises 71–74, find the reference angle θ' and sketch θ and θ' in standard position.

71. $\theta = 264°$ **72.** $\theta = 635°$

73. $\theta = -6\pi/5$ **74.** $\theta = 17\pi/3$

In Exercises 75–80, evaluate the sine, cosine, and tangent of the angle without using a calculator.

75. $\pi/3$ **76.** $\pi/4$

77. $-7\pi/3$ **78.** $-5\pi/4$

79. $495°$ **80.** $-150°$

In Exercises 81–84, use a calculator to evaluate the trigonometric function. Round your answer to four decimal places.

81. $\sin 4$ **82.** $\cot(-4.8)$

83. $\sin(12\pi/5)$ **84.** $\tan(-25\pi/7)$

4.5 In Exercises 85–92, sketch the graph of the function. Include two full periods.

85. $y = \sin 6x$ **86.** $y = -\cos 3x$

87. $f(x) = 5 \sin(2x/5)$ **88.** $f(x) = 8 \cos(-x/4)$

89. $y = 5 + \sin x$ **90.** $y = -4 - \cos \pi x$

91. $g(t) = \frac{5}{2} \sin(t - \pi)$ **92.** $g(t) = 3 \cos(t + \pi)$

93. SOUND WAVES Sound waves can be modeled by sine functions of the form $y = a \sin bx$, where x is measured in seconds.

  (a) Write an equation of a sound wave whose amplitude is 2 and whose period is $\frac{1}{264}$ second.

  (b) What is the frequency of the sound wave described in part (a)?

94. DATA ANALYSIS: METEOROLOGY The times S of sunset (Greenwich Mean Time) at 40° north latitude on the 15th of each month are: 1(16:59), 2(17:35), 3(18:06), 4(18:38), 5(19:08), 6(19:30), 7(19:28), 8(18:57), 9(18:09), 10(17:21), 11(16:44), 12(16:36). The month is represented by t, with $t = 1$ corresponding to January. A model (in which minutes have been converted to the decimal parts of an hour) for the data is $S(t) = 18.09 + 1.41 \sin[(\pi t/6) + 4.60]$.

  (a) Use a graphing utility to graph the data points and the model in the same viewing window.

  (b) What is the period of the model? Is it what you expected? Explain.

  (c) What is the amplitude of the model? What does it represent in the model? Explain.

4.6 In Exercises 95–102, sketch a graph of the function. Include two full periods.

95. $f(x) = 3 \tan 2x$ **96.** $f(t) = \tan\left(t + \dfrac{\pi}{2}\right)$

97. $f(x) = \frac{1}{2} \cot x$ **98.** $g(t) = 2 \cot 2t$

99. $f(x) = 3 \sec x$ **100.** $h(t) = \sec\left(t - \dfrac{\pi}{4}\right)$

101. $f(x) = \dfrac{1}{2} \csc \dfrac{x}{2}$ **102.** $f(t) = 3 \csc\left(2t + \dfrac{\pi}{4}\right)$

In Exercises 103 and 104, use a graphing utility to graph the function and the damping factor of the function in the same viewing window. Describe the behavior of the function as x increases without bound.

103. $f(x) = x \cos x$ **104.** $g(x) = x^4 \cos x$

4.7 In Exercises 105–110, evaluate the expression. If necessary, round your answer to two decimal places.

105. $\arcsin\left(-\frac{1}{2}\right)$ **106.** $\arcsin(-1)$

107. $\arcsin 0.4$ **108.** $\arcsin 0.213$

109. $\sin^{-1}(-0.44)$ **110.** $\sin^{-1} 0.89$

In Exercises 111–114, evaluate the expression without using a calculator.

111. $\arccos\left(-\sqrt{2}/2\right)$

112. $\arccos\left(\sqrt{2}/2\right)$

113. $\cos^{-1}(-1)$

114. $\cos^{-1}\left(\sqrt{3}/2\right)$

 In Exercises 115–118, use a calculator to evaluate the expression. Round your answer to two decimal places.

115. $\arccos 0.324$

116. $\arccos(-0.888)$

117. $\tan^{-1}(-1.5)$

118. $\tan^{-1} 8.2$

 In Exercises 119–122, use a graphing utility to graph the function.

119. $f(x) = 2\arcsin x$

120. $f(x) = 3\arccos x$

121. $f(x) = \arctan(x/2)$

122. $f(x) = -\arcsin 2x$

In Exercises 123–128, find the exact value of the expression.

123. $\cos\left(\arctan \frac{3}{4}\right)$

124. $\tan\left(\arccos \frac{3}{5}\right)$

125. $\sec\left(\tan^{-1} \frac{12}{5}\right)$

126. $\sec\left[\sin^{-1}\left(-\frac{1}{4}\right)\right]$

127. $\cot\left(\arctan \frac{7}{10}\right)$

128. $\cot\left[\arcsin\left(-\frac{12}{13}\right)\right]$

In Exercises 129 and 130, write an algebraic expression that is equivalent to the expression.

129. $\tan[\arccos (x/2)]$

130. $\sec[\arcsin(x - 1)]$

In Exercises 131–134, evaluate each expression without using a calculator.

131. $\operatorname{arccot} \sqrt{3}$

132. $\operatorname{arcsec}(-1)$

133. $\operatorname{arcsec}\left(-\sqrt{2}\right)$

134. $\operatorname{arccsc} 1$

 In Exercises 135–138, use a calculator to approximate the value of the expression. Round your result to two decimal places.

135. $\operatorname{arccot}(10.5)$

136. $\operatorname{arcsec}(-7.5)$

137. $\operatorname{arcsec}\left(-\frac{5}{2}\right)$

138. $\operatorname{arccsc}(-2.01)$

139. ANGLE OF ELEVATION The height of a radio transmission tower is 70 meters, and it casts a shadow of length 30 meters. Draw a diagram and find the angle of elevation of the sun.

140. HEIGHT Your football has landed at the edge of the roof of your school building. When you are 25 feet from the base of the building, the angle of elevation to your football is 21°. How high off the ground is your football?

141. DISTANCE From city A to city B, a plane flies 650 miles at a bearing of 48°. From city B to city C, the plane flies 810 miles at a bearing of 115°. Find the distance from city A to city C and the bearing from city A to city C.

142. WAVE MOTION Your fishing bobber oscillates in simple harmonic motion from the waves in the lake where you fish. Your bobber moves a total of 1.5 inches from its high point to its low point and returns to its high point every 3 seconds. Write an equation modeling the motion of your bobber if it is at its high point at time $t = 0$.

EXPLORATION

TRUE OR FALSE? In Exercises 143 and 144, determine whether the statement is true or false. Justify your answer.

143. $y = \sin \theta$ is not a function because $\sin 30° = \sin 150°$.

144. Because $\tan 3\pi/4 = -1$, $\arctan(-1) = 3\pi/4$.

145. WRITING Describe the behavior of $f(\theta) = \sec \theta$ at the zeros of $g(\theta) = \cos \theta$. Explain your reasoning.

146. CONJECTURE

(a) Use a graphing utility to complete the table.

θ	0.1	0.4	0.7	1.0	1.3
$\tan\left(\theta - \dfrac{\pi}{2}\right)$					
$-\cot \theta$					

(b) Make a conjecture about the relationship between $\tan[\theta - (\pi/2)]$ and $-\cot \theta$.

147. WRITING When graphing the sine and cosine functions, determining the amplitude is part of the analysis. Explain why this is not true for the other four trigonometric functions.

148. OSCILLATION OF A SPRING A weight is suspended from a ceiling by a steel spring. The weight is lifted (positive direction) from the equilibrium position and released. The resulting motion of the weight is modeled by $y = Ae^{-kt}\cos bt = \frac{1}{5}e^{-t/10}\cos 6t$, where y is the distance in feet from equilibrium and t is the time in seconds. The graph of the function is shown in the figure. For each of the following, describe the change in the system without graphing the resulting function.

(a) A is changed from $\frac{1}{5}$ to $\frac{1}{3}$.

(b) k is changed from $\frac{1}{10}$ to $\frac{1}{3}$.

(c) b is changed from 6 to 9.

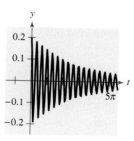

4 CHAPTER TEST

See www.CalcChat.com for worked-out solutions to odd-numbered exercises.

Take this test as you would take a test in class. When you are finished, check your work against the answers given in the back of the book.

1. Consider an angle that measures $\dfrac{5\pi}{4}$ radians.

    (a) Sketch the angle in standard position.

    (b) Determine two coterminal angles (one positive and one negative).

    (c) Convert the angle to degree measure.

2. A truck is moving at a rate of 105 kilometers per hour, and the diameter of its wheels is 1 meter. Find the angular speed of the wheels in radians per minute.

3. A water sprinkler sprays water on a lawn over a distance of 25 feet and rotates through an angle of 130°. Find the area of the lawn watered by the sprinkler.

4. Find the exact values of the six trigonometric functions of the angle θ shown in the figure.

5. Given that $\tan \theta = \frac{3}{2}$, find the other five trigonometric functions of θ.

6. Determine the reference angle θ' for the angle $\theta = 205°$ and sketch θ and θ' in standard position.

7. Determine the quadrant in which θ lies if $\sec \theta < 0$ and $\tan \theta > 0$.

8. Find two exact values of θ in degrees ($0 \le \theta < 360°$) if $\cos \theta = -\sqrt{3}/2$. (Do not use a calculator.)

9. Use a calculator to approximate two values of θ in radians ($0 \le \theta < 2\pi$) if $\csc \theta = 1.030$. Round the results to two decimal places.

In Exercises 10 and 11, find the remaining five trigonometric functions of θ satisfying the conditions.

10. $\cos \theta = \frac{3}{5}$, $\tan \theta < 0$

11. $\sec \theta = -\frac{29}{20}$, $\sin \theta > 0$

In Exercises 12 and 13, sketch the graph of the function. (Include two full periods.)

12. $g(x) = -2 \sin\left(x - \dfrac{\pi}{4}\right)$

13. $f(\alpha) = \dfrac{1}{2} \tan 2\alpha$

In Exercises 14 and 15, use a graphing utility to graph the function. If the function is periodic, find its period.

14. $y = \sin 2\pi x + 2 \cos \pi x$

15. $y = 6e^{-0.12t} \cos(0.25t)$, $0 \le t \le 32$

16. Find a, b, and c for the function $f(x) = a \sin(bx + c)$ such that the graph of f matches the figure.

17. Find the exact value of $\cot\left(\arcsin \frac{3}{8}\right)$ without the aid of a calculator.

18. Graph the function $f(x) = 2 \arcsin\left(\frac{1}{2}x\right)$.

19. A plane is 90 miles south and 110 miles east of London Heathrow Airport. What bearing should be taken to fly directly to the airport?

20. Write the equation for the simple harmonic motion of a ball on a spring that starts at its lowest point of 6 inches below equilibrium, bounces to its maximum height of 6 inches above equilibrium, and returns to its lowest point in a total of 2 seconds.

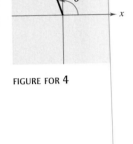

$(-2, 6)$

FIGURE FOR **4**

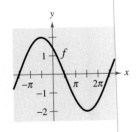

FIGURE FOR **16**

PROOFS IN MATHEMATICS

The Pythagorean Theorem

The Pythagorean Theorem is one of the most famous theorems in mathematics. More than 100 different proofs now exist. James A. Garfield, the twentieth president of the United States, developed a proof of the Pythagorean Theorem in 1876. His proof, shown below, involved the fact that a trapezoid can be formed from two congruent right triangles and an isosceles right triangle.

The Pythagorean Theorem

In a right triangle, the sum of the squares of the lengths of the legs is equal to the square of the length of the hypotenuse, where a and b are the legs and c is the hypotenuse.

$$a^2 + b^2 = c^2$$

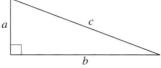

Proof

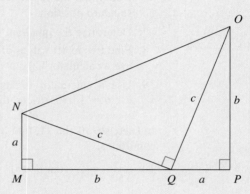

$$\text{Area of trapezoid } MNOP = \text{Area of } \triangle MNQ + \text{Area of } \triangle PQO + \text{Area of } \triangle NOQ$$

$$\frac{1}{2}(a + b)(a + b) = \frac{1}{2}ab + \frac{1}{2}ab + \frac{1}{2}c^2$$

$$\frac{1}{2}(a + b)(a + b) = ab + \frac{1}{2}c^2$$

$$(a + b)(a + b) = 2ab + c^2$$

$$a^2 + 2ab + b^2 = 2ab + c^2$$

$$a^2 + b^2 = c^2$$

PROBLEM SOLVING

This collection of thought-provoking and challenging exercises further explores and expands upon concepts learned in this chapter.

1. The restaurant at the top of the Space Needle in Seattle, Washington is circular and has a radius of 47.25 feet. The dining part of the restaurant revolves, making about one complete revolution every 48 minutes. A dinner party was seated at the edge of the revolving restaurant at 6:45 P.M. and was finished at 8:57 P.M.

   (a) Find the angle through which the dinner party rotated.

   (b) Find the distance the party traveled during dinner.

2. A bicycle's gear ratio is the number of times the freewheel turns for every one turn of the chainwheel (see figure). The table shows the numbers of teeth in the freewheel and chainwheel for the first five gears of an 18-speed touring bicycle. The chainwheel completes one rotation for each gear. Find the angle through which the freewheel turns for each gear. Give your answers in both degrees and radians.

Gear number	Number of teeth in freewheel	Number of teeth in chainwheel
1	32	24
2	26	24
3	22	24
4	32	40
5	19	24

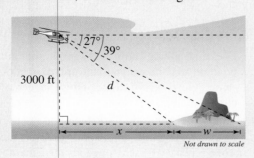

Freewheel

Chainwheel

3. A surveyor in a helicopter is trying to determine the width of an island, as shown in the figure.

3000 ft

d

x

w

Not drawn to scale

(a) What is the shortest distance d the helicopter would have to travel to land on the island?

(b) What is the horizontal distance x that the helicopter would have to travel before it would be directly over the nearer end of the island?

(c) Find the width w of the island. Explain how you obtained your answer.

4. Use the figure below.

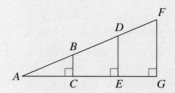

(a) Explain why $\triangle ABC$, $\triangle ADE$, and $\triangle AFG$ are similar triangles.

(b) What does similarity imply about the ratios

$$\frac{BC}{AB}, \frac{DE}{AD}, \text{ and } \frac{FG}{AF}?$$

(c) Does the value of $\sin A$ depend on which triangle from part (a) is used to calculate it? Would the value of $\sin A$ change if it were found using a different right triangle that was similar to the three given triangles?

(d) Do your conclusions from part (c) apply to the other five trigonometric functions? Explain.

5. Use a graphing utility to graph h, and use the graph to decide whether h is even, odd, or neither.

   (a) $h(x) = \cos^2 x$

   (b) $h(x) = \sin^2 x$

6. If f is an even function and g is an odd function, use the results of Exercise 5 to make a conjecture about h, where

   (a) $h(x) = [f(x)]^2$

   (b) $h(x) = [g(x)]^2$.

7. The model for the height h (in feet) of a Ferris wheel car is

   $$h = 50 + 50 \sin 8\pi t$$

   where t is the time (in minutes). (The Ferris wheel has a radius of 50 feet.) This model yields a height of 50 feet when $t = 0$. Alter the model so that the height of the car is 1 foot when $t = 0$.

8. The pressure P (in millimeters of mercury) against the walls of the blood vessels of a patient is modeled by

$$P = 100 - 20 \cos\left(\frac{8\pi}{3}t\right)$$

where t is time (in seconds).

(a) Use a graphing utility to graph the model.

(b) What is the period of the model? What does the period tell you about this situation?

(c) What is the amplitude of the model? What does it tell you about this situation?

(d) If one cycle of this model is equivalent to one heartbeat, what is the pulse of this patient?

(e) If a physician wants this patient's pulse rate to be 64 beats per minute or less, what should the period be? What should the coefficient of t be?

9. A popular theory that attempts to explain the ups and downs of everyday life states that each of us has three cycles, called biorhythms, which begin at birth. These three cycles can be modeled by sine waves.

Physical (23 days): $P = \sin\dfrac{2\pi t}{23}, \quad t \geq 0$

Emotional (28 days): $E = \sin\dfrac{2\pi t}{28}, \quad t \geq 0$

Intellectual (33 days): $I = \sin\dfrac{2\pi t}{33}, \quad t \geq 0$

where t is the number of days since birth. Consider a person who was born on July 20, 1988.

(a) Use a graphing utility to graph the three models in the same viewing window for $7300 \leq t \leq 7380$.

(b) Describe the person's biorhythms during the month of September 2008.

(c) Calculate the person's three energy levels on September 22, 2008.

10. (a) Use a graphing utility to graph the functions given by

$$f(x) = 2 \cos 2x + 3 \sin 3x \quad \text{and}$$

$$g(x) = 2 \cos 2x + 3 \sin 4x.$$

(b) Use the graphs from part (a) to find the period of each function.

(c) If α and β are positive integers, is the function given by $h(x) = A \cos \alpha x + B \sin \beta x$ periodic? Explain your reasoning.

11. Two trigonometric functions f and g have periods of 2, and their graphs intersect at $x = 5.35$.

(a) Give one smaller and one larger positive value of x at which the functions have the same value.

(b) Determine one negative value of x at which the graphs intersect.

(c) Is it true that $f(13.35) = g(-4.65)$? Explain your reasoning.

12. The function f is periodic, with period c. So, $f(t + c) = f(t)$. Are the following equal? Explain.

(a) $f(t - 2c) = f(t)$

(b) $f\left(t + \frac{1}{2}c\right) = f\left(\frac{1}{2}t\right)$

(c) $f\left(\frac{1}{2}(t + c)\right) = f\left(\frac{1}{2}t\right)$

13. If you stand in shallow water and look at an object below the surface of the water, the object will look farther away from you than it really is. This is because when light rays pass between air and water, the water refracts, or bends, the light rays. The index of refraction for water is 1.333. This is the ratio of the sine of θ_1 and the sine of θ_2 (see figure).

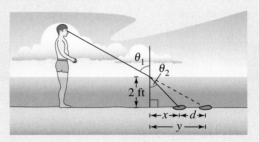

(a) You are standing in water that is 2 feet deep and are looking at a rock at angle $\theta_1 = 60°$ (measured from a line perpendicular to the surface of the water). Find θ_2.

(b) Find the distances x and y.

(c) Find the distance d between where the rock is and where it appears to be.

(d) What happens to d as you move closer to the rock? Explain your reasoning.

14. In calculus, it can be shown that the arctangent function can be approximated by the polynomial

$$\arctan x \approx x - \frac{x^3}{3} + \frac{x^5}{5} - \frac{x^7}{7}$$

where x is in radians.

(a) Use a graphing utility to graph the arctangent function and its polynomial approximation in the same viewing window. How do the graphs compare?

(b) Study the pattern in the polynomial approximation of the arctangent function and guess the next term. Then repeat part (a). How does the accuracy of the approximation change when additional terms are added?

Analytic Trigonometry

In Mathematics

Analytic trigonometry is used to simplify trigonometric expressions and solve trigonometric equations.

In Real Life

Analytic trigonometry is used to model real-life phenomena. For instance, when an airplane travels faster than the speed of sound, the sound waves form a cone behind the airplane. Concepts of trigonometry can be used to describe the apex angle of the cone. (See Exercise 137, page 415.)

Christopher Pasatier/Reuters/Landow

IN CAREERS

There are many careers that use analytic trigonometry. Several are listed below.

- Mechanical Engineer
  Exercise 89, page 396

- Physicist
  Exercise 90, page 403

- Bridge Designer
  Exercise 49, page 423

- Surveyor
  Exercise 43, page 430

5.1 USING FUNDAMENTAL IDENTITIES

Introduction

In Chapter 4, you studied the basic definitions, properties, graphs, and applications of the individual trigonometric functions. In this chapter, you will learn how to use the fundamental identities to do the following.

1. Evaluate trigonometric functions.
2. Simplify trigonometric expressions.
3. Develop additional trigonometric identities.
4. Solve trigonometric equations.

Fundamental Trigonometric Identities

Reciprocal Identities

$$\sin u = \frac{1}{\csc u} \qquad \cos u = \frac{1}{\sec u} \qquad \tan u = \frac{1}{\cot u}$$

$$\csc u = \frac{1}{\sin u} \qquad \sec u = \frac{1}{\cos u} \qquad \cot u = \frac{1}{\tan u}$$

Quotient Identities

$$\tan u = \frac{\sin u}{\cos u} \qquad \cot u = \frac{\cos u}{\sin u}$$

Pythagorean Identities

$$\sin^2 u + \cos^2 u = 1 \qquad 1 + \tan^2 u = \sec^2 u \qquad 1 + \cot^2 u = \csc^2 u$$

Cofunction Identities

$$\sin\left(\frac{\pi}{2} - u\right) = \cos u \qquad \cos\left(\frac{\pi}{2} - u\right) = \sin u$$

$$\tan\left(\frac{\pi}{2} - u\right) = \cot u \qquad \cot\left(\frac{\pi}{2} - u\right) = \tan u$$

$$\sec\left(\frac{\pi}{2} - u\right) = \csc u \qquad \csc\left(\frac{\pi}{2} - u\right) = \sec u$$

Even/Odd Identities

$$\sin(-u) = -\sin u \qquad \cos(-u) = \cos u \qquad \tan(-u) = -\tan u$$

$$\csc(-u) = -\csc u \qquad \sec(-u) = \sec u \qquad \cot(-u) = -\cot u$$

Pythagorean identities are sometimes used in radical form such as

$$\sin u = \pm\sqrt{1 - \cos^2 u}$$

or

$$\tan u = \pm\sqrt{\sec^2 u - 1}$$

where the sign depends on the choice of u.

Using the Fundamental Identities

One common application of trigonometric identities is to use given values of trigonometric functions to evaluate other trigonometric functions.

Example 1   Using Identities to Evaluate a Function

Use the values $\sec u = -\frac{3}{2}$ and $\tan u > 0$ to find the values of all six trigonometric functions.

Solution

Using a reciprocal identity, you have

$$\cos u = \frac{1}{\sec u} = \frac{1}{-3/2} = -\frac{2}{3}.$$

Using a Pythagorean identity, you have

$$\sin^2 u = 1 - \cos^2 u \qquad \text{Pythagorean identity}$$

$$= 1 - \left(-\frac{2}{3}\right)^2 \qquad \text{Substitute } -\frac{2}{3} \text{ for } \cos u.$$

$$= 1 - \frac{4}{9} = \frac{5}{9}. \qquad \text{Simplify.}$$

Because $\sec u < 0$ and $\tan u > 0$, it follows that u lies in Quadrant III. Moreover, because $\sin u$ is negative when u is in Quadrant III, you can choose the negative root and obtain $\sin u = -\sqrt{5}/3$. Now, knowing the values of the sine and cosine, you can find the values of all six trigonometric functions.

$$\sin u = -\frac{\sqrt{5}}{3} \qquad\qquad \csc u = \frac{1}{\sin u} = -\frac{3}{\sqrt{5}} = -\frac{3\sqrt{5}}{5}$$

$$\cos u = -\frac{2}{3} \qquad\qquad \sec u = \frac{1}{\cos u} = -\frac{3}{2}$$

$$\tan u = \frac{\sin u}{\cos u} = \frac{-\sqrt{5}/3}{-2/3} = \frac{\sqrt{5}}{2} \qquad \cot u = \frac{1}{\tan u} = \frac{2}{\sqrt{5}} = \frac{2\sqrt{5}}{5}$$

CHECK**Point**   Now try Exercise 21.

Example 2   Simplifying a Trigonometric Expression

Simplify $\sin x \cos^2 x - \sin x$.

Solution

First factor out a common monomial factor and then use a fundamental identity.

$$\sin x \cos^2 x - \sin x = \sin x(\cos^2 x - 1) \qquad \text{Factor out common monomial factor.}$$

$$= -\sin x(1 - \cos^2 x) \qquad \text{Factor out } -1.$$

$$= -\sin x(\sin^2 x) \qquad \text{Pythagorean identity}$$

$$= -\sin^3 x \qquad \text{Multiply.}$$

CHECK**Point**   Now try Exercise 59.

TECHNOLOGY

You can use a graphing utility to check the result of Example 2. To do this, graph

$$y_1 = \sin x \cos^2 x - \sin x$$

and

$$y_2 = -\sin^3 x$$

in the same viewing window, as shown below. Because Example 2 shows the equivalence algebraically and the two graphs appear to coincide, you can conclude that the expressions are equivalent.

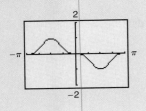

When factoring trigonometric expressions, it is helpful to find a special polynomial factoring form that fits the expression, as shown in Example 3.

Example 3   Factoring Trigonometric Expressions

Factor each expression.

a. $\sec^2 \theta - 1$ **b.** $4 \tan^2 \theta + \tan \theta - 3$

Solution

a. This expression has the form $u^2 - v^2$, which is the difference of two squares. It factors as

$$\sec^2 \theta - 1 = (\sec \theta - 1)(\sec \theta + 1).$$

b. This expression has the polynomial form $ax^2 + bx + c$, and it factors as

$$4 \tan^2 \theta + \tan \theta - 3 = (4 \tan \theta - 3)(\tan \theta + 1).$$

CHECK **Point**   Now try Exercise 61.

> **Algebra Help**
>
> In Example 3, you need to be able to factor the difference of two squares and factor a trinomial. You can review the techniques for factoring in Appendix A.3.

On occasion, factoring or simplifying can best be done by first rewriting the expression in terms of just *one* trigonometric function or in terms of *sine and cosine only*. These strategies are shown in Examples 4 and 5, respectively.

Example 4   Factoring a Trigonometric Expression

Factor $\csc^2 x - \cot x - 3$.

Solution

Use the identity $\csc^2 x = 1 + \cot^2 x$ to rewrite the expression in terms of the cotangent.

$$
\begin{aligned}
\csc^2 x - \cot x - 3 &= (1 + \cot^2 x) - \cot x - 3 \qquad &&\text{Pythagorean identity} \\
&= \cot^2 x - \cot x - 2 \qquad &&\text{Combine like terms.} \\
&= (\cot x - 2)(\cot x + 1) \qquad &&\text{Factor.}
\end{aligned}
$$

CHECK **Point**   Now try Exercise 65.

Example 5   Simplifying a Trigonometric Expression

Simplify $\sin t + \cot t \cos t$.

Solution

Begin by rewriting $\cot t$ in terms of sine and cosine.

$$
\begin{aligned}
\sin t + \cot t \cos t &= \sin t + \left(\frac{\cos t}{\sin t} \right) \cos t \qquad &&\text{Quotient identity} \\
&= \frac{\sin^2 t + \cos^2 t}{\sin t} \qquad &&\text{Add fractions.} \\
&= \frac{1}{\sin t} \qquad &&\text{Pythagorean identity} \\
&= \csc t \qquad &&\text{Reciprocal identity}
\end{aligned}
$$

> **Study Tip**
>
> Remember that when adding rational expressions, you must first find the least common denominator (LCD). In Example 5, the LCD is $\sin t$.

CHECK **Point**   Now try Exercise 71.

Example 6   Adding Trigonometric Expressions

Perform the addition and simplify.

$$\frac{\sin \theta}{1 + \cos \theta} + \frac{\cos \theta}{\sin \theta}$$

Solution

$$\frac{\sin \theta}{1 + \cos \theta} + \frac{\cos \theta}{\sin \theta} = \frac{(\sin \theta)(\sin \theta) + (\cos \theta)(1 + \cos \theta)}{(1 + \cos \theta)(\sin \theta)}$$

$$= \frac{\sin^2 \theta + \cos^2 \theta + \cos \theta}{(1 + \cos \theta)(\sin \theta)} \qquad \text{Multiply.}$$

$$= \frac{1 + \cos \theta}{(1 + \cos \theta)(\sin \theta)} \qquad \begin{array}{l}\text{Pythagorean identity:} \\ \sin^2 \theta + \cos^2 \theta = 1\end{array}$$

$$= \frac{1}{\sin \theta} \qquad \text{Divide out common factor.}$$

$$= \csc \theta \qquad \text{Reciprocal identity}$$

CHECK **Point** Now try Exercise 75.

The next two examples involve techniques for rewriting expressions in forms that are used in calculus.

Example 7   Rewriting a Trigonometric Expression

Rewrite $\dfrac{1}{1 + \sin x}$ so that it is *not* in fractional form.

Solution

From the Pythagorean identity $\cos^2 x = 1 - \sin^2 x = (1 - \sin x)(1 + \sin x)$, you can see that multiplying both the numerator and the denominator by $(1 - \sin x)$ will produce a monomial denominator.

$$\frac{1}{1 + \sin x} = \frac{1}{1 + \sin x} \cdot \frac{1 - \sin x}{1 - \sin x} \qquad \begin{array}{l}\text{Multiply numerator and} \\ \text{denominator by } (1 - \sin x).\end{array}$$

$$= \frac{1 - \sin x}{1 - \sin^2 x} \qquad \text{Multiply.}$$

$$= \frac{1 - \sin x}{\cos^2 x} \qquad \text{Pythagorean identity}$$

$$= \frac{1}{\cos^2 x} - \frac{\sin x}{\cos^2 x} \qquad \text{Write as separate fractions.}$$

$$= \frac{1}{\cos^2 x} - \frac{\sin x}{\cos x} \cdot \frac{1}{\cos x} \qquad \text{Product of fractions}$$

$$= \sec^2 x - \tan x \sec x \qquad \text{Reciprocal and quotient identities}$$

CHECK **Point** Now try Exercise 81.

| **Example 8** | **Trigonometric Substitution** | |

Use the substitution $x = 2 \tan \theta$, $0 < \theta < \pi/2$, to write

$$\sqrt{4 + x^2}$$

as a trigonometric function of θ.

Solution

Begin by letting $x = 2 \tan \theta$. Then, you can obtain

$$\sqrt{4 + x^2} = \sqrt{4 + (2 \tan \theta)^2} \qquad \text{Substitute } 2 \tan \theta \text{ for } x.$$

$$= \sqrt{4 + 4 \tan^2 \theta} \qquad \text{Rule of exponents}$$

$$= \sqrt{4(1 + \tan^2 \theta)} \qquad \text{Factor.}$$

$$= \sqrt{4 \sec^2 \theta} \qquad \text{Pythagorean identity}$$

$$= 2 \sec \theta. \qquad \sec \theta > 0 \text{ for } 0 < \theta < \pi/2$$

CHECK*Point*   Now try Exercise 93.

Figure 5.1 shows the right triangle illustration of the trigonometric substitution $x = 2 \tan \theta$ in Example 8. You can use this triangle to check the solution of Example 8. For $0 < \theta < \pi/2$, you have

$$\text{opp} = x, \quad \text{adj} = 2, \quad \text{and} \quad \text{hyp} = \sqrt{4 + x^2}.$$

With these expressions, you can write the following.

$$\sec \theta = \frac{\text{hyp}}{\text{adj}}$$

$$\sec \theta = \frac{\sqrt{4 + x^2}}{2}$$

$$2 \sec \theta = \sqrt{4 + x^2}$$

So, the solution checks.

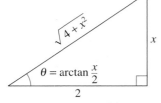

Angle whose tangent is $\pi/2$.
FIGURE 5.1

| **Example 9** | **Rewriting a Logarithmic Expression** |

Rewrite $\ln|\csc \theta| + \ln|\tan \theta|$ as a single logarithm and simplify the result.

Solution

$$\ln|\csc \theta| + \ln|\tan \theta| = \ln|\csc \theta \tan \theta| \qquad \text{Product Property of Logarithms}$$

$$= \ln\left|\frac{1}{\sin \theta} \cdot \frac{\sin \theta}{\cos \theta}\right| \qquad \text{Reciprocal and quotient identities}$$

$$= \ln\left|\frac{1}{\cos \theta}\right| \qquad \text{Simplify.}$$

$$= \ln|\sec \theta| \qquad \text{Reciprocal identity}$$

CHECK*Point*   Now try Exercise 113.

Algebra Help

Recall that for positive real numbers u and v,

$$\ln u + \ln v = \ln(uv).$$

You can review the properties of logarithms in Section 3.3.

5.1   EXERCISES

See www.CalcChat.com for worked-out solutions to odd-numbered exercises.

VOCABULARY: Fill in the blank to complete the trigonometric identity.

1. $\dfrac{\sin u}{\cos u} = $ _____

2. $\dfrac{1}{\csc u} = $ _____

3. $\dfrac{1}{\tan u} = $ _____

4. $\dfrac{1}{\cos u} = $ _____

5. $1 + $ _____ $= \csc^2 u$

6. $1 + \tan^2 u = $ _____

7. $\sin\left(\dfrac{\pi}{2} - u\right) = $ _____

8. $\sec\left(\dfrac{\pi}{2} - u\right) = $ _____

9. $\cos(-u) = $ _____

10. $\tan(-u) = $ _____

SKILLS AND APPLICATIONS

In Exercises 11–24, use the given values to evaluate (if possible) all six trigonometric functions.

11. $\sin x = \dfrac{1}{2}$, $\cos x = \dfrac{\sqrt{3}}{2}$

12. $\tan x = \dfrac{\sqrt{3}}{3}$, $\cos x = -\dfrac{\sqrt{3}}{2}$

13. $\sec \theta = \sqrt{2}$, $\sin \theta = -\dfrac{\sqrt{2}}{2}$

14. $\csc \theta = \dfrac{25}{7}$, $\tan \theta = \dfrac{7}{24}$

15. $\tan x = \dfrac{8}{15}$, $\sec x = -\dfrac{17}{15}$

16. $\cot \phi = -3$, $\sin \phi = \dfrac{\sqrt{10}}{10}$

17. $\sec \phi = \dfrac{3}{2}$, $\csc \phi = -\dfrac{3\sqrt{5}}{5}$

18. $\cos\left(\dfrac{\pi}{2} - x\right) = \dfrac{3}{5}$, $\cos x = \dfrac{4}{5}$

19. $\sin(-x) = -\dfrac{1}{3}$, $\tan x = -\dfrac{\sqrt{2}}{4}$

20. $\sec x = 4$, $\sin x > 0$

21. $\tan \theta = 2$, $\sin \theta < 0$

22. $\csc \theta = -5$, $\cos \theta < 0$

23. $\sin \theta = -1$, $\cot \theta = 0$

24. $\tan \theta$ is undefined, $\sin \theta > 0$

In Exercises 25–30, match the trigonometric expression with one of the following.

(a) $\sec x$ (b) -1 (c) $\cot x$

(d) 1 (e) $-\tan x$ (f) $\sin x$

25. $\sec x \cos x$

26. $\tan x \csc x$

27. $\cot^2 x - \csc^2 x$

28. $(1 - \cos^2 x)(\csc x)$

29. $\dfrac{\sin(-x)}{\cos(-x)}$

30. $\dfrac{\sin[(\pi/2) - x]}{\cos[(\pi/2) - x]}$

In Exercises 31–36, match the trigonometric expression with one of the following.

(a) $\csc x$ (b) $\tan x$ (c) $\sin^2 x$

(d) $\sin x \tan x$ (e) $\sec^2 x$ (f) $\sec^2 x + \tan^2 x$

31. $\sin x \sec x$

32. $\cos^2 x(\sec^2 x - 1)$

33. $\sec^4 x - \tan^4 x$

34. $\cot x \sec x$

35. $\dfrac{\sec^2 x - 1}{\sin^2 x}$

36. $\dfrac{\cos^2[(\pi/2) - x]}{\cos x}$

In Exercises 37–58, use the fundamental identities to simplify the expression. There is more than one correct form of each answer.

37. $\cot \theta \sec \theta$

38. $\cos \beta \tan \beta$

39. $\tan(-x) \cos x$

40. $\sin x \cot(-x)$

41. $\sin \phi(\csc \phi - \sin \phi)$

42. $\sec^2 x(1 - \sin^2 x)$

43. $\dfrac{\cot x}{\csc x}$

44. $\dfrac{\csc \theta}{\sec \theta}$

45. $\dfrac{1 - \sin^2 x}{\csc^2 x - 1}$

46. $\dfrac{1}{\tan^2 x + 1}$

47. $\dfrac{\tan \theta \cot \theta}{\sec \theta}$

48. $\dfrac{\sin \theta \csc \theta}{\tan \theta}$

49. $\sec \alpha \cdot \dfrac{\sin \alpha}{\tan \alpha}$

50. $\dfrac{\tan^2 \theta}{\sec^2 \theta}$

51. $\cos\left(\dfrac{\pi}{2} - x\right) \sec x$

52. $\cot\left(\dfrac{\pi}{2} - x\right) \cos x$

53. $\dfrac{\cos^2 y}{1 - \sin y}$

54. $\cos t(1 + \tan^2 t)$

55. $\sin \beta \tan \beta + \cos \beta$

56. $\csc \phi \tan \phi + \sec \phi$

57. $\cot u \sin u + \tan u \cos u$

58. $\sin \theta \sec \theta + \cos \theta \csc \theta$

In Exercises 59–70, factor the expression and use the fundamental identities to simplify. There is more than one correct form of each answer.

59. $\tan^2 x - \tan^2 x \sin^2 x$

60. $\sin^2 x \csc^2 x - \sin^2 x$

61. $\sin^2 x \sec^2 x - \sin^2 x$

62. $\cos^2 x + \cos^2 x \tan^2 x$

63. $\dfrac{\sec^2 x - 1}{\sec x - 1}$

64. $\dfrac{\cos^2 x - 4}{\cos x - 2}$

65. $\tan^4 x + 2 \tan^2 x + 1$

66. $1 - 2 \cos^2 x + \cos^4 x$

67. $\sin^4 x - \cos^4 x$

68. $\sec^4 x - \tan^4 x$

69. $\csc^3 x - \csc^2 x - \csc x + 1$

70. $\sec^3 x - \sec^2 x - \sec x + 1$

In Exercises 71–74, perform the multiplication and use the fundamental identities to simplify. There is more than one correct form of each answer.

71. $(\sin x + \cos x)^2$

72. $(\cot x + \csc x)(\cot x - \csc x)$

73. $(2 \csc x + 2)(2 \csc x - 2)$

74. $(3 - 3 \sin x)(3 + 3 \sin x)$

In Exercises 75–80, perform the addition or subtraction and use the fundamental identities to simplify. There is more than one correct form of each answer.

75. $\dfrac{1}{1 + \cos x} + \dfrac{1}{1 - \cos x}$

76. $\dfrac{1}{\sec x + 1} - \dfrac{1}{\sec x - 1}$

77. $\dfrac{\cos x}{1 + \sin x} + \dfrac{1 + \sin x}{\cos x}$

78. $\dfrac{\tan x}{1 + \sec x} + \dfrac{1 + \sec x}{\tan x}$

79. $\tan x + \dfrac{\cos x}{1 + \sin x}$

80. $\tan x - \dfrac{\sec^2 x}{\tan x}$

In Exercises 81–84, rewrite the expression so that it is not in fractional form. There is more than one correct form of each answer.

81. $\dfrac{\sin^2 y}{1 - \cos y}$

82. $\dfrac{5}{\tan x + \sec x}$

83. $\dfrac{3}{\sec x - \tan x}$

84. $\dfrac{\tan^2 x}{\csc x + 1}$

NUMERICAL AND GRAPHICAL ANALYSIS In Exercises 85–88, use a graphing utility to complete the table and graph the functions. Make a conjecture about y_1 and y_2.

x	0.2	0.4	0.6	0.8	1.0	1.2	1.4
y_1							
y_2							

85. $y_1 = \cos\left(\dfrac{\pi}{2} - x\right)$, $\quad y_2 = \sin x$

86. $y_1 = \sec x - \cos x$, $\quad y_2 = \sin x \tan x$

87. $y_1 = \dfrac{\cos x}{1 - \sin x}$, $\quad y_2 = \dfrac{1 + \sin x}{\cos x}$

88. $y_1 = \sec^4 x - \sec^2 x$, $\quad y_2 = \tan^2 x + \tan^4 x$

In Exercises 89–92, use a graphing utility to determine which of the six trigonometric functions is equal to the expression. Verify your answer algebraically.

89. $\cos x \cot x + \sin x$

90. $\sec x \csc x - \tan x$

91. $\dfrac{1}{\sin x}\left(\dfrac{1}{\cos x} - \cos x\right)$

92. $\dfrac{1}{2}\left(\dfrac{1 + \sin \theta}{\cos \theta} + \dfrac{\cos \theta}{1 + \sin \theta}\right)$

In Exercises 93–104, use the trigonometric substitution to write the algebraic expression as a trigonometric function of θ, where $0 < \theta < \pi/2$.

93. $\sqrt{9 - x^2}$, $\quad x = 3 \cos \theta$

94. $\sqrt{64 - 16x^2}$, $\quad x = 2 \cos \theta$

95. $\sqrt{16 - x^2}$, $\quad x = 4 \sin \theta$

96. $\sqrt{49 - x^2}$, $\quad x = 7 \sin \theta$

97. $\sqrt{x^2 - 9}$, $\quad x = 3 \sec \theta$

98. $\sqrt{x^2 - 4}$, $\quad x = 2 \sec \theta$

99. $\sqrt{x^2 + 25}$, $\quad x = 5 \tan \theta$

100. $\sqrt{x^2 + 100}$, $\quad x = 10 \tan \theta$

101. $\sqrt{4x^2 + 9}$, $\quad 2x = 3 \tan \theta$

102. $\sqrt{9x^2 + 25}$, $\quad 3x = 5 \tan \theta$

103. $\sqrt{2 - x^2}$, $\quad x = \sqrt{2} \sin \theta$

104. $\sqrt{10 - x^2}$, $\quad x = \sqrt{10} \sin \theta$

In Exercises 105–108, use the trigonometric substitution to write the algebraic equation as a trigonometric equation of θ, where $-\pi/2 < \theta < \pi/2$. Then find $\sin \theta$ and $\cos \theta$.

105. $3 = \sqrt{9 - x^2}$, $\quad x = 3 \sin \theta$

106. $3 = \sqrt{36 - x^2}$, $\quad x = 6 \sin \theta$

107. $2\sqrt{2} = \sqrt{16 - 4x^2}$, $\quad x = 2 \cos \theta$

108. $-5\sqrt{3} = \sqrt{100 - x^2}$, $\quad x = 10 \cos \theta$

In Exercises 109–112, use a graphing utility to solve the equation for θ, where $0 \le \theta < 2\pi$.

109. $\sin \theta = \sqrt{1 - \cos^2 \theta}$

110. $\cos \theta = -\sqrt{1 - \sin^2 \theta}$

111. $\sec \theta = \sqrt{1 + \tan^2 \theta}$

112. $\csc \theta = \sqrt{1 + \cot^2 \theta}$

In Exercises 113–118, rewrite the expression as a single logarithm and simplify the result.

113. $\ln|\cos x| - \ln|\sin x|$ **114.** $\ln|\sec x| + \ln|\sin x|$

115. $\ln|\sin x| + \ln|\cot x|$ **116.** $\ln|\tan x| + \ln|\csc x|$

117. $\ln|\cot t| + \ln(1 + \tan^2 t)$

118. $\ln(\cos^2 t) + \ln(1 + \tan^2 t)$

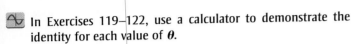 In Exercises 119–122, use a calculator to demonstrate the identity for each value of **θ**.

119. $\csc^2 \theta - \cot^2 \theta = 1$

(a) $\theta = 132°$ (b) $\theta = \dfrac{2\pi}{7}$

120. $\tan^2 \theta + 1 = \sec^2 \theta$

(a) $\theta = 346°$ (b) $\theta = 3.1$

121. $\cos\left(\dfrac{\pi}{2} - \theta\right) = \sin \theta$

(a) $\theta = 80°$ (b) $\theta = 0.8$

122. $\sin(-\theta) = -\sin \theta$

(a) $\theta = 250°$ (b) $\theta = \frac{1}{2}$

123. FRICTION The forces acting on an object weighing W units on an inclined plane positioned at an angle of θ with the horizontal (see figure) are modeled by

$$\mu W \cos \theta = W \sin \theta$$

where μ is the coefficient of friction. Solve the equation for μ and simplify the result.

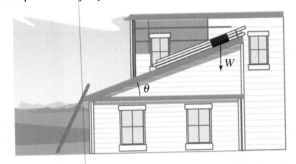

124. RATE OF CHANGE The rate of change of the function $f(x) = -x + \tan x$ is given by the expression $-1 + \sec^2 x$. Show that this expression can also be written as $\tan^2 x$.

125. RATE OF CHANGE The rate of change of the function $f(x) = \sec x + \cos x$ is given by the expression $\sec x \tan x - \sin x$. Show that this expression can also be written as $\sin x \tan^2 x$.

126. RATE OF CHANGE The rate of change of the function $f(x) = -\csc x - \sin x$ is given by the expression $\csc x \cot x - \cos x$. Show that this expression can also be written as $\cos x \cot^2 x$.

EXPLORATION

TRUE OR FALSE? In Exercises 127 and 128, determine whether the statement is true or false. Justify your answer.

127. The even and odd trigonometric identities are helpful for determining whether the value of a trigonometric function is positive or negative.

128. A cofunction identity can be used to transform a tangent function so that it can be represented by a cosecant function.

In Exercises 129–132, fill in the blanks. (*Note:* The notation $x \to c^+$ indicates that x approaches c from the right and $x \to c^-$ indicates that x approaches c from the left.)

129. As $x \to \dfrac{\pi^-}{2}$, $\sin x \to$ ▢ and $\csc x \to$ ▢ .

130. As $x \to 0^+$, $\cos x \to$ ▢ and $\sec x \to$ ▢ .

131. As $x \to \dfrac{\pi^-}{2}$, $\tan x \to$ ▢ and $\cot x \to$ ▢ .

132. As $x \to \pi^+$, $\sin x \to$ ▢ and $\csc x \to$ ▢ .

In Exercises 133–138, determine whether or not the equation is an identity, and give a reason for your answer.

133. $\cos \theta = \sqrt{1 - \sin^2 \theta}$ **134.** $\cot \theta = \sqrt{\csc^2 \theta + 1}$

135. $\dfrac{(\sin k\theta)}{(\cos k\theta)} = \tan \theta$, k is a constant.

136. $\dfrac{1}{(5 \cos \theta)} = 5 \sec \theta$

137. $\sin \theta \csc \theta = 1$ **138.** $\csc^2 \theta = 1$

139. Use the trigonometric substitution $u = a \sin \theta$, where $-\pi/2 < \theta < \pi/2$ and $a > 0$, to simplify the expression $\sqrt{a^2 - u^2}$.

140. Use the trigonometric substitution $u = a \tan \theta$, where $-\pi/2 < \theta < \pi/2$ and $a > 0$, to simplify the expression $\sqrt{a^2 + u^2}$.

141. Use the trigonometric substitution $u = a \sec \theta$, where $0 < \theta < \pi/2$ and $a > 0$, to simplify the expression $\sqrt{u^2 - a^2}$.

142. CAPSTONE

(a) Use the definitions of sine and cosine to derive the Pythagorean identity $\sin^2 \theta + \cos^2 \theta = 1$.

(b) Use the Pythagorean identity $\sin^2 \theta + \cos^2 \theta = 1$ to derive the other Pythagorean identities, $1 + \tan^2 \theta = \sec^2 \theta$ and $1 + \cot^2 \theta = \csc^2 \theta$. Discuss how to remember these identities and other fundamental identities.

5.2 VERIFYING TRIGONOMETRIC IDENTITIES

What you should learn
- Verify trigonometric identities.

Why you should learn it

You can use trigonometric identities to rewrite trigonometric equations that model real-life situations. For instance, in Exercise 70 on page 386, you can use trigonometric identities to simplify the equation that models the length of a shadow cast by a gnomon (a device used to tell time).

Introduction

In this section, you will study techniques for verifying trigonometric identities. In the next section, you will study techniques for solving trigonometric equations. The key to verifying identities *and* solving equations is the ability to use the fundamental identities and the rules of algebra to rewrite trigonometric expressions.

Remember that a *conditional equation* is an equation that is true for only some of the values in its domain. For example, the conditional equation

$$\sin x = 0 \qquad \text{Conditional equation}$$

is true only for $x = n\pi$, where n is an integer. When you find these values, you are *solving* the equation.

On the other hand, an equation that is true for all real values in the domain of the variable is an *identity*. For example, the familiar equation

$$\sin^2 x = 1 - \cos^2 x \qquad \text{Identity}$$

is true for all real numbers x. So, it is an identity.

Verifying Trigonometric Identities

Although there are similarities, verifying that a trigonometric equation is an identity is quite different from solving an equation. There is no well-defined set of rules to follow in verifying trigonometric identities, and the process is best learned by practice.

Guidelines for Verifying Trigonometric Identities

1. Work with one side of the equation at a time. It is often better to work with the more complicated side first.

2. Look for opportunities to factor an expression, add fractions, square a binomial, or create a monomial denominator.

3. Look for opportunities to use the fundamental identities. Note which functions are in the final expression you want. Sines and cosines pair up well, as do secants and tangents, and cosecants and cotangents.

4. If the preceding guidelines do not help, try converting all terms to sines and cosines.

5. Always try *something*. Even paths that lead to dead ends provide insights.

Verifying trigonometric identities is a useful process if you need to convert a trigonometric expression into a form that is more useful algebraically. When you verify an identity, you cannot *assume* that the two sides of the equation are equal because you are trying to verify that they *are* equal. As a result, when verifying identities, you cannot use operations such as adding the same quantity to each side of the equation or cross multiplication.

Example 1   Verifying a Trigonometric Identity

Verify the identity $(\sec^2 \theta - 1)/\sec^2 \theta = \sin^2 \theta$.

Solution

The left side is more complicated, so start with it.

$$\frac{\sec^2 \theta - 1}{\sec^2 \theta} = \frac{(\tan^2 \theta + 1) - 1}{\sec^2 \theta} \qquad \text{Pythagorean identity}$$

$$= \frac{\tan^2 \theta}{\sec^2 \theta} \qquad \text{Simplify.}$$

$$= \tan^2 \theta (\cos^2 \theta) \qquad \text{Reciprocal identity}$$

$$= \frac{\sin^2 \theta}{(\cos^2 \theta)} (\cos^2 \theta) \qquad \text{Quotient identity}$$

$$= \sin^2 \theta \qquad \text{Simplify.}$$

Notice how the identity is verified. You start with the left side of the equation (the more complicated side) and use the fundamental trigonometric identities to simplify it until you obtain the right side.

CHECK*Point*   Now try Exercise 15.

There can be more than one way to verify an identity. Here is another way to verify the identity in Example 1.

$$\frac{\sec^2 \theta - 1}{\sec^2 \theta} = \frac{\sec^2 \theta}{\sec^2 \theta} - \frac{1}{\sec^2 \theta} \qquad \text{Rewrite as the difference of fractions.}$$

$$= 1 - \cos^2 \theta \qquad \text{Reciprocal identity}$$

$$= \sin^2 \theta \qquad \text{Pythagorean identity}$$

> **⚠ WARNING / CAUTION**
>
> Remember that an identity is only true for all real values in the domain of the variable. For instance, in Example 1 the identity is not true when $\theta = \pi/2$ because $\sec^2 \theta$ is not defined when $\theta = \pi/2$.

Example 2   Verifying a Trigonometric Identity

Verify the identity $2 \sec^2 \alpha = \dfrac{1}{1 - \sin \alpha} + \dfrac{1}{1 + \sin \alpha}$.

Algebraic Solution

The right side is more complicated, so start with it.

$$\frac{1}{1 - \sin \alpha} + \frac{1}{1 + \sin \alpha} = \frac{1 + \sin \alpha + 1 - \sin \alpha}{(1 - \sin \alpha)(1 + \sin \alpha)} \qquad \text{Add fractions.}$$

$$= \frac{2}{1 - \sin^2 \alpha} \qquad \text{Simplify.}$$

$$= \frac{2}{\cos^2 \alpha} \qquad \text{Pythagorean identity}$$

$$= 2 \sec^2 \alpha \qquad \text{Reciprocal identity}$$

Numerical Solution

Use the *table* feature of a graphing utility set in *radian* mode to create a table that shows the values of $y_1 = 2/\cos^2 x$ and $y_2 = 1/(1 - \sin x) + 1/(1 + \sin x)$ for different values of x, as shown in Figure 5.2. From the table, you can see that the values appear to be identical, so $2 \sec^2 x = 1/(1 - \sin x) + 1/(1 + \sin x)$ appears to be an identity.

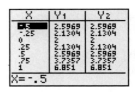

X	Y₁	Y₂
-.5	2.5969	2.5969
-.25	2.1304	2.1304
0	2	2
.25	2.1304	2.1304
.5	2.5969	2.5969
.75	3.7357	3.7357
1	6.851	6.851

X=-.5

FIGURE 5.2

CHECK*Point*   Now try Exercise 31.

Example 3 Verifying a Trigonometric Identity

Verify the identity $(\tan^2 x + 1)(\cos^2 x - 1) = -\tan^2 x$.

Algebraic Solution

By applying identities before multiplying, you obtain the following.

$(\tan^2 x + 1)(\cos^2 x - 1) = (\sec^2 x)(-\sin^2 x)$ Pythagorean identities

$$= -\frac{\sin^2 x}{\cos^2 x} \qquad \text{Reciprocal identity}$$

$$= -\left(\frac{\sin x}{\cos x}\right)^2 \qquad \text{Rule of exponents}$$

$$= -\tan^2 x \qquad \text{Quotient identity}$$

Graphical Solution

Use a graphing utility set in *radian* mode to graph the left side of the identity $y_1 = (\tan^2 x + 1)(\cos^2 x - 1)$ and the right side of the identity $y_2 = -\tan^2 x$ in the same viewing window, as shown in Figure 5.3. (Select the *line* style for y_1 and the *path* style for y_2.) Because the graphs appear to coincide, $(\tan^2 x + 1)(\cos^2 x - 1) = -\tan^2 x$ appears to be an identity.

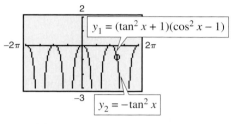

FIGURE **5.3**

CHECK*Point*   Now try Exercise 53.

Example 4 Converting to Sines and Cosines

Verify the identity $\tan x + \cot x = \sec x \csc x$.

Solution

Try converting the left side into sines and cosines.

$$\tan x + \cot x = \frac{\sin x}{\cos x} + \frac{\cos x}{\sin x} \qquad \text{Quotient identities}$$

$$= \frac{\sin^2 x + \cos^2 x}{\cos x \sin x} \qquad \text{Add fractions.}$$

$$= \frac{1}{\cos x \sin x} \qquad \text{Pythagorean identity}$$

$$= \frac{1}{\cos x} \cdot \frac{1}{\sin x} \qquad \text{Product of fractions.}$$

$$= \sec x \csc x \qquad \text{Reciprocal identities}$$

CHECK*Point*   Now try Exercise 25.

> **WARNING / CAUTION**
>
> Although a graphing utility can be useful in helping to verify an identity, you must use algebraic techniques to produce a *valid* proof.

Study Tip

As shown at the right, $\csc^2 x(1 + \cos x)$ is considered a simplified form of $1/(1 - \cos x)$ because the expression does not contain any fractions.

Recall from algebra that *rationalizing the denominator* using conjugates is, on occasion, a powerful simplification technique. A related form of this technique, shown below, works for simplifying trigonometric expressions as well.

$$\frac{1}{1 - \cos x} = \frac{1}{1 - \cos x}\left(\frac{1 + \cos x}{1 + \cos x}\right) = \frac{1 + \cos x}{1 - \cos^2 x} = \frac{1 + \cos x}{\sin^2 x}$$

$$= \csc^2 x(1 + \cos x)$$

This technique is demonstrated in the next example.

Example 5   Verifying a Trigonometric Identity

Verify the identity $\sec x + \tan x = \dfrac{\cos x}{1 - \sin x}$.

Algebraic Solution

Begin with the *right* side because you can create a monomial denominator by multiplying the numerator and denominator by $1 + \sin x$.

$$\frac{\cos x}{1 - \sin x} = \frac{\cos x}{1 - \sin x}\left(\frac{1 + \sin x}{1 + \sin x}\right) \qquad \text{Multiply numerator and denominator by } 1 + \sin x.$$

$$= \frac{\cos x + \cos x \sin x}{1 - \sin^2 x} \qquad \text{Multiply.}$$

$$= \frac{\cos x + \cos x \sin x}{\cos^2 x} \qquad \text{Pythagorean identity}$$

$$= \frac{\cos x}{\cos^2 x} + \frac{\cos x \sin x}{\cos^2 x} \qquad \text{Write as separate fractions.}$$

$$= \frac{1}{\cos x} + \frac{\sin x}{\cos x} \qquad \text{Simplify.}$$

$$= \sec x + \tan x \qquad \text{Identities}$$

CHECK*Point*   Now try Exercise 59.

Graphical Solution

Use a graphing utility set in the *radian* and *dot* modes to graph $y_1 = \sec x + \tan x$ and $y_2 = \cos x/(1 - \sin x)$ in the same viewing window, as shown in Figure 5.4. Because the graphs appear to coincide, $\sec x + \tan x = \cos x/(1 - \sin x)$ appears to be an identity.

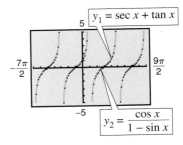

FIGURE **5.4**

In Examples 1 through 5, you have been verifying trigonometric identities by working with one side of the equation and converting to the form given on the other side. On occasion, it is practical to work with each side *separately*, to obtain one common form equivalent to both sides. This is illustrated in Example 6.

Example 6   Working with Each Side Separately

Verify the identity $\dfrac{\cot^2 \theta}{1 + \csc \theta} = \dfrac{1 - \sin \theta}{\sin \theta}$.

Algebraic Solution

Working with the left side, you have

$$\frac{\cot^2 \theta}{1 + \csc \theta} = \frac{\csc^2 \theta - 1}{1 + \csc \theta} \qquad \text{Pythagorean identity}$$

$$= \frac{(\csc \theta - 1)(\csc \theta + 1)}{1 + \csc \theta} \qquad \text{Factor.}$$

$$= \csc \theta - 1. \qquad \text{Simplify.}$$

Now, simplifying the right side, you have

$$\frac{1 - \sin \theta}{\sin \theta} = \frac{1}{\sin \theta} - \frac{\sin \theta}{\sin \theta} \qquad \text{Write as separate fractions.}$$

$$= \csc \theta - 1. \qquad \text{Reciprocal identity}$$

The identity is verified because both sides are equal to $\csc \theta - 1$.

CHECK*Point*   Now try Exercise 19.

Numerical Solution

Use the *table* feature of a graphing utility set in *radian* mode to create a table that shows the values of $y_1 = \cot^2 x/(1 + \csc x)$ and $y_2 = (1 - \sin x)/\sin x$ for different values of x, as shown in Figure 5.5. From the table you can see that the values appear to be identical, so $\cot^2 x/(1 + \csc x) = (1 - \sin x)/\sin x$ appears to be an identity.

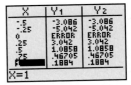

FIGURE **5.5**

In Example 7, powers of trigonometric functions are rewritten as more complicated sums of products of trigonometric functions. This is a common procedure used in calculus.

Example 7 Three Examples from Calculus

Verify each identity.

a. $\tan^4 x = \tan^2 x \sec^2 x - \tan^2 x$

b. $\sin^3 x \cos^4 x = (\cos^4 x - \cos^6 x) \sin x$

c. $\csc^4 x \cot x = \csc^2 x(\cot x + \cot^3 x)$

Solution

a. $\tan^4 x = (\tan^2 x)(\tan^2 x)$ Write as separate factors.

$\qquad = \tan^2 x(\sec^2 x - 1)$ Pythagorean identity

$\qquad = \tan^2 x \sec^2 x - \tan^2 x$ Multiply.

b. $\sin^3 x \cos^4 x = \sin^2 x \cos^4 x \sin x$ Write as separate factors.

$\qquad = (1 - \cos^2 x) \cos^4 x \sin x$ Pythagorean identity

$\qquad = (\cos^4 x - \cos^6 x) \sin x$ Multiply.

c. $\csc^4 x \cot x = \csc^2 x \csc^2 x \cot x$ Write as separate factors.

$\qquad = \csc^2 x(1 + \cot^2 x) \cot x$ Pythagorean identity

$\qquad = \csc^2 x(\cot x + \cot^3 x)$ Multiply.

CHECK *Point*   Now try Exercise 63.

CLASSROOM DISCUSSION

Error Analysis You are tutoring a student in trigonometry. One of the homework problems your student encounters asks whether the following statement is an identity.

$$\tan^2 x \sin^2 x \overset{?}{=} \frac{5}{6} \tan^2 x$$

Your student does not attempt to verify the equivalence algebraically, but mistakenly uses only a graphical approach. Using range settings of

$\text{Xmin} = -3\pi$ $\text{Ymin} = -20$

$\text{Xmax} = 3\pi$ $\text{Ymax} = 20$

$\text{Xscl} = \pi/2$ $\text{Yscl} = 1$

your student graphs both sides of the expression on a graphing utility and concludes that the statement is an identity.

What is wrong with your student's reasoning? Explain. Discuss the limitations of verifying identities graphically.

5.2   EXERCISES

See www.CalcChat.com for worked-out solutions to odd-numbered exercises.

VOCABULARY

In Exercises 1 and 2, fill in the blanks.

1. An equation that is true for all real values in its domain is called an _____.

2. An equation that is true for only some values in its domain is called a _____ _____.

In Exercises 3–8, fill in the blank to complete the trigonometric identity.

3. $\dfrac{1}{\cot u} =$ _____

4. $\dfrac{\cos u}{\sin u} =$ _____

5. $\sin^2 u +$ _____ $= 1$

6. $\cos\left(\dfrac{\pi}{2} - u\right) =$ _____

7. $\csc(-u) =$ _____

8. $\sec(-u) =$ _____

SKILLS AND APPLICATIONS

In Exercises 9–50, verify the identity.

9. $\tan t \cot t = 1$

10. $\sec y \cos y = 1$

11. $\cot^2 y(\sec^2 y - 1) = 1$

12. $\cos x + \sin x \tan x = \sec x$

13. $(1 + \sin \alpha)(1 - \sin \alpha) = \cos^2 \alpha$

14. $\cos^2 \beta - \sin^2 \beta = 2\cos^2 \beta - 1$

15. $\cos^2 \beta - \sin^2 \beta = 1 - 2\sin^2 \beta$

16. $\sin^2 \alpha - \sin^4 \alpha = \cos^2 \alpha - \cos^4 \alpha$

17. $\dfrac{\tan^2 \theta}{\sec \theta} = \sin \theta \tan \theta$

18. $\dfrac{\cot^3 t}{\csc t} = \cos t(\csc^2 t - 1)$

19. $\dfrac{\cot^2 t}{\csc t} = \dfrac{1 - \sin^2 t}{\sin t}$

20. $\dfrac{1}{\tan \beta} + \tan \beta = \dfrac{\sec^2 \beta}{\tan \beta}$

21. $\sin^{1/2} x \cos x - \sin^{5/2} x \cos x = \cos^3 x \sqrt{\sin x}$

22. $\sec^6 x(\sec x \tan x) - \sec^4 x(\sec x \tan x) = \sec^5 x \tan^3 x$

23. $\dfrac{\cot x}{\sec x} = \csc x - \sin x$

24. $\dfrac{\sec \theta - 1}{1 - \cos \theta} = \sec \theta$

25. $\csc x - \sin x = \cos x \cot x$

26. $\sec x - \cos x = \sin x \tan x$

27. $\dfrac{1}{\tan x} + \dfrac{1}{\cot x} = \tan x + \cot x$

28. $\dfrac{1}{\sin x} - \dfrac{1}{\csc x} = \csc x - \sin x$

29. $\dfrac{1 + \sin \theta}{\cos \theta} + \dfrac{\cos \theta}{1 + \sin \theta} = 2\sec \theta$

30. $\dfrac{\cos \theta \cot \theta}{1 - \sin \theta} - 1 = \csc \theta$

31. $\dfrac{1}{\cos x + 1} + \dfrac{1}{\cos x - 1} = -2\csc x \cot x$

32. $\cos x - \dfrac{\cos x}{1 - \tan x} = \dfrac{\sin x \cos x}{\sin x - \cos x}$

33. $\tan\left(\dfrac{\pi}{2} - \theta\right)\tan \theta = 1$

34. $\dfrac{\cos[(\pi/2) - x]}{\sin[(\pi/2) - x]} = \tan x$

35. $\dfrac{\tan x \cot x}{\cos x} = \sec x$

36. $\dfrac{\csc(-x)}{\sec(-x)} = -\cot x$

37. $(1 + \sin y)[1 + \sin(-y)] = \cos^2 y$

38. $\dfrac{\tan x + \tan y}{1 - \tan x \tan y} = \dfrac{\cot x + \cot y}{\cot x \cot y - 1}$

39. $\dfrac{\tan x + \cot y}{\tan x \cot y} = \tan y + \cot x$

40. $\dfrac{\cos x - \cos y}{\sin x + \sin y} + \dfrac{\sin x - \sin y}{\cos x + \cos y} = 0$

41. $\sqrt{\dfrac{1 + \sin \theta}{1 - \sin \theta}} = \dfrac{1 + \sin \theta}{|\cos \theta|}$

42. $\sqrt{\dfrac{1 - \cos \theta}{1 + \cos \theta}} = \dfrac{1 - \cos \theta}{|\sin \theta|}$

43. $\cos^2 \beta + \cos^2\left(\dfrac{\pi}{2} - \beta\right) = 1$

44. $\sec^2 y - \cot^2\left(\dfrac{\pi}{2} - y\right) = 1$

45. $\sin t \csc\left(\dfrac{\pi}{2} - t\right) = \tan t$

46. $\sec^2\left(\dfrac{\pi}{2} - x\right) - 1 = \cot^2 x$

47. $\tan(\sin^{-1} x) = \dfrac{x}{\sqrt{1 - x^2}}$

48. $\cos(\sin^{-1} x) = \sqrt{1 - x^2}$

49. $\tan\left(\sin^{-1}\dfrac{x - 1}{4}\right) = \dfrac{x - 1}{\sqrt{16 - (x - 1)^2}}$

50. $\tan\left(\cos^{-1}\dfrac{x + 1}{2}\right) = \dfrac{\sqrt{4 - (x + 1)^2}}{x + 1}$

ERROR ANALYSIS   In Exercises 51 and 52, describe the error(s).

51. $(1 + \tan x)[1 + \cot(-x)]$
$= (1 + \tan x)(1 + \cot x)$
$= 1 + \cot x + \tan x + \tan x \cot x$
$= 1 + \cot x + \tan x + 1$
$= 2 + \cot x + \tan x$

52. $\dfrac{1 + \sec(-\theta)}{\sin(-\theta) + \tan(-\theta)} = \dfrac{1 - \sec\theta}{\sin\theta - \tan\theta}$

$= \dfrac{1 - \sec\theta}{(\sin\theta)[1 - (1/\cos\theta)]}$

$= \dfrac{1 - \sec\theta}{\sin\theta(1 - \sec\theta)}$

$= \dfrac{1}{\sin\theta} = \csc\theta$

In Exercises 53–60, (a) use a graphing utility to graph each side of the equation to determine whether the equation is an identity, (b) use the *table* feature of a graphing utility to determine whether the equation is an identity, and (c) confirm the results of parts (a) and (b) algebraically.

53. $(1 + \cot^2 x)(\cos^2 x) = \cot^2 x$

54. $\csc x(\csc x - \sin x) + \dfrac{\sin x - \cos x}{\sin x} + \cot x = \csc^2 x$

55. $2 + \cos^2 x - 3\cos^4 x = \sin^2 x(3 + 2\cos^2 x)$

56. $\tan^4 x + \tan^2 x - 3 = \sec^2 x(4\tan^2 x - 3)$

57. $\csc^4 x - 2\csc^2 x + 1 = \cot^4 x$

58. $(\sin^4\beta - 2\sin^2\beta + 1)\cos\beta = \cos^5\beta$

59. $\dfrac{1 + \cos x}{\sin x} = \dfrac{\sin x}{1 - \cos x}$ **60.** $\dfrac{\cot\alpha}{\csc\alpha + 1} = \dfrac{\csc\alpha + 1}{\cot\alpha}$

In Exercises 61–64, verify the identity.

61. $\tan^5 x = \tan^3 x \sec^2 x - \tan^3 x$

62. $\sec^4 x \tan^2 x = (\tan^2 x + \tan^4 x)\sec^2 x$

63. $\cos^3 x \sin^2 x = (\sin^2 x - \sin^4 x)\cos x$

64. $\sin^4 x + \cos^4 x = 1 - 2\cos^2 x + 2\cos^4 x$

In Exercises 65–68, use the cofunction identities to evaluate the expression without using a calculator.

65. $\sin^2 25° + \sin^2 65°$ **66.** $\cos^2 55° + \cos^2 35°$

67. $\cos^2 20° + \cos^2 52° + \cos^2 38° + \cos^2 70°$

68. $\tan^2 63° + \cot^2 16° - \sec^2 74° - \csc^2 27°$

69. RATE OF CHANGE   The rate of change of the function $f(x) = \sin x + \csc x$ with respect to change in the variable x is given by the expression $\cos x - \csc x \cot x$. Show that the expression for the rate of change can also be $-\cos x \cot^2 x$.

70. SHADOW LENGTH   The length s of a shadow cast by a vertical gnomon (a device used to tell time) of height h when the angle of the sun above the horizon is θ (see figure) can be modeled by the equation

$$s = \frac{h\sin(90° - \theta)}{\sin\theta}.$$

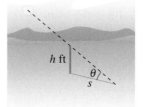

(a) Verify that the equation for s is equal to $h\cot\theta$.

(b) Use a graphing utility to complete the table. Let $h = 5$ feet.

θ	15°	30°	45°	60°	75°	90°
s						

(c) Use your table from part (b) to determine the angles of the sun that result in the maximum and minimum lengths of the shadow.

(d) Based on your results from part (c), what time of day do you think it is when the angle of the sun above the horizon is 90°?

EXPLORATION

TRUE OR FALSE?   In Exercises 71 and 72, determine whether the statement is true or false. Justify your answer.

71. There can be more than one way to verify a trigonometric identity.

72. The equation $\sin^2\theta + \cos^2\theta = 1 + \tan^2\theta$ is an identity because $\sin^2(0) + \cos^2(0) = 1$ and $1 + \tan^2(0) = 1$.

THINK ABOUT IT   In Exercises 73–77, explain why the equation is not an identity and find one value of the variable for which the equation is not true.

73. $\sin\theta = \sqrt{1 - \cos^2\theta}$ **74.** $\tan\theta = \sqrt{\sec^2\theta - 1}$

75. $1 - \cos\theta = \sin\theta$ **76.** $\csc\theta - 1 = \cot\theta$

77. $1 + \tan\theta = \sec\theta$

78. CAPSTONE   Write a short paper in your own words explaining to a classmate the difference between a trigonometric identity and a conditional equation. Include suggestions on how to verify a trigonometric identity.

5.3 SOLVING TRIGONOMETRIC EQUATIONS

What you should learn

- Use standard algebraic techniques to solve trigonometric equations.
- Solve trigonometric equations of quadratic type.
- Solve trigonometric equations involving multiple angles.
- Use inverse trigonometric functions to solve trigonometric equations.

Why you should learn it

You can use trigonometric equations to solve a variety of real-life problems. For instance, in Exercise 92 on page 396, you can solve a trigonometric equation to help answer questions about monthly sales of skiing equipment.

Introduction

To solve a trigonometric equation, use standard algebraic techniques such as collecting like terms and factoring. Your preliminary goal in solving a trigonometric equation is to *isolate* the trigonometric function in the equation. For example, to solve the equation $2 \sin x = 1$, divide each side by 2 to obtain

$$\sin x = \frac{1}{2}.$$

To solve for x, note in Figure 5.6 that the equation $\sin x = \frac{1}{2}$ has solutions $x = \pi/6$ and $x = 5\pi/6$ in the interval $[0, 2\pi)$. Moreover, because $\sin x$ has a period of 2π, there are infinitely many other solutions, which can be written as

$$x = \frac{\pi}{6} + 2n\pi \qquad \text{and} \qquad x = \frac{5\pi}{6} + 2n\pi \qquad \text{General solution}$$

where n is an integer, as shown in Figure 5.6.

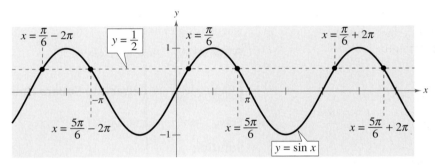

FIGURE 5.6

Another way to show that the equation $\sin x = \frac{1}{2}$ has infinitely many solutions is indicated in Figure 5.7. Any angles that are coterminal with $\pi/6$ or $5\pi/6$ will also be solutions of the equation.

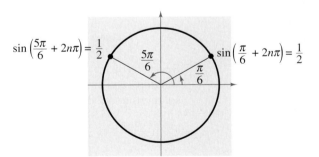

FIGURE 5.7

When solving trigonometric equations, you should write your answer(s) using exact values rather than decimal approximations.

Example 1   Collecting Like Terms

Solve $\sin x + \sqrt{2} = -\sin x$.

Solution

Begin by rewriting the equation so that $\sin x$ is isolated on one side of the equation.

$\sin x + \sqrt{2} = -\sin x$	Write original equation.
$\sin x + \sin x + \sqrt{2} = 0$	Add $\sin x$ to each side.
$\sin x + \sin x = -\sqrt{2}$	Subtract $\sqrt{2}$ from each side.
$2 \sin x = -\sqrt{2}$	Combine like terms.
$\sin x = -\dfrac{\sqrt{2}}{2}$	Divide each side by 2.

Because $\sin x$ has a period of 2π, first find all solutions in the interval $[0, 2\pi)$. These solutions are $x = 5\pi/4$ and $x = 7\pi/4$. Finally, add multiples of 2π to each of these solutions to get the general form

$$x = \frac{5\pi}{4} + 2n\pi \qquad \text{and} \qquad x = \frac{7\pi}{4} + 2n\pi \qquad \text{General solution}$$

where n is an integer.

CHECK*Point*   Now try Exercise 11.

Example 2   Extracting Square Roots

Solve $3 \tan^2 x - 1 = 0$.

Solution

Begin by rewriting the equation so that $\tan x$ is isolated on one side of the equation.

$3 \tan^2 x - 1 = 0$	Write original equation.
$3 \tan^2 x = 1$	Add 1 to each side.
$\tan^2 x = \dfrac{1}{3}$	Divide each side by 3.
$\tan x = \pm\dfrac{1}{\sqrt{3}} = \pm\dfrac{\sqrt{3}}{3}$	Extract square roots.

Because $\tan x$ has a period of π, first find all solutions in the interval $[0, \pi)$. These solutions are $x = \pi/6$ and $x = 5\pi/6$. Finally, add multiples of π to each of these solutions to get the general form

$$x = \frac{\pi}{6} + n\pi \qquad \text{and} \qquad x = \frac{5\pi}{6} + n\pi \qquad \text{General solution}$$

where n is an integer.

CHECK*Point*   Now try Exercise 15.

⚠ **WARNING / CAUTION**

When you extract square roots, make sure you account for both the positive and negative solutions.

The equations in Examples 1 and 2 involved only one trigonometric function. When two or more functions occur in the same equation, collect all terms on one side and try to separate the functions by factoring or by using appropriate identities. This may produce factors that yield no solutions, as illustrated in Example 3.

Example 3   Factoring

Solve $\cot x \cos^2 x = 2 \cot x$.

Solution

Begin by rewriting the equation so that all terms are collected on one side of the equation.

$$\cot x \cos^2 x = 2 \cot x \qquad \text{Write original equation.}$$

$$\cot x \cos^2 x - 2 \cot x = 0 \qquad \text{Subtract 2 cot } x \text{ from each side.}$$

$$\cot x (\cos^2 x - 2) = 0 \qquad \text{Factor.}$$

By setting each of these factors equal to zero, you obtain

$$\cot x = 0 \qquad \text{and} \qquad \cos^2 x - 2 = 0$$

$$x = \frac{\pi}{2} \qquad\qquad\qquad \cos^2 x = 2$$

$$\cos x = \pm\sqrt{2}.$$

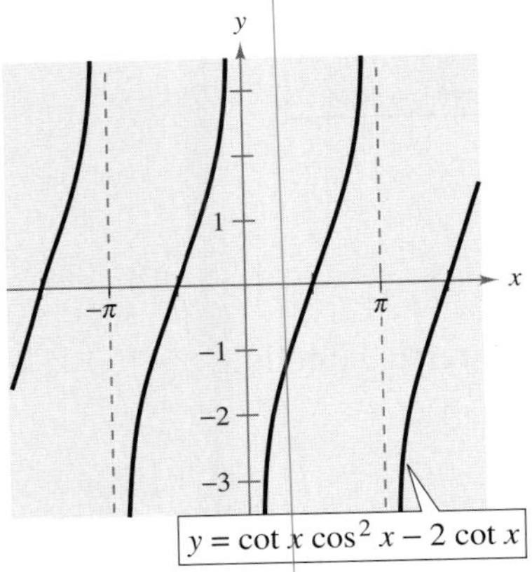

The equation $\cot x = 0$ has the solution $x = \pi/2$ [in the interval $(0, \pi)$]. No solution is obtained for $\cos x = \pm\sqrt{2}$ because $\pm\sqrt{2}$ are outside the range of the cosine function. Because $\cot x$ has a period of π, the general form of the solution is obtained by adding multiples of π to $x = \pi/2$, to get

$$x = \frac{\pi}{2} + n\pi \qquad\qquad \text{General solution}$$

where n is an integer. You can confirm this graphically by sketching the graph of $y = \cot x \cos^2 x - 2 \cot x$, as shown in Figure 5.8. From the graph you can see that the x-intercepts occur at $-3\pi/2$, $-\pi/2$, $\pi/2$, $3\pi/2$, and so on. These x-intercepts correspond to the solutions of $\cot x \cos^2 x - 2 \cot x = 0$.

CHECK *Point*   Now try Exercise 19.

$y = \cot x \cos^2 x - 2 \cot x$

FIGURE 5.8

Equations of Quadratic Type

Many trigonometric equations are of quadratic type $ax^2 + bx + c = 0$. Here are a couple of examples.

Quadratic in sin x	*Quadratic in sec x*
$2\sin^2 x - \sin x - 1 = 0$	$\sec^2 x - 3\sec x - 2 = 0$
$2(\sin x)^2 - \sin x - 1 = 0$	$(\sec x)^2 - 3(\sec x) - 2 = 0$

To solve equations of this type, factor the quadratic or, if this is not possible, use the Quadratic Formula.

Algebra Help

You can review the techniques for solving quadratic equations in Appendix A.5.

Example 4   Factoring an Equation of Quadratic Type

Find all solutions of $2 \sin^2 x - \sin x - 1 = 0$ in the interval $[0, 2\pi)$.

Algebraic Solution

Begin by treating the equation as a quadratic in $\sin x$ and factoring.

$$2 \sin^2 x - \sin x - 1 = 0 \qquad \text{Write original equation.}$$

$$(2 \sin x + 1)(\sin x - 1) = 0 \qquad \text{Factor.}$$

Setting each factor equal to zero, you obtain the following solutions in the interval $[0, 2\pi)$.

$$2 \sin x + 1 = 0 \qquad \text{and} \quad \sin x - 1 = 0$$

$$\sin x = -\frac{1}{2} \qquad\qquad\qquad \sin x = 1$$

$$x = \frac{7\pi}{6}, \frac{11\pi}{6} \qquad\qquad\qquad x = \frac{\pi}{2}$$

Graphical Solution

Use a graphing utility set in *radian* mode to graph $y = 2 \sin^2 x - \sin x - 1$ for $0 \le x < 2\pi$, as shown in Figure 5.9. Use the *zero* or *root* feature or the *zoom* and *trace* features to approximate the x-intercepts to be

$$x \approx 1.571 \approx \frac{\pi}{2}, \quad x \approx 3.665 \approx \frac{7\pi}{6}, \quad \text{and} \quad x \approx 5.760 \approx \frac{11\pi}{6}.$$

These values are the approximate solutions of $2 \sin^2 x - \sin x - 1 = 0$ in the interval $[0, 2\pi)$.

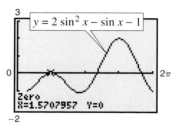

FIGURE **5.9**

CHECK *Point*   Now try Exercise 33.

Example 5   Rewriting with a Single Trigonometric Function

Solve $2 \sin^2 x + 3 \cos x - 3 = 0$.

Solution

This equation contains both sine and cosine functions. You can rewrite the equation so that it has only cosine functions by using the identity $\sin^2 x = 1 - \cos^2 x$.

$$2 \sin^2 x + 3 \cos x - 3 = 0 \qquad \text{Write original equation.}$$

$$2(1 - \cos^2 x) + 3 \cos x - 3 = 0 \qquad \text{Pythagorean identity}$$

$$2 \cos^2 x - 3 \cos x + 1 = 0 \qquad \text{Multiply each side by } -1.$$

$$(2 \cos x - 1)(\cos x - 1) = 0 \qquad \text{Factor.}$$

Set each factor equal to zero to find the solutions in the interval $[0, 2\pi)$.

$$2 \cos x - 1 = 0 \quad \implies \quad \cos x = \frac{1}{2} \quad \implies \quad x = \frac{\pi}{3}, \frac{5\pi}{3}$$

$$\cos x - 1 = 0 \quad \implies \quad \cos x = 1 \quad \implies \quad x = 0$$

Because $\cos x$ has a period of 2π, the general form of the solution is obtained by adding multiples of 2π to get

$$x = 2n\pi, \quad x = \frac{\pi}{3} + 2n\pi, \quad x = \frac{5\pi}{3} + 2n\pi \qquad \text{General solution}$$

where n is an integer.

CHECK *Point*   Now try Exercise 35.

Sometimes you must square each side of an equation to obtain a quadratic, as demonstrated in the next example. Because this procedure can introduce extraneous solutions, you should check any solutions in the original equation to see whether they are valid or extraneous.

Example 6   Squaring and Converting to Quadratic Type

Find all solutions of $\cos x + 1 = \sin x$ in the interval $[0, 2\pi)$.

Solution

It is not clear how to rewrite this equation in terms of a single trigonometric function. Notice what happens when you square each side of the equation.

$$\cos x + 1 = \sin x \qquad \text{Write original equation.}$$

$$\cos^2 x + 2\cos x + 1 = \sin^2 x \qquad \text{Square each side.}$$

$$\cos^2 x + 2\cos x + 1 = 1 - \cos^2 x \qquad \text{Pythagorean identity}$$

$$\cos^2 x + \cos^2 x + 2\cos x + 1 - 1 = 0 \qquad \text{Rewrite equation.}$$

$$2\cos^2 x + 2\cos x = 0 \qquad \text{Combine like terms.}$$

$$2\cos x(\cos x + 1) = 0 \qquad \text{Factor.}$$

Setting each factor equal to zero produces

$$2\cos x = 0 \qquad \text{and} \qquad \cos x + 1 = 0$$

$$\cos x = 0 \qquad\qquad\qquad \cos x = -1$$

$$x = \frac{\pi}{2}, \frac{3\pi}{2} \qquad\qquad\qquad x = \pi.$$

Because you squared the original equation, check for extraneous solutions.

Check $x = \pi/2$

$$\cos\frac{\pi}{2} + 1 \overset{?}{=} \sin\frac{\pi}{2} \qquad \text{Substitute } \pi/2 \text{ for } x.$$

$$0 + 1 = 1 \qquad \text{Solution checks. } \checkmark$$

Check $x = 3\pi/2$

$$\cos\frac{3\pi}{2} + 1 \overset{?}{=} \sin\frac{3\pi}{2} \qquad \text{Substitute } 3\pi/2 \text{ for } x.$$

$$0 + 1 \ne -1 \qquad \text{Solution does not check.}$$

Check $x = \pi$

$$\cos\pi + 1 \overset{?}{=} \sin\pi \qquad \text{Substitute } \pi \text{ for } x.$$

$$-1 + 1 = 0 \qquad \text{Solution checks. } \checkmark$$

Of the three possible solutions, $x = 3\pi/2$ is extraneous. So, in the interval $[0, 2\pi)$, the only two solutions are $x = \pi/2$ and $x = \pi$.

CHECK Point   Now try Exercise 37.

Study Tip

You square each side of the equation in Example 6 because the squares of the sine and cosine functions are related by a Pythagorean identity. The same is true for the squares of the secant and tangent functions and for the squares of the cosecant and cotangent functions.

Functions Involving Multiple Angles

The next two examples involve trigonometric functions of multiple angles of the forms $\sin ku$ and $\cos ku$. To solve equations of these forms, first solve the equation for ku, then divide your result by k.

Example 7 Functions of Multiple Angles

Solve $2 \cos 3t - 1 = 0$.

Solution

$$2 \cos 3t - 1 = 0 \qquad \text{Write original equation.}$$

$$2 \cos 3t = 1 \qquad \text{Add 1 to each side.}$$

$$\cos 3t = \frac{1}{2} \qquad \text{Divide each side by 2.}$$

In the interval $[0, 2\pi)$, you know that $3t = \pi/3$ and $3t = 5\pi/3$ are the only solutions, so, in general, you have

$$3t = \frac{\pi}{3} + 2n\pi \qquad \text{and} \qquad 3t = \frac{5\pi}{3} + 2n\pi.$$

Dividing these results by 3, you obtain the general solution

$$t = \frac{\pi}{9} + \frac{2n\pi}{3} \qquad \text{and} \qquad t = \frac{5\pi}{9} + \frac{2n\pi}{3} \qquad \text{General solution}$$

where n is an integer.

CHECK Point Now try Exercise 39.

Example 8 Functions of Multiple Angles

Solve $3 \tan \dfrac{x}{2} + 3 = 0$.

Solution

$$3 \tan \frac{x}{2} + 3 = 0 \qquad \text{Write original equation.}$$

$$3 \tan \frac{x}{2} = -3 \qquad \text{Subtract 3 from each side.}$$

$$\tan \frac{x}{2} = -1 \qquad \text{Divide each side by 3.}$$

In the interval $[0, \pi)$, you know that $x/2 = 3\pi/4$ is the only solution, so, in general, you have

$$\frac{x}{2} = \frac{3\pi}{4} + n\pi.$$

Multiplying this result by 2, you obtain the general solution

$$x = \frac{3\pi}{2} + 2n\pi \qquad \text{General solution}$$

where n is an integer.

CHECK Point Now try Exercise 43.

Using Inverse Functions

In the next example, you will see how inverse trigonometric functions can be used to solve an equation.

Example 9   Using Inverse Functions

Solve $\sec^2 x - 2 \tan x = 4$.

Solution

$\sec^2 x - 2 \tan x = 4$	Write original equation.
$1 + \tan^2 x - 2 \tan x - 4 = 0$	Pythagorean identity
$\tan^2 x - 2 \tan x - 3 = 0$	Combine like terms.
$(\tan x - 3)(\tan x + 1) = 0$	Factor.

Setting each factor equal to zero, you obtain two solutions in the interval $(-\pi/2, \pi/2)$. [Recall that the range of the inverse tangent function is $(-\pi/2, \pi/2)$.]

$$\tan x - 3 = 0 \qquad \text{and} \qquad \tan x + 1 = 0$$

$$\tan x = 3 \qquad\qquad\qquad \tan x = -1$$

$$x = \arctan 3 \qquad\qquad\qquad x = -\frac{\pi}{4}$$

Finally, because $\tan x$ has a period of π, you obtain the general solution by adding multiples of π

$$x = \arctan 3 + n\pi \qquad \text{and} \qquad x = -\frac{\pi}{4} + n\pi \qquad \text{General solution}$$

where n is an integer. You can use a calculator to approximate the value of arctan 3.

CHECK *Point* Now try Exercise 63.

CLASSROOM DISCUSSION

Equations with No Solutions   One of the following equations has solutions and the other two do not. Which two equations do not have solutions?

a. $\sin^2 x - 5 \sin x + 6 = 0$

b. $\sin^2 x - 4 \sin x + 6 = 0$

c. $\sin^2 x - 5 \sin x - 6 = 0$

Find conditions involving the constants b and c that will guarantee that the equation

$$\sin^2 x + b \sin x + c = 0$$

has at least one solution on some interval of length 2π.

5.3 EXERCISES

See www.CalcChat.com for worked-out solutions to odd-numbered exercises.

VOCABULARY: Fill in the blanks.

1. When solving a trigonometric equation, the preliminary goal is to _____ the trigonometric function involved in the equation.

2. The equation $2 \sin \theta + 1 = 0$ has the solutions $\theta = \dfrac{7\pi}{6} + 2n\pi$ and $\theta = \dfrac{11\pi}{6} + 2n\pi$, which are called _____ solutions.

3. The equation $2 \tan^2 x - 3 \tan x + 1 = 0$ is a trigonometric equation that is of _____ type.

4. A solution of an equation that does not satisfy the original equation is called an _____ solution.

SKILLS AND APPLICATIONS

In Exercises 5–10, verify that the x-values are solutions of the equation.

5. $2 \cos x - 1 = 0$

   (a) $x = \dfrac{\pi}{3}$ (b) $x = \dfrac{5\pi}{3}$

6. $\sec x - 2 = 0$

   (a) $x = \dfrac{\pi}{3}$ (b) $x = \dfrac{5\pi}{3}$

7. $3 \tan^2 2x - 1 = 0$

   (a) $x = \dfrac{\pi}{12}$ (b) $x = \dfrac{5\pi}{12}$

8. $2 \cos^2 4x - 1 = 0$

   (a) $x = \dfrac{\pi}{16}$ (b) $x = \dfrac{3\pi}{16}$

9. $2 \sin^2 x - \sin x - 1 = 0$

   (a) $x = \dfrac{\pi}{2}$ (b) $x = \dfrac{7\pi}{6}$

10. $\csc^4 x - 4 \csc^2 x = 0$

   (a) $x = \dfrac{\pi}{6}$ (b) $x = \dfrac{5\pi}{6}$

In Exercises 11–24, solve the equation.

11. $2 \cos x + 1 = 0$ **12.** $2 \sin x + 1 = 0$

13. $\sqrt{3} \csc x - 2 = 0$ **14.** $\tan x + \sqrt{3} = 0$

15. $3 \sec^2 x - 4 = 0$ **16.** $3 \cot^2 x - 1 = 0$

17. $\sin x(\sin x + 1) = 0$

18. $(3 \tan^2 x - 1)(\tan^2 x - 3) = 0$

19. $4 \cos^2 x - 1 = 0$ **20.** $\sin^2 x = 3 \cos^2 x$

21. $2 \sin^2 2x = 1$ **22.** $\tan^2 3x = 3$

23. $\tan 3x(\tan x - 1) = 0$ **24.** $\cos 2x(2 \cos x + 1) = 0$

In Exercises 25–38, find all solutions of the equation in the interval $[0, 2\pi)$.

25. $\cos^3 x = \cos x$ **26.** $\sec^2 x - 1 = 0$

27. $3 \tan^3 x = \tan x$ **28.** $2 \sin^2 x = 2 + \cos x$

29. $\sec^2 x - \sec x = 2$ **30.** $\sec x \csc x = 2 \csc x$

31. $2 \sin x + \csc x = 0$ **32.** $\sec x + \tan x = 1$

33. $2 \cos^2 x + \cos x - 1 = 0$

34. $2 \sin^2 x + 3 \sin x + 1 = 0$

35. $2 \sec^2 x + \tan^2 x - 3 = 0$

36. $\cos x + \sin x \tan x = 2$

37. $\csc x + \cot x = 1$ **38.** $\sin x - 2 = \cos x - 2$

In Exercises 39–44, solve the multiple-angle equation.

39. $\cos 2x = \dfrac{1}{2}$ **40.** $\sin 2x = -\dfrac{\sqrt{3}}{2}$

41. $\tan 3x = 1$ **42.** $\sec 4x = 2$

43. $\cos \dfrac{x}{2} = \dfrac{\sqrt{2}}{2}$ **44.** $\sin \dfrac{x}{2} = -\dfrac{\sqrt{3}}{2}$

In Exercises 45–48, find the x-intercepts of the graph.

45. $y = \sin \dfrac{\pi x}{2} + 1$ **46.** $y = \sin \pi x + \cos \pi x$

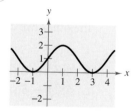
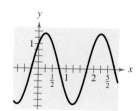

47. $y = \tan^2 \left(\dfrac{\pi x}{6}\right) - 3$ **48.** $y = \sec^4 \left(\dfrac{\pi x}{8}\right) - 4$

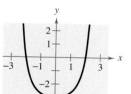
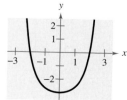

In Exercises 49–58, use a graphing utility to approximate the solutions (to three decimal places) of the equation in the interval $[0, 2\pi)$.

49. $2 \sin x + \cos x = 0$

50. $4 \sin^3 x + 2 \sin^2 x - 2 \sin x - 1 = 0$

51. $\dfrac{1 + \sin x}{\cos x} + \dfrac{\cos x}{1 + \sin x} = 4$ **52.** $\dfrac{\cos x \cot x}{1 - \sin x} = 3$

53. $x \tan x - 1 = 0$ **54.** $x \cos x - 1 = 0$

55. $\sec^2 x + 0.5 \tan x - 1 = 0$

56. $\csc^2 x + 0.5 \cot x - 5 = 0$

57. $2 \tan^2 x + 7 \tan x - 15 = 0$

58. $6 \sin^2 x - 7 \sin x + 2 = 0$

 In Exercises 59–62, use the Quadratic Formula to solve the equation in the interval $[0, 2\pi)$. Then use a graphing utility to approximate the angle x.

59. $12 \sin^2 x - 13 \sin x + 3 = 0$

60. $3 \tan^2 x + 4 \tan x - 4 = 0$

61. $\tan^2 x + 3 \tan x + 1 = 0$

62. $4 \cos^2 x - 4 \cos x - 1 = 0$

In Exercises 63–74, use inverse functions where needed to find all solutions of the equation in the interval $[0, 2\pi)$.

63. $\tan^2 x + \tan x - 12 = 0$

64. $\tan^2 x - \tan x - 2 = 0$

65. $\tan^2 x - 6 \tan x + 5 = 0$

66. $\sec^2 x + \tan x - 3 = 0$

67. $2 \cos^2 x - 5 \cos x + 2 = 0$

68. $2 \sin^2 x - 7 \sin x + 3 = 0$

69. $\cot^2 x - 9 = 0$

70. $\cot^2 x - 6 \cot x + 5 = 0$

71. $\sec^2 x - 4 \sec x = 0$

72. $\sec^2 x + 2 \sec x - 8 = 0$

73. $\csc^2 x + 3 \csc x - 4 = 0$

74. $\csc^2 x - 5 \csc x = 0$

 In Exercises 75–78, use a graphing utility to approximate the solutions (to three decimal places) of the equation in the given interval.

75. $3 \tan^2 x + 5 \tan x - 4 = 0$, $\left[-\dfrac{\pi}{2}, \dfrac{\pi}{2}\right]$

76. $\cos^2 x - 2 \cos x - 1 = 0$, $[0, \pi]$

77. $4 \cos^2 x - 2 \sin x + 1 = 0$, $\left[-\dfrac{\pi}{2}, \dfrac{\pi}{2}\right]$

78. $2 \sec^2 x + \tan x - 6 = 0$, $\left[-\dfrac{\pi}{2}, \dfrac{\pi}{2}\right]$

 In Exercises 79–84, (a) use a graphing utility to graph the function and approximate the maximum and minimum points on the graph in the interval $[0, 2\pi)$, and (b) solve the trigonometric equation and demonstrate that its solutions are the x-coordinates of the maximum and minimum points of f. (Calculus is required to find the trigonometric equation.)

Function	*Trigonometric Equation*
79. $f(x) = \sin^2 x + \cos x$	$2 \sin x \cos x - \sin x = 0$
80. $f(x) = \cos^2 x - \sin x$	$-2 \sin x \cos x - \cos x = 0$
81. $f(x) = \sin x + \cos x$	$\cos x - \sin x = 0$
82. $f(x) = 2 \sin x + \cos 2x$	$2 \cos x - 4 \sin x \cos x = 0$
83. $f(x) = \sin x \cos x$	$-\sin^2 x + \cos^2 x = 0$
84. $f(x) = \sec x + \tan x - x$	

$$\sec x \tan x + \sec^2 x - 1 = 0$$

FIXED POINT   In Exercises 85 and 86, find the smallest positive fixed point of the function f. [A *fixed point* of a function f is a real number c such that $f(c) = c$.]

85. $f(x) = \tan \dfrac{\pi x}{4}$ **86.** $f(x) = \cos x$

87. GRAPHICAL REASONING   Consider the function given by

$$f(x) = \cos \frac{1}{x}$$

and its graph shown in the figure.

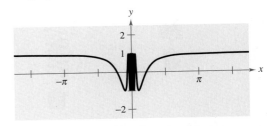

(a) What is the domain of the function?

(b) Identify any symmetry and any asymptotes of the graph.

(c) Describe the behavior of the function as $x \to 0$.

(d) How many solutions does the equation

$$\cos \frac{1}{x} = 0$$

have in the interval $[-1, 1]$? Find the solutions.

(e) Does the equation $\cos(1/x) = 0$ have a greatest solution? If so, approximate the solution. If not, explain why.

88. GRAPHICAL REASONING Consider the function given by $f(x) = (\sin x)/x$ and its graph shown in the figure.

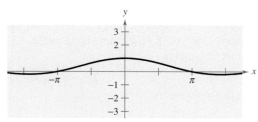

(a) What is the domain of the function?

(b) Identify any symmetry and any asymptotes of the graph.

(c) Describe the behavior of the function as $x \to 0$.

(d) How many solutions does the equation

$$\frac{\sin x}{x} = 0$$

have in the interval $[-8, 8]$? Find the solutions.

89. HARMONIC MOTION A weight is oscillating on the end of a spring (see figure). The position of the weight relative to the point of equilibrium is given by $y = \frac{1}{12}(\cos 8t - 3 \sin 8t)$, where y is the displacement (in meters) and t is the time (in seconds). Find the times when the weight is at the point of equilibrium ($y = 0$) for $0 \le t \le 1$.

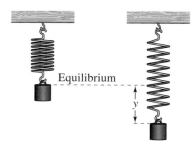

90. DAMPED HARMONIC MOTION The displacement from equilibrium of a weight oscillating on the end of a spring is given by $y = 1.56e^{-0.22t}\cos 4.9t$, where y is the displacement (in feet) and t is the time (in seconds). Use a graphing utility to graph the displacement function for $0 \le t \le 10$. Find the time beyond which the displacement does not exceed 1 foot from equilibrium.

91. SALES The monthly sales S (in thousands of units) of a seasonal product are approximated by

$$S = 74.50 + 43.75 \sin \frac{\pi t}{6}$$

where t is the time (in months), with $t = 1$ corresponding to January. Determine the months in which sales exceed 100,000 units.

92. SALES The monthly sales S (in hundreds of units) of skiing equipment at a sports store are approximated by

$$S = 58.3 + 32.5 \cos \frac{\pi t}{6}$$

where t is the time (in months), with $t = 1$ corresponding to January. Determine the months in which sales exceed 7500 units.

93. PROJECTILE MOTION A batted baseball leaves the bat at an angle of θ with the horizontal and an initial velocity of $v_0 = 100$ feet per second. The ball is caught by an outfielder 300 feet from home plate (see figure). Find θ if the range r of a projectile is given by $r = \frac{1}{32}v_0^2 \sin 2\theta$.

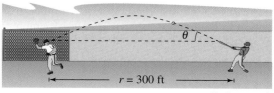

Not drawn to scale

94. PROJECTILE MOTION A sharpshooter intends to hit a target at a distance of 1000 yards with a gun that has a muzzle velocity of 1200 feet per second (see figure). Neglecting air resistance, determine the gun's minimum angle of elevation θ if the range r is given by

$$r = \frac{1}{32}v_0^2 \sin 2\theta.$$

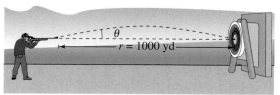

Not drawn to scale

95. FERRIS WHEEL A Ferris wheel is built such that the height h (in feet) above ground of a seat on the wheel at time t (in minutes) can be modeled by

$$h(t) = 53 + 50 \sin\left(\frac{\pi}{16}t - \frac{\pi}{2}\right).$$

The wheel makes one revolution every 32 seconds. The ride begins when $t = 0$.

(a) During the first 32 seconds of the ride, when will a person on the Ferris wheel be 53 feet above ground?

(b) When will a person be at the top of the Ferris wheel for the first time during the ride? If the ride lasts 160 seconds, how many times will a person be at the top of the ride, and at what times?

96. DATA ANALYSIS: METEOROLOGY The table shows the average daily high temperatures in Houston H (in degrees Fahrenheit) for month t, with $t = 1$ corresponding to January. (Source: National Climatic Data Center)

Month, t	Houston, H
1	62.3
2	66.5
3	73.3
4	79.1
5	85.5
6	90.7
7	93.6
8	93.5
9	89.3
10	82.0
11	72.0
12	64.6

(a) Create a scatter plot of the data.

(b) Find a cosine model for the temperatures in Houston.

(c) Use a graphing utility to graph the data points and the model for the temperatures in Houston. How well does the model fit the data?

(d) What is the overall average daily high temperature in Houston?

(e) Use a graphing utility to describe the months during which the average daily high temperature is above 86°F and below 86°F.

97. GEOMETRY The area of a rectangle (see figure) inscribed in one arc of the graph of $y = \cos x$ is given by $A = 2x \cos x, \ 0 < x < \pi/2$.

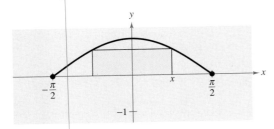

(a) Use a graphing utility to graph the area function, and approximate the area of the largest inscribed rectangle.

(b) Determine the values of x for which $A \geq 1$.

98. QUADRATIC APPROXIMATION Consider the function given by $f(x) = 3 \sin(0.6x - 2)$.

(a) Approximate the zero of the function in the interval $[0, 6]$.

(b) A quadratic approximation agreeing with f at $x = 5$ is $g(x) = -0.45x^2 + 5.52x - 13.70$. Use a graphing utility to graph f and g in the same viewing window. Describe the result.

(c) Use the Quadratic Formula to find the zeros of g. Compare the zero in the interval $[0, 6]$ with the result of part (a).

EXPLORATION

TRUE OR FALSE? In Exercises 99 and 100, determine whether the statement is true or false. Justify your answer.

99. The equation $2 \sin 4t - 1 = 0$ has four times the number of solutions in the interval $[0, 2\pi)$ as the equation $2 \sin t - 1 = 0$.

100. If you correctly solve a trigonometric equation to the statement $\sin x = 3.4$, then you can finish solving the equation by using an inverse function.

101. THINK ABOUT IT Explain what would happen if you divided each side of the equation $\cot x \cos^2 x = 2 \cot x$ by $\cot x$. Is this a correct method to use when solving equations?

102. GRAPHICAL REASONING Use a graphing utility to confirm the solutions found in Example 6 in two different ways.

(a) Graph both sides of the equation and find the x-coordinates of the points at which the graphs intersect.

Left side: $y = \cos x + 1$

Right side: $y = \sin x$

(b) Graph the equation $y = \cos x + 1 - \sin x$ and find the x-intercepts of the graph. Do both methods produce the same x-values? Which method do you prefer? Explain.

103. Explain in your own words how knowledge of algebra is important when solving trigonometric equations.

104. CAPSTONE Consider the equation $2 \sin x - 1 = 0$. Explain the similarities and differences between finding all solutions in the interval $\left[0, \dfrac{\pi}{2} \right)$, finding all solutions in the interval $[0, 2\pi)$, and finding the general solution.

PROJECT: METEOROLOGY To work an extended application analyzing the normal daily high temperatures in Phoenix and in Seattle, visit this text's website at *academic.cengage.com*. (Data Source: NOAA)

5.4   SUM AND DIFFERENCE FORMULAS

What you should learn

- Use sum and difference formulas to evaluate trigonometric functions, verify identities, and solve trigonometric equations.

Why you should learn it

You can use identities to rewrite trigonometric expressions. For instance, in Exercise 89 on page 403, you can use an identity to rewrite a trigonometric expression in a form that helps you analyze a harmonic motion equation.

Richard Megna/Fundamental Photographs

Using Sum and Difference Formulas

In this and the following section, you will study the uses of several trigonometric identities and formulas.

Sum and Difference Formulas

$$\sin(u + v) = \sin u \cos v + \cos u \sin v$$

$$\sin(u - v) = \sin u \cos v - \cos u \sin v$$

$$\cos(u + v) = \cos u \cos v - \sin u \sin v$$

$$\cos(u - v) = \cos u \cos v + \sin u \sin v$$

$$\tan(u + v) = \frac{\tan u + \tan v}{1 - \tan u \tan v} \qquad \tan(u - v) = \frac{\tan u - \tan v}{1 + \tan u \tan v}$$

For a proof of the sum and difference formulas, see Proofs in Mathematics on page 422.

Examples 1 and 2 show how **sum and difference formulas** can be used to find exact values of trigonometric functions involving sums or differences of special angles.

Example 1   Evaluating a Trigonometric Function

Find the exact value of $\sin \dfrac{\pi}{12}$.

Solution

To find the *exact* value of $\sin \dfrac{\pi}{12}$, use the fact that

$$\frac{\pi}{12} = \frac{\pi}{3} - \frac{\pi}{4}.$$

Consequently, the formula for $\sin(u - v)$ yields

$$\sin \frac{\pi}{12} = \sin\left(\frac{\pi}{3} - \frac{\pi}{4}\right)$$

$$= \sin \frac{\pi}{3} \cos \frac{\pi}{4} - \cos \frac{\pi}{3} \sin \frac{\pi}{4}$$

$$= \frac{\sqrt{3}}{2}\left(\frac{\sqrt{2}}{2}\right) - \frac{1}{2}\left(\frac{\sqrt{2}}{2}\right)$$

$$= \frac{\sqrt{6} - \sqrt{2}}{4}.$$

Try checking this result on your calculator. You will find that $\sin \dfrac{\pi}{12} \approx 0.259$.

CHECK Point   Now try Exercise 7.

Study Tip

Another way to solve
Example 2 is to use the fact that
$75° = 120° - 45°$ together with
the formula for $\cos(u - v)$.

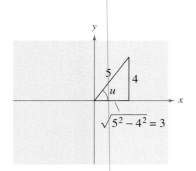

FIGURE **5.10**

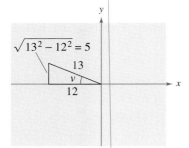

FIGURE **5.11**

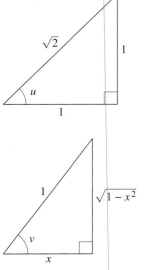

FIGURE **5.12**

Example 2   Evaluating a Trigonometric Function

Find the exact value of $\cos 75°$.

Solution

Using the fact that $75° = 30° + 45°$, together with the formula for $\cos(u + v)$, you obtain

$$\cos 75° = \cos(30° + 45°)$$

$$= \cos 30° \cos 45° - \sin 30° \sin 45°$$

$$= \frac{\sqrt{3}}{2}\left(\frac{\sqrt{2}}{2}\right) - \frac{1}{2}\left(\frac{\sqrt{2}}{2}\right) = \frac{\sqrt{6} - \sqrt{2}}{4}.$$

CHECK *Point*   Now try Exercise 11.

Example 3   Evaluating a Trigonometric Expression

Find the exact value of $\sin(u + v)$ given

$$\sin u = \frac{4}{5}, \text{ where } 0 < u < \frac{\pi}{2}, \quad \text{and} \quad \cos v = -\frac{12}{13}, \text{ where } \frac{\pi}{2} < v < \pi.$$

Solution

Because $\sin u = 4/5$ and u is in Quadrant I, $\cos u = 3/5$, as shown in Figure 5.10. Because $\cos v = -12/13$ and v is in Quadrant II, $\sin v = 5/13$, as shown in Figure 5.11. You can find $\sin(u + v)$ as follows.

$$\sin(u + v) = \sin u \cos v + \cos u \sin v$$

$$= \left(\frac{4}{5}\right)\left(-\frac{12}{13}\right) + \left(\frac{3}{5}\right)\left(\frac{5}{13}\right)$$

$$= -\frac{48}{65} + \frac{15}{65}$$

$$= -\frac{33}{65}$$

CHECK *Point*   Now try Exercise 43.

Example 4   An Application of a Sum Formula

Write $\cos(\arctan 1 + \arccos x)$ as an algebraic expression.

Solution

This expression fits the formula for $\cos(u + v)$. Angles $u = \arctan 1$ and $v = \arccos x$ are shown in Figure 5.12. So

$$\cos(u + v) = \cos(\arctan 1) \cos(\arccos x) - \sin(\arctan 1) \sin(\arccos x)$$

$$= \frac{1}{\sqrt{2}} \cdot x - \frac{1}{\sqrt{2}} \cdot \sqrt{1 - x^2}$$

$$= \frac{x - \sqrt{1 - x^2}}{\sqrt{2}}.$$

CHECK *Point*   Now try Exercise 57.

HISTORICAL NOTE

Hipparchus, considered the most eminent of Greek astronomers, was born about 190 B.C. in Nicaea. He was credited with the invention of trigonometry. He also derived the sum and difference formulas for $\sin(A \pm B)$ and $\cos(A \pm B)$.

Example 5 shows how to use a difference formula to prove the cofunction identity

$$\cos\left(\frac{\pi}{2} - x\right) = \sin x.$$

Example 5   Proving a Cofunction Identity

Prove the cofunction identity $\cos\left(\dfrac{\pi}{2} - x\right) = \sin x.$

Solution

Using the formula for $\cos(u - v)$, you have

$$\cos\left(\frac{\pi}{2} - x\right) = \cos\frac{\pi}{2}\cos x + \sin\frac{\pi}{2}\sin x$$

$$= (0)(\cos x) + (1)(\sin x)$$

$$= \sin x.$$

CHECK Point   Now try Exercise 61.

Sum and difference formulas can be used to rewrite expressions such as

$$\sin\left(\theta + \frac{n\pi}{2}\right) \quad \text{and} \quad \cos\left(\theta + \frac{n\pi}{2}\right), \quad \text{where } n \text{ is an integer}$$

as expressions involving only $\sin\theta$ or $\cos\theta$. The resulting formulas are called **reduction formulas.**

Example 6   Deriving Reduction Formulas

Simplify each expression.

a. $\cos\left(\theta - \dfrac{3\pi}{2}\right)$ **b.** $\tan(\theta + 3\pi)$

Solution

a. Using the formula for $\cos(u - v)$, you have

$$\cos\left(\theta - \frac{3\pi}{2}\right) = \cos\theta\cos\frac{3\pi}{2} + \sin\theta\sin\frac{3\pi}{2}$$

$$= (\cos\theta)(0) + (\sin\theta)(-1)$$

$$= -\sin\theta.$$

b. Using the formula for $\tan(u + v)$, you have

$$\tan(\theta + 3\pi) = \frac{\tan\theta + \tan 3\pi}{1 - \tan\theta\tan 3\pi}$$

$$= \frac{\tan\theta + 0}{1 - (\tan\theta)(0)}$$

$$= \tan\theta.$$

CHECK Point   Now try Exercise 73.

Example 7   **Solving a Trigonometric Equation**

Find all solutions of $\sin\left(x + \dfrac{\pi}{4}\right) + \sin\left(x - \dfrac{\pi}{4}\right) = -1$ in the interval $[0, 2\pi)$.

Algebraic Solution

Using sum and difference formulas, rewrite the equation as

$$\sin x \cos \frac{\pi}{4} + \cos x \sin \frac{\pi}{4} + \sin x \cos \frac{\pi}{4} - \cos x \sin \frac{\pi}{4} = -1$$

$$2 \sin x \cos \frac{\pi}{4} = -1$$

$$2(\sin x)\left(\frac{\sqrt{2}}{2}\right) = \frac{1}{2}$$

$$\sin x = -\frac{1}{\sqrt{2}}$$

$$\sin x = -\frac{\sqrt{2}}{2}.$$

So, the only solutions in the interval $[0, 2\pi)$ are

$$x = \frac{5\pi}{4} \quad \text{and} \quad x = \frac{7\pi}{4}.$$

Graphical Solution

Sketch the graph of

$$y = \sin\left(x + \frac{\pi}{4}\right) + \sin\left(x - \frac{\pi}{4}\right) + 1 \text{ for } 0 \le x < 2\pi.$$

as shown in Figure 5.13. From the graph you can see that the x-intercepts are $5\pi/4$ and $7\pi/4$. So, the solutions in the interval $[0, 2\pi)$ are

$$x = \frac{5\pi}{4} \quad \text{and} \quad x = \frac{7\pi}{4}.$$

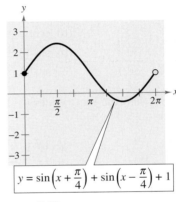

$$y = \sin\left(x + \frac{\pi}{4}\right) + \sin\left(x - \frac{\pi}{4}\right) + 1$$

FIGURE **5.13**

CHECK*Point*   Now try Exercise 79.

The next example was taken from calculus. It is used to derive the derivative of the sine function.

Example 8   **An Application from Calculus**

Verify that $\dfrac{\sin(x + h) - \sin x}{h} = (\cos x)\left(\dfrac{\sin h}{h}\right) - (\sin x)\left(\dfrac{1 - \cos h}{h}\right)$ where $h \ne 0$.

Solution

Using the formula for $\sin(u + v)$, you have

$$\frac{\sin(x + h) - \sin x}{h} = \frac{\sin x \cos h + \cos x \sin h - \sin x}{h}$$

$$= \frac{\cos x \sin h - \sin x(1 - \cos h)}{h}$$

$$= (\cos x)\left(\frac{\sin h}{h}\right) - (\sin x)\left(\frac{1 - \cos h}{h}\right).$$

CHECK*Point*   Now try Exercise 105.

5.4 EXERCISES

See www.CalcChat.com for worked-out solutions to odd-numbered exercises.

VOCABULARY: Fill in the blank.

1. $\sin(u - v) =$ _____

2. $\cos(u + v) =$ _____

3. $\tan(u + v) =$ _____

4. $\sin(u + v) =$ _____

5. $\cos(u - v) =$ _____

6. $\tan(u - v) =$ _____

SKILLS AND APPLICATIONS

In Exercises 7–12, find the exact value of each expression.

7. (a) $\cos\left(\dfrac{\pi}{4} + \dfrac{\pi}{3}\right)$ (b) $\cos\dfrac{\pi}{4} + \cos\dfrac{\pi}{3}$

8. (a) $\sin\left(\dfrac{3\pi}{4} + \dfrac{5\pi}{6}\right)$ (b) $\sin\dfrac{3\pi}{4} + \sin\dfrac{5\pi}{6}$

9. (a) $\sin\left(\dfrac{7\pi}{6} - \dfrac{\pi}{3}\right)$ (b) $\sin\dfrac{7\pi}{6} - \sin\dfrac{\pi}{3}$

10. (a) $\cos(120° + 45°)$ (b) $\cos 120° + \cos 45°$

11. (a) $\sin(135° - 30°)$ (b) $\sin 135° - \cos 30°$

12. (a) $\sin(315° - 60°)$ (b) $\sin 315° - \sin 60°$

In Exercises 13–28, find the exact values of the sine, cosine, and tangent of the angle.

13. $\dfrac{11\pi}{12} = \dfrac{3\pi}{4} + \dfrac{\pi}{6}$

14. $\dfrac{7\pi}{12} = \dfrac{\pi}{3} + \dfrac{\pi}{4}$

15. $\dfrac{17\pi}{12} = \dfrac{9\pi}{4} - \dfrac{5\pi}{6}$

16. $-\dfrac{\pi}{12} = \dfrac{\pi}{6} - \dfrac{\pi}{4}$

17. $105° = 60° + 45°$

18. $165° = 135° + 30°$

19. $195° = 225° - 30°$

20. $255° = 300° - 45°$

21. $\dfrac{13\pi}{12}$

22. $-\dfrac{7\pi}{12}$

23. $-\dfrac{13\pi}{12}$

24. $\dfrac{5\pi}{12}$

25. $285°$

26. $-105°$

27. $-165°$

28. $15°$

In Exercises 29–36, write the expression as the sine, cosine, or tangent of an angle.

29. $\sin 3 \cos 1.2 - \cos 3 \sin 1.2$

30. $\cos\dfrac{\pi}{7}\cos\dfrac{\pi}{5} - \sin\dfrac{\pi}{7}\sin\dfrac{\pi}{5}$

31. $\sin 60° \cos 15° + \cos 60° \sin 15°$

32. $\cos 130° \cos 40° - \sin 130° \sin 40°$

33. $\dfrac{\tan 45° - \tan 30°}{1 + \tan 45° \tan 30°}$

34. $\dfrac{\tan 140° - \tan 60°}{1 + \tan 140° \tan 60°}$

35. $\dfrac{\tan 2x + \tan x}{1 - \tan 2x \tan x}$

36. $\cos 3x \cos 2y + \sin 3x \sin 2y$

In Exercises 37–42, find the exact value of the expression.

37. $\sin\dfrac{\pi}{12}\cos\dfrac{\pi}{4} + \cos\dfrac{\pi}{12}\sin\dfrac{\pi}{4}$

38. $\cos\dfrac{\pi}{16}\cos\dfrac{3\pi}{16} - \sin\dfrac{\pi}{16}\sin\dfrac{3\pi}{16}$

39. $\sin 120° \cos 60° - \cos 120° \sin 60°$

40. $\cos 120° \cos 30° + \sin 120° \sin 30°$

41. $\dfrac{\tan(5\pi/6) - \tan(\pi/6)}{1 + \tan(5\pi/6)\tan(\pi/6)}$

42. $\dfrac{\tan 25° + \tan 110°}{1 - \tan 25° \tan 110°}$

In Exercises 43–50, find the exact value of the trigonometric function given that $\sin u = \frac{5}{13}$ and $\cos v = -\frac{3}{5}$. (Both u and v are in Quadrant II.)

43. $\sin(u + v)$

44. $\cos(u - v)$

45. $\cos(u + v)$

46. $\sin(v - u)$

47. $\tan(u + v)$

48. $\csc(u - v)$

49. $\sec(v - u)$

50. $\cot(u + v)$

In Exercises 51–56, find the exact value of the trigonometric function given that $\sin u = -\frac{7}{25}$ and $\cos v = -\frac{4}{5}$. (Both u and v are in Quadrant III.)

51. $\cos(u + v)$

52. $\sin(u + v)$

53. $\tan(u - v)$

54. $\cot(v - u)$

55. $\csc(u - v)$

56. $\sec(v - u)$

In Exercises 57–60, write the trigonometric expression as an algebraic expression.

57. $\sin(\arcsin x + \arccos x)$

58. $\sin(\arctan 2x - \arccos x)$

59. $\cos(\arccos x + \arcsin x)$

60. $\cos(\arccos x - \arctan x)$

In Exercises 61–70, prove the identity.

61. $\sin\left(\dfrac{\pi}{2} - x\right) = \cos x$ **62.** $\sin\left(\dfrac{\pi}{2} + x\right) = \cos x$

63. $\sin\left(\dfrac{\pi}{6} + x\right) = \dfrac{1}{2}(\cos x + \sqrt{3}\sin x)$

64. $\cos\left(\dfrac{5\pi}{4} - x\right) = -\dfrac{\sqrt{2}}{2}(\cos x + \sin x)$

65. $\cos(\pi - \theta) + \sin\left(\dfrac{\pi}{2} + \theta\right) = 0$

66. $\tan\left(\dfrac{\pi}{4} - \theta\right) = \dfrac{1 - \tan\theta}{1 + \tan\theta}$

67. $\cos(x + y)\cos(x - y) = \cos^2 x - \sin^2 y$
68. $\sin(x + y)\sin(x - y) = \sin^2 x - \sin^2 y$
69. $\sin(x + y) + \sin(x - y) = 2\sin x \cos y$
70. $\cos(x + y) + \cos(x - y) = 2\cos x \cos y$

In Exercises 71–74, simplify the expression algebraically and use a graphing utility to confirm your answer graphically.

71. $\cos\left(\dfrac{3\pi}{2} - x\right)$ **72.** $\cos(\pi + x)$

73. $\sin\left(\dfrac{3\pi}{2} + \theta\right)$ **74.** $\tan(\pi + \theta)$

In Exercises 75–84, find all solutions of the equation in the interval $[0, 2\pi)$.

75. $\sin(x + \pi) - \sin x + 1 = 0$
76. $\sin(x + \pi) - \sin x - 1 = 0$
77. $\cos(x + \pi) - \cos x - 1 = 0$
78. $\cos(x + \pi) - \cos x + 1 = 0$

79. $\sin\left(x + \dfrac{\pi}{6}\right) - \sin\left(x - \dfrac{\pi}{6}\right) = \dfrac{1}{2}$

80. $\sin\left(x + \dfrac{\pi}{3}\right) + \sin\left(x - \dfrac{\pi}{3}\right) = 1$

81. $\cos\left(x + \dfrac{\pi}{4}\right) - \cos\left(x - \dfrac{\pi}{4}\right) = 1$

82. $\tan(x + \pi) + 2\sin(x + \pi) = 0$

83. $\sin\left(x + \dfrac{\pi}{2}\right) - \cos^2 x = 0$

84. $\cos\left(x - \dfrac{\pi}{2}\right) + \sin^2 x = 0$

In Exercises 85–88, use a graphing utility to approximate the solutions in the interval $[0, 2\pi)$.

85. $\cos\left(x + \dfrac{\pi}{4}\right) + \cos\left(x - \dfrac{\pi}{4}\right) = 1$

86. $\tan(x + \pi) - \cos\left(x + \dfrac{\pi}{2}\right) = 0$

87. $\sin\left(x + \dfrac{\pi}{2}\right) + \cos^2 x = 0$

88. $\cos\left(x - \dfrac{\pi}{2}\right) - \sin^2 x = 0$

89. HARMONIC MOTION   A weight is attached to a spring suspended vertically from a ceiling. When a driving force is applied to the system, the weight moves vertically from its equilibrium position, and this motion is modeled by

$$y = \dfrac{1}{3}\sin 2t + \dfrac{1}{4}\cos 2t$$

where y is the distance from equilibrium (in feet) and t is the time (in seconds).

(a) Use the identity

$$a\sin B\theta + b\cos B\theta = \sqrt{a^2 + b^2}\,\sin(B\theta + C)$$

where $C = \arctan(b/a)$, $a > 0$, to write the model in the form $y = \sqrt{a^2 + b^2}\,\sin(Bt + C)$.

(b) Find the amplitude of the oscillations of the weight.

(c) Find the frequency of the oscillations of the weight.

90. STANDING WAVES   The equation of a standing wave is obtained by adding the displacements of two waves traveling in opposite directions (see figure). Assume that each of the waves has amplitude A, period T, and wavelength λ. If the models for these waves are

$$y_1 = A\cos 2\pi\left(\dfrac{t}{T} - \dfrac{x}{\lambda}\right) \quad \text{and} \quad y_2 = A\cos 2\pi\left(\dfrac{t}{T} + \dfrac{x}{\lambda}\right)$$

show that

$$y_1 + y_2 = 2A\cos\dfrac{2\pi t}{T}\cos\dfrac{2\pi x}{\lambda}.$$

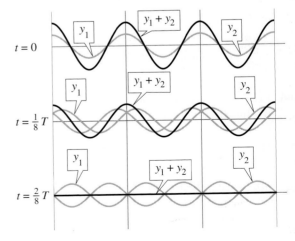

EXPLORATION

TRUE OR FALSE? In Exercises 91–94, determine whether the statement is true or false. Justify your answer.

91. $\sin(u \pm v) = \sin u \cos v \pm \cos u \sin v$

92. $\cos(u \pm v) = \cos u \cos v \pm \sin u \sin v$

93. $\tan\left(x - \dfrac{\pi}{4}\right) = \dfrac{\tan x + 1}{1 - \tan x}$

94. $\sin\left(x - \dfrac{\pi}{2}\right) = -\cos x$

In Exercises 95–98, verify the identity.

95. $\cos(n\pi + \theta) = (-1)^n \cos\theta, \quad n$ is an integer

96. $\sin(n\pi + \theta) = (-1)^n \sin\theta, \quad n$ is an integer

97. $a \sin B\theta + b \cos B\theta = \sqrt{a^2 + b^2} \sin(B\theta + C)$,

where $C = \arctan(b/a)$ and $a > 0$

98. $a \sin B\theta + b \cos B\theta = \sqrt{a^2 + b^2} \cos(B\theta - C)$,

where $C = \arctan(a/b)$ and $b > 0$

In Exercises 99–102, use the formulas given in Exercises 97 and 98 to write the trigonometric expression in the following forms.

(a) $\sqrt{a^2 + b^2}\, \sin(B\theta + C)$ (b) $\sqrt{a^2 + b^2}\, \cos(B\theta - C)$

99. $\sin\theta + \cos\theta$ **100.** $3\sin 2\theta + 4\cos 2\theta$

101. $12\sin 3\theta + 5\cos 3\theta$ **102.** $\sin 2\theta + \cos 2\theta$

In Exercises 103 and 104, use the formulas given in Exercises 97 and 98 to write the trigonometric expression in the form $a \sin B\theta + b \cos B\theta$.

103. $2\sin\left(\theta + \dfrac{\pi}{4}\right)$ **104.** $5\cos\left(\theta - \dfrac{\pi}{4}\right)$

 105. Verify the following identity used in calculus.

$$\frac{\cos(x + h) - \cos x}{h}$$

$$= \frac{\cos x(\cos h - 1)}{h} - \frac{\sin x \sin h}{h}$$

106. Let $x = \pi/6$ in the identity in Exercise 105 and define the functions f and g as follows.

$$f(h) = \frac{\cos[(\pi/6) + h] - \cos(\pi/6)}{h}$$

$$g(h) = \cos\frac{\pi}{6}\left(\frac{\cos h - 1}{h}\right) - \sin\frac{\pi}{6}\left(\frac{\sin h}{h}\right)$$

(a) What are the domains of the functions f and g?

(b) Use a graphing utility to complete the table.

h	0.5	0.2	0.1	0.05	0.02	0.01
$f(h)$						
$g(h)$						

(c) Use a graphing utility to graph the functions f and g.

(d) Use the table and the graphs to make a conjecture about the values of the functions f and g as $h \to 0$.

In Exercises 107 and 108, use the figure, which shows two lines whose equations are $y_1 = m_1x + b_1$ and $y_2 = m_2x + b_2$. Assume that both lines have positive slopes. Derive a formula for the angle between the two lines. Then use your formula to find the angle between the given pair of lines.

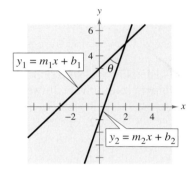

107. $y = x$ and $y = \sqrt{3}x$

108. $y = x$ and $y = \dfrac{1}{\sqrt{3}}x$

In Exercises 109 and 110, use a graphing utility to graph y_1 and y_2 in the same viewing window. Use the graphs to determine whether $y_1 = y_2$. Explain your reasoning.

109. $y_1 = \cos(x + 2), \quad y_2 = \cos x + \cos 2$

110. $y_1 = \sin(x + 4), \quad y_2 = \sin x + \sin 4$

111. PROOF

(a) Write a proof of the formula for $\sin(u + v)$.

(b) Write a proof of the formula for $\sin(u - v)$.

112. CAPSTONE Give an example to justify each statement.

(a) $\sin(u + v) \neq \sin u + \sin v$

(b) $\sin(u - v) \neq \sin u - \sin v$

(c) $\cos(u + v) \neq \cos u + \cos v$

(d) $\cos(u - v) \neq \cos u - \cos v$

(e) $\tan(u + v) \neq \tan u + \tan v$

(f) $\tan(u - v) \neq \tan u - \tan v$

5.5   MULTIPLE-ANGLE AND PRODUCT-TO-SUM FORMULAS

What you should learn

- Use multiple-angle formulas to rewrite and evaluate trigonometric functions.
- Use power-reducing formulas to rewrite and evaluate trigonometric functions.
- Use half-angle formulas to rewrite and evaluate trigonometric functions.
- Use product-to-sum and sum-to-product formulas to rewrite and evaluate trigonometric functions.
- Use trigonometric formulas to rewrite real-life models.

Why you should learn it

You can use a variety of trigonometric formulas to rewrite trigonometric functions in more convenient forms. For instance, in Exercise 135 on page 415, you can use a double-angle formula to determine at what angle an athlete must throw a javelin.

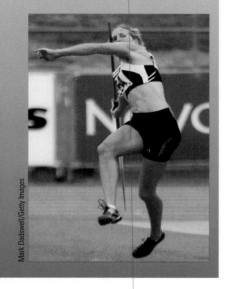

Mark Dadswell/Getty Images

Multiple-Angle Formulas

In this section, you will study four other categories of trigonometric identities.

1. The first category involves *functions of multiple angles* such as $\sin ku$ and $\cos ku$.

2. The second category involves *squares of trigonometric functions* such as $\sin^2 u$.

3. The third category involves *functions of half-angles* such as $\sin(u/2)$.

4. The fourth category involves *products of trigonometric functions* such as $\sin u \cos v$.

You should learn the **double-angle formulas** because they are used often in trigonometry and calculus. For proofs of these formulas, see Proofs in Mathematics on page 423.

Double-Angle Formulas

$$\sin 2u = 2 \sin u \cos u$$

$$\tan 2u = \frac{2 \tan u}{1 - \tan^2 u}$$

$$\cos 2u = \cos^2 u - \sin^2 u$$
$$= 2 \cos^2 u - 1$$
$$= 1 - 2 \sin^2 u$$

Example 1   Solving a Multiple-Angle Equation

Solve $2 \cos x + \sin 2x = 0$.

Solution

Begin by rewriting the equation so that it involves functions of x (rather than $2x$). Then factor and solve.

$2 \cos x + \sin 2x = 0$	Write original equation.
$2 \cos x + 2 \sin x \cos x = 0$	Double-angle formula
$2 \cos x(1 + \sin x) = 0$	Factor.
$2 \cos x = 0$ and $1 + \sin x = 0$	Set factors equal to zero.
$x = \dfrac{\pi}{2}, \dfrac{3\pi}{2}$ $x = \dfrac{3\pi}{2}$	Solutions in $[0, 2\pi)$

So, the general solution is

$$x = \frac{\pi}{2} + 2n\pi \quad \text{and} \quad x = \frac{3\pi}{2} + 2n\pi$$

where n is an integer. Try verifying these solutions graphically.

CHECK*Point*   Now try Exercise 19.

Example 2 Using Double-Angle Formulas to Analyze Graphs

Use a double-angle formula to rewrite the equation

$$y = 4 \cos^2 x - 2.$$

Then sketch the graph of the equation over the interval $[0, 2\pi]$.

Solution

Using the double-angle formula for $\cos 2u$, you can rewrite the original equation as

$y = 4 \cos^2 x - 2$	Write original equation.
$= 2(2 \cos^2 x - 1)$	Factor.
$= 2 \cos 2x.$	Use double-angle formula.

Using the techniques discussed in Section 4.5, you can recognize that the graph of this function has an amplitude of 2 and a period of π. The key points in the interval $[0, \pi]$ are as follows.

Maximum	Intercept	Minimum	Intercept	Maximum
$(0, 2)$	$\left(\dfrac{\pi}{4}, 0\right)$	$\left(\dfrac{\pi}{2}, -2\right)$	$\left(\dfrac{3\pi}{4}, 0\right)$	$(\pi, 2)$

Two cycles of the graph are shown in Figure 5.14.

CHECK Point Now try Exercise 33.

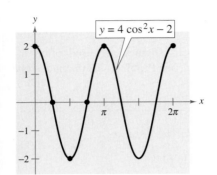

FIGURE 5.14

Example 3 Evaluating Functions Involving Double Angles

Use the following to find $\sin 2\theta$, $\cos 2\theta$, and $\tan 2\theta$.

$$\cos \theta = \frac{5}{13}, \qquad \frac{3\pi}{2} < \theta < 2\pi$$

Solution

From Figure 5.15, you can see that $\sin \theta = y/r = -12/13$. Consequently, using each of the double-angle formulas, you can write

$$\sin 2\theta = 2 \sin \theta \cos \theta = 2\left(-\frac{12}{13}\right)\left(\frac{5}{13}\right) = -\frac{120}{169}$$

$$\cos 2\theta = 2 \cos^2 \theta - 1 = 2\left(\frac{25}{169}\right) - 1 = -\frac{119}{169}$$

$$\tan 2\theta = \frac{\sin 2\theta}{\cos 2\theta} = \frac{120}{119}.$$

CHECK Point Now try Exercise 37.

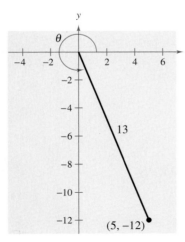

FIGURE 5.15

The double-angle formulas are not restricted to angles 2θ and θ. Other *double* combinations, such as 4θ and 2θ or 6θ and 3θ, are also valid. Here are two examples.

$$\sin 4\theta = 2 \sin 2\theta \cos 2\theta \qquad \text{and} \qquad \cos 6\theta = \cos^2 3\theta - \sin^2 3\theta$$

By using double-angle formulas together with the sum formulas given in the preceding section, you can form other multiple-angle formulas.

| **Example 4**   **Deriving a Triple-Angle Formula** |

$$\sin 3x = \sin(2x + x)$$

$$= \sin 2x \cos x + \cos 2x \sin x$$

$$= 2 \sin x \cos x \cos x + (1 - 2 \sin^2 x) \sin x$$

$$= 2 \sin x \cos^2 x + \sin x - 2 \sin^3 x$$

$$= 2 \sin x(1 - \sin^2 x) + \sin x - 2 \sin^3 x$$

$$= 2 \sin x - 2 \sin^3 x + \sin x - 2 \sin^3 x$$

$$= 3 \sin x - 4 \sin^3 x$$

CHECK *Point*   Now try Exercise 117.

Power-Reducing Formulas

The double-angle formulas can be used to obtain the following **power-reducing formulas.** Example 5 shows a typical power reduction that is used in calculus.

Power-Reducing Formulas

$$\sin^2 u = \frac{1 - \cos 2u}{2} \qquad \cos^2 u = \frac{1 + \cos 2u}{2} \qquad \tan^2 u = \frac{1 - \cos 2u}{1 + \cos 2u}$$

For a proof of the power-reducing formulas, see Proofs in Mathematics on page 423.

| **Example 5**   **Reducing a Power** |

Rewrite $\sin^4 x$ as a sum of first powers of the cosines of multiple angles.

Solution

Note the repeated use of power-reducing formulas.

$$\sin^4 x = (\sin^2 x)^2 \qquad\qquad\qquad \text{Property of exponents}$$

$$= \left(\frac{1 - \cos 2x}{2}\right)^2 \qquad\qquad \text{Power-reducing formula}$$

$$= \frac{1}{4}(1 - 2 \cos 2x + \cos^2 2x) \qquad \text{Expand.}$$

$$= \frac{1}{4}\left(1 - 2 \cos 2x + \frac{1 + \cos 4x}{2}\right) \qquad \text{Power-reducing formula}$$

$$= \frac{1}{4} - \frac{1}{2} \cos 2x + \frac{1}{8} + \frac{1}{8} \cos 4x \qquad \text{Distributive Property}$$

$$= \frac{1}{8}(3 - 4 \cos 2x + \cos 4x) \qquad \text{Factor out common factor.}$$

CHECK *Point*   Now try Exercise 43.

Half-Angle Formulas

You can derive some useful alternative forms of the power-reducing formulas by replacing u with $u/2$. The results are called **half-angle formulas.**

Half-Angle Formulas

$$\sin \frac{u}{2} = \pm \sqrt{\frac{1 - \cos u}{2}}$$

$$\cos \frac{u}{2} = \pm \sqrt{\frac{1 + \cos u}{2}}$$

$$\tan \frac{u}{2} = \frac{1 - \cos u}{\sin u} = \frac{\sin u}{1 + \cos u}$$

The signs of $\sin \dfrac{u}{2}$ and $\cos \dfrac{u}{2}$ depend on the quadrant in which $\dfrac{u}{2}$ lies.

Example 6   Using a Half-Angle Formula

Find the exact value of $\sin 105°$.

Solution

Begin by noting that $105°$ is half of $210°$. Then, using the half-angle formula for $\sin(u/2)$ and the fact that $105°$ lies in Quadrant II, you have

$$\sin 105° = \sqrt{\frac{1 - \cos 210°}{2}}$$

$$= \sqrt{\frac{1 - (-\cos 30°)}{2}}$$

$$= \sqrt{\frac{1 + \left(\sqrt{3}/2\right)}{2}}$$

$$= \frac{\sqrt{2 + \sqrt{3}}}{2}.$$

The positive square root is chosen because $\sin \theta$ is positive in Quadrant II.

CHECK Point   Now try Exercise 59.

Use your calculator to verify the result obtained in Example 6. That is, evaluate $\sin 105°$ and $\left(\sqrt{2 + \sqrt{3}}\right)/2$.

$$\sin 105° \approx 0.9659258$$

$$\frac{\sqrt{2 + \sqrt{3}}}{2} \approx 0.9659258$$

You can see that both values are approximately 0.9659258.

Study Tip

To find the exact value of a trigonometric function with an angle measure in D°M′S″ form using a half-angle formula, first convert the angle measure to decimal degree form. Then multiply the resulting angle measure by 2.

| **Example 7** | **Solving a Trigonometric Equation** |

Find all solutions of $2 - \sin^2 x = 2 \cos^2 \dfrac{x}{2}$ in the interval $[0, 2\pi)$.

Algebraic Solution

$$2 - \sin^2 x = 2 \cos^2 \frac{x}{2} \qquad \text{Write original equation.}$$

$$2 - \sin^2 x = 2\left(\pm \sqrt{\frac{1 + \cos x}{2}}\right)^2 \qquad \text{Half-angle formula}$$

$$2 - \sin^2 x = 2\left(\frac{1 + \cos x}{2}\right) \qquad \text{Simplify.}$$

$$2 - \sin^2 x = 1 + \cos x \qquad \text{Simplify.}$$

$$2 - (1 - \cos^2 x) = 1 + \cos x \qquad \text{Pythagorean identity}$$

$$\cos^2 x - \cos x = 0 \qquad \text{Simplify.}$$

$$\cos x (\cos x - 1) = 0 \qquad \text{Factor.}$$

By setting the factors $\cos x$ and $\cos x - 1$ equal to zero, you find that the solutions in the interval $[0, 2\pi)$ are

$$x = \frac{\pi}{2}, \quad x = \frac{3\pi}{2}, \quad \text{and} \quad x = 0.$$

CHECK *Point*   Now try Exercise 77.

Graphical Solution

Use a graphing utility set in *radian* mode to graph $y = 2 - \sin^2 x - 2 \cos^2(x/2)$, as shown in Figure 5.16. Use the *zero* or *root* feature or the *zoom* and *trace* features to approximate the x-intercepts in the interval $[0, 2\pi)$ to be

$$x = 0, \, x \approx 1.571 \approx \frac{\pi}{2}, \text{ and } x \approx 4.712 \approx \frac{3\pi}{2}.$$

These values are the approximate solutions of $2 - \sin^2 x - 2 \cos^2(x/2) = 0$ in the interval $[0, 2\pi)$.

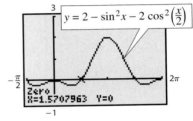

FIGURE 5.16

Product-to-Sum Formulas

Each of the following **product-to-sum formulas** can be verified using the sum and difference formulas discussed in the preceding section.

Product-to-Sum Formulas
$\sin u \sin v = \dfrac{1}{2}[\cos(u - v) - \cos(u + v)]$
$\cos u \cos v = \dfrac{1}{2}[\cos(u - v) + \cos(u + v)]$
$\sin u \cos v = \dfrac{1}{2}[\sin(u + v) + \sin(u - v)]$
$\cos u \sin v = \dfrac{1}{2}[\sin(u + v) - \sin(u - v)]$

Product-to-sum formulas are used in calculus to evaluate integrals involving the products of sines and cosines of two different angles.

Example 8 Writing Products as Sums

Rewrite the product $\cos 5x \sin 4x$ as a sum or difference.

Solution

Using the appropriate product-to-sum formula, you obtain

$$\cos 5x \sin 4x = \tfrac{1}{2}[\sin(5x + 4x) - \sin(5x - 4x)]$$

$$= \tfrac{1}{2}\sin 9x - \tfrac{1}{2}\sin x.$$

CHECK**Point** Now try Exercise 85.

Occasionally, it is useful to reverse the procedure and write a sum of trigonometric functions as a product. This can be accomplished with the following **sum-to-product formulas.**

Sum-to-Product Formulas

$$\sin u + \sin v = 2 \sin\left(\frac{u + v}{2}\right) \cos\left(\frac{u - v}{2}\right)$$

$$\sin u - \sin v = 2 \cos\left(\frac{u + v}{2}\right) \sin\left(\frac{u - v}{2}\right)$$

$$\cos u + \cos v = 2 \cos\left(\frac{u + v}{2}\right) \cos\left(\frac{u - v}{2}\right)$$

$$\cos u - \cos v = -2 \sin\left(\frac{u + v}{2}\right) \sin\left(\frac{u - v}{2}\right)$$

For a proof of the sum-to-product formulas, see Proofs in Mathematics on page 424.

Example 9 Using a Sum-to-Product Formula

Find the exact value of $\cos 195° + \cos 105°$.

Solution

Using the appropriate sum-to-product formula, you obtain

$$\cos 195° + \cos 105° = 2 \cos\left(\frac{195° + 105°}{2}\right) \cos\left(\frac{195° - 105°}{2}\right)$$

$$= 2 \cos 150° \cos 45°$$

$$= 2\left(-\frac{\sqrt{3}}{2}\right)\left(\frac{\sqrt{2}}{2}\right)$$

$$= -\frac{\sqrt{6}}{2}.$$

CHECK**Point** Now try Exercise 99.

Example 10 Solving a Trigonometric Equation

Solve $\sin 5x + \sin 3x = 0$.

Algebraic Solution

$$\sin 5x + \sin 3x = 0 \qquad \text{Write original equation.}$$

$$2 \sin\left(\frac{5x + 3x}{2}\right) \cos\left(\frac{5x - 3x}{2}\right) = 0 \qquad \text{Sum-to-product formula}$$

$$2 \sin 4x \cos x = 0 \qquad \text{Simplify.}$$

By setting the factor $2 \sin 4x$ equal to zero, you can find that the solutions in the interval $[0, 2\pi)$ are

$$x = 0, \frac{\pi}{4}, \frac{\pi}{2}, \frac{3\pi}{4}, \pi, \frac{5\pi}{4}, \frac{3\pi}{2}, \frac{7\pi}{4}.$$

The equation $\cos x = 0$ yields no additional solutions, so you can conclude that the solutions are of the form

$$x = \frac{n\pi}{4}$$

where n is an integer.

Graphical Solution

Sketch the graph of

$$y = \sin 5x + \sin 3x,$$

as shown in Figure 5.17. From the graph you can see that the x-intercepts occur at multiples of $\pi/4$. So, you can conclude that the solutions are of the form

$$x = \frac{n\pi}{4}$$

where n is an integer.

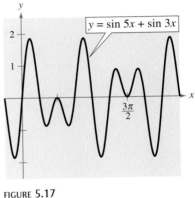

FIGURE 5.17

CHECK *Point* Now try Exercise 103.

Example 11 Verifying a Trigonometric Identity

Verify the identity $\dfrac{\sin 3x - \sin x}{\cos x + \cos 3x} = \tan x$.

Solution

Using appropriate sum-to-product formulas, you have

$$\frac{\sin 3x - \sin x}{\cos x + \cos 3x} = \frac{2 \cos\left(\dfrac{3x + x}{2}\right) \sin\left(\dfrac{3x - x}{2}\right)}{2 \cos\left(\dfrac{x + 3x}{2}\right) \cos\left(\dfrac{x - 3x}{2}\right)}$$

$$= \frac{2 \cos(2x) \sin x}{2 \cos(2x) \cos(-x)}$$

$$= \frac{\sin x}{\cos(-x)}$$

$$= \frac{\sin x}{\cos x} = \tan x.$$

CHECK *Point* Now try Exercise 121.

Application

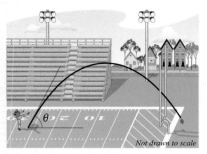

FIGURE 5.18

Example 12   Projectile Motion

Ignoring air resistance, the range of a projectile fired at an angle θ with the horizontal and with an initial velocity of v_0 feet per second is given by

$$r = \frac{1}{16}v_0{}^2 \sin \theta \cos \theta$$

where r is the horizontal distance (in feet) that the projectile will travel. A place kicker for a football team can kick a football from ground level with an initial velocity of 80 feet per second (see Figure 5.18).

a. Write the projectile motion model in a simpler form.

b. At what angle must the player kick the football so that the football travels 200 feet?

c. For what angle is the horizontal distance the football travels a maximum?

Solution

a. You can use a double-angle formula to rewrite the projectile motion model as

$$r = \frac{1}{32}v_0{}^2(2 \sin \theta \cos \theta)$$ Rewrite original projectile motion model.

$$= \frac{1}{32}v_0{}^2 \sin 2\theta.$$ Rewrite model using a double-angle formula.

b. $r = \frac{1}{32}v_0{}^2 \sin 2\theta$ Write projectile motion model.

$$200 = \frac{1}{32}(80)^2 \sin 2\theta$$ Substitute 200 for r and 80 for v_0.

$$200 = 200 \sin 2\theta$$ Simplify.

$$1 = \sin 2\theta$$ Divide each side by 200.

You know that $2\theta = \pi/2$, so dividing this result by 2 produces $\theta = \pi/4$. Because $\pi/4 = 45°$, you can conclude that the player must kick the football at an angle of 45° so that the football will travel 200 feet.

c. From the model $r = 200 \sin 2\theta$ you can see that the amplitude is 200. So the maximum range is $r = 200$ feet. From part (b), you know that this corresponds to an angle of 45°. Therefore, kicking the football at an angle of 45° will produce a maximum horizontal distance of 200 feet.

CHECK**Point**   Now try Exercise 135.

CLASSROOM DISCUSSION

Deriving an Area Formula   Describe how you can use a double-angle formula or a half-angle formula to derive a formula for the area of an isosceles triangle. Use a labeled sketch to illustrate your derivation. Then write two examples that show how your formula can be used.

5.5   EXERCISES

VOCABULARY: Fill in the blank to complete the trigonometric formula.

1. $\sin 2u =$ _____

2. $\dfrac{1 + \cos 2u}{2} =$ _____

3. $\cos 2u =$ _____

4. $\dfrac{1 - \cos 2u}{1 + \cos 2u} =$ _____

5. $\sin \dfrac{u}{2} =$ _____

6. $\tan \dfrac{u}{2} =$ _____

7. $\cos u \cos v =$ _____

8. $\sin u \cos v =$ _____

9. $\sin u + \sin v =$ _____

10. $\cos u - \cos v =$ _____

SKILLS AND APPLICATIONS

In Exercises 11–18, use the figure to find the exact value of the trigonometric function.

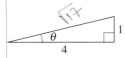

θ 4   1

11. $\cos 2\theta$

12. $\sin 2\theta$

13. $\tan 2\theta$

14. $\sec 2\theta$

15. $\csc 2\theta$

16. $\cot 2\theta$

17. $\sin 4\theta$

18. $\tan 4\theta$

In Exercises 19–28, find the exact solutions of the equation in the interval $[0, 2\pi)$.

19. $\sin 2x - \sin x = 0$

20. $\sin 2x + \cos x = 0$

21. $4 \sin x \cos x = 1$

22. $\sin 2x \sin x = \cos x$

23. $\cos 2x - \cos x = 0$

24. $\cos 2x + \sin x = 0$

25. $\sin 4x = -2 \sin 2x$

26. $(\sin 2x + \cos 2x)^2 = 1$

27. $\tan 2x - \cot x = 0$

28. $\tan 2x - 2 \cos x = 0$

In Exercises 29–36, use a double-angle formula to rewrite the expression.

29. $6 \sin x \cos x$

30. $\sin x \cos x$

31. $6 \cos^2 x - 3$

32. $\cos^2 x - \frac{1}{2}$

33. $4 - 8 \sin^2 x$

34. $10 \sin^2 x - 5$

35. $(\cos x + \sin x)(\cos x - \sin x)$

36. $(\sin x - \cos x)(\sin x + \cos x)$

In Exercises 37–42, find the exact values of $\sin 2u$, $\cos 2u$, and $\tan 2u$ using the double-angle formulas.

37. $\sin u = -\dfrac{3}{5}, \quad \dfrac{3\pi}{2} < u < 2\pi$

38. $\cos u = -\dfrac{4}{5}, \quad \dfrac{\pi}{2} < u < \pi$

39. $\tan u = \dfrac{3}{5}, \quad 0 < u < \dfrac{\pi}{2}$

40. $\cot u = \sqrt{2}, \quad \pi < u < \dfrac{3\pi}{2}$

41. $\sec u = -2, \quad \dfrac{\pi}{2} < u < \pi$

42. $\csc u = 3, \quad \dfrac{\pi}{2} < u < \pi$

In Exercises 43–52, use the power-reducing formulas to rewrite the expression in terms of the first power of the cosine.

43. $\cos^4 x$

44. $\sin^4 2x$

45. $\cos^4 2x$

46. $\sin^8 x$

47. $\tan^4 2x$

48. $\sin^2 x \cos^4 x$

49. $\sin^2 2x \cos^2 2x$

50. $\tan^2 2x \cos^4 2x$

51. $\sin^4 x \cos^2 x$

52. $\sin^4 x \cos^4 x$

In Exercises 53–58, use the figure to find the exact value of the trigonometric function.

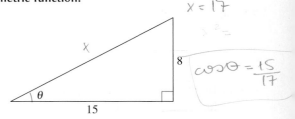

$x = 17$

x 8 $\cos\theta = \dfrac{15}{17}$

θ 15

53. $\cos \dfrac{\theta}{2}$

54. $\sin \dfrac{\theta}{2}$

55. $\tan \dfrac{\theta}{2}$

56. $\sec \dfrac{\theta}{2}$

57. $\csc \dfrac{\theta}{2}$

58. $\cot \dfrac{\theta}{2}$

In Exercises 59–66, use the half-angle formulas to determine the exact values of the sine, cosine, and tangent of the angle.

59. 75°   **60.** 165°

61. 112° 30′   **62.** 67° 30′

63. $\pi/8$ **64.** $\pi/12$

65. $3\pi/8$ **66.** $7\pi/12$

In Exercises 67–72, (a) determine the quadrant in which $u/2$ lies, and (b) find the exact values of $\sin(u/2)$, $\cos(u/2)$, and $\tan(u/2)$ using the half-angle formulas.

67. $\cos u = \dfrac{7}{25}, \quad 0 < u < \dfrac{\pi}{2}$

68. $\sin u = \dfrac{5}{13}, \quad \dfrac{\pi}{2} < u < \pi$

69. $\tan u = -\dfrac{5}{12}, \quad \dfrac{3\pi}{2} < u < 2\pi$

70. $\cot u = 3, \quad \pi < u < \dfrac{3\pi}{2}$

71. $\csc u = -\dfrac{5}{3}, \quad \pi < u < \dfrac{3\pi}{2}$

72. $\sec u = \dfrac{7}{2}, \quad \dfrac{3\pi}{2} < u < 2\pi$

In Exercises 73–76, use the half-angle formulas to simplify the expression.

73. $\sqrt{\dfrac{1 - \cos 6x}{2}}$ **74.** $\sqrt{\dfrac{1 + \cos 4x}{2}}$

75. $-\sqrt{\dfrac{1 - \cos 8x}{1 + \cos 8x}}$ **76.** $-\sqrt{\dfrac{1 - \cos(x - 1)}{2}}$

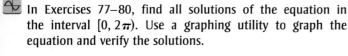 In Exercises 77–80, find all solutions of the equation in the interval $[0, 2\pi)$. Use a graphing utility to graph the equation and verify the solutions.

77. $\sin \dfrac{x}{2} + \cos x = 0$ **78.** $\sin \dfrac{x}{2} + \cos x - 1 = 0$

79. $\cos \dfrac{x}{2} - \sin x = 0$ **80.** $\tan \dfrac{x}{2} - \sin x = 0$

In Exercises 81–90, use the product-to-sum formulas to write the product as a sum or difference.

81. $\sin \dfrac{\pi}{3} \cos \dfrac{\pi}{6}$ **82.** $4 \cos \dfrac{\pi}{3} \sin \dfrac{5\pi}{6}$

83. $10 \cos 75° \cos 15°$ **84.** $6 \sin 45° \cos 15°$

85. $\sin 5\theta \sin 3\theta$ **86.** $3 \sin(-4\alpha) \sin 6\alpha$

87. $7 \cos(-5\beta) \sin 3\beta$ **88.** $\cos 2\theta \cos 4\theta$

89. $\sin(x + y) \sin(x - y)$ **90.** $\sin(x + y) \cos(x - y)$

In Exercises 91–98, use the sum-to-product formulas to write the sum or difference as a product.

91. $\sin 3\theta + \sin \theta$ **92.** $\sin 5\theta - \sin 3\theta$

93. $\cos 6x + \cos 2x$ **94.** $\cos x + \cos 4x$

95. $\sin(\alpha + \beta) - \sin(\alpha - \beta)$ **96.** $\cos(\phi + 2\pi) + \cos \phi$

97. $\cos\left(\theta + \dfrac{\pi}{2} \right) - \cos\left(\theta - \dfrac{\pi}{2} \right)$

98. $\sin\left(x + \dfrac{\pi}{2} \right) + \sin\left(x - \dfrac{\pi}{2} \right)$

In Exercises 99–102, use the sum-to-product formulas to find the exact value of the expression.

99. $\sin 75° + \sin 15°$ **100.** $\cos 120° + \cos 60°$

101. $\cos \dfrac{3\pi}{4} - \cos \dfrac{\pi}{4}$ **102.** $\sin \dfrac{5\pi}{4} - \sin \dfrac{3\pi}{4}$

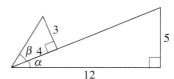 In Exercises 103–106, find all solutions of the equation in the interval $[0, 2\pi)$. Use a graphing utility to graph the equation and verify the solutions.

103. $\sin 6x + \sin 2x = 0$ **104.** $\cos 2x - \cos 6x = 0$

105. $\dfrac{\cos 2x}{\sin 3x - \sin x} - 1 = 0$ **106.** $\sin^2 3x - \sin^2 x = 0$

In Exercises 107–110, use the figure to find the exact value of the trigonometric function.

107. $\sin 2\alpha$ **108.** $\cos 2\beta$

109. $\cos(\beta/2)$ **110.** $\sin(\alpha + \beta)$

In Exercises 111–124, verify the identity.

111. $\csc 2\theta = \dfrac{\csc \theta}{2 \cos \theta}$ **112.** $\sec 2\theta = \dfrac{\sec^2 \theta}{2 - \sec^2 \theta}$

113. $\sin \dfrac{\alpha}{3} \cos \dfrac{\alpha}{3} = \dfrac{1}{2} \sin \dfrac{2\alpha}{3}$ **114.** $\dfrac{\cos 3\beta}{\cos \beta} = 1 - 4 \sin^2 \beta$

115. $1 + \cos 10y = 2 \cos^2 5y$

116. $\cos^4 x - \sin^4 x = \cos 2x$

117. $\cos 4\alpha = \cos^2 2\alpha - \sin^2 2\alpha$

118. $(\sin x + \cos x)^2 = 1 + \sin 2x$

119. $\tan \dfrac{u}{2} = \csc u - \cot u$

120. $\sec \dfrac{u}{2} = \pm \sqrt{\dfrac{2 \tan u}{\tan u + \sin u}}$

121. $\dfrac{\cos 4x + \cos 2x}{\sin 4x + \sin 2x} = \cot 3x$

122. $\dfrac{\sin x \pm \sin y}{\cos x + \cos y} = \tan \dfrac{x \pm y}{2}$

123. $\sin\left(\dfrac{\pi}{6} + x\right) + \sin\left(\dfrac{\pi}{6} - x\right) = \cos x$

124. $\cos\left(\dfrac{\pi}{3} + x\right) + \cos\left(\dfrac{\pi}{3} - x\right) = \cos x$

In Exercises 125–128, use a graphing utility to verify the identity. Confirm that it is an identity algebraically.

125. $\cos 3\beta = \cos^3 \beta - 3 \sin^2 \beta \cos \beta$

126. $\sin 4\beta = 4 \sin \beta \cos \beta (1 - 2 \sin^2 \beta)$

127. $(\cos 4x - \cos 2x)/(2 \sin 3x) = -\sin x$

128. $(\cos 3x - \cos x)/(\sin 3x - \sin x) = -\tan 2x$

In Exercises 129 and 130, graph the function by hand in the interval $[0, 2\pi]$ by using the power-reducing formulas.

129. $f(x) = \sin^2 x$ **130.** $f(x) = \cos^2 x$

In Exercises 131–134, write the trigonometric expression as an algebraic expression.

131. $\sin(2 \arcsin x)$ **132.** $\cos(2 \arccos x)$

133. $\cos(2 \arcsin x)$ **134.** $\sin(2 \arccos x)$

135. PROJECTILE MOTION The range of a projectile fired at an angle θ with the horizontal and with an initial velocity of v_0 feet per second is

$$r = \dfrac{1}{32} v_0^2 \sin 2\theta$$

where r is measured in feet. An athlete throws a javelin at 75 feet per second. At what angle must the athlete throw the javelin so that the javelin travels 130 feet?

136. RAILROAD TRACK When two railroad tracks merge, the overlapping portions of the tracks are in the shapes of circular arcs (see figure). The radius of each arc r (in feet) and the angle θ are related by

$$\dfrac{x}{2} = 2r \sin^2 \dfrac{\theta}{2}.$$

Write a formula for x in terms of $\cos \theta$.

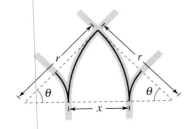

137. MACH NUMBER The mach number M of an airplane is the ratio of its speed to the speed of sound. When an airplane travels faster than the speed of sound, the sound waves form a cone behind the airplane (see figure). The mach number is related to the apex angle θ of the cone by $\sin(\theta/2) = 1/M$.

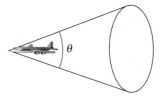

(a) Find the angle θ that corresponds to a mach number of 1.

(b) Find the angle θ that corresponds to a mach number of 4.5.

(c) The speed of sound is about 760 miles per hour. Determine the speed of an object with the mach numbers from parts (a) and (b).

(d) Rewrite the equation in terms of θ.

EXPLORATION

138. CAPSTONE Consider the function given by $f(x) = \sin^4 x + \cos^4 x$.

(a) Use the power-reducing formulas to write the function in terms of cosine to the first power.

(b) Determine another way of rewriting the function. Use a graphing utility to rule out incorrectly rewritten functions.

(c) Add a trigonometric term to the function so that it becomes a perfect square trinomial. Rewrite the function as a perfect square trinomial minus the term that you added. Use a graphing utility to rule out incorrectly rewritten functions.

(d) Rewrite the result of part (c) in terms of the sine of a double angle. Use a graphing utility to rule out incorrectly rewritten functions.

(e) When you rewrite a trigonometric expression, the result may not be the same as a friend's. Does this mean that one of you is wrong? Explain.

TRUE OR FALSE? In Exercises 139 and 140, determine whether the statement is true or false. Justify your answer.

139. Because the sine function is an odd function, for a negative number u, $\sin 2u = -2 \sin u \cos u$.

140. $\sin \dfrac{u}{2} = -\sqrt{\dfrac{1 - \cos u}{2}}$ when u is in the second quadrant.

5.6 LAW OF SINES

Introduction

In Chapter 4, you studied techniques for solving right triangles. In this section and the next, you will solve **oblique triangles**—triangles that have no right angles. As standard notation, the angles of a triangle are labeled A, B, and C, and their opposite sides are labeled a, b, and c, as shown in Figure 5.19.

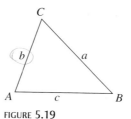

FIGURE **5.19**

To solve an oblique triangle, you need to know the measure of at least one side and any two other measures of the triangle—either two sides, two angles, or one angle and one side. This breaks down into the following four cases.

1. Two angles and any side (AAS or ASA)

2. Two sides and an angle opposite one of them (SSA)

3. Three sides (SSS)

4. Two sides and their included angle (SAS)

The first two cases can be solved using the **Law of Sines,** whereas the last two cases require the Law of Cosines (see Section 5.7).

Law of Sines

If ABC is a triangle with sides a, b, and c, then

$$\frac{a}{\sin A} = \frac{b}{\sin B} = \frac{c}{\sin C}.$$

A is acute. A is obtuse.

The Law of Sines can also be written in the reciprocal form

$$\frac{\sin A}{a} = \frac{\sin B}{b} = \frac{\sin C}{c}.$$

For a proof of the Law of Sines, see Proofs in Mathematics on page 444.

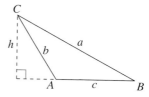

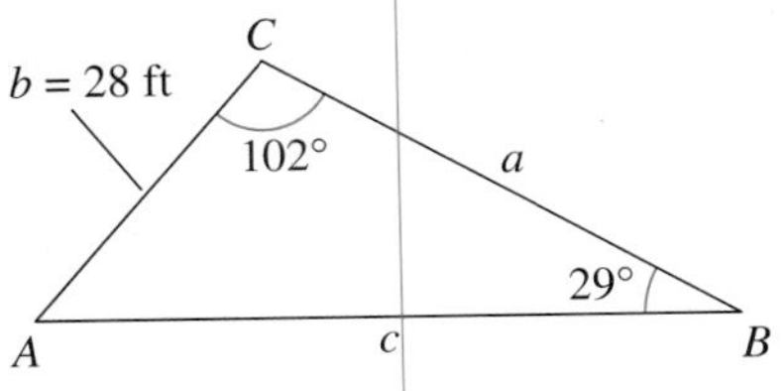

b = 28 ft
C
102°
a
29°
A   c   B

FIGURE **5.20**

Example 1   Given Two Angles and One Side—AAS

For the triangle in Figure 5.20, $C = 102°$, $B = 29°$, and $b = 28$ feet. Find the remaining angle and sides.

Solution

The third angle of the triangle is

$$A = 180° - B - C$$
$$= 180° - 29° - 102°$$
$$= 49°.$$

By the Law of Sines, you have

$$\frac{a}{\sin A} = \frac{b}{\sin B} = \frac{c}{\sin C}.$$

Using $b = 28$ produces

$$a = \frac{b}{\sin B}(\sin A) = \frac{28}{\sin 29°}(\sin 49°) \approx 43.59 \text{ feet}$$

and

$$c = \frac{b}{\sin B}(\sin C) = \frac{28}{\sin 29°}(\sin 102°) \approx 56.49 \text{ feet}.$$

CHECK**Point**   Now try Exercise 5.

Study Tip

When solving triangles, a careful sketch is useful as a quick test for the feasibility of an answer. Remember that the longest side lies opposite the largest angle, and the shortest side lies opposite the smallest angle.

Example 2   Given Two Angles and One Side—ASA

A pole tilts *toward* the sun at an 8° angle from the vertical, and it casts a 22-foot shadow. The angle of elevation from the tip of the shadow to the top of the pole is 43°. How tall is the pole?

Solution

From Figure 5.21, note that $A = 43°$ and $B = 90° + 8° = 98°$. So, the third angle is

$$C = 180° - A - B$$
$$= 180° - 43° - 98°$$
$$= 39°.$$

By the Law of Sines, you have

$$\frac{a}{\sin A} = \frac{c}{\sin C}.$$

Because $c = 22$ feet, the length of the pole is

$$a = \frac{c}{\sin C}(\sin A) = \frac{22}{\sin 39°}(\sin 43°) \approx 23.84 \text{ feet}.$$

CHECK**Point**   Now try Exercise 45.

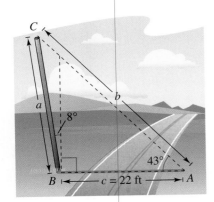

FIGURE **5.21**

For practice, try reworking Example 2 for a pole that tilts *away from* the sun under the same conditions.

The Ambiguous Case (SSA)

In Examples 1 and 2, you saw that two angles and one side determine a unique triangle. However, if two sides and one opposite angle are given, three possible situations can occur: (1) no such triangle exists, (2) one such triangle exists, or (3) two distinct triangles may satisfy the conditions.

The Ambiguous Case (SSA)

Consider a triangle in which you are given a, b, and A.   ($h = b \sin A$)

	A is acute.	A is acute.	A is acute.	A is acute.	A is obtuse.	A is obtuse.
Sketch						
Necessary condition	$a < h$	$a = h$	$a \geq b$	$h < a < b$	$a \leq b$	$a > b$
Triangles possible	None	One	One	Two	None	One

Example 3   Single-Solution Case—SSA

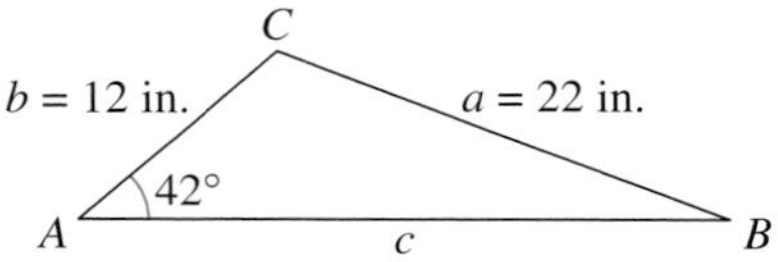

For the triangle in Figure 5.22, $a = 22$ inches, $b = 12$ inches, and $A = 42°$. Find the remaining side and angles.

Solution

By the Law of Sines, you have

$$\frac{\sin B}{b} = \frac{\sin A}{a} \qquad \text{Reciprocal form}$$

$$\sin B = b\left(\frac{\sin A}{a}\right) \qquad \text{Multiply each side by } b.$$

$$\sin B = 12\left(\frac{\sin 42°}{22}\right) \qquad \text{Substitute for } A, a, \text{ and } b.$$

$$B \approx 21.41°. \qquad B \text{ is acute.}$$

Now, you can determine that

$$C \approx 180° - 42° - 21.41° = 116.59°.$$

Then, the remaining side is

$$\frac{c}{\sin C} = \frac{a}{\sin A}$$

$$c = \frac{a}{\sin A}(\sin C) = \frac{22}{\sin 42°}(\sin 116.59°) \approx 29.40 \text{ inches.}$$

CHECKPoint   Now try Exercise 25.

$b = 12$ in. $a = 22$ in. $42°$

One solution: $a \geq b$

FIGURE **5.22**

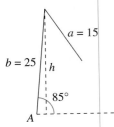

No solution: $a < h$

FIGURE **5.23**

Example 4   No-Solution Case—SSA

Show that there is no triangle for which $a = 15$, $b = 25$, and $A = 85°$.

Solution

Begin by making the sketch shown in Figure 5.23. From this figure it appears that no triangle is formed. You can verify this using the Law of Sines.

$$\frac{\sin B}{b} = \frac{\sin A}{a} \qquad \text{Reciprocal form}$$

$$\sin B = b\left(\frac{\sin A}{a}\right) \qquad \text{Multiply each side by } b.$$

$$\sin B = 25\left(\frac{\sin 85°}{15}\right) \approx 1.660 > 1$$

This contradicts the fact that $|\sin B| \leq 1$. So, no triangle can be formed having sides $a = 15$ and $b = 25$ and an angle of $A = 85°$.

CHECK**Point**   Now try Exercise 27.

Example 5   Two-Solution Case—SSA

Find two triangles for which $a = 12$ meters, $b = 31$ meters, and $A = 20.5°$.

Solution

By the Law of Sines, you have

$$\frac{\sin B}{b} = \frac{\sin A}{a} \qquad \text{Reciprocal form}$$

$$\sin B = b\left(\frac{\sin A}{a}\right) = 31\left(\frac{\sin 20.5°}{12}\right) \approx 0.9047.$$

There are two angles, $B_1 \approx 64.8°$ and $B_2 \approx 180° - 64.8° = 115.2°$, between $0°$ and $180°$ whose sine is 0.9047. For $B_1 \approx 64.8°$, you obtain

$$C \approx 180° - 20.5° - 64.8° = 94.7°$$

$$c = \frac{a}{\sin A}(\sin C) = \frac{12}{\sin 20.5°}(\sin 94.7°) \approx 34.15 \text{ meters.}$$

For $B_2 \approx 115.2°$, you obtain

$$C \approx 180° - 20.5° - 115.2° = 44.3°$$

$$c = \frac{a}{\sin A}(\sin C) = \frac{12}{\sin 20.5°}(\sin 44.3°) \approx 23.93 \text{ meters.}$$

The resulting triangles are shown in Figure 5.24.

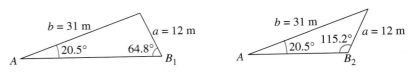

FIGURE **5.24**

CHECK**Point**   Now try Exercise 29.

Area of an Oblique Triangle

The procedure used to prove the Law of Sines leads to a simple formula for the area of an oblique triangle. Referring to Figure 5.25, note that each triangle has a height of $h = b \sin A$. Consequently, the area of each triangle is

$$\text{Area} = \frac{1}{2}(\text{base})(\text{height}) = \frac{1}{2}(c)(b \sin A) = \frac{1}{2}bc \sin A.$$

By similar arguments, you can develop the formulas

$$\text{Area} = \frac{1}{2}ab \sin C = \frac{1}{2}ac \sin B.$$

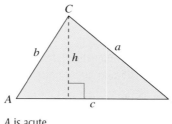
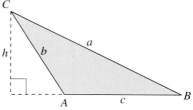

A is acute.   A is obtuse.

FIGURE **5.25**

Area of an Oblique Triangle

The area of any triangle is one-half the product of the lengths of two sides times the sine of their included angle. That is,

$$\text{Area} = \frac{1}{2}bc \sin A = \frac{1}{2}ab \sin C = \frac{1}{2}ac \sin B.$$

Note that if angle A is 90°, the formula gives the area for a right triangle:

$$\text{Area} = \frac{1}{2}bc(\sin 90°) = \frac{1}{2}bc = \frac{1}{2}(\text{base})(\text{height}). \qquad \text{sin } 90° = 1$$

Similar results are obtained for angles C and B equal to 90°.

Example 6   Finding the Area of a Triangular Lot

Find the area of a triangular lot having two sides of lengths 90 meters and 52 meters and an included angle of 102°.

Solution

Consider $a = 90$ meters, $b = 52$ meters, and angle $C = 102°$, as shown in Figure 5.26. Then, the area of the triangle is

$$\text{Area} = \frac{1}{2}ab \sin C = \frac{1}{2}(90)(52)(\sin 102°) \approx 2289 \text{ square meters.}$$

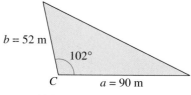

FIGURE **5.26**

CHECK*Point*   Now try Exercise 39.

Application

| Example 7 | **An Application of the Law of Sines** |

The course for a boat race starts at point A in Figure 5.27 and proceeds in the direction S 52° W to point B, then in the direction S 40° E to point C, and finally back to A. Point C lies 8 kilometers directly south of point A. Approximate the total distance of the race course.

Solution

Because lines BD and AC are parallel, it follows that $\angle BCA \cong \angle CBD$. Consequently, triangle ABC has the measures shown in Figure 5.28. The measure of angle B is $180° - 52° - 40° = 88°$. Using the Law of Sines,

$$\frac{a}{\sin 52°} = \frac{b}{\sin 88°} = \frac{c}{\sin 40°}.$$

Because $b = 8$,

$$a = \frac{8}{\sin 88°}(\sin 52°) \approx 6.308$$

and

$$c = \frac{8}{\sin 88°}(\sin 40°) \approx 5.145.$$

The total length of the course is approximately

$$\text{Length} \approx 8 + 6.308 + 5.145$$

$$= 19.453 \text{ kilometers.}$$

CHECK *Point*   Now try Exercise 49.

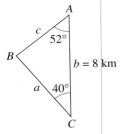

FIGURE 5.27

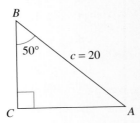

FIGURE 5.28

CLASSROOM DISCUSSION

Using the Law of Sines   In this section, you have been using the Law of Sines to solve *oblique* triangles. Can the Law of Sines also be used to solve a right triangle? If so, write a short paragraph explaining how to use the Law of Sines to solve each triangle. Is there an easier way to solve these triangles?

a. (AAS)   b. (ASA)

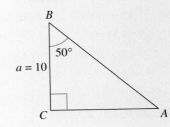

5.6   EXERCISES

See www.CalcChat.com for worked-out solutions to odd-numbered exercises.

VOCABULARY: Fill in the blanks.

1. An _____ triangle is a triangle that has no right angle.

2. For triangle ABC, the Law of Sines is given by $\dfrac{a}{\sin A} = $ _____ $= \dfrac{c}{\sin C}$.

3. Two _____ and one _____ determine a unique triangle.

4. The area of an oblique triangle is given by $\frac{1}{2}bc \sin A = \frac{1}{2}ab \sin C = $ _____ .

SKILLS AND APPLICATIONS

In Exercises 5–24, use the Law of Sines to solve the triangle. Round your answers to two decimal places.

5.

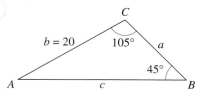

6.

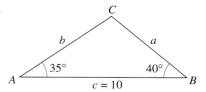

7.

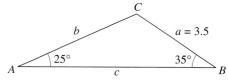

8.

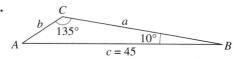

9. $A = 102.4°$, $C = 16.7°$, $a = 21.6$
10. $A = 24.3°$, $C = 54.6°$, $c = 2.68$
11. $A = 83° \, 20'$, $C = 54.6°$, $c = 18.1$
12. $A = 5° \, 40'$, $B = 8° \, 15'$, $b = 4.8$
13. $A = 35°$, $B = 65°$, $c = 10$
14. $A = 120°$, $B = 45°$, $c = 16$
15. $A = 55°$, $B = 42°$, $c = \frac{3}{4}$
16. $B = 28°$, $C = 104°$, $a = 3\frac{5}{8}$
17. $A = 36°$, $a = 8$, $b = 5$
18. $A = 60°$, $a = 9$, $c = 10$
19. $B = 15° \, 30'$, $a = 4.5$, $b = 6.8$

20. $B = 2° \, 45'$, $b = 6.2$, $c = 5.8$
21. $A = 145°$, $a = 14$, $b = 4$
22. $A = 100°$, $a = 125$, $c = 10$
23. $A = 110° \, 15'$, $a = 48$, $b = 16$
24. $C = 95.20°$, $a = 35$, $c = 50$

In Exercises 25–34, use the Law of Sines to solve (if possible) the triangle. If two solutions exist, find both. Round your answers to two decimal places.

25. $A = 110°$, $a = 125$, $b = 100$
26. $A = 110°$, $a = 125$, $b = 200$
27. $A = 76°$, $a = 18$, $b = 20$
28. $A = 76°$, $a = 34$, $b = 21$
29. $A = 58°$, $a = 11.4$, $b = 12.8$
30. $A = 58°$, $a = 4.5$, $b = 12.8$
31. $A = 120°$, $a = b = 25$
32. $A = 120°$, $a = 25$, $b = 24$
33. $A = 45°$, $a = b = 1$
34. $A = 25° \, 4'$, $a = 9.5$, $b = 22$

In Exercises 35–38, find values for b such that the triangle has (a) one solution, (b) two solutions, and (c) no solution.

35. $A = 36°$, $a = 5$
36. $A = 60°$, $a = 10$
37. $A = 10°$, $a = 10.8$
38. $A = 88°$, $a = 315.6$

In Exercises 39–44, find the area of the triangle having the indicated angle and sides.

39. $C = 120°$, $a = 4$, $b = 6$
40. $B = 130°$, $a = 62$, $c = 20$
41. $A = 43° \, 45'$, $b = 57$, $c = 85$
42. $A = 5° \, 15'$, $b = 4.5$, $c = 22$
43. $B = 72° \, 30'$, $a = 105$, $c = 64$
44. $C = 84° \, 30'$, $a = 16$, $b = 20$

45. HEIGHT Because of prevailing winds, a tree grew so that it was leaning 4° from the vertical. At a point 40 meters from the tree, the angle of elevation to the top of the tree is 30° (see figure). Find the height h of the tree.

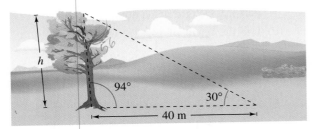

46. HEIGHT A flagpole at a right angle to the horizontal is located on a slope that makes an angle of 12° with the horizontal. The flagpole's shadow is 16 meters long and points directly up the slope. The angle of elevation from the tip of the shadow to the sun is 20°.

(a) Draw a triangle to represent the situation. Show the known quantities on the triangle and use a variable to indicate the height of the flagpole.

(b) Write an equation that can be used to find the height of the flagpole.

(c) Find the height of the flagpole.

47. ANGLE OF ELEVATION A 10-meter utility pole casts a 17-meter shadow directly down a slope when the angle of elevation of the sun is 42° (see figure). Find θ, the angle of elevation of the ground.

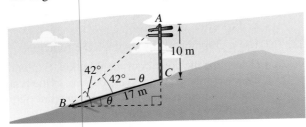

48. FLIGHT PATH A plane flies 500 kilometers with a bearing of 316° from Naples to Elgin (see figure). The plane then flies 720 kilometers from Elgin to Canton (Canton is due west of Naples). Find the bearing of the flight from Elgin to Canton.

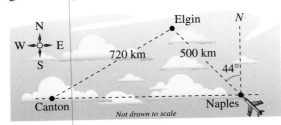

49. BRIDGE DESIGN A bridge is to be built across a small lake from a gazebo to a dock (see figure). The bearing from the gazebo to the dock is S 41° W. From a tree 100 meters from the gazebo, the bearings to the gazebo and the dock are S 74° E and S 28° E, respectively. Find the distance from the gazebo to the dock.

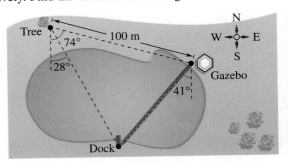

50. RAILROAD TRACK DESIGN The circular arc of a railroad curve has a chord of length 3000 feet corresponding to a central angle of 40°.

(a) Draw a diagram that visually represents the situation. Show the known quantities on the diagram and use the variables r and s to represent the radius of the arc and the length of the arc, respectively.

(b) Find the radius r of the circular arc.

(c) Find the length s of the circular arc.

51. GLIDE PATH A pilot has just started on the glide path for landing at an airport with a runway of length 9000 feet. The angles of depression from the plane to the ends of the runway are 17.5° and 18.8°.

(a) Draw a diagram that visually represents the situation.

(b) Find the air distance the plane must travel until touching down on the near end of the runway.

(c) Find the ground distance the plane must travel until touching down.

(d) Find the altitude of the plane when the pilot begins the descent.

52. LOCATING A FIRE The bearing from the Pine Knob fire tower to the Colt Station fire tower is N 65° E, and the two towers are 30 kilometers apart. A fire spotted by rangers in each tower has a bearing of N 80° E from Pine Knob and S 70° E from Colt Station (see figure). Find the distance of the fire from each tower.

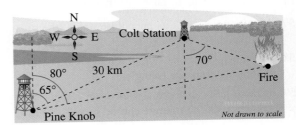

53. DISTANCE A boat is sailing due east parallel to the shoreline at a speed of 10 miles per hour. At a given time, the bearing to the lighthouse is S 70° E, and 15 minutes later the bearing is S 63° E (see figure). The lighthouse is located at the shoreline. What is the distance from the boat to the shoreline?

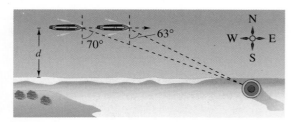

54. DISTANCE A family is traveling due west on a road that passes a famous landmark. At a given time the bearing to the landmark is N 62° W, and after the family travels 5 miles farther the bearing is N 38° W. What is the closest the family will come to the landmark while on the road?

55. ALTITUDE The angles of elevation to an airplane from two points A and B on level ground are 55° and 72°, respectively. The points A and B are 2.2 miles apart, and the airplane is east of both points in the same vertical plane. Find the altitude of the plane.

56. DISTANCE The angles of elevation θ and φ to an airplane from the airport control tower and from an observation post 2 miles away are being continuously monitored (see figure). Write an equation giving the distance d between the plane and observation post in terms of θ and φ.

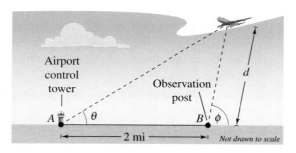

Not drawn to scale

EXPLORATION

TRUE OR FALSE? In Exercises 57–59, determine whether the statement is true or false. Justify your answer.

57. If a triangle contains an obtuse angle, then it must be oblique.

58. Two angles and one side of a triangle do not necessarily determine a unique triangle.

59. If three sides or three angles of an oblique triangle are known, then the triangle can be solved.

60. GRAPHICAL AND NUMERICAL ANALYSIS In the figure, α and β are positive angles.

(a) Write α as a function of β.

(b) Use a graphing utility to graph the function in part (a). Determine its domain and range.

(c) Use the result of part (a) to write c as a function of β.

(d) Use a graphing utility to graph the function in part (c). Determine its domain and range.

(e) Complete the table. What can you infer?

β	0.4	0.8	1.2	1.6	2.0	2.4	2.8
α							
c							

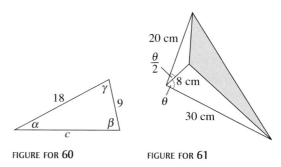

FIGURE FOR **60**   FIGURE FOR **61**

61. GRAPHICAL ANALYSIS

(a) Write the area A of the shaded region in the figure as a function of θ.

(b) Use a graphing utility to graph the function.

(c) Determine the domain of the function. Explain how the area of the region and the domain of the function would change if the eight-centimeter line segment were decreased in length.

62. CAPSTONE In the figure, a triangle is to be formed by drawing a line segment of length a from (4, 3) to the positive x-axis. For what value(s) of a can you form (a) one triangle, (b) two triangles, and (c) no triangles? Explain your reasoning.

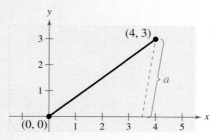

5.7   LAW OF COSINES

Daniel Bendjy/istockphoto.com

Introduction

Two cases remain in the list of conditions needed to solve an oblique triangle—SSS and SAS. If you are given three sides (SSS), or two sides and their included angle (SAS), none of the ratios in the Law of Sines would be complete. In such cases, you can use the **Law of Cosines.**

Law of Cosines

Standard Form

$$a^2 = b^2 + c^2 - 2bc \cos A$$

$$b^2 = a^2 + c^2 - 2ac \cos B$$

$$c^2 = a^2 + b^2 - 2ab \cos C$$

Alternative Form

$$\cos A = \frac{b^2 + c^2 - a^2}{2bc}$$

$$\cos B = \frac{a^2 + c^2 - b^2}{2ac}$$

$$\cos C = \frac{a^2 + b^2 - c^2}{2ab}$$

For a proof of the Law of Cosines, see Proofs in Mathematics on page 445.

Example 1   Three Sides of a Triangle—SSS

Find the three angles of the triangle in Figure 5.29.

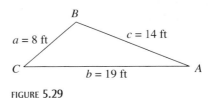

FIGURE **5.29**

Solution

It is a good idea first to find the angle opposite the longest side—side b in this case. Using the alternative form of the Law of Cosines, you find that

$$\cos B = \frac{a^2 + c^2 - b^2}{2ac} = \frac{8^2 + 14^2 - 19^2}{2(8)(14)} \approx -0.45089.$$

Because $\cos B$ is negative, you know that B is an *obtuse* angle given by $B \approx 116.80°$. At this point, it is simpler to use the Law of Sines to determine A.

$$\sin A = a\left(\frac{\sin B}{b}\right) \approx 8\left(\frac{\sin 116.80°}{19}\right) \approx 0.37583$$

You know that A must be acute because B is obtuse, and a triangle can have, at most, one obtuse angle. So, $A \approx 22.08°$ and $C \approx 180° - 22.08° - 116.80° = 41.12°$.

CHECK*Point*   Now try Exercise 5.

Do you see why it was wise to find the largest angle *first* in Example 1? Knowing the cosine of an angle, you can determine whether the angle is acute or obtuse. That is,

$$\cos \theta > 0 \quad \text{for} \quad 0° < \theta < 90° \qquad \text{Acute}$$

$$\cos \theta < 0 \quad \text{for} \quad 90° < \theta < 180°. \qquad \text{Obtuse}$$

So, in Example 1, once you found that angle B was obtuse, you knew that angles A and C were both acute. If the largest angle is acute, the remaining two angles are acute also.

Example 2   Two Sides and the Included Angle—SAS

Find the remaining angles and side of the triangle in Figure 5.30.

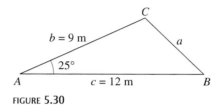

FIGURE **5.30**

Study Tip

When solving an oblique triangle given three sides, you use the alternative form of the Law of Cosines to solve for an angle. When solving an oblique triangle given two sides and their included angle, you use the standard form of the Law of Cosines to solve for an unknown.

Solution

Use the Law of Cosines to find the unknown side a in the figure.

$$a^2 = b^2 + c^2 - 2bc \cos A$$

$$a^2 = 9^2 + 12^2 - 2(9)(12) \cos 25°$$

$$a^2 \approx 29.2375$$

$$a \approx 5.4072$$

Because $a \approx 5.4072$ meters, you now know the ratio $(\sin A)/a$ and you can use the reciprocal form of the Law of Sines to solve for B.

$$\frac{\sin B}{b} = \frac{\sin A}{a}$$

$$\sin B = b\left(\frac{\sin A}{a}\right)$$

$$= 9\left(\frac{\sin 25°}{5.4072}\right)$$

$$\approx 0.7034$$

There are two angles between $0°$ and $180°$ whose sine is 0.7034, $B_1 \approx 44.7°$ and $B_2 \approx 180° - 44.7° = 135.3°$.

For $B_1 \approx 44.7°$,

$$C_1 \approx 180° - 25° - 44.7° = 110.3°.$$

For $B_2 \approx 135.3°$,

$$C_2 \approx 180° - 25° - 135.3° = 19.7°.$$

Because side c is the longest side of the triangle, C must be the largest angle of the triangle. So, $B \approx 44.7°$ and $C \approx 110.3°$.

CHECK *Point*   Now try Exercise 7.

Applications

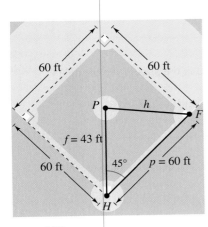

FIGURE **5.31**

| **Example 3** | **An Application of the Law of Cosines** |

The pitcher's mound on a women's softball field is 43 feet from home plate and the distance between the bases is 60 feet, as shown in Figure 5.31. (The pitcher's mound is not halfway between home plate and second base.) How far is the pitcher's mound from first base?

Solution

In triangle *HPF*, $H = 45°$ (line *HP* bisects the right angle at *H*), $f = 43$, and $p = 60$. Using the Law of Cosines for this SAS case, you have

$$h^2 = f^2 + p^2 - 2fp \cos H$$
$$= 43^2 + 60^2 - 2(43)(60) \cos 45° \approx 1800.3.$$

So, the approximate distance from the pitcher's mound to first base is

$$h \approx \sqrt{1800.3} \approx 42.43 \text{ feet.}$$

CHECK*Point*   Now try Exercise 43.

| **Example 4** | **An Application of the Law of Cosines** |

A ship travels 60 miles due east, then adjusts its course northward, as shown in Figure 5.32. After traveling 80 miles in that direction, the ship is 139 miles from its point of departure. Describe the bearing from point *B* to point *C*.

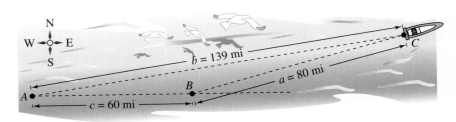

FIGURE **5.32**

Solution

You have $a = 80$, $b = 139$, and $c = 60$. So, using the alternative form of the Law of Cosines, you have

$$\cos B = \frac{a^2 + c^2 - b^2}{2ac}$$

$$= \frac{80^2 + 60^2 - 139^2}{2(80)(60)}$$

$$\approx -0.97094.$$

So, $B \approx \arccos(-0.97094) \approx 166.15°$, and thus the bearing measured from due north from point *B* to point *C* is

$$166.15° - 90° = 76.15°, \text{ or N } 76.15° \text{ E.}$$

CHECK*Point*   Now try Exercise 49.

Heron's Area Formula

The Law of Cosines can be used to establish the following formula for the area of a triangle. This formula is called **Heron's Area Formula** after the Greek mathematician Heron (c. 100 B.C.).

Heron's Area Formula

Given any triangle with sides of lengths a, b, and c, the area of the triangle is

$$\text{Area} = \sqrt{s(s-a)(s-b)(s-c)}$$

where $s = (a+b+c)/2$.

For a proof of Heron's Area Formula, see Proofs in Mathematics on page 446.

Example 5   Using Heron's Area Formula

Find the area of a triangle having sides of lengths $a = 43$ meters, $b = 53$ meters, and $c = 72$ meters.

Solution

Because $s = (a+b+c)/2 = 168/2 = 84$, Heron's Area Formula yields

$$\text{Area} = \sqrt{s(s-a)(s-b)(s-c)}$$
$$= \sqrt{84(41)(31)(12)}$$
$$\approx 1131.89 \text{ square meters.}$$

CHECK Point Now try Exercise 59.

You have now studied three different formulas for the area of a triangle.

Standard Formula: $\text{Area} = \frac{1}{2}bh$

Oblique Triangle: $\text{Area} = \frac{1}{2}bc \sin A = \frac{1}{2}ab \sin C = \frac{1}{2}ac \sin B$

Heron's Area Formula: $\text{Area} = \sqrt{s(s-a)(s-b)(s-c)}$

CLASSROOM DISCUSSION

The Area of a Triangle Use the most appropriate formula to find the area of each triangle below. Show your work and give your reasons for choosing each formula.

a.
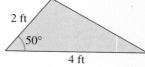
2 ft 50° 4 ft

b.
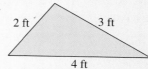
2 ft 3 ft 4 ft

c.
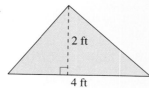
2 ft 4 ft

d.
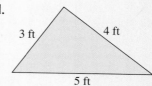
3 ft 4 ft 5 ft

5.7 EXERCISES

See www.CalcChat.com for worked-out solutions to odd-numbered exercises.

VOCABULARY: Fill in the blanks.

1. If you are given three sides of a triangle, you would use the Law of _____ to find the three angles of the triangle.

2. If you are given two angles and any side of a triangle, you would use the Law of _____ to solve the triangle.

3. The standard form of the Law of Cosines for $\cos B = \dfrac{a^2 + c^2 - b^2}{2ac}$ is _____ .

4. The Law of Cosines can be used to establish a formula for finding the area of a triangle called _____ _____ Formula.

SKILLS AND APPLICATIONS

In Exercises 5–20, use the Law of Cosines to solve the triangle. Round your answers to two decimal places.

5.
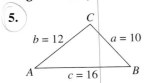

6.
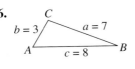

7.
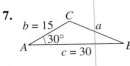

8.
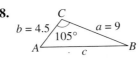

9. $a = 11$, $b = 15$, $c = 21$
10. $a = 55$, $b = 25$, $c = 72$
11. $a = 75.4$, $b = 52$, $c = 52$
12. $a = 1.42$, $b = 0.75$, $c = 1.25$.
13. $A = 120°$, $b = 6$, $c = 7$
14. $A = 48°$, $b = 3$, $c = 14$
15. $B = 10° \, 35'$, $a = 40$, $c = 30$
16. $B = 75° \, 20'$, $a = 6.2$, $c = 9.5$
17. $B = 125° \, 40'$, $a = 37$, $c = 37$
18. $C = 15° \, 15'$, $a = 7.45$, $b = 2.15$
19. $C = 43°$, $a = \frac{4}{9}$, $b = \frac{7}{9}$
20. $C = 101°$, $a = \frac{3}{8}$, $b = \frac{3}{4}$

In Exercises 21–26, complete the table by solving the parallelogram shown in the figure. (The lengths of the diagonals are given by c and d.)

	a	b	c	d	θ	ϕ
21.	5	8			45°	
22.	25	35				120°
23.	10	14	20			
24.	40	60		80		
25.	15		25	20		
26.		25	50	35		

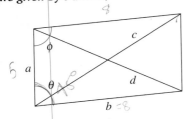

In Exercises 27–32, determine whether the Law of Sines or the Law of Cosines is needed to solve the triangle. Then solve the triangle.

27. $a = 8$, $c = 5$, $B = 40°$
28. $a = 10$, $b = 12$, $C = 70°$
29. $A = 24°$, $a = 4$, $b = 18$
30. $a = 11$, $b = 13$, $c = 7$
31. $A = 42°$, $B = 35°$, $c = 1.2$
32. $a = 160$, $B = 12°$, $C = 7°$

In Exercises 33–40, use Heron's Area Formula to find the area of the triangle.

33. $a = 8$, $b = 12$, $c = 17$
34. $a = 33$, $b = 36$, $c = 25$
35. $a = 2.5$, $b = 10.2$, $c = 9$
36. $a = 75.4$, $b = 52$, $c = 52$
37. $a = 12.32$, $b = 8.46$, $c = 15.05$
38. $a = 3.05$, $b = 0.75$, $c = 2.45$
39. $a = 1$, $b = \frac{1}{2}$, $c = \frac{3}{4}$
40. $a = \frac{3}{5}$, $b = \frac{5}{8}$, $c = \frac{3}{8}$

41. NAVIGATION   A boat race runs along a triangular course marked by buoys *A*, *B*, and *C*. The race starts with the boats headed west for 3700 meters. The other two sides of the course lie to the north of the first side, and their lengths are 1700 meters and 3000 meters. Draw a figure that gives a visual representation of the situation, and find the bearings for the last two legs of the race.

42. NAVIGATION   A plane flies 810 miles from Franklin to Centerville with a bearing of 75°. Then it flies 648 miles from Centerville to Rosemount with a bearing of 32°. Draw a figure that visually represents the situation, and find the straight-line distance and bearing from Franklin to Rosemount.

43. SURVEYING   To approximate the length of a marsh, a surveyor walks 250 meters from point *A* to point *B*, then turns 75° and walks 220 meters to point *C* (see figure). Approximate the length *AC* of the marsh.

44. SURVEYING   A triangular parcel of land has 115 meters of frontage, and the other boundaries have lengths of 76 meters and 92 meters. What angles does the frontage make with the two other boundaries?

45. SURVEYING   A triangular parcel of ground has sides of lengths 725 feet, 650 feet, and 575 feet. Find the measure of the largest angle.

46. STREETLIGHT DESIGN   Determine the angle θ in the design of the streetlight shown in the figure.

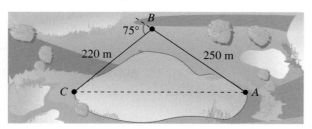

47. DISTANCE   Two ships leave a port at 9 A.M. One travels at a bearing of N 53° W at 12 miles per hour, and the other travels at a bearing of S 67° W at 16 miles per hour. Approximate how far apart they are at noon that day.

48. LENGTH   A 100-foot vertical tower is to be erected on the side of a hill that makes a 6° angle with the horizontal (see figure). Find the length of each of the two guy wires that will be anchored 75 feet uphill and downhill from the base of the tower.

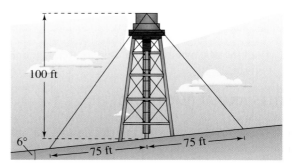

49. NAVIGATION   On a map, Orlando is 178 millimeters due south of Niagara Falls, Denver is 273 millimeters from Orlando, and Denver is 235 millimeters from Niagara Falls (see figure).

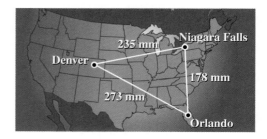

(a) Find the bearing of Denver from Orlando.

(b) Find the bearing of Denver from Niagara Falls.

50. NAVIGATION   On a map, Minneapolis is 165 millimeters due west of Albany, Phoenix is 216 millimeters from Minneapolis, and Phoenix is 368 millimeters from Albany (see figure).

(a) Find the bearing of Minneapolis from Phoenix.

(b) Find the bearing of Albany from Phoenix.

51. BASEBALL   On a baseball diamond with 90-foot sides, the pitcher's mound is 60.5 feet from home plate. How far is it from the pitcher's mound to third base?

52. BASEBALL   The baseball player in center field is playing approximately 330 feet from the television camera that is behind home plate. A batter hits a fly ball that goes to the wall 420 feet from the camera (see figure). The camera turns 8° to follow the play. Approximately how far does the center fielder have to run to make the catch?

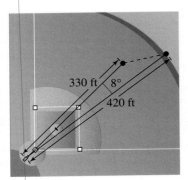

53. AIRCRAFT TRACKING   To determine the distance between two aircraft, a tracking station continuously determines the distance to each aircraft and the angle A between them (see figure). Determine the distance a between the planes when $A = 42°$, $b = 35$ miles, and $c = 20$ miles.

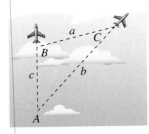

54. AIRCRAFT TRACKING   Use the figure for Exercise 53 to determine the distance a between the planes when $A = 11°$, $b = 20$ miles, and $c = 20$ miles.

55. TRUSSES Q is the midpoint of the line segment $\overline{PR}$ in the truss rafter shown in the figure. What are the lengths of the line segments $\overline{PQ}$, $\overline{QS}$, and $\overline{RS}$?

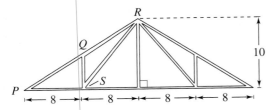

 56. ENGINE DESIGN   An engine has a seven-inch connecting rod fastened to a crank (see figure).

(a) Use the Law of Cosines to write an equation giving the relationship between x and θ.

(b) Write x as a function of θ. (Select the sign that yields positive values of x.)

(c) Use a graphing utility to graph the function in part (b).

(d) Use the graph in part (c) to determine the maximum distance the piston moves in one cycle.

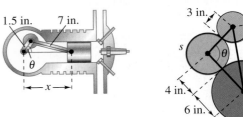

FIGURE FOR 56   FIGURE FOR 57

57. PAPER MANUFACTURING   In a process with continuous paper, the paper passes across three rollers of radii 3 inches, 4 inches, and 6 inches (see figure). The centers of the three-inch and six-inch rollers are d inches apart, and the length of the arc in contact with the paper on the four-inch roller is s inches. Complete the table.

d (inches)	9	10	12	13	14	15	16
θ (degrees)							
s (inches)							

58. AWNING DESIGN   A retractable awning above a patio door lowers at an angle of 50° from the exterior wall at a height of 10 feet above the ground (see figure). No direct sunlight is to enter the door when the angle of elevation of the sun is greater than 70°. What is the length x of the awning?

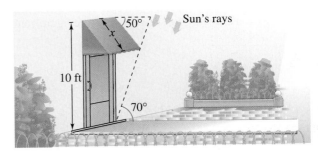

59. GEOMETRY   The lengths of the sides of a triangular parcel of land are approximately 200 feet, 500 feet, and 600 feet. Approximate the area of the parcel.

60. GEOMETRY A parking lot has the shape of a parallelogram (see figure). The lengths of two adjacent sides are 70 meters and 100 meters. The angle between the two sides is 70°. What is the area of the parking lot?

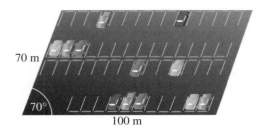

70 m

70°

100 m

61. GEOMETRY You want to buy a triangular lot measuring 510 yards by 840 yards by 1120 yards. The price of the land is $2000 per acre. How much does the land cost? (*Hint:* 1 acre = 4840 square yards)

62. GEOMETRY You want to buy a triangular lot measuring 1350 feet by 1860 feet by 2490 feet. The price of the land is $2200 per acre. How much does the land cost? (*Hint:* 1 acre = 43,560 square feet)

EXPLORATION

TRUE OR FALSE? In Exercises 63 and 64, determine whether the statement is true or false. Justify your answer.

63. In Heron's Area Formula, s is the average of the lengths of the three sides of the triangle.

64. In addition to SSS and SAS, the Law of Cosines can be used to solve triangles with SSA conditions.

65. WRITING A triangle has side lengths of 10 centimeters, 16 centimeters, and 5 centimeters. Can the Law of Cosines be used to solve the triangle? Explain.

66. WRITING Given a triangle with $b = 47$ meters, $A = 87°$, and $C = 110°$, can the Law of Cosines be used to solve the triangle? Explain.

67. CIRCUMSCRIBED AND INSCRIBED CIRCLES Let R and r be the radii of the circumscribed and inscribed circles of a triangle ABC, respectively (see figure), and let
$$s = \frac{a + b + c}{2}.$$

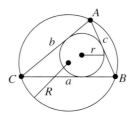

(a) Prove that $2R = \dfrac{a}{\sin A} = \dfrac{b}{\sin B} = \dfrac{c}{\sin C}$.

(b) Prove that $r = \sqrt{\dfrac{(s - a)(s - b)(s - c)}{s}}$.

CIRCUMSCRIBED AND INSCRIBED CIRCLES In Exercises 68 and 69, use the results of Exercise 67.

68. Given a triangle with $a = 25$, $b = 55$, and $c = 72$, find the areas of (a) the triangle, (b) the circumscribed circle, and (c) the inscribed circle.

69. Find the length of the largest circular running track that can be built on a triangular piece of property with sides of lengths 200 feet, 250 feet, and 325 feet.

70. THINK ABOUT IT What familiar formula do you obtain when you use the third form of the Law of Cosines $c^2 = a^2 + b^2 - 2ab \cos C$, and you let $C = 90°$? What is the relationship between the Law of Cosines and this formula?

71. THINK ABOUT IT In Example 2, suppose $A = 115°$. After solving for a, which angle would you solve for next, B or C? Are there two possible solutions for that angle? If so, how can you determine which angle is the correct solution?

72. WRITING Describe how the Law of Cosines can be used to solve the ambiguous case of the oblique triangle ABC, where $a = 12$ feet, $b = 30$ feet, and $A = 20°$. Is the result the same as when the Law of Sines is used to solve the triangle? Describe the advantages and the disadvantages of each method.

73. WRITING In Exercise 72, the Law of Cosines was used to solve a triangle in the two-solution case of SSA. Can the Law of Cosines be used to solve the no-solution and single-solution cases of SSA? Explain.

74. CAPSTONE Determine whether the Law of Sines or the Law of Cosines is needed to solve the triangle.

(a) A, C, and a (b) a, c, and C

(c) b, c, and A (d) A, B, and c

(e) b, c, and C (f) a, b, and c

75. PROOF Use the Law of Cosines to prove that
$$\frac{1}{2}bc(1 + \cos A) = \frac{a + b + c}{2} \cdot \frac{-a + b + c}{2}.$$

76. PROOF Use the Law of Cosines to prove that
$$\frac{1}{2}bc(1 - \cos A) = \frac{a - b + c}{2} \cdot \frac{a + b - c}{2}.$$

5 CHAPTER SUMMARY

What Did You Learn?	Explanation/Examples	Review Exercises

Section 5.1

What Did You Learn?	Explanation/Examples	Review Exercises
Recognize and write the fundamental trigonometric identities *(p. 372)*.	**Reciprocal Identities** $\sin u = 1/\csc u \quad \cos u = 1/\sec u \quad \tan u = 1/\cot u$ $\csc u = 1/\sin u \quad \sec u = 1/\cos u \quad \cot u = 1/\tan u$ **Quotient Identities:** $\tan u = \dfrac{\sin u}{\cos u}, \quad \cot u = \dfrac{\cos u}{\sin u}$ **Pythagorean Identities:** $\sin^2 u + \cos^2 u = 1,$ $1 + \tan^2 u = \sec^2 u, \quad 1 + \cot^2 u = \csc^2 u$ **Cofunction Identities** $\sin[(\pi/2) - u] = \cos u \qquad \cos[(\pi/2) - u] = \sin u$ $\tan[(\pi/2) - u] = \cot u \qquad \cot[(\pi/2) - u] = \tan u$ $\sec[(\pi/2) - u] = \csc u \qquad \csc[(\pi/2) - u] = \sec u$ **Even/Odd Identities** $\sin(-u) = -\sin u \quad \cos(-u) = \cos u \quad \tan(-u) = -\tan u$ $\csc(-u) = -\csc u \quad \sec(-u) = \sec u \quad \cot(-u) = -\cot u$	1–6
Use the fundamental trigonometric identities to evaluate trigonometric functions, and simplify and rewrite trigonometric expressions *(p. 373)*.	In some cases, when factoring or simplifying trigonometric expressions, it is helpful to rewrite the expression in terms of just *one* trigonometric function or in terms of *sine and cosine only*.	7–28

Section 5.2

What Did You Learn?	Explanation/Examples	Review Exercises
Verify trigonometric identities *(p. 380)*.	**Guidelines for Verifying Trigonometric Identities** 1. Work with one side of the equation at a time. 2. Look to factor an expression, add fractions, square a binomial, or create a monomial denominator. 3. Look to use the fundamental identities. Note which functions are in the final expression you want. Sines and cosines pair up well, as do secants and tangents, and cosecants and cotangents. 4. If the preceding guidelines do not help, try converting all terms to sines and cosines. 5. Always try *something*.	29–36

Section 5.3

What Did You Learn?	Explanation/Examples	Review Exercises
Use standard algebraic techniques to solve trigonometric equations *(p. 387)*.	Use standard algebraic techniques such as collecting like terms, extracting square roots, and factoring to solve trigonometric equations.	37–42
Solve trigonometric equations of quadratic type *(p. 389)*.	To solve trigonometric equations of quadratic type $ax^2 + bx + c = 0$, factor the quadratic or, if this is not possible, use the Quadratic Formula.	43–46
Solve trigonometric equations involving multiple angles *(p. 392)*.	To solve equations that contain forms such as $\sin ku$ or $\cos ku$, first solve the equation for ku, then divide your result by k.	47–52
Use inverse trigonometric functions to solve trigonometric equations *(p. 393)*.	After factoring an equation and setting the factors equal to 0, you may get an equation such as $\tan x - 3 = 0$. In this case, use inverse trigonometric functions to solve. (See Example 9.)	53–56

What Did You Learn?	**Explanation/Examples**	**Review Exercises**
Section 5.4 — Use sum and difference formulas to evaluate trigonometric functions, verify identities, and solve trigonometric equations (p. 398).	**Sum and Difference Formulas** $\sin(u + v) = \sin u \cos v + \cos u \sin v$ $\sin(u - v) = \sin u \cos v - \cos u \sin v$ $\cos(u + v) = \cos u \cos v - \sin u \sin v$ $\cos(u - v) = \cos u \cos v + \sin u \sin v$ $\tan(u + v) = \dfrac{\tan u + \tan v}{1 - \tan u \tan v}$ $\tan(u - v) = \dfrac{\tan u - \tan v}{1 + \tan u \tan v}$	57–80
Use multiple-angle formulas to rewrite and evaluate trigonometric functions (p. 405).	**Double-Angle Formulas** $\sin 2u = 2 \sin u \cos u$ $\cos 2u = \cos^2 u - \sin^2 u$ $\tan 2u = \dfrac{2 \tan u}{1 - \tan^2 u}$ $= 2 \cos^2 u - 1$ $= 1 - 2 \sin^2 u$	81–86
Use power-reducing formulas to rewrite and evaluate trigonometric functions (p. 407).	**Power-Reducing Formulas** $\sin^2 u = \dfrac{1 - \cos 2u}{2}, \quad \cos^2 u = \dfrac{1 + \cos 2u}{2}$ $\tan^2 u = \dfrac{1 - \cos 2u}{1 + \cos 2u}$	87–90
Use half-angle formulas to rewrite and evaluate trigonometric functions (p. 408).	**Half-Angle Formulas** $\sin \dfrac{u}{2} = \pm\sqrt{\dfrac{1 - \cos u}{2}}, \quad \cos \dfrac{u}{2} = \pm\sqrt{\dfrac{1 + \cos u}{2}}$ $\tan \dfrac{u}{2} = \dfrac{1 - \cos u}{\sin u} = \dfrac{\sin u}{1 + \cos u}$ The signs of $\sin(u/2)$ and $\cos(u/2)$ depend on the quadrant in which $u/2$ lies.	91–100
Use product-to-sum formulas (p. 409) and sum-to-product formulas (p. 410) to rewrite and evaluate trigonometric functions.	**Product-to-Sum Formulas** $\sin u \sin v = (1/2)[\cos(u - v) - \cos(u + v)]$ $\cos u \cos v = (1/2)[\cos(u - v) + \cos(u + v)]$ $\sin u \cos v = (1/2)[\sin(u + v) + \sin(u - v)]$ $\cos u \sin v = (1/2)[\sin(u + v) - \sin(u - v)]$ **Sum-to-Product Formulas** $\sin u + \sin v = 2 \sin\left(\dfrac{u + v}{2}\right) \cos\left(\dfrac{u - v}{2}\right)$ $\sin u - \sin v = 2 \cos\left(\dfrac{u + v}{2}\right) \sin\left(\dfrac{u - v}{2}\right)$ $\cos u + \cos v = 2 \cos\left(\dfrac{u + v}{2}\right) \cos\left(\dfrac{u - v}{2}\right)$ $\cos u - \cos v = -2 \sin\left(\dfrac{u + v}{2}\right) \sin\left(\dfrac{u - v}{2}\right)$	101–108
Use trigonometric formulas to rewrite real-life models (p. 412).	A trigonometric formula can be used to rewrite the projectile motion model $r = (1/16)v_0^2 \sin \theta \cos \theta$. (See Example 12.)	109–114

Section 5.5

What Did You Learn?	Explanation/Examples	Review Exercises
Section 5.6 — Use the Law of Sines to solve oblique triangles (AAS or ASA) *(p. 416)*.	**Law of Sines** If *ABC* is a triangle with sides *a*, *b*, and *c*, then $$\frac{a}{\sin A} = \frac{b}{\sin B} = \frac{c}{\sin C}.$$ *A* is acute. *A* is obtuse.	115–126
Use the Law of Sines to solve oblique triangles (SSA) *(p. 418)*.	If two sides and one opposite angle are given, three possible situations can occur: (1) no such triangle exists (see Example 4), (2) one such triangle exists (see Example 3), or (3) two distinct triangles may satisfy the conditions (see Example 5).	115–126
Find the areas of oblique triangles *(p. 420)*.	The area of any triangle is one-half the product of the lengths of two sides times the sine of their included angle. That is, $$\text{Area} = \frac{1}{2}bc \sin A = \frac{1}{2}ab \sin C = \frac{1}{2}ac \sin B.$$	127–130
Use the Law of Sines to model and solve real-life problems *(p. 421)*.	The Law of Sines can be used to approximate the total distance of a boat race course. (See Example 7.)	131–134
Section 5.7 — Use the Law of Cosines to solve oblique triangles (SSS or SAS) *(p. 425)*.	**Law of Cosines** *Standard Form* · *Alternative Form* $$a^2 = b^2 + c^2 - 2bc \cos A \qquad \cos A = \frac{b^2 + c^2 - a^2}{2bc}$$ $$b^2 = a^2 + c^2 - 2ac \cos B \qquad \cos B = \frac{a^2 + c^2 - b^2}{2ac}$$ $$c^2 = a^2 + b^2 - 2ab \cos C \qquad \cos C = \frac{a^2 + b^2 - c^2}{2ab}$$	135–144
Use the Law of Cosines to model and solve real-life problems *(p. 427)*.	The Law of Cosines can be used to find the distance between the pitcher's mound and first base on a women's softball field. (See Example 3.)	149–152
Use Heron's Area Formula to find the area of a triangle *(p. 428)*.	**Heron's Area Formula** Given any triangle with sides of lengths *a*, *b*, and *c*, the area of the triangle is $$\text{Area} = \sqrt{s(s - a)(s - b)(s - c)}$$ where $$s = \frac{a + b + c}{2}.$$	153–156

5 REVIEW EXERCISES

See www.CalcChat.com for worked-out solutions to odd-numbered exercises.

5.1 In Exercises 1–6, name the trigonometric function that is equivalent to the expression.

1. $\dfrac{\sin x}{\cos x}$

2. $\dfrac{1}{\sin x}$

3. $\dfrac{1}{\sec x}$

4. $\dfrac{1}{\tan x}$

5. $\sqrt{\cot^2 x + 1}$

6. $\sqrt{1 + \tan^2 x}$

In Exercises 7–10, use the given values and trigonometric identities to evaluate (if possible) all six trigonometric functions.

7. $\sin x = \frac{5}{13}, \quad \cos x = \frac{12}{13}$

8. $\tan \theta = \frac{2}{3}, \quad \sec \theta = \frac{\sqrt{13}}{3}$

9. $\sin\left(\frac{\pi}{2} - x\right) = \frac{\sqrt{2}}{2}, \quad \sin x = -\frac{\sqrt{2}}{2}$

10. $\csc\left(\frac{\pi}{2} - \theta\right) = 9, \quad \sin \theta = \frac{4\sqrt{5}}{9}$

In Exercises 11–24, use the fundamental trigonometric identities to simplify the expression.

11. $\dfrac{1}{\cot^2 x + 1}$

12. $\dfrac{\tan \theta}{1 - \cos^2 \theta}$

13. $\tan^2 x(\csc^2 x - 1)$

14. $\cot^2 x(\sin^2 x)$

15. $\dfrac{\sin\left(\frac{\pi}{2} - \theta\right)}{\sin \theta}$

16. $\dfrac{\cot\left(\frac{\pi}{2} - u\right)}{\cos u}$

17. $\dfrac{\sin^2 \theta + \cos^2 \theta}{\sin \theta}$

18. $\dfrac{\sec^2(-\theta)}{\csc^2 \theta}$

19. $\cos^2 x + \cos^2 x \cot^2 x$

20. $\tan^2 \theta \csc^2 \theta - \tan^2 \theta$

21. $(\tan x + 1)^2 \cos x$

22. $(\sec x - \tan x)^2$

23. $\dfrac{1}{\csc \theta + 1} - \dfrac{1}{\csc \theta - 1}$

24. $\dfrac{\tan^2 x}{1 + \sec x}$

In Exercises 25 and 26, use the trigonometric substitution to write the algebraic expression as a trigonometric function of θ, where $0 < \theta < \pi/2$.

25. $\sqrt{25 - x^2}, x = 5 \sin \theta$

26. $\sqrt{x^2 - 16}, x = 4 \sec \theta$

27. RATE OF CHANGE The rate of change of the function $f(x) = \csc x - \cot x$ is given by the expression $\csc^2 x - \csc x \cot x$. Show that this expression can also be written as

$$\dfrac{1 - \cos x}{\sin^2 x}.$$

28. RATE OF CHANGE The rate of change of the function $f(x) = 2\sqrt{\sin x}$ is given by the expression $\sin^{-1/2} x \cos x$. Show that this expression can also be written as $\cot x \sqrt{\sin x}$.

5.2 In Exercises 29–36, verify the identity.

29. $\cos x(\tan^2 x + 1) = \sec x$

30. $\sec^2 x \cot x - \cot x = \tan x$

31. $\sec\left(\frac{\pi}{2} - \theta\right) = \csc \theta$

32. $\cot\left(\frac{\pi}{2} - x\right) = \tan x$

33. $\dfrac{1}{\tan \theta \csc \theta} = \cos \theta$

34. $\dfrac{1}{\tan x \csc x \sin x} = \cot x$

35. $\sin^5 x \cos^2 x = (\cos^2 x - 2 \cos^4 x + \cos^6 x) \sin x$

36. $\cos^3 x \sin^2 x = (\sin^2 x - \sin^4 x) \cos x$

5.3 In Exercises 37–42, solve the equation.

37. $\sin x = \sqrt{3} - \sin x$

38. $4 \cos \theta = 1 + 2 \cos \theta$

39. $3\sqrt{3} \tan u = 3$

40. $\frac{1}{2} \sec x - 1 = 0$

41. $3 \csc^2 x = 4$

42. $4 \tan^2 u - 1 = \tan^2 u$

In Exercises 43–52, find all solutions of the equation in the interval $[0, 2\pi)$.

43. $2 \cos^2 x - \cos x = 1$

44. $2 \sin^2 x - 3 \sin x = -1$

45. $\cos^2 x + \sin x = 1$

46. $\sin^2 x + 2 \cos x = 2$

47. $2 \sin 2x - \sqrt{2} = 0$

48. $2 \cos \frac{x}{2} + 1 = 0$

49. $3 \tan^2\left(\frac{x}{3}\right) - 1 = 0$

50. $\sqrt{3} \tan 3x = 0$

51. $\cos 4x(\cos x - 1) = 0$

52. $3 \csc^2 5x = -4$

In Exercises 53–56, use inverse functions where needed to find all solutions of the equation in the interval $[0, 2\pi)$.

53. $\sin^2 x - 2 \sin x = 0$

54. $2 \cos^2 x + 3 \cos x = 0$

55. $\tan^2 \theta + \tan \theta - 6 = 0$

56. $\sec^2 x + 6 \tan x + 4 = 0$

5.4 In Exercises 57–60, find the exact values of the sine, cosine, and tangent of the angle.

57. $285° = 315° - 30°$

58. $345° = 300° + 45°$

59. $\dfrac{25\pi}{12} = \dfrac{11\pi}{6} + \dfrac{\pi}{4}$

60. $\dfrac{19\pi}{12} = \dfrac{11\pi}{6} - \dfrac{\pi}{4}$

In Exercises 61–64, write the expression as the sine, cosine, or tangent of an angle.

61. $\sin 60° \cos 45° - \cos 60° \sin 45°$

62. $\cos 45° \cos 120° - \sin 45° \sin 120°$

63. $\dfrac{\tan 25° + \tan 10°}{1 - \tan 25° \tan 10°}$

64. $\dfrac{\tan 68° - \tan 115°}{1 + \tan 68° \tan 115°}$

In Exercises 65–70, find the exact value of the trigonometric function given that $\tan u = \frac{3}{4}$ and $\cos v = -\frac{4}{5}$. (u is in Quadrant I and v is in Quadrant III.)

65. $\sin(u + v)$

66. $\tan(u + v)$

67. $\cos(u - v)$

68. $\sin(u - v)$

69. $\cos(u + v)$

70. $\tan(u - v)$

In Exercises 71–76, verify the identity.

71. $\cos\left(x + \dfrac{\pi}{2}\right) = -\sin x$

72. $\sin\left(x - \dfrac{3\pi}{2}\right) = \cos x$

73. $\tan\left(x - \dfrac{\pi}{2}\right) = -\cot x$

74. $\tan(\pi - x) = -\tan x$

75. $\cos 3x = 4 \cos^3 x - 3 \cos x$

76. $\dfrac{\sin(\alpha - \beta)}{\sin(\alpha + \beta)} = \dfrac{\tan \alpha - \tan \beta}{\tan \alpha + \tan \beta}$

In Exercises 77–80, find all solutions of the equation in the interval $[0, 2\pi)$.

77. $\sin\left(x + \dfrac{\pi}{4}\right) - \sin\left(x - \dfrac{\pi}{4}\right) = 1$

78. $\cos\left(x + \dfrac{\pi}{6}\right) - \cos\left(x - \dfrac{\pi}{6}\right) = 1$

79. $\sin\left(x + \dfrac{\pi}{2}\right) - \sin\left(x - \dfrac{\pi}{2}\right) = \sqrt{3}$

80. $\cos\left(x + \dfrac{3\pi}{4}\right) - \cos\left(x - \dfrac{3\pi}{4}\right) = 0$

5.5 In Exercises 81–84, find the exact values of $\sin 2u$, $\cos 2u$, and $\tan 2u$ using the double-angle formulas.

81. $\sin u = -\dfrac{4}{5}, \quad \pi < u < \dfrac{3\pi}{2}$

82. $\cos u = -\dfrac{2}{\sqrt{5}}, \quad \dfrac{\pi}{2} < u < \pi$

83. $\sec u = -3, \quad \dfrac{\pi}{2} < u < \pi$

84. $\cot u = 2, \quad \pi < u < \dfrac{3\pi}{2}$

In Exercises 85 and 86, use double-angle formulas to verify the identity algebraically and use a graphing utility to confirm your result graphically.

85. $\sin 4x = 8 \cos^3 x \sin x - 4 \cos x \sin x$

86. $\tan^2 x = \dfrac{1 - \cos 2x}{1 + \cos 2x}$

In Exercises 87–90, use the power-reducing formulas to rewrite the expression in terms of the first power of the cosine.

87. $\tan^2 2x$

88. $\cos^2 3x$

89. $\sin^2 x \tan^2 x$

90. $\cos^2 x \tan^2 x$

In Exercises 91–94, use the half-angle formulas to determine the exact values of the sine, cosine, and tangent of the angle.

91. $-75°$

92. $15°$

93. $\dfrac{19\pi}{12}$

94. $-\dfrac{17\pi}{12}$

In Exercises 95–98, (a) determine the quadrant in which $u/2$ lies, and (b) find the exact values of $\sin(u/2)$, $\cos(u/2)$, and $\tan(u/2)$ using the half-angle formulas.

95. $\sin u = \frac{7}{25}$, $0 < u < \pi/2$

96. $\tan u = \frac{4}{3}$, $\pi < u < 3\pi/2$

97. $\cos u = -\frac{2}{7}$, $\pi/2 < u < \pi$

98. $\sec u = -6$, $\pi/2 < u < \pi$

In Exercises 99 and 100, use the half-angle formulas to simplify the expression.

99. $-\sqrt{\dfrac{1 + \cos 10x}{2}}$

100. $\dfrac{\sin 6x}{1 + \cos 6x}$

In Exercises 101–104, use the product-to-sum formulas to write the product as a sum or difference.

101. $\cos \dfrac{\pi}{6} \sin \dfrac{\pi}{6}$

102. $6 \sin 15° \sin 45°$

103. $\cos 4\theta \sin 6\theta$

104. $2 \sin 7\theta \cos 3\theta$

In Exercises 105–108, use the sum-to-product formulas to write the sum or difference as a product.

105. $\sin 4\theta - \sin 8\theta$

106. $\cos 6\theta + \cos 5\theta$

107. $\cos\left(x + \dfrac{\pi}{6}\right) - \cos\left(x - \dfrac{\pi}{6}\right)$

108. $\sin\left(x + \dfrac{\pi}{4}\right) - \sin\left(x - \dfrac{\pi}{4}\right)$

109. PROJECTILE MOTION A baseball leaves the hand of the player at first base at an angle of θ with the horizontal and at an initial velocity of $v_0 = 80$ feet per second. The ball is caught by the player at second base 100 feet away. Find θ if the range r of a projectile is

$$r = \dfrac{1}{32} v_0^2 \sin 2\theta.$$

110. GEOMETRY A trough for feeding cattle is 4 meters long and its cross sections are isosceles triangles with the two equal sides being $\frac{1}{2}$ meter (see figure). The angle between the two sides is θ.

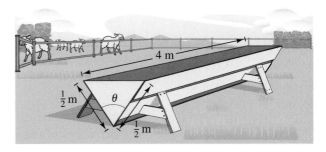

(a) Write the trough's volume as a function of $\theta/2$.

(b) Write the volume of the trough as a function of θ and determine the value of θ such that the volume is maximum.

HARMONIC MOTION In Exercises 111–114, use the following information. A weight is attached to a spring suspended vertically from a ceiling. When a driving force is applied to the system, the weight moves vertically from its equilibrium position, and this motion is described by the model $y = 1.5 \sin 8t - 0.5 \cos 8t$, where y is the distance from equilibrium (in feet) and t is the time (in seconds).

111. Use a graphing utility to graph the model.

112. Write the model in the form

$$y = \sqrt{a^2 + b^2} \sin(Bt + C).$$

113. Find the amplitude of the oscillations of the weight.

114. Find the frequency of the oscillations of the weight.

5.6 In Exercises 115–126, use the Law of Sines to solve (if possible) the triangle. If two solutions exist, find both. Round your answers to two decimal places.

115.

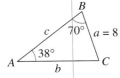

116.

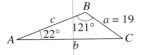

117. $B = 72°$, $C = 82°$, $b = 54$
118. $B = 10°$, $C = 20°$, $c = 33$
119. $A = 16°$, $B = 98°$, $c = 8.4$
120. $A = 95°$, $B = 45°$, $c = 104.8$
121. $A = 24°$, $C = 48°$, $b = 27.5$
122. $B = 64°$, $C = 36°$, $a = 367$
123. $B = 150°$, $b = 30$, $c = 10$
124. $B = 150°$, $a = 10$, $b = 3$
125. $A = 75°$, $a = 51.2$, $b = 33.7$
126. $B = 25°$, $a = 6.2$, $b = 4$

In Exercises 127–130, find the area of the triangle having the indicated angle and sides.

127. $A = 33°$, $b = 7$, $c = 10$
128. $B = 80°$, $a = 4$, $c = 8$
129. $C = 119°$, $a = 18$, $b = 6$
130. $A = 11°$, $b = 22$, $c = 21$

131. HEIGHT From a certain distance, the angle of elevation to the top of a building is 17°. At a point 50 meters closer to the building, the angle of elevation is 31°. Approximate the height of the building.

132. GEOMETRY Find the length of the side w of the parallelogram.

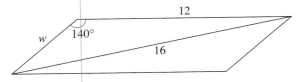

133. HEIGHT A tree stands on a hillside of slope 28° from the horizontal. From a point 75 feet down the hill, the angle of elevation to the top of the tree is 45° (see figure). Find the height of the tree.

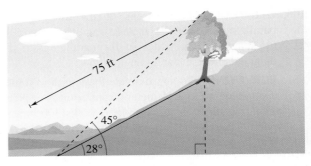

FIGURE FOR 133

134. RIVER WIDTH A surveyor finds that a tree on the opposite bank of a river flowing due east has a bearing of N 22° 30′ E from a certain point and a bearing of N 15° W from a point 400 feet downstream. Find the width of the river.

5.7 In Exercises 135–144, use the Law of Cosines to solve the triangle. Round your answers to two decimal places.

135.

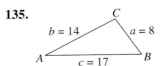

136.

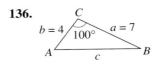

137. $a = 6$, $b = 9$, $c = 14$
138. $a = 75$, $b = 50$, $c = 110$
139. $a = 2.5$, $b = 5.0$, $c = 4.5$
140. $a = 16.4$, $b = 8.8$, $c = 12.2$
141. $B = 108°$, $a = 11$, $c = 11$
142. $B = 150°$, $a = 10$, $c = 20$
143. $C = 43°$, $a = 22.5$, $b = 31.4$
144. $A = 62°$, $b = 11.34$, $c = 19.52$

In Exercises 145–148, determine whether the Law of Sines or the Law of Cosines is needed to solve the triangle. Then solve the triangle.

145. $b = 9$, $c = 13$, $C = 64°$
146. $a = 4$, $c = 5$, $B = 52°$
147. $a = 13$, $b = 15$, $c = 24$
148. $A = 44°$, $B = 31°$, $c = 2.8$

149. GEOMETRY The lengths of the diagonals of a parallelogram are 10 feet and 16 feet. Find the lengths of the sides of the parallelogram if the diagonals intersect at an angle of 28°.

150. GEOMETRY The lengths of the diagonals of a parallelogram are 30 meters and 40 meters. Find the lengths of the sides of the parallelogram if the diagonals intersect at an angle of 34°.

151. SURVEYING To approximate the length of a marsh, a surveyor walks 425 meters from point A to point B. Then the surveyor turns 65° and walks 300 meters to point C (see figure). Approximate the length AC of the marsh.

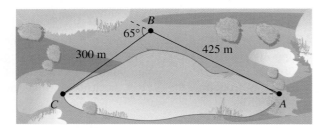

152. NAVIGATION Two planes leave an airport at approximately the same time. One is flying 425 miles per hour at a bearing of 355°, and the other is flying 530 miles per hour at a bearing of 67°. Draw a figure that gives a visual representation of the situation and determine the distance between the planes after they have flown for 2 hours.

In Exercises 153–156, use Heron's Area Formula to find the area of the triangle.

153. $a = 3$, $b = 6$, $c = 8$

154. $a = 15$, $b = 8$, $c = 10$

155. $a = 12.3$, $b = 15.8$, $c = 3.7$

156. $a = \frac{4}{5}$, $b = \frac{3}{4}$, $c = \frac{5}{8}$

EXPLORATION

TRUE OR FALSE? In Exercises 157–162, determine whether the statement is true or false. Justify your answer.

157. If $\frac{\pi}{2} < \theta < \pi$, then $\cos \frac{\theta}{2} < 0$.

158. $\sin(x + y) = \sin x + \sin y$

159. $4 \sin(-x) \cos(-x) = -2 \sin 2x$

160. $4 \sin 45° \cos 15° = 1 + \sqrt{3}$

161. The Law of Sines is true if one of the angles in the triangle is a right angle.

162. When the Law of Sines is used, the solution is always unique.

163. List the reciprocal identities, quotient identities, and Pythagorean identities from memory.

164. State the Law of Sines from memory.

165. State the Law of Cosines from memory.

166. THINK ABOUT IT If a trigonometric equation has an infinite number of solutions, is it true that the equation is an identity? Explain.

167. THINK ABOUT IT Explain why you know from observation that the equation $a \sin x - b = 0$ has no solution if $|a| < |b|$.

168. SURFACE AREA The surface area of a honeycomb is given by the equation

$$S = 6hs + \frac{3}{2}s^2\left(\frac{\sqrt{3} - \cos \theta}{\sin \theta}\right), \quad 0 < \theta \le 90°$$

where $h = 2.4$ inches, $s = 0.75$ inch, and θ is the angle shown in the figure.

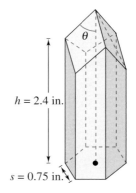

(a) For what value(s) of θ is the surface area 12 square inches?

(b) What value of θ gives the minimum surface area?

In Exercises 169 and 170, use the graphs of y_1 and y_2 to determine how to change one function to form the identity $y_1 = y_2$.

169. $y_1 = \sec^2\left(\frac{\pi}{2} - x\right)$
$y_2 = \cot^2 x$

170. $y_1 = \frac{\cos 3x}{\cos x}$
$y_2 = (2 \sin x)^2$

In Exercises 171 and 172, use the *zero* or *root* feature of a graphing utility to approximate the zeros of the function.

171. $y = \sqrt{x + 3} + 4 \cos x$

172. $y = 2 - \frac{1}{2}x^2 + 3 \sin \frac{\pi x}{2}$

5 CHAPTER TEST

See www.CalcChat.com for worked-out solutions to odd-numbered exercises.

Take this test as you would take a test in class. When you are finished, check your work against the answers given in the back of the book.

1. If $\tan \theta = \frac{6}{5}$ and $\cos \theta < 0$, use the fundamental identities to evaluate all six trigonometric functions of θ.

2. Use the fundamental identities to simplify $\csc^2 \beta(1 - \cos^2 \beta)$.

3. Factor and simplify $\dfrac{\sec^4 x - \tan^4 x}{\sec^2 x + \tan^2 x}$.   4. Add and simplify $\dfrac{\cos \theta}{\sin \theta} + \dfrac{\sin \theta}{\cos \theta}$.

In Exercises 5–10, verify the identity.

5. $\sin \theta \sec \theta = \tan \theta$

6. $\sec^2 x \tan^2 x + \sec^2 x = \sec^4 x$

7. $\dfrac{\csc \alpha + \sec \alpha}{\sin \alpha + \cos \alpha} = \cot \alpha + \tan \alpha$

8. $\tan\left(x + \dfrac{\pi}{2}\right) = -\cot x$

9. $\sin(n\pi + \theta) = (-1)^n \sin \theta$, n is an integer.

10. $(\sin x + \cos x)^2 = 1 + \sin 2x$

11. Rewrite $\sin^4(x/2)$ in terms of the first power of the cosine.

12. Use a half-angle formula to simplify the expression $(\sin 4\theta)/(1 + \cos 4\theta)$.

13. Write $4 \sin 3\theta \cos 2\theta$ as a sum or difference.

14. Write $\cos 3\theta - \cos \theta$ as a product.

In Exercises 15–18, find all solutions of the equation in the interval $[0, 2\pi)$.

15. $\tan^2 x + \tan x = 0$

16. $\sin 2\alpha - \cos \alpha = 0$

17. $4 \cos^2 x - 3 = 0$

18. $\csc^2 x - \csc x - 2 = 0$

19. Find the exact value of $\cos 105°$ using the fact that $105° = 135° - 30°$.

20. Use the figure to find the exact values of $\sin 2u$, $\cos 2u$, and $\tan 2u$.

In Exercises 21–26, use the information to solve (if possible) the triangle. If two solutions exist, find both solutions. Round your answers to two decimal places.

21. $A = 24°$, $B = 68°$, $a = 12.2$

22. $B = 110°$, $C = 28°$, $a = 15.6$

23. $A = 24°$, $a = 11.2$, $b = 13.4$

24. $a = 4.0$, $b = 7.3$, $c = 12.4$

25. $B = 100°$, $a = 15$, $b = 23$

26. $C = 121°$, $a = 34$, $b = 55$

27. Cheyenne, Wyoming has a latitude of $41°$ N. At this latitude, the position of the sun at sunrise can be modeled by

$$D = 31 \sin\left(\frac{2\pi}{365}t - 1.4\right)$$

where t is the time (in days) and $t = 1$ represents January 1. In this model, D represents the number of degrees north or south of due east that the sun rises. Use a graphing utility to determine the days on which the sun is more than $20°$ north of due east at sunrise.

28. A triangular parcel of land has borders of lengths 60 meters, 70 meters, and 82 meters. Find the area of the parcel of land.

29. An airplane flies 370 miles from point A to point B with a bearing of $24°$. It then flies 240 miles from point B to point C with a bearing of $37°$ (see figure). Find the distance and bearing from point A to point C.

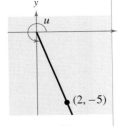

FIGURE FOR 20

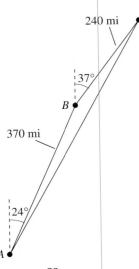

FIGURE FOR 29

PROOFS IN MATHEMATICS

Proof

You can use the figures at the left for the proofs of the formulas for $\cos(u \pm v)$. In the top figure, let A be the point $(1, 0)$ and then use u and v to locate the points $B = (x_1, y_1)$, $C = (x_2, y_2)$, and $D = (x_3, y_3)$ on the unit circle. So, $x_i^2 + y_i^2 = 1$ for $i = 1, 2,$ and 3. For convenience, assume that $0 < v < u < 2\pi$. In the bottom figure, note that arcs AC and BD have the same length. So, line segments AC and BD are also equal in length, which implies that

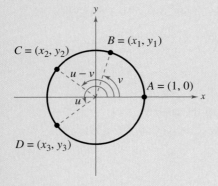

$$\sqrt{(x_2 - 1)^2 + (y_2 - 0)^2} = \sqrt{(x_3 - x_1)^2 + (y_3 - y_1)^2}$$

$$x_2^2 - 2x_2 + 1 + y_2^2 = x_3^2 - 2x_1x_3 + x_1^2 + y_3^2 - 2y_1y_3 + y_1^2$$

$$(x_2^2 + y_2^2) + 1 - 2x_2 = (x_3^2 + y_3^2) + (x_1^2 + y_1^2) - 2x_1x_3 - 2y_1y_3$$

$$1 + 1 - 2x_2 = 1 + 1 - 2x_1x_3 - 2y_1y_3$$

$$x_2 = x_3x_1 + y_3y_1.$$

Finally, by substituting the values $x_2 = \cos(u - v)$, $x_3 = \cos u$, $x_1 = \cos v$, $y_3 = \sin u$, and $y_1 = \sin v$, you obtain $\cos(u - v) = \cos u \cos v + \sin u \sin v$. The formula for $\cos(u + v)$ can be established by considering $u + v = u - (-v)$ and using the formula just derived to obtain

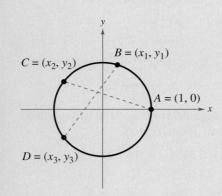

$$\cos(u + v) = \cos[u - (-v)] = \cos u \cos(-v) + \sin u \sin(-v)$$

$$= \cos u \cos v - \sin u \sin v.$$

You can use the sum and difference formulas for sine and cosine to prove the formulas for $\tan(u \pm v)$.

$$\tan(u \pm v) = \frac{\sin(u \pm v)}{\cos(u \pm v)} \qquad \text{Quotient identity}$$

$$= \frac{\sin u \cos v \pm \cos u \sin v}{\cos u \cos v \mp \sin u \sin v} \qquad \text{Sum and difference formulas}$$

$$= \frac{\dfrac{\sin u \cos v \pm \cos u \sin v}{\cos u \cos v}}{\dfrac{\cos u \cos v \mp \sin u \sin v}{\cos u \cos v}} \qquad \text{Divide numerator and denominator by } \cos u \cos v.$$

$$= \frac{\dfrac{\sin u \cos v}{\cos u \cos v} \pm \dfrac{\cos u \sin v}{\cos u \cos v}}{\dfrac{\cos u \cos v}{\cos u \cos v} \mp \dfrac{\sin u \sin v}{\cos u \cos v}}$$

Write as separate fractions.

$$= \frac{\dfrac{\sin u}{\cos u} \pm \dfrac{\sin v}{\cos v}}{1 \mp \dfrac{\sin u}{\cos u} \cdot \dfrac{\sin v}{\cos v}}$$

Product of fractions

$$= \frac{\tan u \pm \tan v}{1 \mp \tan u \tan v}$$

Quotient identity

Trigonometry and Astronomy

Trigonometry was used by early astronomers to calculate measurements in the universe. Trigonometry was used to calculate the circumference of Earth and the distance from Earth to the moon. Another major accomplishment in astronomy using trigonometry was computing distances to stars.

Double-Angle Formulas   (p. 405)

$$\sin 2u = 2 \sin u \cos u \qquad \cos 2u = \cos^2 u - \sin^2 u$$
$$= 2 \cos^2 u - 1 = 1 - 2 \sin^2 u$$
$$\tan 2u = \frac{2 \tan u}{1 - \tan^2 u}$$

Proof

To prove all three formulas, let $v = u$ in the corresponding sum formulas.

$$\sin 2u = \sin(u + u) = \sin u \cos u + \cos u \sin u = 2 \sin u \cos u$$
$$\cos 2u = \cos(u + u) = \cos u \cos u - \sin u \sin u = \cos^2 u - \sin^2 u$$
$$\tan 2u = \tan(u + u) = \frac{\tan u + \tan u}{1 - \tan u \tan u} = \frac{2 \tan u}{1 - \tan^2 u}$$

Power-Reducing Formulas   (p. 407)

$$\sin^2 u = \frac{1 - \cos 2u}{2} \qquad \cos^2 u = \frac{1 + \cos 2u}{2} \qquad \tan^2 u = \frac{1 - \cos 2u}{1 + \cos 2u}$$

Proof

To prove the first formula, solve for $\sin^2 u$ in the double-angle formula $\cos 2u = 1 - 2 \sin^2 u$, as follows.

$$\cos 2u = 1 - 2 \sin^2 u$$

Write double-angle formula.

$$2 \sin^2 u = 1 - \cos 2u$$

Subtract $\cos 2u$ from and add $2 \sin^2 u$ to each side.

$$\sin^2 u = \frac{1 - \cos 2u}{2}$$

Divide each side by 2.

In a similar way you can prove the second formula, by solving for $\cos^2 u$ in the double-angle formula $\cos 2u = 2\cos^2 u - 1$. To prove the third formula, use a quotient identity, as follows.

$$\tan^2 u = \frac{\sin^2 u}{\cos^2 u} = \frac{\dfrac{1 - \cos 2u}{2}}{\dfrac{1 + \cos 2u}{2}} = \frac{1 - \cos 2u}{1 + \cos 2u}$$

Sum-to-Product Formulas   (p. 410)

$$\sin u + \sin v = 2\sin\left(\frac{u + v}{2}\right)\cos\left(\frac{u - v}{2}\right)$$

$$\sin u - \sin v = 2\cos\left(\frac{u + v}{2}\right)\sin\left(\frac{u - v}{2}\right)$$

$$\cos u + \cos v = 2\cos\left(\frac{u + v}{2}\right)\cos\left(\frac{u - v}{2}\right)$$

$$\cos u - \cos v = -2\sin\left(\frac{u + v}{2}\right)\sin\left(\frac{u - v}{2}\right)$$

Proof

To prove the first formula, let $x = u + v$ and $y = u - v$. Then substitute $u = (x + y)/2$ and $v = (x - y)/2$ in the product-to-sum formula.

$$\sin u \cos v = \frac{1}{2}[\sin(u + v) + \sin(u - v)]$$

$$\sin\left(\frac{x + y}{2}\right)\cos\left(\frac{x - y}{2}\right) = \frac{1}{2}(\sin x + \sin y)$$

$$2\sin\left(\frac{x + y}{2}\right)\cos\left(\frac{x - y}{2}\right) = \sin x + \sin y$$

The other sum-to-product formulas can be proved in a similar manner.

Law of Sines   (p. 416)

If ABC is a triangle with sides a, b, and c, then

$$\frac{a}{\sin A} = \frac{b}{\sin B} = \frac{c}{\sin C}.$$

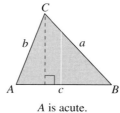

A is acute.

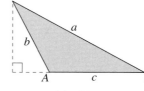

A is obtuse.

Proof

Let h be the altitude of either triangle found in the figure on the previous page. Then you have

$$\sin A = \frac{h}{b} \quad \text{or} \quad h = b \sin A \qquad \text{and} \qquad \sin B = \frac{h}{a} \quad \text{or} \quad h = a \sin B.$$

Equating these two values of h, you have

$$a \sin B = b \sin A \qquad \text{or} \qquad \frac{a}{\sin A} = \frac{b}{\sin B}.$$

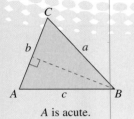

A is acute.

Note that $\sin A \neq 0$ and $\sin B \neq 0$ because no angle of a triangle can have a measure of $0°$ or $180°$. In a similar manner, construct an altitude from vertex B to side AC (extended in the obtuse triangle), as shown at the left. Then you have

$$\sin A = \frac{h}{c} \quad \text{or} \quad h = c \sin A \qquad \text{and} \qquad \sin C = \frac{h}{a} \quad \text{or} \quad h = a \sin C.$$

Equating these two values of h, you have

$$a \sin C = c \sin A \qquad \text{or} \qquad \frac{a}{\sin A} = \frac{c}{\sin C}.$$

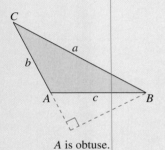

A is obtuse.

By the Transitive Property of Equality you know that

$$\frac{a}{\sin A} = \frac{b}{\sin B} = \frac{c}{\sin C}.$$

So, the Law of Sines is established.

Law of Cosines   *(p. 425)*

Standard Form	Alternative Form
$a^2 = b^2 + c^2 - 2bc \cos A$	$\cos A = \dfrac{b^2 + c^2 - a^2}{2bc}$
$b^2 = a^2 + c^2 - 2ac \cos B$	$\cos B = \dfrac{a^2 + c^2 - b^2}{2ac}$
$c^2 = a^2 + b^2 - 2ab \cos C$	$\cos C = \dfrac{a^2 + b^2 - c^2}{2ab}$

Proof

To prove the first formula, consider the triangle at the left, which has three acute angles. Note that vertex B has coordinates $(c, 0)$. Furthermore, C has coordinates (x, y), where $x = b \cos A$ and $y = b \sin A$. Because a is the distance from vertex C to vertex B, it follows that

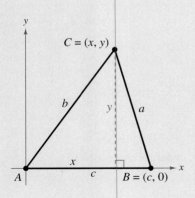

$$a = \sqrt{(x - c)^2 + (y - 0)^2} \qquad \text{Distance Formula}$$

$$a^2 = (x - c)^2 + (y - 0)^2 \qquad \text{Square each side.}$$

$$a^2 = (b \cos A - c)^2 + (b \sin A)^2 \qquad \text{Substitute for } x \text{ and } y.$$

$$a^2 = b^2 \cos^2 A - 2bc \cos A + c^2 + b^2 \sin^2 A \qquad \text{Expand.}$$

$$a^2 = b^2(\sin^2 A + \cos^2 A) + c^2 - 2bc \cos A \qquad \text{Factor out } b^2.$$

$$a^2 = b^2 + c^2 - 2bc \cos A. \qquad \sin^2 A + \cos^2 A = 1$$

Similar arguments can be used to establish the second and third formulas.

Heron's Area Formula   *(p. 428)*

Given any triangle with sides of lengths a, b, and c, the area of the triangle is

$$\text{Area} = \sqrt{s(s - a)(s - b)(s - c)}$$

where $s = \dfrac{a + b + c}{2}$.

Proof

From Section 5.6, you know that

$$\text{Area} = \frac{1}{2}bc \sin A \qquad \text{Formula for the area of an oblique triangle}$$

$$(\text{Area})^2 = \frac{1}{4}b^2 c^2 \sin^2 A \qquad \text{Square each side.}$$

$$\text{Area} = \sqrt{\frac{1}{4}b^2 c^2 \sin^2 A} \qquad \text{Take the square root of each side.}$$

$$= \sqrt{\frac{1}{4}b^2 c^2(1 - \cos^2 A)} \qquad \text{Pythagorean Identity}$$

$$= \sqrt{\left[\frac{1}{2}bc(1 + \cos A)\right]\left[\frac{1}{2}bc(1 - \cos A)\right]}. \qquad \text{Factor.}$$

Using the Law of Cosines, you can show that

$$\frac{1}{2}bc(1 + \cos A) = \frac{a + b + c}{2} \cdot \frac{-a + b + c}{2}$$

and

$$\frac{1}{2}bc(1 - \cos A) = \frac{a - b + c}{2} \cdot \frac{a + b - c}{2}.$$

Letting $s = (a + b + c)/2$, these two equations can be rewritten as

$$\frac{1}{2}bc(1 + \cos A) = s(s - a) \qquad \text{and} \qquad \frac{1}{2}bc(1 - \cos A) = (s - b)(s - c).$$

By substituting into the last formula for area, you can conclude that

$$\text{Area} = \sqrt{s(s - a)(s - b)(s - c)}.$$

PROBLEM SOLVING

This collection of thought-provoking and challenging exercises further explores and expands upon concepts learned in this chapter.

1. (a) Write each of the other trigonometric functions of θ in terms of $\sin \theta$.

    (b) Write each of the other trigonometric functions of θ in terms of $\cos \theta$.

2. Verify that for all integers n,

$$\cos\left[\frac{(2n+1)\pi}{2}\right] = 0.$$

3. Verify that for all integers n,

$$\sin\left[\frac{(12n+1)\pi}{6}\right] = \frac{1}{2}.$$

 4. A particular sound wave is modeled by

$$p(t) = \frac{1}{4\pi}(p_1(t) + 30p_2(t) + p_3(t) + p_5(t) + 30p_6(t))$$

where $p_n(t) = \frac{1}{n}\sin(524n\pi t)$, and t is the time (in seconds).

    (a) Find the sine components $p_n(t)$ and use a graphing utility to graph each component. Then verify the graph of p that is shown.

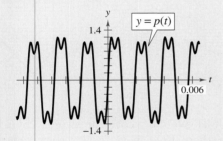

    (b) Find the period of each sine component of p. Is p periodic? If so, what is its period?

    (c) Use the *zero* or *root* feature or the *zoom* and *trace* features of a graphing utility to find the t-intercepts of the graph of p over one cycle.

    (d) Use the *maximum* and *minimum* features of a graphing utility to approximate the absolute maximum and absolute minimum values of p over one cycle.

5. Three squares of side s are placed side by side (see figure). Make a conjecture about the relationship between the sum $u + v$ and w. Prove your conjecture by using the identity for the tangent of the sum of two angles.

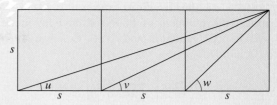

FIGURE FOR 5

6. The path traveled by an object (neglecting air resistance) that is projected at an initial height of h_0 feet, an initial velocity of v_0 feet per second, and an initial angle θ is given by

$$y = -\frac{16}{v_0^2 \cos^2 \theta}x^2 + (\tan \theta)x + h_0$$

where x and y are measured in feet. Find a formula for the maximum height of an object projected from ground level at velocity v_0 and angle θ. To do this, find half of the horizontal distance

$$\frac{1}{32}v_0^2 \sin 2\theta$$

and then substitute it for x in the general model for the path of a projectile (where $h_0 = 0$).

7. Use the figure to derive the formulas for

$$\sin\frac{\theta}{2}, \cos\frac{\theta}{2}, \text{ and } \tan\frac{\theta}{2}$$

where θ is an acute angle.

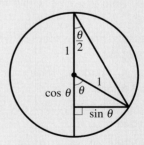

8. The force F (in pounds) on a person's back when he or she bends over at an angle θ is modeled by

$$F = \frac{0.6W \sin(\theta + 90°)}{\sin 12°}$$

where W is the person's weight (in pounds).

    (a) Simplify the model.

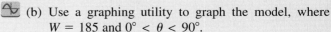   (b) Use a graphing utility to graph the model, where $W = 185$ and $0° < \theta < 90°$.

    (c) At what angle is the force a maximum? At what angle is the force a minimum?

447

9. The number of hours of daylight that occur at any location on Earth depends on the time of year and the latitude of the location. The following equations model the numbers of hours of daylight in Seward, Alaska (60° latitude) and New Orleans, Louisiana (30° latitude).

$$D = 12.2 - 6.4 \cos\left[\frac{\pi(t + 0.2)}{182.6}\right] \quad \text{Seward}$$

$$D = 12.2 - 1.9 \cos\left[\frac{\pi(t + 0.2)}{182.6}\right] \quad \text{New Orleans}$$

In these models, D represents the number of hours of daylight and t represents the day, with $t = 0$ corresponding to January 1.

 (a) Use a graphing utility to graph both models in the same viewing window. Use a viewing window of $0 \le t \le 365$.

(b) Find the days of the year on which both cities receive the same amount of daylight.

(c) Which city has the greater variation in the number of daylight hours? Which constant in each model would you use to determine the difference between the greatest and least numbers of hours of daylight?

(d) Determine the period of each model.

10. The tide, or depth of the ocean near the shore, changes throughout the day. The water depth d (in feet) of a bay can be modeled by

$$d = 35 - 28 \cos\frac{\pi}{6.2}t$$

where t is the time in hours, with $t = 0$ corresponding to 12:00 A.M.

(a) Algebraically find the times at which the high and low tides occur.

(b) Algebraically find the time(s) at which the water depth is 3.5 feet.

 (c) Use a graphing utility to verify your results from parts (a) and (b).

11. Find the solution of each inequality in the interval $[0, 2\pi]$.

(a) $\sin x \ge 0.5$ (b) $\cos x \le -0.5$

(c) $\tan x < \sin x$ (d) $\cos x \ge \sin x$

12. (a) Write a sum formula for $\sin(u + v + w)$.

(b) Write a sum formula for $\tan(u + v + w)$.

13. (a) Derive a formula for $\cos 3\theta$.

(b) Derive a formula for $\cos 4\theta$.

14. The heights h (in inches) of pistons 1 and 2 in an automobile engine can be modeled by

$$h_1 = 3.75 \sin 733t + 7.5$$

and

$$h_2 = 3.75 \sin 733\left(t + \frac{4\pi}{3}\right) + 7.5$$

where t is measured in seconds.

 (a) Use a graphing utility to graph the heights of these two pistons in the same viewing window for $0 \le t \le 1$.

(b) How often are the pistons at the same height?

15. In the figure, a beam of light is directed at the blue mirror, reflected to the red mirror, and then reflected back to the blue mirror. Find the distance PT that the light travels from the red mirror back to the blue mirror.

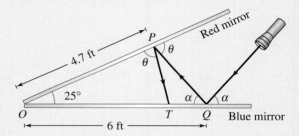

16. A triathlete sets a course to swim S 25° E from a point on shore to a buoy $\frac{3}{4}$ mile away. After swimming 300 yards through a strong current, the triathlete is off course at a bearing of S 35° E. Find the bearing and distance the triathlete needs to swim to correct her course.

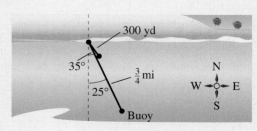

448

ANSWERS TO ODD-NUMBERED EXERCISES AND TESTS

Chapter 1

Section 1.1   *(page 8)*

1. (a) v (b) vi (c) i (d) iv (e) iii (f) ii
3. Distance Formula
5. $A: (2, 6)$, $B: (-6, -2)$, $C: (4, -4)$, $D: (-3, 2)$
7.

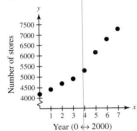

9.

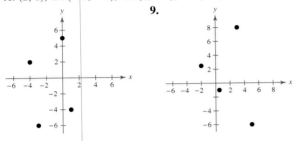

11. $(-3, 4)$ **13.** $(-5, -5)$ **15.** Quadrant IV
17. Quadrant II **19.** Quadrant III or IV **21.** Quadrant III
23. Quadrant I or III
25.

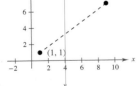

27. 8 **29.** 5 **31.** 13 **33.** $\sqrt{61}$ **35.** $\dfrac{\sqrt{277}}{6}$
37. 8.47 **39.** (a) 4, 3, 5 (b) $4^2 + 3^2 = 5^2$
41. (a) $10, 3, \sqrt{109}$ (b) $10^2 + 3^2 = (\sqrt{109})^2$
43. $(\sqrt{5})^2 + (\sqrt{45})^2 = (\sqrt{50})^2$
45. Distances between the points: $\sqrt{29}, \sqrt{58}, \sqrt{29}$
47. (a)

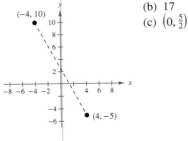

(b) 10
(c) $(5, 4)$

49. (a)

(b) 17
(c) $\left(0, \frac{5}{2}\right)$

51. (a)

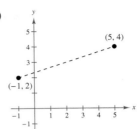

(b) $2\sqrt{10}$
(c) $(2, 3)$

53. (a)

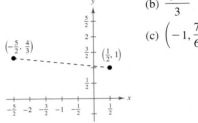

(b) $\dfrac{\sqrt{82}}{3}$
(c) $\left(-1, \frac{7}{6}\right)$

55. (a)

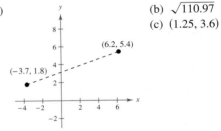

(b) $\sqrt{110.97}$
(c) $(1.25, 3.6)$

57. $30\sqrt{41} \approx 192$ km **59.** \$4415 million
61. $(0, 1), (4, 2), (1, 4)$ **63.** $(-3, 6), (2, 10), (2, 4), (-3, 4)$
65. \$3.87/gal; 2007
67. (a) About 9.6% (b) About 28.6%
69. The number of performers elected each year seems to be nearly steady except for the middle years. Five performers will be elected in 2010.
71. \$24,331 million
73. (a)

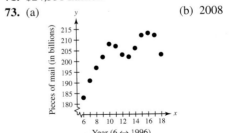

(b) 2008

(c) Answers will vary. Sample answer: Technology now enables us to transport information in many ways other than by mail. The Internet is one example.
75. $(2x_m - x_1, 2y_m - y_1)$
77. $\left(\dfrac{3x_1 + x_2}{4}, \dfrac{3y_1 + y_2}{4}\right), \left(\dfrac{x_1 + x_2}{2}, \dfrac{y_1 + y_2}{2}\right), \left(\dfrac{x_1 + 3x_2}{4}, \dfrac{y_1 + 3y_2}{4}\right)$

79.

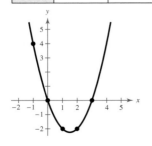

Points plotted: $(-3, 5)$, $(3, 5)$, $(-2, 1)$, $(2, 1)$, $(-7, -3)$, $(7, -3)$

(a) The point is reflected through the y-axis.
(b) The point is reflected through the x-axis.
(c) The point is reflected through the origin.

81. False. The Midpoint Formula would be used 15 times.
83. No. It depends on the magnitudes of the quantities measured.
85. Use the Midpoint Formula to prove that the diagonals of the parallelogram bisect each other.

$$\left(\frac{b+a}{2}, \frac{c+0}{2}\right) = \left(\frac{a+b}{2}, \frac{c}{2}\right)$$

$$\left(\frac{a+b+0}{2}, \frac{c+0}{2}\right) = \left(\frac{a+b}{2}, \frac{c}{2}\right)$$

Section 1.2   *(page 21)*

1. solution or solution point   **3.** intercepts
5. circle; (h, k); r
7. (a) Yes   (b) Yes   **9.** (a) Yes   (b) No
11. (a) Yes   (b) No   **13.** (a) No   (b) Yes
15.

x	-1	0	1	2	$\frac{5}{2}$
y	7	5	3	1	0
(x, y)	$(-1, 7)$	$(0, 5)$	$(1, 3)$	$(2, 1)$	$\left(\frac{5}{2}, 0\right)$

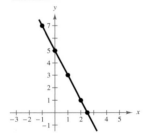

17.

x	-1	0	1	2	3
y	4	0	-2	-2	0
(x, y)	$(-1, 4)$	$(0, 0)$	$(1, -2)$	$(2, -2)$	$(3, 0)$

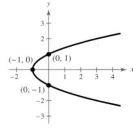

19. x-intercept: $(3, 0)$
y-intercept: $(0, 9)$
23. x-intercept: $\left(\frac{6}{5}, 0\right)$
y-intercept: $(0, -6)$
27. x-intercept: $\left(\frac{7}{3}, 0\right)$
y-intercept: $(0, 7)$
31. x-intercept: $(6, 0)$
y-intercepts: $\left(0, \pm\sqrt{6}\right)$
33. y-axis symmetry
37. Origin symmetry

21. x-intercept: $(-2, 0)$
y-intercept: $(0, 2)$
25. x-intercept: $(-4, 0)$
y-intercept: $(0, 2)$
29. x-intercepts: $(0, 0)$, $(2, 0)$
y-intercept: $(0, 0)$

35. Origin symmetry
39. x-axis symmetry

41.

43.

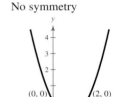

45. x-intercept: $\left(\frac{1}{3}, 0\right)$
y-intercept: $(0, 1)$
No symmetry

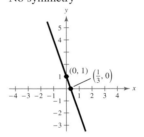

47. x-intercepts: $(0, 0)$, $(2, 0)$
y-intercept: $(0, 0)$
No symmetry

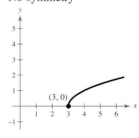

49. x-intercept: $\left(\sqrt[3]{-3}, 0\right)$
y-intercept: $(0, 3)$
No symmetry

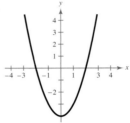

51. x-intercept: $(3, 0)$
y-intercept: None
No symmetry

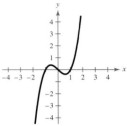

53. x-intercept: $(6, 0)$
y-intercept: $(0, 6)$
No symmetry

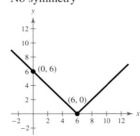

55. x-intercept: $(-1, 0)$
y-intercepts: $(0, \pm 1)$
x-axis symmetry

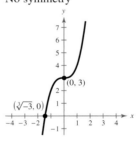

57.
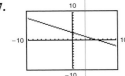
Intercepts: $(6, 0)$, $(0, 3)$

59.
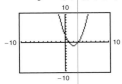
Intercepts: $(3, 0)$, $(1, 0)$, $(0, 3)$

61.
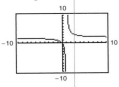
Intercept: $(0, 0)$

63.
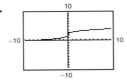
Intercepts: $(-8, 0)$, $(0, 2)$

65.
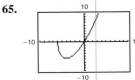
Intercepts: $(0, 0)$, $(-6, 0)$

67.
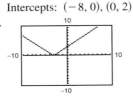
Intercepts: $(-3, 0)$, $(0, 3)$

69. $x^2 + y^2 = 16$ **71.** $(x - 2)^2 + (y + 1)^2 = 16$
73. $(x + 1)^2 + (y - 2)^2 = 5$
75. $(x - 3)^2 + (y - 4)^2 = 25$
77. Center: $(0, 0)$; Radius: 5 **79.** Center: $(1, -3)$; Radius: 3

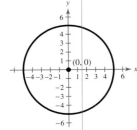
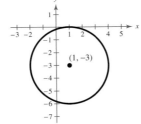

81. Center: $\left(\frac{1}{2}, \frac{1}{2}\right)$; Radius: $\frac{3}{2}$ **83.**

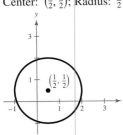
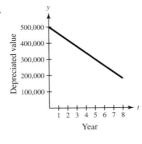

85. (a)
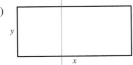
(b) Answers will vary.

(c)
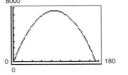
(d) $x = 86\frac{2}{3}$, $y = 86\frac{2}{3}$

(e) A regulation NFL playing field is 120 yards long and $53\frac{1}{3}$ yards wide. The actual area is 6400 square yards.
87. (a)
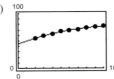
(b) 75.66 yr
(c) 1993

The model fits the data very well.
(d) The projection given by the model, 77.2 years, is less.
(e) Answers will vary.
89. (a) $a = 1, b = 0$ (b) $a = 0, b = 1$

Section 1.3 *(page 33)*

1. linear **3.** parallel **5.** rate or rate of change
7. general **9.** (a) L_2 (b) L_3 (c) L_1
11.
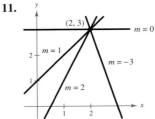

13. $\frac{3}{2}$ **15.** -4
17. $m = 5$
y-intercept: $(0, 3)$
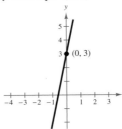

19. $m = -\frac{1}{2}$
y-intercept: $(0, 4)$
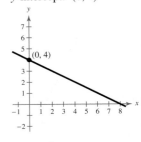

21. m is undefined.
There is no y-intercept.
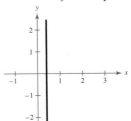

23. $m = -\frac{7}{6}$
y-intercept: $(0, 5)$
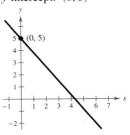

25. $m = 0$

y-intercept: $(0, 3)$

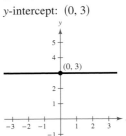

27. m is undefined.

There is no y-intercept.

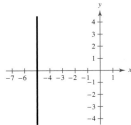

55. $y = -\frac{1}{3}x + \frac{4}{3}$

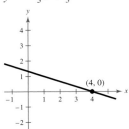

57. $y = -\frac{1}{2}x - 2$

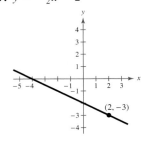

29.

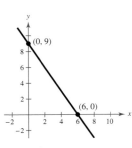

$m = -\frac{3}{2}$

31.

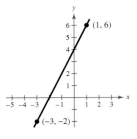

$m = 2$

59. $x = 6$

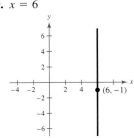

61. $y = \frac{5}{2}$

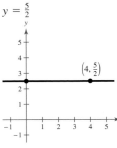

33.

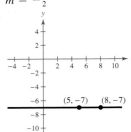

$m = 0$

35.

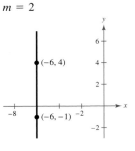

m is undefined.

63. $y = 5x + 27.3$

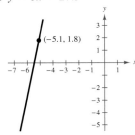

65. $y = -\frac{3}{5}x + 2$

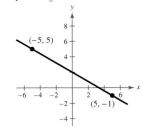

37.

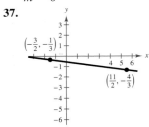

$m = -\frac{1}{7}$

39.

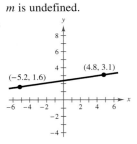

$m = 0.15$

67. $x = -8$

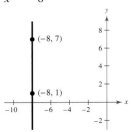

69. $y = -\frac{1}{2}x + \frac{3}{2}$

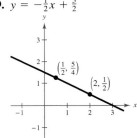

41. $(0, 1), (3, 1), (-1, 1)$

43. $(6, -5), (7, -4), (8, -3)$

45. $(-8, 0), (-8, 2), (-8, 3)$

47. $(-4, 6), (-3, 8), (-2, 10)$

49. $(9, -1), (11, 0), (13, 1)$

51. $y = 3x - 2$

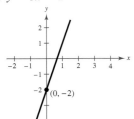

53. $y = -2x$

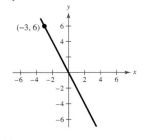

71. $y = -\frac{6}{5}x - \frac{18}{25}$

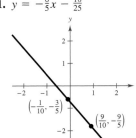

73. $y = 0.4x + 0.2$

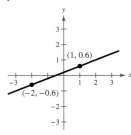

75. $y = -1$

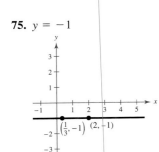

77. $x = \frac{7}{3}$

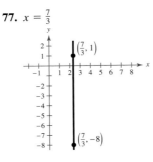

79. Parallel   **81.** Neither   **83.** Perpendicular
85. Parallel   **87.** (a) $y = 2x - 3$ (b) $y = -\frac{1}{2}x + 2$
89. (a) $y = -\frac{3}{4}x + \frac{3}{8}$ (b) $y = \frac{4}{3}x + \frac{127}{72}$
91. (a) $y = 0$ (b) $x = -1$
93. (a) $x = 3$ (b) $y = -2$
95. (a) $y = x + 4.3$ (b) $y = -x + 9.3$
97. $3x + 2y - 6 = 0$ **99.** $12x + 3y + 2 = 0$
101. $x + y - 3 = 0$
103. Line (b) is perpendicular to line (c).

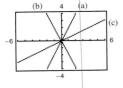

105. Line (a) is parallel to line (b).
Line (c) is perpendicular to line (a) and line (b).

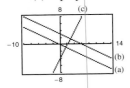

107. $3x - 2y - 1 = 0$ **109.** $80x + 12y + 139 = 0$
111. (a) Sales increasing 135 units/yr
(b) No change in sales
(c) Sales decreasing 40 units/yr
113. (a) The average salary increased the greatest from 2006 to 2008 and increased the least from 2002 to 2004.
(b) $m = 2350.75$
(c) The average salary increased $2350.75 per year over the 12 years between 1996 and 2008.
115. 12 ft   **117.** $V(t) = 3790 - 125t$
119. V-intercept: initial cost; Slope: annual depreciation
121. $V = -175t + 875$ **123.** $S = 0.8L$
125. $W = 0.07S + 2500$
127. $y = 0.03125t + 0.92875$; $y(22) \approx \$1.62$; $y(24) \approx \$1.68$
129. (a) $y(t) = 442.625t + 40,571$
(b) $y(10) = 44,997$; $y(15) = 47,210$
(c) $m = 442.625$; Each year, enrollment increases by about 443 students.
131. (a) $C = 18t + 42,000$ (b) $R = 30t$
(c) $P = 12t - 42,000$ (d) $t = 3500$ h

133. (a)

(b) $y = 8x + 50$

(c)

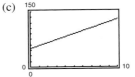

(d) $m = 8$, 8 m

135. (a) and (b)

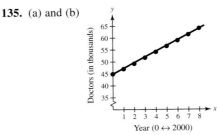

(c) Answers will vary. Sample answer: $y = 2.39x + 44.9$
(d) Answers will vary. Sample answer: The y-intercept indicates that in 2000 there were 44.9 thousand doctors of osteopathic medicine. The slope means that the number of doctors increases by 2.39 thousand each year.
(e) The model is accurate.
(f) Answers will vary. Sample answer: 73.6 thousand
137. False. The slope with the greatest magnitude corresponds to the steepest line.
139. Find the distance between each two points and use the Pythagorean Theorem.
141. No. The slope cannot be determined without knowing the scale on the y-axis. The slopes could be the same.
143. The line $y = 4x$ rises most quickly, and the line $y = -4x$ falls most quickly. The greater the magnitude of the slope (the absolute value of the slope), the faster the line rises or falls.
145. No. The slopes of two perpendicular lines have opposite signs (assume that neither line is vertical or horizontal).

Section 1.4   (page 48)

1. domain; range; function   **3.** independent; dependent
5. implied domain   **7.** Yes   **9.** No
11. Yes, each input value has exactly one output value.
13. No, the input values 7 and 10 each have two different output values.
15. (a) Function
(b) Not a function, because the element 1 in A corresponds to two elements, -2 and 1, in B.
(c) Function
(d) Not a function, because not every element in A is matched with an element in B.
17. Each is a function. For each year there corresponds one and only one circulation.
19. Not a function   **21.** Function   **23.** Function

25. Not a function   **27.** Not a function   **29.** Function
31. Function   **33.** Not a function   **35.** Function
37. (a) -1 (b) -9 (c) $2x - 5$
39. (a) 36π (b) $\frac{9}{2}\pi$ (c) $\frac{32}{3}\pi r^3$
41. (a) 15 (b) $4t^2 - 19t + 27$ (c) $4t^2 - 3t - 10$
43. (a) 1 (b) 2.5 (c) $3 - 2|x|$
45. (a) $-\dfrac{1}{9}$ (b) Undefined   (c) $\dfrac{1}{y^2 + 6y}$
47. (a) 1 (b) -1 (c) $\dfrac{|x - 1|}{x - 1}$
49. (a) -1 (b) 2 (c) 6 **51.** (a) -7 (b) 4 (c) 9

53.

x	-2	-1	0	1	2
$f(x)$	1	-2	-3	-2	1

55.

t	-5	-4	-3	-2	-1
$h(t)$	1	$\frac{1}{2}$	0	$\frac{1}{2}$	1

57.

x	-2	-1	0	1	2
$f(x)$	5	$\frac{9}{2}$	4	1	0

59. 5 **61.** $\frac{4}{3}$ **63.** ± 3 **65.** $0, \pm 1$ **67.** $-1, 2$
69. $0, \pm 2$ **71.** All real numbers x
73. All real numbers t except $t = 0$
75. All real numbers y such that $y \geq 10$
77. All real numbers x except $x = 0, -2$
79. All real numbers s such that $s \geq 1$ except $s = 4$
81. All real numbers x such that $x > 0$
83. $\{(-2, 4), (-1, 1), (0, 0), (1, 1), (2, 4)\}$
85. $\{(-2, 4), (-1, 3), (0, 2), (1, 3), (2, 4)\}$
87. $A = \dfrac{P^2}{16}$
89. (a) The maximum volume is 1024 cubic centimeters.
(b)

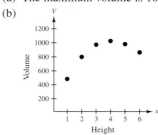

Yes, V is a function of x.
(c) $V = x(24 - 2x)^2$, $0 < x < 12$
91. $A = \dfrac{x^2}{2(x - 2)}$, $x > 2$
93. Yes, the ball will be at a height of 6 feet.
95. 1998: $136,164   2003: $180,419
1999: $140,971   2004: $195,900
2000: $147,800   2005: $216,900
2001: $156,651   2006: $224,000
2002: $167,524   2007: $217,200

97. (a) $C = 12.30x + 98{,}000$ (b) $R = 17.98x$
(c) $P = 5.68x - 98{,}000$
99. (a) $R = \dfrac{240n - n^2}{20}$, $n \geq 80$
(b)

n	90	100	110	120	130	140	150
$R(n)$	$675	$700	$715	$720	$715	$700	$675

The revenue is maximum when 120 people take the trip.
101. (a)

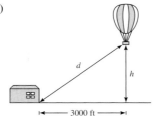

(b) $h = \sqrt{d^2 - 3000^2}$, $d \geq 3000$
103. $3 + h$, $h \neq 0$ **105.** $3x^2 + 3xh + h^2 + 3$, $h \neq 0$
107. $-\dfrac{x + 3}{9x^2}$, $x \neq 3$ **109.** $\dfrac{\sqrt{5x} - 5}{x - 5}$
111. $g(x) = cx^2$; $c = -2$ **113.** $r(x) = \dfrac{c}{x}$; $c = 32$
115. False. A function is a special type of relation.
117. False. The range is $[-1, \infty)$.
119. Domain of $f(x)$: all real numbers $x \geq 1$
Domain of $g(x)$: all real numbers $x > 1$
Notice that the domain of $f(x)$ includes $x = 1$ and the domain of $g(x)$ does not because you cannot divide by 0.
121. No; x is the independent variable, f is the name of the function.
123. (a) Yes. The amount you pay in sales tax will increase as the price of the item purchased increases.
(b) No. The length of time that you study will not necessarily determine how well you do on an exam.

Section 1.5   (page 61)

1. ordered pairs   **3.** zeros   **5.** maximum   **7.** odd
9. Domain: $(-\infty, -1] \cup [1, \infty)$; Range: $[0, \infty)$
11. Domain: $[-4, 4]$; Range: $[0, 4]$
13. Domain: $(-\infty, \infty)$; Range: $[-4, \infty)$
(a) 0 (b) -1 (c) 0 (d) -2
15. Domain: $(-\infty, \infty)$; Range: $(-2, \infty)$
(a) 0 (b) 1 (c) 2 (d) 3
17. Function   **19.** Not a function   **21.** Function
23. $-\frac{5}{2}, 6$ **25.** 0 **27.** $0, \pm\sqrt{2}$ **29.** $\pm\frac{1}{2}, 6$ **31.** $\frac{1}{2}$
33.   **35.**

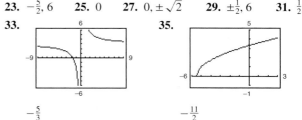

$-\frac{5}{3}$ $-\frac{11}{2}$

37. 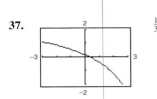 $\frac{1}{3}$

39. Increasing on $(-\infty, \infty)$
41. Increasing on $(-\infty, 0)$ and $(2, \infty)$
Decreasing on $(0, 2)$
43. Increasing on $(1, \infty)$; Decreasing on $(-\infty, -1)$
Constant on $(-1, 1)$
45. Increasing on $(-\infty, 0)$ and $(2, \infty)$; Constant on $(0, 2)$
47. 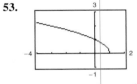 Constant on $(-\infty, \infty)$

49. 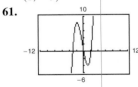   **51.**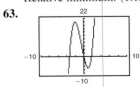
Decreasing on $(-\infty, 0)$ Increasing on $(-\infty, 0)$
Increasing on $(0, \infty)$ Decreasing on $(0, \infty)$

53.   **55.**
Decreasing on $(-\infty, 1)$ Increasing on $(0, \infty)$

57.   **59.**
Relative minimum: Relative maximum:
$(1, -9)$ $(1.5, 0.25)$

61.
Relative maximum: $(-1.79, 8.21)$
Relative minimum: $(1.12, -4.06)$

63.
Relative maximum: $(-2, 20)$
Relative minimum: $(1, -7)$

65.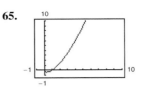
Relative minimum: $(0.33, -0.38)$

67. 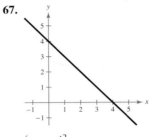   **69.**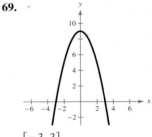
$(-\infty, 4]$ $[-3, 3]$

71. 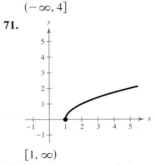   **73.**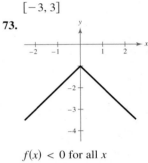
$[1, \infty)$ $f(x) < 0$ for all x

75. The average rate of change from $x_1 = 0$ to $x_2 = 3$ is -2.
77. The average rate of change from $x_1 = 1$ to $x_2 = 5$ is 18.
79. The average rate of change from $x_1 = 1$ to $x_2 = 3$ is 0.
81. The average rate of change from $x_1 = 3$ to $x_2 = 11$ is $-\frac{1}{4}$.
83. Even; y-axis symmetry   **85.** Odd; origin symmetry
87. Neither; no symmetry   **89.** Neither; no symmetry

91.    **93.**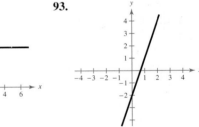
Even   Neither

95. 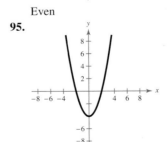   **97.**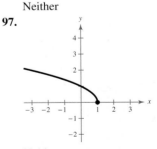
Even   Neither

CHAPTER 1

99.

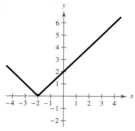

Neither

101. $h = -x^2 + 4x - 3$ **103.** $h = 2x - x^2$

105. $L = \frac{1}{2}y^2$ **107.** $L = 4 - y^2$

109. (a)

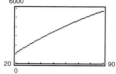

(b) 30 W

111. (a) Ten thousands   (b) Ten millions   (c) Percents

113. (a)

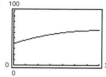

(b) The average rate of change from 1970 to 2005 is 0.705. The enrollment rate of children in preschool has slowly been increasing each year.

115. (a) $s = -16t^2 + 64t + 6$

(b)

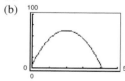

(c) Average rate of change = 16

(d) The slope of the secant line is positive.

(e) Secant line: $16t + 6$

(f)

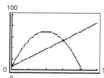

117. (a) $s = -16t^2 + 120t$

(b)

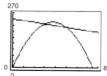

(c) Average rate of change = -8

(d) The slope of the secant line is negative.

(e) Secant line: $-8t + 240$

(f)

119. (a) $s = -16t^2 + 120$

(b)

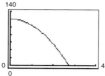

(c) Average rate of change = -32

(d) The slope of the secant line is negative.

(e) Secant line: $-32t + 120$

(f)

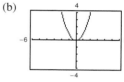

121. False. The function $f(x) = \sqrt{x^2 + 1}$ has a domain of all real numbers.

123. (a) Even. The graph is a reflection in the x-axis.

(b) Even. The graph is a reflection in the y-axis.

(c) Even. The graph is a vertical translation of f.

(d) Neither. The graph is a horizontal translation of f.

125. (a) $\left(\frac{3}{2}, 4\right)$ (b) $\left(\frac{3}{2}, -4\right)$

127. (a) $(-4, 9)$ (b) $(-4, -9)$

129. (a) $(-x, -y)$ (b) $(-x, y)$

131. (a)

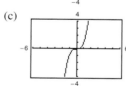

(b)

(c)

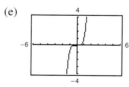

(d)

(e)

(f)

All the graphs pass through the origin. The graphs of the odd powers of x are symmetric with respect to the origin, and the graphs of the even powers are symmetric with respect to the y-axis. As the powers increase, the graphs become flatter in the interval $-1 < x < 1$.

133. 60 ft/sec; As the time traveled increases, the distance increases rapidly, causing the average speed to increase with each time increment. From $t = 0$ to $t = 4$, the average speed is less than from $t = 4$ to $t = 9$. Therefore, the overall average from $t = 0$ to $t = 9$ falls below the average found in part (b).

135. Answers will vary.

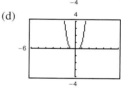

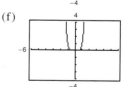

Section 1.6   *(page 71)*

1. g   **2.** i   **3.** h   **4.** a   **5.** b   **6.** e   **7.** f

8. c   **9.** d

11. (a) $f(x) = -2x + 6$

(b)

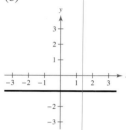

13. (a) $f(x) = -3x + 11$

(b)

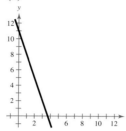

15. (a) $f(x) = -1$

(b)

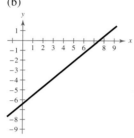

17. (a) $f(x) = \frac{6}{7}x - \frac{45}{7}$

(b)

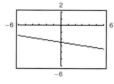

19.

21.

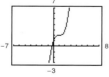

23.

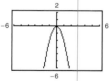

25.

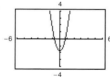

27.

29.

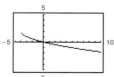

31.

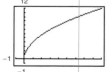

33.

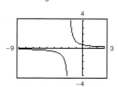

35.

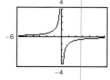

37.

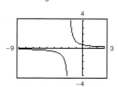

39.

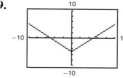

41.

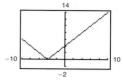

43. (a) 2   (b) 2   (c) -4 (d) 3

45. (a) 1   (b) 3   (c) 7   (d) -19

47. (a) 6   (b) -11 (c) 6   (d) -22

49. (a) -10 (b) -4 (c) -1 (d) 41

51.

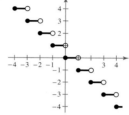

53.

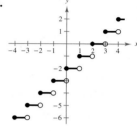

55.

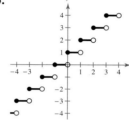

57.

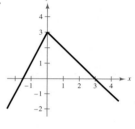

59.

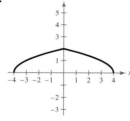

61.

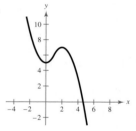

63.

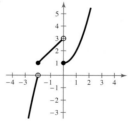

65. (a)

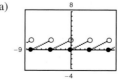

(b) Domain: $(-\infty, \infty)$

Range: $[0, 2)$

(c) Sawtooth pattern

67. (a)
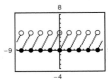
(b) Domain: $(-\infty, \infty)$
   Range: $[0, 4)$
(c) Sawtooth pattern

69. (a)
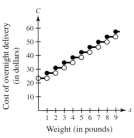
(b) $57.15

71. (a) $W(30) = 420;\; W(40) = 560;$
   $W(45) = 665;\; W(50) = 770$

(b) $W(h) = \begin{cases} 14h, & 0 < h \le 45 \\ 21(h - 45) + 630, & h > 45 \end{cases}$

73. (a)
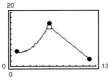

$f(x) = \begin{cases} 0.505x^2 - 1.47x + 6.3, & 1 \le x \le 6 \\ -1.97x + 26.3, & 6 < x \le 12 \end{cases}$

Answers will vary. Sample answer: The domain is determined by inspection of a graph of the data with the two models.

(b) $f(5) = 11.575, f(11) = 4.63$; These values represent the revenue for the months of May and November, respectively.

(c) These values are quite close to the actual data values.

75. False. A linear equation could be a horizontal or vertical line.

Section 1.7   *(page 78)*

1. rigid   **3.** nonrigid   **5.** vertical stretch; vertical shrink

7. (a)   (b)
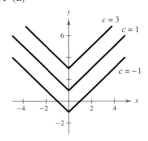
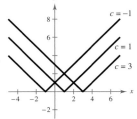

(c)
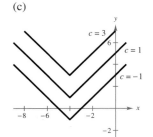

9. (a)   (b)
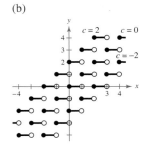

(c)
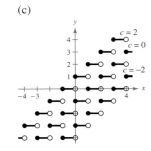

11. (a)   (b)
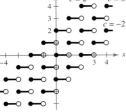

(c)   (d)
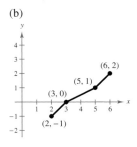

(e)   (f)
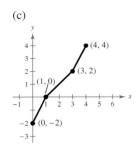
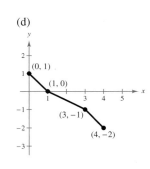
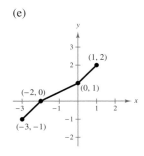
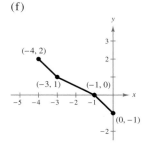

(g)

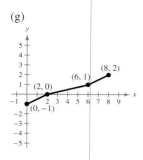

13. (a)

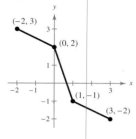

(b)

(c)

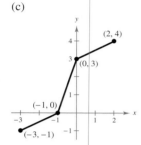

(d)

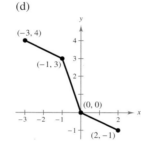

(e)

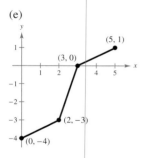

(f)

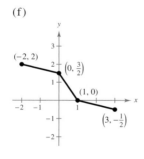

(g)

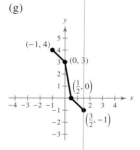

15. (a) $y = x^2 - 1$ (b) $y = 1 - (x + 1)^2$
   (c) $y = -(x - 2)^2 + 6$ (d) $y = (x - 5)^2 - 3$
17. (a) $y = |x| + 5$ (b) $y = -|x + 3|$
   (c) $y = |x - 2| - 4$ (d) $y = -|x - 6| - 1$

19. Horizontal shift of $y = x^3$; $y = (x - 2)^3$
21. Reflection in the x-axis of $y = x^2$; $y = -x^2$
23. Reflection in the x-axis and vertical shift of $y = \sqrt{x}$;
   $y = 1 - \sqrt{x}$
25. (a) $f(x) = x^2$
   (b) Reflection in the x-axis and vertical shift 12 units upward
   (c)

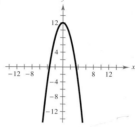

   (d) $g(x) = 12 - f(x)$
27. (a) $f(x) = x^3$
   (b) Vertical shift seven units upward
   (c)   (d) $g(x) = f(x) + 7$

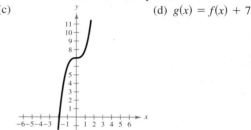

29. (a) $f(x) = x^2$
   (b) Vertical shrink of two-thirds and vertical shift four units
       upward
   (c)   (d) $g(x) = \frac{2}{3}f(x) + 4$

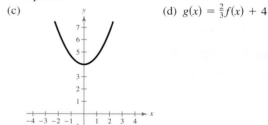

31. (a) $f(x) = x^2$
   (b) Reflection in the x-axis, horizontal shift five units to the
       left, and vertical shift two units upward
   (c)   (d) $g(x) = 2 - f(x + 5)$

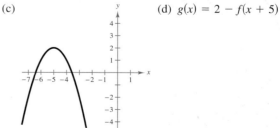

33. (a) $f(x) = x^2$
   (b) Vertical stretch of two, horizontal shift four units to the
       right, and vertical shift three units upward

(c)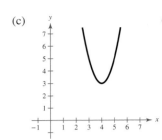

(d) $g(x) = 3 + 2f(x - 4)$

35. (a) $f(x) = \sqrt{x}$
(b) Horizontal shrink of one-third
(c)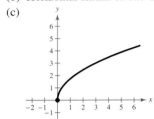

(d) $g(x) = f(3x)$

37. (a) $f(x) = x^3$
(b) Vertical shift two units upward and horizontal shift one unit to the right
(c)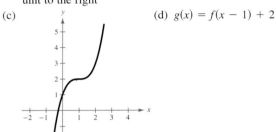

(d) $g(x) = f(x - 1) + 2$

39. (a) $f(x) = x^3$
(b) Vertical stretch of three and horizontal shift two units to the right
(c)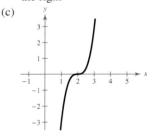

(d) $g(x) = 3f(x - 2)$

41. (a) $f(x) = |x|$
(b) Reflection in the x-axis and vertical shift two units downward
(c)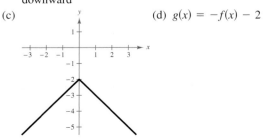

(d) $g(x) = -f(x) - 2$

43. (a) $f(x) = |x|$
(b) Reflection in the x-axis, horizontal shift four units to the left, and vertical shift eight units upward
(c)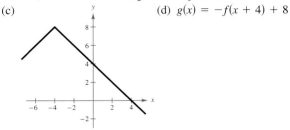

(d) $g(x) = -f(x + 4) + 8$

45. (a) $f(x) = |x|$
(b) Reflection in the x-axis, vertical stretch of two, horizontal shift one unit to the right, and vertical shift four units downward
(c)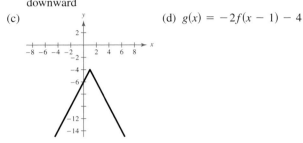

(d) $g(x) = -2f(x - 1) - 4$

47. (a) $f(x) = [\![x]\!]$
(b) Reflection in the x-axis and vertical shift three units upward
(c)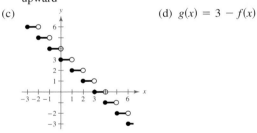

(d) $g(x) = 3 - f(x)$

49. (a) $f(x) = \sqrt{x}$
(b) Horizontal shift nine units to the right
(c)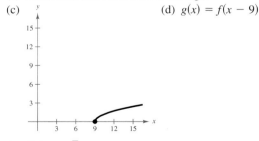

(d) $g(x) = f(x - 9)$

51. (a) $f(x) = \sqrt{x}$
(b) Reflection in the y-axis, horizontal shift seven units to the right, and vertical shift two units downward

(c)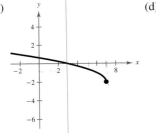

(d) $g(x) = f(7 - x) - 2$

(e)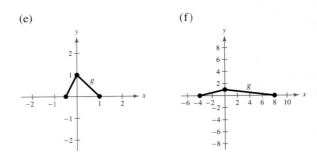

(f)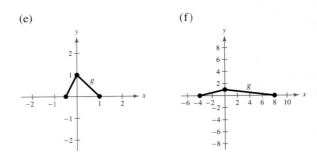

53. (a) $f(x) = \sqrt{x}$
(b) Horizontal stretch and vertical shift four units downward
(c)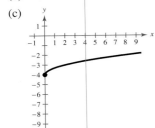

(d) $g(x) = f\left(\frac{1}{2}x\right) - 4$

55. $g(x) = (x - 3)^2 - 7$ **57.** $g(x) = (x - 13)^3$
59. $g(x) = -|x| + 12$ **61.** $g(x) = -\sqrt{-x + 6}$
63. (a) $y = -3x^2$ (b) $y = 4x^2 + 3$
65. (a) $y = -\frac{1}{2}|x|$ (b) $y = 3|x| - 3$
67. Vertical stretch of $y = x^3$; $y = 2x^3$
69. Reflection in the x-axis and vertical shrink of $y = x^2$;
$y = -\frac{1}{2}x^2$
71. Reflection in the y-axis and vertical shrink of $y = \sqrt{x}$;
$y = \frac{1}{2}\sqrt{-x}$
73. $y = -(x - 2)^3 + 2$ **75.** $y = -\sqrt{x} - 3$
77. (a)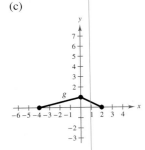

(b)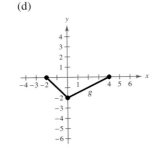

(c)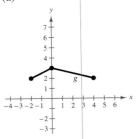

(d)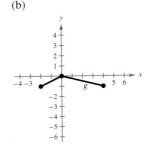

79. (a) Vertical stretch of 128.0 and a vertical shift of 527 units upward
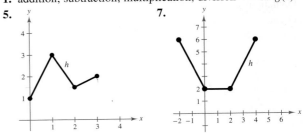

(b) 32; Each year, the total number of miles driven by vans, pick-ups, and SUVs increases by an average of 32 billion miles.
(c) $f(t) = 527 + 128\sqrt{t + 10}$; The graph is shifted 10 units to the left.
(d) 1127 billion miles; Answers will vary. Sample answer: Yes, because the number of miles driven has been steadily increasing.
81. False. The graph of $y = f(-x)$ is a reflection of the graph of $f(x)$ in the y-axis.
83. True. $|-x| = |x|$
85. (a) $g(t) = \frac{3}{4}f(t)$ (b) $g(t) = f(t) + 10,000$
(c) $g(t) = f(t - 2)$
87. $(-2, 0), (-1, 1), (0, 2)$
89. No. $g(x) = -x^4 - 2$. Yes. $h(x) = -(x - 3)^4$.

Section 1.8   *(page 88)*

1. addition; subtraction; multiplication; division   **3.** $g(x)$
5. 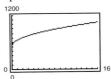   **7.**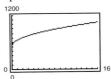

9. (a) $2x$ (b) 4 (c) $x^2 - 4$
(d) $\dfrac{x + 2}{x - 2}$; all real numbers x except $x = 2$
11. (a) $x^2 + 4x - 5$ (b) $x^2 - 4x + 5$ (c) $4x^3 - 5x^2$
(d) $\dfrac{x^2}{4x - 5}$; all real numbers x except $x = \dfrac{5}{4}$
13. (a) $x^2 + 6 + \sqrt{1 - x}$ (b) $x^2 + 6 - \sqrt{1 - x}$
(c) $(x^2 + 6)\sqrt{1 - x}$
(d) $\dfrac{(x^2 + 6)\sqrt{1 - x}}{1 - x}$; all real numbers x such that $x < 1$

15. (a) $\dfrac{x+1}{x^2}$ (b) $\dfrac{x-1}{x^2}$ (c) $\dfrac{1}{x^3}$

(d) x; all real numbers x except $x = 0$

17. 3   **19.** 5   **21.** $9t^2 - 3t + 5$ **23.** 74

25. 26   **27.** $\frac{3}{5}$

29. 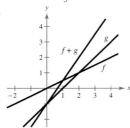   **31.**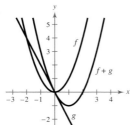

33. 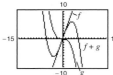   **35.**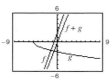

$f(x), g(x)$ $f(x), f(x)$

37. (a) $(x-1)^2$ (b) $x^2 - 1$ (c) $x - 2$

39. (a) x (b) x (c) $x^9 + 3x^6 + 3x^3 + 2$

41. (a) $\sqrt{x^2 + 4}$ (b) $x + 4$

Domains of f and $g \circ f$: all real numbers x such that $x \geq -4$

Domains of g and $f \circ g$: all real numbers x

43. (a) $x + 1$ (b) $\sqrt{x^2 + 1}$

Domains of f and $g \circ f$: all real numbers x

Domains of g and $f \circ g$: all real numbers x such that $x \geq 0$

45. (a) $|x + 6|$ (b) $|x| + 6$

Domains of $f, g, f \circ g$, and $g \circ f$: all real numbers x

47. (a) $\dfrac{1}{x+3}$ (b) $\dfrac{1}{x} + 3$

Domains of f and $g \circ f$: all real numbers x except $x = 0$

Domain of g: all real numbers x

Domain of $f \circ g$: all real numbers x except $x = -3$

49. (a) 3   (b) 0   **51.** (a) 0   (b) 4

53. $f(x) = x^2$, $g(x) = 2x + 1$

55. $f(x) = \sqrt[3]{x}$, $g(x) = x^2 - 4$

57. $f(x) = \dfrac{1}{x}$, $g(x) = x + 2$ **59.** $f(x) = \dfrac{x+3}{4+x}$, $g(x) = -x^2$

61. (a) $T = \frac{3}{4}x + \frac{1}{15}x^2$

(b)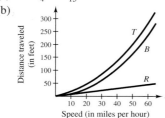

(c) The braking function $B(x)$. As x increases, $B(x)$ increases at a faster rate than $R(x)$.

63. (a) $c(t) = \dfrac{b(t) - d(t)}{p(t)} \times 100$

(b) $c(5)$ is the percent change in the population due to births and deaths in the year 2005.

65. (a) $(N + M)(t) = 0.227t^3 - 4.11t^2 + 14.6t + 544$,   which represents the total number of Navy and Marines personnel combined.

$(N + M)(0) = 544$

$(N + M)(6) \approx 533$

$(N + M)(12) \approx 520$

(b) $(N - M)(t) = 0.157t^3 - 3.65t^2 + 11.2t + 200$,   which represents the difference between the number of Navy personnel and the number of Marines personnel.

$(N - M)(0) = 200$

$(N - M)(6) \approx 170$

$(N - M)(12) \approx 80$

67. $(B - D)(t) = -0.197t^3 + 10.17t^2 - 128.0t + 2043$,   which represents the change in the United States population.

69. (a) For each time t there corresponds one and only one temperature T.

(b) $60°$, $72°$

(c) All the temperature changes occur 1 hour later.

(d) The temperature is decreased by 1 degree.

(e) $T(t) = \begin{cases} 60, & 0 \leq t \leq 6 \\ 12t - 12, & 6 < t < 7 \\ 72, & 7 \leq t \leq 20 \\ -12t + 312, & 20 < t < 21 \\ 60, & 21 \leq t \leq 24 \end{cases}$

71. $(A \circ r)(t) = 0.36\pi t^2$; $(A \circ r)(t)$ represents the area of the circle at time t.

73. (a) $N(T(t)) = 30(3t^2 + 2t + 20)$; This represents the number of bacteria in the food as a function of time.

(b) About 653 bacteria   (c) 2.846 h

75. $g(f(x))$ represents 3 percent of an amount over $500,000.

77. False. $(f \circ g)(x) = 6x + 1$ and $(g \circ f)(x) = 6x + 6$

79. (a) $O(M(Y)) = 2(6 + \frac{1}{2}Y) = 12 + Y$

(b) Middle child is 8 years old; youngest child is 4 years old.

81. Proof

83. (a) Proof

(b) $\frac{1}{2}[f(x) + f(-x)] + \frac{1}{2}[f(x) - f(-x)]$

$= \frac{1}{2}[f(x) + f(-x) + f(x) - f(-x)]$

$= \frac{1}{2}[2f(x)]$

$= f(x)$

(c) $f(x) = (x^2 + 1) + (-2x)$

$k(x) = \dfrac{-1}{(x+1)(x-1)} + \dfrac{x}{(x+1)(x-1)}$

Section 1.9   (page 98)

1. inverse   **3.** range; domain   **5.** one-to-one

7. $f^{-1}(x) = \frac{1}{6}x$ **9.** $f^{-1}(x) = x - 9$ **11.** $f^{-1}(x) = \dfrac{x-1}{3}$

13. $f^{-1}(x) = x^3$ **15.** c   **16.** b   **17.** a   **18.** d

19. $f(g(x)) = f\left(-\dfrac{2x+6}{7}\right) = -\dfrac{7}{2}\left(-\dfrac{2x+6}{7}\right) - 3 = x$

$g(f(x)) = g\left(-\dfrac{7}{2}x - 3\right) = -\dfrac{2\left(-\frac{7}{2}x - 3\right) + 6}{7} = x$

21. $f(g(x)) = f\left(\sqrt[3]{x-5}\right) = \left(\sqrt[3]{x-5}\right)^3 + 5 = x$

$g(f(x)) = g(x^3 + 5) = \sqrt[3]{(x^3+5) - 5} = x$

23. (a) $f(g(x)) = f\left(\dfrac{x}{2}\right) = 2\left(\dfrac{x}{2}\right) = x$

$g(f(x)) = g(2x) = \dfrac{(2x)}{2} = x$

(b)

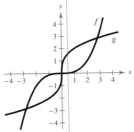

25. (a) $f(g(x)) = f\left(\dfrac{x-1}{7}\right) = 7\left(\dfrac{x-1}{7}\right) + 1 = x$

$g(f(x)) = g(7x + 1) = \dfrac{(7x+1) - 1}{7} = x$

(b)

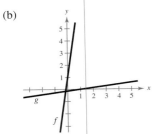

27. (a) $f(g(x)) = f\left(\sqrt[3]{8x}\right) = \dfrac{\left(\sqrt[3]{8x}\right)^3}{8} = x$

$g(f(x)) = g\left(\dfrac{x^3}{8}\right) = \sqrt[3]{8\left(\dfrac{x^3}{8}\right)} = x$

(b)

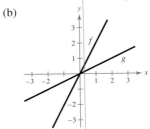

29. (a) $f(g(x)) = f(x^2 + 4), \ x \ge 0$

$= \sqrt{(x^2 + 4) - 4} = x$

$g(f(x)) = g\left(\sqrt{x - 4}\right)$

$= \left(\sqrt{x-4}\right)^2 + 4 = x$

(b)

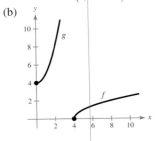

31. (a) $f(g(x)) = f\left(\sqrt{9 - x}\right), \ x \le 9$

$= 9 - \left(\sqrt{9-x}\right)^2 = x$

$g(f(x)) = g(9 - x^2), \ x \ge 0$

$= \sqrt{9 - (9 - x^2)} = x$

(b)

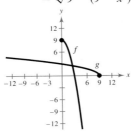

33. (a) $f(g(x)) = f\left(-\dfrac{5x+1}{x-1}\right) = \dfrac{-\left(\dfrac{5x+1}{x-1}\right) - 1}{-\left(\dfrac{5x+1}{x-1}\right) + 5}$

$= \dfrac{-5x - 1 - x + 1}{-5x - 1 + 5x - 5} = x$

$g(f(x)) = g\left(\dfrac{x-1}{x+5}\right) = \dfrac{-5\left(\dfrac{x-1}{x+5}\right) - 1}{\dfrac{x-1}{x+5} - 1}$

$= \dfrac{-5x + 5 - x - 5}{x - 1 - x - 5} = x$

(b)

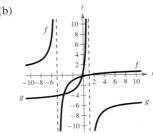

35. No

37.

x	-2	0	2	4	6	8
$f^{-1}(x)$	-2	-1	0	1	2	3

39. Yes **41.** No

43.

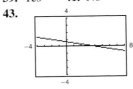

The function has an inverse.

45.

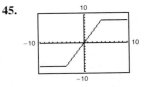

The function does not have an inverse.

47.

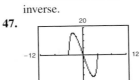

The function does not have an inverse.

49. (a) $f^{-1}(x) = \dfrac{x + 3}{2}$

(b)

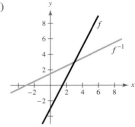

(c) The graph of f^{-1} is the reflection of the graph of f in the line $y = x$.

(d) The domains and ranges of f and f^{-1} are all real numbers.

51. (a) $f^{-1}(x) = \sqrt[5]{x + 2}$

(b)

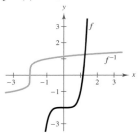

(c) The graph of f^{-1} is the reflection of the graph of f in the line $y = x$.

(d) The domains and ranges of f and f^{-1} are all real numbers.

53. (a) $f^{-1}(x) = \sqrt{4 - x^2}$, $0 \le x \le 2$

(b)

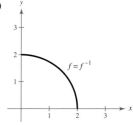

(c) The graph of f^{-1} is the same as the graph of f.

(d) The domains and ranges of f and f^{-1} are all real numbers x such that $0 \le x \le 2$.

55. (a) $f^{-1}(x) = \dfrac{4}{x}$

(b)

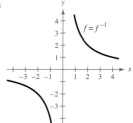

(c) The graph of f^{-1} is the same as the graph of f.

(d) The domains and ranges of f and f^{-1} are all real numbers x except $x = 0$.

57. (a) $f^{-1}(x) = \dfrac{2x + 1}{x - 1}$

(b)

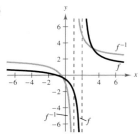

(c) The graph of f^{-1} is the reflection of the graph of f in the line $y = x$.

(d) The domain of f and the range of f^{-1} are all real numbers x except $x = 2$. The domain of f^{-1} and the range of f are all real numbers x except $x = 1$.

59. (a) $f^{-1}(x) = x^3 + 1$

(b)

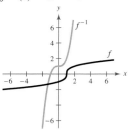

(c) The graph of f^{-1} is the reflection of the graph of f in the line $y = x$.

(d) The domains and ranges of f and f^{-1} are all real numbers.

61. (a) $f^{-1}(x) = \dfrac{5x - 4}{6 - 4x}$

(b)

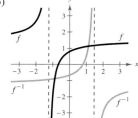

(c) The graph of f^{-1} is the reflection of the graph of f in the line $y = x$.

(d) The domain of f and the range of f^{-1} are all real numbers x except $x = -\frac{5}{4}$. The domain of f^{-1} and the range of f are all real numbers x except $x = \frac{3}{2}$.

63. No inverse   **65.** $g^{-1}(x) = 8x$ **67.** No inverse

69. $f^{-1}(x) = \sqrt{x} - 3$ **71.** No inverse   **73.** No inverse

75. $f^{-1}(x) = \dfrac{x^2 - 3}{2}$, $x \ge 0$

77. $f^{-1}(x) = \sqrt{x} + 2$

The domain of f and the range of f^{-1} are all real numbers x such that $x \ge 2$. The domain of f^{-1} and the range of f are all real numbers x such that $x \ge 0$.

79. $f^{-1}(x) = x - 2$

The domain of f and the range of f^{-1} are all real numbers x such that $x \ge -2$. The domain of f^{-1} and the range of f are all real numbers x such that $x \ge 0$.

81. $f^{-1}(x) = \sqrt{x} - 6$

The domain of f and the range of f^{-1} are all real numbers x such that $x \geq -6$. The domain of f^{-1} and the range of f are all real numbers x such that $x \geq 0$.

83. $f^{-1}(x) = \dfrac{\sqrt{-2(x - 5)}}{2}$

The domain of f and the range of f^{-1} are all real numbers x such that $x \geq 0$. The domain of f^{-1} and the range of f are all real numbers x such that $x \leq 5$.

85. $f^{-1}(x) = x + 3$

The domain of f and the range of f^{-1} are all real numbers x such that $x \geq 4$. The domain of f^{-1} and the range of f are all real numbers x such that $x \geq 1$.

87. 32   **89.** 600   **91.** $2\sqrt[3]{x + 3}$

93. $\dfrac{x + 1}{2}$ **95.** $\dfrac{x + 1}{2}$

97. (a) Yes; each European shoe size corresponds to exactly one U.S. shoe size.
　　(b) 45　(c) 10　(d) 41　(e) 13

99. (a) Yes
　　(b) S^{-1} represents the time in years for a given sales level.
　　(c) $S^{-1}(8430) = 6$
　　(d) No, because then the sales for 2007 and 2009 would be the same, so the function would no longer be one-to-one.

101. (a) $y = \dfrac{x - 10}{0.75}$

x = hourly wage; y = number of units produced
　　(b) 19 units

103. False. $f(x) = x^2$ has no inverse.   **105.** Proof

107.

x	1	3	4	6
y	1	2	6	7

x	1	2	6	7
$f^{-1}(x)$	1	3	4	6

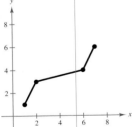

109. This situation could be represented by a one-to-one function if the runner does not stop to rest. The inverse function would represent the time in hours for a given number of miles completed.

111. This function could not be represented by a one-to-one function because it oscillates.

113. $k = \frac{1}{4}$

115.

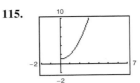

There is an inverse function $f^{-1}(x) = \sqrt{x} - 1$ because the domain of f is equal to the range of f^{-1} and the range of f is equal to the domain of f^{-1}.

Section 1.10   (page 108)

1. variation; regression   **3.** least squares regression
5. directly proportional   **7.** directly proportional
9. combined

11.

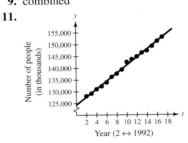

The model is a good fit for the actual data.

13.

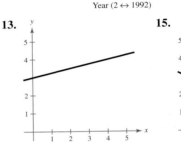

$y = \frac{1}{4}x + 3$

15.

$y = -\frac{1}{2}x + 3$

17. (a) and (b)

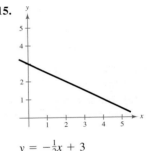

$y \approx t + 130$
　　(c) $y = 1.01t + 130.82$　(d) The models are similar.
　　(e) Part (b): 242 ft; Part (c): 243.94 ft
　　(f) Answers will vary.

19. (a)

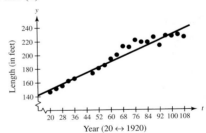

　　(b) $S = 38.3t + 224$

(c)

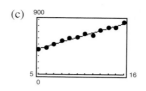

The model is a good fit.

(d) 2007: \$875.1 million; 2009: \$951.7 million

(e) Each year the annual gross ticket sales for Broadway shows in New York City increase by \$38.3 million.

21. Inversely

23.

x	2	4	6	8	10
$y = kx^2$	4	16	36	64	100

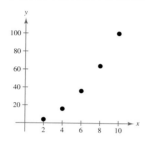

25.

x	2	4	6	8	10
$y = kx^2$	2	8	18	32	50

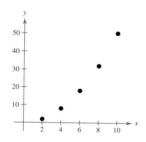

27.

x	2	4	6	8	10
$y = k/x^2$	$\frac{1}{2}$	$\frac{1}{8}$	$\frac{1}{18}$	$\frac{1}{32}$	$\frac{1}{50}$

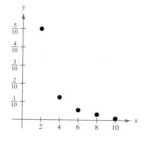

29.

x	2	4	6	8	10
$y = k/x^2$	$\frac{5}{2}$	$\frac{5}{8}$	$\frac{5}{18}$	$\frac{5}{32}$	$\frac{1}{10}$

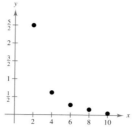

31. $y = \dfrac{5}{x}$ **33.** $y = -\dfrac{7}{10}x$ **35.** $y = \dfrac{12}{5}x$

37. $y = 205x$ **39.** $I = 0.035P$

41. Model: $y = \frac{33}{13}x$; 25.4 cm, 50.8 cm

43. $y = 0.0368x$; \$8280

45. (a) 0.05 m (b) $176\frac{2}{3}$ N **47.** 39.47 lb

49. $A = kr^2$ **51.** $y = \dfrac{k}{x^2}$ **53.** $F = \dfrac{kg}{r^2}$ **55.** $P = \dfrac{k}{V}$

57. $F = \dfrac{km_1 m_2}{r^2}$

59. The area of a triangle is jointly proportional to its base and height.

61. The area of an equilateral triangle varies directly as the square of one of its sides.

63. The volume of a sphere varies directly as the cube of its radius.

65. Average speed is directly proportional to the distance and inversely proportional to the time.

67. $A = \pi r^2$ **69.** $y = \dfrac{28}{x}$ **71.** $F = 14rs^3$

73. $z = \dfrac{2x^2}{3y}$ **75.** About 0.61 mi/h **77.** 506 ft

79. 1470 J **81.** The velocity is increased by one-third.

83. (a)

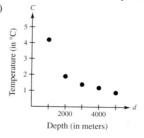

(b) Yes. $k_1 = 4200$, $k_2 = 3800$, $k_3 = 4200$, $k_4 = 4800$, $k_5 = 4500$

(c) $C = \dfrac{4300}{d}$

(d)

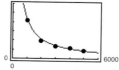

(e) About 1433 m

85. (a)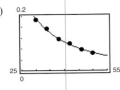

(b) 0.2857 $\mu W/cm^2$

87. False. E is jointly proportional to the mass of an object and the square of its velocity.

89. (a) Good approximation   (b) Poor approximation
(c) Poor approximation   (d) Good approximation

91. As one variable increases, the other variable will also increase.

93. (a) y will change by a factor of one-fourth.
(b) y will change by a factor of four.

Review Exercises   *(page 116)*

1.

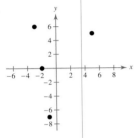

3. Quadrant IV

5. (a) 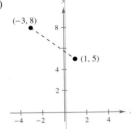   (b) 5
(c) $\left(-1, \frac{13}{2}\right)$

7. (a) 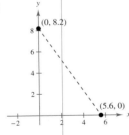   (b) $\sqrt{98.6}$
(c) $(2.8, 4.1)$

9. $(0, 0), (2, 0), (0, -5), (2, -5)$

11. \$6.275 billion

13.

x	-2	-1	0	1	2
y	-11	-8	-5	-2	1

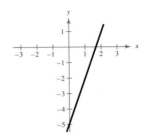

15.

x	-1	0	1	2	3	4
y	4	0	-2	-2	0	4

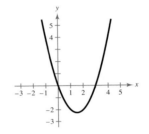

17. 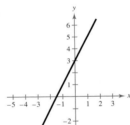   **19.**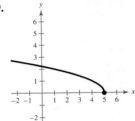

21.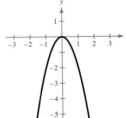

23. x-intercept: $\left(-\frac{7}{2}, 0\right)$ **25.** x-intercepts: $(1, 0), (5, 0)$
y-intercept: $(0, 7)$ y-intercept: $(0, 5)$

CHAPTER 1

27. x-intercept: $\left(\frac{1}{4}, 0\right)$
y-intercept: $(0, 1)$
No symmetry

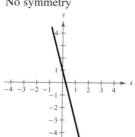

29. x-intercepts: $\left(\pm\sqrt{5}, 0\right)$
y-intercept: $(0, 5)$
y-axis symmetry

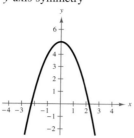

31. x-intercept: $\left(\sqrt[3]{-3}, 0\right)$
y-intercept: $(0, 3)$
No symmetry

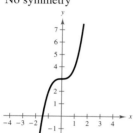

33. x-intercept: $(-5, 0)$
y-intercept: $\left(0, \sqrt{5}\right)$
No symmetry

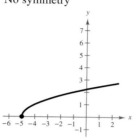

35. Center: $(0, 0)$
Radius: 3

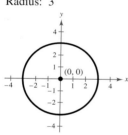

37. Center: $(-2, 0)$
Radius: 4

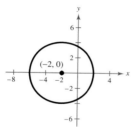

39. Center: $\left(\frac{1}{2}, -1\right)$
Radius: 6

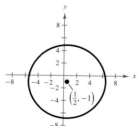

41. $(x - 2)^2 + (y + 3)^2 = 13$

43. (a)   (b) 2008

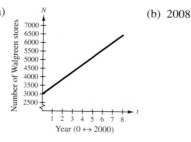

45. Slope: 0
y-intercept: 6

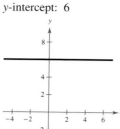

47. Slope: 3
y-intercept: 13

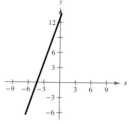

49.

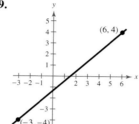

$m = \frac{8}{9}$

51.

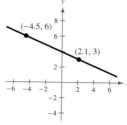

$m = -\frac{5}{11}$

53. $y = \frac{2}{3}x - 2$

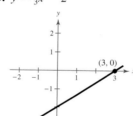

55. $y = -\frac{1}{2}x + 2$

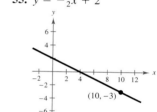

57. $x = 0$ **59.** $y = \frac{2}{7}x + \frac{2}{7}$

61. (a) $y = \frac{5}{4}x - \frac{23}{4}$ (b) $y = -\frac{4}{5}x + \frac{2}{5}$

63. $V = -850t + 21{,}000, \quad 10 \le t \le 15$

65. No   **67.** Yes

69. (a) 5   (b) 17   (c) $t^4 + 1$ (d) $t^2 + 2t + 2$

71. All real numbers x such that $-5 \le x \le 5$

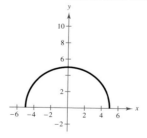

73. All real numbers x except $x = 3, -2$

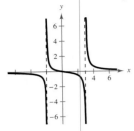

75. (a) 16 ft/sec   (b) 1.5 sec   (c) -16 ft/sec
77. $4x + 2h + 3$, $h \neq 0$ **79.** Function
81. Not a function   **83.** $\frac{7}{3}, 3$ **85.** $-\frac{3}{8}$
87.

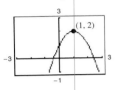

Increasing on $(0, \infty)$
Decreasing on $(-\infty, -1)$
Constant on $(-1, 0)$

89.

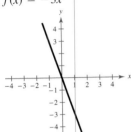

91.

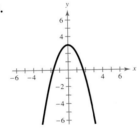

93. 4   **95.** $\dfrac{1 - \sqrt{2}}{2}$ **97.** Neither   **99.** Odd

101. $f(x) = -3x$ **103.**

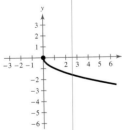

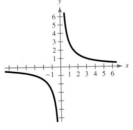

105.

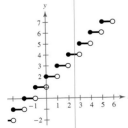

107.

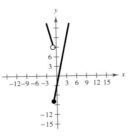

109.

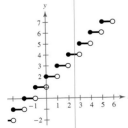

111.

113. $y = x^3$
115. (a) $f(x) = x^2$
   (b) Vertical shift nine units downward
   (c)

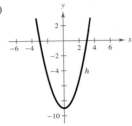

   (d) $h(x) = f(x) - 9$
117. (a) $f(x) = \sqrt{x}$
   (b) Reflection in the x-axis and vertical shift four units upward
   (c)

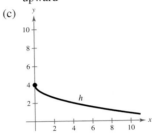

   (d) $h(x) = -f(x) + 4$
119. (a) $f(x) = x^2$
   (b) Reflection in the x-axis, horizontal shift two units to the left, and vertical shift three units upward
   (c)

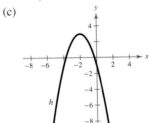

   (d) $h(x) = -f(x + 2) + 3$
121. (a) $f(x) = [\![x]\!]$
   (b) Reflection in the x-axis and vertical shift six units upward
   (c)

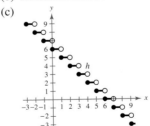

   (d) $h(x) = -f(x) + 6$
123. (a) $f(x) = |x|$
   (b) Reflections in the x-axis and the y-axis, horizontal shift four units to the right, and vertical shift six units upward

(c)

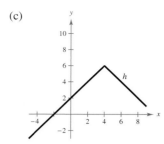

(d) $h(x) = -f(-x + 4) + 6$

125. (a) $f(x) = [\![x]\!]$

  (b) Horizontal shift nine units to the right and vertical stretch

  (c)

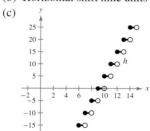

  (d) $h(x) = 5 f(x - 9)$

127. (a) $f(x) = \sqrt{x}$

  (b) Reflection in the x-axis, vertical stretch, and horizontal shift four units to the right

  (c)

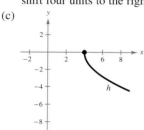

  (d) $h(x) = -2 f(x - 4)$

129. (a) $x^2 + 2x + 2$ (b) $x^2 - 2x + 4$

  (c) $2x^3 - x^2 + 6x - 3$

  (d) $\dfrac{x^2 + 3}{2x - 1}$; all real numbers x except $x = \dfrac{1}{2}$

131. (a) $x - \dfrac{8}{3}$ (b) $x - 8$

  Domains of $f, g, f \circ g$, and $g \circ f$: all real numbers x

133. $f(x) = x^3, g(x) = 1 - 2x$

135. (a) $(r + c)(t) = 178.8t + 856$; This represents the average annual expenditures for both residential and cellular phone services.

  (b)

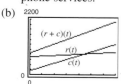

  (c) $(r + c)(13) = \$3180.40$

137. $f^{-1}(x) = \frac{1}{3}(x - 8)$ **139.** The function has an inverse.

141.

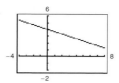

The function has an inverse.

143.

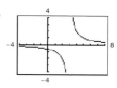

The function has an inverse.

145. (a) $f^{-1}(x) = 2x + 6$

  (b)

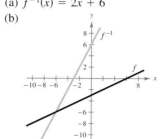

  (c) The graph of f^{-1} is the reflection of the graph of f in the line $y = x$.

  (d) Both f and f^{-1} have domains and ranges that are all real numbers.

147. (a) $f^{-1}(x) = x^2 - 1, \ x \geq 0$

  (b)

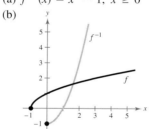

  (c) The graph of f^{-1} is the reflection of the graph of f in the line $y = x$.

  (d) f has a domain of $[-1, \infty)$ and a range of $[0, \infty)$; f^{-1} has a domain of $[0, \infty)$ and a range of $[-1, \infty)$.

149. $x > 4; \ f^{-1}(x) = \sqrt{\dfrac{x}{2} + 4}, x \neq 0$

151. (a)

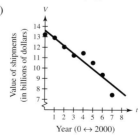

  (b) The model is a good fit for the actual data.

153. Model: $k = \frac{8}{5}m$; 3.2 km, 16 km

155. A factor of 4   **157.** About 2 h, 26 min

159. False. The graph is reflected in the *x*-axis, shifted 9 units to the left, and then shifted 13 units downward.

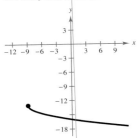

161. The Vertical Line Test is used to determine if the graph of *y* is a function of *x*. The Horizontal Line Test is used to determine if a function has an inverse function.

Chapter Test   (page 121)

1.

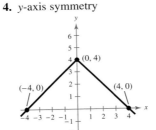

Midpoint: $\left(2, \frac{5}{2}\right)$; Distance: $\sqrt{89}$

2. About 11.937 cm

3. No symmetry

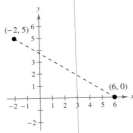

4. *y*-axis symmetry

5. *y*-axis symmetry

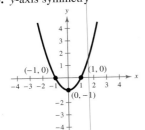

6. $(x - 1)^2 + (y - 3)^2 = 16$

7. $y = -2x + 1$ **8.** $y = -1.7x + 5.9$

9. (a) $5x + 2y - 8 = 0$ (b) $-2x + 5y - 20 = 0$

10. (a) $-\dfrac{1}{8}$ (b) $-\dfrac{1}{28}$ (c) $\dfrac{\sqrt{x}}{x^2 - 18x}$ **11.** $x \le 3$

12. (a) $0, \pm 0.4314$

(b)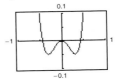

(c) Increasing on $(-0.31, 0)$, $(0.31, \infty)$
Decreasing on $(-\infty, -0.31)$, $(0, 0.31)$

(d) Even

13. (a) $0, 3$

(b)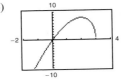

(c) Increasing on $(-\infty, 2)$
Decreasing on $(2, 3)$

(d) Neither

14. (a) -5

(b)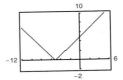

(c) Increasing on $(-5, \infty)$
Decreasing on $(-\infty, -5)$

(d) Neither

15.

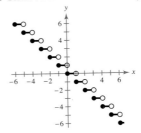

16. Reflection in the *x*-axis of $y = [\![x]\!]$

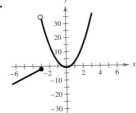

17. Reflection in the x-axis, horizontal shift, and vertical shift of $y = \sqrt{x}$

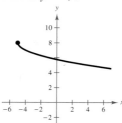

18. Reflection in the x-axis, vertical stretch, horizontal shift, and vertical shift of $y = x^3$

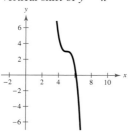

19. (a) $2x^2 - 4x - 2$ (b) $4x^2 + 4x - 12$
    (c) $-3x^4 - 12x^3 + 22x^2 + 28x - 35$

    (d) $\dfrac{3x^2 - 7}{-x^2 - 4x + 5}$, $x \neq -5, 1$

    (e) $3x^4 + 24x^3 + 18x^2 - 120x + 68$
    (f) $-9x^4 + 30x^2 - 16$

20. (a) $\dfrac{1 + 2x^{3/2}}{x}$, $x > 0$ (b) $\dfrac{1 - 2x^{3/2}}{x}$, $x > 0$

    (c) $\dfrac{2\sqrt{x}}{x}$, $x > 0$ (d) $\dfrac{1}{2x^{3/2}}$, $x > 0$

    (e) $\dfrac{\sqrt{x}}{2x}$, $x > 0$ (f) $\dfrac{2\sqrt{x}}{x}$, $x > 0$

21. $f^{-1}(x) = \sqrt[3]{x - 8}$ **22.** No inverse

23. $f^{-1}(x) = \left(\frac{1}{3}x\right)^{2/3}$, $x \geq 0$ **24.** $v = 6\sqrt{s}$

25. $A = \dfrac{25}{6}xy$ **26.** $b = \dfrac{48}{a}$

Problem Solving   (page 123)

1. (a) $W_1 = 2000 + 0.07S$ (b) $W_2 = 2300 + 0.05S$
    (c)

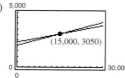

    Both jobs pay the same monthly salary if sales equal $15,000.
    (d) No. Job 1 would pay $3400 and job 2 would pay $3300.

3. (a) The function will be even.
    (b) The function will be odd.
    (c) The function will be neither even nor odd.

5. $f(x) = a_{2n}x^{2n} + a_{2n-2}x^{2n-2} + \cdots + a_2x^2 + a_0$
    $f(-x) = a_{2n}(-x)^{2n} + a_{2n-2}(-x)^{2n-2}$
           $+ \cdots + a_2(-x)^2 + a_0$
           $= f(x)$

7. (a) $81\frac{2}{3}$ h   (b) $25\frac{5}{7}$ mi/h

    (c) $y = -\frac{180}{7}x + 3400$

      Domain: $0 \leq x \leq \frac{1190}{9}$

      Range: $0 \leq y \leq 3400$

    (d)

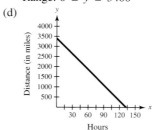

9. (a) $(f \circ g)(x) = 4x + 24$ (b) $(f \circ g)^{-1}(x) = \frac{1}{4}x - 6$
    (c) $f^{-1}(x) = \frac{1}{4}x$; $g^{-1}(x) = x - 6$
    (d) $(g^{-1} \circ f^{-1})(x) = \frac{1}{4}x - 6$; They are the same.
    (e) $(f \circ g)(x) = 8x^3 + 1$; $(f \circ g)^{-1}(x) = \frac{1}{2}\sqrt[3]{x - 1}$;
      $f^{-1}(x) = \sqrt[3]{x - 1}$; $g^{-1}(x) = \frac{1}{2}x$;
      $(g^{-1} \circ f^{-1})(x) = \frac{1}{2}\sqrt[3]{x - 1}$
    (f) Answers will vary.
    (g) $(f \circ g)^{-1}(x) = (g^{-1} \circ f^{-1})(x)$

11. (a)   (b)

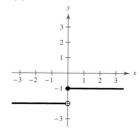

    (c)   (d)

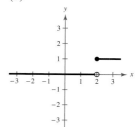

    (e)   (f)

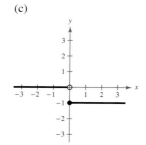
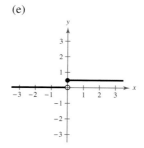
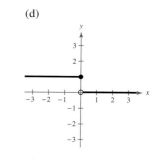
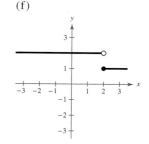

13. Proof

15. (a)

x	-4	-2	0	4
$f(f^{-1}(x))$	-4	-2	0	4

(b)

x	-3	-2	0	1
$(f + f^{-1})(x)$	5	1	-3	-5

(c)

x	-3	-2	0	1
$(f \cdot f^{-1})(x)$	4	0	2	6

(d)

x	-4	-3	0	4
$|f^{-1}(x)|$	2	1	1	3

Chapter 2

Section 2.1   *(page 132)*

1. polynomial   **3.** quadratic; parabola
5. positive; minimum
7. e   **8.** c   **9.** b   **10.** a   **11.** f   **12.** d
13. (a)   (b)

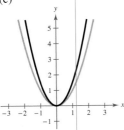

Vertical shrink

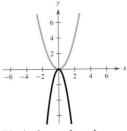

Vertical shrink and
reflection in the x-axis

(c)   (d)

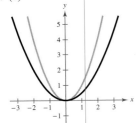

Vertical stretch

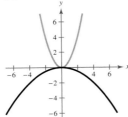

Vertical stretch and
reflection in the x-axis

15. (a)   (b)

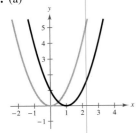

Horizontal shift one unit
to the right

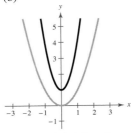

Horizontal shrink and
vertical shift one unit upward

(c)   (d)

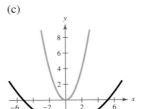

Horizontal stretch and
vertical shift three units
downward

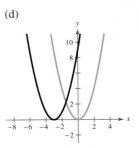

Horizontal shift three units
to the left

17.   **19.**

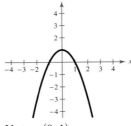

Vertex: $(0, 1)$
Axis of symmetry: y-axis
x-intercepts: $(-1, 0)\ (1, 0)$

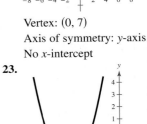

Vertex: $(0, 7)$
Axis of symmetry: y-axis
No x-intercept

21.   **23.**

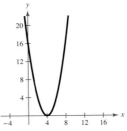

Vertex: $(0, -4)$
Axis of symmetry: y-axis
x-intercepts: $(\pm 2\sqrt{2}, 0)$

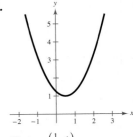

Vertex: $(-4, -3)$
Axis of symmetry: $x = -4$
x-intercepts: $(-4 \pm \sqrt{3}, 0)$

25.   **27.**

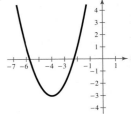

Vertex: $(4, 0)$
Axis of symmetry: $x = 4$
x-intercept: $(4, 0)$

Vertex: $(\frac{1}{2}, 1)$
Axis of symmetry: $x = \frac{1}{2}$
No x-intercept

29.

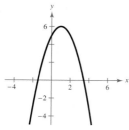

Vertex: $(1, 6)$
Axis of symmetry: $x = 1$
x-intercepts: $\left(1 \pm \sqrt{6}, 0\right)$

31.

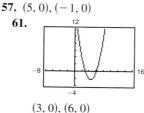

Vertex: $\left(\frac{1}{2}, 20\right)$
Axis of symmetry: $x = \frac{1}{2}$
No x-intercept

33.

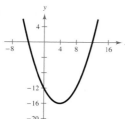

Vertex: $(4, -16)$
Axis of symmetry: $x = 4$
x-intercepts: $(-4, 0), (12, 0)$

35.

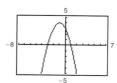

Vertex: $(-1, 4)$
Axis of symmetry: $x = -1$
x-intercepts: $(1, 0), (-3, 0)$

37.

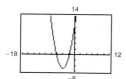

Vertex: $(-4, -5)$
Axis of symmetry: $x = -4$
x-intercepts: $\left(-4 \pm \sqrt{5}, 0\right)$

39.

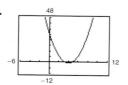

Vertex: $(4, -1)$
Axis of symmetry: $x = 4$
x-intercepts: $\left(4 \pm \frac{1}{2}\sqrt{2}, 0\right)$

41.

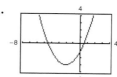

Vertex: $(-2, -3)$
Axis of symmetry: $x = -2$
x-intercepts: $\left(-2 \pm \sqrt{6}, 0\right)$

43. $y = -(x + 1)^2 + 4$
45. $y = -2(x + 2)^2 + 2$
47. $f(x) = (x + 2)^2 + 5$
49. $f(x) = 4(x - 1)^2 - 2$
51. $f(x) = \frac{3}{4}(x - 5)^2 + 12$
53. $f(x) = -\frac{24}{49}\left(x + \frac{1}{4}\right)^2 + \frac{3}{2}$
55. $f(x) = -\frac{16}{3}\left(x + \frac{5}{2}\right)^2$
57. $(5, 0), (-1, 0)$

59.

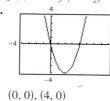

$(0, 0), (4, 0)$

61.

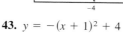

$(3, 0), (6, 0)$

63.

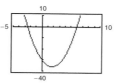

$\left(-\frac{5}{2}, 0\right), (6, 0)$

65. $f(x) = x^2 - 2x - 3$
$\quad\ g(x) = -x^2 + 2x + 3$
67. $f(x) = x^2 - 10x$
$\quad\ g(x) = -x^2 + 10x$
69. $f(x) = 2x^2 + 7x + 3$
$\quad\ g(x) = -2x^2 - 7x - 3$
71. 55, 55 **73.** 12, 6 **75.** 16 ft **77.** 20 fixtures
79. (a) \$14,000,000; \$14,375,000; \$13,500,000
(b) \$24; \$14,400,000
Answers will vary.
81. (a) $A = \dfrac{8x(50 - x)}{3}$

(b)

x	5	10	15	20	25	30
a	600	$1066\frac{2}{3}$	1400	1600	$1666\frac{2}{3}$	1600

$x = 25$ ft, $y = 33\frac{1}{3}$ ft

(c)

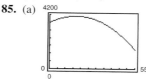

$x = 25$ ft, $y = 33\frac{1}{3}$ ft
(d) $A = -\frac{8}{3}(x - 25)^2 + \frac{5000}{3}$ (e) They are identical.
83. (a) $R = -100x^2 + 3500x, \quad 15 \le x \le 20$
(b) \$17.50; \$30,625
85. (a)

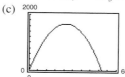

(b) 4075 cigarettes; Yes, the warning had an effect because the maximum consumption occurred in 1966.
(c) 7366 cigarettes per year; 20 cigarettes per day
87. True. The equation has no real solutions, so the graph has no x-intercepts.
89. True. The graph of a quadratic function with a negative leading coefficient will be a downward-opening parabola.
91. $b = \pm 20$ **93.** $b = \pm 8$
95. $f(x) = a\left(x + \dfrac{b}{2a}\right)^2 + \dfrac{4ac - b^2}{4a}$
97. (a)

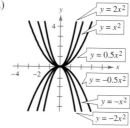

As $|a|$ increases, the parabola becomes narrower. For $a > 0$, the parabola opens upward. For $a < 0$, the parabola opens downward.

(b)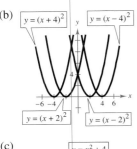

For $h < 0$, the vertex will be on the negative x-axis. For $h > 0$, the vertex will be on the positive x-axis. As $|h|$ increases, the parabola moves away from the origin.

(c)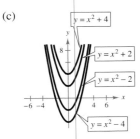

As $|k|$ increases, the vertex moves upward (for $k > 0$) or downward (for $k < 0$), away from the origin.

99. Yes. A graph of a quadratic equation whose vertex is on the x-axis has only one x-intercept.

Section 2.2 (page 145)

1. continuous   **3.** x^n

5. (a) solution; (b) $(x - a)$; (c) x-intercept   **7.** standard

9. c   **10.** g   **11.** h   **12.** f

13. a   **14.** e   **15.** d   **16.** b

17. (a) 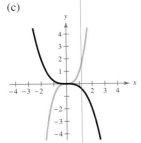   (b)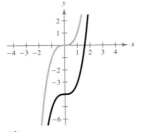

19. (a) 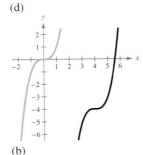   (b)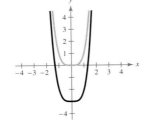

(c) 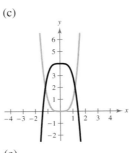   (d)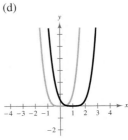

(e) 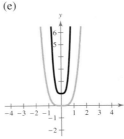   (f)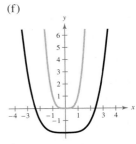

21. Falls to the left, rises to the right

23. Falls to the left, falls to the right

25. Rises to the left, falls to the right

27. Rises to the left, falls to the right

29. Falls to the left, falls to the right

31. 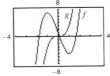   **33.**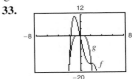

35. (a) ± 6
(b) Odd multiplicity; number of turning points: 1
(c)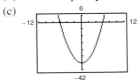

37. (a) 3
(b) Even multiplicity; number of turning points: 1
(c)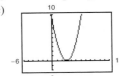

39. (a) $-2, 1$
(b) Odd multiplicity; number of turning points: 1
(c)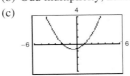

41. (a) $0, 2 \pm \sqrt{3}$
(b) Odd multiplicity; number of turning points: 2

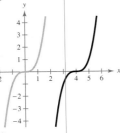
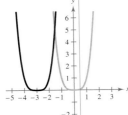

(c)

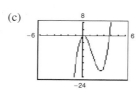

43. (a) 0, 4

(b) 0, odd multiplicity; 4, even multiplicity; number of turning points: 2

(c)

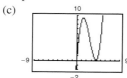

45. (a) $0, \pm\sqrt{3}$

(b) 0, odd multiplicity; $\pm\sqrt{3}$, even multiplicity; number of turning points: 4

(c)

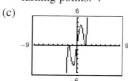

47. (a) No real zeros

(b) Number of turning points: 1

(c)

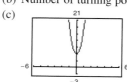

49. (a) $\pm 2, -3$

(b) Odd multiplicity; number of turning points: 2

(c)

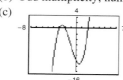

51. (a)

(b) x-intercepts: $(0, 0), \left(\frac{5}{2}, 0\right)$ (c) $x = 0, \frac{5}{2}$

(d) The answers in part (c) match the x-intercepts.

53. (a)

(b) x-intercepts: $(0, 0), (\pm 1, 0), (\pm 2, 0)$

(c) $x = 0, 1, -1, 2, -2$

(d) The answers in part (c) match the x-intercepts.

55. $f(x) = x^2 - 8x$ **57.** $f(x) = x^2 + 4x - 12$

59. $f(x) = x^3 + 9x^2 + 20x$

61. $f(x) = x^4 - 4x^3 - 9x^2 + 36x$ **63.** $f(x) = x^2 - 2x - 2$

65. $f(x) = x^2 + 6x + 9$ **67.** $f(x) = x^3 + 4x^2 - 5x$

69. $f(x) = x^3 - 3x$ **71.** $f(x) = x^4 + x^3 - 15x^2 + 23x - 10$

73. $f(x) = x^5 + 16x^4 + 96x^3 + 256x^2 + 256x$

75. (a) Falls to the left, rises to the right

(b) $0, 5, -5$ (c) Answers will vary.

(d)

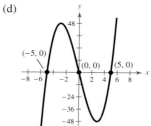

77. (a) Rises to the left, rises to the right

(b) No zeros   (c) Answers will vary.

(d)

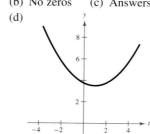

79. (a) Falls to the left, rises to the right

(b) $0, 2$ (c) Answers will vary.

(d)

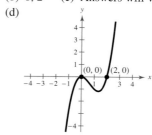

81. (a) Falls to the left, rises to the right

(b) $0, 2, 3$ (c) Answers will vary.

(d)

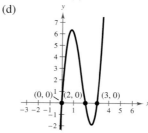

83. (a) Rises to the left, falls to the right

(b) $-5, 0$ (c) Answers will vary.

(d)
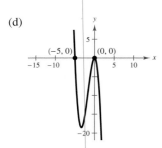

85. (a) Falls to the left, rises to the right
(b) 0, 4   (c) Answers will vary.
(d)
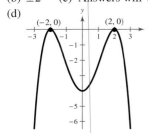

87. (a) Falls to the left, falls to the right
(b) ±2   (c) Answers will vary.
(d)
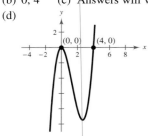

89.

Zeros: 0, ±4,
odd multiplicity

91.
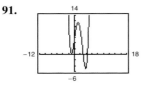

Zeros: −1,
even multiplicity;
3, $\frac{9}{2}$, odd multiplicity

93. $[-1, 0], [1, 2], [2, 3]$; about $-0.879, 1.347, 2.532$
95. $[-2, -1], [0, 1]$; about $-1.585, 0.779$
97. (a) $V(x) = x(36 - 2x)^2$ (b) Domain: $0 < x < 18$
(c)
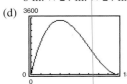

6 in. × 24 in. × 24 in.
(d)
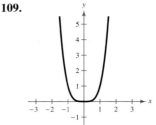

$x = 6$; The results are the same.
99. (a) $A = -2x^2 + 12x$ (b) $V = -384x^2 + 2304x$
(c) 0 in. $< x < 6$ in.

(d)
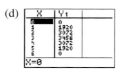

When $x = 3$, the volume is maximum at $V = 3456$;
dimensions of gutter are 3 in. × 6 in. × 3 in.
(e)
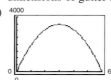

The maximum value is the same.
(f) No. Answers will vary.
101. (a)
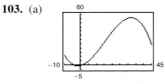

(b) The model fits the data well.
(c) Relative minima: $(0.21, 300.54), (6.62, 410.74)$
    Relative maximum: $(3.62, 681.72)$
(d) Increasing: $(0.21, 3.62), (6.62, 7)$
    Decreasing: $(0, 0.21), (3.62, 6.62)$
(e) Answers will vary.
103. (a)
[graph]
    (b) $t \approx 15$

(c) Vertex: $(15.22, 2.54)$
(d) The results are approximately equal.
105. False. A fifth-degree polynomial can have at most four turning points.
107. True. The degree of the function is odd and its leading coefficient is negative, so the graph rises to the left and falls to the right.
109.
[graph]

(a) Vertical shift two units upward; Even
(b) Horizontal shift two units to the left; Neither
(c) Reflection in the y-axis; Even
(d) Reflection in the x-axis; Even
(e) Horizontal stretch; Even
(f) Vertical shrink; Even
(g) $g(x) = x^3, x \geq 0$; Neither
(h) $g(x) = x^{16}$; Even

CHAPTER 2

111. (a)

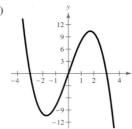

Zeros: 3
Relative minimum: 1
Relative maximum: 1
The number of zeros is the same as the degree, and the number of extrema is one less than the degree.

(b)

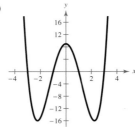

Zeros: 4
Relative minima: 2
Relative maximum: 1
The number of zeros is the same as the degree, and the number of extrema is one less than the degree.

(c)

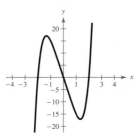

Zeros: 3
Relative minimum: 1
Relative maximum: 1
The number of zeros and the number of extrema are both less than the degree.

Section 2.3 (page 156)

1. $f(x)$: dividend; $d(x)$: divisor; $q(x)$: quotient; $r(x)$: remainder

3. improper   **5.** Factor   **7.** Answers will vary.

9. (a) and (b)

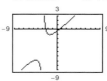

(c) Answers will vary.

11. $2x + 4, \quad x \neq -3$ **13.** $x^2 - 3x + 1, \quad x \neq -\frac{5}{4}$

15. $x^3 + 3x^2 - 1, \quad x \neq -2$ **17.** $x^2 + 3x + 9, \quad x \neq 3$

19. $7 - \dfrac{11}{x + 2}$ **21.** $x - \dfrac{x + 9}{x^2 + 1}$ **23.** $2x - 8 + \dfrac{x - 1}{x^2 + 1}$

25. $x + 3 + \dfrac{6x^2 - 8x + 3}{(x - 1)^3}$ **27.** $3x^2 - 2x + 5, \quad x \neq 5$

29. $6x^2 + 25x + 74 + \dfrac{248}{x - 3}$ **31.** $4x^2 - 9, \quad x \neq -2$

33. $-x^2 + 10x - 25, \quad x \neq -10$

35. $5x^2 + 14x + 56 + \dfrac{232}{x - 4}$

37. $10x^3 + 10x^2 + 60x + 360 + \dfrac{1360}{x - 6}$

39. $x^2 - 8x + 64, \quad x \neq -8$

41. $-3x^3 - 6x^2 - 12x - 24 - \dfrac{48}{x - 2}$

43. $-x^3 - 6x^2 - 36x - 36 - \dfrac{216}{x - 6}$

45. $4x^2 + 14x - 30, \quad x \neq -\frac{1}{2}$

47. $f(x) = (x - 4)(x^2 + 3x - 2) + 3, \quad f(4) = 3$

49. $f(x) = \left(x + \frac{2}{3}\right)(15x^3 - 6x + 4) + \frac{34}{3}, \quad f\left(-\frac{2}{3}\right) = \frac{34}{3}$

51. $f(x) = \left(x - \sqrt{2}\right)\left[x^2 + \left(3 + \sqrt{2}\right)x + 3\sqrt{2}\right] - 8,$
$f\left(\sqrt{2}\right) = -8$

53. $f(x) = \left(x - 1 + \sqrt{3}\right)\left[-4x^2 + \left(2 + 4\sqrt{3}\right)x + \left(2 + 2\sqrt{3}\right)\right],$
$f\left(1 - \sqrt{3}\right) = 0$

55. (a) -2 (b) 1 (c) $-\frac{1}{4}$ (d) 5

57. (a) -35 (b) -22 (c) -10 (d) -211

59. $(x - 2)(x + 3)(x - 1)$; Solutions: $2, -3, 1$

61. $(2x - 1)(x - 5)(x - 2)$; Solutions: $\frac{1}{2}, 5, 2$

63. $\left(x + \sqrt{3}\right)\left(x - \sqrt{3}\right)(x + 2)$; Solutions: $-\sqrt{3}, \sqrt{3}, -2$

65. $(x - 1)\left(x - 1 - \sqrt{3}\right)\left(x - 1 + \sqrt{3}\right)$;
Solutions: $1, 1 + \sqrt{3}, 1 - \sqrt{3}$

67. (a) Answers will vary.   (b) $2x - 1$
(c) $f(x) = (2x - 1)(x + 2)(x - 1)$
(d) $\frac{1}{2}, -2, 1$ (e)

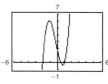

69. (a) Answers will vary.   (b) $(x - 1), (x - 2)$
(c) $f(x) = (x - 1)(x - 2)(x - 5)(x + 4)$
(d) $1, 2, 5, -4$ (e)

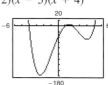

71. (a) Answers will vary.   (b) $x + 7$
(c) $f(x) = (x + 7)(2x + 1)(3x - 2)$
(d) $-7, -\frac{1}{2}, \frac{2}{3}$ (e)

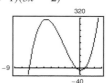

73. (a) Answers will vary.   (b) $x - \sqrt{5}$
(c) $f(x) = \left(x - \sqrt{5}\right)\left(x + \sqrt{5}\right)(2x - 1)$
(d) $\pm\sqrt{5}, \frac{1}{2}$ (e)

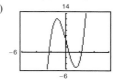

75. (a) Zeros are 2 and about ± 2.236.
(b) $x = 2$ (c) $f(x) = (x - 2)\left(x - \sqrt{5}\right)\left(x + \sqrt{5}\right)$

77. (a) Zeros are -2, about 0.268, and about 3.732.
(b) $t = -2$
(c) $h(t) = (t + 2)\left[t - \left(2 + \sqrt{3}\right)\right]\left[t - \left(2 - \sqrt{3}\right)\right]$

79. (a) Zeros are 0, 3, 4, and about ± 1.414.

(b) $x = 0$

(c) $h(x) = x(x - 4)(x - 3)(x + \sqrt{2})(x - \sqrt{2})$

81. $2x^2 - x - 1$, $x \neq \frac{3}{2}$ **83.** $x^2 + 3x$, $x \neq -2, -1$

85. (a) and (b)

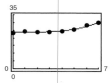

$A = 0.0349t^3 - 0.168t^2 + 0.42t + 23.4$

(c)

t	0	1	2	3
$A(t)$	23.4	23.7	23.8	24.1

t	4	5	6	7
$A(t)$	24.6	25.7	27.4	30.1

(d) \$45.7 billion; No, because the model will approach infinity quickly.

87. False. $-\frac{4}{7}$ is a zero of f.

89. True. The degree of the numerator is greater than the degree of the denominator.

91. $x^{2n} + 6x^n + 9$, $x^n \neq -3$ **93.** The remainder is 0.

95. $c = -210$ **97.** $k = 7$

99. (a) $x + 1$, $x \neq 1$ (b) $x^2 + x + 1$, $x \neq 1$

(c) $x^3 + x^2 + x + 1$, $x \neq 1$

In general, $\dfrac{x^n - 1}{x - 1} = x^{n-1} + x^{n-2} + \cdots + x + 1$, $x \neq 1$

Section 2.4   (page 164)

1. (a) iii   (b) i   (c) ii   **3.** principal square

5. $a = -12$, $b = 7$ **7.** $a = 6$, $b = 5$ **9.** $8 + 5i$

11. $2 - 3\sqrt{3}i$ **13.** $4\sqrt{5}i$ **15.** 14   **17.** $-1 - 10i$

19. $0.3i$ **21.** $10 - 3i$ **23.** 1   **25.** $3 - 3\sqrt{2}i$

27. $-14 + 20i$ **29.** $\frac{1}{6} + \frac{7}{6}i$ **31.** $5 + i$ **33.** $108 + 12i$

35. 24   **37.** $-13 + 84i$ **39.** -10 **41.** $9 - 2i$, 85

43. $-1 + \sqrt{5}i$, 6   **45.** $-2\sqrt{5}i$, 20   **47.** $\sqrt{6}$, 6

49. $-3i$ **51.** $\frac{8}{41} + \frac{10}{41}i$ **53.** $\frac{12}{13} + \frac{5}{13}i$ **55.** $-4 - 9i$

57. $-\frac{120}{1681} - \frac{27}{1681}i$ **59.** $-\frac{1}{2} - \frac{5}{2}i$ **61.** $\frac{62}{949} + \frac{297}{949}i$

63. $-2\sqrt{3}$ **65.** -15

67. $(21 + 5\sqrt{2}) + (7\sqrt{5} - 3\sqrt{10})i$ **69.** $1 \pm i$

71. $-2 \pm \frac{1}{2}i$ **73.** $-\frac{5}{2}, -\frac{3}{2}$ **75.** $2 \pm \sqrt{2}i$

77. $\frac{5}{7} \pm \frac{5\sqrt{15}}{7}$ **79.** $-1 + 6i$ **81.** $-14i$

83. $-432\sqrt{2}i$ **85.** i **87.** 81

89. (a) $z_1 = 9 + 16i$, $z_2 = 20 - 10i$

(b) $z = \dfrac{11,240}{877} + \dfrac{4630}{877}i$

91. (a) 16   (b) 16   (c) 16   (d) 16

93. False. If the complex number is real, the number equals its conjugate.

95. False.

$i^{44} + i^{150} - i^{74} - i^{109} + i^{61} = 1 - 1 + 1 - i + i = 1$

97. $i, -1, -i, 1, i, -1, -i, 1$; The pattern repeats the first four results. Divide the exponent by 4.

If the remainder is 1, the result is i.

If the remainder is 2, the result is -1.

If the remainder is 3, the result is $-i$.

If the remainder is 0, the result is 1.

99. $\sqrt{-6}\sqrt{-6} = \sqrt{6}i\sqrt{6}i = 6i^2 = -6$ **101.** Proof

Section 2.5   (page 176)

1. Fundamental Theorem of Algebra   **3.** Rational Zero

5. linear; quadratic; quadratic   **7.** Descartes's Rule of Signs

9. 0, 6   **11.** $2, -4$ **13.** $-6, \pm i$ **15.** $\pm 1, \pm 2$

17. $\pm 1, \pm 3, \pm 5, \pm 9, \pm 15, \pm 45, \pm\frac{1}{2}, \pm\frac{3}{2}, \pm\frac{5}{2}, \pm\frac{9}{2}, \pm\frac{15}{2}, \pm\frac{45}{2}$

19. 1, 2, 3   **21.** $1, -1, 4$ **23.** $-6, -1$ **25.** $\frac{1}{2}, -1$

27. $-2, 3, \pm\frac{2}{3}$ **29.** $-2, 1$ **31.** $-4, \frac{1}{2}, 1, 1$

33. (a) $\pm 1, \pm 2, \pm 4$

(b)

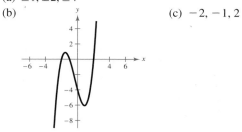

(c) $-2, -1, 2$

35. (a) $\pm 1, \pm 3, \pm\frac{1}{2}, \pm\frac{3}{2}, \pm\frac{1}{4}, \pm\frac{3}{4}$

(b)

(c) $-\frac{1}{4}, 1, 3$

37. (a) $\pm 1, \pm 2, \pm 4, \pm 8, \pm\frac{1}{2}$

(b)

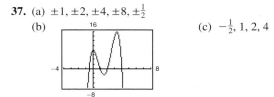

(c) $-\frac{1}{2}, 1, 2, 4$

39. (a) $\pm 1, \pm 3, \pm\frac{1}{2}, \pm\frac{3}{2}, \pm\frac{1}{4}, \pm\frac{3}{4}, \pm\frac{1}{8}, \pm\frac{3}{8}, \pm\frac{1}{16}, \pm\frac{3}{16}, \pm\frac{1}{32}, \pm\frac{3}{32}$

(b)

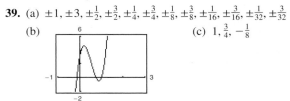

(c) $1, \frac{3}{4}, -\frac{1}{8}$

41. (a) ± 1, about ± 1.414 (b) $\pm 1, \pm\sqrt{2}$

(c) $f(x) = (x + 1)(x - 1)(x + \sqrt{2})(x - \sqrt{2})$

43. (a) 0, 3, 4, about ± 1.414 (b) $0, 3, 4, \pm\sqrt{2}$

(c) $h(x) = x(x - 3)(x - 4)(x + \sqrt{2})(x - \sqrt{2})$

45. $x^3 - x^2 + 25x - 25$ **47.** $x^3 - 12x^2 + 46x - 52$

49. $3x^4 - 17x^3 + 25x^2 + 23x - 22$

51. (a) $(x^2 + 9)(x^2 - 3)$ (b) $(x^2 + 9)(x + \sqrt{3})(x - \sqrt{3})$

(c) $(x + 3i)(x - 3i)(x + \sqrt{3})(x - \sqrt{3})$

CHAPTER 2

53. (a) $(x^2 - 2x - 2)(x^2 - 2x + 3)$

    (b) $\left(x - 1 + \sqrt{3}\right)\left(x - 1 - \sqrt{3}\right)(x^2 - 2x + 3)$

    (c) $\left(x - 1 + \sqrt{3}\right)\left(x - 1 - \sqrt{3}\right)\left(x - 1 + \sqrt{2}i\right)$

       $\left(x - 1 - \sqrt{2}i\right)$

55. $\pm 2i, 1$ **57.** $\pm 5i, -\frac{1}{2}, 1$ **59.** $-3 \pm i, \frac{1}{4}$

61. $2, -3 \pm \sqrt{2}i, 1$ **63.** $\pm 6i; (x + 6i)(x - 6i)$

65. $1 \pm 4i; (x - 1 - 4i)(x - 1 + 4i)$

67. $\pm 2, \pm 2i; (x - 2)(x + 2)(x - 2i)(x + 2i)$

69. $1 \pm i; (z - 1 + i)(z - 1 - i)$

71. $-1, 2 \pm i; (x + 1)(x - 2 + i)(x - 2 - i)$

73. $-2, 1 \pm \sqrt{2}i; (x + 2)\left(x - 1 + \sqrt{2}i\right)\left(x - 1 - \sqrt{2}i\right)$

75. $-\frac{1}{5}, 1 \pm \sqrt{5}i; (5x + 1)\left(x - 1 + \sqrt{5}i\right)\left(x - 1 - \sqrt{5}i\right)$

77. $2, \pm 2i; (x - 2)^2(x + 2i)(x - 2i)$

79. $\pm i, \pm 3i; (x + i)(x - i)(x + 3i)(x - 3i)$

81. $-10, -7 \pm 5i$ **83.** $-\frac{3}{4}, 1 \pm \frac{1}{2}i$ **85.** $-2, -\frac{1}{2}, \pm i$

87. One positive zero   **89.** One negative zero

91. One positive zero, one negative zero

93. One or three positive zeros   **95–97.** Answers will vary.

99. $1, -\frac{1}{2}$ **101.** $-\frac{3}{4}$ **103.** $\pm 2, \pm \frac{3}{2}$ **105.** $\pm 1, \frac{1}{4}$

107. d   **108.** a   **109.** b   **110.** c

111. (a)

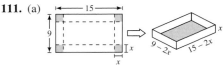

    (b) $V(x) = x(9 - 2x)(15 - 2x)$

       Domain: $0 < x < \frac{9}{2}$

    (c)

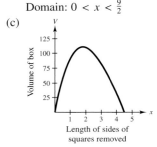

       $1.82 \text{ cm} \times 5.36 \text{ cm} \times 11.36 \text{ cm}$

    (d) $\frac{1}{2}, \frac{7}{2}, 8$; 8 is not in the domain of V.

113. $x \approx 38.4$, or \$384,000

115. (a) $V(x) = x^3 + 9x^2 + 26x + 24 = 120$

    (b) $4 \text{ ft} \times 5 \text{ ft} \times 6 \text{ ft}$

117. $x \approx 40$, or 4000 units

119. No. Setting $p = 9{,}000{,}000$ and solving the resulting equation yields imaginary roots.

121. False. The most complex zeros it can have is two, and the Linear Factorization Theorem guarantees that there are three linear factors, so one zero must be real.

123. r_1, r_2, r_3 **125.** $5 + r_1, 5 + r_2, 5 + r_3$

127. The zeros cannot be determined.

129. Answers will vary. There are infinitely many possible functions for f. Sample equation and graph:

$f(x) = -2x^3 + 3x^2 + 11x - 6$

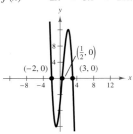

131. Answers will vary. Sample graph:

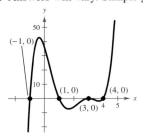

133. $f(x) = x^4 + 5x^2 + 4$ **135.** $f(x) = x^3 - 3x^2 + 4x - 2$

137. (a) $-2, 1, 4$

    (b) The graph touches the x-axis at $x = 1$.

    (c) The least possible degree of the function is 4, because there are at least four real zeros (1 is repeated) and a function can have at most the number of real zeros equal to the degree of the function. The degree cannot be odd by the definition of multiplicity.

    (d) Positive. From the information in the table, it can be concluded that the graph will eventually rise to the left and rise to the right.

    (e) $f(x) = x^4 - 4x^3 - 3x^2 + 14x - 8$

    (f)

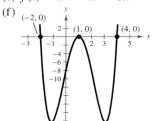

139. (a) Not correct because f has $(0, 0)$ as an intercept.

    (b) Not correct because the function must be at least a fourth-degree polynomial.

    (c) Correct function

    (d) Not correct because k has $(-1, 0)$ as an intercept.

Section 2.6   (*page 190*)

1. rational functions   **3.** horizontal asymptote

5. (a)

x	$f(x)$
0.5	-2
0.9	-10
0.99	-100
0.999	-1000

x	$f(x)$
1.5	2
1.1	10
1.01	100
1.001	1000

x	$f(x)$
5	0.25
10	$0.\overline{1}$
100	$0.\overline{01}$
1000	$0.\overline{001}$

(b) Vertical asymptote: $x = 1$
Horizontal asymptote: $y = 0$
(c) Domain: all real numbers x except $x = 1$

7. (a)

x	$f(x)$
0.5	-1
0.9	-12.79
0.99	-147.8
0.999	-1498

x	$f(x)$
1.5	5.4
1.1	17.29
1.01	152.3
1.001	1502

x	$f(x)$
5	3.125
10	$3.\overline{03}$
100	$3.\overline{0003}$
1000	3

(b) Vertical asymptotes: $x = \pm 1$
Horizontal asymptote: $y = 3$
(c) Domain: all real numbers x except $x = \pm 1$

9. Domain: all real numbers x except $x = 0$
Vertical asymptote: $x = 0$
Horizontal asymptote: $y = 0$

11. Domain: all real numbers x except $x = 5$
Vertical asymptote: $x = 5$
Horizontal asymptote: $y = -1$

13. Domain: all real numbers x except $x = \pm 1$
Vertical asymptotes: $x = \pm 1$

15. Domain: all real numbers x
Horizontal asymptote: $y = 3$

17. d   **18.** a   **19.** c   **20.** b   **21.** 3   **23.** 9

25. Domain: all real numbers x except $x = \pm 4$;
Vertical asymptote: $x = -4$; horizontal asymptote: $y = 0$

27. Domain: all real numbers x except $x = -1, 5$;
Vertical asymptote: $x = -1$; horizontal asymptote: $y = 1$

29. Domain: all real numbers x except $x = -1, \frac{1}{2}$;
Vertical asymptote: $x = \frac{1}{2}$; horizontal asymptote: $y = \frac{1}{2}$

31. (a) Domain: all real numbers x except $x = -2$
(b) y-intercept: $\left(0, \frac{1}{2}\right)$
(c) Vertical asymptote: $x = -2$
Horizontal asymptote: $y = 0$
(d)

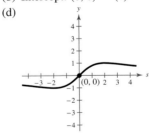

33. (a) Domain: all real numbers x except $x = -4$
(b) y-intercept: $\left(0, -\frac{1}{4}\right)$

(c) Vertical asymptote: $x = -4$
Horizontal asymptote: $y = 0$
(d)

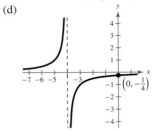

35. (a) Domain: all real numbers x except $x = -2$
(b) x-intercept: $\left(-\frac{7}{2}, 0\right)$
y-intercept: $\left(0, \frac{7}{2}\right)$
(c) Vertical asymptote: $x = -2$
Horizontal asymptote: $y = 2$
(d)

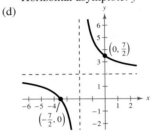

37. (a) Domain: all real numbers x
(b) Intercept: $(0, 0)$
(c) Horizontal asymptote: $y = 1$
(d)

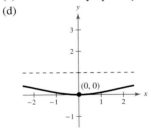

39. (a) Domain: all real numbers s
(b) Intercept: $(0, 0)$ (c) Horizontal asymptote: $y = 0$
(d)

41. (a) Domain: all real numbers x except $x = \pm 2$
(b) x-intercepts: $(1, 0)$ and $(4, 0)$
y-intercept: $(0, -1)$
(c) Vertical asymptotes: $x = \pm 2$
Horizontal asymptote: $y = 1$

(d)
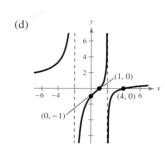

(d)
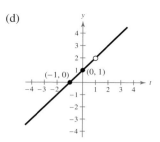

43. (a) Domain: all real numbers x except $x = \pm 1, 2$
(b) x-intercepts: $(3, 0)$, $\left(-\frac{1}{2}, 0\right)$
y-intercept: $\left(0, -\frac{3}{2}\right)$
(c) Vertical asymptotes: $x = 2$, $x = \pm 1$
Horizontal asymptote: $y = 0$
(d)
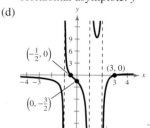

45. (a) Domain: all real numbers x except $x = 2, -3$
(b) Intercept: $(0, 0)$
(c) Vertical asymptote: $x = 2$
Horizontal asymptote: $y = 1$
(d)
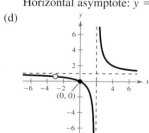

47. (a) Domain: all real numbers x except $x = -\frac{3}{2}, 2$
(b) x-intercept: $\left(\frac{1}{2}, 0\right)$
y-intercept: $\left(0, -\frac{1}{3}\right)$
(c) Vertical asymptote: $x = -\frac{3}{2}$
Horizontal asymptote: $y = 1$
(d)
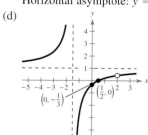

49. (a) Domain: all real numbers t except $t = 1$
(b) t-intercept: $(-1, 0)$
y-intercept: $(0, 1)$
(c) Vertical asymptote: None
Horizontal asymptote: None

51. (a) Domain of f: all real numbers x except $x = -1$
Domain of g: all real numbers x
(b) $x - 1$; Vertical asymptotes: None
(c)

x	-3	-2	-1.5	-1	-0.5	0	1
$f(x)$	-4	-3	-2.5	Undef.	-1.5	-1	0
$g(x)$	-4	-3	-2.5	-2	-1.5	-1	0

(d)
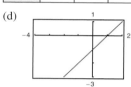

(e) Because there are only a finite number of pixels, the graphing utility may not attempt to evaluate the function where it does not exist.

53. (a) Domain of f: all real numbers x except $x = 0, 2$
Domain of g: all real numbers x except $x = 0$
(b) $\dfrac{1}{x}$; Vertical asymptote: $x = 0$
(c)

x	-0.5	0	0.5	1	1.5	2	3
$f(x)$	-2	Undef.	2	1	$\frac{2}{3}$	Undef.	$\frac{1}{3}$
$g(x)$	-2	Undef.	2	1	$\frac{2}{3}$	$\frac{1}{2}$	$\frac{1}{3}$

(d)
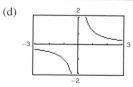

(e) Because there are only a finite number of pixels, the graphing utility may not attempt to evaluate the function where it does not exist.

55. (a) Domain: all real numbers x except $x = 0$
(b) x-intercepts: $(-3, 0)$, $(3, 0)$
(c) Vertical asymptote: $x = 0$
Slant asymptote: $y = x$
(d)

57. (a) Domain: all real numbers x except $x = 0$
  (b) No intercepts
  (c) Vertical asymptote: $x = 0$
    Slant asymptote: $y = 2x$
  (d)

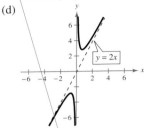

59. (a) Domain: all real numbers x except $x = 0$
  (b) No intercepts
  (c) Vertical asymptote: $x = 0$
    Slant asymptote: $y = x$
  (d)

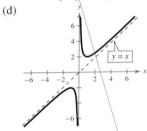

61. (a) Domain: all real numbers t except $t = -5$
  (b) y-intercept: $\left(0, -\frac{1}{5}\right)$
  (c) Vertical asymptote: $t = -5$
    Slant asymptote: $y = -t + 5$
  (d)

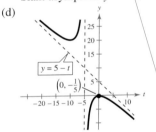

63. (a) Domain: all real numbers x except $x = \pm 2$
  (b) Intercept: $(0, 0)$
  (c) Vertical asymptotes: $x = \pm 2$
    Slant asymptote: $y = x$
  (d)

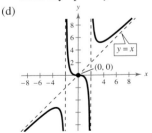

65. (a) Domain: all real numbers x except $x = 1$
  (b) y-intercept: $(0, -1)$
  (c) Vertical asymptote: $x = 1$
    Slant asymptote: $y = x$

  (d)

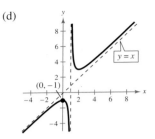

67. (a) Domain: all real numbers x except $x = -1, -2$
  (b) y-intercept: $\left(0, \frac{1}{2}\right)$
    x-intercepts: $\left(\frac{1}{2}, 0\right), (1, 0)$
  (c) Vertical asymptote: $x = -2$
    Slant asymptote: $y = 2x - 7$
  (d)

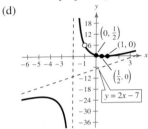

69.

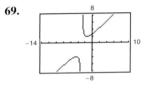

  Domain: all real numbers x except $x = -3$
  Vertical asymptote: $x = -3$
  Slant asymptote: $y = x + 2$
  $y = x + 2$

71.

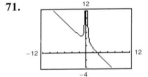

  Domain: all real numbers x except $x = 0$
  Vertical asymptote: $x = 0$
  Slant asymptote: $y = -x + 3$
  $y = -x + 3$

73. (a) $(-1, 0)$ (b) -1
75. (a) $(1, 0), (-1, 0)$ (b) ± 1
77. (a)

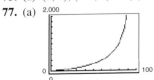

  (b) \$28.33 million; \$170 million; \$765 million
  (c) No. The function is undefined at $p = 100$.
79. (a) 333 deer, 500 deer, 800 deer (b) 1500 deer
81. (a) $A = \dfrac{2x(x + 11)}{x - 4}$ (b) $(4, \infty)$

(c)
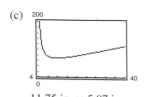

11.75 in. × 5.87 in.

83. (a) Answers will vary.

(b) Vertical asymptote: $x = 25$
Horizontal asymptote: $y = 25$

(c)
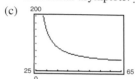

(d)

x	30	35	40	45	50	55	60
y	150	87.5	66.7	56.3	50	45.8	42.9

(e) Sample answer: No. You might expect the average speed for the round trip to be the average of the average speeds for the two parts of the trip.

(f) No. At 20 miles per hour you would use more time in one direction than is required for the round trip at an average speed of 50 miles per hour.

85. False. Polynomials do not have vertical asymptotes.

87. False. If the degree of the numerator is greater than the degree of the denominator, no horizontal asymptote exists. However, a slant asymptote exists only if the degree of the numerator is one greater than the degree of the denominator.

89. c

Section 2.7 *(page 201)*

1. positive; negative   **3.** zeros; undefined values

5. (a) No   (b) Yes   (c) Yes   (d) No

7. (a) Yes   (b) No   (c) No   (d) Yes

9. $-\frac{2}{3}, 1$ **11.** 4, 5

13. $(-3, 3)$
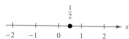

15. $[-7, 3]$

17. $(-\infty, -5] \cup [1, \infty)$

19. $(-3, 2)$

21. $(-3, 1)$

23. $\left(-\infty, -\frac{4}{3}\right) \cup (5, \infty)$

25. $(-\infty, -3) \cup (6, \infty)$

27. $(-1, 1) \cup (3, \infty)$

29. $x = \frac{1}{2}$

31. $(-\infty, 0) \cup \left(0, \frac{3}{2}\right)$ **33.** $[-2, 0] \cup [2, \infty)$ **35.** $[-2, \infty)$

37.
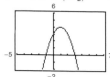

(a) $x \le -1, \ x \ge 3$

(b) $0 \le x \le 2$

39.
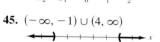

(a) $-2 \le x \le 0,$
$2 \le x < \infty$

(b) $x \le 4$

41. $(-\infty, 0) \cup \left(\frac{1}{4}, \infty\right)$

43. $\left(-\infty, \frac{5}{3}\right] \cup (5, \infty)$

45. $(-\infty, -1) \cup (4, \infty)$

47. $(-5, 3) \cup (11, \infty)$

49. $\left(-\frac{3}{4}, 3\right) \cup [6, \infty)$

51. $(-3, -2) \cup [0, 3)$

53. $(-\infty, -1) \cup (1, \infty)$

55.
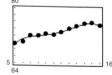

(a) $0 \le x < 2$

(b) $2 < x \le 4$

57.
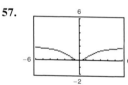

(a) $|x| \ge 2$

(b) $-\infty < x < \infty$

59. $[-2, 2]$ **61.** $(-\infty, 4] \cup [5, \infty)$

63. $(-5, 0] \cup (7, \infty)$ **65.** $(-3.51, 3.51)$

67. $(-0.13, 25.13)$ **69.** $(2.26, 2.39)$

71. (a) $t = 10$ sec   (b) 4 sec $< t < 6$ sec

73. 13.8 m $\le L \le 36.2$ m

75. $40{,}000 \le x \le 50{,}000; \$50.00 \le p \le \$55.00$

77. (a) and (c)

The model fits the data well.

(b) $N = -0.00412t^4 + 0.1705t^3 - 2.538t^2 + 16.55t + 31.5$

(d) 2003 to 2006

(e) No; The model decreases sharply after 2006.

79. $R_1 \ge 2$ ohms

81. True. The test intervals are $(-\infty, -3), (-3, 1), (1, 4),$ and $(4, \infty)$.

83. (a) $(-\infty, -4] \cup [4, \infty)$

(b) If $a > 0$ and $c > 0$, $b \le -2\sqrt{ac}$ or $b \ge 2\sqrt{ac}$.

85. (a) $\left(-\infty, -2\sqrt{30}\right] \cup \left[2\sqrt{30}, \infty\right)$
(b) If $a > 0$ and $c > 0$, $b \leq -2\sqrt{ac}$ or $b \geq 2\sqrt{ac}$.

87.

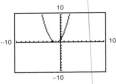

For part (b), the y-values that are less than or equal to 0 occur only at $x = -1$.

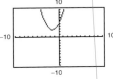

For part (c), there are no y-values that are less than 0.

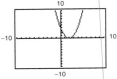

For part (d), the y-values that are greater than 0 occur for all values of x except 2.

Review Exercises   *(page 206)*

1. (a)

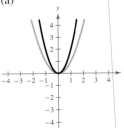

Vertical stretch

(b)

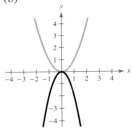

Vertical stretch and reflection in the x-axis

(c)

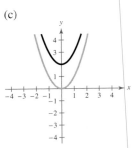

Vertical shift two units upward

(d)

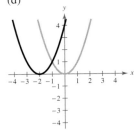

Horizontal shift two units to the left

3. $g(x) = (x - 1)^2 - 1$

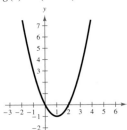

Vertex: $(1, -1)$
Axis of symmetry: $x = 1$
x-intercepts: $(0, 0)$, $(2, 0)$

5. $f(x) = (x + 4)^2 - 6$

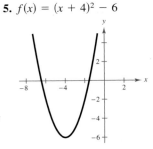

Vertex: $(-4, -6)$
Axis of symmetry: $x = -4$
x-intercepts: $\left(-4 \pm \sqrt{6}, 0\right)$

7. $f(t) = -2(t - 1)^2 + 3$

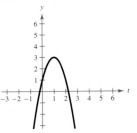

Vertex: $(1, 3)$
Axis of symmetry: $t = 1$
t-intercepts: $\left(1 \pm \dfrac{\sqrt{6}}{2}, 0\right)$

9. $h(x) = 4\left(x + \tfrac{1}{2}\right)^2 + 12$

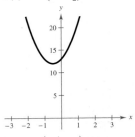

Vertex: $\left(-\tfrac{1}{2}, 12\right)$
Axis of symmetry: $x = -\tfrac{1}{2}$
No x-intercept

11. $h(x) = \left(x + \tfrac{5}{2}\right)^2 - \tfrac{41}{4}$

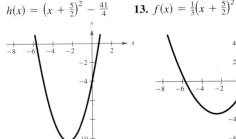

Vertex: $\left(-\tfrac{5}{2}, -\tfrac{41}{4}\right)$
Axis of symmetry: $x = -\tfrac{5}{2}$
x-intercepts: $\left(\dfrac{\pm\sqrt{41} - 5}{2}, 0\right)$

13. $f(x) = \tfrac{1}{3}\left(x + \tfrac{5}{2}\right)^2 - \tfrac{41}{12}$

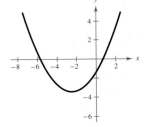

Vertex: $\left(-\tfrac{5}{2}, -\tfrac{41}{12}\right)$
Axis of symmetry: $x = -\tfrac{5}{2}$
x-intercepts: $\left(\dfrac{\pm\sqrt{41} - 5}{2}, 0\right)$

15. $f(x) = -\tfrac{1}{2}(x - 4)^2 + 1$

17. $f(x) = (x - 1)^2 - 4$

19. $y = -\tfrac{11}{36}\left(x + \tfrac{3}{2}\right)^2$

21. (a)

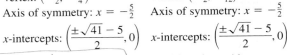

(b) $y = 500 - x$
$A(x) = 500x - x^2$
(c) $x = 250$, $y = 250$

23. 1091 units

CHAPTER 2

25. 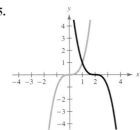 **27.**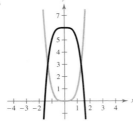

29.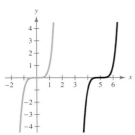

31. Falls to the left, falls to the right
33. Rises to the left, rises to the right
35. $-8, \frac{4}{3}$, odd multiplicity; turning points: 1
37. $0, \pm\sqrt{3}$, odd multiplicity; turning points: 2
39. 0, even multiplicity; $\frac{2}{3}$, odd multiplicity; turning points: 2
41. (a) Rises to the left, falls to the right   (b) -1
(c) Answers will vary.
(d)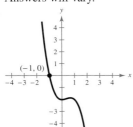

43. (a) Rises to the left, rises to the right   (b) $-3, 0, 1$
(c) Answers will vary.
(d)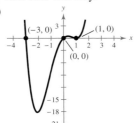

45. (a) $[-1, 0]$ (b) About -0.900
47. (a) $[-1, 0], [1, 2]$ (b) About -0.200, about 1.772
49. $6x + 3 + \dfrac{17}{5x - 3}$ **51.** $5x + 4, \quad x \neq \dfrac{5}{2} \pm \dfrac{\sqrt{29}}{2}$

53. $x^2 - 3x + 2 - \dfrac{1}{x^2 + 2}$

55. $6x^3 + 8x^2 - 11x - 4 - \dfrac{8}{x - 2}$

57. $2x^2 - 9x - 6, \quad x \neq 8$
59. (a) Yes   (b) Yes   (c) Yes   (d) No

61. (a) -421 (b) -9
63. (a) Answers will vary.
(b) $(x + 7), (x + 1)$
(c) $f(x) = (x + 7)(x + 1)(x - 4)$
(d) $-7, -1, 4$
(e)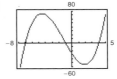

65. (a) Answers will vary.   (b) $(x + 1), (x - 4)$
(c) $f(x) = (x + 1)(x - 4)(x + 2)(x - 3)$
(d) $-2, -1, 3, 4$
(e)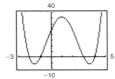

67. $8 + 10i$ **69.** $-1 + 3i$ **71.** $3 + 7i$
73. $63 + 77i$ **75.** $-4 - 46i$ **77.** $39 - 80i$
79. $\dfrac{23}{17} + \dfrac{10}{17}i$ **81.** $\dfrac{21}{13} - \dfrac{1}{13}i$ **83.** $\pm\dfrac{\sqrt{10}}{5}i$ **85.** $1 \pm 3i$
87. $0, 3$ **89.** $2, 9$ **91.** $-4, 6, \pm 2i$
93. $\pm 1, \pm 3, \pm 5, \pm 15, \pm\frac{1}{2}, \pm\frac{3}{2}, \pm\frac{5}{2}, \pm\frac{15}{2}, \pm\frac{1}{4}, \pm\frac{3}{4}, \pm\frac{5}{4}, \pm\frac{15}{4}$
95. $-6, -2, 5$ **97.** $1, 8$ **99.** $-4, 3$
101. $f(x) = 3x^4 - 14x^3 + 17x^2 - 42x + 24$
103. $4, \pm i$ **105.** $-3, \frac{1}{2}, 2 \pm i$
107. $0, 1, -5; f(x) = x(x - 1)(x + 5)$
109. $-4, 2 \pm 3i; g(x) = (x + 4)^2(x - 2 - 3i)(x - 2 + 3i)$
111. Two or no positive zeros, one negative zero
113. Answers will vary.
115. Domain: all real numbers x except $x = -10$
117. Domain: all real numbers x except $x = 6, 4$
119. Vertical asymptote: $x = -3$
Horizontal asymptote: $y = 0$
121. Vertical asymptote: $x = 6$
Horizontal asymptote: $y = 0$
123. (a) Domain: all real numbers x except $x = 0$
(b) No intercepts
(c) Vertical asymptote: $x = 0$
Horizontal asymptote: $y = 0$
(d)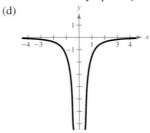

125. (a) Domain: all real numbers x except $x = 1$
(b) x-intercept: $(-2, 0)$
y-intercept: $(0, 2)$

(c) Vertical asymptote: $x = 1$
    Horizontal asymptote: $y = -1$

(d)

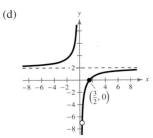

127. (a) Domain: all real numbers x
  (b) Intercept: $(0, 0)$
  (c) Horizontal asymptote: $y = \frac{5}{4}$
  (d)

129. (a) Domain: all real numbers x
  (b) Intercept: $(0, 0)$
  (c) Horizontal asymptote: $y = 0$
  (d)

131. (a) Domain: all real numbers x
  (b) Intercept: $(0, 0)$
  (c) Horizontal asymptote: $y = -6$
  (d)

133. (a) Domain: all real numbers x except $x = 0, \frac{1}{3}$
  (b) x-intercept: $\left(\frac{3}{2}, 0\right)$
  (c) Vertical asymptote: $x = 0$
      Horizontal asymptote: $y = 2$

(d)

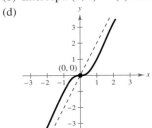

135. (a) Domain: all real numbers x
  (b) Intercept: $(0, 0)$ (c) Slant asymptote: $y = 2x$
  (d)

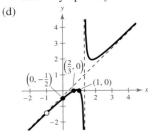

137. (a) Domain: all real numbers x except $x = \frac{4}{3}, -1$
  (b) y-intercept: $\left(0, -\frac{1}{2}\right)$
      x-intercepts: $\left(\frac{2}{3}, 0\right), (1, 0)$
  (c) Vertical asymptote: $x = \frac{4}{3}$
      Slant asymptote: $y = x - \frac{1}{3}$
  (d)

139. $\overline{C} = 0.5 = \$0.50$

141. (a)

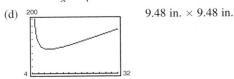

  (b) $A = \dfrac{2x(2x + 7)}{x - 4}$ (c) $4 < x < \infty$

  (d) 9.48 in. $\times$ 9.48 in.

143. $\left(-\frac{2}{3}, \frac{1}{4}\right)$ **145.** $[-4, 0] \cup [4, \infty)$
147. $[-5, -1) \cup (1, \infty)$ **149.** $(-\infty, 0) \cup [4, 5]$
151. 4.9%
153. False. A fourth-degree polynomial can have at most four zeros, and complex zeros occur in conjugate pairs.

CHAPTER 2

155. Find the vertex of the quadratic function and write the function in standard form. If the leading coefficient is positive, the vertex is a minimum. If the leading coefficient is negative, the vertex is a maximum.

157. An asymptote of a graph is a line to which the graph becomes arbitrarily close as x increases or decreases without bound.

Chapter Test   *(page 210)*

1. (a) Reflection in the x-axis followed by a vertical shift two units upward

(b) Horizontal shift $\frac{3}{2}$ units to the right

2. $y = (x - 3)^2 - 6$

3. (a) 50 ft

(b) 5. Yes, changing the constant term results in a vertical translation of the graph and therefore changes the maximum height.

4. Rises to the left, falls to the right

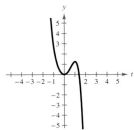

5. $3x + \dfrac{x - 1}{x^2 + 1}$ **6.** $2x^3 + 4x^2 + 3x + 6 + \dfrac{9}{x - 2}$

7. $(2x - 5)(x + \sqrt{3})(x - \sqrt{3})$;

Zeros: $\frac{5}{2}, \pm\sqrt{3}$

8. (a) $-3 + 5i$ (b) 7   **9.** $2 - i$

10. $f(x) = x^4 - 7x^3 + 17x^2 - 15x$

11. $f(x) = x^4 - 6x^3 + 16x^2 - 24x + 16$

12. $-5, -\frac{2}{3}, 1$ **13.** $-2, 4, -1 \pm \sqrt{2}i$

14. x-intercepts: $(-2, 0), (2, 0)$

Vertical asymptote: $x = 0$

Horizontal asymptote: $y = -1$

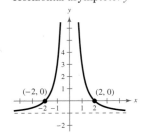

15. x-intercept: $\left(-\frac{3}{2}, 0\right)$

y-intercept: $\left(0, \frac{3}{4}\right)$

Vertical asymptote: $x = -4$

Horizontal asymptote: $y = 2$

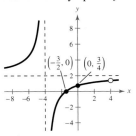

16. y-intercept: $(0, -2)$

Vertical asymptote: $x = 1$

Slant asymptote: $y = x + 1$

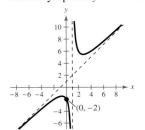

17. $x < -4$ or $x > \frac{3}{2}$ **18.** $x \le -12$ or $-6 < x < 0$

Problem Solving   *(page 213)*

1. Answers will vary.   **3.** 2 in. $\times$ 2 in. $\times$ 5 in.

5. (a) and (b) $y = -x^2 + 5x - 4$

7. (a) $f(x) = (x - 2)x^2 + 5 = x^3 - 2x^2 + 5$

(b) $f(x) = -(x + 3)x^2 + 1 = -x^3 - 3x^2 + 1$

9. $(a + bi)(a - bi) = a^2 + abi - abi - b^2 i^2$
$$= a^2 + b^2$$

11. (a) As $|a|$ increases, the graph stretches vertically. For $a < 0$, the graph is reflected in the x-axis.

(b) As $|b|$ increases, the vertical asymptote is translated. For $b > 0$, the graph is translated to the right. For $b < 0$, the graph is reflected in the x-axis and is translated to the left.

Chapter 3

Section 3.1   *(page 224)*

1. algebraic   **3.** One-to-One   **5.** $A = P\left(1 + \dfrac{r}{n}\right)^{nt}$

7. 0.863   **9.** 0.006   **11.** 1767.767

13. d   **14.** c   **15.** a   **16.** b

17.

x	−2	−1	0	1	2
f(x)	4	2	1	0.5	0.25

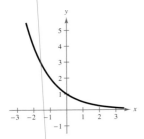

19.

x	−2	−1	0	1	2
f(x)	36	6	1	0.167	0.028

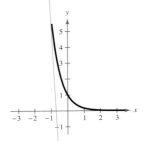

21.

x	−2	−1	0	1	2
f(x)	0.125	0.25	0.5	1	2

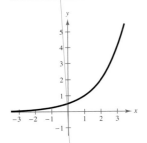

23. Shift the graph of f one unit upward.
25. Reflect the graph of f in the x-axis and shift three units upward.
27. Reflect the graph of f in the origin.

29. 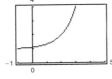   **31.**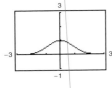

33. 0.472   **35.** 3.857×10^{-22} **37.** 7166.647

39.

x	−2	−1	0	1	2
f(x)	0.135	0.368	1	2.718	7.389

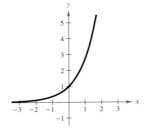

41.

x	−8	−7	−6	−5	−4
f(x)	0.055	0.149	0.406	1.104	3

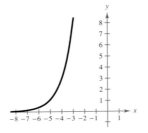

43.

x	−2	−1	0	1	2
f(x)	4.037	4.100	4.271	4.736	6

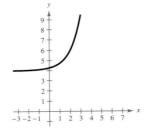

45. 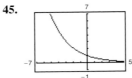   **47.**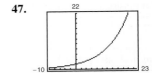

49.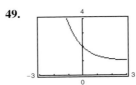

51. $x = 2$ **53.** $x = -5$ **55.** $x = \frac{1}{3}$ **57.** $x = 3, -1$

59.

n	1	2	4	12
A	\$1828.49	\$1830.29	\$1831.19	\$1831.80

n	365	Continuous
A	\$1832.09	\$1832.10

61.

n	1	2	4	12
A	\$5477.81	\$5520.10	\$5541.79	\$5556.46

n	365	Continuous
A	\$5563.61	\$5563.85

63.

t	10	20	30
A	\$17,901.90	\$26,706.49	\$39,841.40

t	40	50
A	\$59,436.39	\$88,668.67

65.

t	10	20	30
A	\$22,986.49	\$44,031.56	\$84,344.25

t	40	50
A	\$161,564.86	\$309,484.08

67. \$104,710.29   **69.** \$35.45

71. (a)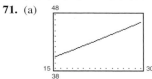

(b)

t	15	16	17	18	19	20
P (in millions)	40.19	40.59	40.99	41.39	41.80	42.21

t	21	22	23	24	25	26
P (in millions)	42.62	43.04	43.47	43.90	44.33	44.77

t	27	28	29	30
P (in millions)	45.21	45.65	46.10	46.56

(c) 2038

73. (a) 16 g   (b) 1.85 g

(c)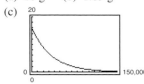

75. (a) $V(t) = 30,500\left(\frac{7}{8}\right)^t$ (b) \$17,878.54

77. True. As $x \to -\infty$, $f(x) \to -2$ but never reaches -2.

79. $f(x) = h(x)$ **81.** $f(x) = g(x) = h(x)$

83. 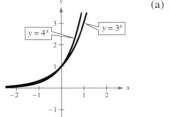   (a) $x < 0$ (b) $x > 0$

85.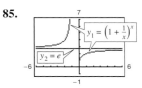

As the x-value increases, y_1 approaches the value of e.

87. (a)   (b)

In both viewing windows, the constant raised to a variable power increases more rapidly than the variable raised to a constant power.

89. (a) $A = \$5466.09$ (b) $A = \$5466.35$

(c) $A = \$5466.36$ (d) $A = \$5466.38$

No. Answers will vary.

Section 3.2   (page 234)

1. logarithmic   **3.** natural; e **5.** $x = y$ **7.** $4^2 = 16$

9. $9^{-2} = \frac{1}{81}$ **11.** $32^{2/5} = 4$ **13.** $64^{1/2} = 8$

15. $\log_5 125 = 3$ **17.** $\log_{81} 3 = \frac{1}{4}$ **19.** $\log_6 \frac{1}{36} = -2$

21. $\log_{24} 1 = 0$ **23.** 6   **25.** 0   **27.** 2

29. -0.058 **31.** 1.097   **33.** 7   **35.** 1

37. 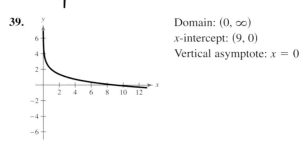   Domain: $(0, \infty)$

x-intercept: $(1, 0)$

Vertical asymptote: $x = 0$

39.   Domain: $(0, \infty)$

x-intercept: $(9, 0)$

Vertical asymptote: $x = 0$

41.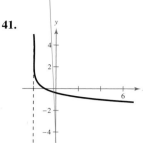
Domain: $(-2, \infty)$
x-intercept: $(-1, 0)$
Vertical asymptote: $x = -2$

43.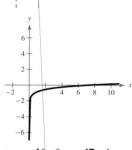
Domain: $(0, \infty)$
x-intercept: $(7, 0)$
Vertical asymptote: $x = 0$

45. c   **46.** f   **47.** d   **48.** e   **49.** b   **50.** a
51. $e^{-0.693...} = \frac{1}{2}$ **53.** $e^{1.945...} = 7$ **55.** $e^{5.521...} = 250$
57. $e^0 = 1$ **59.** $\ln 54.598... = 4$
61. $\ln 1.6487... = \frac{1}{2}$ **63.** $\ln 0.406... = -0.9$
65. $\ln 4 = x$ **67.** 2.913   **69.** -23.966
71. 5   **73.** $-\frac{5}{6}$
75.
Domain: $(4, \infty)$
x-intercept: $(5, 0)$
Vertical asymptote: $x = 4$

77.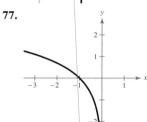
Domain: $(-\infty, 0)$
x-intercept: $(-1, 0)$
Vertical asymptote: $x = 0$

79. 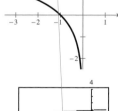   **81.**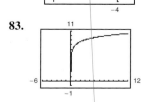

83.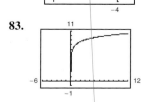

85. $x = 5$ **87.** $x = 7$ **89.** $x = 8$ **91.** $x = -5, 5$
93. (a) 30 yr; 10 yr   (b) \$323,179; \$199,109

(c) \$173,179; \$49,109
(d) $x = 750$; The monthly payment must be greater than \$750.
95. (a)

t	1	2	3	4	5	6
C	10.36	9.94	9.37	8.70	7.96	7.15

(b)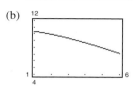

(c) No, the model begins to decrease rapidly, eventually producing negative values.
97. (a)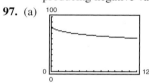

(b) 80   (c) 68.1   (d) 62.3
99. False. Reflecting $g(x)$ about the line $y = x$ will determine the graph of $f(x)$.
101. 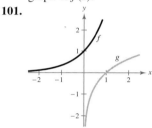   **103.**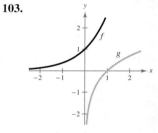

The functions f and g are inverses.

The functions f and g are inverses.

105.

x	-2	-1	0	1	2
$f(x) = 10^x$	$\frac{1}{100}$	$\frac{1}{10}$	1	10	100

x	$\frac{1}{100}$	$\frac{1}{10}$	1	10	100
$f(x) = \log x$	-2	-1	0	1	2

The domain of $f(x) = 10^x$ is equal to the range of $f(x) = \log x$ and vice versa. $f(x) = 10^x$ and $f(x) = \log x$ are inverses of each other.
107. (a)

x	1	5	10	10^2
$f(x)$	0	0.322	0.230	0.046

x	10^4	10^6
$f(x)$	0.00092	0.0000138

(b) 0

(c)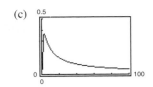

109. Answers will vary.

111. (a) 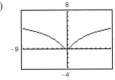   (b) Increasing: $(0, \infty)$
  Decreasing: $(-\infty, 0)$
  (c) Relative minimum: $(0, 0)$

Section 3.3   (page 241)

1. change-of-base   **3.** $\dfrac{1}{\log_b a}$ **4.** c   **5.** a   **6.** b

7. (a) $\dfrac{\log 16}{\log 5}$ (b) $\dfrac{\ln 16}{\ln 5}$ **9.** (a) $\dfrac{\log x}{\log \frac{1}{5}}$ (b) $\dfrac{\ln x}{\ln \frac{1}{5}}$

11. (a) $\dfrac{\log \frac{3}{10}}{\log x}$ (b) $\dfrac{\ln \frac{3}{10}}{\ln x}$ **13.** (a) $\dfrac{\log x}{\log 2.6}$ (b) $\dfrac{\ln x}{\ln 2.6}$

15. 1.771   **17.** -2.000 **19.** -1.048 **21.** 2.633

23. $\frac{3}{2}$ **25.** $-3 - \log_5 2$ **27.** $6 + \ln 5$ **29.** 2

31. $\frac{3}{4}$ **33.** 4   **35.** -2 is not in the domain of $\log_2 x$.

37. 4.5   **39.** $-\frac{1}{2}$ **41.** 7   **43.** 2   **45.** $\ln 4 + \ln x$

47. $4 \log_8 x$ **49.** $1 - \log_5 x$ **51.** $\frac{1}{2} \ln z$

53. $\ln x + \ln y + 2 \ln z$ **55.** $\ln z + 2 \ln(z - 1)$

57. $\frac{1}{2} \log_2(a - 1) - 2 \log_2 3$ **59.** $\frac{1}{3} \ln x - \frac{1}{3} \ln y$

61. $2 \ln x + \frac{1}{2} \ln y - \frac{1}{2} \ln z$

63. $2 \log_5 x - 2 \log_5 y - 3 \log_5 z$

65. $\dfrac{3}{4} \ln x + \dfrac{1}{4} \ln(x^2 + 3)$ **67.** $\ln 2x$ **69.** $\log_4 \dfrac{z}{y}$

71. $\log_2 x^2 y^4$ **73.** $\log_3 \sqrt[4]{5x}$ **75.** $\log \dfrac{x}{(x + 1)^2}$

77. $\log \dfrac{xz^3}{y^2}$ **79.** $\ln \dfrac{x}{(x + 1)(x - 1)}$

81. $\ln \sqrt[3]{\dfrac{x(x + 3)^2}{x^2 - 1}}$ **83.** $\log_8 \dfrac{\sqrt[3]{y(y + 4)^2}}{y - 1}$

85. $\log_2 \frac{32}{4} = \log_2 32 - \log_2 4$; Property 2

87. $\beta = 10(\log I + 12)$; 60 dB   **89.** 70 dB

91. $\ln y = \frac{1}{4} \ln x$ **93.** $\ln y = -\frac{1}{4} \ln x + \ln \frac{5}{2}$

95. $y = 256.24 - 20.8 \ln x$

97. (a) and (b)   (c)

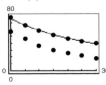   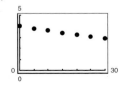

$T = 21 + e^{-0.037t + 3.997}$
The results are similar.

(d)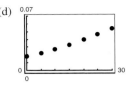

$T = 21 + \dfrac{1}{0.001t + 0.016}$

(e) Answers will vary.

99. Proof

101. False; $\ln 1 = 0$ **103.** False; $\ln(x - 2) \neq \ln x - \ln 2$

105. False; $u = v^2$

107. $f(x) = \dfrac{\log x}{\log 2} = \dfrac{\ln x}{\ln 2}$ **109.** $f(x) = \dfrac{\log x}{\log \frac{1}{2}} = \dfrac{\ln x}{\ln \frac{1}{2}}$

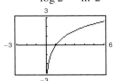   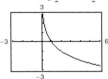

111. $f(x) = \dfrac{\log x}{\log 11.8} = \dfrac{\ln x}{\ln 11.8}$

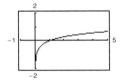

113. $f(x) = h(x)$; Property 2

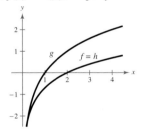

115. $\ln 1 = 0$ $\ln 9 \approx 2.1972$
  $\ln 2 \approx 0.6931$ $\ln 10 \approx 2.3025$
  $\ln 3 \approx 1.0986$ $\ln 12 \approx 2.4848$
  $\ln 4 \approx 1.3862$ $\ln 15 \approx 2.7080$
  $\ln 5 \approx 1.6094$ $\ln 16 \approx 2.7724$
  $\ln 6 \approx 1.7917$ $\ln 18 \approx 2.8903$
  $\ln 8 \approx 2.0793$ $\ln 20 \approx 2.9956$

Section 3.4   (page 251)

1. solve

3. (a) One-to-One   (b) logarithmic; logarithmic
  (c) exponential; exponential

5. (a) Yes   (b) No

7. (a) No   (b) Yes   (c) Yes, approximate

9. (a) Yes, approximate   (b) No   (c) Yes

11. (a) No   (b) Yes   (c) Yes, approximate

13. 2   **15.** -5 **17.** 2   **19.** $\ln 2 \approx 0.693$

21. $e^{-1} \approx 0.368$ **23.** 64 **25.** (3, 8) **27.** (9, 2)

29. 2, −1 **31.** About 1.618, about −0.618

33. $\dfrac{\ln 5}{\ln 3} \approx 1.465$ **35.** $\ln 5 \approx 1.609$ **37.** $\ln 28 \approx 3.332$

39. $\dfrac{\ln 80}{2 \ln 3} \approx 1.994$ **41.** 2 **43.** 4

45. $3 - \dfrac{\ln 565}{\ln 2} \approx -6.142$ **47.** $\dfrac{1}{3} \log\left(\dfrac{3}{2}\right) \approx 0.059$

49. $1 + \dfrac{\ln 7}{\ln 5} \approx 2.209$ **51.** $\dfrac{\ln 12}{3} \approx 0.828$

53. $-\ln \dfrac{3}{5} \approx 0.511$ **55.** 0 **57.** $\dfrac{\ln \frac{8}{3}}{3 \ln 2} + \dfrac{1}{3} \approx 0.805$

59. $\ln 5 \approx 1.609$ **61.** $\ln 4 \approx 1.386$

63. $2 \ln 75 \approx 8.635$ **65.** $\frac{1}{2} \ln 1498 \approx 3.656$

67. $\dfrac{\ln 4}{365 \ln\left(1 + \frac{0.065}{365}\right)} \approx 21.330$ **69.** $\dfrac{\ln 2}{12 \ln\left(1 + \frac{0.10}{12}\right)} \approx 6.960$

71.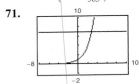

2.807

73.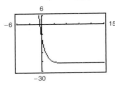

−0.427

75.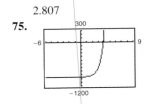

3.847

77.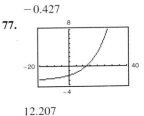

12.207

79.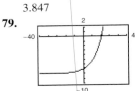

16.636

81. $e^{-3} \approx 0.050$ **83.** $e^7 \approx 1096.633$ **85.** $\dfrac{e^{2.4}}{2} \approx 5.512$

87. 1,000,000 **89.** $\dfrac{e^{10/3}}{5} \approx 5.606$

91. $e^2 - 2 \approx 5.389$ **93.** $e^{-2/3} \approx 0.513$

95. $\dfrac{e^{19/2}}{3} \approx 4453.242$ **97.** $2(3^{11/6}) \approx 14.988$

99. No solution **101.** $1 + \sqrt{1 + e} \approx 2.928$

103. No solution **105.** 7 **107.** $\dfrac{-1 + \sqrt{17}}{2} \approx 1.562$

109. 2 **111.** $\dfrac{725 + 125\sqrt{33}}{8} \approx 180.384$

113.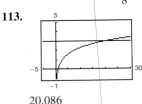

20.086

115.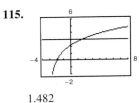

1.482

117. (a) 13.86 yr (b) 21.97 yr

119. (a) 27.73 yr (b) 43.94 yr

121. −1, 0 **123.** 1 **125.** $e^{-1/2} \approx 0.607$

127. $e^{-1} \approx 0.368$ **129.** (a) 210 coins (b) 588 coins

131. (a)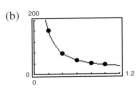

(b) $V = 6.7$; The yield will approach 6.7 million cubic feet per acre.

(c) 29.3 yr

133. 2003

135. (a) $y = 100$ and $y = 0$; The range falls between 0% and 100%.

(b) Males: 69.71 in.   Females: 64.51 in.

137. (a)

x	0.2	0.4	0.6	0.8	1.0
y	162.6	78.5	52.5	40.5	33.9

(b)

The model appears to fit the data well.

(c) 1.2 m

(d) No. According to the model, when the number of g's is less than 23, x is between 2.276 meters and 4.404 meters, which isn't realistic in most vehicles.

139. $\log_b uv = \log_b u + \log_b v$
True by Property 1 in Section 3.3.

141. $\log_b(u - v) = \log_b u - \log_b v$
False
$1.95 \approx \log(100 - 10) \neq \log 100 - \log 10 = 1$

143. Yes. See Exercise 103.

145. Yes. Time to double: $t = \dfrac{\ln 2}{r}$;

Time to quadruple: $t = \dfrac{\ln 4}{r} = 2\left(\dfrac{\ln 2}{r}\right)$

147. (a)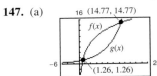

(b) $a = e^{1/e}$
(c) $1 < a < e^{1/e}$

Section 3.5 *(page 262)*

1. $y = ae^{bx}$; $y = ae^{-bx}$ **3.** normally distributed

5. $y = \dfrac{a}{1 + be^{-rx}}$ **7.** c **8.** e **9.** b

10. a **11.** d **12.** f

13. (a) $P = \dfrac{A}{e^{rt}}$ (b) $t = \dfrac{\ln\left(\dfrac{A}{P}\right)}{r}$

Initial Investment	Annual % Rate	Time to Double	Amount After 10 years
15. $1000	3.5%	19.8 yr	$1419.07
17. $750	8.9438%	7.75 yr	$1834.37
19. $500	11.0%	6.3 yr	$1505.00
21. $6376.28	4.5%	15.4 yr	$10,000.00

23. $303,580.52

25. (a) 7.27 yr (b) 6.96 yr (c) 6.93 yr (d) 6.93 yr

27.

r	2%	4%	6%	8%	10%	12%
t	54.93	27.47	18.31	13.73	10.99	9.16

29.

r	2%	4%	6%	8%	10%	12%
t	55.48	28.01	18.85	14.27	11.53	9.69

31.

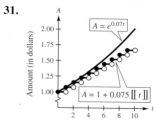

Continuous compounding

	Half-life (years)	Initial Quantity	Amount After 1000 Years
33.	1599	10 g	6.48 g
35.	24,100	2.1 g	2.04 g
37.	5715	2.26 g	2 g

39. $y = e^{0.7675x}$ **41.** $y = 5e^{-0.4024x}$

43. (a)

Year	1970	1980	1990	2000	2007
Population	73.7	103.74	143.56	196.35	243.24

(b) 2014

(c) No; The population will not continue to grow at such a quick rate.

45. $k = 0.2988$; About 5,309,734 hits

47. (a) $k = 0.02603$; The population is increasing because $k > 0$.

    (b) 449,910; 512,447   (c) 2014

49. About 800 bacteria

51. (a) About 12,180 yr old   (b) About 4797 yr old

53. (a) $V = -5400t + 23,300$ (b) $V = 23,300e^{-0.311t}$

    (c)

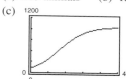

The exponential model depreciates faster.

(d)

t		1 yr	3 yr
$V = -5400t + 23,300$		17,900	7100
$V = 23,300e^{-0.311t}$		17,072	9166

(e) Answers will vary.

55. (a) $S(t) = 100(1 - e^{-0.1625t})$

    (b)

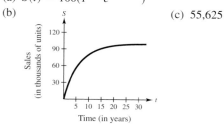

    (c) 55,625

57. (a)

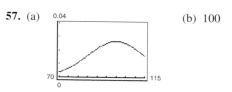

    (b) 100

59. (a) 715; 90,880; 199,043

    (b)

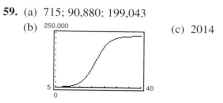

    (c) 2014

    (d) $235,000 = \dfrac{237,101}{1 + 1950e^{-0.355t}}$

    $t \approx 34.63$

61. (a) 203 animals   (b) 13 mo

    (c)

Horizontal asymptotes: $p = 0, p = 1000$. The population size will approach 1000 as time increases.

63. (a) $10^{8.5} \approx 316,227,766$ (b) $10^{5.4} \approx 251,189$

    (c) $10^{6.1} \approx 1,258,925$

65. (a) 20 dB   (b) 70 dB   (c) 40 dB   (d) 120 dB

67. 95%   **69.** 4.64   **71.** 1.58×10^{-6} moles/L

73. $10^{5.1}$ **75.** 3:00 A.M.

77. (a)

    (b) $t \approx 21$ yr; Yes

79. False. The domain can be the set of real numbers for a logistic growth function.

81. False. The graph of $f(x)$ is the graph of $g(x)$ shifted upward five units.

83. Answers will vary.

Review Exercises   *(page 270)*

1. 0.164   **3.** 0.337   **5.** 1456.529

7. Shift the graph of *f* two units downward.

9. Reflect *f* in the *y*-axis and shift two units to the right.

11. Reflect *f* in the *x*-axis and shift one unit upward.

13. Reflect *f* in the *x*-axis and shift two units to the left.

15.

x	−1	0	1	2	3
f(x)	8	5	4.25	4.063	4.016

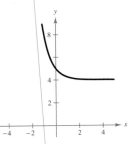

17.

x	−1	0	1	2	3
f(x)	4.008	4.04	4.2	5	9

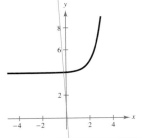

19.

x	−2	−1	0	1	2
f(x)	3.25	3.5	4	5	7

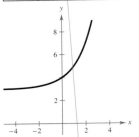

21. $x = 1$ **23.** $x = 4$ **25.** 2980.958   **27.** 0.183

29.

x	−2	−1	0	1	2
h(x)	2.72	1.65	1	0.61	0.37

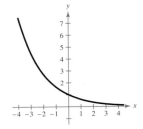

31.

x	−3	−2	−1	0	1
f(x)	0.37	1	2.72	7.39	20.09

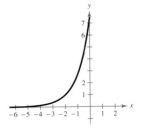

33.

n	1	2	4	12
A	$6719.58	$6734.28	$6741.74	$6746.77

n	365	Continuous
A	$6749.21	$6749.29

35. (a) 0.154   (b) 0.487   (c) 0.811

37. $\log_3 27 = 3$ **39.** $\ln 2.2255 \ldots = 0.8$

41. 3   **43.** −2   **45.** $x = 7$ **47.** $x = -5$

49. Domain: $(0, \infty)$ **51.** Domain: $(-5, \infty)$

　　　 x-intercept: $(1, 0)$ 　　*x*-intercept: $(9995, 0)$

　　　 Vertical asymptote: $x = 0$ 　Vertical asymptote: $x = -5$

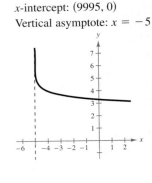

53. (a) 3.118   (b) −0.020

55. Domain: $(0, \infty)$
x-intercept: $(e^{-3}, 0)$
Vertical asymptote: $x = 0$

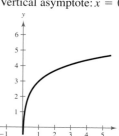

57. Domain: $(-\infty, 0), (0, \infty)$
x-intercept: $(\pm 1, 0)$
Vertical asymptote: $x = 0$

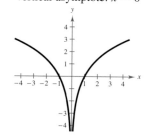

59. 53.4 in.   **61.** 2.585   **63.** -2.322

65. $\log 2 + 2 \log 3 \approx 1.255$ **67.** $2 \ln 2 + \ln 5 \approx 2.996$

69. $1 + 2 \log_5 x$ **71.** $2 - \frac{1}{2} \log_3 x$

73. $2 \ln x + 2 \ln y + \ln z$ **75.** $\log_2 5x$ **77.** $\ln \dfrac{x}{\sqrt[4]{y}}$

79. $\log_3 \dfrac{\sqrt{x}}{(y + 8)^2}$

81. (a) $0 \le h < 18{,}000$

(b)
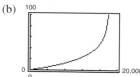

Vertical asymptote: $h = 18{,}000$

(c) The plane is climbing at a slower rate, so the time required increases.

(d) 5.46 min

83. 3   **85.** $\ln 3 \approx 1.099$ **87.** $e^4 \approx 54.598$

89. $x = 1, 3$ **91.** $\dfrac{\ln 32}{\ln 2} = 5$

93.
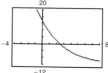
2.447

95. $\frac{1}{3}e^{8.2} \approx 1213.650$ **97.** $3e^2 \approx 22.167$

99. $e^8 \approx 2980.958$ **101.** No solution   **103.** 0.900

105.
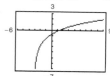
1.482

107.
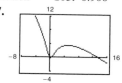
0, 0.416, 13.627

109. 31.4 yr   **111.** e   **112.** b   **113.** f   **114.** d

115. a   **116.** c   **117.** $y = 2e^{0.1014x}$

119. (a)
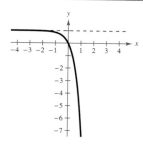

The model fits the data well.

(b) 2022; Answers will vary.

121. (a)
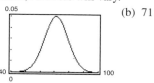
(b) 71

123. (a) 10^{-6} W/m²   (b) $10\sqrt{10}$ W/m²
(c) 1.259×10^{-12} W/m²

125. True by the inverse properties

Chapter Test   *(page 273)*

1. 2.366   **2.** 687.291   **3.** 0.497   **4.** 22.198

5.

x	-1	$-\frac{1}{2}$	0	$\frac{1}{2}$	1
$f(x)$	10	3.162	1	0.316	0.1

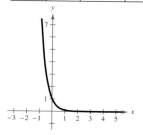

6.

x	-1	0	1	2	3
$f(x)$	-0.005	-0.028	-0.167	-1	-6

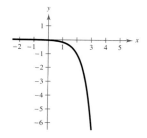

7.

x	-1	$-\frac{1}{2}$	0	$\frac{1}{2}$	1
$f(x)$	0.865	0.632	0	-1.718	-6.389

8. (a) -0.89 (b) 9.2

9.

x	$\frac{1}{2}$	1	$\frac{3}{2}$	2	4
$f(x)$	-5.699	-6	-6.176	-6.301	-6.602

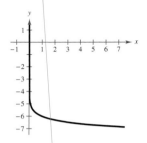

Vertical asymptote: $x = 0$

10.

x	5	7	9	11	13
$f(x)$	0	1.099	1.609	1.946	2.197

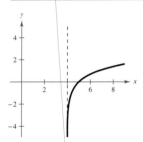

Vertical asymptote: $x = 4$

11.

x	-5	-3	-1	0	1
$f(x)$	1	2.099	2.609	2.792	2.946

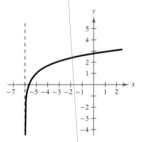

Vertical asymptote: $x = -6$

12. 1.945 **13.** -0.167 **14.** -11.047

15. $\log_2 3 + 4\log_2 |a|$ **16.** $\ln 5 + \frac{1}{2}\ln x - \ln 6$

17. $3\log(x-1) - 2\log y - \log z$ **18.** $\log_3 13y$

19. $\ln \dfrac{x^4}{y^4}$ **20.** $\ln\left(\dfrac{x^3 y^2}{x+3}\right)$ **21.** $x = -2$

22. $x = \dfrac{\ln 44}{-5} \approx -0.757$ **23.** $\dfrac{\ln 197}{4} \approx 1.321$

24. $e^{1/2} \approx 1.649$ **25.** $e^{-11/4} \approx 0.0639$ **26.** 20

27. $y = 2745e^{0.1570t}$ **28.** 55%

29. (a)

x	$\frac{1}{4}$	1	2	4	5	6
H	58.720	75.332	86.828	103.43	110.59	117.38

(b) 103 cm; 103.43 cm

Cumulative Test for Chapters 1–3 (*page 274*)

1.

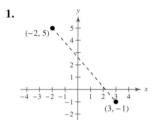

Midpoint: $\left(\frac{1}{2}, 2\right)$; Distance: $\sqrt{61}$

2.

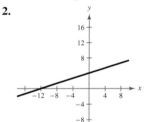

3.

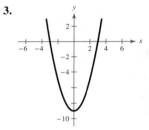

4.

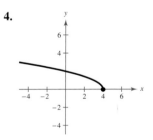

5. $y = 2x + 2$

6. For some values of x there correspond two values of y.

7. (a) $\dfrac{3}{2}$ (b) Division by 0 is undefined. (c) $\dfrac{s+2}{s}$

8. (a) Vertical shrink by $\frac{1}{2}$

(b) Vertical shift two units upward

(c) Horizontal shift two units to the left

9. (a) $5x - 2$ (b) $-3x - 4$ (c) $4x^2 - 11x - 3$

(d) $\dfrac{x-3}{4x+1}$; Domain: all real numbers x except $x = -\dfrac{1}{4}$

10. (a) $\sqrt{x-1} + x^2 + 1$ (b) $\sqrt{x-1} - x^2 - 1$
(c) $x^2\sqrt{x-1} + \sqrt{x-1}$
(d) $\dfrac{\sqrt{x-1}}{x^2+1}$; Domain: all real numbers x such that $x \geq 1$

11. (a) $2x + 12$ (b) $\sqrt{2x^2 + 6}$
Domain of $f \circ g$: all real numbers x such that $x \geq -6$
Domain of $g \circ f$: all real numbers x

12. (a) $|x| - 2$ (b) $|x - 2|$
Domain of $f \circ g$ and $g \circ f$: all real numbers x

13. Yes; $h^{-1}(x) = -\frac{1}{5}(x - 3)$ **14.** 2438.65 kW

15. $y = -\frac{3}{4}(x + 8)^2 + 5$

16.

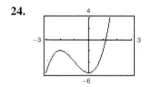

17.

18.

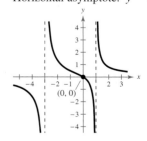

19. $-2, \pm 2i$; $(x + 2)(x + 2i)(x - 2i)$

20. $-7, 0, 3$; $x(x)(x - 3)(x + 7)$

21. $4, -\frac{1}{2}, 1 \pm 3i$; $(x - 4)(2x + 1)(x - 1 + 3i)(x - 1 - 3i)$

22. $3x - 2 - \dfrac{3x - 2}{2x^2 + 1}$ **23.** $3x^3 + 6x^2 + 14x + 23 + \dfrac{49}{x - 2}$

24.

Interval: $[1, 2]$; 1.20

25. Intercept: $(0, 0)$
Vertical asymptotes: $x = -3, x = 1$
Horizontal asymptote: $y = 0$

26. y-intercept: $(0, 2)$
x-intercept: $(2, 0)$
Vertical asymptote: $x = 1$
Horizontal asymptote: $y = 1$

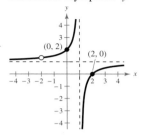

27. y-intercept: $(0, 6)$
x-intercepts: $(2, 0), (3, 0)$
Vertical asymptote: $x = -1$
Slant asymptote: $y = x - 6$

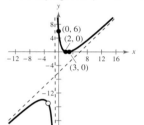

28. $x \leq -3$ or $0 \leq x \leq 3$

29. All real numbers x such that $x < -5$ or $x > -1$

30. Reflect f in the x-axis and y-axis, and shift three units to the right.

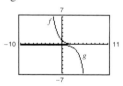

31. Reflect f in the x-axis, and shift four units upward.

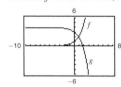

32. 1.991 **33.** -0.067 **34.** 1.717 **35.** 0.281

36. $\ln(x + 4) + \ln(x - 4) - 4 \ln x$, $x > 4$

37. $\ln \dfrac{x^2}{\sqrt{x + 5}}$, $x > 0$ **38.** $x = \dfrac{\ln 12}{2} \approx 1.242$

39. $\ln 6 \approx 1.792$ or $\ln 7 \approx 1.946$ **40.** $e^6 - 2 \approx 401.429$

41. (a)

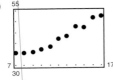

(b) $S = -0.0297t^3 + 1.175t^2 - 12.96t + 79.0$

(c)

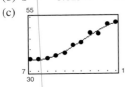

The model is a good fit for the data.

(d) $25.3 billion; Answers will vary. Sample answer: No, this is not reasonable because the model decreases sharply after 2009.

42. 6.3 h

Problem Solving (page 277)

1.

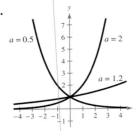

$y = 0.5^x$ and $y = 1.2^x$

$0 < a \le e^{1/e}$

3. As $x \to \infty$, the graph of e^x increases at a greater rate than the graph of x^n.

5. Answers will vary.

7. (a)

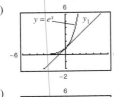

(b)

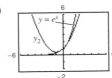

(c)

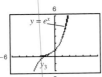

9.

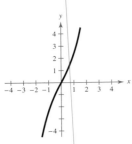

$f^{-1}(x) = \ln\left(\dfrac{x + \sqrt{x^2 + 4}}{2}\right)$

11. c   **13.** $t = \dfrac{\ln c_1 - \ln c_2}{\left(\dfrac{1}{k_2} - \dfrac{1}{k_1}\right)\ln \dfrac{1}{2}}$

15. (a) $y_1 = 252{,}606(1.0310)^t$

(b) $y_2 = 400.88t^2 - 1464.6t + 291{,}782$

(c)

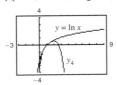

(d) The exponential model is a better fit. No, because the model is rapidly approaching infinity.

17. $1, e^2$

19. $y_4 = (x - 1) - \frac{1}{2}(x - 1)^2 + \frac{1}{3}(x - 1)^3 - \frac{1}{4}(x - 1)^4$

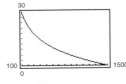

The pattern implies that

$\ln x = (x - 1) - \frac{1}{2}(x - 1)^2 + \frac{1}{3}(x - 1)^3 - \cdots$.

21.

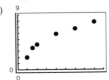

17.7 ft³/min

23. (a)

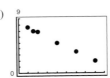

(b)–(e) Answers will vary.

25. (a)

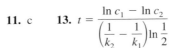

(b)–(e) Answers will vary.

Chapter 4

Section 4.1 (page 288)

1. Trigonometry   **3.** coterminal   **5.** acute; obtuse

7. degree   **9.** linear; angular   **11.** 1 rad   **13.** 5.5 rad

15. -3 rad

17. (a) Quadrant I   (b) Quadrant III

19. (a) Quadrant IV   (b) Quadrant IV

21. (a) Quadrant III   (b) Quadrant II

23. (a)   (b)

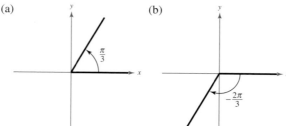

25. (a)   (b)

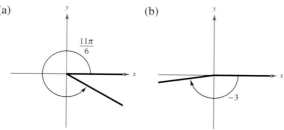

27. Sample answers: (a) $\dfrac{13\pi}{6}, -\dfrac{11\pi}{6}$ (b) $\dfrac{17\pi}{6}, -\dfrac{7\pi}{6}$

29. Sample answers: (a) $\dfrac{8\pi}{3}, -\dfrac{4\pi}{3}$ (b) $\dfrac{25\pi}{12}, -\dfrac{23\pi}{12}$

31. (a) Complement: $\dfrac{\pi}{6}$; Supplement: $\dfrac{2\pi}{3}$

　　(b) Complement: $\dfrac{\pi}{4}$; Supplement: $\dfrac{3\pi}{4}$

33. (a) Complement: $\dfrac{\pi}{2} - 1 \approx 0.57$;

　　Supplement: $\pi - 1 \approx 2.14$

　　(b) Complement: none; Supplement: $\pi - 2 \approx 1.14$

35. $210°$ **37.** $-60°$ **39.** $165°$

41. (a) Quadrant II   (b) Quadrant IV

43. (a) Quadrant III   (b) Quadrant I

45. (a)   (b)

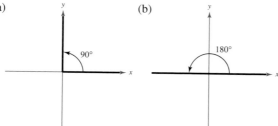

47. (a)   (b)

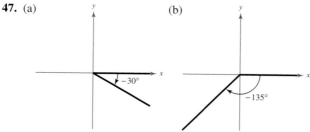

49. Sample answers: (a) $405°, -315°$ (b) $324°, -396°$

51. Sample answers: (a) $600°, -120°$ (b) $180°, -540°$

53. (a) Complement: $72°$; Supplement: $162°$

　　(b) Complement: $5°$; Supplement: $95°$

55. (a) Complement: none; Supplement: $30°$

　　(b) Complement: $11°$; Supplement: $101°$

57. (a) $\dfrac{\pi}{6}$ (b) $\dfrac{\pi}{4}$ **59.** (a) $-\dfrac{\pi}{9}$ (b) $-\dfrac{\pi}{3}$

61. (a) $270°$ (b) $210°$ **63.** (a) $225°$ (b) $-420°$

65. 0.785 **67.** -3.776 **69.** 9.285 **71.** -0.014

73. $25.714°$ **75.** $337.500°$ **77.** $-756.000°$

79. $-114.592°$ **81.** (a) $54.75°$ (b) $-128.5°$

83. (a) $85.308°$ (b) $330.007°$

85. (a) $240°36'$ (b) $-145°48'$

87. (a) $2°30'$ (b) $-3°34'48''$ **89.** 10π in. ≈ 31.42 in.

91. 2.5π m ≈ 7.85 m   **93.** $\dfrac{9}{2}$ rad   **95.** $\dfrac{21}{50}$ rad   **97.** $\dfrac{1}{2}$ rad

99. 4 rad   **101.** 6π in.$^2 \approx 18.85$ in.^{2} **103.** 12.27 ft^2

105. 591.3 mi   **107.** 0.071 rad $\approx 4.04°$ **109.** $\dfrac{5}{12}$ rad

111. (a) $10{,}000\pi$ rad/min $\approx 31{,}415.93$ rad/min

　　(b) 9490.23 ft/min

113. (a) $[400\pi, 1000\pi]$ rad/min   (b) $[2400\pi, 6000\pi]$ cm/min

115. (a) 910.37 revolutions/min   (b) 5720 rad/min

117.

$A = 87.5\pi$ m$^2 \approx 274.89$ m^2

119. (a) $\dfrac{14\pi}{3}$ ft/sec ≈ 10 mi/h   (b) $d = \dfrac{7\pi}{7920}n$

　　(c) $d = \dfrac{7\pi}{7920}t$ (d) The functions are both linear.

121. False. A measurement of 4π radians corresponds to two complete revolutions from the initial side to the terminal side of an angle.

123. False. The terminal side of the angle lies on the x-axis.

125. Radian. 1 rad $\approx 57.3°$

127. Proof

Section 4.2 *(page 297)*

1. unit circle   **3.** period

5. $\sin t = \dfrac{5}{13}$ $\csc t = \dfrac{13}{5}$

　　$\cos t = \dfrac{12}{13}$ $\sec t = \dfrac{13}{12}$

　　$\tan t = \dfrac{5}{12}$ $\cot t = \dfrac{12}{5}$

7. $\sin t = -\dfrac{3}{5}$ $\csc t = -\dfrac{5}{3}$

　　$\cos t = -\dfrac{4}{5}$ $\sec t = -\dfrac{5}{4}$

　　$\tan t = \dfrac{3}{4}$ $\cot t = \dfrac{4}{3}$

9. $(0, 1)$ **11.** $\left(\dfrac{\sqrt{2}}{2}, \dfrac{\sqrt{2}}{2}\right)$ **13.** $\left(-\dfrac{\sqrt{3}}{2}, \dfrac{1}{2}\right)$

15. $\left(-\dfrac{1}{2}, -\dfrac{\sqrt{3}}{2}\right)$

17. $\sin \dfrac{\pi}{4} = \dfrac{\sqrt{2}}{2}$ **19.** $\sin\left(-\dfrac{\pi}{6}\right) = -\dfrac{1}{2}$

　　$\cos \dfrac{\pi}{4} = \dfrac{\sqrt{2}}{2}$ $\cos\left(-\dfrac{\pi}{6}\right) = \dfrac{\sqrt{3}}{2}$

　　$\tan \dfrac{\pi}{4} = 1$ $\tan\left(-\dfrac{\pi}{6}\right) = -\dfrac{\sqrt{3}}{3}$

21. $\sin\left(-\dfrac{7\pi}{4}\right) = \dfrac{\sqrt{2}}{2}$

$\cos\left(-\dfrac{7\pi}{4}\right) = \dfrac{\sqrt{2}}{2}$

$\tan\left(-\dfrac{7\pi}{4}\right) = 1$

23. $\sin\dfrac{11\pi}{6} = -\dfrac{1}{2}$

$\cos\dfrac{11\pi}{6} = \dfrac{\sqrt{3}}{2}$

$\tan\dfrac{11\pi}{6} = -\dfrac{\sqrt{3}}{3}$

25. $\sin\left(-\dfrac{3\pi}{2}\right) = 1$

$\cos\left(-\dfrac{3\pi}{2}\right) = 0$

$\tan\left(-\dfrac{3\pi}{2}\right)$ is undefined.

27. $\sin\dfrac{2\pi}{3} = \dfrac{\sqrt{3}}{2}$ $\csc\dfrac{2\pi}{3} = \dfrac{2\sqrt{3}}{3}$

$\cos\dfrac{2\pi}{3} = -\dfrac{1}{2}$ $\sec\dfrac{2\pi}{3} = -2$

$\tan\dfrac{2\pi}{3} = -\sqrt{3}$ $\cot\dfrac{2\pi}{3} = -\dfrac{\sqrt{3}}{3}$

29. $\sin\dfrac{4\pi}{3} = -\dfrac{\sqrt{3}}{2}$ $\csc\dfrac{4\pi}{3} = -\dfrac{2\sqrt{3}}{3}$

$\cos\dfrac{4\pi}{3} = -\dfrac{1}{2}$ $\sec\dfrac{4\pi}{3} = -2$

$\tan\dfrac{4\pi}{3} = \sqrt{3}$ $\cot\dfrac{4\pi}{3} = \dfrac{\sqrt{3}}{3}$

31. $\sin\dfrac{3\pi}{4} = \dfrac{\sqrt{2}}{2}$ $\csc\dfrac{3\pi}{4} = \sqrt{2}$

$\cos\dfrac{3\pi}{4} = -\dfrac{\sqrt{2}}{2}$ $\sec\dfrac{3\pi}{4} = -\sqrt{2}$

$\tan\dfrac{3\pi}{4} = -1$ $\cot\dfrac{3\pi}{4} = -1$

33. $\sin\left(-\dfrac{\pi}{2}\right) = -1$ $\csc\left(-\dfrac{\pi}{2}\right) = -1$

$\cos\left(-\dfrac{\pi}{2}\right) = 0$ $\sec\left(-\dfrac{\pi}{2}\right)$ is undefined.

$\tan\left(-\dfrac{\pi}{2}\right)$ is undefined. $\cot\left(-\dfrac{\pi}{2}\right) = 0$

35. $\sin 4\pi = \sin 0 = 0$ **37.** $\cos\dfrac{7\pi}{3} = \cos\dfrac{\pi}{3} = \dfrac{1}{2}$

39. $\cos\dfrac{17\pi}{4} = \cos\dfrac{\pi}{4} = \dfrac{\sqrt{2}}{2}$

41. $\sin\left(-\dfrac{8\pi}{3}\right) = \sin\dfrac{4\pi}{3} = -\dfrac{\sqrt{3}}{2}$

43. (a) $-\dfrac{1}{2}$ (b) -2 **45.** (a) $-\dfrac{1}{5}$ (b) -5

47. (a) $\dfrac{4}{5}$ (b) $-\dfrac{4}{5}$ **49.** 0.7071 **51.** 1.0000

53. -0.1288 **55.** 1.3940 **57.** -1.4486

59. (a) 0.25 ft (b) 0.02 ft (c) -0.25 ft

61. False. $\sin(-t) = -\sin(t)$ means that the function is odd, not that the sine of a negative angle is a negative number.

63. False. The real number 0 corresponds to the point $(1, 0)$.

65. (a) y-axis symmetry (b) $\sin t_1 = \sin(\pi - t_1)$
   (c) $\cos(\pi - t_1) = -\cos t_1$

67. Answers will vary. **69.** It is an odd function.

71. (a)

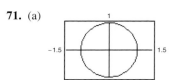

Circle of radius 1 centered at $(0, 0)$

(b) The t-values represent the central angle in radians. The x- and y-values represent the location in the coordinate plane.

(c) $-1 \le x \le 1, -1 \le y \le 1$

Section 4.3 *(page 306)*

1. (a) v (b) iv (c) vi (d) iii (e) i (f) ii

3. complementary

5. $\sin\theta = \dfrac{3}{5}$ $\csc\theta = \dfrac{5}{3}$

$\cos\theta = \dfrac{4}{5}$ $\sec\theta = \dfrac{5}{4}$

$\tan\theta = \dfrac{3}{4}$ $\cot\theta = \dfrac{4}{3}$

7. $\sin\theta = \dfrac{9}{41}$ $\csc\theta = \dfrac{41}{9}$

$\cos\theta = \dfrac{40}{41}$ $\sec\theta = \dfrac{41}{40}$

$\tan\theta = \dfrac{9}{40}$ $\cot\theta = \dfrac{40}{9}$

9. $\sin\theta = \dfrac{8}{17}$ $\csc\theta = \dfrac{17}{8}$

$\cos\theta = \dfrac{15}{17}$ $\sec\theta = \dfrac{17}{15}$

$\tan\theta = \dfrac{8}{15}$ $\cot\theta = \dfrac{15}{8}$

The triangles are similar, and corresponding sides are proportional.

11. $\sin\theta = \dfrac{1}{3}$ $\csc\theta = 3$

$\cos\theta = \dfrac{2\sqrt{2}}{3}$ $\sec\theta = \dfrac{3\sqrt{2}}{4}$

$\tan\theta = \dfrac{\sqrt{2}}{4}$ $\cot\theta = 2\sqrt{2}$

The triangles are similar, and corresponding sides are proportional.

13.

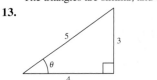

$\sin\theta = \dfrac{3}{5}$ $\csc\theta = \dfrac{5}{3}$

$\cos\theta = \dfrac{4}{5}$ $\sec\theta = \dfrac{5}{4}$

$\cot\theta = \dfrac{4}{3}$

15.

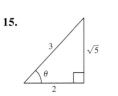

$\sin\theta = \dfrac{\sqrt{5}}{3}$ $\csc\theta = \dfrac{3\sqrt{5}}{5}$

$\cos\theta = \dfrac{2}{3}$

$\tan\theta = \dfrac{\sqrt{5}}{2}$ $\cot\theta = \dfrac{2\sqrt{5}}{5}$

17.

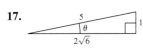

$\csc\theta = 5$

$\cos\theta = \dfrac{2\sqrt{6}}{5}$ $\sec\theta = \dfrac{5\sqrt{6}}{12}$

$\tan\theta = \dfrac{\sqrt{6}}{12}$ $\cot\theta = 2\sqrt{6}$

19.

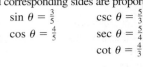

$\sin\theta = \dfrac{\sqrt{10}}{10}$ $\csc\theta = \sqrt{10}$

$\cos\theta = \dfrac{3\sqrt{10}}{10}$ $\sec\theta = \dfrac{\sqrt{10}}{3}$

$\tan\theta = \dfrac{1}{3}$

21. $\dfrac{\pi}{6}; \dfrac{1}{2}$ **23.** $45°; \sqrt{2}$ **25.** $60°; \dfrac{\pi}{3}$ **27.** $30°; 2$

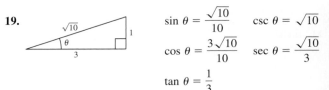

29. $45°; \dfrac{\pi}{4}$ **31.** (a) $\dfrac{1}{2}$ (b) $\dfrac{\sqrt{3}}{2}$ (c) $\sqrt{3}$ (d) $\dfrac{\sqrt{3}}{3}$

33. (a) $\dfrac{2\sqrt{2}}{3}$ (b) $2\sqrt{2}$ (c) 3 (d) 3

35. (a) $\dfrac{1}{5}$ (b) $\sqrt{26}$ (c) $\dfrac{1}{5}$ (d) $\dfrac{5\sqrt{26}}{26}$

37–45. Answers will vary.   **47.** (a) 0.1736   (b) 0.1736

49. (a) 0.2815   (b) 3.5523   **51.** (a) 0.9964   (b) 1.0036

53. (a) 5.0273   (b) 0.1989   **55.** (a) 1.8527   (b) 0.9817

57. (a) $30° = \dfrac{\pi}{6}$ (b) $30° = \dfrac{\pi}{6}$

59. (a) $60° = \dfrac{\pi}{3}$ (b) $45° = \dfrac{\pi}{4}$

61. (a) $60° = \dfrac{\pi}{3}$ (b) $45° = \dfrac{\pi}{4}$

63. $9\sqrt{3}$ **65.** $\dfrac{32\sqrt{3}}{3}$

67. 443.2 m; 323.3 m   **69.** $30° = \pi/6$

71. (a) 219.9 ft   (b) 160.9 ft

73. $(x_1, y_1) = \left(28\sqrt{3},\ 28\right)$
$(x_2, y_2) = \left(28,\ 28\sqrt{3}\right)$

75. $\sin 20° \approx 0.34$, $\cos 20° \approx 0.94$, $\tan 20° \approx 0.36$,
$\csc 20° \approx 2.92$, $\sec 20° \approx 1.06$, $\cot 20° \approx 2.75$

77. True, $\csc x = \dfrac{1}{\sin x}$.   **79.** False, $\dfrac{\sqrt{2}}{2} + \dfrac{\sqrt{2}}{2} \neq 1$.

81. False, $1.7321 \neq 0.0349$.

83. (a)

θ	0.1	0.2	0.3	0.4	0.5
$\sin \theta$	0.0998	0.1987	0.2955	0.3894	0.4794

(b) θ (c) As $\theta \to 0$, $\sin \theta \to 0$ and $\dfrac{\theta}{\sin \theta} \to 1$.

85. Corresponding sides of similar triangles are proportional.

87. Yes, $\tan \theta$ is equal to opp/adj. You can find the value of the hypotenuse by the Pythagorean Theorem, then you can find $\sec \theta$, which is equal to hyp/adj.

Section 4.4   (page 316)

1. $\dfrac{y}{r}$ **3.** $\dfrac{y}{x}$ **5.** $\cos \theta$ **7.** zero; defined

9. (a) $\sin \theta = \dfrac{3}{5}$ ⠀⠀ $\csc \theta = \dfrac{5}{3}$
$\cos \theta = \dfrac{4}{5}$ ⠀⠀ $\sec \theta = \dfrac{5}{4}$
$\tan \theta = \dfrac{3}{4}$ ⠀⠀ $\cot \theta = \dfrac{4}{3}$
(b) $\sin \theta = \dfrac{15}{17}$ ⠀ $\csc \theta = \dfrac{17}{15}$
$\cos \theta = -\dfrac{8}{17}$ ⠀ $\sec \theta = -\dfrac{17}{8}$
$\tan \theta = -\dfrac{15}{8}$ ⠀ $\cot \theta = -\dfrac{8}{15}$

11. (a) $\sin \theta = -\dfrac{1}{2}$ ⠀ $\csc \theta = -2$
$\cos \theta = -\dfrac{\sqrt{3}}{2}$ ⠀ $\sec \theta = -\dfrac{2\sqrt{3}}{3}$
$\tan \theta = \dfrac{\sqrt{3}}{3}$ ⠀ $\cot \theta = \sqrt{3}$

(b) $\sin \theta = -\dfrac{\sqrt{17}}{17}$ ⠀ $\csc \theta = -\sqrt{17}$
$\cos \theta = \dfrac{4\sqrt{17}}{17}$ ⠀ $\sec \theta = \dfrac{\sqrt{17}}{4}$
$\tan \theta = -\dfrac{1}{4}$ ⠀ $\cot \theta = -4$

13. $\sin \theta = \dfrac{12}{13}$ ⠀ $\csc \theta = \dfrac{13}{12}$
$\cos \theta = \dfrac{5}{13}$ ⠀ $\sec \theta = \dfrac{13}{5}$
$\tan \theta = \dfrac{12}{5}$ ⠀ $\cot \theta = \dfrac{5}{12}$

15. $\sin \theta = -\dfrac{2\sqrt{29}}{29}$ ⠀ $\csc \theta = -\dfrac{\sqrt{29}}{2}$
$\cos \theta = -\dfrac{5\sqrt{29}}{29}$ ⠀ $\sec \theta = -\dfrac{\sqrt{29}}{5}$
$\tan \theta = \dfrac{2}{5}$ ⠀ $\cot \theta = \dfrac{5}{2}$

17. $\sin \theta = \dfrac{4}{5}$ ⠀ $\csc \theta = \dfrac{5}{4}$
$\cos \theta = -\dfrac{3}{5}$ ⠀ $\sec \theta = -\dfrac{5}{3}$
$\tan \theta = -\dfrac{4}{3}$ ⠀ $\cot \theta = -\dfrac{3}{4}$

19. Quadrant I   **21.** Quadrant II

23. $\sin \theta = \dfrac{15}{17}$ ⠀ $\csc \theta = \dfrac{17}{15}$
$\cos \theta = -\dfrac{8}{17}$ ⠀ $\sec \theta = -\dfrac{17}{8}$
$\tan \theta = -\dfrac{15}{8}$ ⠀ $\cot \theta = -\dfrac{8}{15}$

25. $\sin \theta = \dfrac{3}{5}$ ⠀ $\csc \theta = \dfrac{5}{3}$
$\cos \theta = -\dfrac{4}{5}$ ⠀ $\sec \theta = -\dfrac{5}{4}$
$\tan \theta = -\dfrac{3}{4}$ ⠀ $\cot \theta = -\dfrac{4}{3}$

27. $\sin \theta = -\dfrac{\sqrt{10}}{10}$ ⠀ $\csc \theta = -\sqrt{10}$
$\cos \theta = \dfrac{3\sqrt{10}}{10}$ ⠀ $\sec \theta = \dfrac{\sqrt{10}}{3}$
$\tan \theta = -\dfrac{1}{3}$ ⠀ $\cot \theta = -3$

29. $\sin \theta = -\dfrac{\sqrt{3}}{2}$ ⠀ $\csc \theta = -\dfrac{2\sqrt{3}}{3}$
$\cos \theta = -\dfrac{1}{2}$ ⠀ $\sec \theta = -2$
$\tan \theta = \sqrt{3}$ ⠀ $\cot \theta = \dfrac{\sqrt{3}}{3}$

31. $\sin \theta = 0$ ⠀ $\csc \theta$ is undefined.
$\cos \theta = -1$ ⠀ $\sec \theta = -1$
$\tan \theta = 0$ ⠀ $\cot \theta$ is undefined.

33. $\sin \theta = \dfrac{\sqrt{2}}{2}$ ⠀ $\csc \theta = \sqrt{2}$
$\cos \theta = -\dfrac{\sqrt{2}}{2}$ ⠀ $\sec \theta = -\sqrt{2}$
$\tan \theta = -1$ ⠀ $\cot \theta = -1$

35. $\sin \theta = -\dfrac{2\sqrt{5}}{5}$ ⠀ $\csc \theta = -\dfrac{\sqrt{5}}{2}$
$\cos \theta = -\dfrac{\sqrt{5}}{5}$ ⠀ $\sec \theta = -\sqrt{5}$
$\tan \theta = 2$ ⠀ $\cot \theta = \dfrac{1}{2}$

37. 0   **39.** Undefined   **41.** 1   **43.** Undefined

45. $\theta' = 20°$

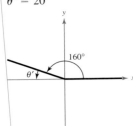

47. $\theta' = 55°$

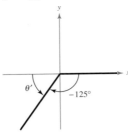

49. $\theta' = \dfrac{\pi}{3}$

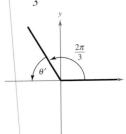

51. $\theta' = 2\pi - 4.8$

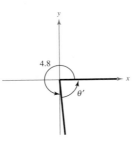

53. $\sin 225° = -\dfrac{\sqrt{2}}{2}$

$\cos 225° = -\dfrac{\sqrt{2}}{2}$

$\tan 225° = 1$

55. $\sin 750° = \dfrac{1}{2}$

$\cos 750° = \dfrac{\sqrt{3}}{2}$

$\tan 750° = \dfrac{\sqrt{3}}{3}$

57. $\sin(-150°) = -\dfrac{1}{2}$

$\cos(-150°) = -\dfrac{\sqrt{3}}{2}$

$\tan(-150°) = \dfrac{\sqrt{3}}{3}$

59. $\sin\dfrac{2\pi}{3} = \dfrac{\sqrt{3}}{2}$

$\cos\dfrac{2\pi}{3} = -\dfrac{1}{2}$

$\tan\dfrac{2\pi}{3} = -\sqrt{3}$

61. $\sin\dfrac{5\pi}{4} = -\dfrac{\sqrt{2}}{2}$

$\cos\dfrac{5\pi}{4} = -\dfrac{\sqrt{2}}{2}$

$\tan\dfrac{5\pi}{4} = 1$

63. $\sin\left(-\dfrac{\pi}{6}\right) = -\dfrac{1}{2}$

$\cos\left(-\dfrac{\pi}{6}\right) = \dfrac{\sqrt{3}}{2}$

$\tan\left(-\dfrac{\pi}{6}\right) = -\dfrac{\sqrt{3}}{3}$

65. $\sin\dfrac{9\pi}{4} = \dfrac{\sqrt{2}}{2}$

$\cos\dfrac{9\pi}{4} = \dfrac{\sqrt{2}}{2}$

$\tan\dfrac{9\pi}{4} = 1$

67. $\sin\left(-\dfrac{3\pi}{2}\right) = 1$

$\cos\left(-\dfrac{3\pi}{2}\right) = 0$

$\tan\left(-\dfrac{3\pi}{2}\right)$ is undefined.

69. $\dfrac{4}{5}$ **71.** $-\dfrac{\sqrt{13}}{2}$ **73.** $\dfrac{8}{5}$

75. 0.1736   **77.** -0.3420 **79.** -1.4826 **81.** 3.2361

83. 4.6373   **85.** 0.3640   **87.** -0.6052 **89.** -0.4142

91. (a) $30° = \dfrac{\pi}{6}$, $150° = \dfrac{5\pi}{6}$ (b) $210° = \dfrac{7\pi}{6}$, $330° = \dfrac{11\pi}{6}$

93. (a) $60° = \dfrac{\pi}{3}$, $120° = \dfrac{2\pi}{3}$ (b) $135° = \dfrac{3\pi}{4}$, $315° = \dfrac{7\pi}{4}$

95. (a) $45° = \dfrac{\pi}{4}$, $225° = \dfrac{5\pi}{4}$ (b) $150° = \dfrac{5\pi}{6}$, $330° = \dfrac{11\pi}{6}$

97. (a) 12 mi (b) 6 mi (c) 6.9 mi

99. (a) $N = 22.099 \sin(0.522t - 2.219) + 55.008$
$F = 36.641 \sin(0.502t - 1.831) + 25.610$

(b) February: $N = 34.6°$, $F = -1.4°$
March: $N = 41.6°$, $F = 13.9°$
May: $N = 63.4°$, $F = 48.6°$
June: $N = 72.5°$, $F = 59.5°$
August: $N = 75.5°$, $F = 55.6°$
September: $N = 68.6°$, $F = 41.7°$
November: $N = 46.8°$, $F = 6.5°$

(c) Answers will vary.

101. (a) 2 cm (b) 0.11 cm (c) -1.2 cm

103. False. In each of the four quadrants, the signs of the secant function and the cosine function will be the same, because these functions are reciprocals of each other.

105. As θ increases from 0° to 90°, x decreases from 12 cm to 0 cm and y increases from 0 cm to 12 cm. Therefore, $\sin \theta = y/12$ increases from 0 to 1 and $\cos \theta = x/12$ decreases from 1 to 0. Thus, $\tan \theta = y/x$ increases without bound. When $\theta = 90°$, the tangent is undefined.

107. (a) $\sin t = y$ (b) $r = 1$ because it is a unit circle.
$\cos t = x$
(c) $\sin \theta = y$ (d) $\sin t = \sin \theta$, and $\cos t = \cos \theta$.
$\cos \theta = x$

Section 4.5 *(page 326)*

1. cycle   **3.** phase shift   **5.** Period: $\dfrac{2\pi}{5}$; Amplitude: 2

7. Period: 4π; Amplitude: $\dfrac{3}{4}$ **9.** Period: 6; Amplitude: $\dfrac{1}{2}$

11. Period: 2π; Amplitude: 4

13. Period: $\dfrac{\pi}{5}$; Amplitude: 3

15. Period: $\dfrac{5\pi}{2}$; Amplitude: $\dfrac{5}{3}$

17. Period: 1; Amplitude: $\dfrac{1}{4}$

19. g is a shift of f π units to the right.

21. g is a reflection of f in the x-axis.

23. The period of f is twice the period of g.

25. g is a shift of f three units upward.

27. The graph of g has twice the amplitude of the graph of f.

29. The graph of g is a horizontal shift of the graph of f π units to the right.

31.

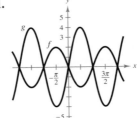

33.

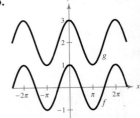

35.

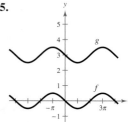

37.

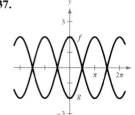

59.

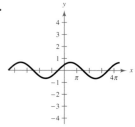

39.

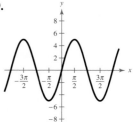

41.

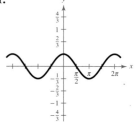

61. (a) $g(x)$ is obtained by a horizontal shrink of four, and one cycle of $g(x)$ corresponds to the interval $[\pi/4, 3\pi/4]$.

(b)

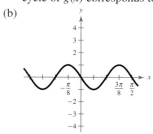

(c) $g(x) = f(4x - \pi)$

63. (a) One cycle of $g(x)$ corresponds to the interval $[\pi, 3\pi]$, and $g(x)$ is obtained by shifting $f(x)$ upward two units.

(b)

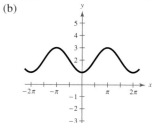

(c) $g(x) = f(x - \pi) + 2$

65. (a) One cycle of $g(x)$ is $[\pi/4, 3\pi/4]$. $g(x)$ is also shifted down three units and has an amplitude of two.

(b)

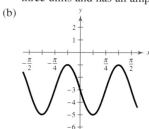

(c) $g(x) = 2f(4x - \pi) - 3$

43.

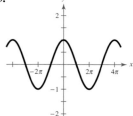

45.

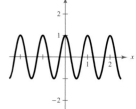

47.

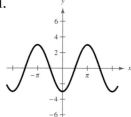

49.

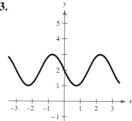

51.

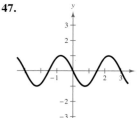

53.

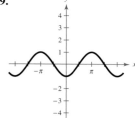

55.

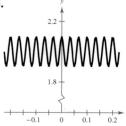

57.

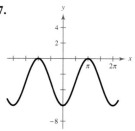

67.

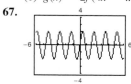

69.

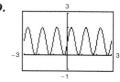

71.

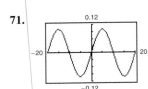

73. $a = 2, d = 1$ **75.** $a = -4, d = 4$

77. $a = -3, b = 2, c = 0$ **79.** $a = 2, b = 1, c = -\dfrac{\pi}{4}$

81.

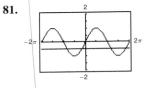

$$x = -\dfrac{\pi}{6}, -\dfrac{5\pi}{6}, \dfrac{7\pi}{6}, \dfrac{11\pi}{6}$$

83. $y = 1 + 2\sin(2x - \pi)$ **85.** $y = \cos(2x + 2\pi) - \dfrac{3}{2}$

87. (a) 6 sec   (b) 10 cycles/min

(c)

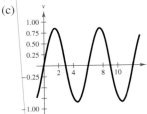

89. (a) $I(t) = 46.2 + 32.4\cos\left(\dfrac{\pi t}{6} - 3.67\right)$

(b)

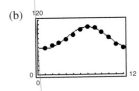

The model fits the data well.

(c)

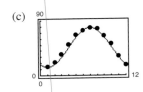

The model fits the data well.

(d) Las Vegas: 80.6°; International Falls: 46.2°
   The constant term gives the annual average temperature.

(e) 12; yes; One full period is one year.

(f) International Falls; amplitude; The greater the amplitude, the greater the variability in temperature.

91. (a) $\frac{1}{440}$ sec   (b) 440 cycles/sec

93. (a) 365; Yes, because there are 365 days in a year.

(b) 30.3 gal; the constant term

(c)

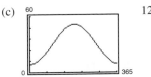

$124 < t < 252$

95. False. The graph of $f(x) = \sin(x + 2\pi)$ translates the graph of $f(x) = \sin x$ exactly one period to the left so that the two graphs look identical.

97. True. Because $\cos x = \sin\left(x + \dfrac{\pi}{2}\right)$, $y = -\cos x$ is a reflection in the x-axis of $y = \sin\left(x + \dfrac{\pi}{2}\right)$.

99.

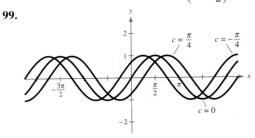

The value of c is a horizontal translation of the graph.

101.

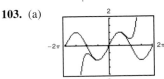

Conjecture:

$$\sin x = \cos\left(x - \dfrac{\pi}{2}\right)$$

103. (a)

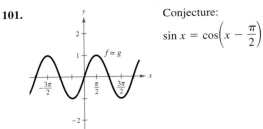

The graphs appear to coincide from $-\dfrac{\pi}{2}$ to $\dfrac{\pi}{2}$.

(b)

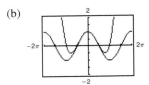

The graphs appear to coincide from $-\dfrac{\pi}{2}$ to $\dfrac{\pi}{2}$.

(c) $-\dfrac{x^7}{7!}, -\dfrac{x^6}{6!}$

The interval of accuracy increased.

Section 4.6   *(page 337)*

1. odd; origin   3. reciprocal   5. π
7. $(-\infty, -1] \cup [1, \infty)$ 9. e, π 10. c, 2π
11. a, 1   12. d, 2π 13. f, 4   14. b, 4

15. 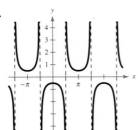   17.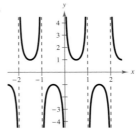

19. 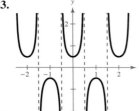   21.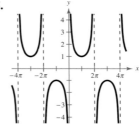

23. 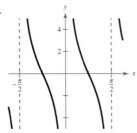   25.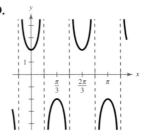

27. 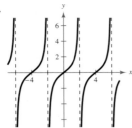   29.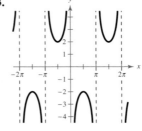

31. 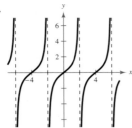   33.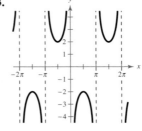

35. 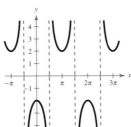   37.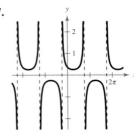

39. 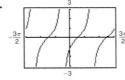   41.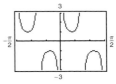

43. 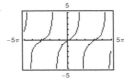   45.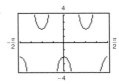

47.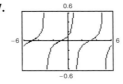

49. $-\dfrac{7\pi}{4}, -\dfrac{3\pi}{4}, \dfrac{\pi}{4}, \dfrac{5\pi}{4}$ 51. $-\dfrac{4\pi}{3}, -\dfrac{\pi}{3}, \dfrac{2\pi}{3}, \dfrac{5\pi}{3}$

53. $-\dfrac{4\pi}{3}, -\dfrac{2\pi}{3}, \dfrac{2\pi}{3}, \dfrac{4\pi}{3}$ 55. $-\dfrac{7\pi}{4}, -\dfrac{5\pi}{4}, \dfrac{\pi}{4}, \dfrac{3\pi}{4}$

57. Even   59. Odd   61. Odd   63. Even

65. (a)    (b) $\dfrac{\pi}{6} < x < \dfrac{5\pi}{6}$

(c) f approaches 0 and g approaches $+\infty$ because the cosecant is the reciprocal of the sine.

67.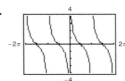

The expressions are equivalent except when $\sin x = 0$, y_1 is undefined.

69.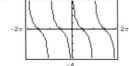

The expressions are equivalent.

71.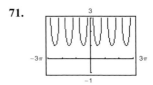

The expressions are equivalent.

73. d, $f \to 0$ as $x \to 0$.   **74.** a, $f \to 0$ as $x \to 0$.
75. b, $g \to 0$ as $x \to 0$.   **76.** c, $g \to 0$ as $x \to 0$.
77.   **79.**

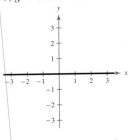

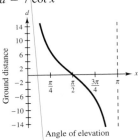

The functions are equal.   The functions are equal.

81.   **83.**

As $x \to \infty$, $g(x) \to 0$.   As $x \to \infty$, $f(x) \to 0$.

85. 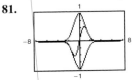   **87.**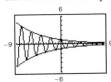

As $x \to 0$, $y \to \infty$.   As $x \to 0$, $g(x) \to 1$.
As $x \to 0$, $f(x)$ oscillates
between 1 and -1.

89.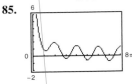

91. $d = 7 \cot x$

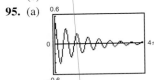

93. (a) Period of $H(t)$: 12 mo
    Period of $L(t)$: 12 mo
    (b) Summer; winter
    (c) About 0.5 mo

95. (a)   (b) y approaches 0 as t increases.

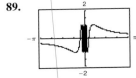

97. True. $y = \sec x$ is equal to $y = 1/\cos x$, and if the reciprocal
of $y = \sin x$ is translated $\pi/2$ units to the left, then

$$\frac{1}{\sin\left(x + \dfrac{\pi}{2}\right)} = \frac{1}{\cos x} = \sec x.$$

99. (a) As $x \to \dfrac{\pi}{2}^{+}$, $f(x) \to -\infty$.

    (b) As $x \to \dfrac{\pi}{2}^{-}$, $f(x) \to \infty$.

    (c) As $x \to -\dfrac{\pi}{2}^{+}$, $f(x) \to -\infty$.

    (d) As $x \to -\dfrac{\pi}{2}^{-}$, $f(x) \to \infty$.

101. (a) As $x \to 0^{+}$, $f(x) \to \infty$.
    (b) As $x \to 0^{-}$, $f(x) \to -\infty$.
    (c) As $x \to \pi^{+}$, $f(x) \to \infty$.
    (d) As $x \to \pi^{-}$, $f(x) \to -\infty$.

103. (a)

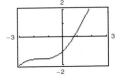

    0.7391
    (b) 1, 0.5403, 0.8576, 0.6543, 0.7935, 0.7014, 0.7640,
    0.7221, 0.7504, 0.7314, . . .; 0.7391

105.   The graphs appear to
                        coincide on the interval
                        $-1.1 \le x \le 1.1$.

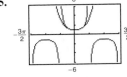

Section 4.7   *(page 347)*

1. $y = \sin^{-1} x$; $-1 \le x \le 1$

3. $y = \tan^{-1} x$; $-\infty < x < \infty$; $-\dfrac{\pi}{2} < y < \dfrac{\pi}{2}$

5. $\dfrac{\pi}{6}$ **7.** $\dfrac{\pi}{3}$ **9.** $\dfrac{\pi}{6}$ **11.** $\dfrac{5\pi}{6}$ **13.** $-\dfrac{\pi}{3}$

15. $\dfrac{2\pi}{3}$ **17.** $-\dfrac{\pi}{3}$ **19.** 0

21.

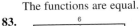

23. 1.19   **25.** -0.85 **27.** -1.25 **29.** 0.32
31. 1.99   **33.** 0.74   **35.** 1.07   **37.** 1.36

39. -1.52 **41.** $-\dfrac{\pi}{3}$, $-\dfrac{\sqrt{3}}{3}$, 1   **43.** $\theta = \arctan \dfrac{x}{4}$

45. $\theta = \arcsin \dfrac{x + 2}{5}$ **47.** $\theta = \arccos \dfrac{x + 3}{2x}$

49. 0.3   **51.** -0.1 **53.** 0   **55.** $\dfrac{3}{5}$ **57.** $\dfrac{\sqrt{5}}{5}$

59. $\dfrac{12}{13}$ **61.** $\dfrac{\sqrt{34}}{5}$ **63.** $\dfrac{\sqrt{5}}{3}$ **65.** 2   **67.** $\dfrac{1}{x}$

69. $\sqrt{1 - 4x^2}$ **71.** $\sqrt{1 - x^2}$ **73.** $\dfrac{\sqrt{9 - x^2}}{x}$

75. $\dfrac{\sqrt{x^2 + 2}}{x}$

77.

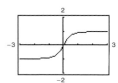

Asymptotes: $y = \pm 1$

79. $\dfrac{9}{\sqrt{x^2 + 81}}, \; x > 0; \; \dfrac{-9}{\sqrt{x^2 + 81}}, \; x < 0$

81. $\dfrac{|x - 1|}{\sqrt{x^2 - 2x + 10}}$

83.

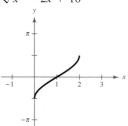

The graph of g is a horizontal shift one unit to the right of f.

85.

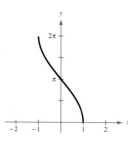

87.

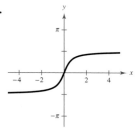

89.

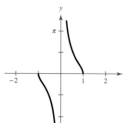

91.

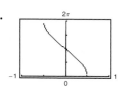

93.

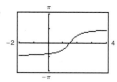

95.

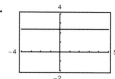

97. $3\sqrt{2} \sin\left(2t + \dfrac{\pi}{4}\right)$

The graph implies that the identity is true.

99. $\dfrac{\pi}{2}$ **101.** $\dfrac{\pi}{2}$ **103.** π

105. (a) $\theta = \arcsin \dfrac{5}{s}$ (b) $0.13, 0.25$

107. (a)

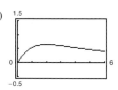

(b) 2 ft   (c) $\beta = 0$; As x increases, β approaches 0.

109. (a) $\theta \approx 26.0°$ (b) 24.4 ft

111. (a) $\theta = \arctan \dfrac{x}{20}$ (b) $14.0°, 31.0°$

113. False. $\dfrac{5\pi}{4}$ is not in the range of the arctangent.

115. Domain: $(-\infty, \infty)$
Range: $(0, \pi)$

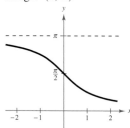

117. Domain: $(-\infty, -1] \cup [1, \infty)$
Range: $[-\pi/2, 0) \cup (0, \pi/2]$

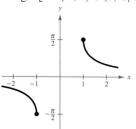

119. $\dfrac{\pi}{4}$ **121.** $\dfrac{3\pi}{4}$ **123.** $\dfrac{\pi}{6}$ **125.** $\dfrac{\pi}{3}$ **127.** 1.17

129. 0.19   **131.** 0.54   **133.** -0.12

135. (a) $\dfrac{\pi}{4}$ (b) $\dfrac{\pi}{2}$ (c) 1.25   (d) 2.03

137. (a) $f \circ f^{-1}$ $f^{-1} \circ f$

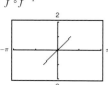 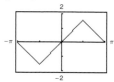

(b) The domains and ranges of the functions are restricted. The graphs of $f \circ f^{-1}$ and $f^{-1} \circ f$ differ because of the domains and ranges of f and f^{-1}.

Section 4.8   (page 357)

1. bearing   **3.** period

5. $a \approx 1.73$ **7.** $a \approx 8.26$ **9.** $c = 5$
$c \approx 3.46$ $c \approx 25.38$ $A \approx 36.87°$
$B = 60°$ $A = 19°$ $B \approx 53.13°$

11. $a \approx 49.48$
$A \approx 72.08°$
$B \approx 17.92°$

13. $a \approx 91.34$
$b \approx 420.70$
$B = 77°45'$

15. 3.00   **17.** 2.50   **19.** 214.45 ft   **21.** 19.7 ft
23. 19.9 ft   **25.** 11.8 km   **27.** 56.3°   **29.** 2.06°

31. (a) $\sqrt{h^2 + 34h + 10{,}289}$ (b) $\theta = \arccos\left(\dfrac{100}{l}\right)$

(c) 53.02 ft

33. (a) $l = 250$ ft, $A \approx 36.87°$, $B \approx 53.13°$
(b) 4.87 sec

35. 554 mi north; 709 mi east

37. (a) 104.95 nautical mi south; 58.18 nautical mi west
(b) S 36.7° W; distance = 130.9 nautical mi

39. N 56.31° W   **41.** (a) N 58° E   (b) 68.82 m
43. 78.7°   **45.** 35.3°   **47.** 29.4 in.   **49.** $y = \sqrt{3}\,r$
51. $a \approx 12.2, b \approx 7$ **53.** $d = 4\sin(\pi t)$

55. $d = 3\cos\left(\dfrac{4\pi t}{3}\right)$

57. (a) 9   (b) $\frac{3}{5}$ (c) 9   (d) $\frac{5}{12}$
59. (a) $\frac{1}{4}$ (b) 3   (c) 0   (d) $\frac{1}{6}$ **61.** $\omega = 528\pi$

63. (a)   (b) $\dfrac{\pi}{8}$ (c) $\dfrac{\pi}{32}$

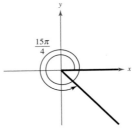

65. (a)

θ	L_1	L_2	$L_1 + L_2$
0.1	$\dfrac{2}{\sin 0.1}$	$\dfrac{3}{\cos 0.1}$	23.0
0.2	$\dfrac{2}{\sin 0.2}$	$\dfrac{3}{\cos 0.2}$	13.1
0.3	$\dfrac{2}{\sin 0.3}$	$\dfrac{3}{\cos 0.3}$	9.9
0.4	$\dfrac{2}{\sin 0.4}$	$\dfrac{3}{\cos 0.4}$	8.4

(b)

θ	L_1	L_2	$L_1 + L_2$
0.5	$\dfrac{2}{\sin 0.5}$	$\dfrac{3}{\cos 0.5}$	7.6
0.6	$\dfrac{2}{\sin 0.6}$	$\dfrac{3}{\cos 0.6}$	7.2
0.7	$\dfrac{2}{\sin 0.7}$	$\dfrac{3}{\cos 0.7}$	7.0
0.8	$\dfrac{2}{\sin 0.8}$	$\dfrac{3}{\cos 0.8}$	7.1

7.0 m

(c) $L = L_1 + L_2 = \dfrac{2}{\sin\theta} + \dfrac{3}{\cos\theta}$

(d)   7.0 m; The answers are the same.

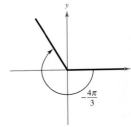

67. (a)

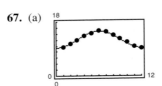

(b) 12; Yes, there are 12 months in a year.
(c) 2.77; The maximum change in the number of hours of daylight

69. False. The scenario does not create a right triangle because the tower is not vertical.

Review Exercises   *(page 364)*

1. (a)

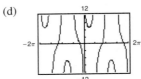

(b) Quadrant IV
(c) $\dfrac{23\pi}{4}, -\dfrac{\pi}{4}$

3. (a)

(b) Quadrant II
(c) $\dfrac{2\pi}{3}, -\dfrac{10\pi}{3}$

5. (a)

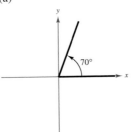

(b) Quadrant I
(c) 430°, −290°

7. (a)

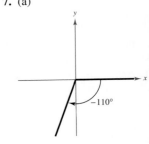

(b) Quadrant III
(c) 250°, −470°

9. 7.854   **11.** −0.589   **13.** 54.000°   **15.** −200.535°
17. 198° 24′   **19.** 0° 39′   **21.** 48.17 in.

23. Area ≈ 339.29 in.^{2} **25.** $\left(-\dfrac{1}{2}, \dfrac{\sqrt{3}}{2}\right)$

27. $\left(-\dfrac{\sqrt{3}}{2}, -\dfrac{1}{2}\right)$

29. $\sin \dfrac{7\pi}{6} = -\dfrac{1}{2}$ $\csc \dfrac{7\pi}{6} = -2$

$\cos \dfrac{7\pi}{6} = -\dfrac{\sqrt{3}}{2}$ $\sec \dfrac{7\pi}{6} = -\dfrac{2\sqrt{3}}{3}$

$\tan \dfrac{7\pi}{6} = \dfrac{\sqrt{3}}{3}$ $\cot \dfrac{7\pi}{6} = \sqrt{3}$

31. $\sin\left(-\dfrac{2\pi}{3}\right) = -\dfrac{\sqrt{3}}{2}$ $\csc\left(-\dfrac{2\pi}{3}\right) = -\dfrac{2\sqrt{3}}{3}$

$\cos\left(-\dfrac{2\pi}{3}\right) = -\dfrac{1}{2}$ $\sec\left(-\dfrac{2\pi}{3}\right) = -2$

$\tan\left(-\dfrac{2\pi}{3}\right) = \sqrt{3}$ $\cot\left(-\dfrac{2\pi}{3}\right) = \dfrac{\sqrt{3}}{3}$

33. $\sin \dfrac{11\pi}{4} = \sin \dfrac{3\pi}{4} = \dfrac{\sqrt{2}}{2}$

35. $\sin\left(-\dfrac{17\pi}{6}\right) = \sin \dfrac{7\pi}{6} = -\dfrac{1}{2}$

37. -75.3130 **39.** 3.2361

41. $\sin \theta = \dfrac{4\sqrt{41}}{41}$ $\csc \theta = \dfrac{\sqrt{41}}{4}$

$\cos \theta = \dfrac{5\sqrt{41}}{41}$ $\sec \theta = \dfrac{\sqrt{41}}{5}$

$\tan \theta = \dfrac{4}{5}$ $\cot \theta = \dfrac{5}{4}$

43. (a) 3 (b) $\dfrac{2\sqrt{2}}{3}$ (c) $\dfrac{3\sqrt{2}}{4}$ (d) $\dfrac{\sqrt{2}}{4}$

45. (a) $\dfrac{1}{4}$ (b) $\dfrac{\sqrt{15}}{4}$ (c) $\dfrac{4\sqrt{15}}{15}$ (d) $\dfrac{\sqrt{15}}{15}$

47. 0.6494 **49.** 0.5621 **51.** 3.6722

53. 0.6104 **55.** 71.3 m

57. $\sin \theta = \frac{4}{5}$ $\csc \theta = \frac{5}{4}$

$\cos \theta = \frac{3}{5}$ $\sec \theta = \frac{5}{3}$

$\tan \theta = \frac{4}{3}$ $\cot \theta = \frac{3}{4}$

59. $\sin \theta = \dfrac{15\sqrt{241}}{241}$ $\csc \theta = \dfrac{\sqrt{241}}{15}$

$\cos \theta = \dfrac{4\sqrt{241}}{241}$ $\sec \theta = \dfrac{\sqrt{241}}{4}$

$\tan \theta = \dfrac{15}{4}$ $\cot \theta = \dfrac{4}{15}$

61. $\sin \theta = \dfrac{9\sqrt{82}}{82}$ $\csc \theta = \dfrac{\sqrt{82}}{9}$

$\cos \theta = \dfrac{-\sqrt{82}}{82}$ $\sec \theta = -\sqrt{82}$

$\tan \theta = -9$ $\cot \theta = -\dfrac{1}{9}$

63. $\sin \theta = \dfrac{4\sqrt{17}}{17}$ $\csc \theta = \dfrac{\sqrt{17}}{4}$

$\cos \theta = \dfrac{\sqrt{17}}{17}$ $\sec \theta = \sqrt{17}$

$\tan \theta = 4$ $\cot \theta = \dfrac{1}{4}$

65. $\sin \theta = -\dfrac{\sqrt{11}}{6}$ $\csc \theta = -\dfrac{6\sqrt{11}}{11}$

$\cos \theta = \dfrac{5}{6}$ $\cot \theta = -\dfrac{5\sqrt{11}}{11}$

$\tan \theta = -\dfrac{\sqrt{11}}{5}$

67. $\cos \theta = -\dfrac{\sqrt{55}}{8}$ $\sec \theta = -\dfrac{8\sqrt{55}}{55}$

$\tan \theta = -\dfrac{3\sqrt{55}}{55}$ $\cot \theta = -\dfrac{\sqrt{55}}{3}$

$\csc \theta = \dfrac{8}{3}$

69. $\sin \theta = \dfrac{\sqrt{21}}{5}$ $\sec \theta = -\dfrac{5}{2}$

$\tan \theta = -\dfrac{\sqrt{21}}{2}$ $\cot \theta = -\dfrac{2\sqrt{21}}{21}$

$\csc \theta = \dfrac{5\sqrt{21}}{21}$

71. $\theta' = 84°$ **73.** $\theta' = \dfrac{\pi}{5}$

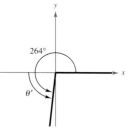 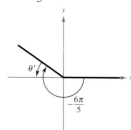

75. $\sin \dfrac{\pi}{3} = \dfrac{\sqrt{3}}{2}$; $\cos \dfrac{\pi}{3} = \dfrac{1}{2}$; $\tan \dfrac{\pi}{3} = \sqrt{3}$

77. $\sin\left(-\dfrac{7\pi}{3}\right) = -\dfrac{\sqrt{3}}{2}$; $\cos\left(-\dfrac{7\pi}{3}\right) = \dfrac{1}{2}$;

$\tan\left(-\dfrac{7\pi}{3}\right) = -\sqrt{3}$

79. $\sin 495° = \dfrac{\sqrt{2}}{2}$; $\cos 495° = -\dfrac{\sqrt{2}}{2}$; $\tan 495° = -1$

81. -0.7568 **83.** 0.9511

85.   **87.**

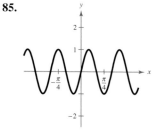 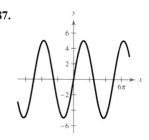

89.   **91.**

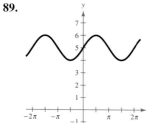 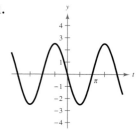

93. (a) $y = 2 \sin 528\pi x$ (b) 264 cycles/sec

95. 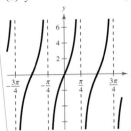 **97.**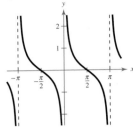

99. 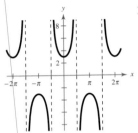 **101.**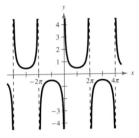

103.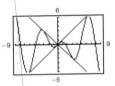

As $x \to +\infty$, $f(x)$ oscillates.

105. $-\dfrac{\pi}{6}$ **107.** 0.41 **109.** -0.46 **111.** $\dfrac{3\pi}{4}$

113. π **115.** 1.24 **117.** -0.98

119. 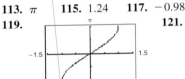 **121.**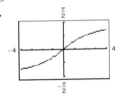

123. $\dfrac{4}{5}$ **125.** $\dfrac{13}{5}$ **127.** $\dfrac{10}{7}$ **129.** $\dfrac{\sqrt{4 - x^2}}{x}$

131. $\dfrac{\pi}{6}$ **133.** $\dfrac{3\pi}{4}$ **135.** 0.09 **137.** 1.98

139.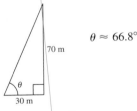

$\theta \approx 66.8°$

70 m

30 m

141. 1221 mi, 85.6°

143. False. For each θ there corresponds exactly one value of y.

145. The function is undefined because $\sec \theta = 1/\cos \theta$.

147. The ranges of the other four trigonometric functions are $(-\infty, \infty)$ or $(-\infty, -1] \cup [1, \infty)$.

Chapter Test *(page 367)*

1. (a)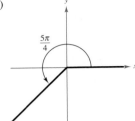

(b) $\dfrac{13\pi}{4}$, $-\dfrac{3\pi}{4}$

(c) 225°

2. 3500 rad/min **3.** About 709.04 ft²

4. $\sin \theta = \dfrac{3\sqrt{10}}{10}$ $\qquad$ $\csc \theta = \dfrac{\sqrt{10}}{3}$

$\cos \theta = -\dfrac{\sqrt{10}}{10}$ $\qquad$ $\sec \theta = -\sqrt{10}$

$\tan \theta = -3$ $\qquad$ $\cot \theta = -\dfrac{1}{3}$

5. For $0 \le \theta < \dfrac{\pi}{2}$: $\qquad$ For $\pi \le \theta < \dfrac{3\pi}{2}$:

$\sin \theta = \dfrac{3\sqrt{13}}{13}$ $\qquad$ $\sin \theta = -\dfrac{3\sqrt{13}}{13}$

$\cos \theta = \dfrac{2\sqrt{13}}{13}$ $\qquad$ $\cos \theta = -\dfrac{2\sqrt{13}}{13}$

$\csc \theta = \dfrac{\sqrt{13}}{3}$ $\qquad$ $\csc \theta = -\dfrac{\sqrt{13}}{3}$

$\sec \theta = \dfrac{\sqrt{13}}{2}$ $\qquad$ $\sec \theta = -\dfrac{\sqrt{13}}{2}$

$\cot \theta = \dfrac{2}{3}$ $\qquad$ $\cot \theta = \dfrac{2}{3}$

6. $\theta' = 25°$

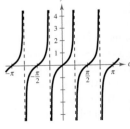

205°

θ'

7. Quadrant III **8.** 150°, 210° **9.** 1.33, 1.81

10. $\sin \theta = -\dfrac{4}{5}$ $\qquad$ **11.** $\sin \theta = \dfrac{21}{29}$

$\tan \theta = -\dfrac{4}{3}$ $\qquad$ $\cos \theta = -\dfrac{20}{29}$

$\csc \theta = -\dfrac{5}{4}$ $\qquad$ $\tan \theta = -\dfrac{21}{20}$

$\sec \theta = \dfrac{5}{3}$ $\qquad$ $\csc \theta = \dfrac{29}{21}$

$\cot \theta = -\dfrac{3}{4}$ $\qquad$ $\cot \theta = -\dfrac{20}{21}$

12. 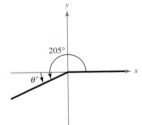 **13.**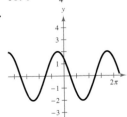

14.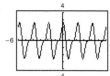

Period: 2

15.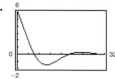

Not periodic

16. $a = -2, b = \dfrac{1}{2}, c = -\dfrac{\pi}{4}$ **17.** $\dfrac{\sqrt{55}}{3}$

18.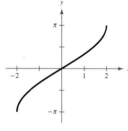

19. 309.3°

20. $d = -6 \cos \pi t$

Problem Solving *(page 369)*

1. (a) $\dfrac{11\pi}{2}$ rad or 990°   (b) About 816.42 ft

3. (a) 4767 ft   (b) 3705 ft

(c) $w = 2183$ ft, $\tan 63° = \dfrac{w + 3705}{3000}$

5. (a) 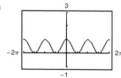   (b)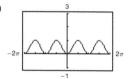

Even   Even

7. $h = 51 - 50 \sin\left(8\pi t + \dfrac{\pi}{2}\right)$

9. (a)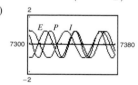

(b)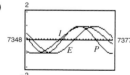

(c) $P(7369) = 0.631$
$E(7369) = 0.901$
$I(7369) = 0.945$

All three drop earlier in the month, then peak toward the middle of the month, and drop again toward the latter part of the month.

11. (a) 3.35, 7.35   (b) -0.65

(c) Yes. There is a difference of nine periods between the values.

13. (a) 40.5°   (b) $x \approx 1.71$ ft; $y \approx 3.46$ ft

(c) About 1.75 ft

(d) As you move closer to the rock, d must get smaller and smaller. The angles θ_1 and θ_2 will decrease along with the distance y, so d will decrease.

Chapter 5

Section 5.1 *(page 377)*

1. $\tan u$ **3.** $\cot u$ **5.** $\cot^2 u$ **7.** $\cos u$ **9.** $\cos u$

11. $\sin x = \dfrac{1}{2}$

$\cos x = \dfrac{\sqrt{3}}{2}$

$\tan x = \dfrac{\sqrt{3}}{3}$

$\csc x = 2$

$\sec x = \dfrac{2\sqrt{3}}{3}$

$\cot x = \sqrt{3}$

13. $\sin \theta = -\dfrac{\sqrt{2}}{2}$

$\cos \theta = \dfrac{\sqrt{2}}{2}$

$\tan \theta = -1$

$\csc \theta = -\sqrt{2}$

$\sec \theta = \sqrt{2}$

$\cot \theta = -1$

15. $\sin x = -\dfrac{8}{17}$

$\cos x = -\dfrac{15}{17}$

$\tan x = \dfrac{8}{15}$

$\csc x = -\dfrac{17}{8}$

$\sec x = -\dfrac{17}{15}$

$\cot x = \dfrac{15}{8}$

17. $\sin \phi = -\dfrac{\sqrt{5}}{3}$

$\cos \phi = \dfrac{2}{3}$

$\tan \phi = -\dfrac{\sqrt{5}}{2}$

$\csc \phi = -\dfrac{3\sqrt{5}}{5}$

$\sec \phi = \dfrac{3}{2}$

$\cot \phi = -\dfrac{2\sqrt{5}}{5}$

19. $\sin x = \dfrac{1}{3}$

$\cos x = -\dfrac{2\sqrt{2}}{3}$

$\tan x = -\dfrac{\sqrt{2}}{4}$

$\csc x = 3$

$\sec x = -\dfrac{3\sqrt{2}}{4}$

$\cot x = -2\sqrt{2}$

21. $\sin \theta = -\dfrac{2\sqrt{5}}{5}$

$\cos \theta = -\dfrac{\sqrt{5}}{5}$

$\tan \theta = 2$

$\csc \theta = -\dfrac{\sqrt{5}}{2}$

$\sec \theta = -\sqrt{5}$

$\cot \theta = \dfrac{1}{2}$

23. $\sin \theta = -1$ $\csc \theta = -1$
$\cos \theta = 0$ $\sec \theta$ is undefined.
$\tan \theta$ is undefined. $\cot \theta = 0$

25. d   **26.** a   **27.** b   **28.** f   **29.** e   **30.** c
31. b   **32.** c   **33.** f   **34.** a   **35.** e   **36.** d
37. $\csc \theta$ **39.** $-\sin x$ **41.** $\cos^2 \phi$ **43.** $\cos x$
45. $\sin^2 x$ **47.** $\cos \theta$ **49.** 1   **51.** $\tan x$
53. $1 + \sin y$ **55.** $\sec \beta$ **57.** $\cos u + \sin u$ **59.** $\sin^2 x$
61. $\sin^2 x \tan^2 x$ **63.** $\sec x + 1$ **65.** $\sec^4 x$
67. $\sin^2 x - \cos^2 x$ **69.** $\cot^2 x(\csc x - 1)$
71. $1 + 2 \sin x \cos x$ **73.** $4 \cot^2 x$ **75.** $2 \csc^2 x$
77. $2 \sec x$ **79.** $\sec x$ **81.** $1 + \cos y$
83. $3(\sec x + \tan x)$

85.

x	0.2	0.4	0.6	0.8	1.0
y_1	0.1987	0.3894	0.5646	0.7174	0.8415
y_2	0.1987	0.3894	0.5646	0.7174	0.8415

x	1.2	1.4
y_1	0.9320	0.9854
y_2	0.9320	0.9854

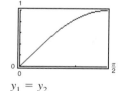

$y_1 = y_2$

87.

x	0.2	0.4	0.6	0.8	1.0
y_1	1.2230	1.5085	1.8958	2.4650	3.4082
y_2	1.2230	1.5085	1.8958	2.4650	3.4082

x	1.2	1.4
y_1	5.3319	11.6814
y_2	5.3319	11.6814

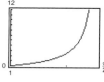

$y_1 = y_2$

89. $\csc x$ **91.** $\tan x$ **93.** $3 \sin \theta$ **95.** $4 \cos \theta$
97. $3 \tan \theta$ **99.** $5 \sec \theta$ **101.** $3 \sec \theta$ **103.** $\sqrt{2} \cos \theta$
105. $3 \cos \theta = 3$; $\sin \theta = 0$; $\cos \theta = 1$
107. $4 \sin \theta = 2\sqrt{2}$; $\sin \theta = \dfrac{\sqrt{2}}{2}$; $\cos \theta = \dfrac{\sqrt{2}}{2}$

109. $0 \le \theta \le \pi$ **111.** $0 \le \theta < \dfrac{\pi}{2}, \dfrac{3\pi}{2} < \theta < 2\pi$

113. $\ln|\cot x|$ **115.** $\ln|\cos x|$ **117.** $\ln|\csc t \sec t|$
119. (a) $\csc^2 132° - \cot^2 132° \approx 1.8107 - 0.8107 = 1$

  (b) $\csc^2 \dfrac{2\pi}{7} - \cot^2 \dfrac{2\pi}{7} \approx 1.6360 - 0.6360 = 1$

121. (a) $\cos(90° - 80°) = \sin 80° \approx 0.9848$

  (b) $\cos\left(\dfrac{\pi}{2} - 0.8\right) = \sin 0.8 \approx 0.7174$

123. $\mu = \tan \theta$ **125.** Answers will vary.
127. True. For example, $\sin(-x) = -\sin x$.
129. 1, 1   **131.** ∞, 0
133. Not an identity because $\cos \theta = \pm\sqrt{1 - \sin^2 \theta}$

135. Not an identity because $\dfrac{\sin k\theta}{\cos k\theta} = \tan k\theta$

137. An identity because $\sin \theta \cdot \dfrac{1}{\sin \theta} = 1$

139. $a \cos \theta$ **141.** $a \tan \theta$

Section 5.2   *(page 385)*

1. identity   **3.** $\tan u$ **5.** $\cos^2 u$ **7.** $-\csc u$
9–49. Answers will vary.
51. In the first line, $\cot(x)$ is substituted for $\cot(-x)$, which is incorrect; $\cot(-x) = -\cot(x)$.

53. (a) 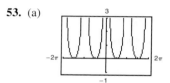   (b)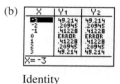

Identity

  (c) Answers will vary.
55. (a) 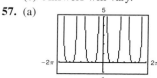   (b)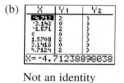

Not an identity

  (c) Answers will vary.
57. (a) 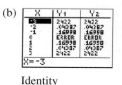   (b)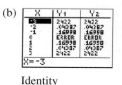

Identity

  (c) Answers will vary.
59. (a) 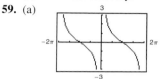   (b)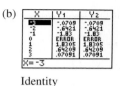

Identity

  (c) Answers will vary.
61–63. Answers will vary.   **65.** 1   **67.** 2
69. Answers will vary.
71. True. Many different techniques can be used to verify identities.
73. The equation is not an identity because $\sin \theta = \pm\sqrt{1 - \cos^2 \theta}$.

  Possible answer: $\dfrac{7\pi}{4}$

75. The equation is not an identity because $1 - \cos^2 \theta = \sin^2 \theta$.

  Possible answer: $-\dfrac{\pi}{2}$

77. The equation is not an identity because $1 + \tan^2 \theta = \sec^2 \theta$.

  Possible answer: $\dfrac{\pi}{6}$

Section 5.3   *(page 394)*

1. isolate   **3.** quadratic   **5–9.** Answers will vary.

11. $\dfrac{2\pi}{3} + 2n\pi, \dfrac{4\pi}{3} + 2n\pi$ **13.** $\dfrac{\pi}{3} + 2n\pi, \dfrac{2\pi}{3} + 2n\pi$

15. $\dfrac{\pi}{6} + n\pi, \dfrac{5\pi}{6} + n\pi$ **17.** $n\pi, \dfrac{3\pi}{2} + 2n\pi$

19. $\dfrac{\pi}{3} + n\pi, \dfrac{2\pi}{3} + n\pi$ **21.** $\dfrac{\pi}{8} + \dfrac{n\pi}{2}, \dfrac{3\pi}{8} + \dfrac{n\pi}{2}$

23. $\dfrac{n\pi}{3}, \dfrac{\pi}{4} + n\pi$ **25.** $0, \dfrac{\pi}{2}, \pi, \dfrac{3\pi}{2}$

27. $0, \pi, \dfrac{\pi}{6}, \dfrac{5\pi}{6}, \dfrac{7\pi}{6}, \dfrac{11\pi}{6}$ **29.** $\dfrac{\pi}{3}, \dfrac{5\pi}{3}, \pi$

31. No solution   **33.** $\pi, \dfrac{\pi}{3}, \dfrac{5\pi}{3}$

35. $\dfrac{\pi}{6}, \dfrac{5\pi}{6}, \dfrac{7\pi}{6}, \dfrac{11\pi}{6}$ **37.** $\dfrac{\pi}{2}$ **39.** $\dfrac{\pi}{6} + n\pi, \dfrac{5\pi}{6} + n\pi$

41. $\dfrac{\pi}{12} + \dfrac{n\pi}{3}$ **43.** $\dfrac{\pi}{2} + 4n\pi, \dfrac{7\pi}{2} + 4n\pi$ **45.** $3 + 4n$

47. $-2 + 6n, 2 + 6n$ **49.** $2.678, 5.820$

51. $1.047, 5.236$ **53.** $0.860, 3.426$

55. $0, 2.678, 3.142, 5.820$ **57.** $0.983, 1.768, 4.124, 4.910$

59. $0.3398, 0.8481, 2.2935, 2.8018$

61. $1.9357, 2.7767, 5.0773, 5.9183$

63. $\arctan(-4) + \pi, \arctan(-4) + 2\pi, \arctan 3, \arctan 3 + \pi$

65. $\dfrac{\pi}{4}, \dfrac{5\pi}{4}, \arctan 5, \arctan 5 + \pi$ **67.** $\dfrac{\pi}{3}, \dfrac{5\pi}{3}$

69. $\arctan\left(\tfrac{1}{3}\right), \arctan\left(\tfrac{1}{3}\right) + \pi, \arctan\left(-\tfrac{1}{3}\right) + \pi, \arctan\left(-\tfrac{1}{3}\right) + 2\pi$

71. $\arccos\left(\tfrac{1}{4}\right), 2\pi - \arccos\left(\tfrac{1}{4}\right)$

73. $\dfrac{\pi}{2}, \arcsin\left(-\dfrac{1}{4}\right) + 2\pi, \arcsin\left(\dfrac{1}{4}\right) + \pi$

75. $-1.154, 0.534$ **77.** 1.110

79. (a) 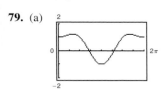   (b) $\dfrac{\pi}{3} \approx 1.0472$

$\dfrac{5\pi}{3} \approx 5.2360$

0

$\pi \approx 3.1416$

Maximum: $(1.0472, 1.25)$

Maximum: $(5.2360, 1.25)$

Minimum: $(0, 1)$

Minimum: $(3.1416, -1)$

81. (a) 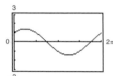   (b) $\dfrac{\pi}{4} \approx 0.7854$

$\dfrac{5\pi}{4} \approx 3.9270$

Maximum: $(0.7854, 1.4142)$

Minimum: $(3.9270, -1.4142)$

83. (a) 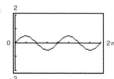   (b) $\dfrac{\pi}{4} \approx 0.7854$

$\dfrac{5\pi}{4} \approx 3.9270$

$\dfrac{3\pi}{4} \approx 2.3562$

$\dfrac{7\pi}{4} \approx 5.4978$

Maximum: $(0.7854, 0.5)$

Maximum: $(3.9270, 0.5)$

Minimum: $(2.3562, -0.5)$

Minimum: $(5.4978, -0.5)$

85. 1

87. (a) All real numbers x except $x = 0$

(b) y-axis symmetry; Horizontal asymptote: $y = 1$

(c) Oscillates   (d) Infinitely many solutions; $\dfrac{2}{2n\pi + \pi}$

(e) Yes, 0.6366

89. 0.04 sec, 0.43 sec, 0.83 sec

91. February, March, and April   **93.** $36.9°, 53.1°$

95. (a) $t = 8$ sec and $t = 24$ sec

(b) 5 times: $t = 16, 48, 80, 112, 144$ sec

97. (a)   (b) $0.6 < x < 1.1$

$A \approx 1.12$

99. True. The first equation has a smaller period than the second equation, so it will have more solutions in the interval $[0, 2\pi)$.

101. The equation would become $\cos^2 x = 2$; this is not the correct method to use when solving equations.

103. Answers will vary.

Section 5.4   (page 402)

1. $\sin u \cos v - \cos u \sin v$ **3.** $\dfrac{\tan u + \tan v}{1 - \tan u \tan v}$

5. $\cos u \cos v + \sin u \sin v$

7. (a) $\dfrac{\sqrt{2} - \sqrt{6}}{4}$ (b) $\dfrac{\sqrt{2} + 1}{2}$

9. (a) $\dfrac{1}{2}$ (b) $\dfrac{-\sqrt{3} - 1}{2}$

11. (a) $\dfrac{\sqrt{6} + \sqrt{2}}{4}$ (b) $\dfrac{\sqrt{2} - \sqrt{3}}{2}$

13. $\sin \dfrac{11\pi}{12} = \dfrac{\sqrt{2}}{4}\left(\sqrt{3} - 1\right)$

$\cos \dfrac{11\pi}{12} = -\dfrac{\sqrt{2}}{4}\left(\sqrt{3} + 1\right)$

$\tan \dfrac{11\pi}{12} = -2 + \sqrt{3}$

15. $\sin \dfrac{17\pi}{12} = -\dfrac{\sqrt{2}}{4}\left(\sqrt{3} + 1\right)$

$\cos \dfrac{17\pi}{12} = \dfrac{\sqrt{2}}{4}\left(1 - \sqrt{3}\right)$

$\tan \dfrac{17\pi}{12} = 2 + \sqrt{3}$

17. $\sin 105° = \dfrac{\sqrt{2}}{4}\left(\sqrt{3} + 1\right)$

$\cos 105° = \dfrac{\sqrt{2}}{4}\left(1 - \sqrt{3}\right)$

$\tan 105° = -2 - \sqrt{3}$

19. $\sin 195° = \dfrac{\sqrt{2}}{4}\left(1 - \sqrt{3}\right)$

$\cos 195° = -\dfrac{\sqrt{2}}{4}\left(\sqrt{3} + 1\right)$

$\tan 195° = 2 - \sqrt{3}$

21. $\sin \dfrac{13\pi}{12} = \dfrac{\sqrt{2}}{4}\left(1 - \sqrt{3}\right)$

$\cos \dfrac{13\pi}{12} = -\dfrac{\sqrt{2}}{4}\left(1 + \sqrt{3}\right)$

$\tan \dfrac{13\pi}{12} = 2 - \sqrt{3}$

23. $\sin\left(-\dfrac{13\pi}{12}\right) = \dfrac{\sqrt{2}}{4}\left(\sqrt{3} - 1\right)$

$\cos\left(-\dfrac{13\pi}{12}\right) = -\dfrac{\sqrt{2}}{4}\left(\sqrt{3} + 1\right)$

$\tan\left(-\dfrac{13\pi}{12}\right) = -2 + \sqrt{3}$

25. $\sin 285° = -\dfrac{\sqrt{2}}{4}\left(\sqrt{3}+1\right)$

$\cos 285° = \dfrac{\sqrt{2}}{4}\left(\sqrt{3}-1\right)$

$\tan 285° = -\left(2+\sqrt{3}\right)$

27. $\sin(-165°) = -\dfrac{\sqrt{2}}{4}\left(\sqrt{3}-1\right)$

$\cos(-165°) = -\dfrac{\sqrt{2}}{4}\left(1+\sqrt{3}\right)$

$\tan(-165°) = 2-\sqrt{3}$

29. $\sin 1.8$ **31.** $\sin 75°$ **33.** $\tan 15°$ **35.** $\tan 3x$

37. $\dfrac{\sqrt{3}}{2}$ **39.** $\dfrac{\sqrt{3}}{2}$ **41.** $-\sqrt{3}$ **43.** $-\dfrac{63}{65}$ **45.** $\dfrac{16}{65}$

47. $-\dfrac{63}{16}$ **49.** $\dfrac{65}{56}$ **51.** $\dfrac{3}{5}$ **53.** $-\dfrac{44}{117}$ **55.** $-\dfrac{125}{44}$

57. 1 **59.** 0 **61–69.** Proofs   **71.** $-\sin x$

73. $-\cos\theta$ **75.** $\dfrac{\pi}{6},\dfrac{5\pi}{6}$ **77.** $\dfrac{2\pi}{3},\dfrac{4\pi}{3}$ **79.** $\dfrac{\pi}{3},\dfrac{5\pi}{3}$

81. $\dfrac{5\pi}{4},\dfrac{7\pi}{4}$ **83.** $0,\dfrac{\pi}{2},\dfrac{3\pi}{2}$ **85.** $\dfrac{\pi}{4},\dfrac{7\pi}{4}$ **87.** $\dfrac{\pi}{2},\pi,\dfrac{3\pi}{2}$

89. (a) $y = \dfrac{5}{12}\sin(2t+0.6435)$ (b) $\dfrac{5}{12}$ ft   (c) $\dfrac{1}{\pi}$ cycle/sec

91. True. $\sin(u\pm v) = \sin u\cos v \pm \cos u\sin v$

93. False. $\tan\left(x-\dfrac{\pi}{4}\right) = \dfrac{\tan x - 1}{1+\tan x}$

95–97. Answers will vary.

99. (a) $\sqrt{2}\sin\left(\theta+\dfrac{\pi}{4}\right)$ (b) $\sqrt{2}\cos\left(\theta-\dfrac{\pi}{4}\right)$

101. (a) $13\sin(3\theta+0.3948)$ (b) $13\cos(3\theta-1.1760)$

103. $\sqrt{2}\sin\theta + \sqrt{2}\cos\theta$ **105.** Answers will vary.

107. $15°$

109. 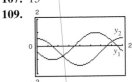   No, $y_1 \neq y_2$ because their graphs are different.

111. (a) and (b) Proofs

Section 5.5   (page 413)

1. $2\sin u\cos u$

3. $\cos^2 u - \sin^2 u = 2\cos^2 u - 1 = 1 - 2\sin^2 u$

5. $\pm\sqrt{\dfrac{1-\cos u}{2}}$ **7.** $\dfrac{1}{2}[\cos(u-v)+\cos(u+v)]$

9. $2\sin\left(\dfrac{u+v}{2}\right)\cos\left(\dfrac{u-v}{2}\right)$ **11.** $\dfrac{15}{17}$ **13.** $\dfrac{8}{15}$

15. $\dfrac{17}{8}$ **17.** $\dfrac{240}{289}$ **19.** $0,\dfrac{\pi}{3},\pi,\dfrac{5\pi}{3}$

21. $\dfrac{\pi}{12},\dfrac{5\pi}{12},\dfrac{13\pi}{12},\dfrac{17\pi}{12}$ **23.** $0,\dfrac{2\pi}{3},\dfrac{4\pi}{3}$ **25.** $0,\dfrac{\pi}{2},\pi,\dfrac{3\pi}{2}$

27. $\dfrac{\pi}{2},\dfrac{\pi}{6},\dfrac{5\pi}{6},\dfrac{7\pi}{6},\dfrac{3\pi}{2},\dfrac{11\pi}{6}$ **29.** $3\sin 2x$ **31.** $3\cos 2x$

33. $4\cos 2x$ **35.** $\cos 2x$

37. $\sin 2u = -\dfrac{24}{25}$, $\cos 2u = \dfrac{7}{25}$, $\tan 2u = -\dfrac{24}{7}$

39. $\sin 2u = \dfrac{15}{17}$, $\cos 2u = \dfrac{8}{17}$, $\tan 2u = \dfrac{15}{8}$

41. $\sin 2u = -\dfrac{\sqrt{3}}{2}$, $\cos 2u = -\dfrac{1}{2}$, $\tan 2u = \sqrt{3}$

43. $\dfrac{1}{8}(3+4\cos 2x+\cos 4x)$ **45.** $\dfrac{1}{8}(3+4\cos 4x+\cos 8x)$

47. $\dfrac{(3-4\cos 4x+\cos 8x)}{(3+4\cos 4x+\cos 8x)}$ **49.** $\dfrac{1}{8}(1-\cos 8x)$

51. $\dfrac{1}{16}(1-\cos 2x-\cos 4x+\cos 2x\cos 4x)$

53. $\dfrac{4\sqrt{17}}{17}$ **55.** $\dfrac{1}{4}$ **57.** $\sqrt{17}$

59. $\sin 75° = \dfrac{1}{2}\sqrt{2+\sqrt{3}}$

$\cos 75° = \dfrac{1}{2}\sqrt{2-\sqrt{3}}$

$\tan 75° = 2+\sqrt{3}$

61. $\sin 112° 30' = \dfrac{1}{2}\sqrt{2+\sqrt{2}}$

$\cos 112° 30' = -\dfrac{1}{2}\sqrt{2-\sqrt{2}}$

$\tan 112° 30' = -1-\sqrt{2}$

63. $\sin\dfrac{\pi}{8} = \dfrac{1}{2}\sqrt{2-\sqrt{2}}$

$\cos\dfrac{\pi}{8} = \dfrac{1}{2}\sqrt{2+\sqrt{2}}$

$\tan\dfrac{\pi}{8} = \sqrt{2}-1$

65. $\sin\dfrac{3\pi}{8} = \dfrac{1}{2}\sqrt{2+\sqrt{2}}$

$\cos\dfrac{3\pi}{8} = \dfrac{1}{2}\sqrt{2-\sqrt{2}}$

$\tan\dfrac{3\pi}{8} = \sqrt{2}+1$

67. (a) Quadrant I

(b) $\sin\dfrac{u}{2} = \dfrac{3}{5}$, $\cos\dfrac{u}{2} = \dfrac{4}{5}$, $\tan\dfrac{u}{2} = \dfrac{3}{4}$

69. (a) Quadrant II

(b) $\sin\dfrac{u}{2} = \dfrac{\sqrt{26}}{26}$, $\cos\dfrac{u}{2} = -\dfrac{5\sqrt{26}}{26}$, $\tan\dfrac{u}{2} = -\dfrac{1}{5}$

71. (a) Quadrant II

(b) $\sin\dfrac{u}{2} = \dfrac{3\sqrt{10}}{10}$, $\cos\dfrac{u}{2} = -\dfrac{\sqrt{10}}{10}$, $\tan\dfrac{u}{2} = -3$

73. $|\sin 3x|$ **75.** $-|\tan 4x|$

77. π **79.** $\dfrac{\pi}{3},\pi,\dfrac{5\pi}{3}$

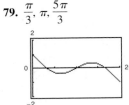

81. $\dfrac{1}{2}\left(\sin\dfrac{\pi}{2}+\sin\dfrac{\pi}{6}\right)$ **83.** $5(\cos 60° + \cos 90°)$

85. $\dfrac{1}{2}(\cos 2\theta - \cos 8\theta)$ **87.** $\dfrac{7}{2}(\sin(-2\beta) - \sin(-8\beta))$

89. $\dfrac{1}{2}(\cos 2y - \cos 2x)$ **91.** $2\sin 2\theta\cos\theta$

93. $2\cos 4x\cos 2x$ **95.** $2\cos\alpha\sin\beta$

97. $-2\sin\theta\sin\dfrac{\pi}{2} = -2\sin\theta$ **99.** $\dfrac{\sqrt{6}}{2}$ **101.** $-\sqrt{2}$

103. $0, \dfrac{\pi}{4}, \dfrac{\pi}{2}, \dfrac{3\pi}{4}, \pi, \dfrac{5\pi}{4}, \dfrac{3\pi}{2}, \dfrac{7\pi}{4}$ **105.** $\dfrac{\pi}{6}, \dfrac{5\pi}{6}$

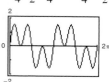

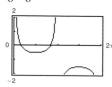

107. $\dfrac{120}{169}$ **109.** $\dfrac{3\sqrt{10}}{10}$ **111–123.** Answers will vary.

125.

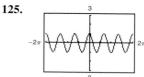

127.

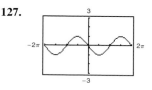

129.

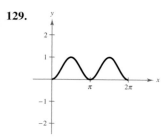

131. $2x\sqrt{1 - x^2}$ **133.** $1 - 2x^2$ **135.** $23.85°$

137. (a) π (b) 0.4482 (c) 760 mi/h; 3420 mi/h

(d) $\theta = 2 \sin^{-1}\left(\dfrac{1}{M}\right)$

139. False. For $u < 0$,

$$\sin 2u = -\sin(-2u)$$
$$= -2\sin(-u)\cos(-u)$$
$$= -2(-\sin u)\cos u$$
$$= 2\sin u \cos u.$$

Section 5.6   *(page 422)*

1. oblique   **3.** angles; side

5. $A = 30°, a \approx 14.14, c \approx 27.32$

7. $C = 120°, b \approx 4.75, c \approx 7.17$

9. $B = 60.9°, b \approx 19.32, c \approx 6.36$

11. $B = 42°4', a \approx 22.05, b \approx 14.88$

13. $C = 80°, a \approx 5.82, b \approx 9.20$

15. $C = 83°, a \approx 0.62, b \approx 0.51$

17. $B \approx 21.55°, C \approx 122.45°, c \approx 11.49$

19. $A \approx 10°11', C \approx 154°19', c \approx 11.03$

21. $B \approx 9.43°, C = 25.57°, c \approx 10.53$

23. $B \approx 18°13', C \approx 51°32', c \approx 40.06$

25. $B \approx 48.74°, C \approx 21.26°, c \approx 48.23$

27. No solution

29. Two solutions:

$B \approx 72.21°, C \approx 49.79°, c \approx 10.27$

$B \approx 107.79°, C \approx 14.21°, c \approx 3.30$

31. No solution   **33.** $B = 45°, C = 90°, c \approx 1.41$

35. (a) $b \le 5, b = \dfrac{5}{\sin 36°}$ (b) $5 < b < \dfrac{5}{\sin 36°}$

(c) $b > \dfrac{5}{\sin 36°}$

37. (a) $b \le 10.8, b = \dfrac{10.8}{\sin 10°}$ (b) $10.8 < b < \dfrac{10.8}{\sin 10°}$

(c) $b > \dfrac{10.8}{\sin 10°}$

39. 10.4 **41.** 1675.2 **43.** 3204.5 **45.** 24.1 m

47. $16.1°$ **49.** 77 m

51. (a)

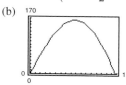

(b) 22.6 mi

(c) 21.4 mi

(d) 7.3 mi

Not drawn to scale

53. 3.2 mi   **55.** 5.86 mi

57. True. If an angle of a triangle is obtuse (greater than $90°$), then the other two angles must be acute and therefore less than $90°$. The triangle is oblique.

59. False. If just three angles are known, the triangle cannot be solved.

61. (a) $A = 20\left(15 \sin\dfrac{3\theta}{2} - 4\sin\dfrac{\theta}{2} - 6\sin\theta\right)$

(b)

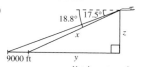

(c) Domain: $0 \le \theta \le 1.6690$

The domain would increase in length and the area would have a greater maximum value.

Section 5.7   *(page 429)*

1. Cosines   **3.** $b^2 = a^2 + c^2 - 2ac \cos B$

5. $A \approx 38.62°, B \approx 48.51°, C \approx 92.87°$

7. $B \approx 23.79°, C \approx 126.21°, a \approx 18.59$

9. $A \approx 30.11°, B \approx 43.16°, C \approx 106.73°$

11. $A \approx 92.94°, B \approx 43.53°, C \approx 43.53°$

13. $B \approx 27.46°, C \approx 32.54°, a \approx 11.27$

15. $A \approx 141°45', C \approx 27°40', b \approx 11.87$

17. $A \approx 27°10', C \approx 27°10', b \approx 65.84$

19. $A \approx 33.80°, B \approx 103.20°, c \approx 0.54$

	a	b	c	d	θ	ϕ
21.	5	8	12.07	5.69	$45°$	$135°$
23.	10	14	20	13.86	$68.2°$	$111.8°$
25.	15	16.96	25	20	$77.2°$	$102.8°$

27. Law of Cosines; $A \approx 102.44°, C \approx 37.56°, b \approx 5.26$

29. Law of Sines; No solution

31. Law of Sines; $C = 103°, a \approx 0.82, b \approx 0.71$

33. 43.52 **35.** 10.4 **37.** 52.11 **39.** 0.18

41.   N $37.1°$ E, S $63.1°$ E

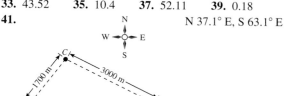

43. 373.3 m   **45.** 72.3°   **47.** 43.3 mi
49. (a) N 58.4° W   (b) S 81.5° W   **51.** 63.7 ft
53. 24.2 mi   **55.** $\overline{PQ} \approx 9.4$, $\overline{QS} = 5$, $\overline{RS} \approx 12.8$
57.

d (inches)	9	10	12	13	14
θ (degrees)	60.9°	69.5°	88.0°	98.2°	109.6°
s (inches)	20.88	20.28	18.99	18.28	17.48

d (inches)	15	16
θ (degrees)	122.9°	139.8°
s (inches)	16.55	15.37

59. 46,837.5 ft²   **61.** $83,336.37
63. False. For s to be the average of the lengths of the three sides of the triangle, s would be equal to $(a + b + c)/3$.
65. No. The three side lengths do not form a triangle.
67. (a) and (b) Proofs   **69.** 405.2 ft
71. Either; Because A is obtuse, there is only one solution for B or C.
73. The Law of Cosines can be used to solve the single-solution case of SSA. There is no method that can solve the no-solution case of SSA.
75. Proof

Review Exercises   *(page 436)*

1. $\tan x$ **3.** $\cos x$ **5.** $|\csc x|$

7. $\tan x = \dfrac{5}{12}$
$\csc x = \dfrac{13}{5}$
$\sec x = \dfrac{13}{12}$
$\cot x = \dfrac{12}{5}$

9. $\cos x = \dfrac{\sqrt{2}}{2}$
$\tan x = -1$
$\csc x = -\sqrt{2}$
$\sec x = \sqrt{2}$
$\cot x = -1$

11. $\sin^2 x$ **13.** 1   **15.** $\cot \theta$ **17.** $\csc \theta$
19. $\cot^2 x$ **21.** $\sec x + 2 \sin x$ **23.** $-2 \tan^2 \theta$
25. $5 \cos \theta$ **27–35.** Answers will vary.

37. $\dfrac{\pi}{3} + 2n\pi, \dfrac{2\pi}{3} + 2n\pi$ **39.** $\dfrac{\pi}{6} + n\pi$

41. $\dfrac{\pi}{3} + n\pi, \dfrac{2\pi}{3} + n\pi$ **43.** $0, \dfrac{2\pi}{3}, \dfrac{4\pi}{3}$ **45.** $0, \dfrac{\pi}{2}, \pi$

47. $\dfrac{\pi}{8}, \dfrac{3\pi}{8}, \dfrac{9\pi}{8}, \dfrac{11\pi}{8}$ **49.** $\dfrac{\pi}{2}$

51. $0, \dfrac{\pi}{8}, \dfrac{3\pi}{8}, \dfrac{5\pi}{8}, \dfrac{7\pi}{8}, \dfrac{9\pi}{8}, \dfrac{11\pi}{8}, \dfrac{13\pi}{8}, \dfrac{15\pi}{8}$ **53.** $0, \pi$

55. $\arctan(-3) + \pi, \arctan(-3) + 2\pi, \arctan 2, \arctan 2 + \pi$

57. $\sin 285° = -\dfrac{\sqrt{2}}{4}(\sqrt{3} + 1)$
$\cos 285° = \dfrac{\sqrt{2}}{4}(\sqrt{3} - 1)$
$\tan 285° = -2 - \sqrt{3}$

59. $\sin \dfrac{25\pi}{12} = \dfrac{\sqrt{2}}{4}(\sqrt{3} - 1)$
$\cos \dfrac{25\pi}{12} = \dfrac{\sqrt{2}}{4}(\sqrt{3} + 1)$
$\tan \dfrac{25\pi}{12} = 2 - \sqrt{3}$

61. $\sin 15°$ **63.** $\tan 35°$ **65.** $-\dfrac{24}{25}$ **67.** -1
69. $-\dfrac{7}{25}$ **71–75.** Answers will vary.

77. $\dfrac{\pi}{4}, \dfrac{7\pi}{4}$ **79.** $\dfrac{\pi}{6}, \dfrac{11\pi}{6}$

81. $\sin 2u = \dfrac{24}{25}$
$\cos 2u = -\dfrac{7}{25}$
$\tan 2u = -\dfrac{24}{7}$

83. $\sin 2u = -\dfrac{4\sqrt{2}}{9}$, $\cos 2u = -\dfrac{7}{9}$, $\tan 2u = \dfrac{4\sqrt{2}}{7}$

85.

87. $\dfrac{1 - \cos 4x}{1 + \cos 4x}$

89. $\dfrac{3 - 4\cos 2x + \cos 4x}{4(1 + \cos 2x)}$

91. $\sin(-75°) = -\dfrac{1}{2}\sqrt{2 + \sqrt{3}}$ **93.** $\sin \dfrac{19\pi}{12} = -\dfrac{1}{2}\sqrt{2 + \sqrt{3}}$
$\cos(-75°) = \dfrac{1}{2}\sqrt{2 - \sqrt{3}}$ $\cos \dfrac{19\pi}{12} = \dfrac{1}{2}\sqrt{2 - \sqrt{3}}$
$\tan(-75°) = -2 - \sqrt{3}$ $\tan \dfrac{19\pi}{12} = -2 - \sqrt{3}$

95. (a) Quadrant I
(b) $\sin \dfrac{u}{2} = \dfrac{\sqrt{2}}{10}$, $\cos \dfrac{u}{2} = \dfrac{7\sqrt{2}}{10}$, $\tan \dfrac{u}{2} = \dfrac{1}{7}$

97. (a) Quadrant I
(b) $\sin \dfrac{u}{2} = \dfrac{3\sqrt{14}}{14}$, $\cos \dfrac{u}{2} = \dfrac{\sqrt{70}}{14}$, $\tan \dfrac{u}{2} = \dfrac{3\sqrt{5}}{5}$

99. $-|\cos 5x|$ **101.** $\dfrac{1}{2}\left(\sin \dfrac{\pi}{3} - \sin 0\right) = \dfrac{1}{2}\sin \dfrac{\pi}{3}$

103. $\dfrac{1}{2}[\sin 10\theta - \sin(-2\theta)]$ **105.** $2 \cos 6\theta \sin(-2\theta)$

107. $-2 \sin x \sin \dfrac{\pi}{6}$ **109.** $\theta = 15°$ or $\dfrac{\pi}{12}$

111.

113. $\dfrac{1}{2}\sqrt{10}$ ft

115. $C = 72°$, $b \approx 12.21$, $c \approx 12.36$
117. $A = 26°$, $a \approx 24.89$, $c \approx 56.23$
119. $C = 66°$, $a \approx 2.53$, $b \approx 9.11$
121. $B = 108°$, $a \approx 11.76$, $c \approx 21.49$

123. $A \approx 20.41°$, $C \approx 9.59°$, $a \approx 20.92$
125. $B \approx 39.48°$, $C \approx 65.52°$, $c \approx 48.24$
127. 19.06   **129.** 47.23   **131.** 31.1 m   **133.** 31.01 ft
135. $A \approx 27.81°$, $B \approx 54.75°$, $C \approx 97.44°$
137. $A \approx 16.99°$, $B \approx 26.00°$, $C \approx 137.01°$
139. $A \approx 29.92°$, $B \approx 86.18°$, $C \approx 63.90°$
141. $A = 36°$, $C = 36°$, $b \approx 17.80$
143. $A \approx 45.76°$, $B \approx 91.24°$, $c \approx 21.42$
145. Law of Sines; $A \approx 77.52°$, $B \approx 38.48°$, $a \approx 14.12$
147. Law of Cosines; $A \approx 28.62°$, $B \approx 33.56°$, $C \approx 117.82°$
149. About 4.3 ft, about 12.6 ft
151. 615.1 m   **153.** 7.64   **155.** 8.36
157. False. If $(\pi/2) < \theta < \pi$, then $\cos(\theta/2) > 0$. The sign of $\cos(\theta/2)$ depends on the quadrant in which $\theta/2$ lies.
159. True. $4 \sin(-x) \cos(-x) = 4(-\sin x) \cos x$
$$= -4 \sin x \cos x$$
$$= -2(2 \sin x \cos x)$$
$$= -2 \sin 2x$$
161. True. $\sin 90°$ is defined in the Law of Sines.
163. Reciprocal identities:

$\sin \theta = \dfrac{1}{\csc \theta}$, $\cos \theta = \dfrac{1}{\sec \theta}$, $\tan \theta = \dfrac{1}{\cot \theta}$,

$\csc \theta = \dfrac{1}{\sin \theta}$, $\sec \theta = \dfrac{1}{\cos \theta}$, $\cot \theta = \dfrac{1}{\tan \theta}$

Quotient identities: $\tan \theta = \dfrac{\sin \theta}{\cos \theta}$, $\cot \theta = \dfrac{\cos \theta}{\sin \theta}$

Pythagorean identities: $\sin^2 \theta + \cos^2 \theta = 1$,
$1 + \tan^2 \theta = \sec^2 \theta$, $1 + \cot^2 \theta = \csc^2 \theta$
165. $a^2 = b^2 + c^2 - 2bc \cos A$, $b^2 = a^2 + c^2 - 2ac \cos B$,
$c^2 = a^2 + b^2 - 2ab \cos C$
167. $-1 \le \sin x \le 1$ for all x **169.** $y_1 = y_2 + 1$
171. -1.8431, 2.1758, 3.9903, 8.8935, 9.8820

Chapter Test   (page 441)

1. $\sin \theta = -\dfrac{6\sqrt{61}}{61}$ $\csc \theta = -\dfrac{\sqrt{61}}{6}$

$\cos \theta = -\dfrac{5\sqrt{61}}{61}$ $\sec \theta = -\dfrac{\sqrt{61}}{5}$

$\tan \theta = \dfrac{6}{5}$ $\cot \theta = \dfrac{5}{6}$

2. 1   **3.** 1   **4.** $\csc \theta \sec \theta$
5–10. Answers will vary.   **11.** $\frac{1}{8}(3 - 4 \cos x + \cos 2x)$
12. $\tan 2\theta$ **13.** $2(\sin 5\theta + \sin \theta)$
14. $-2 \sin 2\theta \sin \theta$ **15.** $0, \dfrac{3\pi}{4}, \pi, \dfrac{7\pi}{4}$
16. $\dfrac{\pi}{6}, \dfrac{\pi}{2}, \dfrac{5\pi}{6}, \dfrac{3\pi}{2}$ **17.** $\dfrac{\pi}{6}, \dfrac{5\pi}{6}, \dfrac{7\pi}{6}, \dfrac{11\pi}{6}$
18. $\dfrac{\pi}{6}, \dfrac{5\pi}{6}, \dfrac{3\pi}{2}$ **19.** $\dfrac{\sqrt{2} - \sqrt{6}}{4}$
20. $\sin 2u = -\frac{20}{29}$, $\cos 2u = -\frac{21}{29}$, $\tan 2u = \frac{20}{21}$
21. $C = 88°$, $b \approx 27.81$, $c \approx 29.98$
22. $A = 42°$, $b \approx 21.91$, $c \approx 10.95$
23. Two solutions:
$B \approx 29.12°$, $C \approx 126.88°$, $c \approx 22.03$
$B \approx 150.88°$, $C \approx 5.12°$, $c \approx 2.46$

24. No solution   **25.** $A \approx 39.96°$, $C \approx 40.04°$, $c \approx 15.02$
26. $A \approx 21.90°$, $B \approx 37.10°$, $c \approx 78.15$
27. Day 123 to day 223
28. 2052.5 m^2 **29.** 606.3 mi; 29.1°

Problem Solving   (page 447)

1. (a) $\cos \theta = \pm\sqrt{1 - \sin^2 \theta}$ (b) $\sin \theta = \pm\sqrt{1 - \cos^2 \theta}$

$\tan \theta = \pm\dfrac{\sin \theta}{\sqrt{1 - \sin^2 \theta}}$ $\tan \theta = \pm\dfrac{\sqrt{1 - \cos^2 \theta}}{\cos \theta}$

$\cot \theta = \pm\dfrac{\sqrt{1 - \sin^2 \theta}}{\sin \theta}$ $\csc \theta = \pm\dfrac{1}{\sqrt{1 - \cos^2 \theta}}$

$\sec \theta = \pm\dfrac{1}{\sqrt{1 - \sin^2 \theta}}$ $\sec \theta = \dfrac{1}{\cos \theta}$

$\csc \theta = \dfrac{1}{\sin \theta}$ $\cot \theta = \pm\dfrac{\cos \theta}{\sqrt{1 - \cos^2 \theta}}$

3. Answers will vary.   **5.** $u + v = w$

7. $\sin \dfrac{\theta}{2} = \sqrt{\dfrac{1 - \cos \theta}{2}}$

$\cos \dfrac{\theta}{2} = \sqrt{\dfrac{1 + \cos \theta}{2}}$

$\tan \dfrac{\theta}{2} = \dfrac{\sin \theta}{1 + \cos \theta}$

9. (a)

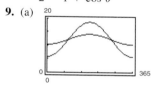

(b) $t \approx 91$ (April 1), $t \approx 274$ (October 1)
(c) Seward; The amplitudes: 6.4 and 1.9
(d) 365.2 days

11. (a) $\dfrac{\pi}{6} \le x \le \dfrac{5\pi}{6}$ (b) $\dfrac{2\pi}{3} \le x \le \dfrac{4\pi}{3}$

(c) $\dfrac{\pi}{2} < x < \pi$, $\dfrac{3\pi}{2} < x < 2\pi$

(d) $0 \le x \le \dfrac{\pi}{4}$, $\dfrac{5\pi}{4} \le x \le 2\pi$

13. (a) $\cos 3\theta = \cos \theta - 4 \sin^2 \theta \cos \theta$
(b) $\cos 4\theta = \cos^4 \theta - 6 \sin^2 \theta \cos^2 \theta + \sin^4 \theta$
15. 2.01 ft

Chapter 6

Section 6.1   (page 454)

1. inclination   **3.** $\left|\dfrac{m_2 - m_1}{1 + m_1 m_2}\right|$ **5.** $\dfrac{\sqrt{3}}{3}$ **7.** -1

9. $\sqrt{3}$ **11.** 3.2236   **13.** $\dfrac{3\pi}{4}$ rad, 135°

15. $\dfrac{\pi}{4}$ rad, 45°   **17.** 0.6435 rad, 36.9°   **19.** $\dfrac{\pi}{6}$ rad, 30°

21. $\dfrac{5\pi}{6}$ rad, 150°   **23.** 1.0517 rad, 60.3°

25. 2.1112 rad, 121.0°   **27.** $\dfrac{3\pi}{4}$ rad, 135°

29. $\frac{\pi}{4}$ rad, 45°   **31.** $\frac{5\pi}{6}$ rad, 150°   **33.** 1.2490 rad, 71.6°

35. 2.1112 rad, 121.0°   **37.** 1.1071 rad, 63.4°

39. 0.1974 rad, 11.3°   **41.** 1.4289 rad, 81.9°

43. 0.9273 rad, 53.1°   **45.** 0.8187 rad, 46.9°

47. $(1, 5) \leftrightarrow (4, 5)$: slope $= 0$
$(4, 5) \leftrightarrow (3, 8)$: slope $= -3$
$(3, 8) \leftrightarrow (1, 5)$: slope $= \frac{3}{2}$
$(1, 5)$: 56.3°; $(4, 5)$: 71.6°; $(3, 8)$: 52.1°

49. $(-4, -1) \leftrightarrow (3, 2)$: slope $= \frac{3}{7}$
$(3, 2) \leftrightarrow (1, 0)$: slope $= 1$
$(1, 0) \leftrightarrow (-4, -1)$: slope $= \frac{1}{5}$
$(-4, -1)$: 11.9°; $(3, 2)$: 21.8°; $(1, 0)$: 146.3°

51. 0   **53.** $\frac{4\sqrt{10}}{5} \approx 2.5298$ **55.** 7

57. $\frac{8\sqrt{37}}{37} \approx 1.3152$

59. (a)   (b) 4   (c) 8

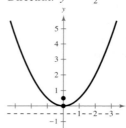

61. (a)   (b) $\frac{35\sqrt{37}}{74}$ (c) $\frac{35}{8}$

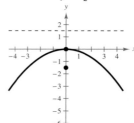

63. $2\sqrt{2}$ **65.** 0.1003, 1054 ft   **67.** 31.0°

69. $\alpha \approx 33.69°$; $\beta \approx 56.31°$

71. True. The inclination of a line is related to its slope by $m = \tan\theta$. If the angle is greater than $\pi/2$ but less than π, then the angle is in the second quadrant, where the tangent function is negative.

73. (a) $d = \dfrac{4}{\sqrt{m^2 + 1}}$ (c) $m = 0$

(b)

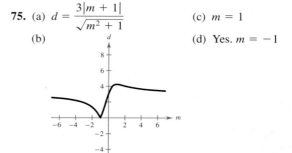

(d) The graph has a horizontal asymptote of $d = 0$. As the slope becomes larger, the distance between the origin and the line, $y = mx + 4$, becomes smaller and approaches 0.

75. (a) $d = \dfrac{3|m + 1|}{\sqrt{m^2 + 1}}$ (c) $m = 1$

(b)   (d) Yes. $m = -1$

(e) $d = 3$. As the line approaches the vertical, the distance approaches 3.

Section 6.2   *(page 462)*

1. conic   **3.** locus   **5.** axis   **7.** focal chord

9. A circle is formed when a plane intersects the top or bottom half of a double-napped cone and is perpendicular to the axis of the cone.

11. A parabola is formed when a plane intersects the top or bottom half of a double-napped cone, is parallel to the side of the cone, and does not intersect the vertex.

13. e   **14.** b   **15.** d   **16.** f   **17.** a   **18.** c

19. $x^2 = \frac{3}{2}y$ **21.** $x^2 = 2y$ **23.** $y^2 = -8x$

25. $x^2 = -4y$ **27.** $y^2 = 4x$ **29.** $x^2 = \frac{8}{3}y$

31. $y^2 = -\frac{25}{2}x$

33. Vertex: $(0, 0)$ **35.** Vertex: $(0, 0)$
Focus: $(0, \frac{1}{2})$ Focus: $(-\frac{3}{2}, 0)$
Directrix: $y = -\frac{1}{2}$ Directrix: $x = \frac{3}{2}$

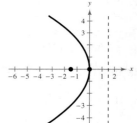

37. Vertex: $(0, 0)$ **39.** Vertex: $(1, -2)$
Focus: $(0, -\frac{3}{2})$ Focus: $(1, -4)$
Directrix: $y = \frac{3}{2}$ Directrix: $y = 0$

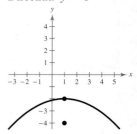

41. Vertex: $\left(-3, \frac{3}{2}\right)$
Focus: $\left(-3, \frac{5}{2}\right)$
Directrix: $y = \frac{1}{2}$

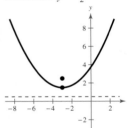

43. Vertex: $(1, 1)$
Focus: $(1, 2)$
Directrix: $y = 0$

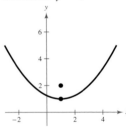

45. Vertex: $(-2, -3)$
Focus: $(-4, -3)$
Directrix: $x = 0$

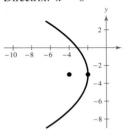

47. Vertex: $(-2, 1)$
Focus: $\left(-2, -\frac{1}{2}\right)$
Directrix: $y = \frac{5}{2}$

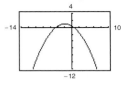

49. Vertex: $\left(\frac{1}{4}, -\frac{1}{2}\right)$
Focus: $\left(0, -\frac{1}{2}\right)$
Directrix: $x = \frac{1}{2}$

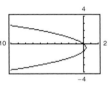

51. $(x - 3)^2 = -(y - 1)$ **53.** $y^2 = 4(x + 4)$
55. $(y - 3)^2 = 8(x - 4)$ **57.** $x^2 = -8(y - 2)$
59. $(y - 2)^2 = 8x$ **61.** $y = \sqrt{6(x + 1)} + 3$
63.

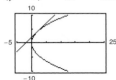

$(2, 4)$

65. $4x - y - 8 = 0; (2, 0)$ **67.** $4x - y + 2 = 0; \left(-\frac{1}{2}, 0\right)$
69.

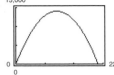

$x = 106$ units

71. (a)

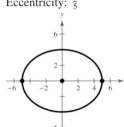

(−640, 152)   (640, 152)

(b) $y = \dfrac{19x^2}{51,200}$

(c)

Distance, x	0	100	250	400	500
Height, y	0	3.71	23.19	59.38	92.77

73. (a) $y = -\frac{1}{640}x^2$ (b) 8 ft
75. (a) $x^2 = 180,000y$ (b) $300\sqrt{2}$ cm ≈ 424.26 cm
77. $x^2 = -\frac{25}{4}(y - 48)$
79. (a) $17,500\sqrt{2}$ mi/h $\approx 24,750$ mi/h
(b) $x^2 = -16,400(y - 4100)$
81. (a) $x^2 = -49(y - 100)$ (b) 70 ft
83. False. If the graph crossed the directrix, there would exist points closer to the directrix than the focus.
85. $m = \dfrac{x_1}{2p}$
87. (a)

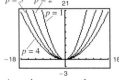

As p increases, the graph becomes wider.
(b) $(0, 1), (0, 2), (0, 3), (0, 4)$ (c) $4, 8, 12, 16; 4|p|$
(d) It is an easy way to determine two additional points on the graph.

Section 6.3   *(page 472)*

1. ellipse; foci   **3.** minor axis
5. b   **6.** c   **7.** d   **8.** f   **9.** a   **10.** e
11. $\dfrac{x^2}{4} + \dfrac{y^2}{16} = 1$ **13.** $\dfrac{x^2}{49} + \dfrac{y^2}{45} = 1$ **15.** $\dfrac{x^2}{49} + \dfrac{y^2}{24} = 1$
17. $\dfrac{21x^2}{400} + \dfrac{y^2}{25} = 1$ **19.** $\dfrac{(x - 2)^2}{1} + \dfrac{(y - 3)^2}{9} = 1$
21. $\dfrac{(x - 4)^2}{16} + \dfrac{(y - 2)^2}{1} = 1$ **23.** $\dfrac{x^2}{48} + \dfrac{(y - 4)^2}{64} = 1$
25. $\dfrac{x^2}{16} + \dfrac{(y - 4)^2}{12} = 1$ **27.** $\dfrac{(x - 2)^2}{4} + \dfrac{(y - 2)^2}{1} = 1$
29. Ellipse
Center: $(0, 0)$
Vertices: $(\pm 5, 0)$
Foci: $(\pm 3, 0)$
Eccentricity: $\frac{3}{5}$

31. Circle
Center: $(0, 0)$
Radius: 5

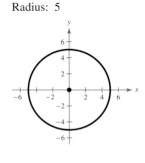

33. Ellipse
Center: $(0, 0)$
Vertices: $(0, \pm 3)$
Foci: $(0, \pm 2)$
Eccentricity: $\frac{2}{3}$

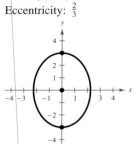

35. Ellipse
Center: $(4, -1)$
Vertices: $(4, -6), (4, 4)$
Foci: $(4, 2), (4, -4)$
Eccentricity: $\frac{3}{5}$

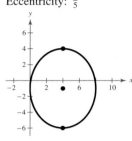

37. Circle
Center: $(0, -1)$
Radius: $\frac{2}{3}$

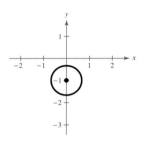

39. Ellipse
Center: $(-2, -4)$
Vertices: $(-3, -4), (-1, -4)$
Foci: $\left(\dfrac{-4 \pm \sqrt{3}}{2}, -4 \right)$
Eccentricity: $\dfrac{\sqrt{3}}{2}$

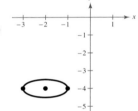

41. Ellipse
Center: $(-2, 3)$
Vertices: $(-2, 6), (-2, 0)$
Foci: $\left(-2, 3 \pm \sqrt{5}\right)$
Eccentricity: $\dfrac{\sqrt{5}}{3}$

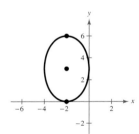

43. Circle
Center: $(1, -2)$
Radius: 6

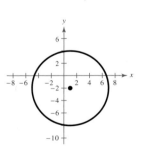

45. Ellipse
Center: $(-3, 1)$
Vertices: $(-3, 7), (-3, -5)$
Foci: $\left(-3, 1 \pm 2\sqrt{6}\right)$
Eccentricity: $\dfrac{\sqrt{6}}{3}$

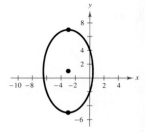

47. Ellipse
Center: $\left(3, -\dfrac{5}{2}\right)$
Vertices: $\left(9, -\dfrac{5}{2}\right), \left(-3, -\dfrac{5}{2}\right)$
Foci: $\left(3 \pm 3\sqrt{3}, -\dfrac{5}{2}\right)$
Eccentricity: $\dfrac{\sqrt{3}}{2}$

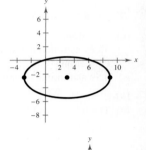
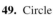

49. Circle
Center: $(-1, 1)$
Radius: $\frac{2}{3}$

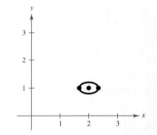

51. Ellipse
Center: $(2, 1)$
Vertices: $\left(\frac{7}{3}, 1\right), \left(\frac{5}{3}, 1\right)$
Foci: $\left(\frac{34}{15}, 1\right), \left(\frac{26}{15}, 1\right)$
Eccentricity: $\frac{4}{5}$

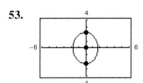

53.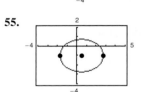
Center: $(0, 0)$
Vertices: $\left(0, \pm\sqrt{5}\right)$
Foci: $\left(0, \pm\sqrt{2}\right)$

55.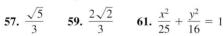
Center: $\left(\frac{1}{2}, -1\right)$
Vertices: $\left(\frac{1}{2} \pm \sqrt{5}, -1\right)$
Foci: $\left(\frac{1}{2} \pm \sqrt{2}, -1\right)$

57. $\dfrac{\sqrt{5}}{3}$ **59.** $\dfrac{2\sqrt{2}}{3}$ **61.** $\dfrac{x^2}{25} + \dfrac{y^2}{16} = 1$

CHAPTER 6

63. (a)

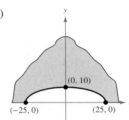

(b) $\dfrac{x^2}{625} + \dfrac{y^2}{100} = 1$

(c) Yes

65. (a) $\dfrac{x^2}{321.84} + \dfrac{y^2}{20.89} = 1$ (b)

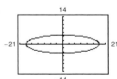

(c) Aphelion:
35.29 astronomical units
Perihelion:
0.59 astronomical unit

67. (a) $y = -8\sqrt{0.04 - \theta^2}$ (b)
(c) The bottom half

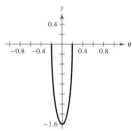

69.

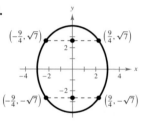

71.

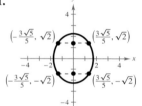

73. False. The graph of $(x^2/4) + y^4 = 1$ is not an ellipse. The degree of y is 4, not 2.

75. (a) $A = \pi a(20 - a)$ (b) $\dfrac{x^2}{196} + \dfrac{y^2}{36} = 1$

(c)

a	8	9	10	11	12	13
A	301.6	311.0	314.2	311.0	301.6	285.9

$a = 10$, circle

(d)

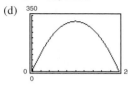

The shape of an ellipse with a maximum area is a circle. The maximum area is found when $a = 10$ (verified in part c) and therefore $b = 10$, so the equation produces a circle.

77. $\dfrac{(x - 6)^2}{324} + \dfrac{(y - 2)^2}{308} = 1$ **79.** Proof

Section 6.4   *(page 482)*

1. hyperbola; foci   **3.** transverse axis; center
5. b   **6.** c   **7.** a   **8.** d

9. Center: $(0, 0)$
Vertices: $(\pm 1, 0)$
Foci: $(\pm \sqrt{2}, 0)$
Asymptotes: $y = \pm x$

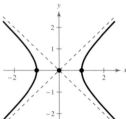

11. Center: $(0, 0)$
Vertices: $(0, \pm 5)$
Foci: $\left(0, \pm \sqrt{106}\right)$
Asymptotes: $y = \pm \dfrac{5}{9}x$

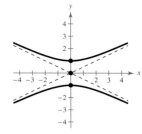

13. Center: $(0, 0)$
Vertices: $(0, \pm 1)$
Foci: $\left(0, \pm \sqrt{5}\right)$
Asymptotes: $y = \pm \dfrac{1}{2}x$

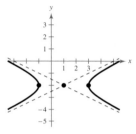

15. Center: $(1, -2)$
Vertices: $(3, -2), (-1, -2)$
Foci: $\left(1 \pm \sqrt{5}, -2\right)$
Asymptotes:
$y = -2 \pm \dfrac{1}{2}(x - 1)$

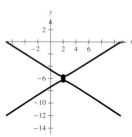

17. Center: $(2, -6)$
Vertices:
$\left(2, -\dfrac{17}{3}\right), \left(2, -\dfrac{19}{3}\right)$
Foci: $\left(2, -6 \pm \dfrac{\sqrt{13}}{6}\right)$
Asymptotes:
$y = -6 \pm \dfrac{2}{3}(x - 2)$

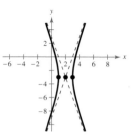

19. Center: $(2, -3)$
Vertices: $(3, -3), (1, -3)$
Foci: $\left(2 \pm \sqrt{10}, -3\right)$
Asymptotes:
$y = -3 \pm 3(x - 2)$

21. The graph of this equation is two lines intersecting at $(-1, -3)$.

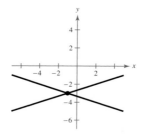

23. Center: $(0, 0)$
Vertices: $(\pm\sqrt{3}, 0)$
Foci: $(\pm\sqrt{5}, 0)$
Asymptotes: $y = \pm\dfrac{\sqrt{6}}{3}x$

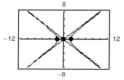

25. Center: $(0, 0)$
Vertices: $(\pm 3, 0)$
Foci: $(\pm\sqrt{13}, 0)$
Asymptotes: $y = \pm\frac{2}{3}x$

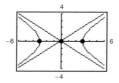

27. Center: $(1, -3)$
Vertices: $(1, -3 \pm \sqrt{2})$
Foci: $(1, -3 \pm 2\sqrt{5})$
Asymptotes:
$y = -3 \pm \frac{1}{3}(x - 1)$

29. $\dfrac{y^2}{4} - \dfrac{x^2}{12} = 1$ **31.** $\dfrac{x^2}{1} - \dfrac{y^2}{25} = 1$

33. $\dfrac{17y^2}{1024} - \dfrac{17x^2}{64} = 1$ **35.** $\dfrac{(x-4)^2}{4} - \dfrac{y^2}{12} = 1$

37. $\dfrac{(y-5)^2}{16} - \dfrac{(x-4)^2}{9} = 1$ **39.** $\dfrac{y^2}{9} - \dfrac{4(x-2)^2}{9} = 1$

41. $\dfrac{(y-2)^2}{4} - \dfrac{x^2}{4} = 1$ **43.** $\dfrac{(x-2)^2}{1} - \dfrac{(y-2)^2}{1} = 1$

45. $\dfrac{(x-3)^2}{9} - \dfrac{(y-2)^2}{4} = 1$ **47.** $\dfrac{y^2}{9} - \dfrac{x^2}{9/4} = 1$

49. $\dfrac{(x-3)^2}{4} - \dfrac{(y-2)^2}{16/5} = 1$

51. (a) $\dfrac{x^2}{1} - \dfrac{y^2}{169/3} = 1$ (b) About 2.403 ft

53. $(3300, -2750)$

55. (a) $\dfrac{x^2}{1} - \dfrac{y^2}{27} = 1$; $-9 \le y \le 9$ (b) 1.89 ft

57. Ellipse   **59.** Hyperbola   **61.** Hyperbola
63. Parabola   **65.** Ellipse   **67.** Parabola
69. Parabola   **71.** Circle
73. True. For a hyperbola, $c^2 = a^2 + b^2$. The larger the ratio of b to a, the larger the eccentricity of the hyperbola, $e = c/a$.
75. False. When $D = -E$, the graph is two intersecting lines.
77. Answers will vary.
79. $y = 1 - 3\sqrt{\dfrac{(x-3)^2}{4} - 1}$

81.

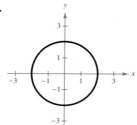

The equation $y = x^2 + C$ is a parabola that could intersect the circle in zero, one, two, three, or four places depending on its location on the y-axis.

(a) $C > 2$ and $C < -\frac{17}{4}$ (b) $C = 2$
(c) $-2 < C < 2$, $C = -\frac{17}{4}$ (d) $C = -2$
(e) $-\frac{17}{4} < C < -2$

Section 6.5   (page 490)

1. plane curve   **3.** eliminating; parameter
5. (a)

t	0	1	2	3	4
x	0	1	$\sqrt{2}$	$\sqrt{3}$	2
y	3	2	1	0	-1

(b)

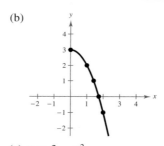

(c) $y = 3 - x^2$

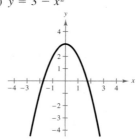

The graph of the rectangular equation shows the entire parabola rather than just the right half.

7. (a)

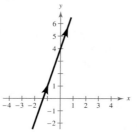

(b) $y = 3x + 4$

9. (a)

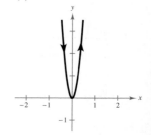

(b) $y = 16x^2$

11. (a)

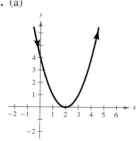

(b) $y = x^2 - 4x + 4$

13. (a)

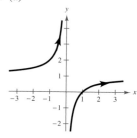

(b) $y = \dfrac{(x - 1)}{x}$

15. (a)

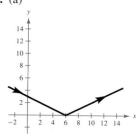

(b) $y = \left| \dfrac{x}{2} - 3 \right|$

17. (a)

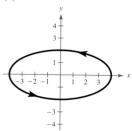

(b) $\dfrac{x^2}{16} + \dfrac{y^2}{4} = 1$

19. (a)

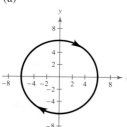

(b) $\dfrac{x^2}{36} + \dfrac{y^2}{36} = 1$

21. (a)

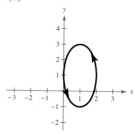

(b) $\dfrac{(x - 1)^2}{1} + \dfrac{(y - 1)^2}{4} = 1$

23. (a)

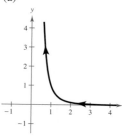

(b) $y = \dfrac{1}{x^3}, \quad x > 0$

25. (a)

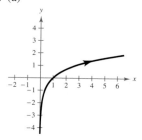

(b) $y = \ln x$

27. Each curve represents a portion of the line $y = 2x + 1$.

Domain	Orientation
(a) $(-\infty, \infty)$	Left to right
(b) $[-1, 1]$	Depends on θ
(c) $(0, \infty)$	Right to left
(d) $(0, \infty)$	Left to right

29. $y - y_1 = m(x - x_1)$

31. $\dfrac{(x - h)^2}{a^2} + \dfrac{(y - k)^2}{b^2} = 1$

33. $x = 3t$
$y = 6t$

35. $x = 3 + 4 \cos \theta$
$y = 2 + 4 \sin \theta$

37. $x = 5 \cos \theta$
$y = 3 \sin \theta$

39. $x = 4 \sec \theta$
$y = 3 \tan \theta$

41. (a) $x = t, \ y = 3t - 2$ (b) $x = -t + 2, \ y = -3t + 4$

43. (a) $x = t, y = 2 - t$ (b) $x = -t + 2, y = t$

45. (a) $x = t, y = t^2 - 3$ (b) $x = 2 - t, y = t^2 - 4t + 1$

47. (a) $x = t, y = \dfrac{1}{t}$ (b) $x = -t + 2, y = -\dfrac{1}{t - 2}$

49.

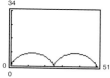

51.

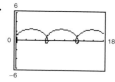

53.

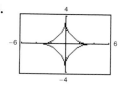

55.

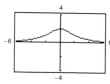

57. b
Domain: $[-2, 2]$
Range: $[-1, 1]$

58. c
Domain: $[-4, 4]$
Range: $[-6, 6]$

59. d
Domain: $(-\infty, \infty)$
Range: $(-\infty, \infty)$

60. a
Domain: $(-\infty, \infty)$
Range: $[-2, 2]$
Maximum height: 90.7 ft
Range: 209.6 ft

61. (a)

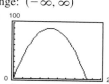

(b)

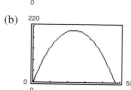

Maximum height: 204.2 ft
Range: 471.6 ft

(c)

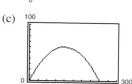

Maximum height: 60.5 ft
Range: 242.0 ft

(d)

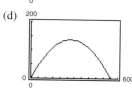

Maximum height: 136.1 ft
Range: 544.5 ft

63. (a) $x = (146.67 \cos \theta)t$
$y = 3 + (146.67 \sin \theta)t - 16t^2$

(b)

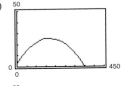

No

(c)

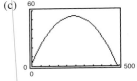

Yes

(d) $19.3°$

65. Answers will vary.

67. $x = a\theta - b \sin \theta$
$y = a - b \cos \theta$

69. True
$x = t$
$y = t^2 + 1 \Rightarrow y = x^2 + 1$
$x = 3t$
$y = 9t^2 + 1 \Rightarrow y = x^2 + 1$

71. Parametric equations are useful when graphing two functions simultaneously on the same coordinate system. For example, they are useful when tracking the path of an object so that the position and the time associated with that position can be determined.

73. $-1 < t < \infty$

Section 6.6 *(page 497)*

1. pole   **3.** polar

5.

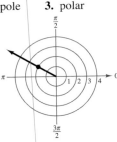

$\left(2, -\dfrac{7\pi}{6}\right), \left(-2, -\dfrac{\pi}{6}\right)$

7.

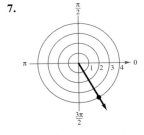

$\left(4, \dfrac{5\pi}{3}\right), \left(-4, -\dfrac{4\pi}{3}\right)$

9.

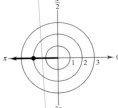

$(2, \pi), (-2, 0)$

11.

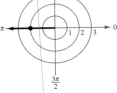

$\left(-2, -\dfrac{4\pi}{3}\right), \left(2, \dfrac{5\pi}{3}\right)$

13.

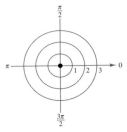

$\left(0, \dfrac{5\pi}{6}\right), \left(0, -\dfrac{\pi}{6}\right)$

15.

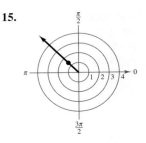

$\left(\sqrt{2}, -3.92\right), \left(-\sqrt{2}, -0.78\right)$
$(-3, 4.71), (3, 1.57)$

17.

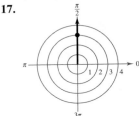

19. $(0, 3)$ **21.** $\left(\dfrac{\sqrt{2}}{2}, \dfrac{\sqrt{2}}{2}\right)$ **23.** $\left(-\sqrt{2}, \sqrt{2}\right)$

25. $\left(\sqrt{3}, 1\right)$ **27.** $(-1.1, -2.2)$ **29.** $(1.53, 1.29)$

31. $(-1.20, -4.34)$ **33.** $(-0.02, 2.50)$

35. $(-3.60, 1.97)$ **37.** $\left(\sqrt{2}, \dfrac{\pi}{4}\right)$ **39.** $\left(3\sqrt{2}, \dfrac{5\pi}{4}\right)$

41. $(6, \pi)$ **43.** $\left(5, \dfrac{3\pi}{2}\right)$ **45.** $(5, 2.21)$

47. $\left(\sqrt{6}, \dfrac{5\pi}{4}\right)$ **49.** $\left(2, \dfrac{11\pi}{6}\right)$ **51.** $\left(3\sqrt{13}, 0.98\right)$

53. $(13, 1.18)$ **55.** $\left(\sqrt{13}, 5.70\right)$ **57.** $\left(\sqrt{29}, 2.76\right)$

59. $\left(\sqrt{7}, 0.86\right)$ **61.** $\left(\dfrac{17}{6}, 0.49\right)$ **63.** $\left(\dfrac{\sqrt{85}}{4}, 0.71\right)$

65. $r = 3$ **67.** $r = 4 \csc \theta$ **69.** $r = 10 \sec \theta$

71. $r = -2 \csc \theta$ **73.** $r = \dfrac{-2}{3 \cos \theta - \sin \theta}$

75. $r^2 = 16 \sec \theta \csc \theta = 32 \csc 2\theta$

77. $r = \dfrac{4}{1 - \cos \theta}$ or $-\dfrac{4}{1 + \cos \theta}$ **79.** $r = a$

81. $r = 2a \cos \theta$ **83.** $r = \cot^2 \theta \csc \theta$

85. $x^2 + y^2 - 4y = 0$ **87.** $x^2 + y^2 + 2x = 0$

89. $\sqrt{3}x + y = 0$ **91.** $\dfrac{\sqrt{3}}{3}x + y = 0$

93. $x^2 + y^2 = 16$ **95.** $y = 4$ **97.** $x = -3$

99. $x^2 + y^2 - x^{2/3} = 0$ **101.** $(x^2 + y^2)^2 = 2xy$

103. $(x^2 + y^2)^2 = 6x^2y - 2y^3$ **105.** $x^2 + 4y - 4 = 0$

107. $4x^2 - 5y^2 - 36y - 36 = 0$

109. The graph of the polar equation consists of all points that are six units from the pole.
$x^2 + y^2 = 36$

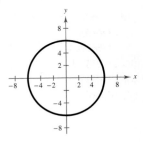

111. The graph of the polar equation consists of all points on the line that makes an angle of $\pi/6$ with the positive polar axis.
$- \sqrt{3}x + 3y = 0$

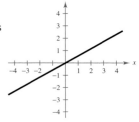

113. The graph of the polar equation is not evident by simple inspection, so convert to rectangular form.
$x^2 + (y - 1)^2 = 1$

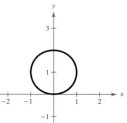

115. The graph of the polar equation is not evident by simple inspection, so convert to rectangular form.
$(x + 3)^2 + y^2 = 9$

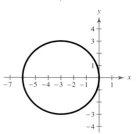

117. The graph of the polar equation is not evident by simple inspection, so convert to rectangular form.
$x - 3 = 0$

119. True. Because r is a directed distance, the point (r, θ) can be represented as $(r, \theta \pm 2\pi n)$.

121. $(x - h)^2 + (y - k)^2 = h^2 + k^2$
Radius: $\sqrt{h^2 + k^2}$
Center: (h, k)

123. (a) Answers will vary.
(b) $(r_1, \theta_1), (r_2, \theta_2)$ and the pole are collinear.
$d = \sqrt{r_1^2 + r_2^2 - 2r_1r_2} = |r_1 - r_2|$
This represents the distance between two points on the line $\theta = \theta_1 = \theta_2$.
(c) $d = \sqrt{r_1^2 + r_2^2}$
This is the result of the Pythagorean Theorem.
(d) Answers will vary. For example:
Points: $(3, \pi/6), (4, \pi/3)$
Distance: 2.053
Points: $(-3, 7\pi/6), (-4, 4\pi/3)$
Distance: 2.053

125. (a)

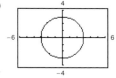

(b) Yes. $\theta \approx 3.927, x \approx -2.121,$
$y \approx -2.121$
(c) Yes. Answers will vary.

Section 6.7   (*page 505*)

1. $\theta = \dfrac{\pi}{2}$ **3.** convex limaçon   **5.** lemniscate

7. Rose curve with 4 petals   **9.** Limaçon with inner loop
11. Rose curve with 3 petals   **13.** Polar axis

15. $\theta = \dfrac{\pi}{2}$ **17.** $\theta = \dfrac{\pi}{2}$, polar axis, pole

19. Maximum: $|r| = 20$ when $\theta = \dfrac{3\pi}{2}$

Zero: $r = 0$ when $\theta = \dfrac{\pi}{2}$

21. Maximum: $|r| = 4$ when $\theta = 0, \dfrac{\pi}{3}, \dfrac{2\pi}{3}$

Zeros: $r = 0$ when $\theta = \dfrac{\pi}{6}, \dfrac{\pi}{2}, \dfrac{5\pi}{6}$

23.

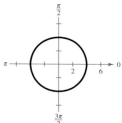

25.

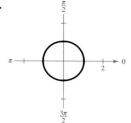

27.

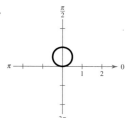

29.

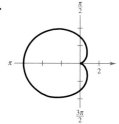

31.

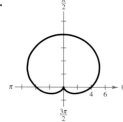

33.

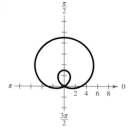

35.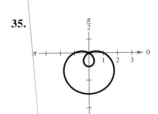

37.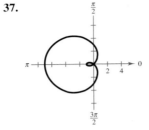

39.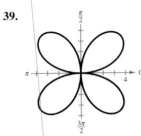

41.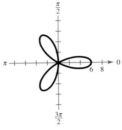

43.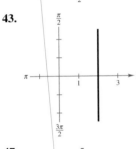

45.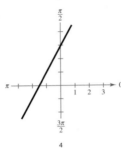

47.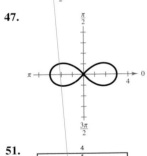

49.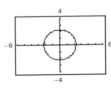

51.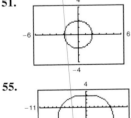

53.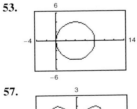

55.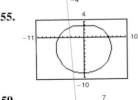

57.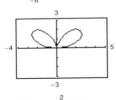

59.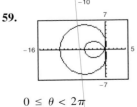

$0 \le \theta < 2\pi$

61.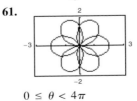

$0 \le \theta < 4\pi$

63.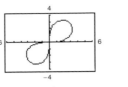

$0 \le \theta < \pi/2$

65.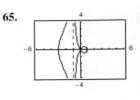

67.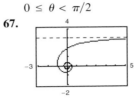

69. True. For a graph to have polar axis symmetry, replace (r, θ) by $(r, -\theta)$ or $(-r, \pi - \theta)$.

71. (a) 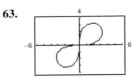   (b)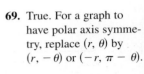

Upper half of circle   Lower half of circle

(c) 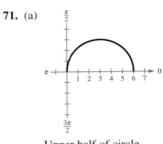   (d)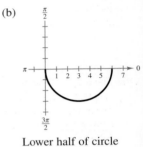

Full circle   Left half of circle

73. Answers will vary.

75. (a) $r = 2 - \dfrac{\sqrt{2}}{2}(\sin \theta - \cos \theta)$ (b) $r = 2 + \cos \theta$

(c) $r = 2 + \sin \theta$ (d) $r = 2 - \cos \theta$

77. (a) 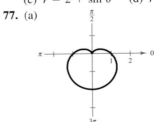   (b)

79. 8 petals; 3 petals; For $r = 2 \cos n\theta$ and $r = 2 \sin n\theta$, there are n petals if n is odd, $2n$ petals if n is even.

81. (a) 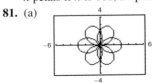   (b)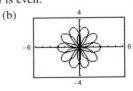

$0 \le \theta < 4\pi$ $0 \le \theta < 4\pi$

(c) Yes. Explanations will vary.

CHAPTER 6

Section 6.8 (page 511)

1. conic   **3.** vertical; right

5. $e = 1$: $r = \dfrac{2}{1 + \cos\theta}$, parabola

$e = 0.5$: $r = \dfrac{1}{1 + 0.5\cos\theta}$, ellipse

$e = 1.5$: $r = \dfrac{3}{1 + 1.5\cos\theta}$, hyperbola

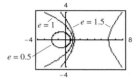

7. $e = 1$: $r = \dfrac{2}{1 - \sin\theta}$, parabola

$e = 0.5$: $r = \dfrac{1}{1 - 0.5\sin\theta}$, ellipse

$e = 1.5$: $r = \dfrac{3}{1 - 1.5\sin\theta}$, hyperbola

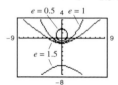

9. e   **10.** c   **11.** d   **12.** f   **13.** a   **14.** b
15. Parabola   **17.** Parabola

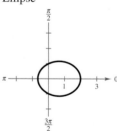

19. Ellipse   **21.** Ellipse

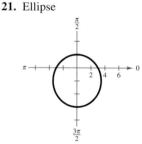

23. Hyperbola   **25.** Hyperbola

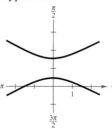
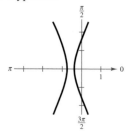

27. Ellipse   **29.**

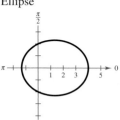
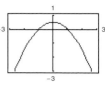

Parabola

31.   **33.**

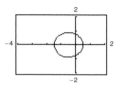
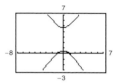

Ellipse   Hyperbola

35.   **37.**

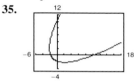
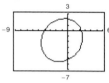

39. $r = \dfrac{1}{1 - \cos\theta}$ **41.** $r = \dfrac{1}{2 + \sin\theta}$

43. $r = \dfrac{2}{1 + 2\cos\theta}$ **45.** $r = \dfrac{2}{1 - \sin\theta}$

47. $r = \dfrac{10}{1 - \cos\theta}$ **49.** $r = \dfrac{10}{3 + 2\cos\theta}$

51. $r = \dfrac{20}{3 - 2\cos\theta}$ **53.** $r = \dfrac{9}{4 - 5\sin\theta}$

55. Answers will vary.

57. $r = \dfrac{9.5929 \times 10^7}{1 - 0.0167\cos\theta}$

Perihelion: 9.4354×10^7 mi

Aphelion: 9.7558×10^7 mi

59. $r = \dfrac{1.0820 \times 10^8}{1 - 0.0068\cos\theta}$

Perihelion: 1.0747×10^8 km

Aphelion: 1.0894×10^8 km

61. $r = \dfrac{1.4039 \times 10^8}{1 - 0.0934\cos\theta}$

Perihelion: 1.2840×10^8 mi

Aphelion: 1.5486×10^8 mi

63. $r = \dfrac{0.624}{1 + 0.847\sin\theta}$; $r \approx 0.338$ astronomical unit

65. (a) $r = \dfrac{8200}{1 + \sin\theta}$

(b)

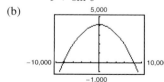

(c) 1467 mi   (d) 394 mi

67. True. The graphs represent the same hyperbola.

69. True. The conic is an ellipse because the eccentricity is less than 1.

71. The original equation graphs as a parabola that opens downward.
   (a) The parabola opens to the right.
   (b) The parabola opens up.
   (c) The parabola opens to the left.
   (d) The parabola has been rotated.

73. Answers will vary.

75. $r^2 = \dfrac{24{,}336}{169 - 25\cos^2\theta}$ **77.** $r^2 = \dfrac{144}{25\cos^2\theta - 9}$

79. $r^2 = \dfrac{144}{25\cos^2\theta - 16}$

81. (a) Ellipse
   (b) The given polar equation, r, has a vertical directrix to the left of the pole. The equation r_1 has a vertical directrix to the right of the pole, and the equation r_2 has a horizontal directrix below the pole.
   (c)

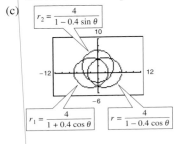

$r_2 = \dfrac{4}{1 - 0.4\sin\theta}$

$r_1 = \dfrac{4}{1 + 0.4\cos\theta}$ $r = \dfrac{4}{1 - 0.4\cos\theta}$

Review Exercises   *(page 516)*

1. $\dfrac{\pi}{4}$ rad, $45°$ **3.** 1.1071 rad, $63.43°$

5. 0.4424 rad, $25.35°$ **7.** 0.6588 rad, $37.75°$

9. $4\sqrt{2}$ **11.** Hyperbola

13. $y^2 = 16x$

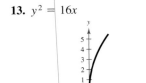

15. $(y - 2)^2 = 12x$

17. $y = -4x - 2;\ \left(-\tfrac{1}{2}, 0\right)$ **19.** $8\sqrt{6}$ m

21. $\dfrac{(x - 3)^2}{25} + \dfrac{y^2}{16} = 1$ **23.** $\dfrac{(x - 2)^2}{4} + \dfrac{(y - 1)^2}{1} = 1$

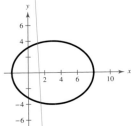

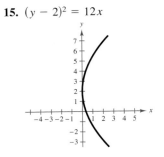

25. The foci occur 3 feet from the center of the arch on a line connecting the tops of the pillars.

27. Center: $(-1, 2)$
   Vertices: $(-1, 9), (-1, -5)$
   Foci: $\left(-1, 2 \pm 2\sqrt{6}\right)$
   Eccentricity: $\dfrac{2\sqrt{6}}{7}$

29. Center:
   Vertices: $(1, 0), (1, -8)$
   Foci: $\left(1, -4 \pm \sqrt{7}\right)$
   Eccentricity: $\dfrac{\sqrt{7}}{4}$

31. $\dfrac{y^2}{1} - \dfrac{x^2}{3} = 1$ **33.** $\dfrac{5(x - 4)^2}{16} - \dfrac{5y^2}{64} = 1$

35. Center: $(5, -3)$
   Vertices: $(11, -3), (-1, -3)$
   Foci: $\left(5 \pm 2\sqrt{13}, -3\right)$
   Asymptotes:
     $y = -3 \pm \tfrac{2}{3}(x - 5)$

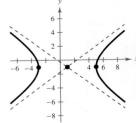

37. Center: $(1, -1)$
   Vertices: $(5, -1), (-3, -1)$
   Foci: $(6, -1), (-4, -1)$
   Asymptotes:
     $y = -1 \pm \tfrac{3}{4}(x - 1)$

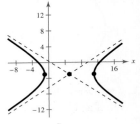

39. 72 mi   **41.** Hyperbola   **43.** Ellipse

45. (a)

t	-2	-1	0	1	2
x	-8	-5	-2	1	4
y	15	11	7	3	-1

   (b)

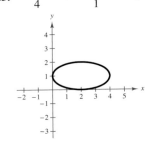

47. (a)

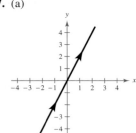

(b) $y = 2x$

49. (a)

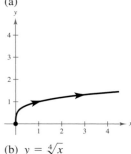

(b) $y = \sqrt[4]{x}$
(b) $x^2 + y^2 = 9$

51. (a)

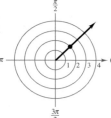

53. $x = -4 + 13t$
       $y = 4 - 14t$

55. $x = -3 + 4\cos\theta$
       $y = 4 + 3\sin\theta$

57.

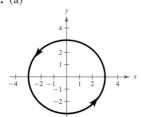

$(2, -7\pi/4), (-2, 5\pi/4)$

59.

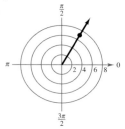

$(7, 1.05), (-7, -2.09)$

61. $\left(-\dfrac{1}{2}, -\dfrac{\sqrt{3}}{2}\right)$ **63.** $\left(-\dfrac{3\sqrt{2}}{2}, \dfrac{3\sqrt{2}}{2}\right)$ **65.** $\left(1, \dfrac{\pi}{2}\right)$

67. $\left(2\sqrt{13}, 0.9828\right)$ **69.** $r = 9$ **71.** $r = 6\sin\theta$

73. $r^2 = 10\csc 2\theta$ **75.** $x^2 + y^2 = 25$ **77.** $x^2 + y^2 = 3x$

79. $x^2 + y^2 = y^{2/3}$

81. Symmetry: $\theta = \dfrac{\pi}{2}$, polar axis, pole

Maximum value of $|r|$: $|r| = 6$ for all values of θ
No zeros of r

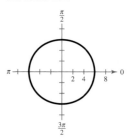

83. Symmetry: $\theta = \dfrac{\pi}{2}$, polar axis, pole

Maximum value of $|r|$: $|r| = 4$ when $\theta = \dfrac{\pi}{4}, \dfrac{3\pi}{4}, \dfrac{5\pi}{4}, \dfrac{7\pi}{4}$

Zeros of r: $r = 0$ when
$$\theta = 0, \dfrac{\pi}{2}, \pi, \dfrac{3\pi}{2}$$

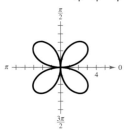

85. Symmetry: polar axis
Maximum value of $|r|$:
$|r| = 4$ when $\theta = 0$
Zeros of r: $r = 0$
when $\theta = \pi$

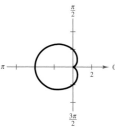

87. Symmetry: $\theta = \dfrac{\pi}{2}$

Maximum value of $|r|$: $|r| = 8$ when $\theta = \dfrac{\pi}{2}$

Zeros of r: $r = 0$ when $\theta = 3.4814, 5.9433$

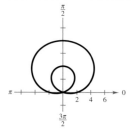

89. Symmetry: $\theta = \dfrac{\pi}{2}$, polar axis, pole

Maximum value of $|r|$: $|r| = 3$ when $\theta = 0, \dfrac{\pi}{2}, \pi, \dfrac{3\pi}{2}$

Zeros of r: $r = 0$ when $\theta = \dfrac{\pi}{4}, \dfrac{3\pi}{4}, \dfrac{5\pi}{4}, \dfrac{7\pi}{4}$

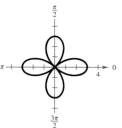

91. Limaçon

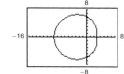

93. Rose curve

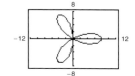

95. Hyperbola

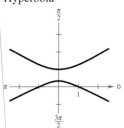

97. Ellipse

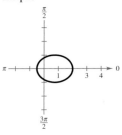

99. $r = \dfrac{4}{1 - \cos\theta}$ **101.** $r = \dfrac{5}{3 - 2\cos\theta}$

103. $r = \dfrac{7978.81}{1 - 0.937\cos\theta}$; 11,011.87 mi

105. False. The equation of a hyperbola is a second-degree equation.

107. False. $(2, \pi/4), (-2, 5\pi/4)$, and $(2, 9\pi/4)$ all represent the same point.

109. (a) The graphs are the same.   (b) The graphs are the same.

Chapter Test   (page 519)

1. 0.3805 rad, 21.8°   **2.** 0.8330 rad, 47.7°

3. $\dfrac{7\sqrt{2}}{2}$

4. Parabola: $y^2 = 2(x - 1)$
Vertex: $(1, 0)$
Focus: $\left(\frac{3}{2}, 0\right)$

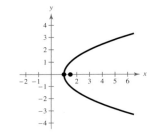

5. Hyperbola: $\dfrac{(x-2)^2}{4} - y^2 = 1$

Center: $(2, 0)$
Vertices: $(0, 0), (4, 0)$
Foci: $\left(2 \pm \sqrt{5}, 0\right)$

Asymptotes: $y = \pm\dfrac{1}{2}(x - 2)$

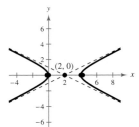

6. Ellipse: $\dfrac{(x+3)^2}{16} + \dfrac{(y-1)^2}{9} = 1$

Center: $(-3, 1)$
Vertices: $(1, 1), (-7, 1)$
Foci: $\left(-3 \pm \sqrt{7}, 1\right)$

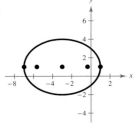

7. Circle: $(x - 2)^2 + (y - 1)^2 = \frac{1}{2}$
Center: $(2, 1)$

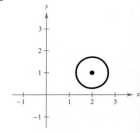

8. $(x - 2)^2 = \dfrac{4}{3}(y + 3)$ **9.** $\dfrac{5(y - 2)^2}{4} - \dfrac{5x^2}{16} = 1$

10.

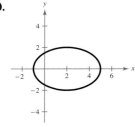

$\dfrac{(x - 2)^2}{9} + \dfrac{y^2}{4} = 1$

11. $x = 6 + 4t$
$y = 4 + 7t$

12. $\left(\sqrt{3}, -1\right)$ **13.** $\left(2\sqrt{2}, \frac{7\pi}{4}\right), \left(-2\sqrt{2}, \frac{3\pi}{4}\right), \left(2\sqrt{2}, -\frac{\pi}{4}\right)$

14. $r = 3\cos\theta$

15.

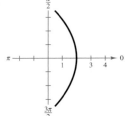

Parabola

16.

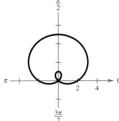

Ellipse

17.

Limaçon with inner loop

18.

Rose curve

19. Answers will vary. For example: $r = \dfrac{1}{1 + 0.25\sin\theta}$

20. Slope: 0.1511; Change in elevation: 789 ft

21. No; Yes

CHAPTER 6

Cumulative Test for Chapters 4–6   *(page 520)*

1. (a)

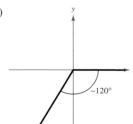

     (b) 240°

     (c) $-\dfrac{2\pi}{3}$

     (d) 60°

     (e) $\sin(-120°) = -\dfrac{\sqrt{3}}{2}$ $\csc(-120°) = -\dfrac{2\sqrt{3}}{3}$

        $\cos(-120°) = -\dfrac{1}{2}$ $\sec(-120°) = -2$

        $\tan(-120°) = \sqrt{3}$ $\cot(-120°) = \dfrac{\sqrt{3}}{3}$

2. $-83.1°$ **3.** $\dfrac{20}{29}$

4.

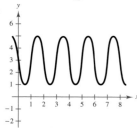

5.

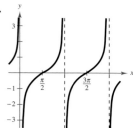

6.

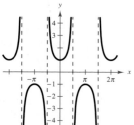

7. $a = -3, b = \pi, c = 0$

8.

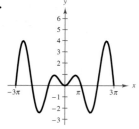

9. 4.9   **10.** $\dfrac{3}{4}$

11. $\sqrt{1 - 4x^2}$ **12.** 1   **13.** $2\tan\theta$

14–16. Answers will vary.   **17.** $\dfrac{\pi}{3}, \dfrac{\pi}{2}, \dfrac{3\pi}{2}, \dfrac{5\pi}{3}$

18. $\dfrac{\pi}{6}, \dfrac{5\pi}{6}, \dfrac{7\pi}{6}, \dfrac{11\pi}{6}$ **19.** $\dfrac{3\pi}{2}$ **20.** $\dfrac{16}{63}$ **21.** $\dfrac{4}{3}$

22. $\dfrac{\sqrt{5}}{5}, \dfrac{2\sqrt{5}}{5}$ **23.** $\dfrac{5}{2}\left(\sin\dfrac{5\pi}{2} - \sin\pi\right)$

24. $-2\sin 8x \sin x$ **25.** $B \approx 26.39°, C \approx 123.61°, c \approx 14.99$

26. $B \approx 52.48°, C \approx 97.52°, a \approx 5.04$

27. $B = 60°, a \approx 5.77, c \approx 11.55$

28. $A \approx 26.28°, B \approx 49.74°, C \approx 103.98°$

29. Law of Sines; $C = 109°, a \approx 14.96, b \approx 9.27$

30. Law of Cosines; $A \approx 6.75°, B \approx 93.25°, c \approx 9.86$

31. 41.48 in.^{2} **32.** 599.09 m^2

33. Ellipse   **34.** Circle

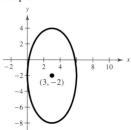   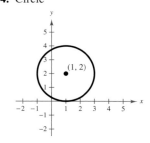

35. $\dfrac{x^2}{1} + \dfrac{(y-5)^2}{25} = 1$

36.

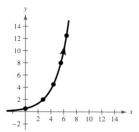

The corresponding rectangular equation is $y = \dfrac{\sqrt{e^x}}{2}$.

37.

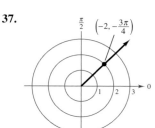

$\left(-2, \dfrac{5\pi}{4}\right), \left(2, -\dfrac{7\pi}{4}\right), \left(2, \dfrac{\pi}{4}\right)$

38. $-8r\cos\theta - 3r\sin\theta + 5 = 0$

39. $9x^2 + 20x - 16y^2 + 4 = 0$

40.   **41.**

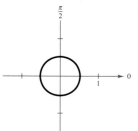   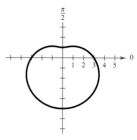

Circle   Dimpled limaçon

42.

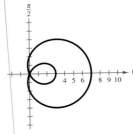

Limaçon with an open loop

43. About 395.8 rad/min; about 8312.7 in./min
44. 42π yd^2 ≈ 131.95 yd^2 **45.** 5 ft   **46.** 22.6°

47. $d = 4 \cos \dfrac{\pi}{4} t$

Problem Solving   *(page 525)*

1. (a) 1.2016 rad   (b) 2420 ft, 5971 ft
3. $y^2 = 4p(x + p)$ **5.** Answers will vary.

7. $\dfrac{(x - 6)^2}{9} - \dfrac{(y - 2)^2}{7} = 1$

9. (a) The first set of parametric equations models projectile motion along a straight line. The second set of parametric equations models projectile motion of an object launched at a height of h units above the ground that will eventually fall back to the ground.

  (b) $y = (\tan \theta)x$; $y = h + x \tan \theta - \dfrac{16x^2 \sec^2 \theta}{v_0{}^2}$

  (c) In the first case, the path of the moving object is not affected by a change in the velocity because eliminating the parameter removes v_0.

11. (a)

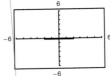

(b)

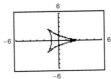

The graph is a line between -2 and 2 on the x-axis.

The graph is a three-sided figure with counterclockwise orientation.

(c)

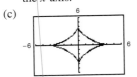

(d)

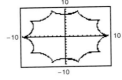

The graph is a four-sided figure with counterclockwise orientation.

The graph is a 10-sided figure with counterclockwise orientation.

(e)

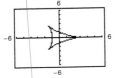

(f)

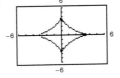

The graph is a three-sided figure with clockwise orientation.

The graph is a four-sided figure with clockwise orientation.

13.

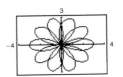 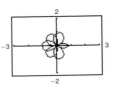

$r = 3 \sin\left(\dfrac{5\theta}{2}\right)$ $r = -\cos(\sqrt{2}\theta)$, $-2\pi \le \theta \le 2\pi$

Sample answer: If n is a rational number, then the curve has a finite number of petals. If n is an irrational number, then the curve has an infinite number of petals.

15. (a) No. Because of the exponential, the graph will continue to trace the butterfly curve at larger values of r.

  (b) $r \approx 4.1$. This value will increase if θ is increased.

17. (a) $r_{\text{Neptune}} = \dfrac{4.4947 \times 10^9}{1 - 0.0086 \cos \theta}$

      $r_{\text{Pluto}} = \dfrac{5.54 \times 10^9}{1 - 0.2488 \cos \theta}$

  (b) Neptune: Aphelion = 4.534×10^9 km
              Perihelion = 4.456×10^9 km

     Pluto: Aphelion = 7.375×10^9 km
             Perihelion = 4.437×10^9 km

  (c)

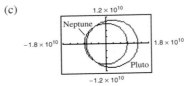

  (d) Yes, at times Pluto can be closer to the sun than Neptune. Pluto was called the ninth planet because it has the longest orbit around the sun and therefore also reaches the furthest distance away from the sun.

  (e) If the orbits were in the same plane, then they would intersect. Furthermore, since the orbital periods differ (Neptune = 164.79 years, Pluto = 247.68 years), then the two planets would ultimately collide if the orbits intersect. The orbital inclination of Pluto is significantly larger than that of Neptune (17.16° vs. 1.769°), so further analysis is required to determine if the orbits intersect.

INDEX

Definition of the Six Trigonometric Functions

Right triangle definitions, where $0 < \theta < \pi/2$

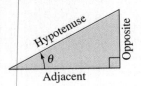

$$\sin \theta = \frac{\text{opp.}}{\text{hyp.}} \qquad \csc \theta = \frac{\text{hyp.}}{\text{opp.}}$$

$$\cos \theta = \frac{\text{adj.}}{\text{hyp.}} \qquad \sec \theta = \frac{\text{hyp.}}{\text{adj.}}$$

$$\tan \theta = \frac{\text{opp.}}{\text{adj.}} \qquad \cot \theta = \frac{\text{adj.}}{\text{opp.}}$$

Circular function definitions, where θ *is any angle*

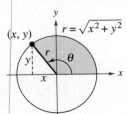

$$\sin \theta = \frac{y}{r} \qquad \csc \theta = \frac{r}{y}$$

$$\cos \theta = \frac{x}{r} \qquad \sec \theta = \frac{r}{x}$$

$$\tan \theta = \frac{y}{x} \qquad \cot \theta = \frac{x}{y}$$

Reciprocal Identities

$$\sin u = \frac{1}{\csc u} \qquad \cos u = \frac{1}{\sec u} \qquad \tan u = \frac{1}{\cot u}$$

$$\csc u = \frac{1}{\sin u} \qquad \sec u = \frac{1}{\cos u} \qquad \cot u = \frac{1}{\tan u}$$

Quotient Identities

$$\tan u = \frac{\sin u}{\cos u} \qquad \cot u = \frac{\cos u}{\sin u}$$

Pythagorean Identities

$$\sin^2 u + \cos^2 u = 1$$

$$1 + \tan^2 u = \sec^2 u \qquad 1 + \cot^2 u = \csc^2 u$$

Cofunction Identities

$$\sin\left(\frac{\pi}{2} - u\right) = \cos u \qquad \cot\left(\frac{\pi}{2} - u\right) = \tan u$$

$$\cos\left(\frac{\pi}{2} - u\right) = \sin u \qquad \sec\left(\frac{\pi}{2} - u\right) = \csc u$$

$$\tan\left(\frac{\pi}{2} - u\right) = \cot u \qquad \csc\left(\frac{\pi}{2} - u\right) = \sec u$$

Even/Odd Identities

$$\sin(-u) = -\sin u \qquad \cot(-u) = -\cot u$$
$$\cos(-u) = \cos u \qquad \sec(-u) = \sec u$$
$$\tan(-u) = -\tan u \qquad \csc(-u) = -\csc u$$

Sum and Difference Formulas

$$\sin(u \pm v) = \sin u \cos v \pm \cos u \sin v$$

$$\cos(u \pm v) = \cos u \cos v \mp \sin u \sin v$$

$$\tan(u \pm v) = \frac{\tan u \pm \tan v}{1 \mp \tan u \tan v}$$

Double-Angle Formulas

$$\sin 2u = 2 \sin u \cos u$$

$$\cos 2u = \cos^2 u - \sin^2 u = 2 \cos^2 u - 1 = 1 - 2 \sin^2 u$$

$$\tan 2u = \frac{2 \tan u}{1 - \tan^2 u}$$

Power-Reducing Formulas

$$\sin^2 u = \frac{1 - \cos 2u}{2}$$

$$\cos^2 u = \frac{1 + \cos 2u}{2}$$

$$\tan^2 u = \frac{1 - \cos 2u}{1 + \cos 2u}$$

Sum-to-Product Formulas

$$\sin u + \sin v = 2 \sin\left(\frac{u + v}{2}\right) \cos\left(\frac{u - v}{2}\right)$$

$$\sin u - \sin v = 2 \cos\left(\frac{u + v}{2}\right) \sin\left(\frac{u - v}{2}\right)$$

$$\cos u + \cos v = 2 \cos\left(\frac{u + v}{2}\right) \cos\left(\frac{u - v}{2}\right)$$

$$\cos u - \cos v = -2 \sin\left(\frac{u + v}{2}\right) \sin\left(\frac{u - v}{2}\right)$$

Product-to-Sum Formulas

$$\sin u \sin v = \frac{1}{2}[\cos(u - v) - \cos(u + v)]$$

$$\cos u \cos v = \frac{1}{2}[\cos(u - v) + \cos(u + v)]$$

$$\sin u \cos v = \frac{1}{2}[\sin(u + v) + \sin(u - v)]$$

$$\cos u \sin v = \frac{1}{2}[\sin(u + v) - \sin(u - v)]$$

FORMULAS FROM GEOMETRY

Triangle:

$h = a \sin \theta$

$\text{Area} = \dfrac{1}{2}bh$

$c^2 = a^2 + b^2 - 2ab \cos \theta$ (Law of Cosines)

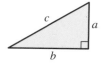

Right Triangle:

Pythagorean Theorem
$c^2 = a^2 + b^2$

Equilateral Triangle:

$h = \dfrac{\sqrt{3}s}{2}$

$\text{Area} = \dfrac{\sqrt{3}s^2}{4}$

Parallelogram:

$\text{Area} = bh$

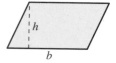

Trapezoid:

$\text{Area} = \dfrac{h}{2}(a + b)$

Circle:

$\text{Area} = \pi r^2$

$\text{Circumference} = 2\pi r$

Sector of Circle:

$\text{Area} = \dfrac{\theta r^2}{2}$

$s = r\theta$

θ in radians

Circular Ring:

$\text{Area} = \pi(R^2 - r^2)$

$\qquad = 2\pi pw$

p = average radius,

w = width of ring

Sector of Circular Ring:

$\text{Area} = \theta pw$

p = average radius,

w = width of ring,

θ in radians

Ellipse:

$\text{Area} = \pi ab$

$\text{Circumference} \approx 2\pi\sqrt{\dfrac{a^2 + b^2}{2}}$

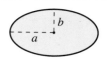

Cone:

$\text{Volume} = \dfrac{Ah}{3}$

A = area of base

Right Circular Cone:

$\text{Volume} = \dfrac{\pi r^2 h}{3}$

$\text{Lateral Surface Area} = \pi r\sqrt{r^2 + h^2}$

Frustum of Right Circular Cone:

$\text{Volume} = \dfrac{\pi(r^2 + rR + R^2)h}{3}$

$\text{Lateral Surface Area} = \pi s(R + r)$

Right Circular Cylinder:

$\text{Volume} = \pi r^2 h$

$\text{Lateral Surface Area} = 2\pi rh$

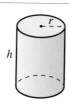

Sphere:

$\text{Volume} = \dfrac{4}{3}\pi r^3$

$\text{Surface Area} = 4\pi r^2$

Wedge:

$A = B \sec \theta$

A = area of upper face,

B = area of base

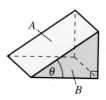